2011 13th International Symposium on Integrated Circuits

(ISIC 2011)

Singapore
12 – 14 December 2011

IEEE Catalog Number: CFP1176B-PRT
ISBN: 978-1-61284-863-1

**Copyright © 2011 by the Institute of Electrical and Electronic Engineers, Inc
All Rights Reserved**

Copyright and Reprint Permissions: Abstracting is permitted with credit to the source. Libraries are permitted to photocopy beyond the limit of U.S. copyright law for private use of patrons those articles in this volume that carry a code at the bottom of the first page, provided the per-copy fee indicated in the code is paid through Copyright Clearance Center, 222 Rosewood Drive, Danvers, MA 01923.

For other copying, reprint or republication permission, write to IEEE Copyrights Manager, IEEE Service Center, 445 Hoes Lane, Piscataway, NJ 08854. All rights reserved.

***This publication is a representation of what appears in the IEEE Digital Libraries. Some format issues inherent in the e-media version may also appear in this print version.**

IEEE Catalog Number: CFP1176B-PRT
ISBN 13:   978-1-61284-863-1

Additional Copies of This Publication Are Available From:

Curran Associates, Inc
57 Morehouse Lane
Red Hook, NY 12571 USA
Phone: (845) 758-0400
Fax: (845) 758-2633
E-mail: curran@proceedings.com
Web: www.proceedings.com

2011 International Symposium on
Integrated Circuits

12 – 14 December 2011
Suntec International Convention and Exhibition Centre
Singapore

Technical support & inquiries

Research Publishing Services
t:+65-6492 1137; f:+65-6747 4355
e:enquiries@rpsonline.com.sg

Organized by

Technically Co-Sponsored by

Message from the General Chair

On behalf of the Organizing Committee, I welcome you to the 2011 International Symposium on Integrated Circuits (ISIC-2011).

ISIC-2011 is organized by the School of Electrical & Electronic Engineering of Nanyang Technological University (NTU) and the Singapore Section of IEEE. It is supported by the IEEE Solid-State Circuits Society and sponsored by Lee Foundation and Agilent Technologies. Since its first founding in 1984, ISIC has become one of the leading international conferences for researchers from universities, research institutes and industries to share the latest progress and development in integrated circuit designs.

An inaugural Chip Design Competition is also held in conjunction with ISIC-2011. The competition aims to promote novel integrated circuits and systems design methodologies to make the chips more energy efficient as well as to circumvent variations in process technology. The competition has received overwhelming response and after reviewing all the submitted entries, 13 finalists have been shortlisted to present their designs before a panel of judges for the final competition on 13 December 2011. With the generous sponsorships from Contact Singapore, Mediatek and Infineon, the top 10 winners will receive very attractive cash prizes.

ISIC-2011 received close to 200 papers from 23 countries, making it a truly international conference. We thank the international technical program committee for their rigorous review of the papers. The accepted quality papers will be organized into 4 special sessions and 26 regular sessions. Besides oral presentations of these research papers, we are honored to have 3 internationally renowned experts to deliver the keynotes. Professor Kaushik Roy from the School of Electrical and Computer Engineering, Purdue University, USA, will talk about *Post-Silicon Technologies for Logic and Memory: Prospects and Perspective;* Professor Simon Wong from the Electrical Engineering Department of Stanford University, USA, will share with us the *Prospect of 3D Integrated Circuits;* and Dr Kiyoo Itoh, an Honorary Fellow of Hitachi Ltd, Japan, will address us from the industrial perspective on *0.5-V High-Speed Circuit Designs for Nanoscale SoCs*. We have also lined up 2 highly sought after tutorials, one on *Error Resilient and Low-Energy Computing* by Professor Kaushik Roy and another on *Advanced ESD Protection Design in CMOS Integrated Circuits* by Professor Ming-Dou Ker from Institute of Electronics, National Chiao-Tung University, Taiwan.

I want to take this opportunity to thank all my colleagues in the Organizing Committee, who have worked extremely hard to prepare a rich and exciting conference programme for the delegates. I must thank all sponsors, distinguished keynote and tutorial speakers for their support to ISIC-2011. I also like to express my appreciation to the ISIC-2011 Secretariat, my technical staff and student helpers for doing an excellent job.

Finally, I wish all delegates a rewarding experience at the Symposium and a pleasant stay in the beautiful garden city of Singapore.

Prof Kye Yak See
General Chair, ISIC 2011 Organizing Committee

Organizers

About ISIC

The International Symposium on Integrated Circuits (ISIC) dates back to 1985 and is now recognized as one of the major conferences in the highly important field of theory, design, and implementation of integrated circuits and systems. The ISIC-2011 will offer a rich program of the highest quality with distinguished invited speakers from all over the world and provide a broad forum of exchanges for researchers and IC designers.

The history of the symposium series is listed below:

- Year 2009:   Singapore   180 delegates
- Year 2007:   Singapore   150 delegates
- Year 2005:   Singapore   130 delegates
- Year 2003:   Singapore   120 delegates
- Year 2001:   Singapore   110 delegates

About the Organizers

 NTU is an internationally reputed research-intensive tertiary institution. Our broad-based education covers science and technology, business and the arts, entrepreneurial and leadership skills to prepare students for the global working world.

 IEEE Singapore Section Society was legally registered with the Singapore Registry of Societies as a trade association in Singapore in 1978 and, at the same time, recognized by the IEEE as a geographical entity in Region 10 named as IEEE Singapore Section.

Organizing Committee

Advisors
Kam Chan Hin, *Nanyang Technological University, Singapore*
Cheng Tee Hiang, *Nanyang Technological University, Singapore*

General Chair
See Kye Yak, *Nanyang Technological University, Singapore*

Vice General Chair
Do Manh Anh, *Nanyang Technological University, Singapore*

Secretary
Lau Kim Teen, *Nanyang Technological University, Singapore*

Technical Program Co-Chair
Gwee Bah Hwee, *Nanyang Technological University, Singapore*
Zheng Yuanjin, *Nanyang Technological University, Singapore*

Poster & Special Session Chair
Joseph Chang, *Nanyang Technological University, Singapore*

Publication & Publicity Co-Chair
Siek Liter, *Nanyang Technological University, Singapore*
Jong Ching Chuen, *Nanyang Technological University, Singapore*

Finance Chair
Yvonne Lam, *Nanyang Technological University, Singapore*

Web & Local Arrangement Chair
Ng Lian Soon, *Nanyang Technological University, Singapore*

Technical Program Committee

Basu Arindam, *Nanyang Technological University, Singapore*
Chang Kok Keong, *Institute of Materials Research & Engineering, A*STAR, Singapore*
Chen Shoushun, *Nanyang Technological University, Singapore*
Chong Kwen Siong, *Temasek Laboratory, Singapore*
Chua Chin Wang, *National Sun Yat-Sen University, Taiwan*
Gao Yuan, *Institute of Microelectronics, A*STAR, Singapore*
Ge Tong, *Temasek Laboratory, Singapore*
He Jin, *Peking University, China*
Howard Luong, *Hong Kong University of Science & Technology, Hong Kong*
Hyunchol Shin, *Kwangwoon University, Korea*
Je Minkyu, *Institute of Microelectronics, A*STAR, Singapore*
Jong Ching Chuen, *Nanyang Technological University, Singapore*
Kim Tae Hyoung, Tony, *Nanyang Technological University, Singapore*
Li Er Ping, *Zhe Jiang University, China*
Li Huiyun, *Chinese Academy of Sciences, China*
Lin Tong, *Nanyang Technological University, Singapore*
Patrick Yue, *UC Santa Barbara, USA*
RanjitGharpurey, *University of Texas at Austin, USA*
Ting Chan Wai, *Temasek Laboratory, Singapore*
Shaojun Wei, *Tsinghua University, China*
Shi Yiqiong, *Temasek Laboratory, Singapore*
Shu Wei, *Temasek Laboratory, Singapore*
Sun PinPing, *IBM Fishkill, USA*
Sungyong Jung, *University of Texas at Arlington, USA*
Tiew Kei Tee, *Institute of Microelectronics, A*STAR, Singapore*
Victor Adrian, *Temasek Laboratory, Singapore*
Yang Xu, *Illinois Institute of Technology, USA*
Zhao Yiqiang, *Tian Jing University, China*

International Advisory Committee

Frank Chang, *University of California at Los Angeles, USA*
Chua-Chin Wang, *National Sun Yat-Sen University, Taiwan*
Bernard Courtois, *Circuits Multi Project (CMP), France*
Morris (Ming-Dou) Ker, *National Chiao-Tung University, Taiwan*
Akira Matsuzawa, *Tokyo Institute of Technology, Japan*
Bram Nauta, *University of Twente, The Netherlands*
Harjani Ramesh, *University of Minnesota, USA*
Ulf Schlichtmann, *Technical University of Munich, Germany*
David Skellern, *National ICT, Australia*
Myung H. Sunwoo, *Ajou University, Korea*
Edgar Sanchez-Sinencio, *Texas A&M University, USA*

Tutorials

Tutorial 1

Professor Kaushik Roy
Purdue University, Indiana, USA
https://engineering.purdue.edu/ECE/People/profile?resource_id=3085

Error Resilient and Low-Energy Computing

The scaling of technology and the limitations of photolithography has led to large variations of device parameters, worse device electrostatics, increased leakage current, power density, and reduced yield. Traditional design techniques usually takes a-worst case design approach. However, such an approach may lead to very high power consumption and over-design in sub 50nm process technology. Hence, there is a need for change in design paradigm to address the above issues. In this tutorial I address various techniques (design time and run-time) to simultaneously improve power consumption and error resiliency under severe parameter variations for both logic and memory. Finally, I will present recent technology developments and approaches to energy scavenging techniques – in particular, I will focus on thermo-electric devices and solar cells.

Tutorial 2

Professor Ming-Dou Ker
IEEE Fellow
Institute of Electronics, National Chiao-Tung University, Hsinchu, Taiwan

Ming-Dou Ker received the Ph.D. degree from the Institute of Electronics, National Chiao-Tung University, Hsinchu, Taiwan, in 1993. He ever worked as the Department Manager with the VLSI Design Division, Computer and Communication Research Laboratories, Industrial Technology Research Institute (ITRI), Hsinchu, Taiwan. Since 2004, he has been a Full Professor with the Department of Electronics Engineering, National Chiao-Tung University, Hsinchu, Taiwan. During 2008 ~ 2011, he was rotated to be Chair Professor and Vice President of I-Shou University, Kaohsiung, Taiwan. Now, he has been the Distinguished Professor in the Department of Electronics Engineering, National Chiao-Tung University, Taiwan. He ever served as the Executive Director of National Science and Technology Program on System-on-Chip (NSoC) in Taiwan during 2010 ~ 2011. Since 2011, he is working as the Executive Director of National Science and Technology Program on Nano Technology (NPNT) in Taiwan. In the technical field of reliability and quality design for microelectronic circuits and systems, he has published over 400 technical papers in international journals and conferences. He has proposed many solutions to improve the reliability of integrated circuits and/or microelectronic systems, which have been granted with 185 U.S. patents and 157 Taiwan patents. He had been invited to teach and/or to consult the reliability and quality design for integrated circuits by hundreds of design houses and semiconductor companies in the worldwide IC industry. His current research interests include reliability and quality design for nanoelectronics and gigascale systems, high-speed and mixed-voltage I/O interface circuits, on-glass circuits for system-on-panel applications, and biomimetic circuits and systems for intelligent prosthesis. Prof. Ker has served as the member of Technical Program Committee and the Session Chair of numerous international conferences for many years. He ever served as the Associate Editor for the IEEE Transactions on VLSI Systems, 2006-2007. He was selected as the Distinguished Lecturer in the IEEE Circuits and Systems Society (2006–2007) and in the IEEE Electron Devices Society (2008-2012). He was the President of Foundation in Taiwan ESD Association. In 2008, he has been an IEEE Fellow "for the contributions to the electrostatic protection in integrated circuits and the performance optimization of VLSI Microsystems". In 2009, he was awarded as one of the top ten Distinguished Inventors in Taiwan.

Advanced ESD Protection Design in CMOS Integrated Circuits

As the increase of applications with more integrated circuits (ICs) products in our life, the reliability of IC products has become one of the most important issues. To reduce the weight of electronic products, to integrate more functions into the electronic products, as well as to reduce the power consumption of electronic products, the CMOS technology has been continually scaled into nanometer scale to realize VLSI/SoC for microelectronic systems. With the transistors in the nano-scale dimension, the gate-oxide thickness of MOSFET is only 10~15Å for operating with sub-1V power supply. Such thinner gate oxide is very easily ruptured by electrostatic discharge (ESD) events, which frequently happen in our environments with the voltage level of hundreds or even thousands volts. The electronics products are typically weaker to sustain such ESD stresses during the assembly, testing, package, and the applications.

To verify the ESD reliability of IC products for safe applications, there are already some industry standards developed, such as Human Body Model (HBM), Machine Model (MM), and Charged Device Model (CDM), to investigate the chip-level ESD robustness of IC products. Typically, for safe production of ICs, the ESD robustness for commercial IC products has been requested to sustain the ESD levels of ±2kV in the HBM ESD test and ±1kV in the CDM ESD test. Besides, in the IEC 61000-4-2 standard of system-level ESD test, the electronic products are zapped by the ESD gun with ESD voltage of up to 15kV (level 4) in the air-discharge mode. ESD protection for CMOS ICs is not only the process issue but also highly dependent to the design issue. How to design the effective ESD protection circuits to meet the ESD specifications is a quite difficult challenge, which has been an important topic that the IC designers need to know.

In this Tutorial, a brief introduction on ESD issue and ESD standards is presented with some failure analysis pictures to demonstrate the impact of ESD on IC products. The basic design concept for on-chip ESD protection will be presented. Some state-of-the-art ESD protection techniques, gate-driven design and substrate-triggered design, will be shown with circuit implementation in silicon. The power-rail ESD clamp circuit plays an important role to effectively improve ESD robustness of IC products. Some new design consideration of power-rail ESD clamp circuit in the nanosclae CMOS processes will be discussed. After the chip-level ESD protection, the system-level ESD protection will be presented. The solutions (board-level and chip-level) to meet the system-level ESD test will be included.

vii

Keynote Lectures

Keynote Lectures 1

Professor Kiyoo Itoh
Honorary Fellow of Hitachi Ltd
http://www.hitachi.com/rd/fellow_itoh.html

0.5-V High-Speed Circuit Designs for Nanoscale SoCs

0.5-V high-speed circuit designs for nanoscale SoCs are described. First, after clarifying the importance of repair techniques and fully-depleted (FD-) MOSFETs to reduce the V_t-variation (DVt) of MOSFETs, a planar ultra-thin BOX FD-MOSFET (SOTB) and FinFETs are compared in terms of DV_t and thus voltage scalability. Second, various dual-V_{DD} (0.5/0.25V) dual-V_t (0.3/0.05V) static logic circuits utilizing the gate-source reverse-biasing scheme are proposed and evaluated with a 25-nm SOTB. It is consequenly found that, for example, the resulting bus driver reduces the power of a conventional 0.5-V high-V_t CMOS driver to about one-fourth with halving the total MOS channel width. Third, 0.5-V circuits and devices for 1-Gb embedded-SRAMs/DRAMs are proposed and evaluated in terms of low-voltage potential. One of highlights here is a boosted word-voltage scheme for the six-transistor SRAM cell to cope with rapidly reduced voltage margin as VDD is reduced. Fourth, the state-of-the-art compensation circuits for process, voltage, and temperature variations are investigated. Finally, it is shown that 0.5-V high-speed 25-nm memory-rich SoCs are possible while reducing the power to one-tenth that of a conventional 1-V 32-nm SoCs, if the above devices and circuits are used and the within-wafer V_t-variation is compensated for.

Keynote Lectures 2

Professor Kaushik Roy
Purdue University, Indiana, USA
https://engineering.purdue.edu/ECE/People/profile?resource_id=3085

Post-Silicon Technologies for Logic and Memory: Prospects and Perspective

Scaling of technology has adverse effects on stability of on-chip memories and leads to increased delay variation in logic. Due to increased leakage current and parameter variations and the need for minimal sized transistors for high density, the standard 6T SRAM cells show high failure rate at low supply voltages. In this talk I will explore different technologies for logic and future on-chip caches. In particular, I will focus on design and optimization of spin-transfer torque magnetic memories (spin as a state variable) which has the possibility of replacing high level on-chip caches/ main memory in future processors. Finally, I will consider the possibilities of logic design using spin as a state variable and compare the pros and cons of spin and charge-based electronics.

Keynote Lectures 3

Professor Simon Wong
IEEE Fellow
http://marco.stanford.edu/swong/

Prospect of 3D Integrated Circuits

Various advancements in 3-dimensional integrated circuits (3D-IC) will be discussed. The performance advantages of 3D-IC will be illustrated with specific examples, including 3D-memories and 3D-FPGA. Through strategic modification of the architectures and circuits to take advantage of 3D, significant improvement in speed and reduction in power consumption can be achieved.

Technical Programme

Chip Design Competition Finalist Presentation 1

A 0.5V 25Mpixels/s SVGA 30fps H.264 Video Decoder Chip 1
Jia-Wei Chen, Pei-Yao Chang, Keng-Jui Chang, Tzu-Yuan Kuo, Wei-Han Hsu, Jinn-Shyan Wang, Cheng-An Chien, Hsiu-Cheng Chang and Jiun-In Guo

A 6.72-GB/S 8PJ/BIT/Iteration IEEE 802.15.3C LDPC Decoder Chip 7
Zhixiang Chen, Xiao Peng, Xiongxin Zhao, Qian Xie, Leona Okamura, Dajiang Zhou and Satoshi Goto

Using Built-In Fine Resolution Clipping Technique for High-Speed Testing by Using Low-Speed Wireless Tester 13
Ching-Hwa Cheng

A 0.35 V, 100 MHz, 0.19 µW/MHz, 3-Locking-Cycle All Digital Delay Locked Loop with Asynchronous-Deskewing Technology in 55 nm CMOS Technology 19
Chun-Yuan Cheng, Jinn-Shyan Wang and Cheng-Tai Yeh

A Wideband Fully Integrated +30dBm Class-D Outphasing RF PA in 65nm CMOS 25
Jonas Fritzin, Christer Svensson and Atila Alvandpour

A 77-135GHz Down-Conversion IQ Mixer for 10Gbps Multiband Applications 29
Sanming Hu, Yong-Zhong Xiong, Lei Wang, Jinglin Shi and Teck-Guan Lim

Analog IC (1)

Supply Voltage and Temperature Insensitive Current Reference for the 4 MHz Oscillator 35
Chi-Hsiung Wang, Cheng-Feng Lin, Wei-Bin Yang and Yu-Lung Lo

DDCCs based Voltage-Mode One Input Five outputs Biquadratic Filter with High Input Impedance 39
Wei-Yuan Chiu, Jiun-Wei Horng, Yi-Sing Guo and Ching-Yao Tseng

Design of High-Speed Laser Diode Driver with Fast Switching and Reaction Time 43
Heng-Shou Hsu, Don-Gey Liu and Shih-Chi Liu

High-Speed Laser Diode Driver with Low Sensitivity to Process Variation and Improvement on Overshoot Performance 47
Heng-Shou Hsu, Don-Gey Liu and Chia-Ming Chuang

Digital IC (1)

Image Classifying Algorithm and its VLSI Implementation based on The Directional Features 51
Dongfang Wang, Ningmei Yu, Yvonne Lam Y.H. and Yuanjin Zheng

A High-performance Configurable VLSI Architecture for Integer Motion Estimation in H.264 55
Ningmei Yu, Wenhua Jia, Meihua Gu, Dongfang Wang, Gang Xi and Yuanjin Zheng

ROM-less DDFS using Non-equal Division Parabolic Polynomial Interpolation Method 59
Chia-Hao Hsu, Yun-Chi Chen and Chua-Chin Wang

Accelerated Evaluation Method for the SRAM Cell Write Margin using Word Line Voltage Shift 63
Hiroshi Makino, Shunji Nakata, Hirotsugu Suzuki, Hiroki Morimura, Shin'ichiro Mutoh, Masayuki Miyama, Tsutomu Yoshimura, Shuhei Iwade and Yoshio Matsuda

Reconfigurable Back Propagation based Neural Network Architecture 67
Gin-Der Wu, Zhen-Wei Zhu and Bo-Wei Lin

Memory-Bank based Radix-2^2 Fast Fourier Transform 71
Gin-Der Wu, Zhen-Wei Zhu and Hung-Yi Chang

RF/MM-Wave IC (1)

CMOS Ku-Band LNB with High Image Suppression Capability for Satellite Application 75
Lin Jia and M. Annamalai Arasu

A 0.68-1.65GHz CMOS LC Voltage-Controlled Oscillator with Small VCO-Gain and Step Variation 79
Liheng Lou, Lingling Sun, Haijun Gao and Jincai Wen

Low Power *S*-band Receiver using GaAs pHEMT Technology 83
Yangyang Peng, Xiaoying Wang, Fangyue Ma and Wenquan Sui

Design of an X-band Board-band Lumped-element Quadrature Hybird 87
Fangyue Ma, Yangyang Peng, Xiaoying Wang, Saier Liu, Jin Lan and Wenquan Sui

A Ka-band MMIC Doherty Power Amplifier using GaAs pHEMT Technology 91
Xiaoying Wang, Yangyang Peng, Fangyue Ma and Wenquan Sui

Special Session: News from Munich - Dealing with Variations and Aging; MPSoC Thread Assignment and FPGA Signal Processing

An Energy-Efficient Supply Voltage Scheme using In-Situ Pre-Error Detection for on-the-fly Voltage Adaptation to PVT Variations 94
Martin Wirnshofer, Leonhard Heiß, Georg Georgakos and Doris Schmitt-Landsiedel

Variability-Aware Automated Sizing of Analog Circuits Considering Discrete Design Parameters 98
Michael Pehl, Michael Zwerger and Helmut Graeb

Program-Aware Circuit Level Timing Analysis 102
Veit B. Kleeberger, Sebastian Kiesel, Ulf Schlichtmann and Samarjit Chakraborty

Hardware Assisted Thread Assignment for RISC based MPSoCs in Invasive Computing 106
Ravi Kumar Pujari, Thomas Wild, Andreas Herkersdorf, Benjamin Vogel and Jörg Henkel

Real-Time Signal Processing on Low-Cost-FPGAs using Dynamic Partial Reconfiguration 110
Michael Feilen, Matthias Ihmig, Anton Zahlheimer and Walter Stechele

Chip Design Competition Presentation 2

Versatile Ultra Low Noise Low Power Analog Signal Conditioning Chip With Integrated Drivers　114
Sanjay Joshi, Viral Thaker and Maryam Shojaei Baghini

A Novel Digital PLL With Good Performance and Very Small Area　118
Luo Zhihong, Au Yeung On, Benjamin Lau and Henry Law

A 900MHz RFID Reader Chip with RC Calibration　124
Mou Shouxian, Ma Kaixue and Yeo Kiat Seng

Energy Efficient Integrated Gas Sensor System with Post CMOS Functionalization　130
Pramod M

Analog IC (2)

Micro Energy Management for Energy Harvesting at Maximum Power Point　136
Sewan Heo, Yil Suk Yang, Jaewoo Lee, Sang-kyun Lee and Jongdae Kim

Low-Power, Low-offset Stacked Analog Latch using an Offset Cancellation Technique　140
Minehiko Tateno, Hiroki Date and Kenichi Ohhata

Wireless Powering and Bidirectional Telemetry Front-End for Implantable Biomedical Devices　144
Rui-Feng Xue, Hyouk-Kyu Cha, Jia Hao Cheong, Pradeep Basappa K, Minkyu Je and Yuanjin Zheng

Multiple Output Switched Capacitor DC-DC Converter for Low Power Applications　148
Ravinder Pal Singh and Minkyu Je

High Voltage Electrostatic Driving of MEMS Micromirrors　152
Ravinder Pal Singh, Tal Langer and Minkyu Je

Visualization of Intrinsic Harmonic Distortion for Closed-loop Class D Amplifier　156
Jun Yu, Wang Ling Goh and Meng Tong Tan

Digital IC (2)

A Voltage Management Technique for Low-Power Domino Circuits　160
Ching-Hwa Cheng

A 30K 2.5Gb/s Decision-Eased Soft RS (224, 216) Decoder for Wireless Systems　164
Yi-Min Lin, Yu-Chun Huang, Chi-Heng Yang and Hsie-Chia Chang

A 0.16nJ/bit/iteration 3.38mm 2 Turbo Decoder Chip for WiMAX/LTE Standards　168
Cheng-Hung Lin, Chun-Yu Chen, En-Jui Chang and An-Yeu (Andy) Wu

Comparative Design of Floating-Point Arithmetic Units Using the Balsa Synthesis System　172
Ren-Der Chen, Yu-Cheng Chou and Wan-Chen Liu

Investigating the FIFO Design Styles Based on the Balsa Synthesis System　176
Ren-Der Chen, Che-An Lee and Pei-Hua Hsieh

Digital System for Low Power Wireless Neural Recording System   180
Peng Li, Xin Liu, Bin Zhao and Minkyu Je

RF/MM-Wave IC (2)

A Wideband 0.6dB Insertion Loss +20.5dBm P1dB CMOS T/R Switch   184
Xuesong Chen and M. Kumarasamy Raja

60-GHz SP4T Switch with ESD Protection   188
Jin He, Yong-Zhong Xiong and Yue Ping Zhang

A Dual-Band LC Voltage-Controlled Oscillator in 0.13μm CMOS Technology   192
Siti Maisurah M. H., Nazif Emran F., Norman Fadhil Idham M. and A. I. Abdul Rahim

A 5/6-bit Multi-Modulus Frequency Divider in 0.13μm CMOS Technology   196
Siti Maisurah M. H., Nazif Emran F., Norman Fadhil Idham M. and A. I. Abdul Rahim

Optimizing Gain of 5 GHz RF Amplifier keeping Minimum Deviation in Center Frequency and Noise   200
Figure
Jai Narayan Tripathi, Prakash R. Apte and Jayanta Mukherjee

A 5mA 2.4GHz 2-Point Modulator with QVCO for Zigbee Tranceiver in 0.18-μm CMOS   204
Dan Lei Yan, Bin Zhao, M. Kumarasamy Raja and Yuan Xiaojun

Special Session: CMOS Circuit Techniques for Flexible Wideband Radio Front-Ends

A 2GHz Tx LO Generation Circuit with Active PPF and 3/2 Divider in 65nm CMOS   208
Markus Törmänen, Andreas Axholt, Jonas Lindstrand and Henrik Sjöland

RF Receiver System for Cognitive Radio Application   212
Yuanzhong Xue, Yuanjin Zheng and Yeo Kiat Seng

A Divide-by-Two Injection-Locked Frequency Divider with 13-GHz Locking Range in 0.18-μm CMOS   216
Technology
Xiang Yi, Chirn Chye Boon, Jia Fu Lin, Manh Anh Do, Kiat Seng Yeo and Wei Meng Lim

Wideband RF Frontend Design for Flexible Radio Receiver   220
Fahad Qazi, Quoc-Tai Duong and Jerzy J. Dabrowski

A Wideband Fully Integrated +30dBm Class-D Outphasing RF PA in 65nm CMOS   N/A
Jonas Fritzin, Christer Svensson and Atila Alvandpour

Chip Design Competition Presentation 3 & Analog IC (4)

Ultra-Low-Power Wireless Implantable Blood Flow Sensing Microsystem for Vascular Graft Applications   224
Rui-Feng Xue, Jia Hao Cheong, Hyouk-Kyu Cha, Xin Liu, Peng Li, Huey Jen Lim, Li Shiah Lim, Ming-Yuan Cheng, Cairan He, Woo-Tae Park and Minkyu Je

An 80-dB SNR 4th-Order Discrete-Time Sigma-Delta Modulator   230
Chan-Keun Kwon, Chan-Hui Jeong, Young-Jae Min, Young-Mok Jung and Soo-Won Kim

A CMOS Single Stage Fully Differential Folded Cascode Amplifier Employing Gain Boosting Technique 234
S. A. Enche Ab Rahim and I. M. Azmi

A Novel Ripple Controlled Modulation for High Efficiency DC-DC Converters 238
Zhuochao Sun and Liter Siek

Gm-Enhanced Differential Colpitts VCO 242
Xue-Fei Xiao, Wang Ling Goh, Minkyu Je and Jae-Hong Chang

32kHz MEMS-Based Oscillator for Implantable Medical Devices 246
Jae-Hong Chang, ShengXi Diao, Raja Muthusamy Kumarasamy and Minkyu Je

Analog IC (3)

Pseudo Differential Operational Transconductance Amplifier using Common Mode Feed Forward and HD3 250
Feed Forward
Yen-Shuo Chang, Hong-Chong Wu, Miin-Shyue Shiau, Don-Gey Liu and Heng-shou Hsu

High Gain, High Speed OTA for S/H Circuit in 14-b 100-MS/s Pipeline ADC 254
Shuai Chen, Lenian He and Lu Zhang

A 10 Gb/s 4-PAM Transceiver with Adaptive Pre-Emphasis 258
Sungmin Yoo, Daeho Yun, Bongsub Song, Jinwook Burm, Jinil Chung and Jun Hyun Chun

Processing *N*-ary Trees in Hardware Circuits 262
Valery Sklyarov, Iouliia Skliarova, Dmitri Mihhailov and Alexander Sudnitson

Simulation Environment for Visual Prototyping of Circuits and Systems 266
Iouliia Skliarova and Valery Sklyarov

A Compact Millimeter-Wave CMOS Bandpass Filter Using a Dual-Mode Ring Resonator 270
Sha Luo, Aaron V. Do, Chirn Chye Boon, Lei Zhu and Manh Anh Do

Computer-Aided Design, Logic and System Synthesis

Transfer Function Analysis for Model Topology Determination of On-Chip Transmission Lines 273
Huang Wang, Lingling Sun, Jun Liu, Jincai Wen and Zhiping Yu

Design Optimization of MOS Operational Amplifiers using Finite Difference Sensitivity 277
Binbin Weng and Guoyong Shi

Automated Synthesis Design Flow of Power Converter Circuits Aimed at SOC Applications 281
Hsin-Yu Luo, Hsiu-Wen Li, Long-Ching Yeh and Chien-Nan Jimmy Liu

A Resource Binding Technique for TSV Number Minimization in High-Level Synthesis of 3D ICs 285
Wei-Kai Cheng and Yi-Chun Yen

Low Power Digital Type ADC 289
Richard Wee Tar Ng and Liter Siek

System-on-Chip (SoC)

Ultra Low Power SOC for Portable Health Monitoring Platforms 293
Richard Wee Tar Ng, Attard Laurent and Sie Boo Chiang

A Scalable Strategy for Runtime Resource Management on NoC based Manycore Systems 297
Xiongfei Liao and Thambipillai Srikanthan

Hybrid Non-preemptive/Cooperative Multitasking on NoC Based Manycore Systems 301
Xiongfei Liao, Thambipillai Srikanthan and Xiao He

An Autonomous Vehicle Using a Multi-Thread and Event-Driven Processor 305
Touta Hayashi and Kenji Ohmori

A Dynamic Comparator with Analog Offset Calibration for Biomedical SAR ADC Applications 309
M. M. J. Herath and P. K. Chan

Testing and Yield Enhancement

Automated Wafer Defect Map Generation for Process Yield Improvement 313
Cher Ming Tan and Kheng Tuan Lau

Delay Defect Diagnosis Methodology using Path Delay Measurements 317
Eun Jung Jang, Jaeyong Chung and Jacob A. Abraham

A Compact Model of AlGaN/GaN on Silicon Schottky Diode and its Application 321
Yihu Li, Lei Wang, S. Arulkumaran, Yong-Zhong Xiong, Geok Ing Ng, Wang Ling Goh, Shane Todd and Patrick Lo

Prototyping A Bidirectional Processor Design Based on Reversible Principles 325
Dilip Vasudevan, Michel Schellekens, Nasim Zeinolabedini and Emanuel Popovici

Efficient Pipelined VLSI Architectre with Dual Scanning Method for 2-D Lifting-Based Discrete Wavelet Transform 329
Anand Darji , S. N. Merchant and A. N. Chandorkar

Mixed-Signal IC (1)

A Fast-locking Clock and Data Recovery Circuit with A Lock Detector Loop 332
Chih-Lin Chen, Chua-Chin Wang and Chun-Ying Juan

A 5-bit 500-MS/s Time-Domain Flash ADC in 0.18-µm CMOS 336
Young-Jae Min, Ammar Abdullah, Hoon-Ki Kim and Soo-Won Kim

Parallel Background Calibration with Signal-Shifted Correlation for Pipelined ADC 340
Kexu Sun, Xuan Wang and Lenian He

A 3 bit 36 GS/s Flash ADC in 65 nm Low Power CMOS Technology 344
Damir Ferenci, Simon Mauch, Markus Grözing, Felix Lang and Manfred Berroth

Adaptive Spread Spectrum Clock Tracking for Interpolator-based Clock and Data Recovery   348
Chuan-Thim Khor and Alan Chai

Analog IC (5)

Temperature Behavior Mismatch of Halo Implanted Short Channel Transistors and its Influence on PUF   352
Circuits
Maximilian Hofer, Christoph Böhm and Wolfgang Pribyl

Low-Power Wireless Receivers for Healthcare Applications   356
Alper Cabuk, Yuan Gao, Shengxi Diao, Yuanjin Zheng, Minkyu Je and Chun Huat Heng

Versatile MIMO Voltage-Mode OTA-C Universal Biquadratic Filter   360
Montree Kumngern

New Current-Mode First-Order Allpass Filter using a Single CCCDTA   364
Montree Kumngern

A 15nV/√Hz Noise 0.2μV Offset Chopper Conditioning Amplifier for Monolithic Infrared Sensing Systems   368
Juanda, Wei Shu, Joseph Chang and Wenfeng Yu

A CMOS Circuit Design of a Loss of Signal and the Application in Optical Receivers   372
Feiyan Qin, Guoqing Xu and Huiyun Li

Digital IC (3)

A Power-Efficient Integrated Input/Output Completion Detection Circuit for Asynchronous-Logic Quasi-   376
Delay-Insensitive Pre-Charged Half-Buffer
Weng-Geng Ho, Kwen-Siong Chong, Bah-Hwee Gwee, Joseph. S. Chang and Ming-Fatt Yee

A Low-Cost and High-Throughput Architecture for H.264/AVC Integer Transform by Using Four   380
Computation Streams
Yuan-Ho Chen, Tsin-Yuan Chang and Chih-Wen Lu

Design of Support Vector Machine Circuit for Real-time Classification   384
Soojin Kim, Seonyoung Lee, Kyoungwon Min and Kyeongsoon Cho

An All-Digital DLL with Dual-Loop Control for Multiphase Clock Generator   388
Yu-Lung Lo, Pei-Yuan Chou, Hsiang-Hui Cheng, Shu-Fen Tsai and Wei-Bin Yang

Design and Fabrication of Configurable Digital Controller Interface for Micro Mirror Projector ASIC   392
Jianwen Luo, Peng Li, Chin Yann Pang, Pradeep Kumar Gopalakrishnan, Tal Langer and Minkyu Je

A Low Power JPEG Image Compression IC for Wireless Ingestible Endoscopy   396
Wei-Da Toh, Bin Zhao, Yuan Gao, Yuanjin Zheng, Minkyu Je and Chun-Huat Heng

Sensor Systems

Design and Implementation of a Bio Sensor Array (BSA) for Cancer Cell Detection   400
Lim Lay Keng, Antoine Jalabert and Roshan Weerasekera

Closed Loop Wireless Power Transmission for Implantable Medical Devices 404
Luis Andia, Rui-Feng Xue, Kuang-Wei Cheng and Minkyu Je

High-voltage Pulser for Ultrasound Medical Imaging Applications 408
Dongning Zhao, Meng Tong Tan, Hyouk-Kyu Cha, Jinli Qu, Yan Mei, Hao Yu, Arindam Basu and Minkyu Je

Design of a Radiation Tolerant CMOS Image Sensor 412
Xinyuan Qian, Hang Yu, Bo Zhao, Shoushun Chen and Kay Soon Low

A Review of CMOS Multimodal Neuromonitoring Sensors and Systems 416
Wai Pan Chan and Minkyu Je

On-Chip RF Energy Harvesting Circuit for Image Sensor 420
Jun Wu Zhang, Xiang Yu Zhang, Zhuang Liang Chen, Kye Yak See, Cher Ming Tan and Shou Shun Chen

Special Session: Silicon Based High Performance ICs at RF and Millimeter Wave Range-Part 1

Scalable Modeling Based on Fill Ratio for Planar Spiral Inductors 424
Lin Zhong, Lingling Sun, Jun Liu and Huang Wang

Model of On-Chip VGP-CPW with P+ Implant in CMOS Process 428
Jincai Wen, Jia Lou and Lingling Sun

A Novel Accurate dB-Linear Control Circuit Topology for Variable Gain Amplifiers in BiCMOS 432
Technology
Zhenghao Lu, C. H. Hu, X. P. Yu, W. M. Lim, Y. Liu and K. S. Yeo

A 6-GHz dual-modulus prescaler using 180nm SiGe technology 436
C. Z. Nan, X. P. Yu, B. Y. Hu, Z. H. Lu, W. M. Lim, Y. Liu, K. S. Yeo and ChangHui Hu

Self-demodulated Receiver at MM-wave Range using SiGe Technology 440
X. P. Yu, B. Y Hu, X. L. Yan, Z. H. Lu, W. M. Lim, Y. Liu, K. S. Yeo and C. H. Hu

Mixed-Signal IC (2)

Programmable Low-Dithering-Jitter Interpolator-based CDR 444
Lip-Kai Soh, Wai-Tat Wong, Swee-Wah Lee and Chuan-Thim Khor

A Power Efficient ΣΔ Modulator Based on CBSC IIR Filter in 0.18µm CMOS 448
Mehdi Taghizadeh, Majid Zamani, Payman Goodarzi and Ammar Rahimi Kazerooni

Low-Power Design Techniques with Process Tagging and Dynamic Power Management 452
Daniel Cooley, Yuwono Rahman, Jin Ruan, Xun Yu, Lei Chen and Jianyuan Deng

CMOS Based 16-Channel Neural/Muscular Stimulation System with Arbitrary Waveform and Active 456
Charge Balancing Circuit
Lei Yao and Minkyu Je

Physical Design Exploration of 3DIC Wireless Transceiver using Through-Si-Vias   460
Mini Jayakrishnan, Xin Liu, Hong Yu Li, Jingjing Lan and Wang Ling Goh

A PLL with a VCO of Improved PVT Tolerance   464
Kok-Foong Chong, Liter Siek and Benjamin Lau

Analog IC (6)

The Phase Locked Loop for MEMS Horizontal Scanning Control of Micro-Laser Projection ASIC   468
Dan Lei Yan, Luo Jian Wen, Li Peng, Ravinder Pal Singh, Duy-Dong Pham, Tal Langer and Minkyu Je

A Flipped Voltage Follower based Low-Dropout Regulator with Composite Power Transistor   **472**
S. S. Chong and P. K. Chan

20 MHz Accurate Peak Detector for FPW Allergy Biosensor With Digital Calibration   476
Tzung-Je Lee, Wei-Chih Hsiao and Chua-Chin Wang

2.45 GHz ZigBee Receiver Frontend for HAN With Smart Meter   480
Tzung-Je Lee, Wayne Luo, Shang-Hsien Yang, Ming-Hung Shih, Ko-Chi Kuo and Chua-Chin Wang

High Frequency Tow-Thomas Tunable Filter using OTA based Voltage Op-Amp   484
Walid Zemouri, Eman A. Soliman and Soliman A. Mahmoud

Cascaded Third-Order Tunable Low-Pass Filter using Low Voltage Low Power OTA   488
Sondos H. Ismail, Eman A. Soliman and Soliman A. Mahmoud

Reconfigurable Systems (1)

Modularized Development Platform for Hardware/Software Design   492
Kai-Chao Yang, Yu-Tsang Chang, Chien-Ming Wu and Chun-Ming Huang

An Improved Dynamic-Biased CMOS Operational Amplifier for Biomedical Circuit Applications   496
H. L. Tan, G. T. Ong and P. K. Chan

An Angina Diagnosing System using Fuzzy Clustering and Correlation in FPGA   **500**
Evaldo R. F. Cintra, Tales C. Pimenta and Robson L. Moreno

Secret Sharing based Countermeasure for AES S-Box   504
Yi Wang, Zheng Yuan, Zhican Li and Renfa Li

FPGA based Optimized SHA-3 Finalist in Reconfigurable Hardware   508
Qian Song, Yi Wang, Zhican Li, Quan Zhou, Wufei Wu, Demin Han, Wenlong Xu, Zuo Chen and Renfa Li

A Code Reuse Method for Many-Core Coarse-Grained Reconfigurable Architecture Function Library   512
Development
Shuo Li, Guo Chen and Ahmed Hemani

Semiconductor Devices, Fabrication and Assembly

Charge Collection Probability: Normal-Collector Configuration   516
Chee Chin Tan, Vincent K. S. Ong and K. Radhakrishnan

A Study of the Effect of Shallow Trench Isolation Technology on MOSFET DC Characteristic   520
Xia Fang, Lingling Sun, Jun Liu and Huang Wang

Low Frequency Noise Investigation of AlGaN/GaNOn Silicon Schottky Diode   524
Yihu Li, Yong-Zhong Xiong, S. Arulkumaran, Lei Wang, Wang Ling Goh, Geok Ing Ng, Shane Todd and Patrick Lo

Optimization of Vertical Silicon Nanowire based Solar Cell using 3D TCAD Simulation   528
Jitendra Kumar, S. K. Manhas, Dharmendra Singh and Ramesh Vaddi

Amorphous Carbon Step Coverage Improvement applied to Advanced Hard Mask for Lithographic Application   532
Zitu-Tin Lin, Chun-Chi Chen, Hung-Ju Chien and Hiroshi Matsuo

Special Session: Silicon Based High Performance ICs at RF and Millimeter Wave Range-Part 2

Wideband Receiver for Software Defined Radio in GHz range using Standard 40nm CMOS technology   535
F. Yang, Y. Liu, X. L. Zhang, Z. H. Lu, X. P. Yu, W. M. Lim and C. H. Hu

A $\Delta\Sigma$ Fractional-N PLL with Fast Auto-Frequency Calibration for CMMB Tuners   539
Jing Jin, Xiaoming Liu, Peng Qin and Jianjun Zhou

A 1mW 5GHz Current Reuse CMOS VCO with Low Phase Noise and Balanced Differential Outputs   543
Wenrong Ying, Peng Qin, Jing Jin and Tingting Mo

A Curvature Compensated Bandgap Reference with low Drift and low Noise   547
Junmin Jiang, Zhihua Ning and Lenian He

GSM/EDGE Power Amplifier Module with Improved Low-Power Efficiency   551
Jinbo Li, Tingting Mo and Feng Xu

Analog IC (8)

Comparative Study and Analysis of Noise Reduction Techniques for Front-End Amplifiers   555
Lei Liu, Xiaodan Zou, Wang Ling Goh and Minkyu Je

Temperature Insensitive Current Reference for the 6.27 MHz Oscillator   559
Wei-Bin Yang, Zheng-Yi Huang, Ching-Tsan Cheng and Yu-Lung Lo

A Novel Mode Switching Power Management System IC Design for Implantable Biomedical Instrumentations   563
Arnold C. Paglinawan, Charmaine C. Paglinawan, Glenn O. Avendaño, Ying-Hsiang Wang and Wen-Yaw Chung

Analog IC (7)

A 19-nW Sub-Bandgap Reference with 15ppm/°C Temperature Coefficient   567
Jia Hao Cheong and Minkyu Je

Double Regulated Voltage Supply for High Precision MEMS Accelerometers   571
Huey Jen Lim, Ravinder Pal Singh, Kevin Chai Tshun Chuan, David Nuttman and Minkyu Je

Design and Optimzation of High Precision CMOS Voltage Reference Using Taguchi Orthogonal Array   575
Technique
Hande Vinayak, Maryam Shojaei Baghini and Prakash Apte

Reconfigurable Systems (2)

A 2.72GOPS/11mW Low Power Reconfigurable Accelerator with a Highly Parallel Datapath Consisting of   579
Combinatorial Circuits in 65nm CMOS
N. Ozaki, Y. Yasuda, Y. Saito, D. Ikebuchi, M. Kimura, H. Amano, H. Nakamura, K. Usami, M. Namiki and M. Kondo

A New Source of Secure Pseudorandom Numbers Exploiting IMCGs Implemented in an FPGA   585
Mieczyslaw Jessa and Michal Jaworski

A Reed-Solomon Architecture for Soft-Core Implementation   589
Thullyo D. C. R. Ferreira, Luís H. C. Ferreira, Robson L. Moreno and Tales C. Pimenta

Simulation, Verification and Testability (1)

Debugging Methodology for A Synthesizable Testbench FPGA Emulator   593
A. W. Ruan, H. C. Huang, C. Q. Li, Z. J. Song, Y. B. Liao and W. Tang

A Novel Methodology for Hardware Acceleration and Emulation   597
Y. B. Liao, C. Q. Li, H. C. Huang, C. Y. Xiang, A. W. Ruan and W. Tang

Ultra-low Power High Efficient Rectifiers with 3T/4T Double-gate MOSFETs for RFID Applications   601
Ramesh Vaddi and Tony T. Kim

Simulation, Verification and Testability (2)

Low-Power 4-Bit Flash ADC For Digitally Controlled DC-DC Converter   605
Guolei Yu and Liter Siek

Synthesizable Verification IP to Stress Test System- On-Chip Emulation and Prototyping Platforms   609
Subramanian Shiva Shankar and Jayaratnam Siva Shankar

1-Bit Heuristic Adaptive Quantizer (HAQ) for on Chip Image Compression in CMOS Image Sensors   613
Michael Barrow, Amine Bermak and Shoushun Chen

0.5-V High-Speed Circuit Designs for Nanoscale SoCs –Challenges and Solutions-   **617**
Kiyoo Itoh, Akira Kotabe, Dai Hisamoto, Ryuta Tsuchiya, Riichiro Takemura

A 0.5V 25Mpixels/s SVGA 30fps H.264 Video Decoder Chip

Jia-Wei Chen, Pei-Yao Chang, Keng-Jui Chang, Tzu-Yuan Kuo, Wei-Han Hsu, and Jinn-Shyan Wang
Department of Electrical Engineering
National Chung Cheng University
Chiayi, Taiwan
92jiawei@vlsi.ee.ccu.edu.tw

Cheng-An Chien, Hsiu-Cheng Chang, and Jiun-In Guo
Department of Computer Science and Information Engineering
National Chung Cheng University
Chiayi, Taiwan
jiguo@cs.ccu.edu.tw

Abstract—**A sub-threshold voltage, high throughput H.264 video decoder design is proposed for portable applications in this paper. To improve the performance for high throughput rate applications, the computational complexity in H.264 video decoding is optimized in the proposed design. To reduce complexity, both a shared adder-based hardware sharing scheme and an advanced data management scheme are presented. Moreover, to reduce power consumption, both a low area sub-threshold voltage SRAM and a high performance sub-threshold voltage CMOS circuit design scheme are presented. Exploiting all the design techniques, the proposed 90nm 0.5V 25Mpixels/s SVGA 30fps H.264 decoder outperforms the 65nm design at 0.5V through a 31x improvement in throughput.**

Keywords-Video codecs, H.264/AVC, CMOS digital integratd circuits, Sub-threshold circuit design, Low-power electronics, SRAM chip.

I. INTRODUCTION

H.264/AVC video coding standard [1] is an important evolution that Scalable Video Coding (SVC) and Multi-view video coding standard inherit all coding tools to make them processing higher compression ratio. The H.264/AVC adopts new coding tools to improve its data compression efficiency. In addition to the common coding tools like Context Adaptive Variable Length Coding (CAVLC), Variable Block-Size Motion Estimation (VBSME) and Motion Compensation (MC), Integer Transform (IT), and In-Loop de-blocking Filter (ILF), it contains the Main Profile (MP) coding tools like Context Adaptive Binary Arithmetic Coding (CABAC), Picture Adaptive Frame Field (PAFF) coding, and MacroBlock Adaptive Frame Field (MBAFF) coding as well as the High Profile (HP) coding tools like 8x8 transform. However, high coding efficiency of H.264/AVC induces high computational complexity as well, which is more than 4 times complexity as compared to that of MPEG-2 [2] in decoders. Moreover, due to the advanced prediction techniques adopted in the coding tools of H.264/AVC, more and more prediction data need to be stored in internal SRAM or transferred to off-chip memory during video coding. These facts induce high complexity and large internal memory in H.264/AVC, which tend to consume the majority of power consumption in portable devices.

Voltage scaling is an effective technique that reduces power consumption by a quadratic factor. When the requirement of supply voltage for a design is scaled down to 0.5V and below, the solar cells [3, 4] have better translation efficiency to generate the energy. Thus, aggressive voltage scaling to 0.5V and below is beneficial to using solar cells as the power supply on portable devices.

However, the voltage scaling causes the problem of decreasing the driving capability of transistors. Decreasing the circuit speed is a severe challenge for real-time video applications like the designs [5, 6] needed to be operated at frequencies in the range of tens of MHz. Thus, the challenge in the proposed design is to enable the design operating at low voltage with high processing performance. In addition, for low voltage (LV) video designs, the design of embedded SRAM modules is another challenge due to the increased sensitivity to variations.

To overcome above problems, we propose a 0.5V 25Mpixels/s SVGA 30fps H.264 MP/HP video decoder in 90nm CMOS for portable multimedia by reducing computational complexity, reducing internal SRAM size, and encapsulating both low area sub-threshold voltage SRAM and the in-house LV standard cell logics with high performance sub-threshold voltage CMOS circuits. Compared to the designs [7, 8], the proposed design owns 4x and 100x improvement in workable frequency, respectively, which implies a 31x improvement in throughput compared to the design [8] in decoding QCIF 17fps H.264 video at 0.5V.

The rest of this paper is organized as follows. Section II first describes the architecture of the proposed sub-threshold voltage high throughput H.264 MP/HP decoder and then illustrates the presented techniques to reduce the complexity in the proposed design. Section III presents a sub-threshold voltage circuit design technique to improve the circuit processing performance. A low area sub-threshold voltage SRAM is proposed in Section IV. Section V shows the implementation results of the proposed design and performance comparison of the proposed design with others. Finally, a simple conclusion is given in Section VI.

978-1-61284-863-1/11 $26.00 © 2011 IEEE

II. ARCHITECTURE DESIGN FOR THE PROPOSED H.264 VIDEO DECODER

Fig. 1 shows the block diagram of the proposed LV H.264 video decoder connected to a RISC processor and an external memory through 32-bit and 64-bit buses, respectively. It contains three domains for low power management, including level converters in the first domain, the proposed Adaptive Power Controller (APC)/Power switches (PSW)/Adaptive Body Bias (ABB) circuits in the second domain, and the proposed LV H.264 decoding core in the third domain. Three power saving modes (i.e., active, idle, and sleep modes) are designed in the proposed LV decoding core containing high throughput Bit-Stream Decoder (BSD), and Texture Decoder (TD) both with zero-block skipping, as well as 4-pixel parallel Predicted Pixel Compensator (PPC) and In-Loop de-blocking Filter (ILF). In addition, we optimize the functional blocks through both reducing algorithmic complexity and sharing data-path hardware to carry out computation for supporting H.264 HP coding tools. Moreover, based on the proposed sub-threshold voltage high throughput CMOS circuits, the APC configures the proposed LV decoding core to achieve high throughput in active mode and reduce power consumption in idle/sleep modes.

In the following, we will first describe the encapsulated design techniques to reduce the computational complexity of the proposed design.

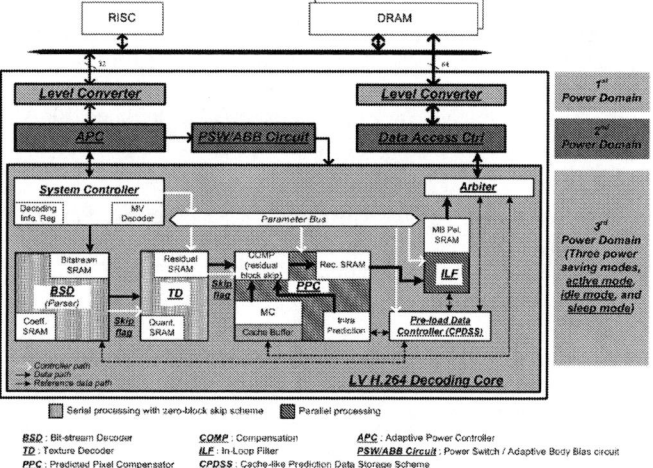

Figure 1. Block diagram of the proposed low voltage (LV) H.264 video decoder

A. Low Cost Adder-based Sharing Technique

For reducing the complexity, we propose a low cost adder-based sharing filter structure to realize these filter-like operations efficiently by calculating the common intermediate results first without using multipliers and then sharing them in computing the final filter results. The similar concept is also adopted in the design [9] to realize both the 4x4 transform and 4x4 ILF operations. To further reduce the complexity for H.264 HP video decoder, the common terms in realizing both the 4x4-block filters and 8x8-block filters are generated and then some adder trees are used to compute the output values. To give an example in transform operations, the architecture of

1-D inverse transform is shown in Fig. 2. One shifter-adder filter is used to realize the even-point transform for both the 4x4 and 8x8 blocks, and then one adder-based filter with post-processing unit is used to realize the odd-point transform for 8x8 blocks, in which the even-point 8x8 IDCT and 4x4 IDCT could share the same architecture to generate the transform outputs. In addition to transform filter, the operation in intra prediction, inter prediction, and in-loop filter are also realized by the design concept of shared adder-based technique for low hardware cost. Therefore, a single 4x4 block filter is designed to support the filtering operations of both 4x4-block and 8x8-block of luminance pixels. Using the adder-based sharing filter structure contributes 35% of area reduction in average for both supporting the 4x4/8x8-block filter in transform, intra prediction filter, MC six-tape filter, and ILF.

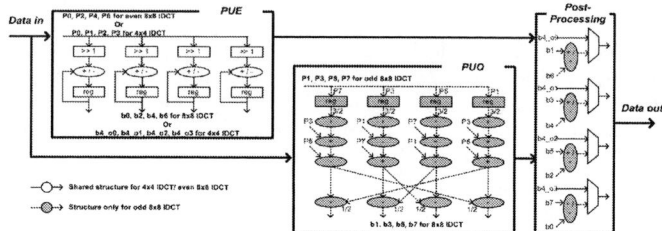

Figure 2. Low cost adder-based sharing structure for different filter-like operations (1-D inverse transform)

B. Low Complexity Techniques for Realizing MBAFF Coding

To reduce the control complexity, we adopt a *Frame-based Prediction Data Management* that all the neighboring prediction data are stored in a frame-based organization to simplify the 36 cases of prediction direction into two cases and reduce a half of internal memory requirement as well. To further reduce the internal memory size, we implement a *Cache-like Prediction Data Storage Scheme* in the pre-load data controller unit to access the required data efficiently between the proposed design and the off-chip memory, as shown in Fig. 3. The 3 MB-pair block (i.e., 1.5kB) of prediction data cache stores the rearranged prediction data for different frame/field prediction modes, which reduces 98% of internal memory as compared to the direct implementation to store all the prediction data internally.

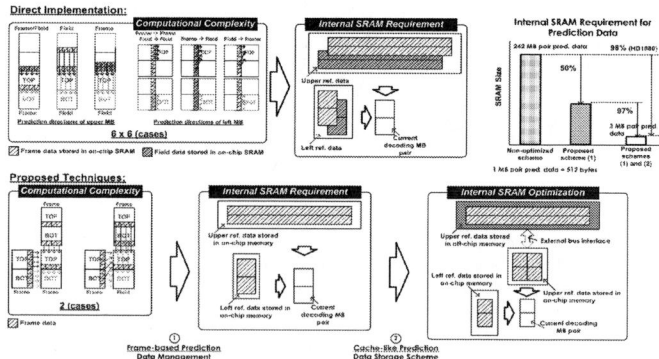

Figure 3. Proposed techniques to overcome the complexity of H.264 MBAFF coding tool

978-1-61284-863-1/11 $26.00 © 2011 IEEE

III. PROPOSED HIGH PERFORMANCE SUB-THRESHOLD VOLTAGE CMOS CIRCUIT

Fig. 4 shows the architecture of high performance sub-threshold voltage CMOS circuits. The body bias is useful to help increasing workable frequency. To ensure high throughput at low Vdd, through adaptive body bias (ABB) circuit, we support dynamic NP-swappable body bias scheme and generate the body biases (i.e., VNW and VPW) for the in-house standard cell circuits. To support controllable body bias, we design multi-rail standard cells that adopt multiple power-rail architecture with one M3 rail for Vdd, one M1 rail for VNW, one M3 rail for GND and one M1 rail for VPW. Through an advanced T-well CMOS process, the body biases of the MOSFET could be separated and directly implemented in a standard cell design flow based on the in-house cell library in 90nm CMOS technology.

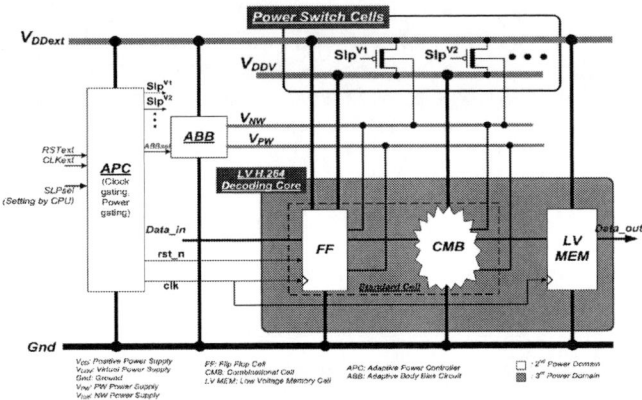

Figure 4.   High performance sub-threshold voltage CMOS circuits

IV. PROPOSED LOW AREA SUB-THRESHOLD VOLTAGE SRAM

Fig. 5 shows the SRAM architecture, which is designed to support enough driving capability for multi-MHz operations, low area overhead and high immunity to process variation. We propose a 7T cell SRAM with asymmetric active VDD/GND gating (called asymmetric 7T, i.e., a7T cell) for lower VDD. The proposed SRAM cell and its operation waveforms are shown at the bottom of Fig. 5. Transistors M3 and M4 are connected to the virtual VDD (VVDD) and the real VDD, respectively. Transistors M1 and M2 have a common virtual ground (VGND). This is called the asymmetric active VDD/GND gating because it is applied in the active operation and is asymmetric between VDD and GND and between M3 and M4. Cell writing is executed by differential operations, but cell reading is performed only through the right port. When the clock signal CK is low, bit lines are pre-charged, and VGND is boosted ΔV1 above GND for reducing leakage power. During writing and reading phases, transistor M7 is disabled to break the butterfly loop. During writing, VGND is maintained at ΔV1 for increasing write margin (WRM). During reading, only VGND of the accessed row is pulled down to GND for reading "0" and others are maintained at ΔV1 for decreasing undesirable BL voltage drop. VVDD becomes floating to conquer the problem of destructive read during reading.

Fig. 6 shows the performance and comparison of the proposed a7T SRAM. The proposed LV SRAM supports a wide operating voltage range from 0.25V to 1.0V and allows the operation frequency from 1 MHz to 250 MHz with low area a7T design. Compared to the LV SRAM designs [10-11] with over 60% cell overhead (compared to 6T cell SRAM), the proposed LV SRAM respectively owns 27.4x and 6x improvement in workable frequency with only 24.9% in cell overhead.

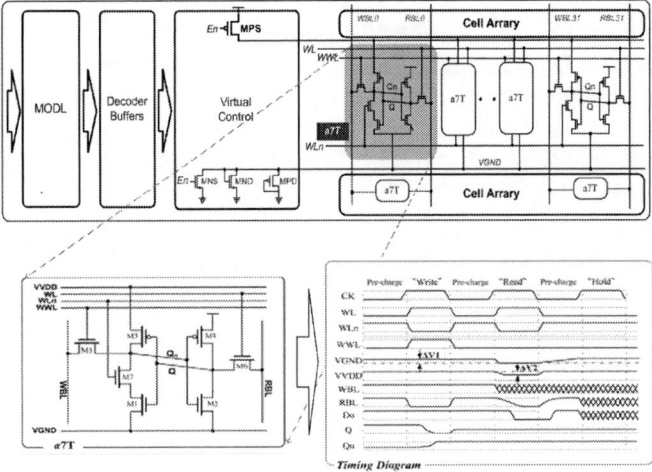

Figure 5.   The proposed low area sub-threshold voltage SRAM design

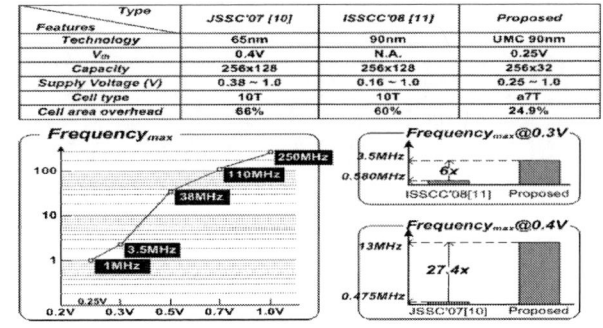

Type / Features	JSSC'07 [10]	ISSCC'08 [11]	Proposed
Technology	65nm	90nm	UMC 90nm
V_{th}	0.4V	N.A.	0.25V
Capacity	256x128	256x128	256x32
Supply Voltage (V)	0.38 ~ 1.0	0.16 ~ 1.0	0.25 ~ 1.0
Cell type	10T	10T	a7T
Cell area overhead	66%	60%	24.9%

Figure 6.   Performance evaluation of the proposed LV SRAM

V. DESIGN IMPLEMENTATION AND PERFORMANCE COMPARISON

The proposed LV H.264 video decoder was implemented in 90-nm CMOS low-Vt technology and designed with three power domains, as shown in Fig. 7. To test the functionality of the implemented chip, we adopt an ARM-based platform, which includes a CPU mother-board and a FPGA daughter-board connected together through AMBA on-chip-bus. Fig. 8 and Fig. 9 both show the comparison of the proposed design with the existing LV video designs [7-9]. Fig. 8 shows the workable frequency comparison under different operation voltages. Compared to the JPEG design [7], it owns 4x/1.6x improvement in workable frequency at 0.5V/0.7V, respectively. Fig. 9 shows the throughput rate comparison under different operation voltages. Compared to the LV H.264 video decoder

978-1-61284-863-1/11 $26.00 © 2011 IEEE

[8] to support QCIF@17fps at 0.5V, it owns a 31x improvement in throughput to support SVGA@30fps video decoding at 0.5V with high profile tools, which is twice more complex than the baseline profile tools supported in [8]. Compared to the H.264 decoder [9] to support CIF@30fps baseline video decoding at 0.7V, the proposed design owns an 9x improvement in throughput to support HD720@30fps video decoding at 0.7V.

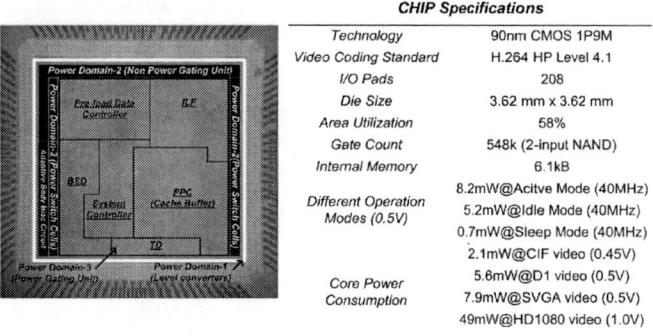

CHIP Specifications	
Technology	90nm CMOS 1P9M
Video Coding Standard	H.264 HP Level 4.1
I/O Pads	208
Die Size	3.62 mm x 3.62 mm
Area Utilization	58%
Gate Count	548k (2-input NAND)
Internal Memory	6.1kB
Different Operation Modes (0.5V)	8.2mW@Acitve Mode (40MHz)
	5.2mW@Idle Mode (40MHz)
	0.7mW@Sleep Mode (40MHz)
Core Power Consumption	2.1mW@CIF video (0.45V)
	5.6mW@D1 video (0.5V)
	7.9mW@SVGA video (0.5V)
	49mW@HD1080 video (1.0V)

Figure 7. Die photo showing the different domains

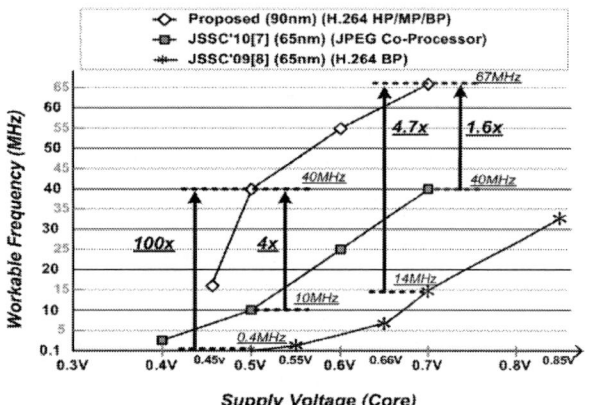

Figure 8. Workable frequecy comparison under different operation voltages with the existing LV designs

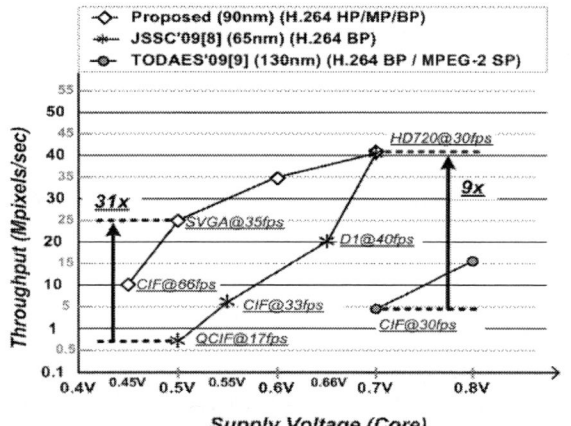

Figure 9. Throughput rate comparison under different operation voltages with the exiting LV designs

VI. CONCLUSION

In this paper an H.264 video decoder is proposed to demonstrate high throughput video decoding to support real-time SVGA video when operated at 0.5V for portable applications. To support high throughput rate, the complexity and the required memory bandwidth are reduced by the proposed low complexity and low memory bandwidth design techniques. To support sub-threshold voltage operation with both high performance and low leakage power consumption, a sub-threshold voltage circuit design technique with a low area sub-threshold voltage SRAM are proposed to enable the proposed design implemented by a 90nm low-Vt CMOS technology to be operated at 0.5V for real-time SVGA video decoding.

ACKNOWLEDGMENT

The authors are grateful to the National Science Council and Ministry of Economical Affairs of Taiwan to provide the financial support under grant NSC98-2220-E-194-001, and 98-EC-17-A-01-S1-040, respectively. The authors are also grateful to the Chip Implementation Center (CIC) of Taiwan for the support to the test chip implementation.

REFERENCES

[1] ITU-T Recommendation and International Standard of Joint Video Specification, ITU-T H.264, Mar. 2009.

[2] J. Ostermann, et al., "Video coding with H.264/AVC: tools, performance, and complexity," *IEEE Circuits Syst. Mag.*, vol.4, no.1, pp. 7- 28, First Quarter 2004.

[3] Poly-crystalline I-Cell provided by MOTECH. [Online]. Available: http://www.motech.com.tw/Files/ProductDoc/IM156B3-101124.pdf

[4] Multicrystalline Silicon Solar Cell provided by GINTECH. [Online]. Available: http://www.gintech.com.tw/thumb/8de7a336f2fe51d88208cbf58c5ae077.pdf

[5] C. A. Chien, et al., "A H.264/MPEG-2 dual mode video decoder chip supporting temporal/spatial scalable video," in *Proc. ASP-DAC*, 2011.

[6] T. D. Chuang, et al., "A 59.5mW scalable/multi-view video decoder chip for Quad/3D Full HDTV and video streaming applications," in *Proc. ISSCC*, 2010, pp. 330-331.

[7] Y. Pu, J. P. de Gyvez, H. Corporaal, and Y. Ha, "An ultra-low-energy multi-standard JPEG co-processor in 65 nm CMOS with sub/near threshold supply voltage," *IEEE J. Solid-State Circuits*, vol.45, no.3, pp.668-680, Mar. 2010.

[8] V. Sze, D. F. Finchelstein, M. E. Sinangil, and A. P. Chandrakasan, "A 0.7-V 1.8-mW H.264/AVC 720p video decoder," *IEEE J. Solid-State Circuits*, vol. 44, no. 11, pp. 2943-2956, Nov. 2009.

[9] C. D. Chien, C. A. Chien, J. C. Chu, J. I. Guo, and C. H. Cheng, "A 252Kgate/4.9Kbyte SRAM/71mW multi-standard video decoder for high definition video applications," *ACM Trans. on Design Automation of Electronic Systems (TODAES)*, vol. 14, no. 1, Jan. 2009.

[10] B. H. Calhoun and A. P. Chandrakasan, "A 256-kb 65-nm sub-threshold SRAM design for ultra-low-voltage operation," *IEEE J. Solid-State Circuits*, vol.42, no.3, pp. 680-688, Mar. 2007.

[11] I. J. Chang, J. J. Kim, S. P. Park, and K. Roy, "A 32kb 10T subthreshold SRAM array with bit-interleaving and differential read scheme in 90nm CMOS," in *Proc. ISSCC*, 2008, pp. 388-622.

Appedix

● **The in-house standard cell circuits with the proposed ABB feature**

Fig. A1 shows the proposed ABB circuits to generate the body biases (i.e., VNW and VPW) for the in-house standard cell circuits that adopt multiple power-rail architecture with one M3 rail for Vdd, one M1 rail for VNW, one M3 rail for GND and one M1 rail for VPW. Through an advanced T-well CMOS process, the body biases of the MOSFET could be separated and directly implemented in a standard cell design flow based on the in-house cell library in 90nm CMOS technology.

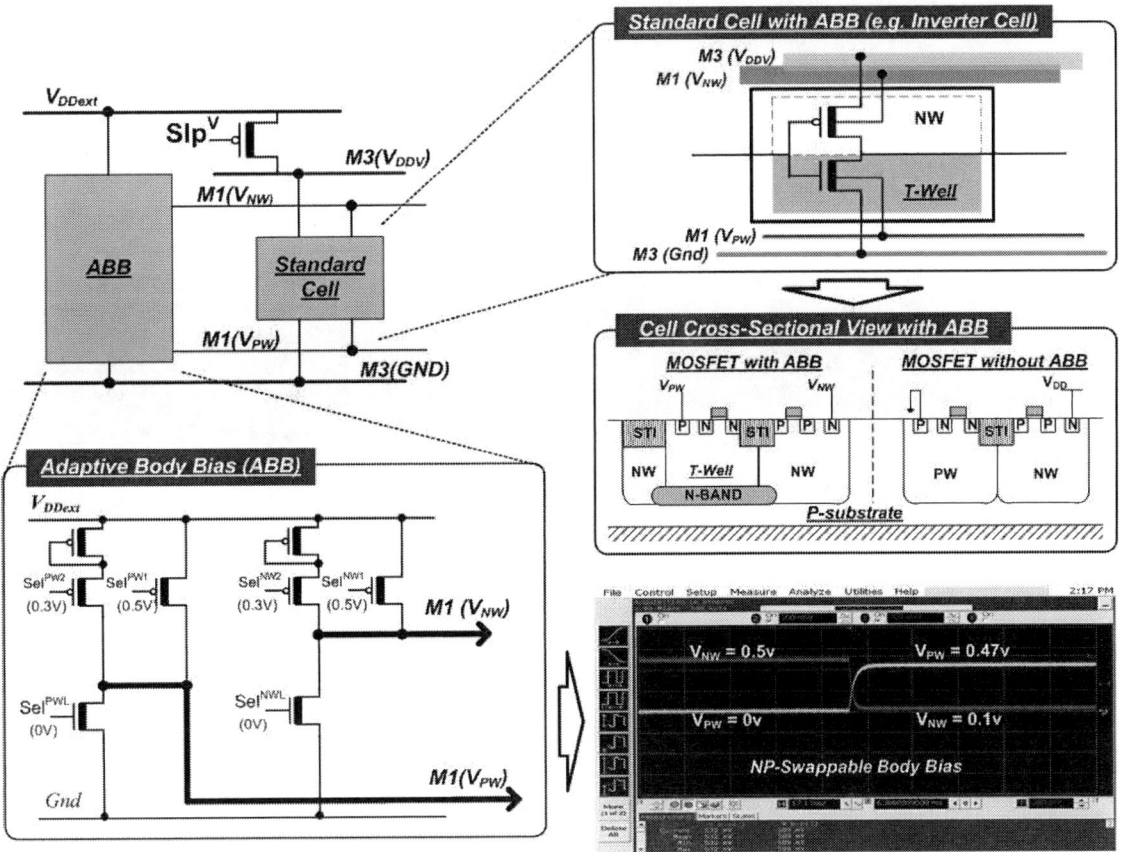

Figure A1. The in-house standard cell circuits with the ABB in the proposed LV H.264 video decoding core

● **Measured results of the adopted low area sub-threshold voltage SRAM**

The experimental results of the adopted low area sub-thresold voltage SRAM test chip are summarized in the following Fig. A2. The measured VDD_{min} is 0.25V, and the SRAM runs at 1.0MHz. The SRAM has a multi-MHz performance when VDD is above 0.3V with the lower part of Fig. A2 showing the write patterns and read-out results from the logic analyzer.

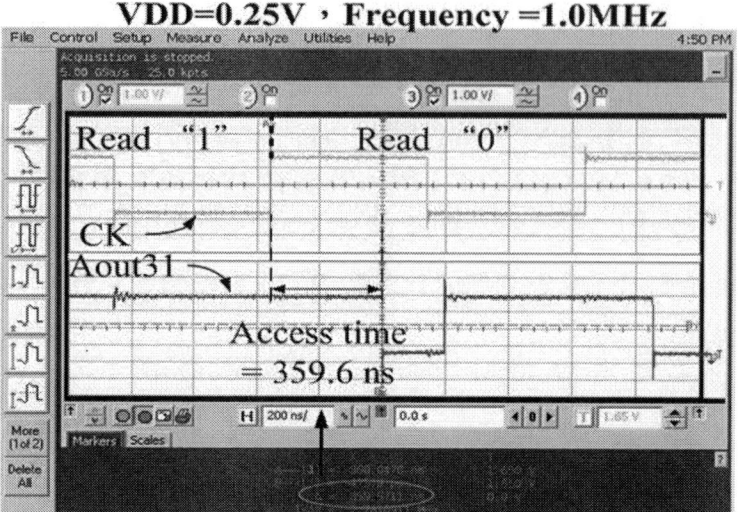

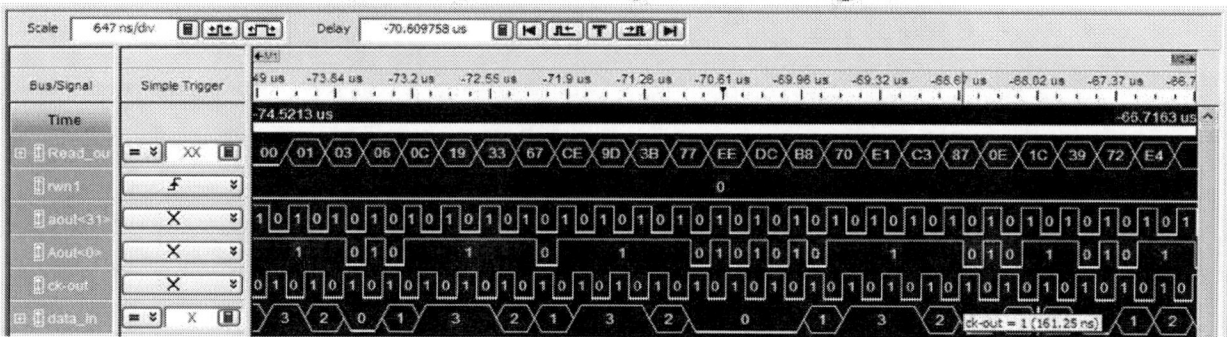

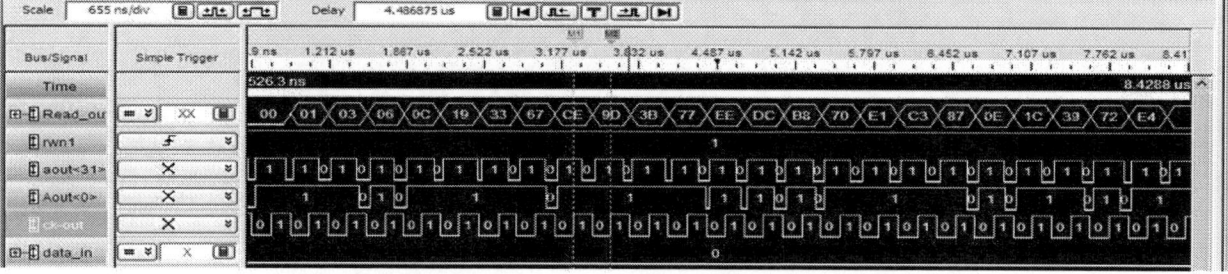

Figure A2. Measured operation waveforms of the LV SRAM test chip

A 6.72-GB/S 8PJ/BIT/ITERATION IEEE 802.15.3C LDPC DECODER CHIP

Zhixiang CHEN[*], Xiao PENG, Xiongxin ZHAO, Qian XIE, Leona OKAMURA, Dajiang ZHOU and Satoshi GOTO

Graduate School of Information, Production and Systems, Waseda University, Japan

2-7, Hibikino, Wakamatsu, City of Kitakyushu, 808-0135, Fukuoka, Japan

[*]Email: zhixiangchen@ruri.waseda.jp

ABSTRACT

In this paper, we introduce an LDPC decoder design for decoding length-672 code adopted in IEEE 802.15.3c standard. The proposed decoder features high performance in both data rate and power efficiency. A macro-layer level fully parallel layered decoding architecture is proposed to support the throughput requirement in the standard. The decoder takes only 4 clock cycles to process one decoding iteration. While parallelism increases, the chip routing congestion problem becomes more severe because of the more complicated interconnection network used for message passing. This problem is nicely solved by our proposed efficient message permutation scheme utilizing the parity check matrix features. The proposed message permutation network features high compatibility and zero-logic-gate VLSI implementation, which contribute to the remarkable improvements in both area utilization ratio and total gate count. To verify the above techniques, the proposed decoder is implemented on a chip fabricated using Fujitsu 65nm 1P12L LVT CMOS process. The chip occupies a core area of 1.30mm^2 with area utilization ratio 86.3%. According to the measurement results, working at 1.2V, 400 MHz and 10 iterations the proposed decoder delivers a 6.72Gb/s data throughput and dissipates a power of 537.6mW, resulting in an energy efficiency 8.0pJ/bit/iteration.

I. INTRODUCTION

Originally proposed in 1960s and rediscovered in 1990s, low-density parity-check (LDPC) code shows its excellent error correcting performance in communication systems. With proposals on hardware optimized decoding algorithm [1] and on architecture aware (AA) code with its decoder [2], real application of LDPC code became possible and popular. Recently Quasi-Cyclic (QC) LDPC code, a kind of AA LDPC codes was adopted by several modern wireless communication standards such as IEEE 802.11n, IEEE 802.16e and IEEE 802.15.3c. [3]

IEEE 802.15.3c as a part of wireless personal area networks (WPAN) standard aims at high-rate wireless data transfer like low-latency bidirectional data transfer and delivery of uncompressed, high-definition video. According to the standard, data rate with the highest 5.7 Gb/s is required. However, it is challenging to design a decoder supporting such kind of high throughput for an LDPC code with length 672 only, not mentioning that 4 different code rates should also be compatible in the decoder. Recently, a kind of architecture called partially parallel decoder [4] was proposed to decode QC-LDPC code and its relevant optimizations can be found in [5][6]. However, even for the state of the art work [6], this architecture seems impossible to fulfill the WPAN high throughput requirement. By increasing the decoding parallelism, Sha et al. proposed a WPAN decoder which can deliver a 3.6 Gb/s throughput at 500 MHz clock frequency in [7]. Still it has gap in data rate for the latest-version WPAN standard. Therefore, new decoder architecture is expected.

In this paper, we propose a macro-layer level fully parallel decoder for length 672 code defined in WPAN standard. For the proposed architecture, only 4 clock cycles are needed for processing one layered decoding iteration. The complex-interconnection problem due to highly parallel architecture is solved by implementing two simple message permutation networks which consist no logic gates but wires only based on the exploitation and utilization of the parity check matrix (PCM) features. All four code rates are supported by the proposed decoder. A frame level pipeline is integrated in the design to reduce the length of the critical path. A 65nm CMOS chip is fabricated to verify the proposed architecture. Measured at 1.2V, 400MHz and 10 iterations the proposed decoder achieves a data throughput 6.72Gb/s and consumes a power of 537.6mw.

The rest of this paper is organized as follows. In section II, the PCM of length 672 LDPC code in WPAN is introduced and exploited with some definitions given. Section III gives the details of our proposed decoder architecture. Chip implementation result is presented in section IV with comparison to the state of the art work. Finally, in section V we draw the conclusion.

II. OBSERVATIONS AND IMPLICATIONS ON PCM

The length-672 LDPC code defined in WPAN is a kind of QC-LDPC code. The PCM of such code is obtained by expanding a so called base matrix, an example of which is provided in our previous work [5]. For the length-672 WPAN code, the size of sub-block (sub-matrix) is fixed 21 and four code rates, 1/2, 5/8, 3/4 and 7/8 are supported. The base matrices of rate 5/8 and 7/8 are given in Fig. 1.

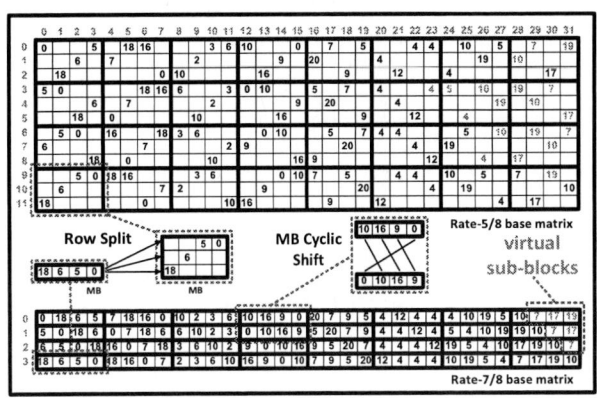

FIGURE 1. PCMS OF WPAN LDPC CODE

A. Row Split Feature

Unlike other codes, for WPAN length-672 code, base matrices of rate 1/2, 5/8 and 3/4 are generated by splitting the rows of rate 7/8. Take rate 5/8 as an example, 12 rows are obtained by splitting each row of rate 7/8 into 3. From the leftmost one, 4 consecutive sub-blocks in one row of rate-7/8 base matrix form a part defined as a **Macro Block (MB)** which is enclosed in bold black rectangle in Fig. 1. In this paper MB is also used to define the correspondingly

---

This research is supported by "Ambient SoC Global COE program of Waseda University" of MEXT, Japan. The chip fabrication is supported by the Semiconductor Technology Academic Research Center (STARC), Japan.

generated part from rate 7/8 in another three rates. Lower left MBs of rate 5/8 and rate 7/8 are shown in Fig. 1. In this example, by row splitting four nonempty sub-blocks 18, 6, 5 and 0 in rate-7/8 MB are allocated to four positions of rate-5/8 MB with adding empty sub-blocks to another 8. MB can be treated as an element constructing the base matrix that every 8 MBs in horizontal form a *Macro Layer* (**ML**) and 4 MLs in vertical form a base matrix.

B. MB Cyclic Shift Feature

Row split feature mentioned above indicates the relations between two MBs from two base matrices while the relations between MBs in the same base matrix is so called cyclic shift. Except for the redundant bit part, generally, MB in a ML can be attained by cyclically right shift one sub-block in the MB above. The left fourth MBs in first and second MLs of rate-7/8 base matrix are used as an example to show this feature in Fig. 1.

C. Virtual Sub-blocks and Implications

Shown in Fig. 1 row split and MB cyclic shift features will be satisfied for the entire base matrix after some virtual sub-blocks, highlighted in red, are added into the redundant bit part. These virtual sub-blocks are used to help you understand the two important features mentioned above. During the real decoding process, they are actually dismissed.

More importantly, the row split feature implies that the message interconnection patterns between check nodes and variable nodes inside ML for different code rates are the same. The MB cyclic shift feature implies that the message interconnection patterns for different MLs in one rate are the same if we cyclically shift appropriate number of sub-blocks in each MB. Therefore, based on above two implications, it is possible for a simple network, particularly designed for a single ML, to be reused for all MLs with small modification.

III. PROPOSED ML FULLY PARALLEL DECODER

In this section, firstly we introduce the proposed decoder architecture by showing its block diagram. Then two important modules of this decoder, macro layer processing engine (MLPE) and message permutation network will be analyzed in detail.

A. Decoder Architecture

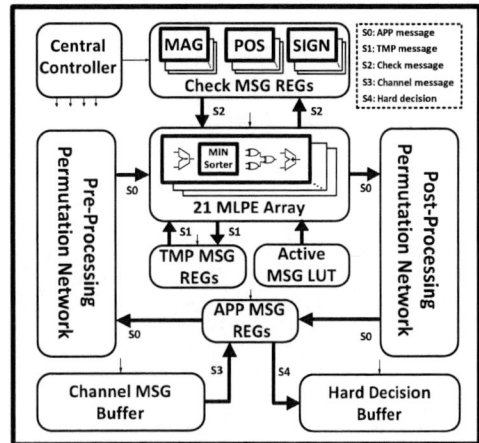

FIGURE 2. DECODER ARCHITECTURE BLOCK DIAGRAM

Block diagram of proposed decoder architecture is given in Fig. 2. Decoding process is started from initializing the registers (REGs) for storing a posterior probability (APP) messages (MSG) with channel messages in the buffer. APP messages are then sent through a pre-processing permutation network into a message processing array. Each macro layer processing engine (MLPE) in the array is capable of simultaneously updating up to 32 APP messages which forms one row according to layered decoding algorithm. Thus an entire ML (672 messages at most) can be updated concurrently by implementing 21 copies of MLPE. To reduce the length of critical path, a code-frame level pipeline processing is introduced in the design. Two code frames are decoded alternatively in the decoder. The temporary (TMP) messages and check (CHK) messages generated in the decoding process are also stored in the dedicated registers. An active message look up table (LUT) is used to indicate which messages among 32 are active or involved in the current ML updating, that is to say their corresponding sub-blocks are non-empty. After being updated, APP messages are reordered by a post-processing permutation network and stored back into registers. When the decoding is finished, hard decisions, the most significant bit of APP messages, are written into the hard decision buffer for output.

B. Macro Layer Processing Engine

MLPE is used to update APP messages according to the layered message updating scheme, the detail of which can be found in [8]. Algorithm applied for message calculation is so called normalized min-sum algorithm [1] with factor 0.75. A register stage is added to reduce the length of critical path, thus the entire message updating process is divided into two phases, namely, temporary message and APP message updating. Two code frames are updated alternatively in these two phases resulting in a frame-level pipelining. Detailed data flow and architecture diagram are shown in Fig. 3 with data bus, messages of different code frames distinguished by different colors.

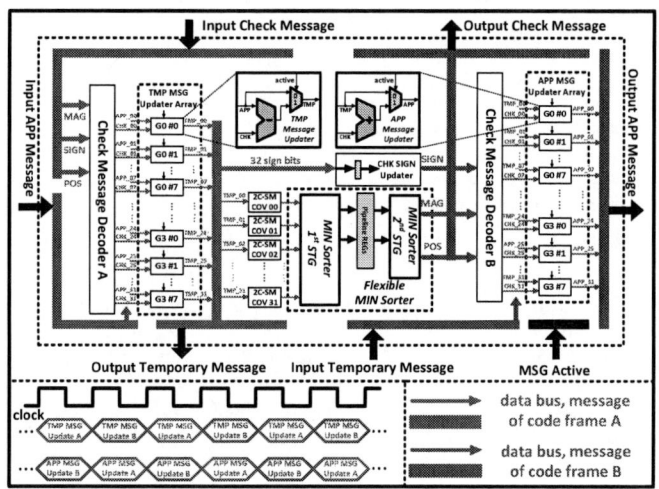

FIGURE 3. DATA FLOW OF MLPE AND TIMING DIAGRAM

Take rate 7/8 as an example, in the temporary message updating phase 32 APP messages related to 32 sub-blocks are inputted into 32 temporary (TMP) message (MSG) updaters. Meanwhile, the corresponding 32 check messages are generated by a check message decoder according to the compressed check information composed of magnitudes (MAG) of first and second minimum values, 32 signs and the position (POS) of first minimum value. If an APP message is involved in this ML updating like left 29 sub-blocks in first ML, rate 7/8, the output TMP MSG equals to the variable message which is APP MSG subtracted by CHK MSG. Otherwise the original APP

MSG is bypassed to output as a TMP MSG. These 32 TMP MSGs are then outputted and stored in the TMP MSG REGs waiting for next-phase process. At the same time, 32 converters (2C-SM COV) covert TMP messages in 6-bit two's complement form to 5- bit sign-magnitude numbers for min sorter use. For an inactive APP MSG, it is tied to the maximum magnitude to avoid its effect on min sorter.

The 32-input flexible min sorter is constructed based on a tree structure approach which was proposed in [9]. Small modifications are added to make it support up to 4 group (8 inputs in one group) independent sorting for cases of other rates. A pipeline register stage divides it into two parts, called first and second stages of the min sorter.

For the other frame involved in the APP message updating phase, new check information generated by flexible min sorter and CHK sign updater are outputted and stored in the check message registers in a compressed way. Meanwhile, after being decoded, 32 CHK MSGs and together with 32 TMP MSGs generated in the previous phase are sent into 32 APP MSG updaters for APP MSG updating. Similar with TMP MSG updater, APP MSG can be either original TMP MSG (old APP MSG) or the sum of TMP (variable message) and CHK MSG according to the message active control. After two-phase processing, updating of all the APP MSG in one ML is finished in two clock cycles. Considering two frame pipeline decoding and each containing 4 MLs, averagely, it takes only 4 clock cycles to finish one decoding iteration for one code frame which is shown in the timing diagram in Fig. 3.

For code rates other than 7/8, the total 32 messages, their related updaters and min sorter are divided or configured into several groups, two for rate 3/4, three for 5/8 and four for 1/2. Each group deals with one row in a ML independently and an entire ML is processed simultaneously. It is worth mentioning that the APP messages are grouped outside the MLPE by pre-processing permutation network which will be discussed in the following.

C. Permutation Network

For LDPC decoder design, implementation of interconnection between check and variable nodes in the Tanner Graph of the code is always a critical problem, especially in case of high parallelism decoder. However, for our proposed WPAN decoder only two simple networks, namely, pre and post processing permutation networks are implemented to solve the interconnection problem. The interconnection complexity is greatly reduced by making these two simple fixed networks be reusable for all the MLs and code rates based on the two features observed in section II. For both of the two networks, they contain no logic circuits but only 672 6-bit wide wires connecting 672 pairs of message input and output.

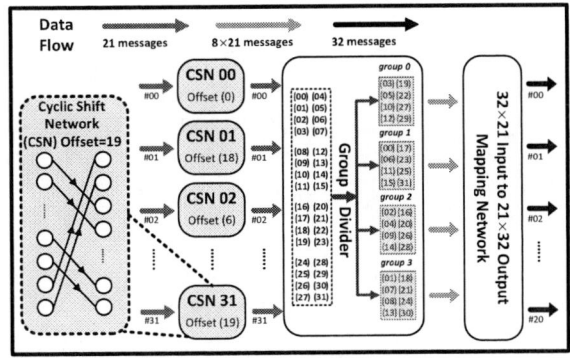

FIGURE 4. PRE-PROCESSING PERMUTATION NETWORK

Shown in Fig. 4, pre-processing permutation network mainly contains three parts, 32 fixed cyclic shift networks (CSN), one group divider and one mapping network. 32 clusters of APP messages (21 in each cluster) corresponding to 32 sub-blocks are inputted into 32 CSNs. Each CSN performs fixed-offset-cyclic shift for 21 messages according to the offsets of sub-blocks, including the virtual sub-blocks defined before, in first ML of rate-7/8 base matrix. An example of CSN with offset 19 is given in the figure. After cyclic shift within each cluster, 32 clusters of messages are then fed into a module, namely, group divider. According to the distribution of sub-blocks in four rows of the first ML in rate-1/2 base matrix, these 32 clusters are divided into 4 groups. The expression (*i*) in the figure represents the message cluster related to sub-block with column index *i*, where *i* is from 0 to 31. For rate 1/2, these four groups of APP messages are updated independently in MLPE which is configured to four-row process mode. For rate 5/8, group 0 and 1 are merged and updated together in MLPE with the other two remaining independent. Similarly, when processing rate-3/4 code, MLPE is configured to two-row-process mode with each row relating to two groups. After final mapping, 32 clusters of APP messages are distributed into 21 clusters, each of which consists 32 messages and will be directly sent to a MLPE. An APP message in cluster *j* with inner offset *k* at the input side of this mapping network is mapped into cluster *k* with inner offset *j* at the output side.

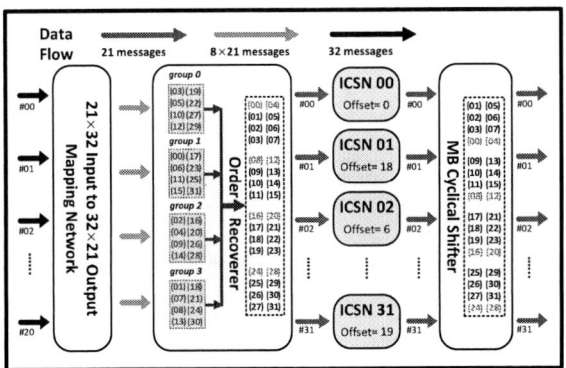

FIGURE 5. POST-PROCESSING PERMUTATION NETWORK

Figure 5 shows the details of post-processing permutation network. Firstly, updated APP messages from 21 MLPEs are remapped to 32 clusters. 32 clusters in different 4 groups are merged together in its original order, from 0 to 31, through a module called order recoverer. 32 fixed inverse cyclic shift networks (ICSN) cyclically permute the messages in an inverse direction of CSN to recover the original order, 0 to 21, within a sub-block. Cooperating with the second feature mentioned in section II, MB cyclic shifter is used to cyclically permute each MB by one sub-block. In the figure, the first sub-block of each MB is marked red for your notice.

Taking the virtual sub-blocks into consideration, all 16 MLs of four code rates share the same check to variable interconnection patterns if the sub-blocks in the MBs are cyclically shifted, which is actually realized by the MB cyclical shifter inside the post-processing permutation network. The effect of virtual sub-blocks on message updating is avoided by setting the corresponding APP messages inactive in the active message LUT. Therefore, these two permutation networks can be reused for processing all MLs in our proposed decoder. Moreover, since all the connections between signals inside or outside the modules in these two permutation networks are fixed wires, electrically, both of them can be treated as 672 input to 672 output fixed mapping network only consisting 672 pairs of wires. Thus the wire congestion overhead introduced by them is relatively small which can be revealed by the post layout implementation result shown in the next section.

IV. IMPLEMENTATION RESULT

To verify the proposed architecture, it is implemented on a **65nm LVT CMOS** chip. The die photo and chip features are shown in Fig. 6. At clock frequency 400MHz, supply voltage 1.2V the on-chip decoder achieves a data throughput 6.72Gb/s with power consumption 537.6mW which turns to an energy efficiency 8pJ/bit·iter. For the low-demand scenario, we set the supply voltage 1.0V and clock frequency 200MHz, delivering 3.36Gb/s data throughput, the chip consumes 209.8mW power resulting energy efficiency 6.2 pJ/bit·iter. The details on chip measurement and more performance results are available in the appendix attached.

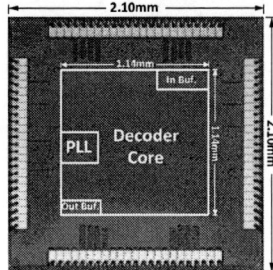

Technology	Fujitsu 65nm LVT CMOS	
Die Area	2.1mm×2.1mm	
Core Area	1.14mm×1.14mm	
Density	86.3%	
Iteration	10	
Frequency	400MHz	200MHz
Supply	1.2V	1.0V
Power	537.6mW	209.8mW
Throughput	6.72Gb/s	3.36Gb/s
Energy Eff.	8pJ/bit·iter	6.2pJ/bit·iter

FIGURE 6. CHIP DIE PHOTO AND MAIN CHARACTERISTICS

TABLE I. COMPARISONS PUBLISHED LDPC DECODER ASICs

Publication	this work	A-SSCC'10 Hung [10]	ESSCIRC'09 Chen [11]	VLSI'10 Xiang [6]
Technology (nm)	65	65	90	130
Supply voltage (V)	1.2	1.0	0.8	1.2
Code length	672	672	2048	576-2304
Code rate	1/2, 5/8 3/4, 7/8	1/2, 3/4 5/8, 7/8	15/16	1/2, 2/3 3/4, 5/6
Quantization bits	6 bits	6 bits	6 bits	6 bits
Algorithm	layered	layered	layered	layered
Chip core area (mm²)	1.30	1.56	3.84	3.03
Decoder core gate count	430K[1]	647K	708K	470K
Clock frequency (MHz)	400	197	120	214
Iterations per frame	10	5	4	10
Data throughput (Gb/s)	6.72	6.6[2]	6.15[2]	0.955
Power consumption (mW)	537.6	361	191.2	397
Energy efficiency (pJ/bit/iteration)	8.0	12.5	8.3	42
Normalized energy [3] efficiency (pJ/bit/iteration)	8.0	18	12.5	16.8

1) 1.92μm² for each two-input one-output NAND gate;

2) Original information throughput is scaled to data throughput;

3) Normalized to 65nm, 1.2V technology with $P_{65} = P_x \cdot \frac{C_{65}}{C_x} \cdot \left(\frac{V_{65}}{V_x}\right)^2$ assuming $\frac{C_{65}}{C_{90}} = 0.67$, $\frac{C_{65}}{C_{130}} = 0.4$;

According to the post-layout report, the decoder core occupies an equivalent gate[*] 430.8K, where 140K is used as registers for pipeline and message storage and the rest 290.8K consist of logic and clock tree. As mentioned before, by using our proposed message permutation networks the wire congestion overhead of the decoder chip is greatly reduced, which can be revealed by the facts that compared to the synthesis result, the degradation on timing performance after placement and route is less than 7% and that the chip density achieves 86.3%. Performance comparisons with the previous work aiming at high throughput are summarized in Tab. I.

The proposed decoder achieves highest power efficiency compared to LDPC decoders for different applications with technology scaling taken into consideration. Compared to the state-of-art design for WPAN standard [10], our proposed decoder improves 16.7%, 33.5% and 44.4% in area, gate count and power efficiency.

V. CONCLUSION

This paper discusses the design of an LDPC decoder for IEEE 802.15.3c application. To achieve the data rate required in the standard, macro-layer level fully parallel decoder architecture is proposed. To mitigate the interconnection complexity raised by high parallelism processing, a matrix-exploited message permutation scheme is applied. A decoder with pipeline for high clock frequency is implemented in 65nm CMOS. Measurement result shows that it achieves 6.72Gb/s at 400MHz, 10 decoding iterations with energy efficiency 8.0pJ/bit/iteration. The proposed decoder achieves highest power efficiency compared to LDPC decoders for different applications with technology scaling taken into consideration. Compared to the state-of-art design for WPAN standard [10], our proposed decoder improves 16.7%, 33.5% and 44.4% in area, gate count and power efficiency.

REFERENCE

[1] J. Chen, "Decoding low-density parity-check codes with normalized APP-based algorithm," *IEEE GLOBECOM 2001 Conf.*, vol. 2, pp. 1026–1030.

[2] M. Mansour, "A 640-Mb/s 2048-Bit Programmable LDPC Decoder Chip," *IEEE Journal of Solid-State Circuit*, vol. 41, No. 3, pp. 684-698, 2006.

[3] *Part 15.3: wireless medium access control (MAC) and physical layer specifications for high rate wireless personal area networks (WPANs)*, IEEE Std. P802.15.3c-DF8, 2009.

[4] Y. Sun, "VLSI Decoder Architecture for High Throughput, Variable Block-size and Multi-rate LDPC Codes," *IEEE ISCAS 2007 Conf.*, pp. 2104-2107.

[5] Z. Chen, "A High Parallelism LDPC Decoder with an Early Stopping Criterion for WiMax and WiFi Application," *IPSJ Trans. TSLDM*, Vol. 3, pp. 292-302, 2010.

[6] B. Xiang, "A 4.84 mm² 847-955 Mb/s 397 mW Dual-Path Fully-Overlapped QC-LDPC Decoder for the WiMAX System in 0.13 μm CMOS," *IEEE VLSI Symp. 2010*, pp. 211-212.

[7] J. Sha, "LDPC Decoder Design for High Rate Wireless Personal Area Networks," *IEEE Trans. Consumer Electronics*, vol. 55, issue 2, pp. 488-460, 2009.

[8] E. Sharon, "Efficient Serial Message-Passing Schedules for LDPC Decoding," *IEEE Trans. Inform. Theory*, vol. 53, pp. 4076-4091, 2007.

[9] C. WEY, "Algorithms of Finding the First Two Minimum Values and Their Hardware Implementation," *IEEE Trans. Circuit and Systems I*, vol. 55, No. 11, pp. 3430-3437, 2008.

[10] S. Hung, "A 5.7Gbps Row-Based Layered Scheduling LDPC Decoder for IEEE 802.15.3c Applications," *IEEE ASSCC 2010 Conf.*, pp. 12-2.

[11] C. Chen, "A 11.5-Gbps LDPC Decoder Based on CP-PEG Code Construction," *IEEE ESSCIRC 2009 Conf.*, pp. 412-415.

[*] A two-input-one-output **NAND** gate in this process occupies 1.92μm² area.

Appendix

CHIP MEASUREMENT: EXPERIMENT AND RESULTS

ABSTRACT

Chip measurement work mainly includes chip logic verification and power measurement, both of which rely on a verification system developed based on Mitsubishi-MU300 LSI evaluation board. For the logic verification, chip responses were compared with the results generated by on-FPGA decoder core with the same test pattern input. In condition of environment temperature 300K, 1.2V supply voltage, the observed maximum clock frequency under which the chip works correctly is 450MHz. (Working correctly here means the chip response of one million test code frames are exactly the same as the on-FPGA core.) The maximum throughput at frequency 450MHz measured is 7.56Gb/s. For additive white Gaussian noise (AWGN) channel decoder's bit error ratio (BER) performance is evaluated via the measurement system. For the power measurement, chip power consumption in different scenarios like, working or halt, supply voltage, clock frequency and code rate are measured.

I. LOGIC VERIFICATION

In this section, we firstly introduce our self-designed chip verification system based on an FPGA board. With this system chip logic verification is achieved by compared the chip responses to the same design implemented on FPGA using the same test pattern. Maximum frequency is observed as well as the evaluation of the decoder BER performance.

A. Verification system

To verify the logic function of the chip, a verification system based on Mitsubishi MU300 LSI evaluation board is developed. Unlike the recommended test flow provided by vender, the data flow and verification mechanism are redesigned by ourselves taking advantage of platforms and tools provided by Altera company. The entire verification system includes PC, MU300 FPGA board, software and hardware interface between PC and board, DUT board as an interface to connect FPGA and ASIC chip. The dataflow of the verification system is shown in Appdx-Fig. 1.

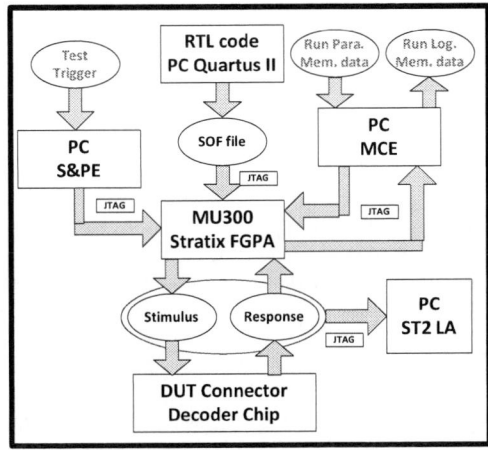

APPENDIX-FIGURE 1. DATA FLOW OF THE VERIFICATION SYSTEM

In this system, an Altera Stratix II FPGA (EP2S130F1508C4) is used to implement the verification core which generates the test pattern and analyze the test response. Quartus 10.0 is used to compile the RTL code and generate the FPGA configuration file. Several Altera provided tools, Signal Tap II Logic Analyzer (ST2LA), In System Source and Probe Editor (S&PE) and Memory Content Editor (MCE), are interfaced with FPGA for data collection and debugging. For each verification test, run parameters will be generated and sent to the pre-specified block memory in FPGA via MCE and the run results are also recorded in one block memory which will be read back to PC through MCE. S&PE is used to give trigger signal and write the on-the-fly changeable register. DUT connector provided by vender is used to connect the FPGA with the chip tested.

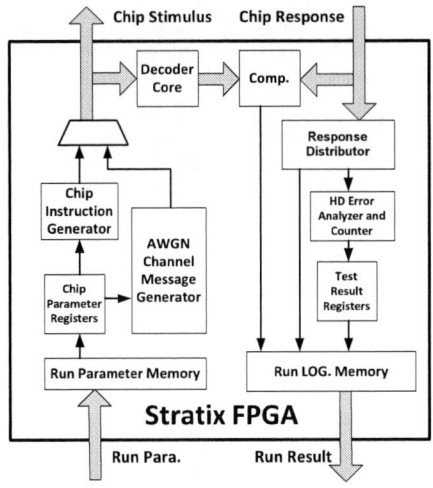

APPENDIX-FIGURE 2. VERIFICATION CORE ON FPGA

The block diagram of the verification core is shown in Appdx-Fig. 2. According to the run parameters set in this test, chip instructions such as, setting the code rate, iteration number and workload are automatically generated. Meanwhile, test code frames BPSK-AWGN log likelihood ratio (LLR) channel messages are generated by an array (3 replicas) of AWGN noise generators, which is implemented according to the proposals in [1]. This noise generator features high speed and low H/W cost with moderate noise quality, which is far than enough for our case. The stimulus or test patterns are then sent into two sinks, the ASIC chip and the decoder core implemented on FPGA. The chip responses including chip status and decoded code frame, the hard decision data, are sent back to verification core. A comparator is used to compare the responses from the chip with the on-FPGA decoder core to evaluate the correctness of the chip logic function. The chip is concluded working correctly if there is no difference between the responses from two decoders among a preset fixed number of test code frames. Moreover, the BER performance of the ASIC decoder is evaluated by analyzing the hard decision data and the corresponding channel status. To do so, some counters are used. Finally, a run log is generated and written into the specified memory for further evaluation.

On chip clock is adjusted by on-chip PLL and the clock system in FPGA can be easily designed using the Quartus platform. According to the Quartus report, FPGA hardware overhead is listed in Appendix-Table I for your reference. A photo showing the real verification system is given in Appdx-Fig. 3.

978-1-61284-863-1/11 $26.00 © 2011 IEEE

APPENDIX-TALBE I. HARDWARE OVERHEAD OF VERIFICATION SYSTEM

FPGA Resource	EP2S130F1508C4
Combinational ALUTs	53,022/106,032 (50%)
Total registers	25,083/106,032 (24%)
Pins	99/1,127 (9%)
Block memory bits	833,632/6,747,840 (13%)
DSP block elements	72/504 (15%)
PLL	1/12 (9%)

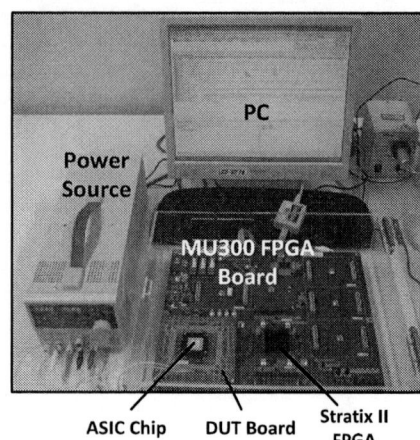

APPENDIX-FIGURE 3. A PHOTO OF THE VERIFICATION SYSTEM

B. Logic verification results

The logic verification includes two parts, measuring the maximum clock frequency of the chip and evaluating the decoding performance in terms of BER.

For the maximum clock frequency, since the chip is not formally tested using the professional tester and we could not ensure all paths of the circuit being covered by our test patterns. Thus to evaluate the correctness of logic function under a particular frequency, we set the following criterion.

If and only if the decoding results of one million random code frames, 672 hard decision bits for each frame, are exactly the same as the results from a design replica implemented on FPGA with the same test patterns, the logic function is then considered as correct.

Appendix-Table II lists the comparison result at different clock frequencies. The results are observed in condition of code rate 1/2, signal to noise ratio (SNR) 2.8dB, environment temperature 300K and 1.2V VDD supply voltage. Accordingly, the maximum throughput at maximum frequency 450MHz is 7.56Gb/s as observed.

APPENDIX-TALBE II. LOGIC COMPARISON RESULT

Frequency (MHz)	# of Difference
360	0
420	0
450	0
465	225,152
480	657,987

According to the measured results the BER vs. SNR curves are given in Appdx-Fig. 4. From the figure one can observe an approximate 0.3dB quantization-raised performance degradation when it is compared to the floating number based software simulation.

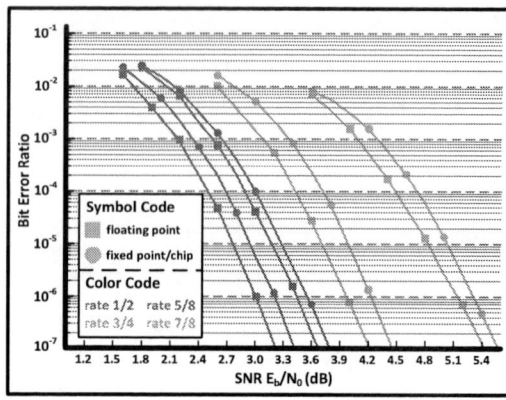

APPENDIX-FIGURE 4. BER PERFORMANCE CURVES OF THE DECODER

II. POWER MEASUREMENT

One of the most important reasons that LDPC codes are widely applied in modern communication systems is due to its high energy efficiency. To show the advancement of our proposal, a precise power consumption measurement is necessary. In the measurement process, we use TEXIO PW18 power supply as the power source, which supports mA level current measurement. The observed data are shown in Appendix-Table III. To give an evaluation on energy efficiency (EE), a commonly used metric, the energy consumed for decoding one code bit of one iteration is also given in the talbe. The data are acquired in condition of code rate 1/2, iteration number 10, SNR 2.8dB.

APPENDIX-TALBE III. POWER MEASUREMENT RESULTS

Frequency (MHz)	200	240	240	320	400
VDD Voltage (V)	1.0	1.0	1.2	1.2	1.2
VDD Current (mA)	209	244	297	374	448
Power (mW)	209	244	356.4	448.8	537.6
EE. (pJ/bit/iter.)	6.22	6.05	8.84	8.35	8.00

III. REMARKS

This appendix report describes an FPGA board based verification system for simplified chip performance evaluation. An LDPC decoder chip design is verified and evaluated. The detailed measurement results are listed. The logic function of the chip is correct and the maximum frequency observed is 450MHz. The maximum throughput at frequency 450MHz measured is 7.56Gb/s The decoder can achieve energy efficiency 8.00pJ/bit/iteration when working at 400MHz, 1.2V supply.

ACKNOWLEDGEMENT

The authors would like to show sincere gratitude to the Semiconductor Technology Academic Research Center (STARC), Japan for providing us with this valuable opportunity to implement and verify our novel idea.

REFERENCE

[1] Dong-U LEE, "A Hardware Gaussian Noise Generator Using the Wallace Method," *IEEE Trans. VLSI Syst.*, vol. 13, No. 8, pp. 911-920 2005.

Using Built-In Fine Resolution Clipping Technique for High-Speed Testing by Using Low-Speed Wireless Tester

Ching-Hwa Cheng { chengch@fcu.edu.tw}
Department of Electronic Engineering, Feng-Chia University, Taiwan, R.O.C.

Abstract

The delay testing of high-speed system integrated circuits is highly complex. Many test challenges are generated from performance testing requirements. The BIST circuit can help solve traditionally slower ATE tester limitations. In this paper, a double edge clipping (DEC) technique is proposed for high-speed performance testing by utilizing a low-price slow-speed ATE. DEC differs from traditional circuit delay testing techniques by changing the clock rate using external ATE. DEC technique uses a lower-speed input clock frequency, then applies internal clipping and a BIST mechanism to adjust clock edges for high-speed circuit functional testing. The postlayout simulations show that the wide-range (26.5%~76%), fine-scale (16ps) duty cycle adjustment technique with high-precision (28ps) calibration circuit is effective for binning performance and high-speed circuit performance testing. Test chips with wireless test system integration are fully validated.
Keywords: high-speed test, performance binning.

1. Introduction

The increasing complexities of SoC systems limit the accessibility of internal circuit modules. The circuit performance test is a common requirement, e.g, Shmoo test, and this requirement can be achieved by using Automatic Test Equipment (ATE). The delay effects from process, voltage, and temperature variations also become serious in very deep submicron technologies. Circuit performance testing and failure debugging are important requirements after chip manufacturing. However, these challenges will become more difficult when facing future high-speed complex designs. For example, high resolution ATE cannot measure the small circuit path delays internal to the chip.

The largest delay time for a circuit can be identified by propagating a group of transition patterns through primary inputs to the primary outputs. A circuit has lower performance when the delay time of any of its paths exceeds a specified clock cycle time. Therefore, a circuit performance can be binned after applying a large number of functional test pattern and observing whether the outputs are correct within a specific clock cycle.

The scan chain can be used as the high-speed test mechanism. Two well known scan-based high-speed test methods are launch-off shift (LOS) and launch-off capture (LOC). With LOS, it is difficult to estimate functional performance accurately during the test phase.

For the LOC delay test, the second vector V2 is the CUT's response to V1. LOC scheme needs precision control of the launch and capture signals in normal mode during the scan delay test. LOC is shown in Fig. 1. For the LOC test scheme, the scan chain can be used to capture the combinational circuit output responses and to launch test patterns into the combinational circuitry.

The path delay represents the sum of delays from the gates, interconnections on that path, and the setup time of input flip-flop. The general clock strategy is designed to obtain circuit output transition from output flip-flop of the second clock cycle. Two clock positive edges are used to trigger the flip-flops in the synchronous circuit operation. The shortest time of the two clock edges is equal to the circuit's highest performance.

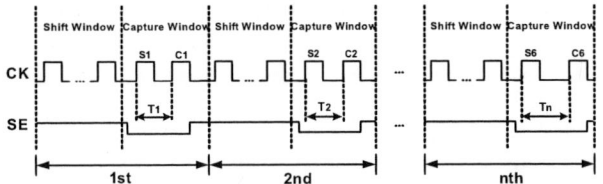

Fig. 1. The launch-off capture (LOC) scan delay testing.

The conventional circuit delay testing technique is carried out by adjusting the clock pulse edge from ATE, then evaluating the circuit's correct operation. The clock rates and scan enabled control mechanism both need to be adapted for testing circuit performance. T2 must be adjusted to be larger than T1. Traditional LOC techniques propose that the clock edge adjustable technique can be well executed using external ATE. The LOC scan delay test strategy covers the circuit functionality test and can detect the inter/intra clock domain delay faults. In this way, the system clock is speed up in the scan delay test phase. However, such a high-speed clock cannot be obtained nor applied to the circuit from using low-speed ATE.

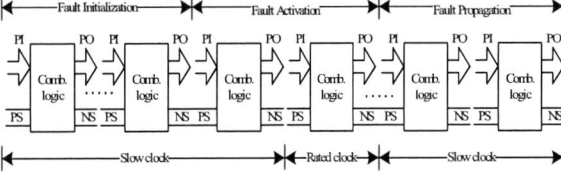

Fig. 2. The conventional slow-fast-slow high-speed testing technique.

Due to the limited length of this paper, we review only a small part of the current high-speed test literature. Datta et. al. [2] presented the modified vernier delay line (VDL) technique to provide high-resolution capability, as a long chain is needed to implement monitoring of the process variation. Fig. 2 shows the slow-fast-slow delay testing technique that uses two clock frequencies during a test operation [11]. A slow speed clock is used to fault initialization and propagation. The delay tester uses a low speed clock to send the pattern to initialize the circuit. After the test pattern preparation phase, a high speed clock is applied to activate the circuit high-speed operation. By using this technique, the circuit delay fault can be determined.

[12] proposed apply two clocks for pulse trigger flip-flop by using the adjustable cycle. As most circuits are designed by using a single clock, this technique needs to replace the flip-flops and increases design complexity.

As the high redesign efforts and limited clock rates make the above methods are not generally employed high-speed clock signals to the chip. It is also difficult to apply above techniques to future complex SoC chips with a multi-cycle clock domain.

978-1-61284-863-1/11 $26.00 © 2011 IEEE

Due to the IO pad's large load inability to correctly estimate the interconnection signal propagation time accurately, as well as pad quantity limitations, the insert test points for outside chip delay debugging are not practical. As the generic design uses a specified clock rate, designed circuit is hard to adopt arbitrary clock rates from external of the chip for the whole circuit. Thus, the internal chip built-in mechanism for resolving circuit delay and performance measurement related testing issues become essential.

2. The Proposed Double Edge Clipping Method

Scan-based high-speed testing for delay faults requires patterns that launch a transition from a scan cell or primary input (PI), and then capture the transition at a scan cell or primary output (PO). The key to high-speed testing is to generate a pair of clock pulses for the launch and capture events.

We proposed a built in high-speed testing BIST by precisely controlling the launch and capture edges (two trigger signals) during test operation. Our high-speed test scheme is based on the LOC scheme. Two built-in duty cycle adjust clocks (DGCK1 & DGCK2) were adopted for the CUT performance test, as shown in Fig. 3. We named this built-in high-speed test with calibration mechanism as the Double Edge Clipping (DEC) technique. The measured circuit delay time can be computed by using the difference in time within the launch and capture edges of the DEC. DEC has highly accurate test results with low area overhead that can be applied to generic chips to support the built in high-speed functional testing requirements of high-speed chips.

Using LOC scan delay test scheme, the capture operation is dependent on the first launch and second capture edge. The DEC differs from the circuit performance in that is uses two adjustable positive clock edges to execute the capture operation from the scan chain operation for a circuit. High-speed testing, speed binning and speed diagnosis were performed by decreasing the duration of two clock positive edges of the circuit until function failure. The test timing diagrams are shown in Fig. 3(a).

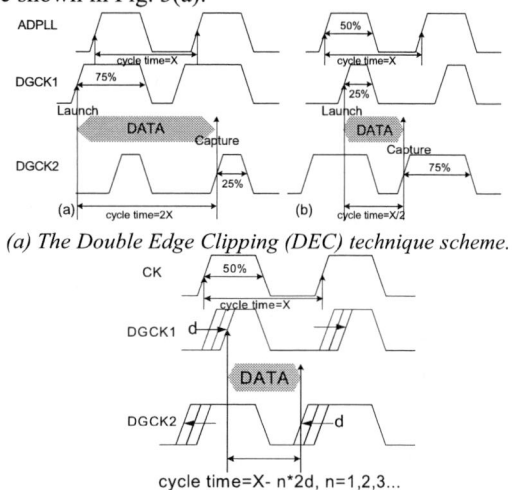

(a) The Double Edge Clipping (DEC) technique scheme.

(b) The clipped cycle time computations.

Fig. 3. The proposed Double Edge Clipping technique.

DEC technique adjusts the first and second trigger edges for circuit delay testing. The varied widths of the two positive trigger edges can be used to decide the circuit's real performance. From the duration time of two positive edges, we can compute the circuit's highest

working clock frequency. This is equivalent to increasing the CUT input external clock frequency.

In the DEC technique, two DCPG (duty clock pulse generators) generate progressive closed clock edges (cycle time reduction) after starting the next new testing cycle. Two clock edge widths (cycle time) represent the circuit delay time. The progressive closed edge adjustment method is used to push the circuit toward high-speed operation during the next new test. In Fig. 3(b), one DCPG adapting time is *d* and two DCPG clipped time is *2d*. The cycle time is X and the clipped cycle times become *X3/2-2d*, *X3/2-4d*, *X3/2-6d*...(*X3/2-n*2d, n=1,2,3...*). In Fig. 3(b), if we assume the original clock frequency is *f* (equals 1/X), and the duty cycle adapting minimum and maximum ranges are 25% and 75%, respectively, the highest and lowest frequencies are *2f* and *2f/3*, respectively.

This DEC test scheme can be implemented both for the launch-off-shift and launch-off-capture of two traditional delay test schemes. For DEC, it is best to use LOC scheme rather than a LOS scheme to test the high speed circuit.

2.1 Duty Cycle Adjustment Circuit Design

Fig. 4(a) shows the circuit structure of the DCPG. The DCPG will adjust the duty cycle from 26.47%~76.16%. To maintain the accuracy and the stability of the design, we adopted a delay line to design this circuit. There are 2-bit coarse and 5-bit fine adjustable mechanisms, and 128 stages that can be identified to adjust the edge. The coarse stage adjusting technique is controlled by the delay line length. The fine stage adjusting technique is controlled by the MOS capacitance (source-drain connected). The average of a fine adjustment edge duration time is 16ps, and the average of a coarse stage duration is 262ps.

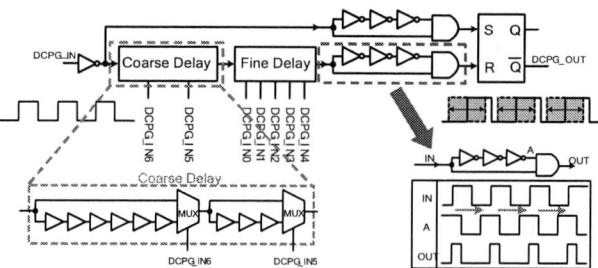

(a) A DCPG circuit with duty cycle adaption.

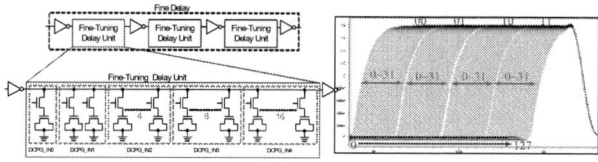

(b) The fine edge adaption simulation waveform.

Fig. 4. Duty cycle pulse generator circuit.

The DCPG fine adapting edge's ability and prelayout simulation is shown in Fig. 4(b). The simulation waveforms were obtained using the HSPICE sweep function. The areas of overlapping waveform are from coarse control mode transformation. This phenomenon might lead to adjusted pulse width duplication from the overlapping region. DCPG has a fine pulse adjusting ability. From the TT, FF, and SS model simulations, the MEAN values show that the DCPG circuit has very good stability with little variation. The TT(1.8V,25°C), FF (1.92 V, 0°C) and SS (1.62V, 125°C) models are

978-1-61284-863-1/11 $26.00 © 2011 IEEE

incorporated in the design simulation. From the SS model slope, there are two regions with discontinuous lines during the coarse adjustment.

Two DCPG are used to adapt the internal clock duty cycle and trigger edges to support the high speed test requirement internal to the chip. DGCK1 and DGCK2 represent the edge adjustment by using DCPG1 and DCPG2, respectively. In this manner, the internal high-speed testing and speed binning requirements can be satisfied. The DCPG fine turning edge has good resolution. These high accurate testing results allow a quick response to the internal chip's physical status.

The double edge trigger signal is simple generated, as shown in Fig. 5. The DGCK1 will be sent to the CUT to trigger the circuit launch operation. When DGCK1 captures the logic-0 from the scan enable signal, it allows the multiplexer to select the DGCK2 signal. The scan enable signal transitions from high to low before the CUT goes to double edge test mode. The DGCK1 output transition to high signal will capture logic-0 from the scan enable signal, allowing the multiplexer to choose DGCK2 signal for the CUT capturing operation. There is a 20ps delay in the transmission of the DGCK1 and DGCK2 pulse edges to the CUT.

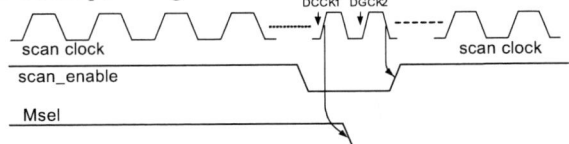

Fig. 5. The clipping edges generation timing diagram.

In Table 1, when the internal chip clock generator ADPLL produces two types of clock signals (250MHz or 500MHz) sent to the DCPG, which generates the adjustable clock duty cycle (equivalent to clock trigger edge adjustment). That is equivalent to the frequency band of 197MHz to 922.5MHz.

Table 1. The double edge adjustment range.

Double Edge Clipping Technique		
Input Freq.	Output (DGCK1 & DGCK2)	
250MHz	Period	5.063ns~2.962ns
	Frequency	197MHz~337.6MHz
500MHz	Period	3.063ns~1.084ns
	Frequency	326.5MHz~922.5MHz

3. Built-in Self Clipping Mechanism

The proposed DEC technique cooperates with the LOC test timing diagram is shown in Fig. 6(a). It applies a launch-followed-by-capture pulse for high-speed testing for an intra clock domain. This approach is conducted by a synchronized adjustment of the launch and capture edges to test all clock domains.

To quickly understand the circuit's highest performance, a quick locate mechanism is used in this adjustable mechanism as shown in Fig. 6(b). The clipping edge adjustment is activated dependent on a functional BIST checking result. If the BIST checking result is correct, the progressive clipping action will proceed. If BIST checking result is incorrect, the clipping mechanism will be released. These progressive clipping actions follow the binary search indication.

The Binary-Search circuit is used to control two DCPG, quickly reaching the CUT ratio clock frequency. From the BIST comparisons, the circuit delay time testing results can be rapidly obtained by quickly applying the suitable two trigger (clipped) edges to CUT.

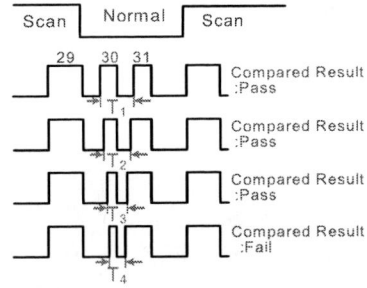

(a) The DEC LOC test timing diagram.

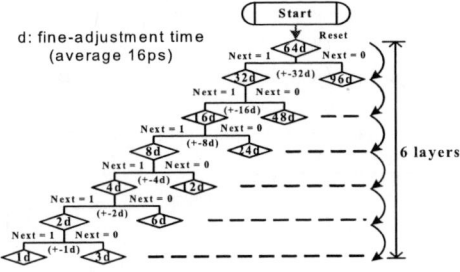

(b) The quick clipped edge binary search technique.
Fig. 6. The DEC quick clipping mechanism.

Fig. 7 shows from Binary_value=64, the original two pulse width (cycle time) is 2ns (500MHz). At Binary_value=96, the two pulse width (cycle time) is 1.54ns (649.3MHz). When the fine clipping result for Binary_value=126 is a two pulse width (cycle time) of 1.35ns (740.7MHz). For Binary_value=127, the two pulse width (cycle time) is 1.33ns (751.9MHz). Test chips postlayout simulation results show that the proposed clipping technique can accurately adjust the clipped edges. The binary search supports fine clipped edge adjustment ability, with a time difference (fine clipped edge adjustment ability) of 16ps. The circuit small delay time can be effectively identified by this mechanism.

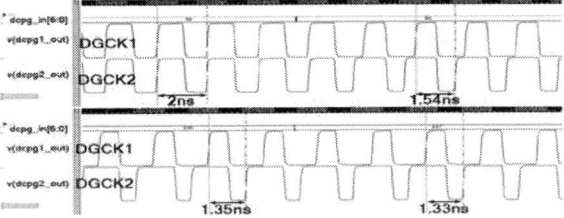

Fig. 7. DEC DGCK1 & DGCK2 postlayout simulations

4. BIDM Calibration Module in DEC Technique

An all-digital high resolution built-in delay measurement (BIDM) circuit that applies Vernier Delay Line (VDL) is proposed to calibrate the DEC technique. The circuit structure and physical design are shown in Fig. 8(a)(b). BIDM circuit utilizes the VDL concept to provide a high precision measurement when compared to previous techniques and can be easily embedded within the CUT. The VDL stages are kept short for maintaining measurement accuracy. We adopted VDL design to transfer measuring time into digital values, as a time-to-digital transformer. The VDL is composed of a positive edge trigger D-type Flip-Flops (DFF) and buffers. The wide OR (WOR) dynamic circuits are used to decrease the circuit delay and improve measurement accuracy with less area.

978-1-61284-863-1/11 $26.00 © 2011 IEEE

The BIDM measures the time within two clock pulses (DGCK1 and DGCK2), BIDM is used to calibrate the DCPG cycle time adjustment results. After the post-layout simulation, the average measured delay times for one fine stage and one coarse stage are 28ps and 140ps, respectively. BIDM mechanisms can support the BDBC test chip clock measurement maximum range of 2800ps.

BIDM circuit has high timing precision and has been proven in silicon by real chips [3]. Thus, our design makes it possible to calibrate small time differences inside two positive clock edges, and uses scan chains to scan out the measurement results for external chip observation. BIDM is digitally designed with less area overhead. The fully cell-based designed circuit has higher capability to tolerate process variations than other analog circuits.

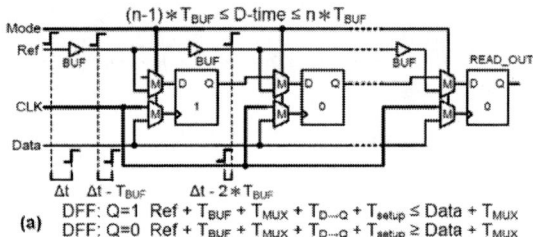

(a)
$$DFF: Q=1 \quad Ref + T_{BUF} + T_{MUX} + T_{D \to Q} + T_{setup} \leq Data + T_{MUX}$$
$$DFF: Q=0 \quad Ref + T_{BUF} + T_{MUX} + T_{D \to Q} + T_{setup} \geq Data + T_{MUX}$$

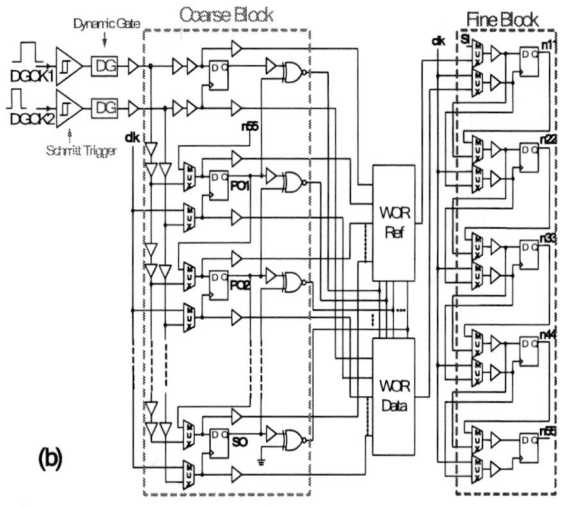

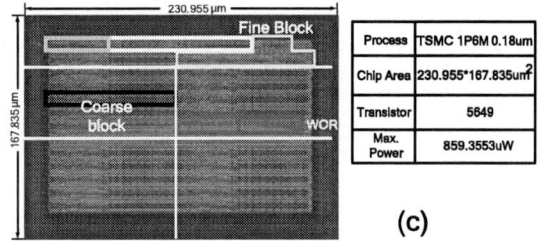

Fig. 8. The BIDM circuit structure with layout.

The BIDM circuit, layout and specification are shown in Fig. 8(c). BIDM captures two clock pulse rising edges (DGCK1 and DGCK2) from DCPG and measures the width time within these two pulses. BIDM measurement values are used to calibrate the DCPG adjusting results (i.e. to compare with DCPG control bits). The BIDM measurement results and the DCPG control bits are stored and scanned out externally of the chip for computation and comparison.

5. DEC Test Chip Implementation

Two test chips were implemented. Chip1 with HOY interface was used for the designed CUT connection to the HOY wireless test system. Chip2 was used as the hard IP, implemented within the CUT for connecting to a slow clock.

The test Chip1 integrated double edge clipping test scheme with HOY interface is shown in Fig. 9. The DGCK1 and DGCK2 are applied to the CUT during the time the test circuit goes into normal mode.

The DEC technique includes two digital controlled pulse generator circuits (DCPG), the built-in delay time measurement circuit (BIDM) with several functional BIST circuitries. Two DCPG are used to adjust the launch and capture trigger clock edges to push the circuit to its highest working frequency. The DCPG design target is to design a fine-scale, large-range clock duty cycle adjustable mechanism for circuit high-speed testing purposes. The other circuit parts not mentioned here support the communication and control signals with the HOY wireless test systems.

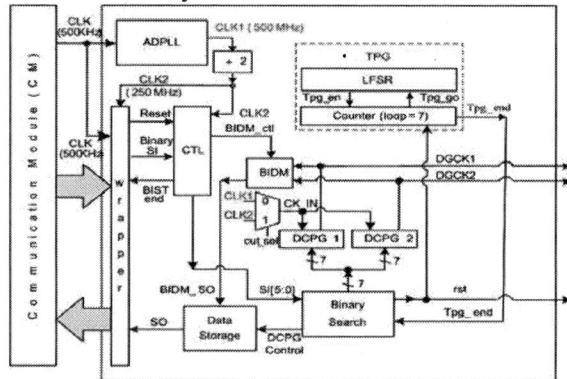

Fig. 9. The test Chip1 circuit block diagram.

The ADPLL is the circuit's original clock signal generator. Two DCPGs use this clock source to generate the two pulse edge adapted signals to test the CUT. Our built in testing mechanisms can automatically implement the clock edge adjustable technique.

The test Chip1 is fabricated by the National Chip Implementation Center (CIC). The full chip view and specifications are shown in Fig. 10.

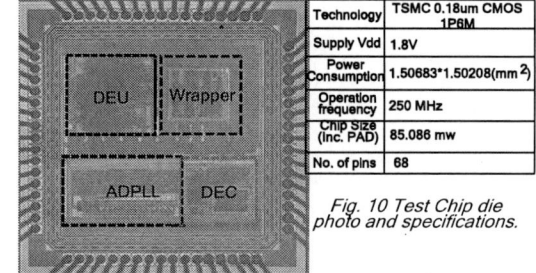

Technology	TSMC 0.18um CMOS 1P6M
Supply Vdd	1.8V
Power Consumption	1.50683*1.50208(mm^2)
Operation frequency	250 MHz
Chip Size (inc. PAD)	85.086 mw
No. of pins	68

Fig. 10 Test Chip die photo and specifications.

6. Conclusion

The DEC test chip design was effectively validated by a wireless (HOY) test system. The characteristics of the proposed DEC technique are proven: 1. On chip functional performance binning; 2. Quick built-in high-speed testing mechanism; 3. Support of LOC scan delay test strategy; 4. With circuit BIST functional test; 5. Support of low-speed external clock speed transformation transfer to high-speed internal pulse clock ability; 6. High-precision delay time measurement (BIDM) circuit for self-calibration mechanism design.

978-1-61284-863-1/11 $26.00 © 2011 IEEE

References:

[1] C. W. Wu, C. T. Huang, S. Y. Huang, P. C. Huang, T. Y. Chang, Y. T. Hsing, "The HOY Tester-Can IC Testing Go Wireless?," IEEE International Symposium on VLSI Design, Automation and Test, pp. 183–186, 2006.

[2] R. Datta, A. Sebastine, A. Raghunathan, J. A. Abraham. "On-chip Delay Measurement for Silicon Debug," IEEE Great Lakes Symposium on VLSI, pp.145-148, 2004.

[3] M. C. Tsai, C. H. Cheng, C. M. Yang, "An All-Digital High-Precision Built-In Delay Time Measurement Circuit," IEEE VLSI Test Symposium, 2008.

[4] Wei-Ming Lin; Shen-Iuan Liu, "An all-digital reused-SAR delay-locked loop with adjustable duty cycle", IEEE Asian Solid-State Circuits Conference, pp:312 – 315, 2007.

[5] K. Minami, M. Mizuno, H. Yamaguchi, T. Nakano, Y. Matsushima, Y. Sumi, T. Sato, H. Yamashida, M. Yamashina, "A 1GHz portable digital delay-locked loop with infinite phase capture ranges," in IEEE Int. Solid-State Circuits Conf. Dig. Tech. Papers, Feb. 2000, pp. 350-351.

[6] J. T. Kwak, C. Ki Kwon, K. W. Kim, S. H. Lee, J. S. Kih, "A low cost high performance register-controlled digital DLL for 1Gbps x32 DDR SDRAM," Symp. VLSI Circuits Dig. 17, pp. 283-284, Jun. 2003.

[7] T. Hamamoto, K. Furutani, T. Kubo, S. Kawasaki, H. Iga, T. Kono, Y. Konishi and T. Yoshihara, "A 667-Mb/s operating digital DLL architecture for 512-Mb DDR SDRAM," IEEE J. Solid-State Circuits, vol. 39, no. 1, pp. 194-206, Jan. 2004.

[8] Y. J. Jeon, J. H. Lee, H. C. Lee, K. W. Jin, K. S. Min, J. Y. Chung, and H. J. Park, "A 66–333-MHz 12-mW register-controlled DLL with a single delay line and adaptive-duty-cycle clock dividers for production DDR SDRAMs," IEEE J. Solid-State Circuits, vol. 39, no. 11, pp. 2087-2092, Nov. 2004.

[9] J. S. Wang, Y. M. Wang, C. H. Chen, Y. C. Liu, "An ultra-low-power fast-lock-in small-jitter all-digital DLL," in IEEE Int. Solid-State Circuit Conf. Dig. Tech. Papers, Feb. 2005, pp. 422-423 and 607.

[10] Chen-I Chung, Jyun-Sian Jhou, Ching-Hwa Cheng, and Sih-Yan Li, "Functional Built-In Delay Binning and Calibration Mechanism for On-Chip at-Speed Self Test", IEEE Asia Test Symposium, 2009.

[11] Angela Krstić, Kwang-Ting Cheng, "Delay Faults Testing for VLSI Circuits", Kluwer Academic Publishers, 1998.

[12] V.D. Agrawal, T. J. Chakraborty,"High-performance circuit testing with slow-speed testers", IEEE International Test Conference, 1995.

Appendix: Test Chip Measurement Analysis

The test chip measurement results from using oscillator scope and logic analyzer are shown in Fig. A1.

Table A1 shows comparisons with existing designs. The DEC considers the area and power consumption overhead, and has a tradeoff between internal delay testing accuracy with less area and power increases. Our design has less area with a fine duty cycle adjustment capability. However, there is a large peak to peak jitter in the same DCPG. This impaction is less for DEC delay testing accuracy, as two edges are sampled from different DCPG during the test phase.

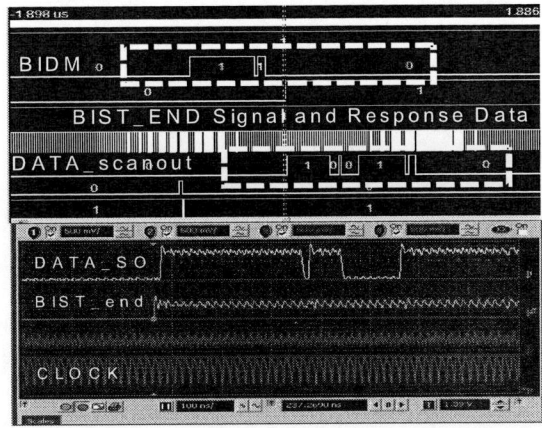

Fig. A1. Test chip validations use scope & logic analyzer.
Table A1. The DCPG comparisons with existent designs.

	[5]	[6]	[7]	[8]	[9]	[4]	Our DCPG
Technology	CMOS 0.15um	CMOS 0.13um	CMOS 0.13um	CMOS 0.15um	CMOS 0.25um	CMOS 0.18um	CMOS 0.18um
Supply	1.6~2V	1.8V	X	2.1V	1V	1.8V	1.8V
No. of output clocks	1	1	1	1	1	2(Expandable)	2
Max. / Min. operating frequency	1.2GHz/ 0.8GHz	500MHz/ 66MHz	333MHz	333MHz/ 66MHz	100MHz	800MHz/ 300MHz	500MHz
Output duty cycle	X	50%±2%	50%	50%	X	16.7%~83.2% @300MHz	26.5%~76% @500MHz (922.5MHz~ 326.5MHz)
Jitter (pp)	29ps	<25ps	88ps (2 clocks lock) 117ps (3 clock lock) @200MHz	X	30ps @100MHz	16ps @300MHz, 9.78ps @800MHz	89ps@500MHz
Power	36mW @1.8V	24mW @400MHz	12.6mA @250MHz	12mW @200MHz	2.43mW @100MHz	0.9mW @300MHz 2.7mW @800MHz	1.7mW @500MHz
Area	0.32mm²	X	0.254mm²	0.16mm²	0.096mm²	0.054mm²	0.004mm²

A1. HOY Wireless Test System Platform

The HOY wireless test system, shown in Fig.A2, uses wireless testing mechanisms to solve the testing challenges of advanced process technologies of current ATE [1]. This methodology decreases testing costs and is likely to become the main trend for future SoC testing. The ATE controller is a notebook using Linux OS. The test software programs are developed using the C++ and Python languages.

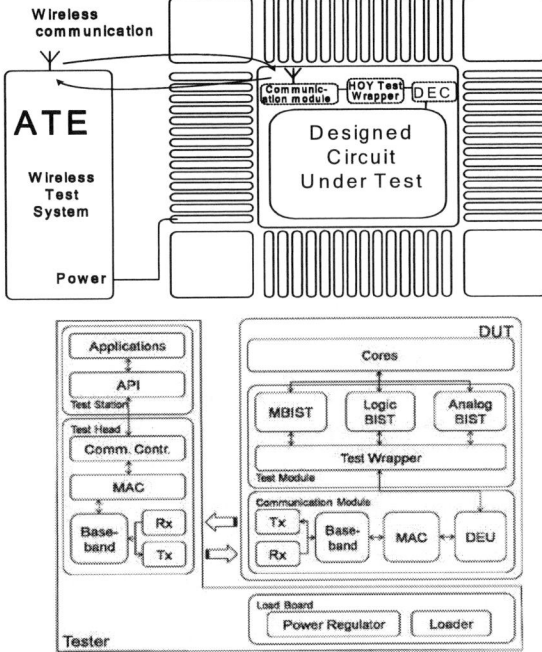

Fig. A2. The DEC integrated within HOY wireless test system.

As HOY is a low-price slow-speed tester, for HOY testing of higher speed circuit, the Built-in Delay Binning and Calibration Mechanism (BDBC) [10] has been proposed for scan based delay testing techniques. BDBC testing mechanism uses the single edge adjusting technique, which can raise the chip's lower internal system clock (from ADPLL) to a higher speed clock pulse during the testing mode. However, the BDBC has sufficient test frequency for low speed circuits.

By apply the DEC, the HOY test system uses a slow clock and can test the high-speed CUT (nearly 1GHz). In this paper, the DEC proposes an internal chip self-test technique to satisfy: high-speed testing, efficient speed binning, and delay diagnosis requirements. This technique can be used in the HOY wireless test system and other low-price slow-speed ATEs.

A.3 HOY Wireless Test System Validation

The proposed DEC algorithm can be adopted for the conventional slow-speed ATE. In this paper, the system validation is applied by using the HOY wireless test system.

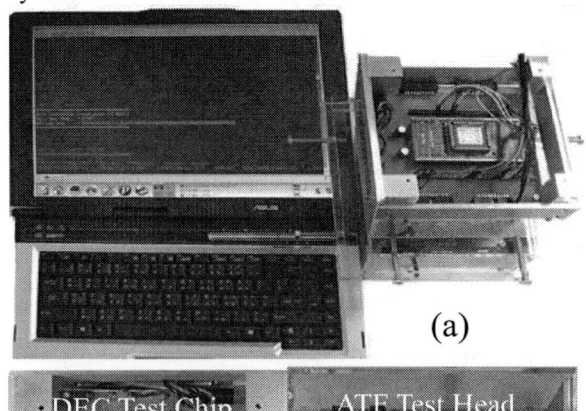

(a)

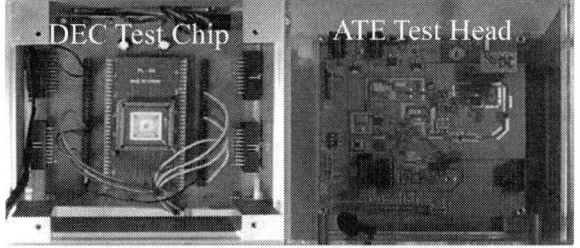

(b). The ATE response screen

Fig. A3. Chip1 validation uses HOY test system.

Test chips with wireless test system proved that DEC technique is fully functional and can accurately measure small path delay times from internal chip. Reporting of the timing measurement results for outside observation by scan chain, DEC chip test mechanism can be easily enabled from the generic test system and HOY wireless test system, as shown in Figs. A3(a).

The BIDM and DCPG results are scanned out after the built-in test mechanism is executed 6 times. For one CUT, a total of 33 bits are scanned out. BIDM scan out bits are used to measure the DCPG adjusting clock edges, and supports the high precision delay time measurement circuit with resolution capabilities of 140ps and 28ps for coarse and fine stages, respectively. The Binary-Search and BIDM bits can be used to mutually calibrate the

adapting pulse edges of DCPG. Fig. A3(b) shows the clipping control bits with BIDM measurement results that can be scanned out for outside calculation.

Table A2. The comparisons of DEC with the existent ATE.

Capture - Clocking Scheme	Intra - Structural	Intra - Delay	Inter - Structural	Inter - Delay	Sync. Design	Async. Design
One - hot single-capture	√	-	√	-	√	√
Staggered single- capture	√	-	√	√	√	√
One - hot skewed-load	√	√	√	-	√	√
Aligned skewed - load	√	√	√	√	√	-
Staggered skewed - load	√	√	√	√	√	√
One - hot double-capture	√	√	√	-	√	√
Aligned double-capture	√	√	√	√	√	-
Staggered double-capture	√	√	√	√	√	√
Our method	√	√	√	√	√	-

Table A2 summarizes the comparisons of the DEC technique with the existent ATE solutions. The clipping technique functionality is suitable for all synchronous type CUT. It should be noted that DEC technique is a convenient built-in high-speed testing technique. This technique makes automatic circuit at speed delay testing feasible. The proposed DEC technique affords high-speed with fine adjustment clock signal to test chip using a low-price slow-speed ATE.

The DEC has highly accurate test results with a low area overhead in generic chips. From the comparisons of the DEC technique with existing ATE solutions, the clipping technique functionality is suitable for all synchronous type CUT.

The DEC technique is a convenient built-in high-speed testing technique. The DEC serves as an internal design–for-test hard macro circuit to support the high speed internal test requirement. The DEC technique might also be used for external test application (on board) requirement.

A.3 DEC for Multi-CUT Test System Design

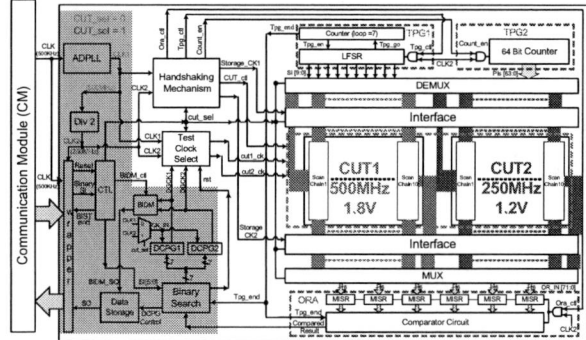

Fig. A4. Built in self high-speed test integrated design.

The test chip focus on the implementation of the shaded part, as shown in Fig. 9. The shaded regions are used to generate clipped pulse clock signals for applying the DEC test technique for two CUTs and the HOY wireless test wrapper design. The wholly integrated built-in high-speed test circuitry for multi-CUT (multi voltage domain) is shown in Fig. A4. The test assisted circuits include a test pattern generation and test compression modules, as well as a BIST executing mechanism. The BIST module supports functional testing to the CUT1 and CUT2 by using counter, linear feedback shift register (LFSR) and multiple input shift register (MISR) circuits for CUT1 and CUT2 functional testing internal to the chip.

A 0.35 V, 100 MHz, 0.19 μW/MHz, 3-Locking-Cycle All Digital Delay Locked Loop with Asynchronous-Deskewing Technology in 55 nm CMOS Technology

Chun-Yuan Cheng, Jinn-Shyan Wang, and Cheng-Tai Yeh

Dept. of Electrical Engineering/SOC Research Center

National Chung-Cheng University, Taiwan

Abstract—This paper presents an all digital delay-locked loop (*ADDLL*) that uses asynchronous-deskewing technology and achieves low power/voltage, small jitter, fast locking, and high process, voltage, and temperature (PVT)-variation tolerance. The measurement results show that the maximum frequency is 100 MHz at 0.35 V with 19μW power dissipation, 62 ps peak-to-peak jitter, and 3 locking cycles. When operated at 0.5 V, the measured maximal operating clock frequency is 450 MHz with 12 ps peak-to-peak jitter, 6 locking cycles, and 119μW power dissipation. The *ADDLL* is fabricated with 55nm CMOS technology, and the active area is only 0.019 mm^2.

I. INTRODUCTION

The delay locked loop (*DLL*) [1]-[5] has become an important component for safely clocking intellectual property (*IP*) blocks of future system-on-chips (*SoCs*) that employ the block-based power-down design [6]. Thus, low power, small jitter, and fast lock-in are the three basic design goals of the *DLL* for *SoC* applications. To achieve highly energy-efficient designs, scaling voltage to near or sub-threshold voltage regions is an important technique [7]. When the supply voltage is degraded to a sub-threshold region, the ultra-low-voltage (*ULV*) designs operate on the order of tens of megahertz. However, energy-saving circuits are sensitive to PVT variations in the ultra-low-voltage region. Thus, low voltage, high frequency, and robustness to PVT variations are other important design criterions for a *DLL* designed for energy efficient *SoC* systems.

Recently, a 55 nm *ADDLL* [5] showed very good performance in terms of power consumption, jitter, and lock-in speed by adopting a second-generation half-delay-line skew-compensation circuitry (*HDSC-II*) and an improved SAR (*iSAR*) controller. The *HDSC-II* is a low-power de-skewing circuit, and it serves as the coarse-operation engine to help the *ADDLL* get locked with a phase

error of about the delay time of a coarse delay cell in two clock cycles. The *iSAR* is used to achieve fast fine lock-in and has the fine-control-word bound-detection capability for intimate PVT tracking during operation. However, because of the use of the *HDSC-II*, the external clock signal must propagate though at least one coarse delay cell, and the highest frequency of the *ADDLL* decreases. The other bottleneck of the *HDSC*-based *DLL* at high frequencies is delayed feedback control, which occurs in the *iSAR* counter. Traditional *DLLs* [2] often use the clock division technique to lower the operating frequency of the counter so that the loop counter can operate correctly. Nevertheless, the technique induces longer locking cycles.

To pursue all the design goals, several design techniques are adopted in this work. First, a new coarse-fine architecture, which is constructed on top of a modified open-loop skew compensation circuit, and an asynchronous SAR controller are used to achieve low power, small jitter, and fast lock-in. Second, a low-power up/down (U/D) controller with a high-speed phase detector is added to track run-time variations after lock-in. Third, a fine delay cell with high delay resolution for achieving small jitter is designed using differential digital techniques [5] to increase the ability of the cell to deal with PVT variations in advanced CMOS technology.

The proposed 55 nm *ADDLL* is designed with a supply voltage of 0.5 V with a maximum frequency of 450 MHz. When the supply voltage is reduced to 0.35 V, the proposed *ADDLL* still shows good performance in terms of power efficiency, jitter, and locking speed. The rest of this paper is organized as follows. The architecture of the proposed *DLL* is described in section II. Section III details the circuit implementations, and section IV provides the experimental results. Finally, conclusions are drawn in section V.

II. ARCHITECTURE AND LOCKING OPERATION

The block diagram of the proposed *ADDLL* is shown in Fig. 1.

978-1-61284-863-1/11 $26.00 © 2011 IEEE

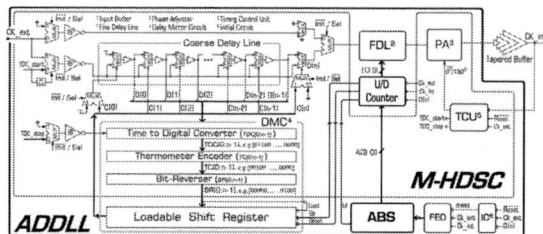

Fig. 1 The block diagram of the proposed *ADDLL*

This design integrates a modified half-delay-line de-skewing circuit (*M-HDSC*), a loadable shift register, an asynchronous binary search circuit (*ABS*), a first-edge detector (*FED*), an initial circuit (*IC*), a U/D counter, and a fine delay line (*FDL*). The HDSC-II only uses a half delay line to compensate the clock skew and can tolerate a seriously distorted duty cycle; thus, it can be easily applied to clock synchronization in a *SoC*. The *M-HDSC* evolved from the *HDSC-II* with an additional selection path inserted after the coarse delay line (*CDL*) because the clock signal can propagate though the added path without passing along the *CDL* when the clock skew is smaller than the delay of the coarse delay cell. This small design change increases the maximum frequency of the *ADDLL*. In addition, the *M-HDSC* removes the dummy circuits, including the *FDL* and phase adjuster (*PA*) in the *HDSC-II*, and it reduces the hardware overhead. The *IC* circuit controls the *ABS* controller whenever the *DLL* is to be activated. The *FED* circuit generates the required signals for the *ABS* loop counter. The *FDL* is a binary weighted delay line with a control code tuned by the U/D counter.

Fig. 2 shows the timing diagram of the proposed *ADDLL*. There are three phases operated in sequence: coarse locking, fine locking, and maintenance. During the first phase, the initial circuit, the first-edge detector, and the *ABS* are disabled. In this case, the *ADDLL* behaves as a modified *HDSC* with the *FDL* inserted along the clock paths and the shift register added below the bit reverser. The delay mirror circuit (*DMC*) is responsible for generating the control code for the coarse delay line. At the beginning of operation, the time-to-digital converter is reset, and all the output bits of the bit reverser are cleared. After obtaining the control code with a particular bit being set to logic 1, a large OR gate is used to generate the load signal automatically for the loadable shift register. Then, the control code is downloaded into the shift register and fed back to the coarse delay line. Because of the non-zero response time of the coarse control word produced by the *DMC* circuit, the

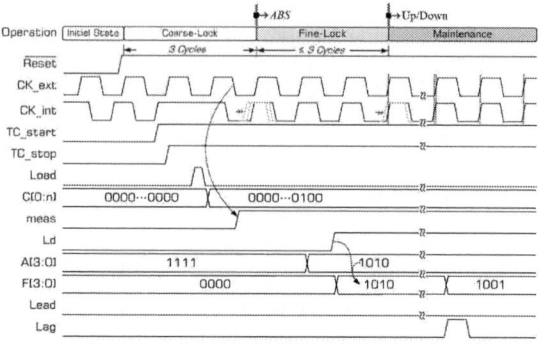

Fig. 2 Operation waveforms of the proposed *ADDLL*

ADDLL still requires one cycle to stabilize during high frequency operation. Therefore, it is easy to show the *ADDLL* can become coarse locked after three clock cycles.

After coarse locking, the phase error between *CK_ext* and *CK_int* is no more than the delay time of a coarse delay cell, and the *DLL* enters the second phase to further reduce the phase error. The *ABS* controller is activated by the *IC* circuit to perform an asynchronous binary successive approximation search to find the control word for the *FDL*. In this design, we adopt a 4-bit binary weighted *FDL*, and therefore, it takes at most three clock cycles when operating at the highest frequency. As a whole, the proposed *ADDLL* takes at most six clock cycles to get locked after releasing the reset signal. After fine locking, the control code generated from the *ABS* controller is loaded into the U/D controller, and the *ADDLL* enters the maintenance phase. Thus, the output of the up/down counter can be adjusted based on the outputs of the phase detector inserted in the up/down counter, and then, the delay time of the fine delay line can be fine-tuned to track the environmental variations.

III. CIRCUIT IMPLEMENTATION

A. Asynchronous binary search circuit

To increase the locking speed, an asynchronous-binary-search circuit has been presented [4]. However, large power consumption is a serious problem because it uses several phase detectors and delay lines in the controller, and the clock signal continues to pass through the sub-circuits even after the lock-in operation. Therefore, the *ABS* counter adopted in this work is only activated in the initial fine lock-in to achieve low power dissipation. The 3-stage *ABS* circuit is shown in Fig. 3. The proposed *ABS* is modified from the traditional asynchronous binary circuit [1]. The circuit comprises

978-1-61284-863-1/11 $26.00 © 2011 IEEE

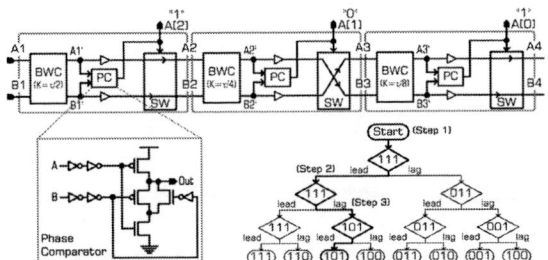

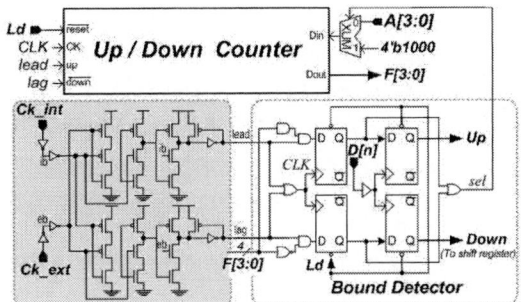

Fig. 3 A 3bit *ABS* counter and the operation principle

Fig. 4 The implementation of the U/D counter

path balancing buffers, binary-weighted cells (*BWC*), phase comparators (*PC*), and switches (*SW*). For fast locking and robustness in *ULV* designs, we adopt a simple *PC* with input buffers to increase the driving capability, and the asynchronous binary algorithm is different from the traditional algorithm.

After the coarse locking operation, the *ABS* circuit begins to convert the phase delay between the internal and external clocks into a binary-weighted control code. Only the first pulses of the external clock and the internal clock are extracted to be fed to the *ABS* controller. The signal A1 represents the rising edge of the external clock, and the signal B1 represents that of the internal clock. To accomplish the binary search, B1 first passes through the first *BWC* with a delay time of 0.5τ. The factor τ is the unit delay of the coarse delay line. Then, the switch determines the paths for A2 and B2 and is controlled by the output of the *PC*. If B1' leads to A1', the outputs of the *PC* are pulled to high, and A1' and B1' will be passed directly to become A2 and B2, respectively. If B2' lags behind A2', the output of the phase comparator will be low, and A2' and B2' become B3 and A3, respectively. This action mimics the action from step 2 to step 3. The phase procedure continues in the following stages until all control bits are obtained.

B. U/D counter

The U/D counter is shown in Fig. 4. It consists of an up/down counter, a bound detector, and a phase detector. After fine locking, the control code of the *FDL* is loaded into the counter, and the *ADDLL* enters the maintenance phase. Now, the output of the counter can be adjusted based on the output of the *PD*. Thus, the delay time of the *FDL* can be fine-tuned to track the environmental change. In addition, the clock source of the up/down counter and the bound detector is the output of the *PD*. Therefore, the circuits are disabled to save power if the clock skew is smaller then the dead zone of the phase detector.

The adopted *PD* is derived from a conventional high-speed dynamic PD [5]. NC²MOS latches are added, and input buffers are used to reduce the effect of a long rise or fall time of the input clock on the performance. Therefore, the *PD* is utilized for high operating speed to drive a heavy load and maintain high PVT-variation tolerance. In addition, a suitable dead zone is helpful for suppressing the output jitter caused by the dithering phenomenon [2]; the *PD* in this work is designed to be four times as large as the delay step of the fine delay line.

C. Fine delay line

The primary design goals of a fine delay line include high delay resolution and the ability to tolerate PVT variations. In this work, we used a difference delay cell [5], as shown in Fig. 5, to construct the *FDL*. This design is low power because only the components in one path of the delay cell are activated. The relationship between the control code and the delay time under the typical operating conditions (0.5 V, 25°C, TT) is shown in Fig. 5. The average delay resolution is designed to be 23.3 ps. The delay difference between two successive codes of the FDL is shown in Fig. 5. The delays range from 22 ps to 25 ps.

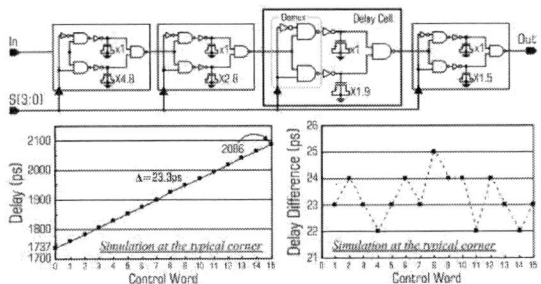

Fig. 5 The fine delay line

978-1-61284-863-1/11 $26.00 © 2011 IEEE

IV. EXPERIMENTAL RESULTS

The proposed *ADDLL* is implemented in 55nm 1P7M CMOS technology. The chip microphotograph is shown in Fig. 6, and the active area is only 0.019 mm². Performance comparisons are shown in Table I. When the *ADDLL* is operated at the maximum clock frequency of 100 MHz with 0.35 V, the peak-to-peak jitter is 62 ps, the lock-in time is 3 clock cycles, and the power dissipation is 19 µW. When operated at 0.5 V, the measured maximal operating clock frequency is 450 MHz with 12 ps peak-to-peak jitter, 6 locking cycles, and a 0.119mW power dissipation. The proposed *ADDLL* shows good performance in terms of power, locking speed, and jitter. It can also be used for voltage scaling.

V. CONCLUSIONS

For low power/voltage, high frequency, small jitter, fast lock-in, and robustness to PVT variation, several design techniques, including a new coarse-fine architecture constructed on top of the *M-HDSC* and the *ABS* controller, a low power U/D counter with a high-speed phase detector, and a variation-tolerable high-resolution fine delay cell, are proposed to design a new *DLL*. To the best of our knowledge, this is the first all-digital *DLL* operating at an ultra-low voltage of 0.35 V that maintains a wide range, small jitter, and sub-0.2 µW/MHz active power.

Fig. 6 The chip micrograph

Table I Performance comparisons

	[3] ('09)	[5] ('10)	This Work	
Design	Analog	Digital	Digital	
Process	0.18 µm	55 nm	55 nm	
Voltage	0.6 V	0.65 V	0.35 V	0.5 V
Area (mm²)	0.258	0.007	0.019	
f_{max}	550 MHz	300 MHz	100 MHz	450 MHz
f_{min}	85 MHz	80 MHz	4 MHz	20 MHz
f_{max}/f_{min}	6.5	3.8	25	22.5
Lock Time	n.a.	2-6 cycles	3-6 cycles	
p-p Jitter	25.6 ps @f_{max}	9 ps @f_{max}	62 ps @f_{max}	12 ps @f_{max}
Power @ f_{max}	4.2 mW	0.16 mW	0.019 mW	0.119 mW
Power Index (µW/MHz)	7.64	0.53	0.19	0.26

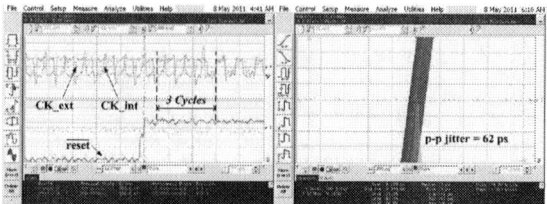

Fig. 7 Measurement results of the 0.35 V *ADDLL* at 100MHz

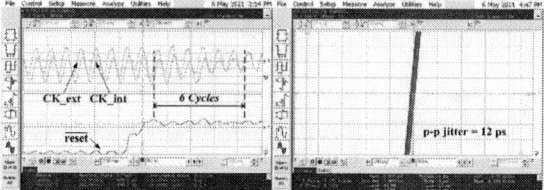

Fig. 8 Measurement results of the 0.5 V *ADDLL* at 450MHz

ACKNOWLEDGEMENT

The authors thank the National Science Council, the Ministry of Economic Affairs, and the NSoC Project of Taiwan for funding, and also thank UMC for supporting the chip fabrication.

REFERENCES

[1] J.-S. Wang, et al., "An ultra low power, fast lock-in, small jitter, all digital delay locked loop," *ISSCC Dig. Tech. Papers,* 22.7, pp. 422-423, Feb. 2005.

[2] R.-J. Yang, et al., "A 2.5 GHz all-digital delay- locked loop in 0.13µm CMOS technology," *IEEE J. Solid-State Circuits*, vol. 42, no. 11, pp. 2338-2347, Nov. 2007.

[3] C.-T. Lu, et al., "A 0.6 V low-power wide-range delay- locked loop in 0.18 µm CMOS," *IEEE Microwave and Wireless Components Letters*, vol. 19, no. 10, pp. 662- 664, Oct. 2009.

[4] D. Shin, et al., "A 7 ps 0.053 mm² fast-lock all-digital DLL with wide-range and high resolution all digital DCC," *IEEE J. Solid-State Circuits*, vol. 44, no. 9, pp. 2437-2451, Sep. 2009.

[5] J.-S.. Wang, et al., "A duty-cycle-distortion-tolerant half-delay-line low-power fast-lock-in all digital delay-locked loop," *IEEE J. Solid-State Circuits*, vol. 45, no. 5, pp. 1036-1047, May 2010.

[6] A. Allen, et al., "Dynamic frequency-switching clock system on a quad-core Itanium processor," *ISSCC Dig. Tech. Papers,* 3.4, pp. 62-63, Feb. 2009.

[7] M. Seok, et al., "A 0.27V 30MHz 17.7nJ/transform 1024-pt complex FFT core with super-pipelining," *ISSCC Dig. Tech. Papers,* 19.6, pp. 342-343, Feb. 2011.

978-1-61284-863-1/11 $26.00 © 2011 IEEE

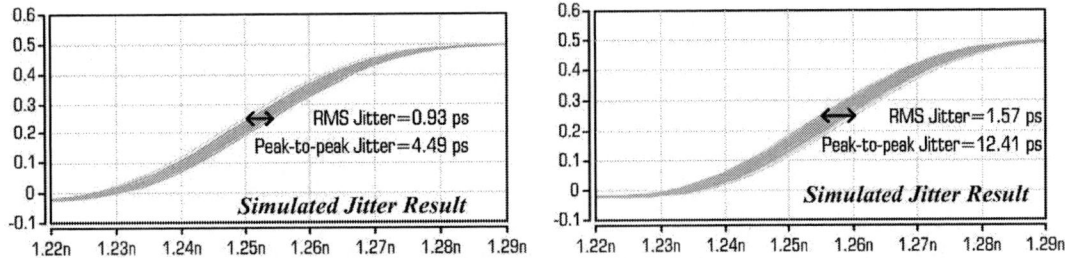

The left figure shows the simulated results at the supply voltage of 0.5 V. When operated at the clock frequency of 400 MHz, the root-mean-square (RMS) jitter is 0.93 ps and the peak-to-peak jitter is 4.49 ps. The right figure shows the simulated results with a 50 mV$_{pp}$ sinusoidal power noise. The RMS and the peak-to-peak jitters increase to 1.57 ps and 12.41 ps, respectively.

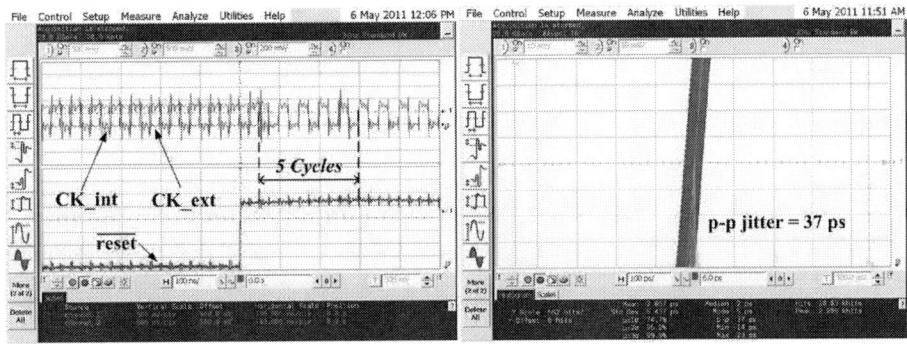

The figure shows the measurement results at the supply voltage of 0.5 V. For the sake of simplicity, the input duty cycle was set to 50% for these measurements. When operated at the minimal clock frequency of 20 MHz, the lock-in time is 5 clock cycles and the peak-to-peak jitter is 37 ps.

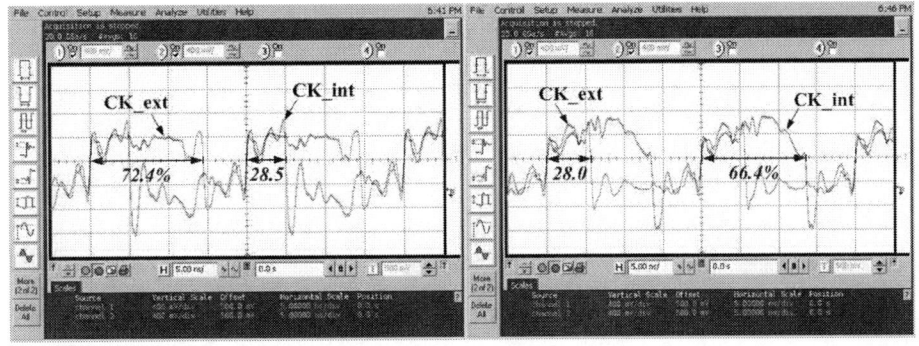

The measurement results for verifying the duty-cycle-variation tolerance, where the 0.5 V 50 MHz input clock has a duty cycle of 72.4% and 28.0%, respectively.

978-1-61284-863-1/11 $26.00 © 2011 IEEE

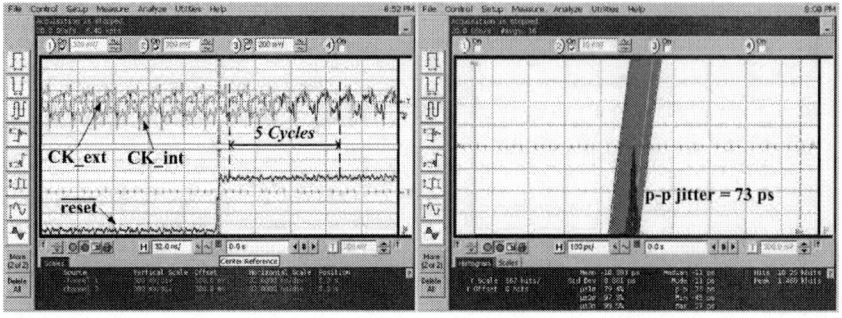

The figure shows the measurement results at the supply voltage of 0.35 V. When operated at the clock frequency of 50 MHz, the lock-in time is 5 clock cycles and the peak-to-peak jitter is 73 ps.

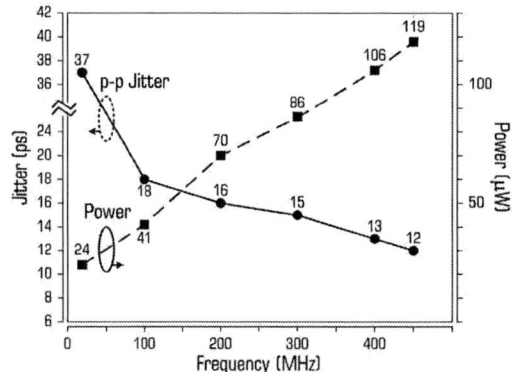

Measured jitters and power consumption of the ADDLL with a supply voltage of 0.5 V.

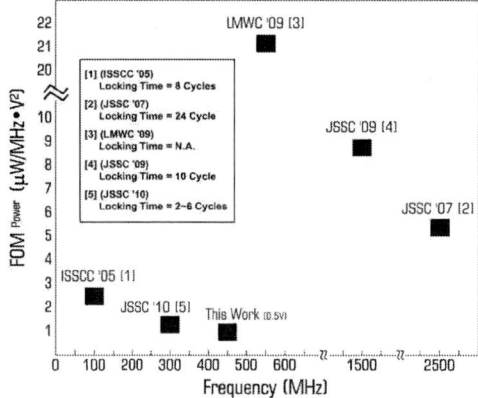

The figures-of-merit (FOM) for power consumption is displayed in the figure. FOM_{power} is defined as

$$FOM_{power} = \frac{Power\ Consumption(\mu W)}{Operating\ Frequency(MHz) \times Voltage(V^2)}.$$

A Wideband Fully Integrated +30dBm Class-D Outphasing RF PA in 65nm CMOS

Jonas Fritzin, Christer Svensson and Atila Alvandpour

Division of Electronic Devices, Department of Electrical Engineering, Linköping University

SE-581 83 Linköping, Sweden, Email: {fritzin, christer, atila}@isy.liu.se

Abstract—This paper presents a Class-D outphasing RF Power Amplifier (PA) which can operate at a 5.5 V supply and deliver +29.7 dBm with 26.6 % PAE at 1.95 GHz in a standard 65nm CMOS technology. The PA utilizes two on-chip transformers to combine the outputs of four Class-D stages. The Class-D stages utilize a cascode configuration, driven by an AC-coupled low-voltage driver, to allow a 5.5 V supply without excessive device voltage stress. The measured 3 dB bandwidth was 1.6 GHz (1.2-2.8 GHz). The PA was continuously operated for 168 hours (1 week) without any performance degradation. To evaluate the linearity of the outphasing PA, a WCDMA and an LTE signal (20 MHz, 16-QAM) were used. At +26.0 dBm channel power for the WCDMA signal, the measured ACLR at 5 MHz and 10 MHz offset were -35.6 dBc and -48.4 dBc, respectively. At +22.9 dBm channel power for the LTE signal, the measured ACLR at 20 MHz offset was -35.9 dBc.

Index Terms—outphasing, CMOS, power amplifier.

I. INTRODUCTION

With the scaling of CMOS transistors, the speed of the transistors have increased while being operated at lower supply voltages, making it more challenging to meet the requirement on output power and efficiency in Power Amplifiers (PA). With the improved speed of CMOS transistors, highly efficient switched PAs, like Class-D and Class-E, have gained increased interest in polar modulation and outphasing [1]–[4]. In the outphasing amplifier, an input signal, $s(t)$, containing both amplitude and phase modulation is divided into two constant-envelope phase-modulated signals, $s_1(t)$ and $s_2(t)$, as in Fig. 1(a). Fig. 1(b) shows how the two signals are separately amplified by efficient switched amplifiers, A_1 and A_2, and connected to a power combiner, whose output, $y(t)$, is an amplified replica of the input signal.

The output power of CMOS Class-D RF PAs has, until recently [1], been lower than +30 dBm [2]–[4]. To achieve higher output power, either a high supply voltage can be used to obtain high voltage swing or a low load impedance, i.e. a high impedance transformation ratio, is needed. Using a high impedance transformation ratio can result in low efficiency, especially with on-chip matching networks, and bandwidth reduction [5]. A higher voltage swing can be achieved in Class-D PAs by utilizing cascoding techniques and two supply voltages, $2 \times V_{DD}$ and V_{DD}, in the output stage and the drivers [2]–[4]. In that way, a higher voltage swing is achieved, but the device voltage stress is limited to the nominal supply voltage.

This paper presents a Class-D outphasing RF PA which can operate at a 5.5 V supply and deliver +29.7 dBm at 1.95 GHz in

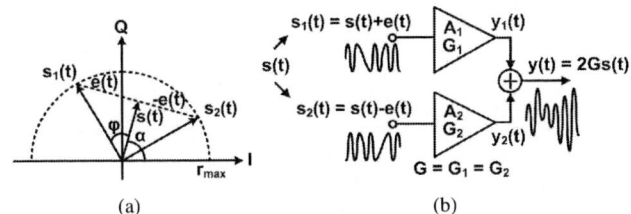

Fig. 1. (a) Outphasing concept and signal decomposition. (b) Ideal power combining of the two constant-envelope signals.

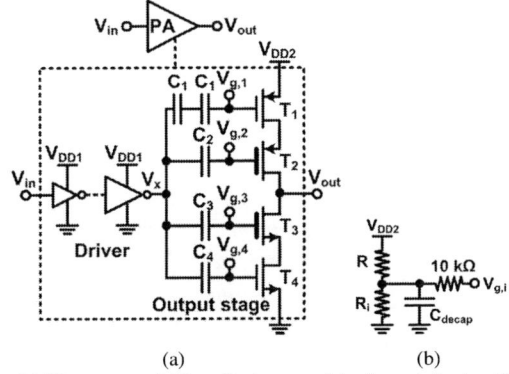

Fig. 2. (a) The proposed Class-D stage used in the outphasing PA. C_1-C_4 are MIM capacitors. T_4 is also biased due to reliability considerations. (b) Off-chip biasing resistors, R and R_i.

a standard 65nm CMOS technology. For a 6.0 V supply, the output power was +30.5 dBm. The PA utilizes two on-chip transformers to combine the outputs of four Class-D stages as shown in Fig. 5. The Class-D stages utilize a cascode configuration [1] illustrated in Fig. 2(a), driven by an AC-coupled low-voltage driver operating at 1.3 V, V_{DD1}, to allow a 5.5 V, V_{DD2}, supply without excessive device voltage stress. By driving all transistors in the cascode configuration it is possible to achieve a low on-resistance in the on-state, and distribute the voltage stress on the devices in the off-state, enabling the use of a high supply voltage in the output stage to achieve a high output power.

The outline of the paper is as follows. In Section II, the design of the Class-D stage, the outphasing RF PA and the transformer are presented. Moreover, the reliability considerations and the stress on the devices are discussed. In Section III, the measured RF performance and the performance for modulated signals are presented and compared with other work. In Section IV, the conclusions are provided.

978-1-61284-863-1/11 $26.00 © 2011 IEEE

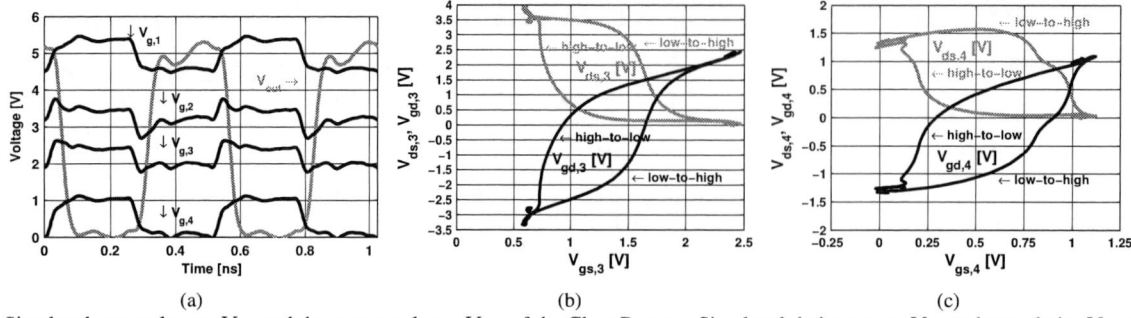

Fig. 3. (a) Simulated gate voltages, $V_{g,i}$, and the output voltage, V_{out}, of the Class-D stage. Simulated drain-source, V_{ds}, and gate-drain, V_{gd}, voltages of the (b) thick-oxide, T_3, and the (c) thin-oxide, T_4, devices.

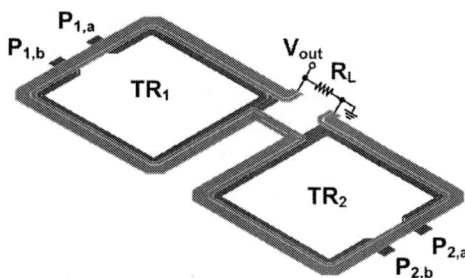

Fig. 4. Layout of the transformers. The two ports of the primary winding (in black) are indicated by $P_{i,a}$ and $P_{i,b}$. The equivalent secondary winding is drawn in grey. The floating metal shields are excluded for clarity.

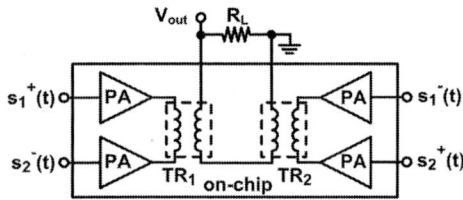

Fig. 5. The implemented Class-D outphasing RF PA using two transformers to combine the outputs of four amplifier stages.

II. DESIGN OF THE CLASS-D STAGE AND THE OUTPHASING RF POWER AMPLIFIER

A. Design and Operation of the Class-D Stage

The Class-D stages, denoted PA in Fig. 2(a) and Fig. 5, operate with a high supply voltage of 5.5 V, V_{DD2}, and utilize cascoded devices. The transistors used are 1.2 V thin-oxide devices, T_1 and T_4, and 2.5 V thick-oxide devices, T_2 and T_3, with oxide thickness of 1.8 nm and 5.0 nm, respectively. The gates of T_1-T_4 are separated from the driver stage by AC coupling capacitors, C_1-C_4, and individually biased via two off-chip resistors as in Fig. 2(b). All transistors are driven by an AC-coupled low-voltage driver operating at 1.3 V, V_{DD1}. The gate bias levels of T_1-T_4 are assumed to be $V_{DD2}-V_{DD1}/2$, $V_{DD2}/2+V_{DD1}/2$, $V_{DD2}/2-V_{DD1}/2$, and $V_{DD1}/2$, respectively.

When the output signal from the driver, V_x in Fig. 2(a), is high, the gate voltage of T_3 is raised above the bias level, which reduces the on-resistance of T_3. When the output signal from the driver, V_x, is low, the gate voltage of T_3 is lowered below the bias level. This also lowers the gate-drain, $V_{gd,4}$,

and drain-source, $V_{ds,4}$, voltages of T_4, but also increases gate-drain, $V_{gd,3}$, and drain-source, $V_{ds,3}$, voltages of T_3. The operation of T_1 and T_2 is the same, but they are in their on-state (off-state) when T_3 and T_4 are in their off-state (on-state). Consequently, by choosing suitable bias points and driving all transistors with a low-voltage driver, the voltage stress on the devices can be suitably distributed during the whole RF-cycle and a high supply voltage can be used. If only T_1 and T_4 are driven by the low-voltage driver, either a lower V_{DD2} or fixed bias levels $> V_{DD2}/2 + V_{DD1}/2$ for T_2 (or $< V_{DD2}/2 - V_{DD1}/2$ for T_3) must be used, which would reduce the output power.

To decouple V_{DD2}, MIM capacitors with a maximum allowed voltage of 5.5 V were used. The top/bottom plates of the MIM capacitors were connected to V_{DD2}/V_{DD1}. The low-voltage driver of the output stage is a tapered buffer with tapering factor $\lambda = 2.0$. The transistor widths of T_1-T_4 in Fig. 2(a) were 2.85 mm, 3.6 mm, 2.55 mm, and 2.23 mm, respectively. The channel lengths of the thin-oxide, T_1 and T_4, and thick-oxide, T_2 and T_3, devices are 0.06 μm and 0.28 μm, respectively. The bulks/N-wells are connected to G_{ND}/V_{DD2}.

B. Reliability Considerations

The reliability of CMOS transistors due to oxide degradation is especially important to consider in circuits with large voltage swings, like PAs. As discussed in [6], [7], the impact of RF stress is not as damaging as DC stress. During RF operation, the Time-Dependent Dielectric Breakdown (TDDB) is proportional to the root mean square (rms) value of the electric field applied to the gate oxide [8], [9]. In [8], the transistors had similar time to failure when rms RF and DC stress experiments were compared. Fig. 3(a) show the gate voltages, $V_{g,i}$, of T_1-T_4 in Fig. 2(a). Fig. 3(b) and Fig. 3(c) show the simulated drain-source, V_{ds}, and gate-drain, V_{gd}, voltages of the NMOS thick-oxide, T_3, and the thin-oxide, T_4, devices for the low-to-high and high-to-low transitions of the gate voltage, $V_{g,i}$, in the Class-D stage. The simulated rms electric fields between gate-drain, gate-source, and gate-bulk are < 0.7 V/nm gate oxide, which gives an expected lifetime of more than 10 years [10].

As V_{ds} is high ($\approx 1.5 \times V_{DD,nominal}$) when the transistors are in their off-state and V_{gs} is small, the HC stress is minimized [9]. The simulated V_{ds} is smaller compared to Class-AB PAs,

978-1-61284-863-1/11 $26.00 © 2011 IEEE

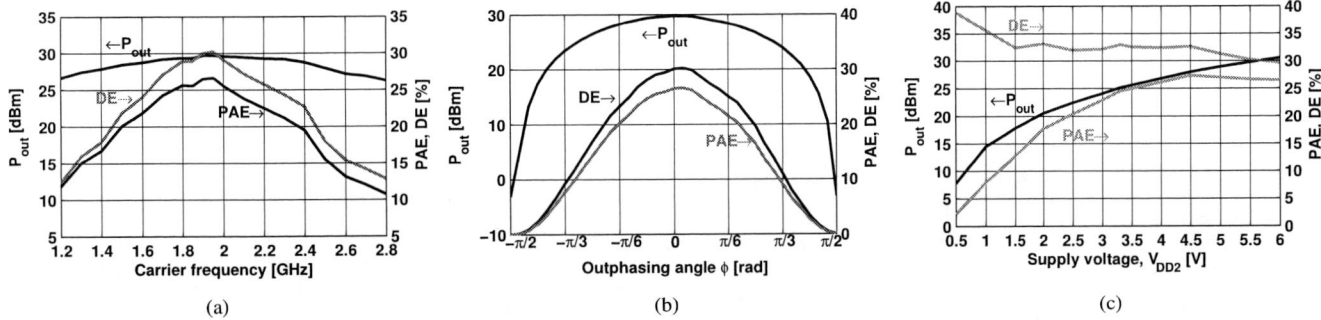

Fig. 6. Measured P_{out}, DE, and PAE for $V_{DD1} = 1.3$ V and $V_{DD2} = 5.5$ V: (a) over carrier frequency, and (b) over outphasing angle, ϕ, at 1.95 GHz. (c) Measured P_{out}, DE, and PAE over V_{DD2} for $V_{DD1} = 1.3$ V at 1.95 GHz.

where the cascode device is typically not driven by the driver and V_{ds} approaches $2 \times V_{DD,nominal}$ [9].

The drain/well breakdown of the 65nm process is 10 V. As the drain voltage of the proposed Class-D stage never exceeds V_{DD2}, this breakdown voltage is never exceeded.

C. Design of the Outphasing RF PA

Fig. 5 showed the implemented outphasing PA. In order to achieve a high voltage swing and high output power, the outputs of four amplifiers are combined by using two transformers, TR_1 and TR_2. One primary winding connect outputs between $s_1^+(t)$ and $s_2^-(t)$, and the other primary winding connect outputs between $s_1^-(t)$ and $s_2^+(t)$ [3]. With the secondary windings connected in series, the load impedance seen at the primary side of each transformer becomes $R_L/2$ (at maximum output power). At power back-off, the load impedance increase and the matching network losses are reduced [3]. No tuning capacitors were needed to optimize performance at 1.95 GHz.

D. Transformer Design

To get a good coupling factor and large bandwidth in the transformers, an overlay structure (stacked conductors) was used. The eight-metal stack of the 65nm process, does not contain thick metal layers ($> 2.5\,\mu$m) and limits the quality factors, Q, of the transformer windings. To reduce the losses in the windings, the traces were made wide. However, wide traces increase the parasitic capacitances and limit the number of turns in the windings. Therefore only transformers with a 1:1 turns ratio were used as shown in Fig. 4. Under TR_1 and TR_2, floating metal shields were placed in M_1 and M_2 to reduce the losses [11].

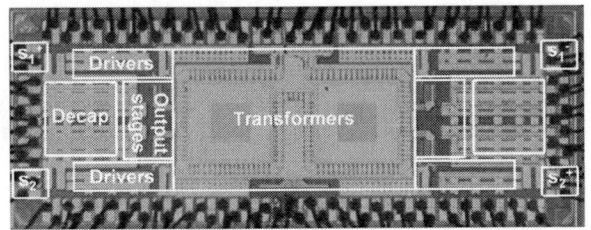

Fig. 7. Photo of the chip with size 2.5×1.0 mm². The photo has the same orientation as the simplified PA schematic in Fig. 5.

TABLE I
COMPARISON OF CMOS CLASS-D RF PAS - PEAK VALUES

Ref., Year	P_{out} [dBm]	V_{DD} [V]	DE [%]	PAE [%]	f [GHz]	Tech. [nm]	BW [GHz]
[3] 2010	+25.1	2.0	-	40.6	2.40	32	1.0[a]
[4] 2011	+25.2	2.5	-	55.2	2.25	90	1.0[b]
[2] 2009	+28.1	2.4	-	19.7	2.25	45	0.6[b]
[1] 2011	+32.0	5.5	20.1	15.3	1.85	130	0.9[b]
This work	**+29.7**	**5.5**	**30.2**	**26.6**	**1.95**	**65**	**1.6[b]**

(a) 1 dB and (b) 3 dB bandwidth (BW)

TABLE II
MEASURED SPECTRAL AND MODULATION PERFORMANCE AT 1.95 GHz

Standard	Parameter	Measured	Required
WCDMA	ACLR @ 5 MHz [dBc]	-35.6	-33
	ACLR @ 10 MHz [dBc]	-48.4	-43
	EVM [%]	< 1.0	17.5
	Channel power [dBm]	+26.0	
LTE	ACLR @ 20 MHz [dBc]	-35.9	-30
	EVM [%]	< 3.0	12.5[a]
	Channel power [dBm]	+22.9	

(a) 16-QAM modulation

The primary windings of TR_1 and TR_2 were implemented in M_5 and M_6 with total thickness and width of 1.1 μm and 45.0 μm, respectively. The secondary winding was placed in M_7 and M_8 with thickness and width of 1.3 μm and 45.0 μm, respectively. The self-inductances of the galvanically isolated primary windings, L_p, and equivalent secondary winding, L_s, (two series-connected secondary windings) were 0.8 nH and 2.0 nH, respectively. At 2.0 GHz, the quality factors, Q_p and Q_s, were approximately 5 and 5, respectively. In EM simulations, the loss of a single transformer was 1.0 dB.

III. MEASUREMENT RESULTS

A. Measured RF Performance

Fig. 7 shows the chip photo of the PA implemented in a 65nm CMOS process. The chip was attached to an FR4 PCB and connected with bond-wires. Two R&S SMBV100A signal generators with phase-coherent RF outputs and maximum IQ sample rate of 150 MHz were used in the measurements.

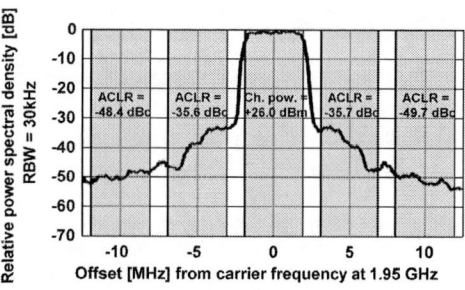

Fig. 8. Measured WCDMA spectrum for a channel power of +26 dBm.

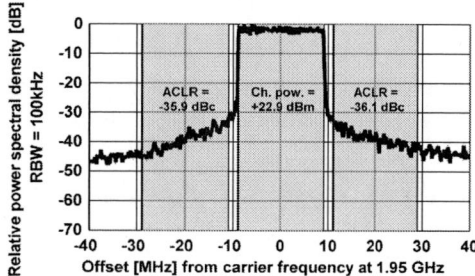

Fig. 9. Measured LTE spectrum for a 20 MHz signal with 16-QAM modulation for a channel power of +22.9 dBm.

Fig. 6(a) and Fig. 6(b) show the measured output power (P_{out}), drain efficiency (DE), and power-added efficiency (PAE) over frequency and outphasing angle, ϕ, for V_{DD1} = 1.3 V and V_{DD2} = 5.5 V. At 1.95 GHz, the output power was +29.7 dBm with a DE and PAE of 30.2 % and 26.6 % (including all drivers), respectively. The gain was 26 dB from the drivers to the output. The DC power consumption of the smallest drivers was considered as input power. For V_{DD2} = 6.0 V, the output power was +30.5 dBm with a DE and PAE of 29.6 % and 26.5 %, respectively. The 3 dB bandwidth was 1.6 GHz (1.2-2.8 GHz).

To obtain an initial assessment of the reliability of the Class-D stage and the PA, the PA was continuously operated for 168 hours for a V_{DD2} = 5.5 V without any performance degradation. Thus, no noticeable degradation of the devices has occured.

The PA is also suitable for polar modulation as shown in Fig. 6(c). By lowering V_{DD2}, the power consumption due to switching of the drain capacitances in the output stages is reduced. With a power efficient driver based on thin-oxide devices, the PAE is still as high as 17 % at 10 dB back-off (+20 dBm). An amplitude modulator has not been implemented in this work.

In Table I, published CMOS Class-D RF PAs are listed. Compared to previous works, the PA presented in this work has the largest 3 dB bandwidth. The efficiency is the highest of the Class-D RF PAs with an output power larger than +27 dBm (0.5 W) [1], [2].

B. Measured Performance of Modulated Signals

Table II presents the measured and required performance when uplink WCDMA and LTE signals were applied to the outphasing PA at 1.95 GHz. The channel powers of the WCDMA and LTE signals were +26.0 dBm and +22.9 dBm,

respectively. The PAPR of the WCDMA and the LTE signal were 3.5 dB and 6.6 dB, respectively. The measured spectrums are shown in Fig. 8 and Fig. 9. Spectral and modulation requirements were met without requiring predistortion. The PAE was 7.5 % when amplifying the LTE signal.

IV. CONCLUSIONS

This paper has presented a Class-D outphasing RF PA which can operate at a 5.5 V supply and deliver +29.7 dBm with 26.6 % PAE in a standard 65nm CMOS technology. The PA utilizes two on-chip transformers to combine the outputs of four Class-D stages. The Class-D stages utilize a cascode configuration to allow a 5.5 V supply without excessive device voltage stress. All transistors in the cascode configuration are driven by an AC-coupled 1.3 V driver. The measured 3 dB bandwidth was 1.6 GHz (1.2-2.8 GHz). The PA was continuously operated for 168 hours (1 week), with a 5.5 V supply, without any performance degradation. WCDMA and LTE (20 MHz, 16-QAM) signals were applied to the PA and for channel powers of +26.0 dBm and +22.9 dBm, respectively, the PA successfully met the spectral and modulation requirements. To the authors' best knowledge, the PA presented in this work has the largest 3 dB bandwidth of all Class-D RF PAs.

REFERENCES

[1] J. Fritzin, C. Svensson, and A. Alvandpour, "A +32dBm 1.85GHz Class-D Outphasing RF PA in 130nm CMOS for WCDMA/LTE," in *IEEE Europ. Solid-State Circ. Conf.*, Sep. 2011, to be presented.

[2] H. Xu, Y. Palaskas, A. Ravi, M. Sajadieh, M. Elmala, and K. Soumyanath, "A 28.1dBm class-D Outphasing Power Amplifier in 45nm LP Digital CMOS," in *VLSI Symp.*, Jun. 2009, pp. 206–207.

[3] H. Xu, Y. Palaskas, A. Ravi, and K. Soumyanath, "A Highly Linear 25dBm Outphasing Power Amplifier in 32nm CMOS for WLAN Application," in *IEEE Europ. Solid-State Circ. Conf.*, Sep. 2010, pp. 306–309.

[4] S.-M. Yoo, J. Walling, E. Woo, and D. Allstot, "A Switched-Capacitor Power Amplifier for EER/Polar transmitters," in *ISSCC Dig. Tech. Papers*, Feb. 2011, pp. 427–428.

[5] I. Aoki, S. Kee, D. Rutledge, and A. Hajimiri, "Distributed Active Transformer - New Power-Combining and Impedance-Transformation Technique," *IEEE Trans. Microw. Theory Techn.*, vol. 50, no. 1, pp. 316–331, Jan. 2002.

[6] C. Yu and J. Yuan, "MOS RF Reliability Subject to Dynamic Voltage Stress - Modeling and Analysis," *IEEE Trans. Electron Devices*, vol. 52, no. 8, pp. 1751–1758, Aug. 2005.

[7] J. Yuan and J. Ma, "Evaluation of RF-stress Effect on Class-E MOS Power Amplifier Efficiency," *IEEE Trans. Electron Devices*, vol. 55, no. 1, pp. 430–434, Jan. 2008.

[8] L. Larcher, D. Sanzogni, R. Brama, A. Mazzanti, and F. Svelto, "Oxide Breakdown After RF Stress: Experimental Analysis and Effects on Power Amplifier Operation," in *IEEE Reliability Physics Symposium Proc.*, Mar. 2006, pp. 283–288.

[9] M. Ruberto, O. Degani, S. Wail, A. Tendler, A. Fridman, and G. Goltman, "A Reliability-Aware Power Amplifier Design for CMOS Radio Chip Integration," in *IEEE Reliability Physics Symposium Proc.*, Jul. 2008, pp. 536–540.

[10] W. Chan and J. Long, "A 58-65 GHz Neutralized CMOS Power Amplifier With PAE Above 10% at 1-V Supply," *IEEE J. Solid-State Circuits*, vol. 45, no. 3, pp. 554–564, Mar. 2010.

[11] T. Cheung and J. Long, "Shielded Passive Devices for Silicon-Based Monolithic Microwave and Millimeter-Wave Integrated Circuits," *IEEE J. Solid-State Circuits*, vol. 41, no. 5, pp. 1183–1200, May 2006.

A 77-135GHz Down-Conversion IQ Mixer for 10Gbps Multiband Applications

Sanming Hu, Yong-Zhong Xiong, Lei Wang, Jinglin Shi, and Teck-Guan Lim

Institute of Microelectronics, A*STAR (Agency for Science, Technology, and Research), Singapore
11 Science Park Road, Science Park II, 117685, Singapore
Email: husm@ime.a-star.edu.sg

Abstract—This paper presents a down-conversion IQ mixer in 0.13-μm SiGe BiCMOS. The mixer exhibits a measured conversion gain of 1.6 dB at f_{LO} = 135 GHz and P_{LO} = - 4.2 dBm. It can directly down-convert 10-Gbps modulated millimeter-wave (mmWave) signals to base-band data. Without any design change or DC bias adjustment, the IQ mixer can operate at any frequency between 77 and 135 GHz. The mixer is therefore universal for many mmWave applications such as 77GHz automotive radars, 94GHz imaging, and 122GHz communications.

Keywords-Down-conversion; IQ mixer; millimeter-wave; multi-band; wideband

I. INTRODUCTION

The emerging applications such as automotive radars, high-resolution imaging, and high-date-rate communications are creating a steadily growing market for the silicon-based millimeter-wave (mmWave) systems [1].

As a key block in the above systems, down-conversion mixers have been reported in [2]-[9]. Nevertheless, many reported mixers do not have the IQ function and only operate at a narrow frequency band. Therefore, they are not suitable for multiband high-speed communications or reconfigurable applications.

In this paper, a direct down-conversion IQ mixer is described for 10-Gbps demodulation. Moreover, the designed mixer can efficiently operate at any frequency between 77 and 135 GHz without any physical change.

II. CIRCUIT DESIGN

The architecture diagram of the IQ mixer is illustrated in Fig. 1. It consists of a Wilkinson power divider, two identical single-ended mixers, and a Lange coupler which has 90° phase difference between the coupled and through output ports.

As shown in Fig. 2, the single-ended mixer adopts a Lange coupler to combine the RF and LO signals. A short-circuited stub (TL1) and capacitor (C1) are employed for input matching [10]. The short-circuited stub (TL1) also works as an electrostatic discharge (ESD) protection circuit because it can conduct the static charges to the ground directly. The mixing mechanism is enhanced by the nonlinear characteristic of

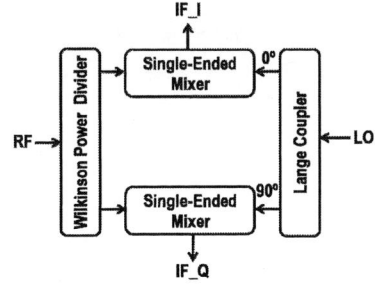

Fig. 1. The architecture diagram of the IQ mixer.

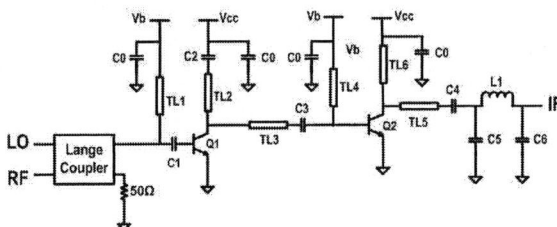

Fig. 2. The circuit schematic of the single-ended mixer.

transistor Q1 when the bias voltage at the base (Vb) is slightly higher than the threshold voltage of the transistor.

To suppress the RF and LO signal leakage at the IF port, a transmission line (TL2) is connected to the ground via a capacitor (C2) which also facilitate DC biasing.

An IF amplifier (Q2) is then integrated within the mixer to reduce conversion loss. It also works as a part of the matching network. A Π-network consisting C5, L1, and C6 is used for low-pass filtering of the LO and RF leakage at the IF port.

Simulated performance of this single-ended mixer is shown in Fig. 3. The results show that the 3-dB conversion-gain bandwidths are 13 GHz, 12 GHz, 12 GHz, and 13 GHz in the case of the carrier frequencies are 77 GHz, 94 GHz, 122 GHz, and 135 GHz, respectively. These 3-dB bandwidths are mainly limited by the low-pass filtering network. In addition, the conversion gain variation at these four frequency bands is within 3 dB.

978-1-61284-863-1/11 $26.00 © 2011 IEEE

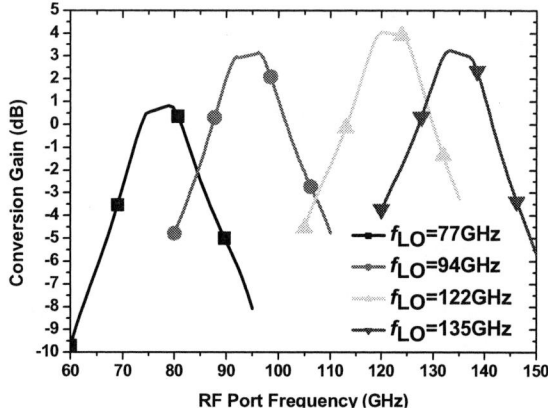

Fig. 3. Simulated conversion gain of the single-ended mixer integrated with an IF amplifier (Vcc = 1.6 V, Vb= 0.92 V, Ic = 2 mA, P_{RF} = -19.7 dBm, and P_{LO} = -7.9 dBm).

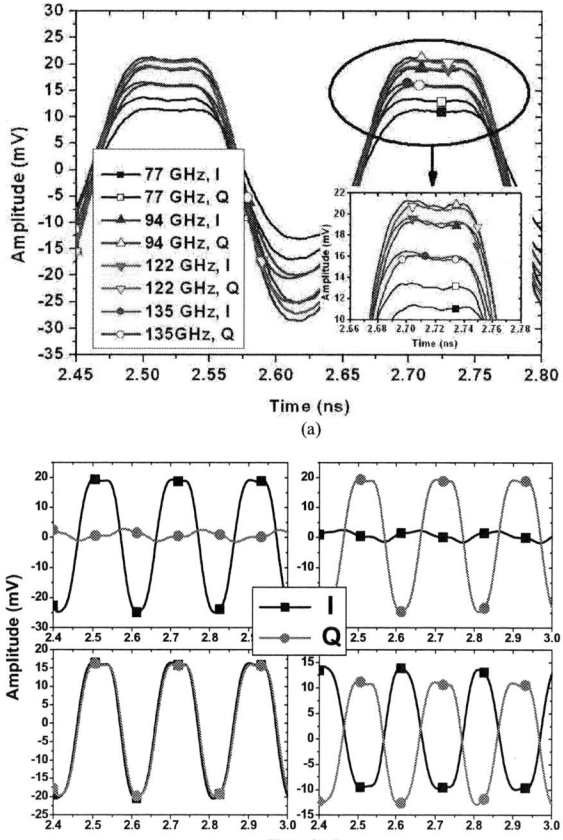

Fig. 4. Time-domain characteristics of the direct down-conversion IQ mixer (Vcc = 1.6 V, Vb = 0.92 V, Ic = 4 mA, P_{RF} = -16 dBm, and P_{LO} = - 4.2 dBm). (a) I and Q output with different operating frequencies when the LO and RF signals have phase difference of 45°. (b) I and Q output with different LO-RF phase difference (from left-top to right-bottom: 0°, 90°, 45°, and 135°.

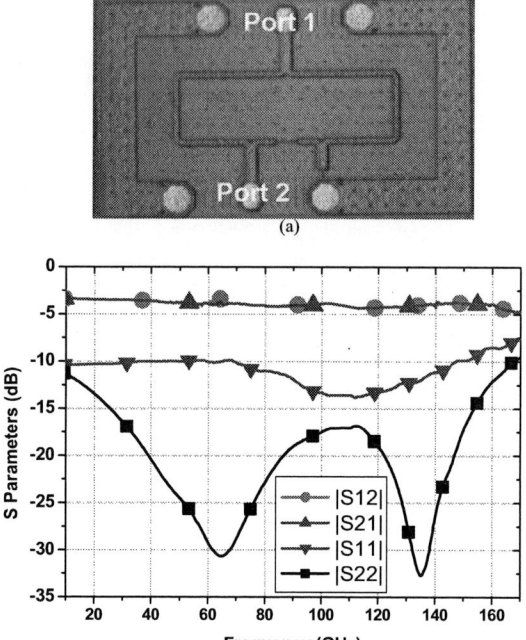

Fig. 5. Wilkinson power divider. (a) Die photograph. (b) S parameters.

To confirm wideband operation of the IQ mixer, a time-domain analysis was carried out when the RF signal is on-off keying (OOK) modulated. The time-domain waveforms for I and Q channels, when the LO and RF frequencies are the same and simultaneously changed, are depicted in Fig. 4 (a). The amplitude deviation is within 3 dB, covering these four frequencies. The 10-Gbps demodulated time-domain output for 135 GHz is shown in Fig. 4 (b).

III. MEASURED RESULTS AND DISCUSSIONS

The fabricated Wilkinson power divider is depicted in Fig. 5 (a). The die area of the Wilkinson power divider (including pads) is 0.5 × 0.3 mm².

After de-embedding the measured raw data, the two-port S parameters are illustrated in Fig. 5 (b). It shows that the measured 10-dB return-loss bandwidths of the two ports are 70.7-146.3 GHz, and 110-170 GHz, respectively. From 70 to 146 GHz, the total insertion loss of the Wilkinson power divider is about 0.76 dB.

The Lange coupler exhibits a measured insertion loss of 0.7 dB and a simulated phase difference of 89° between the through and coupled ports at 135GHz. More details are available in [11].

The IQ mixer fabricated in 0.13-μm SiGe BiCMOS technology is depicted in Fig. 6. The chip occupies 0.8 × 0.9 mm² including pads. A symmetrical layout with respect to the

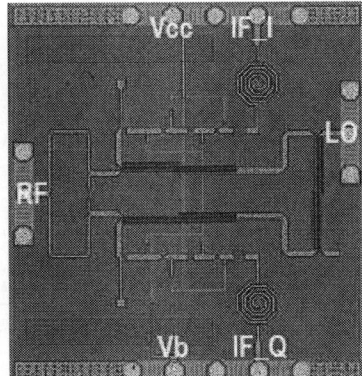

Fig. 6. Die photograph of the IQ mixer.

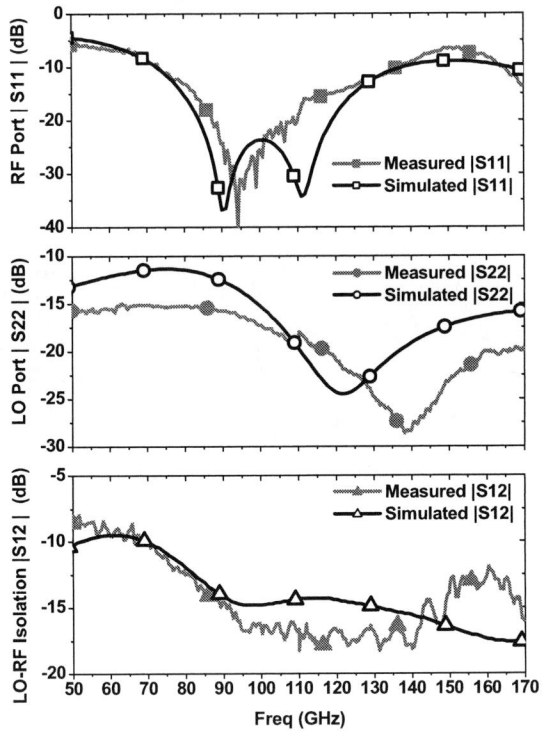

Fig. 7. Measured and simulated S parameters of the IQ mixer (Vcc = 1.6 V, Vb= 0.92 V, Ic = 4 mA, P_{RF} = -16 dBm, and P_{LO} = - 4.2 dBm).

RF-to-LO line is employed for proper IQ operation of the mixer.

S parameters of the IQ mixer are illustrated in Fig. 7. The 10-dB return-loss bandwidth of the IQ mixer covers the frequency range of 73 – 138 GHz at both RF and LO ports. The LO-RF isolation is higher than 11 dB. A complete circuit

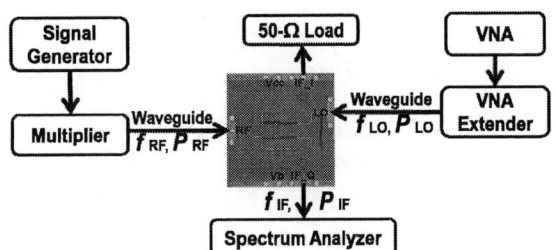

Fig. 8. On-wafer conversion-gain measurement setup.

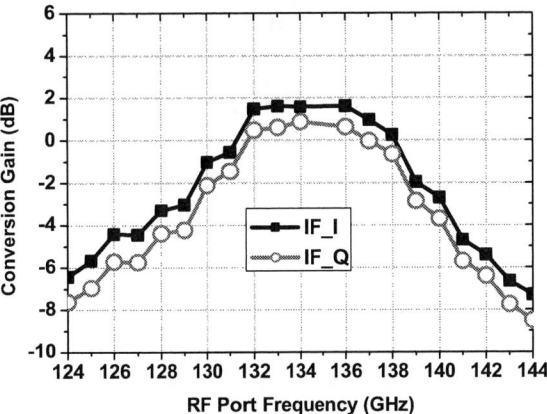

Fig. 9. Measured conversion gain of the IQ ports (Vcc = 1.6 V, Vb= 0.92 V, Ic = 4 mA, P_{RF} = -16 dBm, P_{LO} = -4.2 dBm, and f_{LO} = 135 GHz).

including LO buffer and LNA stages would further improve the LO-RF isolation.

The setup illustrated in Fig. 8 was adopted to measure conversion gain of the IQ mixer. Due to frequency limitation (120-170 GHz) of the measurement system, only the 135 GHz band measurement was carried out. As shown in Fig. 9, the conversion gain is 1.6 dB when the LO power is -4.2 dBm. The 3-dB conversion-gain bandwidth of the IQ mixer is from 130 to 138 GHz. The IQ amplitude imbalance is within 1 dB covering this frequency band. These frequency-domain results show that the designed mixer has wideband IQ performance when the carrier frequency is as high as 135 GHz.

The performances of this IQ mixer and other state-of-the-art designs are tabulated in Table I. The comparison shows that the IQ mixer in this brief exhibits the most multiple frequency bands together with a high conversion gain and low DC power consumption.

IV. CONCLUSIONS

An IQ mixer for ultra-high-speed communications is presented in this brief. This SiGe mixer can operate at

978-1-61284-863-1/11 $26.00 © 2011 IEEE

TABLE I

COMPARISON OF REPORTED MILLIMETER-WAVE MIXERS

Ref.	Tech.	Type	IQ	Freq (GHz)	P_{DC} (mW)	Conversion Gain (dB)	Area (mm²)
[2]	130 nm CMOS	Single-Balanced	×	58-73	7.5	0.5	0.8 x 0.7
[3]	130 nm CMOS	Single-Balanced	×	57-63	2.4	-2	1.6 x 1.7
[5]	250 nm SiGe HBT	Micromixer	×	77/79	175.5	13.4/7	0.5 x 0.55
[8]	SiGe HBT	Double-Balanced	×	160-180	49.5	-24	0.65 × 0.7
[9]	65 nm CMOS	Double-Balanced	√	85-100	208	Max 10.5 (whole Rx)	0.51 × 0.7
This Work	**130 nm SiGe HBT**	**Single-Balanced**	√	**77/94/122/135**	**6.4**	**1.6 (at 135GHz)**	**0.8 x 0.9**

frequencies including 77, 94, 122 and 135 GHz to directly down-convert 10-Gbps OOK modulated millimeter-wave signal. The developed mixer is a good candidate for the multiband communication and universal millimeter-wave system applications.

ACKNOWLEDGMENT

This work was supported by the A*STAR SERC under Grant 082-141-0040.

REFERENCES

[1] B. Razavi, " Design of millimeter-wave CMOS radios: A tutorial," *IEEE Trans. Circuits and Systems – I: Regular Papers*, vol. 56, no.1, pp. 4-16, Jan. 2009.

[2] C. H. Lien, C. H. Wang, C. S. Lin, P. S. Wu, K. Y. Lin, and H. Wang, "Analysis and design of reduced-size Marchand rat-race hybrid for millimeter-wave compact balanced mixers in 130-nm CMOS process," *IEEE Trans. Microw. Theory. Tech.*, vol. 57, no. 8, pp. 1966-1977. Aug. 2009.

[3] S. Emami, C. H. Doan, A. M. Niknejad, and R. W. Broderson, "A 60-GHz down-conversion CMOS single-gate mixer," in *IEEE Radio Frequency Integrated Circuits (RFIC) Symp. Dig. Tech. Papers*, Jun. 2005, pp.163-166.

[4] A. Parsa and B. Razavi, "A new transceiver architecture for the 60-GHz band," *IEEE J. Solid-State Circuits*, vol. 44, no. 3, pp. 751-762, Mar. 2009.

[5] L. Wang, S. Glisic, J. Borngaeber, W. Winkler, and J. C. Scheytt, "A single-ended fully integrated SiGe 77/79 GHz receiver for automotive radar," *IEEE J. Solid-State Circuits*, vol. 43, no. 9, pp. 1897-1908, Sep. 2008.

[6] S. K. Reynolds and J. D. Powell, "77 and 94-GHz downcoversion mixers in SiGe BiCMOS," in *IEEE Asian Solid-State Circuits Conference (A-SSCC) Dig. Tech. Papers*, pp. 191-194, Nov. 2006.

[7] K. Schmalz, W. Winkler, J. Borngräber, W. Debski, B. Heinemann, and C. Scheytt, "A 122 GHz Receiver in SiGe Technology," in *IEEE Bipolar/BiCMOS Circuits and Technology Meeting (BCTM)*, pp. 182-185, Oct. 2009.

[8] E. Laskin, P. Chevalier, A. Chantre, B. Sautreuil, and S. P. Voinigescu, "165-GHz transceiver in SiGe Technology," *IEEE J. Solid-State Circuits*, vol. 43, no. 5, pp. 1087-1100, May 2008.

[9] E. Laskin, M. Khanpour, S. T. Nicolson, A. Tomkins, P. Garcia, A. Cathelin, D. Belot, and S. P. Voinigescu, "Nanoscale CMOS transceiver design in the 90-170-GHz range," *IEEE Trans. Microw. Theory. Tech.*, vol. 57, no. 12, pp. 3477-3490. Dec. 2009.

[10] B. Zhang, Y. Z. Xiong, L. Wang, S. Hu, T. G. Lim, Y. Q. Zhuang, L. W. Li, and X. Yuan, "130-GHz gain-enhanced SiGe low noise amplifier," in *IEEE Asian Solid-State Circuits Conference (A-SSCC) Dig. Tech. Papers*, 11-2, Nov. 2010.

[11] L. Wang, Y. Z. Xiong, B. Zhang, S. Hu, T. G. Lim, and X. Yuan, "0.7-dB insertion-loss D-band Lange coupler design and characterization in 0.13um SiGe BiCMOS technology," in *Journal of Infrared, Millimeter, and Terahertz Waves*, vol.31, no. 10, pp. 1136-1145, Aug. 2010.

77-135GHz IQ Mixer

A*STAR IME

FEATURES

- Frequency: 77-135 GHz
- Operating Voltage: 1.6 V
- Biasing Voltage: 0.92 V
- Conversion Gain: 1.6 dB @135 GHz
- DC Power Consumption: 6.4 mW
- Process: 0.13 um IHP SiGe BiCMOS

APPLICATIONS

- 77/79GHz automotive radars
- 94GHz high-resolution imaging
- 122GHz high-data-rate communications

DESCRIPTION

The IQ mixer can directly down-convert 10-Gbps modulated millimeter-wave (mmWave) signals to base-band data. Without any design change or DC bias adjustment, the IQ mixer can operate at any frequency between 77 and 135 GHz. Therefore, the mixer is universal for many mmWave applications such as 77GHz automotive radars, 94GHz imaging, and 122GHz communications.

BLOCK SCHEMATIC

As illustrated in Fig.1, the IQ mixer mainly consists of a Wilkinson power divider, two identical single-ended mixers, and a Lange coupler. The mixing mechanism is enhanced by the nonlinear characteristic of a HBT when Vb is slightly higher than the threshold voltage of the transistor. An IF amplifier is integrated within the single-ended mixer to increase the conversion gain.

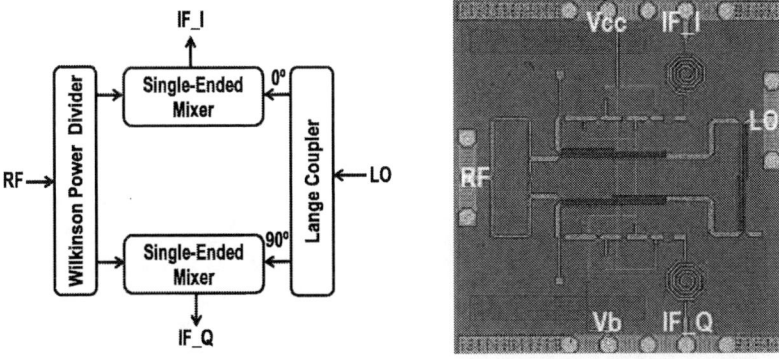

Fig. 1. Block schematic (left) and die photograph (right) of the IQ mixer.

77-135GHz IQ Mixer　　　　　　　　　　　　　　　　**A*STAR IME**

TYPICAL PERFORMANCE CURVES

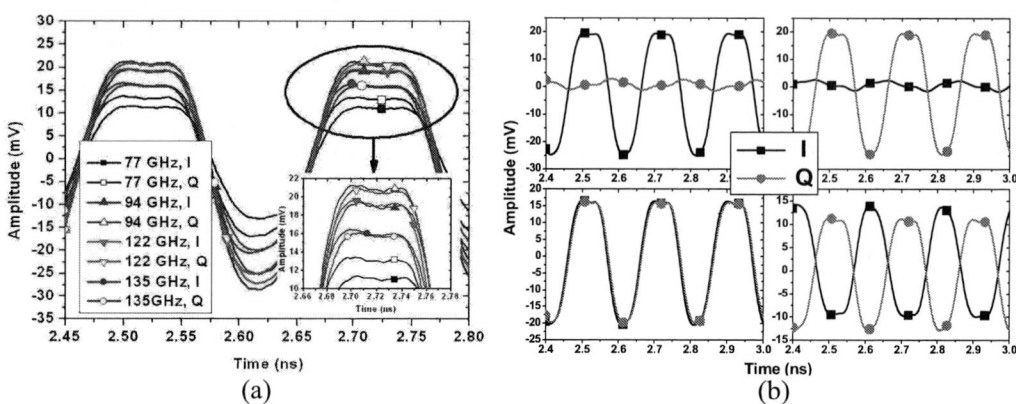

Fig. 2.　Time-domain characteristics of the IQ mixer (Vcc = 1.6 V, Vb = 0.92 V, Ic = 4 mA, P_{RF} = -16 dBm, and P_{LO} = - 4.2 dBm). (a) I and Q output with different operating frequencies when the LO and RF signals have phase difference of 45°. (b) I and Q output with different LO-RF phase difference (from left-top to right-bottom: 0°, 90°, 45°, and 135°.

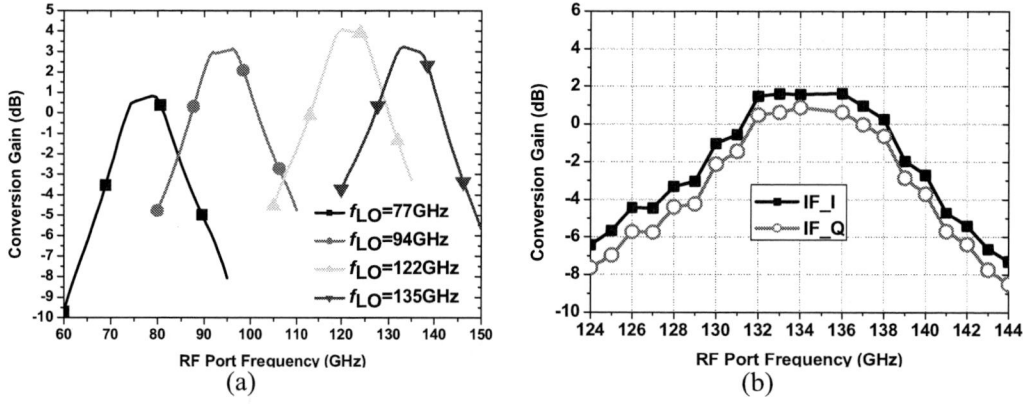

Fig. 3.　Conversion gain of the mixer.

978-1-61284-863-1/11 $26.00 © 2011 IEEE

Supply Voltage and Temperature Insensitive Current Reference for the 4 MHz Oscillator

Chi-Hsiung Wang	Cheng-Feng Lin	Wei-Bin Yang	Yu-Lung Lo
Dept. of E. E.	Dept. of E. E.	Dept. of E. E.	Dept. of E. E.
Tamkang University	Tamkang University	Tamkang University	National Kaohsiung Normal
New Taipei City, Taiwan	New Taipei City, Taiwan	New Taipei City, Taiwan	University
chwang0327@gmail.com	792350257@s92.tku.edu.tw	robin@ee.tku.edu.tw	Kaohsiung, Taiwan
			yllo@nknu.edu.tw

Abstract—This paper presents a 4 MHz current control ring oscillator with a new temperature and supply voltage immune current reference implemented by 0.35μm CMOS technology. Compared to the conventional oscillator with current reference techniques, the proposed approach shows a significant improvement for the sensitivities of temperature and supply voltage. The current reference is designed by combining positive and negative temperature effect circuits, such that it can exempt from the temperature and supply voltage variations. By HSPICE simulation, this new current reference is insensitive to the supply voltage with variations of −0.47%~0.67% over the supply voltage range of 2.97V to 3.63V, and it is also insensitive to the temperature with variation of 366 ppm/°C over the temperature range of −40°C to 100°C. The proposed oscillator frequency is insensitive to the supply voltage with variations of −15%~20% over the supply voltage range of 2.97V to 3.63V, and it is insensitive to temperature with variation of 404 ppm/°C over the temperature range of −40°C to 100°C.

Keywords- *current reference, supply voltage, ring oscillator*

I. INTRODUCTION

Current reference is an essential block in many analog circuits, such as the bias sources for oscillators, amplifiers, and phase lock loops. For those applications, the current references must be insensitive to supply voltage and temperature variations. Since it is easy to implement a voltage reference by bandgap circuit, the current references can be derived from voltage references by applying Ohm's law for voltage to current conversion [1]. However, this kind of current reference may take more silicon area since it contains some bipolar transistors, operational amplifier, and resistors. The current reference proposed by Sansen et al. [2] is a circuit without resistor. With an absolute temperature T, the current is proportional to T0.5 with 3.5V minimum supply voltage. The current reference proposed by Oquey et al. [3] is a low voltage approach using the PTAT-like technique. Here the resistor R is replaced by an n-channel MOSFET working in the triode region. Like [2], the output current is proportional to T0.5. It can work with a power supply as low as 1.2V and produce 1-100 nA output current. However, this current reference is strongly dependent on the supply voltage.

In this research work, we present a new current reference. The proposed current reference consists of two components.

The first one is adopted from the well-known circuit given by Razavi's text book [4], which uses two PMOS transistors as current mirror and a resistor to define the output current. This circuit exhibits a positive supply voltage coefficient and the current is also proportional to the absolute temperature (PTAT). The second component has the characteristics of negative supply voltage coefficient and negative temperature coefficient. The two circuits can compensate each other such that it provides a current insensitive to supply voltage and temperature. We apply the proposed current reference further to the current control ring oscillator and make the ring oscillator have low sensitivity to temperature and supply voltage variations.

The rest of this paper is organized as follows. Section 2 describes the circuit descriptions and analyses. The simulation results are addressed in Section 3 and a brief conclusion is given in Section 4.

II. CIRCUIT DESCRIPTION AND ANALYSIS

In this section, the circuit architecture and associated operating principles of temperature and supply voltage compensated ring oscillator are analyzed and discussed. The system block diagram of the current control ring oscillator with temperature and supply voltage compensated scheme is shown in Fig. 1(a); the system block diagram of the temperature and supply voltage compensated current reference is shown in Fig. 1(b). The current reference is comprised by two parts, positive temperature current reference circuit and negative temperature current reference circuit, and the details of the current references are described in the following subsection.

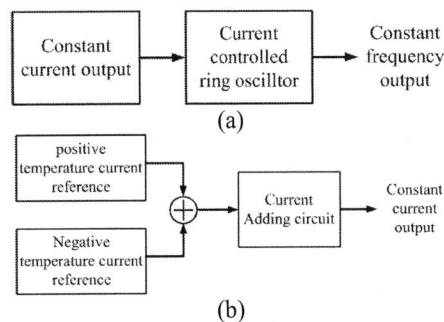

Figure 1. (a) Temperature and supply voltage compensated ring oscillator
(b) Temperature and supply voltage compensated current reference

978-1-61284-863-1/11 $26.00 © 2011 IEEE

A. Positive temperature current reference circuit

A well-known conventional current reference given by Razavi [4] is shown in Fig. 2. According to Fig. 2, the output current is given as follows:

$$I_{10} = \frac{2}{\mu_n C_{ox} \left(W/L \right)} \frac{1}{R^2} \left(1 - \frac{1}{\sqrt{K}} \right)^2, \tag{1}$$

where K is the ratio of the size of M_{N17} to M_{N18}. From (1), the current reference is insensitive to the supply voltage. In fact, based on the simulation results depicted in Fig. 3 with a supply voltage range of 2.97V to 3.63V, the current reference still depends on the supply voltage slightly for the channel length modulation of transistors M_{P12} and M_{P13}.

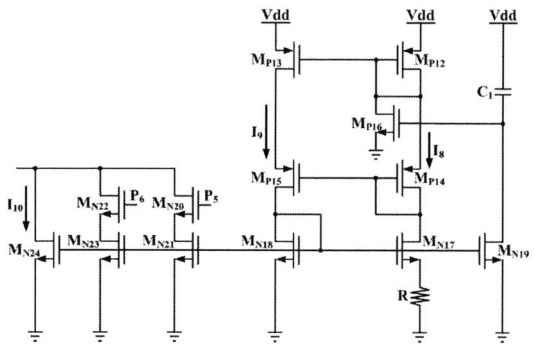

Figure 2. Positive temperature current reference circuit

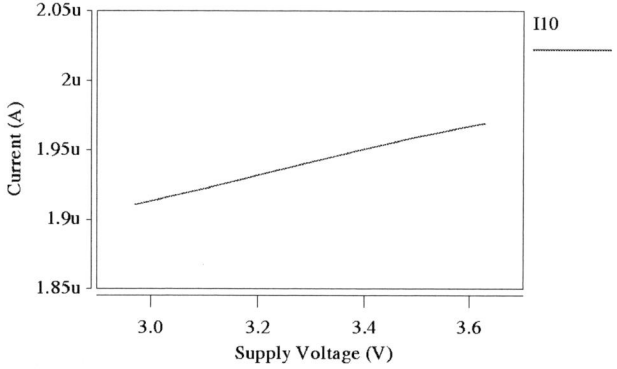

Figure 3. The simulation of currents versus supply voltage for the positive temperature current reference

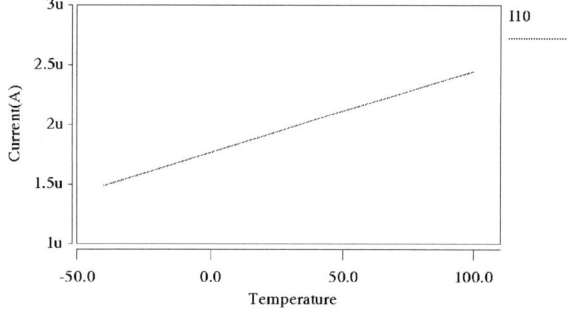

Figure 4. The simulation of currents versus temperature for the positive temperature current reference

Moreover, due to the mobility in the denominator of (1), this current reference has the effect of positive temperature coefficient as shown in Fig. 4.

B. Negative temperature current reference circuit

The negative temperature current reference circuit proposed here is to compensate the positive temperature current reference. At first, we propose two different types of current references as the basic current operational component.

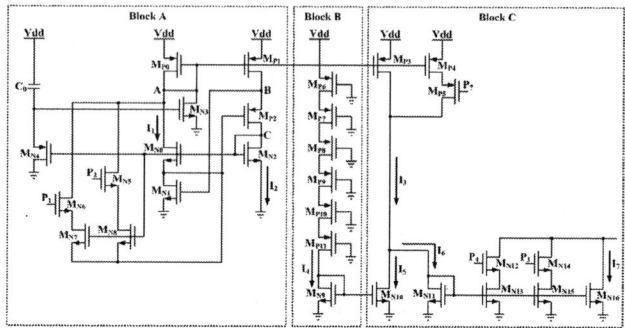

Figure 5. Negative temperature current reference circuit

The first component shown in block A of Fig. 5 is almost the same as the known current reference depicted in Fig. 2 except that the resistor is replaced by transistor M_{N1} which works in the triode region. Besides, transistor M_{P2} is placed in between M_{P1} and M_{N2} to provide a bias voltage for M_{N1}. As a result, M_{N1} works in the triode region, and I_1 can be expressed as:

$$
\begin{aligned}
I_1 &= \mu_n C_{ox} \left(\frac{W}{L} \right)_{N1} \left(V_{GS} - V_{TH} \right) V_{DS} \\
&= \mu_n C_{ox} \gamma V_{DS} \left(\frac{W}{L} \right)_{N1} \left(V_{DD} - \frac{V_{TH}}{\gamma} \right) \\
&= \alpha_1 \left(\frac{W}{L} \right)_{N1} \left(V_{DD} - \beta_1 \right)
\end{aligned}
\tag{2}
$$

where $VGS = \gamma \times VDD$, $\alpha 1 = \mu n Cox \gamma VDS$, and $\beta 1 = VTH/\gamma$. The second component of the proposed current reference shown in block B of Fig. 5 composes of M_{P6} - M_{P11} and M_{N9}. Transistors M_{P6} - M_{P11} all work in the triode region. Based on HSPICE simulation, the current reference of these two components will be almost linearly dependent on the supply voltage. Therefore, I_4 can be expressed as:

$$
\begin{aligned}
I_4 &= \mu_n C_{ox} \left(\frac{W}{L} \right)_{N9} \left[\left(V_{GS} - V_{TH} \right) V_{DS} - \frac{1}{2} V_{DS}^{\;2} \right] \\
&= \alpha_2 \left(\frac{W}{L} \right)_{N9} \left(V_{DD} - \beta_2 \right)
\end{aligned}
\tag{3}
$$

where $VGS = VDD$, $\alpha 2 = \mu n Cox VDS$, and $\beta 2 = VTH + 1/2(VDS)$. According to blocks A and B of Fig. 4, M_{P0} and M_{P3} are a set of current mirror; M_{N9} and M_{N10} are another set of current mirror. By adjusting the ratio of $(W/L)_{P3}$ and $(W/L)_{N10}$, we can make $\alpha_1 (W/L)_{P3}$ to be smaller than $\alpha_2 (W/L)_{N10}$. According to

978-1-61284-863-1/11 $26.00 © 2011 IEEE

the operation of current subtraction ($I_6 = I_3–I_5$), current I_6 presents the negative supply voltage coefficient due to currents I_5 behaving a larger supply voltage coefficient than I_3. Moreover, current I_6 also presents the negative temperature coefficient due to the mobility.

C. Current Adding circuit

According to block C of Fig. 5 and Fig. 2, I_7 and I_{10} are the duplicated currents of the negative temperature current reference and the positive temperature current reference, respectively. By adding currents of I_7 and I_{10}, we can derive a new current reference I_{11} which has low sensitivity to temperature and supply voltage variations. Fig. 6 shows the combination circuit to generate the temperature and supply voltage immune current I_{11} and the currents (I_{12} and I_{13}) required for the current control ring oscillator.

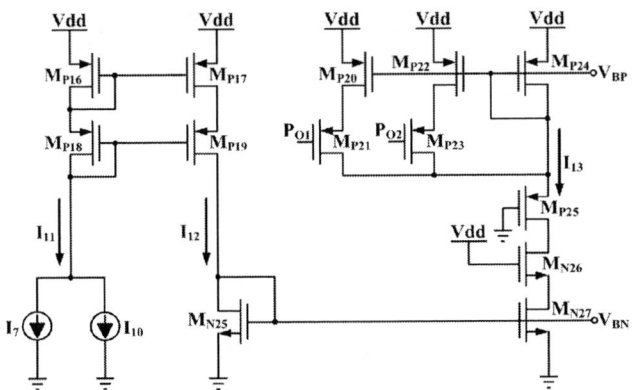

Figure 6.   Current Adding circuit

D. Current Controlled Oscillator

Fig 7 shows the current control oscillator (CCO). In Fig. 7, INV is a dummy circuit. The required references are provided from the circuit described in the previous subsection. The buffer can drive the capacitance loading as high as 30pF, and the designed output frequency is 4MHz.

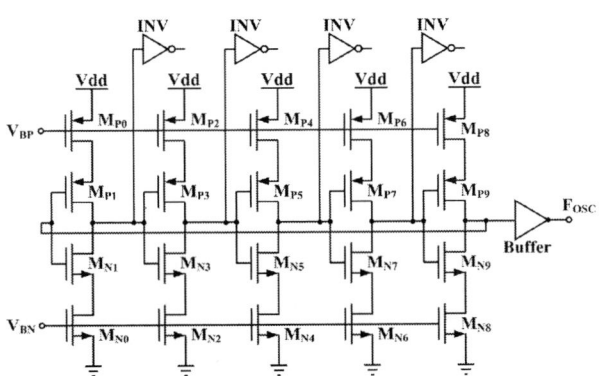

Figure 7.   Current Controlled Oscillator (CCO)

III.   SIMULATION RESULTS

The HSPICE simulator has been performed on the temperature and supply voltage compensated ring oscillator using TSMC 0.35um models for the MOS device. Fig. 8 shows the current reference versus voltage variation simulation results. I_7 and I_{10} exhibit linear relationship with the supply voltage from 2.97V to 3.63V. It is clear that current I_{11}, $I_{11}=I_7+I_{10}$, is weakly dependent on the supply voltage from 2.97V to 3.63V. Fig. 9 shows the current reference versus temperature variation simulation results. According to the simulation, I_7 has the effect of negative temperature coefficient, and I_{10} has the effect of positive temperature coefficient from −40°C to 100°C. Moreover, it is clear that current I_{11}, $I_{11}=I_7+I_{10}$, shows the insensitivity to the temperature variations.

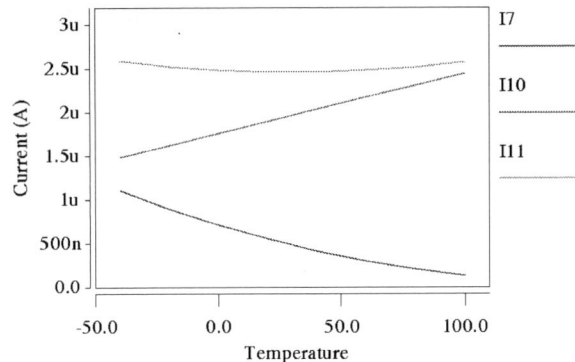

Figure 8.   The simulation results of the currents versus supply voltages

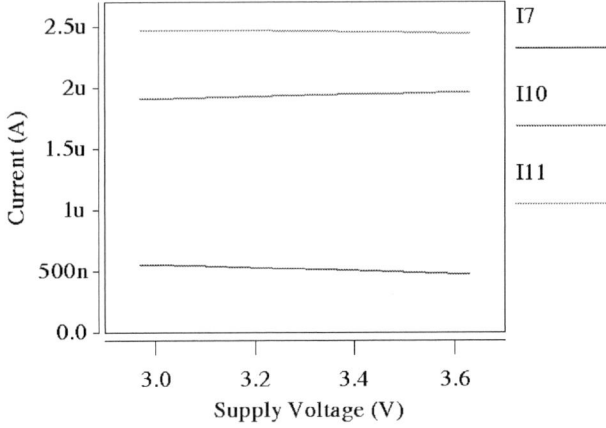

Figure 9.   The simulation results of the currents versus different temperature

The simulation results of the output frequency of the temperature and supply voltage compensated ring oscillator is shown in Figs. 10 and 11. Because the current reference I_{13} possesses low sensitivity to temperature and supply voltage variations according to Fig. 6, the oscillator frequency is proved to be insensitive to supply voltage with variations of −15%~20% over the supply voltage range of 2.97V to 3.63V and insensitive to temperature with a variation of 404 ppm/°C over a temperature range of −40°C to 100°C. The performance specification and comparisons with other research works of the ring oscillator are shown in Table. 1.

978-1-61284-863-1/11 $26.00 © 2011 IEEE

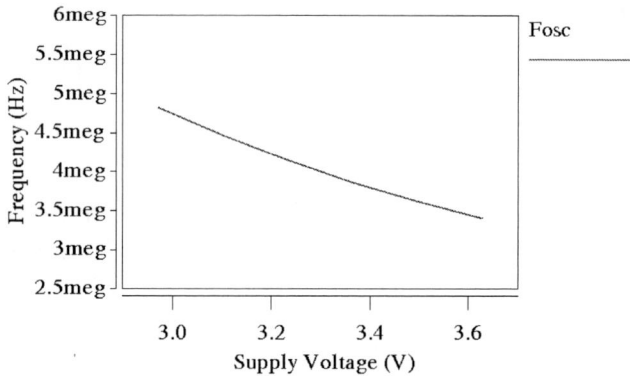

Figure 10. The simulation results of the CCO frequency versus supply voltages

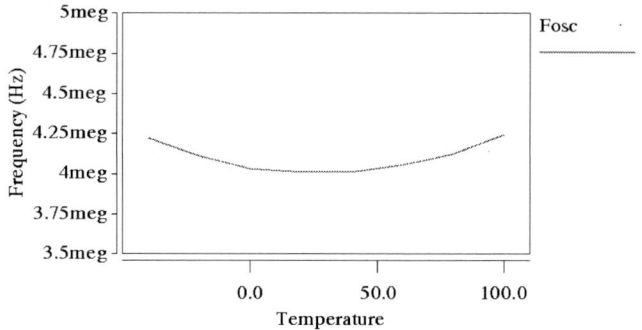

Figure 11. The simulation results of the CCO frequency versus different temperature

TABLE I. PERFORMANCE SUMMARY AND COMPARISONS

Parameters	[5]	[6]	[7]	[8]	[9]	This work
Process (μm)	0.5	0.25	0.28	0.25	0.5	0.35
Frequency (Hz)	80k	7M	2.4G	800M	12.8M	4M
Supply Voltage (V)	1	2.5	2.5	2.5	3	3.3
Consume current (μA)	1.14	600	7680	7580	133	67
Temperature (℃)	0 ~80	-40 ~125	-40 ~120	N/A	-40 ~125	-40 ~100
Area (mm2)	0.24	1.6	0.0121	N/A	0.1848	0.02

IV. CONCLUSIONS

In this paper we propose a new negative temperature current reference circuit to compensate the positive temperature current reference and combine these two current references to form a new current reference which can exempt from the effects of temperature and supply voltage variations. This temperature and supply voltage immune reference is further applied to design a 4MHz current control oscillator. The simulation results prove that the proposed current reference is insensitive to the supply voltage with a variation of −0.47%~0.67% over a supply voltage range of 2.97V to 3.63V, and it is also insensitive to temperature with variation of 366 ppm/°C over the temperature range of −40°C to 100°C. The simulations also show that the 4MHz current control oscillator whose output frequency is insensitive to the supply voltage with variation of −15%~20% over the supply voltage range of 2.97V to 3.63V, and it is insensitive to temperature with variation of 404 ppm/°C over the temperature range of −40°C to 100°C.

V. ACKNOWLEDGMENTS

This research is partially supported by National Science Council (NSC), Taiwan. The authors also would like to thank National Chip Implementation Center (CIC), Taiwan, for their chip fabrication supporting.

REFERENCES

[1] E. Vittoz, "The design of high performance analog circuits on digital CMOS chips," *Solid-State Circuits, IEEE Journal of*, vol. 20, no. 3, pp. 657-665, 1985.

[2] W. M. Sansen, F. Op't Eynde, and M. Steyaert, "A CMOS temperature-compensated current reference," *Solid-State Circuits, IEEE Journal of*, vol. 23, no. 3, pp. 821-824, 1988.

[3] H. J. Oquey and D. Aebischer, "CMOS current reference without resistance," *Solid-State Circuits, IEEE Journal of*, vol. 32, no. 7, pp. 1132-1135, 1997.

[4] B. Razavi, *Design of Analog CMOS Integrated Circuits*, McGRAW–Hill Publication Co., 2001.

[5] G. D. Vita, F. Marraccini and G. Iannaccone, "Low-voltage low-power CMOS oscillator with low temperature and process sensitivity," *Circuits and Systems, 2007. ISCAS 2007. IEEE International Symposium on*, Issue 27-30, pp. 2152-2155, May 2007.

[6] K. Sundaresan, P. E. Allen and F. Ayazi, "Process and temperature compensation in a 7-MHz CMOS clock oscillator," *Solid-State Circuits, IEEE Journal of*, vol. 41, no. 2, pp.433-442, Feb. 2006.

[7] W. Rahajandraibe, L. Zaid, V. Cheynet de Beaupre and G. Bas , "Temperature Compensated 2.45 GHz Ring Oscillator with Double Frequency Control," in proc. IEEE Radio Frequency Integrated Circuits Symposium, Issue 3-5, pp. 409-412, June 2007.

[8] S. S. Lee, T. G. Kim, J. T. and S. W. Kim, "Process-and-temperature compensated CMOS voltage-controlled oscillator for clock generators," ELECTRONICS LETTERS Vol. 39 No. 21, pp. 1481-1485, oct. 2003.

[9] A. Olmos, "A temperature compensated fully trimmable on-chip IC oscillator," *Integrated Circuits and Systems Design, 2003. SBCCI 2003. Proceedings. 16th Symposium on*, Issue 8-11, pp. 181-186, May 2003.

DDCCs Based Voltage-Mode One Input Five outputs Biquadratic Filter With High Input Impedance

Wei-Yuan Chiu, Jiun-Wei Horng[+], Yi-Sing Guo and Ching-Yao Tseng

Department of Electronic Engineering,
Chung Yuan Christian University,
Chung-Li, 32023, Taiwan
[+]e-mail: jwhorng@cycu.edu.tw

Abstract—**A new high input impedance voltage-mode universal biquadratic filter with one input terminals and five output terminals is presented. The proposed circuit uses three plus-type differential difference current conveyors, two resistors and two grounded capacitors. The proposed circuit can realize all the standard filter functions: lowpass, bandpass, highpass, notch and allpass, simultaneously, without component matching conditions. The proposed circuit offers the features of high input impedance and the use of only grounded capacitors.**

Keywords-active filter; current conveyor; analog circuit

I. INTRODUCTION

To obtain more various filter functions simultaneously in the same circuit topology will increase the usefulness of the circuit. Several voltage-mode universal biquads with one input terminal and five output terminals were presented in [1-5]. These circuits can realize all the standard filter functions; namely highpass, bandpass, lowpass, notch and allpass, simultaneously. The one input and five outputs universal biquad in [1] uses five current feedback amplifiers (CFAs), six resistors and two grounded capacitors. Since a CFA is equivalent to a plus-type second-generation current conveyor (CCII) with a voltage follower, so the circuit in [1] needs five CCIIs and five voltage followers in fact. The one input and five outputs universal biquads in [2]-[3] use multiple output second-generation current conveyors (MOCCIIs) as active components. The circuit at Fig. 4 of [3] is of special interest because it uses only two MOCCIIs, five resistors and two grounded capacitors. The high input impedance one input and five outputs universal biquad in [4] uses three differential voltage current conveyors (DVCCs), four resistors and two grounded capacitors. The circuit in [5] uses two DVCCs, three resistors and two grounded capacitors.

In this paper, a new voltage-mode universal biquadratic filter with one input terminal and five output terminals is presented. The proposed filter circuit employs three plus-type differential difference current conveyors (DDCCs), two resistors and two grounded capacitors and can realize all the standard filter functions; highpass, bandpass, lowpass, notch and allpass, simultaneously, without critical component matching conditions. Voltage followers are needed while cascaded the proposed circuit to the next stages.

The proposed circuit uses only grounded capacitors which are attractive for integrated circuit implementation [6]. The circuit enjoys high input impedance so that it can be directly connected in cascade to implement higher-order filters. Since the implementation configuration of the plus-type DDCC is simpler than that of the minus-type DDCC, the proposed circuit employs the plus-type DDCCs only.

II. CIRCUIT DESCRIPTION

Using standard notation, the port relations of an ideal DDCC can be characterized by [7]

$$
\begin{bmatrix} V_x \\ I_{y1} \\ I_{y2} \\ I_{y3} \\ I_z \end{bmatrix} = \begin{bmatrix} 1 & -1 & 1 & 0 \\ 0 & 0 & 0 & 0 \\ 0 & 0 & 0 & 0 \\ 0 & 0 & 0 & 0 \\ 0 & 0 & 0 & \pm 1 \end{bmatrix} \begin{bmatrix} V_{y1} \\ V_{y2} \\ V_{y3} \\ I_x \end{bmatrix} \tag{1}
$$

where the plus and minus signs indicate whether the conveyor is configured as a non-inverting or inverting circuit, termed DDCC+ or DDCC–.

The proposed configuration is shown in Fig. 1. The transfer functions can be expressed as

$$
\frac{V_{out1}}{V_{in}} = \frac{G_1 G_2}{s^2 C_1 C_2 + s C_1 G_2 + G_1 G_2} \tag{2}
$$

$$
\frac{V_{out2}}{V_{in}} = \frac{s C_1 G_2}{s^2 C_1 C_2 + s C_1 G_2 + G_1 G_2} \tag{3}
$$

$$
\frac{V_{out3}}{V_{in}} = -\frac{s^2 C_1 C_2 - s C_1 G_2 + G_1 G_2}{s^2 C_1 C_2 + s C_1 G_2 + G_1 G_2} \tag{4}
$$

978-1-61284-863-1/11 $26.00 © 2011 IEEE

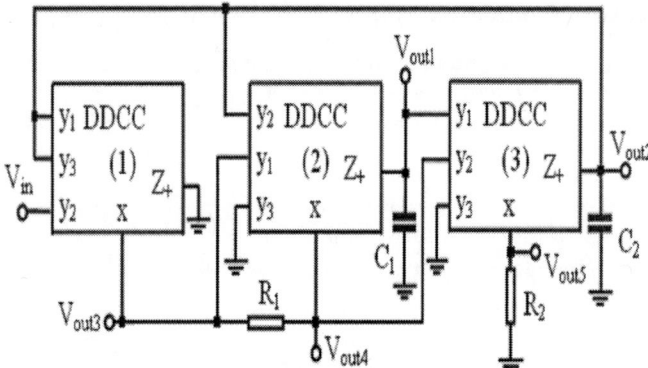

Fig. 1 The proposed high-input impedance voltage-mode biquadratic filter

$$\frac{V_{out4}}{V_{in}} = -\frac{s^2 C_1 C_2 + G_1 G_2}{s^2 C_1 C_2 + s C_1 G_2 + G_1 G_2} \qquad (5)$$

$$\frac{V_{out5}}{V_{in}} = \frac{s^2 C_1 C_2}{s^2 C_1 C_2 + s C_1 G_2 + G_1 G_2} \qquad (6)$$

From (2-6) it can be seen that a lowpass response is obtained from V_{out1}, a bandpass response is obtained from V_{out2}, a notch response is obtained from V_{out3}, a highpass response is obtained from V_{out4} and an allpass response is obtained from V_{out5}. Due to the input voltage signal is connected directly to the y2 port of the DDCC(1) and the input current to the y2 port is zero ($I_{y2} = 0$), the circuit has the feature of high input impedance.

Moreover, the proposed circuit has the following features: employs the minimum number of passive components, uses only plus-type DDCCs that simplify the circuit configuration and using only grounded capacitors that are attractive for integrated circuit implementation.

III. SENSITIVITIES ANALYSIS

Taking the non-idealities of the DDCC into account, the relationship of the terminal voltages and currents can be rewritten as

$$\begin{bmatrix} V_x \\ I_{y1} \\ I_{y2} \\ I_{y3} \\ I_z \end{bmatrix} = \begin{bmatrix} \alpha_{k1} & -\alpha_{k2} & \alpha_{k3} & 0 \\ 0 & 0 & 0 & 0 \\ 0 & 0 & 0 & 0 \\ 0 & 0 & 0 & 0 \\ 0 & 0 & 0 & \pm\beta_k \end{bmatrix} \begin{bmatrix} V_{y1} \\ V_{y2} \\ V_{y3} \\ I_x \end{bmatrix} \qquad (7)$$

where $\alpha_{k1} = 1 - \varepsilon_{k1v}$ and ε_{k1v} ($|\varepsilon_{k1v}| \ll 1$) denotes the voltage tracking error from V_{y1} terminal to V_x terminal of the k-th DDCC, $\alpha_{k2} = 1 - \varepsilon_{k2v}$ and ε_{k2v} ($|\varepsilon_{k2v}| \ll 1$) denotes the

voltage tracking error from V_{y2} terminal to V_x terminal of the k-th DDCC, $\alpha_{k3} = 1 - \varepsilon_{k3v}$ and ε_{k3v} ($|\varepsilon_{k3v}| \ll 1$) denotes the voltage tracking error from V_{y3} terminal to V_x terminal of the k-th DDCC and $\beta_k = 1 - \varepsilon_{ki}$ and ε_{ki} ($|\varepsilon_{ki}| \ll 1$) denotes the current tracking error of the k-th DDCC. The denominator of non-ideal voltage transfer function in Fig. 1 becomes

$$D(s) = s^2 C_1 C_2 + s C_1 G_2 \, \alpha_{32} \beta_3 [(\alpha_{11} + \alpha_{13}) \alpha_{21} - \alpha_{22}]$$
$$+ G_1 G_2 \, \alpha_{31} \beta_2 \beta_3 [(\alpha_{11} + \alpha_{13})(1 - \alpha_{21}) + \alpha_{22}] \qquad (8)$$

The resonance angular frequency ω_o and quality factor Q are obtained by

$$\omega_o = \sqrt{\frac{G_1 G_2 \, \alpha_{31} \beta_2 \beta_3 [(\alpha_{11} + \alpha_{13})(1 - \alpha_{21}) + \alpha_{22}]}{C_1 C_2}} \qquad (9)$$

$$Q = \frac{\sqrt{C_2 G_1 \, \alpha_{31} \beta_2 [(\alpha_{11} + \alpha_{13})(1 - \alpha_{21}) + \alpha_{22}]}}{\alpha_{32}[(\alpha_{11} + \alpha_{13}) \alpha_{21} - \alpha_{22}] \sqrt{C_1 G_2 \beta_3}} \qquad (10)$$

The active and passive sensitivities of ω_o and Q are shown as

$$S^{\omega_o}_{\alpha_{31}, \beta_2, \beta_3} = \frac{1}{2} \; ; \; S^{\omega_o}_{G_1, G_2} = -S^{\omega_o}_{C_1, C_2} = \frac{1}{2} \; ;$$

$$S^{\omega_o}_{\alpha_{11}} = \frac{1}{2} \frac{\alpha_{11}(1 - \alpha_{21})}{(\alpha_{11} + \alpha_{13})(1 - \alpha_{21}) + \alpha_{22}} \; ;$$

$$S^{\omega_o}_{\alpha_{13}} = \frac{1}{2} \frac{\alpha_{13}(1 - \alpha_{21})}{(\alpha_{11} + \alpha_{13})(1 - \alpha_{21}) + \alpha_{22}} \; ;$$

$$S^{\omega_o}_{\alpha_{21}} = -\frac{1}{2} \frac{\alpha_{21}(\alpha_{11} + \alpha_{13})}{(\alpha_{11} + \alpha_{13})(1 - \alpha_{21}) + \alpha_{22}} \; ;$$

$$S^{\omega_o}_{\alpha_{22}} = \frac{1}{2} \frac{\alpha_{22}}{(\alpha_{11} + \alpha_{13})(1 - \alpha_{21}) + \alpha_{22}} \; ;$$

$$S^{Q}_{\alpha_{31}, \beta_2} = -S^{Q}_{\beta_3} = \frac{1}{2} \; ; \; S^{Q}_{C_2, G_1} = -S^{Q}_{C_1, G_2} = \frac{1}{2} \; ;$$

$$S^{Q}_{\alpha_{32}} = -1 \; ;$$

$$S^{Q}_{\alpha_{11}} =$$
$$\frac{-\alpha_{11}\alpha_{21}}{(\alpha_{11} + \alpha_{13})\alpha_{21} - \alpha_{22}} + \frac{\alpha_{11}}{2} \frac{1 - \alpha_{21}}{(\alpha_{11} + \alpha_{13})(1 - \alpha_{21}) + \alpha_{22}} \; ;$$

$$S^{Q}_{\alpha_{13}} =$$
$$\frac{-\alpha_{13}\alpha_{21}}{(\alpha_{11} + \alpha_{13})\alpha_{21} - \alpha_{22}} + \frac{\alpha_{13}}{2} \frac{1 - \alpha_{21}}{(\alpha_{11} + \alpha_{13})(1 - \alpha_{21}) + \alpha_{22}} \; ;$$

$$S^Q_{\alpha_{21}} =$$

$$\frac{-\alpha_{21}(\alpha_{11} + \alpha_{13})}{(\alpha_{11} + \alpha_{13})\,\alpha_{21} - \alpha_{22}} - \frac{\alpha_{21}}{2}\frac{\alpha_{11} + \alpha_{13}}{(\alpha_{11} + \alpha_{13})(1 - \alpha_{21}) + \alpha_{22}};$$

$$S^Q_{\alpha_{22}} =$$

$$\frac{\alpha_{22}}{(\alpha_{11} + \alpha_{13})\,\alpha_{21} - \alpha_{22}} + \frac{\alpha_{22}}{2}\frac{1}{(\alpha_{11} + \alpha_{13})(1 - \alpha_{21}) + \alpha_{22}}$$

IV. SIMULATION RESULTS

The proposed circuit was simulated using HSPICE. The DDCC was realized by the CMOS implementation in Fig. 2 [8] using TSMC 0.25 μm CMOS technology process parameters. The aspect ratios of the MOS transistors were chosen in Table 1 and the power supply was ±1.25V. The biasing voltage V_b was taken as -0.75V. The total power dissipation is 215.1 uW. Fig. 3(a), (b), (c), (d) and (e) represents the simulated amplitude-frequency responses and phase-frequency responses for the lowpass, bandpass, allpass, notch and highpass filters of Fig. 1, respectively, designed with f_o =1.59MHz, $C_1 = C_2 =$ 10pF and $R_1 = R_2$ =10kΩ.

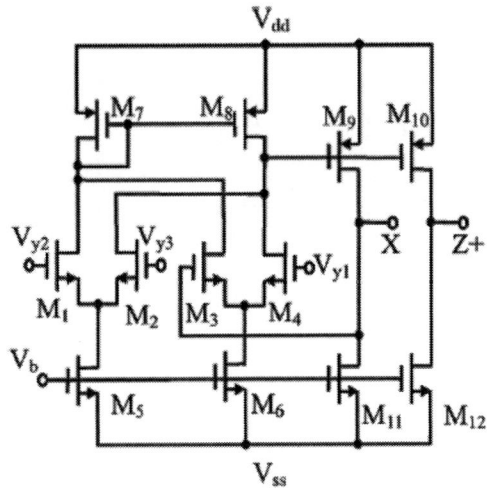

Fig. 2 CMOS realization of the plus-type DDCC

TABLE I
ASPECT RATIOS OF THE MOS IN FIG. 2

MOS transistors	W / L
M1, M2, M3, M4	5 / 0.5
M7, M8, M9, M10	10 / 0.5
M5, M6, M11, M12	5 / 0.5

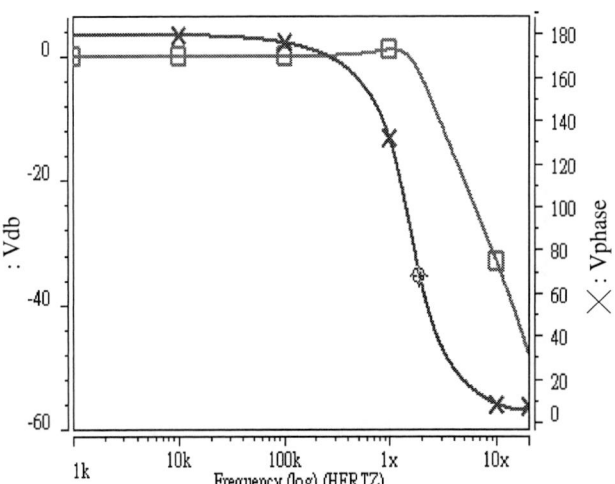

Fig. 3 (a) Simulated frequency responses of the proposed lowpass filter

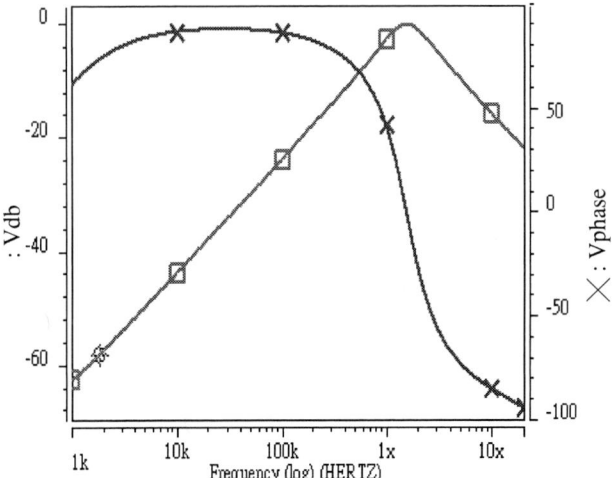

Fig. 3 (b) Simulated frequency responses of the proposed bandpass filter

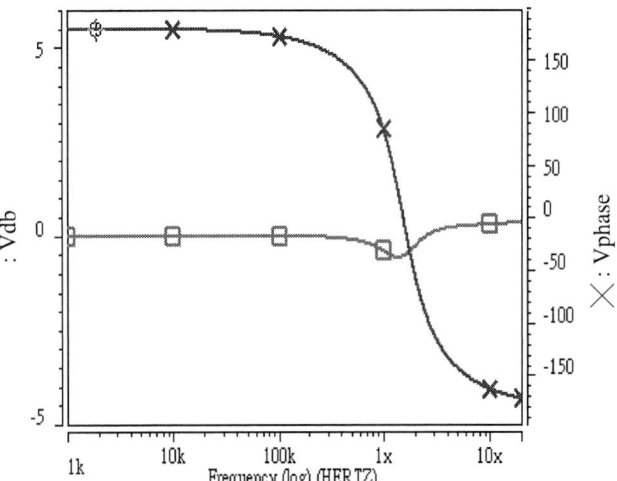

Fig. 3 (c) Simulated frequency responses of the proposed allpass filter

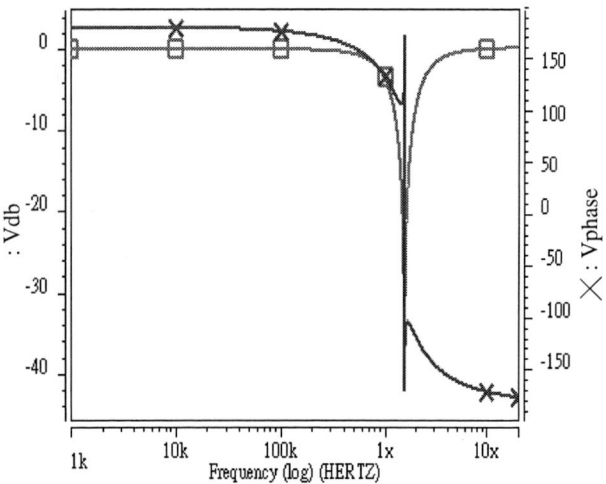

Fig. 3 (d)   Simulated frequency responses of the proposed notch filter

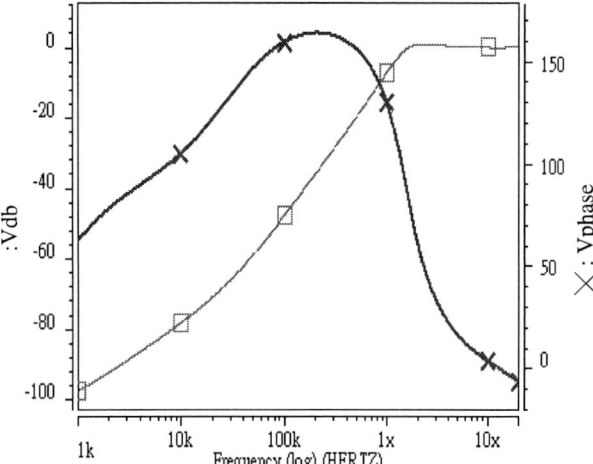

Fig. 3 (e)   Simulated frequency responses of the proposed highpass filter

V.   CONCLUSION

In this paper, a new single input and five outputs voltage-mode universal biquadratic filter with high input impedance is presented. The proposed circuit uses three plus-type DDCCs, two grounded capacitors and two resistors and offers following advantages: high input impedance, the use of only three plus-type DDCCs, the use of only grounded capacitors, realization of highpass, bandpass, lowpass, notch, and allpass filters, simultaneously, without component matching conditions.

ACKNOWLEDGEMENTS

The National Science Council, Republic of China supported this work under grant number NSC 100-2221-E-033-053.

REFERENCES

[1]   M . T . A buelm a' atti and H . A . A l-Zaher, "N ew un iversal filter with one input and five outputs using current-feedback am plifiers", *Analog Integrated Circuits Signal Process.*, vol. 16, no. 3, pp. 239 -244, 1998.

[2]   J. W. Horng, C. L. Hou, C. M. Chang, W. Y. Chung, and H. Y. Wei, "V oltage-mode universal biquadratic filters with one input and five outputs using M OCC IIs", *Computers and Electrical Engineering*, vol. 31, no. 3, pp. 190 -202, 2005.

[3]   J.W .H omg, C .L .H ou, C .M .Chang and W .Y .Chung, "Voltage-mode universal biquadratic filters with one input and five outputs", *Analog Integrated Circuits and Signal Processing*, vol. 47, no.1, pp. 73 -83, 2006.

[4]   J. W. Horng, C. L. Hou, C. M. Chang, H. P. Chou and C .T . L in , "High input impedance voltage-mode universal biquadratic filter with one input and five outputs using current conveyors", *Circuits Systems Signal Processing*, vol. 25, no. 6, pp. 767 -777, 2006.

[5]   J. W. Horng, "Lossless inductance simulation and voltage-mode universal biquadratic filter with one input and five outputs using DVCCs", *Analog Integrated Circuits Signal Process.*, vol. 62, no. 3, pp. 407 -413, 2010.

[6]   M. Bhushan and R . W . N ewcom b, "G rounding of capacitors in integrated circuits," *Electronics Letters*, vol. 3, no. 4, pp. 148 -149, 1967.

[7]   W. Chiu, S. I. Liu, H. W. Tsao and J. J. Chen , "CMOS differential difference current conveyors and their applications", *IEE Proceedings-Circuits Devices and Systems*, vol. 143, pp. 91-96, 1996.

[8]   H. O. Elwan and A.M. Soliman, "Novel CMOS differential voltage current conveyor and its applications", IEE Proceedings, Part G, vol. 144, no. 3, pp. 195-200, 1997.

Design of High-Speed Laser Diode Driver with Fast Switching and Reaction Time

Heng-Shou Hsu
Department of Electronic Engineering
Feng Chia University
Taichung, Taiwan, R.O.C.
hshsu@fcu.edu.tw

Don-Gey Liu
Department of Electronic Engineering
Feng Chia University
Taichung, Taiwan, R.O.C.
dgliu@fcu.edu.tw

Shih-Chi Liu
Department of Electronic Engineering
Feng Chia University
Taichung, Taiwan, R.O.C.
m9836650@fcu.edu.tw

Abstract—We proposed the design and implementation of the laser diode driver (LDD) circuitry with large driving current capability and high speed. The novel LDD circuitry with SFM (Source Follower Mode) architecture has been developed and carefully simulated. Besides, we also compared previous SFM architecture with the proposed novel SFM architecture. It can be shown that the novel SFM LDD faster speed performance than that of previous architecture. And novel SFM LDD is better than previous including Slew-Rate, reaction time, overshoot /Undershoot and corner case. The novel LDD is designed and implemented in TSMC 0.35-μ m 5V CMOS process.

Keywords- DVD/CD, DVD Player, LDD (Laser Diode Driver)

I. INTRODUCTION

Laser Diode Driver (LDD) can be find in many applications optical disks such as CD, DVD, Blue-ray disc, and fiber-optical communication....etc. Today the data is bigger and more complex than before. Therefore, a high-speed LDD is very important with high capacity optical store devices and high-speed optical communication. We main target enhance to Slew-Rate (rising time and falling time), reaction time, overshoot/Undershoot (<10%) and corner case (FF SS FS SF). In this paper, we focus on design of high speed and large current Laser Diode Driver for DVD/CD Read/Write [1-6].

II. DESIGN SOURCE FOLLOWER MODE LDD

A. Previous SFM Architecture [7]

The LDD output transistor size is very large. The added source follower buffer then isolated the original feedback path by directly shorting gate node and drain node of driver transistor. Therefore, large parasitic capacitances formed by driver transistor and output transistor are isolated with the drain node of driver transistor which makes the closed-loop response faster and with good settling performance. The SFM LDD had advantage of fast rising time and falling, but it had not enough fast reaction time, and disadvantage of corner case.

B. Novel SFM Architecture

The novel circuit used an Operational Amplifier and eight LDD to produce 200mA driving Laser Diode. The new LDD system Architecture is shown in Fig.1.

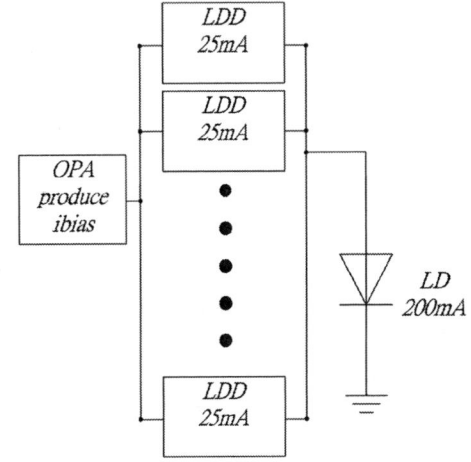

Figure 1.   LDD system Architecture

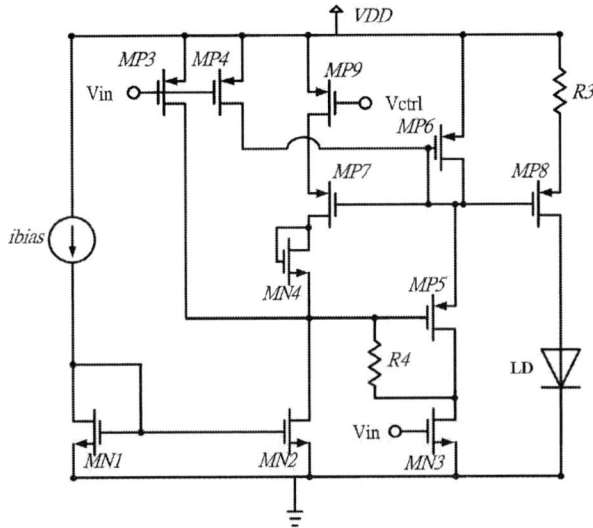

Figure 2.   Novel SFM Architecture

Fig. 2 Show the novel SFM LDD circuit. We used Operational Amplifier (OPA) [8] to generate I_{bias}. This current mirror is used to amplify I_{ref} (1) for large output driving current I_{out}(3). If channel-length modulation effect is neglected, we get

$$I_{ref} = \frac{\mu_p C_{ox}}{2} \left(\frac{W}{L}\right)_{MN2} (V_{GS} - V_{tn})^2 + \left(\frac{V_{DSMP5}}{R4}\right) \quad (1)$$

$$I_{out} = \frac{\left(\frac{W}{L}\right)_{MP4}}{\left(\frac{W}{L}\right)_{MP3}} I_{ref} \quad (2)$$

Besides, the laser diode (LD) equivalent circuit is shown in Fig. 3.

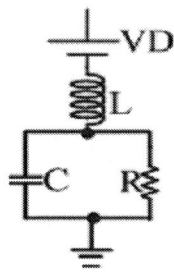

Figure 3.   LD (Laser Diode) equivalent circuit

Firstly, we use the diode connected transistors (MP6) as source follower's current source when Vin voltage is high and MP6 is cutoff, when Vin is low. It effectively decreases rising time and falling time as shown in Fig.4

Secondly, we add MN4 in I_{ref} to decrease overshoot and undershoot influence. Thirdly, we used the Parallel Switch Mode [9] in the circuit. This transistor size is smaller than Series Switch Mode. It can decrease parasitic capacitances. During the discharge mode the Vin set low. The I_{ref} current through MP3 to MN2. The MP4 provide a discharge path, it cause to quickly close output transistor MP8. Fourthly, the switch transistor (MN3) provided extra current and quickly reaction time. The resistor's (R4) is big when we design. SO extra current (V_{ds}MP5/R4) can be ignored. The equation (1) can be rewritten as equation (3).

$$I_{ref} = \frac{\mu_p C_{ox}}{2} \left(\frac{W}{L}\right)_{MN2} (V_{GSMN2} - V_{mMN2})^2 \quad (3)$$

Finally, the resistor (R3) is small resistor. Although bigger (R3) have better corner case, but it produce voltage drop to reduce output current. The transistor (MP9) working in deep triode like a variable resistor. According to the simulation results find the applicable V_{ctrl} voltage. Let MP3 source voltage be equal to MP4 source voltage. We also used V_{ctrl} to adjust the output current.

III.   SIMULATION RESULTS

In this section, we demonstrate the simulation results of the novel SFM architecture. The novel SFM architecture have been implemented in TSMC 0.35-μm 5V CMOS process and simulated by HSPICE with supply voltage "VDD" to be 5V.

The novel SFM architecture LDD simulation waveforms are demonstrated from Fig.5 to Fig.11. The signal integrity in novel SFM LDD is better than previous work SFM LDD. While the overshoot also in specifications. Although the novel SFM LDD undershoot is bigger than previous, the novel circuit can be quickly stablized at 0mA as shown Fig.5 and Fig.6.

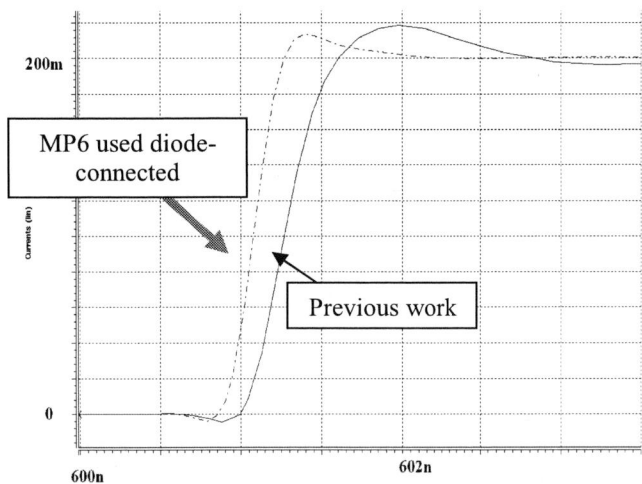

Figure 4.   Rising time comparison with previous work.

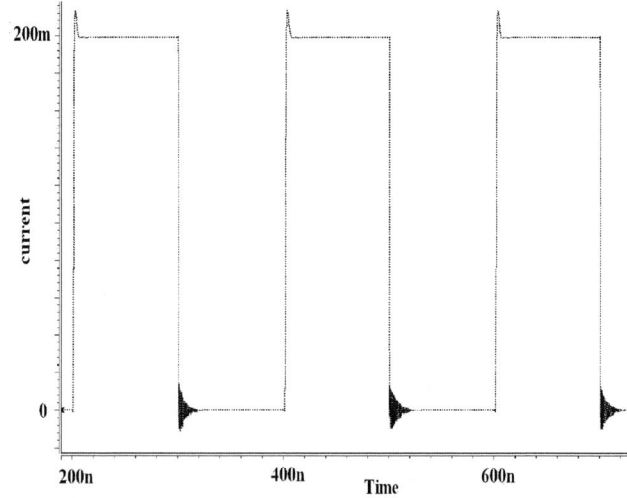

Figure 5.   Response of Previous work during high-speed recording

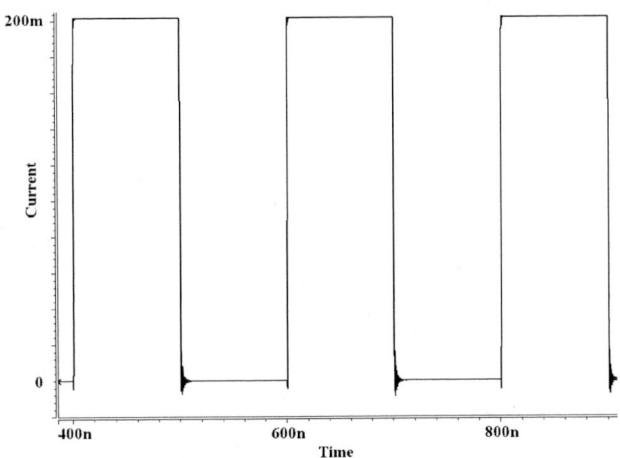

Figure 6. Response of novel SFM LDD during high-speed recording.

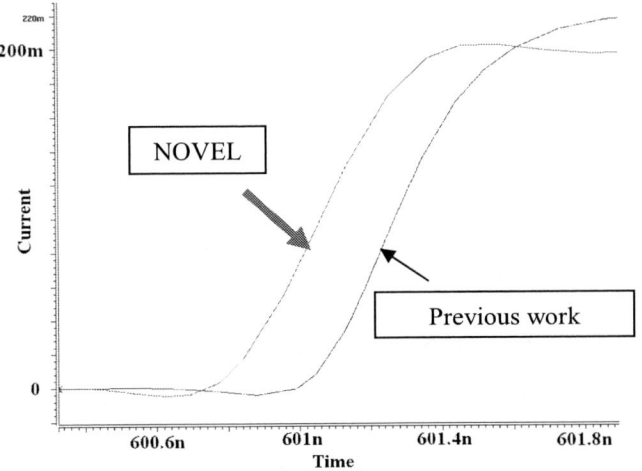

Figure 7. The rising time and reaction time of novel SFM LDD.

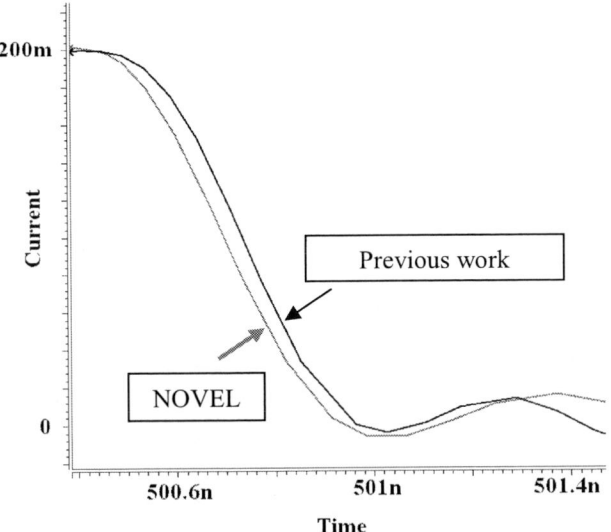

Figure 8. The falling time of Novel SFM LDD.

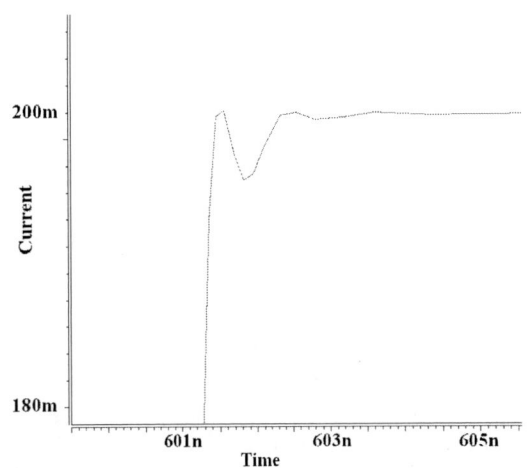

Figure 9. The overshoot of Novel SFM LDD

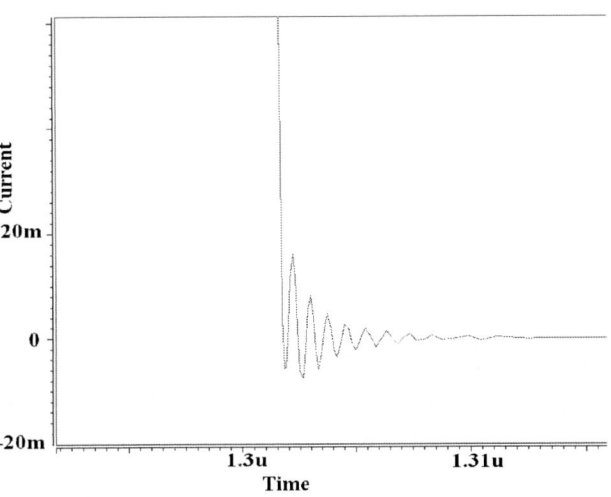

Figure 10. The undershoot of Novel SFM LDD

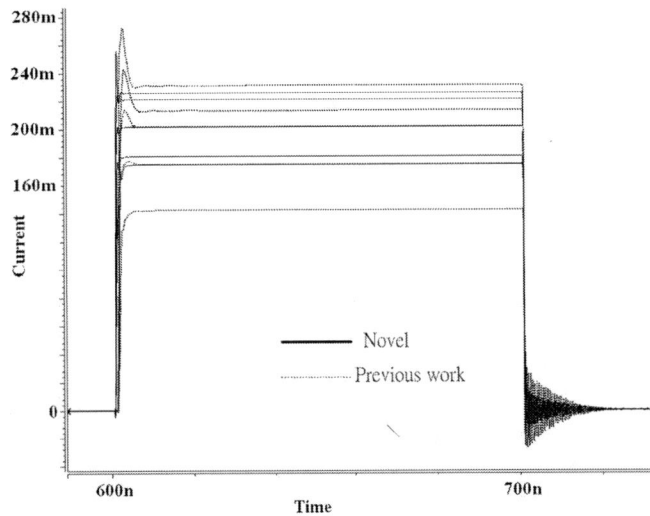

Figure 11. Response of SFM LDD in Corner case(TT SS FF SF FS) at novel circuit without adjust V_{ctrl}.

978-1-61284-863-1/11 $26.00 © 2011 IEEE   45

TABLE I. COMPARING PREVIOUS CORNER CASE WITH NOVEL VERSION

	Previous work in [7]	Novel
TT 25℃	200mA	200mA
FF 0℃	231mA	229mA →213mA
SS 100℃	142mA	174mA →191mA

※ → means after adjusting V_{ctrl}.

IV. CONCLUSION

The proposed novel SFM LDD has been implemented by 0.35-μm 5V CMOS process and simulated by HSPICE. The rising time is 0.5ns and falling time is about 0.4ns which is fast enough for high speed DVD recording. Besides, the overshoot and undershoot is small enough to maintain the disk quality during high speed recording. And the reaction time is faster than previous work. The layout of the novel SFM LDD has been fabricated with the pattern illustrated as Fig.12. The comparison between previous LDD and novel version duting corner cases has been summarized as TABLE I whereas the overall performances of this novel LDD have been listed as TABLE II.

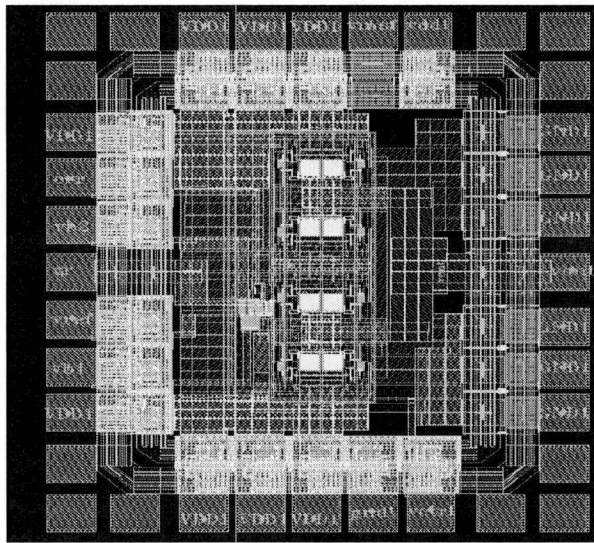

Figure 12. Layout of the novel SFM LDD.

TABLE II. OVERALL PERFORMANCE SUMMARY OF THE PROPOSE NOVEL SFM LDD

	Previous work in [7]	Novel
Technology	TSMC 0.35um	
Driver current	200mA	
Power Supply	5V	
Rising Time	0.59ns	0.49ns
Falling Time	0.33ns	0.39ns
Reaction Time	Faster than 0.9ns	
Undershoot	13mA	15mA
Overshoot	14mA	1.2mA
SS	231mA	229mA
FF	142mA	174mA

REFERENCES

[1] Koichiro Nishimura, Shinrou Inui, Masaaki Kurebayashi, Toshimitsu Kaku, and Akihiro Asada "High-Speed DVD-Multi Drive System", *IEEE Transaction on Consumer Electronics*, vol. 50, NO. 1, pp. 198-203, February 2004

[2] Udo Karthaus, Stefan Schabel "Write Pulse Generator for 16x DVD Recording With Symmetric CMOS Inverter Ring Oscillator"*IEEE Journal of Solid-State Circuits*, Vol. 40, NO. 11, pp. 2286- 2295 November 2005.

[3] Johannes Sturm, Martin Leifhelm, Harald Schatzmayr, Stefan Groib, and Horst Zimmermann, "Optical Receiver IC for CD/DVD/Blue-Laser Application" *IEEE Journal of Solid-State Circuits*, vol. 40 NO. 7, pp. 1406-1413, July 2005.

[4] Mehrdad Nourani, Amir R. Attarha "Detecting Signal-Overshoots for Reliability Analysis in High-Speed System-on-Chips" *IEEE Transactions on Reliability*, vol. 51, NO. 4, pp. 494-504, December 2002.

[5] Woogeun Rhee, Bang-Sup Song, Akbar Ali "A 1.1-GHz CMOS Fractional-N Frequency Synthesizer with a 3-b Third-Order ΔΣ Modulator" *IEEE Journal of Solid-State Circuits*, vol. 35, NO. 10, pp. 1453-1460, October 2000.

[6] Hoi Lee, Philip K. T. Mok, Ka Nang Leung "Design of Low-Power Analog Drivers Based on Slew-Rate Enhancement Circuits for CMOS Low-Dropout Regulators" *IEEE Transactions on Circuits and Systems-II Express Briefs*, vol. 52, NO. 9, pp. 563-537 September 2005

[7] Heng Shou Hsu, Hung Chieh Chung, "High Speed Laser Diode Driver" IEEE International Symposium on Integrated Circuits, pp. 570-573, December 2009.

[8] Heng Shou Hsu, Hung Chieh Chung, "Design of High Speed Laser Diode Driver" Workshop on Consumer Electronics, pp. 730-733, Nov. 2010.

[9] Behzad Razavi, Design of Analog CMOS Integrated Circuits, 1st edition, New York, NY: McGraw-Hill, 2001.

[10] Yun Yang, Jia Guo, Yasuaki Inoue and Hong Yu, Graduate School of Information, Production and Systems, Wasada University "A Large Current and High Speed Laser Diode Driver Using Switch Position Modification" *IEEE Communications Circuits and Systems Proceedings*, vol. 4, pp. 2319-2323, June 2006.

High-Speed Laser Diode Driver with Low Sensitivity to Process Variation and Improvement on Overshoot Performance

Heng-Shou Hsu
Department of Electronic Engineering
Feng Chia University
Taichung, Taiwan, R.O.C.
hshsu@fcu.edu.tw

Don-Gey Liu
Department of Electronic Engineering
Feng Chia University
Taichung, Taiwan, R.O.C.
dgliu@fcu.edu.tw

Chia-Ming Chuang
Department of Electronic Engineering
Feng Chia University
Taichung, Taiwan, R.O.C.
m9833019@fcu.edu.tw

Abstract—**In this paper, we proposed a high speed and large current laser diode driving circuit with low sensitivity to process variation. It is commonly applied in the DVD player. In a large current and high speed output LDD circuit, topics such as overshoot, undershoot, rising time, falling time and slew-rate are necessary to be considered. Therefore, to improve the signal integrity of Combination Switch Mode (CSM) LDD Architecture in [5], modification is made. The simulation results show that the output current turned slightly slower in speed, whereas the performance in overshoot and undershoot are significantly improved. Finally, we proposed the Source Degeneration Type with Improve Overshoot Performance (CSM) LDD Architecture. The circuit has been implemented in 0.35-μ m 5V CMOS process and simulated by HSPICE.**

Keywords- DVD Player, Laser Diode Driver

I. INTRODUCTION

Laser Diode Driver (LDD) is widely used in high speed data transmission, and several studies have been made to derive large and high speed drive current. In the proposed paper, the LDD is employed in the DVD Player Pick-Up Head. Fig. 1 shows the system architecture of laser diode driver on DVD Player. Being transmitted by LVDS, the date is then red, write and rewrite by the control circuit. The input signal is used to generate the driving current for the Laser Diode. In this thesis, we describe the design of high speed laser diode driver (LDD), with the feature of providing large driving current and high speed rate at output terminal to activate the laser diode. [1-3].

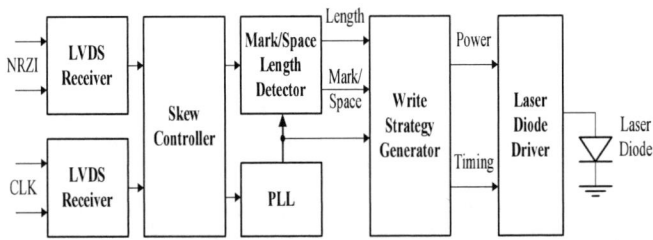

Figure 1. System architecture of laser diode driver

II. LASER DIODE DRIVER ARCHITECTURE

A laser diode that connected to LDD circuit needs a large current to ensure the output laser functionality, therefore a large output and high speed current is needed to attain high signal integrity. In the large current and high speed output of the LDD circuit, topics such as overshoot, undershoot and slew-rate (rising time and falling time) are necessary to be considered. Because of the desired large current outputs, the laser diode driver circuit design is achieved by current mirror architecture. [4]

In this paper, the Combination Switch Mode (CSM) LDD Architecture is [5] shown in Fig. 2. The schematic is composed of current mirror architecture, and transistors MP1, MP2 and MN1 receive control signal and act as switches.

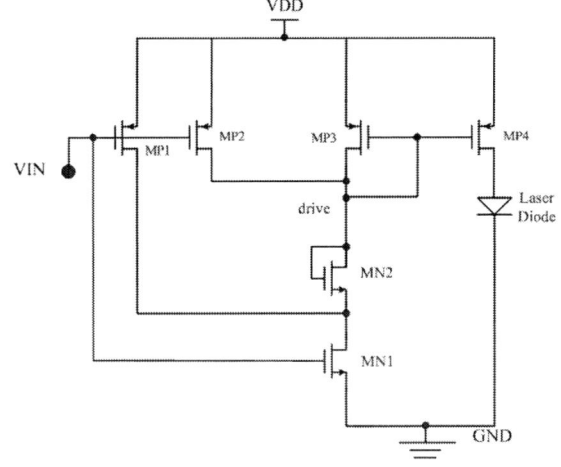

Figure 2. Combination Switch Mode (CSM) LDD Architecture [5]

In the above CSM architecture, transistors MP3 and MP4 form the current mirror. The current mirror is used to amplify I_{ref}, and get the output drive current I_{out}. We can acquire the large output current by using this technique. Despite several relevant studies were made according to [6-7], there are still issues needed to be considered, therefore, we utilize some techniques to improve the architecture in [5].

978-1-61284-863-1/11 $26.00 © 2011 IEEE

III. CIRCUIT DESIGN

In this paper, we introduce the laser diode drive circuit architecture as shown in the Fig. 2 [5]. The circuit architecture is modified to have better the signal integrity performance. And the circuit affected by temperature, voltage, process and other environmental influences can be reduced.

A. Source Degeneration Type LDD Architecture

In Fig. 2, the source of transistor MP3 and MP4 are connected to VDD, thus, variation in power supply could cause the output voltage instable, moreover, the PVT variation may cause the circuit not generate the exact current value than we expected. In order to reduce the mismatch between the ideal case and actual cases, we utilized three resistances R1, R2 and R3 to make the voltage that across the VDS of MP4 and MP3 more stable. Fig. 3 illustrates the proposed the Source Degeneration Type LDD Architecture.

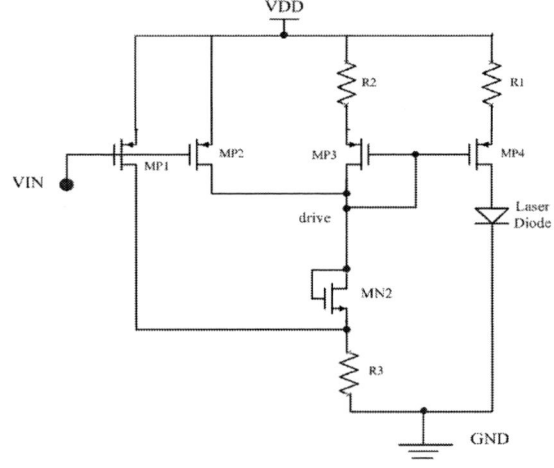

Figure 3. CSM LDD with Source-Degeneration

In this circuit, NMOS MN2 is to generate current I $_{ref}$. The current source is given by equation (1). The VGS2 and VSG3 are described in equation (2) and equation (3).

$$I_{ref} = \frac{\mu_n C_{ox}}{2} \left(\frac{W}{L}\right)_{MN2} (V_{GS2} - V_{tn})^2 (1 + \lambda V_{DS2}) \quad (1)$$

$$V_{GS2} = V_{DD} - V_{SG3} \quad (2)$$

$$V_{SG3} = |V_{tp}| + \sqrt{\frac{2I_{ref}}{\mu_P C_{ox} \left(\frac{W}{L}\right)_{MP3}}} \quad (3)$$

Furthermore, we also removed the control switch transistor MN1, only two transistors MP1and MP2 work as control switches.

B. Source Degeneration Type with Improved Overshoot Performance forLDD Architecture

In the Source Degeneration Type (CSM) LDD Architecture, we added three resistors, though the simulation results show the output current mismatch between worst case and ideal case has been reduced to nearly ten percent, the results revealed that the output voltage of LDD is unstable during charging and discharging moment which would lead to overshoot and undershoot. Being incompatible with the specification, high speed recording can't ensure the disk quality. To overcome this problem, an extra transistor MP5 is applied as shown in Fig.4 with gate and drain connected to drive point to form a feedback mechanism, thereby enhance the bias drive voltage stability. After modification, the simulated results showed the overshoot and undershoot are smaller. Despite the slightly slower discharge rate in the results, the transmission speed is still compatible with the specification. Compared to Combination Switch Mode (CSM) LDD Architecture, the results were improved. Fig 4 gives the Source-Degeneration type with improve overshoot performance for CSM LDD architecture.

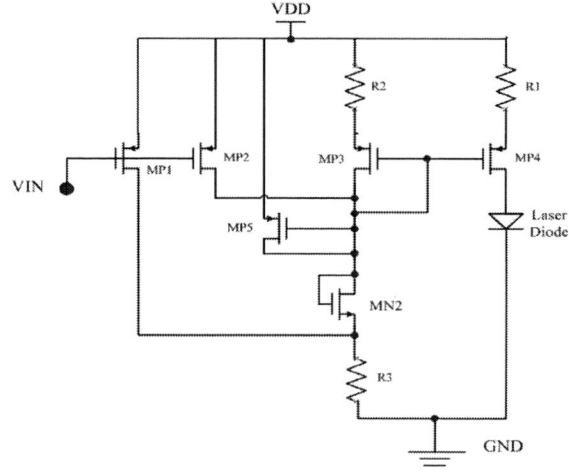

Figure 4. CSM LDD with Source-Degeneration and improved overshoot performance.

IV. SIMULATION REsULTS

In this section, we demonstrate the simulation results of the Source Degeneration Type (CSM) LDD Architecture and Source Degeneration Type with Improved Overshoot Performance (CSM) LDD Architecture, and the comparison of CSM LDD architecture and the proposed LDD architecture have been made. The Source Degeneration Type (CSM) LDD Architecture and Source Degeneration Type with Improved Overshoot Performance (CSM) LDD Architecture was implemented in 0.35-μ m 5V CMOS process and simulated by HSPICE with supply voltage 5V.

To drive the laser diode, we demand four driving circuits to obtain a current of 200 mA. Each drive circuit provides 50mA for the laser diode, and the circuit achieves a total amount of 200mA to turn on the laser diode. Fig. 5 describes the drive circuit block diagram.

978-1-61284-863-1/11 $26.00 © 2011 IEEE

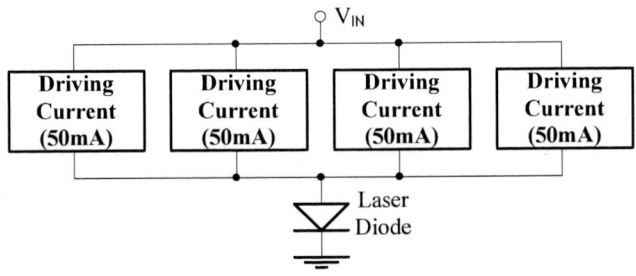

Figure 5. The drive circuit block diagram

Fig. 6 presents the laser diode equivalent model that was to be drove by the LDD circuit proposed in this thesis.

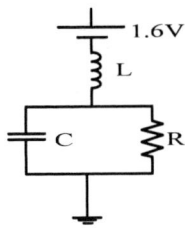

Figure 6. The Laser Diode equivalent circuit

The Combination Switch Mode (CSM) LDD Architecture simulation waveform is show in Fig. 7. The Source Degeneration Type (CSM) LDD Architecture and Source Degeneration Type with Improved Overshoot Performance (CSM) LDD Architecture simulation waveforms are demonstrated from Fig. 8 to Fig. 14. Comparison of these three LDD architectures is presented in the following results. Fig. 7 and Fig. 8 show the output current mismatch between the worst case and ideal case is decreased to nearly ten percent. Described in Fig. 9 and Fig. 10, the rising time and falling time are slightly slower, whereas the overshoot and undershoot performances are significantly improved. From Fig. 11 to Fig. 14, the simulation results with overshoot and undershoot improvement are shown.

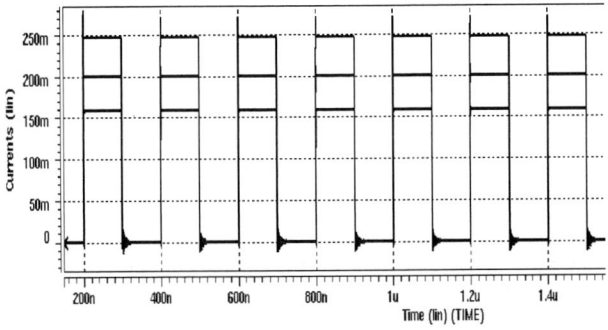

Figure 7. The corner-case of Combination Switch Mode (CSM) LDD

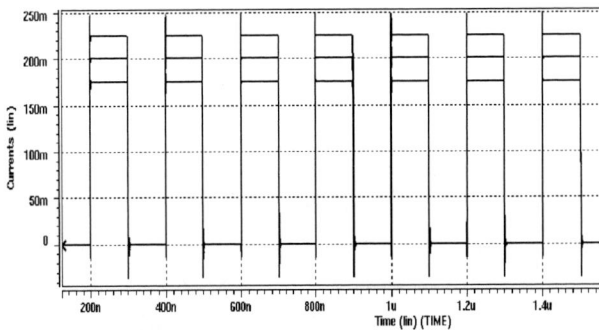

Figure 8. The corner-case of CSM LDD with Source Degeneration

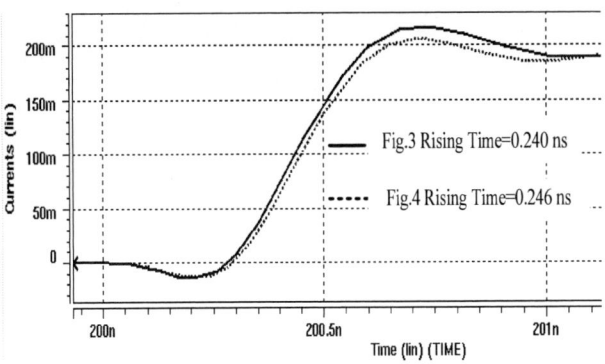

Figure 9. The comparison of rising time between Fig. 3 and Fig. 4

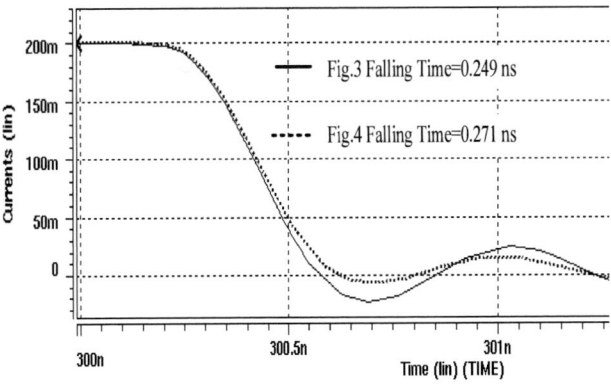

Figure 10. The comparison of falling time between Fig. 3 and Fig. 4.

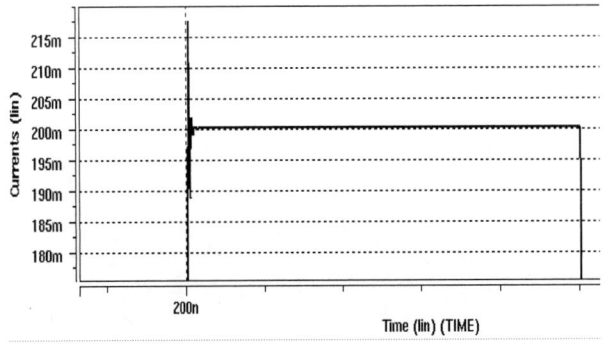

Figure 11. The overshoot of CSM LDD with Source-Degeneration.

978-1-61284-863-1/11 $26.00 © 2011 IEEE

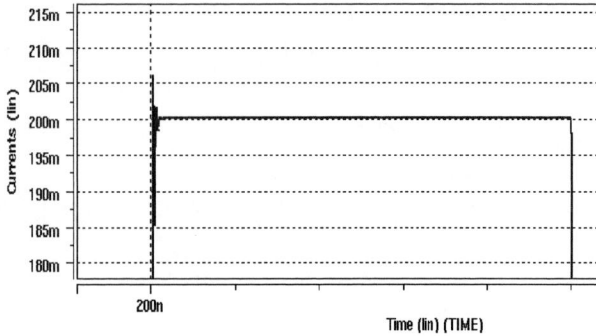

Figure 12. The overshoot of CSM LDD with Source-Degeneration and improved overshoot response.

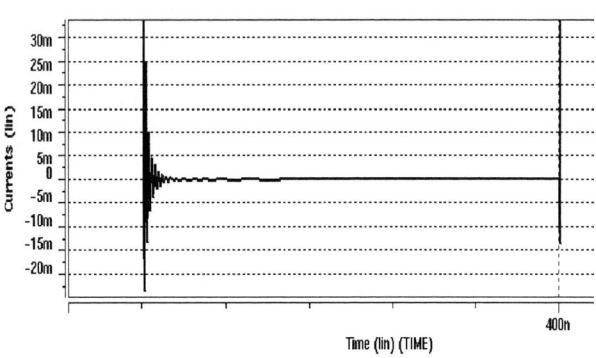

Figure 13. The undershoot of CSM LDD with Source-Degeneration

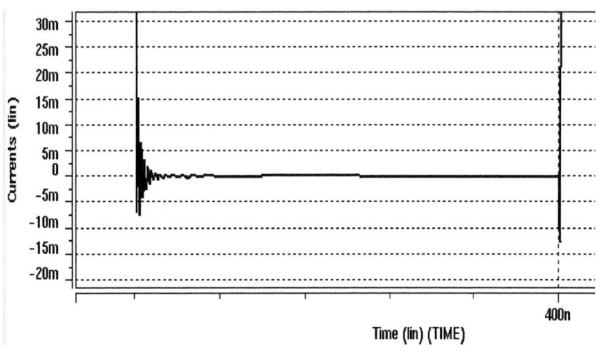

Figure 14. The undershoot of CSM LDD with Source-Degeneration and improved undershoot response.

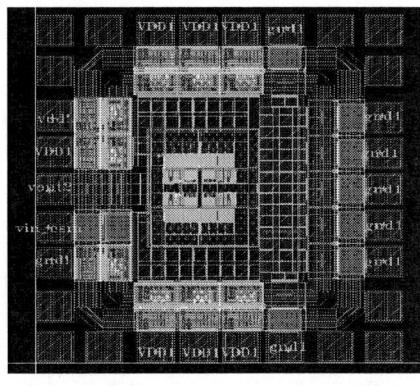

Figure 15. LDD Chip Layout

V. CONCLUSIONS

The proposed Source Degeneration Type with Improved Overshoot Performance (CSM) LDD Architecture. The proposed LDD circuit has been implemented by 0.35-μm 5V CMOS process and simulated by HSPICE. The rising time is 0.24ns and falling time is about 0.27ns that is fast enough for high speed DVD recording with speed rate up to 16X. As for signal integrity concerns, the overshoot is 206mA, and undershoot is -7.3mA. The overshoot and undershoot are small enough to ensure the disk quality during high speed recording.

TABLE I. PERFORMANCE COMPARISON OF SIMULATION RESULT

	Reference [5]	CSM) with Source Degeneration	Source Degeneration Type with Improve Overshoot Performance (CSM)
Output current	200mA	200mA	200mA
Rising/Falling	0.2/0.25ns	0.24/0.25ns	0.24/0.27ns
Overshoot	215mA	218mA	206mA
Undershoot	-8mA	-23.5mA	-7.3mA
FF 0° C	247 mA	224.6 mA	223 mA
TT 25° C	200 mA	200 mA	200 mA
SS 100° C	158 mA	175.5 mA	176 mA

REFERENCES

[1] Koichiro Nishimura, Shinrou Inui, Masaaki Kurebayashi, Toshimitsu Kaku, and Akihiro Asada "High-Speed DVD-Multi Drive System", *IEEE Transaction on Consumer Electronics*, vol. 50, NO. 1, pp. 198-203, February 2004.

[2] Udo Karthaus, Stefan Schabel "Write Pulse Generator for 16x DVD Recording With Symmetric CMOS Inverter Ring Oscillator"*IEEE Journal of Solid-State Circuits*, Vol. 40, NO. 11, pp. 2286- 2295 November 2005.

[3] Johannes Sturm, Martin Leifhelm, Harald Schatzmayr, Stefan Groib, and Horst Zimmermann, "Optical Receiver IC for CD/DVD/Blue-Laser Application" *IEEE Journal of Solid-State Circuits*, vol. 40 NO. 7, pp. 1406-1413, July 2005.

[4] Mehrdad Nourani, Amir R. Attarha "Detecting Signal-Overshoots for Reliability Analysis in High-Speed System-on-Chips" *IEEE Transactions on Reliability*, vol. 51, NO. 4, pp. 494-504, December 2002.

[5] Heng Shou Hsu, Hung Chieh Chung, "Design of High Speed Laser Diode Driver" Workshop on Consumer Electronics, pp. 730-733, Nov. 2010.

[6] Yun Yang, Jia Guo, Yasuaki Inoue and Hong Yu, Graduate School of Information, Production and Systems, Wasada University "A Large Current and High Speed Laser Diode Driver Using Switch Position Modification" *IEEE Communications Circuits and Systems Proceedings*, vol. 4, pp. 2319-2323, June 2006.

[7] Heng Shou Hsu, Hung Chieh Chung, "High Speed Laser Diode Driver" IEEE International Symposium on Integrated Circuits, pp. 570-573, December200

Image Classifying Algorithm And Its VLSI Implementation Based On The Directional Features

Dongfang Wang, Ningmei Yu
Dept. Electronics Engineering
Xi'an University of Technology
Xi'an ,China
wangdong@xaut.edu.cn

Yvonne Lam Y.H. , Yuanjin Zheng
School of Electrical and Electronic Engineering
Nanyang Technological University
Singapore
YJZHENG@ntu.edu.sg

Abstract—**vector quantization (VQ) is widely used in the field of image coding because of its simple decoding algorithm and high compression rate. But owing to its high complexity of image codeword matching process ,it is often limited out of some place requisite of real time. In order to accelerate this process, a novel direction classifying criterion is presented in this paper, it is based on the good image directional selectivity of wavelet transform, and deduced and simplified from the relationship of the coefficients of quadric wavelet transform. This criterion was used to improve the efficiency of codewords matching in VQ algorithm. Simulation results show that, in average, this algorithm can reduce time passing to 38.4% within 1.8% PSNR lost in contrast with the common VQ. At the same time, this criterion is simple and easy to be implemented using VLSI technology, and its pipe-line VLSI architecture can be easily integrated in other image encoding chips due to its regularity and modularity. In this paper, it was integrated into PDVQ chip fabricated with 0.35μm CMOS process, to speed up its image encoding. Test results show that, the PDVQ chip can determine the image direction correctly, and at 3.0V power supply, PDVQ chip can operate steadily under 100MHz, and which can support real-time encoding application for 512×512 gray images at 30 frame/s.**

Keyword: image coding; vector quantization; VLSI

I. INTRODUCTION

Image communication over wireless transmission channels is an emerging technology, and there are several challenging issues to overcome current technology limitations. The first requirement for wireless image communication is low delay transmission. Conventional standards for video compression, such as MPEG4 and H.264, have good coding efficiency but still take a longer delay time than that of wireless communication channels to be currently supported. Secondly, high compression efficiency to satisfy the limited bandwidth of wireless channels is essential to the wireless image communication, but the popular standards mentioned above have good coding efficiency but still require a much larger bandwidth than that of common wireless communication channels to be currently supported. Therefore, it is desirable to design a high efficient image-coding technique, which has a low delay and a high compression ratio over a wireless communication channel. Finally, we should consider image compression algorithms and their VLSI architectures which allow portable decoders with small size, low-power consumption, and acceptable reconstructed image quality.

Vector quantization (VQ)[1] is an important technique for data compression, and has been widely used in speech coding, image coding and video coding because of its high compression ratio and simple decoding algorithm. One of the major barriers for using VQ in real applications is the large computational complexity (huge time cost and storage cost) involved in VQ encoding process. But if the codebook used in common VQ is constrained with various conditions, the matching complexity will be reduced [2]. Two kinds of constraints for the codebooks are often used, one is based on the content of codebook, in which codewords are often transformed using various transform algorithms such as pyramids architecture[3] and Wavelet Transform [4], and then replaced by its coefficients sequentially according to the importance to distinguish it, but generally these algorithms are not easy to be implemented by hardware. The other constraint is based on the structure of codebooks, in which the sequences of codewords are related with its matching algorithm, so the codewords often are allocated into fixed areas by the algorithm such as classified VQ (CVQ) [5], tree VQ[6], Multistage VQ[7], to reduce the matching computation.

A novel and interesting direction classifying criterion based on the good image directional selectivity of wavelet transform is presented in this paper, which can be applied in VQ algorithm as CVQ. In this criterion, the images can be classified into smooth, horizontal, vertical and irregular categories, it is easier to be implemented than the classification of edge angle [8], and is more efficient than the classification only including edge and smooth direction [9]. In next section, the classifying criterion will be introduced.

II. ALOGRITHM INTRODUCTION

A. Direction Classifying Criterion

In fact, for the human vision, the image direction is often as a first criterion to match the image blocks. For example, in Fig. 1, through judging the gray direction, we can know that image block 1 is more similar to input image block than image block 2. So, the match efficiency can be improved highly, if input image block is firstly classified by directional feature based on the human vision, which show that, for our eyes, it is more

---

Project supported by Natural Science Basic Research Plan in Shaanxi Province of China (No.2011JQ8032) and Natural Science Research Plan in Xi'an university of Technology of China (105-210907).

sensitive to the low-frequency smooth regions than to the high-frequency edges, and to image edges, it is more sensitive to horizontal and vertical direction changes than other angle direction changes. So, the most sensitive direction change is smooth, followed by horizontal and vertical, the last is irregular.

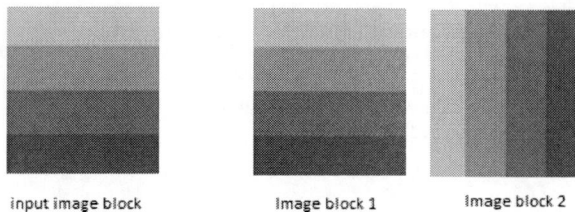

input image block Image block 1 Image block 2

Figure 1. Directional features of image blocks

As we know, wavelet transform has good image directional selectivity[10], it can decompose an image to *HL* (horizontal direction), *LH* (vertical direction) and *HH* (diagonal direction), furthermore, after quadric wavelet transform, a image block can be decomposed to low frequency, horizontal, vertical and diagonal wavelet coefficients, and the maximum in *HL2*, *LH2* and *HH2* determines its direction [10]. So, its coefficients can be used to analyze image directional features.

Generally, for the smooth images, the difference between these 3 coefficients is very small. For vertical images, $th\_v=|ch2/cv2|>4$. For horizontal images, $th\_h=|cv2/ch2|>4$. For diagonal images, *HH2* is much larger or smaller than other two coefficients. If we select the appropriate th_v and th_h, the images can be classified into the corresponding categories. Table 1 shows the relationship between th_h, th_v and image determinative accuracy. But it is complex for hardware circuit to calculate th_v and th_h and not easy to be implemented. So, in literature [10], we proposed a novel method to simple the calculation of the coefficients of quadric wavelet transform by using the sums of the four 2×2 sub-blocks to replace them.

For a 4×4 image block, each pixel represented by the variable ai ($i = 1,2 \ldots \ldots, 16$) is shown in Fig. 2, and *Aj* ($j=1,2,3,4$) denote the gray sum of each 2×2 sub-image block. In literature [10], the relations of *Aj* ($j=1,2,3,4$) are used to determine the directions of images in place of using *ch2* and *cv2* .

TABLE I. THRESHOLD VALUE AND DETERMINATIVE ACCURACY

Accuracy	Threshold			
	7	6	5	4
Vertical	87.06%	88.51%	86.98%	85.44%
Horizontal	86.63%	86.21%	85.29%	83.33%

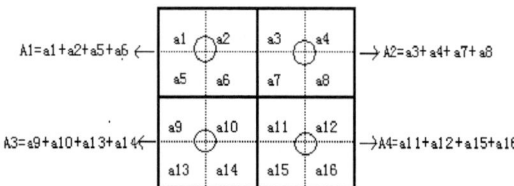

Figure 2. Partition of son image block

In addition, in order to simple the classification, we merge the diagonal category into irregular category. So, the images can be classified into four corresponding categories, they are

smooth, horizontal, vertical and irregular. And the threshold values were selected by iterative algorithm in terms of the impacts on testing images qualities. The details of the iterative algorithm can be referred to the literature [10], but the original criterion expression in it is very complex. In literature [11], the simple criterion was deduced, and the final direction classifying criterion is depicted as follows:

- If *max(A1,A2,A3,A4)-min(A1,A2,A3,A4)<64*, this image block can be classified into the smooth category.

- Else, if *|A1-A2|<32 && |A3-A4|<32*, it can be classified into the vertical category.

- Else, if *|A1-A3|<32 && |A2-A4|<32*, it can be classified into the horizontal category.

- Else, it can be classified into the irregular category.

B. Criterion Test

The criterion presented in this paper was tested through designing a CVQ encoding algorithm. Firstly, codewords in the codebook were classified into four direction sub-codebooks offline. When encoding, the CVQ encoding system will judge the direction of input image blocks to choose corresponding direction sub-codebook, and then search the most matching codeword in selected sub codebook.

The test results of the 10 standard test images are shown in Fig. 3. In average, for the codebook with 1024×16 codebook, this CVQ algorithm can reduce time passing to 38.4% within 1.8% PSNR lost in contrast with the common VQ. At the same time, the reconstructive images 'Lena' Using category codebook and using no-category codebook are also shown in Fig. 4.

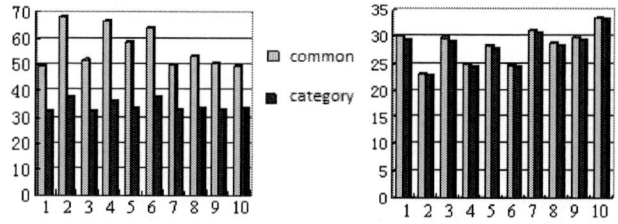

Figure 3. (a) Encoding times (b) PSNR

(a) Using category codebook (b) Using no-category codebook

Figure 4. Reconstructive image 'Lena'

C. Criterion Application

In order to speed up our PDVQ (partition dynamically VQ) image coding system, this direction judging criterion was integrated into it, and its VLSI architecture were also designed

978-1-61284-863-1/11 $26.00 © 2011 IEEE

with the whole encoding system. The PDVQ encoding state flow is illustrated in Fig. 5, in which, the task of **Sort Function** is to classify the direction of the input images, the architecture and test results of the Sort Function unit in PDVQ chip will be described in followed sections.

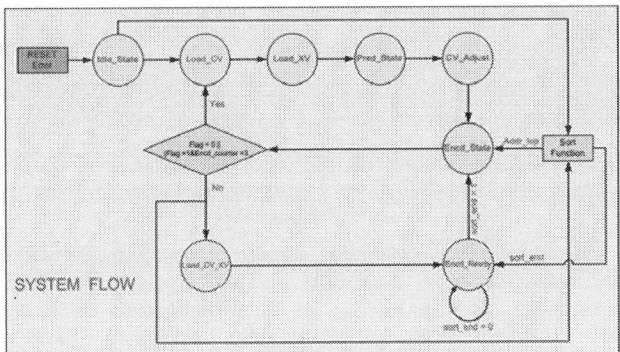

Figure 5.   State flow of PDVQ encoding system

III.   VLSI ARCHITECTURE OF THE SORT FUNCTION

In the PDVQ chip, the function of the sort function unit is constituted with two modules. One is CMPU_SUM module to calculate the sum of each 2×2 sub-image block; the other is TYPE_JUDGE module to classify the direction of the input image block using direction-classifying criteria.

A.   CMPU_SUM Module

In this module, the main work is to calculate the sums including the sum of each 2×2 sub image block and the sum of the whole input image block. The Fig. 6 illustrates the sums needed to be computed.

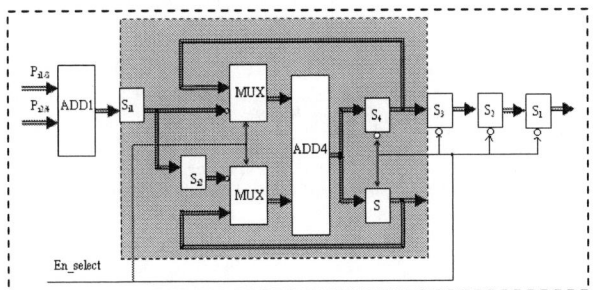

Figure 6.   Sum values needed computing

In order to improve the speed and reduce the chip area, pipeline and time-multiplexing technologies are designed in this module, and its circuit architecture is shown in Fig. 7, in which, the adder ADD1 is designed to calculate the sums of the two pixels, and the adder ADD4 is used to compute both the sums of 2×2 sub image blocks and the sum of the whole input image blocks. Four shift registers Si (i=1,2,3,4) and register S are used to store the computing results.

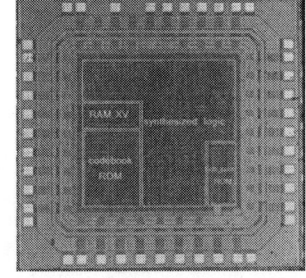

Figure 7.   Structure of sum computer

B.   TYPE_JUDGE Module

This module is designed to determine the direction of the image block based on the sums gotten from the CMPU_SUM module. According to the direction-determining criteria described in section II, the main judge conditions are $max(A1,A2,A3,A4)-min(A1,A2,A3,A4)<64$, $|S1-S3| <32$ and $|S2-S4|<32$. By analyzing these expression and determinative criteria, we determined that combination circuits would be large-scale, which would constrain the system clock frequency. So, we adopted pipeline technology to improve the speed of this module circuit. The block diagrams of smooth and vertical direction judging circuit are shown in Fig. 8 and Fig. 9.

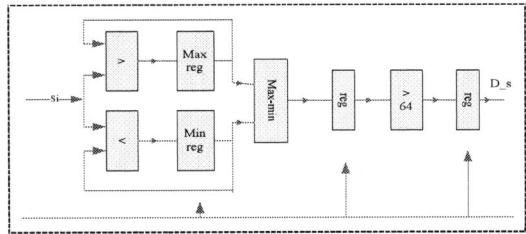

Figure 8.   Block diagram of smooth direction judging

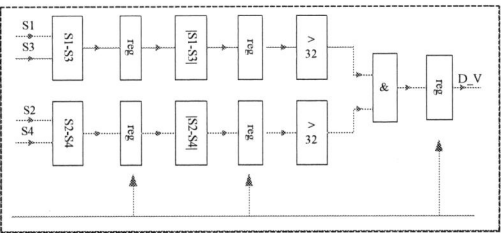

Figure 9.   Block diagram of vertical direction judging

IV.   CHIP TEST

Because the sort function circuit was implemented as a part of PDVQ image encoding system, so its test results can be gotten from testing the PDVQ chip. The PDVQ chip shown in Fig. 10 was fabricated based on charter 0.35 micron CMOS standard cell technology, and the size of chip is 2.08mm×2.08mm. The test results show that the chip can work stably at 100MHz, and support 512×512 images real-time application under 30fps when the operation frequency is 60MHz. In order to test its ability to determine the direction of the image block, the directions of 8×8 image blocks were firstly simulated by MATLAB, then these image blocks were written into a FPGA testing chip as test data, the final output results of PDVQ chip can show whether its direction determining is right or not. We used ten 512×512 test images to test the PDVQ chip, but in order to demonstrate it clearly, we selected two 8×8 image blocks including 4 direction types to clarify that the PDVQ chip can classify the direction of the image block correctly.

Figure 10. Microphotograph of PDVQ chip

A. Image Block 1

Gray values of the first 8× 8 image block are shown in Fig. 11(a), and are re-displayed as an image in Fig. 11(b). Two sub image blocks on the left side, L1 and L2, can be directly determined horizontal using our eyes, and two blocks on the right side, R1 and R2, are vertical. The MATLAB simulation results of these four sub image blocks are given in hexadecimal form, where L1= DAH, R1= B3H ,L2=DAH ,R2=B0H .

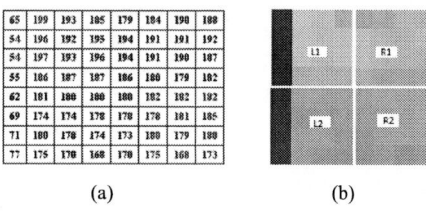

(a)   (b)

Figure 11. Image block 1

The chip output results were tested using logic analyzer and shown in Fig. 12, in which, *clk* denotes the system clock signal (60MHz), *rst* denotes the reset signal, *data_in*(16bit) denotes the input image data signal, *Encd_end* denotes the encoding end signal, and the *addr* (8bit) denotes the encoding result, it is also the output signal of the PDVQ encoder.

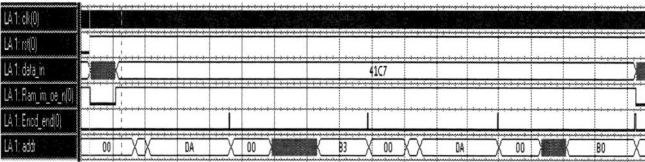

Figure 12. Output waveforms of image block 1

As shown in Fig. 12, when *Encd_end* signal is available (high level), this illustrates that the PDVQ chip has finished a 8× 8 image block encoding process and the output signals (*addr*) can be sampled and sent to the logic analyzer. From the Fig. 12, we can know that the output encoding results of the image block 1 are DA,B3,DA,B0, they are consistent with the results of simulation, which can prove that the PDVQ chip can encode the horizontal and smooth direction image block correctly.

B. Image Block 2

The gray matrix of image block 2 is shown in Fig. 13. The top sub blocks are smooth, the bottom left block is irregular and the bottom right block is vertical. The MATLAB simulation results in hexadecimal form are A9H, 6BH, 4BH, FBH. The Fig. 14 illustrates the output waveform sampled by logic analyzer, and the output encoding results of the image block 2 are A9, 6B, 4B, FB, they are consistent with the results of simulation, which can prove that the PDVQ chip can encode the vertical and irregular direction image block correctly.

143	144	144	156	175	184	171	193
151	152	146	150	176	189	193	198
152	149	156	165	195	200	204	208
153	144	146	152	193	206	207	211
150	142	145	147	169	192	188	191
144	142	156	158	167	184	176	150
131	132	142	145	150	166	183	167
99	101	103	113	111	111	111	106

Figure 13. Image block 2

The test image blocks 1 and 2 include all four directions: irregular, horizontal, Vertical and smooth, and from the experimental results of ten 512×512   test images, it can be concluded that the encoding results of the PDVQ chip are consistent with the simulation results, and the PDVQ chip can determine the image direction correctly.

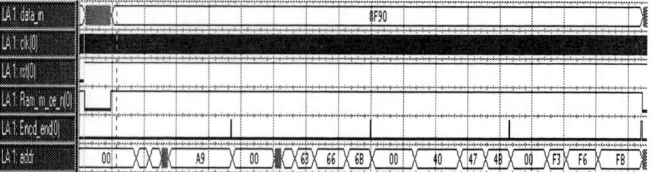

Figure 14. Output waveforms of image block 2

V.   CONCLUSION

In this paper, we presented a direction classifying algorithm based on the wavelet transform, and its direction classifying criteria were simplified using the sums of the 2×2 sub image blocks, this criterion can reduce time passing to 38.4% in contrast with the common VQ, at the same time, the direction-classifying criteria are simple and easy to be implemented by VLSI technology, so it was integrated into our PDVQ image encoding system to improve codewords matching efficiency. Its VLSI architecture was implemented as a part of PDVQ chip, which was fabricated based on charter 0.35 micron CMOS standard cell technology, and it can support 512×512 images real-time application under 30fps when the operation frequency is 60MHz. The PDVQ chip testing results show that it can determine the image direction correctly.

REFERENCES

[1]   Y. Linde, A. Buzo, R. M. Gray. "An Algorithm for Vector Quantizer Design." IEEE Transactions on Communications, vol. 28, pp. 84-95, 1980.

[2]   S. Sun,Z. Lu,Vector Quantization Technology and Application, Beijing: Science Press,2002.

[3]   Byung Cheol Song, Jong Beom Ra." A Fast Search Algorithm for Vector Quantization Using L2-norm Pyramid of Codewords." IEEE Transactions on Image Processing. vol. 11, pp. 10-15: 2002.

[4]   PK. Meher , BK. Mohanty , JC. Patra ," Hardware-efficient Systolic-like Modular Design forTwo- dimensional Discrete Wavelet Transform." IEEE Trans. on Circuits and Systems II-Express Briefs, vol. 55 , pp. 151-155, 2008.

[5]   Ho, Y.-S.Gersho, A. "Classified Transform Coding of Image Using Interpolative Vector Quantization." IEEE international conference on Acoustics, Speech and signal processing, pp.1890-1893, May 1989.

[6]   Channa, Arshad Hussain; Hussain, Syed Afaq ." Image Coder using Ant Tree Vector Quantization Algorithm." 9th International Multitopic Conference. pp.1 - 6. Dec. 2005,

[7]   Ho, Y.-S.Gersho, A. "Variable-rate Multistage Vector Quantization for Image Coding. "IEEE International Conferrence on Acoustics,speech and signal Procession, pp. 1156—1159,1988.

[8]   B.Ramamurthi, A.Gersho," Classified Vector Quantization of Images," IEEE Transactions on Communications. Vol.34 , 11, pp. 1105 - 1115, 1986.

[9]   X. Yang,D. Xu,and Y. Qi," Bag-of-words image representation based on classified vector quantization," 2010 International Conference on Machine Learning and Cybernetics (ICMLC), , Vol. 2, pp. 708 - 712, 2010.

[10] L. Yin, N. Yu, W. Ma,and D. Wang,"VQ Category Encoding Algorithm Based on Direction," Computer Engineering and Applications, vol. 14 , pp. 44-47, 2005.

[11] D. Wang,Research on Real-time Image Encoding chip Based on Vector Quantization Technology.The doctoral thesis of Xi'an university of technology,2009.

A High-performance Configurable VLSI Architecture for Integer Motion Estimation in H.264

Ningmei Yu, Wenhua Jia, Meihua Gu,

Dongfang Wang, Gang Xi
Dept. Electronics Engineering
Xi'an University of Technology
Xi'an ,China
yunm@xaut.edu.cn

Yuanjin Zheng
School of Electrical and Electronic Engineering
Nanyang Technological University
Singapore
YJZHENG@ntu.edu.sg

Abstract—**A high-performance configurable integer motion estimation VLSI architecture based on parallelogram data matching pattern for H.264 is proposed in this paper. Through rational design for the data flow and processing module array, the memory traffic is reduced; data reusability in vertical direction is improved. Furthermore, the number of processing element is configured according to the area-speed requirement, data reusability in horizontal direction is controlled, and fast matching in large searching window is realized. The design is described with Verilog HDL, and is logic synthesized with Synopsys DC under SMIC 0.13μm process. With 300MHz clock frequency, when the PE number is the configured to 5, the search window size is 65x65, the speed can reach 36 fps, which can meet the speed requirements of real-time high-definition video encoding（1920×1088@30fps）.**

Keywords-H.264; integer motion estimation; VLSI; configurable architecture

I. INTRODUCTION

H.264/AVC is the newest video coding standard [1]. The coding efficiency is improved remarkably by adopting various complex predictive coding technologies. However, the computational complexity and workload of H.264 encoder are also increased, where the most time and resources consuming part is integer motion estimation (IME), which is also called Variable Block Size Motion Estimation (VBSME). Therefore, hardware acceleration of IME module is very important for the speed improving of H.264 encoder.

Data re-usable rate, search window size and matching speed are the important parameters for measuring the performance of IME architecture. There are four levels in data re-usable scheme, which are denoted as A, B, C and D respectively [2]. Level A has the lowest data reusability, only the overlapping reference pixels in the neighbor modules are used, the on-chip memory is very small, and vast amounts of reference data must be read from off-chip memory. In comparison, level D increases the data re-usable rate by increasing the on-chip memory to store the temporary data. Therefore, contradictions-balancing between the storage size and the amount of memory access is one of the key problems in designing IME architecture. The size of search window will affect the encoding quality, the larger the size is, the higher the encoding quality will be, while the encoding time is increased, and the encoding speed is descended. Therefore, with an unchanged encoding speed, the larger the search window size is, the higher the matching accuracy will be.

Reference [3] proposed a kind of B-level data-reusable structure. The data re-usability of the structure proposed in reference [4] achieved level-C by making use of on-chip memory. In order to improve the data re-usability and reduce the computational complexity, most of IME architectures produced larger size SADs by accumulating the small size SADs. Reference [5] proposed a kind of 2-D array with 16x16 PEs, in which 1-D data broadcasting, 1-D part results re-using. The high-performance reconfigurable VLSI structure proposed in reference [6] supported "zigzag" scan format and high data re-usability. Although the calculation array can get 100% PE utilization, the search area was only [-16, +16]. Reference [7] presented a scalable H.264/AVC ME VLSI structure based on the full search algorithm. Although this structure was flexible, and could be graded by the search window and PE array, the search area was only [-16, + 15].

This paper presents a high-performance integer motion estimation VLSI architecture aiming at the large computation and memory access, the circuit structure of the crucial module will be represent in detail in the following sections.

II. INTEGER MOTION ESTIMATION ALGORITHM

In VBSME, the macroblock (MB) can be segmented into sub-blocks smaller than 16x16, and each sub-blocks has motion vector respectively, which can improve the rate distortion(RD) performance of H.264 encoder. The block sizes of 4×4, 4×8, 8×4, 8×8, 16×8, 8×16 and 16×16 are supported in H.264, as shown in Fig.1, where the former 4 types belong to MB type, there are also four sub-modes belong to sub-MB type under the 8x8 mode. There are total 41 sub-blocks in one MB.

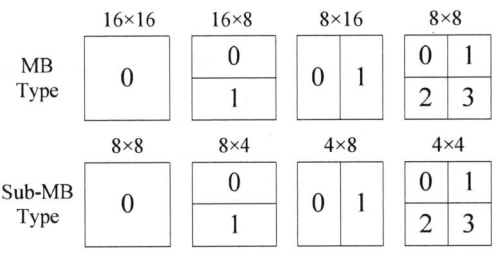

Figure 1.   Segmentations of MB

Project supported by Natural Science Basic Research Plan in Shaanxi Province of China(No.2011JQ8032) and Natural Science Research Plan in Xi'an university of Technology of China.

For the FS-VBS-IME(Full Search Variable Block Search Integer Motion Estimation) used in H.264, the search range of one MB is the region centered by the position of the predicted MV pointing to, as shown in Fig.2. There are total $(2p+1)\times(2p+1)$ search points in the region, and the reference window size is $(2p+16)\times(2p+16)$, FS-VBS-IME takes all the $(2p+1)\times(2p+1)$ search points into consider. Therefore, the value of p is in proportion to the precision of matching, and p is set to be 32 here, the corresponding search window size is 65×65. For a specific search position, the current MB has a corresponding reference MB, which includes each reference blocks of the sub-blocks into this MB.

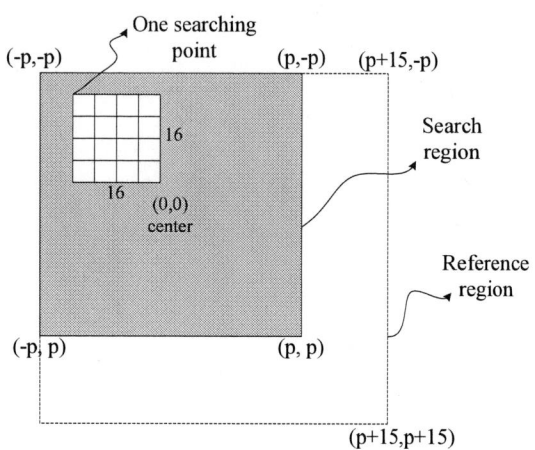

Figure 2.   Search region and reference window

The cost function used in FS-VBS-IME is sum of absolute difference (SAD). For a M×N (M=4,8,16; N=4,8,16) block, the calculation is as (1).

$$SAD(s, c(m)) = \sum_{x=1}^{M} \sum_{y=1}^{N} \left| s(x, y) - c(x - m_x, y - m_y) \right| \quad (1)$$

Where s is the original block, c is the predict block pointed by m.

For each search point, 41 SADs can be obtained by summing up 16 4×4SADs which has been obtained before. The minimum motion costs and the best integer motion vecters (IMVs) can be obtained after traversing all the 65×65 searching points, there are total 41 IMVs corresponding to 41 sub-blocks in one MB.

III.   THE PROPOSED ARCHITECTURE

IME is the most computation consuming module in H.264 encoder, as well as the focus part of the whole ME. VBS-IME has the highest computational complexity, the largest proportion of area and delay. How to improve the data utilization rate, reduce the memory access, and ensure the high performance of the time-area efficiency is the key problem for the IME architecture design.

A.   General Structure

Full search scheme is adopted in the proposed IME architecture, a configurable PE unit is also used, as shown in Fig.3. Reference data in the search window is read line by line

from the on-chip SRAM, and sent to the parallel I-PE-Arrays, each PE unit accepts the reference data in the same column, completes the matching between the current MB and the reference data of one list, and stores the best position information and cost of the column, N lines of data can be processed by N PEs simultaneously. Therefore, when 65 PEs(N=65) are adopted, the circuit area is the largest, and the processing speed is also the fastest. For simplicity, 65 I-PE-Arrays parallel processing are assumed in the following description of each circuit block.

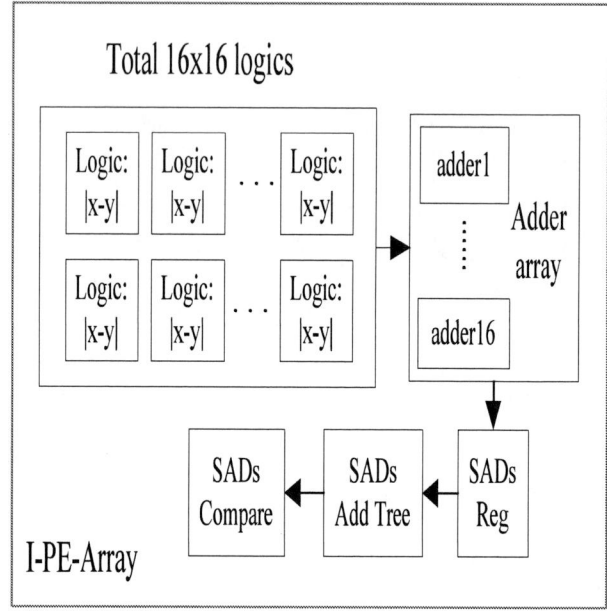

Figure 3.   The whole architecture of IME

B.   Data Reading and Matching

The reference window pixels determined by MB's predictive MV are stored in SRAM, where the searching points are 65×65, the corresponding reference pixels are 80×80. By expanding the original searching window (for sub-pixel interpolation), 86x86 reference pixels should be stored. In IME, valid reference data (4-83 lines) are read line by line from SRAM, N data are extracted, and sent into N PEs. The number of data reading times are determined by the number of PE units. With 65 PEs, reading once from SRAM can match all the points. Taking the balance of area and speed into account, 5 PEs is used in the proposed architecture.

Matching calculation module is responsible for the entire pixels motion search, the best integer pixel location in the 65×65 searching range can be obtained by the matching computation of the current MB and the reference pixel. Each PE accepts 16 reference data corresponding to one location once, and matches the 16 lines data in MB, the calculation results are saved by the data control unit.

C.   Integer-pixel Process Element I-PE-Arrays

I-PE-Array is in charge of the calculation of the absolute value of the residual, accumulation of SADs, piece jointing, costs saving, and outputting the costs and the position information. I-PE-Array includes 4 components: residua absolute-value calculation unit, adder array and data allocation unit, SADs adder tree, and comparison module, as shown in Fig.4. Multiple PEs in parallel can receive the adjacent position pixel data from SRAM, the common data in horizontal can be shared, and the data reusability can be improved efficiently.

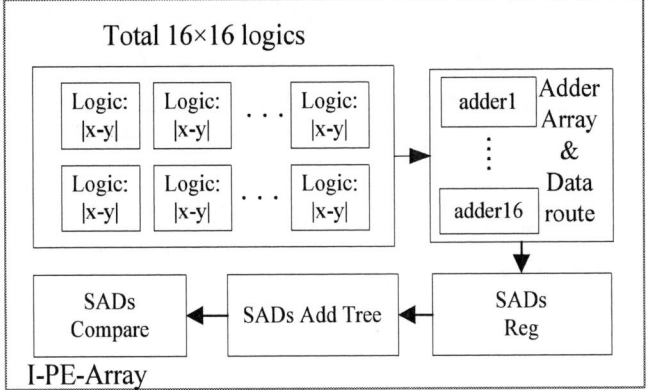

Figure 4.   Block diagram structure of IME

1) Residua absolute-value calculation unit

Residua absolute-value calculation unit can realize the absolute computation of the residua between the input current line 16 reference pixels and the 256 pixels of 16 lines of MB. There are total 256 residua absolute-value calculation units in each PE, according to the short-distance pipeline structure, one line of data is accepted in one cycle, and 256 absolute-values are produced.

In order to reduce the memory traffic, and improve the encoding speed, the re-usability of the data read from SRAM should be improved. Fig. 5 shows the reference window with 80×80 pixels, where the corresponding pixel-height is the subsequent 16-pixel, and the corresponding pixel-height of the 65 search positions in line 2 is the subsequent 16-pixel, there are 15 common lines of reference pixels corresponding to these two lines of search positions, as shown in Fig.5. The corresponding reference pixels for 65 index position lines are cross and common, so after data is read from SRAM, 256 residua absolute-value calculation units process them at the same time, the residua absolute-value for this line data in different position calculation in different position is computed, which avoids repeatedly reading the same date from SRAM, so the data reusability in vertical direction can be improved.

For 80×80 reference data, from the first line to the sixteenth line, the matching row-number of each line with the reference rows in one MB are on the increase (date in line 1 only participate in the calculation for position 1, date in line 2 participate in the calculation for the position 2, and so forth). From row 16 to row 66, the matching row-number retains 16; from row 66 to row 80, the matching row-number is on the decrease, as shown in Fig. 6. The rows of MB are 16, and the rows of reference window (RW) are 80. Parallelogram in Fig.6 reflects the corresponding matching-row relationship and the change of the matching row-number.

2) Addition Array and the Data Allocation

The addition array uses several small SADs to compose larger SADs, which avoids the recalculation from scratch. 256 result-data from the residua absolute-value calculation unit are summed in the addition array, which is composed with 16 adder unit. Each adder unit is in charge of the addition of 16 different calculations, 64 local 1×4 SADs for 16 positions are produced simultaneously. Then the 1x4 SAD results in the corresponding position are added respectively. The final result is calibrated by the control signal to judge the production of the 4×4 SADs. If 4×4 SAD is produced, it will be sent to the SAD tree module and the addition registration will be cleared to receive the next local SAD.

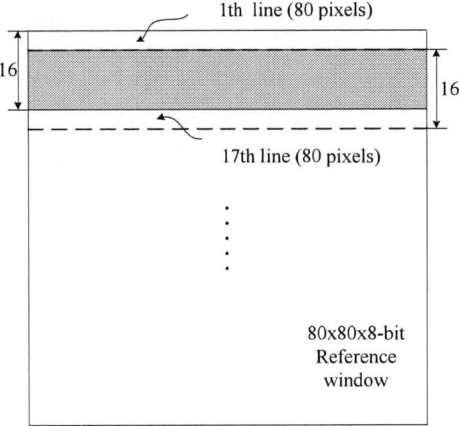

Figure 5.   Adjacent search positions share the reference region

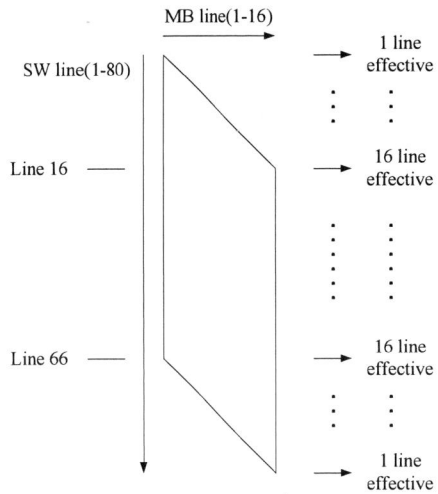

Figure 6.   Matching law

3) SAD addition tree

One MB can be partitioned into 16 blocks of 4×4 array. When 16 4×4 SADs in a certain position are sent into SADs registor, larger SADs are produced immediately by splicing those 16 4×4 SADs. There are 8 SADs of 4x8, 8 SADs of 8×4, 4 SADs of 8×8, 2 SADs of 16×8 , 2 SADs of 8×16 and one SADs of 16×16, as shown in Fig.7. Therefore, 41 SADs responding to one position can be obtained in one cycle.

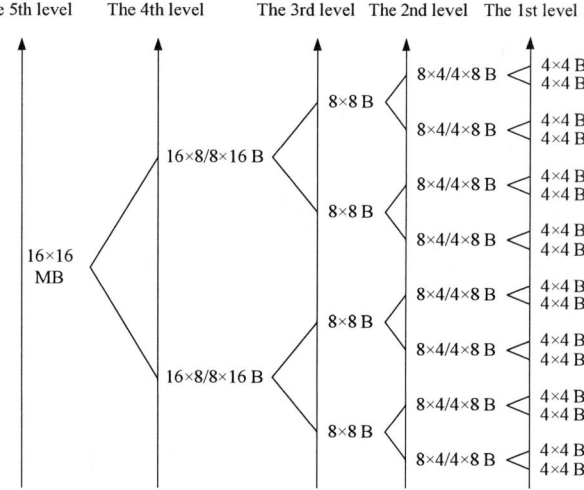

| The 5th level | The 4th level | The 3rd level | The 2nd level | The 1st level |

Figure 7. SADs adder tree

TABLE I. PERFORMANCE OF IME ARCHITECTURE

Algorithm	Full Search VBSME
Search range	65×65
Block size	4×4,4×8,8×4,8×8,16×8,8×16,16×16
Technology	SMIC 0.13 μm CMOS
Gate counts	26K× n (n=1, 5, 13, or 65)
SRAM on chip	86×86×8 bits
Max frequency	300MHz
Cycles/MB	80cycle×m （m=65,13,5，or 1）

TABLE II. COMPARISON OF THE ARCHITECTURES

	Ref.[6]	Ref.[7]	Proposed
Process	0.18 μm	0.18 μm	0.13 μm
SR	33×33	32×32	65×65
Block size	4×4 to 16×16	4×4 to 16×16	4×4 to 16×16
Frequency	200MHz	416MHz	300MHz
Throughput	1920×1088 @19fps	1920×1088 @12fps	1920×1088 @36fps
Gate counts	160k	39k	130k
Cycles/MB	1234	4267	1040

4) Compare Modue

There are total 41 units in comparison module, which is responsible for the cost comparison of 41 SADs, finding the minimum cost for different dimensions of sub-blocks, and the corresponding optimal 41 of MVs. Once 41 SADs are produced, and sent to the comparison module immediately, 41 comparison unit are in charge of the processing of 41 SADs respectively. Each unit receives the new data and compares with the minimum data stored before, updates the minimum data in memory, which ensures the output data is the minimum one. In the end, 41 minimum SADS, and the corresponding best positions (41 best IMVs) for one search position are exported.

IV. IMPLEMENTATION AND RESULTS ANALYSIS

The proposed full search IME architecture are described with Verilog HDL language, simulated under VCS, and synthesized by the synopsis DC with 0.13μm standards technology library. The search range is 65×65(-32,+32), each image block size from 4×4 to 16×16 can be processed. As the structure is configurable, the I-PE-array number can be configured, while the I-PE-array number affects the area of circuits, speed, and even power dissipation. The performance of IME circuit architecture is shown in table 1. There are total 26k×n (n=1, 5, 13, or 65) gate counts besides SRAM, where n represents the number of I-PE-array, SRAM size is 7936 bytes. Processing time required for each MB is 80×m (m=65,13,5, or 1), where m is the serial number, and n satisfies the relationship of m×n=65.

According to the speed requirements, 5 I-PE-array in parallel are selected, the comparison of the proposed IME performance with the other architecture from the references is shown in table 2. The proposed IME circuit supports the maximum search range, and has the best encoding performance. The processing speed for HD 1920×1088 video image is 36 fps under 300MHz working frequency, which can meet the HDTV (1920×1088) @30fps video coding motion estimation speed requirement.

V. CONCLUSITONS

According to calculation features of H.264/AVC's integer motion estimation, a configurable IME architecture is presented in this paper. Through rational configuration, the hardware utilization, data reuse rate, storage access times, speed and area of IME module can be adjusted. Finally, the proposed IME architecture is functional simulated with Verilog HDL, and logical synthesized with Synopsys DC and SMIC 0.18μm process under the clock frequency of 300MHz. Experimental results show that the processing speed can meet the requirement of HDTV real-time encoding (HDTV1920×1088@30fps). Compared with the other structures, the proposed structure has greater throughput capacity, less storage bandwidth, and is very suitable for high-resolution video motion estimation.

REFERENCES

[1] Joint Video Team. Draft ITU-T Recommendation and Final Draft International Standard of Joint Video Specification[S], 2005.

[2] J.C. Tuan, T.S. Chang, C.W. Jen. On the Data Reuse and Memory Bandwidth Analysis for Full-search Block-matching VLSI Architecture[J]. IEEE Transactions on Circuits and Systems for Video Technology, 2002, 12(1):61-72.

[3] W. Cao, Z.G. Mao. A Novel VLSI Architecture for VBSME in MPEG-4 AVC/H.264[C]. IEEE International Symposium on Circuits and Systems, 2005,1:1794-1797.

[4] C.M. Ou, C.F. Le, W.J. Hwang. An Efficient VLSI Architecture for H.264 Variable Block Size Motion Estimation[J]. IEEE Transactions on Consumer Electronics, 2005,51(4): 1291-1299.

[5] Y.W.Huang, T.C.Wang, B.Y.Hsieh, and L.G. Chen. Hardware Architecture Design for Variable Block Size Motion Estimation in MPEG-4AVC/JVT/ITU-T H.264[C]. Proceedings of IEEE International Symposium Circuits System, 2003, II: 796-799.

[6] C. Wei, H. Hui, T. Jiarong, and M. Hao. A High-Performance Reconfigurable VLSI Architecture for VBSME in H.264[J]. IEEE Transactions on Consumer Electronics,2008, 54(3): 1338-1345.

[7] J. Kim and T. Park. A Novel VLSI Architecture for Full-Search Variable Block-Size Motion Estimation[J]. IEEE Transactions on Consumer Electronics, 2009, 55(2): 728-733.

ROM-less DDFS Using Non-equal Division Parabolic Polynomial Interpolation Method

Chia-Hao Hsu, Yun-Chi Chen, and Chua-Chin Wang[†], *Senior Member, IEEE*

Department of Electrical Engineering
National Sun Yat-Sen University
Kaohsiung, Taiwan 80424
Email: ccwang@ee.nsysu.edu.tw

Abstract—A direct digital frequency synthesizer (DDFS) based on a non-equal division parabolic polynomial interpolation method is proposed in this paper. To attain high spurious free dynamic range (SDRF) and reduce area cost, a parabolic polynomial interpolation method is adopted in the proposed design to replace conventional ROM-based phase-to-sine mapper methods. Particularly, the left 1/4 of the phase range is approximated using a low-curvature parabolic curve. The proposed design is manufactured using a standard 0.18 μm CMOS technology. The maximum output frequency is 50 MHz, the core area is 1.4528 mm^2, and the spurious free dynamic range (SFDR) is 68.67 dBc. The proposed DDFS outperforms prior works' SFDR and energy efficiency.

Keywords—Direct digital frequency synthesizer (DDFS), parabolic polynomial interpolation, spurious free dynamic range (SFDR).

I. INTRODUCTION

Frequency synthesizer is a well-known technique to generate signals with a selective frequency, which is an important component for many communication systems, e.g., digital radios, mobile telephones, GPSs (Global Positioning Systems), etc. Traditionally, a signal with a selective frequency is mostly generated by phase-locked loop (PLL) [1]- [2] circuits. However, two major drawbacks of the the PLL-based frequency generators are slow frequency switching speed and poor spectral purity [3]. The deficiencies of the PLL-based frequency generators are then inadequately used in the modern wireless communication systems to meet fast frequency switching demand. By contrast, the direct digital frequency synthesizer (DDFS) has been considered as a better alternative than the PLL-based frequency generators, because it is proved to provide fast frequency switching and excellent spectral purity.

The DDFS technique was first proposed in 1971's [4], which utilized digital data processing technique to generate a tunable frequency signal driven by a precise clock signal. Fig. 1 shows the architecture of the conventional DDFS. The phase accumulator is used to generate the digital phase as well as the frequency. The samples of sine wave amplitude are

[†]Prof. C.-C. Wang is with the Department of Electrical Engineering, National Sun Yat-Sen University, Kaohsiung, Taiwan 80424. (e-mail: ccwang@ee.nsysu.edu.tw). He is the contact author.

stored in the ROM-based look-up table, which are possibly derived from MATLAB simulation results. The samples are converted into digital sine wave signals by the amplitude complementor. Finally, the digital-to-analog converter (DAC) converts the digital sine wave signals into an analog sine wave signal. Unfortunately, the conventional DDFS has a major intrinsic difficulty: it demands very large ROM as the storage of the sine wave amplitude samples. Therefore, the conventional DDFS will suffer from three inherent drawbacks of the ROM, which are large power consumption, large chip area, and slow operating speed. Although the ROM size can be significantly reduced by truncating the output bit-wide of the phase accumulator, the unwanted spurious noise caused by the truncation will then degrade the spectral purity performance of the generated sine wave.

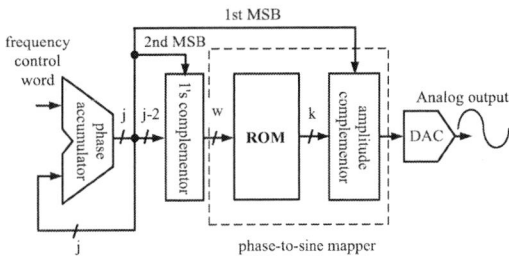

Fig. 1. Conventional DDFS structure

Lately, ROM-less DDFS [5]- [9] researches have been proposed to attain hardware reduction advantage, which utilized different high-order (probably more than three order) polynomial algorithms instead of the ROM-based look-up tables to realize the phase-to-sine mappers. Due to the intrinsic massive computation complexity, any algorithm based on high-order polynomials will be hard to meet the high speed requirement of modern wireless communication applications. Moreover, according to the reports of prior works [12], any polynomial whose order is larger than three may be inefficient to improve the spurious free dynamic range (SFDR) of the sine wave. This paper proposes a novel DDFS based on a 2nd-order non-equal division parabolic polynomial interpolation method to resolve all of the mentioned difficulties. Compared with several prior works, the proposed DDFS shows a satisfactory speed and an exceptional spectral purity.

II. THE PROPOSED DDFS ARCHITECTURE

A. Parabolic Polynomial Interpolation

To implement the phase-to-sine mapper, the parabolic polynomial could be the most convenient choice to provide superior SFDR. A quadrant of sinusoid is partitioned into M segments, where every segment is approximated by the 2nd-order parabolic polynomial, as shown in Eqn. (1).

$$y(x) = a_i x^2 + c_i, \quad i = 1 \sim M. \tag{1}$$

where the parameters, a_i and c_i, can be derived by the least square method. However, large M will result in divergence when performing the least square method, though the SFDR of this DDFS will be enhanced as the increase of M. The parabolic polynomial method has difficulty in fitting the curvature of the sinusoid. The curvature at x, $k(x)$, of a given curve $y = f(x)$ can be given by

$$k(x) = \frac{y''}{(1 + y'^2)^{3/2}} \tag{2}$$

where the 1st-order derivative, y', is equal to $2a_i x$ and the 2nd-derivative, y'', is equal to $2a_i$. Obviously, y' is proportional to y''. Thus, the fitting procedure is difficult to determine both the slope and the curvature of the curve if only one parameter a_i is allowed.

To resolve the problem of Eqn. (2), this paper adopts another parabolic polynomial method. We propose to add a coefficient, x_i, in Eqn. (2). Then, Eqn. (2) is rewritten as Eqn. (3) to enhance the adjustability of the parabolic polynomial.

$$y(x) = a_i(x + x_i)^2 + c_i, \quad i = 1 \sim M \tag{3}$$

Therefore, the new 1st-order derivative formula is given by $y' = 2a_i(x + x_i)$. Regarding the computation complexity, Eqn. (3) needs only one more addition compared with Eqn. (1).

The coefficient, x_i, is derived by the variation between an ideal sine wave and the generated sine wave, which is used to adjust the characteristics of the curve in each segment to carry out the phase-to-sine mapping function. Notably, x_i, is given an initial value, "0", at the start state. Then, a_i and c_i are, respectively, derived by the least square method. Therefore, Eqn. (3) can be re-formulated into a matrix expression as follows.

$$
\begin{bmatrix} y_1 \\ y_2 \\ y_3 \\ \vdots \\ y_M \end{bmatrix}
=
\begin{bmatrix} (x + x_1)^2 & 1 \\ (x + x_2)^2 & 1 \\ (x + x_3)^2 & 1 \\ \vdots & \vdots \\ (x + x_M)^2 & 1 \end{bmatrix}
\begin{bmatrix} a_i \\ c_i \end{bmatrix}
\tag{4}
$$

The transpose of the first matrix at the righthand side of Eqn. (4) is used to be multiplied at both sides of Eqn. (4) to attain the following equation.

$$
\begin{bmatrix} (x + x_1)^2 & 1 \\ (x + x_2)^2 & 1 \\ (x + x_3)^2 & 1 \\ \vdots & \vdots \\ (x + x_M)^2 & 1 \end{bmatrix}^{\mathrm{T}}
\begin{bmatrix} y_1 \\ y_2 \\ y_2 \\ \vdots \\ y_M \end{bmatrix}
$$
$$
=
\begin{bmatrix} (x + x_1)^2 & 1 \\ (x + x_2)^2 & 1 \\ (x + x_3)^2 & 1 \\ \vdots & \vdots \\ (x + x_M)^2 & 1 \end{bmatrix}^{\mathrm{T}}
\begin{bmatrix} (x + x_1)^2 & 1 \\ (x + x_2)^2 & 1 \\ (x + x_3)^2 & 1 \\ \vdots & \vdots \\ (x + x_M)^2 & 1 \end{bmatrix}
\begin{bmatrix} a_i \\ c_i \end{bmatrix}
\tag{5}
$$

Eqn. (5) is then reorganized as follows.

$$
\begin{bmatrix} a_i \\ c_i \end{bmatrix}
=
\left(
\begin{bmatrix} (x + x_1)^2 & 1 \\ (x + x_2)^2 & 1 \\ (x + x_3)^2 & 1 \\ \vdots & \vdots \\ (x + x_M)^2 & 1 \end{bmatrix}^{\mathrm{T}}
\begin{bmatrix} (x + x_1)^2 & 1 \\ (x + x_2)^2 & 1 \\ (x + x_3)^2 & 1 \\ \vdots & \vdots \\ (x + x_M)^2 & 1 \end{bmatrix}
\right)^{-1}
$$
$$
\begin{bmatrix} (x + x_1)^2 & 1 \\ (x + x_2)^2 & 1 \\ (x + x_3)^2 & 1 \\ \vdots & \vdots \\ (x + x_M)^2 & 1 \end{bmatrix}^{\mathrm{T}}
\begin{bmatrix} y_1 \\ y_2 \\ y_3 \\ \vdots \\ y_M \end{bmatrix}
\tag{6}
$$

Thus, the coefficients, a_i and c_i, can be derived based on Eqn. (6). Furthermore, the approximation value of the sine wave amplitude can be easily deduced according to Eqn. (4) and Eqn. (6), which is expressed as Eqn. (7).

$$
\begin{bmatrix} y_1' \\ y_2' \\ y_3' \\ \vdots \\ y_M' \end{bmatrix} = \begin{bmatrix} (x+x_1)^2 & 1 \\ (x+x_2)^2 & 1 \\ (x+x_3)^2 & 1 \\ \vdots & \vdots \\ (x+x_M)^2 & 1 \end{bmatrix} \left(\begin{bmatrix} (x+x_1)^2 & 1 \\ (x+x_2)^2 & 1 \\ (x+x_3)^2 & 1 \\ \vdots & \vdots \\ (x+x_M)^2 & 1 \end{bmatrix}^{\mathrm{T}} \right.
$$
$$
\left. \begin{bmatrix} (x+x_1)^2 & 1 \\ (x+x_2)^2 & 1 \\ (x+x_3)^2 & 1 \\ \vdots & \vdots \\ (x+x_M)^2 & 1 \end{bmatrix} \right)^{-1} \begin{bmatrix} (x+x_1)^2 & 1 \\ (x+x_2)^2 & 1 \\ (x+x_3)^2 & 1 \\ \vdots & \vdots \\ (x+x_M)^2 & 1 \end{bmatrix}^{\mathrm{T}} \begin{bmatrix} y_1 \\ y_2 \\ y_3 \\ \vdots \\ y_M \end{bmatrix}
\tag{7}
$$

Therefore, the least square error, x_i, is found according to Eqn. (7). Finally, we then back-substitute x_i into Eqn. (6) such that a_i and c_i can be derived, respectively.

In this study, a quadrant of sinusoid is divided into 16 segments ($M = 16$). To attain a better spectral purity, we analyze two different segment division methods to choose a better one so as to decide the range of each segment. As shown in Fig. 2(a), each segment is equally distributed over the entire phase range, which is called equal-segment division method. By contrast, Fig. 2(b) shows the proposed non-equal division method.

Fig. 3 depicts the absolute errors of equal-segment division method and non-equal division method, which are, respectively, derived by the difference between an ideal sine wave and the generated sine waves. Notably, the left 1/4 quadrant of the sinusoid ($\sin \theta$, $0 \leq \theta < \pi/8$) is approximate by a low-curvature parabolic curve. By contrast, the right 3/4 quadrant of the sinusoid ($\sin \theta$, $\pi/8 \leq \theta < \pi/2$) is approximated using a high-curvature parabolic curve with intensive samples to reduce the error. In short, the 1st segment, S1, contains 1/4 quadrant of the sinusoid at left side, while the remaining 3/4 quadrant of the sinusoid is equally divided into the other 15 segments. Therefore, the proposed non-equal division method will synthesize a better sine wave without paying too much hardware overhead.

Fig. 4 shows the schematic of the proposed DDFS, which utilizes a 32-bit frequency control word (FCW) to attain a wide frequency tuning range. The output of the squarer through the selectors is shifted by the shift registers, a_{i-j}, where the symbol a_{i-j} represents the j-th bit of the sequence for the i-th segment. Then, the outputs of the shift registers are selected by multiplexers (MUXs), respectively. The summation of the total 16 multiplexer's outputs is realized by an adder.

III. Implementation and Simulation

The proposed DDFS is realized on silicon using TSMC (Taiwan Semiconductor Manufacturing Company) standard 0.35 μm CMOS technology. All of the PVT corner simulations ([0°C, +25°C, +75°C, +100°C], (VDD, (1±10%)VDD), and SS, FS, TT, SF, FF models), have been carried out to justify its robustness. Fig. 5 illustrates the chip layout including I/O

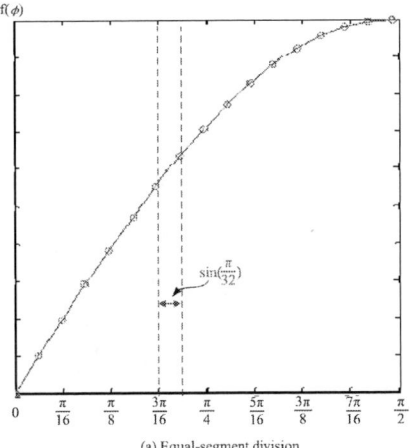

(a) Equal-segment division

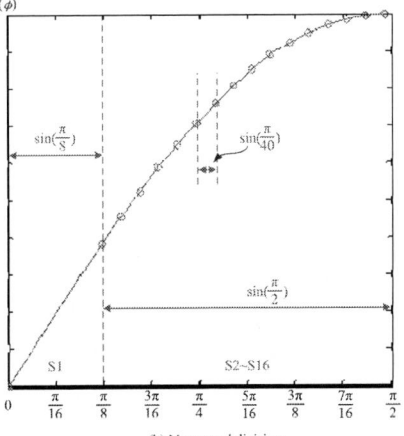

(b) Non-equal division

Fig. 2. Segment division methods

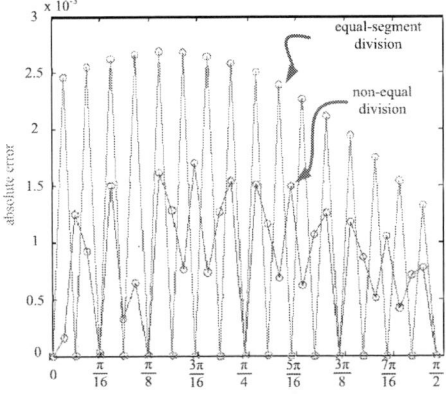

Fig. 3. Absolute error of two different segment division methods

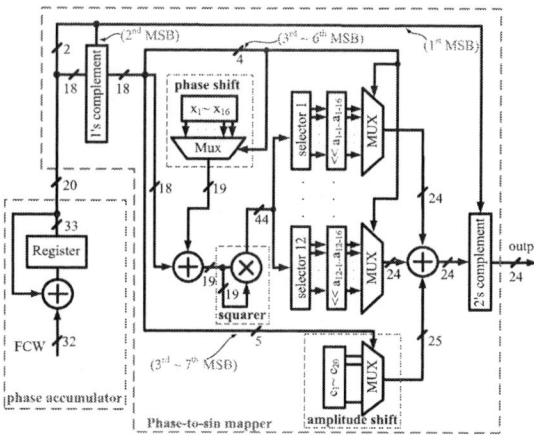

Fig. 4. Schematic of the proposed DDFS

TABLE I
COMPARISON OF DDFS DESIGNS

	[10]	[11]	[12]	ours
Process (μm)	0.35	*	0.13	0.35
Type	sine	quad	sine	sine
SFDR (dBc)	35	50.89	64.72	68.67
Phase accumulator bits	8	15	20	20
Output resolution (bits)	8	14	24	24
Power Efficiency (mW/MHz)	410000	1044.35	0.35	0.31
Max. Clock (MHz)	2000	11.25	161	50
Area (core) mm^2	N/A	N/A	0.33	1.4528
Area (with pad) mm^2	3.99	N/A	2.015	4.5109
Year	2005	2008	2009	2011

*The DDFS design was implemented by FPGA.

by the Southern Taiwan Science Park Administration (STSPA), Taiwan, R.O.C. under contract no. EZ-10-09-44-98. It is also partially supported by Ministry of Economic Affairs, Taiwan, under grant 99-EC-17-A-01-S1-104 and 99-EC-17-A-19-S1-133. The authors would like to express their deepest gratefulness to Chip Implementation Center of National Applied Research Laboratories, Taiwan, for their thoughtful chip fabrication service.

REFERENCES

[1] Y. Sun, X. Yu, W. Rhee, D. Wang, and Z. Wang, "A fast settling dual-path fractional-N PLL with hybrid-mode dynamic bandwidth control," *IEE Microwave and Wireless Components Letters*, vol. 20, no. 8, pp. 462-464, Aug. 2010.

[2] W.-Y. Shin, M. Kim, G.-M. Hong, and S. Kim, "A fast-acquisition PLL using split half-duty sampled feedforward loop filter," *IEEE Transactions on Consumer Electronics*, vol. 56, no. 3, pp. 1856-1859, Aug. 2010.

[3] M. Kesoulis, D. Soudris, C. Koukourlis, and A. Thanailakis, "Systematic methodology for designing low power direct digital frequency synthesizers," *IET Circuits, Devices & Systems*, vol. 1, no. 4, pp. 293-304, Aug. 2007.

[4] J. Tierney, C. Rader, and B. Gold, "A digital frequency synthesizer," *IEEE Trans. on Audio and Electroacoustics*, vol. 19, no. 1, pp. 48-57, Mar. 1971.

[5] A. Ashrafi and R. Adhami, "A direct digital frequency synthesizer utilizing quasi-linear interpolation method," in Proc. of *37th IEEE Southeastern Symposium on System Theory*, pp. 144-148, Mar. 2005.

[6] A. Ashrafi and R. Adhami, "An optimized direct digital frequency synthesizer based on even fourth order polynomial interpolation," in Proc. of *38th IEEE Southeastern Symposium on System Theory*, pp. 109-113, Mar. 2006.

[7] H. Jafari, A. Ayatollahi, and S. Mirzakuchaki, "A low power, high SFDR, ROM-less direct digital frequency synthesize," in Proc. of *2005 IEEE Conf. on Electron Devices and Solid-State Circuits*, pp. 50-54, Dec. 2005.

[8] D. De Caro and A. G. M. Strollo, "High-performance direct digital frequency synthesizers in 0.25 μm CMOS using dual-slope approximation," *IEEE J. of Solid-State Circuits*, vol. 40, no. 11, pp. 2220-2227, Nov. 2005.

[9] Y. Song and B. Kim, "A 250 MHz direct digital frequency synthesizer with delta-sigma noise shaping," in Proc. of *2003 IEEE Int. Solid-State Circuits Conf. (ISSCC 2003)* vol. 1, pp. 472-509, Jan. 2003.

[10] X. Yu, F. F. Dai, Y. Shi, and R. Zhu, "2 GHz 8-bit CMOS ROM-less direct digital frequency synthesizer," in Proc. of *IEEE Inter. Symp. on Circuits and Systems, 2005. (ISCAS 2005)*, vol. 5, pp. 4397-4400, May 2005.

[11] S. S. Jeng, H. C. Lin, and C. Y. Wu, "DDFS design using the equi-section division method for SDR transceiver," in Proc. of *IEEE 19th International Symposium on Personal, Indoor and Mobile Radio Communications*, pp. 1-5, Sept. 2008.

[12] C.-C. Wang, C.-H. Hsu, C.-C. Lee, and J.-M. Huang, "A ROM-less DDFS based on a parabolic polynomial interpolation method with an offset," *Journal of Signal Processing Systems* vol. 61, pp.1-9, May 2010.

PADs of the proposed DDFS design. The comparison between the proposed DDFS and prior DDFS designs is tabulated in Table I. The proposed DDFS attains the largest SFDR (68.67 dBc) and the best energy efficiency. The power consumption is 15 mW at a 50 MHz clock.

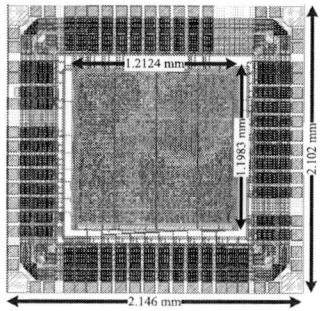

Fig. 5. Layout view of the proposed DDFS

IV. CONCLUSION

This paper proposes a ROM-less DDFS using a 16-segment non-equal division parabolic polynomial to realize the phase-to-sine mapping function. The proposed DDFS demonstrates a significant improvement in SFDR compared to prior works. Besides, the proposed DDFS is carried out as a multiplier-less design. The proposed DDFS attains a SFDR of 68.67 dBc, a 0.31 mW/MHz energy efficiency and the maximum clock rate is 50 MHz.

ACKNOWLEDGEMENT

This investigation is partially supported by National Science Council under grant NSC 99-2221-E-110-081-MY3 and NSC 99-2923-E-110-0 02-MY2. Besides, this research is supported

978-1-61284-863-1/11 $26.00 © 2011 IEEE

Accelerated Evaluation Method for the SRAM Cell Write Margin using Word Line Voltage Shift

Hiroshi Makino[1], Shunji Nakata[2], Hirotsugu Suzuki[3], Hiroki Morimura[2], Shin'ichiro Mutoh[2], Masayuki Miyama[3], Tsutomu Yoshimura[4], Shuhei Iwade[1] and Yoshio Matsuda[3]

[1]Faculty of Information Science and Technology/[4]Faculty of Engineering, Osaka Institute of Technology, [1]Hirakata/[4]Osaka, Japan
[2] NTT Microsystem Integration Laboratories, Nippon Telegraph and Telephone Corporation, Atsugi, Japan
[3]Graduate School of Natural Science, Kanazawa University, Kanazawa, Japan

Abstract— **An accelerated evaluation method for the SRAM cell write margin is proposed based on the conventional Write Noise Margin (WNM) definition. The WNM is measured under a lower word line voltage than the power supply voltage VDD. A lower word line voltage is used because the access transistor operates in the saturation mode over a wide range of threshold voltage variation. The final WNM at the VDD word line voltage, the Accelerated Write Noise Margin (AWNM), is obtained by shifting the measured WNM at the lower word line voltage. The amount of WNM shift is determined from the WNM dependence on the word line voltage. As a result, the cumulative frequency of the AWNM displays a normal distribution. A normal distribution of the AWNM drastically improves development efficiency, because the write failure probability can be estimated by a small number of samples. Effectiveness of the proposed method is verified using the Monte Carlo simulation.**

Keywords-SRAM; write noise margin; Vth fluctuation; word line voltage; write margin distribution

I. INTRODUCTION

The recent progress of process technology has caused various fluctuation problems in the device characteristics due to transistor area reduction. The Vth fluctuation caused by dopant fluctuation has the greatest influence on device characteristics [1-2]. Generally, this dopant induced Vth fluctuation is random and obeys the normal distribution.

The stability of the SRAM cell is greatly affected by Vth fluctuation, because the SRAM cell is usually designed using minimum design rules. Vth fluctuation degrades both the read and write operation stabilities. It has been said that the read operation is usually less stable than the write operation under Vth fluctuation. However, the write operation is also affected by a large Vth fluctuation. In addition, a recent paper reports that write operation failure is more dominant than read operation failure in low power supply voltages [3]. Therefore, an accurate evaluation of the write operation stability is as important as the evaluation of the read operation stability.

Conventionally, the Write Noise Margin (WNM) is used as a metric of write operation stability [4]. Although the WNM is easy to measure, it does not obey a normal distribution because it is not sensitive to Vth variation when the WNM is large [5]. If the write margin obeys the normal distribution, the write margin distribution can be easily estimated by a small number of samples. This drastically improves

development efficiency.

In this paper, we propose an accelerated method for evaluating the SRAM cell write margin based on the conventional WNM definition. The WNM is measured at a lower word line voltage than the VDD of the power supply voltage and is calibrated to the WNM of the VDD word line voltage. In the proposed method, the write margin obeys the normal distribution even though it uses the conventional WNM definition.

II. CONVENTIONAL WRITE NOISE MARGIN

The circuit of the SRAM write operation is shown in Fig. 1(a). Let us assume that the inverted data are written to the SRAM cell where "1" is stored on the internal node V1 and "0" on the V2. Then, the data "0" and "1" are given on the bit lines BL and /BL, respectively, under the activated word line WL. If the voltages of nodes V1 and V2 are inverted, the write operation is successful. Hereupon, the V1, V2, BL, /BL and WL are also used as the voltages of their nodes. The definition of the conventional Write Noise Margin (WNM) is shown in Fig. 1(b). There are DC characteristic curves of the inverter A (InvA) and the inverter B (InvB) under WL=VDD, BL=0 V and /BL=VDD. The VDD is the power supply voltage. The WNM is defined as the width of the smallest embedded square between the two DC characteristic curves.

Generally, the write margin is a function of the threshold voltage Vth's of the SRAM cell. If the write margin is linear for the Vth's, it is expected to obey the normal distribution, allowing us to predict the write margin distribution accurately from a small number of samples. If the write margin distribution follows the normal distribution, the write yield can also be easily estimated [6].

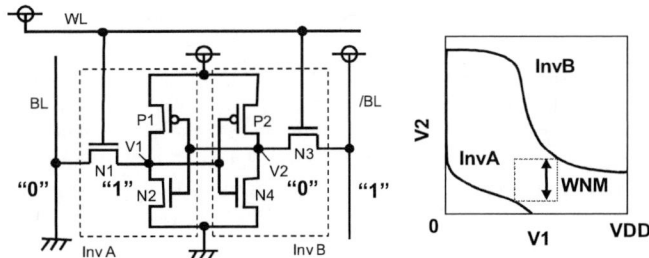

Fig.1. (a) SRAM cell circuit in the write operation (b) Definition of the Write Noise Margin.

978-1-61284-863-1/11 $26.00 © 2011 IEEE

The dependence of the WNM on the Vth is examined using the SPICE simulation. The transistor parameter of 45-nm process technology [7] is used with a power supply voltage of VDD=1.0 V. The typical values of threshold voltages are Vthn=0.404 V for the NMOS transistors and Vthp=-0.384 for the PMOS transistors. The transistor sizes are W=55 nm, 83 nm, and 55 nm with L=45 nm for the access, driver, and load transistors, respectively.

The simulation results are shown in Fig. 2. We set ΔVth=0 when the threshold voltages are typical. While the WNM is not linear on the Vth of the access transistor N1, the WNM is almost linear on the Vth's of the other transistors. Non-linearity on the N1 causes the WNM to deviate from the normal distribution [6]. In the lower ΔVth region, the load transistor P1 determines WNM=0. In the higher Vth region, the access transistor N1 determines WNM=0. The slope of the WNM for the N1 notably changes around ΔVth=0.1 V. The WNM is completely linear for ΔVth>0.1V. We call this line the linear section of the WNM for the N1. The point WNM=0 is on this straight line. When ΔVth<0.1V, the slope of the WNM is nearly equal to 0. This means that the WNM is not sensitive to the Vth fluctuation of the N1 when the WNM is large. This is consistent with a previous work [5]. The access transistors affect the WNM only when a large Vth fluctuation occurs. In other words, the WNM distribution has a tail at the side of the small margin. In this case, a large number of samples is needed in order to estimate the distribution. If we estimate the distribution with a small number of samples, almost every sample appears around the WNM for ΔVth=0. This creates a very sharp predicted distribution, resulting in an overestimation of the ΔVth for WNM=0, because the slope of the WNM is nearly equal to 0 around ΔVth=0.

The reason why the WNM has different slopes around ΔVth = 0.1 V can be explained by whether the access transistors are in the saturation mode or in the linear mode when the WNM is evaluated. The dependence of the V1 on the ΔVth of N1 is examined using the SPICE simulation at V2=0 V. The results are shown, together with the WNM, in Fig.3. The dashed line, which is around ΔVth=0.1 V, represents the changing point of the WNM slope. The V1 rapidly increases from that point. This means the operation mode of the N1 changes from a linear mode to a saturation mode around the dashed line. Therefore, a change in the slope of the WNM is related to a change in the operation mode of the access transistor N1.

In the AC write operation of a SRAM cell, the access transistor N1 is always in the saturation mode at the beginning of the operation because the V1 is not less than the WL. The write failure occurs when the N1 stays in the saturation mode during the write operation. Therefore, the write margin should be evaluated in the saturation mode of the access transistor N1. Contrary to the actual AC write operation, the conventional WNM is evaluated in the linear mode of the access transistor when the write margin is large. The conventional WNM definition is not considered adequate to evaluate the stability of a SRAM cell.

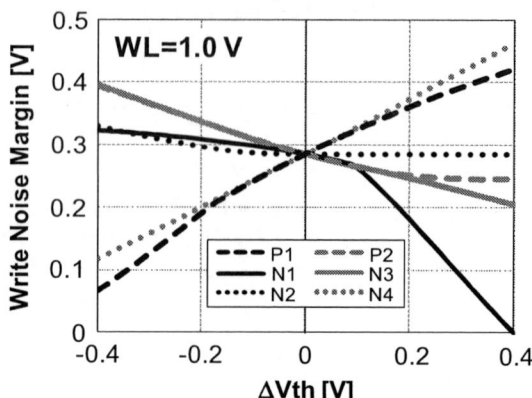

Fig. 2. Dependence of the WNM on the Vth in the conventional definition at WL=VDD.

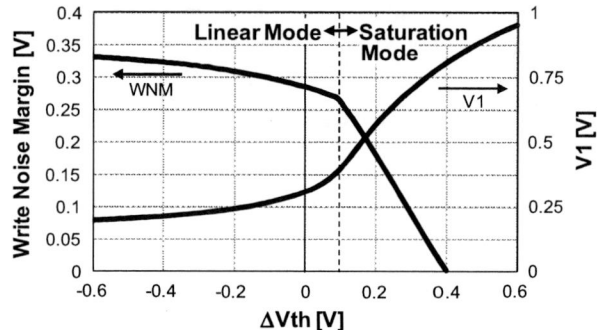

Fig. 3. The voltage of V1 at the fixed voltage of V2=0 V for the threshold voltage variation of the access transistor in the write operation.

III. ACCELERATED EVALUATION METHOD

In this section, we propose an accelerated evaluation method for the SRAM cell write margin based on the conventional WNM definition. In the proposed method, the access transistor is forced to operate in the saturation mode by lowering the word line voltage from the VDD. The WNM is measured under a lower word line voltage. The word line voltage is chosen within a range where the access transistor operates in the saturation mode. The WNM at the word line voltage of the VDD is calibrated from the WNM at a lower word line voltage. This calibrated WNM is called the Accelerated WNM (AWNM).

First, we measure the dependence of the WNM on the word line voltage. The WNM given by the SPICE simulation is shown in Fig. 4. The power supply voltage is VDD=1.0 V. The solid line represents the simulation results. The WNM is linear for word line voltages of less than 0.9 V. The slope change of the WNM at the word line voltage of 0.9V corresponds to the change in the operation mode of the access transistor N1 around the threshold voltage of ΔVth=0.1 V in Fig. 3. The N1 operates in a saturation mode in WL<0.9 V. The dashed line represents the extrapolated line. The extrapolated value of the WNM is 0.35 V which, in the proposed method, is the WNM at WL=1.0 V. This is the AWNM. The slope of the WNM for the WL voltage in the

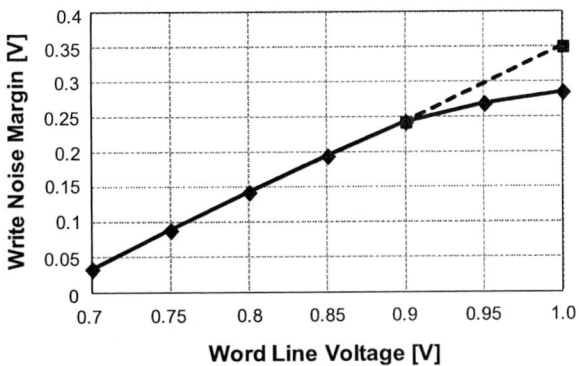

Fig. 4. Dependence of the WNM on the word line voltage.

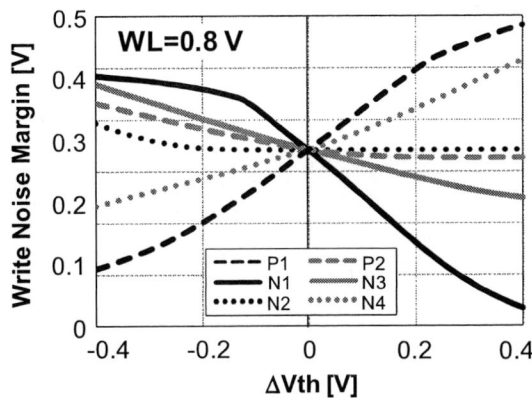

Fig. 5. Dependence of the WNM on ΔVth under WL=0.8 V.

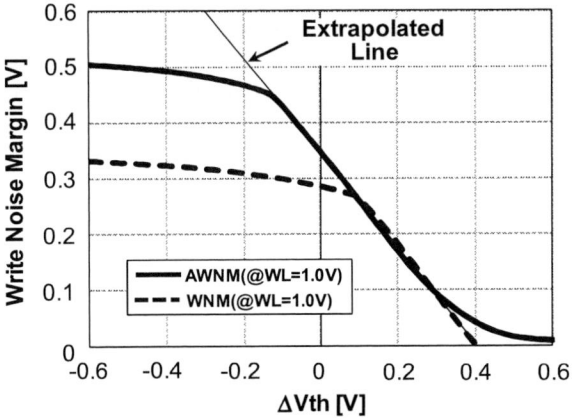

Fig. 6. The AWNM and the WNM at WL=1.0V for the access transistor N1.

linear section is used in the calibration of the measured WNM, as shown later.

Although the accelerated evaluation method gives a good linearity for the WNM, the value of the WNM itself is small when compared to the WNM at WL=1.0V. This value is calibrated to the WNM at WL=1.0 V. In the accelerated evaluation method, the AWNM at WL=1.0V, the AWNM (WL$_{1.0}$), is obtained from the WNM at a low word line voltage, WNM(WL$_m$), as:

$$AWNM(WL_{1.0}) = WNM(WL_m) + \alpha(WL_{1.0} - WL_m), \quad (1)$$

where α is the slope of the WNM for the WL voltage in the linear section.

Fig. 5 shows the dependence of the WNM on ΔVth at the word line voltage of 0.8V. A negative value is defined as the maximum length of an embedded square in the crossed butterfly curves. This means that the data are not inverted. In Fig. 5, the WNM for the N1 is linear around ΔVth=0. In Fig. 6, the dependences of the AWNM and the WNM on the Vth's are shown for the transistor N1. The solid line represents the AWNM and the dashed line represents the WNM. The thin solid line is an extrapolated line from the slope of AWNM at ΔVth =0V. The most important point is that the extrapolated line gives WNM=0 correctly. The threshold voltage ΔVth's giving AWNM=0 and WNM=0 are the same for the most influential transistor, N1.

IV. MONTE CARLO SIMULATION

The proposed method is verified using the Monte Carlo simulation. The Vth's are assumed to obey the normal distribution with a variance of σ_{Vth} =50 mV and the means of Vthn=0.404 V and Vthp=-0.384 V. In the Monte Carlo simulation, we make the Vth's of six SRAM cell transistors independently change at random. The number of samples is 100,000. For simplicity, we set the same variance for all of the transistors.

In this simulation, we use Vth's of typical values, but the actual threshold voltages of SRAM cell transistors are not known when real devices are measured. By measuring the dependence of the WNM on the word line voltage in several samples, it is possible to determine a word line voltage that causes the access transistor to operate in the saturation mode. We should exclude samples with a very low Vth, where the access transistor operates in the linear mode in the WNM

measurements, when determining the word line voltage. The excluded samples do not influence the measurements, because the probability of encountering such devices is very small.

The dependence of the WNM on the word line voltage is shown for the first ten samples in Fig. 7. The slopes α=ΔWNM/ΔWL are almost the same for the ten samples in each linear section. The word line voltage of 0.8 V was chosen from these data. The cumulative frequency scaled by the variance σ is shown in Fig. 8. The straight line of the cumulative frequency means a normal distribution of the write margin. Lines I, II, and III are the WNM at WL=0.7 V, 0.8 V, and 1.0 V, respectively. For reference, the data at WL=0.7 are also shown. Line IV is the AWNM corresponding to WL=1.0 V calibrated from line I, that is, the WSNM at WL=0.7 V. Line V is the AWNM corresponding to WL=1.0 V calibrated from line II, the AWNM at WL=0.8 V. For the calibration, we used α=1.060, which is the mean value of the ten samples. The mean values of the write margins are summarized in Table I. The two means of the AWNM at WL=1.0 V, μ_{AWNM}'s, calibrated from the WNM at WL=0.7 V and WL=0.8 V match well. The AWNMs at WL=1.0 V calibrated from I and II almost overlap, which shows the universality of the proposed method.

The slope of the cumulative frequency of the WNM at WL=1.0 V changes at about WNM=0.2 V. Obviously, the

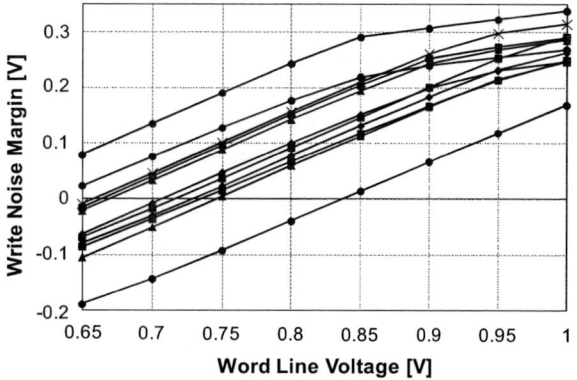

Fig. 7. Dependence of the WNM on the word line voltage for the first ten samples in the Monte Carlo simulation.

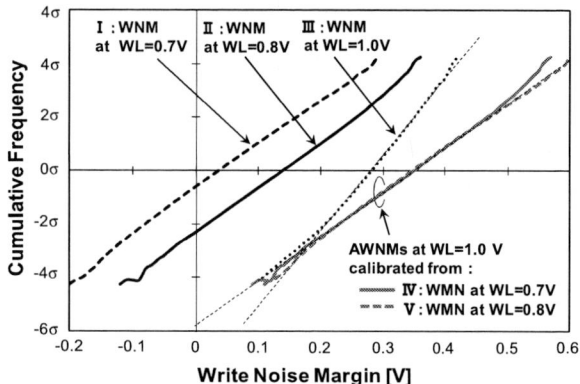

Fig. 8. The cumulative frequency of the AWNM and the WNM. Lines I, II and III are the WNM at WL=0.7 V, 0.8 V, and 1.0 V, respectively. Line IV and V are the AWNMs at WL=1.0 V calibrated from lines I and II, respectively.

Table I. Mean values of the WNM and AWNM.

	μ_{WNM}	μ_{AWNM}
WL=0.7 V	0.036	0.35
WL=0.8 V	0.14	0.35
WL=1.0 V	0.28	

Table II. Extrapolated values of $-\mu_{WM}/\sigma_{WM}$ in Fig. 8.

	$-\mu_{WNM}/\sigma_{WNM}$
WNM	-8.58* / -6.08**
AWNM at WL=1.0 V from WL=0.7 V	-5.72
AWNM at WL=1.0 V from WL=0.8 V	-5.78

*) The extrapolated value of a straight line with the slope of the WNM around the mean value of 0.28 V.

**) The extrapolated value of a straight line with the slope of the WNM around the mean value of 0.15 V.

definition. The WNM is measured under a lower word line voltage than the VDD of the power supply voltage and the access transistor is forced to operate in the saturation mode. The Accelerated Write Noise Margin (AWNM), the WNM at WL=VDD in this method, is obtained by shifting the WNM at the lower word line voltage. The degree of shift is determined from the pre-measured WNM dependence on the word line voltage. The cumulative frequency of the AWNM is linear for the AWNM, which means a normal distribution of the AWNM. A normal distribution of the AWNM dramatically improves development efficiency. The effectiveness of the proposed accelerated evaluation method for the write margin is verified using the Monte Carlo simulation.

conventional WNM does not obey the normal distribution. Therefore, the WNM at WL=1.0 V gives a small write failure probability if the probability is estimated from the slope in the neighborhood of $\Delta V_{th}=0$. On the other hand, the cumulative frequency of the WNM at WL=0.7 V and WL=0.8V is a straight line. As a result, the AWNM is also straight and obeys the normal distribution.

The extrapolated $-\mu_{WNM}/\sigma_{WNM}$'s are shown in Table II. Although the two values of AWNMs agree well, the value of WNM does not match them in this simulation range. We can estimate the write failure probability precisely not from the WNM but from the AWNM. The write failure probability P_{WF} is obtained when considering both the cases of "0" writing and "1" writing as [6]:

$$ P_{WF} = 2 \int_{\infty}^{-\mu_{WM} / \sigma_{WM}} \frac{1}{\sqrt{2\pi}} \exp\left(-\frac{x^2}{2}\right) dx \quad (2) $$

The value $-\mu_{WM}/\sigma_{WM}$ is the cumulative frequency at AWNM=0 in the scale of σ, corresponding to the $-\mu_{WNM}/\sigma_{WNM}$ in Table II.

V. CONCLUSION

We have proposed an accelerated evaluation method for the SRAM cell write margin based on the conventional WNM

REFERENCES

[1] M. J. M. Pelgrom, A. C. J. Duinmaijer, and A. P. G. Welbers, "Matching properties of MOS transistors," *IEEE J. Solid-State Circuits*, vol. 24, no. 5, pp. 1433–1440, Oct. 1989.

[2] P. A. Stolk, F. P. Widdershoven and D. B. M. Klaassen, "Modeling statistical dopant fluctuations in MOS transistors", *IEEE Trans. on Electron Devices*, vol. 45, no. 9, pp. 1960-1971, 1998.

[3] O. Hirabayashi, A. Kawasumi, A. Suzuki, Y. Takeyama, K. Kushida, T. Sasaki, A. Katayama, G. Fukano, Y. Fujimura, T. Nakazato, Y. Shizuki, N. Kushiyama, and T. Yabe, "A process-variation-tolerant dual-power-supply SRAM with 0.179um² cell in 40 nm CMOS using level programmable wordline driver," *ISSCC Dig. Tech. Papers*, pp. 458-459, Feb. 2009.

[4] A. Bhavnagarwala, S. Kosonocky, C. Radens, K. Stawiasz, R. Mann, Q. Ye and K. Chin, "Fluctuation limits & scaling opportunities for CMOS SRAM cells," *IEEE IEDM Dig. Tech. Papers*, pp. 659-662, Nov., 2005.

[5] K. Takeda, H. Ikeda, Y. Hagihara, M. Nomura, and H. Kobatake, "Redefinition of write margin for next-generation SRAM and write-margin monitoring circuit," *IEEE ISSCC, Digest of Tech. Papers*, pp.630-631, Feb. 2006.

[6] H. Makino, S. Nakata, H. Suzuki, S. Mutoh, M. Miyama, T. Yoshimura, S. Iwade, and Y. Matsuda, "Re-examination of SRAM cell write margin definitions in view of predicting distribution," *Trans. on Circuits and Systems II: Brief Express*, vol. 58, issue 4, pp.230-234, April 2011.

[7] Visit http://www.eas.asu.edu/~ptm/ for "45nm PTM HP model: V2.1," Predictive Technology Model.

Reconfigurable Back Propagation Based Neural Network Architecture

Gin-Der Wu
Department of Electrical Engineering
National Chi Nan University
Puli, Taiwan, R.O. C.
ginderwu@ncnu.edu.tw

Zhen-Wei Zhu and Bo-Wei Lin
Department of Electrical Engineering
National Chi Nan University
Puli, Taiwan, R.O. C.
s96323910@ncnu.edu.tw

Abstract- **Since the topology of neural networks is very crucial to the performance, the reconfigurable ability of the neural network hardware is very important. Therefore, this paper proposes an efficient architecture to implement the reconfigurable back propagation based neural network (BPNN). To further reduce the hardware, this paper adopts the resource sharing method. Finally, Xilinx – ISE is used to synthesize BPNN into the field-programmable gate arrays (FPGA) in experiments.**

Keywords: neural networks, reconfigurable ability, BPNN, FPGA.

I. INTRODUCTION

Artificial neural networks (ANNs) are used in many applications such as pattern recognition, classification, prediction and control. The most common training method for ANN is back propagation (BP) algorithm [1]. Reconfiguration is defined as the ability to modify the topology of the neuron network [2]. It changes the interconnection weights among neurons. Reconfigurable neural network can be implemented by analog and digital technology [2]-[4]. The analog method is attractive in speed and chip-area. However, digital implementation has better flexibility in contrast to the analog method [5].

To consider the representation of data [6], floating-point and fixed-point data format are two commonly numerical methods. In this paper, fixed-point type is adopted. A special data format called "dual fixed-point" is used in [7]-[8]. This format is similar to the fixed-point type. By using "exponent" bit, it can select two different fixed-point scaling. This method reduces the cost and power consumption [7]. To implement the activation function in neural network, some approximate methods are proposed. They are piecewise linear (PWL), piecewise linear approximation of a nonlinear function (PLAN) [9], centre recursive interpolation (CRI) [10] and look up table (LUT) methods. In these methods, PWL approximation has small maximum-error and small average-error. Its circuit design has low complexity. PLAN uses digital gates to do direct transform from x to y, where x is the input, and y is the approximated output. The corresponding gate-level logic has the small area. CRI method improves the accuracy, but it needs more than one clock cycle. Hence CRI is a slow approximation method [10]-

[11]. Since PWL has low complexity in the circuit design, this paper adopts PWL method to implement the sigmoid function.

Based on the above discussion, this paper proposes a reconfigurable BPNN architecture in FPGA. The proposed architecture can be configured by the 31-bit configuration instruction from the program memory (PM). The number of hidden layers or neurons can be modified by this instruction which includes four kinds of information. The proposed architecture can reduce the hardware resource of FPGA. The design is described in verilog hardware description language (Verilog HDL). The circuit is synthesized by using Xilinx-ISE design tool.

II. BACK PROPAGATION NEURAL NETWORK

In this paper, the number of hidden layers or neurons is reconfigurable. In [12], more than one hidden layers are applied especially in discontinuous mapping. Fig. 1 shows K-1 hidden layers where the number of the corresponding neurons is L. Fig. 2 shows the mathematical model of a neuron. This neuron executes multiplication and accumulation first. Then it applies the sigmoid function $y_i = f(\sum w_{im} \cdot x_m)$ in the output.

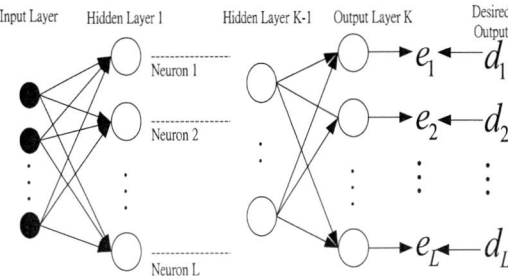

Figure 1. N-layers architecture of the neural network.

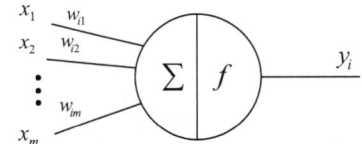

Figure 2. Mathematical model of an artificial neuron.

In the training phase, the patterns are repeatedly fed into the BPNN. The learning algorithm BP updates connecting weights to minimize the cost function until convergence. If it does not converge, the BP stops while the iteration reaches some threshold. The learning steps are described as follows [13].

Step 0: A set of training pairs (x, d) are prepared as input and desired output vectors. Initialize the weights to small random numbers. Set a learning rate ($\eta>0$) for training.

Step 1: Propagate the signal forward through the network.

$$O_j^{(t)} = f(T_j^{(t)}) = f(\sum_k w_{jk}^{(t)} \cdot o_k^{t-1}) \qquad (1)$$

Step 2: Compute the error signal δ_j^T from the output layer.

$$E = \frac{1}{2}\sum_{j=1}^{n}(d_j^{(t)} - o_j^T) + E \qquad (2)$$

$$\delta_j^T = (d_j^{(t)} - o_j^T) \times o(1-o_i) \qquad (3)$$

Step 3: Propagate the error signal backward to the previous layers and update the weights w_{jk}^t.

$$\delta_j^{t-1} = o^{t-1} \cdot (1-o^{t-1})\sum_k w_{jk}^t \cdot \delta_k^t, \qquad (4)$$

$$for \ t = T, T-1,...,2$$

$$w_{jk}^{t(new)} = w_{jk}^{t(old)} + \Delta w_{jk}^t \qquad (5)$$

$$\Delta w_{jk}^t = \eta^t \delta^{t-1} o_k \qquad (6)$$

Step 4: If all training patterns have been processed, go to step5. Otherwise go to step1.

Step 5: If the total error is acceptable, terminate the training process. Otherwise, initialize the new epoch and go to step1.

III. RECONFIGURABLE ARCHITECTURE

A reconfigurable BPNN architecture is proposed in this section. Fig. 3 shows the dataflow and control signal. There are five function blocks: (1) Program Memory (2) Forward Propagation (3) Back Propagation (4) Weight Array (5) Control Unit. In this figure, control unit (CU) receives the 31-bit configuration instruction from the program memory (PM). The number of hidden layers or neurons can be modified by this instruction which includes four kinds of information: (1) Learn/Recall (2) Number of Layer (3) Net_Configure (4) Iteration_Times.

Table I shows the control bit of Learn/Recall mode. This bit determines the status of BPNN to be learn-mode or recall-mode. If this bit is 1, then BPNN runs the learn-mode. Otherwise, the recall- mode is selected. Table II shows the control bits about the Number of Layer (NOL). NOL is 2-bit to set the number of layers in BPNN. Table III shows the configuration of 16-bit Net_Configure. It includes input-layer, output-layer and two hidden layers. When it runs in one hidden layer architecture, label [23:20] become output layer configuration. Since it is limited by 4 bits capacity, the maximum number of nodes in each layer is 15.

Table I. Learn/Recall configure list

Learn/Recall	Bit [30]
Learn mode	1'b1
Recall mode	1'b0

Table II. Number of layer configure list

Number of Layer	Bit [29:28]
1 output	2'b00
1 output & 1 hidden layer	2'b01
1 output & 2 hidden layers	2'b10

Table III. Configuration of neural network.

Net_Configure	[27:24]	[23:20]	[19:16]	[15:12]
Definition	Output Layer	Hidden Layer2 / Output Layer	Hidden Layer1	Input Layer

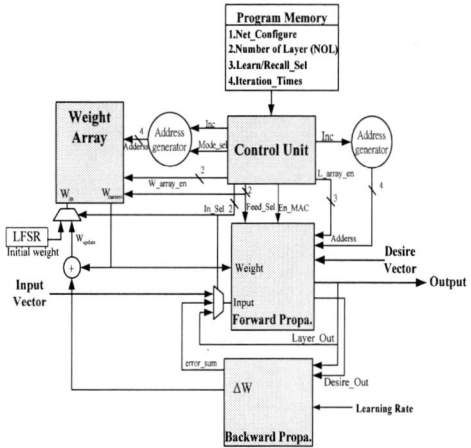

Figure 3. The proposed reconfigurable BPNN architecture.

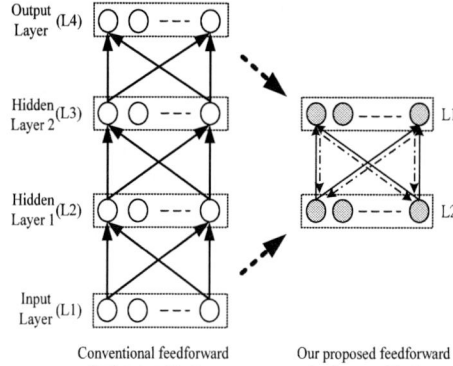

Conventional feedforward hardware architecture   Our proposed feedforward hardware architecture

Figure 4. The proposed feedforward architecture.

978-1-61284-863-1/11 $26.00 © 2011 IEEE

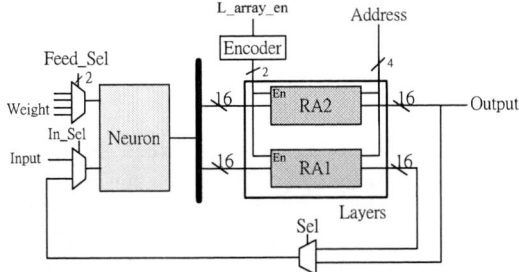

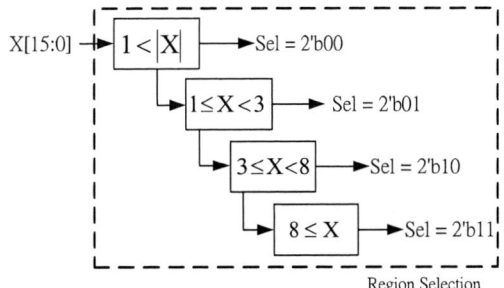

Figure 5. The proposed hardware architecture for neural network

Figure 7. Flowchart of the region-selector.

In Fig. 4, the left-hand side illustrates the conventional four layers feedforward neural network where L1, L2, L3 and L4 are used. To reduce the hardware cost, the right-hand side illustrates our proposed method. Its characteristic is that the forward propagation can be merged into only two layers. Compared with conventional type, the proposed design can reduce slices resource utilization in FPGA. After removing redundant Register Arrays (RAs), our proposed hardware architecture is shown in Fig. 5.

To implement the sigmoid function, PWL approximation uses linear equations to model the sigmoid function. Its function is shown in (7). Between -8 and 8, it has five linear segments. The circuit design of PWL has low complexity. Besides, it has small maximum-error and small average-error.

$$f(x) = \begin{cases} 0 & \text{Region1} & x \leq -8 \\ 1 - \dfrac{|x| + 120}{128} & \text{Region2} & -8 < x \leq -3 \\ 1 - \dfrac{|x| + 5}{8} & \text{Region3} & -3 \leq x < 1 \\ \dfrac{x + 2}{4} & \text{Region4} & |x| < 1 \\ \dfrac{x + 5}{8} & \text{Region5} & 1 \leq x < 3 \\ \dfrac{x + 120}{128} & \text{Region6} & 3 \leq x < 8 \\ 1 & \text{Region7} & x \geq 8 \end{cases} \quad (7)$$

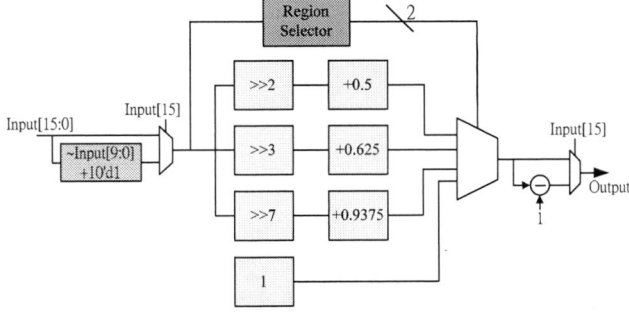

Figure 6. Hardware architecture of PWL.

The hardware implementation of PWL is shown in Fig. 6. In contrast to LUT, PWL needs smaller hardware resource. To fit FPGA hardware, all dividends are the power of two. To reduce the hardware, the arithmetic shift is used to replace the multiplier operator. Fig. 7 shows the flowchart of the region-selector. This selector is used to choose the correct data path for output.

IV. EXPERIMENTS

This paper adopts Xilinx Virtex-4 XC4VLX60 to implement the proposed neural network. The clock frequency is set to be 37.946MHz. The electrical design automation (EDA) software tools are ISE 8.1i. To test the performance, the neural network is applied to predict two kinds of nonlinear function. One is sine function, and the other is logarithm function. In the training phase, the learning rate η is 0.75, and the iteration number is 1000. In the testing phase, the total execution time is 0.25 sec (9,728,000 cycles $\times$ 26.353nS). Fig. 8 shows the prediction of sine function with different hidden nodes. Fig. 9 shows the prediction of logarithm with different hidden nodes. Finally, Fig. 10 shows the average error of these two functions under different hidden nodes. Although increasing hidden nodes results in good precision, the chip-area increases. To balance the chip-area and precision, the architecture is set to be 2-14-2 (Input-Hidden-Output).

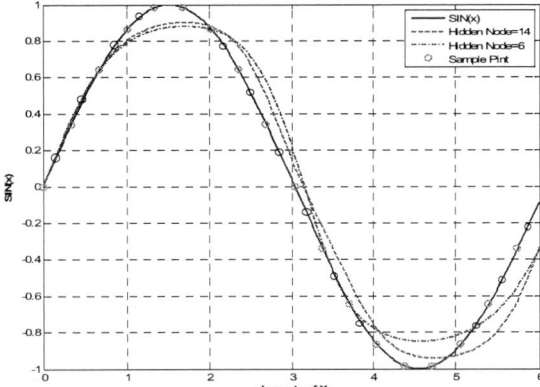

Figure 8. Prediction of sine function with different hidden nodes.

978-1-61284-863-1/11 $26.00 © 2011 IEEE

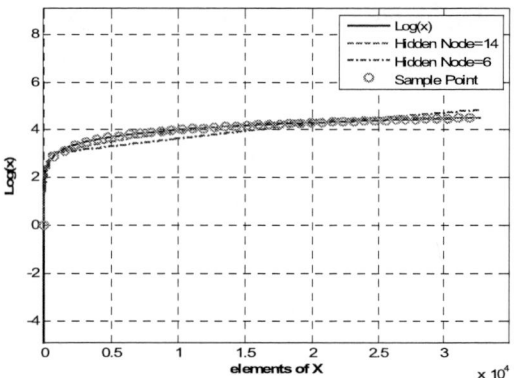

Figure 9. Prediction of logarithm function with different hidden nodes.

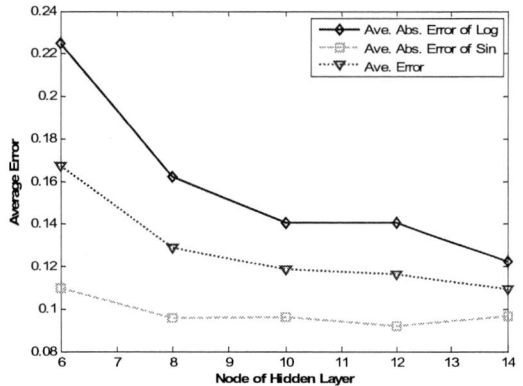

Figure 10. Average error with difference nodes in two nonlinear functions.

Table IV. Analysis of the resource utilization in different architecture.

	A. Muth. [15]	Proposed architecture	
Process Type	Parallel	Parallel	Serial
Network Architecture	8-5-5-3	8-5-5-3	8-5-5-3
Reconfigurable	No	Yes	Yes
Slices	2993	2560	1771
Slices Flip Flops	2558	2452	2495
LUTs with 4 input	5460	2382	2304
Mult18x18	0	8	1
BRAMs	2	0	0
GCLKs	1	2	2

Table V. Analysis of the resource utilization in different architecture.

	M. Moussa [6]		Proposed Architecture	
Reconfigurable	No		Yes	
Network Architecture	Slices	MULT18x18	Slices	MULT18x18
5-5-2	4917	34	4494	5
10-5-2	4915	34	4494	5
10-10-2	4915	34	4494	5
10-10-5	8783	58	4494	5

The analysis of the resource utilization in different feedforward neural network architecture is shown in Table IV

and Table V. Based on these two tables, the proposed reconfigurable BPNN architecture can reduce the utilization of hardware resource in FPGA.

V. CONCLUSION

A reconfigurable BPNN architecture is proposed in this paper. The architecture can be configured by the instruction. Therefore, the parameters such as Learn/Recall, number of layer, network configuration and iteration times can be modified immediately. It is very convenient to be applied in different applications. Since PWL has low complexity in circuit design, this paper adopts PWL approximation method to implement the sigmoid function. In experiments, the design is described in Verilog hardware description language (Verilog HDL). The circuit is synthesized by using Xilinx-ISE design tool. The experimental results show that the proposed architecture can reduce the utilization of hardware resource in FPGA.

REFERENCE

[1] D. E. Rumelhart, G. E. and R. J, "Learning internal representations by error propagation" *Parallel Distributed Processing, Cambrifge, MA: MIT Press* vol. 1, chap 8, 1988.

[2] S. Satyanarayana, Y.P. Tsividis, H.P. Graf, "A reconfigurable VLSI neural network" *IEEE Trans. Solid-State Circuits*, vol.27, pp.67-81, 1992

[3] L. M. Reyneri, "Implementation issues of neuro-fuzzy hardware: going toward HW/SW codesign" *IEEE Trans. Neural Networks*, vol.14, pp. 176-194, 2003.

[4] Y. Sungjoon, E. Oruklu, J. Saniie, "Dynamically reconfigurable neural network hardware design for ultrasonic target detection" *IEEE Ultrasonics Symposium*, pp.1377-1380, 2006.

[5] J. Y. Chang, L. R. Dung, "VLSI design of back propagation networks with on-chip learning" *Master Thesis, National Chiao Tung University, Hsinchu, Taiwan. R.O.C.*, 2001.

[6] A. W. Savich, M. Moussa, "The Impact of arithmetic representation on implementing MLP-BP on FPGAs: A Study" *IEEE Trans. on Neural Networks*, vol. 11, pp.240-252, 2007.

[7] C. T. Ewe, "Dual fixed-point: an efficient alternative to floating-point computation for DSP applications" *Field Programmable Logic and Applications*, pp. 715-716, 2005.

[8] C. T. Ewe, P. V. K. Cheung, G. A. Constantinides "Error modeling of dual fixed-point arithmetic and its application in field programmable logic" *Field Programmable Logic and Applications*, pp.124-129, 2005.

[9] H. Amin, K. M Curtus, and B. R. Hayes-Gill, "Piecewise linear approximation applied to nonlinear function of a neural network", *IEE Circuits, Devices and Systems*, vol.144, pp.313-317, 1997.

[10] M. T. Tommiska, "Efficient digital implementation of the sigmoid function for reprogrammable logic" *IEE Computers and Digital Techniques*, vol.150, pp.403-411, 2003.

[11] K. Basterretxea, J. M. Tarela, I. del Campo, "Approximation of sigmoid function and the derivative for hardware implementation of artificial neurons" *IEE Circuits, Devices and Systems*, vol.151 , pp.18-24, 2004.

[12] E. D. Sontag, "Feedback stabilization using two-hidden-layer nets" *IEEE Trans. on Neural Networks*, vol.3, pp.981-990, 1992.

[13] C. T. Lin and C. S. George Lee, "Neural fuzzy system", *NJ: Prentice Hall*, 1996.

[14] Xilinx, Virtex-4I Platform FPGAs: XST user guide. [Online]. available: www.xilinx.com, 2004.

[15] S. Himavathi, D. Anitha, A. Muthuramalingam, "Feedforward neural network implementation in FPGA using layer multiplexing for effective resource utilization" *IEEE Trans. on Neural Networks*, vol. 18, pp.880-888, 2007

Memory-Bank Based Radix-2^2 Fast Fourier Transform

Gin-Der Wu
Department of Electrical Engineering
National Chi Nan University
Puli, 545 Taiwan, R.O.C.
ginderwu@ncnu.edu.tw

Zhen-Wei Zhu and Hung-Yi Chang
Department of Electrical Engineering
National Chi Nan University
Puli, 545 Taiwan, R.O.C.
s96323910@ncnu.edu.tw

*Abstract-***A memory-bank based reconfigurable radix-** 2^2 **Fast Fourier Transform (FFT) is proposed in this paper. It can be configured to different size which ranges between 16 and 1024 points. Compared with the other methods, our experiments show that the proposed FFT processor can reduce more power consumption. It is very suitable to be applied in the speech processing, image processing, and communication system.**

Keywords: Fast Fourier Transform, radix-2^2 FFT, Memory-base archirecture.

I. INTRODUCTION

Fast Fourier Transform (FFT) is usually applied in digital signal processing, such as speech signal [1], image signal, and communication system. To implement FFT, there are two methods. One is pipeline-based design [2], such as multiple-path delay commutator pipeline architecture (MDC) and single-path delay feedback pipeline architecture (SDF). The other one is memory-based design [3], such as single-port and dual-port memory-based architecture. For pipeline architecture, radix-2 MDC (R2MDC) [4] is the classic example of the pipeline-based FFT design. It can provide high performance. However, the hardware architecture is quite complex, and the memory size is very large. Therefore, radix-2 SDF (R2SDF) [5] is proposed to overcome these disadvantages. Although the pipeline architecture provides high performance, the bit-reversed output order needs extra buffer to reorder the output data. This architecture still needs much hardware resource. For memory-based architecture, this architecture usually uses one butterfly unit (BF) to compute FFT. Unlike SDF and MDC architecture, it usually has one or two memories to store the results which are computed by the butterfly unit. In addition, it just needs one ROM to store the coefficients. In contrast to SDF and MDC, it needs lower circuit-area. However, the low throughput rate is its fault.

In [6] and [7], a reconfigurable FFT processor adopts an asynchronous register array to reduce power consumption. In [8], power saving is achieved by adjusting the FFT size in wireless receivers. In [9], clock gating technique is adopted to disable the memory modules which are not in use in order to reduce the power. In [10], the CORDIC algorithm is adopted to design a low power FFT processor. In [11], a novel low power reconfigurable FFT processor is modeled in SystemC. This paper proposes a memory-bank based reconfigurable radix-2^2

fast fourier transform to get the optimal balance between flexibility and power consumption. Power saving is achieved by using the appropriate FFT size instead of a fixed large FFT size. Besides, the memory-based architecture is adopted to design the reconfigurable FFT processor. Unlike the conventional memory-based architecture, this paper adopts the pipeline architecture to improve the performance.

II. FFT ARCHITECTURE

A. Radix-4

To analyze the digital signal, the discrete fourier transform (DFT) is usually applied as

$$X_{4k} = \sum_{n=0}^{N-1} x_n W_N^{4nk} \qquad (1)$$

According to the radix-4 concept, all digital samples can be decomposed into four parts. The first part is

$$X_{4k} = \sum_{n=0}^{N/4-1} x_n W_N^{4nk} + \sum_{n=N/4}^{2N/4-1} x_n W_N^{4nk} + \sum_{n=2N/4}^{3N/4-1} x_n W_N^{4nk} + \sum_{n=3N/4}^{N-1} x_n W_N^{4nk} \qquad (2)$$

$$X_{4k} = \sum_{n=0}^{N/4-1} x_n W_N^{4nk} + \sum_{n=0}^{N/4-1} x_{n+N/4} W_N^{4(n+N/4)k} + \sum_{n=0}^{N/4-1} x_{n+2N/4} W_N^{4(n+2N/4)k} + \sum_{n=0}^{N/4-1} x_{n+3N/4} W_N^{4(n+3N/4)k} \qquad (3)$$

$$X_{4k} = \sum_{n=0}^{N/4-1} (x_n + x_{n+N/4} W_N^{Nk} + x_{n+2N/4} W_N^{2Nk} + x_{n+3N/4} W_N^{3Nk}) W_N^{4nk} \qquad (4)$$

$$X_{4k} = \sum_{n=0}^{N/4-1} (x_n + x_{n+N/4} + x_{n+2N/4} + x_{n+3N/4}) W_N^{4nk} \qquad (5)$$

By the same method, the other three parts are shown as follows.

$$X_{4k+1} = \sum_{n=0}^{N/4-1} (x_n - j x_{n+N/4} - x_{n+2N/4} + j\, x_{n+3N/4}) W_N^{nk} W_N^{4nk} \qquad (6)$$

$$X_{4k+2} = \sum_{n=0}^{N/4-1} (x_n - x_{n+N/4} + x_{n+2N/4} - x_{n+3N/4}) W_N^{2nk} W_N^{4nk} \qquad (7)$$

$$X_{4k+3} =$$
$$\sum_{n=0}^{N/4-1}(x_n + jx_{n+N/4} - x_{n+2N/4} - j\,x_{n+3N/4})W_N^{3nk}W_N^{4nk} \quad (8)$$

where W_N^n, W_N^{2n}, and W_N^{3n} represent three twiddle factors. The decomposition can be mapped to a butterfly graph in Fig 1.

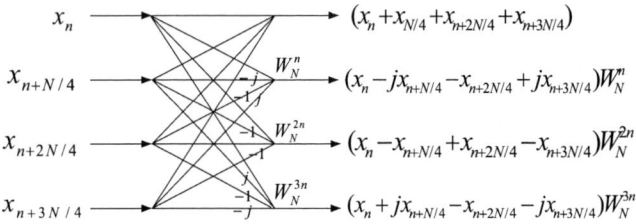

Figure 1. The butterfly graph of the radix-4 FFT.

B. Radix-2^2

To consider the radix-2^2 algorithm, the previous equations are rewritten as follows.

$$X_{4k} = \sum_{n=0}^{N/4-1}(T1 + T2)W_{N/4}^{nk} \quad (9)$$
$$X_{4k+1} = \sum_{n=0}^{N/4-1}(T3 - jT4)W_N^n W_{N/4}^{nk} \quad (10)$$
$$X_{4k+2} = \sum_{n=0}^{N/4-1}(T1 - T2)W_N^{2n} W_{N/4}^{nk} \quad (11)$$
$$X_{4k+3} = \sum_{n=0}^{N/4-1}(T3 + jT4)W_N^{3n} W_{N/4}^{nk} \quad (12)$$

where

$$T1 = x_n + x_{n+2N/4} \quad (13)$$

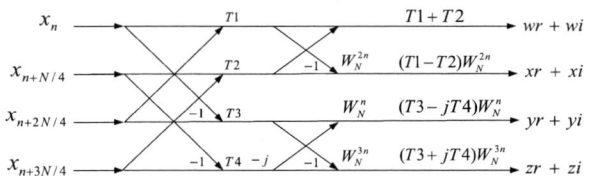

Figure 2 The butterfly graph of the radix-2^2 FFT.

$$T2 = x_{n+N/4} + x_{n+3N/4} \quad (14)$$
$$T3 = x_n - x_{n+2N/4} \quad (15)$$
$$T4 = x_{n+N/4} - x_{n+3N/4} \quad (16)$$

Fig. 2 shows the radix-2^2 butterfly. Compared with the radix-4 butterfly, it reduces the multiplication complexity. Besides, the radix-2^2 FFT is faster than the conventional radix-2 FFT. Furthermore, the input data x of each butterfly in radix-2^2 FFT can be rearranged as real parts and imaginary parts.

$$x_n = ar + jai \quad (17)$$
$$x_{n+N/4} = br + jbi \quad (18)$$
$$x_{n+2N/4} = cr + jci \quad (19)$$
$$x_{n+3N/4} = dr + jdi \quad (20)$$

The corresponding outputs have four real parts and four imaginary parts.

$$T1 = (ar + cr) + j(ai + ci) \quad (21)$$
$$T2 = (br + dr) + j(bi + di) \quad (22)$$
$$T3 = (ar - cr) + j(ai - ci) \quad (23)$$
$$T4 = (br - dr) + j(bi - di) \quad (24)$$

Finally, the corresponding radix-2^2 butterfly architecture is shown in Fig. 3.

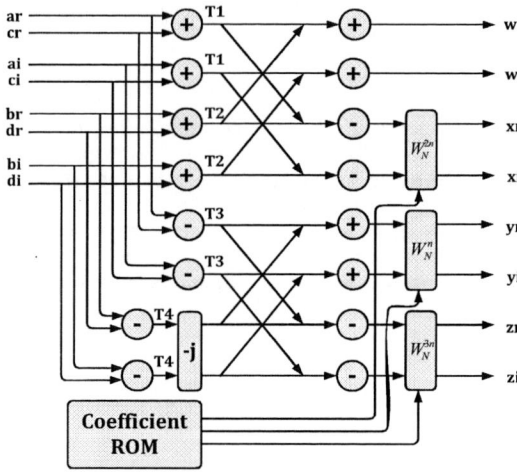

Figure 3. The radix-2^2 butterfly architecture.

To design the reconfigurable radix-22 FFT, a memory-bank based architecture is proposed. The overall architecture is shown in Fig. 4. The dual-port SRAM bank (DPSB) is used to store the intermediate data. The coefficient ROM supplies cosine and sine function for butterfly unit. The FFT/IFFT selector can choose FFT or Inverse FFT mode to be executed. Besides, two multiplexer groups, MG1 and MG2, are designed. MG1 decides which data are written into DPSB; MG2 decides which data are read form DPSB.

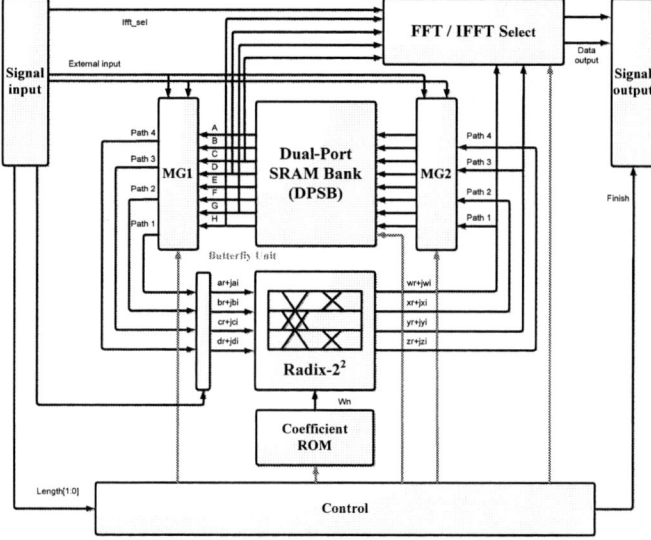

Figure 4. The architecture of the proposed reconfigurable FFT.

978-1-61284-863-1/11 $26.00 © 2011 IEEE

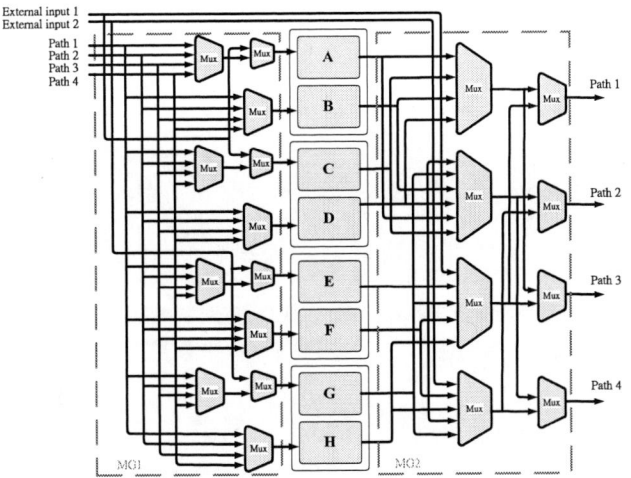

Figure 5. The memory-bank architecture of the reconfigurable FFT.

Table I. The mode of the reconfigurable FFT processor

	16-point	64-point	256-point	1024-point
Control[1:0]	2'b00	2'b01	2'b10	2'b11

The control unit can control all modules. The operation mode of the hardware is shown in TABLE I. To process N-point FFT, the flowchart can be decomposed into $(\log_4 N) + 2$ stages.

Total stage: $(\log_4 N) + 2$ stages.
where   N = {16, 64, 256, 1024}.

These stages are introduced as follows.

Catch stage: Store input data.
Deferred stage: Final data output.
$\log_4 N$ stage: FFT computation and final data output.
Other stages: FFT computation.

For an example of 1024-point FFT, the number of total stages is 7. First of all, the data is stored into the memory-bank for FFT computation. In execution, the block of butterfly unit (BF) is executed for 256 times, and each BF produces eight outputs. Four dual-port memories are adopted to save these outputs.

Each dual port memory has 256 words. The memory-bank needs two groups of multiplexer to deliver correct data into butterfly unit. Fig. 5 illustrates the structure of the memory-bank.

For another example of 16-point FFT, the number of total stages is 4. Similar to the previous example, the detail steps are described as follows.

1) In catch stage, input data are stored into memory-bank.
2) Then, the data are read form memory-bank and external-input.
3) The algorithm of FFT is performed.
4) In deferred stage, finish the final results.

Fig. 6 illustrates the detail signal flow of 16-point FFT. According to above discussion, we can find a regular rule. The rule is summarized in TABLE II.

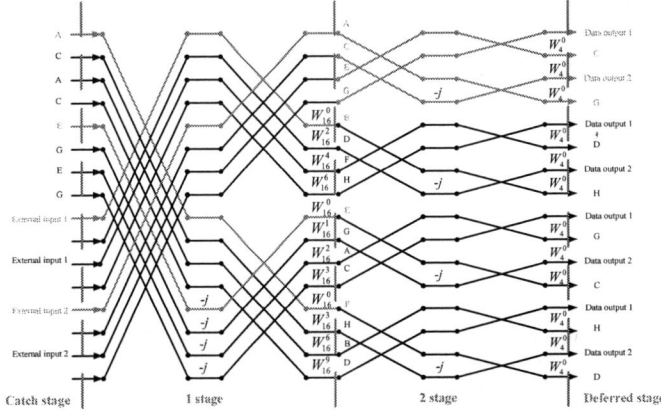

Figure 6. The signal flow of 16-point.

Table II. The rule of the memory bank

N-point	Path 1	Path 2	Path 3	Path 4
Catch stage	AC	EG		
1 stage	AC	EG	External input 1	External input 2
Other stage	ACBDEGFH	BDACFHEG	EGFHACBD	FHEGBDAC
Log4(N) stage	ACEG‖ABEF	BDFH‖CDGH	EGAC‖EFAB	FHBD‖GHCD
Deferred stage	Data output 1	CDGH	Data output 2	GHCD

Table III. The comparison of several recent researches of conventional FFT processor

	Hasan[8]	Zhao[9]	Liu[10]	Ahmadinia[11]	Chen[12]	Wu[6]	Wu[7]	Propoesd Mehod
Reconfigurable	Yes	Yes	Yes	Yes	Yes	No	Yes	Yes
Technology	UMC 0.18μm	UMC 0.18μm	UMC 0.18μm	0.13μm	UMC 0.13μm	TSMC 0.18μm	TSMC 0.13μm	UMC 90nm
FFT radix	4	2	2/4/8	2	$2^3/2^2$	2	2^2	2^2
Power consumption (mW/MHz)	5.496	1.305	4.309	1.305	0.75	0.88	0.56	0.12

III. ANALYSIS

To test the performance, the proposed reconfigurable FFT processor is described in synthesizable Verilog HDL. Then it is synthesized and estimated by UMC 90nm CMOS standard cell technology library with Synopsys Design Compiler. The total gate-count of the FFT processor is about 318818 (includes four SRAM and one coefficient ROM). The maximal clock frequency of this FFT processor is 50 MHz. The format of input data is 16-bit. To consider 1024-point FFT, the total execution time needs 6664 clock cycles. Hence, the latency is about 66.64μs (6664×20ns). Finally, the comparison of several recent researches is shown in Table III. Compared with the other methods, our proposed architecture can reduce more power consumption.

IV. CONCLUSIONS

This paper proposes a memory-bank based radix-2^2 FFT processor. It can be configured form 16 to 1024 points FFT/IFFT. Power saving is achieved by using the appropriate FFT size instead of a fixed large FFT size. The memory-based architecture is adopted to design the reconfigurable FFT processor. Unlike the conventional memory-based architecture, this paper adopts the pipeline architecture to improve the performance. Based on experiments, the proposed FFT processor has low power consumption.

REFERENCES

[1] O. B. Tuzun, M. Demirekler, and K. B. Nakiboglu, "Comparison of parametric and non-parametric representations of speech for recognition," *in Proceedings of the 7th Mediterranean Electrotechnical,* vol. 1, pp. 65-68, 1994.

[2] S. S. He and M. Torkelson, "Design and implementation of a 1024-point FFT processor," *in Proceedings of the IEEE 1998 Custom Integrated Circuit,* pp. 131-134, 1998.

[3] C. H. Chang, C. L. Wang, and Y. T. Chang, "Efficient VLSI architectures for fast computation of the discrete Fourier transform and its inverse," *IEEE Transactions on Signal Processing,* vol. 48, pp. 3206-3216, 2000.

[4] L. R. Rabiner and B. Gold, "Theory and Application of Digital Processing", *Prentice-Hall, Inc.,* 1975.

[5] E. H. Wold, and A. M. Despain, "Pipeline and parallel-pipeline FFT Processor for VLSI implementation," *IEEE Transactions on Computers,* vol. C-33, pp.414-426, 1984.

[6] G. D. Wu, and Y. Lei, "A register array based low power FFT processor for speech recognition," *Journal of Information Science and Engineering,* vol. 24, no. 3, pp. 981-991, 2008.

[7] G. D. Wu, and Y. M. Liu, "Radix-2^2 Based Low Power Reconfigurable FFT Processor," *IEEE International Symposium on Industrial Electronics,* pp. 1134~1138, July 5-8, 2009.

[8] M. Hasan, T. Arslan, and J. S. Thompson, "A delay spread based low power reconfigurable FFT processor architecture for wireless receiver," *IEEE International Symposium on System-on-Chip,* pp. 135-138, 2003.

[9] Y. Zhao, A. T. Erdogan, and T. Arslan, "A Low-Power and Domain-Specific Reconfigurable FFT Fabric for System-on-Chip Applications," *IEEE International Symposium on Parallel and Distributed,* pp. 4-8, 2005.

[10] G. Liu, and Q. Feng, "ASIC Design of Low-power Reconfigurable FFT Processor," *IEEE International Conference on ASIC,* pp. 44-47, 2007.

[11] A. Ahmadinia, B.Ahmad, T. Arslan, "System Level Modelling of Reconfigurable FFT Architecture for System-on-Chip Design," *Second NASA/ESA Conference on Adaptive Hardware and Systems,* pp. 169-175, 2007.

[12] Y. Chen, Y. W. Lin, and C. Y. Lee, "A Block Scaling FFT/IFFT Processor for WiMAX Applications," *in Proc. 2nd IEEE Asian Solid-State Circuits Conference,* pp. 203–206, 2007.

CMOS Ku-Band LNB with High Image Suppression Capability for Satellite Application

Lin Jia and M. Annamalai Arasu

Institute of Microelectronics, A*STAR (Agency for Science, Technology and Research), Singapore

Abstract— **This paper presents a fully integrated Ku-band low noise block (LNB) front-end receiver with high image suppression capability. It translates the RF-band (10.7-12.75 GHz) to an IF ranging in the L-band (0.95-2.15 GHz). The receiver exhibits a conversion gain of 50 dB, a single-sideband noise figure of 3.8 dB and image suppression level of 40 dB. The integrated receiver draws 80mA from 1.1 V supply voltage. This paper demonstrates the feasibility of a Ku-band heterodyne receiver implemented on a 65nm CMOS technology.**

Index Terms— **Direct Broadcast Satellites, low noise amplifier, image rejected load, mixer, amplifier, receiver**

I. INTRODUCTION

Growing demand for new high-data rate digital services via digital broadcast satellite (DBS), such as high definition television (HDTV), and rising demand for one-way access high speed Internet services will have a profound impact on the future of telecommunications. Moreover, the new two-way digital video broadcast return channel via satellite standard (DVB-RCS), has already reached maturity and is attracting an increasing number of users [1]-[4].

A low-noise block (LNB) down-converter is a critical block in a digital broadcast satellite (DBS) receiver. It is usually installed outdoors with a dish antenna. The orthogonal-linear polarized signals in Ku-band (10.7-12.75 GHz) from satellite are transmitted to earth and picked up by a parabolic dish antenna, then an LNB down-converts the received RF signals to an IF frequencies, ranging of the L-band (0.95-2.15 GHz). The IF signals is sent to an integrated receiver decoder (IDR) which is located indoor for tuning and digital demodulation. Ku-band LNB covers the whole RF band, Low Band (LB): 10.7-11.7 GHz / High Band (HB): 11.7-12.75 GHz, and detects both Vertical (V) and Horizontal (H) polarized signals. Up till now, commercial LNBs have all been fabricated by discrete solution and III-V technology. Thus, the cost is quite high. To lower the cost, several integrated LNB ICs have been reported in latest years [5]-[7]. However, they are beyond in partial integration level because of the two difficult factors: 1) stringent noise figure (NF) <0.6dB, is quite challenge for CMOS technology due to nature noise substrate. Two stages external HEMTs have to be used as given in Fig.1. Moreover, CMOS and HEMT are not comparable technologies. 2) Image

rejection filter (IRF) shown in Fig.1 is necessary for application. Hence, LNB requirements have posed the principal obstacle for a monolithic solution.

This work demonstrates an implementation of a fully integrated down-converter front-end with high image suppression capability in a low-cost 65nm CMOS technology. The design covers Ku-band frequency range, and the noise figure of the proposed front end IC is 3.8dB. NF of a modern LNB down-converter can be reduced to 0.6dB or even lower by applied one or two stages discrete HEMT amplifier in front of a CMOS down-converter as usual. A small gain variation (<6dB) along the whole frequency range is achieved since the received signals always contain information from all channels simultaneously. And high OIP3 and P1dB are obtained to meet linearity requirements.

The paper is organized as follows. The down-converter front-end architecture is described in Section II. Designs of circuit blocks are discussed in Section III and Section IV. Experimental results are summarized in Section V.

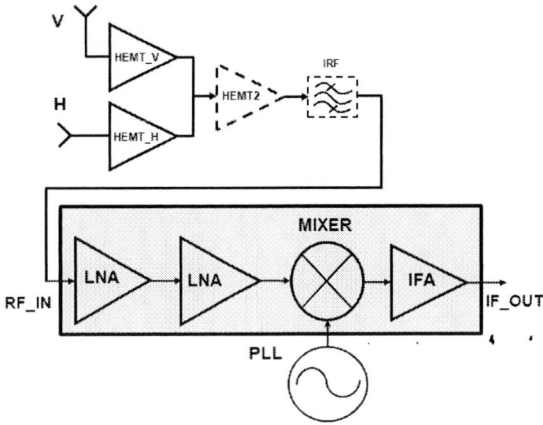

Fig. 1. Architecture of Universal LNB for DBS application.

II. PROPOSED LNA WITH IMAGE REJECTED LOAD

The two-stage LNA has a single-ended input and a single-ended output. The first stage LNA adopts a common source (CS) design since it has superior noise performance over other topologies, while the second stage based on a cascode configuration provides higher power gain. In order to achieve wideband input matching as well as flat gain over a wide operation frequency, capacitive feedback approach is used. The schematic of the LNA with detail matching network is shown in Fig.2

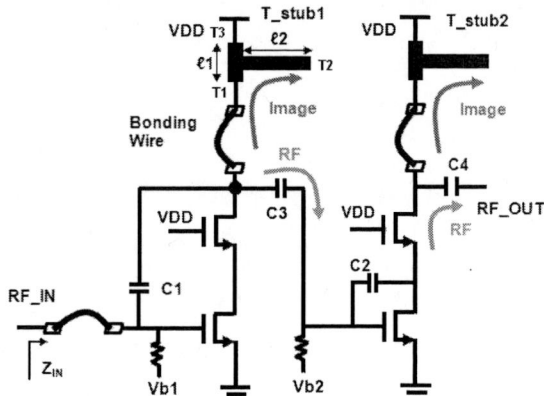

Fig. 2. Two-stage LNA with image rejected load.

A. Input Matching

The capacitive feedback components, C_1 and C_2, are used to provide the desired input impedance matching without adding any extra noise comparing with the conventional resistive shunt-shunt feedback. The equivalent of the small-signal model for capacitive feedback LNA is illustrated in Fig.3. R_g is the gate resistance of the input transistor M_1, which is lower bounded by the channel induced gate resistance $(R_g > 1/(5g_m))$, consider the simplified transistor noise model, assume $R_g = 0$ for analysis. R_L is the load impedance when L_L and C_L resonate at the frequency of interest. Thus we have a small-signal analysis of the input impedance as given:

$$Z_{in} = sL_g + \left(\cfrac{1}{sC_{gs} + sC_1 \cfrac{1 + g_m R_L}{1 + sC_1 R_L}} \right)$$

$$= sL_g + \cfrac{\cfrac{1}{sC_1}(1 + g_m R_L + \cfrac{C_{gs}}{C_1} - C_{gs} C_1 R_L^2 s^2)}{(1 + g_m R_L + \cfrac{C_{gs}}{C_1})^2 + (R_L C_{gs})^2} \quad (1)$$

$$+ \cfrac{R_L (1 + g_m R_L)}{(1 + g_m R_L + \cfrac{C_{gs}}{C_1})^2 + (R_L C_{gs})^2}$$

To simple (1) while $g_m R_L \gg 1$, $s = j\, \omega_0^2$, ω_0 is input matching resonant frequency. Z_{in} is approximated to:

$$Z_{in} \approx \cfrac{g_m R_L^2}{(g_m R_L + \cfrac{C_{gs}}{C_1})^2 + (R_L C_{gs})^2} \quad (2)$$

The imaginary terms in (1) are dependent on R_L, C_1, L_G, C_{gs}

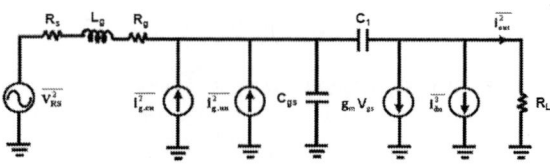

Fig. 3. Small-Signal model of the capacitive feedback LNA for noise analysis.

and g_m. hence, the capacitive feedback provides more freedom for the choice of the single inductance L_G whose quality factor is crucial for high noise performance design.

B. Noise Matching

The small-signal model of Fig.3 is employed to perform the noise analysis. The noise contribution of the cascade transistor, M_2, is neglected due to the noise cancellation mechanism of the cascade transistor when the output impedance of the input device is much higher than the inverse of the transconductance of the cascade transistor. The model include two internal independent noise source $\overline{i_g^2}$ and $\overline{i_d^2}$. The noise factor of amplifier can be calculated as:

$$F = 1 + \frac{R_L}{R_s} + \frac{R_g}{R_s} + \gamma g_{do} R_s \left(\frac{\omega_0}{\omega_t} \right)^2 \quad (3)$$

The short-channel excess noise factor γ is bias dependent but measurements show that it ranges between 1~2. A factor α is usually used to account for the difference for short channel device, thus, $g_{do} = g_m / \alpha$. R_s is source resistance of 50ohm.

At a given frequency, given lossless feedback and matching networks, selection of the optimum device width and optimum bias voltage at each frequency results in an input match and an overall noise figure of F_{min}. In practical circuit design, the matching inductor has finite quality factor, requiring a careful trade-off between input matching and noise figure.

C. Image Signal Rejected Load

Two quarter wave guides fabricated on Printed Circuit Board (PCB) are used to suppress image signals, which is located at the frequency range of 7.8-8.8GHz (LB) and 8.4-9.4 GHz (HB). T-stub can be divided into two parts: the first part is a symmetrical waveguide with the length of l_1, the width of w_1 and the thickness of h_1. h is the distance

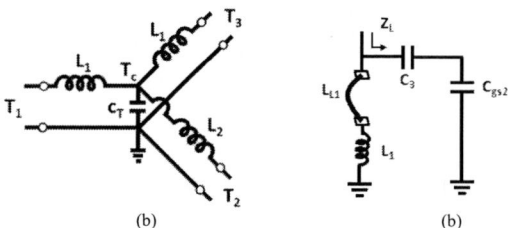

(b)   (b)

Fig. 4. (a) Model of T_Stub; (b) Load resonator of LNA.

between the symmetrical waveguide and the ground plane. The second part is comprised by a waveguide connecting the symmetrical one with an open end. The shape is sized as the length of l_2, and the width of w as well as the thickness of h. The equivalent circuit for T_stub is depicted in Fig.4 (a). Two terminals (T_1 and T_3) of T-stub is the symmetrical waveguide terminals, which is series with LNA load inductor. T2 with open end forms T-stub of waveguide, which decides the image frequency (ω_i) when L2 and C_T resonate at the image interest as given in (4). Therefore, the quarter waveguide is shorted to ground at the image interest of ω_i. Meanwhile, the image signal is suppressed. Due to the wide image signal frequency, two T-stubs are cascaded to apply in the design and to resonate at different location of the image band to extend desired image suppression.

$$\omega_i = \frac{1}{\sqrt{L_2 C_T}} \qquad (4)$$

The equivalent circuit for RF load is given in Fig.5, and the resonation frequency of the RF output is given in (5):

$$\omega_0 = \frac{1}{\sqrt{(L_{L1} + L_1)\dfrac{C_{gs2}C_3}{C_{gs2} + C_3}}} \approx \frac{1}{\sqrt{L_{L1}\dfrac{C_{gs2}C_3}{C_{gs2} + C_3}}} \qquad (5)$$

Because $L_1 \ll L_{L1}$, and the proposed LNAs are broadband application. Hence, T-stubs will not cause the load resonate frequency shift, but it will impact the RF signal gain.

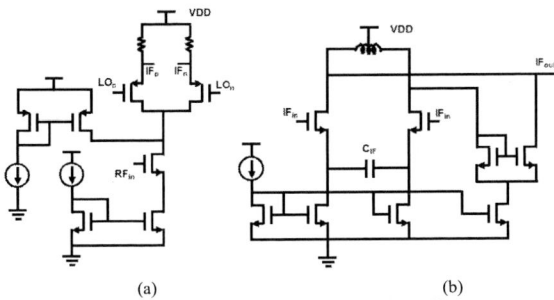

(a)   (b)

Fig. 5. (a) Schematic of (a) Mixer; (b) IF amplifier with buffer.

III. MIXER AND IF AMPLIFIER

The mixer consisted of a V-I converter and a single ended Gilbert-cell topology, as illustrated in Fig.5 (a). The RF input transistor Mn1 acts as the input stage of a cascade LNA. The gate bias voltage is set to 0.55V with a corresponding f_T of 153GHz. Current bleeding circuit is used to improve conversion gain as well as NF [8].

The schematic of the IF amplifier (IFA) is shown in Fig.5 (b). The IFA implemented a fully integrated balanced to unbalanaced converter, avoiding the need for an external balun. Capacitors C_{IF} is set to provide power gain. A bandwidth is required as large as the DBS IF band, and the output is matching at 75 Ω.

IV. EXPERIMENT RESULTS

The circuits were fabricated in 65nm CMOS process with

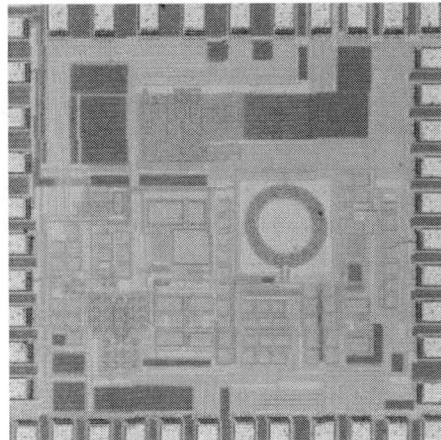

Fig. 6. Die photograph of Receiver.

standard 10-layer Cu back-end. The f_t of 153GHz and f_{max} of 186GHz for an n-MOSFET with 32 gate fingers and 0.5μm finger width, contacted on one side of the gate, were measured on the same die with receiver. The receiver occupies 1.2mmX1.2mm, including all pads, and the die photograph is shown in Fig.6.

In Fig.7, the input and output reflection coefficients of the

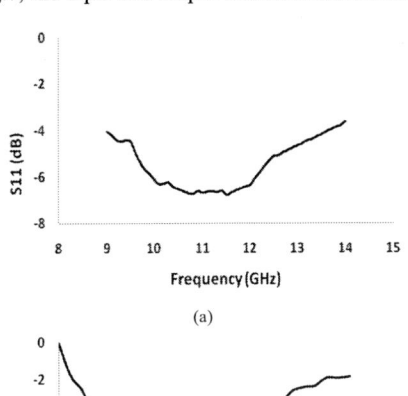

(a)

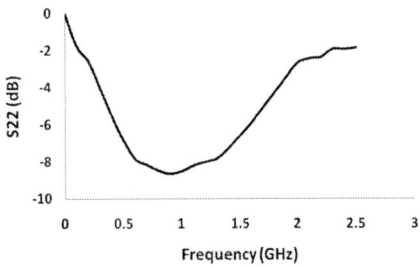

(b)

Fig. 7. Measured Input and Output Matching.

978-1-61284-863-1/11 $26.00 © 2011 IEEE   77

receiver (S_{11} and S_{22}) are reported versus the RF and IF band respectively. The measured conversion gains of around 50dB versus the LB and HB RF frequencies are presented in Fig.8, the gain flatness is less than 6 dB. The maximum image rejection levels of 40 dB can be observed. The measured noise figures of around 3.5dB of the LB and HB front end receiver are demonstrated in Fig. 9.

To meet the noise figure requirement of the LNB, an external HEMT with low noise figure of 0.45dB and conversion gain of 10dB, is necessary to apply for the first stage. The total cascade noise figure of 0.6 dB was achieved by the noise figure cascade principle [11].

Comparing with measurement and simulation, the absolute value of gain is lower by 10 dB, the noise figure are worse

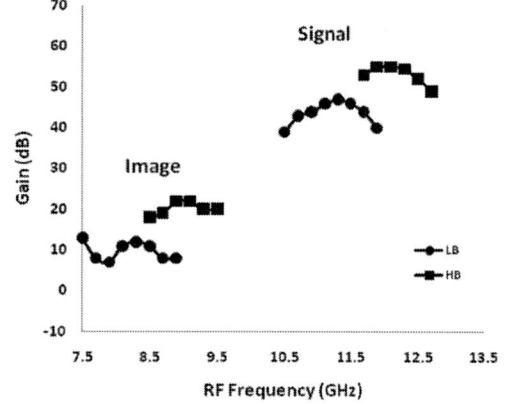

Fig. 8. Measured Conversion Gain.

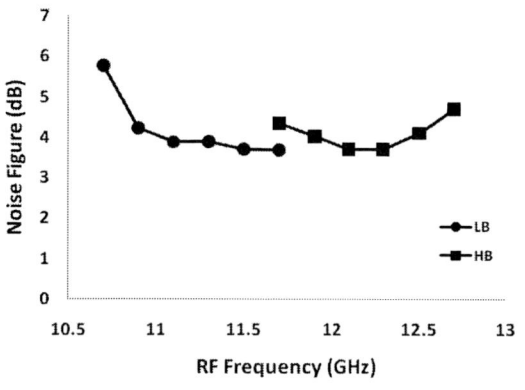

Fig. 8. Measured Noise Figure.

2dB at least, and the input and output matching is lower by 10 dB also. Inaccurate transistor model could be accounted for difference between measured and simulated results. The power consumption is 80mA with 1.1 voltage supply. Table.1 gives the comparison of the proposed work and the art design.

V. CONCLUSION

This work demonstrates a CMOS front-end receiver for application of Universal Ku-band LNB down conversion. It achieves a gain of 50dB with ±3dB variation. The in-band NF is between 3.0 dB to 4 dB. The image rejection level of 40dB is obtained. The power consumption is 80mW with 1.2V power supply. The receiver was implemented in a low cost CMOS technology. To compare with the art design, the proposed LNB satisfies the requirement of Ku-band application and lowers the cost.

TABLE.1 COMPARISON BETWEEN THIS WORK AND THE ART DESIGN.

Ref.	Band (GHz)	f_{LO} (GHz)	Size (mm²)	Gain (dB)	NF (dB)	Power (V_{cc}/I_{cc})	Tech.
[5]	LB HB	4.875 5.3	10	36	6	3.3V/180mA	Silicon bipolar
[6]	LB HB	4.875 5.3	NA	38	7	3.3V/160mA	Silicon bipolar
[7]	LB HB	9.75 10.6	NA	32	9	3.3V/102mA	SiGe
[9]	LB HB	9.75 10.75 (Ex)	1.62	50	4.2	75mA/1.8V	0.18µm CMOS
This Work	LB HB	9.75 10.6	1.44	50	3.6	110mA/1.1V	65nm CMOS

LB: 10.7-11.70 / 0.95-1.95 (GHz);HB: 11.7-12.75 /1.1-2.15 (GHz)

ACKNOWLEDGMENT

The authors gratefully acknowledge contributions of their colleagues, in particular RFIC group of IME. The authors also wish to thank the assistance received from both IME and UMC staffs.

REFERENCES

[1] Satellite Earth Stations and System (SES); Television Receive-Only (TVRO) Satellite Earth Station Operating in 11/12 GHz Frequency Bands, *ETSI, ETS 300 784*, Jul. 1997.

[2] Digital Video Broadcasting (DVB); DVB Framing Structure, Channel Coding and Modulation for 11/12 GHz Satellite Services, *ETSI, EN 300421*.

[3] Overview of DVB-S—Annex B, *EUTELSAT*, Jun. 1999.

[4] Y. Konishi and Y. Fukuoka, "Satellite receivers technologies," IEEE *Trans. Broadcasting*, vol. 34, pp. 449–456, Dec. 1988.

[5] T. Copani, S. A. Smerzi, G. Girlando, and G. Palmisano, "A 12-GHz Silicon Bipolar Dual-Conversion Receiver for Digital Satellite Applications," *IEEE JSSC., Vol. 40, No. 6*, June 2005, PP.1278-1287.

[6] G. Girlando, S. A. Smerzi, T. Copani, and G. Palmisano, "A Monolithic 12-GHz Heterodtne Receiver for DVB-S Applications in Silicon Bipolar Technology," *IEEE Trans. on TMTT., Vol. 53, No. 3*, Mar. 2005, PP.952-959.

[7] Z. Deng, J. Chen, J. Tsai, and A. Niknejad, " A CMOS Ku-Band Single-Conversion Low-Noise Block Front-End for Satellite Receivers," *IEEE Proc. On RFIC Sym.*, 2009, pp.135-137.

[8] J. Park, C.H. Lee, B.S.Kim, J.Laskar, "Design and Analsysis of Low Flicker Noise CMOS Mixers for Direct-Conversion Receivers," *IEEE Trans. MTT, Vol. 54, No. 12, Dec.* 2006, PP.4372-4380.

[9] Manual of *TFF1004HN_N1_1_NXP*, 2006.

[10] D.K. Shaeffer, et al, "A 1.5-V 1.5-GHz CMOS low noise amplifier," *IEEE J. Solid-State Circuits, Vol.32*, No. 05, pp.745-759, May 1997.

[11] B. Razavi, RF Microelectronics, Prentice hall PTR, Upper Saddle River, NJ 07458.

A 0.68-1.65GHz CMOS LC Voltage-Controlled Oscillator with Small VCO-Gain and Step Variation

Liheng Lou[1,2], Lingling Sun[2,*], Haijun Gao[2] and Jincai Wen[2]

[1] Institute of VLSI, Zhejiang University, Hangzhou, China

[2] Key Laboratory of RF Circuits and Systems of Ministry of Education, Hangzhou Dianzi University, Hangzhou, China

[*] Corresponding Author-Email: sunll@hdu.edu.cn

Abstract—A wideband CMOS LC voltage controlled oscillator (VCO) with small VCO gain (K_{VCO}) and band step variation was developed. In order to get small K_{VCO} and step variation across different sub-bands, 8-unit capacitor array was optimized and an additional array of varactors was introduced into the LC-tank, both of which are temperature weighted. **Implemented in a 65 nm CMOS RF technology, post-layout simulation shows that the proposed VCO can be tuned from 681 MHz to1656 MHz. VCO gain is around 170 MHz/V with 17% variation across the tuning range, while the step between two sub-bands is 99 MHz. Under a 1.2 V supply, the VCO exhibits phase noise of -73dBc/Hz at 10 kHz offset and phase noise -124.3 dBc/Hz at 1 MHz offset from the carrier. VCO figure-of-merit (FoM) of -184.0dBc/Hz is achieved.**

Keywords- CMOS; LC; voltage controlled oscillator (VCO); VCO gain (K_{VCO}); wide tuning range

I. INTRODUCTION

For the better phase noise performance compared to a ring oscillator, a LC-tank voltage controlled oscillator (VCO) is preferable in phase-locked loop (PLL) of radio frequency (RF) applications. A VCO with wide tuning range is indispensable for broadband and multi-band RF transceivers. As one of the most effective technology, switched-capacitor array is widely used to extend tuning range while keeping the tuning sensitivity low [1]-[3]. Adopted in a phase-locked loop (PLL), this kind of VCO firstly have coarse tuning process by a digital block known as auto frequency calibration (AFC), and subsequently fine tuning process by an analog tuning voltage.

Conventionally, the switched capacitor array is binary weighted, which maximizes its switching capacitor number under a given control bit. However, this kind of structure always causes large variations in tuning sensitivity across different sub-bands. These huge variations are not desirable for the VCO itself as well as PLL which adopts this VCO. Some technology [4], [5] was proposed to reduce K_{VCO} by varactor size optimization. However, the improvement is still limited due to the employment of conventional binary weighted switched capacitor array. The other problem is the significant difference of the step between two adjacent sub-bands across the whole tuning range, which makes the AFC of look-up table scheme less robust considering PVT.

In this paper, a switched varactor-capacitor array structure is proposed. Instead of the conventional binary weighted coding, this structure employs 8-bit temperature weighted coding, making both K_{VCO} and step almost constant.

II. DESIGN CONSIDERATION

A. Voltage controlled oscillator gain (K_{VCO})

There is a trade-off between K_{VCO} and the number of switched capacitors in conventional VCO which employs binary weighted capacitor array. The more switched capacitors we use, the smaller the K_{VCO} we have, which leads to better phase noise performance (of course, K_{VCO} is not the only condition to insure low phase noise, while circuit structure also has great impacts to phase noise performance [6]). But we can not increase the switched capacitors with no limit. On one hand, more capacitors mean higher complexity which degrades the circuit stability. On the other hand, lower K_{VCO} is not the only insurance as mentioned above. It is reported that the binary weighted control bit number would be acceptable less than 6. A typical schematic of a conventional VCO which employs 3-bit binary weighted switched capacitor array and corresponding frequency tuning characteristics are presented in Fig.1.

As we can see in Fig1.(b), simulation shows that the top sub-band of maximum frequency has the largest K_{VCO}, 210 MHz/V with the worst phase noise performance, -63 dBc/Hz at 10 kHz offset from the carrier and -120dBc/Hz at 1 MHz offset. The bottom sub-band of minimum frequency has the smallest K_{VCO}, 54 MHz/V with the best phase noise performance, -74 dBc/Hz at 10 kHz offset from the carrier and -126 dBc/Hz at 1 MHz offset. There is 118.2% in K_{VCO} variation between the top sub-band and the bottom one. This huge K_{VCO} variation among different sub-bands would alter the loop bandwidth and deteriorate phase noise performance.

B. Sub-band step

To ensure that the F-V curves cover all the tuning range, the step between two adjacent sub-bands is determined by K_{VCO} and required overlap. In conventional VCO, the step between two adjacent sub-bands of higher frequency is always larger than the one between two adjacent sub-bands of lower frequency, see Fig.1(b). The reason is that the equal

---

This work was supported by the National Natural Science Foundation of China (No.61001066).

capacitance is switched in or out from the tank every time the switching happens in binary weighted switched capacitor array, while the frequency is not directly proportional to capacitance. The step variation is obvious especially in a wide-band VCO, and is inevitable as long as the binary weighted switched capacitor array is employed.

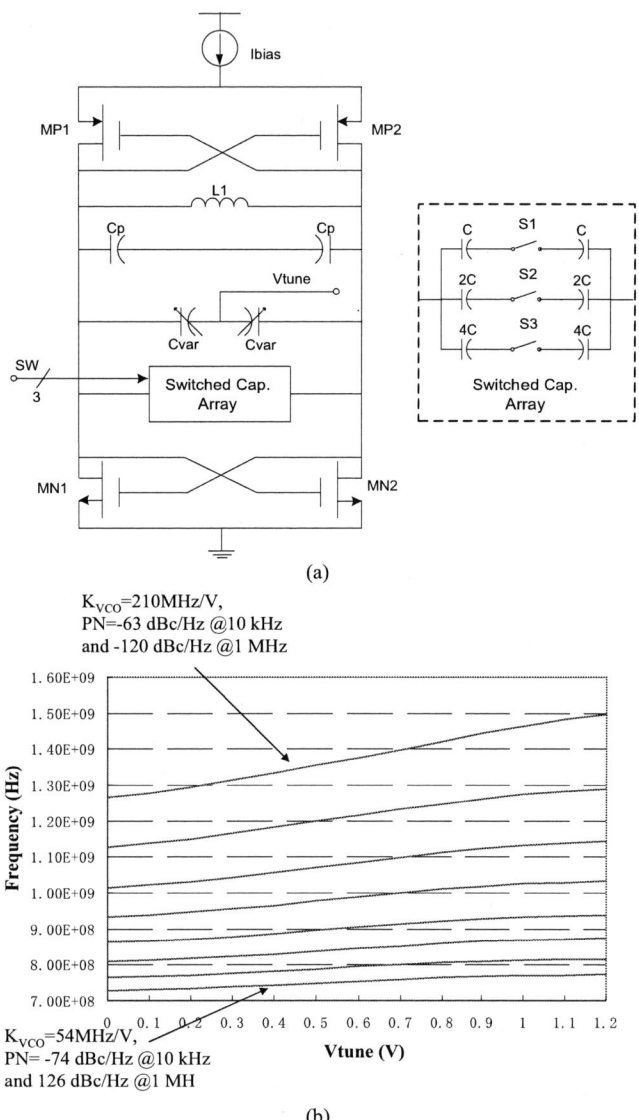

K_{VCO}=210MHz/V,
PN=-63 dBc/Hz @10 kHz
and -120 dBc/Hz @1 MHz

K_{VCO}=54MHz/V,
PN= -74 dBc/Hz @10 kHz
and 126 dBc/Hz @1 MH

(b)

Fig. 1. Conventional VCO: (a) Typical PMOS biased complementary structure with 3-bit binary weighted switched capacitor array, (b) Frequency tuning (F-V) characteristics.

III. PROPOSED WIDEBAND VCO

As discussed in the previous section, there is huge variation in K_{VCO} and step in a conventional VCO. The F-V curves distribute non-uniformly across the tuning range using the binary weighted switched capacitance array (see Fig.1(b)). There are too many curves of small K_{VCO} and step crowding in the bottom region, leaving the top part sparse. In order to lower the maximum K_{VCO}, more switched capacitors are needed.

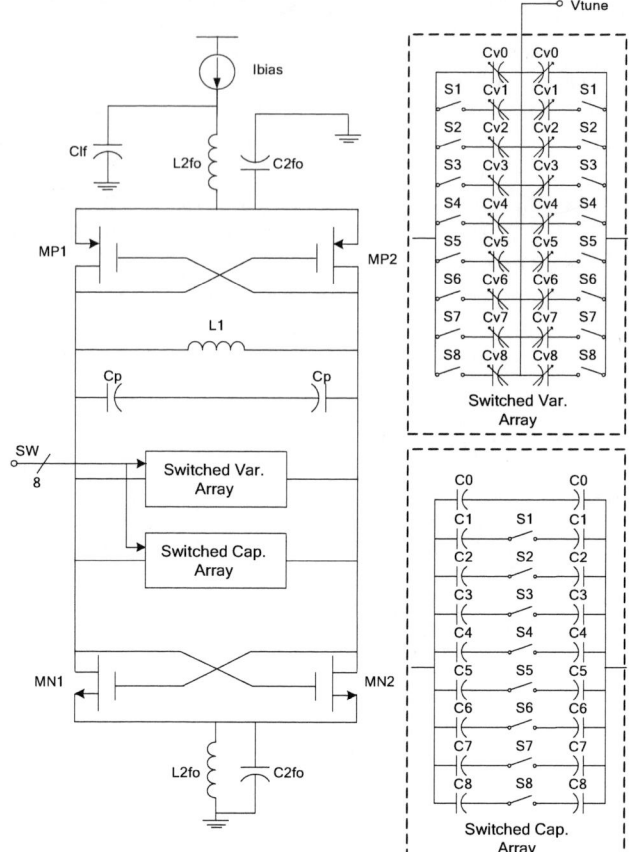

Fig. 2. Proposed wideband VCO.

Generally, more switched capacitors introduce longer AFC coarse tuning phase. Another problem is the variation in K_{VCO} makes it hard to predict the corresponding frequency range considering the effect of PVT, which means a more complex AFC algorithm is needed.

To overcome the disadvantages inherent in the binary weighted capacitor array scheme, we propose the temperature weighted scheme, using both switched capacitor array and switched varactor array, as depicted in Fig.2(a).

The 8-unit switched varactor array makes sure more varactors are switched into the tank to keep K_{VCO} constant, while the step between two adjacent sub-bands is mainly determined by the 8-unit switched capacitor array. All switches in the proposed wideband VCO is realized using NMOS. There are explicit expressions to derive the capacitance of both capacitors and varactors. Obviously, the bottom sub-band frequency F_{MIN} can be expressed as a function of top sub-band frequency F_{MAX}, the number of total number of switched units N and step frequency between two adjacent sub-bands f_{step} as

$$F_{MIN} = F_{MAX} - N \cdot f_{step} . \qquad (1)$$

Assuming every switch in the switched varactor array has equal parasitic capacitance C_{gd}, the relationship between oscillation frequency and the corresponding capacitance of a varactor in array, $Cv(n)$, and capacitance of a capacitor in array, $C(n)$, can be expressed as

$$f_{OSC}(n) = F_{MAX} - nf_{step} = \frac{1}{2\pi\sqrt{L}}$$
$$\cdot \left(Cp + \sum_{n=0}^{N}(C(n)+Cv(n)) + 2(N-n)C_{gd} \right)^{-0.5} \quad (2)$$

where, Cp, n and $f_{OSC}(n)$ refers to the capacitance parallel with the tank, the nth sub-band blow the top one(the 0th sub-band refers to the top one), and the oscillation frequency of the nth sub-band, respectively. Here, $C(n)$ and $Cv(n)$ refers to the corresponding n^{th} capacitor and varactor(see Fig.2). So, K_{VCO} of each band is calculated as

$$K_{VCO}(n) = \frac{\partial f_{OSC}}{\partial Vtune} = -\frac{1}{4\pi\sqrt{L}}$$
$$\cdot \frac{\sum_{n=0}^{N}\left(\frac{Cv(n)}{Cv(0)}\right)}{\left(Cp + \sum_{n=0}^{N}(C(n)+Cv(n)) + 2(N-n)C_{gd} \right)^{1.5}} \cdot \frac{\partial Cv}{\partial Vtune} \quad (3)$$

where $\partial Cv/\partial Vtune$ represents the capacitance change ratio of $Cv(0)$ against Vtune.

Since the span between F_{MAX} and F_{MIN} must cover the required tuning range, they are determined by the tuning range. And once N and f_{step} is chosen by (1), Cp is firstly determined by(2). Then, K_{VCO} is determined according to N and f_{step}. Note that the target is to keep K_{VCO} of different sub-band as a

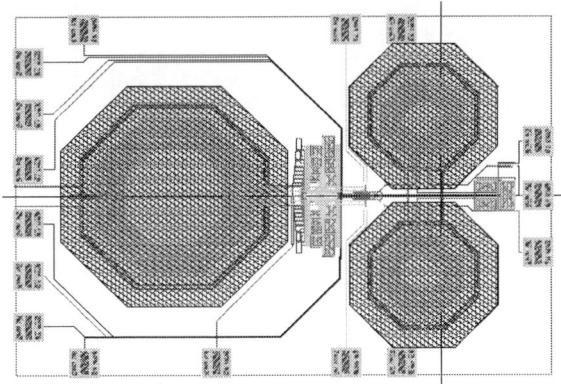

Fig. 3. Layout of proposed wideband VCO.

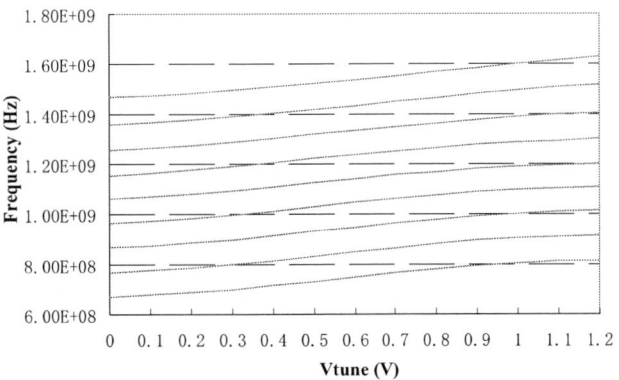

Fig. 4. Frequency tuning characteristics of proposed wideband VCO.

constant, so $C(n)$ and $Cv(n)$ of the nth sub-band can be derived from (2) and (3).

IV. SIMULATION RESULE

Proposed wideband VCO with little K_{VCO} and step variation has been designed and delivered taping-out in a 65nm 1-poly, 8-metal CMOS RF technology. The VCO layout is given in Fig.3.

Fig.4 shows the frequency tuning characteristics, and it is observed that K_{VCO} and step are almost constant. Fig.5 shows K_{VCO} variation of both proposed VCO and a conventional one (as Fig.1 shows) across different sub-bands when the tuning voltage Vtune is fixed at 0.6V. As Fig.5 presents, the variation of K_{VCO} is less than 17% while the one of conventional VCO is as large as 118%. Fig.6 gives phase noise of top sub-band (SW=00000000) and bottom sub-band (SW=11111111). Fig.7 shows phase noise of proposed VCO across different sub-bands when Vtune is fixed at 0.6V, in which phase noise is also of little variation. Besides, the proposed VCO exhibits good phase noise performance: -72.4 dBc/Hz at 10 kHz offset and -122.2 dBc/Hz at 1 MHz offset from the carrier, while consuming 2mA from a 1.2 V supply.

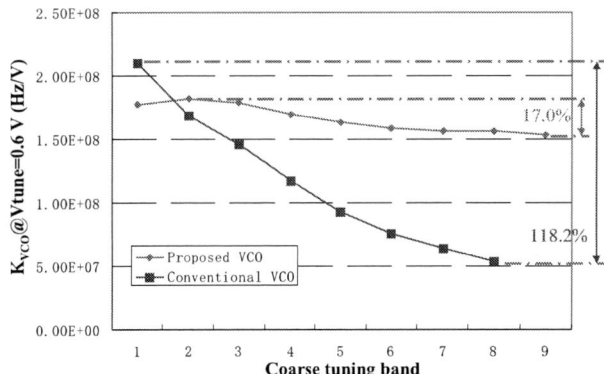

Fig. 5. K_{VCO} variation of proposed VCO and conventional VCO across different sub-bands when Vtune is fixed at 0.6V.

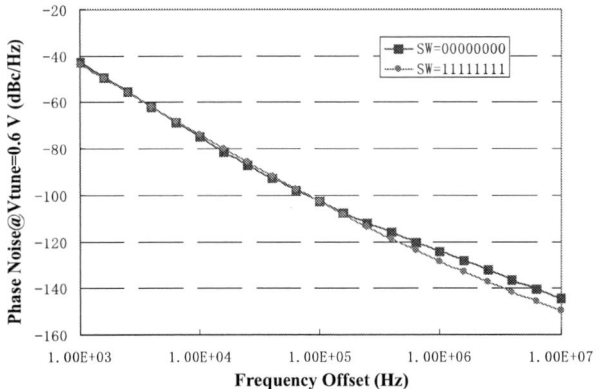

Fig. 6. Phase noise of top and bottom sub-band of proposed VCO when Vtune is fixed at 0.6V.

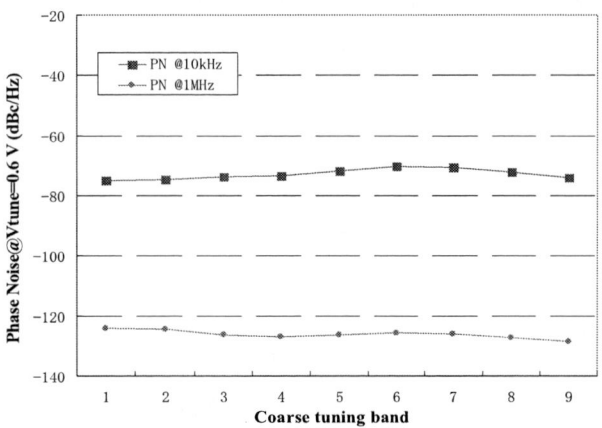

Fig. 7. Phase noise of each sub-band of proposed VCO when Vtune is fixed at 0.6V.

Some previous works in K_{VCO} variation control as well as this work are summarized in Table I [7]–[10], in which FoM represents the widely used VCO figure-of-merit given by

$$FoM = L(\Delta f) - 20 \cdot \log\left(\frac{f_{OSC}}{\Delta f}\right) + 10 \cdot \log\left(\frac{P_{diss}}{1 \text{ mW}}\right) \quad (4)$$

where L is phase noise, Δf is the offset frequency, f_{OSC} is the oscillation frequency, and P_{diss} is the power dissipation.

TABLE I
PERFORMANCE COMPARISON WITH OTHER WIDE-BAND VCOS

Ref.	[7]	[8]	[9]	[10]	This work*
fosc [GHz]	5.2	1.7	1.85	6.0	1.5
ΔK_{VCO} [%]	9.56	69.5	54.4	57.5	17.0
Tuning Range [%]	18.0	63.1	66.7	5.1	83.4
Phase Noise [dBc/Hz]	-113.7 @1M	-128 @1.25M	-127.1 @1M	-115.2 @1M	-124.3 @1M
Power Diss. [mW]	9.7	14.0	10.8	12.5	2.4
FoM [dBc/Hz]	-180.0	-179.2	-181.6	-179.8	-184.0
Tech. [nm]	180	180	180	130	65

* Based on post-layout simulation.

V. CONCLUSION

A wideband CMOS LC VCO employing a switched capacitor-varactor array controlled by temperature weighted coding has been proposed. With additional switched varactor array, small variation in VCO gain (K_{VCO}) across different sub-bands is achieved. Implemented in 65 nm RF CMOS technology, the proposed VCO exhibits about 17% variation in the K_{VCO} while the frequency tuning range is from 681 MHz to 1656 MHz. By optimizing the switched capacitor array, step between two adjacent sub-bands is kept equal, which helps to improve AFC efficiency. The VCO phase noise is -73 dBc/Hz at 10 kHz offset and -124.3 dBc/Hz at 1 MHz offset from the carrier while consuming 2 mA from a 1.2V supply. And the VCO figure-of-merit (FoM) of -184.0dBc/Hz is achieved.

REFERENCES

[1] J. W. M. Rogers, J. A. Macedo, and C. Plett, "The effect of varactor nonlinearity on the phase noise of completely integrated VCOs," *IEEE J. Solid-State Circuits*, vol. 35, no. 9, pp. 1360–1367, Sep. 2000

[2] K. Manetakis, D. Jessie and C. Narathong, "A CMOS VCO with 48% tuning range for modern broadband systems," *IEEE Custom Integrated Circuits Conference 2004*.

[3] Z. Li and K. K. O, "A low-phase-noise and low-power multiband CMOS voltage-controlled oscillator," *IEEE J. Solid-State Circuits*, Vol. 40, No. 6, pp. 1296 – 1302, June 2005.

[4] D. Hauspie, E.-C. Park, and J. Craninckx, "Wide-band VCO with simultaneous switching of frequency band, active core, and varactor size," *IEEE J. Solid-State Circuits*, vol. 42, no. 7, pp. 1472–1480, Jul. 2007.

[5] S. S. Broussev, T. A. Lehtonen, and N. T. Tchamov, "A wide-band low phase-noise LC-VCO with programmable Kvco," *IEEE Microw. Wireless Compon. Lett.*, vol. 17, no. 4, pp. 274–276, Apr. 2007.

[6] Ali Hajimiri and Thomas H. Lee, "Design issues in CMOS differential LC oscillators," *IEEE J. Solid-State Circuits*, vol. 34, no. 5, pp. 717–724, May. 1999.

[7] Y. J. Moon, Y. S. Roh, C. Y. Jeong, and C. Yoo, "A 4.39–5.26 GHz LC-tank CMOS voltage-controlled oscillator with small VCO-gain variation," *IEEE Microw. Wireless Compon. Lett.*, vol. 19, no. 8, pp. 524–526, Aug. 2009.

[8] E. Y. Sung, K. S. Lee, D. H. Baek, Y. J. Kim, and B. H. Park, "A wideband 0.18-um CMOS ΣΔ fractional-N frequency synthesizer with a single VCO for DVB-T," in *IEEE Asian Solid-State Circuits Conf. Dig.*, Nov. 2005, pp. 193–196.

[9] J. Kim, J. Shin, S. Kim, and H. Shin, "A wide-band CMOS LC VCO with linearized coarse tuning characteristics," *IEEE Trans. Circuits and Systems—II: Express Briefs*, vol.55, pp. 399–403, May. 2008.

[10] L. Jia, Y. B. Choi, and W. G. Yeoh, "A 5.8-GHz VCO with precision gain control," in *IEEE RFIC Symp. Dig.*, Jun. 2007, pp. 701–704.

978-1-61284-863-1/11 $26.00 © 2011 IEEE

A Low Power S-band Receiver Using GaAs pHEMT Technology

Yangyang Peng[1], Xiaoying Wang[2], Fangyue Ma[3] and Wenquan Sui[4*]

Nanoelectronic Platform, Zhejiang-California Nanosystems Institute, Zhejiang University.

[1]*yy.peng@live.com*   [2]*xiaoying8581@gmail.com*
[3]*marilu@yahoo.cn*   [4]*goldenglobe99@hotmail.com*

Abstract—**A two-chip S-band receiver has been designed and demonstrated in this paper. The proposed receiver is composed by a two stage low-noise amplifier (LNA) and a resistive type mixer. To achieve wide operation bandwidth as well as low noise figure, the LNA uses wideband matching network and negative feedback technique. Measured results show the LNA obtains a minimum noise figure of 2.4 dB with 17 dB gain. The input and output return loss all exceed -10 dB across the working frequency. To obtain high linearity performance, a resistive type mixer is adopted in this design. With simulation, the mixer shows a conversion loss of 7.2 dB. The isolations among RF, LO and IF ports are all greater than 20 dB. The power consumption of this receiver is 33 mW.**

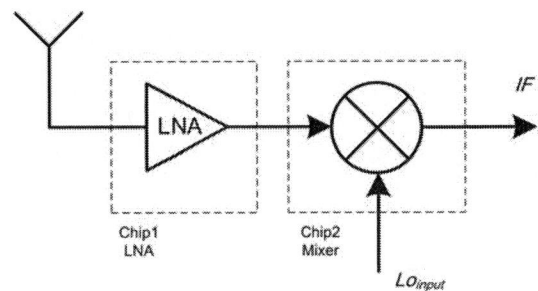

Fig. 1.   Schematic of the S-band receiver.

I. INTRODUCTION

Receivers are critical components in wireless system. It is used to amplify the weak signal that received by the antenna and convert it to a low frequency. They are the first blocks of the systems in most of communication architectures. The S-band has long been an application space servicing both commercial and military radars, which requires high performance and reliability of the circuits. Therefore, high-performance microwave receivers with low noise, low power and high linearity are required.

The high yield and reliability make GaAs MMIC a reliable technology and has been consequently widely adopted for realization of circuits and subsystems for microwave and millimeter wave applications. Though some new processes and technology of III-V compounds have been developed to provide better performance, the yield is still a problem and not reliable for commercial use [1][2]. Although higher gain GaN devices are starting to make a significant impact in wireless applications with better power performance, the cost and power consumption are still problems for its wide appication. The silicon substrate technology such as CMOS and BiCMOS process have achieved cut-off frequencies over 300 GHz. However the critical components of the systems, such as the ultra low noise wideband amplifier, high power amplifier and switch, still need to be developed with GaAs technology due to its good performance and high reliability.

There are some previous woks published to design low noise amplifiers and mixers around S-band and C-band in recent publications targeting for low noise or low voltage applications [4-10]. In this paper, a low power, high gain LNA MMIC is implemented in a commercial 0.5-μm AlGaAs/GaAs pHEMT

technology and achieves 2.4-3 dB noise figure in 2.5-5 GHz frequency range. The gain of the LNA is greater than 17 dB through the working frequency. The output P_{1dB} is 2.3 dBm. The resistive type mixer is also designed using pHEMT technology. Simulation results shows the mixer obtains a conversion loss of 7.2 dB with isolation among RF, LO and IF port greater than 20 dB. The return loss of RF and IF ports is also designed to be better than -10 dB to eliminate the mismatch problems between the mixer and LNA.

II. CIRCUIT DESIGN

A. Low-Noise Amplifier

The schematic of the LNA is shown in Fig. 2. The first stage has been designed to obtain the low noise figure of the circuit. Utilizing source degeneration technical with common source topology, the input return loss and noise match can be achieved simultaneously. The central frequencies of the two stages have been designed to have a little different to get a wideband frequency response. The input and output are all matched to 50 Ω through the matching network. Lossy matches are placed at the output to make a stable gain as well as not to deteriorate the noise figure.

To design a low power amplifier, devices of smaller size are preferred because of the less power consumptions. Although smaller devices also have another advantage of higher maximum available gain, the linearity will be degraded. In this work, the total gate width of the first stage is 200 μm and it is biased at $V_{DD} = 1.5\ V, V_G = 0.7\ V$, with $I_{DD} = 11\ mA$. To make the common source stage more stable and easily achieve noise match and input match simul-

978-1-61284-863-1/11 $26.00 © 2011 IEEE

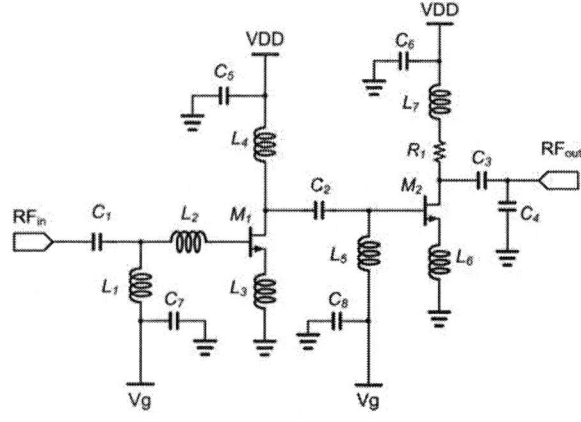

Fig. 2.   Schematic of the LNA.

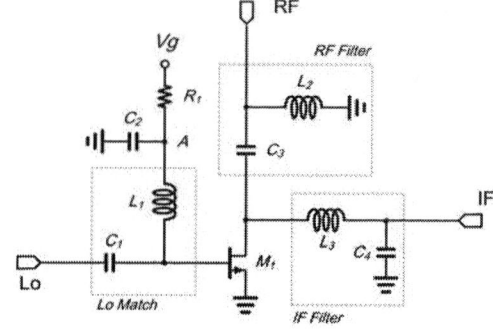

Fig. 4.   Schematic of the mixer.

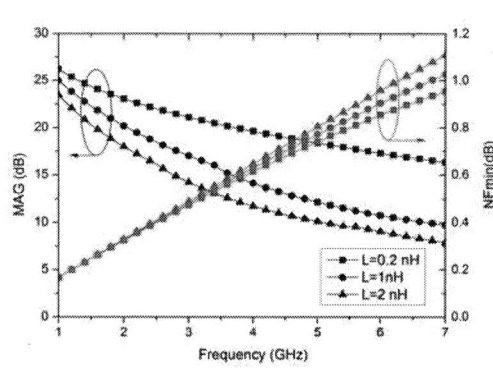

Fig. 3.   Simulation of MAG/MSG and NF_{min} of the first stage with various inductor value..

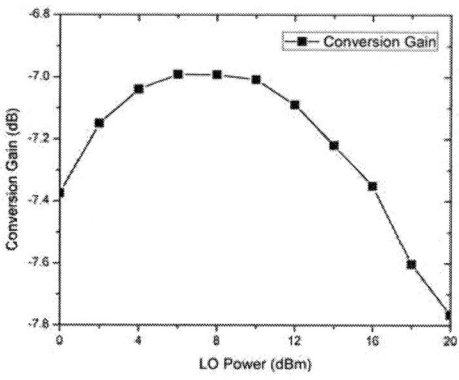

Fig. 5.   Conversion gain *vs.* LO power. (LO=3 GHz, RF=3.5 GHz.)

taneously, the source degeneration inductor must be selected carefully. With proper selection of L_s, the matching of S_{opt} and S_{11} can be achieved simultaneously. Besides, the stability factor, maximum available gain and minimum noise figure are also needed to be taken into considerations when selecting the inductor value. Fig. 3 shows the simulation results of MAG/MSG and NF_{min} of the first stage with various source degeneration inductor values. When $L_s = 2\ nH$, the amplifier cell is stable above 1 GHz and the minimum noise figure is also reduced, but the MAG/MSG is decreased by 10 dB comparing with simple common source cell. In this design, the source degeneration inductor is selected to be 1 nH. The amplifier cell is conditionally stable in the working frequency, matching networking are selected to keep the amplifier away from the unstable region.

The noise contribution of the second stage is minimized by matching the FET input near optimum noise match Γ_{opt}. And source degeneration inductor also placed at the source to move the conjugate of the input reflection coefficient towards Γ_{opt}. The total gate width of the second stage is 200 μm and it is biased at $V_{DD} = 1.5\ V, V_G = 0.7\ V$, with $I_{DD} = 11\ mA$. To further ensure the stability of the amplifier, lossy matches are utilized at the output of the third stage. The output port is

matched to 50 Ω for the on chip measurement.

B. Mixer

There are many types of mixers in microwave technology. The resistive mixer has good linearity and is capable of high output power at moderate LO levels. In spite of having conversion loss instead of conversion gain, resistive mixer have many advantages such as high linearity and superior intermodulation properties [11][12]. In this design, a resistive FET mixer topology is adopted.

The schematic of the mixer is shown in Fig. 4. The FET M_1 is operated at zero DC drain voltage. The LO power is applied at the gate. With the help of C_1 and L_1, a good input match at LO port is obtained. C_2 is a bypass capacitor, it makes the node A as a RF ground point. The value of the gate bias resistor R_1 is 2 KΩ. L_2 and C_3 compose the high-pass RF filter at RF port. Meanwhile, L_3 and C_4 compose the low-pass IF filter at IF port. The isolations can be enhanced with these filters.

The performance of the mixer has good relationship with the power level of the LO. As shown in Fig. 5. When the LO power is 8 dBm, the mixer gets a peak conversion gain of -7.0 dBm. It is a moderate LO power level and can be easily obtained by LO driver.

978-1-61284-863-1/11 $26.00 © 2011 IEEE

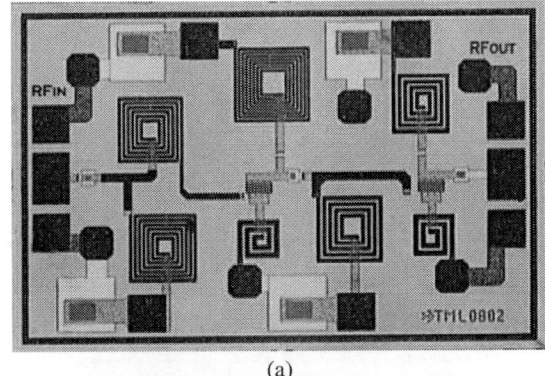

(a)

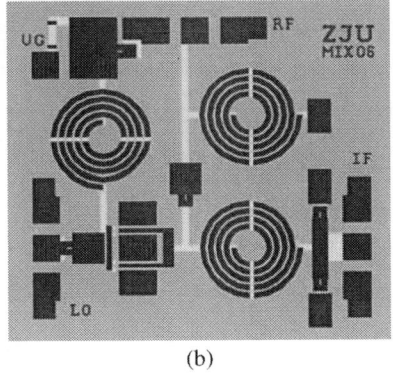

(b)

Fig. 6. Chip photo and layout of the LNA and mixer.

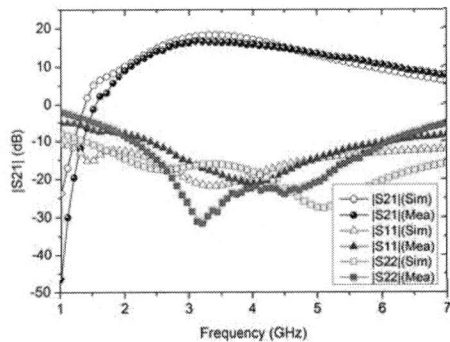

Fig. 7. Simulated and measured small signal gain and return losses of the LNA.

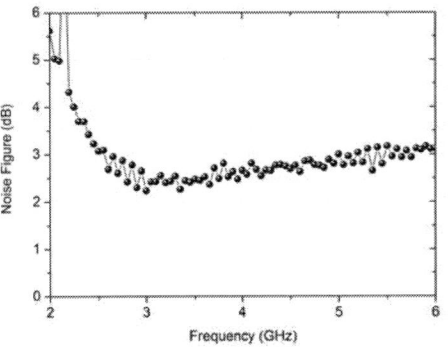

Fig. 8. Measured noise figure of the LNA.

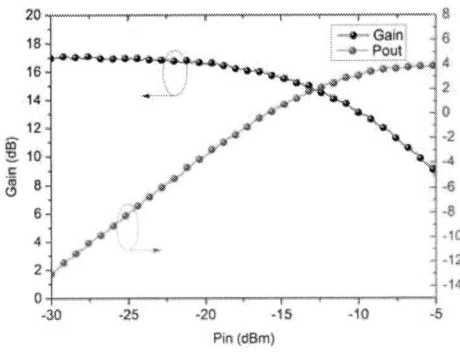

Fig. 9. Measured power performance of the LNA at 4 GHz.

III. LAYOUT CONSIDERATIONS

For the LNA, the DC biases are fed through RF chock inductors. The RF chock inductor values are properly selected to share with the matching networks. The bypass capacitors are carefully designed as well. For the mixer, the inductors are separated enough to make sure there is little coupling among them. To realize the parasitic and coupling effects of the circuits, a commercial electromagnetic (EM) simulator is utilized to analysis the circuit. And the characters of the inductors in this design have been checked with measurement data to design the circuits accurately. Yield analysis was also utilized to keep the circuit not sensitive to process variation. Ground-Signal-Ground (GSG) pads are used for on-wafer measurement. The dimensions of the LNA and mixer chips are $1.5 \times 1 \ mm^2$ and $0.9 \times 0.9 \ mm^2$ individually. The layout of the chips are shown in Fig. 6 .

IV. RESULTS AND DISCUSSION

Performance of the LNA was measured on-wafer using Anritsu 37397D VNA and Agilent 8975A noise measurement equipment. Wafer-level calibration was performed to take cable and probe losses into account. The DC biases are fed using bond wires connected to a PCB test board.

The measured small signal gain and return losses of the LNA are shown in Fig. 7. The measured data fitted the simulated one very well. The average gain is 17 dB with flatness of 1.6 dB from 2.5 GHz to 5 GHz. Input and output return losses over this band are less than 10 dB. The DC

bias of this condition is $V_{DD} = 1.5 \ V, V_G = 0.7 \ V$, with $I_{DD} = 22 \ mA$. Fig. 8 illustrates the measured noise figure. The LNA has a noise figure of 2.4 to 3 dB from 2.5 to 5 GHz when the bias current is 33 mA.The power performances of the LNA are shown in Fig. 9 . The LNA has an output P_{1dB} of 2.3 dBm with total power consumption of 33 mW.

The simulation results of the mixers are shown as bellows. Fig. 10 is the conversion gain under different RF frequencies.

978-1-61284-863-1/11 $26.00 © 2011 IEEE

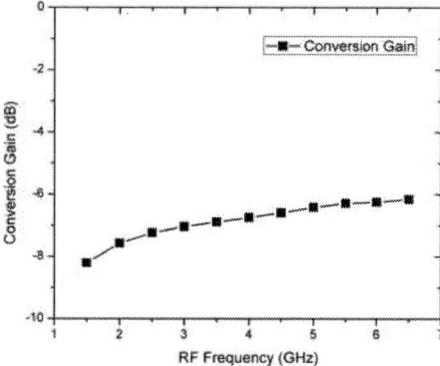

Fig. 10. Simulated conversion gain of the mixer. (Fix IF=0.5 GHz, Sweep LO and RF.)

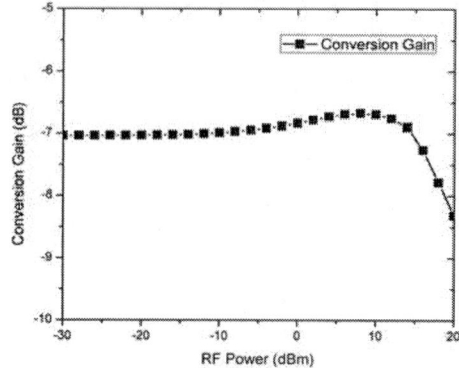

Fig. 12. Simulated power performance of the mixer. (IF=0.5 GHz, RF=3 GHz.)

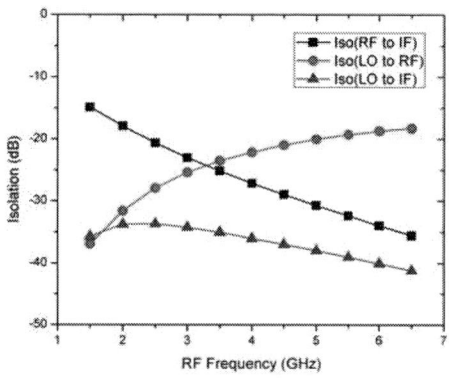

Fig. 11. Simulated isolation of the mixer. (Fix IF=0.5 GHz, Sweep LO and RF.)

The conversion gain is bigger than -8 dB when RF frequency sweep from 2 GHz to 6 GHz. As shown in Fig. 12, the isolation is smaller than -20 dB across the working band. The input P_{1dB} of the mixer is 18 dBm.

V. CONCLUSION

An S-band receiver has been designed with a LNA and mixer chip set. The LNA has been fabricated using 0.5-μm enhanced mode AlGaAs/GaAs pHEMT technology. It achieves good performance with 17 dB gain and 2.4 to 3 dB noise figure from 2.5 to 5 GHz. The chip occupies 1.5 mm^2 and dissipates 33 mW with 1.5 V power supply. The mixer is based on FET resistive topology. With a LO power of 8 dBm, simulated results show the mixer has a conversion gain greater than -8 dB when IF fixed at 0.5 GHz and RF swept from 2 GHz to 7 GHz. The isolation among RF, LO and IF ports are greater than 20 dB. The size of the mixer layout is 0.8 mm^2.

ACKNOWLEDGMENT

This work is partially supported by the Natural Science Foundation under grant number 60971058.

REFERENCES

[1] M.S. Heins, *et al*, "X-band GaAs mHEMT LNAs with 0.5 dB noise figure. ", *Microwave Symposium Digest, 2004 IEEE MTT-S International*, pp. 149-152, 2004.

[2] V.G. Mokerov, *et al*, " X-band MMIC Low-Noise Amplifier Based on 0.15 um GaAs Phemt Technology", *International Crimeam Conference*, pp.77-78, 2007.

[3] S.E. Rosenbaum, *et al*, "A 2-GHz three-stage AlInAs-GaInAs-InP HEMT MMIC low-noise amplifier", *IEEE Microwave and Guided Wave Letters*, vol. 3, no. 8, pp. 265-267, 1993.

[4] M. Soyuer, J.-O. Plouchart, H. Ainspan, and J. Burghartz, "A 5.8 GHz 1-V low noise amplieer in SiGe bipolar technology, *IEEE Radio Frequency Integrated Circuits Symp. Dig.*, pp. 19C22, 1997.

[5] B.G. Choi, Y.S. Lee, C.S. Park, K.S. Yoon, "A low noise on-chip matched MMIC LNA of 0.76 dB noise figure at 5 GHz for high speed wireless LAN applications", *Gallium Arsenide Integrated Circuit*, pp. 143-146, 2000.

[6] H.-A. Zulfa, Y.H. CHow, Y.W. Eng, "A low-voltage, fully-integrated 1.5-6 Ghz low noise amplifier in e-mode pHEMT technology for multiband, multimode applications", *European Microwave Intetrated Circuit Conference*, pp. 306-309, 2008.

[7] H. Huang, H.Y. Zhang, J.J. Yin, T.C. Ye, "Enhancement Mode pHEMT LNA with Super Low Noise and High Gain for S Band Application", *Solid-State and Integrated Circuit Technology*, pp. 947-976, 2007.

[8] B.G. Choi, Y.S. Lee, K.S. Yoon, H.C. Seo, C.S. Park, "Low noise pHEMT and its MMIC LNA implementation for C-band applications". *Microwave and Millimeter Wave Technology, International Conference on*, pp. 56-59, 2000.

[9] F. Ellinger, U. Lott, W. Bachtold, "Ultra low power GaAs MMIC low noise amplifier for smart antenna combining at 5.2 GHz", *Radio Frequency Integrated Circuits (RFIC) Symposium*, pp. 157-159, 2000.

[10] J. Wang, Y. Cen, X. Chen, "S-band PHEMT monolithic frequency variable receiver front-end". *International Conference on Microwave and Millimeter Wave Technology Proceedings*, pp. 234-237, 1998.

[11] S. A. Maas, "A GaAs MESFET Mixer with Very Low Intermodulation". *IEEE Transactions on Microwave Theory and Techniques*. pp. 425-429, vol. 35, no. 4. 1987.

[12] B. M. Motlagh, S. E. Gunnarsson, M. Ferndahl, and H. Zirath, "Fully Integrated 60-GHz Single-Ended Resistive Mixer in 90-nm CMOS Technology". *IEEE Microwave and Wireless Componets Letters*. pp. 25-28, vol. 16, no.1. 2006.

[13] Y. Peng, K. Lu, W. Sui, "A 7-to 14-GHz GaAs pHEMT LNA with 1.1 dB noise figure and 26 dB gain". *Microwave and Optical Technology Letters*. pp. 2615-2617, vol. 52, issue. 11. 2010.

Design of an X-band Broad-band Lumped-element Quadrature Hybrid

Fangyue Ma, Yangyang Peng, Xiaoying Wang, Saier Liu, Jin Lan, and Wenquan Sui*

Zhejiang-California International Nanosystem Institute, Electronic Platform, Zhejiang University

Hangzhou, 310029 P.R. China

goldenglobe99@hotmail.com

Abstract—A method for synthesizing a two-section inductively-coupled lumped-element quadrature hybrid for broad-band applications is first proposed. Based on this analysis, a two-section inductively-coupled lumped-element quadrature hybrid operated at X-band is presented. The measured magnitude imbalance and phase difference are 2.4dB and 90°± 4° over the frequency band of 8-12GHz, respectively. Results show good agreement with the theoretical predictions.

Keywords-Lumped-element circuit, SiGe, broad-band quadrature hybrid.

I. INTRODUCTION

QUADRATURE hybrids are a special case of the usual directional coupler whose coupling is 3dB. In many radio frequency (RF) and microwave circuits such as balance amplifiers and image rejection mixers, quadrature hybrids are used to obtain the desired circuit performance. They are typically implemented by using resonant quarter-wavelength transmission lines as their building block. Examples are branch-line couplers and Lange couplers [1], [2]. However, at frequencies below 20GHz, a transmission line structure will occupy too much valuable chip area [3]. Therefore, various lumped-element quadrature hybrids are proposed for microwave integrated circuit (MIC) or monolithic microwave integrated circuit (MMIC) applications.

Lumped-element quadrature hybrid is a symmetric four-ports network whose port orders are shown in Fig. 1. As usual, lumped-element quadrature hybrids are implemented by single section capacitively-coupled (co-directional) or inductively-coupled (contra-directional) structures, as shown in Fig. 2 [4]-[9]. Although the single section structures achieve a good isolation and voltage standing-wave ratio(VSWR), the relatively narrow-band characteristics limit their scope of applications. This paper presents a new set of design equations which are proposed to realize the hybrid with good balance in

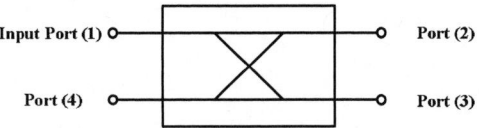

Fig. 1. Lumped-element quadrature hybrid symbol showing port order.

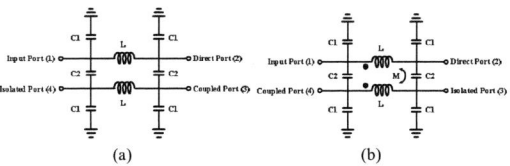

Fig. 2. Conventional single section lumped-element quadrature hybrid: (a) capacitively-coupled hybrid, and (b) inductively-coupled hybrid.

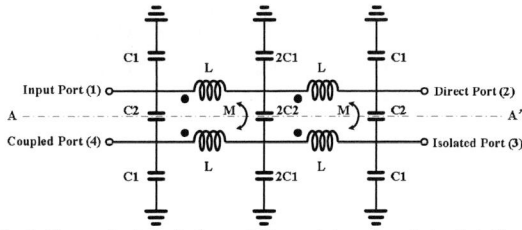

Fig. 3. Two-section inductively-coupled lumped-element quadrature hybrid.

magnitude and phase for the first time. The commercial electromagnetic (EM) simulator (Momentum) is utilized to analyze the frequency properties of the designed hybrid network. The simulation results show less than 1.6dB magnitude imbalance and 90°± 1.16° phase shift between the direct and coupled ports over frequency band of 8-12GHz. In order to verify the design concept, the lumped topology is implemented using the JAZZ SiGe SBC18QTD process. Due to the parasitic effects, the measurement results show 2.4dB magnitude imbalance and 90°± 4° phase shift but still match the simulation results well.

978-1-61284-863-1/11 $26.00 © 2011 IEEE

II. ANALYSIS METHODOLOGY

The circuit of the two-section inductively-coupled lumped hybrid in Fig. 3 is symmetrical about the AA'-plane, so it can be analyzed by the conventional even- and odd-mode analysis method. If all the ports of hybrid are assumed to be terminated at impedance Z_0 (usually $Z_0 = 50\Omega$), the transmission wave matrix of the even- and odd-mode half-circuits is given by

$$\left[T_{e,o}\right] = \begin{bmatrix} T_{11e,o} & T_{12e,o} \\ T_{21e,o} & T_{22e,o} \end{bmatrix}$$

and the elements of the transmission matrix are

$$T_{11e} = 1 - 2X_e B_e \left(2 - X_e B_e\right) + j\left(1 - X_e B_e\right)$$
$$\times \left[X_e Y_0 + B_e Z_0 \left(2 - X_e B_e\right)\right] \tag{1a}$$

$$T_{11o} = 1 - 2X_o B_o \left(2 - X_o B_o\right) + j\left(1 - X_o B_o\right)$$
$$\times \left[X_o Y_0 + B_o Z_0 \left(2 - X_o B_o\right)\right] \tag{1b}$$

$$T_{12e} = -j\left(1 - X_e B_e\right)\left[X_e Y_0 - B_e Z_0 \left(2 - X_e B_e\right)\right] \tag{2a}$$

$$T_{12o} = -j\left(1 - X_o B_o\right)\left[X_o Y_0 - B_o Z_0 \left(2 - X_o B_o\right)\right] \tag{2b}$$

$$T_{21e,o} = -T_{12e,o} \tag{3}$$

$$T_{22e} = 1 - 2X_e B_e \left(2 - X_e B_e\right) - j\left(1 - X_e B_e\right)$$
$$\times \left[X_e Y_0 + B_e Z_0 \left(2 - X_e B_e\right)\right], \tag{4a}$$

$$T_{22o} = 1 - 2X_o B_o \left(2 - X_o B_o\right) - j\left(1 - X_o B_o\right)$$
$$\times \left[X_o Y_0 + B_o Z_0 \left(2 - X_o B_o\right)\right], \tag{4b}$$

where Y_0, B_e, and B_o equal $1/Z_0$, ωC_1, and $\omega\left(C_1 + 2C_2\right)$, respectively; X_e and X_o are equivalent to $\omega\left(L + M\right)$ and $\omega\left(L - M\right)$, respectively [10].

According to the method of synthesis proposed in [4], the following conditions must be satisfied for constructing a contra-directional ($S_{31} = 0$) quadrature hybrid:

$$T_{11e} = T_{11o} \tag{5}$$

$$T_{12e} = -T_{12o} \tag{6}$$

$$\mathrm{Re}\left(T_{12e}\right) = \mathrm{Re}\left(T_{12o}\right) = 0. \tag{7}$$

In respect that T_{12e} and T_{12o} are pure imaginary numbers, condition (7) is satisfied definitely here. Substituting (1a) and (1b) into (5) yields the following conditions:

$$X_e B_e \left(2 - X_e B_e\right) = X_o B_o \left(2 - X_o B_o\right) \tag{8}$$

$$\left(1 - X_e B_e\right)\left[X_e Y_0 + B_e Z_0 \left(2 - X_e B_e\right)\right]$$
$$= \left(1 - X_o B_o\right)\left[X_o Y_0 + B_o Z_0 \left(2 - X_o B_o\right)\right] \tag{9}$$

and another equation obtained from (2a), (2b), and (6) can be

express as

$$\left(1 - X_e B_e\right)\left[X_e Y_0 - B_e Z_0 \left(2 - X_e B_e\right)\right]$$
$$= -\left(1 - X_o B_o\right)\left[X_o Y_0 - B_o Z_0 \left(2 - X_o B_o\right)\right]. \tag{10}$$

Adding (9) to (10) gives

$$\left(1 - X_e B_e\right)X_e Y_0 = \left(1 - X_o B_o\right)B_o Z_0 \left(2 - X_o B_o\right) \tag{11}$$

and subtracting (9) from (10) gives

$$\left(1 - X_e B_e\right)B_e Z_0 \left(2 - X_e B_e\right) = \left(1 - X_o B_o\right)X_o Y_0. \tag{12}$$

Cross multiply (11) and (12) together and we can get

$$X_e Y_0^2 X_o = B_o Z_0^2 B_e \left(2 - X_e B_e\right)\left(2 - X_o B_o\right). \tag{13}$$

Thus

$$X_e^2 Y_0^2 X_o = X_e Y_0^2 X_o \times X_e$$
$$= B_o Z_0^2 X_e B_e \left(2 - X_e B_e\right)\left(2 - X_o B_o\right)$$
$$= Z_0^2 B_o^2 \left(2 - X_o B_o\right)^2 X_o \tag{14}$$

which results in

$$X_e Y_0 = \pm Z_0 B_o \left(2 - X_o B_o\right). \tag{15}$$

When $X_e Y_0 = -Z_0 B_o \left(2 - X_o B_o\right)$, based on (8), (11), and (12), the equations about B_e, B_o, X_e, and X_o can be obtained as

$$X_e Y_0 = -Z_0 B_o \left(2 - X_o B_o\right) \tag{16}$$

$$X_o Y_0 = -Z_0 B_e \left(2 - X_e B_e\right) \tag{17}$$

and

$$X_e B_e + X_o B_o = 2. \tag{18}$$

Since B_e, B_o, X_e, and X_o must be greater than zero,

$$2 - X_o B_o = X_e B_e > 0 \tag{19}$$

and

$$-Z_0 B_o \left(2 - X_o B_o\right) < 0. \tag{20}$$

However

$$X_e Y_0 > 0. \tag{21}$$

As a result, it's impossible $X_e Y_0 = -Z_0 B_o \left(2 - X_o B_o\right)$. The reasonable solution can only come from the situation when $X_e Y_0 = Z_0 B_o \left(2 - X_o B_o\right)$. Based on (8), (11), and (12) again, we can gain the relationship between B_e, B_o, X_e, and X_o as

$$X_e Y_0 = Z_0 B_o \left(2 - X_o B_o\right) \tag{22}$$

$$X_o Y_0 = Z_0 B_e \left(2 - X_e B_e\right) \tag{23}$$

and

$$X_e B_e = X_o B_o, \tag{24}$$

where only $X_e B_e < 2$, $X_o B_o < 2$, and $X_e X_o < Z_0^2$ can be used to guarantee that B_e, B_o, X_e, and X_o will be real numbers and greater than zero.

Obviously, four unknowns must be determined from the above equations. Hence, the values of elements shown in Fig. 3 can be acquired for arbitrary given values of X_e and X_o. Then, a microwave circuit simulator can be used to evaluate the characteristics of the hybrid constructed by the values of the elements obtained above.

Five sets of 10GHz quadrature hybrid elements' values and their performances corresponding to various values using ideal lumped elements are listed in Table I. Rows 7 and 8 of the Table I show the maximum magnitude imbalance and phase difference

TABLE I
PERFORMANCE AND ELEMENTS' VALUES FOR CONSTRUCTING 10GHz
IDEAL TWO-SECTION QUADRATURE HYBRID

X_e	90	90	85	60	60
X_o	20	15	20	20	10
L (nH)	0.875	0.836	0.836	0.637	0.557
M (nH)	0.557	0.597	0.518	0.318	0.398
C_1 (fF)	83.3	56.9	81.2	74.0	34.0
C_2 (fF)	146	142	132	74.6	84.9
Magnitude Imbalance (dB)	6	0.7	5.2	5.13	2.6
Phase Difference (Degree)	90 ± 0.151	90 ± 0.036	90 ± 0.119	90 ± 0.028	90 ± 0.003

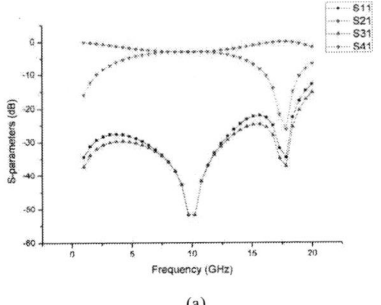

(a)

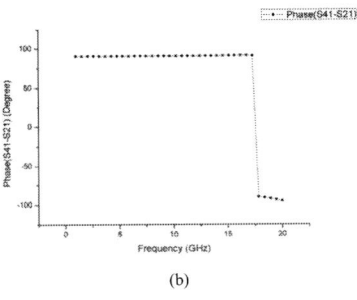

(b)

Fig. 4. Simulated characteristics of the quadrature hybrid constructed by ideal lumped elements: (a) S-parameters, and (b) phase difference between ports 2 and 4.

between the output ports at the frequency range of 8-12GHz, respectively. It is shown that the hybrid constructed by elements listed in Column 3 of Table I has the most broad-band performance. Fig. 4 presents the simulated characteristics of the quadrature hybrid consisting of ideal components with the values in Column 3 of Table I. As it shown, the hybrid displays very good performance at the frequency range from 8-12GHz. Therefore, the proposed analysis methodology is verified very well.

III. DESIGN AND SIMULATION

The hybrid performances in Table I reveal that the hybrid constructed by the elements listed in Column 3 yields most excellent characteristics. Based on the values of the elements in Column 3 of Table I, a two-section inductively-coupled quadrature hybrid with center frequency and frequency band of 10GHz and 8-12GHz respectively is designed and fabricated to verify the design concept. The lumped topology is implemented using the JAZZ SiGe SBC18QTD process with 4 metal layers. This technology supports metal-insulator-metal (MIM) capacitors and the inductively-coupled inductors can be made by the parallel

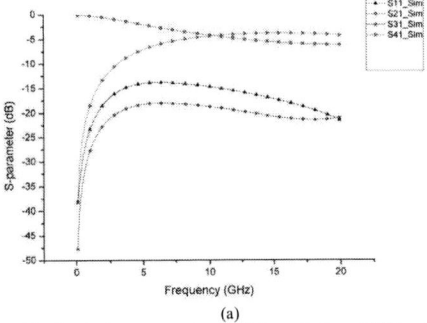

(a)

Fig. 5. Simulated frequency characteristics of the 10GHz two-section inductively-coupled lumped-element quadrature hybrid.

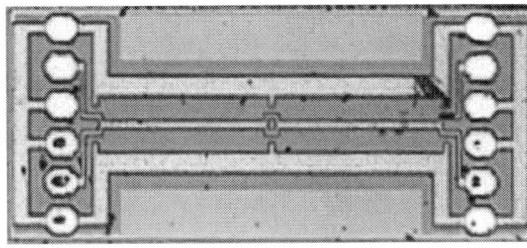

Fig. 6. Photograph of the 10GHz two-section inductively-coupled lumped- element quadrature hybrid. Circuit size (including ground ring and probe pads) is 1.69mm×0.52mm.

978-1-61284-863-1/11 $26.00 © 2011 IEEE

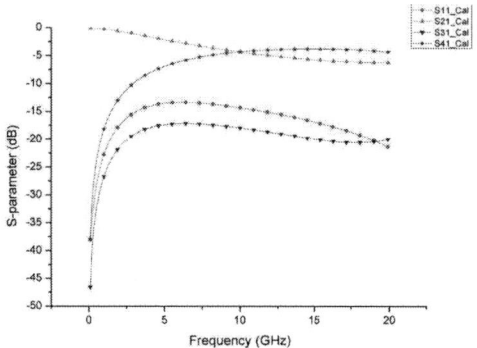

Fig. 7. Measured S-parameters of the 10GHz two-section inductively- coupled lumped-element quadrature hybrid.

coupled top metal microstrip lines.

To realize the desired inductively-coupled inductors, a commercial electromagnetic (EM) simulator (Momentum) is utilized to analyze various lengths and spaces of the parallel-coupled microstrip lines to get the precise self-inductances and mutual inductances. Then, we design the layout of quadrature hybrid using supported capacitors and simulation-obtained inductively-couple inductors. Finally, the entire layout is simulated by Momentum. Due to the parasitic effect of lumped elements, ground ring, and probe pads, the lumped elements usually need be slightly adjusted to improve the circuit performance.

Fig. 5 shows the final simulated characteristics of the two-section inductively-coupled quadrature hybrid. Over the frequency band 8-12GHz, the results show less than 1.6dB magnitude imbalance and 90°± 1.16° phase shift between the direct and coupled ports. The return loss is better than 13.5dB and the insert loss is less than 1.5dB. Fig. 6 shows the photograph of the circuit, which has a die size of 1.69mm × 0.52mm. The measured results of the quadrature hybrid are presented in Fig.7. The magnitude imbalance is 2.4dB and phase shift is 90°± 4°, while the insert loss and isolation remain the same as simulation results. Comparing Fig.5 and 7 reveals a very good agreement between the measurements and the predictions. The little deviation is because the parasitic parameter.

IV. CONCLUSION

This paper has first proposed a method for synthesizing a two-section inductively-coupled lumped-element quadrature hybrid for broad-band applications. Based on this analysis, a high-performance quadrature hybrid using at X-band is designed, simulated and fabricated. The measurement of the circuit shows the presented quadrature hybrid does have wider bandwidth and better performance than the conventional single section lumped-element hybrid. The proposed quadrature hybrid should be able to find applications in future MMICs.

ACKNOWLEDGEMENT

This work is partially supported by the National Science Foundation under grant number 60971058.

REFERENCES

[1] D. M. Pozar, *Microwave Engineering.* New York: Wiley, 1998, ch. 7.
[2] J. Lange, "Interdigitated stripline quadrature coupler," *IEEE Trans. Microwave Theory Tech.,* vol. MTT-17, pp. 1150-1151, Dec. 1969.
[3] R. W. Vogel, "Analysis and design of lumped- and lumped-distributed element directional couplers for MIC and MMIC applications," *IEEE Trans. Microwave Theory and Techniques,* vol. 40, pp. 253-262, Feb. 1992.
[4] F. Ellinger, R. Vogt, W. Bächtold, "Ultracompact reflective-type phase shifter MMIC at C-band with 360° phase-control range for smart antenna combining," *IEEE J. Solid-State Circuits,* vol. 37, pp. 481-486, April. 2002.
[5] R. C. Frye, S. Kapur, and R. C. Melville, "A 2-GHz quasrature hybrid implemented in CMOS," *IEEE J. Solid-State Circuits,* vol. 38, pp. 550-555, March. 2003.
[6] D. Ozis, J. Paramesh, and D. J. Allstot, "Analysis and design of lumped-element quadrature couplers with lossy passive elements," in *Conf. Rec. 2006 IEEE Int. Conf. Circuits and Systems,* pp. 2317-2320.
[7] Y. J. Lee, and J. Y. Park, "Fully embedded CDMA Cellular-band lumped LC-quadrature hybrid coupler into organic package substrate," in *2007 Proc. Microelectronics Conf.,* pp. 81-84.
[8] Y. Zhou, and Y. Chen, "Lumped-element equivalent circuit models for distributed microwave directional couplers," in *Conf. Rec. 2008 IEEE Int. Conf. Microwave and Millimeter Wave technology,* pp. 131-134.
[9] I. Sakagami, M. Tahara, and M. Fuji, "Lumped-element type D branch couplers," in *2009 Proc. Microwave Conf.,* pp. 2656-2659.
[10] Y. C. Chiang, and C. Y. Chen, "Design of a wide-band lumped-element 3-dB quadrature coupler," *IEEE Trans. Microwave Theory and Techniques,* vol. 49, pp. 476-179, March. 2001.
[11] S. Liu, J. Lan, and W. Sui, "An X-band broad-band lumped-element quadrature hybrid implemented in SiGe technology," in *Conf. Rec. 2010 IEEE Int. Conf. on Communications, Circuits and Systems,* pp. 718-721.

A Ka-band MMIC Doherty Power Amplifier using GaAs pHEMT Technology

Xiaoying Wang[1], Yangyang Peng[2], Fangyue Ma[3] and Wenquan Sui[4*]
Nanoelectronic Platform, Zhejiang-California Nanosystems Institute, Zhejiang University.
[1]*xiaoying8581@163.com* [2]*yy.peng@live.com*
[3]*marilu@yahoo.cn* [4]*goldenglobe99@hotmail.com*

Abstract—A fully integrated Ka-band Monolithic microwave integrated circuit (MMIC) Doherty Power Amplifier (PA) is designed and demonstrated in this paper. The proposed Doherty PA maily consists of a Lange coupler, carrier amplifier branch, peaking amplifier branch and impedance transformer network. offset lines are introduced in each branch to overcome the inherent defects of conventional Doherty PA. Electromagnetic (EM) simulated results show the proposed Doherty PA obtains a small signal gain over 5.5dB from 31GHz to 35GHz with a compact die size of 2mm×1.7mm. Power added efficiency (PAE) is over 19.8% at 6dB back-off with saturated output power over 26dBm.

I. INTRODUCTION

In recent years, millimeter-wave (mm-Wave) wireless communication has attracted a great deal of attentions, although it is relatively new in the world of wireless communication [1][2]. Power amplifiers (PAs) with high efficiency and linearity play a great role in wireless communication systems [3]. At low frequency, Doherty PAs are commonly used to improve efficiency at high output power back-off, but most of them are hybrid [4]-[8]. Study of monolithic Doherty PAs at higher frequency is relatively less reported [9][10].

In this paper, a fully integrated Ka-band monolithic microwave integrated circuit (MMIC) Doherty PA is proposed. Fig.1 shows the schematic of the Doherty PA which mainly consists of a Lange coupler, carrier amplifier branch, peaking amplifier branch and impedance transformer network.

The carrier amplifier operates in Class AB while the peaking amplifier operates in Class C. The impedance transformer network is used to transform the 50Ω load to 25Ω, which is necessary for following Doherty PA design.

II. DESIGN OF MMIC KA-BAND DOHERTY PA

As shown in Fig.1, in the low power region, the peaking amplifier does not function for the small input signal, the carrier amplifier branch sees $2Ropt$ ($Ropt$: optimum load impedance, $Ropt = 50\Omega$ in this design) load, thus the carrier amplifier is saturated at a lower input power compared with a normal Class AB amplifier. Efficiency reaches maximum while output power is not saturated. In medium power region, the peaking amplifier begins to function, and the load impedance sees from the carrier branch decreases from $2Ropt$ to $Ropt$ because of the current supplied by the peaking amplifier. Efficiency is still in a high level while output power increases. In high power

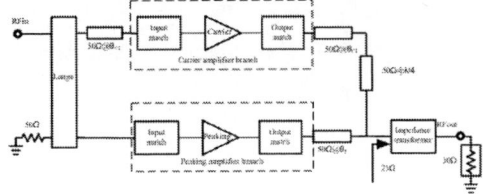

Fig. 1. Schematic of the Ka-band MMIC Doherty PA.

region, both the carrier amplifier and the peaking amplifier function, see load impedance of $Ropt$, output power of the whole circuit reaches its peak.

The active load modulation is the key point of the Doherty PA for obtaining high efficiency at output power back-off. The load of the carrier amplifier is modulated by the peaking amplifier through the λ/4 transmission line with 50Ω characteristic impedance which also compensates the 90° phase difference caused by the Lange coupler.

For conventional Doherty PA, there is a defect in the low power region. The output impedance of the peaking amplifier branch is not ideal open to the carrier branch. It causes output power of carrier branch leaking into the peaking branch. To solve this problem, an offset line with 50Ω characteristic impedance is introduced in many papers [4]-[7]. And in order to compensate the phase difference caused by this offset line, another offset line with the same length needs to be introduced in the carrier branch. In this paper, the offset line of carrier branch is divided into two lines, $\theta c1$ and $\theta c2$, as shown in Fig.1.

The proposed Ka-band MMIC Doherty PA is designed with $0.15um$ GaAs pHEMT MMIC process of Win Semiconductor Corp.. Transistors employed in carrier and peaking branch have identical sizes of 4finger×75um. To ensure stability of the Doherty PA, a paralleled resistor connected to ground structure is adopted in each branch. An appropriate resistance is selected for PA stability, but the gain of the amplifier will be decreased.

Both output match networks are designed to deliver maximum power with load impedance of $Ropt$. For the high operation frequency, the coupling effect and passive circuits were simulated by an Electromagnetic (EM) simulator. Momentum

978-1-61284-863-1/11 $26.00 © 2011 IEEE

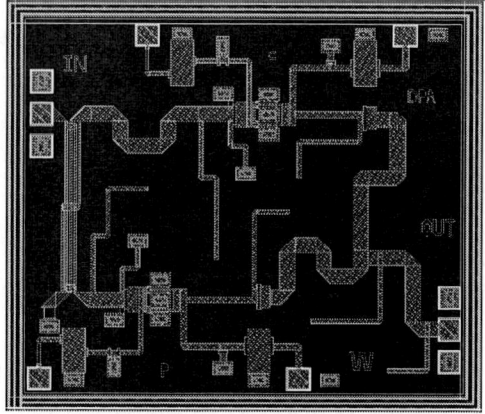

Fig. 2. Layout of the Doherty PA with size of 2mm×1.7mm.

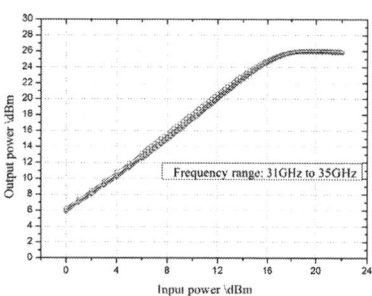

Fig. 3. Output power from 31GHz to 35GHz.

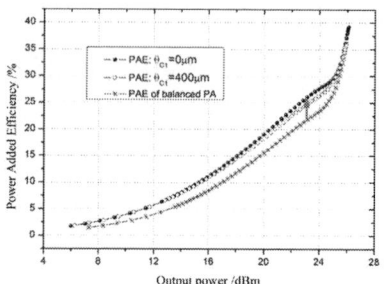

Fig. 4. PAE at 33GHz.

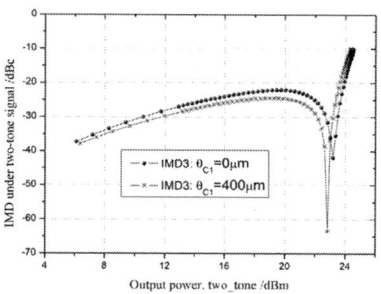

Fig. 5. IMD3 and IMD5 performance.

of ADS (Advanced Design System of Agilent Technologies) is used in this design. Fig.2 is the layout of the Doherty PA with size of 2mm×1.7mm.

III. SIMULATION AND RESULTS

Fig.3 shows the EM simulated output power versus input power from 31GHz to 35GHz. The saturated powers are over 26dBm.

In order to illustrate efficiency improvement of Doherty PA, a balanced PA is also simulated as a comparison. The balanced PA consists of a Lange coupler, two branches with the same structure of the carrier amplifier branch, and bias condition is same in all cases, another Lange coupler is used to combine these two branches. The PAE of the proposed Doherty PA and balanced PA is simulated and results are shown in Fig.4.

As can be seen from Fig.4, PAE of the Doherty PA with $\theta c1=0\mu m$ is higher than that of balanced PA and Doherty PA with $\theta c1=400\mu m$ at the same output power back-off. Although the theoretical efficiency plateau does not appear, the Doherty PA with $\theta c1=0\mu m$ has a PAE over 19.8% at 6dB back-off, and has its peak value 39.2% at saturated output power. Reference [4] discussed the reason of PAE

plateau disappearance. The conduction angle chosen for the peaking amplifier can significantly affect efficiency due to soft turn-on effect of peaking amplifier. And this choice also affects linearity of the Doherty PA. A trade-off between PAE plateau and linearity must be made though changing gate bias condition of peaking amplifier.

Fig. 5 and Fig.6 illustrate the linearity of the proposed Doherty PA. Gain of Doherty PA is less than that of balanced PA due to the nonfunction of peaking amplifier at low input signal. But the gain compression of Doherty PA occurs at higher output power. Third-order inter-modulation distortion (IMD3) characteristic using a two-tone signal with 1MHz tone spacing is simulated. IMD3 of Doherty PA with $\theta c1=400\mu m$ is 30dB better than Doherty PA with $\theta c1=0\mu m$ at output power 22.8 dBm, but the flatness of small signal gain from 31GHz to 35GHz are worse, as can been from Fig.7.The PAE of Doherty PA with $\theta c1=0\mu m$ is also better than Doherty PA with $\theta c1=400\mu m$ (Fig.4).

IV. CONCLUSION

A Ka-band MMIC Doherty PA is designed and demonstrated. The EM simulated results show it obtain a small signal gain over 5.5dB from 31GHz to 35GHz, and the saturated output power is about 26dBm. Simulation results show the

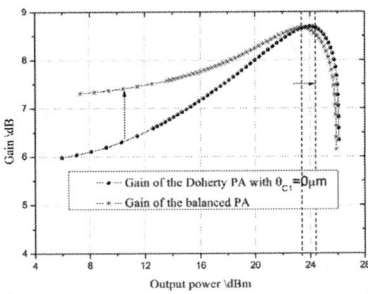

Fig. 6.   Gain versus output power at 33GHz.

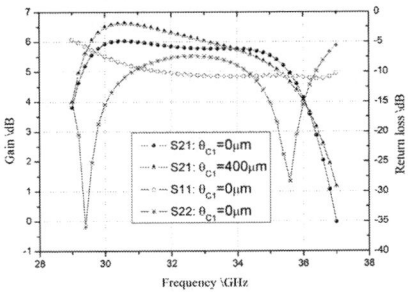

Fig. 7.   Gain and return loss.

[6] K.-J. Cho, W.-J. Kim, J.-H Kim and S. P. Stapleton, "Linearity optimization of a high power Doherty amplifier based on post-distortion compensation,", *IEEE Microw. Wireless Compon. Lett.*, vol. 15, no. 11, pp. 748-750, Nov., 2005

[7] B., Kim, J., Kim, I., Kim, J., Cha and S., Hong, "Microwave Doherty power amplifier for high efficiency and linearity,", *International Workshop on Integrated Nonlinear Microwave and Millimeter-Wave Circuits*, 2006.

[8] Cho, K. J., *et al*, "Gallium-nitride microwave Doherty power amplifier with 40*W* PEP and 68% PAE,", *Electronic Letters*, vol. 42, no. 12, pp. 704-705, June, 2006.

[9] C. F. Campbell, "A fully integrated Ku-band Doherty amplifier MMIC,", *IEEE Microw. Wireless Compon. Lett.*, vol. 9, no. 3, pp. 114-116, Mar. 1999.

[10] J. Tsai and T.Huang, "A 38-46GHz MMIC Doherty power amplifier using post-distortion linearization,", *IEEE Microw. Wireless Compon. Lett.*, vol.17, no.5, pp.388-390, May 2007.

PAE is higher than the balanced PA at the same output power back-off, over 19.8% at 6dB back-off with saturated output power over 26dBm.IMD3 decreased rapidly at output power back-off. The size of the Doherty PA layout is 2mm×1.7mm.

ACKNOWLEDGMENT

This work is partially supported by the Natural Science Foundation under grant number 60971058.

REFERENCES

[1] S.K. Yong and C.-C. Chong, "An overview of multigigabit wireless through millimeter wave technology: potentials and technical challenges,", *EURASIP Journal on Wireless Communications and Networking*, 2004.

[2] C.-C. Chong, K. Hamaguchi, P. F. M. Smulders, and S.-K. Yong, " Millimeter-wave wireless communication sysytems: theory and applications,", *EURASIP Journal on Wireless Communications and Networking*, 2007.

[3] S.C.Cripps, "RF Power Amplifier for Wireless Communications,", *Norwood, MA: Artech House*, 1999.

[4] Jangheon, Kim, Fehri, B, Boumaiza, S and Wood, J, "Power efficiency and linearity enhancment using optimized asymmetrical Doherty power amplifiers, *IEEE Trans. Microw. Theory Tech.*, vol.59, no.2, pp425-434, February, 2011.

[5] Darraji, R, Ghannouchi, F M and Hammi, O, "A dual-input digitally driven Doherty amplifier for performance enhancment of Doherty tansmitter,", *IEEE Trans. Microw. Theory Tech.*, vol.59, no.5, pp1284-1293, May, 2011.

An Energy-Efficient Supply Voltage Scheme using In-Situ Pre-Error Detection for on-the-fly Voltage Adaptation to PVT Variations

Martin Wirnshofer*, Leonhard Heiß* Georg Georgakos[†], Doris Schmitt-Landsiedel*,
* Technische Universität München, Munich, Germany
Email: wirnshofer@tum.de
[†]Infineon Technologies AG, Neubiberg, Germany

Abstract—**The presented Pre-Error voltage scheme dynamically tunes the supply voltage of digital circuits, according to PVT variations. By exploiting unused timing margin, produced by state-of-the-art worst-case designs, power consumption is minimized. Pre-Error flip-flops detect late-arriving signals in critical paths for the pre-error rate driven voltage adaptation. We use a Markov chain model to describe the voltage scheme analytically and analyze the effect of global and local variations on the closed-loop control. For an arithmetic circuit, synthesized in an industrial 65nm design-flow, an average power saving of 23% is achieved for very low error rates below 1E-11.**

I. INTRODUCTION

The trend to more and more functionality in solid-state circuits has made power consumption a major design criterion in digital circuits, especially in the context of battery powered devices [1]. In digital design flows, the modules are typically described on register transfer level (RTL) and then synthesized into hardware by electronic design automation (EDA) tools. These tools minimize power consumption by selecting an optimal set of logic gates to realize the desired functionality. The optimal kind and size of logic gates depends on the given constraints, mainly the specified clock period. To guarantee correct timing for the defined temperature range and to account for process variations, the worst-case is considered when synthesizing circuits. This leads to excessively large safety margins in most of the other operating conditions. Shrinking technology nodes are enlarging these timing margins due to ever increasing process variation [2]. To cut unused timing margins and save power, several approaches were proposed to adapt the supply voltage dynamically to the actual operating condition of the chip.

Ring-oscillators [3] or more sophisticated delay lines [4], [5] aim to replicate the timing of critical paths and are well suited for monitoring global variations on the chip. However, within-die variations of process, voltage and temperature (PVT), as well as aging, affect the timing of these replica circuits differently from the real circuit, causing once more considerable safety margins. Tuning the replica circuits during test improves their tracking but adds test time and complexity [5] and therefore increases cost.

Recent approaches [6]–[9] aim to monitor the timing within the real circuit, i.e in-situ. Special latches [6], [7] or flip-flops [8] are used to detect timing errors. The voltage adaptation is based on the occurrence of timing errors. As errors have to be accepted, micro-architectural recovery circuits are necessary, that repeat single computations after malfunction. Besides this excessive additional complexity, these techniques are not viable for real-time applications.

In contrast, the adaptive voltage scaling (AVS) approach in [9] restricts the reduction of the supply voltage to a lower limit where errors do not yet occur. Therefore, no additional hardware effort and complexity for the recovery circuitry is needed. Here, Pre-Error flip-flops emit warnings - pre-errors - when the timing slack in critical

paths drops below a certain value, i.e. late data transitions occur. The pre-error rate indicates the circuit speed and is used for on-the-fly variation and aging aware voltage adaptation. For zero or few pre-errors, the supply voltage is reduced, whereas an increased pre-error rate triggers a voltage increment.

This paper provides an in-depth analysis of the Pre-Error AVS technique regarding feasibility and efficiency. Section II will give a brief overview of the Pre-Error AVS. In section III a robust circuit implementation of the Pre-Error flip-flop is discussed. We elaborate a model for the closed-loop voltage adaptation in section IV and demonstrate the power saving potential of the Pre-Error AVS scheme in section V.

II. OVERVIEW OF PRE-ERROR AVS

In the Pre-Error AVS scheme the timing information is provided by Pre-Error flip-flops, that detect late data transitions in critical paths. Late data transitions are defined by the pre-error detection window, i.e. a defined time interval before the triggering edge of the clock. The speed of digital circuits is influenced by PVT variations and so is the frequency of pre-errors. The pre-error rate, indicating the timing slack, is thus used to control the supply voltage on-the-fly in an adaptive manner. After each N clock cycles, forming an observation interval, the control logic decides whether to change the voltage. For a pre-error count, during the previous interval, under a lower threshold $n_{limit\downarrow}$ the voltage is decreased, for counts above the upper limit $n_{limit\uparrow}$ it is increased. For counts inside the limits the voltage is maintained.

For analyzing the Pre-Error AVS technique, a 16-bit multiplier test circuit is used. It was synthesized in a commercial 65nm CMOS low-power technology with state-of-the-art EDA tools. The clock frequency was set to $f = 500\,\text{MHz}$ ($T = 2\,\text{ns}$) and we equipped the three most critical outputs with Pre-Error flip-flops, i.e. 9.4% of all 32 outputs.

III. PRE-ERROR FLIP-FLOP IMPLEMENTATION

Fig. 1 shows our proposed implementation for the necessary Pre-Error flip-flop. Besides the regular flip-flop (FF1), the Pre-Error detector latches the input D with the falling clock-edge (FF2). The stored value is compared with input data D by an XOR-gate. The output X1 is then latched with the next rising clock edge to detect transitions during the low clock phase. The detection window, denoted by ΔT, is defined by the falling clock edge. By changing the duty cycle, e.g. by a pulse-width-control loop (PWCL) [10], ΔT can be adjusted.

The transition monitor in the lower part of the schematic in Fig. 1 signals transitions from one rising clock edge to the next. With this additional information about transitions we can distinguish

978-1-61284-863-1/11 $26.00 © 2011 IEEE

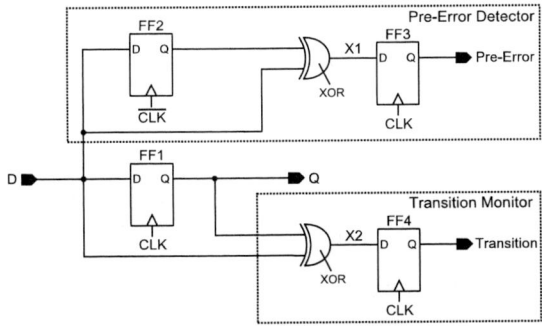

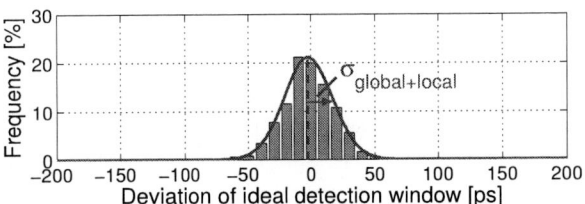

Fig. 2. Histogram showing the effect of local and global process variations on the length of the pre-error detection window at $V_{DD} = 0.8\,\text{V}$. The standard deviation reads as $\sigma = 19\,\text{ps}$. (For local variations only: $\sigma_{local} = 16\,\text{ps}$)

tion window is only slightly affected by variations ($\sigma_{0.8\,\text{V}} = 19\,\text{ps}$) for higher voltages deviations diminish ($\sigma_{1.2\,\text{V}} = 6\,\text{ps}$). Note, that the deviation is independent of the detection window length and that mainly local process variations affect the detection window. Global variations have a share of only 16% of the standard deviation.

For very high operating frequencies, where distributing a constant duty-cycle is rather complex, the detection window can also be realized by a delay-element [9], which is, however, also affected by variations and consumes somewhat more power and area.

IV. MODEL OF THE PRE-ERROR AVS

For a certain output, the signal delay strongly varies in dependence on the applied input patterns. This is due to changing signal propagation paths inside the logic with varying input vectors. Fig. 3 illustrates this effect for the most critical output of the multiplier test circuit. The histogram is obtained by simulating random input patterns. The left dashed line denotes a detection window of $\Delta T = 600\,\text{ps}$. Delays longer than $t_d = T - \Delta T = 1.4\,\text{ns}$ will result in a pre-error. The delay distribution and consequently the pre-error probability depend on the operating condition of the circuit, e.g. a voltage reduction shifts the distribution to the right and increases the pre-error rate.

Fig. 1. Schematic of the Pre-Error flip-flop, consisting of Pre-Error detector and transition monitor. The lower part shows the corresponding simulated timing diagram in the case of a pre-error followed by a cycle without pre-error.

between clock cycles with and without activity. An observation interval consists of N active clock cycles, being cycles in which the data toggle. Otherwise, phases with low activity would bias the pre-error rate. In the extreme case, for an activity rate of zero during an observation interval, no pre-errors occur and the voltage would be lowered independent of the operating condition of the circuit. This will probably lead to timing errors afterwards.

The pre-errors as well as the transitions of the observed outputs are then combined by an OR-tree and fed into the digital AVS unit, consisting of a pre-error counter, a transition counter and a control logic that triggers the voltage regulator to increase, decrease or maintain the voltage after every observation interval. The digital AVS unit was synthesized in a way to safely operate for the defined voltage range of the AVS scheme from 1.2 V down to 0.8 V.

For the AVS scheme it is important to define a detection window as accurate and robust as possible. To determine the sensitivity of ΔT towards variations we applied Monte-Carlo simulations. Besides global and local variations of the Pre-Error flip-flop itself, we also included the variations of the clock tree. This is important as the detection window is defined by the duty-cycle of the clock. We used a conventional H tree with three buffer stages. The histogram in Fig. 2 shows the resulting deviation compared to the ideal detection window length at $V_{DD} = 0.8\,\text{V}$. Even at this most critical voltage, the detec-

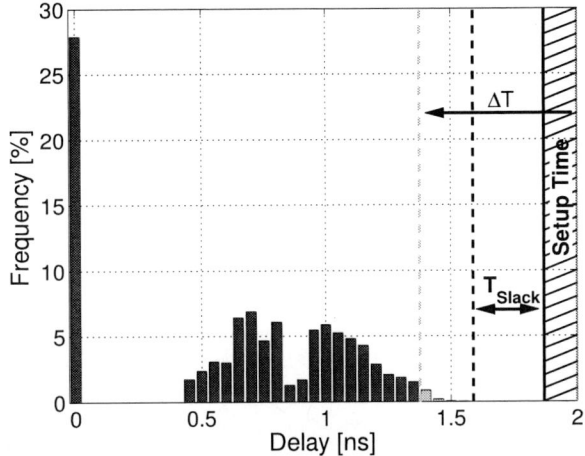

Fig. 3. Delay histogram for the most critical output, when applying random patterns to the synthesized multiplier test circuit.

The pre-error rate is thus used as indicator for the operating condition or remaining timing slack, respectively. Note that the occurrence of pre-errors is a stochastic process and thus the pre-error count during a given time span is a random variable.

978-1-61284-863-1/11 $26.00 © 2011 IEEE

The probability for a pre-error at an observed output in one clock cycle is the number of delays longer than $T - \Delta T$ divided by the total number of delays. The pre-errors of the observed outputs are OR'ed together to get the overall pre-error rate P_{pre}. If the pre-error count n_{pre} during an observation interval of N active clock cycles stays below $n_{limit\downarrow}$, the digital AVS unit decides to reduce the supply voltage. The corresponding probability $P_{V_{DD}\downarrow}$ is given by

$$P\left[n_{pre} < n_{limit\downarrow}\right] = P_{V_{DD}\downarrow} =$$

$$= \sum_{n_{pre}=0}^{n_{limit\downarrow}-1} \binom{N}{n_{pre}} \cdot (P_{pre})^{n_{pre}} \cdot (1 - P_{pre})^{N-n_{pre}} \quad (1)$$

For a given delay distribution, this probability is dependent on the clock cycles N of the observation interval and the threshold $n_{limit\downarrow}$, see Eq. 1. Moreover, it is dependent on the length of the detection window, as ΔT directly affects the pre-error rate P_{pre}, see Fig. 3.

The corresponding probability $P_{V_{DD}\uparrow}$ to increment the supply voltage, in a situation where the circuit slows down (e.g. due to a change in temperature or aging), reads as

$$P\left[n_{pre} > n_{limit\uparrow}\right] = P_{V_{DD}\uparrow} =$$

$$= 1 - \sum_{n_{pre}=0}^{n_{limit\uparrow}} \binom{N}{n_{pre}} \cdot (P_{pre})^{n_{pre}} \cdot (1 - P_{pre})^{N-n_{pre}} \quad (2)$$

For counts inside $n_{limit\downarrow}$ and $n_{limit\uparrow}$ the voltage is maintained. $P_{V_{DD}\rightarrow}$ denotes the probability for this event.

As the voltage is reduced or increased only with a specific probability, the Pre-Error AVS can be modeled by a Markov chain as depicted in Fig. 4.

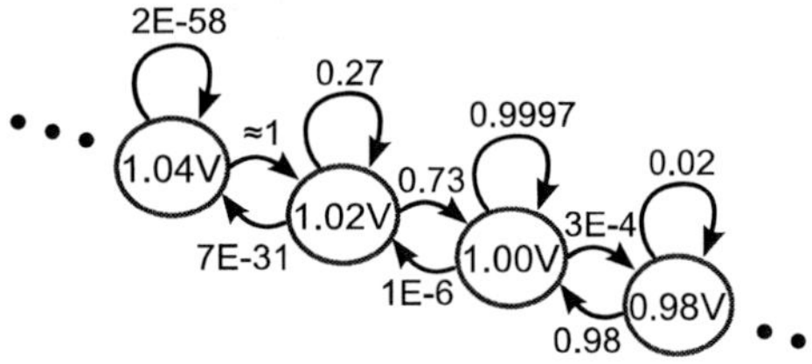

Fig. 4. Markov chain with transition probabilities for the individual voltage levels as model for the Pre-Error AVS. The shown values were obtained for the used 16-bit multiplier design at nominal process and temperature.

The figure shows the transition probabilities, i.e. the probabilities to reduce, increase or maintain the voltage, for nominal process and temperature. The used voltage step size is $20\,\text{mV}$. The particular delay distributions of the three observed outputs result in the statistics of the pre-error rate and so determine the transition probabilities of the Markov chain. The lower the voltage, the smaller the timing slack and consequently the lower the probability $P_{V_{DD}\downarrow}$ to further reduce the voltage and vice versa. The Markov chain, as depicted in Fig. 4, is characterized by the corresponding Markov or transition matrix P, containing the transition probabilities as elements. For a maximum voltage of $V_{DD} = 1.20\,\text{V}$ the Markov matrix reads as

$$P = \begin{pmatrix} P_{1.20\rightarrow} & P_{1.20\downarrow} & 0 & 0 & 0 & \cdots \\ P_{1.18\uparrow} & P_{1.18\rightarrow} & P_{1.18\downarrow} & 0 & 0 & \cdots \\ 0 & P_{1.16\uparrow} & P_{1.16\rightarrow} & P_{1.16\downarrow} & 0 & \cdots \\ \ddots & \ddots & \ddots & \ddots & \ddots \end{pmatrix}$$

$$(3)$$

For a fixed operating condition, the matrix is stationary, i.e. its elements or transition probabilities, respectively, are constant. The corresponding stationary probability vector π, containing the probabilities $P_{V_{DD}}$ to be at a certain voltage level V_{DD}

$$\pi = \begin{bmatrix} P_{1.20} & P_{1.18} & P_{1.16} & \cdots \end{bmatrix} \quad (4)$$

is defined as left eigenvector of the Markov matrix, associated with eigenvalue 1

$$\pi P = \pi \quad (5)$$

This means π is invariant under application of the transition matrix. When starting the AVS at any voltage level, it converges to the state defined by π after a short time. Accordingly, the probability to be at a certain voltage level remains constant then. For changing operating conditions, e.g. for a temperature decrease, the Markov matrix and thus the probability vector automatically adapt to the new condition.

To calculate the dynamic power consumption, we have to account for the probabilities to be at a certain voltage level $P_{V_{DD}}$, obtained by Eq. 5. Each of these likelihoods is multiplied with the dynamic power consumption $P_{dyn,V_{DD}}$ at the corresponding voltage level. All products are finally summed up

$$\overline{P_{dyn}} = \sum_{V_{DD}} P_{V_{DD}} \cdot P_{dyn,V_{DD}} \quad (6)$$

Since the pre-error count is based upon a stochastic process, an overcritical reduction can never be completely ruled out and it is essential to quantify the risk of timing errors. Therefore, we have to multiply the error probability for each voltage level $P_{err,V_{DD}}$ with the frequency $P_{V_{DD}}$ to be at that level. The error rate is given by summing up all products

$$P_{err} = \sum_{V_{DD}} P_{V_{DD}} \cdot P_{err,V_{DD}} \quad (7)$$

For the error probabilities $P_{err,V_{DD}}$, timing failures at all outputs are considered, of course. Timing errors occur for delays violating the setup time constraint.

V. RESULTS

To evaluate the power saving potential of the described AVS technique we use the Markov chain model, described above. This way we can determine the probabilities $P_{V_{DD}}$ (elements of vector π) very fast and accurate. Determining π by circuit simulations is impracticable due to extremely long run times.

Due to the statistical behavior of the pre-error count a long observation interval is advantageous to the Pre-Error AVS. Our analysis showed that an observation interval longer than $N = 1000$ cycles only slightly improves voltage scaling while increasing area and power overhead of the digital AVS unit. Consequently, for the following results, N was set to 1000.

With tuning the AVS settings (ΔT, $n_{limit\downarrow}$ and $n_{limit\uparrow}$) a desired error probability can be achieved. With this fixed setting, the voltage is then adapted automatically to the actual circuit speed, affected by process variations and aging. Temperature variation is a slowly changing effect compared to the response time of the control scheme and is thus also covered. The AVS also compensates for the conservative timing estimate of the synthesis tool as well as for inaccuracies of the voltage regulator, e.g. due to an offset in the bandgap reference. For fast changing effects, such as voltage drops due to activation of neighboring logic blocks, we added the same amount of safety margin to the worst-case design and the AVS scheme. This means

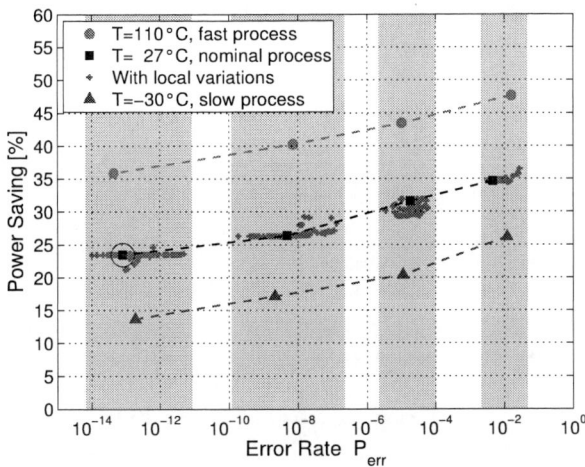

Fig. 5. Correlation between error rate P_{err} and power saving of the Pre-Error AVS scheme for different corners and under local process variations.

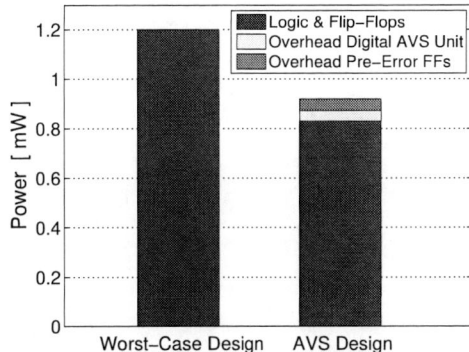

Fig. 6. Comparison of power consumption for worst-case design and AVS scheme, including the power overhead of the AVS scheme.

the synthesis as well as the error estimation for the AVS approach are performed at nominal voltage minus this additional safety margin.

Fig. 5 shows the relation between error rate and power saving. The power saving was computed by Eq. 6. The power consumption at each voltage level $P_{dyn,V_{DD}}$, as well as the error probabilities $P_{err,V_{DD}}$ for Eq. 7, were determined by SPICE simulations. The colored regions in Fig. 5 mark fixed AVS settings. For the leftmost region, the error rate is set to 1E-12 in nominal case. In the presence of local variations, the error rate varies from 1E-14 to 1E-11. Here, local variations of the Pre-Error flip-flops and clock-tree - affecting the detection window - and local variations of the multiplier circuit itself are considered.

For the fast corner ($T = 110\,^\circ$C, fast process)[1] lower voltages are more likely leading to higher power savings. As the AVS exploits the unused timing margin produced by the conservative timing estimation of the synthesis tool, there is considerable power saving even in the slow corner. Under global variations, the AVS keeps the error rate almost constant by tuning the supply voltage accordingly (compare slow and fast corner in Fig. 5). Global fluctuations hardly shift the detection window length and affect all multiplier outputs similarly.

In case of the conservative settings (leftmost region), for the nominal case ($T = 27\,^\circ$C, nominal process) we obtain a power saving of 23% compared to state-of-the-art guardbanding. The error rate can be adjusted by changing the AVS settings. For the different regions in Fig. 5, this is done by tuning ΔT. (ΔT is gradually decrased from leftmost region to rightmost region.) Thus, if higher error rates can be accepted, the power saving can be further increased.

For the power saving we also considered the power overhead of the digital AVS unit and the Pre-Error flip-flops. Fig. 6 illustrates the power consumption for the encircled data point of Fig. 5 in detail.

Including the Pre-Error flip-flops plus OR-tree results in an overhead of 3.9%. The digital AVS unit adds another 3.6% in power. Thus the power saving in logic operation of 30.5% is reduced to 23% as shown in Fig. 5.

---

[1]Note that typically the fast corner is at fast process and low temperature. In this example, however, the voltage is scaled to a point where the circuit is operated at temperature inversion. Here, the effect of decreasing threshold voltage with temperature exceeds the mobility degradation. Consequently, the circuit exhibits an inverted temperature characteristic, as it speeds up with increased temperature and vice versa.

VI. CONCLUSION

In this work, all relevant aspects for implementing the Pre-Error adaptive voltage scaling (AVS) scheme, including a robust implementation of the necessary Pre-Error flip-flop, are covered. We derived an analytical model to evaluate the dynamic voltage scaling and proved the feasibility to set a defined error rate. The error rate stays well within a narrow range under global as well as local variations. Considerable power saving potential, by eliminating unused safety margin, was demonstrated for a synthesized design. For the widely used multiplier circuit, an average power reduction of 23% is achieved, while keeping the error rate below 1E-11. The reliability (error rate) can be traded for further power savings. Digital circuits can thus operate at peak efficiency for a desired error rate.

REFERENCES

[1] M. Horowitz, E. Alon, D. Patil, S. Naffziger, R. Kumar, and K. Bernstein, "Scaling, power, and the future of CMOS," in *IEEE International Electron Devices Meeting (IEDM) Technical Digest*, 2005, pp. 7–13.

[2] S. Saxena *et al.*, "Variation in transistor performance and leakage in nanometer-scale technologies," *IEEE Transactions on Electron Devices*, vol. 55, no. 1, pp. 131–144, Jan. 2008.

[3] T. Burd, T. Pering, A. Stratakos, and R. Brodersen, "A dynamic voltage scaled microprocessor system," in *IEEE International Solid-State Circuits Conference (ISSCC) Digest of Technical Papers*, 2000, pp. 294–295.

[4] T. Fischer, F. Anderson, B. Patella, and S. Naffziger, "A 90nm variable-frequency clock system for a power-managed Itanium-family processor," in *IEEE International Solid-State Circuits Conference (ISSCC) Digest of Technical Papers*, 2005, pp. 294–599.

[5] A. Drake *et al.*, "A distributed critical-path timing monitor for a 65nm high-performance microprocessor," in *IEEE International Solid-State Circuits Conference (ISSCC) Digest of Technical Papers*, Feb. 11–15, 2007, pp. 398–399.

[6] K. A. Bowman *et al.*, "Energy-efficient and metastability-immune resilient circuits for dynamic variation tolerance," *IEEE Journal of Solid-State Circuits*, vol. 44, no. 1, pp. 49–63, Jan. 2009.

[7] S. Das *et al.*, "RazorII: In situ error detection and correction for PVT and SER tolerance," *IEEE Journal of Solid-State Circuits*, vol. 44, no. 1, pp. 32–48, Jan. 2009.

[8] D. Bull, S. Das, K. Shivashankar, G. S. Dasika, K. Flautner, and D. Blaauw, "A power-efficient 32 bit ARM processor using timing-error detection and correction for transient-error tolerance and adaptation to PVT variation," *IEEE Journal of Solid-State Circuits*, vol. 46, no. 1, pp. 18–31, 2011.

[9] M. Eireiner, S. Henzler, G. Georgakos, J. Berthold, and D. Schmitt-Landsiedel, "In-situ delay characterization and local supply voltage adjustment for compensation of local parametric variations," *IEEE Journal of Solid-State Circuits*, vol. 42, no. 7, pp. 1583–1592, July 2007.

[10] Y.-J. Wang, S.-K. Kao, and S.-I. Liu, "All-digital delay-locked loop/pulsewidth-control loop with adjustable duty cycles," *IEEE Journal of Solid-State Circuits*, vol. 41, no. 6, pp. 1262–1274, 2006.

Variability-Aware Automated Sizing of Analog Circuits Considering Discrete Design Parameters

Michael Pehl, Michael Zwerger, and Helmut Graeb
Technische Universitaet Muenchen
Munich, Germany, Email: christian.michael.pehl@mytum.de

Abstract—During analog design, the values of circuit parameters like transistor lengths and widths must be assigned, such that performance specifications are fulfilled for various operating conditions. Additionally, the design should be robust against process variations. The sizing must consider discrete design parameters, e.g., to model multipliers or manufacturing grids. Currently, no tools are available to consider both, process variations and discrete design parameters. This paper presents a new method for this task, which is based on SQP, Branch-and-Bound, and Realistic Worst-Case-Analysis. Simulation results in comparison with Monte Carlo analyses for two amplifiers illustrate the efficiency of the method and show the robustness of the results against variations in operating conditions and process parameters.

I. Introduction

In the step of analog sizing, design parameter values – e.g., transistors dimensions – are assigned, such that performance specifications are fulfilled. The performances are typically specified for a certain operation region – e.g., temperatures between 0 and 80°C – and should consider variability in the process. Additionally, to avoid problems in the sub-sequent layout and manufacturing step, the sizing must consider discrete design parameters, e.g., multipliers to model the number of transistors connected in parallel or manufacturing grids [1]. To support the analog designer in the highly complex task of sizing the circuit, various approaches have been presented. The approaches can roughly be subdivided into a group of stochastic and evolutionary approaches and a group of deterministic approaches. Stochastic approaches (e.g., [2], [3]) are suitable for global optimization and can consider discrete parameters in some cases. However, the runtime of the algorithms is typically high and no stochastic approach can be found, which considers variability in the process during the circuit sizing step. In contrast, deterministic gradient-based approaches can solve the sizing task very fast and efficiently, if a good initial sizing can be provided. Additionally, there are some approaches in this field which allow the designer to consider manufacturing tolerances and operation regions or to optimize the yield of a circuit [4]. However, only a few deterministic gradient-based approaches consider discrete parameters (e.g., [1], [5]) and no approach can be found considering discrete parameters as well as operation regions and manufacturing variabilities. Based on an approach to consider the operation region and variability during sizing [6], this paper presents a new method to compute such a solution under consideration of discrete design parameters. As simple rounding of the solution in the continuous space does not solve the problem in general [7], this work connects the approach in [6] with the SQP and Branch-and-Bound based approach described in [1].

The paper is structured as follows: In Section II, a formal description and mathematical formulation of the sizing task are given. The new approach is presented in Section III. In Section IV, the efficiency and efficacy of the algorithm is shown and the results are evaluated by Monte Carlo analysis. Section V concludes.

II. Problem Definition

A. Parameters

During the step of analog sizing, design parameters – e.g., transistor lengths and widths – must be assigned. For design parameters in this paper it is assumed that they are bounded by lower and upper values $\mathbf{d}_l$ and $\mathbf{d}_u$, respectively, and that the performances can be computed for each design parameter point $\mathbf{d}_{rel}$ in the (relaxed) domain

$$\mathbf{d}_{rel} \in \mathbb{D}_{rel}^N = \{\mathbf{d}_{rel} \mid \mathbf{d}_l \leq \mathbf{d}_{rel} \leq \mathbf{d}_u\} \tag{1}$$

However, some design parameters d_i – e.g., multipliers – are discrete and must be element of a discrete ordered set after sizing:

$$d_i \in \mathbb{D}_i := (\mathbb{D}, <) = \{d_1, ..., d_k, ..., d_{n_i}\}$$
$$\underset{k \in 1, ..., n_i-1}{\forall} d_k < d_{k+1} \tag{2}$$

The resulting mixed discrete and continuous design space is assigned as $\mathbb{D}_{disc}^N$ in the following.

To consider circuit performances for different operating conditions, e.g., for different temperatures, operating parameters

$$\mathbf{o} \in \mathbb{T}_o = \{\mathbf{o} \mid \mathbf{o}_l \leq \mathbf{o} \leq \mathbf{o}_u\} \tag{3}$$

are used. The specifications of lower parameter bounds $\mathbf{o}_l$ and upper parameter bounds $\mathbf{o}_u$ define a tolerance region $\mathbb{T}_o$, where the specifications for the circuit performances must be fulfilled.

To model variabilities in the manufacturing step, process parameters $\mathbf{s}$ are used. Without loss of generality, the N_s process parameters are considered to be normal distributed parameters with

$$pdf(\mathbf{s}) = \frac{1}{\sqrt{2\pi}^{N_s} \sqrt{\det(\mathbf{C})}} \cdot \exp\left(-\frac{1}{2}\beta^2(\mathbf{s})\right) \tag{4}$$

978-1-61284-863-1/11 $26.00 © 2011 IEEE

$\mathbf{C}$ is the covariance matrix. For mean values $\mathbf{s}_0$, $\beta(\mathbf{s})$ is given by

$$\beta^2(\mathbf{s}) = (\mathbf{s} - \mathbf{s}_0)^T \cdot \mathbf{C}^{-1} \cdot (\mathbf{s} - \mathbf{s}_0) \qquad (5)$$

The ellipsoid defined by (5) and a certain value $\beta = \beta_w$ is called the tolerance ellipsoid which defines the tolerance region

$$\mathbb{T}_s = \left\{ \mathbf{s} \,\middle|\, (\mathbf{s} - \mathbf{s}_0) \cdot \mathbf{C}^{-1} \cdot (\mathbf{s} - \mathbf{s}_0) \leq \beta_w^2(\mathbf{s}) \right\} \qquad (6)$$

Thus, if the performance specifications for a circuit are fulfilled for each process parameter point in $\mathbb{T}_s$, $\beta_w \cdot \sigma$ is a measure for the quality of the design and the parametric yield which can be expected [4]:

$$Y = \int_{-\infty}^{\beta_W} \frac{1}{\sqrt{2\pi}} \cdot \exp\left(-\frac{1}{2}\beta^2\right) d\beta \qquad (7)$$

E.g., $\beta_w = 3$ corresponds to a 3σ-design and a parametric yield of 99.9%.

B. Sizing Task

Design parameters should be computed such that the performance specifications are fulfilled for the complete operation region $\mathbb{T}_o$ and requiring a minimum (approximated) yield Y_{min} related to β_w in (7). Assuming exactly one specified upper bound $f_{U,i}$ for each performance $\hat{f}_i(\mathbf{d}, \mathbf{o}, \mathbf{s})$ and transforming $\hat{f}_i(\mathbf{d}, \mathbf{o}, \mathbf{s})$ into

$$f_i(\mathbf{d}, \mathbf{o}, \mathbf{s}) = \frac{\hat{f}_i(\mathbf{d}, \mathbf{o}, \mathbf{s}) - f_{U,i}}{|f_{U,i}|} \qquad (8)$$

this claim is fulfilled for a certain design parameter vector $\mathbf{d}^*$ and for N_f performances if

$$\bigvee_{i=1,...,N_f} \left(\max_{\mathbf{o} \in \mathbb{T}_o \wedge \mathbf{s} \in \mathbb{T}_s} f_i\left(\mathbf{d}^*, \mathbf{o}, \mathbf{s}\right) \right) \leq 0 \qquad (9)$$

Due to (8), the parameter points $\mathbf{o}_{w,i}$ and $\mathbf{s}_{w,i}$, respectively, which solve the maximizations in (9) correspond to the points $\mathbf{o} \in \mathbb{T}_o$ and $\mathbf{s} \in \mathbb{T}_s$ with the worst values for performance i. Each of these points is called the worst-case operation point and, respectively, worst-case process point for performance i. The sizing task can now be formulated as an optimization over all performances at the worst-case points. Transforming this multiobjective program into a single objective one by the least squares approach

$$\varphi(\mathbf{d}) := \sum_{i=1}^{N_f} \max\left(0, \ f_i\left(\mathbf{d}, \mathbf{o}_{w,i}, \mathbf{s}_{w,i}\right)\right)^2 \qquad (10)$$

the task can be written as

$$\min_{\mathbf{d} \in \mathbb{D}_{disc}^N} \varphi(\mathbf{d}) \quad \text{s.t.} \quad \mathbf{c}(\mathbf{d}) \geq \mathbf{0} \qquad (11)$$

The sizing constraints $\mathbf{c}(\mathbf{d})$ [8] are evaluated at the nominal point of the operating and the mean value of the process parameters. The sizing task is solved at $\mathbf{d}^*$, if (9) and (11) are satisfied, i.e., all constraints are fulfilled at the nominal point and performances at the worst-case points are within the specifications:

$$\varphi(\mathbf{d}^*) = 0 \wedge \mathbf{c}(\mathbf{d}^*) \geq \mathbf{0} \qquad (12)$$

An even more robust design might be achievable. This can be obtained by tightening the values of $f_{U,i}$ and Y_{min}.

III. Variability-Aware Sizing Approach

The approach presented in this paper is based on Branch-and-Bound. It requires to solve the sizing task on a relaxed, i.e., continuous, domain (1). For this purpose, Feasible Sequential Quadratic Programming (FSQP) [9] is used.

A. Variability-Aware FSQP

In Sequential Quadratic Programming, the optimization task is formulated using the Lagrangian function

$$\mathcal{L}(\mathbf{d}, \lambda) = \varphi(\mathbf{d}) - \sum_i \lambda_i c_i(\mathbf{d}) \qquad (13)$$

$\varphi(\mathbf{d})$ is the objective function in (10) and λ_i is the Lagrangian multiplier for constraint $c_i(\mathbf{d})$. In each iteration μ, the gradient of $\mathcal{L}$ is linearly approximated and – following the Karush Kuhn Tucker conditions – set to zero. Thus, in each step the equation system

$$\begin{pmatrix} \mathbf{H}^{(\mu)} & -\mathbf{J}^{(\mu)T} \\ -\mathbf{J}^{(\mu)} & \mathbf{0} \end{pmatrix} \begin{pmatrix} \Delta\mathbf{d}^{(\mu)} \\ \Delta\lambda^{(\mu+1)} \end{pmatrix} = \begin{pmatrix} -\mathbf{g}^{(\mu)} \\ \mathbf{c}^{(\mu)} \end{pmatrix} \qquad (14)$$

must be solved. $\Delta\lambda^{(\mu)}$ and $\Delta\mathbf{d}^{(\mu)}$ are the changes in Lagrangian multipliers and design parameters, $\mathbf{c}^{(\mu)}$ is the current constraint value, $\mathbf{J}^{(\mu)} = \nabla_{\mathbf{d}}\mathbf{c}(\mathbf{d})|_{\mathbf{d}^{(\mu)}}$ is the Jacobian matrix of the constraints, and $\mathbf{g}^{(\mu)} = \nabla_{\mathbf{d}}\varphi(\mathbf{d})|_{\mathbf{d}^{(\mu)}} - \mathbf{J}^{(\mu)}\lambda^{(\mu)}$ is the gradient of the Lagrangian function with respect to $\mathbf{d}$. The Hessian matrix of the Lagrangian function $\mathbf{H}^{(\mu)} = \nabla_{\mathbf{dd}}\mathcal{L}(\mathbf{d}, \lambda)|_{\mathbf{d}^{(\mu)}\lambda^{(\mu)}}$, is approximated using, e.g., a BFGS (Broyden Fletcher Goldfarb Shanno) algorithm. In contrast to other SQP approaches, FSQP guarantees that each step computed during the algorithm leads to a feasible point. Whereas the constraints in this work should be evaluated at the nominal point for operating and process parameters, circuit performances are considered at their worst-case points. I.e., the gradient of the objective function $\nabla_{\mathbf{d}}\varphi(\mathbf{d})$ must be computed at the worst-case points.

To reduce the effort, the worst-case points are approximated by a classical and a realistic worst-case analysis [4]. I.e., the worst-case point in $\mathbb{T}_o$ and $\mathbb{T}_s$ is computed for each performance based on a linear model w.r.t. operating and process parameters. To improve the quality of the model, the linearization is done at the newest available worst-case operating and process point [6]. The algorithm in Fig. 1 shows the proceeding for calculating the worst-case points (line 2 to 9) and computing the gradient of the objective function (line 10 to 16). Using this approach for the gradient computation, the FSQP approach results in a variability-aware continuous sizing for the circuit.

B. Branch-and-Bound for Analog Sizing

The algorithm in Fig. 2 shows a recursive Branch-and-Bound approach [1]. In each recursion, FSQP is used to solve the sizing task in a continuous sub-domain $\mathbb{D}_{sub}^N$ (line 5), which is initially given by (1). If $\mathbb{D}_{sub}^N$ does not contain a solution, there is also no solution in a sub-domain of $\mathbb{D}_{sub}^N$ and the branch can be pruned (line 1 and 6). If the solution of FSQP is discrete (line 9), a variability-aware discrete sizing is found and the algorithm terminates.

978-1-61284-863-1/11 $26.00 © 2011 IEEE

Algorithm: Worst_Case

Require: $\mathbf{d}^{(\mu)}, \mathbf{o}_{w,1}^{(\mu)}, ..., \mathbf{o}_{w,N_f}^{(\mu)}, \mathbf{s}_{w,1}^{(\mu)}, ..., \mathbf{s}_{w,N_f}^{(\mu)}$

1: **for** each performance f_i with $i = 1, ..., N_f$ **do**
2:     // gradients w.r.t. operating parameters
3:     $\mathbf{g}_{o,i} = \nabla_{\mathbf{o}} f_i(\mathbf{d}, \mathbf{o}, \mathbf{s})|_{\mathbf{d}=\mathbf{d}^{(\mu)}, \mathbf{o}=\mathbf{o}_{w,i}^{(\mu)}, \mathbf{s}=\mathbf{s}_{w,i}^{(\mu)}}$
4:     // compute worst-case operating point
5:     $\mathbf{o}_{w,i}^{(\mu+1)} = \arg\max\limits_{\mathbf{o} \in \mathbb{T}_o} \mathbf{g}_{o,i}^{(\mu)^T} \cdot \mathbf{o}$
6:     // gradients w.r.t. process parameters
7:     $\mathbf{g}_{s,i} = \nabla_{\mathbf{s}} f_i(\mathbf{d}, \mathbf{o}, \mathbf{s})|_{\mathbf{d}=\mathbf{d}^{(\mu)}, \mathbf{o}=\mathbf{o}_{w,i}^{(\mu+1)}, \mathbf{s}=\mathbf{s}_{w,i}^{(\mu)}}$
8:     // compute worst-case process point
9:     $\mathbf{s}_{w,i}^{(\mu+1)} = \arg\max\limits_{\mathbf{s} \in \mathbb{T}_s} \mathbf{g}_{s,i}^{(\mu)^T} \mathbf{s}$
10:    // gradients w.r.t. design parameters
11:    $\mathbf{g}_{d,i} = \nabla_{\mathbf{d}} f_i(\mathbf{d}, \mathbf{o}, \mathbf{s})|_{\mathbf{d}=\mathbf{d}^{(\mu)}, \mathbf{o}=\mathbf{o}_{w,i}^{(\mu+1)}, \mathbf{s}=\mathbf{s}_{w,i}^{(\mu+1)}}$
12:    // error value for least squres approach (10)
13:    $\epsilon_i = \max\left(0, f_i(\mathbf{d}^{(\mu)}, \mathbf{o}_{w,i}^{(\mu+1)}, \mathbf{s}_{w,i}^{(\mu+1)})\right)$
14: **end for**
15: // gradient of objective function (10)
16: $\mathbf{g}_{\varphi}^{(\mu+1)} = 2 \cdot \sum\limits_{i=1}^{N_f} \epsilon_i \cdot \mathbf{g}_{d,i}$
17: **return** $\mathbf{g}_{\varphi}^{(\mu+1)}, \mathbf{o}_{w,1}^{(\mu+1)}, ..., \mathbf{o}_{w,N_f}^{(\mu+1)}, \mathbf{s}_{w,1}^{(\mu+1)}, ..., \mathbf{s}_{w,N_f}^{(\mu+1)}$

Fig. 1. Algorithm for the computation of the gradient at the worst-case point

Otherwise, a parameter $d_i = d_{i,0}$ which should be discretized is chosen for branching (Section III-C) and the algorithm is recursively started for two new sub-domains:

$$\mathbb{D}_{sub,1}^N = \left\{ \mathbf{d} \in \mathbb{D}_{sub}^N \,|\, d_i \geq \lceil d_{i,0} \rceil \right\} \tag{15}$$

$$\mathbb{D}_{sub,2}^N = \left\{ \mathbf{d} \in \mathbb{D}_{sub}^N \,|\, d_i \leq \lfloor d_{i,0} \rfloor \right\} \tag{16}$$

$\lceil \bullet \rceil$ and $\lfloor \bullet \rfloor$ round the parameter to the next higher and lower discrete value (line 14 to 16). The algorithm ends, if any solution for the sizing task has been found (line 1).

C. Implementation

The algorithm in Section III-B solves the sizing task under consideration of an operation region and process variations. However, it can be further improved by the following methods:
Branching heuristics:

For the presented algorithm, a branching in direction of greatest gradient is used. I.e., branching is done in direction of a design parameter d_i with index:

$$i = \arg\max\limits_{i \in \mathbb{I}} |g_i| \tag{17}$$

g_i is the i-th component of the last Lagrangian gradient, and $\mathbb{I}$ is an index set including the indexes of all parameters which must be discretized in the current Branch-and-Bound recursion. The sub-problem, which lies in direction of $-g_i$ is always considered first and $\mathbb{D}_{sub,1}^N$ and $\mathbb{D}_{sub,2}^N$ in Fig. 2 are exchanged accordingly.
Approximation of the solution

Although the rounded result of the FSQP algorithm is not a valid solution in general, it solves the sizing task in certain cases. To avoid additional computational cost the rounded

Algorithm: Branch-and-Bound

Require: Incumbent $\mathbf{d}_{inc}$, relaxed sub-domain $\mathbb{D}_{sub}^N$

1: **if** $\varphi(\mathbf{d}_{inc}) = 0 \vee \left\{ \mathbf{d} \in \mathbb{D}_{sub}^N \,|\, \mathbf{c}(\mathbf{d}) \geq 0 \right\} = \emptyset$ **then**
2:    // incumbent solves task or no feasible point in $\mathbb{D}_{sub}^N$
3:    **return** $\mathbf{d}_{inc}$
4: **end if**
5: Solve $\mathbf{d}_0 = \arg\min\limits_{\mathbf{d} \in \mathbb{D}_{sub}^N} \varphi(\mathbf{d})$ s.t. $\mathbf{c}(\mathbf{d}) \geq 0$ by FSQP
6: **if** $\varphi(\mathbf{d}_0) \neq 0$ **then**
7:    // no solution in current domain
8:    **return** $\mathbf{d}_{inc}$
9: **else if** $\mathbf{d}_0$ is discrete **then**
10:   // solution found
11:   **return** $\mathbf{d}_0$
12: **else**
13:   // branch
14:   Compute sub-domains $\mathbb{D}_{sub,1}^N, \mathbb{D}_{sub,2}^N$ by (15),(16)
15:   $\mathbf{d}_{inc} = $**Branch-and-Bound**$\left(\mathbf{d}_{inc}, \mathbb{D}_{sub,1}^N\right)$
16:   $\mathbf{d}_{inc} = $**Branch-and-Bound**$\left(\mathbf{d}_{inc}, \mathbb{D}_{sub,2}^N\right)$
17:   **return** $\mathbf{d}_{inc}$
18: **end if**

Fig. 2. Branch-and-Bound algorithm

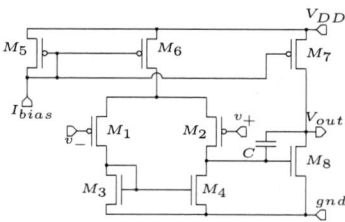

Fig. 3. Miller amplifier

point can be considered in the Algorithm above.
Additionally, (14) is solved in each FSQP step. This problem can be interpreted as the quadratic optimization problem

$$\min \frac{1}{2} \Delta\mathbf{d}^T H \Delta\mathbf{d} + \mathbf{g}^T \Delta\mathbf{d} \quad \text{s.t.} \quad \mathbf{J}\Delta\mathbf{d} + \mathbf{c} \geq 0 \tag{18}$$

Solving the problem from the last FSQP step over the discrete sub-domain $\mathbb{D}_{disc}^N \cap \mathbb{D}_{sub}^N$ in a Branch-and-Bound recursion, results in an approximation of the discrete solution of the underlying problem. As this approximation solves the sizing task in many cases and can be computed fast, it can be used to speed up the algorithm in Fig. 2 [1].

IV. SIMULATION RESULTS

In this section, the results gained for the Miller amplifier (MA) in Fig. 3 and the low voltage amplifier (LVA) in Fig. 4 using the Variability-Aware Discrete Sizing Approach (VADSA) from Section III are discussed. For both, MA and LVA, a $180nm$ technology is used. Widths, length, and multipliers of all transistors, compensation capacitances, and – in case of the LVA – the bias voltages are used as design

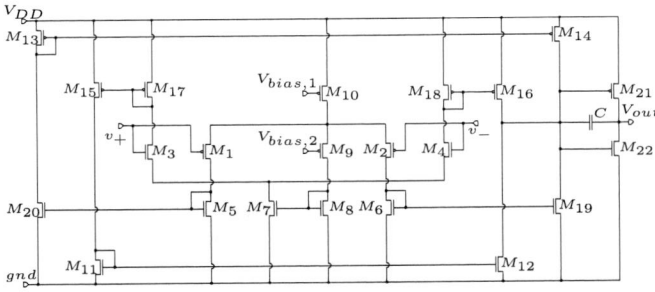

Fig. 4. Low voltage amplifier [10]

TABLE I

OPERATING PARAMETERS FOR MILLER (MA) AND LOW VOLTAGE
AMPLIFIER (LVA)

MA	min	nominal	max		LVA	min	nominal	max
$V_{DD}[V]$	2.3	2.5	2.7		$V_{DD}[V]$	1.9	2.0	2.1
$C_{load}[pF]$	15	20	25		$C_{load}[pF]$	1.9	2	2.1
$T[°C]$	-40	27	125		$T[°C]$	-40	27	125
$I_{bias}[\mu A]$	9	10	11					

TABLE II

SPECIFIED GAIN, PSRR, CMRR, PHASE MARGIN (PM), TRANSIT
FREQUENCY (f_t), RISING AND FALLING SLEW-RATE (SR+, SR-), AND
POWER FOR MILLER (MA) AND LOW VOLTAGE AMPLIFIER (LVA)

	Gain [dB]	PSRR [dB]	CMRR [dB]	PM [°]	f_T [MHz]	SR+ [V/μs]	SR- [V/μs]	Power [mW]
MA	> 75	> 140	> 120	> 60	10	> 8	< −8	< 5
LVA	> 75	> 90	> 90	> 60	20	> 12	< −12	< 7

TABLE III

YIELD AT WORST-CASE OPERATING POINTS FOR NOMINAL POINT (NOM),
ROUNDED CONTINUOUS SIZING (ROUND), AND SIZING WITH VADSA
FOR MILLER (MA) AND LOW VOLTAGE AMPLIFIER (LVA) APPROXIMATED
BY MONTE CARLO WITH 2500 SAMPLES

	NOM	ROUND	VADSA
MA	0%	86.0%	99.8%
LVA	0%	98.0%	99.4%

parameters. After sizing, all transistor lengths and widths should lie on a $10nm$ grid and multipliers must be integers. Capacitances and bias voltages were considered as continuous parameters.

In case of the LVA, all transistor lengths, in case of the MA, all lengths of PMOS and, respectively, NMOS transistors were set equal. For both amplifiers, widths and multipliers of transistors were equalized with respect to symmetry demands and constraints defined in [8]. Additionally, widths of the transistors in the same current mirror were set equal, i.e., the relation of the drain currents was determined by the multipliers. Thus, for the MA, 12 discrete and one continuous, for the LVA, 26 discrete and three continuous design parameters were used.

Operating parameters and specified operation region for both amplifiers are given in Table I. For both circuits, global variations of the oxide thickness, electron mobility, and thresh-

old voltages were considered. Additionally, for the MA local variations of electron mobility and threshold voltages were assumed. Thus, a total of 21 process parameters for the MA and 5 for the LVA were used. The yield was claimed to be at least 99.9% per performance, i.e., the overall yield should be at least 98.9%.

The specifications for both amplifiers are given in Table II. The algorithm was started at a discrete design parameter point which fulfills the constraints and performance specifications for the nominal value of process and operating parameters.

The approximated yield for both amplifiers at the initial point and after the sizing with VADSA is shown in Table III. Additionally the yield after continuous sizing with FSQP and sub-sequent rounding is shown.

The approximated yield at the initial point is 0%. Continuous optimization with sub-sequent rounding does not solve the yield requirements. In contrast, the sizings computed with VADSA fulfill the yield requirements and result in a significant higher robustness of the circuits.

The efficiency of VADSA can be seen from the runtime of the algorithm, which takes with 49 minutes for MA and 122 minutes for LVA (16 times parallelized on an Intel R Xeon R X5500 2.67 GHz CPU) only 5 and, respectively, 4 minutes more than continuous sizing and sub-sequent rounding. The additional runtime is herein dominated by the evaluation of the results using the classical and realistic worst-case analyses.

V. CONCLUSION

In this paper, a new approach for sizing of analog circuits under consideration of operating conditions, process parameters, and mixed continuous and discrete design parameters was presented. The approach is based on FSQP, Branch-and-Bound, and classical and realistic worst-case analysis. Experimental results show that the algorithm is highly efficient and computes a sizing which does not only fulfill specifications for the whole operation region, but is also robust against process variations.

REFERENCES

[1] M. Pehl and H. Graeb, *An SQP and Branch-and-Bound Based Approach for Discrete Sizing of Analog Circuits.* InTech, Feb. 2011, pp. 297–316.

[2] R. Phelps, M. Krasnicki, R. Rutenbar, L. Carley, and J. Hellums, "Anaconda: Simulation-Based Synthesis of Analog Circuits via Stochastic Pattern Search," *IEEE TCAD*, vol. 19, no. 6, Jun. 2000.

[3] A. Somani, P. Chakrabarti, and A. Patra, "An Evolutionary Algorithm-Based Approach to Automated Design of Analog and RF Circuits Using Adaptive Normalized Cost Functions," *IEEE TEC*, vol. 11, no. 3, Jun. 2007.

[4] H. Graeb, *Analog Design Centering and Sizing.* Springer, 2007.

[5] M. Pehl and H. Graeb, "**RaGAzi**: A **Ra**ndom and **G**radient-Based Approach to **A**nalog **Siz**ing for Mixed Discrete and Continuous Parameters," in *IEEE ISIC*, Dec. 2009.

[6] R. Schwencker, F. Schenkel, M. Pronath, and H. Graeb, "Analog Circuit Sizing using Adaptive Worst-Case Parameter Sets," in *DATE*, 2002.

[7] D. Li and X. Sun, *Nonlinear Integer Programming.* Springer, 2006.

[8] T. Massier, H. Graeb, and U. Schlichtmann, "The Sizing Rules Method for CMOS and Bipolar Analog Integrated Circuit Synthesis," *IEEE TCAD*, vol. 27, no. 12, pp. 2209–2222, Dec. 2008.

[9] C. Lawrence and A. Tits, "A Computationally Efficient Feasible Quadratic Programming Algorithm," *SIAM J. on optimization*, vol. 11, no. 4, 2000.

[10] R. Martins, "On the Design of Very Low Power Integrated Circuits," Ph.D. dissertation, Vienna University of Technology, Dec. 1998.

978-1-61284-863-1/11 $26.00 © 2011 IEEE

Program-Aware Circuit Level Timing Analysis

Veit B. Kleeberger, Sebastian Kiesel, Ulf Schlichtmann
Institute for Electronic Design Automation
Technische Universität München
Munich, Germany
http://www.eda.ei.tum.de

Samarjit Chakraborty
Institute for Real-Time Computer Systems
Technische Universität München
Munich, Germany
http://www.rcs.ei.tum.de

Abstract—There is an increasing need for accurate timing information in nanoscale CMOS integrated circuit design. As delay variability increases design margins tend to be overly pessimistic in worst case circuit design. Conventional timing characterization methods do not take the function of the circuit into account. The use of program information tightens the result of circuit level timing analysis and reduces overestimation of worst case circuit delay. The method presented in this paper is able to analyze the impact of program specific details on the delay in the presence of process variations. Additionally we show how the gained information can be used to analyze different instruction sequences towards robustness in a microprocessor.

I. INTRODUCTION AND BACKGROUND

In the design of integrated circuits one important step is the gate level timing analysis to determine the maximum possible clock frequency, at which the circuit operates without errors. Traditionally timing verification is a purely static process, thus ignoring the functionality of the circuit and computing its worst-case topological delay. This can be very inaccurate as false paths, which arise from logic dependencies, redundancies, etc., are typically not recognized automatically by this analysis [1].

The timing of a circuit is not only influenced by the circuit topology, but also by its functionality, i.e., the logic values it computes in one clock cycle. The benefit of including functional information in nominal timing analysis has already been reported in literature [2], [3]. In data path circuits the worst case circuit delay is strongly dependent on the selected values at the data inputs. Fig. 1 shows a simple example of how circuit timing can depend on selected functionality.

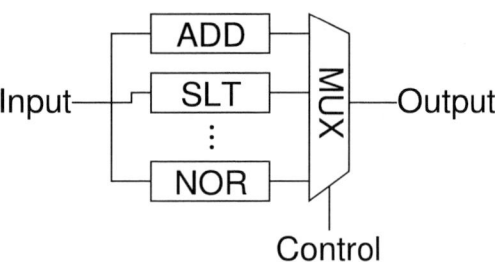

Fig. 1: Simple illustration of dependency of timing on circuit functionality

In the example in Fig. 1 the computation of an addition in the ADD block is clearly more complex than the conjunction of two signals in the NOR block and thus needs more time. As the control signal selects one specific block, it primarily

determines the timing requirements of the whole circuit depending on the selected subblock. Thus, the selected circuit function determines the maximum clock frequency at which the circuit can operate correctly.

While it has been well studied that the nominal worst case circuit delay can be influenced by the selected circuit functionality, the effect of functional information on delay variability has not been studied. As delay variability is also dependent on circuit topology it is obvious that it will also be dependent on the selected function of the circuit. In the example of Fig. 1 the influence of process variations on the delay variation will differ in each subblock. Thus, the control signal influences the nominal delay, as well as its variability depending on process variations.

As timing variability in nanometer technologies increases different functions have different robustness characteristics. The difference in variability and nominal delay leads to different safety margins for different functions for a given clock frequency.

Timing analysis that is aware of functional information shows which instructions are more vulnerable to process variations. This gives an insight into the robustness of a given program running on a microprocessor against variations.

This paper introduces enhancements to statistical static timing analysis that can include given functional information from a program that will be executed on the circuit. The inclusion of functional information into timing analysis gives better estimates for the timing characteristics and process dependent variability of functional units in the chip. The remainder of the paper is structured as follows: Section II introduces the principles of Statistical Static Timing Analysis (SSTA). Section III describes the proposed enhancements to SSTA to incorporate given functional information. Section IV shows the application of program-aware SSTA and the effect of instruction ordering on the timing of a microprocessor ALU. Section V summarizes our contributions and outlines directions for future work.

II. STATISTICAL TIMING ANALYSIS

Although our enhancements are not restricted to a special type of timing analysis, we explain them together with the statistical timing analysis method introduced by Visweswariah et al. [4]. This timing analysis has been shown to be applicable to large scale designs and provides sufficient accuracy for analyzing large scale designs with process variations.

978-1-61284-863-1/11 $26.00 © 2011 IEEE

All delays and arrival times in the circuit are expressed in the same canonical form:

$$a = a_0 + \sum_{i=1}^{n} a_i \Delta X_i + a_{n+1} \Delta R_a \qquad (1)$$

where a_0 is the nominal value, ΔX_i the variation source and a_i the sensitivity of a to the variation source. The last part $a_{n+1} \Delta R_a$ is an independent random variable and its corresponding sensitivity. Without loss of generality all variation sources ΔX_i and ΔR_a are assumed to be Gaussian $N(0, 1)$ distributed.

To compute the maximum circuit delay between the primary inputs and outputs, and so the maximum clock frequency, the circuit graph (Fig. 2) is traversed topologically.

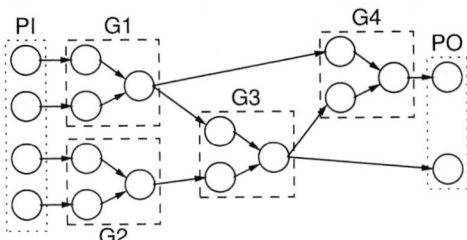

Fig. 2: Timing graph of a circuit consisting of primary inputs (PI), cells (G_i), and primary outputs (PO).

At each node in the graph, for all fan-in edges an edge delay (e.g. for cell pin-to-pin delay) is added first, and then the resulting arrival time is computed using a maximum operation.

As all timing quantities are represented in the linear form (1), the edge delay can be added directly without any approximation. The maximum of two canonical sums can be computed using the concept of tightness probabilities. The tightness probability T_A (3) gives the probability that the canonical sum A is larger than the canonical sum B, as shown below:

$$\theta = \sqrt{Var\{A + B\}} \qquad (2)$$

$$T_A = \Phi\left(\frac{a_0 - b_0}{\theta}\right) \qquad (3)$$

$$E\{max(A, B)\} = a_0 T_A + b_0 (1 - T_A) + \theta\phi\left(\frac{a_0 - b_0}{\theta}\right) \qquad (4)$$

$$Var\{max(A, B)\} = \left(Var\{A\} + a_0^2\right) T_A + \left(Var\{B\} + b_0^2\right)(1 - T_A) + (a_0 + b_0)\theta\phi\left(\frac{a_0 - b_0}{\theta}\right) - E\{max(A, B)\}^2 \qquad (5)$$

ϕ is the Gaussian probability density function and Φ is its cumulative distribution function. The maximum $g = max(a, b)$ in the form of (1) is then computed by first setting g_0 equal to $E\{max(A, B)\}$ from (4). The sensitivities g_i are computed by weighted addition

$$g_i = T_A a_i + (1 - T_A) b_i \qquad i = 1, 2, \ldots, n \qquad (6)$$

and the random part g_{n+1} is computed by equating the variance of g to $Var\{max(A, B)\}$ from (5).

III. Instruction Aware SSTA

In circuit operation, the delay of the combinational logic part in every clock cycle is a function of the input signals. Just switching signals lead to a signal delay in the circuit and determine the maximum clock frequency for this operation. If in contrast the signal stays constant, the clock will always capture the right signal value, no matter at which frequency the circuit is operated. Incorporating signal values (see Tab. I) can thus give tighter bounds on the estimated delay.

TABLE I: Possible values of a signal

Signal Value	Meaning
0	The signal is constant zero
1	The signal is constant one
r	The signal is rising
f	The signal is falling
x	The signal is unknown, i.e., it can be rising or falling

We propose an extended timing analysis which computes, besides the arrival time at every point in the timing graph, a signal value according to Tab. I. Every point in the timing graph (Fig. 2) is therefore annotated with a tuple (a, v). The tuple consists of the arrival time a of the signal given as a canonical sum (1) and its value v according to Tab. I.

The signal propagation in timing analysis must now take into account the arrival time and its signal value to compute a circuit delay which is dependent on the specified signal values. Without loss of generality, the modified arrival time computation is explained in the following for a two input logic cell.

A. Boolean Sensitivity

Before the arrival time of a signal at the output of a cell in the circuit is computed we check if the input conditions lead to a constant output signal. This case happens if some inputs make the cell insensitive to switching signals on the other inputs. For example if for a NAND cell one input is constantly at *0*, the output will be constantly *1*, no matter what value the other input signals have.

This check is performed by inserting constant values from the inputs into the boolean equation that describes the logic cell. A subsequent tautology and contradiction check reveals if the output has to be constant logic zero or constant logic one. If this is not the case, i.e., the output is switching, we continue the delay computation by excluding irrelevant timing arcs[1] from the timing graph.

B. Timing Arc Reduction

After we have verified that the output of the logic cell is not constant, we can proceed with further checks and simplifications in timing propagation.

Therefore, we check for every combination of two input signals if one input definitely switches before the other does. This can be checked by computing the tightness probability T given by (3). If the tightness probability is larger than a user-defined threshold p signal A is considered to switch earlier than signal B.

[1] Timing arc denotes any possible pin-to-pin delay of a logic cell

Let us assume that signal A switches earlier than signal B. Then, depending on the boolean function the cell implements, several simplifications for the input signals are possible.

1) Both inputs are known to be rising or falling: In this case we have three possibilities for the output signal. The first possibility is that the signal change at input A will not cause a change at the output but will make the cell insensitive to input changes at B (Fig. 3a). As the output does not change for this case we can skip further computations and set the output to the constant value.

Another possibility for this case is that the signal change at input A will lead to a signal change at the output and make the cell insensitive to input changes from B. Thus, the signal change at input B will have no effect and we can eliminate all timing arcs for input B.

Fig. 3b shows the last possibility where the input combination causes a glitch at the output. As we are just interested in the latest possible signal, we can exclude all timing arcs from input A by setting them to constant 1 or 0 without compromising the safety of the estimate of worst case circuit delay.

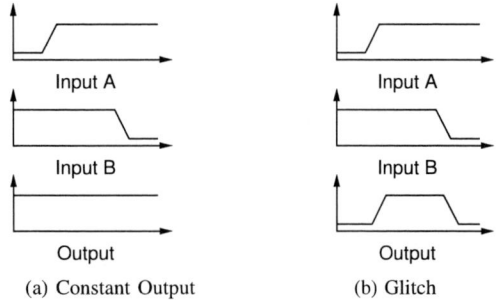

(a) Constant Output   (b) Glitch

Fig. 3: Possible cases checked for input $A=r/f$ and input $B=r/f$

2) Input A is unknown (x) and input B is rising or falling: If we do not know anything about the signal at input A, but know that input B is switching in one direction we can make the following tests.

If both signal changes at input A lead to a rising (falling) output signal, we do not consider input A any longer as x but as rising (falling) according to the signal change that leads to the later arrival time at the output (Fig. 4a).

If one signal transition at input A does not change the output, but the other one leads to a rising or falling edge, we consider input A as rising or falling according to the signal change at input A that causes a change at the output.

The last possibility is that one signal change at input A does not change the output and the other one leads to a glitch (Fig. 4b). In this case we can eliminate all timing arcs for input A.

3) Input A is rising or falling and input B is unknown (x): In this case Fig. 5a shows the first possibility. The signal change at input A leads to transition at the output or leaves the output unchanged, depending on the transition at input B. Therefore, we can eliminate all timing arcs from input B as none of them contribute to a transition at the output.

The second possibility is the same as shown in Fig. 4a. Like before we restrict input B from unknown to rising or falling,

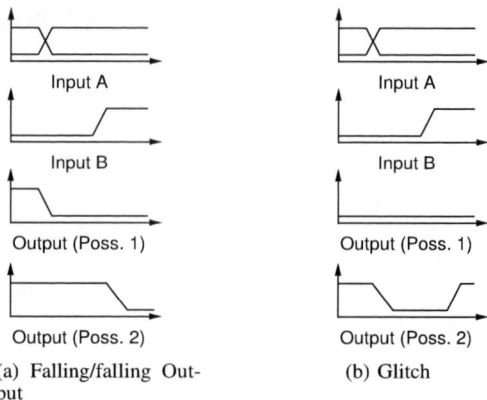

(a) Falling/falling Output   (b) Glitch

Fig. 4: Possible cases checked for input $A=x$ and input $B=r/f$

depending on which transition causes the later transition at the output.

Fig. 5b shows the third possibility. As just one transition of input B is causing a transition at the output, we restrict input B to this case and eliminate the other one, which causes a constant output.

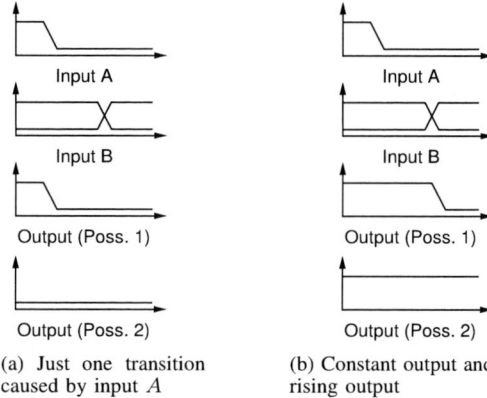

(a) Just one transition caused by input A (b) Constant output and rising output

Fig. 5: Possible cases checked for input $A=r/f$ and input $B=x$

A fourth possibility is again the same as shown in Fig. 4b, where input B causes a glitch in one case. Again, we restrict input B to the signal transition that causes the glitch and set input A to 0 or 1, so that only the second edge of the glitch is generated.

C. Delay Computation

After we have completed the input simplification we iterate over all timing arcs that were not excluded previously. For every timing arc we check if it can appear with the modified input conditions generated in the previous step. If the timing arc cannot appear we exclude it from the max operation in timing analysis. As timing arcs appear directly in the timing graph, this reduces the timing graph. For example a cell in the timing graph shown in Fig. 2 (marked by the dashed rectangle) consists of two timing arcs. If one of these two timing arcs cannot appear, e.g. the lower in the rectangle, the timing at the cell output is solely determined by the timing of the upper input.

The delay computation is demonstrated by the example shown in Fig. 6.

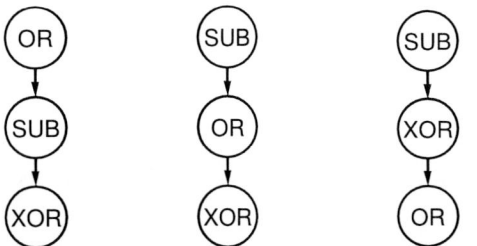

(a) NAND cell with annotated signal values and arrival times

(b) Representation in the timing graph

Fig. 6: Example for the modified delay computation algorithm

First we check if the output can be constant, which is not the case as none of the inputs is constant. From the other cases described above case 3 matches the example. After checking all three possibilities for this case, we can conclude that possibility 2, as shown in Fig. 5a, matches here. Thus, we can ignore input B and compute the arrival time at the output by just using the timing arc $A \to Z$. Thus, we set input B from originally x to f.

IV. RESULTS

To show the applicability of our approach we synthesized a MIPS ALU with Synopsys Design Compiler using the Nangate 45 nm Cell Library [5]. The ALU realizes 8 possible instructions, which are AND, OR, ADD, SUB, SLT, XOR, NOR, LUI.

Following [6] we assume for the scale of process variations a standard deviation of 15% for transistor length, 5.33% for oxide thickness, 4.44% for threshold voltage, and 10% for channel resistance.

We tested different control input combinations and their effect on the worst case circuit delay of the ALU. The time for the analysis of one instruction pair took about 1.5 seconds and the extensive analysis of all 56 possible instruction pairs about 1.5 minutes.

After determining the timing requirements of all possible instruction pairs we conducted an exploratory study on the effect of instruction ordering on the resulting worst case circuit delay distribution. In every clock cycle the instruction from the previous cycle together with the current instruction determines which control inputs in the circuit change, i.e., which stay constant at zero, which inputs rise, etc. This in turn determines the worst case circuit delay which gives the maximum clock frequency for the circuit.

Fig. 7 shows an example where the order of instructions impacts the worst case circuit delay distribution. We start with the original function order shown in Fig. 7a. To improve the timing requirements we exchange the OR with the SUB function (Fig. 7b. As shown in Fig. 7d, this change improves the worst case circuit delay distribution of the ALU significantly. To further improve the timing, we try in the next step to exchange the OR and the XOR function (Fig. 7c). This degrades the timing requirements (Fig. 7d) compared to the order shown in Fig. 7b. Thus, the instruction order in Fig. 7b is in this experiment the best function order in terms of circuit timing. For a given clock frequency it is also the most robust one as it shows the smallest variability of the three tested orderings in Fig.7.

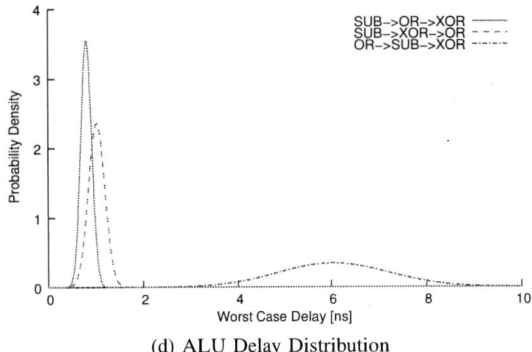

(a) Original Graph (b) Reordered Graph 1 (c) Reordered Graph 2

(d) ALU Delay Distribution

Fig. 7: Instruction Ordering and its effect on the worst case circuit delay distribution of an ALU

V. CONCLUSION AND OUTLOOK

We extended the statistical timing analysis of circuits on gate level to include given functional information. An exploratory study showed that the function order in the operation of a circuit impacts the worst case delay and its variability. As a part of future work, we will study how the obtained information can be used to efficiently guide program optimization towards robustness against process variations. Another topic we want to investigate is the effect of instruction ordering on possible dynamic frequency scaling in a microprocessor.

ACKNOWLEDGMENT

This work was partly supported by the German Research Foundation (DFG) as part of the priority program "Dependable Embedded Systems" (SPP 1500, http://spp1500.itec.kit.edu)

REFERENCES

[1] P. McGeer and R. Brayton, *Integrating functional and temporal domains in logic design: the false path problem and its implications.* Kluwer Academic Publishers, 1991.

[2] H. Yalcin, M. Mortazavi, R. Palermo, C. Bamji, K. A. Sakallah, and J. P. Hayes, "Fast and accurate timing characterization using functional information," *IEEE Transactions on Computer-Aided Design of Integrated Circuits and Systems*, vol. 20, no. 2, pp. 315–331, Feb. 2001.

[3] H. Yalcin, J. P. Hayes, and K. A. Sakallah, "An approximate timing analysis method for datapath circuits," in *IEEE/ACM International Conference on Computer-Aided Design (ICCAD)*, Nov. 1996, pp. 114–118.

[4] C. Visweswariah, K. Ravindran, K. Kalafala, S. G. Walker, and S. Narayan, "First-Order Incremental Block-Based Statistical Timing Analysis," in *ACM/IEEE Design Automation Conference (DAC)*, 2004, pp. 331–336.

[5] *Nangate 45nm Open Cell Library*, Nangate Inc., http://www.nangate.com, 2009.

[6] S. R. Nassif, "Modeling and Analysis of Manufacturing Variations," in *IEEE Custom Integrated Circuits Conference (CICC)*, 2001.

978-1-61284-863-1/11 $26.00 © 2011 IEEE

Hardware Assisted Thread Assignment for RISC based MPSoCs in Invasive Computing

Ravi Kumar Pujari, Thomas Wild, Andreas Herkersdorf
Institute for Integrated Systems
Technical University of Munich, Germany
{ravi.kumar, thomas.wild, herkersdorf}@tum.de

Benjamin Vogel, Jörg Henkel
Chair for Embedded Systems, Dept. of CS
Karlsruhe Institute of Technology, Germany
{benjamin.vogel, henkel}@kit.edu

Abstract—**One of the major challenges of future many-core architectures is the efficient utilization of the abundance of computing power. Invasive computing provides a computing paradigm wherein applications can economically use the available compute resources. Applications can expand and shrink on demand depending on their thread level parallelism and resource availability. In this paper we present an analytical justification for performing a hardware-software co-optimization of the thread assignment in a resource aware programming environment. We propose a dedicated hardware block to support thread assignments as an architectural extension to standard MPSoC designs.**

Index Terms—**thread assignment, resource aware programming, multiprocessor system on a chip, invasive computing**

I. INTRODUCTION

Future MPSoCs (Multi Processor System on a Chip) with up to hundreds of cores provide an abundance of computing power to the applications [1]. To efficiently use the huge amount of computing resources, applications have to be specially adapted to the underlying architecture. Within the invasive computing paradigm proposed in [2], applications may dynamically adapt their run-time behavior to the amount of available hardware resources. Applications can claim resources, use them in parallel and release them.

This new computing paradigm is envisioned to be applied on a tile based, heterogeneous many-core architecture comprising of loosely coupled RISC cores, tightly coupled processor arrays (TCPA [3]), special purpose hardware accelerators and RISC cores with reconfigurable logic blocks (iCores). A hierarchical communication infrastructure to interconnect these numerous processing elements (PE) will be provided. The tile-internal communication is achieved using standard busses while the tile-external communication is performed via an invasive network on chip (iNoC). Moreover, special memory tiles, which can be accessed via the iNoC are included in the architecture.

Applications exhibiting thread level parallelism can make optimum use of such a heterogeneous MPSoC provided an efficient thread assignment policy is used. The contribution of this paper is to analyze the challenges faced during a thread

This work was partly supported by the German Research Foundation (DFG) as part of the Transregional Collaborative Research Centre "Invasive Computing" (SFB/TR 89)

assignment in invasive computing and propose a hardware-software co-optimized hierarchical solution to it.

The remainder of this paper is structured as follows: in Section II, we introduce the novel computing paradigm of invasive computing. In Section III, an overview of related work is presented. In Section IV, we give an analytical justification for performing a hardware-software co-optimization for thread assignment. In Section V, we present our hardware block to assist in thread assignment and finally conclude in Section VI.

II. INVASIVE COMPUTING

The central idea of invasive computing is resource aware programming. Applications dynamically decide which resources to use by making queries for hardware state parameters (e.g. speed, energy consumption, temperature, reliability). If resources could be allocated, which we denote as successful *invasion*, the program adapts itself to the available resources and then *infects* the claimed resources with its execution threads (also called *i*lets). After the execution of *i*lets is finished, the resources are *retreated*, i.e. given back to the middle-ware referred as invasive run-time support system (*i*RTSS). Figure 1 depicts this feedback-loop between the applications, the *i*RTSS, and the underlying hardware.

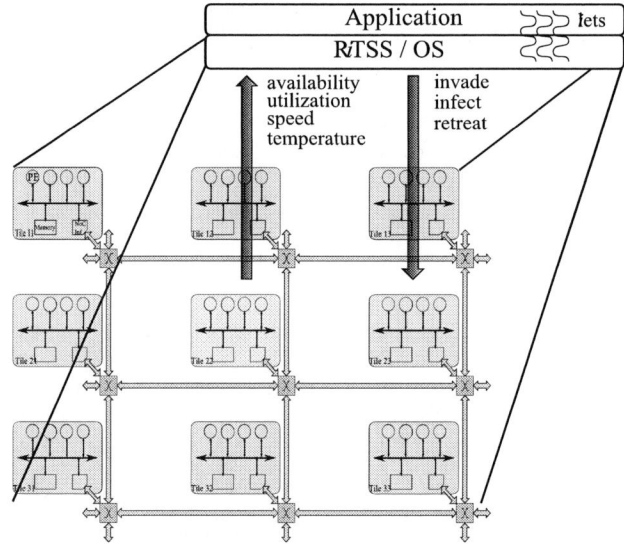

Fig. 1.   Resource Aware Programming

To further illustrate this concept, an example of invasive programs running on our envisaged tiled MPSoC is depicted in Figure 2. Initially, two applications A1 and A2 have already invaded tiles 11, 12, 21 and tiles 31, 32 respectively. Later on, another application A3 starts its invasion from a PE in the right top corner of tile 13 in south-west direction. This invasion might fail due to possibly failing constraints such as unavailability of resources, too high temperature values, or energy and reliability aspects. If the invasion succeeds, A3 infects the claimed PEs with *i*lets and starts executing them in parallel. Once A3 finishes or its level of parallelism decreases, the acquired PEs are (partly) retreated to the *i*RTSS so that other applications can invade them. Applications form dynamically expanding and contracting domains or regions of execution thus exhibiting a sense of "breathing" when running over the complete architecture.

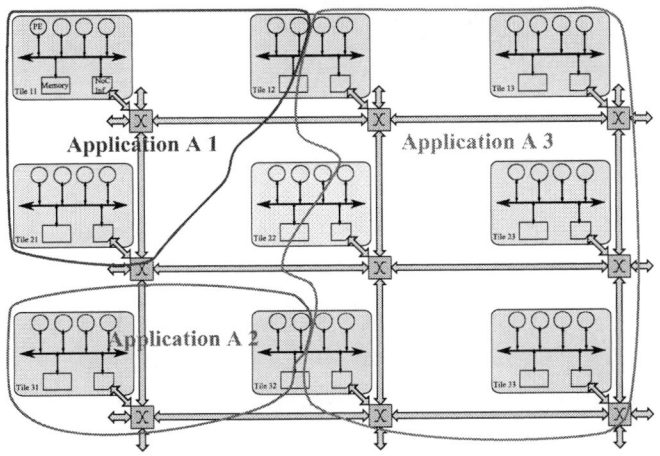

Fig. 2.   Invasive Applications running on a Tiled MPSoC

III. THREAD ASSIGNMENT

A. Related Work

The temporal and spatial mapping of threads is managed by software schedulers [4] in conventional multiprocessing systems. The cell broadband engine design [5] proposes such a solution using a special core to perform the thread assignments to all the available PEs. A centrally located power PE (PPE) takes care of assigning threads to synergistic PEs (SPE). Unfortunately, this approach of having a server (PPE) - client (SPE) model does not relieve the thread assignment decision from the programmer or OS running on the PPE. One has to know a-priori when a task would like to expand thus requiring more SPEs. The reaction time to dynamics in application requirements and chips resource availability would be large which might become a bottleneck in the thread mapping decisions. The scheduler would waste a considerable amount of time in gathering complete status of all the PEs suffering non-uniform access delays in case of our envisioned distributed architecture. This approach is not scalable as getting the exact load situations and other aforementioned hardware status information would be exorbitantly time consuming in the context of hundreds of PEs.

A more scalable approach is used in distributed OS [6], where the scheduler instances running on different PEs make thread assignment decisions based on either incomplete or outdated status information. A part of the scheduler responsible for monitoring resources propagates any changes in the status. This propagation is usually done using interrupt service routines (ISR) or, in a timely scheduled fashion, by a software instance. Thus, the closed loop reaction time to changes in resource status is quite high owing to the software overhead of scheduler or ISRs.

Many specific type of MPSoCs e.g. network [7] or graphic processing units (NPU, GPUs) tend to eliminate the job of thread mapping partially [8] from software making use of dedicated hardwares. They are designed to execute the same task but on different data sets (e.g. network packets, images or video frames). These special purpose PUs are more suitable for single instruction multiple data (SIMD) processing.

A pure software controlled thread assignment policy would not scale in the context of invasive computing given the large multi-criteria search space aggravated by the dynamics in resource conditions. In case of generic MPSoCs, neither there are any architectural supports in the form of dedicated hardware to build in the resource awareness nor do the applications take any of the chips hardware monitoring conditions (temperature, energy consumption, reliability aspects etc.) into consideration while spawning new threads.

B. Requirements for Thread Assignment

To cater a wide range of application classes ranging from high performance computing (HPC), streaming, embedded to networking applications, we identified the following requirements that have to be fulfilled in invasive computing paradigm:

- **Need for Performance**
  The computing power of PEs should be used for running the application's threads (*i*lets) instead of wasting considerable time on deciding where to execute them.
- **Dynamics in computation requirements**
  The application's run-time behavior differs over time [9] as an application can expand or shrink at unpredictable times. Assignment strategies need to be very adaptive.
- **Changing global and local status in hardware**
  The availability of PEs, memory, communication links as well as the temperature values and the power budgets change rapidly. The changes in these parameters have to be incorporated in the thread assignment strategy instantly to cope with potential hardware failures.

IV. HARDWARE ASSISTED THREAD ASSIGNMENT

Thread assignment is a dynamically changing, multi-criteria search problem. It is the task of finding the best PE on the invasive chip to run the next *i*let with minimum search latency while meeting applications' timing constraints. We argue that this task cannot be solved solely either in software or in hardware.

978-1-61284-863-1/11 $26.00 © 2011 IEEE   107

A. Quantitave Analysis of Thread Assignment

Consider an application running on a PE of the center tile (refer Figure 2) wants to expand and run an $i$let in one of the PEs in its neighboring tiles. Even if we contemplate only one constraint for the decision, e.g. the availability of a free PE, the iRTSS/OS running on the $i$let generator has to gather availability status of many other PEs: n_l local PEs within its own tile and n_r remote PEs from neighboring tiles in its vicinity located at h_r hops distance over the iNoC. As the availability status of PEs varies dynamically, a naive approach would be to fetch on every assignment request. On every access, this would suffer σ_{bus} cycles of delay for the bus arbitration and response time and σ_{NoC} cycles for packetization and depacketization as well as round trip request-response time for one hop over the iNoC. Realistically, we need to consider m constraints at a time, often with associated minimum, maximum and average of the monitor values. It is to be noted that the decision process is not only scaled by the number of constraints, but also due to multiple PEs fulfilling the requirements within the bound of constraints, enforcing us to prioritize the comparison. An example set of constraints would be:

```
if((avg_temp(PE[i]) < 50 C) AND
    (max_load(PE[i]) < 20 %))
        select PE[i]
else if((max_hops(PE[j]) < 5 hops) OR
        (energy_budget(PE[j]) > 20 Wh))
            ...

...
```

After fetching the m monitor values the iRTSS still needs $CPI_{compare}$ clock cycles for the less/greater than compare instructions. Even ignoring the clock cycles required for prioritization, the total number of cycles for the thread assignment decision logic (τ_d) can be given by

$$\tau_d = [m \cdot n_l \cdot \sigma_{bus}] + [m \cdot n_r \cdot (\sigma_{bus} + (h_r \cdot \sigma_{NoC}))] + [m \cdot (n_l + n_r) \cdot CPI_{compare}] \quad (1)$$

For our example of selecting another PE, the search space is as follows: 4 local and 16 remote PEs are in the vicinity of one single hop distance over the iNoC from our $i$let generator's tile. Each of the currently 8 assumed monitors will deliver minimum, maximum and average values. If we assume an ideal CPI of 1, 20 CPU cycles access time for the tile internal bus and 100 cycles for the iNoC, we need about 48480 cycles to decide on which PE to run the next thread on. As mentioned earlier, an invasion not necessarily succeeds. In that case, the search space has to be expanded further over many hops causing the invasion to be prohibitively slow. Even after finding a successful PE, we need to take into account the software overhead: the scheduling overhead, the memory overhead to load the $i$let into the target tile's memory, the context switching overhead, and the delay until the $i$let finally starts its execution on the target PE.

B. Hierarchical Decision Making

As both SW and HW are involved in the decision making, a HW-SW co-optimization is a viable solution. An approach to minimize the delay in thread assignment would be to implement (i) global, coarse level mapping decisions in SW for flexibility and scalability reasons, and (ii) finer, localized monitor aggregation, and thread assignment decisions in HW and hide these latencies (τ_d) from the applications.

This is justified in Figure 3 where the overhead due to thread assignment decision τ_d from equation (1) for a total of 20 PEs is depicted with respect to threads runtime τ_{thread}. We modeled different constellations of PEs in remote and local tiles in single hop vicinity. The overhead of the assignment decision for a thread running for 100K cycles is about 32% at point 'A' (n_l=4; n_r=16), which reduces to under 10% at point 'B' (n_l=20; n_r=0) provided we can move all the monitor values of all the PEs from remote tiles into the local tile.

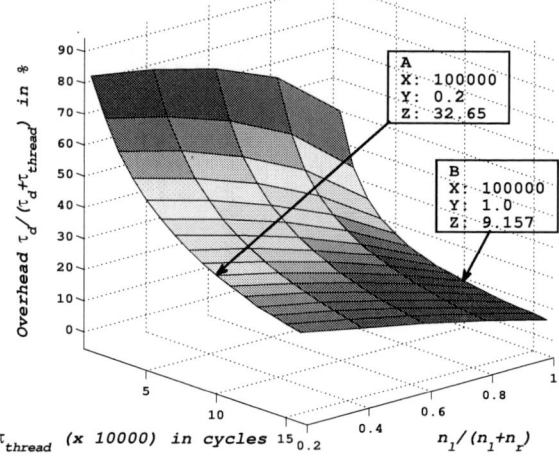

Fig. 3. Thread assignment decision overhead

To mitigate this high overhead for short threads during thread assignment, we propose a hierarchically partitioned thread assignment strategy as follows:

- In the invasion phase, the iRTSS identifies a region or a target tile and not the exact PE. The iRTSS on the source tile uses abstracted status information of the neighborhood region, e.g. average temperature or mean load to take its decision on where to orient its invasion direction. This abstraction causes reduction of factor n_r.
- As the tile's averaged and abstracted monitor data values do not change as rapidly as the individual PE's status, we propose to pro-actively pre-fetch and distribute these abstracted monitor data using a dedicated hardware block, thereby hiding the latency due to σ_{NoC}.
- During infection, the exact mapping and assignment of incoming $i$lets to PEs is done by having a dedicated hardware in the target tile. This hardware block can adapt its assignment decisions to a large set of changing monitor values much faster and without any bus accesses (σ_{bus}), thus supporting a higher infection rate.

By incorporating the above suggestions into the invasive computing paradigm and by providing suitable hardware extensions, we expect to reduce the overhead significantly.

V. Core *i*let Controller

We propose a dynamic many-Core *i*let Controller (C*i*C) as micro-architectural hardware extension to assist the *i*RTSS. The C*i*C assists and accelerates the infection of PEs. Our approach is to delegate those functions to C*i*Cs that can be implemented with much lower latency in hardware, resulting in higher performance of the MPSoC. Figure 4 depicts the following functional blocks that we envisaged in the C*i*C:

A. Monitor Aggregator-Distributor

This block collects detailed and abstracted monitor status of local and remote tiles' IP blocks and provides it to the *i*let mapper. It also forwards its abstracted status to neighboring tiles over the *i*NoC either pro-actively or on demand depending on the configuration by the *i*RTSS.

B. *i*let Mapper

The *i*let mapper selects and triggers the execution of an *i*let on a PE via interrupts or using message passing mechanisms. The incoming *i*lets are evaluated and prioritized based on their processing needs using the local status information from the monitor aggregator. To complete the evaluation within the inter *i*let arrival time, this block is envisioned to be implemented in hardware as a classifier [10] or hash tables [11], configurable by the *i*RTSS.

C. Rule Configurator

It holds the rule base and other adjustable parameters that can be configured at run-time via the *i*RTSS.

D. Energy Manager

This block collects selected, energy-related monitor data (e.g. load/temperature values) from the PEs to influence the local binding as well as to initiate migration decisions due to temperature and power threshold overshoots. Besides, it generates sleep signals to power gate the individual PEs.

E. Infection Support

The infection support accelerates loading and dynamic relocation of the *i*let binary into the local tile memory without additional software intervention.

F. Communication Interface

It provides interfaces to the tile local bus and the network, thus facilitates updating the rule base, reception of *i*lets, as well as the distribution of monitor data over *i*NoC.

VI. Conclusions

In this paper, an analysis of impacts of a novel, resource-aware computing model on the thread assignment strategy from a tile-based MPSoC architecture perspective has been introduced. To mitigate the overhead in thread assignment, we presented our hierarchical, HW-SW co-optimized approach using IP blocks called C*i*C.

Future work will include thorough studies on architecture prototyping based on open source RISC IP cores to measure HW cost figures in terms of area and power consumption. The

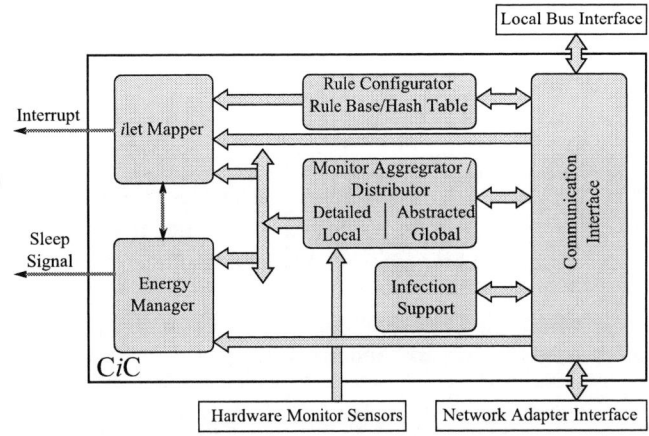

Fig. 4.   Core *i*let Controller

necessary extensions to the programming model to incorporate the resource awareness, the performance benefits as well as its implications on application development, portability, ease of programming etc. are some of the on-going research topics being investigated.

References

[1] S. Borkar and A. A. Chien, "The future of microprocessors," *Commun. ACM*, vol. 54, pp. 67–77, May 2011.

[2] J. Teich, J. Henkel, A. Herkersdorf, D. Schmitt-Landsiedel, W. Schröder-Preikschat, and G. Snelting, "Invasive Computing: An Overview," in *Multiprocessor System-on-Chip – Hardware Design and Tool Integration* (M. Hübner and J. Becker, eds.), pp. 241–268, Springer, Berlin, Heidelberg, 2011.

[3] D. Kissler, F. Hannig, A. Kupriyanov, and J. Teich, "A highly parameterizable parallel processor array architecture," in *Field Programmable Technology, 2006. FPT 2006. IEEE International Conference on*, pp. 105–112, dec. 2006.

[4] A. S. Tanenbaum, *Modern Operating Systems*. Upper Saddle River, NJ, USA: Prentice Hall Press, 3rd ed., 2007.

[5] J. A. Kahle, M. N. Day, H. P. Hofstee, C. R. Johns, T. R. Maeurer, and D. Shippy, "Introduction to the cell multiprocessor," *IBM J. Res. Dev.*, vol. 49, pp. 589–604, July 2005.

[6] T. Casavant and J. Kuhl, "A taxonomy of scheduling in general-purpose distributed computing systems," *IEEE Transactions on Software Engineering*, vol. 14, pp. 141–154, 1988.

[7] N. Shah, "Understanding network processors," Master's thesis, University of California, Berkeley, Sep 2001.

[8] V. W. Lee, C. Kim, J. Chhugani, M. Deisher, D. Kim, A. D. Nguyen, N. Satish, M. Smelyanskiy, S. Chennupaty, P. Hammarlund, R. Singhal, and P. Dubey, "Debunking the 100x gpu vs. cpu myth: an evaluation of throughput computing on cpu and gpu," *SIGARCH Comput. Archit. News*, vol. 38, pp. 451–460, June 2010.

[9] T. Sherwood, E. Perelman, G. Hamerly, S. Sair, and B. Calder, "Discovering and exploiting program phases," *IEEE Micro*, vol. 23, pp. 84–93, November 2003.

[10] J. Zeppenfeld, A. Bouajila, W. Stechele, and A. Herkersdorf, "Learning classifier tables for autonomic systems on chip." Lecture Notes in Informatics,Springer,Vol. 134, p. 771-778, Sep 2008.

[11] R. Ohlendorf, M. Meitinger, T. Wild, and A. Herkersdorf, "A processing path dispatcher in network processor mpsocs," *IEEE Trans. Very Large Scale Integr. Syst.*, vol. 16, pp. 1335–1345, October 2008.

Real-Time Signal Processing on Low-Cost-FPGAs using Dynamic Partial Reconfiguration

Michael Feilen, Matthias Ihmig, Anton Zahlheimer, Walter Stechele

Lehrstuhl für Integrierte Systeme, Technische Universität München

michael.feilen@tum.de, matthias.ihmig@tum.de, anton.zahlheimer@mytum.de, walter.stechele@tum.de

Abstract—The limited number of logic elements makes the implementation of signal processing chains on low-cost FPGAs a challenging task. In order to allow the implementation of complex designs on devices with limited resources, we propose the subpartitioning of a processing chain into several modules, which are loaded and executed in a round-robin fashion using dynamic partial reconfiguration (DPR) of FPGAs. The DPR architecture requires input data pre-buffering which introduces delay. These circumstances are first considered in a theoretical analysis and later applied to a broadcast receiver chain, where the benefits and drawbacks of the architecture are highlighted.

Index Terms—**FPGAs, Signal Processing, Low-Cost, DSP, Xilinx Spartan-6, Dynamic Partial Reconfiguration (DPR)**

I. INTRODUCTION

Partial reconfiguration of FPGA logic elements can be used to increase the utilization and might enable the system designer to realize a complex application on a cheaper FPGA with fewer logic resources. We apply the idea of time-multiplexing FPGA logic resources by executing different blocks of a signal processing chain from one submodule of the processing chain to the next submodule using DPR. This way, the active module of the processing chain can use *all* available FPGA resources in the reconfigurable partition. This makes it possible to trade-off the number of modules, i.e. the quantitative reusability of the hardware, against the processing delay of the system.

Although the FPGA market is shared by different vendors, our work targets the Spartan-6 FPGA line of the vendor Xilinx. However, the general idea of the architecture is applicable to other reconfigurable FPGAs.

II. PRIOR ART

Research related to dynamic partial reconfiguration of FP-GAs mainly focusses on reconfigurable CPUs [1], video-processing systems [2] and security applications [3]. A partitioning approach for the implementation of a sequential DAB receiver on a Virtex platform has been presented in [4]. Different methods for using DPR together with the low-cost Xilinx Spartan-6 platform have been published in [5] and [6].

In our work we will use DPR of FPGAs to realize complex signal processing chains which inherently consumes a high amount of logic. We will derive parameters for the design of a a general-purpose reconfiguration architecture with focus on real-time signal processing, which can be designed and adapted with respect to common system parameters, such as processing delay and sample rate.

III. PROPOSED SYSTEM ARCHITECTURE

We propose a differential reconfiguration approach and a sub-partitioning of the FPGA into one static part and one reconfigurable part. The static part of the system includes a data preprocessing engine, a multi-port memory controller to access external memory, a reconfiguration controller and proxy logic to interface the reconfigurable partition as shown in Figure 1. The reconfiguration controller includes a bitstream loader statemachine that sequentially loads the configuration bitstreams of the processing chain from the external SDRAM. The differential FPGA configuration bitstreams for the different signal processing modules are loaded and executed one after the other in the reconfigurable partition.

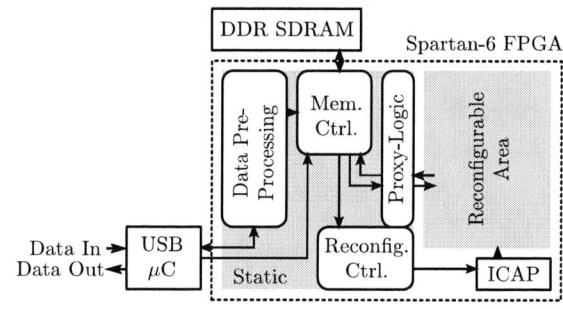

Fig. 1. Schematic of proposed reconfigurable system architecture.

After having briefly introduced the system architecture, we will derive the key parameters which determine the realizability of such a real-time signal processing chain and explain the DPR approach in further detail.

IV. ANALYSIS OF THE ARCHITECTURE

Given is a signal processing chain which is subdivided into M different modules as shown in Figure 2. Each module reflects one configuration of the reconfigurable part of the DPR system. The execution scheme of the modules is assumed to be round-robin, i.e. the processing sequence is $m_1, m_2, m_3, ... m_M, m_1$ and so on. This scheduling scheme is depicted in Figure 3.

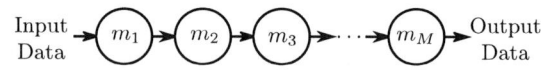

Fig. 2. Processing pipeline consisting of M reconfigurable modules.

978-1-61284-863-1/11 $26.00 © 2011 IEEE

The processing duration for all modules of the pipeline can be expressed by:

$$T_{\text{CYCLE}} = \sum_{m=1}^{M} T_{m,\text{RECO}} + T_{m,\text{EXEC}} + T_{m,\text{CTSW}} \qquad (1)$$

where $T_{m,\text{RECO}}$ is the time needed for reconfiguration of each module, $T_{m,\text{EXEC}}$ is the processing time including memory transfer delays and $T_{m,\text{CTSW}}$ is the time for storing and recovering the module context. The reconfiguration time $T_{m,\text{RECO}}$ is defined by the throughput of the configuration interface (ICAP) and by the size of the bitstream as follows:

$$T_{m,\text{RECO}} = \frac{N_{m,\text{BITSTR}}}{B_{\text{ICAP}} \cdot f_{\text{ICAP}}} \qquad (2)$$

where $N_{m,\text{BITSTR}}$ denotes the number of bytes for the configuration bitstream of module m, B_{ICAP} denotes the width of the ICAP in multiples of bytes and f_{ICAP} states the ICAP clock frequency.

The time for the context switch $T_{m,\text{CTSW}}$ depends on the memory throughput and on the amount of data that needs to be written and restored per module.

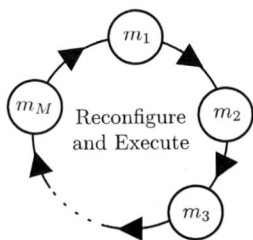

Fig. 3. Round-robin reconfiguration scheduling for the M different modules of the processing pipeline inside the reconfigurable system.

The number of modules M in Eq. (1) is an important parameter of the system since it determines the overall system processing time T_{CYCLE}.

A. Bounds on the Frame-Throughput per Module

In this section we discuss the necessity for input buffering in our architecture and derive a lower and upper bound on the frame-throughput per module given the input buffer size T_{INBUF} and the system cycle time T_{CYCLE}.

The processing of the input data is assumed to be performed framewise. One frame corresponds to a fixed duration in time, further denoted as T_{FRAME}, which is a design-parameter of the processing chain. In order for the system to be real-time-capable each module must process a certain number of input frames during the time it is active. In other words, for $M > 1$, each module must process one frame in less than T_{FRAME} seconds. Hence, processing N_F frames per module execution requires reading the input data faster than new data arrives, which requires to pre-buffer a certain amount of incoming data in an input buffer.

Claim: Since it takes T_{CYCLE} seconds for the first module m_1 to be executed again, the input buffer of the DPR system must be sized to store at least T_{CYCLE} seconds of input data.

Proof: An input buffer *overrun* occurs if the modules can not process the data fast enough referenced to the incoming stream datarate. Hence, each module must be designed to achieve a minimum throughput of

$$N_F \geq \left\lceil \frac{T_{\text{CYCLE}}}{T_{\text{FRAME}}} \right\rceil \qquad (3)$$

input frames per module. In order for the input buffer not to *underrun*, the modules can not read more than a certain number of frames. The maximum number of frames the pipeline can process is limited by the input buffer storage duration T_{INBUF} as follows :

$$N_F \leq \left\lfloor \frac{T_{\text{INBUF}}}{T_{\text{FRAME}}} \right\rfloor \qquad (4)$$

and hence

$$T_{\text{INBUF}} \geq T_{\text{CYCLE}} \qquad (5)$$

In the next section the design of the input buffer and the overall system delay of the DPR architecture are derived.

B. Input-to-Output Delay and Input-Buffer Considerations

The input-to-output delay T_{DELAY} is defined as the time between the arrival of the first input sample and the time the last module of the processing pipeline has finished writing the first set of output data to the memory. The input-to-output delay is an important parameter of real-time systems and furthermore determines the performance of our DPR design.

As already mentioned in Section IV-A the pre-buffered data is continuously read from the input buffer by the processing pipeline. The input buffer was defined to hold T_{INBUF} seconds of data, or at least N_F input frames. Since the first module is required to process all N_F frames at once, the system must wait $N_F \cdot T_{\text{FRAME}}$ seconds initially to pre-buffer the data until it can begin with the execution of the first module in the processing pipeline.

In the following analysis we will focus on input buffer designs with two buffers A and B, where each buffer has the capacity to store

$$T_{\text{INBUF}} = N_F \cdot T_{\text{FRAME}} \qquad (6)$$

seconds of data. While the processing pipeline reads the data from buffer A with capacity T_{INBUF}, the incoming samples are written to another buffer B with the same capacity T_{INBUF}. After T_{INBUF} seconds of incoming data have been written to buffer B, the buffers A and B are swapped. This buffering scheme is also known as *double buffering* or *circular buffering*. Using the buffering constraint in Eq. (6), the required input buffer capacity becomes $2 \cdot T_{\text{INBUF}}$ and the input-to-output delay T_{DELAY} becomes:

$$T_{\text{DELAY}} = T_{\text{INBUF}} + T_{\text{CYCLE}} \qquad (7)$$

Figure 4 summarizes the timing constraints and the framing of the proposed architecture. In this drawing the system idle time T_{IDLE} is defined as the time between the end of the processing of all modules and between the completion of the current input buffer for pre-buffering.

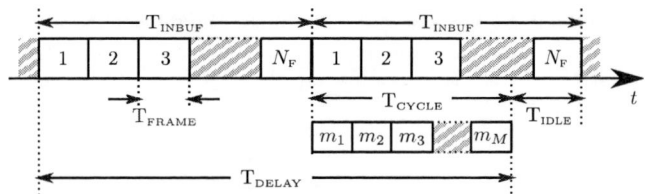

Fig. 4. Continuous stream of input samples sub-partitioned into N_F frames per reconfigurable module with a duration of T_{FRAME} seconds per frame.

C. Upper Bounds on the Number of Modules

Bearing in mind the architectural constraints derived in the last sections we will further derive two upper bounds on the number of modules that can be used for the realization of the processing chain.

1) Upper Bound by Delay Requirement: The throughput N_F in number of frames is upper bounded by the input buffersize of the DPR system and lower bounded by the cycle time. In order for the system to function properly, the input buffer constraints discussed in the last sections must be satisfied. From these constraints we can derive a limit for the maximum number of reconfigurable modules that can be used for the DPR pipeline.

Let the average processing time per module be $\overline{T_M}$. In this case the cycle time T_{CYCLE} can be expressed by:

$$T_{CYCLE} = M \cdot \overline{T_M} \tag{8}$$

Given this approximation, the minimum number of frames per module becomes:

$$N_F = \left\lceil \frac{\overline{T_M} \cdot M}{T_{FRAME}} \right\rceil \tag{9}$$

Furthermore, by putting Eq. (4) in Eq. (9) an upper bound on the number of modules M with respect to the system parameters can be formulated as:

$$M \leq \left\lceil \frac{T_{INBUF}}{\overline{T_M}} \right\rceil \tag{10}$$

From the last inequality we can derive a constraint on the number of modules with respect to the input-to-output delay stated in Eq. (7) as follows:

$$M \leq \left\lceil \frac{T_{DELAY} - T_{CYCLE}}{\overline{T_M}} \right\rceil = \left\lceil \frac{T_{DELAY}}{2 \cdot \overline{T_M}} \right\rceil \tag{11}$$

Eq. (11) states that the number of modules increases with the delay in the DPR system. Additionally, the equation states that the maximum number of modules can be increased by decreasing the processing time per module.

2) Upper Bound by Memory Throughput: Another upper bound on M can be formulated by considering the memory throughput of the system.

The maximum frame-throughput of each module is determined by the throughput of the external memory or, more specifically, by the memory controller subsystem on the FPGA. The memory throughput rate r_{MEM} can be calculated by:

$$r_{MEM} = B_{MEM} \cdot f_{MEM} \tag{12}$$

where B_{MEM} states the number of bytes per transfer and f_{MEM} is the transfer clock frequency. The memory controller is assumed to function full-duplex bidirectional, i.e. the same rates can be achieved for read and write accesses. The incoming data is provided with a sample frequency f_{IN} and with a resolution of B_{IN} bytes per sample at a rate of:

$$r_{IN} = B_{IN} \cdot f_{IN} \tag{13}$$

Usually, the data rate does not increase during the signal processing inside the different modules but rather stays constant or decreases. Given this assumption, in a worst-case scenario all modules use the same memory throughput r_{MEM}. Since the memory controller throughput limits the amount of data a module m can process in one clock cycle, on average the number of modules is constraint to:

$$M \leq \frac{r_{MEM}}{r_{IN}} \tag{14}$$

When the bound in the last equation is tighter than the bound in Eq. (11), the memory throughput limits the processing speed of each module. Otherwise, the delay requirement is the limiting factor for the number of modules.

D. Discussion

For a fixed memory throughput f_{MEM} in Eq. (14) the number of modules M is upper bounded by the input sample rate f_{IN}. This bound is further referred to as *throughput bound*. For a fixed average module processing time $\overline{T_M}$ in Eq. (11) the number of modules is upper bounded by the maximum tolerable input-to-output delay T_{DELAY}, which is why we will refer to this bound as *delay bound*.

Given the constraints presented in the Sections IV-A and IV-C, the chance of meeting realtime requirements of the system increases with an increasing distance to both of the bounds in direction of using fewer modules and lower sample rates.

V. EXAMPLE ANALYSIS FOR OFDM BROADCAST RECEIVERS

An OFDM broadcast receiver requires the implementation of a variety of different signal processing blocks. On FPGAs with limited logic resources the designer might not be able to integrate all components necessary to decode the OFDM signal.

In our prototype system the configuration data is written and read using an ICAP frequency of $f_{ICAP} = 48$ MHz. Since the Spartan-6 ICAP has a wordwidth of 16 bit, the reconfiguration delay per module can be calculated using Eq. (2):

$$T_{m,RECO} = \frac{601.125 \text{ kBytes}}{2 \text{ Bytes} \cdot 48 \text{ MHz}} = 6.26 \text{ ms}$$

Although the size of the differential bitstreams for the reconfigurable partition may be less than 601.125 kBytes, under the condition that we reconfigure almost all resources of the reconfigurable partition, the size of the bitstream asymptotically reaches this value.

1) Number of Modules: In an OFDM receiver it is beneficial to set the frame duration equal to the OFDM symbol duration in time. For the subsequent examination the OFDM symbol duration is defined to be $T_{FRAME} = 2.5$ ms with 500 subcarriers per symbol and the input sample rate shall be equal to $f_{IN} = 4$ MSamples/s. The number of bytes per sample B_{IN} shall be defined to be 4 Bytes, i.e. one complex sample with 16 bit per I and Q component.

Since all receiver types utilize time interleaving schemes, which inevitably apply a constant decoding delay, the delay introduced by the DPR block processing mostly can be neglected. In our example, the maximum delay of the DPR processing shall be defined to be $T_{DELAY} = 500$ ms. In this case, according to the delay bound, the maximum number of modules that can be processed is

$$ M \leq \left\lfloor \frac{500 \text{ ms}}{2 \cdot 60 \text{ ms}} \right\rfloor = 4 $$

Due to the high memory bandwidth in our system, the delay bound with $M = 4$ is tighter than the throughput bound where $M = 12$.

Finally, it is important to check whether the system can handle the frame throughput as this also determines the realizability of the DPR chain. The average number of frames N_F per module can be calculated using Eq. (9). Using the values defined in this section, the number of frames per module N_F must be at least

$$ N_F = \left\lceil \frac{60 \text{ ms} \cdot 4}{2.5 \text{ ms}} \right\rceil = 96 $$

In conclusion, each module must be able to cope with a throughput of 96 frames per execution cycle, which means all processing logic must fit into one reconfigurable partition. Given the FPGA clock frequency of 48 MHz and neglecting the context switch, for the broadcast systems mentioned previously, a throughput of 98 OFDM symbols within a module execution time of roughly $60 - 6.26 = 54$ ms yields the following number of clock cycles per symbol:

$$ \frac{54 \text{ ms} \cdot 48 \text{ MHz}}{96 \text{ symbols}} = 27 \text{ kHz per symbol} $$

With 500 subcarriers per symbol the number of cycles for the processing of each carrier can be calculated to:

$$ \frac{27 \text{ kHz per sym.}}{500 \text{ carriers per sym.}} = 54 \text{ Hz per carrier} $$

In summary, within each of the four modules the designer can utilize 54 clock cycles for the processing of one carrier.

2) OFDM Receiver Partitioning: An important question is how to split the receiver chain into different parts of approximately the same size. For the broadcasting standards mentioned, the DFT logic and the Viterbi decoder logic consume most of the resources [7]. Usually, the channel estimation and equalization stages may require a high amount of logic, depending on the performance of the algorithm. An example for the partitioning of the 4 submodules of an OFDM receiver chain could be:

m_1: Preprocessing and Synchronization
m_2: Demodulation and Demapping
m_3: Channel Equalization and Decoding
m_4: Demultiplexing and Postprocessing

VI. CONCLUSION

The presented DPR architecture enables the design of complex signal processing systems on FPGAs with limited logic resources. For systems comprising many processing blocks, the architecture can enable the use of smaller FPGAs by time-multiplexing the logic resources. However, since the presented method inherently introduces additional delay for context switch and reconfiguration, the designer has to consider the delay requirement of the application when realizing a DPR-based processing chain. As the original pipelined application is broken up into sequentially processed modules, constraints due to additional memory transfers must be taken into account as well.

VII. ACKNOWLEDGEMENTS

The authors would like to thank Dirk Koch and Philipp Schmidbauer for valuable contributions and insightful discussions on the topic of partial reconfiguration with focus on Spartan-6 FPGAs, as well as the Bundesministerium für Wirtschaft und Technologie for supporting this project under Grant 10 P 8012B.

REFERENCES

[1] L. Bauer, M. Shafique, and J. Henkel, "Efficient resource utilization for an extensible processor through dynamic instruction set adaptation," *IEEE Trans. Very Large Scale Integr. Syst.*, vol. 16, pp. 1295–1308, October 2008. [Online]. Available: http://portal.acm.org/citation.cfm?id=1515843.1515848

[2] B. Krill, A. Amira, A. Ahmad, and H. Rabah, "A new fpga-based dynamic partial reconfiguration design flow and environment for image processing applications," in *Visual Information Processing (EUVIP), 2010 2nd European Workshop on*, july 2010, pp. 226 –231.

[3] Z. E. A. A. Ismaili and A. Moussa, "Self-partial and dynamic reconfiguration implementation for aes using fpga," *CoRR*, vol. abs/0909.2369, 2009.

[4] M. Ihmig, N. Alt, and A. Herkersdorf, "Resource-efficient Sequential Architecture for FPGA-based DAB Receiver," in *Proceedings of the 5th Karlsruhe Workshop on Software Radios*, March 5/6 2008, pp. 101–107.

[5] D. Koch, C. Beckhoff, and J. Tørrison, "Advanced partial run-time reconfiguration on spartan-6 fpgas," in *Field-Programmable Technology (FPT), 2010 International Conference on*, dec. 2010, pp. 361 –364.

[6] J. Meyer, J. Noguera, M. Hübner, L. Braun, O. Sander, R. Gil, R. Stewart, and J. Becker, "Fast start-up for spartan-6 fpgas using dynamic partial reconfiguration," in *Design, Automation Test in Europe Conference Exhibition (DATE), 2011*, march 2011, pp. 1 –6.

[7] M. Ihmig, N. Alt, and A. Herkersdorf, "Implementation and Fine-grain partitioning of a DAB SDR receiver on an FPGA-DSP platform," in *Proceedings of the 6th Karlsruhe Workshop on Software Radios*, March 3/4 2010.

Versatile Ultra Low Noise Low Power Analog Signal Conditioning Chip With Integrated Drivers

Sanjay Joshi, Viral Thaker and Maryam Shojaei Baghini, *Senior Member, IEEE*
Department of Electrical Engineering,
Indian Institute of Technology(IIT)-Bombay, Mumbai, India.
Email: sanjayjoshi@ee.iitb.ac.in, viralpthaker@ee.iitb.ac.in, mshojaei@ee.iitb.ac.in

Abstract—An ultra low-noise, low-power and miniature area analog signal conditioning chip in 180nm UMC Mixed-Mode CMOS process with 1.8V supply is presented. The design targets applications like sensors, biomedical and energy-efficient hand-held devices. The test chip features Instrumentation Amplifier (INA) with chopper modulation at the first stage. The 2nd stage is a novel area efficient Spike Removal Filter (SRF). On-chip clock generators with frequency of 4kHz (non-overlapping) and 8kHz for SRF stage is implemented. The last stage is a differential active RC filter to adjust gain and bandwidth of the forward channel. The chip features a reconfigurable Driven-Right-Leg Circuit (DRLC) and Shield Drive Amplifier (SDA) in the feedback path specifically for Bio-medical applications. The DRLC is re-configurable with Operational amplifier (Op-Amp) and Operational Transconductance Amplifier (OTA). The measurement results show that INA achieves input-referred noise density of $28nV/\sqrt{Hz}$ and DC current of $5.9\mu A$ maintaining minimum CMRR of 109dB at 1.91kHz. Moreover, 34dB of 50Hz interference reduction is achieved with DRLC. Wide range of specifications along with reconfigurable modules and interconnections enables the chip for broad range of signal conditioning applications.

Index Terms—Chopper modulated Instrumentation Amplifier, Driven Right Leg Circuit, Shield Drive Amplifier, Spike Removal Filter.

I. INTRODUCTION

There is an enormous demand for portable low-cost health-care devices for monitoring vital signals having amplitudes as low as a few μV and bandwidths up to a 1 kHz [1]–[3]. Furthermore, there are many emerging applications that involve network of sensors and their associated precise instrumentation. For example pollution detectors and sensors for measuring quality of soil or water are such daily life applications. Low power, miniaturized and low cost monitoring/sensing devices are the key components in such systems. The performance of these devices directly depends on analog processing front end, which must extract and amplify extremely small signals amidst a noisy environment. Depending on the usage scenario the key specifications involved in extracting such signals are immunity to interference, high CMRR and low noise performance, high-pass filtering characteristic to compensate for base line drift, configurable gain and filter characteristics and high input impedance. Besides the system must consume minimal power for long power autonomy.

The distinguishing features of the presented work, as compared to earlier works [1]–[5], are on-chip re-configurable Driven Right Leg Circuit (DRLC) using OTA and op-amp

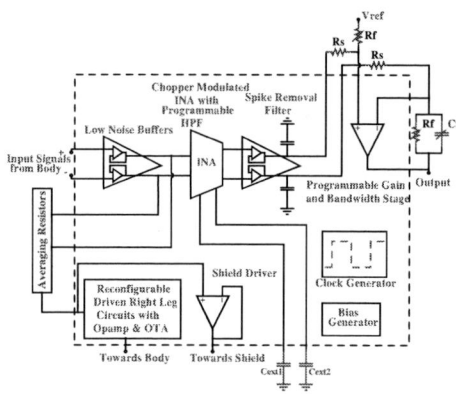

Fig. 1. Block Diagram of Analog Signal Conditioning Chip

plus ultra low-power, ultra low-noise Instrumentation amplifier (INA) with chopper modulation, followed by an area efficient Spike Removal Filter (SRF) while preserving the base line level. The test chip also features simple non-overlapping on-chip clock generators for chopper and SRF stage. All required drivers such as Shield Drive Amplifier (SDA) for further reduction of interference (optional) and buffers are implemented on the chip. This enables the chip as a stand alone general-purpose chip on many target products for low-cost health care. The implemented INA exhibits at least 40% reduction in current consumption, area and input referred noise while maintaining very high CMRR as compared to reported INA. Unlike most of the reported INAs, we have used a scaled technology for implementation of the entire signal conditioning module. This enables integration of entire circuit modules with wireless transceivers all in one single chip.

This paper is organized as follows. Section II presents architecture and internal modules of the test chip. Section III explains the design methodologies. Section IV shows measurement results of the chip and test outputs. Finally Section V concludes the paper.

II. SYSTEM AND IC ARCHITECTURE

Fig. 1 shows the architecture of implemented analog conditioning chip. It consists of Instrumentation Amplifier, Spike Removal Filter (SRF) and a rail-to-rail programmable differential active RC filter in the forward path. Two ultra low-noise

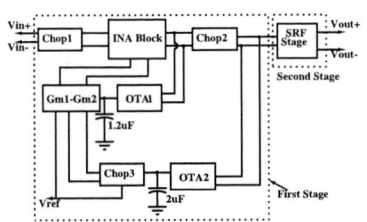
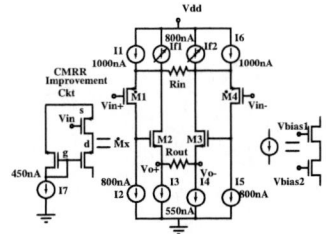
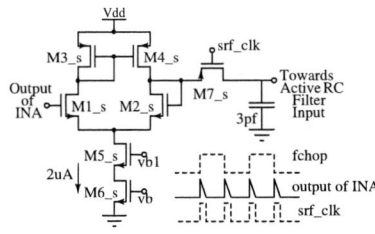

Fig. 2. Block Diagram of Instrumentation Amplifier   Fig. 3. Schematic of Instrumentation Amplifier   Fig. 4. Schematic of Spike Removal Filter

on-chip buffers isolate signal conditioning modules from noisy environment. Other functional modules of the chip are shield drive, bias generator, clock generator frequency of 4kHz (non-overlapping) for chopping and another 8kHz clock for SRF stage. On-chip DRLC is re-configurable to an OTA-based or Op-Amp based driver. The entire signal conditioning channel is able to drive up to 40pF capacitive load for external ICs like data converter and Micro-controller.

III. ELECTRICAL AND PHYSICAL DESIGN

A. Instrumentation Amplifier (INA)

INA is mainly divided into two stages as shown in Fig. 2. First stage [1] amplifies the Input signal by gain of 10V/V maintaining the noise level much below the amplitude of acquired signal by chopper modulation. Second stage is novel area efficient Spike Removal Filter (SRF) to reduce spikes generated by chopper. The main features of implemented INA are as follows.

1) Ultra Low noise performance: To achieve ultra low noise the well-known method of chopper modulation, as shown in Fig. 2, is used [1]–[3], [5]–[7]. The chopper, operating at 4kHz, up-converts the flicker noise. This modulated noise is then filtered out from the base-band signal. Moreover, PMOS input transistors operating in weak inversion are used to reduce the INA flicker noise.

2) High-pass filtering: High pass filtering has been established by using the two gm-C (OTA_2,C=2μF and OTA_1,1.2= μF) in feedback as shown in Fig. 2. It removes unwanted signals from the base-band signal. Therefore, INA works in AC coupled mode of operation.

3) High CMRR: By using chopper modulation technique and AC coupled mode of operation CMRR is improved. Moreover, input transistors are made up by cascode structure (shown in Fig. 3) for keeping V_{ds} of the input transistor constant with change in input DC voltage [1]. This structure increases the impedance of the input transistor. In addition all the current sources have large impedance by using large L which eventually enhances CMRR. All the important transistors are laid down in common centroid technique that reduces mismatch and improves CMRR.

4) Optimal design for area, noise and power performance: Input referred noise expression 1 shows that as we decrease the R_1 resistance the noise and area decreases but, at the same time the power consumption increases. Because input transistor G_m is inversely proportional to R_1. Therefore, by

increasing the R_1 power reduces but the noise level and area increases as the gain is defined as the ratio of R_2/R_1. Therefore, if R_1 increases then R_2 increases by 10 times.

$$V_{in}^2 = 2V_m^2 + V_{R1}^2 + \frac{2gm_{I1}^2}{g1^2}V_{I1}^2 + \frac{2gm_{I2}^2}{g1^2}V_{I2}^2 + \frac{2gm_{I3}^2}{g1^2}V_{I3}^2 \tag{1}$$

Where V_m is noise due to input transistors M1 and M4, V_{R1} is noise due to resistor R1, $gm_{I1}, gm_{I2}, gm_{I3}$ are transconductance of current sources I_1, I_2 and I_3 respectively.

B. Spike Removal Filter(SRF)

The main problem of chopping circuits is the switching noise generated at the output due to charge injection and clock feed-through. To attenuate switching noise, SRF stage is used. Generally, SRF stage is made up of PMOS buffer that has the drawback of shifting the DC level of the signal. Therefore, it will limit the voltage swing at the input of the following DC-coupled stage. In the case of high supply voltage, the swing might not be a concern. However, as the supply voltage scales effect of threshold voltage of the transistor is more. As a result it will be very difficult to have SRF DC-coupled to the next stage. Otherwise SRF needs additional capacitors which add to the silicon area. To overcome this problem we propose a novel solution in the presented chip. Here SRF constitutes of switches series with differential buffers having high input capacitance. This configuration enables attenuating spikes while maintaining the DC level of the signal at the output. The proposed area-efficient SRF stage is shown in Fig. 4.

C. Programmable Gain and Bandwidth stage

To suit gain and bandwidth requirement of different signals, first order differential Active RC filter is designed. The resistor to realize variable gain and capacitor to realize filter are off-chip components. Gain of 20 is kept in order to record ECG of human body.

D. Driven Right Leg Circuit

Traditionally, the common mode voltage V_{cm} can be reduced by employing Driven Right Leg circuit. It senses the common-mode voltage on the body by the averaging resistors R_2 and R_3 and feeds back to the right leg. The negative feedback drives the common mode voltage to a low value [8].

978-1-61284-863-1/11 $26.00 © 2011 IEEE   115

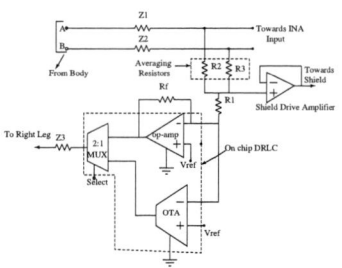
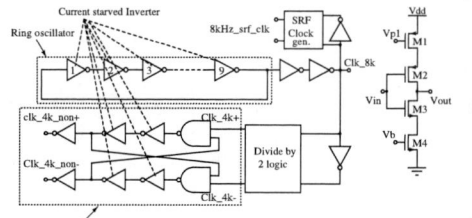
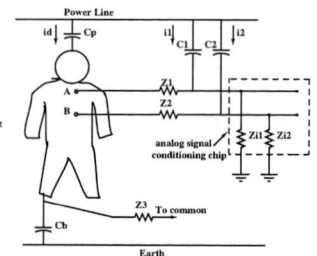

Fig. 5. Re-configurable Driven Right Leg Circuit and Shield Drive Amplifier

Fig. 6. Schematic of Clock Generator circuit and current starved inverter

Fig. 7. Interference model for Biomedical acquisition system

In the case of op-amp based DRLC more reduction in V_{cm} can be achieved by increasing the gain as seen from Eq. 2 [9], [10].

$$V_{cm} = Z_3 \frac{i_d}{\left(1 + \frac{R_f}{R_1}\right)} \quad (2)$$

On the other hand, using trans-conduction amplifier to drive patient's body results in far more advantages then using conventional op-amp. Transfer function of the OTA consists of only 2 effective poles compared to three effective poles in classical Op-Amp [11]. Thus the requirement of on-chip compensation capacitor is relaxed which will in turn reduce area of the chip as well as power. Moreover, external resisters R_f and R_1 are not required which makes DRLC fully on chip with acceptable level of common mode voltage on the body. The DRLC is re-configurable with OTA and Op-Amp as shown is the Fig. 5.

E. On Chip Clock Generator

For modulating and demodulating the signal 4kHz Non overlapping clock and for controlling the SRF stage 8kHz clock having 10% duty cycle is needed. First 8kHz Clock is generated by ring oscillator based on current starved inverter as shown in Fig. 6. Current starved inverter has the benefit of controlling the current flowing through the inverter by control voltage. The frequency of ring oscillator can be controlled by two knobs. One of the knob is current flowing through the inverter and another knob is the parasitic capacitance of the inverter. The relationship for frequency of oscillation is shown by Eq. 3.

$$f_{osc} = \frac{id}{N.C_{tot}V_{dd}} \quad (3)$$

Where i_d is current flowing through inverter, C_{tot} is the total capacitance as a load for each current starved inverter stage,N is the number of stages,V_{dd} is the supply voltage.

In this case we have used N=9 for generating the 8kHz clock. Buffer is connected at the output of ring oscillator that are used to sharpen the clock edges. 4kHz non overlapping clock and 8kHz clock with 10% duty cycle is generated from 8kHz clock by non over lapping clock generation circuit as shown in Fig. 6. For generating the non overlapping 4kHz

clock, 8kHz clock is divided by two and the divide by two clocks output is applied to the non overlapping generation circuit. Delay element for non overlapping clock generation circuit is also made up of current starved inverter. The second clock, Clock SRF is also generated by using similar logic as used for non overlapping clock generation [10].

IV. MEASUREMENT RESULTS

The test chip is fabricated in 180nm UMC Mixed Mode CMOS process. It occupies an area of 655 μm x 365μm, as shown in the chip micro-photograph in Fig. 8. The chip is powered by supply voltage of 1.8V, using two 1.5V batteries. Fig. 9 shows measured noise spectral density of the INA with

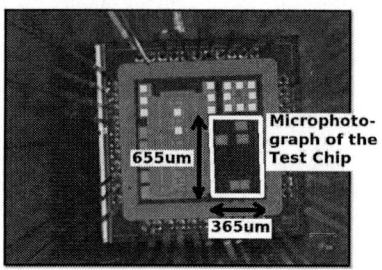

Fig. 8. Micro-photograph of the chip

chopper and without chopper. The measured noise density of INA with chopper is 28nV/$\sqrt{Hz}$. SR 530 Lock-in amplifier is used for precise noise measurement. Fig. 10 shows the

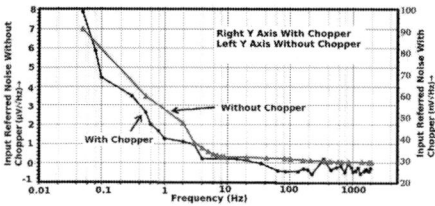

Fig. 9. Input Referred Noise density of INA with and without chopper

induced 50Hz interference voltage on the body. (A) shows the interference voltage on the body without DRLC. The induced peak to peak voltage is 1.72V which may saturate

978-1-61284-863-1/11 $26.00 © 2011 IEEE   116

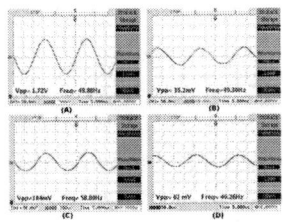

Fig. 10.   Interference voltage on the body (A) Without DRLC (B) DRLC with OTA (C) DRLC with op-amp, gain=1 (D) DRLC with op-amp, gain=20

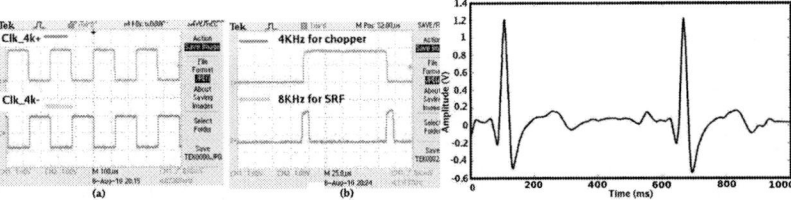

Fig. 11.   Measurement results of clock generator (a) 4kHz clock for chopping (b) 8kHz clock for SRF

Fig. 12.   Recorded ECG with analog conditioning chip

TABLE I
PERFORMANCE COMPARISON OF THE IMPLEMENTED WORK WITH EARLIER WORKS

	This Work	[1]	[3]	[6]	[12]
Technology (nm)	180	500	90	800	350
Supply voltage (V)	1.8	2	3	1.7-3.3	3.3
Differential input range (mV)	40	N/A	N/A	N/A	12
Maximum channel gain	5000	2500	2500	N/A	6
Input common mode range (V)	0.7-1.3	N/A	1.05-2.3	N/A	0.3-2.3
High frequency cutoff (Hz)	Adjustable	0.34	0.4	Adjustable	N/A
Input referred noise(nV/$\sqrt{Hz}$)	28	60	51.4	N/A	N/A
CMRR(dB)	109@1.91 kHz	120	140	80	100@60 Hz
THD(%)	0.016	N/A	0.1	0.52	N/A

or damage INA. As shown in (B) by using OTA-based DRLC the voltage is reduced to 35mV. The interference is reduced by 34dB. (C) and (D) shows the interference reduction by using Op-Amp-based DRLC in unity gain configuration and with gain on 20 respectively. Fig. 11 show the output waveform of 4kHz non-overlapping clock generator (a) and 8kHz SRF clock generator (b), respectively. They perfectly match the desired frequency and duty cycle. Fig. 12 shows the recorded ECG from the human body. Table I shows performance comparison of implemented test chip with earlier works. The Differential voltage generated on the body (V_{ab}) due to potential divider effect is 500μV without applying DRLC and SDA. The V_{ab} reduces to 5μV by applying only DRLC. Moreover, it reduces to 0.5μV by applying both DRLC and SDA. 5μV differential voltage is generated on the body without SDA might not be the problem in acquiring ECG but will be troublesome for the EEG acquisition. Therefore, the differential voltage has further reduced upto to 0.5μV by SDA.

V. CONCLUSION

A low power and ultra low noise on-chip signal conditioning circuit with re-configurable DRLC with OTA and Op-Amp, appropriate for different Bio-potential signals is presented. The proposed circuit eliminates 1/f noise effectively in its current feedback INA using chopper modulation. The INA achieves worst case CMRR of 109dB at 1.91kHz and input referred voltage noise density of 28nV/$\sqrt{Hz}$. DRLC with either OTA or op-amp effectively attenuates large common-mode voltage on the body. However DRLC using OTA has advantage of significant reduction of the area and number of external components. The operation of the test chip has been demonstrated by recording ECG from human body using a battery operated test board. Since all modules of the test

chip are re-configurable, chip can be used in for many sensor applications.

ACKNOWLEDGMENT

The Authors would like to acknowledge support from SMDP Government of India and VLSI lab IIT-Bombay. Authors also would like to thank TCS for financial support for fabrication of test chip. 180nm UMC model files have been provided by Europractice. Authors also would like to thank users of VLSI lab IIT-Bombay for insightful discussions.

REFERENCES

[1] R. F. Yazicioglu, P. Merken, R. Puers, and C. V. Hoof, "A 60 μW 60 nV/ $\sqrt{Hz}$ Readout Front-End for Portable Biopotential Acquisition Systems," *IEEE Journal of Solid-State Circuits*, vol. 42, No.5, pp. 1100–1110, May 2007.

[2] R. F. Yazicioglu, T. Torfs, P. Merken, J. Penders, R. P. Vladimir Leonov, B. Gyselinckx, and C. V. Hoof, "Ultra-low-power biopotential interfaces and their applications in wearable and implantable systems," *Microelectronics Journal*, vol. 40, pp. 1313–1321, 2009.

[3] C.-T. Ma, P.-I. Mak, M.-I. Vai, P.-U. Mak, S.-H. Pun, W. Feng, and R. P. Martins, "A 90nm CMOS Bio-Potential Signal Readout Front-End with Improved Powerline Interference Rejection," *IEEE Circuits and Systems*, vol. 40, pp. 1313–1321, 2009.

[4] L. Fay, V. Misra, and R. Sarpeshkar, "A micropower electrocardiogram amplifier," *IEEE Tran. on Biomedical Circuits and Systems*, vol. 3, NO. 5, pp. 312–320, October 2009.

[5] C. Menolfi and Q. Huang, "A fully integrated, untrimmed CMOS instrumentation amplifier with submicrovolt offset," *IEEE Journal of Solid-State Circuits*, vol. 34, NO. 3, pp. 415–420, March 1999.

[6] T. Denison, K. Consoer, W. Santa, A.-T. Avestruz, J. Cooley, and A. Kelly, "A 2 μW 100 nv/rtHz Chopper-Stabilized Instrumentation Amplifier for Chronic Measurement of Neural Field Potentials," *IEEE Journal of Solid-State Circuits*, vol. 42, NO. 12, pp. 2934–2945, December 2007.

[7] R. F. Yazicioglu, S. Kim, T. Torfs, H. Kim, and C. V. Hoof, "A 30 μW Analog Signal Processor ASIC for Portable Biopotential Signal Monitoring," *IEEE Journal of Solid-State Circuits*, vol. 46, No.1, pp. 209–223, January 2011.

[8] B. Winter and J. Webster, "Driven-Right-Leg Circuit Design," *IEEE Tran. on Biomedical Engineering*, vol. 30, NO. 1, pp. 62–66, January 1983.

[9] A. Wong, K.-P. Pun, Y.-T. Zhang, and C.-S. Choy, "An ECG measurement IC using driven-right-leg circuit," *ISCAS*, 2006.

[10] J. G. Webster, *Medical Instrumentation Application and Design*. Wiley India Pvt. Ltd., New Delhi: Wiley India Pvt. Ltd., 2007.

[11] E. M. Spinelli, N. H. M. nez, and M. A. Mayosky, "A Transconductance Driven-Right-Leg Circuit," *IEEE Tran. on Biomedical Engineering*, vol. 46, NO. 12, pp. 1466–1470, December 1999.

[12] M. S. Baghini, S. Nag, R. K. Lal, and D. K. Sharma, "An Ultra-Low-Power Current-Mode Integrated CMOS Instrumentation Amplifier For Personal ECG Recorders," *Journal of Circuits, Systems, and Computers*, vol. 17, No. 6, pp. 1053–1067, 2008.

A Novel Digital PLL With Good Performance And Very Small Area

Luo Zhihong, Au Yeung On, Benjamin Lau, Henry Law
Design Enablement, GLOBALFOUNDRIES.
60 Woodlands Industrial Park D Street 2 Singapore 738406

Abstract—**A novel digital PLL(Phase Locked Loop) is presented in this paper. It uses three types of digital delay control methods, including delay cell number adjust, delay cell load adjust and cycle control to digitally control the DCO(Digitally Controlled Oscillator) output clock frequency, so as to get wider frequency range and smaller jitter. This PLL uses NAND gate as the basic delay cell, which can completely reset DCO in a very short time, and prevent the jitter accumulation. It uses binary search to achieve fast lock and uses shift chain to get better input clock jitter tolerance. This digital PLL has been silicon validated in GLOBALFOUNDRIES 65nmG process. Its chip area is only 5255um², DCO's frequency have a wide range between 550MHz to 2.45GHz. Its total power is around 1.0mW when DCO's frequency is 1.0GHz. This PLL can be locked very fast in 25 divided reference clock cycles, and its output clock jitter is smaller than 40ps.**

Keywords-PLL; DCO; DPFD; Cycle control; Jitter.

I. INTRODUCTION

PLL(Phase Lock Loop) is a control system used to generate an output signal whose phase is related to the phase of the input "reference" signal. It is widely used in radio, communications, computers and other electronic applications.

Traditionally, PLL is an analog block, it includes the basic components of VCO(voltage control oscillator), PFD(phase and frequency detector), charge pump, LPF(low pass filter), feedback divider and etc. However, analog PLL uses a lot of capacitors which make the chip area very big. It also has some other shortcomings, such as noise sensitivity, not being able to be directly converted to different process, and etc.

In recent years, digital PLL has become more and more popular because of its some special features, i.e. small chip area, noise insensitivity, and easy process conversion. Normally digital PLL uses DCO (Digitally Controlled Oscillator), instead of VCO, to generate the digitally controlled clock. There are many different digital control methods, such as digitally adjusting current source, digitally adjusting delay cell number, digitally adjusting load of delay cells, and etc. The digitally controlled signals for DCO are also generated by some different ways. Recently some different digital PLLs have been launched[1],[2],[3], and they all have demonstrated good performance in some but not all aspects, including chip area, lock time, jitter, power, and DCO frequency range.

II. PROPOSED PLL

Different from any existing PLLs, the PLL presented in this paper can achieve better performance with much smaller chip area. It outperforms the conventional analog PLL with its chip area much smaller than the latter. This PLL is purely composed of logic cells, partitioned into five blocks: DCO, DPFD, Divider(2R), Divider(P), Counter(F). Figure 1 shows the top level structure of this PLL.

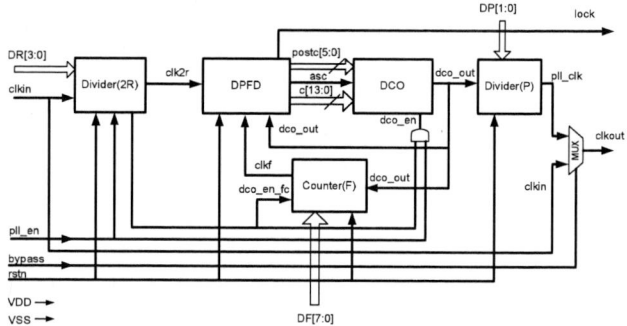

Figure 1. Top level structure of this PLL

When "pll_en" is enabled, this PLL starts to work. Divider(2R) divides "clkin" by 2DR to produce an output clock signal "clk2r" with 50% duty cycle. "dco_en_fc" which is also generated by Divider(2R) is a small pulse(around 200ps), just before the beginning of every clk2r cycle. It is used to reset DCO and Counter(F). DCO generates the high frequency clock "dco_out". Counter(F) starts to count the clock "dco_out" from 0 at the beginning of every clk2r cycle. When it arrives DF which is set by user, the Counter(F)'s output "clkf" turns high. At each falling edge of "clk2r", DPFD detects "clkf" is high or low, if "clkf" is high(low), it means the DCO output clock frequency is higher(lower) than expected, then DPFD will adjust the digital control bits "c[13:0]", "asc", "postc[5:0]" accordingly, which will change the DCO output frequency, until in a certain clk2r cycle, the "clk2r" falling edge and "clkf" rising edge are perfectly match. Then the DCO output clock "dco_out" frequency will equal DF/DR times of the frequency of "clkin". The frequency of the Divider(P) output clock "pll_clk" will equal to (DF/DR)/ 2^{DP} times of the frequency of "clkin".

A. DCO (Digitally Control Oscillator)

DCO is a ring oscillator used to generate the output clock with digitally adjustable frequency. It is necessary to design this DCO in analog way to tune some of the transistor parameters. Figure 2 shows the DCO schematic.

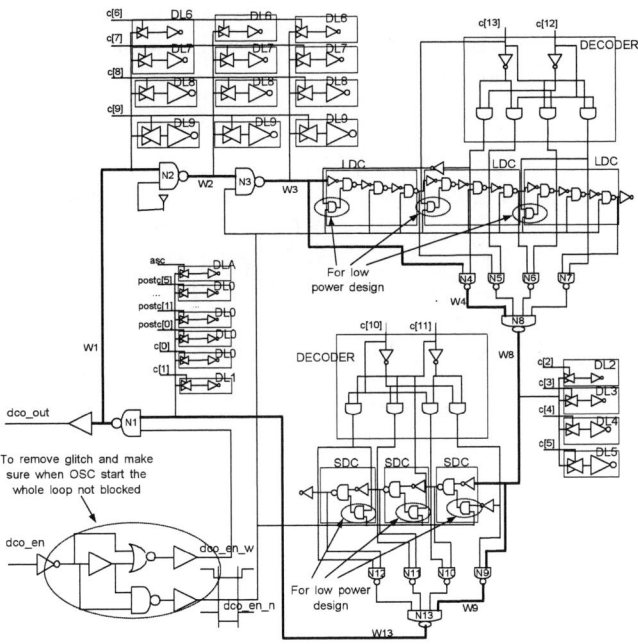

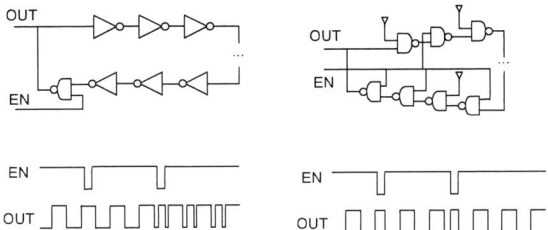

Figure 2. DCO schematic of this PLL

The basic delay cell of this ring oscillator is NAND gates. When c[13:10] are 0, N1, N2, N3, N4, N8, N9, N13 make up the ring oscillator(thick line in Figure 2). A small negative pulse (around 200ps) on "dco_en" at the beginning of clk2r cycle can completely reset DCO to a known state(W1=1, W2=0, W3=1, W4=0,W8=1,W9=0,W13=1), so the jitter will not accumulate to the next clk2r cycle. On the other hand, as shown in Figure 3, if inverters are used as the basic delay cell, the "EN" pulse width must be larger than the whole loop delay, otherwise, the ring oscillator cannot start from a known state after "EN" pulse.

Figure 3. "EN" for inverter ring oscillator and NAND ring oscillator

There are 3 LDCs(Large Delay Cell) and 3 SDCs(Small Delay Cell) in DCO and they can be selected to be included in the ring. "c[13:12]" controls the number of LDCs included in the ring; "c[11:10]" controls the number of SDCs included in the ring. With different number of LDCs and SDCs in the ring, the period of the DCO output clock will be changed accordingly. The DCO in Figure 2 also includes load cells that are used to vary the amount of load on NAND gates N1, N2, N3, N8 and N13. Switching on or off the control signals on these load cells changes the load of these NAND gates so as to change period of the ring oscillator output clock.

The 6 AND cells circled in Figure 2 are used to switch off the cells which is not included in the ring in some certain situation so as to avoid unnecessary power consumption when DCO is oscillating. Two pulse signals "dco_en_w" and "dco_en_n" are generated from dco_en. "dco_en_w" turns low ahead of "dco_en_n" and turns high behind of "dco_en_n". "dco_en_w" is connected to NAND gate N1, "dco_en_n" is connected to other NAND gate in the loop. Therefore it can remove the output clock glitch and make sure that the whole loop is not blocked when DCO starts to oscillate.

B. DPFD (Digital Phase and Frequency Detector)

DPFD mainly includes 7 modules: ud generator, c generator, postc generateor, plus_num generator, asc generator, lock detector and unlock detector. Figure 4 shows the structure of DPFD.

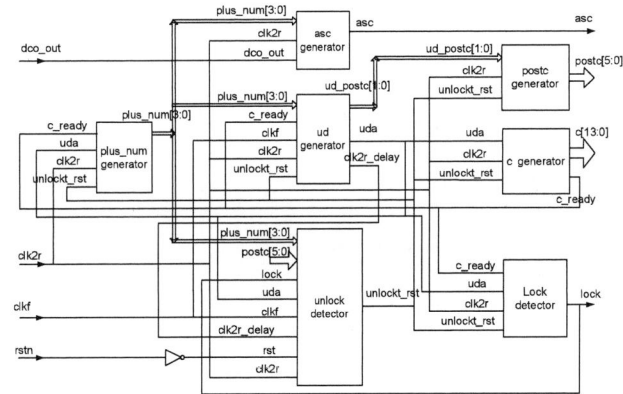

Figure 4. Structure of DPFD

1) ud generator

"ud generator" is used to generate the signal "uda" and "ud_postc". Figure 5 shows the schematic of "ud generator". At each falling edge of clk2r, If "clkf" is high, output signal "uda" will be '1', otherwise "uda" will be '0'. To avoid the meta-stability state of DFF, one Schmitt trigger buffer is inserted before producing the output of uda. The value of the "ud_postc[1:0]" signal in turn varies based on the combination of "plus_num[3:0]" and uda. When "plus_num[3:0]" equals to 4'b1011, and uda is '0', ud_postc[1:0] will be 2'b01, when "plus_num[3:0]" equals to 4'b0000, and "uda" is '1', "ud_postc[1:0]" will be 2'b10, otherwise, "ud_postc[1:0]" will be 2'b00. In Figure 1, when "pll_en" is '1', DCO enable pin "dco_en" will be negative pulsed before each clk2r's rising edge.

978-1-61284-863-1/11 $26.00 © 2011 IEEE   119

When "dco_en" is high, "dco_out" will turn to '0' immediately, and there is a fixed delay 't1' between clk2r's rising edge and dco_out's first falling edge. When Counter(F) reaches DF, "clkf" will turn high at the falling edge of "dco_out". There is a fixed delay 't2' between the dco_out's DF(th) falling edge and clkf's rising edge. Obviously, when clk2r's falling edge and clkf's rising edge match, $t1+DF*T_{dco}+t2 = T_{clk2r}/2$, which is not the expected equation $DF*T_{dco}= T_{clk2r}/2$. To eliminate the effect of delay "t1" and "t2", a block "delay_clk2r" is inserted between clk2r and DFF, "clk2rnd" is generated from clk2r with a compensation delay 't3' equals t1+t2, when clk2rnd's rising edge and clkf's rising edge match, $DF*T_{dco}= T_{clk2r}/2$, which makes the DCO exactly oscillate at the expected frequency. Figure 6 shows the waveform of delay compensation.

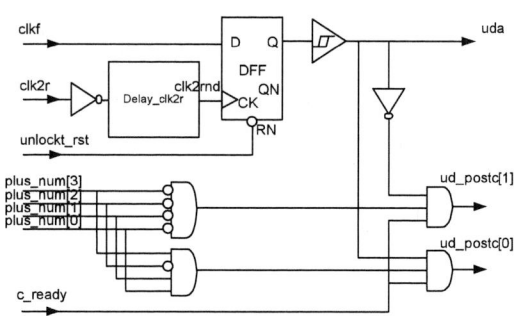

Figure 5. Schematic of "ud generator"

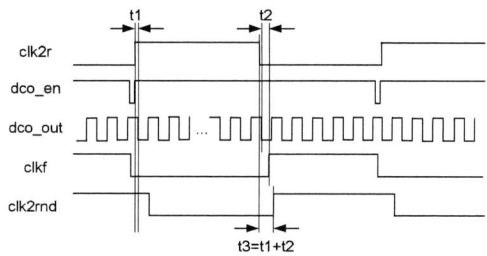

Figure 6. Waveform of delay compensation

2) c generator

Figure 7 is a flow chart illustrating how the "c generator" works, and Figure 8 shows the waveform of "c generator". The "c generator" uses binary search to determine the values of c[13:0] and achieve a fast lock. When binary search is initiated, c[13:0] and c_ready are both set to 0, then at the rising edge of clk2r, set c[13] to '1', at the falling edge of clk2r, "uda" will update its value. At the next rising edge of clk2r, if uda is '1', fix c[13] to '1', else fix c[13] to '0', at the same time, set c[12] to '1'; repeat the above steps to fix the value for c[12] and the remaining c[11:0] until c[0] is fixed, at this point c_ready will turn high. From c[13] to c[0], each control pin has different weight for adjusting the DCO output clock period, c[13] has the biggest and c[0] has the smallest. To get the fastest search, the weight of

c[12] should equal 50% of c[13], c[11] should equals 50% of c[12], ...so c[0] will equal to $1/2^{13}$ of c[13]. However, considering the PVT variation, the actual ratio may not always exactly equal to 50%. If it is smaller than 50%, some frequency in the spec frequency range will not be able to be covered by binary search. So we adjust the ratio to around 60% which can guarantee that actual delay ration still larger than 50% in any variation, and all frequency in the frequency range can be searched.(c[12] to c[13] and c[10] to c[11] are simply delay cell number change, they don't need to follow this rule)

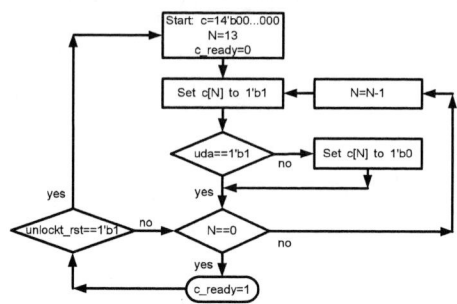

Figure 7. Working procedure of c generator

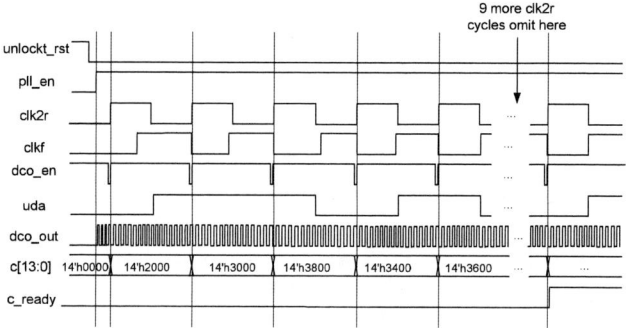

Figure 8. Waveform of c generator

3) Plus_num generator & asc generator(For Cycle control)

After the value of "c[13:0]" is fixed, a coarse capturing of the expected frequency is established. However, this frequency is still not accurate enough and further tuning has to be done. Although the DCO can be designed so that changing the value of "c[0]" allows an adjustment of as small as 1.4ps to be applied to the DCO output clock cycle time, jitter is multiplied after passing through the feedback counter Counter(F) and can be accumulated to cause a jitter of 2*DF*1.4ps, which is significant, at the last dco_out cycle within a clk2r period.

A new method, which we call cycle control, is used to reduce the DCO output clock jitter. With the "asc" signal being given different values within a single clk2r time period, load settings for DCO will also change to adjust the "dco_out" time period, in response to feedback from the "ud generator" which determines whether the dco_out frequency is faster or slower than the target value. This allows for greater precision in tuning the DCO to the required frequency. The dco_out cycles within a clk2r period are grouped into subsets comprising 12 dco_out cycles each. The value of the plus_num[3:0] determines the number of dco_out cycles in which "asc" is equal to high within each subset of 12. The "plus_num generator" module generates the plus_num[3:0] signal at rising edge of the "clk2r" signal. If "uda" equals '1' plus_num[3:0] will increase by 1, which means that asc will be "1" for an additional dco_out cycle within each subset of 12, otherwise plus_num[3:0] will decrease by 1 within each subset of 12. Supposing that "asc" adjustment to a dco_out cycle time is 2ps, the maximum jitter may be reduced to (2*DF/12)*2ps by using cycle control.

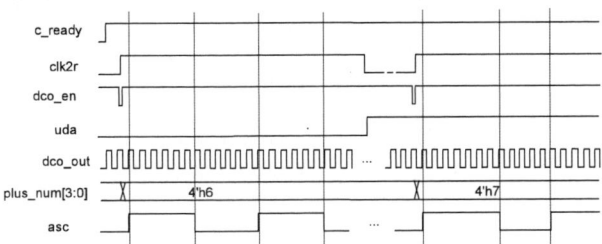

Figure 9. Waveform of cycle control

4) Postc Generator

After PLL is locked, the reference clock may have some minor changes, and "postc generator" consequently will start to work to make DCO output clock still follow the reference clock. Figure 10 shows the schematic of postc generator. As shown in the DCO schematic in Figure 2, each digit in the postc[5:0] binary string is associated with a respective digital load DL0. The value of the "postc" signal is changeable at each clk2r rising edge based on the value of the "ud_postc[1:0]" signal from the "ud generator".

ud_postc=2'b01 => Right shift, with '1' fill in the left
ud_postc=2'b10 => Left shift, with '0' fill in the right
ud_postc=2'b00 or 2'b11 => Keep current value

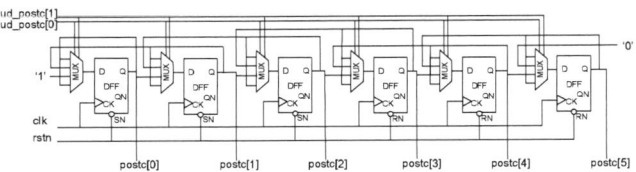

Figure 10. Schematic of postc generator

5) Lock detector & Unlock Detector

After the binary search, c[13:0] are fixed, cycle control procedure starts, module "lock detector" starts to detect "uda" toggling. If "uda" start to change to the opposite value(1->0, or 0->1) for three continuously clk2r cycles, the PLL is regarded as get locked, and output signal "lock" will turn high.

After PLL is locked, the module "unlock detect" will detect whether the skew between clk2r and clkf is larger than the threshold value or not. If it is, a short positive pulse "unlockt_rst"

will be generated to reset the whole PLL, "lock" will turn low, and PLL will restart the procedure of getting locked, normally it can be locked again in maximum 25 divided reference clock cycles(clk2r).

III. SILICON VALIDATION RESULT

This digital PLL has been silicon validated with GLOBAL-FOUNDRIES's 65nmG process. Figure 11 shows the layout for this digital PLL. TABLE I list the performance comparison. It shows that this PLL's area is very small, while its performance of jitter, power, lock time, frequency range are very good comparing with other digital PLLs launched recently.

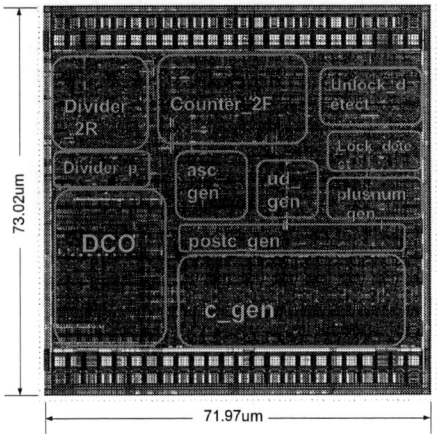

Figure 11. This PLL's layout and floorplan

TABLE I. PERFORMANCE COMPARISON

	*Proposed PLL	[1]	[2]
Data source	Silicon test	Simulation data	Silicon test
Process	65nm G CMOS	130nm CMOS	0.35um CMOS
Chip area	0.005255mm^2	0.012 mm^2	0.71mm^2
DCO Freq	550M ~ 2.45G	224M~1.06G	45M~510M
Lock time	<25 div cycles	<15 cycles	<46 cycles
Power	1.03mW@1GHz	2.9mW@1GHz	100mW@500MHz
Jitter(RMS)	25.11ps@0.9GHz 23.10ps@1.8GHz	48.5ps@1GHz 120ps@300MHz	70ps

*Test condition: TT wafer, VDD =1.0v, T=25C, F$_{clkin}$=50MHz, DR=1.

REFERENCES

[1] M.H. Chang, Z.X. Yang and W. Hwang, "A 1.9mW Portable ADPLL-based Frequency Synthesizer for High Speed Clock Generation", Proceedings of ISCAS '2007. pp.1137~1140.

[2] C.C. Chung and C.Y. Lee, "An All-Digital Phase-Locked Loop for High-Speed Clock Generation", IEEE Journal of Solid-state Circuit, Vol.38, No.2, February 2003.

[3] N. D. Dalt, E. Thaller, P. Gregorius, and L. Gazsi, "A compact tripleband low-jitter digital LC PLL with programmable coil in 130-nm CMOS," IEEE J. Solid-State Circuits, vol. 40, no. 7, pp. 1482–1490, Jul. 2005.

[4] V. Kratyuk, P.K. Hanumolu, U.K. Moon and K. Mayaram, "A design procedure for all-digital phase-locked Loops Based on a charge-pump phased-loop-loop analogy", IEEE Transactions on Circuits and Systems, Vol.54, No.3, March 2007.

Appendix: Silicon test results and performance analysis of this digital PLL

I. DESIGN AND TEST PROCEDURE

The whole design and test flow of this digital PLL shows in the following Figure A1.

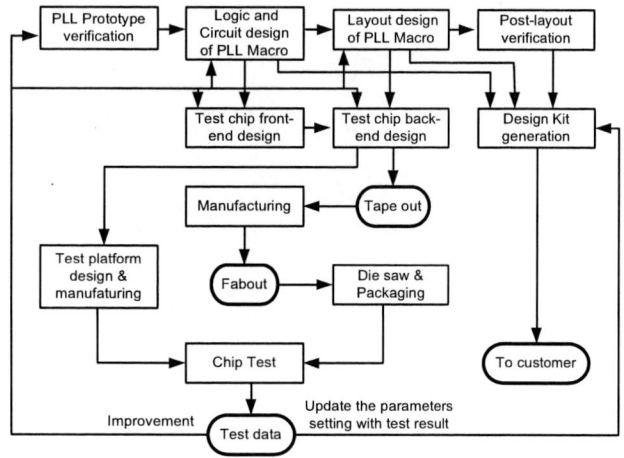

Figure A1. Design and test procedure for this digital PLL

II. TEST CHIP AND TEST PLATFORM DESCRIPTION

In order to test the performance of this PLL macro, a test chip is necessary to design and implement the PLL macro. A test platform is also needed to be designed and manufactured for the test of this test chip.

A good test chip should make all the input pins controllable, all the output pins and some critical internal nets are observable. For the test chip pad number limitation, it is impossible to directly connect all the ports and internal nets of this PLL macro to the test chip's pad. A shift chain is designed to shift in the PLL configuration pins and some internal input signals for debug. The macro output ports and some critical internal signals for observing will share the test chip's output pads by using some MUX cells.

The output clock of this digital PLL is in very high frequency, it may up to 3GHz. Normal GPIO can not transmit such high frequency signal. A high speed pad LVDS is used specially for this Digital PLL's output clock. To guarantee that the PLL's output clock can be measured even the LVDS pad cannot work,

an additional divider is used to get a lower frequency clock, so that this divided clock can directly pass through a GPIO pad.

Since many applications use crystal oscillator as the reference clock input, one oscillator pad is used in this test chip. The reference clock can either come from tester or from crystal oscillator.

III. SILICON TEST RESULT

Six packaged devices (typical wafer) were selected for the functional test and performance test. The following list the test result of one device on typical test condition. (Power voltage: 1.0V, Temperature: 25°C)

A. Functional test

The functionality of this PLL with reference clock from tester and from crystal oscillator are both tested and passed. The test results with reference clock from tester and crystal oscillator are shown in TABLE A1 and TABLE A2 respectively.

TABLE A1. FUNCTIONAL TEST RESULT WITH REFERENCE CLOCK FROM TESTER

	Fin=50MHz, DR=2, DP=0			
DF setting	DF=36	DF=60	DF=72	DF=88
Expect Frequency	0.9GHz	1.5GHz	1.8GHz	2.2GHz
Measured Frequency	0.899996 GHz	1.500004 GHz	1.800034 GHz	2.200050 GHz
Pass/Fail	Pass	Pass	Pass	Pass

TABLE A2. FUNCTIONAL TEST RESULT WITH REFERENCE CLOCK FROM CRYSTAL OSCILLATOR

	Fin=24MHz, DR=2, DP=0			
DF setting	DF=38	DF=41	DF=75	DF=86
Expect Frequency	0.912GHz	0.984GHz	1.8GHz	2.064GHz
Measured Frequency	0.912283 GHz	0.984398 GHz	1.802240 GHz	2.066320 GHz
Pass/Fail	Pass	Pass	Pass	Pass

B. DCO Fmax and Fmin test

TABLE A3 shows the test result of DCO's maximum frequency Fmax and minimum frequency Fmin. From this table, Fmax=2.45GHz and Fmin=0.55GHz.

TABLE A3. FMAX AND FMIN TEST RESULT

	Fin=50MHz, DR=1, DP=0					
DF setting	DF=10	DF=11	DF=12	DF=48	DF=49	DF=50
Frequency	0.5GHz	0.55GHz	0.6GHz	2.4GHz	2.45GHz	2.50GHz
Lock or not	No	Yes	Yes	Yes	Yes	No

C. Lock time test

The following TABLE A4 shows the lock time test result of this PLL. When DR=1, the divided reference clock are 25MHz. When DR=2, the divided reference clock are 12.5MHz.

TABLE A4. LOCK TIME TEST RESULT

	Fin=50MHz, DP=0			
DR setting	DR=1		DR=2	
DF setting	DF=18	DF=24	DF=36	DF=72
Frequency	0.9GHz	1.8GHz	0.9GHz	1.8GHz
Lock Time	0.655us	0.651us	1.250us	1.390us

D. VDDmax and VDDmin test

The following TABLE A5 shows the test result of power supply voltage range within which this PLL can work correctly. To avoid damaging this macro, the maximum supply voltage is set to Fmax=1.5v. The measured minimum voltage Fmin=0.73v.

TABLE A5. VDDMAX AND VDDMIN TEST RESULT

	Fin=50MHz, DR=2, DF=40, DP=0						
VDD	1.5v	1.2v	1.0v	0.8v	0.73v	0.725v	0.72v
Measured Frequency	1.0000 GHz	1.0000 GHz	1.0001 GHz	1.0002 GHz	1.0002 GHz	0.9365 GHz	0.9337 GHz
Pass/Fail	Pass	Pass	Pass	Pass	Pass	Fail	Fail

E. Jitter test

The following TABLE A6 shows the RMS period jitter and RMS cycle to cycle jitter test result of this PLL. The test result is based on DP=1, which means the output clock is generated from the DCO output clock with a divide factor of 2. (For the speed limitation of the LVDS pad, the DCO output need to be divided by 2 before the jitter measurement. Generally, the actual DCO clock jitter is smaller than the result shows in TABLE. A6.)

TABLE A6. JITTER MEASUREMENT RESULT

	Fin=50MHz, DR=2, DP=1			
DF setting	DF=36	DF=60	DF=72	DF=88
DCO Frequency	0.9GHz	1.5GHz	1.8GHz	2.2GHz
Period jitter(RMS)	16.98 ps	19.31 ps	17.71 ps	20.76 ps
Cycle-cycle jitter(RMS)	25.11 ps	24.96 ps	23.10 ps	22.52 ps

F. Duty cycle test

The following TABLE A7 shows the duty cycle test result of this PLL with DCO output divided by 1, 2, 8.

TABLE A7. DUTY CYCLE TEST RESULT

	Fin=50MHz, DR=2, DP=0					
DP setting	DP=0 (/1)		DP=1 (/2)		DP=3 (/8)	
DF setting	DF=36	DF=72	DF=36	DF=72	DF=36	DF=72
Frequency	0.9GHz	1.8GHz	0.9GHz	1.8GHz	0.9GHz	1.8GHz
Duty cycle	48.08%	47.82%	51.57%	51.35%	50.41%	51.05%

G. Power test

The following TABLE A8 shows the power test result of this PLL. The static current is 0.163mA, and the working current is 0.953mA@0.9GHz, 1.572mA@1.8GHz.

TABLE A8. POWER TEST RESULT

	Fin=50MHz, DR=2, DP=0, VDD=1.0v					
DF setting	DF=36	DF=40	DF=48	DF=56	DF=64	DF=72
Frequency	0.9GHz	1.0GHz	1.2GHz	1.4GHZ	1.6GHz	1.8GHz
Static current	0.163mA					
Working current	0.953 mA	1.036 mA	1.224 mA	1.358 mA	1.512 mA	1.572 mA

IV. PERFORMANCE ANALYSIS

The above silicon test result shows this digital PLL has very good performance of lock time, power, jitter and frequency range, while it only uses $5255um^2$ chip area. The following Figure A2 is the performance comparison diagram(For any value over 100 only indicate 100 in this diagram) between this PLL and other digital PLLs launched recently.

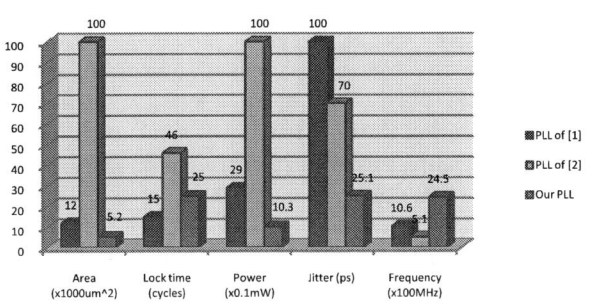

Figure A2. Performance comparison

A 900MHz RFID Reader Chip with RC Calibration

Mou Shouxian [#1], Ma Kaixue [#2], Yeo Kiat Seng [#3]

Circuits and Systems Division, EEE, Nanyang Technological University

Singapore

[1]sxmou@ntu.edu.sg; [2]kxma@ntu.edu.sg; [3]eksyeo@ntu.edu.sg;

Tel: (65) 6790 6387; Fax: [3](65) 6791 2687

Abstract— **An RFID scanner chip targeted to operation frequency range of 860MHz to 960MHz is designed and fabricated. To reduce chip performance degradation due to process and temperature variation, resistor and capacitor calibration is adopted. Resistor calibration output codes are used to adjust main circuit blocks' biasing current. Capacitor calibration output codes are used to fine tune filter operating range and DAC conversion accuracy. The reader is implemented with a 90nm standard CMOS process and has a chip area around 3.1mm × 3.3mm. The chip is packaged with QFN48 and mounted on PCB. The proposed RFID reader consumes around 90mW power, which is possible to be integrated on a mobile phone or for other applications.**

Keywords —**RFID, resistor calibration, RC calibration, CMOS.**

I. INTRODUCTION

In recent years radio frequency identification (RFID) procedures have become very popular in many service industries, such as purchasing and distribution logistics, industry, manufacturing companies and material flow systems. Automatic identification procedures exist to provide information about people, animals, goods and products in transit [1]. Fig. 1 presents an example of information system by using micro RFID to digitally track various commodities, from their manufacturing to the scrapping, which makes it easy to control their quality, security and cost.

Compared to the near-field inductive coupled RFID systems operating at the low frequency of 125 kHz or 134 kHz and the high frequency of 13.56 MHz, the ultra high frequency (UHF) RFID systems are more promising owing to longer read distance and faster data rate [2-7]. UHF RFID specification is based on the UHF RFID standard including ISO/IEC 18000-6 and some other standards from individual companies as illustrated in Table I [1, 8]. The overall frequency range for UHF RFID is from 860MHz to 960MHz. Detailed reader operation frequency is determined by the regulation of each region within the 860–960 MHz band.

The existing UHF RFID readers based on discrete components have been fully developed, but they are bulky, expensive and power hungry [1]. There are also some integrated or partially-integrated UHF RFID readers based on CMOS technology in literature [2-7]. However, most of them still consume high power [3-7]: 540mW for [3] and

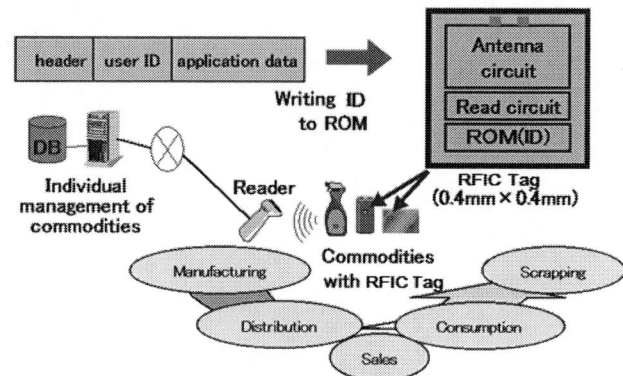

Fig. 1 Digital Tracking Information System using micro RFID.

TABLE I

VARIETY OF RFID TAG AT 900MHZ FREQUENCY RANGE

Product	Memory (b)	Supplier	Protocol	Power for operation
XRA00	64	ST Micro	EPC C1G1	
Quark, Omega	64	Alien	EPC C1G1	
Lepton	96	Alien	EPC C1G1	
Monza	256	Impinj	EPC C1G2	-9dBm (R); -6dBm (W)
XRAG2	432	ST Micro	EPC C1G2	-9dBm (R); -6dBm (W)
UCode EPC G2	512	Philips	EPC C1G2	-9dBm (R); -6dBm (W)
UCode EPC 1.19	256	Philips	ISO 18000-6B	
UCode HSL	1728	Philips	ISO 18000-6B	
u CHIP Hibiki	512	HITACHI	original	-10dBm (R); -2dBm (W)
TagFront	2048	Fujitsu	ISO 18000-6	
T-Junction	1024	Toppan	original	

1.5W for [4], which is not quite practical for mobile applications. This paper describes a single-chip CMOS UHF RFID reader transceiver, which achieves low power consumption (90 mW) and high integration. In the presence of 5dBm transmitter power, it realizes a sensitivity of -60dBm.

978-1-61284-863-1/11 $26.00 © 2011 IEEE

Resistor calibration and resistor-capacitor calibration are adopted in the chip. The main purpose of R-calibration and RC-calibration is to calibrate resistor and resistor/capacitor values, so as to compensate their variation through process and temperature.

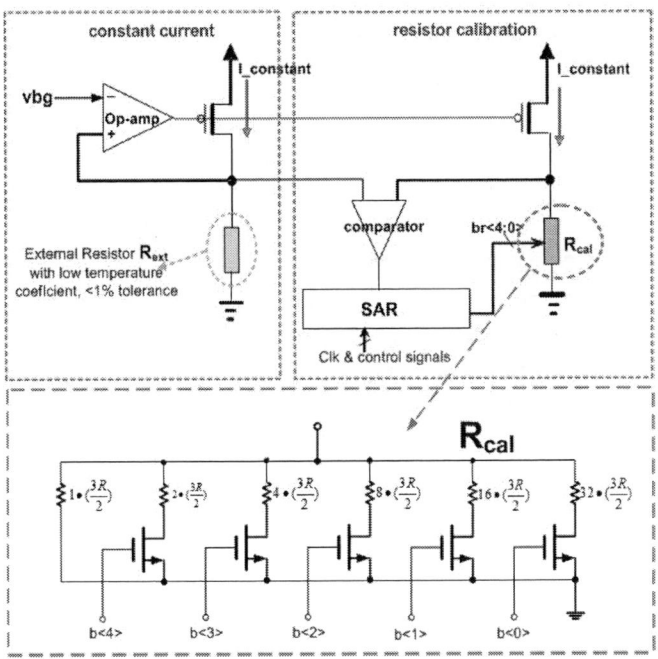

Fig. 2   R-calibration block diagram.

II. RESISTOR CALIBRATION AND RC-CALIBRATION

A. R-calibration

Reference voltage and reference current are necessary in circuit designs. Practically in IC designs, reference voltages are generated based on bandgap voltage. However, in order to generate a current, these voltages are presented across a resistor to generate the bias current. Any current generated using this scheme would inherently possess the process and temperature variations of this resistor. The foundries typically specify a process variation of ±30% [9]. This would indicate that even if the reference voltage is perfect and has no variations over process, the generated bias current would have a variation of ±30%. This is a very large variation and unacceptable for most of the applications.

One solution would be to reference the on-chip generated voltage on a tight-tolerance external (on-board) resistor to generate the desired bias current. This scheme has the limitation that every such current would require a separate pin and an external resistor. Alternatively, a calibration scheme can be used where the generated current based on an on-chip resistor is compared against the current generated with the same reference voltage but with an external tight-tolerance resistor as shown in Fig. 2. The calibration algorithm can then attempt to equalize the currents by switching the proper fractional on-chip resistors until the currents are equal. It is implemented by SAR block. The obtained calibration code br<4:0>, namely (b_4, b_3, b_2, b_1, b_0) can then be applied to all the current generation circuitry

on chip. Utilizing this scheme, bias currents with variations in the few percent range can easily be generated. According to Fig. 2, some equations can be derived:

$$R_{cal} = 1.5R / (1 + \frac{b_4}{2} + \frac{b_3}{4} + \frac{b_2}{8} + \frac{b_1}{16} + \frac{b_0}{32})$$

$$= 48R / (32 + 16b_4 + 8b_3 + 4b_2 + 2b_1 + b_0) \tag{1}$$

$$= 48R / (32 + b_r < 4:0 > in - decimal)$$

$$I_{const}R_{ext} = I_{const}R_{cal} \rightarrow R_{ext} = R_{cal} \tag{2}$$

$$R_{ext} = 48R / (32 + b_r < 4:0 > in - decimal) \tag{3}$$

$$R_{ext} / R = 48 / (32 + b_r < 4:0 > in - decimal) \tag{4}$$

Practically, on-chip resistor R is set to the same value as R_{ext} and the initial calibration code br<4:0> is <1,0,0,0,0> (16 in decimal). In the case that there is no R-calibration block, when on-chip resistor R variation tends to R_{min} as process varies, the generated current based on $V_{bandgap}/R_{on-chip}$ increases. On the contrary, with R-calibration block, according to equation (4), SAR block would generate R-calibration code br<4:0> less than 16 (in decimal). Then the corresponding code, which is less than 16 such as (0,1,0,1) as an example, can be used to tune the resistance of the on-chip resistor block to be higher. Therefore, the final generated current would remain almost constant.

B. RC-calibration

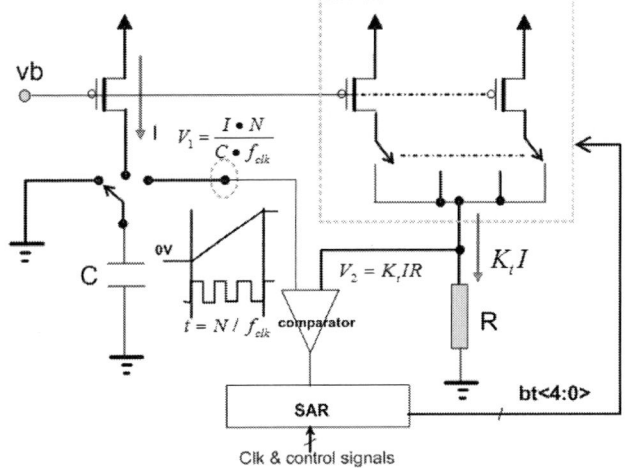

Fig. 3   RC-calibration block diagram.

Most of the on-chip filters are composed of resistors and capacitors. The variation of the resistors and capacitors leads to filter bandwidth variation, which would degrade system performance. Filter time-constant calibration technique can be used to overcome this issue. The filter is calibrated at startup or periodic intervals when it is not in use. Further the technique relies on a replica master RC block that is calibrated. The filter's slave block is built with the same type of resistors and capacitors as the master block. Once the master block is calibrated and the tuning code is obtained, the same tuning code is applied to the slave block. One such approach is to generate a master RC-based relaxation oscillator. The number of cycles "counted" by this RC oscillator is then compared to a certain number of clock cycles of a known high accuracy clock, such

978-1-61284-863-1/11 $26.00 © 2011 IEEE

as the on-PCB crystal oscillator used for the whole chip. Switched capacitors or switched resistors are then turned on or off in order to calibrate the RC oscillator cycle counts against that of the crystal oscillator. The calibration code is then achieved and applied to the slave filter blocks. The RC-calibration block diagram is illustrated in Fig. 3. The calibration block consumes no current after calibration is done.

Both R-calibration and RC-calibration have their calibration accuracy, which depends on the number of calibration code bits. More bits guarantee higher accuracy, while consume longer calibration time. Hence tradeoff must be done according to the system requirement.

III. CIRCUIT IMPLEMENTATION OF THE RFID READER

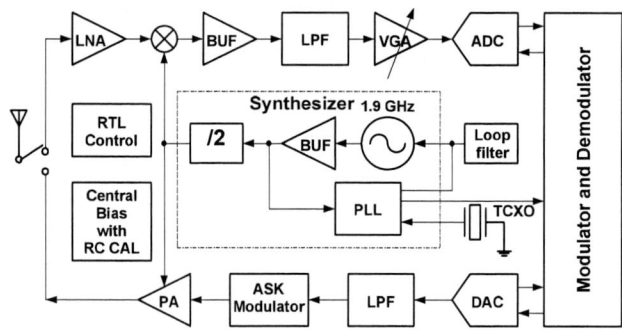

Fig. 4   The RFID reader block diagram.

A UHF RFID system consists of a reader, tag(s) and a computer (optional). The reader(s) sends information to one or more tags by modulating an RF carrier using amplitude-shift-keying (ASK) modulation at a bit rate ranging from 26.7 to 128 kbps [3]. The tag-to-reader data rate can be extended to 640 kb/s by modifying the receiver bandwidth of the reader [7]. Fig. 4 illustrates the RFID reader block diagram, which includes the receiver chain, the transmitter chain, frequency synthesizer, RTL digital control block, central bias block and modulator/demodulator portion. The loop filter and TCXO use off chip solutions.

A. Low Noise Amplifier

In an UHF RFID system, the reader transmits high power to activate and communicate with the RF tags. A passive tag receives the transmitted signal, converts it to DC to build up its voltage supply, and then modulates its antenna impedance to reflect the modulated signal back to the reader with an offset of 40kHz to 250kHz from the reader transmitter carrier [4]. Due to the simultaneous TX/RX operation and limited isolation (around 20dB), the TX output leakage results in an in-band blocker at the RX input, which has a much higher power than the useful signal to be received. It is difficult to provide a highly selective filter to reject a blocker which is only a few hundred kHz away from the desired signal.

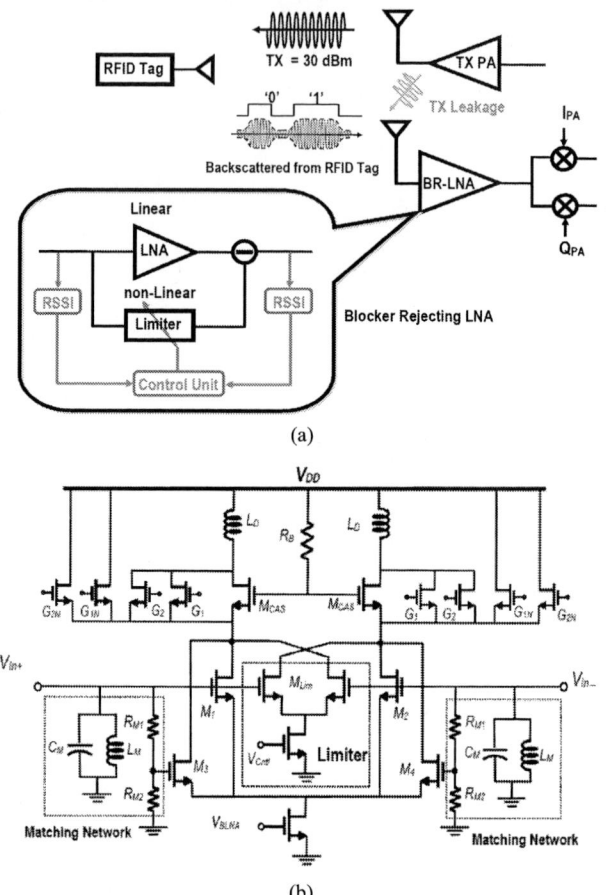

Fig. 5   (a) Block diagram and (b) schematic of the blocker-rejection LNA [4].

A blocker rejection LNA based on [4] is adopted as shown in Fig. 5. TX blocker rejection is achieved through a combination of two RF paths. A linear path amplifies both the desired signal and blocker equally through an LNA and a nonlinear path limits both blocker and the desired signal. The limiting function only preserves the frequency and phase of the stronger signal which is the blocker. The blocker is then rejected by subtracting the outputs of the linear and nonlinear paths. Therefore, the blocker is cancelled out but the desired signal is amplified through the linear path.

The LNA has a main g_m (that of transistors M_1 and M_2) that corresponds to high-gain mode and a secondary g_m (that of transistors M_3 and M_4) that corresponds to low-gain mode. Input transistors of the limiter, M_{Lim}, are highly nonlinear (switching mode) and driven by the large blockers rather than the weak desired signal. The current subtraction proceeds at low-impedance node of cascoded devices M_{CAS}.

B. Frequency Synthesizer

The synthesized frequencies is provided by a Fractional-N synthesizer, which composes a 3-bit swallow counter, 5-bit programmable counter, prescaler and a 1.6GHz to 1.92GHz VCO.

978-1-61284-863-1/11 $26.00 © 2011 IEEE   126

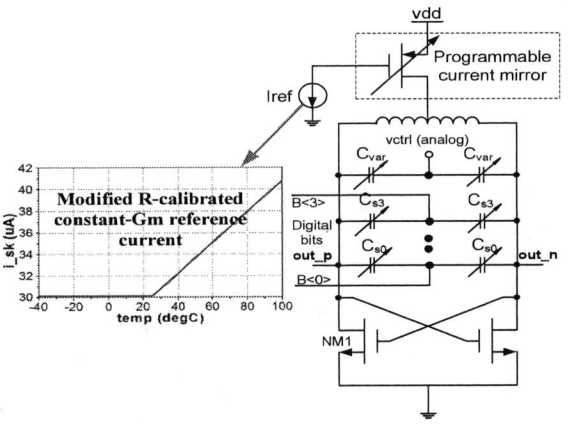

Fig. 6   The 1.6GHz to 1.92GHz VCO with modified constant-Gm biasing current.

The cross-coupled NMOS VCO is illustrated in Fig. 6. A 4-bits varactor bank is for VCO frequency calibration with a minimum step size around 10MHz. The analog-controlled varactor pair is adopted to provide a K_v only around 30MHz/V for better phase noise performance. Owing to degraded g_m of NMOS transistors at high temperature, resistor-calibrated constant-Gm current are adopted as VCO supply current for compensation. To guarantee the performance of low temperature, we generate a modified constant-Gm biasing current as shown on the left side of Fig. 6, which is flat rather than slide down when temperature is lower than room temperature.

C. Transmitter

The transmitter chain is with an 8-bit current-steering DAC, LPF, ASK modulator and PA. Actually, ASK modulator and PA are combined together, which includes a baseband amplifier and an RF amplifier with modulation function.

IV.   EXPERIMENT RESULTS

The chip is fabricated with a 90nm standard CMOS process and packaged with QFN 48pins. It has a layout size of 3.1mm × 3.3mm as shown in Fig. 7. The layouts of the LNA and the PA are placed at the bottom corner of the chip to obtain good RF grounding by bonding wire to the base plate of the QFN package directly. Fig. 8 presents the UHF RFID reader evaluation PCB board. The chip is measured under various conditions: different samples; different sample corners – typical (TT), slow-slow[Note1] (SS), fast-fast (FF), slow-fast (SF) and fast-slow (FS); various temperature from -40°C to 90°C; and also with ±10% VDD variations. The detailed measurement results of the chip are illustrated in the appendix in Fig. 9 to Fig. 18.

Fig. 9 to Fig. 10 demonstrates the R-calibrated reference biasing current and biasing voltage testing results versus temperature. It can be seen by resistor calibration, the generated biasing current and voltage have a variation of less than ±1.5% over 130°C temperature range nearly for all the corners. The modified constant-Gm current I_{gm} shown in Fig. 11 has some shifting to the left and top side. The deviation can be reduced by adjusting the tuning bits of bias generation block. On the chip,

VCO is biased with these modified I_{gm}. Fig. 12 and Fig. 13 show its phase noise at temperature of 25 °C and 80 °C respectively. It can be found that with the proposed I_{gm} biasing current, the VCO achieves -98.6dBc/Hz phase noise at 80 °C with 100kHz offset, which degrades only 2dBc than that (-100.7dBc/Hz) of room temperature. Fig. 14 and Fig. 15 compare the gain of Rx-chain under the conditions with and without resistor calibration. It indicates that with resistor calibration, the gain flatness all over the 130°C temperature range improves.

The PLL has a locking time around 300us as described in Fig. 16. Fig. 17 shows the 1dB gain compression point for Tx chain, which output P_{1dB} is around 0.6dBm. Fig. 18 describes the measured transmitter spectrum with overall mask. For RBW around 3kHz, the transmitter has an output power around 0dBm. For RBW of 100kHz, it is 15dB down from the mask. The measured side-band image rejection is around 56dBc. Detailed performance of the chip is summarized and compared with some designs in literatures in Table II. The results indicate the advantages of the proposed RFID reader: low power and small chip area.

V.   CONCLUSION

A CMOS RFID reader with operating frequency of 860MHz to 960MHz is presented in this paper. On the chip, resistor and capacitor calibration is used to compensate the chip performance degradation due to process and temperature variation. As the results, the gain flatness through the wide temperature variation range improves, and the phase noise performance at high temperature maintains. The reader is implemented with a 90nm standard CMOS process and packaged with QFN48. The proposed RFID reader has compact chip area and low power consumption, which make it possible to be integrated with other systems at various application scenarios.

REFERENCES

[1]   K. Finkenzeller, *RFID Handbook: Fundamentals and Applications in Contactless Smart Cards and Identification.* 2nd edition, Wiley, 2003.

[2]   X.G. Sun, B.Y. Chi, C. Zhang etl., "A 1.8V 74mW UHF RFID reader receiver with 18.5dBm IIP3 and −77dBm sensitivity in 0.18μm CMOS," in *IEEE RFIC Symp. Dig.*, 2010, pp. 597-600.

[3]   P.B. Khannur etl., "A universal UHF RFID reader IC in 0.18-μm CMOS technology," *IEEE J. Solid-State Circuits*, vol. 43, no. 5, pp. 1146–1155, May 2008.

[4]   I. Kipnis, S.Chiu, M. Loyer etl., "A 900MHz UHF RFID reader transceiver IC," in *IEEE ISSCC Dig.*, 2007, pp. 214-215.

[5]   A. Safarian, A. Shameli, A. Rofougaran, M. Rofougaran and F. Flaviis, "An integrated RFID reader," in *IEEE ISSCC Dig.*, 2007, pp. 218-219.

[6]   L. Ye, H. Liao, F. Song etl., "A single-chip CMOS UHF RFID reader transceiver for Chinese mobile applications," *IEEE J. Solid-State Circuits*, vol. 45, no. 7, pp. 1316–1329, Jul. 2010.

[7]   I. Kwon, Y. Eo, H. Bang etl., "A single-chip CMOS transceiver for UHF mobile RFID reader," *IEEE J. Solid-State Circuits*, vol. 43, no. 3, pp. 729–738, Mar. 2008.

[8]   Radio-Frequency Identification for Item Management—Part 6: Parameters for Air Interface Communications at 860 MHz to 960 MHz. ISO/IEC 18000-6:2004/FPDAM 1, 2005.

[9]   A. Behzad, *Wireless LAN Radios System Definition to Transistor Design: Calibration Techniques*, IEEE book chapter, pp. 161 – 177, 2008.

APPENDIX: TESTING RESULTS

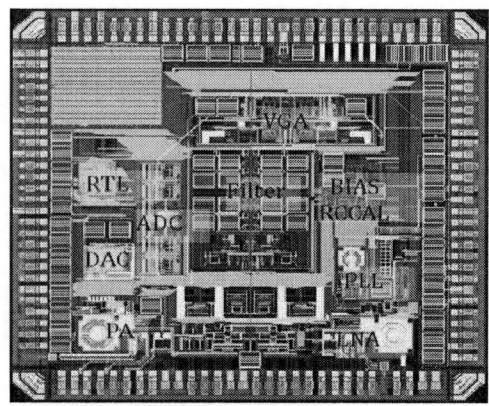

Fig. 7   Layout of the proposed UHF RFID chip 3.1mm × 3.3mm.

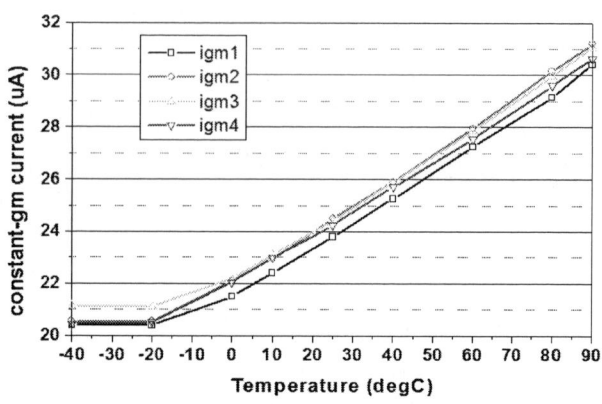

Fig. 11   Generated R-calibrated modified 20uA constant-Gm biasing current.

Fig. 8 Evaluation board.

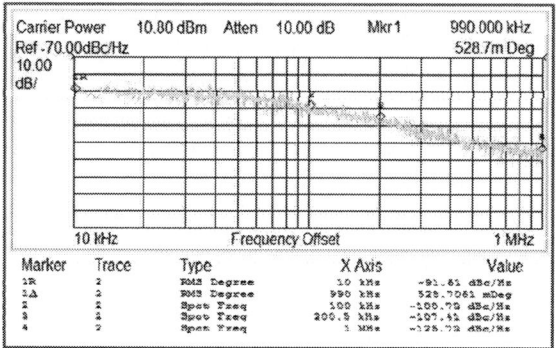

Fig. 12 VCO phase noise performance at 960MHz at room temperature.

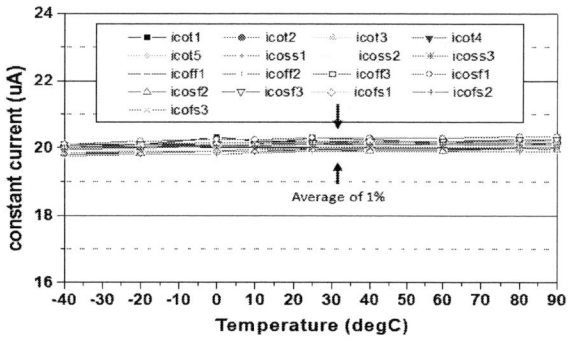

Fig. 9 Generated 20uA constant biasing current for samples of all the corners.

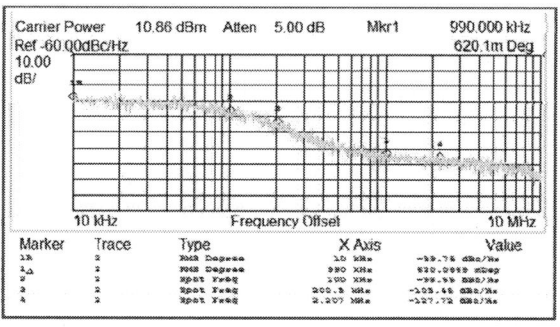

Fig. 13 VCO phase noise performance at 960MHz at 80 °C with R-calibrated constant-Gm current.

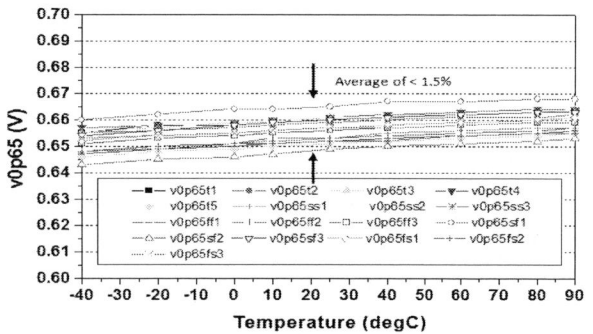

Fig. 10   Generated 0.65V biasing voltage for samples of all the corners.

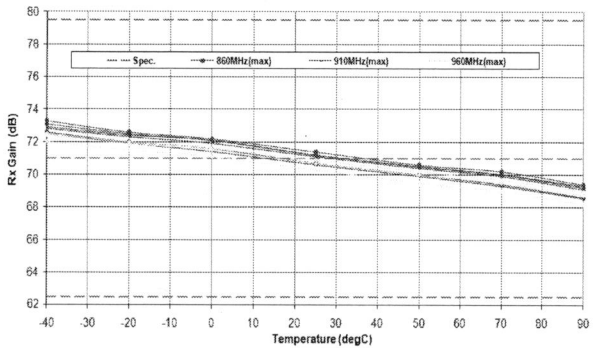

Fig. 14 Rx chain maximum gain versus temperature with non-R-calibrated constant-Gm current.

978-1-61284-863-1/11 $26.00 © 2011 IEEE   128

TABLE II
SUMMARY OF MEASUREMENT RESULTS

Parameters	This Work: Measured @ 25°C					[3]	[4]	[6]	Units
	TT	SS	FF	SF	FS				
Transmit power[Note2]	5 – 7	4 – 6	7.1 – 8.1	4.1 – 6.1	6.9 – 7.8	10	-9 – 4	22	dBm
VCO frequency	1300 – 1950	1310 – 1960	1315 – 1990	1305 – 1960	1315 – 1985				MHz
Phase noise @100kHz	-100.7	-101.5	-101	-98	-99	-101	-87	-103	dBc/Hz
OP_{1dB}@Tx	3 – 5.5	1.5	2.5	5.8	4.3				dBm
BB LPF bandwidth	0.3	0.32	0.33	0.33	0.33				MHz
Rx maximum gain	70 – 75	69	73	72	72		82.5		dB
Rx minimum gain	25 – 33	27	29	29	30		10		dB
Power consumption	90	101	103	93	95	540	160	203/660	mW
Power down power consumption	5	2	10	6	8				uW
Voltage supply	1.2					1.8	1.8	1.8	Volt
Operating frequency	860 – 960					860 – 960	902 – 928	840 – 925	MHz
Chip area	10.3					36	23.8	13.5	mm²
Process	90nm standard CMOS					0.18µm	0.18µm	0.18µm	CMOS

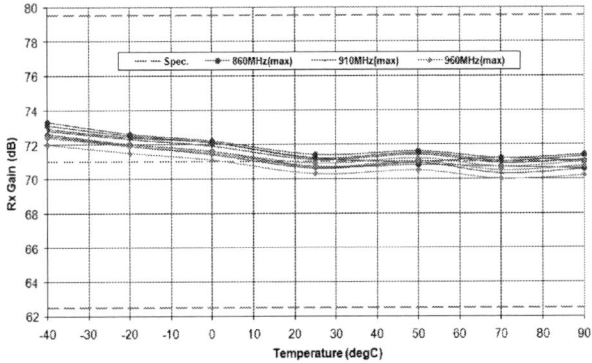

Fig. 15 Rx chain maximum gain versus temperature with R-calibrated constant-Gm current.

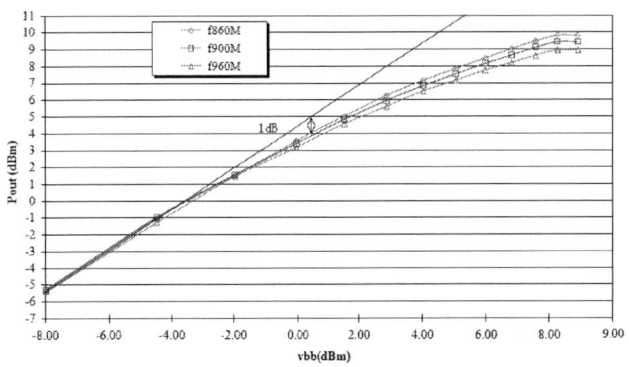

Fig. 17 Tx chain output power versus baseband input signal power (after baseband amplifier).

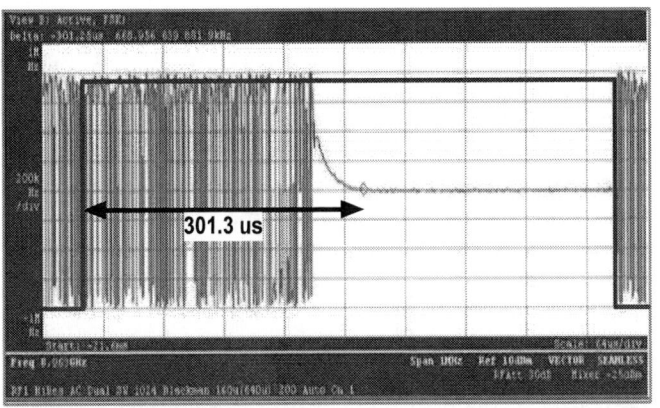

Fig. 16 PLL locking time.

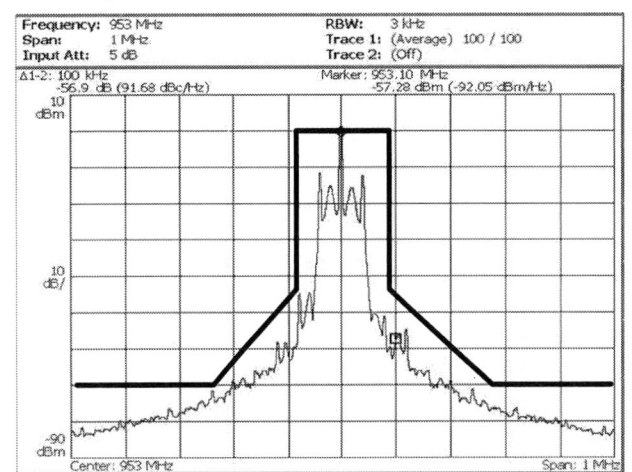

Fig. 18 Tx output spectrum mask.

Note1: slow-slow means NMOS transistor is with slow model and PMOS transistor is with slow model.

Note2: with cable loss and circulator loss compensation.

Energy Efficient Integrated Gas Sensor System with Post CMOS Functionalization*

Pramod M

M.E. (Microelectronics Systems) Student, Department of Electrical Communication Engineering.
Indian Institute of Science, Bangalore, 560012.
Email : pramodm@ece.iisc.ernet.in

Abstract—An energy efficient gas sensor array is presented in 0.35 μm CMOS technology. It contains an array of 27 sensor pixels which can be functionalized with a distinct gas sensing material. A sensor pixel is capable of extracting the impedance of the sensor upto 10 KHz. Each pixel contains digitally programmable gain stages for sensor signal amplification. The amplified signal is subsequently digitized using a second order continuous time delta sigma (CT-$\Delta\Sigma$) modulator. CT-$\Delta\Sigma$ modulators are low power alternatives to discrete time (DT) $\Delta\Sigma$ modulators. Fully differential approach of signal amplification and digitization reduces the effect of process variations. The chip also contains a reference pixel to compensate for the phase shift introduced by the signal processing circuits. An proportional to absolute temperature sensor with digital readout is provided on chip to measure the ambient temperature. A sensor pixel of the chip is functionalized with polycarbazole conducting polymer to sense volatile organic gases at room temperature. Measurement results on exposure to toluene are presented. The chip consumes only 57 mW from 3.3 V supply. An graphical user interface and a microcontroller interface is designed to establish communication between the chip and a computer for chip calibration, data acquisition and analysis.

Index Terms—Gas sensor array, CMOS, Low power, Impedance spectroscopy

I. INTRODUCTION

A gas sensors systems typically contain transducers which sense the analyte gas and respond by change in a physical quantity such as electrical conductivity. These transducers are interfaced to signal processing circuits for signal amplification, digitization etc. In earlier gas sensor systems, discrete gas sensing elements were used which were power hungry, bulky and unreliable. This limited their use only in niche applications such as in hazardous gas detection, gas leaks detection and process control in industries etc. These systems could not be integrated with hand held sensor systems. In recent years, CMOS based gas sensors have created tremendous interest. These chips are low cost, reliable and energy efficient alternatives to gas sensor systems with discrete sensing elements [1].

In gas sensing CMOS chips, the gas sensing materials are integrated with the CMOS signal processing circuits. Materials such as inorganic metal oxides like SnO_2, conducting polymers are commonly used for gas sensing. Inorganic metal

oxides based CMOS gas sensors need to operate around 200-400°C. Therefore, these materials are typically coated on microheaters realized using poly layer of the CMOS process. The microheaters consume considerable power during operation and are not reliable due to high operating temperatures. To increase the efficiency of the microheaters, bulk micromachining is employed, further adding to the complexity of post CMOS processing. Energy inefficient post CMOS process such as RF sputtering, chemical vapour deposition etc. are required for integration of such materials. On the contrary, use of Conducting polymers (CPs) as sensing materials as several advantages over inorganic metal oxides. These materials sense gases at room temperature [2], [3]. CPs can be easily integrated with CMOS process with only fewer number of post processing steps. Most conducting polymers show cross sensitivity to gases. An approach to overcome this is to have an array of sensors along with pattern classification techniques to improve discrimination among gasses [4], [5].

This paper introduces a digital readout interface chip for CMOS gas sensor array fabricated in AMS 0.35 μm 4 metal 2 poly CMOS process. The chip contains 27 sensor pixels which can be functionalized with distinct conducting polymers. The front end signal conditioning circuit at each pixel amplifies the sensor signal through fully differential digitally programmable gain stages and subsequently digitizes it using a fully differential continuous time delta sigma (CT-$\Delta\Sigma$) modulator. Digital data is read out using memory addressing approach for ease of interfacing with an external DSP. An on-chip reference pixel to decouple the phase change of the sensor from the phase shift introduced by signal processing blocks is presented. A proportional to absolute temperature (PTAT) sensor with digital readout is also integrated on chip. The sensing ability of the chip is validated by functionalizing a sensor pixel with polycarbazole conducting polymer. Measurement results on exposure to toluene is presented. An interface circuit is designed to communicate between the chip and a computer using a microcontroller. A graphical user interface (GUI) is also designed for ease of chip calibration, data acquisition and analysis. The rest of the paper is organized as follows. Sec. II details the architecture of the chip. Architecture of each sensor pixel is described in Sec. III. The need for a reference pixel is detailed in Sec. IV. Sec. V describes the design of a PTAT sensor with digital readout. The microcontroller based interface between the chip and the computer is described in

---

*This work is done as part of M.E. project thesis in Microelectronics Systems at Department of Electrical Communication Engineering of Indian Institute of Science, Bangalore, 560012.

978-1-61284-863-1/11 $26.00 © 2011 IEEE

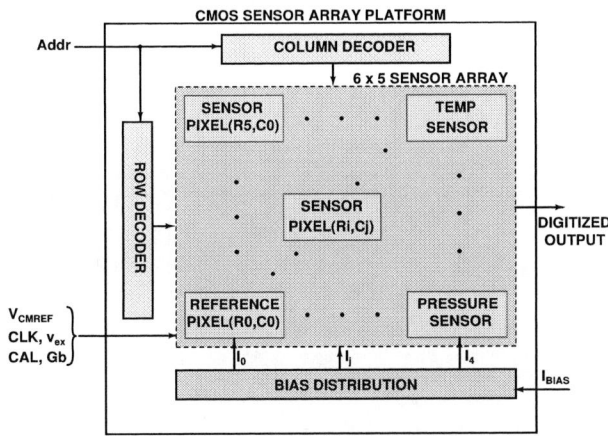

Fig. 1. Architecture of the chip.

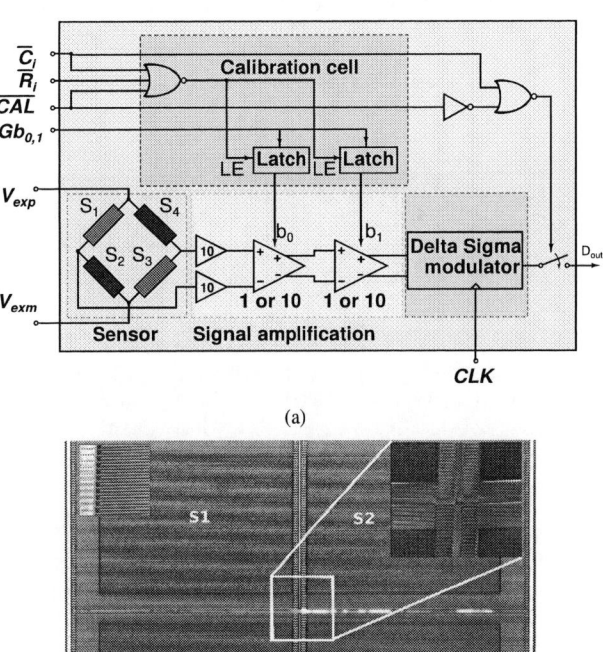

(a)

(b)

Fig. 2. (a) Architecture of each sensor pixel. (b) Sensor structure.

Sec. VI. Sec. VII concludes this paper. The measurement results and performance analysis are presented in appendices.

II. CHIP ARCHITECTURE

The architecture of the chip is shown in Fig. 1. The chip contains an array of 6×5 pixels of which there are 27 sensor pixels, a reference pixel and a temperature sensing pixel which are introduced in this paper. The chip also contains a pressure sensor with digital readout which will be reported elsewhere.

In an array of sensor pixels, the performance of each pixel differs due to intra-die process variations across the chip. An external reference bias current I_{BIAS} is distributed to all the pixels by a current distribution network. This reference current is used to generate the bias voltages at each sensor pixel separately. This approach reduces the effect of intra-die process variations and the impact of ambient noise on bias voltages.

III. SENSOR PIXEL

The sensor pixel contains a sensing element connected to fully differential digitally programmable gain stages for sensor signal amplification. This is followed by a fully differential second order CT-$\Delta\Sigma$ modulator. A calibration cell at each pixel stores the calibration information. Schematic of the sensor pixel is shown in Fig. 2(a). The design of individual modules of the sensor pixel is detailed in the following sections.

A. Sensor

The electrical impedance of the conducting polymer changes on exposure to the analyte gas. It is required to measure the relative change in impedance over its base value to determine the type and concentration of the gas. This is done using a Wheatstone bridge configuration as shown in Fig 2(a). The bridge comprises of four sensor elements S_1 to S_4 of which only two sensors S_1 and S_3 respond to the analyte while S_2 and S_4 are passivated.

Interdigitated capacitor (IDC) structures are used to realize each sensor element as shown in Fig. 2(b). The IDCs are

designed with Metal 4 (top most metal) of the CMOS process with width and gap of 0.6 μm. PAD layer is defined over these IDCs to etch away the passivation layers. Post processing steps required for creating wells to confine the sensing material is avoided by this approach.

B. Sensor Signal Amplification

For a given gas concentration, the response of the wheatstone bridge varies with the type of analyte gas. The signal processing circuits following the wheatstone bridge should account for such variations. The is done by using digitally programmable gain stages (DPGS).

The circuit schematic of DPGS is shown in Fig. 3. The first gain stage uses single ended two stage miller compensated opamp with NMOS input differential pair. The second and third gain stages of DPGS are realized using fully differential two stage miller compensated opamps. Two common mode feedback circuits set the common mode voltage at the outputs of each stage of these opamps. DGPS can be digitally programmed to give overall gain of 10, 100 or 1000.

The gain setting of second and third gain stages of DPGS is done using calibration setup shown in Fig. 2(a). It consists of two latches to store the gain settings. The contents of the latches are controlled by external pins CAL, Gb_0 and Gb_1. Inputs R_i and C_i are row and column select lines from the on-chip address decoders.

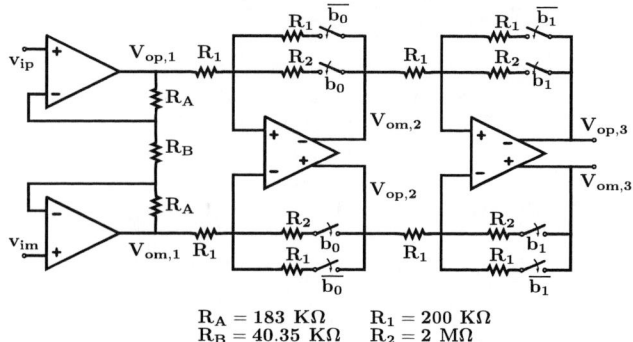

$R_A = 183 \ K\Omega \quad R_1 = 200 \ K\Omega$
$R_B = 40.35 \ K\Omega \quad R_2 = 2 \ M\Omega$

Fig. 3. Circuit schematic of DPGS for sensor signal amplification.

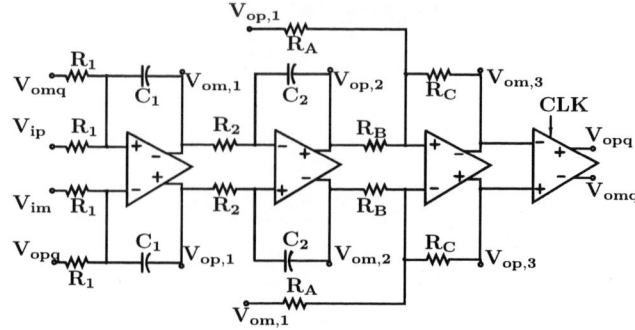

Fig. 4. Second order fully differential low pass CT-$\Delta\Sigma$M employing CIFF loop filter realized using Active-RC integrators.

C. Conversion to Digital Data

In an array of sensors, the amplified sensor signal can be interfaced to a read out channel using analog switches realized using transmission gates. This approach leads to loss of signal integrity due to noise pick up and distortion. To avoid these sources of errors, the amplified sensor signal has to be digitized at each pixel. Among different A/D converter architectures, $\Delta\Sigma$ modulators are particularly attractive because they provide higher conversion accuracy for lower precision requirements from components, easier data readout from an array of modulators.

In particular, having a CT loop filter in the $\Delta\Sigma$ modulator provides the following advantages over discrete time counterparts. There is significant reduction in switching noise, which is important here because the sensor element is placed directly above these circuts. In DT loop filters, the outputs need to settle to within a specified resolution in a given time period, requiring large opamp bandwidth and hence higher power consumption. Whereas in CT loop filters the bandwidth requirements of opamps are relaxed and hence they **consume lower power**. CT-$\Delta\Sigma$ modulators offer inherent anti-alias filtering, obviating the need for an low pass filter before the modulator.

In this work a low pass second order CT-$\Delta\Sigma$ modulator is designed for 10 KHz bandwidth (BW), oversampling ratio (OSR) of 32 and clock frequency (f_{CLK}) of 640 KHz is chosen. The CT loop filter is realized using Cascade of Integrators with Feed Forward (CIFF) architecture. The circuit schematic of the CT-$\Delta\Sigma$ modulator is shown in Fig. 4. The differential dynamic comparator circuit introduced in [6] is used for our design.

IV. REFERENCE PIXEL

The signal processing front-end circuits following the sensing element add additional phase shift Ψ to the phase change $\Delta\phi$ of the sensor. Ψ drifts with time, it is a function of ambient temperature and also varies with the frequency of measurement. In this section we present a novel on-chip solution to decouple Ψ from $\Delta\phi$.

Let Z denote the base impedance of sensors S_1 to S_4 shown in Fig. 2(b). Assume that S_2 and S_4 are passivated and that

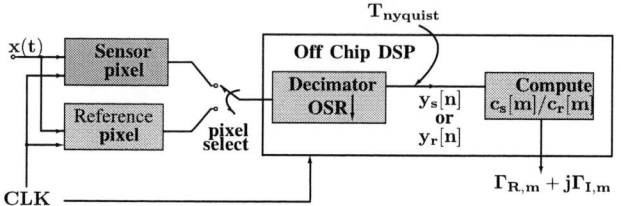

Fig. 5. Setup to determine the real and imaginary component of the sensor impedance.

they do not respond to the analyte. On exposure to the analyte, let the change in impedance of S_1 and S_3 be ΔZ. Also let $Z \gg \Delta Z$ and $\Delta\phi = \angle(\Delta Z/2Z)$.

Consider that an input $x(t) = cos(2\pi f_0 t)$ is fed both to the reference pixel and a sensor pixel as shown in Fig. 5. The output of the sensor pixel and reference pixel after off-chip decimation filtering is given by (also refer to Fig. 2(a)).

$$y_s[n] = \left|\frac{\Delta Z}{2Z}\right| A_s cos(\omega_0 n + \Delta\phi + \boldsymbol{\Psi}) \quad (1)$$

$$y_r[n] = A_r cos(\omega_0 n + \boldsymbol{\Psi}) \quad (2)$$

Where $\omega_0 = 2\pi f_0 T_{nyquist}$, $T_{nyquist} = OSR/f_{CLK}$, A_s and A_r are the gain settings of the senor and reference pixel respectively (Sec. III-B). Let $f_0 T_{nyq} = m/N \quad m < N$ for prime integers m and N so that $y_s[n]$ and $y_r[n]$ are periodic signals with period N. The discrete time fourier series coefficients of $y_s[n]$ and $y_r[n]$ will be,

$$c_s[k] = \frac{1}{N}\sum_{n=0}^{N-1}\left|\frac{\Delta Z}{2Z}\right| A_s cos(2\pi\frac{m}{N}n + \Delta\phi + \boldsymbol{\Psi})e^{-j\frac{2\pi nk}{N}}(3)$$

$$c_r[k] = \frac{1}{N}\sum_{n=0}^{N-1} A_r cos(2\pi\frac{m}{N}n + \boldsymbol{\Psi})e^{-j\frac{2\pi nk}{N}} \quad (4)$$

$$k = 0, 1, 2, \ldots N-1.$$

It can be shown that the summation in Eqn. 3 and Eqn. 4 will be non zero only for $k = m$. Dividing Eqn. 3 by Eqn. 4 we get.

$$\frac{c_s[m]}{c_r[m]} = \left|\frac{\Delta Z}{2Z}\right|\frac{A_s}{A_r}.e^{j\Delta\phi} = \Gamma_{R,m} + j\Gamma_{I,m} \quad (5)$$

Eqn. 5, shows that the real component $\Gamma_{R,m}$ and the imaginary component $\Gamma_{I,m}$ of the sensor impedance change can be

978-1-61284-863-1/11 $26.00 © 2011 IEEE

obtained for frequency f_0 independent of Ψ. This approach is validated through measurements and it is found that the normalized RMS phase error is less than 0.12%.

V. TEMPERATURE PIXEL

The sensitivity of sensing materials is prone to variations in temperature. It is essential to have an on-chip temperature sensor to mitigate the effect of these variations. One pixel of the sensor array is dedicated to measure the temperature. A PTAT based sensing element is used for our design.

Consider two identical diodes D_1 and D_2 of junction area A_1 and A_2 respectively. If the diodes are sufficiently forward biased with identical current I_{Bias} then we have.

$$I_{Bias} = J_0(T)A_1e^{V_1/\phi_t} \quad (6)$$

$$I_{Bias} = J_0(T)A_2e^{V_2/\phi_t} \quad (7)$$

Where, T is the absolute temperature, $J_0(T)$ is the temperature dependent reverse saturation current density of the diode, V_1 is the voltage across D_1 and V_2 is the voltage across D_2. $\phi_t = KT/q$ is the voltage equivalent of temperature. From Eqn. 6 and Eqn. 7 we get,

$$V_{Temp} = V_1 - V_2 = \frac{KT}{q}.ln\,(A_2/A_1) \quad (8)$$

Eqn. 8 shows that the voltage difference V_{Temp} is linearly proportional to absolute temperature. V_{Temp} is connected to the input of a pixel by replacing the IDC structure. I_{Bias} of 5 μA is used to forward bias two diodes with $A_2 : A_1 = 8 : 1$. With this approach, the absolute temperature of the chip can be read out digitally.

VI. CHIP INTERFACE

The chip provides $\Delta\Sigma$ modulated output which has to be acquired and processed off-chip. Also, the individual calibration settings for each pixel has to be stored on the latches. This is done with the help of a microcontroller (μC) interfaced to a computer.

A. Microcontroller Interface

The computer sends the control words to the μC which inturn uses these signals to configure the chip. dsPIC30F4011 μC is chosen for this purpose. The communication between the μC and computer is established through RS232 standard. This μC operates with a supply of 5 V, while the chip operates at 3.3 V. SN74LVC4245A Octal bus transceiver and 3.3 V to 5 V shifter is used for level voltage level translation. Pulse width modulator output of the μC provides the clock signal of 400 KHz to the chip. This μC has 2 KB RAM which is used to store the output of the chip. Data read from the entire chip takes only 2.7 s, which is sufficient for gas sensing applications.

B. Graphical User Interface for chip

Matlab® is used to create a graphical user interface (GUI). With this GUI, a user can set the gain of DGPS of each sensor pixel, reset gain settings of all the pixels to 10, acquire data from the chip and plot the impedance change of the sensor for different frequencies and also view the $\Delta\Sigma$ modulator outputs.

VII. CONCLUSION

In conclusion, a CMOS sensor array platform with digital readout has been fabricated in 0.35 μm CMOS process. There are 27 sensor pixels and a pixel for temperature sensing. The signal conditioning fronted at each pixels performs sensor signal amplification and digitization using CT-$\Delta\Sigma$ modulator. Impedance spectroscopy of the sensor impedance can be performed upto 10 KHz. A novel on-chip technique is presented to decouple phase change of sensor impedance from phase shift introduced by signal processing blocks. A sensor pixel is functionalized with polycarbazole and measurement results on exposure to different concentrations of toluene are shown. A microcontroller interface and a GUI is designed for ease of calibration of the chip and also for data aquisition and analysis.

ACKNOWLEDGMENTS

The author likes thank DST, DAE and MCIT for supporting this project. He also thanks Prof. Navakanta Bhat, Prof. Gaurab Banerjee, Prof. K N Bhat, Prof. Bharadwaj Amrutur, Prof. Praveen C Ramamurthy, Prof. P Subbanna Bhat, Dr. T Laxminidhi and Prof. Anantha Suresh for their guidance and support.

REFERENCES

[1] J. Gardner, , P. Guha, F. Udrea, and J. Covington, "CMOS Interfacing for Integrated Gas Sensors: A Review," *IEEE Sensors J.*, no. 99, p. 1, 2010.

[2] N. Lewis, "A Vlsi Compatible Conducting Polymer Composite Based Electronic Nose Chip," in *Electron Devices Meeting, 2002. IEDM'02. Digest. International.* IEEE, 2002, pp. 485–487.

[3] A. Kukla, A. Pavluchenko, Y. Shirshov, N. Konoshchuk, and O. Posudievsky, "Application of Sensor Arrays based on Thin Films of Conducting Polymers for Chemical Recognition of Volatile Organic Solvents," *Sensors and Actuators B: Chemical*, vol. 135, no. 2, pp. 541–551, 2009.

[4] U. Lange, N. Roznyatovskaya, and V. Mirsky, "Conducting Polymers in Chemical Sensors and Arrays," *Analytica chimica acta*, vol. 614, no. 1, pp. 1–26, 2008.

[5] M. Kermit and O. Tomic, "Independent Component Analysis Applied on Gas Sensor Array Measurement Data," *Sensors Journal, IEEE*, vol. 3, no. 2, pp. 218–228, 2003.

[6] S. Pavan, N. Krishnapura, R. Pandarinathan, and P. Sankar, "A Power Optimized Continuous-Time $\Delta\Sigma$ ADC for Audio Applications," *Solid-State Circuits, IEEE Journal of*, vol. 43, no. 2, pp. 351–360, 2008.

[7] M. Malfatti, D. Stoppa, A. Simoni, L. Lorenzetti, A. Adami, and A. Baschirotto, "A CMOS Interface for a Gas Sensor Array with a 0.5%-Linearity over 500k-to-1G Range and±2.5 C Temperature Control Accuracy," in *IEEE ISSCC proceedings*, 2006, pp. 294–295.

[8] D. Barrettino, M. Graf, S. Taschini, S. Hafizovic, C. Hagleitner, and A. Hierlemann, "CMOS Monolithic Metal–Oxide Gas Sensor Microsystems," *Sensors Journal, IEEE*, vol. 6, no. 2, pp. 276–286, 2006.

[9] B. Guo, A. Bermak, P. Chan, and G. Yan, "A Monolithic Integrated 4× 4 Tin Oxide Gas Sensor Array with On-chip Multiplexing and Differential Readout Circuits," *Solid State Electronics*, vol. 51, pp. 69–76, 2007.

[10] T. Koickal, A. Hamilton, S. Tan, J. Covington, J. Gardner, and T. Pearce, "Analog VLSI Circuit Implementation of an Adaptive Neuromorphic Olfaction Chip," *Circuits and Systems I: Regular Papers, IEEE Transactions on*, vol. 54, no. 1, pp. 60–73, 2007.

[11] C. Wu and K. Tang, "A Polymer-based Gas Sensor Array and its Adaptive Interface Circuit," in *VLSI Design Automation and Test (VLSI-DAT), 2010 International Symposium on.* IEEE, 2010, pp. 355–358.

[12] K. Arshak, V. Velusamy, O. Korostynska, K. Oliwa-Stasiak, and C. Adley, "A CMOS Single-Chip Gas Recognition Circuit for Metal Oxide Gas Sensor Arrays," *Circuits and Systems, IEEE Transactions on*, vol. IEEE Early Access, 2011.

APPENDIX A
MEASUREMENTS

The chip is fabricated in AMS 0.35 μm 4 metal CMOS process through Europractice. The micrograph of the chip is shown in Fig. 6. The chip occupies an area of 3.3×3.3 mm^2.

The reference pixel and sensor pixels are characterized by manually shorting the opposite arms of the Wheatstone bridge using a micromanipulator. The pixels are excited with a differential sinusiod of 3.125 KHz and clock of 640 KHz using Agilent 81150A function generator. All the external digital signals (Gb_0, Gb_1 and $\overline{CAL}$ and address lines) are controlled manually. The gain is set to 10 for all pixels. The chip outputs are logged for 0.5 s using Lecroy MSP-500 Mixed Signal Oscilloscope and later analyzed on a PC.

Fig. 7 shows the power spectral density (PSD) of the pixels for -24.3 dBFS and 3.125 KHz input. The PSD is obtained by averaging 32 sets of 8 K point Hann windowed FFT of the output of the pixel. It can be seen that the noise shaping closely resembles the ideal spectrum. Fig. 8 shows the variation of signal to noise ratio (SNR) and signal to distortion and quantization noise ratio (SNDR) for gain setting of 10 and 100 across different pixels. Tab. I summarizes the measurement results.

TABLE I
MEASURED PERFORMANCE SUMMARY

Technology	0.35 μm CMOS
Supply /**Power**	3.3 V / **57 mW**
Number of sensor pixels	27
Die size	3.3×3.3 mm^2
Sensor signal bandwidth	10 KHz
f_{CLK} / CT-$\Delta\Sigma$M OSR	640 KHz / 32
Loop filter Order / Type	2 / Active-RC CIFF
Peak SNR / SNDR (gain 10)	55.2 / 49.6 dB
Peak SNR / SNDR (gain 100)	47.2 / 45.6 dB
Mean DR Gain 10 / Gain 100	49.6 / 48.9 dB

A sensor pixel of the chip is functionalized with a conducting polymer to sense volatile organic gasses (VOCs). Polycarbazole dissolved in N-Methyl-2-pyrrolidone (NMP) is drop coated on all the four IDCs of the sensor using microinjector and air dried to remove residual solvent. Opposite arms of the Wheatstone bridge are passivated with polydimethylsiloxane (PDMS). The sensor pixel is exposed to different concentrations of toluene diluted with synthetic air using Kintek 491MB Gas Standard Generator. The response of the pixel to different concentration of toluene is shown in Fig. 9 for 3.125 KHz sinusoidal excitation. Drift in sensor response is seen which may be due to other impurities.

The on-chip temperature sensor pixel is characterized by exposing the chip to ramping ambient temperature. The response of this pixel is shown Fig. 10. Here, each data point is the mean value of the 8192 samples of the temperature sensor pixel output. The ambient temperature is measured using a thermocouple. Tab. II compares this work with other CMOS gas sensor array architectures.

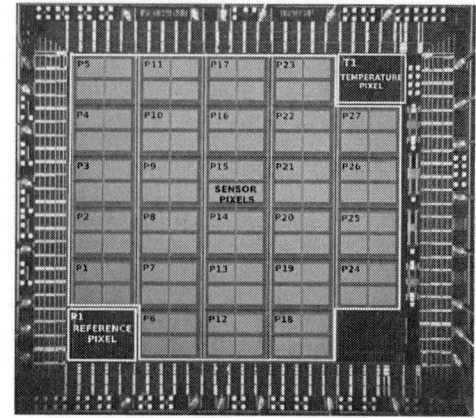

Fig. 6.   Micrograph of the chip.

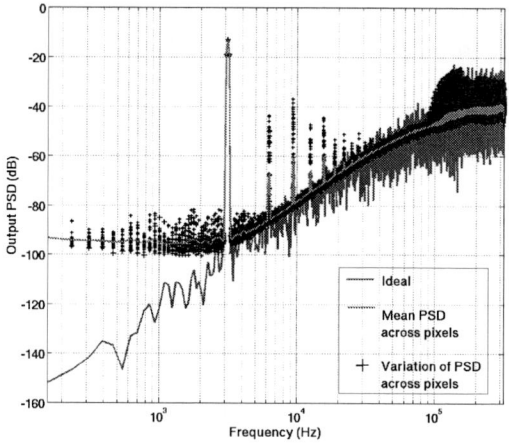

Fig. 7.   Output power spectral density. Distortion peaks present at harmonics of the fundamental are also introduced by gain stages preceding the CT-$\Delta\Sigma$ modulator. Degradation of in-band SNR can be due to non ideal opamps and clock jitter.

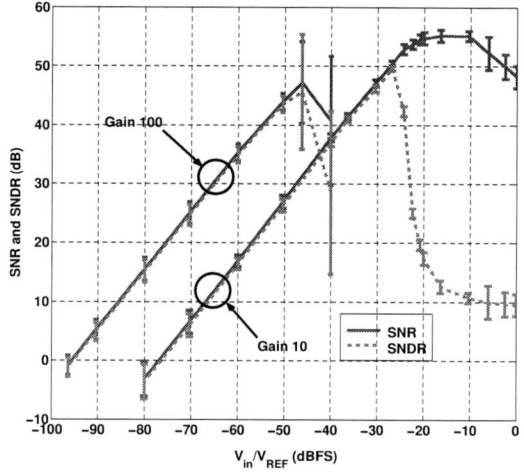

Fig. 8.   Measured SNR and SNDR of 28 pixel outputs. Gain of each pixel is set to 10 or 100. The error bars show the variation of SNR and SNDR across different pixels. Peak SNR of 55.2 dB is obtained for -16.4 dBFS input while Peak SNDR of 49.6 dB occurs at -26.8 dBFS input for gain setting of 10. The measured dynamic range (DR) is 49.6 dB and 48.9 dB is obtained for gain setting of 10 and 100 respectively.

978-1-61284-863-1/11 $26.00 © 2011 IEEE

	[7], 2006	[8], 2006	[9], 2007	[10], 2007	[11], 2010	[12], 2011	**This work**
Impedance Spectroscopy	No	No	No	No	No	No	**Yes**
Digital readout	Yes	Yes	No	No	Yes	Yes	**Yes**
On-chip digital calibration	Capacitor array	Log converter	No	Prog. current	No	No	**Prog. gain**
Number of sensors	8	3	16	70	6	16	**27**
Read out channels	8	3	1	70	1	4	**27**
On-chip intergration of sensor	No	Yes	Yes	Yes	No	No	**Yes**
Sensing material	WO_3	SnO_2	SnO_2	CP	CP	SnO_2	**CP**
Operating temperature (°C)	100-400	200-400	300	Room	Room	300	**Room**
CMOS process (μm)	0.35	0.8	5	0.6	0.18	0.35	**0.35**
On-chip temperature sensor	No	Yes	No	No	No	No	**Yes**

TABLE II

COMPARISON WITH OTHER CMOS GAS SENSOR ARRAY ARCHITECTURES.

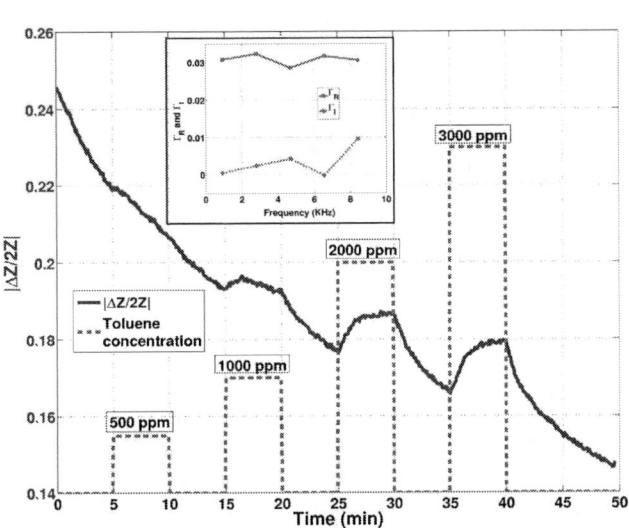

Fig. 9. Response of the sensor on exposure to different concentrations of Toluene. Inset shows the Γ_R and Γ_I of the sensor when it is exposed to 1000 ppm of Toluene and excited with a square wave of 937.5 Hz. The impedance is extracted upto ninth harmonic.

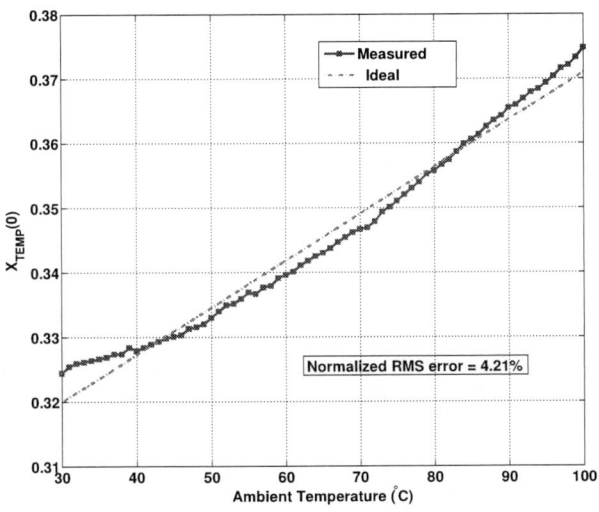

Fig. 10. Output of temperature sensor.

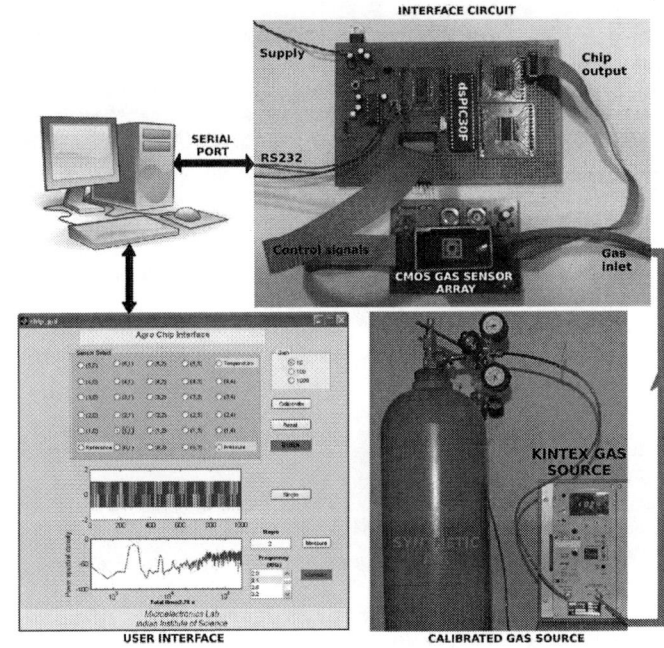

Fig. 11. CMOS sensor array calibration and measurement system.

APPENDIX B

CALIBRATION AND MEASUREMENT SYSTEM

A system for calibration and measurement of CMOS gas sensor array is shown in Fig. 11. Kintek 491MB Gas Standard Generator uses liquid source of a volatile organic gas along with a diluent gas such as synthetic air to generate calibrated concentrations of an analyte gas.

The output of Kintek is connected to a small aluminium chamber covered with glass lid. This aluminium chamber confines the analyte gas over the sensor array. The figure also shows the front end of the GUI designed in Matlab® for ease of calibration of the chip. Data from array of sensors can be acquired and viewed in with this GUI.

Micro Energy Management for Energy Harvesting at Maximum Power Point

Sewan Heo, Yil Suk Yang, Jaewoo Lee, Sang-kyun Lee, and Jongdae Kim
NT Convergence Components Research Department
Electronics and Telecommunications Research Institute
Daejeon, Korea

Abstract—This paper describes an efficient technique for maximum power point tracking of an energy harvesting device. The proposed technique is controlling the operating voltage of maximum power by transferring energy from the harvesting device. With the technique, an energy-aware architecture of the energy management IC maintains the maximum point with low power consumption that is appropriate for low energy generation from the harvesting. An experiment shows that the technique with the IC fabricated in a 0.18μm process maximizes the energy transfer power with negligible error. With a comparison of the energy transfers between a direct transfer from a source to a battery with no conversion loss and the maximum power transfer with the IC, the proposed technique is verified to be more efficient for the low energy harvesting.

Keywords-energy harvesting, maximum power point tracking, energy management

I. INTRODUCTION

Energy harvesting techniques are developed for supplying power to wireless devices such as sensor nodes in a network [1] which cannot obtain stable power. In the wireless situation, the devices should be autonomous and self-powered with an energy source. Since the amount of energy is limited, however, an additional energy is necessary to extend the lifetime of the devices. The energy harvesting generates electrical energy from the surrounding environment or from renewable sources such as solar, wind, vibrational, or thermal energy, thus the devices operate for a longer lifetime with the supplements of the generated energy. However, the amount of the generated energy is very small. Moreover, they are not consistent but instead vary depending on the condition under which they operate. Therefore, an efficient energy management is required to obtain maximum amount of energy from the harvesting considering the operating condition.

The maximum power from an energy harvesting device can be obtained with the maximum power point tracking (MPPT) technique. It maintains the operating voltage of the harvesting device to maximum power point so that the harvesting device generates energy with maximum power. Even though the condition of the device is changed as time passes or by other reasons, the technique searches and maintains a new operating voltage of maximum power generation. There are many techniques of the maximum power point tracking of the harvesting devices, such as solar cells[2][3]; piezoelectric

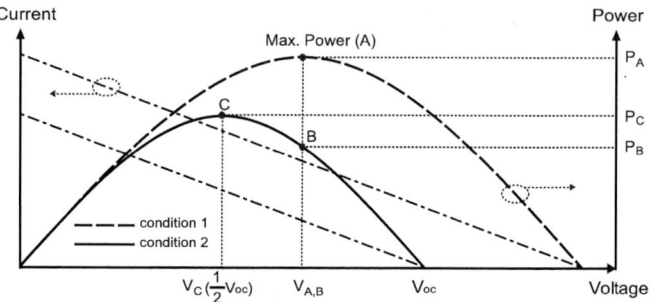

Fig. 1. I-V and P-V Characteristics of the Energy Harvester.

generators (PEG) [4][5]; and thermoelectric generators (TEG) [6][7]. Although the energy is generated at maximum power with the techniques, they are not appropriate to be applied when the amount of harvested energy is relatively small compared to that of consumed energy for the operation of the techniques. Since the real time calculation of the generated power [8] from voltage and current of the harvesting device always consumes considerable energy, a simple and low power technique [9] is required to save energy from the continuous power monitoring.

II. A SIMPLE TECHNIQUE FOR MAXIMUM POWER POINT TRACKING WITH LOW POWER CONSUMPTION

It is necessary for the maximum power point tracking technique to adjust operating voltage to maximum power generation when the operating condition is changed. It is shown in Fig. 1, which is current and power characteristics of a piezoelectric generator or a thermoelectric generator, not a solar cell. The current decreases linearly due to internal resistance when the voltage increases. On the other hand, the power is parabolic depending on the operating voltage, and it is maximized at one point of the voltage. It is important that the one point of the maximum power varies under different condition. Under *condition 1*, V_A is the operating voltage of maximum power and P_A is the generated power. When the condition is changed to *condition 2*, the generated power is deceased to P_B. In this case, the power can be maximized to P_C by adjusting the operating voltage from $V_{A,B}$ to V_C.

Most techniques generally search the voltage of maximum power by a calculation and repeat the procedure for adjusting

978-1-61284-863-1/11 $26.00 © 2011 IEEE

to the operating condition, consequently, inducing very accurate tracking. However, they consume considerable energy for the continuous calculation of power. To get a new solution for maximizing the generated power while minimizing the consumed power, it is necessary for the energy harvesting devices such as piezoelectric or thermoelectric generators to be modeled by a simple and accurate numerical model. By the ideal voltage (or open circuit voltage) source V_{oc} and the internal output resistor R_s in series, the devices can be modeled. In this model, since the output voltage V_o is determined by a voltage drop caused by a current flow through the resistor, the power delivered to the output changes as the current is varied. The output power P_o and the voltage of maximum power V_{opt} are calculated and derived below from the model:

$$P_o = V_o \times \frac{V_{oc} - V_o}{R_s}, \tag{1}$$

$$V_{opt} = \frac{1}{2} V_{oc}. \tag{2}$$

According to (2), which is derived from (1), the output power P_o is maximized when the output voltage V_o is half of the ideal voltage source V_{oc}. As shown in Fig. 1, the voltage of maximum power is V_A or V_C which is half of V_{oc} under each condition. Because V_{oc} varies depending on the device conditions while R_s is a fixed intrinsic parameter, the point of maximum power can be tracked when V_o is controlled so that it is equal to half of V_{oc}. Thus, without the calculation of power from both voltage and current, the maximum power can be obtained by only controlling voltage. Consequently, the power consumption from the calculation of harvested power can be reduced by eliminating use of the digital module such as a multiplier. Moreover, it can be further reduced by using only a comparator for the voltage control rather than the frequent sensing of the voltage and the current via an analog-to-digital converter (ADC). Therefore, based on the model of the harvesting device such as a piezoelectric or a thermoelectric generator, we propose a technique that attains the maximum power from an energy harvesting device with low power consumption.

III. Energy Management

Energy transfer from the harvesting device to a battery requires a DC-DC conversion, which is capable of inducing the maximum power generation by the voltage control and reducing the transfer loss with a high efficiency. Thus, an inductive DC-DC converter is proposed for the energy management with the input voltage regulation. This is an outstanding difference compared to a general DC-DC converter, which regulates output voltage to a fixed value. The energy management IC (EMIC) consists of input source switches, a switched mode converter, and a comparator that enables the input voltage to be regulated to the external reference voltage. It operates in both the pulse frequency modulation (PFM) mode and, simultaneously, discontinuous conduction mode (DCM), reducing power consumption by the sparse operation and by the use of only essential switching power, respectively, because the amount of the input energy is small.

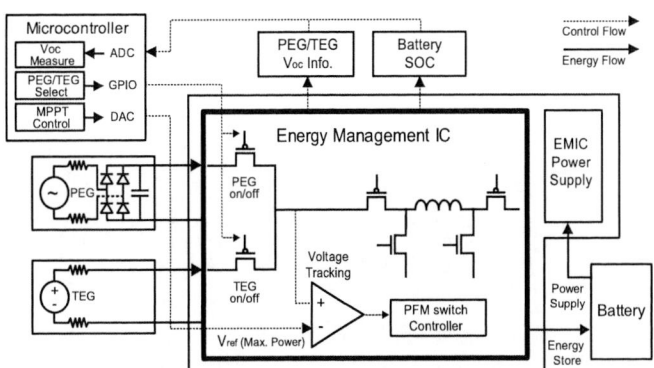

Fig. 2. Energy Harvesting System.

The energy harvesting system is shown in Fig. 2. The harvested energy supplied from two devices is transferred to the battery through the energy management IC controlled by a microcontroller, which consists of an ADC, a general-purpose input and output device (GPIO), and a digital-to-analog converter (DAC). The control step is as follows. Initially, the open circuit voltage is measured from each harvesting devices by ADC. According to the open circuit voltages, the device with higher energy is selected by GPIO and the reference voltage for controlling the input voltage at the maximum power point is determined to half of the open circuit voltage. Finally, the DAC generates and provides the reference voltage to a comparator that regulates the input voltage by controlling the switches.

As an energy store and supply device, the battery can perform diverse roles such as supplying high power for a long time as well as storing much energy with a high capacity. Since power from the energy harvesting device is small and not consistent, the device cannot supply enough power directly to the energy management IC. However, from the small but accumulated energy, the battery can supply instantly higher power to the switching operation of the IC by an additional power converter that provides stable power at a constant voltage from the battery. Moreover, even though the battery cannot obtain enough energy for a long time by little harvested energy, it can sustain power for the time with a high capacity. On the other hand, since the harvested energy is normally larger than the consumed one from the operation of the IC which is designed with low power consumption, the harvested energy in excess of that consumed can accumulate. To protect the battery from charging to a maximum limit, the microcontroller periodically gets the information of the state of charge (SOC) from the battery. When it reaches the limit, the operation of the energy management IC is halted by the microcontroller.

The energy management IC uses two energy harvesting devices, a piezoelectric and a thermoelectric generator. Since these two devices generate the maximum power at half of the open circuit voltage, periodically measuring the voltage and simply maintaining the voltage control lead to generating and tracking the maximum power. In spite of the slightly lowered accuracy of the regulation, the unnecessary power consumption from the calculation of the harvested power can be eliminated. This enables the harvesting system to operate semi-

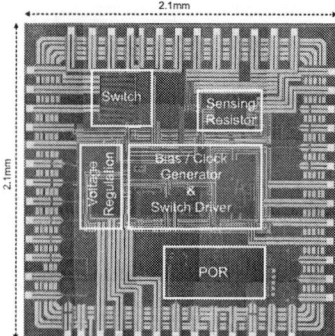

Fig. 3. Chip Photograph of the Energy Management IC.

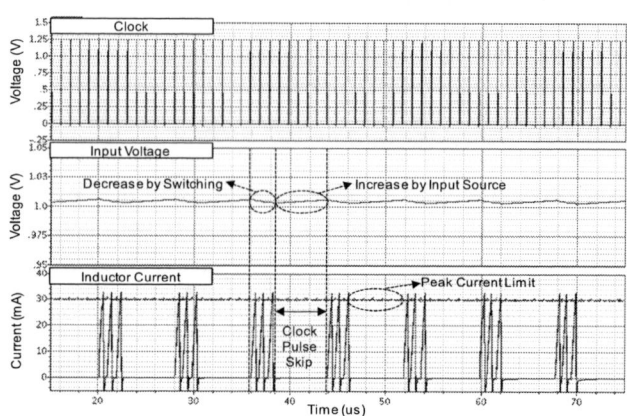

Fig. 4. PFM Mode Operation.

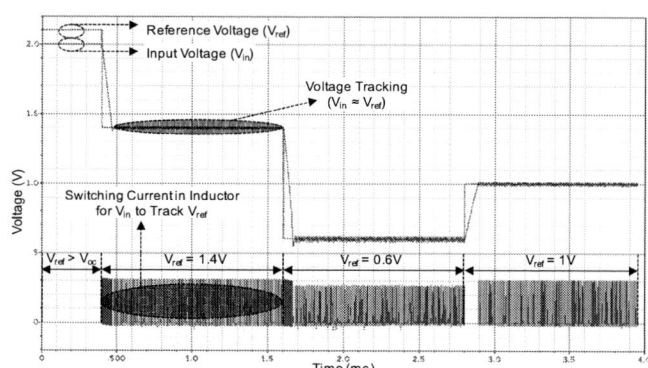

Fig. 5. Voltage Tracking to the Reference Voltage.

permanently with only a small amount of the harvested energy. Thus, we propose the energy management IC with the low power technique for the maximum power point tracking when transferring energy from the harvesting devices.

IV. EXPERIMENTAL RESULTS

The energy management IC was implemented in a 0.18μm BCDMOS process occupying 4.4mm² of a die area as shown in Fig. 3, to evaluate the proposed technique of low power energy harvesting. The DC-DC energy conversion in the IC operates at 1MHz in a PFM mode by the clock pulse skip when the input energy is low. A 10μH inductor is used with 32mA peak current limit. The input voltage range from the energy harvesting device is 0.9V~3.0V and the output voltage range to the battery is 2.6V~3.8V while the IC supply voltage is 1.2V and 3.3V for the core and the IO, respectively. For the accurate evaluation, the IC and the technique was simulated by the accurate Cadence Spectre simulator in the 0.18μm process including the physical parameters, was measured for the actual value, and was compared based on the ideal calculation data. To exclude the influence of the input and output devices, the input energy harvesting source was modeled by a various voltage source with an internal resistance of 500 Ω and the output battery was modeled by a 3.3V super-capacitor.

Above all, the fundamental operation of DC-DC conversion in the energy management IC is simulated as shown in Fig. 4. The input voltage is regulated around 1V, as the reference voltage indicates, by the switching current through the inductor from the input to the output. In the PFM mode, the inductor current is induced by a clock pulse at 1MHz when the input voltage exceeds the reference. On the other hand, when the input voltage is lower than the reference, the inductor current is not induced by the clock pulse skip until the input voltage increases by the input source. Since the inductor current increases to the peak current limit, and then decreases to zero, it always operates in a DCM, which as well as the PFM mode is the most appropriate switching operation for the low input energy source like the energy harvesting.

The input voltage tracks the reference voltage by the voltage comparator. The performance of the tracking is evaluated by a simulation with the reference voltage variation from 2.1V to 1V via 1.4V and 0.6V by 1.2ms, as shown in Fig. 5. When the reference voltage exceeds the maximum value of

the input voltage the IC is in an idle state without transferring energy. When the reference voltage is changed to be lower than the input voltage, however, the IC tries to transfer energy by the switching current in the inductor so that the input voltage tracks the reference voltage. Whereas, when the reference voltage is changed to be higher than the input voltage, the IC waits again in idle state until the input voltage increases to the reference voltage, and then tracks it. The tracking error is less than 20mV, which is determined by the sensitivity of the voltage comparator.

When the amount of generated energy is small, a voltage boosting device is necessary to store the energy to a battery. Moreover, the device is required to get maximum power. The micro energy transfer by the IC with the technique of maximum power point tracking is shown in Fig. 6, comparing with a direct charging to the battery from the input source with a 2kΩ. When the open circuit voltage of the source is lower than that of the battery as normal, the energy can be transferred by only the IC, not by the direct charging. As the open circuit voltage increases, the energy transfer that operated at half of the open circuit voltage becomes greater with the IC rather than the direct charging though it shows rapider increase by no conversion loss.

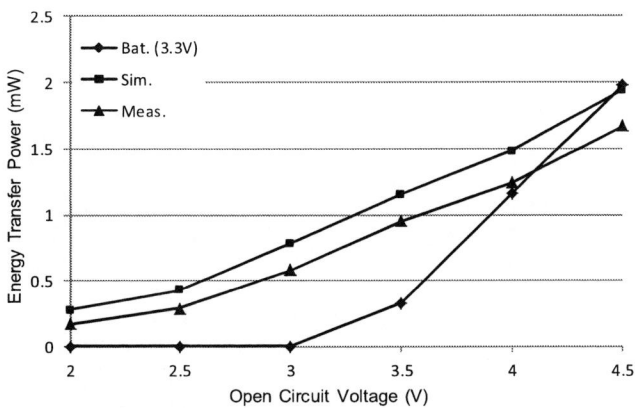

Fig. 6. Energy Transfer with MPPT and Direct Charging.

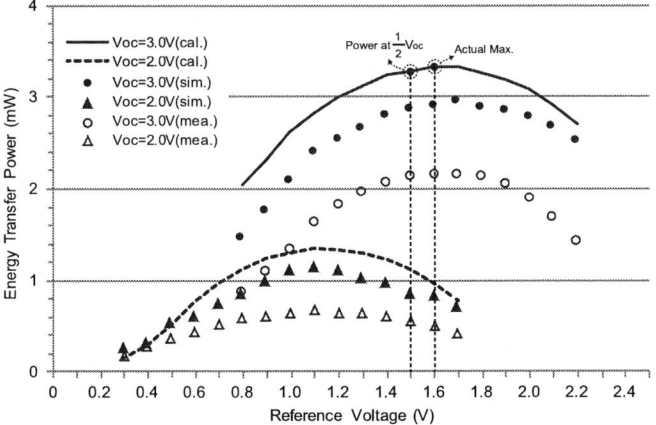

Fig. 7. Variation of the Energy Transfer Power by Reference Voltages.

The energy transfer power varies by different reference voltages and is maximized when it is half of the open circuit voltage as shown in Fig. 7. The power is compared among the calculation, the simulation, and the measurement by estimation from the extracted parameters, the accurate simulator, and the implemented IC, respectively, for two open circuit voltages representing different amounts of energy. There is only a small amount of error, approximately 0.1V, between the half of the open circuit voltage and the voltage for the actual maximum power. Although the data of simulation are lower by 10.7% on average than that of calculation due to the reverse current loss at zero current sensing in the DCM, and that of measurement are much lower, the measured power at half of the open circuit voltage is almost maximal with at most 3.7% of error.

Therefore, energy transfer at maximum power by tracking half of the open circuit voltage is verified.

V. CONCLUSION

This paper proposes a technique for obtaining maximum power transfer from the energy harvesting device by controlling the operating voltage. Based on a simple method and an energy-aware architecture of the energy management IC, it maintains the maximum point with low power consumption that is appropriate for low energy from the harvesting. An experiment verified that the implemented energy management IC with the technique maximizes the energy transfer power so that it is almost close to the actual maximum.

ACKNOWLEDGMENT

This work was supported by the IT R&D program of MKE/KEIT, Rep. of Korea [KI002077, EPMIC Based on Self-Chargeable Power Supply Module].

REFERENCES

[1] V. Raghunathan, C. Schurgers, S. Park, and M. B. Srivastava, "Energy-Aware Wireless Microsensor Networks," IEEE Signal Processing Magazine., vol. 19, no. 2, pp. 40-50, Mar. 2002.

[2] T. Esram, and P.L. Chapman, "Comparison of Photovoltaic Array Maximum Power Point Tracking Techniques," IEEE Trans. Energy Conv., vol. 22, no. 2, pp. 439-449, Jun. 2007.

[3] M. A. S. Masoum, H. Dehbonei, and E. F. Fuchs, "Theoretical and Experimental Analyses of Photovoltaic Systems With Voltage- and Current-Based Maximum Power-Point Tracking," IEEE Trans. Energy Conv., vol. 17, no. 4, pp. 514-522, Dec. 2002.

[4] G.K. Ottman, H.F. Hofmann, A.C. Bhatt, and G.A. Lesieutre, "Adaptive Piezoelectric Energy Harvesting Circuit for Wireless Remote Power Supply," IEEE Trans. Power Electron., vol. 17, no. 5, pp. 669-676, Sep. 2002.

[5] G. K. Ottman, H. F. Hofmann, and G. A. Lesieutre, "Optimized Piezoelectric Energy Harvesting Circuit Using Step-Down Converter in Discontinuous Conduction Mode," IEEE Trans. Power Electron., vol. 18, no. 2, pp. 696-703, Mar. 2003.

[6] H. Nagayoshi, T. Kajikawa, and T. Sugiyama, "Comparison of Maximum Power Point Control Methods for Thermoelectric Power Generator," Proc. Thermoelectronics '02, 2002, p. 450.

[7] H. Lhermet, C. Condemine, M. Plissonnier, R. Salot, P. Audebert, and M. Rosset, "Efficient Power Management Circuit: From Thermal Energy Harvesting to Above-IC Microbattery Energy Storage," IEEE Journal of Solid-State Circuits, vol. 43, no. 1, pp. 246-255, Jan. 2008.

[8] E. Koutroulis, K. Kalaitzakis, and N.C. Voulgaris, "Development of a Microcontroller-Based, Photovoltaic Maximum Power Point Tracking Control System," IEEE Trans. Power Electron., vol.16, no. 1, pp. 46-54, Jan. 2001.

[9] I. Doms, P. Merken, R. P. Mertens, C. V. Hoof, "Capacitive Power-Management Circuit for Micropower Thermoelectric Generators with a 2.1μW Controller," in Proc. IEEE ISSCC, 2008, pp. 300–615.

Low-Power, Low-offset Stacked Analog Latch Using an Offset Cancellation Technique

Minehiko Tateno, Hiroki Date, and Kenichi Ohhata

Department of Electrical and Electronics Engineering, Kagoshima Univ.
1-21-40 Korimoto Kagoshima 890-0065, Japan
Phone: +81-99-285-8420 E-mail: k2026764@kadai.jp

Abstract—**In this paper, a low-power, low-offset analog latch is proposed. A stacked analog latch and a novel offset cancellation technique were combined to achieve low-power and low-offset analog latch. Circuit simulations demonstrated that power dissipation could be reduced by 35% in comparison with using a conventional offset cancellation technique. A test chip fabricated by using 180-nm CMOS technology showed that the proposed circuit reduced the standard deviation of the offset voltage from 15.5 to 3.3 mV. Moreover, the proposed circuit could maintain offset low in the range of the input common level from 0.6 to 1.6 V.**

I. INTRODUCTION

An analog latch is a circuit that outputs a digital signal according to the sign of the input analog signal and is an important basic circuit widely used in analog-to-digital converters and memory circuits [1-4]. In particular, a dynamic analog latch, which consumes no DC power, is a key component in analog-to-digital converters that use parallel architecture [5, 6]. Recently, the offset voltage of the analog latch has begun to cause serious problems because the scaling of MOS transistors decrease not only the supply voltage but also the signal voltage. Applying the offset cancellation technique to the analog latch was proposed to overcome this problem [7]. This circuit measured the threshold voltages of the input differential pair in the reset phase, and then the mismatch of the threshold voltage was canceled in the comparison phase by using the measured threshold voltage. It was reported that the offset voltage was canceled by 70% with this technique. However, this circuit consumed a relatively large amount of power because a double-tail analog latch [2] was used in the circuit resulting in large load capacitance for the 1st stage circuit. Moreover, a clock pulse with a very narrow width was required. Therefore, a large power clock buffer should be used for the small rising and falling times.

In this paper, main problems of the conventional circuit are discussed. We then propose a stacked analog latch that uses a novel offset cancellation technique to overcome these problems. Next, the circuit performances are described by using a circuit simulator. Finally, the experimental results of the fabricated test chip are discussed.

II. CONVENTIONAL ANALOG LATCH

Figure 1 shows a schematic of a conventional analog latch [7]. This circuit is a double-tail analog latch [2] that applies the

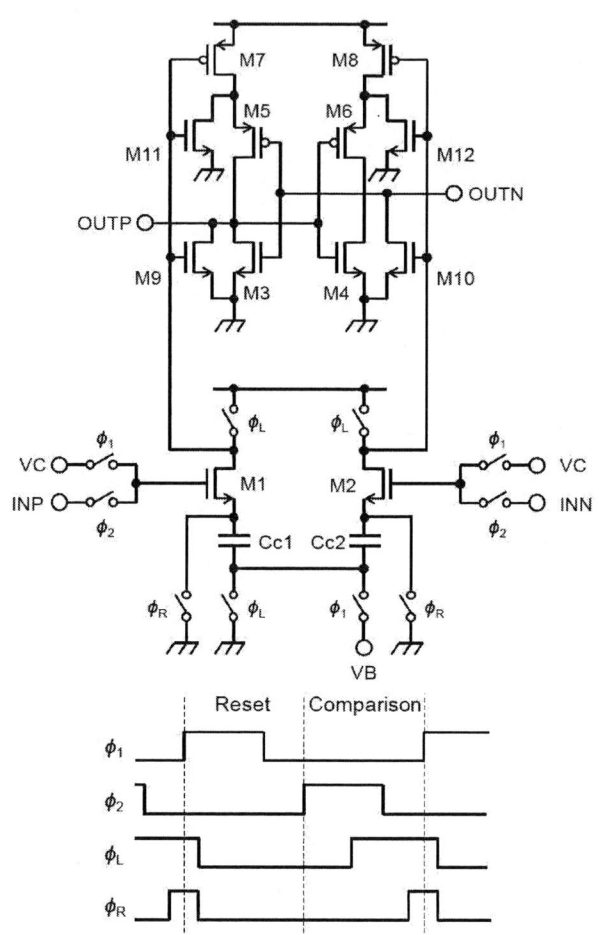

Fig. 1. Schematic of the conventional analog latch.

offset cancellation technique. The 1st stage containing an input differential pair (M1 and M2) amplifies the input signal, and the 2nd stage (M3-M12) latches and outputs the digital signal. The offset voltage due to the differential pair in the 1st stage determines the total offset voltage because the offset contribution of the 2nd stage is reduced by the gain of the 1st stage. Using this fact, the circuit reduces the offset voltage by cancelling only the mismatch in the differential pair. In the reset phase, the threshold voltages of the M1 and M2 are stored

978-1-61284-863-1/11 $26.00 © 2011 IEEE

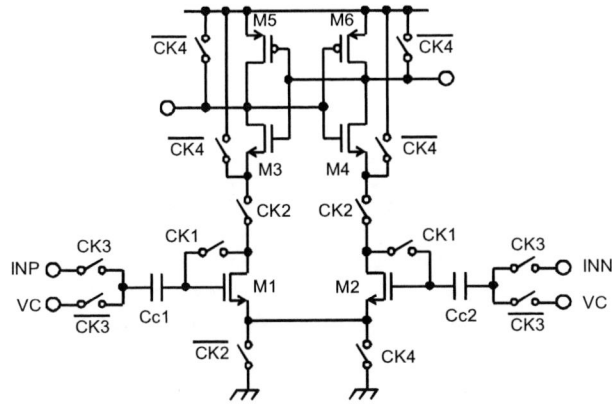

Fig. 2. Schematic of the proposed circuit.

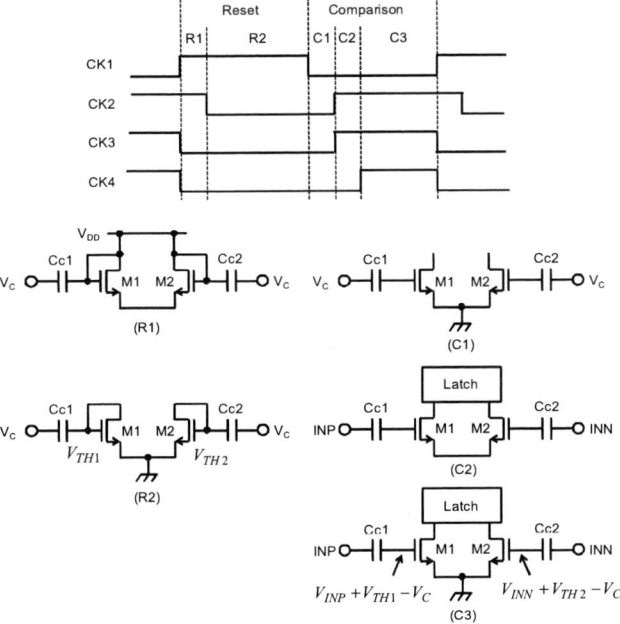

Fig. 3. Clock sequence and corresponding circuits of the proposed circuit.

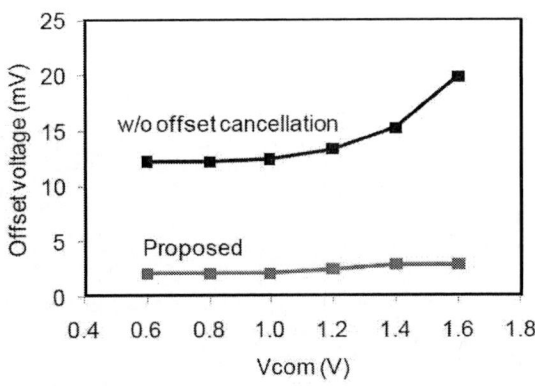

Fig. 4. Standard deviation of the offset voltage dependence on the input common level.

comparison operation. The clock buffer consumes a large amount of power distributing a narrow pulse like this.

III. STACKED ANALOG LATCH USING AN OFFSET CANCELLATION TECHNIQUE

We propose a stacked analog latch that uses the offset cancellation technique shown in Fig. 2 to solve problems of the conventional circuit. The stacked analog latch [1] is comprised of an amplifier part (M1 and M2) which amplifies an input signal and a latch part (M3-M6), which outputs a digital signal. The amplifier and latch parts are stacked; therefore, the number of MOS transistors is small, resulting in low power dissipation. This circuit as well as the conventional circuit shown in Fig. 1 can cancel out the mismatch in the input differential pair. In the reset phase, the threshold voltages of M1 and M2 are stored to Cc1 and Cc2, and then the mismatch in the threshold voltage is cancelled by Cc1 and Cc2 in the comparison phase. Next, the operation of the proposed circuit is discussed by referring to Fig. 3. The reset phase is divided into R1 and R2 periods for each clock state. Capacitors Cc1 and Cc2 are precharged to the V_{DD} by connecting the gates and drains of the M1 and M2 to the V_{DD} in the R1 period. Then, the gates and drains of M1 and M2 are disconnected from the V_{DD}, and the sources of M1 and M2 are connected to the GND in the R2 period. Therefore, capacitors Cc1 and Cc2 are discharged by M1 and M2. M1 and M2 turn off when the gate voltages decrease to the threshold voltages (V_{TH1} and V_{TH2}); therefore, the threshold voltages are stored to capacitors Cc1 and Cc2. The gates and drains are disconnected from each other in the C1 period when the phase shifts to the comparison phase. Then, the drains are connected to the latch circuit in the C2 period. Finally, the sources are pulled down to the GND, and the comparison is performed. At this time, the gate voltages become $V_{INP} + V_{TH1} - V_C$ and $V_{INN} + V_{TH2} - V_C$ because the input signals pass through capacitors Cc1 and Cc2. The overdrive voltages of M1 and M2 are $V_{INP} - V_C$ and $V_{INN} - V_C$; thus, the offset due to these transistors are cancelled. The narrowest clock pulse is $CK4$ for period C3; however, its pulse width is sufficiently wide of 40% of the cycle time. The point that should be noted is that only the offset due to the input differential pair is cancelled by this circuit.

to capacitors Cc1 and Cc2. Then, in the comparison phase, the 1st stage is activated by the $\Box_R$, and the input signal is amplified. At this time, the mismatch in the threshold voltage of M1 and M2 is cancelled because Cc1 and Cc2 are connected to the sources of M1 and M2. It was reported that the test chip fabricated by 90-nm CMOS technology cancelled about 70% of the offset voltage. The power dissipation was 39 µW at an operation frequency of 500 MHz. The conventional circuit consumes relatively large power because three MOS transistors are connected to the output of the 1st stage, resulting in a large load capacitance. Moreover, the width of the reset pulse $\Box_R$ should be about 1/10 of the cycle time because the upper electrode of Cc1 and Cc2 are reset to GND after the

978-1-61284-863-1/11 $26.00 © 2011 IEEE   141

56 x 12.4 µm for an analog latch

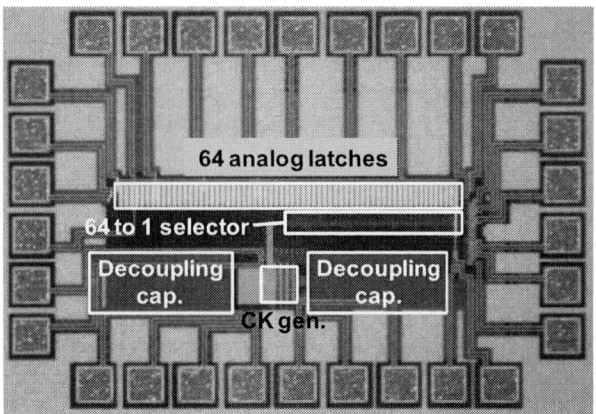

Fig. 5. Photomicrograph of a test chip.

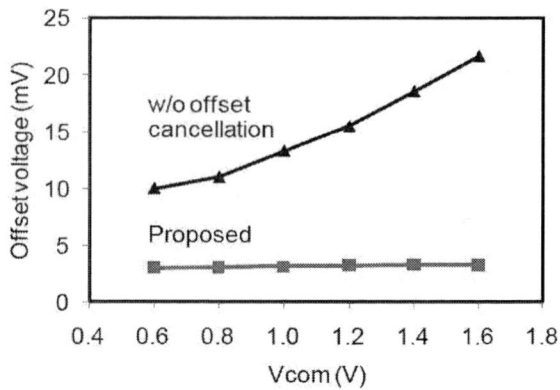

Fig. 7. Measured standard deviation of the offset voltage dependence on the input common level (*Vcom*).

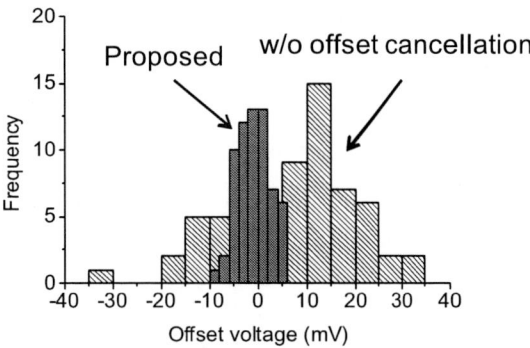

Fig. 6. Measured offset distribution of 64 analog latches.

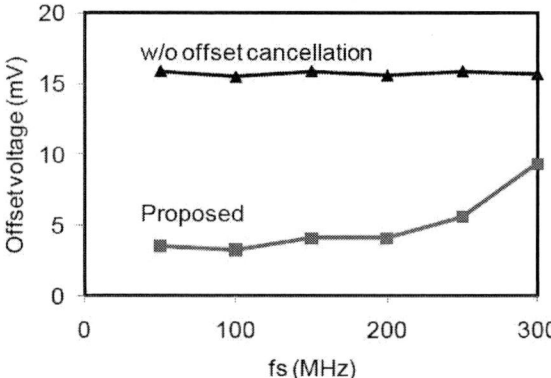

Fig. 8. Measured standard deviation of the offset voltage dependence on the operating frequency (*fs*).

The power dissipation could be reduced by about 35% with the proposed circuit.

Therefore, the gain of the amplifier part should be large to suppress the offset contribution from the latch part. The overdrive voltage of the input differential pair should be small as long as the desired operation speed can be ensured because a lower overdrive voltage results in a higher gain [1]. The overdrive voltage can be adjusted arbitrarily by the V_C in this circuit.

The effectiveness of the proposed circuit was demonstrated by using a circuit simulator with 180-nm CMOS technology. The aspect ratio of the input differential pair was 2 µm/0.18 µm and that of the others was 1 µm/0.18 µm. The supply voltage was 1.8 V, and the operating frequency was 200 MHz. The standard deviation of the offset voltage dependence on the input common level (V_{com}) is shown in Fig. 4. The standard deviation of the offset voltage without using offset cancellation was about 12 mV and increased with V_{com}. This is because the overdrive voltage of the input differential pair became larger with V_{com}. The offset voltage of the proposed circuit could be reduced to 2.5 mV and was independent of V_{com} because the overdrive voltage could maintain a constant by adjusting V_C. The power dissipations of the proposed circuit and the circuit not using offset cancellation were 58 and 88 µW, respectively.

IV. EXPERIMENTAL RESULTS

Figure 5 shows a photomicrograph of a test chip fabricated using 180-nm CMOS technology. The test chip was comprised of 64 analog latches with SR latches, a 64 to 1 selector, and a clock generator. The offset voltage of each analog latch can be measured with the 64 to 1 selector. The area for each analog latch was 56 × 12.4 µm. The supply voltage was 1.8 V, and each analog latch consumed 50 µW at an operating frequency of 200 MHz. The measured offset distribution of the 64 analog latches is shown in Fig. 6. The operating frequency was 100 MHz, and the input common level was 1.2 V. The standard deviation of the offset voltage without offset cancellation was 15.5 mV. In the proposed circuit, the offset voltage was only 3.3 mV. About 80% of offset was reduced by using the proposed circuit. The standard deviation of the offset voltage dependence on the input common level is shown in Fig. 7. The operating frequency was 100 MHz. The standard deviation of the offset voltage without using offset cancellation was 10 mV for when V_{com} = 0.6 V and increased with V_{com}. The offset voltage reached to 21.7 mV for when V_{com} = 1.6 V. The offset voltage of the proposed circuit was 3.2 mV and was independent of V_{com}. The standard deviation of the offset voltage dependence on the operating frequency is shown in Fig. 8. The input common level was 1.2 V. Both circuits could operate up to 300 MHz. The offset voltage of the proposed circuit remained constant approximately up to 200 MHz. However, the offset voltage increased quickly at more than 250 MHz. This is because a large overdrive voltage is required for the input differential pair to operate at more than 250 MHz; therefore, the gain of the 1st stage was reduced, resulting in a large offset contribution of the 2nd stage.

V. Conclusion

A low-power, low-offset stacked analog latch using the offset cancellation technique was proposed. A stacked analog latch with a small parasitic capacitance was used to reduce the power dissipation and a novel offset cancellation technique was applied to reduce the mismatch in the input differential pair. Circuit simulations demonstrated that the power dissipation could be reduced by 35% in comparison with the conventional offset cancellation technique. A test chip fabricated by 180-nm CMOS technology showed that the proposed offset cancellation technique reduced the standard deviation of the offset voltage from 15.5 to 3.3 mV. Moreover, the proposed circuit could maintain a low offset over the input common level of 0.6 to 1.6 V because it could determine the overdrive voltage of the input differential pair independently of the input common level.

Acknowledgment

This research was funded by the Semiconductor Technology Academic Research Center (STARC). The VLSI chip in this study was fabricated in the chip fabrication program of the VLSI Design and Education Center (VDEC), the University of Tokyo in collaboration with Rohm Corporation and Toppan Printing Corporation. The authors would like to thank N. Kano, K. Ono, J. Naka, and O. Kobayashi for their valuable discussions.

References

[1] B. Wicht, T. Nirschl, and D. Schmitt-Landsiedel, "Yield and Speed Optimization of a Latch-Type Voltage Sense Amplifier," IEEE Journal of Solid-State Circuits, vol. 39, no. 7, pp. 1148-1158, July 2004.

[2] D. Schinkel, E, Mensink, E. Klumperink, E. van Tuijl, and B. Nauta, "A Double-Tail Latch-Type Voltage Sense Amplifier with 18 ps Setup + Hold Time," in ISSCC Dig. Tech. Papers, pp. 314-315, Feb. 2007.

[3] B. Razavi, and B. A. Wooley, "Design Techniques for High-Speed, High-Resolusion Comparators," IEEE Journal of Solid-State Circuits, vol. 27, no. 12, pp. 1916-1926, Dec. 1992.

[4] M. Miyahara, Y. Asada, D. Paik, and A. Matsuzawa, "A low-Noise Self-Calibrating Dynamic Comparator for High-Speed ADCs," in Proc. A-SSCC, pp. 269-272, Nov. 2008.

[5] G. V. Plas, S. Decoutere, and S. Donnay, "A 0.16 pJ/Conversion-Step 2.5 mW 1.25 GS/s 4 b ADC in a 90 nm Digital CMOS Process," in ISSCC Dig. Tech. Papers, pp. 566-567, Feb. 2006.

[6] B. Verbruggen, P. Wambacq, M. Kuijk, and G. V. Plas, "A 7.6 mW 1.75 GS/s 5 bit Flash A/D Converter in 90 nm Digital CMOS," in Dig. Symp. VLSI Circuits, pp. 14-15, June 2008.

[7] M. Miyahara, and A. Matsuzawa, "A Low-Offset Latched Comparator Using Zero-Static Power Dynamic Offest Cancellation Technique," in Proc. A-SSCC, pp. 233-236, Nov. 2009.

Wireless Powering and Bidirectional Telemetry Front-End for Implantable Biomedical Devices

Rui-Feng Xue[1], Hyouk-Kyu Cha[1], Jia Hao Cheong[1], Pradeep Basappa K[1], Minkyu Je[1], and Yuanjin Zheng[2]

[1]Institute of Microelectronics, A*STAR (Agency for Science, Technology and Research)
11 Science Park Road, Singapore Science Park II, Singapore 117685
[2]Nanyang Technological University, 50 Nanyang Avenue, Singapore 639798
Email: xuerf@ime.a-star.edu.sg

Abstract—**Wireless powering to implantable biomedical devices is highly desirable due to obviation of batteries or piercing wirings. This paper presents an implantable inductively powered front-end operating at 13.56MHz carrier frequency for biomedical applications, with the capabilities of power transfer, clock extraction and bidirectional command/data communication. The system consideration including the inductive coupling and the circuit building blocks of power management are given and high efficiency is highlighted as well. The ASIC has been fabricated in 0.18μm CMOS process. The rectifier achieves an efficiency of 66% and the ASK-demodulated command, extracted clock and LSK back-telemetry are verified through measurement. This design can be applied to diverse implantable biomedical applications where wireless powering is needed.**

Keywords-implantable biomedical devices; high efficiency; inductive coupling; RF powering; rectifiers; regulators; wireless power transfer.

I. INTRODUCTION

Fueled by the scaling of integrated circuits and the development of microsensors and micropackaging technologies, implantable biomedical devices play more and more vital roles in a myriad of well-known and emerging areas such as monitoring, diagnostic, therapeutic, and interventional applications [1]. A miniature, lightweight and biocompatible implantable microsystem with wireless powering and data telemetry capability is highly desirable. Transcutaneous wireless powering can obviate the need for replacement or recharging of batteries, the risk of infection as well as the possibility of signal distortion induced by penetrating wires, and can also minimize the post-implant trauma of patients.

Fig. 1 shows an example of wireless powering system for biomedical implants. Inductive coupling is commonly employed, where the power and bidirectional data telemetry is achieved using a primary external coil and a secondary implant coil [1-4]. The extracorporeal transmitter typically consists of a class-E power amplifier owing to the high efficiency and ability to generate large voltages across the primary coil from a relatively low-voltage source such as a battery. The induced power at the secondary coil is then rectified and regulated to provide a clean supply for the on-chip electronics. Forward data telemetry from external device unit towards the implant is usually implemented by modulating the envelope of the power

carrier to create detectable changes in the secondary coil, ASK modulation for instance. Reverse telemetry or backtelemetry from the implant toward the external unit is generally based on the load modulation technique known as load shift keying (LSK), wherein the reflected impedance in the primary coil is modulated by changing the impedance seen by the secondary coil.

The constraints in implant size, power dissipation and system functionality are the common requirements imposed on biomedical implants. For example, the neural recording implant microsystem for long-term uninterrupted operation requires about 10 mW power supply, the size of which is confined by the anatomical space available to an area of several cm^2 and a thickness of less than 1 mm. Additionally, tissue-induced loss and the living tissue exposure to electromagnetic fields are also strong concerns since considerable power is transferred transcutaneously. The loss caused by tissue directly affects the power reception. And the temperature rise should be strictly limited to below 1~2 °C to prevent the occurrence of irreversible hurt of the brain. With the above considerations of size, power and tissue effect, high efficiency is the key specification of the wireless powering front-end.

This paper presents an implantable wireless powering front-end operating at 13.56 MHz for biomedical application, including the inductive coupling and the ASIC of power management, clock extraction and bidirectional data telemetry. The system architecture and design considerations are discussed in section II. The circuit building blocks are elaborated in section III. Section IV is dedicated to measurement and verification of the fabricated ASIC, followed by concluding remarks in section V.

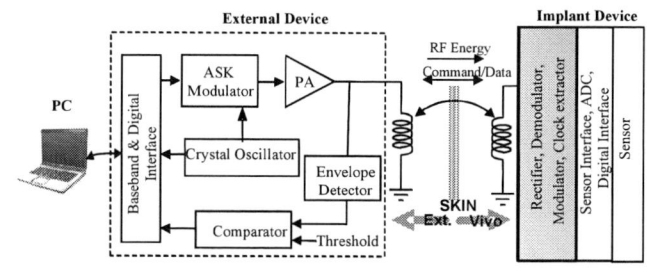

Fig. 1. Simplified block diagram of the wireless powering system.

This work was supported by the Science and Engineering Research Council of A*STAR (Agency for Science, Technology and Research), Singapore. The grant number for the project is 092 148 0069.

978-1-61284-863-1/11 $26.00 © 2011 IEEE

II. FRONT-END ARCHITECTURE

The schematic diagram of the implantable front-end, illustrated in Fig. 2, consists of the coupling coils connected with the ASIC including a power conversion and regulation block, a clock extractor, a ASK demodulator, as well as a backtelemetry modulator. The inductive coupling delivers power and command data forward to the implant ASIC and backscatters the digitized data to external reader. The external device unit configures the implant ASIC by sending the command through ASK-modulated power carrier. The clock is extracted from the incoming power carrier and used further to generate clock for other parts such as the ADC sampling clock. The sensor data in implant device are digitized and converted to a desired format in the digital baseband and sent to the external device by backscattering the incoming RF carrier through a load modulation.

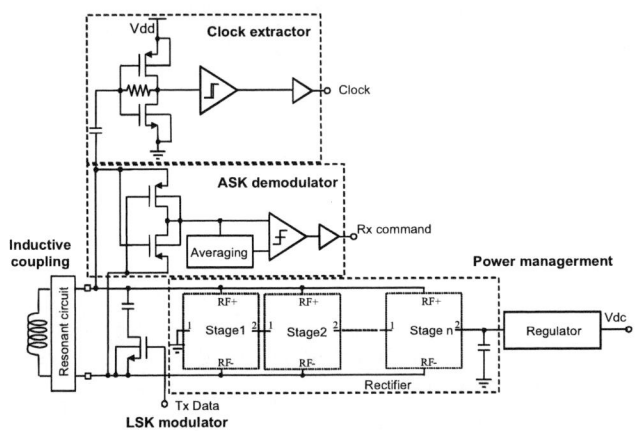

Fig. 2. Schematic diagram of the wireless powering front-end.

The state machine block controls the state transition of the ASIC, illustrated in Fig. 3. After the extracted clock is ready and the power-on-reset signal goes to high, the ASIC goes to "Receive" mode and starts to search the start-of-frame (SOF) and receive the data frame. After all the information bits are received, the ASIC goes to "Processing" mode. After that, all the processed data are read to the 11 bits register and the ASIC goes to "Transmit" mode. All the sampled data are spread to 7 chips and backscatter to the external device unit.

As for the overall efficiency of the wireless powering front-end, it is utmost important how efficiently the RF energy is coupled from the external device as well as how efficiently the power management blocks function, both of which should be designed elaborately. The transcutaneous inductive coupling is the bottle neck so far which is briefed below and the high-efficiency power management is discussed in the next section.

The design of high-efficiency inductive coupling need consider power carrier frequency, tissue effect, geometry size and location of coils. The interplay among these factors adds the design complexity. The choice of the frequency is related to each part of the whole powering link and contains many tradeoffs. First, high frequency is better for smaller size and

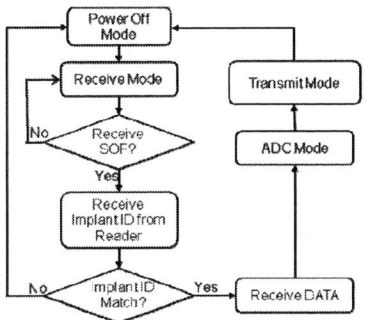

Fig. 3. State machine for control of the front-end ASIC.

higher quality factor of coils, whereas it leads to sensitivity of parasitics, increased requirement of source driving and difficult implementation of rectifier. Second, tissue effect is also a function of operating frequency, which includes tissue-induced loss and living tissue safety considerations. The optimal frequency trades off between tissue-induced loss and power propagation [5]. The safety level with respect to human exposure to electromagnetic fields is regulated in the guidelines in terms of maximum permission exposures (MPEs) and specific absorption ratio (SAR) [6-7]. Taking account of the above factors, 13.56 MHz is chosen as the power carrier frequency.

III. BUILDING CIRCUIT BLOCKS

The power required by the implant device is transferred transcutaneously onto the input port of the ASIC through inductive coupling. The subsequent power management blocks should be designed efficiently in order to achieve high efficiency of the overall link, mainly the rectifier and the on-chip low dropout (LDO) regulator.

A. Rectifier and LDO Regulator

The power conversion efficiency (PCE) of the rectifier is one of the most important parameters [8]. To increase PCE of the rectifier, the dropout voltage and substrate leakage current needs to be minimized. Low dropout voltage can be realized by either increasing the W/L ratio of transistors or using Vth cancellation technique. The dynamical Vth cancellation technique is adopted in this design. In the meanwhile, latch-up needs to be avoided. For converting AC energy to DC energy, an eight-stage differential-drive rectifier is used. The rectifier core has a cross-coupled bridge configuration shown in Fig. 4 (a). A differential-drive active gate bias mechanism enables to achieve both low ON-resistance and small reverse leakage of diode-connected MOS transistors at the same time, resulting in a high PCE. Unit stages are serially stacked along the DC path and connected in parallel to the input RF terminals. By using this multi-stage configuration, appropriate DC output voltage is obtained at the optimal operating point where the PCE is maximized. As shown in Fig. 4 (b), the simulated maximum PCE is higher than 70% with a load resistor of 20 kΩ, and the output voltage increases with input voltage as expected. Additionally, a limiter is inserted for protection of subsequent circuits.

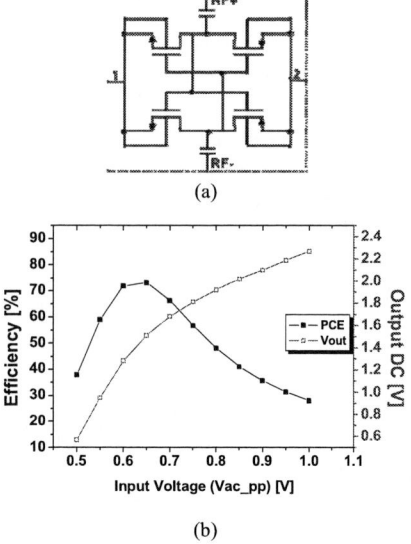

(a)

(b)

Fig. 4. (a) Rectifier core structure and (b) simulated power conversion efficiency and output voltage.

The unregulated DC voltage from the rectifier output is further regulated by the low-power LDO regulator. The simulated performances are shown in Fig. 5. This power management block also generates the desired reference voltages for the other blocks like SAR ADC and sensor interface.

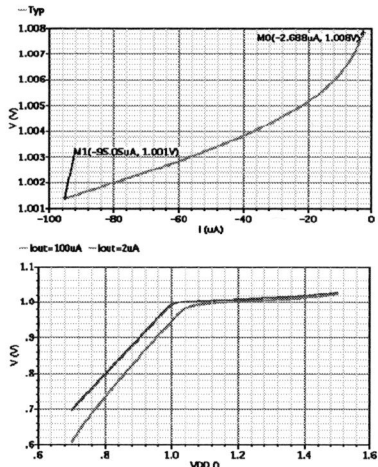

Fig. 5. Simulated load regulation and line regulation of the LDO.

B. Clock Extractor and ASK Demodulator

The clock extractor consists of an input AC-coupled amplifier and a Schmitt trigger which edge-triggers a D flip-flop with its inverting output tied to its input [9]. The clock is divided by two, buffered and fed to the digital core as its reference clock.

Amplitude modulation is a common method used for sending configuration data into implantable devices at fairly low data rates. The envelope of the received ASK-modulated

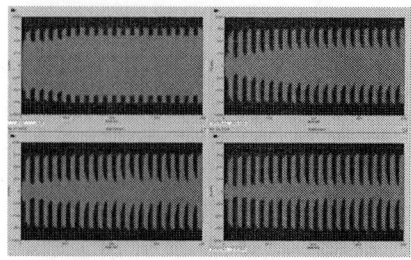

Control bit	Modulation depth
<00>	11%
<01>	31.3%
<10>	48.7%
<11>	67.6%

Fig. 6. Simulated load regulation and line regulation of the LDO.

signal is compared with the average value of the envelope to obtain the command from the external device.

C. Back Telemetry

Backtelemetry from the implant toward the external unit is realized by load shift keying (LSK). A MOS transistor, in parallel connection with the secondary coil, is switched on or off based on data value to be transmitted. This modulates the impedance seen by the secondary coil, which is reflected and read out on the impedance variation of the primary coil. A resistor can be connected in series with the switch as tradeoff between sensitivity in detecting the change at the external device unit and power dissipation on the resistor [10]. The simulated back telemetry with programmable modulation depth from 10% to 70% is shown in Fig. 6.

Moreover, backtelemetry can also be utilized for inductive powering in a closed loop. Any changes in the distance and misalignment between the primary coil and the secondary coil may cause significant change in the received power, which can cause either a malfunction or excessive heat dissipation. The backtelemetry capability can be employed to stabilize the

Fig. 7. Microphotograph of the ASIC.

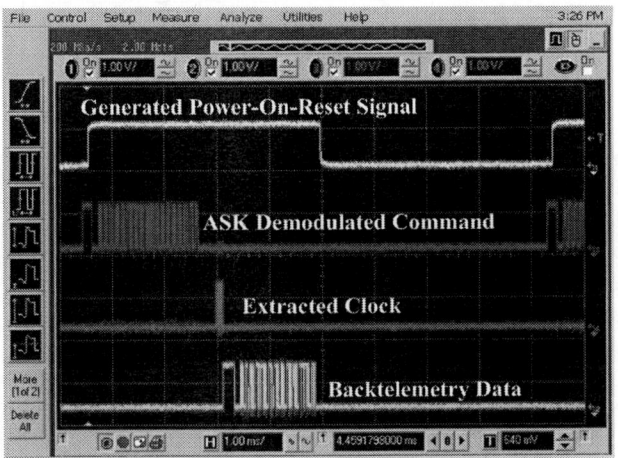

Fig. 8. Measured waveforms of the power-on-reset, ASK demodulated command, extracted clock and backtelemetry data.

received voltage on the implant [11].

IV. MEASURED RESULTS

The wireless powering front-end chip has been fabricated in 0.18μm CMOS process. The ASIC occupies a total active area of $1.5 \times 1.78 mm^2$, together with ADC and sensor interface of the implant microsystem shown in Fig. 7.

The received power from the inductive coupling is converted to DC supply to power the ASIC. As shown in Fig. 8, the configuration command sent from the external device is demodulated, the clock is extracted from the incoming carrier and further divided for ADC sampling in implant ASIC, and the power-on-reset signal is generated to reset the digital baseband. The analog information from the implanted sensor is converted to digital and processed in digital baseband, and then fed to the LSK modulator to backscatter the unmodulated carrier of the external device.

TABLE I
MEASURED PERFORMANCES OF THE FRONT-END

Parameter	Measured Result
Power carrier frequency	13.56 MHz
Forward command data demodulation	ASK (programmable modulation depth from 10% to 90% in steps of 10%)
Communication protocol	Modified and simplified from ISO 14443 RFID standard
Rectifier efficiency	66%
Regulator efficiency and power consumption	56% 12.8 μW
Regulated dc voltage and output current	1 V 10 mA
Backtelemetry	LSK (programmable modulation depth from 10% to 70% in four steps)
Implant coil size	Area = 25×10 mm² Thickness = 0.5 mm
Inductive powering distance	10 mm (nominal)
The safety guidelines for specific absorption rate (SAR)	2 W / kg for any 10 g of tissue

The inductive coupling operates at 10 mm nominal distance within the constraints of implant size. The rectifier achieves an efficiency of 66%. The LDO consumes a power of 12.8μW and generates 1V regulated voltage output with 56% efficiency. The measured front-end performances are summarized in Table I.

V. CONCLUSIONS

An implantable wireless powering front-end with bidirectional telemetry capability is presented in this paper. Connected with the inductive coupling coils working at 13.56 MHz carrier frequency, the ASIC can generate 1 V regulated voltage supply with high efficiency, and extract the command and clock for the implant devices. Backtelemetry can transmit the implant sensor information to external device and provide close-loop power control capability. The design is applicable to diverse implantable biomedical applications.

ACKNOWLEDGMENT

The authors would like to thank the colleagues at IME for technical discussion, IC layout, PCB fabrication and measurement support. The authors also appreciate the valuable comments of the reviewers.

REFERENCES

[1] R. Bashirullah, "Wireless implants," *IEEE Microw. Magazines*, pp. S14-S23, Dec. 2010.

[2] P. B. Khannur, K. L. Chan, J. H. cheong, et al., "A 21.6uW inductively powered implantable IC for blood flow measurement," *IEEE Asian Solid-State Circuits Conference*, Session 9-5, Nov. 8-10, 2010, Beijing, China.

[3] K. Kang, P. B. Khannur, R.-F. Xue, et al., "The RF front-end of a blood flow sensor for vascular graft applications," *Proceedings of the second APSIPA annual Summit and Conference*, pp. 741-744, Biopolis, Singapore, 14-17 Dec. 2010.

[4] R.-F. Xue, P. B. Khannur, K. Kang, et al., "The RF front-end of a blood flow sensor for vascular graft applications," *Proceedings of the second APSIPA annual Summit and Conference*, pp. 761-764, Biopolis, Singapore, 14-17 Dec. 2010.

[5] A. S. Y. Poon, S. O'Driscoll, and T. H. Meng, "Optimal frequency for wireless power transmission into dispersive Tissue," *IEEE Trans. Antennas Propag.*, vol. 58, no. 5, pp. 1739-1750, May 2010.

[6] *IEEE Standard for Safety Levels with Respect to Human Exposure to Radio Frequency Electromagnetic Fields, 3kHz to 300GHz*, IEEE Standard C95.1, 1999.

[7] Federal Communication Commission, Wireless Medical Telemetry [online]. Available:
http://wireless.fcc.gov/services/index.htm?job=service_home&id=wireless_medical_telemetry

[8] Koji Kotani, Atushi Sasaki and Takashi Ito, "High-efficiency differential-drive CMOS rectifier for UHF RFIDs," *IEEE Journal of Solid-State Circuits*, vol. 44, no. 11, pp. 3011-3018, November 2009.

[9] D. J. Black and R. R. Harrison, "Power, clock, and data recovery in a wireless neural recording device" *IEEE Int. Symp. Circuits Syst.*, ISCAS, pp. 5083-5086, 2006.

[10] C. Sauer, M. Stanacevic, G. Cauwenberghs, and N. Thankor, "Power harvesting and telemetry in CMOS for implanted devices," *IEEE Trans. Circuits Syst. –I*, vol. 52, no. 12, pp. 2605-2613, Dec. 2005.

[11] M. Kiani and M. Ghovanloo, "An RFID-based closed-loop wireless power transmission system for biomedical applications," *IEEE Trans. Circuits Syst. – II*, vol. 57, no. 4, pp. 260-264, Apr. 2010.

978-1-61284-863-1/11 $26.00 © 2011 IEEE

Multiple Output Switched Capacitor DC-DC Converter for Low Power Applications

Ravinder Pal Singh and Minkyu Je

Institute of Microelectronics, A*STAR (Agency for Science, Technology and Research),
11, Science Park Road, Singapore, Singapore. 117685
Email: ravinderps@ime.a-star.edu.sg

Abstract— Switched capacitor (SC) DC-DC voltage converters have been widely used for low-power applications. These converters play a role of voltage doublers or are used to obtain a dc voltage with reverse polarity. They may also be configured to implement a particular voltage conversion ratio. However, there are applications which require multiple output voltages. If multiple output voltages are required, such converters are duplicated and tuned to obtain the desired output voltages. Employing a separate switched capacitor DC-DC converter for each output increases the silicon area and results in higher cost. Thus, a simple, low-complexity SC DC-DC conversion scheme is proposed, which obtains multiple output voltages from a single input source. The scheme is scalable in nature and is well suited for integration. The scheme is demonstrated using a test chip fabricated in 0.18μm CMOS process.

I. INTRODUCTION

The modern system-on-chip (SoC) converge different types of devices and functions, onto the single chip. It may contain digital, analog, mixed-signal, and often radio-frequency(RF) functions all integrated on one chip. These sub-blocks have different supply requirements, and thus, a SoC may require more than one power supply for its normal functionality. For example, the processors (and its peripheries), SDRAM, etc operate at a voltage rail of 1.8V whereas RF functional blocks may require 2.8V. Such a portable system is typically powered by a single input source, say a single-cell-type Lithium battery. Such a battery has a nominal voltage output of 3V, but the voltage could reach up to 3.5V at the beginning of discharging and down to 2V near the end of its life. Thus a suitable voltage regulation scheme is required to provide the desired supply voltage to the constituent blocks.

The existing solutions for obtaining multiple output voltages are based on conventional buck converter topology which requires an external inductor. Either one inductor has to be added for every output voltage or a single inductor can be multiplexed among various outputs [1]-[5]. The required inductance value can be of the order of micro-henries, which cannot be integrated on silicon. On the other hand, a switched capacitor voltage conversion is used which does not require external components.

Switched capacitor circuits are employed to act as voltage doublers [6] or are used to obtain a dc voltage with reverse polarity. Alternately, they may also be configured to implement a particular conversion ratio. However, if the input voltage changes, the converter has to be re-designed to adjust the

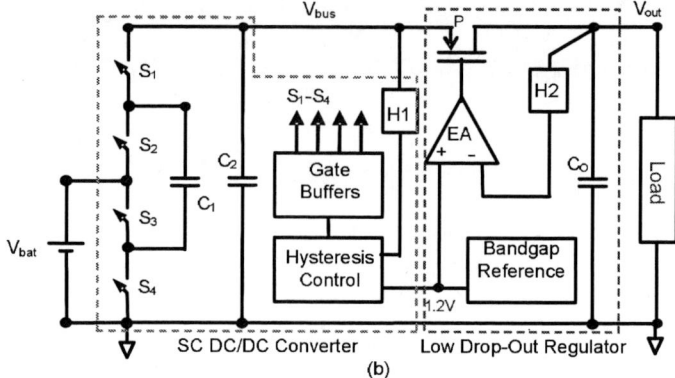

Fig. 1. A switched capacitor DC-DC converter employed with an LDO to filter out the output voltage ripple.

conversion ratio. In order to overcome this limitation, a configurable SC DC-DC converter is proposed [7], which changes its topology based on the input voltage by using 9 switches and 2 capacitors to adjust the step-down voltage ratio. Similarly, a reconfigurable architecture is proposed for stepping the voltage up [8], and it requires 18 switches and 4 capacitors. On the other hand, two different switched-capacitor converters can be used [9] to obtain two different output voltages, one of which is higher and the other is lower than the input voltage. It proposes to use 11 switches and 3 capacitors to obtain two dual voltages, which is area intensive. A multiple voltage-gain SC DC-DC converter has also been proposed [10]. However, it requires 20 switches and 3 capacitors to configure the gain.

The past approaches either use inductive elements which are not well suited for integration or they rely on increased number of switches and capacitors to obtain the given voltage conversion ratio. Thus, a scheme is required which,

1) does not require inductive elements,
2) requires fewer number of switches,
3) is well suited for integration, and
4) is scalable in nature for obtaining multiple outputs.

The proposed scheme is based on the basic SC converter topology. The circuit schematic of the basic SC converter is shown in Fig. 1. It comprises of 4 switches, a flying capacitor C_1 and an output capacitor C_2. In order to reduce the output ripple arising due to switching action, a low drop-out regulator (LDO) is generally used, as shown in Fig. 1[11]. Its operation

978-1-61284-863-1/11 $26.00 © 2011 IEEE

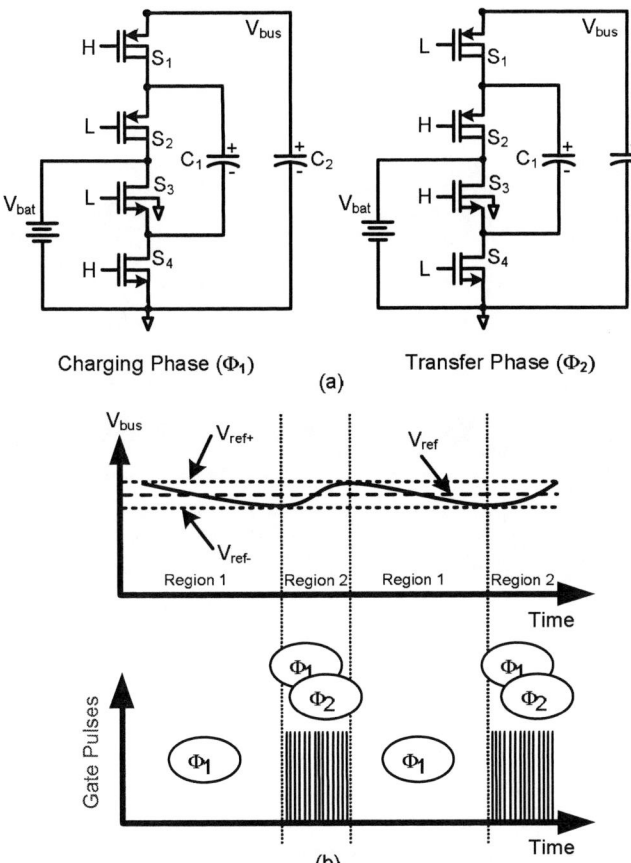

Fig. 2. (a) A basic switched capacitor DC-DC converter; (b) Switching waveforms for a voltage controller DC-DC converter.

can be divided into two phases: In the charging phase Φ_1, the switches S_2 and S_4 are turned on (S_1 and S_3 are off) and the flying capacitor C_1 is charged to V_{bat}. During this phase, the output capacitor C_2 continues to supply the load current and is being discharged. In the transfer phase Φ_2, S_1 and S_3 are turned on (S_2 and S_4 are off). C_1 which was charged to V_{bat} is now in series with the input source. These two voltage sources charge up C_2. Thus, by controlling the frequency and duration of transfer phase operation, the voltage on C_2 (bus voltage V_{bus}) can be controlled.

The charge and transfer phase pulses normally have a duty cycle of 50%. To transfer the energy from input to output, these phases are periodically repeated. For this a small hysteresis band is chosen around the reference voltage, as shown in Fig. 2(b). We denote the upper threshold as V_{ref+} and the lower threshold as V_{ref-}. When the output voltage falls below V_{ref-}, the pulses are repeated alternating Φ_1 and Φ_2, which charges the output capacitor. Once the output voltage exceeds V_{ref+}, the pulses are blocked and no energy transfer takes place. The flying capacitor is kept charged up for the next cycle. In this mode, the output capacitor also continues to provide the load current, which causes the output voltage to drop. The cycle is repeated, once the voltage falls below the

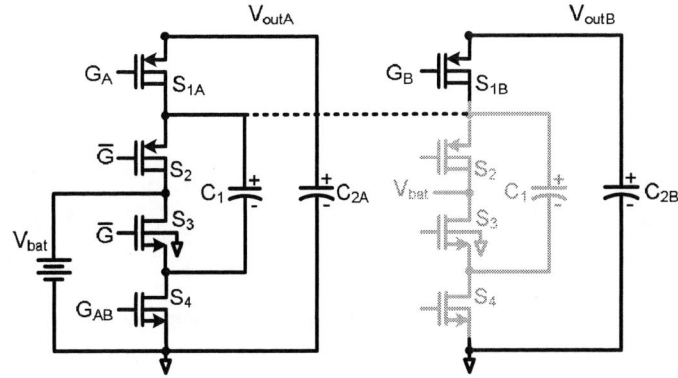

Fig. 3. A dual output switched capacitor DC-DC converter realized using same flying capacitor and the charging phase circuit.

V_{ref-}. Doing so, the output voltage is regulated around the required level V_{ref}, but it introduces a ripple due to the nature of the controller operation. In order to filter out the voltage ripple resulting from the hysteresis voltage control, an LDO is used as a post regulator.

As noted above, these converters normally provide a single output, which may be higher or lower than the input voltage. In order to obtain multiple outputs, the SC DC-DC can be replicated, but it will require 4N switches and 2N capacitors. The existing solutions based on multiple outputs are optimized for given voltage conversion ratio and thus require increased number of switches. We propose a solution which can achieve this with only a single additional switch for every additional output.

II. MULTIPLE OUTPUT SC DC-DC CONVERTER

In the above description of SC DC-DC, during the charging phase the output capacitor supplies the load current and is continuously getting discharged. As seen from Fig. 2, the actual energy transfer takes place only during the charge/transfer phases (region 2). For low power applications, the small load current will discharge the output capacitor slowly. As a result, the periodicity of phase pulses will be small. While the converter is idle (region 1), the charge on the flying capacitor can be used to power up another output capacitor, which will generate an additional output voltage.

Thus, we combine the two converters and share their charging phase circuit and the flying capacitors. An additional switch is added to control the transfer phase for the second output. This is depicted in Fig. 3. Here S_{1B} is added for obtaining a second output V_{outB}. It does not require an additional flying capacitor. It uses the charge on the same flying capacitor C_1 and charges the second output capacitor C_{2B}. The proposed SC DC-DC also works on the principle of the hysteresis voltage controller as described above. The transfer phase pulses depend upon the respective output voltages. While S_{1A} continues to transfer energy to C_{2A} for obtaining V_{outA}, S_{1B} transfers energy to C_{2B} for obtaining V_{outB}. By controlling the pulses to S_{1A} and S_{1B}, the output voltages V_{outA} and V_{outB} can be controlled.

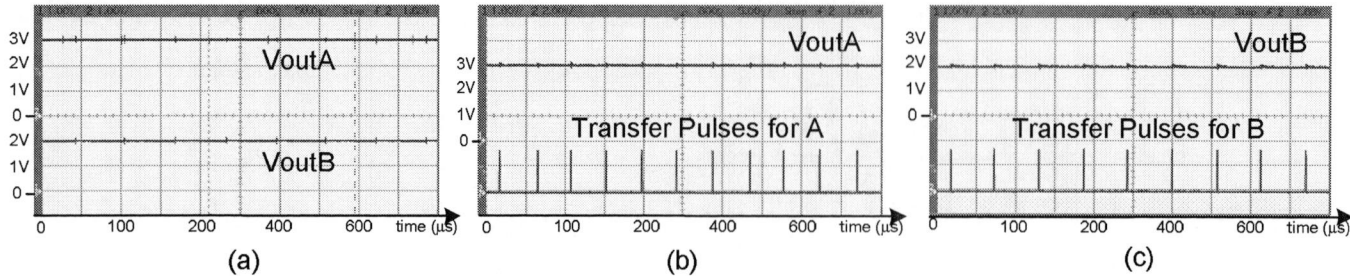

Fig. 4. Experimental results showing the transfer phase pulses for the two outputs in the proposed switched capacitor DC-DC converter.

Fig. 5. Chip photograph of a dual output DC-DC converter. The chip also features four low drop-out regulators.

Thus, by using an additional switch S_{1B}, two outputs can be obtained. This scheme is scalable in nature. By using additional switches, a single-input multiple-output (SIMO) SC DC-DC converter can be realized. However, due to increase in number of output voltages, the flying capacitor will discharge fast. As a result, there may be a need to increase the size of the flying capacitor if the load currents are increased.

III. EXPERIMENTAL RESULTS

The chip was fabricated in a 0.18μm CMOS process. Fig. 5 shows the die micrograph of the test chip which measures 1.58mm x 1.58mm. The SC DC-DC converter has an active area of 360μm x 460μm. Two LDOs were also integrated into the same die, occupying an active area of 450μm x 690μm. This test chip allows testing the LDO regulators and SC DC-DC converter independently. The clock frequency to the SC DC-DC converter was set at 5MHz. The output voltages were preset at $V_{outA} = 3$ V and $V_{outB} = 2$ V, which can be controlled by tuning the external feedback resistors. The gate pulses of the switches S_{1A} and S_{2A} were taken out for monitoring purpose. These signals can be monitored to observe the transfer phase pulses to the two output capacitors. Fig. 4 shows the two output voltages and also the transfer phase

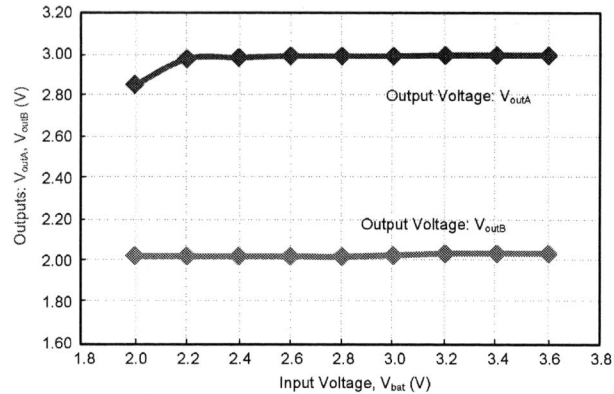

Fig. 6. Experimental results showing the variation of output voltages with input voltage ($f_s = 5$MHz).

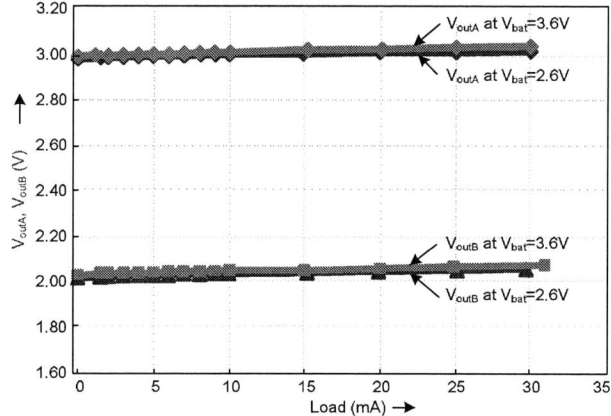

Fig. 7. Experimental results showing the output voltages with varying load.

pulses. The two output voltages are maintained at 2 V and 3 V respectively. These results were obtained at an input voltage of 3.3 V. However, in reality the input voltage may vary between 2.6 V to 3.6 V. Thus, the input voltage is varied to observe the response of the system. The two outputs of the SC DC-DC converter are maintained at 2 V and 3 V respectively, despite the variations in input voltage as shown in Fig. 6.

The output voltages are also monitored with the converter system subjected to varying load. Fig. 7 shows the response of the system when the load current is varied. The response is obtained at two extreme input voltages, that is, $V_{bat} = 2.6$ V

978-1-61284-863-1/11 $26.00 © 2011 IEEE

TABLE I
SUMMARY OF PERFORMANCE COMPARISON

Parameter	[6]	[9]	[8]	This work
Technology	0.6μ m	0.13μ m	0.35μ m	0.18μ m
Chip Area (mm^2)	3.92	4.00	0.65	2.49
Input Voltage	1.5V to 2.5V	1.2V	2.0V to 4.0V	2.4V to 3.6V
Output Voltage	3.0V to 5.0V	0.7V and 2.1V	16V	3V
Conversion Ratio	2	2, 0.66	$8-4$	$0.8-1.2$
Switching Freq.	500kHz	23MHz	50kHz	30MHz
Flying Capacitor	1μ F	1.2nF	0.47μ F	1μ F
Output Capacitor	2.2μ F	N.A.	0.47μ F	1μ F
Maximum Current	<50mA	2.3mA	11.2mA	30mA
Quiescent Current	1.63mA	N.A.	0.49mA	0.212mA
Maximum Efficiency	>90%	66% @1.4mW	N.A.	58.5% @6mW

and $V_{bat} = 3.6$ V. The load is varied by varying the off chip load resistance. It is seen that the output is regulated upto 30mA of load current.

The previous work on multiple output converters is based on inductive solutions [1]-[5]. Some of the work on switched capacitor DC-DC converter is mentioned here and compared with the proposed scheme. Table I shows the performance comparison with other switched capacitor DC-DC converters. The earlier schemes on multiple output switched capacitor converters use increased number of switches. The proposed scheme uses only one additional switch and one output capacitor to obtain another output voltage. Such a scheme provides another design freedom to obtain multiple output voltages while using fewer number of switches and capacitors.

IV. CONCLUSION

A simple and efficient method is presented for obtaining multiple outputs in a switched capacitor DC-DC converter. The SIMO SC DC-DC is proposed which takes the advantage of the characteristic nature of charge-transfer pulses in the simple SC DC-DC converter. The method can be implemented with very little hardware overhead and is suitable for use with low power applications which require multiple supply voltages. The proposed scheme is experimentally verified on a dual output SC DC-DC converter fabricated in 0.18μm CMOS technology. A voltage mode hysteresis controller is used to regulate the output voltages. It is seen that the output voltages are stable for load currents upto 30mA.

ACKNOWLEDGMENT

The authors would like to thank HealthSTATS Pte. Ltd., Singapore for the support.

REFERENCES

[1] D. Ma and et al, "A pseudo-ccm/dcm simo switching converter with freewheel switching," *IEEE Journal of Solid-Stage Circuits*, 2003.
[2] A. Sharma and et al, "A single inductor multiple output converter with adaptive delta current mode control," in *IEEE ISCAS*, 2006.
[3] H.-P. Le and et al, "A single-inductor switching dc-dc converter with 5 outputs and ordered power-distributive control," in *IEEE ISSCC*, 2007.
[4] M. Belloni and et al, "A 4-output single inductor dc-dc buck converter with self-boosted switch drivers and 1.2a total output current," in *IEEE ISSCC*, 2008.
[5] K.-S. Seol and et al, "Multiple-output step-up/down switching dc-dc converter with vestigial current control," in *IEEE ISSCC*, 2009.
[6] H. Lee and P. K. T. Mok, "Switching noise and shoot-through current reduction techniques for switched-capacitor voltage doubler," *IEEE Journal of Solid-State Circuits*, 2005.
[7] C. Jia and et al, "Integrated power management circuit for piezoelectronic generator in wireless monitoring system of orthopaedic implants," *IET Circuit, Devices and Systems*, 2008.
[8] F. Su and W.-H. Ki, "An integrated reconfigurable sc power converter with hybrid gate control scheme for mobile display driver applications," in *IEEE ASSCC*, 2008.
[9] M. Seeman and et al, "An ultra-low-power power management ic for wireless sensor nodes," in *IEEE Custom Integrated Circuits Conference(CICC)*, 2007.
[10] J. Kotowski, "Capacitor dc-dc converter with pfm and gain hopping," in *U.S. Patent 6055168*, 2000.
[11] R. P. Singh and et al, "A buck-boost hybrid dc-dc converter for wearable health monitoring devices," in *Proceedings of APSIPA*, 2010.

High Voltage Electrostatic Driving of MEMS Micromirrors

Ravinder Pal Singh, Tal Langer* and Minkyu Je

Integrated Circuits and Systems Laboratory

Institute of Microelectronics, A*STAR (Agency for Science, Technology and Research)

Singapore 117685.

e-mail: ravinderps@ime.a-star.edu.sg

Abstract—**High Voltage pulses are required to actuate the MEMS for obtaining the desired performance. Charge pump based circuits are used for obtaining the required high-voltage conversion. However, many such stages need to be connected in series for achieving the required voltage gain. Thus, a two-stage boost converter is used to achieve 3.3V to 80V conversion. The PWM controllers are designed in 0.18 μm CMOS process. The pair of nMOS are used to generate the required MEMS actuation pulses.**

I. Introduction

Switched capacitor based circuits or inductor based boost converters are commonly used to generate an output voltage which is higher than the input voltage. While the switched capacitor based circuits are used for low power applications, the latter may be used for medium to high power applications. For example, the switched capacitor based circuits are typically applied in flash memories for read/write operation. It may also be used in low voltage designs to improve the transient response [1]. Recently, such circuits have also been used for MEMS applications, which may require voltages of 40V-80V for actuation [2],[3],[4]. The MEMS actuators could be represented by variable capacitive loads (1 to 10pF), while the high-voltage interface drives the MEMS actuators. In all these applications, a rechargeable lithium-ion battery (3.3V) supplies the required power to the system for practical portable use. Thus, the voltage gain is required to achieve the desired actuation voltage.

The voltage gain provided by a switched capacitor circuit or a charge pump based circuit is a function of the input voltage and the number of stages used to realize the circuit. The ideal voltage for an n-stage charge pump is $n \cdot V_{in}$ where V_{in} is the input voltage. In order to increase the voltage gain, a Dickson charge pump is used. An n-stage ideal Dickson charge pump will have a gain of $(n+1) \cdot V_{in}$. However, in practice, the output voltage of both these circuits will decrease by 0.7V per diode.

*Tal Langer is with Maradin Ltd., Yoqneam, Israel 20692.

e-mail: tal.langer@maradin.co.il

SOI based body diode may be used to reduce the reduction in voltage gain [5]. For the MEMS actuation requiring high-voltage driving, say upto 80V, the required voltage gain is 24. A large number of charge pump stages will be required to achieve this voltage gain. Furthermore, the number of stages will increase when the diode drops from each stage is considered. Given the size of capacitance required to implement these stages, such design approach may not be a viable solution. In order to avoid the large number of stages, a combination of boost converter and a charge Dickson pump is used [6]. 3V to 36V conversion is achieved using a conventional boost converter operating at 500kHz and the Dickson charge pump provides a further voltage gain of 10. However, the drawback of this approach is the inability to regulate the output voltage. While the output of boost converter is regulated using closed-loop Pulse-Width Modulation (PWM) operation, the Dickson charge pump is essentially operating in open-loop. This may result in poor output voltage regulation, especially when the MEMS is actuated. To this end, an inductor based boost converter is designed to generate the required high voltage. The output voltage is tightly regulated by using the closed-loop operation.

II. DC-DC Boost Converter

An inductor based boost converter topology is selected for this purpose. If a single stage boost converter is used to achieve 3.3V to 80V voltage conversion, it will require a duty ratio of 95.8%. Given the rise-time and fall-time considerations, this will be difficult to achieve. Thus, a two stage boost converter is used to achieve the required voltage gain. While the first stage provides a voltage gain to an intermediate voltage of (15-20V), the second stage obtains the 80V for MEMS actuation. The block schematic of the boost converter is given in Fig. 1. In both these stages, controller comprises of an error amplifier and a Pulse Width Modulating (PWM) switching at 1MHz. The detailed specifications of the two stage boost converter are given in Table I.

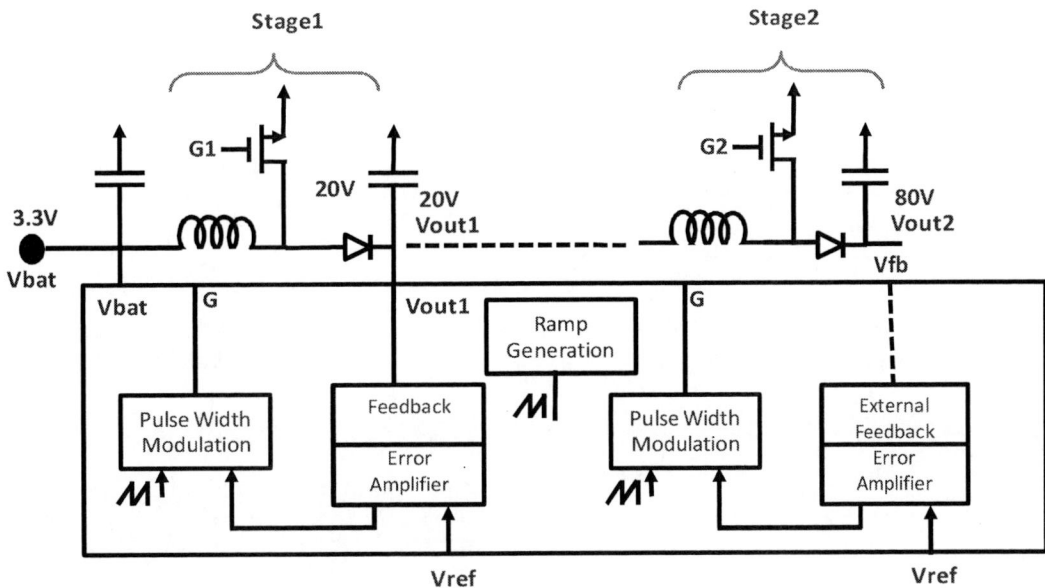

Fig. 1. 2-stage boost converter topology for High-Voltage generation.

TABLE I

SPECIFICATION OF THE HIGH VOLTAGE GENERATION SCHEME

First Stage		
Input Voltage	3.3	Volts
Switching Frequency	1	MHz
Inductance	100	μH
Capacitance	1	μF
Output Voltage	20	Volts
Second Stage		
Input Voltage	20	Volts
Switching Frequency	1	MHz
Inductance	20	μH
Capacitance	0.1	μF
Output Voltage	80	Volts
Load Current	<200	μA
MEMS Driving Interface		
Driving Frequency	10	kHz
Duty Cycle	25-50	%

In order to obtain the closed-loop regulation of the output voltages, a voltage mode controller is used. As the double pole frequency falls within the bandwidth of the system, appropriate lead-lag compensators are used to provide the phase boosting at the resonating frequency. The bode response of the open-loop system is shown in Fig. 2(a). In order to filter out the output voltage effectively, the closed-loop bandwidth is designed as one-tenth of the switching frequency. Thus, the lead-lag controller for this stage is obtained as

$$C_1(s) = 14 \cdot \frac{1 + 7 \times 10^{-6} s}{1 + 7 \times 10^{-8} s} \quad (1)$$

The bode plot of the compensated system is shown in Fig. 2(b). It shows the bandwidth of 130kHz and the phase margin of 60°.

Similarly, the controller for second stage is obtained as

$$C_2(s) = 5 \cdot \frac{1 + 8 \times 10^{-6} s}{1 + 1.7 \times 10^{-8} s} \quad (2)$$

The bode plot of the compensated system is shown in Fig. 3. It shows the bandwidth of 160kHz and the phase margin of 70°.

Both these controllers are realized using an error amplifier and passive components. The compensators and the PWM controller are integrated on-chip. In-order to reduce the dependency of High Voltage process, the external MOSFETs were used. The gate signals were brought out for monitoring as well as for driving purpose. Sufficient gate drivers were used in the chip so as to increase the driving strength of these signals. The two output voltages can be adjusted with the help of external feedback resistors.

The output of the second stage is used with the pulse generation circuit to obtain the actuation pulses. This is realized using a pair of nMOS and the gate drivers as shown in Fig. 4. In drivers provide the required deadtime between the gate pulses, which reduces the output current requirement to a great extent. If the MEMS require only one high voltage(HV) pulse, either D1A or D2A may be used. However, there are applications which require complementary driving pulses [3], [7] both D1A and D2A can be used.

III. EXPERIMENTAL RESULTS

The controller chip for obtaining the 3.3V to 80V conversion was designed in 0.18 μm CMOS process. The chip micrograph is shown in Fig. 5. It occupies an active area of 850μm $\times$ 1150μm. Extra care was taken in the layout while placing the high-voltage metal lines. These lines were shielded

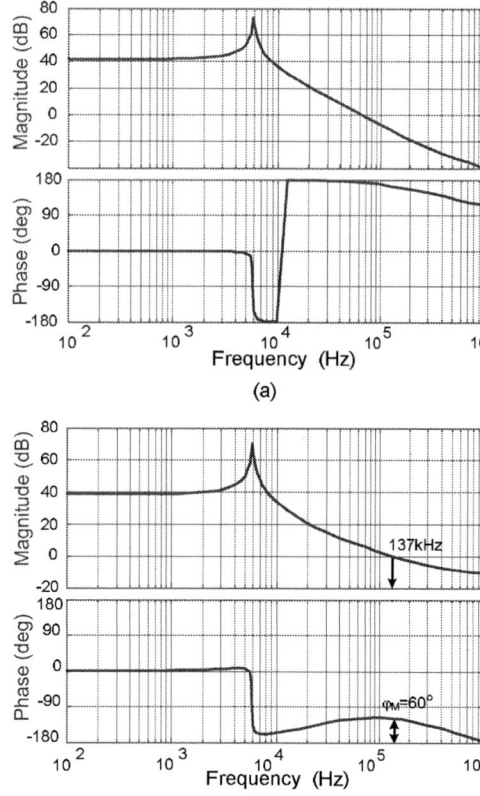

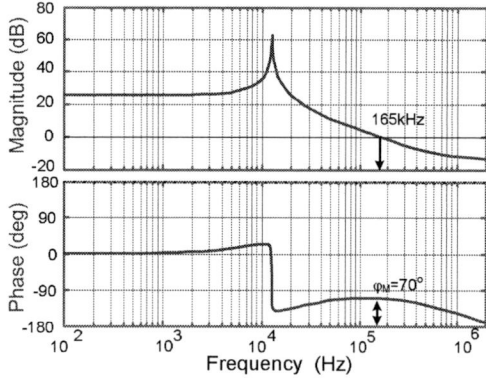

Fig. 2. (a) Bode plot of open loop stage 1; (b) Bode plot of the compensated stage 1 boost converter

Fig. 3. (a) Bode plot of open loop stage 2; (b) Bode plot of the compensated stage 2 boost converter

with ground lines so as to prevent the substrate from getting inverted.

The above controller ASIC is first tested for its Input-Output (I/O) characteristics. The ASIC is used with external inductors and capacitors to realize the two stage boost converter as shown in Fig. 1. The no load output voltages of the two stage boost converters are set as 20V and 80V respectively. The I/O characteristics are obtained by using a variable load at the output of the second stage. The load current and the two output voltages are measured. The typical I/O characteristics

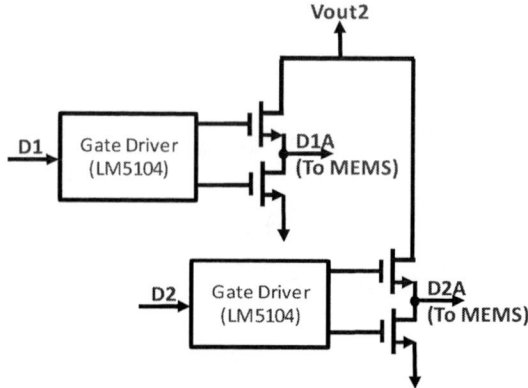

Fig. 4. Interface for generating the high-voltage pulses.

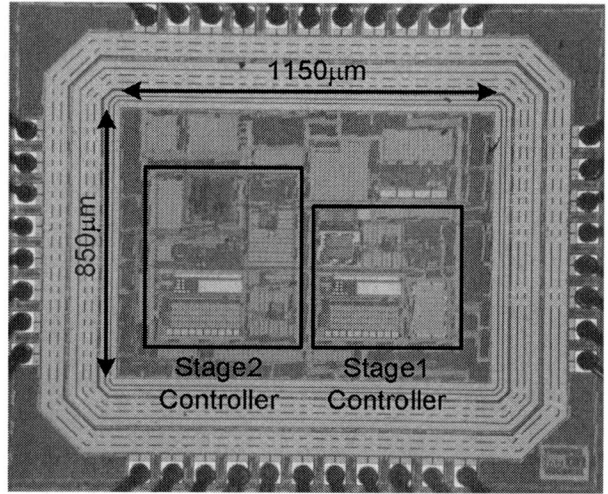

Fig. 5. Die micrograph of high-voltage controller ASIC.

are shown in Fig. 6. It shows the output voltage starts to drop when the load current exceeds 400μA. Since the actual MEMS actuation is capacitive drive, this limit on the load current is acceptable.

The above test was to verify the functionality of the two stage operation. The HV pulse generation was also used with the HV generation scheme. The control pulses were obtained from a signal generator and were used as an input to the driver circuit. The typical waveforms are shown in Fig. 7. The output voltages from the two boost stages is also shown in the figure.

In order to test the two sided driving of the MEMS, two control signals $D1$ and $D2$ are used. These signals have the same duty cycle and the frequency, but differ in phase by 180°. The two control signals and the corresponding pulses are shown in Fig. 8 and the corresponding output voltage is shown in Fig. 9.

The MEMS driving circuit was also tested with MEMS. The MEMS was driven with 60V 10kHz pulses and a laser was pointed at MEMS to observe the deflection on the screen. The resulting projected image is shown in Fig. 10. The projected image measures $\pm 8^{o}$ mechanical deflection.

978-1-61284-863-1/11 $26.00 © 2011 IEEE

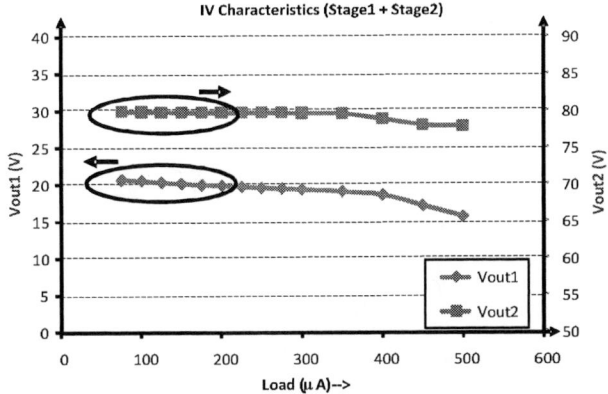

Fig. 6. Input-Output characteristics of two-stage boost conversion. No load output voltages are set as 20V and 80V.

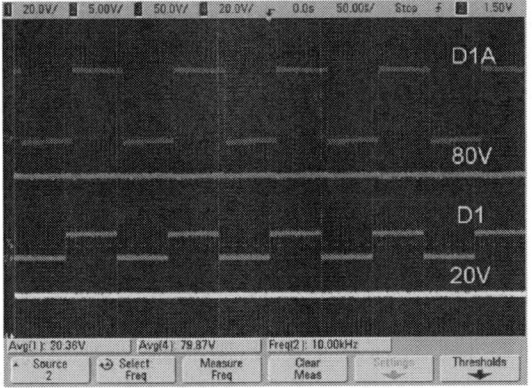

Fig. 7. Waveforms showing the two output voltages and the MEMS actuation signals.

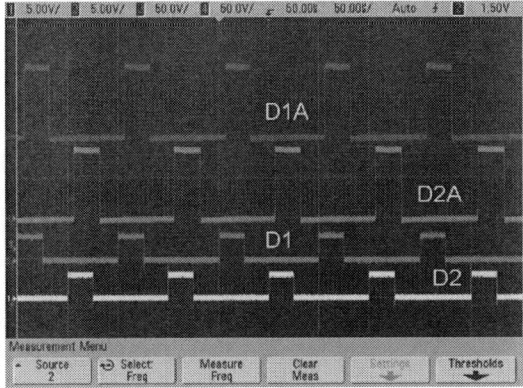

Fig. 8. Waveforms showing the two control signals and the corresponding MEMS actuation signals.

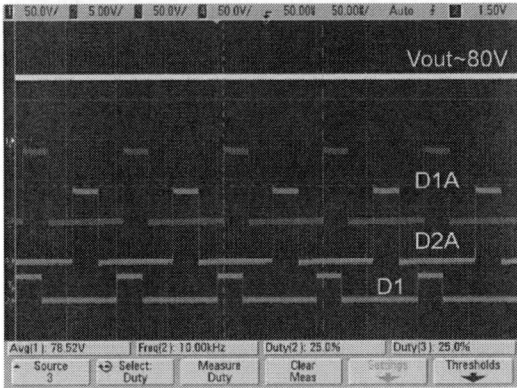

Fig. 9. Waveforms showing the two actuation pulses along with the second stage output.

IV. CONCLUSION

A two-stage boost converter is used to achieve the high-voltage generation required for MEMS actuation. The PWM controllers are designed in 0.18 μm CMOS process and are used to provide 3.3V to 80V conversion. The pair of nMOS are used to generate the required MEMS actuation pulses. The presence of deadtime in the HV interface reduces the output power requirements of such a scheme. The closed loop regulation of the HV generation and actuation scheme is verified with MEMS operating at 10kHz and providing over 16° mechanical deflection.

REFERENCES

[1] Ravinder Pal Singh and A.M. Khambadkone, "Input voltage switched dc-dc converter with improved transient performance," in *IEEE Applied Power Electronics Conference and Exposition*, 2008, pp. 824–830.

[2] M. H. Kiang, O. Solgaard, K. Y. Lau, and Richard S. Muller, "Electrostatic combdrive-actuated micromirrors for laser-beam scanning and positioning," *Journal of Microelectromechanical Systems*, vol. 7, no. 1, pp. 27–37, Mar 1998.

[3] Lubianiker Y. and et. al., "Gimbaled scanning micro–mirror apparatus," *WIPO*, , no. WO 2009147654A1, Dec 2009.

[4] C. Ataman and H. Urey, "Nonlinear frequency response of comb-driven microscanners," in *Proceedings of SPIE*, 2004.

[5] M. R. Hoque and T. McNutt et. al., "A high voltage dickson charge pump in SOI CMOS," in *Proceedings of IEEE Custom Integrated Circuits Conference (CICC)*, 2003.

[6] Jean-Francois Richard and Yvon Savaria, "High voltage charge pump using standard SOI CMOS technology," in *Proceedings of IEEE Northeast Workshop on Circuits adn Systems (NEWCAS)*, 2004.

[7] Duy Dong Pham and Ravinder Pal Singh et. al., "Position sensing and electrostatic actuation circuits for 2-D scanning MEMS micromirror," in *Proceedings of Defence, Science and Research (DSR) Conference*, 2011.

Fig. 10. Projected Image resulting from the MEMS driving.

978-1-61284-863-1/11 $26.00 © 2011 IEEE

Visualization of Intrinsic Harmonic Distortion for Closed-loop Class D Amplifier

Jun Yu and Wang Ling Goh
School of Electrical and Electronic Engineering
Nanyang Technological University
Singapore
yu0002un@e.ntu.edu.sg; ewlgoh@ntu.edu.sg

Meng Tong Tan
Institute of Microelectronics
Agency for Science, Technology and Research (A*STAR)
Singapore
tanmt@ime.a-star.edu.sg

Abstract—**This paper presents an intuitive approach for analyzing the intrinsic harmonic distortion of a closed-loop Class D amplifier. The cubic spline interpolation was deployed for time domain waveform presentation of the phase and duty cycle errors, to provide an insight view of the distortion that are contained in the output pulse width modulation signal. Furthermore, the relationship between the phase and duty cycle errors and the distortion on the demodulated output signal has been shaped and verified through simulation and comparison with our previous analytical results, to further strengthen our conclusion on their effects on the 3rd-order harmonic distortion.**

Keywords-Class D amplifier; intrinsic harmonic distortion; interpolation; phase error; duty cycle error; 2nd-order loop filter

I. INTRODUCTION

With the ever-increasing performance of modern power MOSFETs, the significance of the power stage errors is noted to have diminished [1]. As a result, the distortion generated by the Pulse Width Modulation (PWM) and control process itself has become visible and is widely referred to as the intrinsic harmonic distortion of the closed-loop Class D amplifiers (amps) [1-5]. Although the phenomenon has been well studied, based on the phase and duty cycle errors of the output PWM signal in [1, 4], none of them is able to visualize the pattern of the error sources when a varying input signal is presented.

According to our investigation, the lack of description on the error signals is probably due to two reasons: firstly, the size of the phase and duty cycle errors are small, which makes error detection on the fast changing pulse signal difficult; secondly and most importantly, the timing information of the phase and duty cycle errors are unknown, as they were initially defined as a scalar corresponding to each carrier period.

In this paper, we provide an insightful methodology to investigate the intrinsic harmonic distortion of the closed-loop Class D amplifier by a simple means of comparing the output PWM signals between the closed-loop and open-loop Class D amplifiers. The advantage of this method is that it permits direct visualization of the error signals, allowing the circuit designer to envisage the source of the intrinsic harmonic distortion. Furthermore, the relationship between the error sources and the distortions that are contained in the demodulated output signal is also investigated.

This paper is organized as follows. In Section II, the methodology for plotting the phase and duty cycle errors' time domain waveforms is demonstrated. The analysis on the phase and duty cycle errors, corresponding to a sinusoidal input signal, is illustrated in Section III. The relationship between the error signals and the distortions within the demodulated output signal is then investigated in Section IV. Finally, we draw our conclusions in Section V.

II. THE PHASE AND DUTY CYCLE ERROR WAVEFORMS

The intrinsic harmonic distortion is caused by the residual high frequency components (i.e. the ripple signal) inside the loop filter output signal of a closed-loop Class D amp. As a result, both the rising-edge and the falling-edge switching times of the output PWM signal varied, causing unwanted odd-order harmonics in the demodulated output signal. Since an ideal open-loop PWM-based Class D amp is distortion-free, we had deployed it as a reference to examine the errors caused by the ripple signal.

A. Define the phase and duty cycle errors

The analysis on the intrinsic harmonic distortion begins with a comparison of the output PWM waveforms between the closed-loop and open-loop Class D amps with synchronized input signal and triangular carrier signal. Fig. 1 shows the block diagram that was engaged in our MATLAB simulation work. Due to the negative feedback topology, the output signal of the closed-loop amp is inverted as compared to the input signal. For the output signal of the open-loop amp to be in-phase with the closed-loop amp, the input signal to the open-loop amp is inverted via an unity-gain inverter. The typical PWM waveforms with the inverted input signal and loop filter output signal for the respective open-loop and closed-loop amps are illustrated in Fig. 2. As can be seen, due to the residual high frequency ripples at the output of the loop filter (see solid line in Fig. 2(a)), the switching times of the PWM signal for the closed-loop amp arrives earlier than that of the open-loop amp. The time advancements are unequal at the rising and falling edges of the PWM signals, i.e. t_1 is ahead of t_2, and t_3 is ahead of t_4. Note that the input signal frequency is set to 20 kHz for easy viewing of the mismatch.

The timing errors between the closed-loop and open-loop PWM signals at the rising and falling edges are defined as

978-1-61284-863-1/11 $26.00 © 2011 IEEE

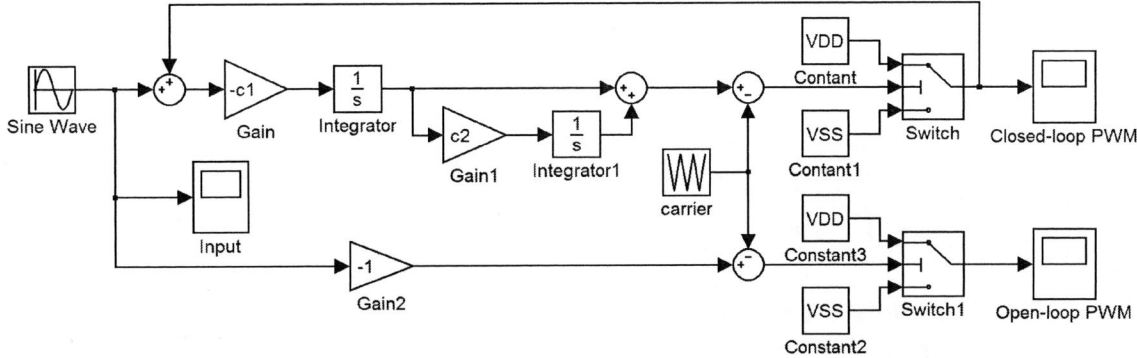

Figure 1. Simulink Model for comparying the PWM output signals between the closed-loop and open-loop designs.

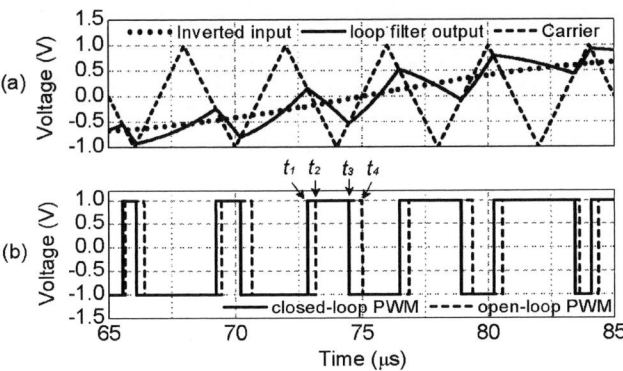

Figure 2. Typical signal waveforms: (a) inverted input signal for the open-loop amp, loop filter ouptut signal of the closed-loop amp and carrier signal; (b) PWM signals for open-loop and closed-loop amps.

$$T_{re} = t_2 - t_1 \quad (1) \quad \text{and} \quad T_{fe} = t_4 - t_3 \quad (2)$$

Alternatively, the mismatch between the two PWM signals can be described through phase and duty cycle errors, as presented in [4]. In this work, the phase error is defined based on the average of the rising-edge and falling-edge timing errors whereas the duty cycle error is defined as the ON duration of the closed-loop output PWM signal minus the ON duration of the open-loop output PWM signal, which is in fact proportional to the difference between the rising-edge and falling-edge timing errors, and can be expressed as follows:

$$\theta_e = \pi \cdot \left(T_{re} + T_{fe}\right)/T \quad (3) \quad \text{and} \quad D_e = \left(T_{re} - T_{fe}\right)/T \quad (4)$$

B. Plot the phase and duty cycle error waveforms

In order to analyze the effect of the phase and duty cycle errors on the output signal, we need to plot the time-domain waveforms of these two errors corresponding to a sinusoidal input signal. However, this is not a straightforward task, as the phase and duty cycle errors are scaled values that are defined in each carrier period, but without a proper time reference. Based on our analyses, the location of the phase and duty cycle errors in time domain is extremely important. For example, by simply placing them at the centre of each period will lead to artificial frequency components in the waveforms created.

In order to circumvent this obstacle, we first locate the rising-edge and falling-edge timing errors at the respective switching time of the ideal open-loop PWM signal (i.e. t_2 and t_4). This will yield two timing error signals, as depicted in Figs. 3(a) and 3(b), where both are discrete in time, with unequal time intervals, and unsynchronized with each other.

In order to calculate the phase and duty cycle errors, we need to align the rising-edge and falling-edge timing errors in the time axis. This is achieved by converting the non-uniform sampled data to uniform sampled data through interpolation. With the assumption that the two timing error signals contained only low frequency components as compared to the carrier signal, cubic spline interpolation is applied to formulate a piecewise function $S(x)$ that contains a group of third degree polynomial defined by

$$s_i\left(x\right) = a_i\left(x - x_i\right)^3 + b_i\left(x - x_i\right)^2 + c_i\left(x - x_i\right) + d_i \quad (5)$$

where a_i, b_i, c_i and d_i are the coefficients of the fitting curve for the interval starting from data input (x_i, y_i) and hence $y_i = s_i(x_i)$. The coefficients are calculated such that the first and second derivative of the piecewise function, i.e. $S'(x)$ and $S''(x)$, are continuous across the entire data interval, to allow the resulting interpolated curve to smoothly pass through all data points. Details on the calculation procedures can be found in [6].

The interpolated uniform-sampled timing error signals are depicted in Figs. 3(c) and 3(d). Note that in the time domain, the interpolated samples for the rising and falling edge timing errors do synchronize with each other. For the purpose of illustration, only two samples were considered at each carrier period. And in the actual application, more samples will be inserted to guarantee a smooth curve.

With the uniform-sampled timing error data, we are able to plot the time domain waveforms of the phase and duty cycle errors by extending the definitions of the phase and duty cycle errors to as follows:

$$\theta_e\left(n \cdot \tau\right) = \pi \cdot \left[T_{re}\left(n \cdot \tau\right) + T_{fe}\left(n \cdot \tau\right)\right]/T \quad (6)$$

$$D_e\left(n \cdot \tau\right) = \left[T_{re}\left(n \cdot \tau\right) - T_{fe}\left(n \cdot \tau\right)\right]/T . \quad (7)$$

where τ is the time interval of the interpolated error data. The computed phase and duty cycle errors are plotted in Figs. 3(e) and 3(f). The complete process was programmed in MATLAB.

978-1-61284-863-1/11 $26.00 © 2011 IEEE

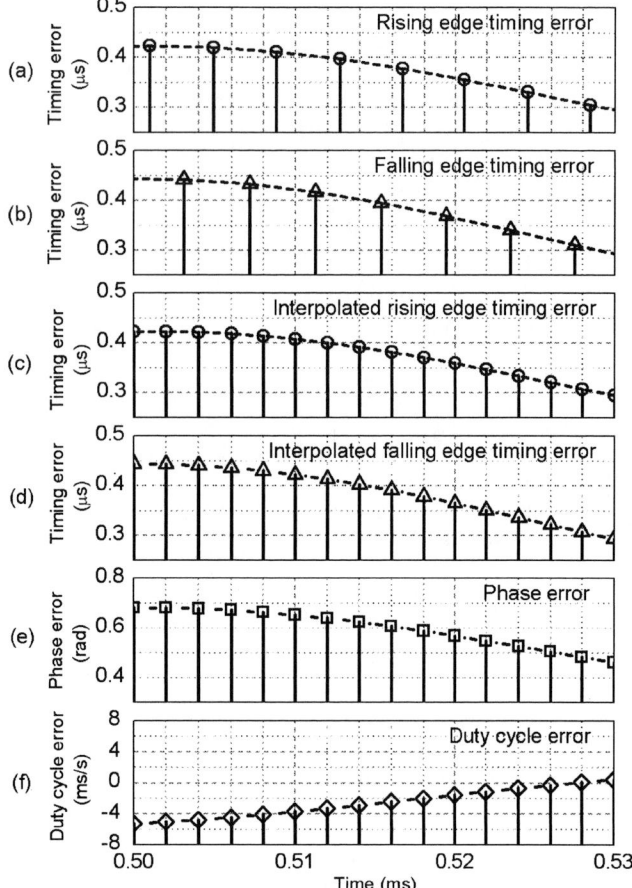

Figure 3. Procedures for generating phase and duty cycle error waveforms: (a) non-uniform sampled rising-edge timing error; (b) non-uniform sampled falling-edge timing error; (c) interpolated rising-edge timing error; (d) interpolated falling-edge timing error; (e) phase error and (f) duty cycle error.

III. ANALYSIS ON PHASE AND DUTY CYCLE ERRORS

In this section, we will demonstrate the phase and duty cycle errors of a 2nd-order Class D amplifier through waveform simulations. The amp's loop filter has a 2nd-order integrator transfer function described in [7]. Fig. 4 shows the resulting phase and duty cycle error waveforms based on sinusoidal input signals that were simulated at different frequencies, i.e. 1 kHz, 2 kHz and 5 kHz. The modulation index of the input signal was set to 0.7. The horizontal axis is the period of the input signal which allows overlapping of error signals with different input signal frequencies. The characteristics of the phase and duty cycle errors can be summarized as follows:

- As shown in Fig. 4(a), the phase error is the lowest when the input signal is at its peak or valley. On the other hand, when the input signal crosses the zero reference, the phase error is at its maximum. As a result, the phase error contains frequency components that are twice the frequency of the input signal.

- In addition, the magnitude of the phase error remained unchanged with different input signal frequencies. This although is a very interesting phenomenon, it does not

imply that the resulting distortion of the output signal is independent of the frequency of the input signal. The relationship between the phase error and the resulting output distortion is given in (8) in Section IV.

- The duty cycle error seen in Fig. 4(b) reaches its peak when the input signal in Fig. 4(a) crosses the zero reference. This indicates that the mismatch between the rising-edge and falling-edge timing errors is at its maximum when the input signal crosses the zero reference. The magnitude of the duty cycle error increases with larger input signal frequency. This is consistent with the fact that the intrinsic harmonic distortion becomes more significant for high frequency input signal.

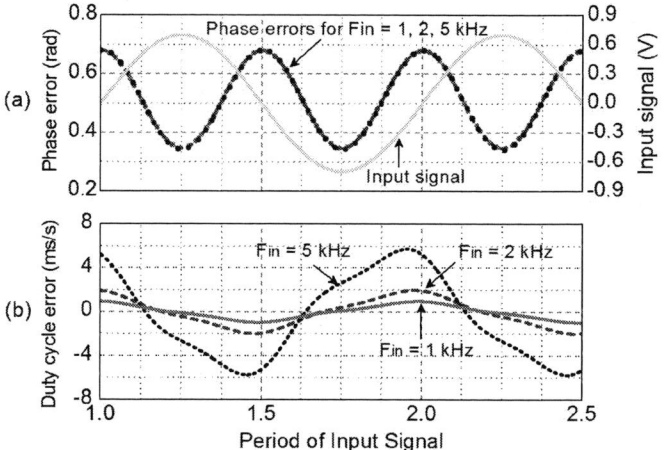

Figure 4. Phase and duty cycle errors with respect to different input frequencies: (a) phase error with input signal; and (b) duty cycle error.

The key frequency components of the phase and duty cycle errors corresponding to a 5 kHz input signal (as shown in Fig. 4) are listed in Table I. The table clearly reflects that the phase error comprises even order harmonics whereas the duty cycle error involves the fundamental component and odd order harmonics.

TABLE I. FREQUENCY COMPONENTS IN PHASE AND DUTY CYCLE ERRORS.

Error source	Frequency components	Magnitude	Phase (Degree)
Phase error (rad)	2nd-harmonic	0.16956	-1.83537
	4th-harmonic	0.00106	-77.38772
Duty cycle error (s/s)	Fundamental	0.005315	-147.7
	3rd-harmonic	7.7466E-4	-166.17815

The variations of the phase and duty cycle errors at different input magnitudes are shown in Fig. 5. The input signal frequency was set to 5 kHz and the modulation index was varied from 0.1 to 0.9, and with a 0.2 step. From the comparison, we note that when the input magnitude decreased, the lower peak of the phase error rose toward to the higher peak and hence the magnitude of the 2nd-order harmonic inside the phase error diminished. At the same time, the duty cycle error became smaller and approached a sinusoidal wave. This implied that both the fundamental and 3rd-order harmonic frequency components would quickly vanish. These phenomena are consistent with the fact that the intrinsic harmonic distortion should become insignificant when the magnitude of the input signal decreases.

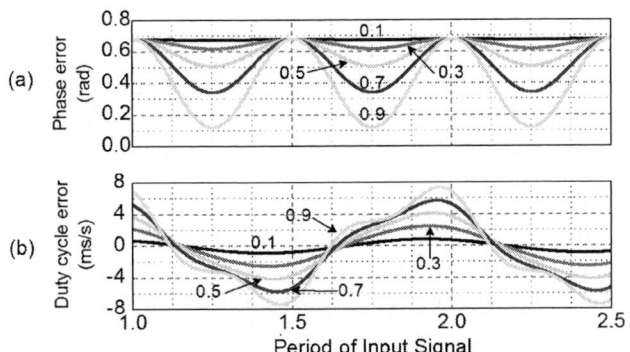

Figure 5.   Phase and duty cycle errors with respect to different input signal magnitudes: (a) Phase error; (b) Duty cycle error.

IV.   RELATIONSHIP BETWEEN PHASE AND DUTY CYCLE ERRORS, AND THE DISTORTION OF THE OUTPUT SIGNAL

Having established the phase and duty cycle errors, we are then able to calculate the major distortion components contained in the demodulated output signal. The phase error behaves like a phase modulation on the ideal output PWM signal. Taking reference from [1], the distortion created by this mechanism can be approximated by the slope of the ideal output signal times the time displacement, which is a linear approximation to the phase modulation. The duty cycle error directly modulates the magnitude of the output signal and can therefore be translated to output distortion by multiplying by 2. Consequently, the generated distortion within the audible output signal can be calculated as follows:

$$E(t) = \theta_e(t) \cdot \frac{T}{2\pi} \cdot \frac{d}{dt} V_{out\_ideal}(t) + 2D_e(t) \qquad (8)$$

From Table I, the most significant frequency component of the phase error is the 2nd-order harmonic of the input signal. Hence, the resulting error signal due to the phase error (i.e. 1st-term at the right hand side of Eqn. (8)) is a 3rd-order harmonic. In addition, since the duty cycle error comprises mainly the fundamental and 3rd-order harmonic of the input signal, the resulting error signal (i.e. 2nd-term at the right hand side of Eqn. (8)) introduces both the 3rd-order harmonic and the distortion on the fundamental frequency component. For instance, based on the data tabulated in Table I, the 3rd-order harmonic distortion of the demodulated output signal can be calculated as follows:

$$H_{3rd} = \left| \begin{array}{l} 0.1696\cos\left(2\omega_{in}t - \frac{1.835}{180}\pi\right) \cdot \frac{T}{2\pi} \cdot \frac{d}{dt}\sin(\omega_{in}t) \\ +2\times0.0007747\cos\left(3\omega_{in}t - \frac{166.2}{180}\pi\right) \end{array} \right|_{@3\omega_{in}} \qquad (9)$$

$$\approx 0.000517$$

In addition, the magnitude of the distorted fundamental component of the output signal is calculated to be equal to 0.70576. A comparison of the MATLAB simulation results, analytical results based on [5], and this work is compiled in Table II. It is evident that the results derived from this work matches well with simulation and our previous analytical work [5] that was based on time-domain modeling technology. It is worthwhile to highlight that the 3rd-order harmonic distortion within the demodulated output signal is contributed by the phase and duty cycle errors, where both are equally important. This finding provides more insight than the previous works [1, 4] which reported that the 3rd-order harmonic distortion of a 2nd-order Class D amp is mainly due to the duty cycle error.

TABLE II.   COMPARISON ON OUTPUT FREQUENCY COMPONENTS

	MATLAB Simulation	Time-domain analysis [5]	This work
Fundamental	0.7055	0.7058	0.70576
3rd-harmonic	0.0005266	0.0005078	0.000517

V.   CONCLUTION

The time domain waveforms of the phase and duty cycle errors were produced by comparing the closed-loop and open-loop PWM signals and with the help of the cubic spline interpolation. The waveforms created provide an insight to the mechanism of the intrinsic harmonic distortion of a closed-loop Class D amplifier. The relationship between the error signals and the distortion inside the output signal is investigated. The results show that both the phase and duty cycle errors play collective roles in the 3rd-order harmonic, which is the most significant distortion component within the demodulated output signal of a 2nd-order Class D amplifier.

ACKNOWLEDGMENT

The authors wish to thank Professor Yajun Yu for her kind discussion on the interpolation techniques. Special appreciation is also extended to Mr Liao Lei for his advice on the use of MATLAB to realize the interpolation function.

REFERENCES

[1]   L. Risbo and C. Neesgaard, "PWM Amplifier Control Loops with Minimum Aliasing Distortion," in *AES 120th Convention*, Paris, France, 2006.

[2]   B. H. Candy and S. M. Cox, "Improved analogue class-D amplifier with carrier symmetry modulation," in *AES 117th Convention*, San Francisco, CA, USA, 2004.

[3]   S. M. Cox and B. H. Candy, "Class-D audio amplifiers with negative feedback," *SIAM J. Appl. Math*, vol. 66, pp. 468-488, 2006.

[4]   W. Shu and J. S. Chang, "THD of Closed-Loop Analog PWM Class-D Amplifiers," *IEEE Trans. Circuits and Systems I: Reg. Papers*, vol. 55, pp. 1769-1777, 2008.

[5]   S. M. Cox, M. T. Tan, and J. Yu, "A second-order Class-D audio amplifier," *SIAM J. Appl. Math*, vol. 71, pp. 270-287, 2011.

[6]   C. D. Boor, A Practical Guide to Splines: Springer-Verlag, 1978.

[7]   M. Berkhout, "An integrated 200-W class-D audio amplifier," *IEEE J. Solid-State Circuits*, vol. 38, pp. 1198-1206, 2003.

A Voltage Management Technique for Low-Power Domino Circuits

Ching-Hwa Cheng {chengch@fcu.edu.tw}

Department of ECE, Feng-Chia University, TaiChung, Taiwan, R.O.C.

ABSTRACT

A low power voltage management technique is proposed to reduce power consumption Exploiting a clock control and charge sharing mechanisms, rising voltage and scalable voltages allow domino circuits have low power consumption with performance management ability. A test chip uses TSMC 0.13um CMOS technology has been successfully validated to achieve 68% dynamic power consumption and 15% static power consumption, respectively.

I. Introduction

The power consumption of conventional CMOS circuits is composed of dynamic and static parts. Most power consumption comes from dynamic power consumption. The generic dynamic power model is given by the equation $P \doteq CV^2 f$, which represents that power consumption is proportional to circuit loading (C), supplied voltage (V) and circuit clock frequency (f). Dynamic power consumption is large because the Vdd charge supplies the circuit's parasitic capacitance during CMOS circuit switching and clock frequency boosting. In most low power techniques, power consumption can be limited by scaling down V and f to the circuit. The V and f can be adjusted as two independent variables.

As domino logic design offers smaller area and higher speed than complementary CMOS design, it has been very commonly used for high-performance processors. CMOS domino gates have two basic operating phases: pre-charge phase when the clock **Phi** is low and evaluation phase when the clock **Phi** is high, as shown in Figure 1. During the pre-charge phase, the capacitance C_o will be charged to high. During the evaluation phase, if all of the inputs In_i to N-transistors are high, the voltage V_x is pulled to ground.

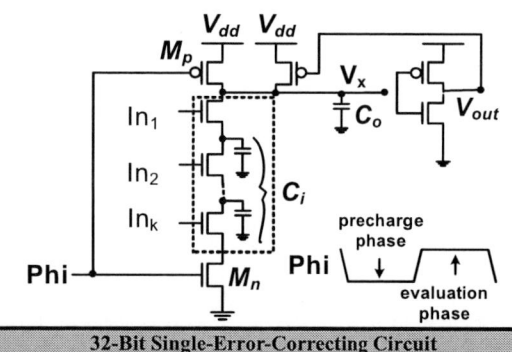

32-Bit Single-Error-Correcting Circuit			
	Delay Time (ns)	**Power** (uW)	**Peak Current** (mA)
Static	3.07	1424	6.9
Domino	1.77(-42%)	3634(+155%)	55.8 (+708%)

Figure 1. The domino gate and design comparisons.

From small-sized test circuit simulation results as shown in Figure 1, the domino circuit has a 42% delay time reduction in comparison with the static circuit. However, the average power consumption of the domino circuit is 155% larger than that of the static circuit. Moreover, the peak current of dynamic circuit is 708% larger than that of static circuit. This power dissipation problem needs to be solved for the domino circuit.

In this paper, a Low Power with Power Management (LPM) mechanism is proposed to design power-performance manageable high speed dynamic circuits. The objective of the proposed low power manageable design is to maintain high performance while decreasing the power consumption of the domino circuits.

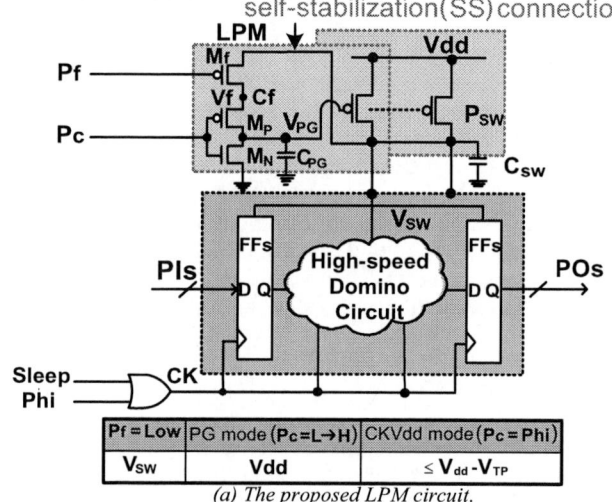

(a) The proposed LPM circuit.

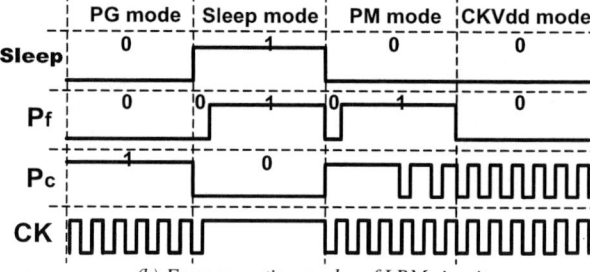

(b) Four operation modes of LPM circuit.

Figure 2. The proposed low power management (LPM) design.

The proposed LPM uses extended power gate techniques to reduce dynamic power during normal operation mode and leakage power during sleep mode. The proposed LPM domino circuit is shown in Figure 2(a). The LPM module controls PMOS transistors (P_{SW}) that connect the power supply to the circuit blocks, to degrade the voltage level during circuit operation. Numerous PMOS transistors (P_{SW}) ensure sufficient current supply to the circuit. The P_{SW} power switches

978-1-61284-863-1/11 $26.00 © 2011 IEEE

are used to shut off the leakage current when the circuits enter sleep mode. The P_{SW} either work in saturation, linear or cutoff region. In linear region, the Vdd voltage level (V_{SW}) to the circuit is scalable. The proposed LPM combines the advantages of rising Vdd and managing Vdd techniques. There are four modes (i.e. Clock Vdd-CKVdd, power management of P_{SW}, and sleep modes) in the proposed LPM of the designed circuits. The PG and sleep mode functionalities are the same as those of the conventional power gate design. The PG mode is shown in the diagram of Figure 2(b). CKVdd mode has greater dynamic power reduction operation mode. This simple LPM can be broadly used as a generic low power design technique.

In this paper, the LPM is used to design a power-performance manageable high speed domino multiplier. Test chip validations show that the LPM expands the domino circuit for high-performance and low-power (slow circuit speed) applications.

In Figure 3, when a rising voltage with rise time Δt is applied to an RC network, the dissipated energy of resistor R can be reduced. As Δt increases to become much larger than the RC time constant, the energy dissipated in the resistor can be approximated as

$$\text{Energy} = \frac{2RC}{\Delta t}(\frac{1}{\Delta t}\frac{1}{2}CV_{SW}^2)$$

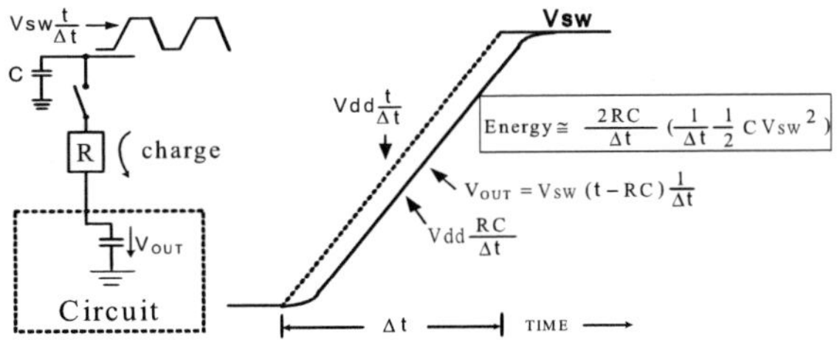

Figure 3. The circuit uses rising voltage to save energy.

II. LPM Low Power Technique

In the normal operation mode, denoted as PG mode, the P_C is set to 1 and Pf is set to 0 or 1. The capacitance Cf and C_{PG} can be used as de-capacitance to filter out the noisy voltage. There is only a small voltage drop for the designed circuits, so that circuit operation is more stable than traditional power gate circuits. This mechanism can control the V_{SW}, to avoid the voltage drop from suddenly turning on the P_{SW}.

The LPM can prevent leakage power when the circuit enters sleep mode. There are two methods to control the designed circuits in the snooze state. The first is to set Pf=0 and P_C=0, V_{PG} nearly equivalent to V_{SW}, so that the P_{SW} are turned off. The second is that when the circuit operation is in PG mode (i.e. Pc=1 and Pf=0), the Cf is charged by Vsw. When control the Pf signal transfers to logic high and P_C signal goes from high to low, the charge sharing from Cf to C_{PG} allows the V_{PG} voltage level to increase, and turns off the P_{SW} transistor.

By regulating the P_{SW} gate input voltage the Vdd supplied current to the designed circuit can be managed. The circuit delay time can be adjusted by controlling the Vdd current supplied to the designed circuit. Moreover, LPM can be used to manage the tradeoff of circuit delay vs. power consumption. The V_{SW} levels under different operation modes are shown in the table of Figure 2(a). The CKVdd and power management modes are described in detail in the following sections.

III. Low-Power CKVdd Mode

The LPM low power mode is CKVdd mode for low power consumption. In CKVdd mode, progressively rising voltage is provided to the circuit during circuit switching to achieve low power consumption. Thus, the designed circuit saves large dynamic power consumption.

When the circuits operate in CKVdd mode, the signal Pf is set to logic low (i.e. Pf=0) and P_C signal is the same as Phi signal. Then, the P_{SW} follow the CK signal cyclically turning on and off. The P_{SW} are fully turned on only for half of a clock cycle, which means that outer power is fully supplied to the circuits for half of every clock cycle. P_{SW} are turned on when the clock (CK) signal is high, and act as active resistors with nonlinear resistance when the CK signal is low. Then, the circuits in CKVdd mode are alternatively charged and discharged. When the clock signals switch the P_{SW} to work in the saturation region or active resistor region, the progressive rising Vdd of the V_{SW} is cyclically generated in synchronous domino circuits.

When both the Pf signal and P_C signal are low, the current of the P_{SW} is determined by gate to source voltage difference (Vgs) of PMOS transistors. The V_{SW} is equal to or less than Vdd-V_{TP}. The IVdd and ICKVdd represent current consumption of stable Vdd and V_{SW}, respectively. The ICKVdd is less than IVdd resulting in less power consumption in the CKVdd mode.

V. Management (PM) Mode of LPM

In LPM designed circuit, the power management (PM) mode is used to manage the P_{SW} working state. The charge sharing stable voltage controls the P_{SW}, which go from working in linear region to working in saturation region. The PM mode provides power-performance trade-off functionality. In addition to the saturation and cutoff regions, the PM mode allows the P_{SW} to work in linear region. One can use the linear region to manage the power consumption with circuit delay time. This mechanism can control the P_{SW}, so that they do not suddenly turn on to prevent voltage drop problem.

In this mode, the performance and power consumption of the designed circuits can be traded off. When the P_C signal is high, M_N is turned on, V_{PG} is discharged to zero voltage level, and P_{SW} are turned on. First, the Mf needs to be turned on for a period of time. The Vf is precharged by Vsw. After that, Mf is turned off, and the pulse signals are applied to P_C. When signal P_C is low, M_N is off, and Mp is on. Cf and C_{PG} will undergo charge sharing, and the voltage level of V_{PG} is equal to $\frac{C_f}{C_f + C_{PG}} \times V_{SW}$. Figure 4 shows the

978-1-61284-863-1/11 $26.00 © 2011 IEEE

test circuit simulation waveforms.

In PM mode, the P_{SW} working region will depend on the V_{PG} voltage level. By controlling the number of P_C pulse signals, the V_{PG} voltage level can be adjusted to lower voltage level, which allows the P_{SW} to be turned on completely. By controlling the number of P_C pulses, the P_{SW} working region can be controlled from cutoff to linear and situation (turned on) regions. Then, the supplied current from Vdd to the designed circuits can be adjusted so that the power consumption and performance of the designed circuits become manageable. The associated circuit simulations of Psw turned on (V_{PG} voltage decreased to 0) in PM mode are shown in Figure 4.

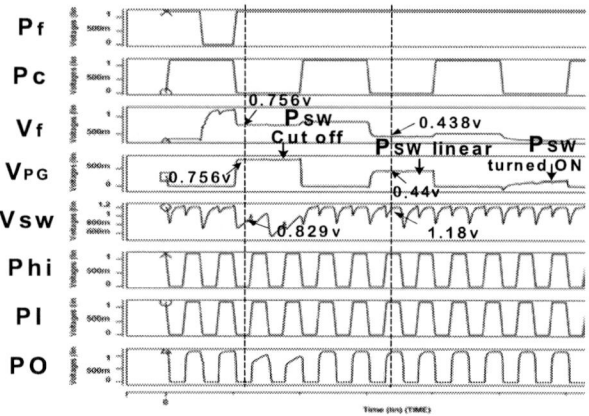

Figure 4. The circuit simulation in PM mode.

VI. Low-Power Domino Circuit Implementation
(1) CAD Design Flow

Most domino circuit research reports have neglected to discuss physical implementation of their designs. In this paper, we attempt to construct a stable mixed static-dynamic high-speed and low-power circuit synthesized environment. The CAD flow aims to partition the circuit into static and dynamic circuit parts, which can decrease the circuit delay time and power consumption, as well as performance issues within this synthesis flow. Some physical test chips have been implemented to validate our design CAD flow. The automatic cell-based design flow takes low power, high performance and skew tolerant issues into consideration. Moreover, a mixed type dynamic cell library was developed, which combines with static circuit cell library to provide a mixed static-dynamic circuit. The mixed circuit partition, synthesis and optimizing techniques are all adopted in this circuit synthesizer.

The automatic low-power with high performance mixed static-dynamic circuit synthesis flow includes four process steps. The first is front-end process for the mixed static-dynamic circuit gate level transform. The second is validation process, which is used to confirm the circuit functionality after circuit transformation of the previous step. The third is chip tapeout process and the fourth is test pattern generation for use in the test chip measurement.

The main achievement of this automatic circuit design flow is that static-dynamic high-performance and low-power circuit designs are quickly obtained. The proposed flow solves the skew-tolerant issue by using current CAD tools including complete tape-out validation flow. The noise-alleviation (charge sharing, crosstalk) domino cell libraries are generated to support the cell-based synthesis CAD design tools.

(2) Test Chip Design

Two 16*16 bit multiplier cores were implemented in a single test chip using TSMC 0.13um technology. There are eight LPM modules for this design.

For comparisons, the two 16*16 bit fast multipliers (MULs) were designed in the same die for verifying the proposed low power technique. MUL *Ori* is a regular domino circuit using outer Vdd with a constant power supply. MUL *New* includes eight LPM modules.

From postlayout simulation, compared to generic domino circuit, the delay increases are 11% and 29% for PG mode and CKVdd mode, respectively. In addition, power reductions are 44% and 53% for PG mode and CKVdd mode, respectively, and the peak current value reductions are 93% and 95%, respectively. CKVdd mode has significantly smaller power consumption with peak current, and saves large amounts of power, especially when working in high speed clock. The circuit delay time also increases but not significantly. The benefits of power saving gain outweigh the disadvantage of performance loss.

Figure 5 shows the die photo of the test chip. As the LPM modules are attached to the power rails, MUL *New* occupies a slightly larger area than MUL *Ori*.

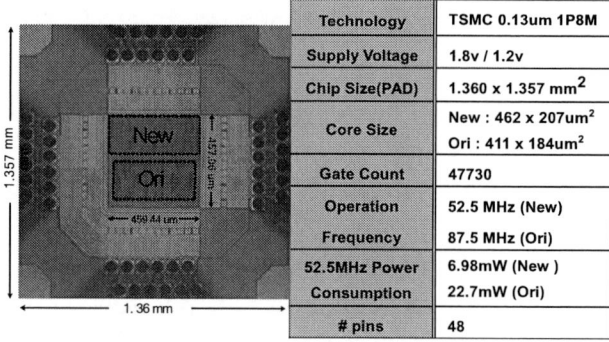

Technology	TSMC 0.13um 1P8M
Supply Voltage	1.8v / 1.2v
Chip Size(PAD)	1.360 x 1.357 mm²
Core Size	New : 462 x 207um² Ori : 411 x 184um²
Gate Count	47730
Operation Frequency	52.5 MHz (New) 87.5 MHz (Ori)
52.5MHz Power Consumption	6.98mW (New) 22.7mW (Ori)
# pins	48

Figure 5. The die photo and test chip specifications.

VII. Conclusion

The proposed PGM is an effective, simple and low power design technique. The LPM allows the designed circuit to have efficient power-performance trade-off capability with a small delay penalty in the output signals. The main contribution of this paper is the use of a feasible Vdd management technique for expanding the domino circuits for both low-power and high-performance applications.

VIII. References

[1] J. Stallman, and E. Habekotte, "Several Driving Configurations with Low-voltage Input Control for a Planar Power Switch", IEEE Journal Solid-State Circuits, vol. 19, pp. 147-154, Feb., 1984.

[2] Peiyi Zhao, J. McNeely, M. Bayoumi, G. Pradeep, Kuang Weidong, "A Low Power Domino with Differential-Controlled-Keeper", IEEE International Symposium on Circuits and Systems, 2007.

[3] S. J. Shieh, J. S. Wang, "Design of low-power domino circuits using multiple supply voltages", IEEE International Conference on Electronics, Circuits and System, 2001.

[4] P. Patra, U. Narayanan, "Automated phase assignment for the synthesis of low power domino circuits", ACM/IEEE Design Automation Conference, 1999.

Appendix: Test Chip Measurements

Figures A1, A2 show the oscillator scope waveforms for MUL *New* working in all modes.

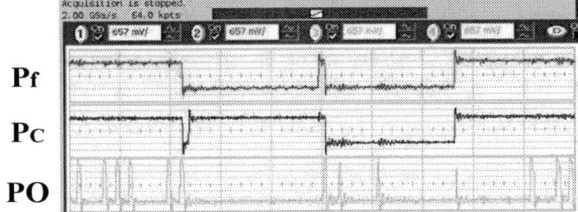

Figure A1. All modes of LPM test chip measurement.

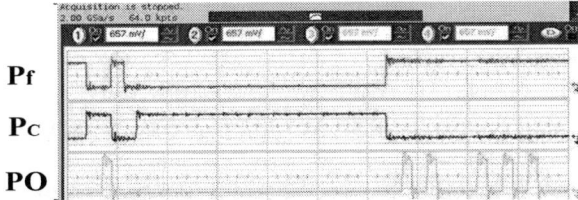

Figure A2. Sleep mode of LPM test chip measurement.

The design of MUL *New* (in CKVdd mode) at fixed voltages of 1.8V and 1.2V was applied to MUL *Ori* for comparing power consumptions. In the following figures the degrading ratios for MUL *New* in different operation modes are based on the comparison basis of MUL *Ori* operated at 1.8v.

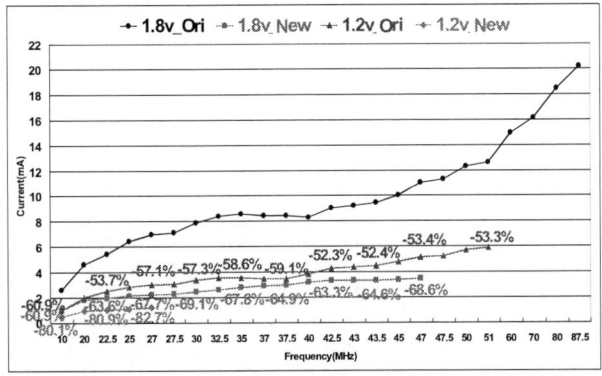

Figure A3. Comparison of power consumption in the proposed test chip.

Figure A3 shows the measurement results by varying clock frequency. At clock frequency of 47MHz, MUL *New* reduces power by 68% and 15% (68%-53%) as compared to MUL *Ori* working at 1.8V and 1.2V, respectively. MUL *New* has less peak current, while

both MUL *New* and MUL *Ori* increase power as frequency increases. It is interesting to note that the power increase rate was lower for MUL *New* than for MUL *Ori*. *The curve flattens when f increases.* This implies that CKVdd mode offers useful power savings in high speed clock designs.

MUL *New* consistently consumes less power than MUL *Ori*. However, MUL *Ori* can work at higher frequency than MUL *New*. As there is no power consumption in sleep mode, PM consumes less power than CKVdd and PG modes.

In addition to the lower power consumption in normal operation mode, LPM is useful for managing the circuit performance. Figure A4 shows the comparison of two PM modes at fixed voltage of 1.8V. PM2 applies one more clock pulse signal than PM1. At a clock frequency of 47MHz, MUL *New* in PM1 and PM2 reduces power by 70.3% and 70.7%, respectively, as compared to MUL *Ori*. However, MUL *New* in PM2 works at a higher frequency and has higher power savings than in PM1. The PM3 applies one more pulse signal than PM2 and the working frequencies are also higher. This results show the LPM fine power-performance manageable mechanism.

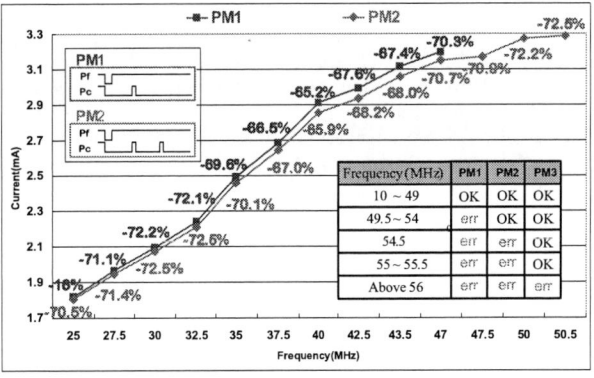

Figure A4. Comparison of power-performance management modes.

The MUL *New* operation in CKVdd mode has lower power consumption and operation frequency than in PG mode as shown in Figure A5.

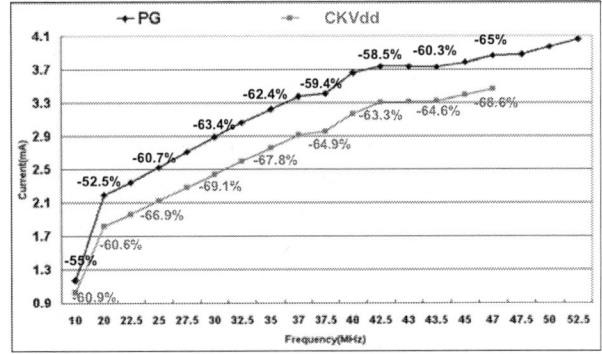

Figure A5. The power consumption comparison of CKVdd and PG modes.

978-1-61284-863-1/11 $26.00 © 2011 IEEE

A 30K 2.5Gb/s Decision-Eased Soft RS (224, 216) Decoder for Wireless Systems

Yi-Min Lin, Yu-Chun Huang, Chi-Heng Yang, and Hsie-Chia Chang

Department of Electronics Engineering & Institute of Electronics
National Chiao Tung University
Hsinchu, Taiwan 300, R.O.C.
E-mail: ymlin@si2lab.org

Abstract—In this paper[1], an area efficient (224, 216; 4) soft Reed-Solomon (RS) decoder is provided for wireless systems. According to the Chase-II algorithm, 3 least reliable positions (LRPs) are chosen, and the decoded codeword is determined from 8 ($= 2^3$) candidate codewords, leading to 0.5 dB coding gain at 10^{-5} BER over the hard RS decoder. Instead of using 8 hard decoders to generate all candidate codewords, our design utilizes only one by reschedule the decoding scheme among all candidate codewords. Therefore, the hardware complexity is significantly reduced. Moreover, the complex decision making unit is eased by using Hamming distance calculation to replace Euclidean distance calculation among all candidate codewords with negligible performance degradation. As compared to low complexity Chase (LCC) soft decoder, our proposed decoder can save around 59.8% complexity. According to the post-layout simulations, the proposed design can achieve 2.5 Gb/s throughput with gate count of 30 K, which is less than one half the normalized area of the hard RS decoder.

Index Terms—Error correction coding, Soft-decidsion, Reed-Solomon (RS) codes, Wireless system.

I. INTRODUCTION

The millimeter-wave (mmWave) system can promise throughput in the order of multi-Gpbs because of the huge bandwidth ranging from 57-64 GHz Hence it has aroused much interest, such as Ecma International TC32-TG20 Task Group [1] and IEEE 802.15.3c Task Group [2]. To support the uncompressed high-definition (HD) video transmission, which can enhance picture quality, the mmWave system is required to provide 3.0 Gb/s throughput if the color depth of a single uncompressed HD (1080p) stream is 8-bit and to achieve 4.5 Gb/s for the future applications with 12-bit color depth. According to [3], two RS (224, 216; 4) decoders are applied for the 4 MSB and 4 LSB bitstreams respectively, implying a RS decoder is demanded to provide 2.25 Gb/s throughput for future video applications. However, the high throughput will damage the reliability of the transmitted data. For higher error correcting performance with the same code rate, the soft decoding algorithms are carried out to improve error correcting ability by making use of the reliability information.

In 1966, Forney firstly developed a generalized-minimum-distance (GMD) method [4] by generating a list of candidate

[1]This work was supported by NSC and MOEA of Taiwan, R.O.C., under grant NSC 100-2220-E-009-062 and 100-EC-17-A-01-S1-124 respectively.

codewords with algebraic algorithms. With the similar concept of candidate list, other algorithms such as Chase [5] and Chase-GMD [6], are also widely used in many applications. Another approach, the Koetter-Vardy algorithm [7], utilizes soft information to interpolate each symbol, leading to larger coding gain. However, the complex soft decoding computations make soft RS decoders more hardware complexity and decoding latency. Hence, the VLSI architecture of the low complexity chase (LCC) based soft RS decoder was provided in 2009 [8], but it still has about three times area in contrast to the hard decoder.

In the above algorithms, the final step is exploited a decision making unit to select the candidate codeword with the smallest Euclidean distance as the decoded codeword. For the long length codeword, however, the Euclidean distance calculation requires large number of multiplication and square root calculations, which are too complex for practical implementation. In this paper, a low complexity and high throughput soft RS decoder with simplified decision making unit is designed for mmWave system. Instead of Euclidean distance calculation, the Hamming distance calculation is exploited with the negligible performance loss. Moreover, with gray code based bit-flipping method, the hardware complexity can be significantly reduced by rescheduling the decoding scheme among all candidate codewords.

This paper is organized as follows. Section II and III describe the decoding process and the proposed area efficient architectures for the decision-eased RS decoding respectively. Based on the proposed method, Section IV demonstrates the implementation results of the proposed soft RS decoder. Section V gives a conclusion of the paper.

II. PROPOSED DECODING SCHEME

An $(N, K; t)$ RS code has block length of N symbols and information length of K symbols, where each symbol consists of m bits while the code is operated under $GF(2^m)$. The error correcting ability t is up to $\lfloor \frac{N-K}{2} \rfloor$, the half minimal distance of the code. To correct an error number larger than t, Fig. 1 shows our proposed decision-eased soft RS (224, 216; 4) decoder, which includes seven major steps: *syndrome calculator, reliability evaluator, syndrome updater,*

978-1-61284-863-1/11 $26.00 © 2011 IEEE

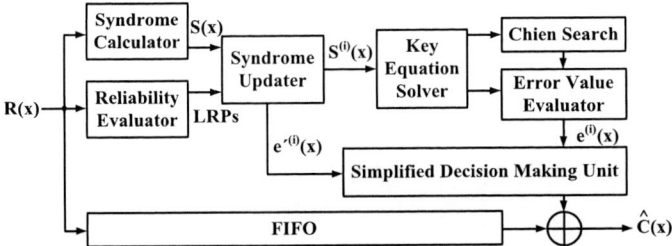

Fig. 1.   Block Diagram of Proposed Decision-Eased Soft RS Decoder

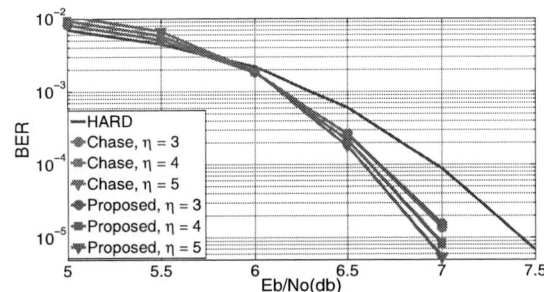

Fig. 2.   Simulation Results of Decision-Eased Algorithm for RS (224, 216; 4) Codes under BPSK Modulation and AWGN Channel.

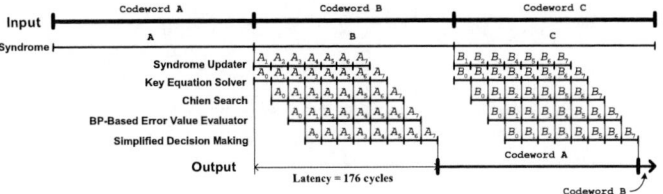

Fig. 3.   Timing schedule of Proposed Decision-Eased Soft RS Decoder

key equation solver, Chien search, error value evaluator, and *simplified decision making unit*. The received data is fed into the syndrome calculator for calculating the syndrome polynomial $S(x)$. In the meantime, the reliability evaluator determines η least reliable positions (LRPs) $\underline{L} = [l_1, l_2, \ldots, l_\eta]$ according to the received data. The corresponding error values: $\underline{E'} = [e'_1, e'_2, \ldots, e'_\eta]$ are also computed. Instead of evaluating of 2^η syndrome polynomials using syndrome calculator for 2^η candidate sequences, a syndrome updater is applied to obtain the i-th syndrome polynomial $S^{(i)}(x)$ from the syndrome polynomial $S(x)$. There are only η uncommon points for all the candidate sequences; therefore, the i-th candidate sequence can be constructed by adding the error pattern $e_f^{(i)}(x)$ induced by the bit flipping procedure to the received data. The $e_f^{(i)}(x)$ is of the form:

$$e_f^{(i)}(x) = a_1^{(i)} e_1^{(i)} x^{l_1} + a_2^{(i)} e_2^{(i)} x^{l_2} + \cdots + a_\eta^{(i)} e_\eta^{(i)} x^{l_\eta}, \quad (1)$$

where $a_j^{(i)} \in GF(2)$ for $i = 0 \sim 2^\eta - 1$ and $j = 1 \sim \eta$. The i-th syndrome polynomial can be derived as

$$S^{(i)}(x) = \sum_{j=1}^{2t} (S_j + e_f^{(i)}(\alpha^j)) x^{j-1}. \quad (2)$$

To further improve the complexity, $S^{(i)}(x)$ can be obtained from the previous syndrome polynomial $S^{(i-1)}(x)$. With gray code based bit flipping order, there is only one different bit between successive two candidate sequences. Hence, $S^{(i)}(x)$ can be simplified as

$$S^{(i)}(x) = \sum_{j=1}^{2t} (S_j^{(i-1)} + e_f'^{(i)}(\alpha^j)) x^{j-1}, \quad (3)$$

where $e_f'^{(i)}(x) = e'_\kappa x^{l_\kappa}$ is difference polynomial between $e_f^{(i-1)}$ and $e_f^{(i)}(x)$ and the only one different bit is at l_κ-th position. After 2^η syndrome polynomials are generated from syndrome updater, the key equation solver, Chien search and error value evaluator are applied to form 2^η candidate codewords. Finally, the decision making unit selects the most possible one from the candidate list as decoded codeword. In the decision making unit, however, all the Euclidean distances are applied only for choosing the most likely codeword from the candidate list. The exact values of all Euclidean distances are not demanded as long as their relations are still held, implying that the square root calculation can be eliminated.

Hence, the Hamming distance calculation is utilized in the proposed decoder for replacing the Euclidean distance calculation due to its much simpler computations. In addition, the i-th candidate codeword $\hat{C}^{(i)}(x)$ in the Chase algorithm can be viewed as

$$\hat{C}^{(i)}(x) = R(x) + e^{(i)}(x) + e_f^{(i)}(x), \quad (4)$$

where $e^{(i)}(x)$ is the i-th estimated error pattern by decoding $R(x) + e_f^{(i)}(x)$. Notice that from (4),

$$\hat{C}^{(i)}(x) + R(x) = e^{(i)}(x) + e_f^{(i)}(x). \quad (5)$$

Therefore, we can computing the weight of $e^{(i)}(x) + e_f^{(i)}(x)$ instead of storing each candidate codeword and directly calculating the Hamming distance between $\hat{C}^{(i)}(x)$ and $R(x)$, leading to significantly reduced memory requirement.

Fig. 2 shows the simulation results for proposed decision-eased decoding with η LRPs, where $\eta = 3 \sim 5$. The performance gain at 10^{-5} BER is 0.6 dB while $\eta = 5$ and is 0.5 dB while $\eta = 3$ in contrast to the hard decoding. Notice that, our proposal almost has no performance loss as compared to the Chase algorithm.

III. PROPOSED VLSI ARCHITECTURES

In our proposed decision-eased soft RS (224, 216; 4) decoder, total 8 candidate codewords will be generated since η equals 3. Fig. 3 is the timing schedule of our design and each block is well-arranged to operate within 16 cycles.

A. Syndrome Updater

Instead of recalculating the i-th syndrome polynomial $S^{(i)}(x)$ with syndrome calculator, the syndrome updater calculates it by updating the $(i-1)$-th syndrome polynomial $S^{(i-1)}(x)$, leading to further hardware reduction. According

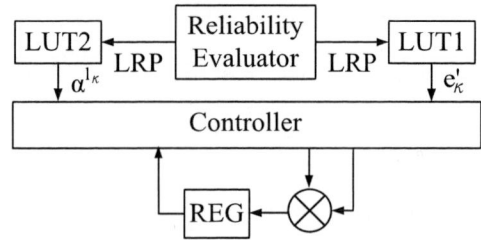

Fig. 4. Syndrome Updater

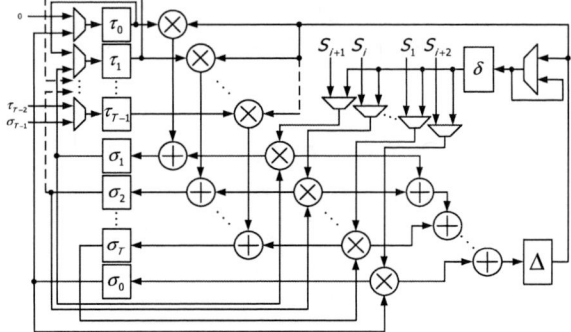

Fig. 5. BM-Based Key Equation Solver

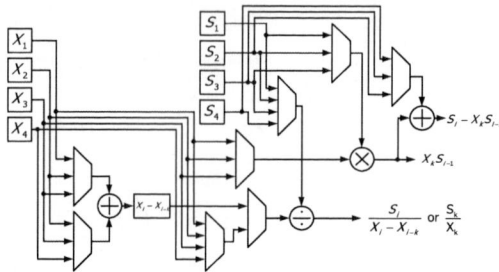

Fig. 6. BP-Based Error Value Evaluator

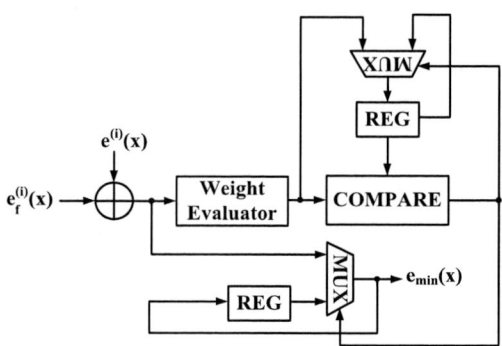

Fig. 7. Decision Making Unit Architecture

to (3), $S^{(i)}(x)$ can be obtained with $e'_\kappa \times x^{l_\kappa}$, where the only one different bit between successive two candidate sequences is at l_κ-th position. The proposed syndrome updater shown in Fig. 4 utilizes look up table 1 (LUT1) and LUT2 to find the e'_κ and α_{l_κ} based on the results of reliability evaluator. Since the syndrome updater tkes 16 cycles to update a syndrome polynomial having 8 syndrome values, the finite field multipliers (FFMs) for updating syndrome polynomials can be shared, resulting in only 1 FFM and 2 LUTs requirements.

B. BM-Based Key Equation Solver

To reduce the hardware complexity for high parallelism error values calculation, the error values are calculated with BP algorithm instead of Forey's algorithm. Therefore, error evaluator polynomial $\Omega(x)$ is not demanded to be generated by the key equation solver. In addition, the key equation solver has 16 decoding cycles to cooperate with other decoding blocks, implying that the number of multipliers can be decreased from $3t+3$ to $2t+1$ with a trade off on decoding cycles from $2t$ to $4t$. Fig. 5 shows our proposed BM-based key equation solver. The proposed decoder takes two cycles in each iteration, where the first cycle is used to update the error location polynomial $\sigma(x)$ and the other cycle is used to calculate discrepancy Δ.

C. BP-Based Error Value Evaluator

The Björck-Pereyra (BP) [9] based method is applied to evaluate error values with the relation between syndrome values S_i and error locators X_i:

$$
\begin{bmatrix}
X_1 & X_2 & X_3 & X_4 \\
X_1^2 & X_2^2 & X_3^2 & X_4^2 \\
X_1^3 & X_2^3 & X_3^3 & X_4^3 \\
X_1^4 & X_2^4 & X_3^4 & X_4^4
\end{bmatrix}
\begin{bmatrix}
e_1 \\
e_2 \\
e_3 \\
e_4
\end{bmatrix}
=
\begin{bmatrix}
S_1 \\
S_2 \\
S_3 \\
S_4
\end{bmatrix}, \quad (6)
$$

which can avoid to use large parallelism architecture for using Forney's algorithm to compute error values within 224 locations in 16 cycles.

The proposed BP-based error value evaluator shown in Fig. 6 has only 1 divider, 1 multiplier and 2 adders. The variable S_i, which initially represents the i-th syndrome value, is updated iteratively. Each calculation of the syndrome represents a row operation in (6). The control logic determines the computation order of the S_i and X_i, and the computation results will be used to update each S_i value. After all computations, S_i indicates the i-th error value.

D. Decision Making Unit

From (5), the weight of $e^{(i)}(x) + e_f^{(i)}(x)$ stands for the Hamming distance between $\hat{C}^{(i)}(x)$ and $R(x)$. That means at most $t+\eta$ symbols are different while counting the distance of candidates codewords. Also, in the decoding process, only the smallest weight error pattern $e^{(\phi)}(x) + e_f^{(\phi)}(x)$ will be added with the received data $R(x)$ for generating the decoded codeword, where

$$
\phi = argmin\{weight(e^{(i)}(x)+e_f^{(i)}(x)) : i = 0 \sim 2^\eta - 1\} \quad (7)
$$

This indicates that we do not have to store all the $e^{(i)}(x) + e_f^{(i)}(x)$ with $i = 0 \sim 2^\eta - 1$.

Fig. 7 illustrates the proposed simplified decision making unit. The estimated error pattern $e^{(i)}(x)$ and the flipped error pattern $e_f^{(i)}(x)$ are added at first and the corresponding weight is calculated. Then the minimum weight error pattern $e_{min}(x)$ is determined by comparing the current minimum weight with the weight of $e^{(i)}(x)+e_f^{(i)}(x)$. If the weight of $e^{(i)}(x)+e_f^{(i)}(x)$

TABLE I
COMPARISON TABLE FOR SOFT RS (224, 216; 4) DECODER

Architecture	$GF(2^8)$ CFFM	$GF(2^8)$ FFM	$GF(2^8)$ FFA	2-to1 MUX (Bit)	Register (Bit)	ROM (Byte)	RAM (Byte)	Latency (Cycle)	Normalized [*] Complexity
Proposed with $\eta = 3$	72	12	86	792	632	456	$0+224 \times 2$	224	13,248
LCC with $\eta = 3$ [8]	4	50	72	997	1430	896	$68+224 \times 8$	464	32,991

[*] The complexity ratio over $GF(2^8)$ among XOR, CFFM, FFM, FFA, MUX, Register, ROM (byte) and RAM (byte) is 1 : 20 : 100 : 8 : 3: 1 : 8 : 8.

TABLE II
IMPLEMENTATION RESULTS AND COMPARISION

	Proposed with $\eta = 3$	pRiBM [10]	pDCME [11]	DCME [12]
Code Type	Soft RS (224, 216; 4)	Hard RS (255, 239; 8)	Hard RS (255, 239; 8)	Hard RS (255, 239; 8)
Technology	90 nm	90 nm	90 nm	0.25 μm
Operation Frequency	312.5MHz (Post Layout)	690MHz (Synthesis)	660MHz (Synthesis)	200MHz (Synthesis)
Gate Count (Normalized) [*]	30.0 K (1.42)	43.6 K(1.03)	53.2 K (1.26)	42.2 K (1)
Throughput	2.5 Gb/s	5.52 Gb/s	5.28 Gb/s	1.6 Gb/s
Coding Gain	0.5 dB @ 10^{-5} BER	-	-	-

[*] The normalized gate count is calculate as: $\dfrac{(gate\ count) \times (\frac{8}{4})}{(gate\ count\ of\ DCME)}$.

is smaller than the stored value, $e^{(i)}(x) + e_f^{(i)}(x)$ becomes the minimum weight error pattern $e_{min}(x)$. While 2^η candidate codewords are all processed, the results in registers are the determined output.

IV. IMPLEMENTATION RESULTS

The proposed decision-eased soft RS decoder with 2-stage pipeline architecture is compared with the LCC soft RS decoder with 4-stage pipeline architecture as shown in TABLE I. Both soft decoders evaluates 3 LRPs for generating candidate sequences. Notice that the complexity of a LCC soft RS (224, 216; 4) decoder is estimated according to the LCC soft RS (255, 239; 8) decoder in [8]. While the complexity of these designs is normalized to XOR gate, the proposed decision-eased soft RS decoder is around 13,248 XOR gates and the LCC soft RS decoder is about 32,991 XOR gates. Due to fewer number of FFMs and storage elements, our proposed decoder can save around 59.8% complexity as compared to LCC decoder, even though the LCC decoder excludes the decision making unit.

TABLE II compares the implementation result among our soft RS decoder and three hard RS decoders. From the post-layout simulation, our design can achieve 2.5 Gb/s throughput with gate count of 30 K in 90nm CMOS process. Moreover, it can fit well for 2.25 Gb/s throughput requirement in mmWave system with 0.5 dB coding gain over hard decoders at 10^{-5} BER. Note that, as considered the difference of the error correcting ability among these designs, a normalized area is calculated for a fair comparison. Our proposed decision-eased soft RS decoder is only 1.38 and 1.42 times as compared to pRiBM and DCME based hard decoders respectively.

V. CONCLUSION

This paper provides a decision-eased soft RS decoder with a simplified decision making unit, which determines the output codeword with the Hamming distance calculation for avoiding the complex computation of Euclidean distance calculation.

Moreover, the rescheduled decoding scheme allows for generating 8 candidate codewords with only one suit hardware, leading to significantly hardware complexity reduction. From the simulation, the proposed soft decoder with 3 LRPs has 0.5 dB coding gain at 10^{-5} BER as compared with the conventional hard decoder. According to the post-layout simulation, the proposed soft RS (224, 216; 4) decoder can achieve 2.5 Gb/s throughput to meet the mmWave system requirement with gate count of 30 K in CMOS 90nm technology, which is less than one half the normalized area of the hard RS decoder.

REFERENCES

[1] High Rate Short Range Wireless Communication, ECMA TC32-TG20 Std., [online]. Available:http://www.ecma-international.org/memento/TC-32-TG20-M.htm/.

[2] IEEE 802.15 WPAN Millimeter-wave Alternative PHY Task Group 3c (TG3c)., IEEE Std. Draft Document, 2009.

[3] H. Singh, H. Niu, X. Qin, H. Shao, C. Kwon, G. Fan, S. Kim, and C. Ngo, "Supporting uncompressed HD video streaming without retransmissions over 60GHz wireless networks," in IEEE Wireless Communications and Networking Conference, WCNC, 2008, pp. 1939–1944.

[4] G. D. Forney, "Generalized Minimum Distance Decoding," IEEE Trans. Inform. Theory, vol. 12, p. 125V131, Apr. 1966.

[5] D. Chase, "A Class of Algorithms for Decoding Block Codes with Channel Measurement Information," IEEE Trans. Inform. Theory, vol. IT-18, p. 170V182, Jan. 1972.

[6] H. Tang, Y. Liu, M. Fossorier, and S. Lin, "On combining chase-2 and gmd decoding algorithms for nonbinary block codes," IEEE Communications Letters, vol. 5, no. 5, pp. 209 –211, May 2001.

[7] R. Koetter and A. Vardy, "Algebraic Soft-Decision Decoding of Reed-Solomon Codes," IEEE Trans Inform. Theory, vol. 49, no. 11, pp. 2809–2825, 2003.

[8] X. Zhang, "High-speed VLSI architecture for low-complexity Chase soft-decision Reed-Solomon decoding," IEEE Inform. Theory and Application Workshop, pp. 422–430, Feb. 2009.

[9] A. Björck and V. Pereyra, "Solutions of Vandermonde Systems of Equations," Math. Comp., vol. 24, pp. 893–903, 1970.

[10] J.-I. Park, K. Lee, C.-S. Choi, and H. Lee, "High-speed low-complexity reed-solomon decoder using pipelined berlekamp-massey algorithm," in IEEE Int. SoC Design Conference (ISOCC), 2009, pp. 452 –455.

[11] S. Lee, H. Lee, J. Shin, and J.-S. Ko, "A High-Speed Pipelined Degree-Computationless Modified Euclidean Algorithm Architecture for Reed-Solomon Decoders," in IEEE Int. Symp. on Circuits and Systems (ISCAS)., May 2007, pp. 901 –904.

[12] J. Baek and M. Sunwoo, "New Degree Computationless Modified Euclid Algorithm and Architecture for Reed-Solomon Decoder," IEEE Trans. VLSI Syst., vol. 14, no. 8, pp. 915 –920, 2006.

A 0.16nJ/bit/iteration 3.38mm^2 Turbo Decoder Chip for WiMAX/LTE Standards

Cheng-Hung Lin
Department of Electrical Engineering
Yuan Ze University
Jungli 320, Taiwan

Chun-Yu Chen
Silicon Motion Technology Corp.
Sindian 231, Taiwan

En-Jui Chang, and
An-Yeu (Andy) Wu
Graduate Institute of
Electronics Engineering,
National Taiwan University,
Taipei 106, Taiwan

Abstract—**This paper presents a turbo decoder chip design supporting distinct convolutional turbo code schemes in WiMAX and LTE systems. A contention-free vectorizable dual-standard interleaver is proposed to enhance the hardware utilization. Moreover, a warm-up free parallel MAP decoding is proposed to improve the throughput rate. The overall VLSI architecture of the proposed CTC decoder is presented for supporting the WiMAX/LTE systems. This chip fabricated in a core area of 3.38 mm^2 by 90nm CMOS process is measured at 152 MHz with a power consumption of 148.1 mW and a throughput rate of 186.1 Mbps. This chip achieves a high area efficiency of 0.36 bit/mm^2 and a low energy efficiency 0.16 nJ/bit/iteration.**

Keywords- WiMAX, LTE, Multi-standard, Turbo Decoder.

I. INTRODUCTION

Recently, convolutional turbo codes (CTCs) have been regular schemes for wireless communications in order to have a reliable transmission over noisy channels. The single-binary (SB) CTC proposed in 1993 [1] was adopted in LTE because of its coding gain close to the Shannon limit. The non-binary CTC proposed in 1999 [2] was introduced to have a superior coding performance than the SB-CTC. The double-binary (DB) CTC was adopted in WiMAX.

To deal with the different data rates and CTC schemes, the multi-standard CTC decoder which works across the multiple standards of wireless WANs can enable smooth migration for different multimedia applications within a same device. The prevalent sliding-window (SW) maximum *a posteriori* algorithm (MAP) deals with variant CTC blocks with intrinsic low throughputs [3]. Thus, to support high-throughput CTC decoding for future wireless WAN systems, a parallel decoding architecture which embeds multiple MAP decoding kernels is inevitable. The parallel decoding architecture may encounter memory contention for parallel interleaved data access. Especially for the multiple-standard CTC interleaving, the memory contention occurs frequently and needs to be solved to make the parallel MAP decoding realizable.

In this paper, a contention-free parallel decoding architecture for WiMAX/LTE CTC decoding is proposed in Section II. Contention-free parallelisms for WiMAX/LTE CTC interleaving are analyzed and a parallel vectorizable WiMAX/LTE CTC interleaver is proposed. In addition, a warm-up free parallel MAP decoding is proposed to achieve

the throughput rates of WiMAX and LTE systems. In Section III, the proposed CTC decoder for the WiMAX/LTE standards has been fabricated in a core size of 3.38 mm^2 by using UMC 90nm CMOS technology. This prototyping chip supports all 17 modes of the WiMAX CTC scheme and selected 18 modes of the LTE CTC scheme. The throughput rate of 186.1 Mbps is maximally measured at 152 MHz with a power consumption of 148.1 mW. Finally, the concluding remarks are given in Section IV.

II. PROPOSED CTC DECODER FOR WiMAX/LTE STANDARDS

A. Contention-free Vectorizable Dual-standard Parallel Interleaver

For parallel decoding, multiple MAP kernels may read/write a same memory bank simultaneously. Since the port of a memory is finite, this simultaneous memory access needs to be prohibited. Without a cautious analysis of the dual-standard CTC interleaving, the memory contention occurs frequently and makes the parallel MAP decoding unrealizable for an unexpected parallelism.

For a hardware design of the parallel interleaving, the solution to memory contention has been discussed in [4]. The interleaver is contention-free when the interleaving sequence $\Pi(t)$ satisfies

$$\left\lfloor \frac{\Pi(t+jW)}{W} \right\rfloor \neq \left\lfloor \frac{\Pi(t+kW)}{W} \right\rfloor , \qquad (1)$$

where N is the decoding block size, W is the basic windows size, P is the parallelism of the parallel MAP decoding ($N = PW$), $0 \leq t < W$, and $0 \leq j, k < P$. The terms on both sides of (1) are indices of the memory banks that are accessed by the jth and k^{th} MAP kernels at the tth time instant. This inequality need to be true for any time instant t for no memory contention.

For an interleaver design, the complexity of the interleaving address generation is also critical. Each physical memory bank requires an address decoder to transform the global interleaving address to the local address for each memory bank. As the parallelism P increases, the duplication of address decoder leads to hardware inefficiency. A better solution is to use the same address for all memory banks. This property requires the

This work was supported by National Science Council under grants NSC 99-2218-E-155-011 and 95-2219-E-002-020.

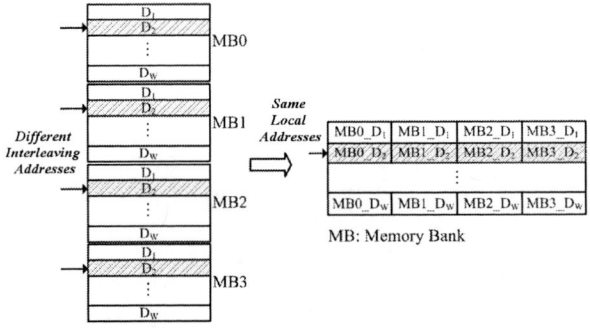

Figure 1. Vectorizable interleaving address for memory banks.

TABLE I. AVAILABLE PARALLELISM FOR WiMAX /LTE SYSTEMS.

LTE (Selected 18 modes)		WiMAX (all 17 modes)	
N	Available parallelism	N	Available parallelism
48	1 2	24	1
72	1 2 3	36	1
96	1 2 3 4	48	1 2
144	1 2 3 4 6	72	1 2 3
192	1 2 3 4 6 8	96	1 2 3 4
216	1 2 3 4 6 8 9	108	1 3
240	1 2 3 4 5 6 8 10	120	1 2 3 4 5
288	1 2 3 4 6 8 9 12	144	1 2 3 4 6
360	1 2 3 4 5 6 8 9 10 12 15	180	1 2 3 4 5 6
384	1 2 3 4 6 8 12 16	192	1 2 3 4 6 8
432	1 2 3 4 6 8 9 12 16 18	216	1 2 3 4 6 8 9
480	1 2 3 4 5 6 8 10 12 15 16 20	240	1 2 3 4 5 6 8 10
960	1 2 3 4 5 6 8 10 12 15 16 20	480	1 2 3 4 5 6 8 10 12
1920	1 2 3 4 5 6 8 10 12 15 16 20	960	1 2 3 4 5 6 8 10 12
2880	1 2 3 4 5 6 8 9 10 12 15 16 18 20	1440	1 2 3 4 5 6 8 9 10 12
3840	1 2 3 4 5 6 8 10 12 15 16 20	1920	1 2 3 4 5 6 8 10 12
4800	1 2 3 4 5 6 8 10 12 15 16 20	2400	1 2 3 4 5 6 8 10 12
6144	1 2 3 4 5 6 8 12 16	-	-

interleaving address to satisfy

$$\Pi(t + jW) \bmod W = \Pi(t) \bmod W, \qquad (2)$$

where $0 \le t < W$, and $0 \le j < P$. The equality implies that each MAP kernel access data based on the same local address. Based on this vectorizable property, only one set of address generator is required. All memory banks can merge into a single physical memory with data stored and fetched as vectors as shown in Fig. 1.

A high-level simulation model for (1) and (2) is used to analyze the available parallelism for the LTE and WiMAX standards with $24 \le W \le 36$. The details of WiMAX and LTE CTC interleaving can be referred to in [5] and [6], respectively. The available parallelism achieving the contention-free and vectorizable interleaving address is shown in Table I and we choose $P = 8$ for the proposed VLSI architecture of WiMAX/LTE CTC decoder. Fig. 2 shows the proposed contention-free vectorizable dual-standard parallel interleaver and the CTC controller. The CTC controller provides control signals and initial parameters. To perform the radix-4 SB/DB EML-MAP decoding [7], the proposed dual-standard address generators generate the WiMAX addresses or the radix-4 LTE even addresses by adopting a hardware sharing technique. The additional LTE address generators generate the radix-4 LTE odd addresses in the LTE modes. The LTE and WiMAX interleaving parameters ($P0$, $TP1$, $TP2$, $TP2$, $P(0)$, $H(0)$, $J(0)$, and f_2) can be implemented in a look-up table. The address

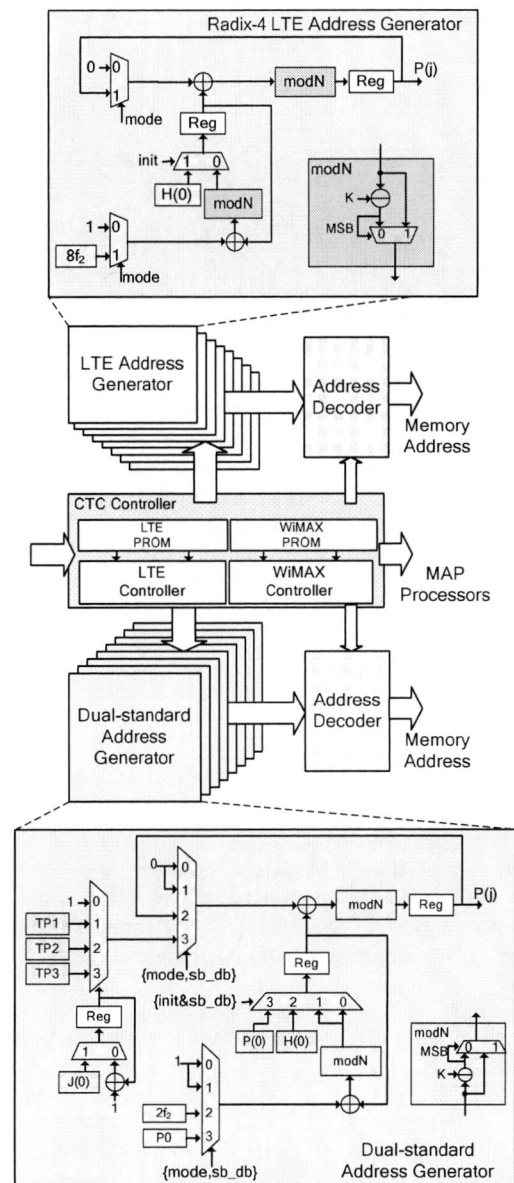

Figure 2. Block diagram of the WiMAX/LTE parallel interleaver and CTC controller.

decoder transforms the interleaving addresses into the contention-free addresses of the memory subbanks.

B. High-throughput Warm-up Free Parallel MAP Decoding

The SW MAP decoding was proposed to facilitate the memory cost of CTC decoders. However, the SW MAP decoding deals with any CTC frame size but has intrinsic low throughputs. The hybrid-window (HW) MAP decoding described in [7] applies parallel MAP kernels to decode one received block. Because of the warm-up processes of forward states metrics, the decoding latency is prolonged to $4W$. Nevertheless, the HW MAP decoding can shorten the decoding cycles to $N/P + 4W$ by working with several sub-blocks simultaneously.

978-1-61284-863-1/11 $26.00 © 2011 IEEE

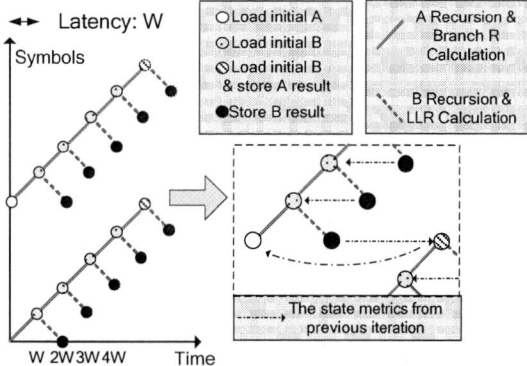

Figure 3. The warm-up free hybrid-window (HW) MAP decoding.

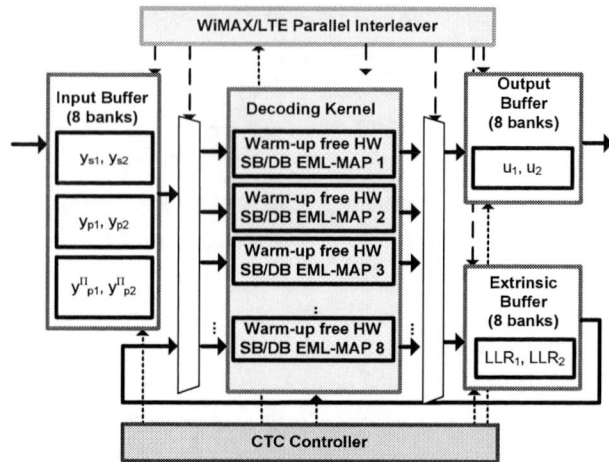

Figure 4. Block diagram of the proposed CTC decoder for WiMAX/LTE standards.

In this paper, a warm-up free HW MAP decoding is proposed to further reduce the decoding latency. The timing chart is shown in Fig. 3. Instead of performing a warm-up recursion, each MAP kernel achieves the initial state metrics (both A and B in Fig. 3) by fetching the state metrics of previous iteration from the border metrics memory (BMM). The loss of coding gain is less than 0.03 dB at BER of 10^{-5}. The decoding latency can be reduced to one W. Compared to the HW MAP decoding in [9] with $P = 8$, $W = 24$, and the radix-4 decoding, the throughput rates can be improved 60% for $N = 48$ and 15% for decoding the $N = 6144$.

C. Proposed CTC Decoder for WiMAX/LTE Systems

Fig. 4 shows the VLSI architecture of the proposed CTC decoder for WiMAX/LTE systems with $P = 8$. For verifying the proposed design techniques, this prototyping decoder supports all 17 modes of the WiMAX CTC scheme and selected 18 modes of the LTE CTC scheme. The proposed WiMAX/LTE parallel interleaver generates addresses for the input memories and extrinsic memories so that the normal-order or interleaved-order values can be processed between the 8 warm-up free HW MAP kernels and the memory subbanks. The number of active MAP kernels with the distinct design modes is shown in Table II. The computational modules and storages of the radix-4 SB/DB EML-MAP decoding in [8] ware applied to the MAP kernels to achieve a high area usage. Moreover, the radix-4 traceback structure in [7] was adopted to reduce the area and power of the state metric cache. Because of the radix-4 SB MAP decoding, two extrinsic information values are accessed in parallel. The bit-level extrinsic information exchange [9] was also adopted to make the extrinsic buffer access only two extrinsic information values for the radix-4 DB MAP decoding. Thus, the extrinsic buffer for the radix-4 SB/DB MAP decoding can be fully shared. When the targeted iteration number is reached or the hard bits of two half iteration are the same, the CTC decoder finishes the decoding procedure and outputs the hard bits from the output memories.

III. EXPERIMENTAL RESULTS

The proposed CTC decoder has been implemented in an ASIC by using Verilog HDL codes synthesized with the standard cell library of UMC 90nm CMOS process. The design of the CTC decoder is simulated using C-to-RTL flow. The

TABLE II. DESIGN MODES OF THE PROPOSED CTC DECODER.

Mode		Info. Bits	Active MAPs	Mode		Info. Bits	Active MAPs
WiMAX	LTE			WiMAX	LTE		
1	18	48	1	10	27	384	8
2	19	72	1	11	28	432	8
3	20	96	2	12	29	480	8
4	21	144	2	13	30	960	8
5	22	192	4	14	31	1920	8
6	23	216	3	15	32	2880	8
7	24	240	4	16	33	3840	8
8	25	288	4	17	34	4800	8
9	26	360	6	-	35	6144	8

quantization parameters are the same to Table III in [8] met the targeted BER of 10^{-5} for WiMAX and of 10^{-3} for LTE. Fig. 5 shows the BER performance of the distinct CTC schemes decoded by the CTC decoder based on AWGN channels and 6 iterations. The proposed CTC decoder chip has been fabricated in a die area of 7.18 mm^2 and a core area of 3.38 mm^2. The total gate count of the proposed contention-free parallel interleaver and CTC controller is 30.4 Kgates. The decoder contains 232.8 Kb RAM. Fig. 6 shows the chip layout and die photo of the prototyping CTC decoder chip. The prototyping chip is fabricated and maximally measured at 152 MHz operating frequency at core supply voltage of 1.1 V. In order to consider reduction of power consumption, the core supply voltage is reduced to form 1.1 V to 0.9 V. The measured maximal operating frequencies and power consumptions are shown in Fig. 7.

For the maximum block size in WiMAX (Mode 17 in Table II), the a throughput rate is 178.4 Mbps (@152 MHz, 6 iterations). For the maximum block size in LTE (Mode 35 in Table II), the chip achieves a throughput rate of 186.1 Mbps with a power consumption of 148.1 mW (@152 MHz, 6 iterations). The comparison of our decoder and other works are shown in Table III. For supporting the both WiMAX and LTE CTC decoding, this chip achieves a high area efficiency of 0.36 bit/mm^2 with a low energy efficiency of 0.16 nJ/bit/iteration.

IV. CONCLUSION

The fabricated CTC decoder chip for WiMAX/LTE standards employs the proposed contention-free vectorizable

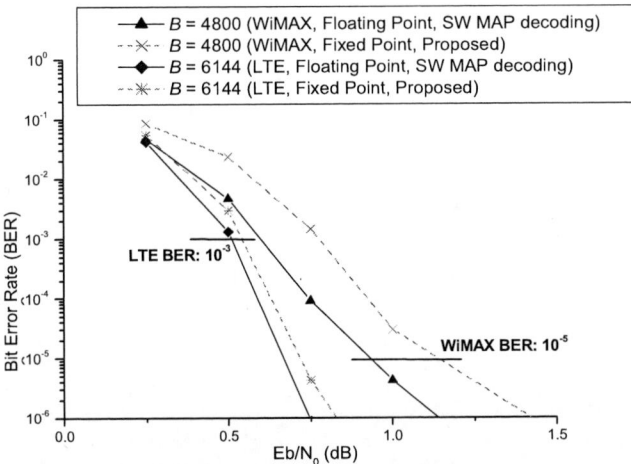

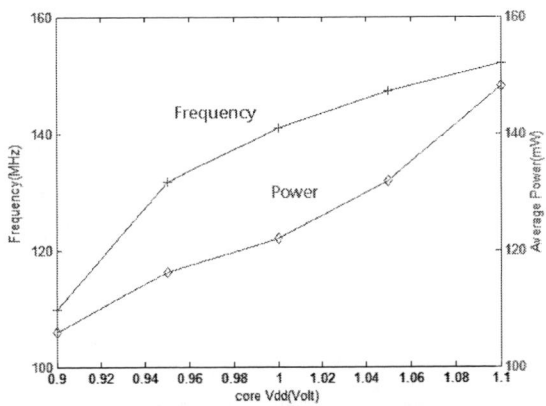

Figure 5. BER performance of the coderate-1/3 CTC decoding by using the prototyping CTC decoder chip.

Figure 7. Measured frequency and power of the proposed CTC decoder chip.

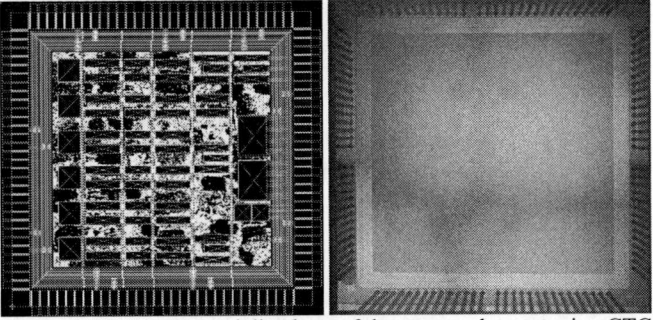

Figure 6. Chip layout and die photo of the proposed prototyping CTC decoder chip.

dual-standard CTC parallel interleaver and high-throughput warm-up free HW MAP decoding. This chip is the dual-standard CTC decoder to meet the both WiMAX and LTE data-rate requirements with a high area efficiency of 0.36 bit/mm^2 and a low energy efficiency 0.16 nJ/bit/iteration.

TABLE III. COMPARISONS OF THE CTC DECODER CHIPS.

	[3]	[10]	[6]	[11]	Proposed
Technology	0.13 μm	0.13 μm	90 nm	0.13 μm	90 nm
Core Voltage	1.2 V	1.2 V	1 V	1.2 V	1.1 V
Standard	UMTS, HSDPA	Mobile WiMAX	LTE	LTE, WiMAX	LTE, WiMAX
Max. info. bits	5144	4800	6144	6144	6144
MAP	Radix-2 SB	Radix-4 DB	Radix-2 SB	Radix-4 SB/DB	Radix-4 SB/DB
Parallel MAP #	1	1	8	8	8
Max. Frequency (MHz)	246	200	275	250	152
Core Area (mm^2)	1.20	2.24	2.10	10.7	3.38
Max. Throughput T (Mbps)	20.2	48.5	129	187.5	186.1
Power (mW) @ T	57.8	N/A	2.0	N/A	148.1
EE [1] (nJ/bit/iter.)	0.70	N/A	0.14	0.61	0.16
AE [2] (bit/mm^2)	0.06	0.11	0.22	0.11	0.36

[1] Energy efficiency (EE) = Power / (Throughput Rate × Iteration).
[2] Area efficiency (AE) = Throughput Rate / (Core Area × Frequency).

ACKNOWLEDGMENT

The authors would like to thank Chip Implementation Center (CIC) for the support of chip fabrication and measurement.

REFERENCES

[1] C. Berrou, A. Glavieux, and P. Thitimajshima, "Near Shannon limit error-correcting coding and decoding: Turbo Codes," in *Proc. Int. Conf. Commun. (ICC)*, 1993, pp. 1064-1070.

[2] C. Berrou, and M Jezequel, "Non binary convolutional codes for turbo coding," *Electron. Lett.*, vol. 35, no. 1, pp. 39-40, Jan. 1999.

[3] C. Benkser *et al.*, "A 58mW 1.22mm^2 HSDPA turbo decoder ASIC in 0.13um CMOS," in *Proc. IEEE Int. Solid-State Circuits Conf. (ISSCC)*, 2008, pp. 264-266.

[4] A. Nimbalker *et al.*, "ARP and QPP Interleavers for LTE Turbo Coding," in *Proc. IEEE Wireless Commun. and Networking Conf. (WCNC)*, 2008, pp. 1032-1037.

[5] C.-H. Lin, C.-Y. Chen, and A.-Y. Wu, "High-throughput 12-Mode CTC decoder for WiMAX standard," in *Proc. IEEE Int. Symp. VLSI Design, Automation, and Test* (VLSI-DAT), pp. 216-219, April, 2008.

[6] C.-C. Wong, Y.-Y. Lee, and H.-C. Chang, "A 188-size 2.1mm^2 reconfigurable turbo decoder chip with parallel architecture for 3GPP LTE system," in *Proc. Int. Symp. VLSI Circuits (VLSIC)*, 2009, pp. 288-289.

[7] C.-H. Lin *et al.*, "Low-power memory-reduced traceback MAP decoding for double-binary convolutional turbo decoder," *IEEE Trans. Circuits. Syst. I: Regular Paper*, vol. 56, no. 5, pp. 1005-1016, May 2009.

[8] C.-H. Lin, C.-Y. Chen, and A.-Y. Wu, "Area-efficient scalable MAP processor design for high-throughput multistandard convolutional turbo decoding," *IEEE Trans. VLSI Syst.*, vol. 19, no. 2, pp. 305-318, Feb. 2011.

[9] J.-H. Kim and I.-C. Park, "Bit-level extrinsic information exchange method for double-binary turbo codes," *IEEE Trans. Circuits. Syst. II: Exp. Briefs*, vol. 56, no. 1, pp. 81–85, Jan. 2009.

[10] J.-H. Kim and I.-C. Park, "A 50Mbps double-binary turbo decoder for WiMAX based on bit-level extrinsic information exchange," in *Proc. IEEE Asian Solid-State Circuits Conf. (A-SSCC)*, 2008, pp. 305-308.

[11] J.-H Lim and I.C. Park, "A unified parallel radix-4 turbo decoder for mobile WiMAX and 3GPP-LTE," in *Proc. IEEE Custom Integrated Circuits Conf. (CICC)*, 2009, pp. 487-490.

978-1-61284-863-1/11 $26.00 © 2011 IEEE

Comparative Design of Floating-Point Arithmetic Units Using the Balsa Synthesis System

Ren-Der Chen   Yu-Cheng Chou   Wan-Chen Liu

Department of Computer Science and Information Engineering
National Changhua University of Education
Changhua, Taiwan, R.O.C.
E-mail: rdchen@cc.ncue.edu.tw

Abstract—**In this paper, the asynchronous floating-point arithmetic units consisting of adders/subtractors and multipliers are designed and compared based on the Balsa synthesis system. For the critical mantissa multiplication in the multiplier, the modified Booth algorithm (radix 2, 4, and 8) is adopted. A pipelined design of the multiplier is also presented to increase performance. Since the Balsa language is compiled using syntax-directed translation, for the two different if statements and one case statement supported by Balsa, three different description styles have been made for each design. It can be seen from the experimental results how the style affects the area cost and simulation time of the resulting circuit. This gives us a guide to choose appropriate control statements for designing Balsa-based asynchronous circuits.**

Keywords-Balsa, asynchronous, floating-point adder/subtractor, multiplier, and modified Booth algorithm.

I. INTRODUCTION

Asynchronous circuits use a local handshaking protocol to control the transfer of data between components. To facilitate the design of large-scale asynchronous circuits, high-level synthesis tools are required [5, 8]. Balsa, an open-source system developed at the University of Manchester [1], is the name for both the synthesis system for designing asynchronous circuits and the language used to describe the circuit behavior [3]. Its compilation process, called syntax-directed translation, performs a one-to-one mapping of each language construct into a network of handshake components [10] that implement it. The SPA processor [6] is a successful design example using Balsa. In this paper, the Balsa language is used as the description language for designing floating-point (FP) adders/subtractors and multipliers. The IEEE 754 [2] is the industry standard for FP number representation. It specifies how binary FP numbers are represented and how to carry out arithmetic operations on them. Typical multiplication of two FP numbers involves several operations such as exponent addition, mantissa multiplication, normalization, and rounding. To reduce the number of partial products to be added during mantissa multiplication, the modified Booth algorithm (radix 2, 4, and 8) [7] is adopted for the multiplier design. The whole FP multiplier is also partitioned into a four-stage pipeline to increase performance.

To further investigate the syntax-directed property of the Balsa synthesis system, three designs with different description styles have been made for each adder/subtractor and multiplier.

These different styles refer to the three types of control statements supported by Balsa, i.e., the if-else, if-|-else, and case-else statements. For the behavioral description of a design unit, only one type of the control statements is used for the whole design. The research on the description styles and language constructs for more concurrent and faster circuits can be found in [4, 9]. The work in [5] also creates a prototype tool on top of the Balsa framework for behavioral synthesis of asynchronous circuits.

This paper is organized as follows. Section II gives an overview of the Balsa synthesis system. Section III explains the IEEE 754 industry standard for FP representation, and the FP addition/subtraction and multiplication. Section IV includes the experimental results and discussions, and finally Section V concludes the paper.

II. BALSA SYNTHESIS SYSTEM

The syntax-directed synthesis of asynchronous circuits is based on the compilation of a high-level description into a communicating network of predesigned modules. Each language construct is mapped one-to-one into a network of components that implement it. The structure of the resulting circuit is hence directly related to the description style, giving designers a high degree of flexibility to optimize circuits at the description language level. The Balsa synthesis system uses the syntax-directed translation to generate a handshake circuit [10] from a description written in the Balsa language. A handshake circuit is a communicating network of handshake components connected point-to-point using handshake channels. Channels carry data from one component to another under the control of a request-acknowledge handshake protocol. Each handshake component has a parameterized gate-level implementation, and a gate-level netlist can be directly generated for a handshake circuit by applying the appropriate parameters, e.g., data width, to each component instance. A large-scale handshake circuit can then be constructed by the composition of small handshake components that are implemented in isolation.

Fig. 1 shows the three different Balsa descriptions for the same behavior. Fig. 1(a) and (c) use the if-else and case-else statements commonly used in high-level languages, whereas Fig. 1(b) uses the if-|-else statement exclusively supported by Balsa. The partial handshake graph generated by Balsa for each description is given in Fig. 2. The graph is depicted only for the "run" handshake component and those relating to it.

978-1-61284-863-1/11 $26.00 © 2011 IEEE

```
begin
  a := 1 || b := 0;
  if (a=0 and b=0) then continue
  else if (a=0 and b=1) then continue
  else if (a=1 and b=0) then continue
  else continue
        end --if
      end --if
    end --if
end --begin
```

(a)

```
begin
  a := 1 || b := 0;
  if (a=0 and b=0) then continue
   | (a=0 and b=1) then continue
   | (a=1 and b=0) then continue
  else continue
  end --if
end --begin
```

(b)

```
begin
  a := 1 || b := 0;
  case ((#b @ #a) as 2 bits) of
    0b00 then continue
   | 0b01 then continue
   | 0b10 then continue
  else continue
  end --case
end --begin
```

(c)

Figure 1. Three different Balsa descriptions for the same behavior. (a) if-else, (b) if-|-else, and (c) case-else.

This run component corresponds to the "continue" statement in the original description. Even though the three different descriptions in Fig. 1(a)-(c) describe the same behavior, the Balsa system generates three different partial handshake graphs for them in Fig. 2(a)-(c), respectively, implying different area cost and simulation time. It can be seen that the graph in Fig. 2(a) uses the longest path of handshake components to control the four run components, whereas the graph in Fig. 2(c) uses the shortest path. We will see in this paper how the difference in the length of the handshake component path affects the area cost and simulation time of the resulting circuit by the various designs of FP adder/subtractor and multiplier.

III. FLOATING-POINT ADDER/SUBTRACTOR AND MULTIPLIER

A. Floating-point number

The IEEE 754 [2] is the industry standard for floating-point (FP) number representation. It specifies how binary FP numbers are represented and how to carry out arithmetic operations on them. A single-precision FP number consists of 32 bits: 1 bit for sign, 8 bits for biased exponent, and 23 bits for mantissa. Since the exponent field needs to represent both positive and negative exponents, it is represented in excess-127 code to facilitate exponent comparison for arithmetic operations. The mantissa field represents the precision of an FP number. To maximize the quantity of representable numbers, FP numbers are typically stored in a normalized

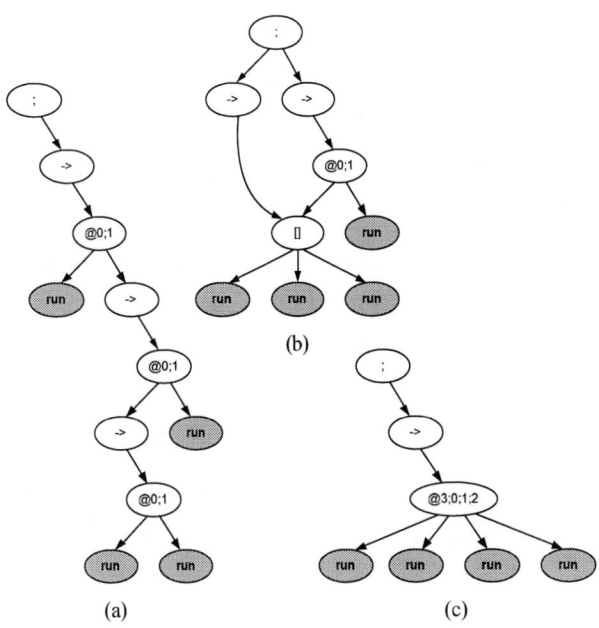

Figure 2. Partial handshake graphs for the three descriptions in Fig. 1. (a) if-else, (b) if-|-else, and (c) case-else.

form. An FP number is normalized when it is adjusted by putting the binary point after the first non-zero bit. This implied bit can be dropped upon storage to allow for increased accuracy and then retrieved when performing arithmetic operations on the mantissa. Similarly, for a double-precision FP number, 64 bits are used for its representation, consisting of 1-bit sign, 11-bit exponent with a bias value 1023, and 52-bit mantissa. In the IEEE 754 format, zero is not directly representable due to the assumption of a leading 1. It is a special value denoted with both exponent and mantissa of zero. If the exponent is all 0's, but the mantissa is nonzero, the FP number is a de-normalized number. It does not have an assumed leading 1 before the binary point. In this paper, the FP addition/subtraction and multiplication are applied only to normalized numbers or zero. De-normalized numbers are generally considered rare and may not justify the use of complicated hardware to handle them.

B. Floating-point addition/subtraction

For two FP numbers to be added or subtracted, they must have equal exponents for their mantissas to be calculated correctly. Shown in Fig. 3(a) is the design flow for FP addition/subtraction. In the zero-detection step, if any of the input operands is zero, the calculation can be avoided and the output will be the nonzero input with its sign complemented if necessary. When both inputs are zero, a zero output is generated. In the exponent-subtraction step, the exponents of the two inputs are subtracted. The difference is used as the amount of shift in the following pre-normalization step, where the mantissa of the input with smaller exponent is shifted right. In the addition/subtraction step, the effective operation on the two mantissas is performed. The implied bit '1' of the resulting mantissa is then located and dropped, and the exponent is adjusted accordingly in the post-normalization step. In the final

978-1-61284-863-1/11 $26.00 © 2011 IEEE

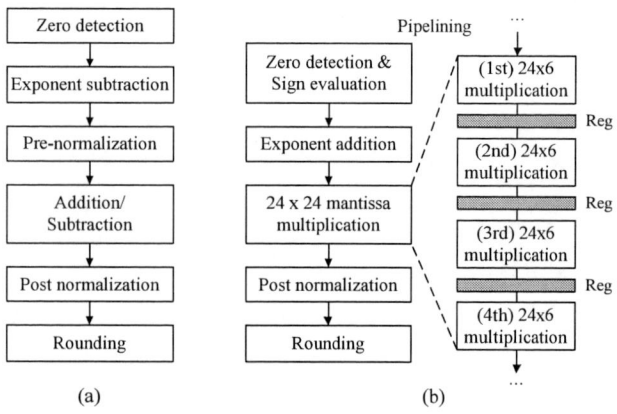

Figure 3. Design flow for the single-precision FP operation. (a) Addition /Subtraction. (b) Multiplication.

TABLE I. MODIFIED BOOTH ENCODING

Radix-2			
Multiplier bits $x_i x_{i-1}$	Multiples	Multiplier bits $x_i x_{i-1}$	Multiples
00	+0M	10	-1M
01	+1M	11	-0M
Radix-4			
Multiplier bits $x_{i+1} x_i x_{i-1}$	Multiples	Multiplier bits $x_{i+1} x_i x_{i-1}$	Multiples
000	+0M	100	-2M
001	+1M	101	-1M
010	+1M	110	-1M
011	+2M	111	-0M
Radix-8			
Multiplier bits $x_{i+2} x_{i+1} x_i x_{i-1}$	Multiples	Multiplier bits $x_{i+2} x_{i+1} x_i x_{i-1}$	Multiples
0000	+0M	1000	-4M
0001	+1M	1001	-3M
0010	+1M	1010	-3M
0011	+2M	1011	-2M
0100	+2M	1100	-2M
0101	+3M	1101	-1M
0110	+3M	1110	-1M
0111	+4M	1111	-0M

rounding step, the mantissa is rounded using the rounding-to-nearest technique, and the exponent is checked for possible overflow.

C. Floating-point multiplication

Typical multiplication of two FP numbers involves several operations such as exponent addition, mantissa multiplication, normalization, and rounding. Shown in Fig. 3(b) is the design flow for single-precision FP multiplication. In the zero-detection step, a zero output is generated when any of the input operands is zero. In the sign-evaluation step, the sign of the final product is determined by a binary XOR operation of the two input signs. In the exponent-addition step, the two exponents are added and a bias of 127 for single precision (1023 for double precision) is subtracted to determine the resultant exponent. This bias subtraction compensates for the bias that has been added in both exponents. The resultant exponent has to be checked for possible overflow or underflow. In the mantissa-multiplication step, the dropped implied bit '1' of each mantissa is retrieved, and a 24x24 binary integer multiplier is designed to calculate the resulting 48-bit mantissa. To make the multiplication process more efficient, the modified Booth algorithm (radix 2, 4, and 8) is adopted here for multiplier design. The work in the post-normalization and rounding steps is the same as that for the FP addition/subtraction.

D. Pipelining

To increase the performance of the FP multiplier, three additional registers are inserted into the mantissa multiplication to make the whole FP multiplier a four-stage pipeline, as shown in Fig. 3(b). To keep the workload of each stage balanced, the 24x24 mantissa multiplication is partitioned into four 24x6 multiplications. The first pipeline stage is composed of some pre-processing of the FP operation and the first 24x6 multiplication. The partial product generated in this stage is fed into the second stage to be added to the new partial product generated there. In the same way, after the partial product from the third stage is added to the new partial product generated in the fourth stage, the result of mantissa multiplication can be obtained. The whole FP

multiplication will then be completed after performing the post-normalization and rounding operations.

E. Modified Booth Algorithm

For the mantissa multiplication, the modified Booth algorithm with high-radix encoding [7] is adopted. It is based on encoding the multiplier operand to reduce the number of partial products to be added. The modified radix-2, 4, and 8 Booth algorithms are based on partitioning the multiplier into overlapping groups of 2, 3, and 4 bits, respectively, with one bit overlap. For radix-8 encoding, the multiplier is scanned three bits at a time, so reducing the number of partial products by a factor of three. The modified Booth encoding for these three radixes is shown in Table I. The partial product is generated through multiplying the multiplicand by a certain coefficient as indicated in the table.

IV. EXPERIMENTAL RESULTS

The asynchronous FP adders/subtractors and multipliers have been designed using the Balsa synthesis system. Classified by single or double precision, and control statement used (if-else, if-|-else, or case-else), six FP adders have been designed. Similarly, classified by single or double precision, nonpipelined or pipelined, radix number (2, 4, or 8), and control statement used, thirty-six FP multipliers have been designed. After applying 200 test patterns to each design, the relative area cost and simulation time of all the adders are given in Table II. Similarly, Table III and Table IV show the relative area cost and simulation time of all the multipliers, respectively. The area cost of each design is estimated by Balsa assuming a particular back-end implementation, and the time is obtained from the Balsa simulation system. For the convenience of comparison, the area cost of the multipliers is shown as the ratio over the smallest area cost of all the designs,

978-1-61284-863-1/11 $26.00 © 2011 IEEE

TABLE II. Relative Area Cost and Simulation Time of the Adders/Subtractors

Precision	Area cost			Simulation time		
	if-else	if-|-else	case-else	if-else	if-|-else	case-else
Single	1.63	1.63	1.64	1.14	1.13	1.04
Double	2.89	2.89	2.91	1.64	1.59	1.50

which is "29006" of the single-precision, nonpipelined, and radix-2 multiplier, described by the case-else statement (labeled with * in Table III). Similarly, the simulation time of the multipliers is shown as the ratio over the smallest time of all the designs, which is "11770600 ns" of the single-precision, pipelined, and radix-8 multiplier, also described by the case-else statement (labeled with * in Table IV).

Some interesting observations can be obtained by simple calculations of the data in Table III and Table IV. When the designs are classified into two categories based on if pipelined, the average speedup of all pipelined designs over their nonpipelined counterparts is 3.35. However, this is achieved at the expense of some additional area overhead for pipelining, and the average area cost ratio of the pipelined over the nonpipelined is 3.66. If the designs are classified into three categories by the radix adopted, the average speedup of all radix-4 and radix-8 designs over the radix-2 counterpart are 1.61 and 1.96, respectively. This is achieved by reducing the number of partial products to be added in the high-radix designs. Additional area overhead, however, cannot be avoided. Compared with the radix-2 design, the average area cost ratio of radix-4 and radix-8 designs are 1.37 and 2.37, respectively.

For each adder/subtractor and multiplier, three different designs have been made based on the three different control statements supported by Balsa. The average time ratio of the designs using the if-else and if-\|-else statements are 1.43 and 1.21, respectively, over the designs using the case-else statement. As far as the operation speed is concerned, the designs adopting the case-else statement are superior to those using the other two if statements. This result can be applied to the adder/subtractor design as well. Furthermore, this does not seem to impose any area overhead on the resulting designs. The average area cost ratio of the designs using the two if statements are 1.01 and 1.00, respectively, over the designs using the case statement.

V. Conclusions

We have designed the asynchronous FP adders/subtractors and multipliers based on the Balsa synthesis system. To increase the performance of the multiplier, the modified Booth algorithm (radix 2, 4, and 8) has been adopted for the mantissa multiplication. The multiplier has also been partitioned into a four-stage pipeline for further improvement in throughput. Since Balsa is a syntax-directed synthesis system, three versions using different control statements have been designed for each adder/subtractor and multiplier. The experimental results show how each description style affects the resulting circuit in area cost and simulation time. This allows us to make more circuit optimizations at the description language level by choosing appropriate control statements for the design.

TABLE III. Relative Area Cost of the Multipliers

Precision	Pipelining	Radix	if-else	if-|-else	case-else
Single	No	2	1.02	1.01	1.00*
		4	1.36	1.35	1.33
		8	2.26	2.24	2.23
	Yes	2	3.55	3.54	3.55
		4	4.91	4.89	4.89
		8	8.53	8.47	8.47
Double	No	2	1.90	1.89	1.89
		4	2.56	2.55	2.58
		8	4.47	4.45	4.43
	Yes	2	6.68	6.67	6.68
		4	9.48	9.46	9.46
		8	16.93	16.86	16.86

TABLE IV. Relative Simulation Time of the Multipliers

Precision	Pipelining	Radix	if-else	if-|-else	case-else
Single	No	2	7.84	7.80	6.50
		4	5.43	4.62	3.70
		8	4.98	3.44	2.77
	Yes	2	2.22	2.24	1.89
		4	1.69	1.48	1.21
		8	1.52	1.20	1.00*
Double	No	2	16.04	16.12	14.05
		4	10.49	9.17	7.62
		8	10.49	6.80	5.49
	Yes	2	4.35	4.41	3.71
		4	2.94	2.66	2.16
		8	2.91	2.15	1.77

References

[1] Balsa website, *http://apt.cs.manchester.ac.uk/projects/tools/balsa*.

[2] IEEE standard for floating-foint arithmetic. IEEE Std 754-2008, pp.1-58, Aug. 2008.

[3] A. Bardsley, "Implementing Balsa handshake circuits," Ph.D. dissertation, Department of Computer Science, University of Manchester, 2000.

[4] J. Hansen and M. Singh, "Concurrency-enhancing transformations for asynchronous behavioral specifications: a data-driven approach," in *Proc. 14th IEEE Int. Symp. Asynchronous Circuits and Systems*, 2008, pp. 15-25.

[5] S. F. Nielsen, J. Sparsø, and J. Madsen, "Behavioral synthesis of asynchronous circuits using syntax directed translation as backend," *IEEE Trans. Very Large Scale Integration (VLSI) Systems*, vol. 17, no. 2, pp. 248-261, Feb. 2009.

[6] L. A. Plana, P. A. Riocreux, W. J. Bainbridge, A. Bardsley, J. D. Garside, and S. Temple, "SPA - a synthesisable Amulet core for smartcard applications," in *Proc. 8th Int. Symp. Asynchronous Circuits and Systems*, 2002, pp. 201-210.

[7] H. Sam and A. Gupta, "A generalized multibit recoding of two's complement binary numbers and its proof with application in multiplier implementations," *IEEE Trans. Computers*, vol. 39, no. 8, pp. 1006-1015, Aug. 1990.

[8] J. Sparsø, "Current trends in high-level synthesis of asynchronous circuits," in *Proc. 16th IEEE Int. Conf. Electronics, Circuits, and Systems (ICECS)*, 2009, pp. 347-350.

[9] L. A. Tarazona, D. A. Edwards, A. Bardsley, and L. A. Plana, "Description-level optimisation of synthesisable asynchronous circuits," in *Proc. 13th Euromicro Conf. Digital System Design: Architectures, Methods and Tools (DSD)*, 2010, pp. 441-448.

[10] K. van Berkel, *Handshake Circuits - An asynchronous architecture for VLSI programming*. Cambridge University Press, 1994.

978-1-61284-863-1/11 $26.00 © 2011 IEEE

Investigating the FIFO Design Styles Based on the Balsa Synthesis System

Ren-Der Chen   Che-An Lee   Pei-Hua Hsieh

Department of Computer Science and Information Engineering
National Changhua University of Education
Changhua, Taiwan, R.O.C.
E-mail: rdchen@cc.ncue.edu.tw

Abstract—**In this paper, three asynchronous FIFO design styles, linear, square, and cubic, are investigated based on the Balsa synthesis system. These styles are designed with the key difference being the path by which data travels through the FIFO. The design with shorter path should result in lower latency and higher throughput, but will require more complicated control. All the FIFOs are designed using the Balsa language, and the area cost and simulation time are compared for each FIFO with varying sizes. A tool is also presented for automatic generation of Balsa code for each FIFO.**

Keywords-Balsa, asynchronous, and FIFO.

I. INTRODUCTION

Asynchronous circuits use a local handshaking protocol to control the transfer of data between components. To facilitate the design of large-scale asynchronous circuits, high-level synthesis tools are required [7, 9]. Balsa, an open-source system developed at the University of Manchester [1], is the name for both the synthesis system for designing asynchronous circuits and the language used to describe the circuit behavior [2, 4]. Its compilation process, called syntax-directed translation, performs a one-to-one mapping of each language construct into a network of handshake components [10] that implement it. The SPA processor [8] is a successful design example using Balsa.

The FIFO is an implementation of a first-in-first-out data structure. Traditionally, the linear FIFO is implemented as a linear array of cells. The work on reducing forward latency in an asynchronous linear FIFO can be found in [6]. The square FIFO [3, 5] is a low-latency, high-throughput implementation consisting of a two-dimensional array of cells. Similarly, the cubic FIFO is a three-dimensional array of cells, featuring even lower latency and higher throughput. In the design of square or cubic FIFO, the data items are distributed over the FIFO cells and must be collected in such a way that the first-in-first-out behavior is satisfied. In this paper, the Balsa synthesis system is used as the framework for designing the three styles of FIFOs, and an automatic tool is also presented to generate the Balsa code for each FIFO with varying sizes.

This paper is organized as follows. Section II gives an overview of the Balsa synthesis system. Section III explains the design of a cubic FIFO. Section IV includes the experimental results and discussions, and finally Section V concludes the paper.

This work was supported by the National Science Council of Taiwan, R.O.C. under the project NSC 99-2221-E-018-027.

II. BALSA SYNTHESIS SYSTEM

The syntax-directed synthesis of asynchronous circuits is based on the compilation of a high-level description into a communicating network of predesigned modules. Each language construct is mapped one-to-one into a corresponding network of components that implement it. The structure of the resulting circuit is hence directly related to the description style, giving designers a high degree of flexibility to optimize circuits at the description language level. The Balsa synthesis system uses the syntax-directed translation to generate a handshake circuit [10] from a description written in the Balsa language. A handshake circuit is a communicating network of handshake components connected point-to-point using handshake channels. Channels carry data from one component to another under the control of a request-acknowledge handshake protocol. Each handshake component has a parameterized gate-level implementation, and a gate-level netlist can be directly generated for a handshake circuit by applying the appropriate parameters, e.g., data width, to each component instance. A large-scale handshake circuit can then be constructed by the composition of small handshake components that are implemented in isolation.

Fig. 1 illustrates the Balsa description for a simple one-place buffer. The buffer has an input channel *in1* and an output channel *out1*. The variable *buf* stores the input data and the operation consists of an infinite repetition (loop..end) of two actions: input data from channel *in1* into *buf* (*in1 -> buf*), sequenced (;) with the second action, output the data in *buf* to channel *out1* (*out1 <- buf*). The handshake circuit generated by Balsa for the description is shown in Fig. 2, where the loop statement in Fig. 1 is implemented by the handshake component labeled with "*".

```
import [balsa.types.basic]
type DataType is 8 bits
procedure buffer
(
        input    in1  : DataType;
        output   out1 : DataType
) is
        variable buf  : DataType
begin
        loop
                in1->buf
                ;
                out1<-buf
        end
end
```

Figure 1. Balsa description for a one-place buffer.

978-1-61284-863-1/11 $26.00 © 2011 IEEE

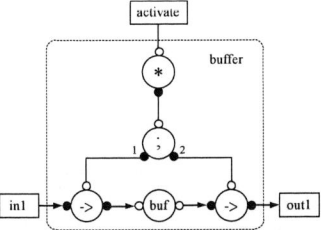

Figure 2. Handshake circuit for the Balsa description of a one-place buffer.

III. CUBIC FIFO DESIGN

The FIFO is an implementation of a first-in-first-out data structure. In this paper, the asynchronous FIFOs with three different structures, linear, square and cubic, are designed and compared using the Balsa synthesis system.

A. Cubic FIFO

The cubic FIFO adopted here is a general cube structure with different length (L), width (W), and height (H), which refer to the number of cells on each edge of the cube. The discussions of linear and square FIFOs are omitted here since the square FIFO is a special case of a cubic FIFO with height equal to one, and the linear FIFO is also a special case of a square FIFO with length equal to one. For the convenience of explanation, each cell in the cubic FIFO is represented as a point (i,j,k) in the three-dimensional coordinate system, where i, j, and k are zeros or positive integers less than L, W, and H, respectively. A cell c with coordinates (i,j,k), denoted as $c(i,j,k)$, is interpreted as being located in row i, column j, and layer k of the cube. Its three adjacent cells $c(i-1,j,k)$, $c(i,j-1,k)$, and $c(i,j,k+1)$ are called its up-row cell, left-column cell, and up-layer cell, respectively. Cell $c(0,0,H-1)$ is the first cell for data items to enter the FIFO, and these data items leave the FIFO from the last cell $c(L-1,W-1,0)$. Besides, a virtual input cell is defined as a cell that writes data to the first cell of the FIFO, and a virtual output cell is a cell that reads data from the FIFO's last cell.

Fig. 3 illustrates the structure of a 4x4x4 cubic FIFO with 64 cells. This FIFO consists of a top layer of cells, a number

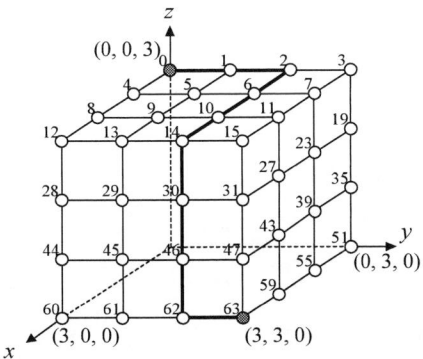

Figure 3. A 4x4x4 cubic FIFO.

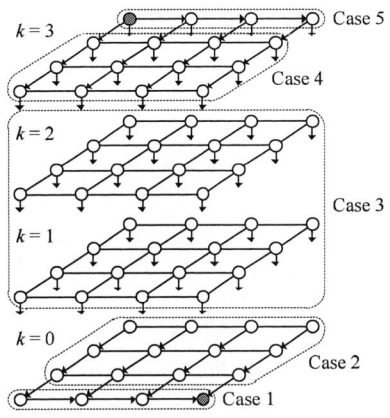

Figure 4. Classification of cells for a 4x4x4 cubic FIFO.

of vertical FIFOs between the layers, and a bottom layer of cells. Both the top and bottom layers are similar to square FIFOs, consisting of a top row of cells, a number of the middle rows of cells, and a bottom row of cells. The vertical FIFOs are simple linear FIFOs for transferring data items between layers. In the top layer, the data items entering the FIFO are distributed to each row of cells. They are then dropped down to the bottom layer through the vertical FIFOs. In the bottom layer, the bottom row collects the data items from each row of cells, and then sends them out in the same order they entered the FIFO. Each data item follows one of several possible paths to travel through the FIFO. In Fig. 3, the path composed of bold edges shows a possible path for data items to go through the FIFO from the first cell $c(0,0,3)$ to the last cell $c(3,3,0)$. It can be seen that only 10 cells in the path are visited by the data items, whereas in a linear or 8x8 square FIFO also with 64 cells, the number of cells to be visited are 64 and 15, respectively.

B. Classification of cells

For a linear FIFO, a cell gets data only from its left-column cell. For a square FIFO, a cell may get data from its up-row cell or left-column cell. For a cubic FIFO, however, a cell may also get data from its up-layer cell, besides its up-row and left-column cells. To preserve the order of data items entering the FIFO, each cell has its specific source cells to get data from. In our method, all the cells of a cubic FIFO are classified into five cases by the type of source cells they can get data from.

The cells in the bottom row of the bottom layer are classified as Case 1. Each cell in this case may get data from its up-layer cell, up-row cell, or left-column cell. The last cell of the FIFO belongs to this case. It is the cell for the virtual output cell in the outside environment to get data from. The cells in the other rows of the bottom layer are classified as Case 2. Each cell in this case may get data from its up-layer cell or up-row cell. The cells that get data only from their up-layer cells are classified as Case 3, existing in the middle layers of a cube. The cells in the middle and bottom rows of the top layer are classified as Case 4. Each cell in this case gets data from its up-row cell only. The remaining cells in the top row of the top layer get data only from their left-column

978-1-61284-863-1/11 $26.00 © 2011 IEEE   177

TABLE I. EVENT SEQUENCE FOR EACH CLASSIFICATION OF CELLS

Case	Event sequence
Case 1: $i = L\text{-}1$, $j = 0..(W\text{-}1)$, $k = 0$	$\overbrace{\alpha, \gamma, \gamma, ..., \gamma}^{j}, \overbrace{\delta, \delta, ..., \delta}^{L\text{-}1}$, where $\delta = \beta, \overbrace{\gamma, \gamma, ..., \gamma}^{j}$
Case 2: $i = 0..(L\text{-}2)$, $j = 0..(W\text{-}1)$, $k = 0$	$\alpha, \overbrace{\beta, \beta, ..., \beta}^{i}$
Case 3: $i = 0..(L\text{-}1)$, $j = 0..(W\text{-}1)$, $k = 0..(H\text{-}2)$	α
Case 4: $i = 1..(L\text{-}1)$, $j = 0..(W\text{-}1)$, $k = H\text{-}1$	β
Case 5: $i = 0$, $j = 0..(W\text{-}1)$, $k = H\text{-}1$	γ

cells. These cells are classified as Case 5. The first cell of the FIFO belongs to this case, and since it has no left-column cell, it gets data from a virtual input cell in the outside environment. Fig. 4 illustrates the classification of cells for the cubic FIFO in Fig. 3.

C. Event sequence

Since a cell in a cubic FIFO may get data from several source cells, some events about data movement need to be defined. For a cell $c(i,j,k)$, the event α is defined as getting data from its up-layer cell $c(i,j,k+1)$ and setting $c(i,j,k+1)$ empty, the event β is defined as getting data from its up-row cell $c(i-1,j,k)$ and setting $c(i-1,j,k)$ empty, and the event γ is defined as getting data from its left-column cell $c(i,j-1,k)$ and setting $c(i,j-1,k)$ empty. Since a cell does not always contain data, each of these three events is activated only when its corresponding source cell contains data; otherwise, the event is still waiting for activation. The event sequence of a cell $c(i,j,k)$, denoted as $seq(i,j,k)$, is defined as a finite ordered list of events to be activated on $c(i,j,k)$. The activation of an event sequence $seq(i,j,k)$ starts from the first event in $seq(i,j,k)$. When an event is activated, the event following it will be the next event waiting to be activated, and when the last event is activated, the next event for activation will be restarted from the first event in $seq(i,j,k)$. The operation of a cubic FIFO can then be modeled as repeatedly checking the empties of each cell $c(i,j,k)$ in the FIFO and then activating a new event in $seq(i,j,k)$ if $c(i,j,k)$ is empty.

In our method, to preserve the order of data items entering the FIFO, each cell $c(i,j,k)$ is associated with an appropriate event sequence. Table I shows the event sequence for each case of cells. In Case 1, each cell gets data first from its up-layer cell once and then from its left-column cell for j times. After that, the following action is repeated for $(L\text{-}1)$ times: getting data from its up-row cell once and then from its left-column cell for j times. For the leftmost cell $c(L\text{-}1,0,0)$, which has no left-column cell, it gets data from its up-layer cell once and then from its up-row cell for $(L\text{-}1)$ times. For example, the event sequence of cell $c(3,2,0)$ in Fig. 3 is (α, γ, γ, β, γ, γ, β, γ, γ, β, γ, γ), and for the leftmost cell $c(3,0,0)$, it is (α, β, β, β). In Case 2, each cell gets data from its up-layer cell once and then from its up-row cell for i times. For the cells in row 0, which have no up-row cells, they get data from their up-layer cells only. For example, the event sequences of cells $c(2,2,0)$ and $c(0,2,0)$ in Fig. 3 are (α, β, β) and (α), respectively. In Case 3,

Case 4, and Case 5, each cell gets data from its up-layer cell, up-row cell, and left-column cell, respectively. However, the left-column cell of $c(0,0,3)$, the first cell of the FIFO, is the virtual input cell that provides the FIFO with new data items.

D. Algorithm

Since Balsa is a control-driven design framework, either the read or write operation of the FIFO must be activated by the outside environment. If it is a read operation, the data in the last cell is read out, and then the data items in the other cells are shifted between cells. If it is a write operation, the shift operation is performed, and then the new data is written into the first cell of the FIFO. For each cell $c(i,j,k)$, this shift operation is performed by getting data from an appropriate cell. The source cell for $c(i,j,k)$ to get data from must be carefully arranged so that the data items are read out in the same order they were written into the FIFO.

The algorithm for the cubic FIFO design is given in Fig. 5. The FIFO behavior is enclosed in an endless loop for receiving any read or write request from the outside environment. When the FIFO is activated, the read event of the virtual output cell is first activated to get data from the last cell of the FIFO. Then the data shift operation is performed on each cell of the FIFO by activating an event in the event sequence of the cell. This activation process starts from the bottom layer to the top layer. In each layer, it starts from the bottom row to the top row, and in each row, it starts from the right column to the left column. In Fig. 3, for example, the number associated with each cell shows the activation order (in decreasing order) of the FIFO. The activation process starts from number 63 to 0, and then restarts again. Since the data items enter the FIFO in the top layer, the events of the cells in layer $H\text{-}2$ will be activated only when n_{rw} is a multiple of the area ($L*W$) of the layer, where n_{rw} is counted as the number of read or written operations performed on the FIFO. This is to ensure that the data items will be correctly distributed to the top layer before going down to the next layer. Similarly in the top layer, since the data items enter the FIFO in the top row, the events of the cells in the second row can be activated only when n_{rw} is a multiple of the width (W) of the layer. This is also to ensure that the data items can be

```
while ( 1 ) {
    if ( read_flag == 1 )
        Activate the read event of the virtual output cell;
    for ( k=0; k <= H-1; k++ ) {
        // n₁ and n₂ are arbitrary positive integers.
        if ( k == H-2  &&  n_rw != n₁ * (L*W) )
            continue;
        for ( i = L-1; i >= 0; i-- ) {
            if ( k == H-1  &&  i == 1  &&  n_rw != n₂* W )
                continue;
            for ( j = W-1; j >= 0; j-- ) {
                if ( c(i,j,k) is empty )
                    Activate an event in seq(i,j,k);
            }
        }
    }
}
```

Figure 5. The cubic FIFO design algorithm.

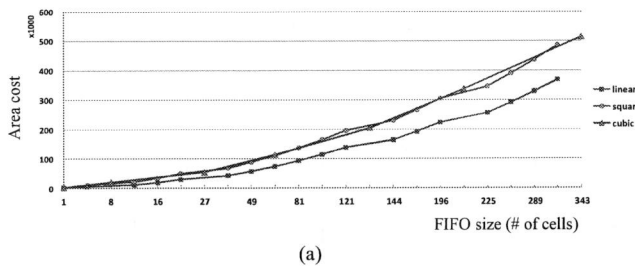

(a)

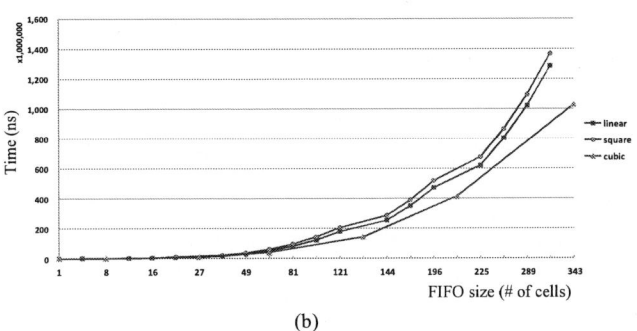

(b)

Figure 6. (a) Comparison of area cost. (b) Comparison of simulation time.

correctly distributed to the top row before going down to the next row in the top layer.

IV. EXPERIMENTAL RESULTS

The three FIFO structures, linear, square, and cubic, have been designed using the Balsa synthesis system, and then tested with varying FIFO sizes. For the linear and square FIFOs, the size varies from 1^2, 2^2, to 18^2, and for the cubic FIFO, it varies from 1^3, 2^3, to 7^3. The data width of the data items entering the FIFO is assumed to be eight bits. Each FIFO is simulated with data items entering it consecutively until it is full and then leaving the FIFO also consecutively until it is empty again. The area cost of each design is estimated by Balsa assuming a particular back-end implementation, and the time for data transferring is also obtained from the Balsa simulation system. For the three FIFO design styles with varying sizes, Fig. 6(a) shows the comparison of area cost. It can be observed that the cost of the linear FIFO is the lowest, whereas there is no big difference in the cost of square and cubic FIFOs. This is reasonable since these two designs are intentionally designed to use more control to shorten the path each data items uses to travel through the FIFO. The comparison of simulation time is given in Fig. 6(b). It is expectable that the cubic FIFO is the fastest for each FIFO size under test. But for the square FIFO, it requires more transfer time than the linear FIFO. This contradicts our original belief that a square FIFO, with shorter path visited by data items, should have a lower latency than a linear FIFO. However, this can be explained by the fact that since Balsa is a control-driven design framework, the delay caused by the additional control overhead here is larger than the time saved by a shorter data transfer path.

To facilitate the design of FIFOs, a tool has been developed to automatically generate the Balsa code for each

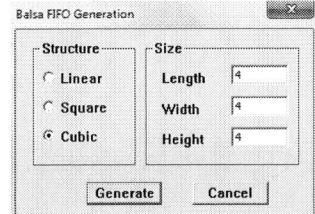

Figure 7. Screenshot of our Balsa FIFO generation tool.

FIFO with varying sizes. The screenshot of the tool is given in Fig. 7. After choosing the FIFO structure, linear, square, or cubic, a synthesizable Balsa code will be generated for a FIFO with the size specified by length, width, and height. To make the design more general, the FIFO may have different values in length, width, and height. However, for two FIFOs of the same structure and with the same size, their resulting circuits may have different area cost and simulation time if they have different length, width, or height.

V. CONCLUSIONS

In this paper, three asynchronous FIFOs, linear, square and cubic, have been designed using the Balsa synthesis system. The resulting circuits have been compared for varying FIFO sizes. It can be seen that the cubic FIFO is superior to the other two FIFOs in transfer time, due to its shorter data path for data items to travel through the FIFO. However, this is achieved at the expense of more complicated control, and hence higher area cost. A tool for automatic generation of each FIFO in Balsa code has also been presented.

REFERENCES

[1] Balsa website, *http://apt.cs.manchester.ac.uk/projects/tools/balsa.*

[2] A. Bardsley, "Implementing Balsa handshake circuits," Ph.D. dissertation, Department of Computer Science, University of Manchester, 2000.

[3] J. Ebergen, "Squaring the FIFO in GasP," in *Proc. 7th Int. Symp. Asynchronous Circuits and Systems*, 2001, pp. 194-205.

[4] D. Edwards and A. Bardsley, "Balsa: an asynchronous hardware synthesis language," *The Computer Journal*, vol. 45, no. 1, pp. 12-18, 2002.

[5] X. Kong and R. Negulescu, "Bolstering faith in GasP circuits through formal verification," in *Proc. 10th IEEE Int. Symp. Asynchronous Circuits and Systems*, 2004, pp. 113-124.

[6] J.-G. Lee, S.-J. Kim, J.-A. Lee, and K. Kim, "A low latency asynchronous FIFO combining a wave pipeline with a handshake scheme," *IEICE Trans. Fundamentals*, vol. E88-A, no. 4, pp. 1031-1037, Apr. 2005.

[7] S. F. Nielsen, J. Sparsø, and J. Madsen, "Behavioral synthesis of asynchronous circuits using syntax directed translation as backend," *IEEE Trans. Very Large Scale Integration (VLSI) Systems*, vol. 17, no. 2, pp. 248-261, Feb. 2009.

[8] L. A. Plana, P. A. Riocreux, W. J. Bainbridge, A. Bardsley, J. D. Garside, and S. Temple, "SPA - a synthesisable Amulet core for smartcard applications," in *Proc. 8th Int. Symp. Asynchronous Circuits and Systems*, 2002, pp. 201-210.

[9] J. Sparsø, "Current trends in high-level synthesis of asynchronous circuits," in *Proc. 16th IEEE Int. Conf. Electronics, Circuits, and Systems (ICECS)*, 2009, pp. 347-350.

[10] K. van Berkel, *Handshake Circuits - An asynchronous architecture for VLSI programming.* Cambridge University Press, 1994.

978-1-61284-863-1/11 $26.00 © 2011 IEEE

Digital System for Low Power Wireless Neural Recording System

Peng Li, Xin Liu, Bin Zhao and Minkyu Je

Institute of Microelectronics, A*STAR (Agency for Science, Technology and Research)
11 Science Park Road, Singapore Science Park II, Singapore 117685
Email: lip@ime.a-star.edu.sg

Abstract— **In this paper, a digital system is proposed for 100-channel low power wireless neural recording system which consists of 100 amplifiers, 10 multiplexers, 10 analog-to-digital converters (ADC), digital baseband and RF transceiver for data link and command link. The proposed system is designed in a 0.18 μm CMOS process and the total power of the digital system of two chips is approximate 800 μW with supplying by 1.8V voltage.**

Keywords-transceiver; hamming code; cross-correlation; synchronization;

I. INTRODUCTION

Recently, the research of multi-channel neural recording system is more and more popular. The multi-channel neural recording systems are used in neuroscience experiments to study complex neural networks of freely behaving animals and it is also a critical component in brain-computer interface used for cortical-controlled neural prosthetics, which has a wide range of applications such as upper and lower limb prostheses, bladder and bowel movement control for spinal cord injury patients [1]. Some multi-channel neural recording systems have been developed [1], [2]. A neural recoding system should meet challenging requirements imposed by the environment, such as recoding a large number of channels, wireless transmission, on-the-fly processing, programmable specifications and power consumption and chip area [1]. To meet these challenging requirements, we propose a 100-channel low power wireless neural recording system.

The digital system is an important part in this 100-channel low power wireless neural recording system. It consists of probe interface control module and baseband processing module. This paper covers majorly the digital system and it is organized as follows. Section II introduces the system architecture of our proposed 100-channel low power wireless neural recoding system. Section III focuses on the digital system architecture and some critical blocks. Section IV discusses the simulation results and the conclusion is summarized in section V.

II. SYSTEM ARCHITECTURE

The proposed system consists of three chips: neural probe interface IC (IC1), communication IC for data transmission and command receiving (IC2) and communication IC for data receiving and command transmission and interface to central processing unit (IC3).The block diagram of whole system is shown in Figure 1.

IC1 contains a 10 x 10 neural probe array which is planted in brain to monitor neuron's signals and convert the amplified analog signal to the digital signal. Then the digitalized signal is processed by digital system in IC2 which is placed between skull and skin. The main function of IC2 is baseband processing and RF modules to enable to communication to IC3. IC2 is connected to IC1 by flexible cables. This wire communication is reliable and allows the IC2 to process the neural data in a real-time. The wireless communication is between IC2 and IC3 in order to avoid the significant restrictions which are imposed by wires. There are two links for data and commands, respectively. The data link has high rate for data throughput and the command link has low rate for command communication where the low power RF can be applied.

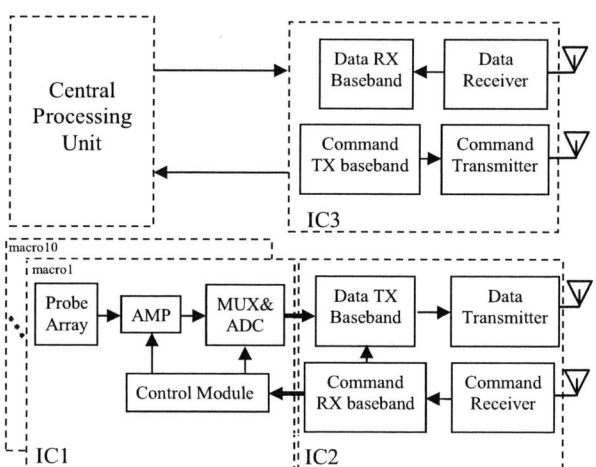

Figure 1. Block diagram of 100-channel wireless neural recording system

III. DIGITAL SYSTEM ARCHITECTURE

A. Control Module

In order to improve the performance of neural devices, a large number of channels should be recorded. In our design, we adopt a 10 x 10 neural probe array. Thus, there are 10 macros in IC1. Within every macro, there are 10 probes, 10 amplifiers,

1 multiplexer, 1 ADC and 1 control module. The block diagram of IC1 is shown in Figure 2.

To accommodate the requirements of various different applications, the specifications, such as the selected channels, the gains and bandwidths of amplifiers should be programmed by control module through command link. We can choose any 1, 2, 5 or 10 channels in one macro. To ease the baseband data process smoothly, channel selection setting should be the same for all 10 macros. The channels which are not used will be disabled to reduce the power consumption.

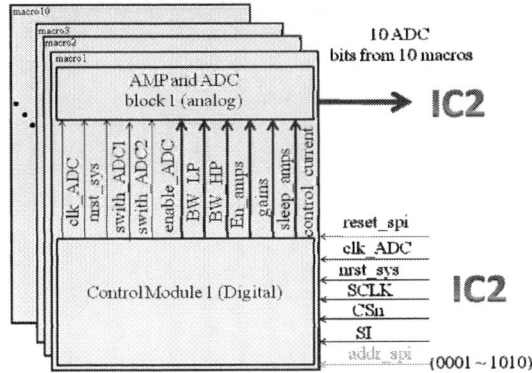

Figure 2. Block diagram of IC1

B. Data TX Baseband

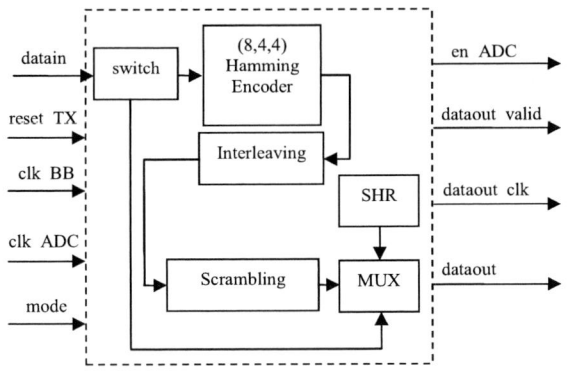

Figure 3. Block diagram of data TX baseband

In data TX baseband block which is shown in Figure 3, we allow the user to disable the channel coding to save power further in high signal-to-noise-ratio (SNR) environment. For balance between performance and complexity in channel coding mode, we adopt the (8, 4, 4) Hamming code because it is relative simple but still have 1-error-correcting and 2-error-detecting capability. The (8, 4, 4) Hamming code is extended from the (7, 4, 3) Hamming code by adding an overall parity check [3], [4]. The minimum distance of (8, 4, 4) Hamming code is then increased to 4. The generator matrix of the (8, 4, 4) Hamming code is defined as

$$ G = \begin{bmatrix} 1 & 0 & 0 & 0 & 1 & 0 & 1 & 1 \\ 0 & 1 & 0 & 0 & 1 & 1 & 1 & 0 \\ 0 & 0 & 1 & 0 & 1 & 1 & 0 & 1 \\ 0 & 0 & 0 & 1 & 0 & 1 & 1 & 1 \end{bmatrix}. $$

From the nature of the (8, 4, 4) Hamming coder, it can be implemented by using a 4-bit shift register and 4 modulo-2 adders tied to the appropriate stages of the shift register. In our design, the hardware implementation of (8, 4, 4) Hamming coder can be illustrated in Figure 4.

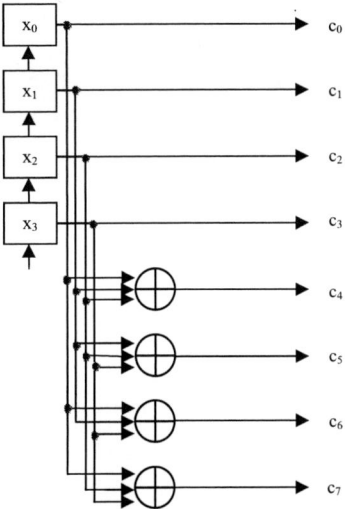

Figure 4. A linear shift register of (8, 4, 4) Hamming coder

Due to the fact that the (8, 4, 4) Hamming encoder can only correct one error and errors normally occur in bursts, it is very necessary to add interleaving block followed by (8, 4, 4) Hamming encoder to improve the performance of Hamming coding. Interleaving can create a more uniform distribution of errors by reordering the coded data. This will be help to spread continuous errors and make full use of Hamming coding. In our design, we use 4x8 matrix interleaver. The interleaving is achieved by filling a matrix with the input row by row and then sending the matrix contents to the output port column by column. The 4x8 matrix interleaver is illustrated in Figure 5. The (8, 4, 4) Hamming encoder combined with 4x8 matrix interleaver can correct up to four continuous burst errors at most. The performance of packet error rate is then greatly improved.

According to the Shannon's theorem, random code is good for transmission. Scrambling can generate pseudo-noise (PN) sequences which are deterministically generated but almost like random sequences to an observer. In our design, we use linear feedback shift register (LFSR) to generate PN sequences. The block diagram of LFSR is shown in Figure 6. The initial

978-1-61284-863-1/11 $26.00 © 2011 IEEE

state of the shift register is 1000. It is made up of four flip-flops and one modulo-2 adder [3].

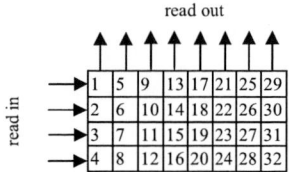

Figure 5. 4x8 matrix interleaver

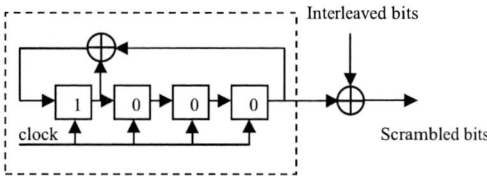

Figure 6. Block diagram of LFSR

Then, the final output which is either scrambled output or the original bitstream is multiplexed. The synchronization header (SHR) will be added to each frame and then go to the RF module.

C. Data RX baseband

The block diagram of data RX baseband is shown in Figure 7.

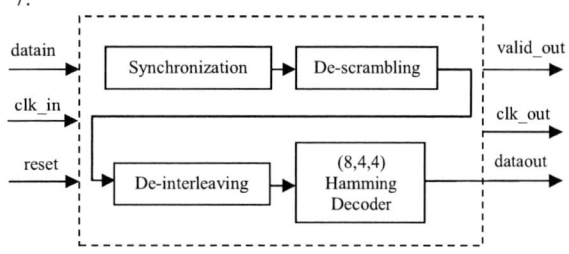

Figure 7. Block diagram of data RX baseband

In data receiving process, timing recovery is one of the most critical functions. In our system, the clock is recovered in the RF receiver block. In other words, symbol synchronization is addressed by RF. For digital RX baseband, we need to handle the frame synchronization. We use the SHR which is proposed by [5]. The SHR consists of four 32 bits preamble sequence [11011001110000110101001000101110] and a 64 bits start-of-frame delimiter (SFD) [00101110110110 011100001101010010110000110101001000101011101101100].

We employ the property of correlation to detect the synchronization header in the receiver. The discrete cross-correlation [6] is defined as

$$p(m) = \sum_{k=0}^{n-1} r(m+k) \cdot w(k). \tag{1}$$

where $r(\cdot)$ is received bitstream, $w(\cdot)$ is SHR sequence and n is the length of window. Here n is 32 for preamble sequence or 64 for SDF. As well known, auto-correlation will always generate a peak correlation value. The cross-correlation should be computed using formula (1) for every clock. Here, we set the preamble threshold T_1=27 and the SFD threshold T_2=51. If the value of computed the cross-correlation between received bitstream and preamble sequence is above the threshold T_1, one preamble is detected. If four continuous peaks (>=27) of preamble cross-correlation are detected, the preamble sequence is confirmed. Then the SFD checking will be enabled. If the cross-correlation value is above the SFD threshold T_2, the SFD is detected and data frame start bit is also determined.

De-scrambling uses the same LFSR as scrambling. The initial state of the shift register is also 1000. De-interleaving restores the data in similar way as interleaving, but it is read out row-wise instead.

The check matrix of (8, 4, 4) Hamming code is

$$H = \begin{bmatrix} 1 & 1 & 1 & 0 & 1 & 0 & 0 & 0 \\ 0 & 1 & 1 & 1 & 0 & 1 & 0 & 0 \\ 1 & 1 & 0 & 1 & 0 & 0 & 1 & 0 \\ 1 & 1 & 1 & 1 & 1 & 1 & 1 & 1 \end{bmatrix}.$$

If R=(r_0, r_1, r_2, r_3, r_4, r_5, r_6, r_7) represents received Hamming code word, syndrome S is defined as

$$S = HR^T. \tag{2}$$

If the values of syndrome are all zeros, then R is the same as the code word which was sent; otherwise the distortion is occurred in the transmission. If one error is occurred, the syndrome will match to one column of H. It represents the corresponding bit is wrong and reversed that bit, the R is then corrected. In our design, the hardware implementation of (8, 4, 4) Hamming decoder can be illustrated in Figure 8.

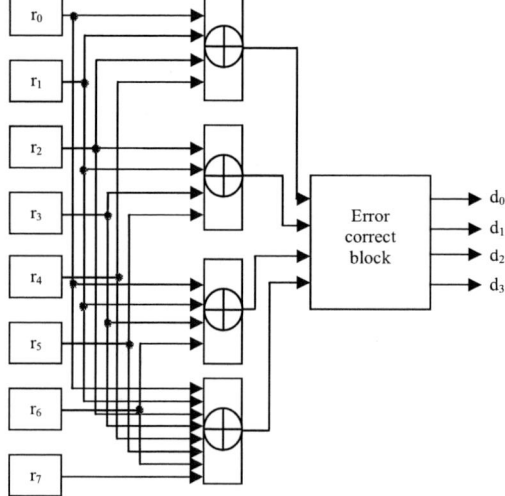

Figure 8. A linear shift register of (8, 4, 4) Hamming decoder

978-1-61284-863-1/11 $26.00 © 2011 IEEE

IV. SIMULATION RESULTS

We use Matlab for system level performance and optimization. Then we convert the Matlab design into Verilog-HDL. The standard digital design flow is then applied, which includes RTL coding, synthesis, place and route and verification.

The simulation results of data TX baseband block are shown in Figure 9 and Figure 10. Figure 9 is for using channel coding and Figure 10 shows a result of bypass mode (In this mode, the channel coding is disabled).

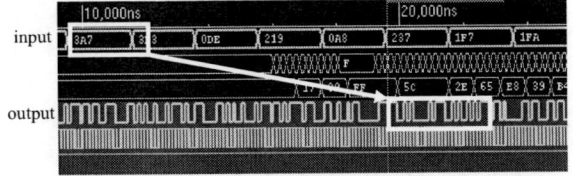

Figure 9. The results of data TX baseband in channel coding mode

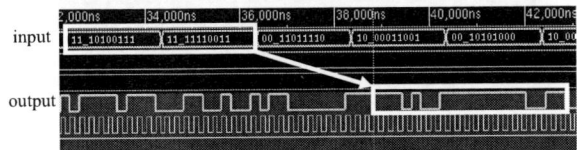

Figure 10. The results of data TX baseband in bypass mode

To simulate the real environment, we add 80 uniform distributed bits before SHR and white Gaussian noise to the outputs of ADC at SNR=3dB. The Matlab and RTL correlation curves for detecting the SHR are shown in Figure 11 and Figure 12.

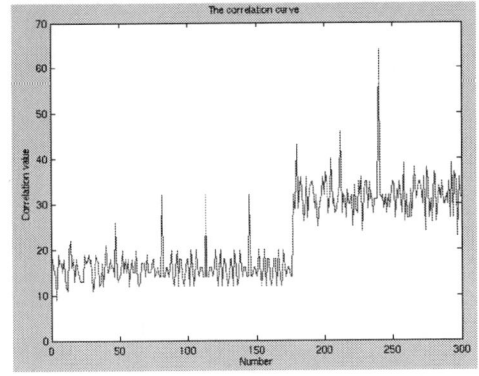

Figure 11. The Matlab correlation curve

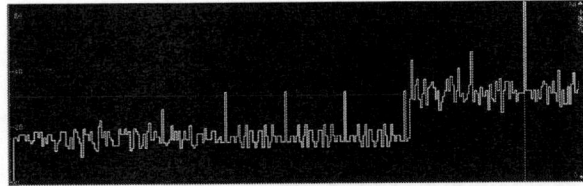

Figure 12. The RTL correlation curve

To verify our performance, we compare the original bitstream and the decoded bitstream. The result tells us that our decoder can correct errors although we added white Gaussian noise at SNR=3dB in the channel.

V. CONCLUSION AND FUTURE WORK

In this paper, we have presented a digital system for our 100-channel low power wireless neural recoding system. It is implemented in 0.18-μm CMOS technology. The core area of digital IC1 is about 230-μm by 230-μm and the core area of digital IC2 is about 250-μm by 900-μm. The test chip of digital IC1 and IC2 are shown in Figure 13. The IC3 is under development now.

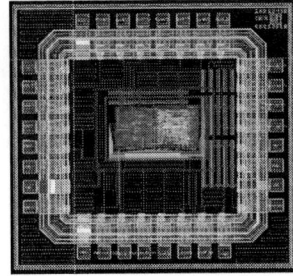

Figure 13. The test chips of digital IC1 (left) and IC2 (right)

ACKNOWLEDGMENT

The authors would like to thank the team members of neurodevices project and ICS support team. This work was supported by the Science and Engineering Research Council of A*STAR (Agency for Science, Technology and Research), Singapore under Neurodevices program. The grant number for the project is 102-171-0163.

REFERENCES

[1] Moo Sung Chae, Zhi Yang, Mehmet R. Yuce, Linh Hoang, and Wentai Liu, "A 128-Channel 6 mW Wireless Neural Recording IC With Spike Feature Extraction and UWB Transmitter," IEEE Trans. Neural Syst. Rehabil. Eng., vol. 17, no. 4, pp.312-321. , August 2009.

[2] Reid R. Harrison, Paul T. Watkins, Ryan J. Kier, Robert O. Lovejoy, Daniel J.Black, Bradley Greger and Florian Solzbacher, "A Low-Power Integrated Circuit for a Wireless 100-Electrode Neural Recoding System," IEEE J. Solid-State Circuits, vol. 42, no. 1, pp. 123-133., January 2007.

[3] John G. Proakis, Digital Communications, fourth Edition, McGraw-Hill, 2001.

[4] Simon Haykin, Communication Systems, 4th Edition, John Wiley & Sons, Inc., 2001.

[5] Xin Liu, Yuanjin Zheng, Bin Zhao, Yisheng Wang and Myint Wai Phyu, "An Ultra Low Power Baseband Transceiver IC for Wireless Body Area Network in 0.18μm CMOS Technology," IEEE TVLSI Systems, unpublished.

[6] Olav E. Liseth, Daniel Mo, Hakon A. Hjortland, Tor Sverre "Bassen" Lande, and Dag T. Wisland, "Power-Efficient Cross-Correlation Beat Detection in Electrocardiogram Analysis Using Bitstreams," IEEE Trans.Biomedical circuits and systems, vol. 4, no. 6, December 2010.

978-1-61284-863-1/11 $26.00 © 2011 IEEE

A Wideband 0.6dB Insertion Loss +20.5dBm P1dB CMOS T/R Switch

Xuesong Chen and M.Kumarasamy Raja

Institute of Microelectronics, A*STAR (Agency for Science, Technology and Research)
11 Science Park Road, Singapore Science Park II, Singapore 117685
E-mail: chenxs@ime.a-star.edu.sg

Abstract—**This paper provides a detailed analysis of performance limitations in RF CMOS T/R switches, and proposes a novel method to improve the insertion loss and power handling capability by dynamically biasing the body and the gate of the switch transistors to make the device operating in linear region under large signals. The concept is implemented with circuits only and no inductor is used. A prototype was designed and implemented in standard 0.18-μm CMOS technology. The experimental results show 0.6dB insertion loss at 2.4GHz, 20.5dBm P1dB for the transmitter path of the T/R Switch with a single 1.8V control voltage. The die area excluding IO pads is 0.14mm by 0.12mm. The performance meets the requirements of most mobile wireless communication systems.**

Keywords-T/R switch; CMOS ; low insertion loss; wideband

I. INTRODUCTION

T/R switch is employed in various wireless communication systems for time multiplexing between transmitter and receiver. With this topology, transmitter and receiver are powered on alternatively which saves the overall system power consumption and reduces the interference between the transmitter and the receiver. It has been adopted in most of the mobile wireless communication systems such as wireless LAN, Bluetooth, cellular PCS/CDMA/WCDMA/TDMA/GSM, and so on. A high quality RF switch in wireless communication systems is required to have low insertion loss, high power handling capability and high isolation. Low insertion loss minimizes the receiver sensitivity degradation and transmitter's output power loss induced by the switch. Power handling capability of a RF switch is specified by the transmitter's maximum output power in a system. High isolation is to minimize the interference from the switch-on path to the switch-off path and vice versa. For mobile wireless communication devices, insertion loss and power handling are the most challenging requirements for a CMOS T/R switch.

MESFET and PIN diodes fabricated in GaAs, InP and As heterostructures are the majority devices used for RF switches due to their high performance. PIN diodes provide high power handling capability but are not suitable for low power applications due to the high conducting current needed for low switch–on resistance. And these devices are generally used as discrete components in a RF wireless system. However, to integrate the T/R switch with RF transceiver and baseband signal processing into a single chip offers a solution of lower cost, higher reliability and smaller form-factor. These benefits are especially attractive for mobile wireless communication

applications. Therefore, CMOS RF T/R switches are promising and potentially great demanding in the market. Comparing to MESFET in GaAs processes, conventional CMOS RF switches have lower power handling capabilities due to lower break-down voltage and parasitic junction diodes. Larger on-resistance of CMOS transistors causes higher insertion loss for CMOS T/R switches. The isolation of CMOS T/R switch is poor due to high conductive silicon substrate. These disadvantages limit the application of CMOS T/R switches. With the advancement of CMOS process technologies and circuit innovations, it is possible for CMOS switches to provide competitive performance to be integrated into wireless communication devices.

In the past decade, CMOS T/R switches have been studied and designed for different applications. In 2001, a conventional CMOS T/R switch was designed and integrated in a Bluetooth transceiver chip using standard 0.35-μm CMOS technology [1]. It reported 2dB insertion loss and >15dB isolation. The performance was not good but affordable for Bluetooth applications. In 2003, Takahiro reported a T/R switch with 0.8dB insertion loss, 23dB isolation and 17.4dBm P1dB at 5GHz [2], in which the transistors are built based on an extended depletion layer in 0.18-μm CMOS process. The idea is to increase the substrate impedance to increase the linearity. This concept can be modified and applied to standard CMOS processes by floating the body [3] or bias the body with a tuned LC tank [4].

In this paper, we will analyze the theoretical operation of CMOS T/R switches and the performance limitations in section II. In section III, we propose a novel method to overcome the limitations and improve the performance. The characterization results of the T/R switch is reported in section IV. Finally, we concluded the work in section V.

II. ANALYSIS OF CMOS T/R SWITCH

MOSFET has been used as a switch in many CMOS circuits like mixer or multiplier, digital logic circuits, etc. For RF T/R switches, NMOS transistors are the most commonly used due to higher mobility than PMOS transistors, which means lower on-resistance at the same transistor size. Fig. 1 shows the circuit diagram of a common CMOS Single Pole Double Throw (SPDT) T/R switch. When the control signal "switch_ctrl" is logic "1", the transistor for the receiver path is at the "on" status and passes through the signal from the antenna to the receiver front-end. While the transistor at the transmit path is at the "off" status, the signal from the transmitter is isolated to the antenna port. If the control signal "switch_ctrl" is logic "0", the receiver path is isolated and the transmitter path is through.

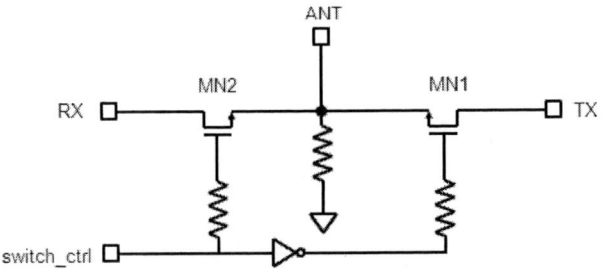

Figure 1.   Schematic of a conventional SPDT T/R switch.

The main concerns of a T/R switch are the performance of insertion loss, power handling capability and isolation. To understand the limitation of the performance in theory, we first analyze the operation of a single NMOS transistor as a switch. Fig. 2 shows a NMOS transistor model and the equivalent circuit of a single RF switch using a NMOS transistor.

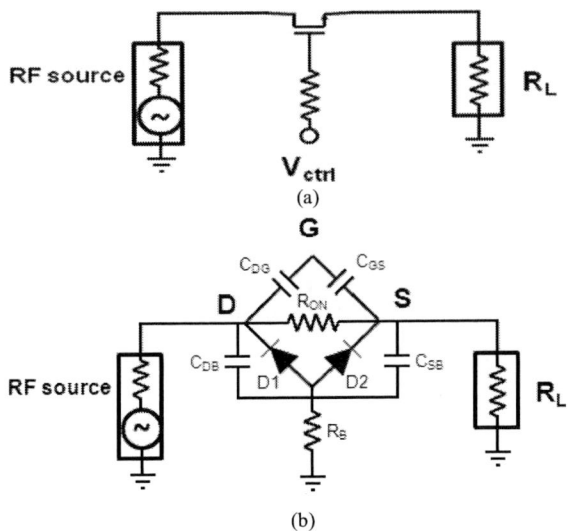

Figure 2.   (a) A single NMOS transistor as a RF switch. (b) Equivalent circuit model of the single RF switch.

From the equivalent circuit model of a single switch, the insertion loss is calculated in [5] as:

$$IL = \frac{(R_{ON} + 2Z_0)^2 + \omega^2 C_T^2 [(R_{ON} + 2Z_0)R_B + (R_{ON} + Z_0)Z_0]^2}{(2Z_0)^2(1 + \omega^2 C_T^2 R_B^2)} \quad (1)$$

Equation (1) tells that at low frequency, the insertion loss is mainly determined by the on-resistance of the transistor which is inversely proportional to the transistor's size and gate voltage. While at RF frequency, the parasitic capacitance of the transistor takes effect in RF signal loss. Therefore, the transistor size has to be optimized for the best insertion loss.

Power handling capability of a switch is basically limited by the non-linearity factors in the switch. For a single NMOS transistor as a RF switch, two different types of non-linear

devices are found in the equivalent circuit model shown in Fig. 2 (b). The obvious one is the two diodes D1 and D2 which are reverse biased junction diodes from source/drain to the body. When a high power signal comes, the voltage swing is large enough so that at some portion of the negative cycle the two diodes are conducted and the RF signal is lost through this path. The second non-linear device in the model is actually the "on-resistor". It is not a passive resistor but an active one whose resistance depends on its bias voltage. Equation (2) shows the on-resistance Ron of a NMOS transistor for CMOS process technology.

$$R_{ON} = \frac{1}{\mu_n C_{OX} \dfrac{W}{L}(V_{GS} - V_{TH})} \quad (2)$$

At small signal, Ron is almost constant as VGS is mainly contributed by the DC bias voltage of the transistor's gate. At large signal, VGS is dynamically changing which causes non-linearity.   To minimize the non-linearity effects, the two junction diodes should be kept reversed biased as much as possible, and the Vgs remain constant despite large signal. One solution is to have more reversed bias voltage for the body and higher control voltage for the gate [6]. It requires additional circuits to generate multiple bias voltage or high control voltage, but still face the issue of oxide breakdown at large RF signals. Another solution is to employ floating gate and body technique, which couples the source and drain voltage to the gate and body to maintain a constant potential dynamically for the gate-source and source-body with no oxide breakdown issues.

For a T/R switch as shown in Fig. 1, the power handling capabilities become more complicated with one switch on and the other one off in series. The linearity for a switch at on status has been analyzed above as a single RF switch. When it is connected to a RF switch at off state, the linearity is further limited due to the non-ideal off state of the "off" switch. In Fig. 1, MN1 is switched on for the transmitter and MN2 is switched off to isolate the transmit signal to the receiver. The gate, source, drain and body of MN2 are all biased at low voltage in DC operating point. With a single supply voltage all are biased at ground level. The switch is "off" for small RF signals. The drain of MN2 is connected to 50 Ohm load, whose voltage is zero when there is no leakage from the transmitter. When the RF signal is large, Vgs of MN2 is positive biased during some portion at the negative RF cycle as the gate's voltage is higher than the source's. It happens with floating gate technique due to the drain's voltage is zero at the "off" state. The gate is not able to track exactly the source voltage since it also couples the signal from the drain. Therefore, more RF signal is leaked to the receiver at higher transmit power which causes nonlinearity.

III.   PROPOSED METHOD AND DESIGN IMPLEMENTATION

Fig. 3 shows the proposed schematic diagram of an asymmetric T/R switch for high power handling. Since insertion loss is more critical for the transmitter, an asymmetric topology is chosen for the T/R switch for optimized system

978-1-61284-863-1/11 $26.00 © 2011 IEEE

performance. MN1 the switch for the transmitter path, MN2 and MN3 are in series for the receiver path. Deep-N well NMOS transistors are chosen for the three switch devices with a few advantages. First, it offers better isolation with reduced substrate coupling. Secondly, the body can be floating to ground which reduces RF leakage. With the floating bias technique to the body, the parasitic capacitance of the body is not ignorable comparing to the coupling capacitance from the drain to the body. Hence the coupled signal at the body is smaller than the RF signal at the drain of the switch transistor. In Fig. 3, the RF signal track blocks track the large RF signal from the transmitter's output to the body of the switch transistor MN1, which is more effective than floating bias of the body. The gate of MN1 is also floating biased with a big resistor to the DC control voltage. Therefore, MN1 is ensured to be in high linear operation.

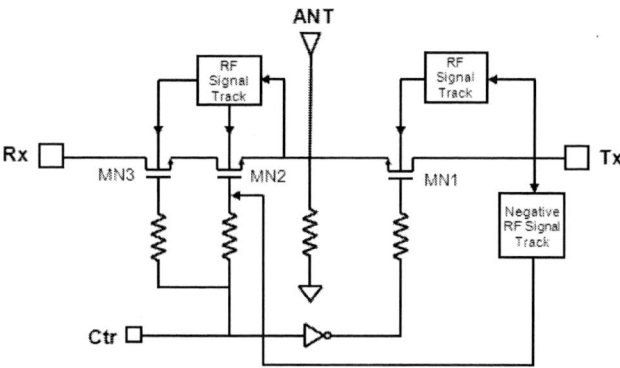

Figure 3. An asymmetric CMOS T/R switch with high linearity for the transmitter at a single control voltage.

Analyzed in section II, another factor limiting the linearity is the leakage through the receiver path, i.e., the series transistor MN2 and MN3. The most RF leakage happens at the negative peak of the large signal when the gate-source voltage of MN2 reaches the maximum. Therefore, a negative RF signal tracking circuits are proposed to add on the gate of MN2, which lowers down the gate voltage during the negative cycle and maintains the switch in a deep "off" status. With this mechanism, the RF leakage is reduced and the linearity of the transmit switch is increased effectively.

Fig. 4 shows the complete circuits for the proposed T/R switch. MN4 is diode connected to track the RF signal during the negative cycle, MN5 is a switch to pass the tracked signal to transistor MN2 at the transmit mode. At the receive mode, MN5 is "off" to isolate the signal from the receiver to the transmitter. R2 is a large resistor to bias the drain of MN5 to ground. The body of MN1 and MN2 track the RF signal through a resistor connected to the transmitter's output. The phase of the tracked signal is a bit delayed due to the parasitic capacitance of the body but in an acceptable range. It can be improved by using an inductor to couple the signal, with the tradeoff of bigger die area. For optimized performance, the transmit switch is designed to be low insertion loss and high power handling capability. Therefore, the size of MN1 is much bigger than the sizes of MN2 and MN3. All the transistors are deep-N well NMOS devices. The deep-N well is connected to

the supply voltage through a large resistor which limits any leakage current.

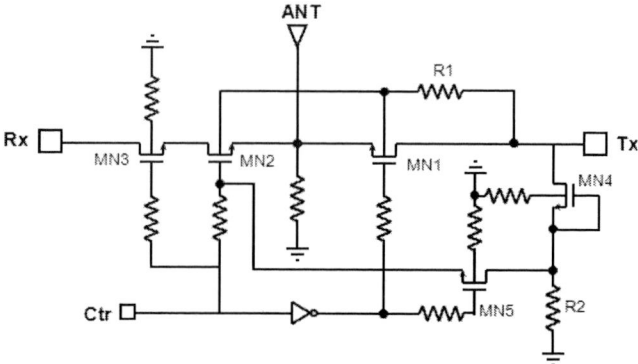

Figure 4. Schematic of the proposed high power CMOS T/R Switch.

With this method, the insertion loss for the transmitter path of the T/R switch is optimized with minimum leakage through the junction diode and the receiver path. Meanwhile, the linearity of the transmitter path is increased as well.

IV. MEASUREMENT AND DISCUSSIONS

The CMOS T/R switch was designed in 0.18 μm standard CMOS process with a single 1.8V DC supply voltage. Fig. 5 is the fabricated chip microphotograph. The die area of the chip excluding I/O pads is 0.14mm by 0.12mm. The chip was bonded to a QFN32 package and assembled to a FR4 PCB for testing.

Fig. 6 shows the return loss at the antenna port and the transmitter port for receive/transmit mode. From 200MHz to 2.4GHz, both the return loss are less than -14dB. For a T/R switch, isolation at the transmit mode is the main concern as the transmitted signal is usually very large while the received signal is very small. Fig. 7 shows the measured isolation at the transmit mode, which is from 25dB to 40dB in the frequency band of 200MHz to 2.4GHz.

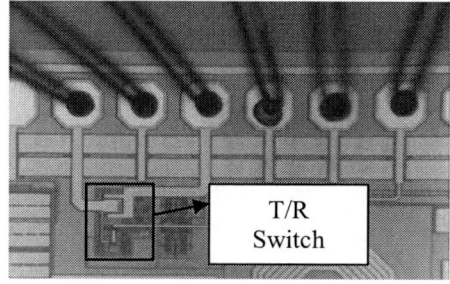

Figure 5. Chip microphotograph of the T/R switch.

The measured insertion loss is 0.3dB/0.4dB/0.6dB at 400MHz/900MHz/2.4GHz respectively for the transmitter path, and 1.4dB/1.45dB/1.6dB at 400MHz/900MHz/2.4GHz respectively for the receiver path. These measured results are very close to the simulated results. Fig. 8 shows the measured P1dB of the transmit path at 2.4GHz, which is 20.5dBm. The transmitter path's P1dB is not as good as expected due to the

978-1-61284-863-1/11 $26.00 © 2011 IEEE

TABLE I.   COMPARISON OF THIS WORK WITH PRIOR ART.

Ref.	Frequency	Insertion Loss (dB)	Input P1dB (dBm)	Isolation (dB)	Control V. (v)	Process Technology
[2]	2.4GHz	0.58	17.4	28.9	2.8/1	0.18μm CMOS (with modified process)
[3]	20GHz	Tx: 3.4 Rx: 3.5	Tx:28.7	22	1.2	90nm CMOS
[4]	2.4GHz	Tx: 1.5, Rx: 1.6	Tx:28.5, Rx:12.5	Tx: 32, Rx: 17	1.8	0.18μm CMOS
[6]	2.4GHz	Tx: 0.8, Rx: 1.2	Tx:28	>24	3.3	0.13μm CMOS technology
This Work	0.2GHz~4GHz	Tx:0.6 Rx:1.6 @2.4GHz	Tx: 20.5	25~38.8	1.8	0.18μm CMOS

diode-connected NMOS transistor "MN4" in Fig. 4 which is not able to withstand large signals. It can be further improved by changing the device to other diode models with high breakdown voltage. For the receiver path, the measured P1dB is 13.3dBm which satisfies the requirement of most receivers. Table I is a summary of the measured results comparing to the other published work.

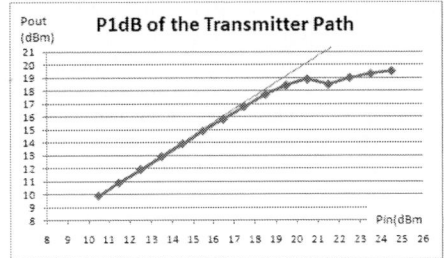

Figure 8.   Measured P1dB for the transmitter path.

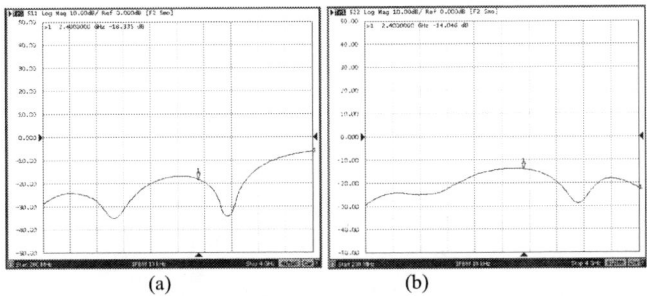

(a)   (b)

Figure 6.   Return loss of the T/R switch: (a) Antenna port, (b) Transmitter port.

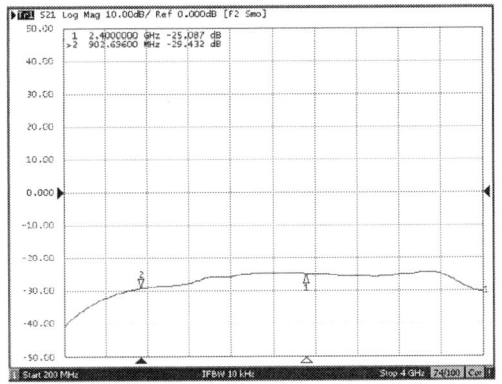

Figure 7.   Isolation between the transmitter path and the receiver path.

V.   CONCLUSION

In this paper, a novel technique was proposed to improve the insertion loss and linearity of CMOS T/R switches. The non-linearity caused by leakage at large RF signals is minimized by dynamic biasing the body and the gate of the switch transistors to maintain a constant bias point even in large signals. The design was implemented in 0.18 μm standard CMOS process with deep-N well option. The T/R switch occupies very small die area and shows good performance. Integration of CMOS T/R switch in a RF transceiver is promising and offers a competitive low cost SoC solution.

REFERENCES

[1]   Aruna Ajjikuttira, et al., "A Fully-Integrated CMOS RFIC for Bluetooth Applications", IEEE International Solid State Circuit Conference Dig. Tech. Papers, Feb. 2001.

[2]   Takahiro Ohnakado, et al., "A 0.8-dB Insertion-Loss, 17.4-dBm Power-Handling, 5-GHz Transmit/Receive Switch With DETs in a 0.18-m CMOS Process", IEEE Electron Device Letters, vol. 24, no. 3, March, 2003, pp. 192~194.

[3]   Piljae Park, et al., "High-Linearity CMOS T/R Switch Design Above 20GHz Using Asymmetrical Topology and AC-Floating Bias", in IEEE Transactions on Microwave Theory and Techniques, vol. 57, no. 4, April 2009, pp. 948-956.

[4]   Niranjan A. Talwalkar, et al., "Integrated CMOS Transmit-Receive Switch Using LC-Tuned Substrate Bias for 2.4GHz and 5.2GHz Applications," in IEEE Journal of Solid State Circuits, vol. 39, no. 6, June 2004, pp. 863-870.

[5]   Feng-Jung Huang and Kenneth O, "A 0.5-um CMOS T/R Switch for 900-MHz Wireless Applications", in IEEE Journal of Solid State Circuits, vol. 36, no. 3, March 2001, pp. 486-492.

[6]   Xu Haifeng, and O. Kenneth K., "High Power T/R Switch using Stacked Transistors", US Patent WO 2008/133620 A1, 6 November 2008.

60-GHz SP4T Switch with ESD Protection

Jin He and Yong-Zhong Xiong
Institute of Microelectronics, A*STAR (Agency for
Science, Technology and Research), Singapore

Yue Ping Zhang
Nanyang Technological University
Singapore

Abstract—**This paper proposes a new π-network-based single-pole-four-throw (SP4T) switch with inherent ESD protection for 60-GHz applications, which achieves a small core area of 274 μm × 262 μm with a 1.2-V 65-nm bulk CMOS RF process. At 60 GHz, the proposed switch exhibits a simulated performance of 3.4-dB insertion loss, 25.7-dB isolation between Ant 1 and Ant 2/3/4, and 43-dB isolation between Ant 2 and Ant 3/4. At an operating frequency of 60 GHz, the simulated input and output return losses are both around 13 dB; the input P_{1dB} of the power-handling capability is 11.1 dBm; and the switching speed is around 2.4 ns. RF ESD protection for the SP4T switch has been verified to pass the 2-kV HBM test by simulation.**

Keywords-CMOS; millimeter-wave integrated cirtuit (MMIC); SP4T switch; ESD;

I. INTRODUCTION

Millimeter-wave (mm-wave) band, especially the worldwide 7-GHz unlicensed bandwidth around 60 GHz, makes it feasible for the wireless applications featuring the high data rate of over gigabits as well as the short range of several meters, such as wireless high definition multimedia interface (HDMI), wireless personal area network (WPAN), and wireless sensor network (WSN). Attributed to constantly shrinking dimensions of devices, CMOS technology becomes a great competitor of III–V technologies like GaAs, Inp, pHEMT, etc. to implement the mm-wave ICs owing to its improving RF performance with low power, low cost, as well as high integration.

As a key building block of time-division-duplexing (TDD)-based front end, the T/R switch is required with low insertion loss, high isolation, and large power-handling capability. Up to present, there have been many single-pole-double-throw (SPDT) switches demonstrated in CMOS for microwave applications [1]–[3]. Only a few attempts have been made on SP4T switches in CMOS for mm-wave applications, such as beam-forming antennas and multiple-input-multiple-output (MIMO) systems. Atesal, *et al.* demonstrated a SP4T 60-GHz switch with the on-chip λ/4 transmission lines in 130-nm CMOS [4]. The SP4T switch had a measured insertion loss of 2.3 dB with an isolation of greater than 22 dB at 60 GHz. Cetinoneri, *et al.* presented a series-shunt SP4T switch manifesting a measured insertion loss of less than 3.5 dB up to 67 GHz with the isolation better than 25 dB [5].

In this paper, a new π-network-based SP4T switch is designed for 60-GHz applications using a 65-nm bulk CMOS RF process. The paper is organized as follows: Section II introduces the fundamental π-network-based SPST switch topology and describes the topology of the proposed SP4T

switch and simulated results. RF ESD protection for the SP4T switch is illustrated in section III. Finally, the paper is concluded in Section IV.

II. THE DESIGN OF 60-GHz SP4T SWITCH

A. Fundamental 60-GHz π-Network-Based SPST Switch

The asymmetric π-network-based topology of the 60-GHz SPST switch is proposed in Fig. 1(a), which is developed from the symmetric one demonstrated in [6].

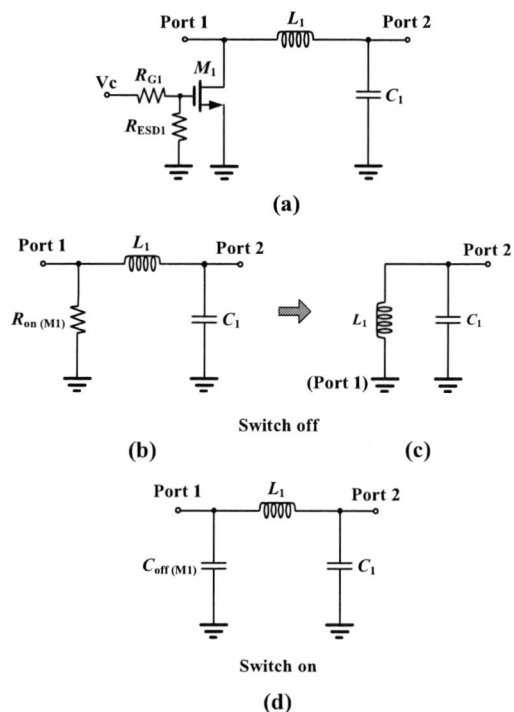

Figure 1. The 60-GHz π-network-based SPST switch: (a) the topology; (b) the small-signal equivalent circuit and (c) its simplification in the off state of (a); (d) the small-signal equivalent circuit in the on state of (a).

When Vc is configured to be high and applied to the gate of M_1 through large bias resistors R_{G1}, M_1 is turned on to perform the function of isolation between Port 1 and Port 2. The switch is turned off in this case so as not to transmit any RF signal. The small-signal equivalent circuit for the proposed SPST switch in the off state is shown in Fig. 1(b). Since the on-resistance $R_{on(M1)}$ of M_1 features a low value of a few ohms, Port1 can be regarded as being directly shortened to ground so that L_1 and C_1 form a resonator which is further simplified in

978-1-61284-863-1/11 $26.00 © 2011 IEEE   188

Fig. 1(c). Obviously, the isolation of the proposed topology is primarily determined by the shunt transistor M_1.

When the control voltage Vc is set to be low voltage, M_1 is turned off and can be simplified as only one shunt off-capacitance $C_{off(M1)}$. Combining L_1 and C_1, they form a π-impedance-matching network between Port 1 and Port 2. The small-signal equivalent circuit for the switch in the on state is presented in Fig. 1(d). In this case, the SPST switch is turned on to transmit the signal. At high frequencies, parasitic series resistance accompanying with L_1 features frequency dependent owing to the skin effect, this phenomenon will degrade the Q of L_1 and thus cause the excessive loss for the switch. Normally, the Q higher than 10 will not influence the performance of the SPST switch significantly.

Note that the shunt transistor M_1 can inherently serve as an ESD protection device to allow ESD event at ports. To further enhance the breakdown characteristics of M_1, a resistor R_{ESD1} of around 100 kΩ is added at the gate of M_1 for the purpose of forcing a uniform gate-assisted bipolar breakdown in case of an ESD event [7].

B. The 60-GHz SP4T Switch

1) The topology of the 60-GHz SP4T switch: The topology of the proposed 60-GHz SP4T switch is shown in Fig. 2(a), which is constructed by four SPST switches with the back-to-back connection. Each SPST switch can be used as a Tx or Rx path. The common node of four SPST switches is labeled as Port. Here the capacitors C_1–C_4 represented by dash lines are combined into one shunt capacitor C for their parallel connection.

As shown in Fig. 2(a), in the Tx or Rx mode, for example, Vc1 is configured to be low voltage whereas Vc2, Vc3, and Vc4 are set to be high voltage synchronously. Consequently, transistor M_1 is turned off to be one off-capacitance $C_{off(M1)}$ acting as a part of the matching network, which forms the Tx path from Port to Ant 1 or Rx path from Ant 1 to Port. Correspondingly, transistors M_2, M_3, and M_4 are turned on to achieve the small on-resistances $R_{on(M2)}$, $R_{on(M3)}$, and $R_{on(M4)}$, which not only separate Ant 2, Ant 3, and Ant 4 from Port and Ant 1, but also isolate Ant 2, Ant 3, and Ant 4 from each other. Fig. 2(b) presents the small-signal equivalent circuit for the SP4T switch in the Tx or Rx mode, where $C_{off(M1)} = C_{off(M2)} = C_{off(M3)} = C_{off(M4)}$, $R_{on(M1)} = R_{on(M2)} = R_{on(M3)} = R_{on(M4)}$, and $L_1 = L_2 = L_3 = L_4$.

Due to the small on-resistances $R_{on(M2)}$, $R_{on(M3)}$, and $R_{on(M4)}$ simultaneously grounding Ant 2, Ant 3, and Ant 4 so as to make inductances L_2, L_3, and L_4 grounding, Fig. 2(b) can be further approximated to be Fig. 2(c), where inductances L_2, L_3, and L_4 are represented with only one inductance $L_2/3$ because they are in parallel connection and the final value becomes 1/3 of each one. It is evident that the matching network between Ant 1 and Port is constructed by a π network and a resonator, they connect in series and share the shunt capacitance C. Table I lists the active device dimensions and passive element values of the SP4T switch for 60-GHz applications.

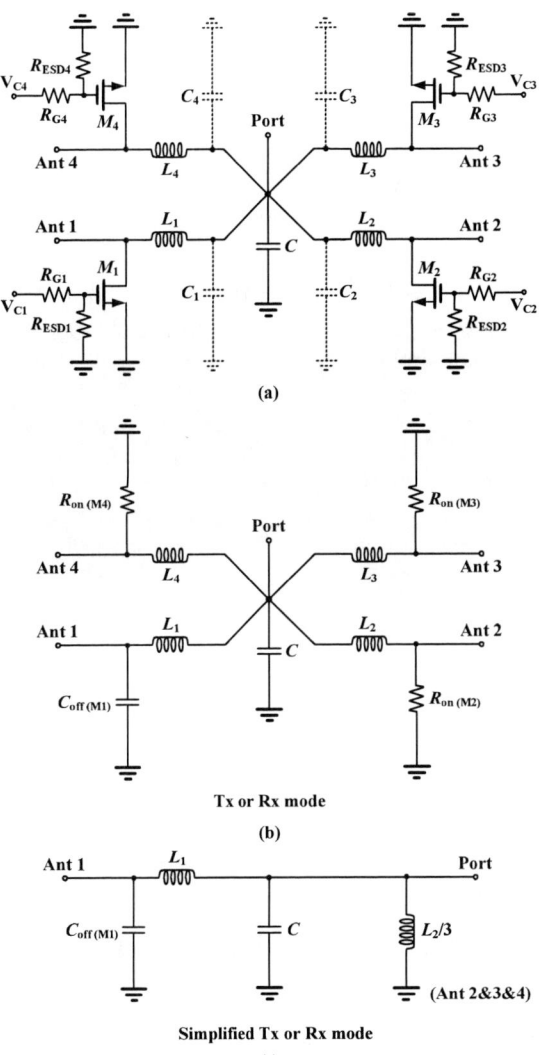

Figure 2. The proposed SP4T switch: (a) the topology, (b) its small-signal equivalent circuit in the Tx or Rx mode, and (c) the further approximation.

TABLE I.   CIRCUIT ELEMENT VALUES OF THE 60-GHz SP4T SWITCH

Circuit Element	Element Value
M_1, M_2, M_3, M_4	130 μm/0.06 μm (R_{on} =2.9 Ω, C_{off} = 76 fF)
R_{G1}, R_{G2}, R_{G3}, R_{G4}	10 kΩ
R_{ESD1}, R_{ESD2}, R_{ESD3}, R_{ESD4}	100 kΩ
L_1, L_2, L_3, L_4	92 pH
C	157 fF
V_{bias} for Tx and Rx nodes	0 V
Vc	1.3 V (on), 0 V (off)
$\overline{Vc}$	0 V (on), 1.3 V (off)

2) Simulated results: The proposed SP4T switch was designed in a 1.2-V 65-nm bulk CMOS RF process with a 7-metal back end and thick-metal inductors. Transistors for low-power (LP) wireless applications achieve a cut-off frequency f_T of around 200 GHz and a maximum oscillation frequency f_{MAX}

of around 250 GHz, respectively. The layout of the proposed SP4T switch is shown in Fig. 3. The whole area including all testing pads and the active area are 465 µm × 456 µm and 274 µm × 262 µm, respectively. Note that in Fig. 3, Ant 3 and Ant 4 are terminated internally with 50-Ω resistances for convenience of measurements since they have identical results to those of Ant 1 and Ant 2 due to the symmetry of the switch topology.

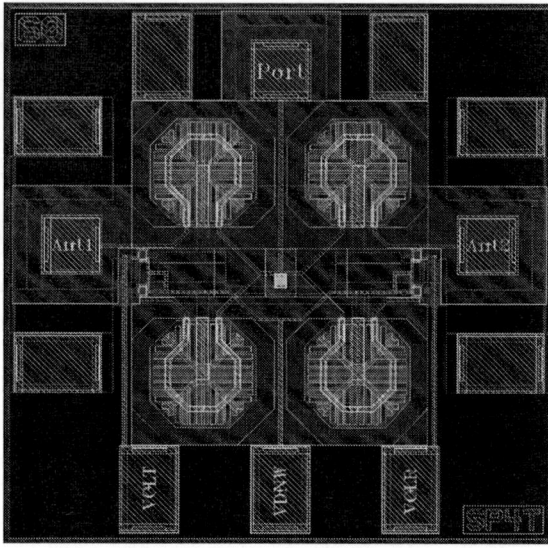

Figure 3.   The layout of the proposed SP4T switch in Fig. 2(a).

With the Cadence SpectreRF simulator, the post-layout simulations of the insertion loss and isolations for the SP4T switch are shown in Fig. 4. It is observed that the switch has an insertion loss of < 4.2 dB over the bandwidth of 57–66 GHz and 3.4 dB at 60 GHz. An isolation of > 25 dB between Ant 1 and Ant 2/3/4 is achieved for 57–66-GHz frequency band and the value of 25.7 dB at the 60 GHz; the isolation between Ant 2 and Ant 3/4 is > 42 dB over the bandwidth of 57–66 GHz and 43 dB at 60 GHz. The simulated return losses are plotted in Fig. 5. It is seen that the input and output return losses are both > 8 dB from 57 to 66 GHz and are both around 13 dB at 60 GHz. Note that the simulated performance of the proposed switch includes the parasitic effects of the testing pads and interconnections, which should be de-embedded from measurements since the switch will be internally integrated with other circuit blocks in practical applications. A thru method discussed in [8] can be adopted for de-embedding. If these parasitic effects are de-embedded from measurements of the real chip, the performance of the proposed SP4T switch can be further improved.

The simulated power-handling capability for the switch is presented in Fig. 6, where the input P_{1dB} exhibits 11.1 dBm at 60 GHz. The switching speed of the switch is simulated in Fig. 7. The rise time (from 10% up to 90% of the maximum output swing) and fall time (from 90% down to 10% of the maximum output swing) of the output signal are 1.2 ns and 1.17 ns, respectively. The switching speed could be improved with gate bias resistances R_{G1}–R_{G4} decreasing, however the insertion loss could be degraded slightly.

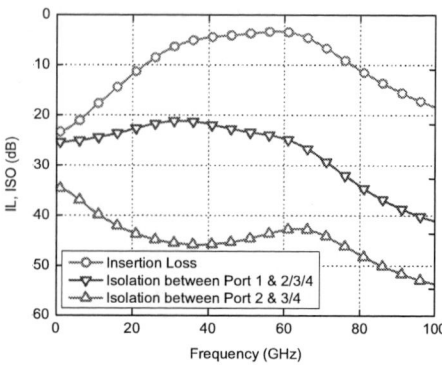

Figure 4.   Simulated insertion loss and isolations for the SP4T switch.

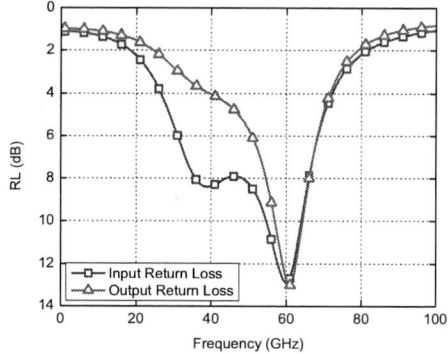

Figure 5.   Simulated return losses for the SP4T switch.

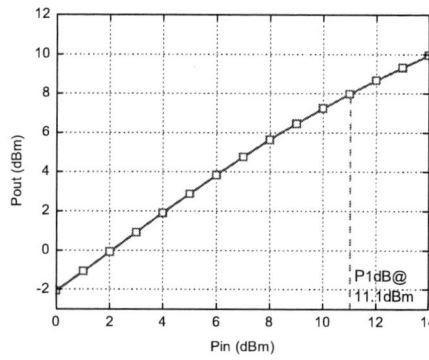

Figure 6.   Simulated input P_{1dB} for the SP4T switch.

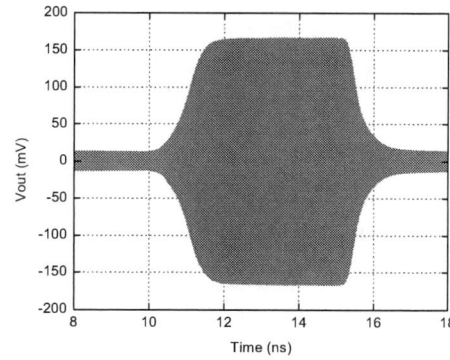

Figure 7.   Simulated switching speed for the SP4T switch.

III. RF ESD PROTECTION FOR THE SP4T SWITCH

Considering the fact that compared with Port of the switch integrated internally, Ant 1–Ant 4 are usually connected to the outside world through the antennas or off-chip antenna filters so that they easily suffer from the ESD events. Therefore, it is very important for us to consider the RF ESD protection in a T/R switch. Note that each shunt nMOS transistor itself of the SP4T switch is the inherent ESD device, thus no additional ESD circuits are needed so as to significantly reduce the unwanted parasitic capacitances, which will degrade the performance of the switch. Moreover, the large resistor R_{ESD} can assist to uniformly trigger all the parasitic parallel npns and further improve the breakdown behaviour of the nMOS ESD device [7].

Generally, Human Body Model (HBM) defined by the JEDEC standard [9] is the widely accepted model to evaluate the performance of the ESD events that a charged human body discharges towards an IC [10]. A typical equivalent schematic adopted for HBM test is shown in Fig. 8 [9], [10], where the human-body-modeled capacitance C_{ESD} of 100 pF is charged up when the switch S_1 is tuned on while the switch S_2 is tuned off and then discharge into the ESD device through a discharging network formed by a inductor L of 7.5 μH and a resistor R of 1.5 kΩ when the switch S_1 is tuned off whereas the switch S_2 is turned on.

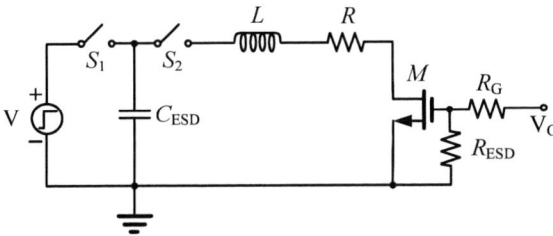

Figure 8.   Typical equivalent schematic for HBM test.

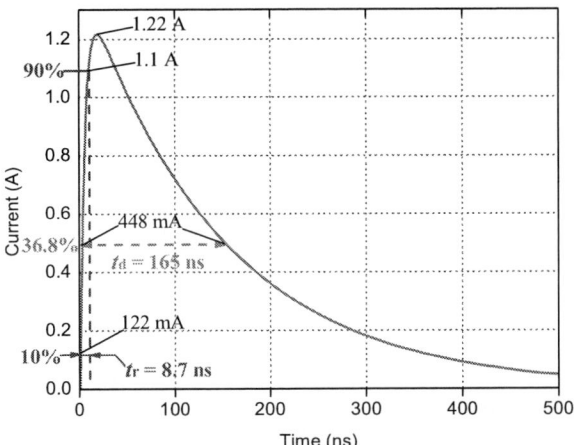

Figure 9.   Simulated discharge current of 2-kV HBM test.

Transient simulation based on Fig. 8 is used to investigate the characteristic of the nMOS ESD device realized with the total width of 130 μm and the fingers of 20. Note that the gate

is floated with no any bias voltage. Fig. 9 shows the simulated discharge current resulting from a 2-kV HBM test between input and ground. The peak current I_{ps} flows through the transistor M is around 1.22 A, the rise time t_r defined from 10% (122 mA) up to 90% (1.1 A) of the peak current I_{ps} achieves 8.7 ns, and the decay time t_d defined as the time interval of 36.8% (448 mA) of the I_{ps} exhibits 165 ns. The simulated I_{ps}, t_r, and t_d meet the wave specification of 2-kV HBM voltage level required in [9]. Actually, the nMOS ESD device implemented in a 0.18-μm CMOS process has been verified to pass an ESD test of 4 kV by measurements [7].

IV. CONCLUSION

A 60-GHz SP4T switch with ESD protection has been proposed based on π-matching networks and simulated in a 1.2-V 65-nm bulk CMOS RF process. The simulation results showed that an insertion loss of 3.4 dB, an isolation of 25.7 dB between Ant 1 and Ant 2/3/4, and an isolation of 43 dB between Ant 2 and Ant 3/4 at 60 GHz were achieved. One can conclude that the proposed SP4T switch is suitable for multi-channel wireless communications.

ACKNOWLEDGMENT

This work is supported by the project titled "Silicon based millimeter wave identification system for wireless communication applications" (A*STAR SERC Grant No.: 1021290051).

REFERENCES

[1] Y. P. Zhang, Q. Li, W. Fan, C. -H. Ang, and H. Li, "A differential CMOS T/R switch for multistandard applications," IEEE Trans. Circuits Syst. II, Express Briefs, vol. 53, no. 8, pp. 782–786, Aug. 2006.

[2] Q. Li and Y. P. Zhang, "CMOS T/R switch design: Towards ultra-wideband and higher frequency," IEEE J. Solid-State Circuits, vol. 42, no. 3. pp. 563–570, Mar. 2007.

[3] Q. Li, Y. P. Zhang, K. S. Yeo, and W. M. Lim, "16.6- and 28-GHz fully integrated CMOS RF switches with improved body floating," IEEE Trans. Microw. Theory Tech., vol. 56, no. 2, pp. 339–345, Feb. 2008.

[4] Y. A. Atesal, B. Cetinoneri, and G. M. Rebeiz, "Low-loss CMOS 50 – 70 GHz SPDT and SP4T switches," in IEEE RFIC Symp., Jun. 2009, pp. 43–46.

[5] B. Cetinoneri, Y. A. Atesal, and G. M. Rebeiz, "A miniature DC–70 GHz SP4T switch in 0.13-μm CMOS," in IEEE MTT-S Int. Microw. Symp. Dig., Jun. 2009, pp. 1093–1096.

[6] J. He and Y. P. Zhang, "Design of SPST/SPDT switches in 65 nm CMOS for 60GHz applications," in Asia-Pacific Microw. Conf., Dec. 2008, pp. 1–4.

[7] N. A. Talwalkar, C. P. Yue, H. Gan, and S. S. Wong, "Integrated CMOS transmit-receive switch using LC-tuned substrate bias for 2.4-GHz and 5.2-GHz applications," IEEE J. Solid-State Circuits, vol. 39, no. 6, pp. 863–870, Jul. 2004.

[8] A. Issaoun, Y.-Z. Xiong, J. Shi, J. Brinkhoff, and F. Lin, "On the deembedding issue of CMOS multigigahertz measurements," IEEE Trans. Microw. Theory Tech., vol. 55, no. 9, pp. 1813–1823, Sept. 2007.

[9] Electrostatic Discharge (ESD) Sensitivity Testing Human Body Model (HBM), International Standard, JEDEC JESD22-A114F, Dec. 2008.

[10] H. Feng, G. Chen, R. Zhan, Q. Wu, X. Guan, H. Xie, A. Z. H. Wang, and R. Gafiteanu, "A mixed-mode ESD protection circuit simulation-design methodology," IEEE J. Solid-State Circuits, vol. 38, no. 6, pp. 995–1006, Jul. 2003.

A Dual-Band LC Voltage-Controlled Oscillator in 0.13µm CMOS Technology

Siti Maisurah M. H., *Member*, *IEEE*, Nazif Emran F., Norman Fadhil Idham M. and A. I. Abdul Rahim

Advance Physical Technology Lab,
TM Research & Development,
TM Innovation Centre, Cyberjaya, Malaysia.
maisurah@tmrnd.com.my

Abstract—A dual-band, low phase-noise LC voltage-controlled oscillator (VCO) has been demonstrated in a 0.13µm CMOS process. The operating frequency of the dual-band VCO covers from 1.51GHz to 1.92GHz and from 2.13GHz to 2.73GHz. The proposed VCO features phase-noise of -116.4dBc/Hz and -121.5dBc/Hz at 1MHz offset frequency for both low corner and high corner end of the low-band operation. For high-band operation, phase-noise performance of -109.6dBC/Hz and -118.2dBc/Hz at 1MHz offset frequency are achieved. At 3V supply voltage, the power dissipation is 54.3mW for the low-band operation and 51.5mW for the high-band operation.

Index Terms – Dual-band VCO, LC-tank, reconfigurable VCO, CMOS RFIC.

I. INTRODUCTION

As the demand for wireless system increases, the idea of having multiple standards or multiple bands in a single transceiver is widely implemented. To realize such concept, the use of a tunable or programmable transceiver provides an option to designers who wish to avoid multiple transceivers design in supporting various standards. In line with this trend, tunable circuit blocks such tunable low-noise amplifier (LNA) and reconfigurable voltage-controlled oscillator (VCO) has been greatly promoted to support the implementation of tunable transceivers.

Several implementations of a dual-band VCO can be found in the literature. This includes the use of switching devices in the resonator tank [1], a wide range tuning VCO that can cover dual bands [2], frequency multiplication method [3] and a set of parallel-connected single-band VCOs [4].

This paper presents a dual-band LC VCO based on the use of two parallel-connected single VCO in a 0.13µm CMOS technology. It can be configured to generate output signals at two frequency bands. The low frequency band covers from 1.51GHz to 1.92GHz while the high frequency band covers from 2.13GHz to 2.73GHz.

II. ARCHITECTURE DESCRIPTION

The proposed architecture of the dual-band VCO is shown in Fig. 1. The dual-band VCO has 2 main sections; which are the VCO Core Circuit and the Band Selection Circuit. The VCO Core circuit is constructed using two individual LC VCO

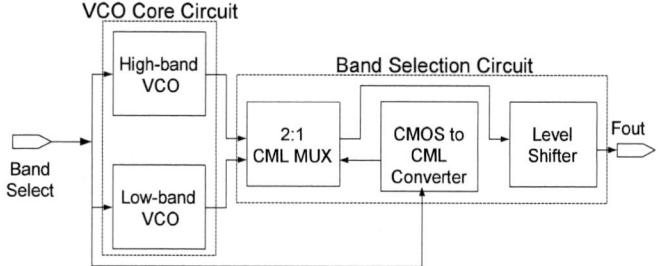

Figure 1. Proposed architecture of the dual-band VCO

that operates in two frequency bands: 1.51GHz to 1.92GHz (Low-band) and 2.13GHz to 2.73GHz (High-band). The selection of these frequency bands is controlled by the Band Selection Circuit which is made up of a 2:1 Current Mode Logic (CML) Multiplexer (MUX) circuit, a CMOS to CML Converter and a Level Shifter Circuit.

A. VCO Core Circuit Design

In this work, the cross-coupled LC-VCO structure is used as the topology for the VCO Core Circuit. Both High-band and Low-band VCOs have the same identical structure except for a control circuitry. The control circuitry provides a mechanism to either turning the VCO ON or OFF depending on the input from the Band Select port. The schematic of the High-band VCO is depicted in Fig. 2.

The VCO can be divided into 5 sections, namely the LC-tank network, negative resistance circuit, current source circuit, output stage circuit and the control circuit. Spiral inductor L1 and junction varactors Var1 and Var2 form the tank circuit of the VCO. Such tank circuit has parasitic resistance associated with the LC components. In this topology, this loss will be compensated by a negative resistance circuit constructed using the NMOS and PMOS transistor pairs Mn1, Mn2, Mp1 and Mp2 [5]. The spectral response of the VCO depends on the Q of the resonator instead of the negative resistance circuit since the frequency response of the negative resistance circuit is very wide.

Transistor Mp3 and Mp4 form the tail current source for the VCO core. The bias current for the core circuit, Iref1 is

978-1-61284-863-1/11 $26.00 © 2011 IEEE

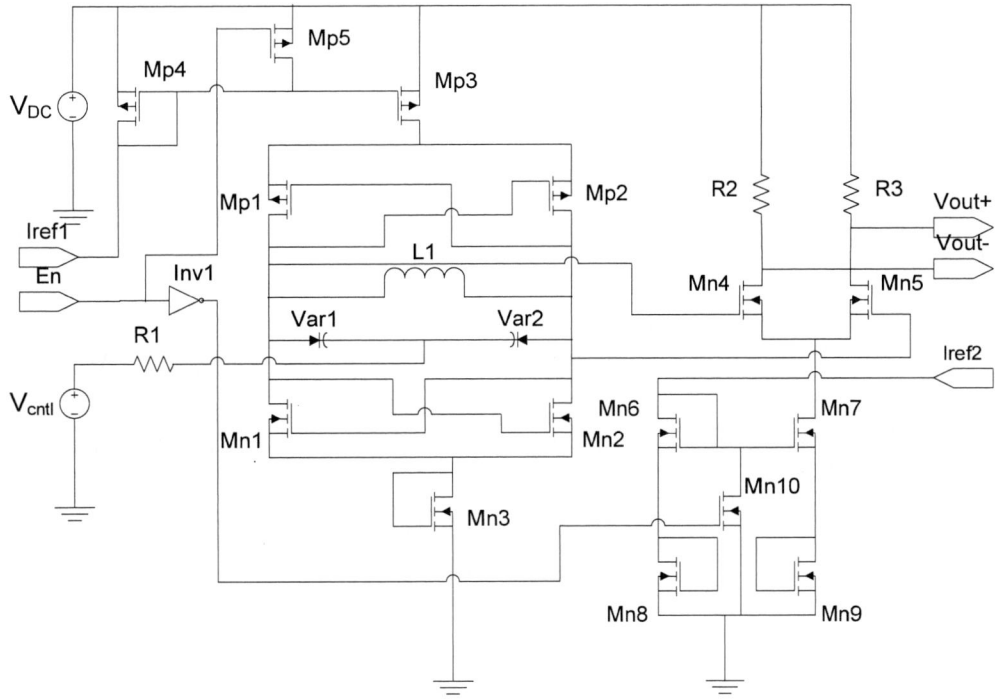

Figure 2.   Schematic of the high-band VCO

connected to a current sink of a bandgap reference circuit. Mn3 is a diode connected transistor which is added to the core circuit to protect and limit the voltage of the NMOS transistor pair. Telescopic cascode configuration is employed in the output stage of the VCO using transistor Mn6 to Mn9 to provide high gain . Transistor Mn4, Mn5 and resistor R2 and R3 forms the output buffer of the VCO circuit.

As mentioned earlier, a control circuitry is included in the VCO to control the operation of the VCO core. At any time, only one VCO core can be turned ON, while the other remain OFF. This is realized by using inverter Inv1, Mp5 and Mn10, as shown in Fig. 2. For the low-band VCO, the connection for the input and output of the Inv1 is reversed whereby the En port is connected to the input of Inv1 and Mn10, while the output of Inv1 is connected to Mp5. This is done so that any input to the High-band VCO will be reversed in the Low-band VCO and vice versa. For example, when the En port is set to 'high', in this case 3V, the high-band VCO will be turned ON and the low-band VCO will be turned OFF. On the other hand, when the En port is set to 0V, the high-band VCO will be turned OFF and while the low-band VCO will be turned ON.

B.   Band Selection Circuit

The Band Selection Circuit is made up of a 2:1 Current Mode Logic (CML) Multiplexer (MUX) circuit, a CMOS to CML Converter and a Level Shifter Circuit. Fig. 3 shows the schematic of the 2:1 CML MUX circuit. The 2:1 CML MUX consists of two symmetrical differential transistor pairs Mn1, Mn2 and Mn3, Mn4 that sample the inputs VCO1, VCO1bar and VCO2, VCO2bar. Differential output signal from the VCO core circuit will be fed into these inputs. The operation of these differential pairs is controlled by the lower-level differential

transistor pair, Mn5 and Mn6 [6]. This transistor pair is a current steering circuit driven by Sel and Selbar input ports. The output voltage swing is determined by resistors R1 and R2.

The 2:1 CML MUX receives its input signal from the CMOS to CML Converter. The schematic of the CMOS to CML Converter is shown in Fig. 4. This converter is needed to convert a single ended CMOS signal to a differential signal with CML voltage level [7]. The CMOS input signal is fed to the circuit through the BandSel port, which is also connected to the En port in the High-band and Low-band VCO. Transistors Mn3, Mp3, Mp4, CML inverter pairs Mn1-Mp1 and Mn2-

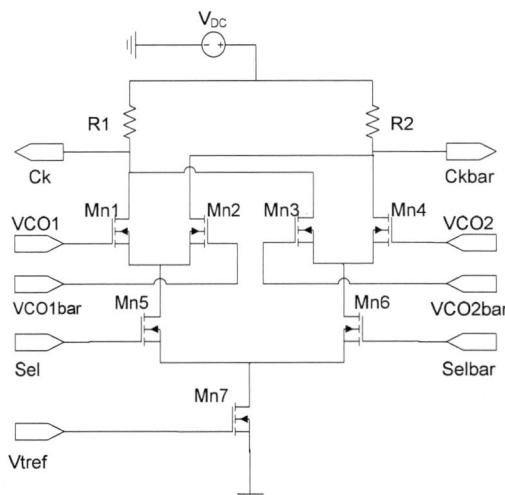

Figure 3.   Schematic of the 2:1 CML MUX

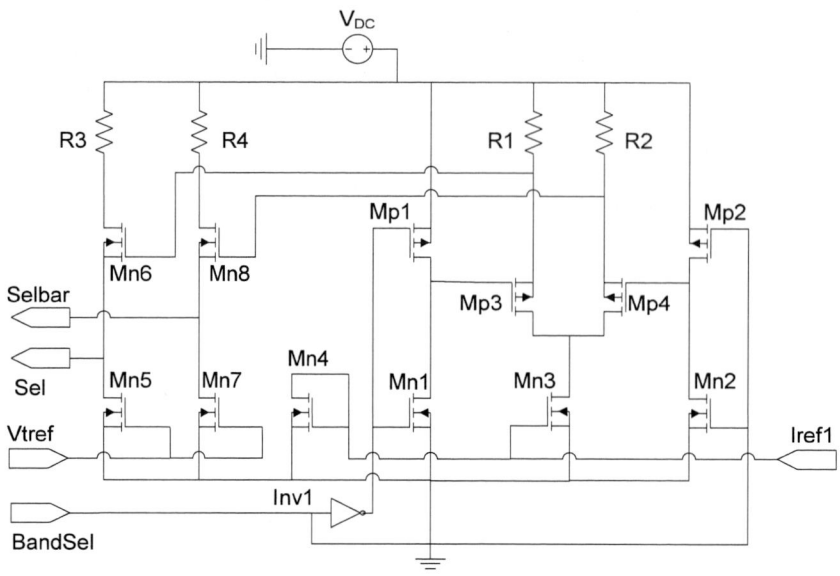

Figure 4. Schematic of the CMOS to CML Converter

Mp2, resistors R1 and R2 together with CMOS inverter, Inv1, are the components needed to construct the CMOS to CML converter. Differential output signal from this converter will then go through to a level shifter circuit which is made up from Mn5, Mn6, Mn7, Mn8, R3 and R4 [8]. The logic levels for the output of the CMOS to CML Converter circuit is different from the input of the 2:1 CML MUX circuit, thus level shifting is needed in order to cascade these two circuit blocks.

The output of the 2:1 CML MUX is connected to a Level Shifter circuit. Fig. 5 shows the schematic of the Level Shifter circuit. Two level shifting is needed to bring the signal down to the desired logic levels. In between the shifter, a CML buffer is used to drive the later level shifter stage. The CML buffer is constructed using transistors Mn5, Mn6, Mn7 and resistors R5

and R6. The rest of the components are parts of the level shifter circuits. The differential output signal of the complete dual-band VCO can be obtained through port Fo and Fobar.

III. SIMULATION RESULTS

Simulation for the proposed dual-band VCO is carried out using Cadence Virtuoso Analog Design Environment software. Periodic Steady State and Periodic Noise Analyses are carried out to determine the tuning range and phase noise performance of the proposed VCO. As shown in Fig. 6, the simulated oscillation frequency for the low-band operation covers from 1.51GHz to 1.92GHz while the high-band operation covers from 2.13GHz to 2.73GHz. These translate to 410MHz of tuning ranges for the low-band VCO and 600MHz for the high-band VCO. The control voltages used to tune to these frequencies are from 0V to 2V.

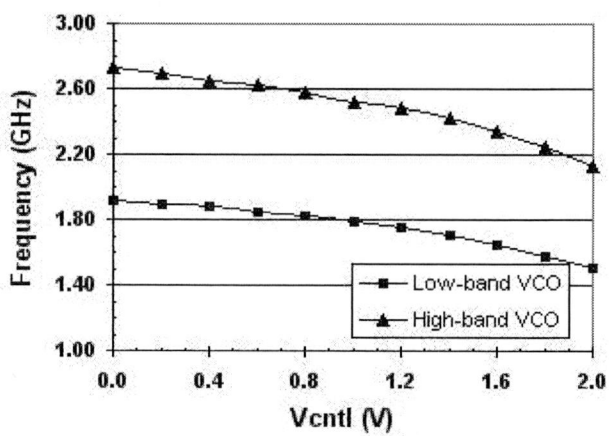

Figure 6. Tuning characteristic of the dual-band VCO

Figure 5. Schematic of the Level Shifter Circuit

978-1-61284-863-1/11 $26.00 © 2011 IEEE

The simulated phase-noise for the low-band operation at both low corner and high corner frequencies taken at 1MHz offset frequency are -116.4dBc/Hz and -121.5dBc/Hz. This can be observed in Fig. 7. For the high-band operation, phase-noise performance of -109.6dBC/Hz and -118.2dBc/Hz at 1MHz offset frequency are achieved for both low corner and high corner frequency, as shown in Fig. 8.

The proposed dual-band VCO demonstrated successful reconfiguration characteristic between the two bands. This can be observed through the simulated output spectrum plotted on the same scale as depicted in Fig. 9. In this plot, Vcntl is set to 0V. The low-band VCO creates a fundamental signal at 1.92GHz while the high-band VCO generates fundamental components at 2.72GHz. The spurious tones for both the bands can also be observed along the frequency.

CONCLUSION

A fully integrated, dual-band LC VCO has been developed based on the parallel-connected VCO method. The operation frequency of the proposed VCO covers from 1.51GHz to 1.92GHz and from 2.13GHz to 2.73GHz. The VCO, implemented in a 0.13μm CMOS technology, shows a good phase noise performance across the tuning frequency in both low-band and high-band operation. Phase-noise performance of -116.46dBC/Hz (at 1.51GHz) and -121.5dBc/Hz (at 1.92GHz) at 1MHz offset frequency are achieved for the low-band operation, while for high-band operation, phase-noise of -109.6 (at 2.33GHz) and -118.2dBc/Hz (at 2.73GHz) are achieved. This VCO is well suited for multi-standard, fully-integrated RF transceiver designs.

ACKNOWLEDGMENT

The authors would like to thank Mr. Anurag Nigam from Nattel Microsystems Private Limited for his advice and technical discussions.

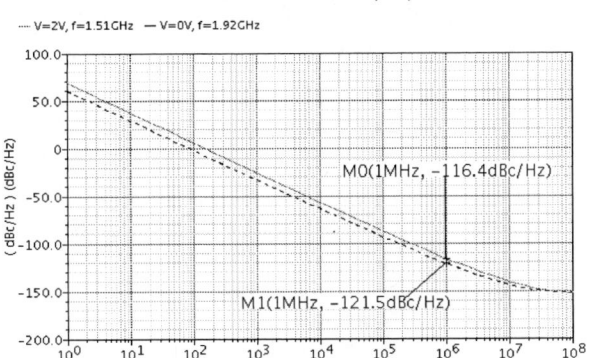

Figure 7.   Phase noise of the low-band VCO at 1.51GHz and 1.92GHz

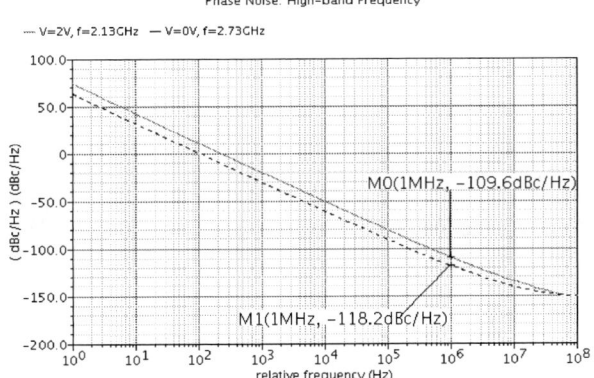

Figure 8.   Phase noise of the high-band VCO at 2.13GHz and 2.73GHz

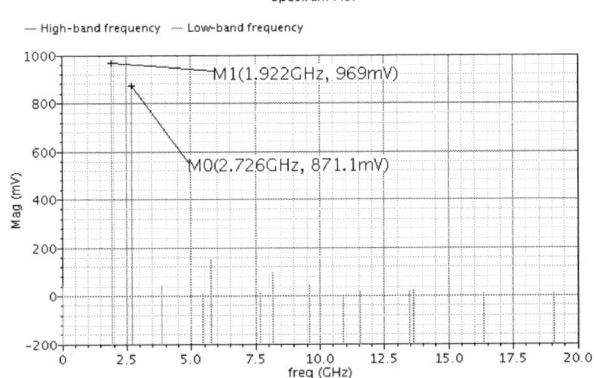

Figure 9.   Output spectrum of the dual-band VCO

REFERENCES

[1]   S. M. Yim and K. O. Kenneth, "Demonstration of a switched resonator concept in a dual-band monolithic CMOS LC-tuned VCO", IEEE Conference on Custom Integrated Circuits, 2001, pp. 205-208.

[2]   J. Tham, M. A. Margarit, B. Pregardier, C. D. Hull, R. Magoon and F. Car, " A 2.7V 900MHz/1.9GHz dual-band transceiver IC for digital wireless communication", Proceeding of the IEEE Custom Integrated Circuits Conference, 1998, pp. 559 - 562.

[3]   J. Lia, G. M. Jian, S. Y. Kiat, P. Y. Xiao, A. D. Manh and M. L. Wei, " A 1.8V 2.4/5.15GHz Dual-Band LC VCO in 0.18um CMOS Technology", IEEE Microwave and Wireless Component Letters, 2006, pp. 194 - 196.

[4]   A. Jayaraman, B. Terry, B. Fransis, P. Sullivan, M. Lindstrom and J. O'Connor, "A fully integrated broadband direct-conversion receiver for DBS applications", IEEE International Solid-State Circuits Conference, Digest of Technical Papers, 2000, pp. 140-141.

[5]   Siti Maisurah M. H., Nazif Emran F., Norman Fadhil Idham M., A. I. Abdul Rahim, "A Low Phase Noise and Large Tuning Range 2.4GHz LC Voltage-Controlled Oscillator", Asia Pacific Conference on Circuit and Systems, 2010, pp. 931-934.

[6]   C. H. Hsiao, M. S. Kao, C. H. Jen, Y. S. Hsu, P. L. Yang, C. T. Chiu, J. M. wu, S. H. Hsu and Y. S. Hsu, "A 3.2Gb/s 20:1 CML Transmitter in 0.18um CMOS Technology", Proceedings of the International COnference on Mixed Design of Integrated Circuits and Systems, 2006, pp. 179 - 183.

[7]   W. Souder, 2009. A Low Power 10GHz Phase Locked Loop for Radar Applications Implemented in 0.13um SiGe Technology. Thesis, (Master). Auburn University.

[8]   M. Ratcliff, 2008. Phase Locked Loop Analysis and Design. Thesis, (Master). Auburn University.

A 5/6-bit Multi-Modulus Frequency Divider in 0.13μm CMOS Technology

Siti Maisurah M. H., *Member*, *IEEE*, Nazif Emran F., Norman Fadhil Idham M. and A. I. Abdul Rahim

Advance Physical Technology Lab,
TM Research & Development,
TM Innovation Centre, Cyberjaya, Malaysia.
maisurah@tmrnd.com.my

Abstract—**A multi-modulus frequency divider with 5-bit or 6-bit operation mode is presented. It can work at clock frequency as high as 2.56GHz. The division ratio of this frequency divider covers from 32 to 63 (5-bit operation mode) and from 64 to 127 (6-bit operation mode) with unit step increment. Key features of the proposed architecture are the 2/3 divider cells which have common logic and almost the same circuit cells implementation.**

Index Terms – Multi-modulus frequency divider, 2/3 divider cell, CML circuit, CMOS RFIC.

I. INTRODUCTION

As wireless system proliferates, the need for multi-standard radios has led to the production of various types of multiband and programmable circuits. In a transceiver, the trend in designing programmable circuits can be observed by looking to the increasing number of designs in tunable amplifiers, multi-band voltage-controlled oscillators (VCO), programmable phase-locked loop (PLL) and others. A major challenge in implementing these circuits lies on the requirement for faster speed, higher frequency operation and low-power circuit.

In a PLL-based frequency synthesizer as shown in Fig. 1, there are two circuit blocks that operate at the highest frequency of the PLL system; namely the frequency divider and the voltage-controlled oscillator [1]. The output frequency range of the frequency synthesizer is determined by the output frequency of the VCO, and also the division ratio of the frequency divider. Therefore, in designing multiband transceivers, wider division ratio of a frequency divider is really desirable.

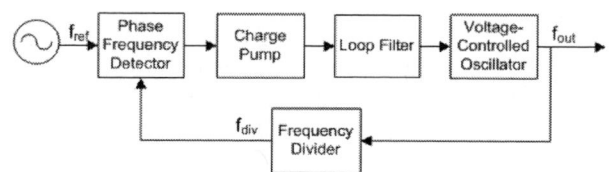

Figure 1. Building block in a PLL

In this paper, a 5/6-bit multi-modulus frequency divider (MMFD) that works up to 2.56GHz clock signal is presented. The proposed MMFD has wide division ratio, which is from 32 to 63 (5-bit operation) and from 64 to 127 (6-bit operation).

II. ARCHITECTURE DESCRIPTION

The proposed architecture of the MMFD is shown in Fig. 2. It consists of six cascaded 2/3 divider cells, two current mode logic (CML) multiplexer (MUX), a CML to CMOS Converter and a Modulus Controller. The Modulus Controller is constructed using combination of D-latch circuits, adders and CMOS to CML converters. In this paper, details of the Modulus Controller will not be discussed.

The MMFD can be controlled to divide input frequency by any programmable ratio from 32 to 127. It can work at clock frequency up to 2.56GHz. The 'Select' bit at the Modulus Controller determines the operation of the MMFD whether to work in 5 bit or 6 bit operation. In this way, users will have an option to use either 20MHz or 40MHz reference frequency for the PLL.

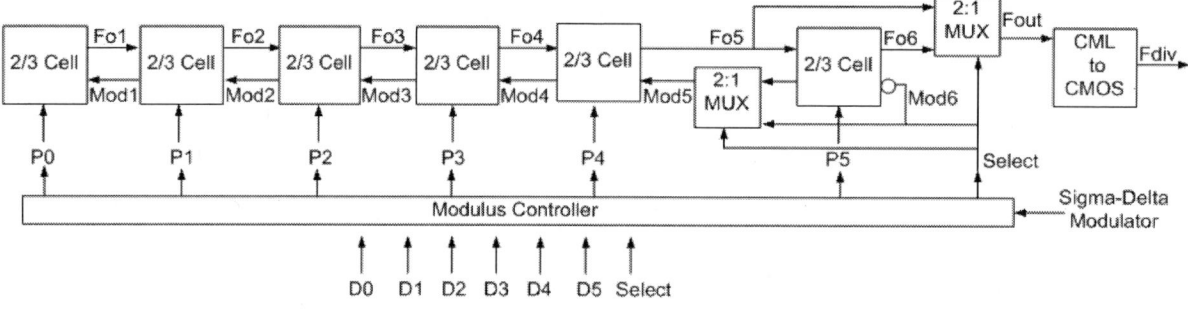

Figure 2. Proposed architecture of the dual-band VCO

978-1-61284-863-1/11 $26.00 © 2011 IEEE

As shown in Fig. 2, the proposed MMFD employed a common ripple fashion architecture based on the cascading of 2/3 dual-modulus cells [2] [3]. Each 2/3 cells divides its input by 2 or 3 depending on its control signal, P, and feedback signal, mod. In every division period, the last 2/3 cell generates a feedback signal, mod_{n-1}, which will propagates up to the preceding cells [4]. An active mod signal will enable the cell to divide by 3 once in a division cycle, with a condition of P set to high. Using this principle, the complete division ratio with unit step increment can be programmable. Setting n as the number of cells, a cascade of n divider cells will be able to produce division ratio between 2^n to 2^{n+1} -1. Thus, in this work, the division ratio vary from 32 to 63 for 5 bit operation, while for 6 bit operation, the division ratio are from 64 to 127.

III. IMPLEMENTATION OF THE MMFD

A. 2/3 Divider Cell

The 2/3 divider cell is constructed using four D-latches and three AND gates as depicted in Fig. 3. As mentioned earlier, the 2/3 divider cells divides the input frequency, Fin, by either 2 or 3 depending on the control signal, P, and feedback signal, Modin. Upon division, the 2/3 divider cell sends the divided output signal to the next cell in the chain. Modin signal is active once in a division cycle. The control signal, P, is checked at the moment Modin is active. The upper part lathes commonly operates at divide-by-2 mode. However, if P is high at the moment Modin is active, latch 3 will force the upper latches to swallow one extra period of the input signal, thus operating at divide-by-3 mode. Latch 4 will incept Modin signal from the subsequent cell and provides Modout signal to the preceding cell [5]. In this work, the logic levels of the input and output of the 2/3 divider cells are different, thus, cascading the 2/3 divider cells directly is not possible. Therefore, level-shifting in between the 2/3 divider cells is required and this is realized using a level shifter circuit.

For the benefits of high operating frequencies, Current Mode Logic (CML) circuits is used to implement the D-latches, AND gates and level shifters. Fig. 4 shows the CML D-Latch. Differential transistor pairs Mn1, Mn2 and Mn3, Mn4 share the same load, R1 and R2, but at different times in clock cycle. When CLK is high, differential pair Mn1 and Mn2 is enabled. All current is passed through Mn3 while Mn4 is off. The output node at Q and Qbar will tract the input D and Dbar. When CLK is low, differential pair Mn1 and Mn2 is turned off while differential pair Mn3 and Mn4 will be enabled. Mn3 and Mn4 are connected in positive feedback, thus the output will

latch on the previous state irrespective of the input [6]. As mentioned earlier, output of one 2/3 divider cell drives the clock input of the next subsequent 2/3 divider cell. This means that the input frequency for each cell is scaled down by the previous one indicating that the maximum allowed delay increases as the cell goes down [7]. Delay in the cell is inversely proportional to the cell's current consumption, thus current in the subsequent cell can be scaled down to half for half the speed and same output swing. This is implemented in our proposed MMFD for power optimization.

For simplicity, the schematic of the AND gate is combined together with the D-latch for latch 1, 3 and 4. This is shown in Fig. 5. An additional differential pair, Mn8, Mn9, is used to implement an AND gate on the D and Dbar input of the existing D-latch circuit. The inputs D and Dbar needs to be level shifted to have same input voltage swing across the AND gate input, A and Abar. This is done using source followers, Mn10 and Mn11. Current to the source followers is mirrored using Mn12 and Mn13.

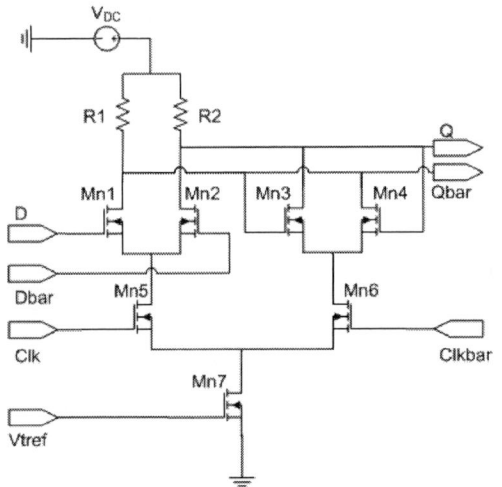

Figure 4. CML D-latch

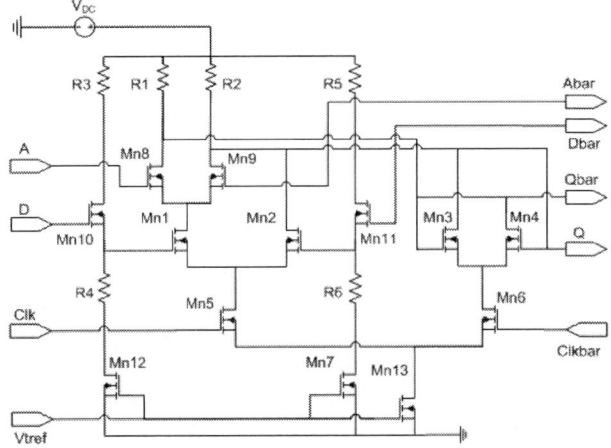

Figure 5. CML D-latch with AND gate input

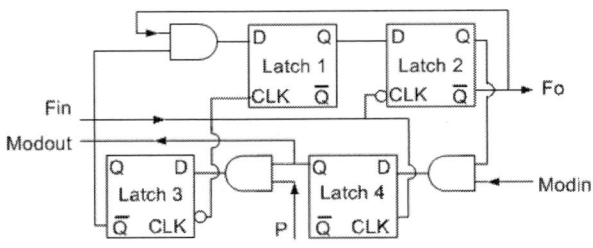

Figure 3. 2/3 divider cell

978-1-61284-863-1/11 $26.00 © 2011 IEEE   197

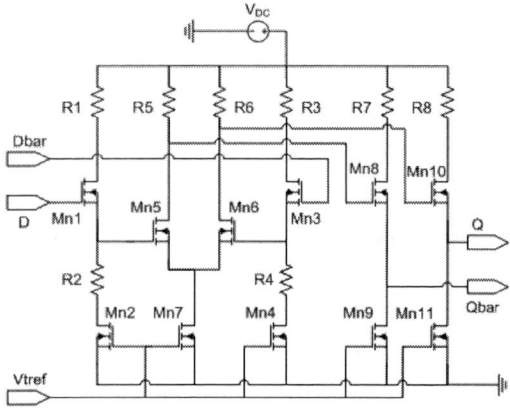

Figure 6. Level shifter

The schematic of the level shifter in between the 2 divider cells is shown in Fig. 6. Two level shifting is needed to bring the signal down to the desired logic levels. In between the shifter, a CML buffer is used to drive the later level shifter stage. The CML buffer is constructed using transistors Mn5, Mn6, Mn7 and resistors R5 and R6. The rest of the components are parts of the level shifter circuits.

B. 2:1 CML MUX

As shown in Fig. 2, there are 2 CML MUX needed to implement the proposed 5/6-bit MMFD. The Select signal from the Modulus Controller controls the operation of these MUX. When Select is set to low, the first MUX will select Modout signal from the sixth divider cell to be fed into the fifth divider cell. Modin signal in the sixth divider cell (Mod6) is held high. At the same time, the second MUX will select the output signal, Fo6, from the sixth 2/3 divider cell to Fout. Thus, in this case, 6-bit operation mode is executed. When Select is set to high, the first MUX will select the input directly from the Select signal, which means Modin in the fifth divider cell (Mod5) is held high. MUX 2 on the other hand will select output signal, Fo5, from the fifth divider cell to Fout. Schematic of the CML MUX is shown in Fig. 7.

The 2:1 CML MUX consists of two symmetrical differential transistor pairs Mn5, Mn6 and Mn8, Mn9 that sample the inputs A, Abar and B, Bbar. The operation of these differential pairs is controlled by the lower-level differential transistor pair, Mn7 and Mn10 [8]. This transistor pair is a current steering circuit driven by S and Sbar input ports. The output voltage swing is determined by resistors R1 and R2. Output Q is the same as input A if S is high and B if S is low.

C. CML to CMOS Converter

The CML output from the divider cells and MUX will be converted into CMOS logics to be further processed by other blocks inside the PLL system. The schematic of the CML to CMOS converter is depicted in Fig. 8. It consists of high gain differential stage, Mn5 and Mn6, whose output currents are mirrored using transistors Mp1 and Mp2. This produces an output swing comparator that provides a CMOS level signal [9].

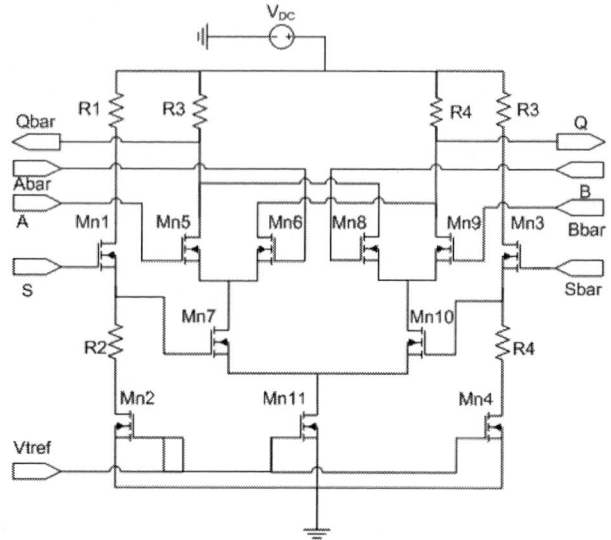

Figure 7. 2:1 CML MUX

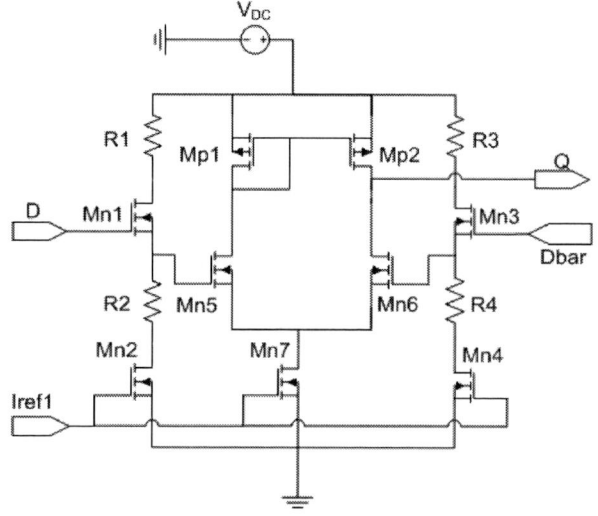

Figure 8. CML to CMOS Converter

IV. SIMULATION RESULTS

Simulation for the proposed MMFD is carried out using Cadence Virtuoso Analog Design Environment software. Fig. 9 shows the MMFD output waveform with a clock signal of 1.28GHz and a division ratio of 32 (5-bit operation mode). It can be calculated that the period of the output signal achieved is around 25.04ns which correspond to 39.94MHz. This is close to the ideal value which is 40MHz. Fig. 10 shows the output waveform with a clock signal of 2.56GHz and a division ratio of 127 (6-bit operation mode). It can be calculated that the period of the output signal achieved is around 49.65ns which correspond to 20.14MHz. This is close to the ideal value which is 20.16MHz. Fig. 11 shows the output waveform of the individual 2/3 divider cell with the same clock signal and division ratio. It can be seen that the period of the output waveform is doubled by half, as the clock signal propagates from the first divider cell to the last divider cell.

978-1-61284-863-1/11 $26.00 © 2011 IEEE   198

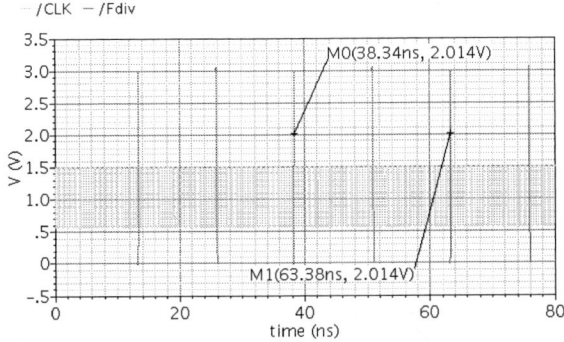

Figure 9. MMFD output signal with clock signal of 1.28GHz and division ratio of 32

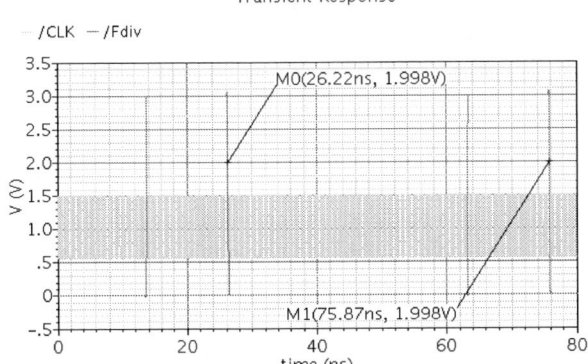

Figure 10. MMFD output signal with clock signal of 2.56GHz and division ratio of 127

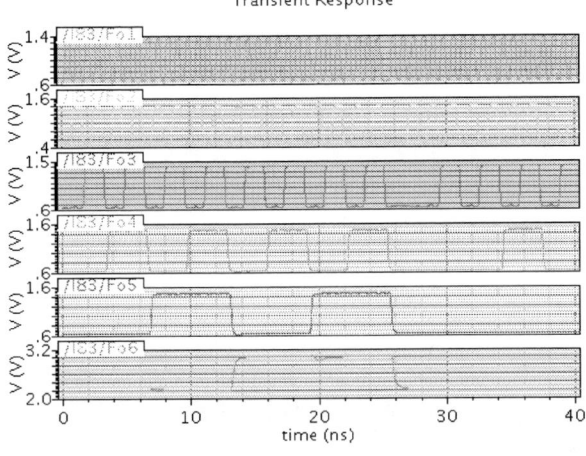

Figure 11. MMFD output signal of each 2/3 divider cell with clock signal of 2.56GHz and division ratio of 127

CONCLUSION

A 5/6 bit multi-modulus frequency divider has been demonstrated in 0.13um CMOS technology. It can reach up to 2.56GHz operating frequency with 47mA current consumption from a DC supply voltage of 3V. The division ratio can be varied from 32 to 63 (5-bit operation) and from 64 to 127 (6-bit operation), with incremental step of 1. The proposed multi-modulus frequency divider is suitable for sigma-delta fractional-N frequency synthesizer implementation.

ACKNOWLEDGMENT

The authors would like to thank Mr. Anurag Nigam from Nattel Microsystems Private Limited for his advice and technical discussions.

REFERENCES

[1] L. Wenguan, C. Honglin and Y. Ruohe, "A 5.5GHz Multi-Modulus Frequency Divider in 0.35um SiGe BiCMOS Technology for Delta-Sigma Fractional-N Frequency Synthesizers", International Conference on Microwave and Milimeter Wave Technology, pp. 1937 - 1940, May 2010.

[2] W. Hongyu, B. Paul and J. Dai, "A Generic Multi-Modulus Divider Architecture for Fractional-N Frequency Synthesizers", IEEE International Frequency Control Symposium, pp. 261-265, June 2007.

[3] Ray M., Souder W., Ratcliff M. Dai, F. and Irwin, J. D.,"A 13GHz Low Power Multi-Modulus Divider Implemented in 0.13um SiGe Technology", IEEE Topical Meeting on Silicon Monolithic Integrated Circuit in RF Systems, pp. 1-4, Jan 2009.

[4] B. S. Anqiao Hu, 2007. Multi-Modulus Divider in Fractional-N Frequency Synthesizer for Direct Conversion DVB-H Receiver, (Master). Ohio State University.

[5] S. Y. Wang, X. L. Wu, J. H. Wu and M. Zhang, "Low power design of multi-modulus programmable frequency divider", IEEE Electronics Letters, Vol. 45, Issue 20, pp. 1017 - 1019, September 2009.

[6] R. John, P. Calvin and D. Foster, "Integrated Circuit Design for High-Speed Frequency Synthesis", Artech House Publishers, 2006.

[7] Vaucher C. S.; Ferencic, I.; Locher, M.; Sedvallson, S.; Voegeli, U. and Wang, Z., "A family of low-power truly modular programmable dividers in standard 0.35um CMOS technology", IEEE Journal of Solid-State Circuits, Vol. 35, Issue 7, pp. 1039 - 1045, 2000.

[8] C. H. Hsiao, M. S. Kao, C. H. Jen, Y. S. Hsu, P. L. Yang, C. T. Chiu, J. M. wu, S. H. Hsu and Y. S. Hsu, "A 3.2Gb/s 20:1 CML Transmitter in 0.18um CMOS Technology", Proceedings of the International COnference on Mixed Design of Integrated Circuits and Systems, 2006, pp. 179 - 183.

[9] Kossel, M.; Morf, T.; Baumberger, W.; Biber, A.; Menolfi, C.; Toifl, T. and Schmatz, M., "A multiphase PLL for 10Gb/s links in SOI CMOS technology, IEEE Radio Frequency Integrated Circuits (RFIC) Symposium, pp. 207-210, 2004.

Optimizing Gain of 5 GHz RF Amplifier keeping Minimum Deviation in Center Frequency and Noise Figure

Jai Narayan Tripathi   Prakash R. Apte   Jayanta Mukherjee

Department of EE, IIT Bombay,
Mumbai, 400076, INDIA.
Email: {jai,jayanta,apte}@ee.iitb.ac.in

Abstract—**Optimizing Gain of an RF Amplifier using Design of Experiments (DOE), is presented. The constraints of keeping minimum deviations in both frequency and Noise Figure (NF) is taken in to account by ANOVA and nominal-the-best approach of Taguchi Methods. Gain of the amplifier is improved by 2.4 dB with frequency and NF deviations of 0% each.**

Key-words : Optimization, Amplifier, Gain, Noise Figure, DOE, ANOVA, Taguchi Methods.

I. INTRODUCTION

In Analog/RF circuits, there are many design constraints to be satisfied simultaneously e.g. power, noise, area, linearity etc. While taking all these constraints into the account, selecting sizes of the active elements is one of the most difficult steps of RF circuit design. There is no specific method for sizing of RF circuits, and even after the analytical calculations to obtain the best sizing for a output response, the results may vary in simulations. So intuition and design iterations are used by the designers for sizing. Design automation (using certain optimization methods) is generally adopted for sizing of transistors and active elements. Using this technique, the design with desired output response can be found. There are various optimization methods (deterministic and stochastic) used for this, such as Geometric Programming, Genetic Optimization, Particle Swarm Optimization etc [1] [2] [3]. Based on such design flow, circuits are designed for particular output but these methods do not provide the solution when an output is required at a target value with minimum variations due to noise. This can only be done by Taguchi methods using DOE. This work presents Design of Experiments (DOE) and Analysis of Variance (ANOVA) for RF circuit design.

II. RF AMPLIFIER AND DOE

The amplifier to be optimized is 5 GHz RF amplifier (Fig. 1). This has gain of 8 dB with noise figure NF of 2.096. This type of amplifier is widely used in many applications in ISM band. The output, to be optimized is gain of amplifier.

Our previous work [6] [7], presents optimization of RF circuits by DOE. In [8], a 3 step method of optimization by DOE, is presented and optimization of an oscillator with one constraint is described. This paper presents optimization of an RF amplifier with two constraints. Maximization of a 5 GHz RF amplifier with almost 0 % deviation in frequency and NF,

using DOE and ANOVA, is presented. The challenge lies in the fact that if the gain of amplifier is increased by varying the physical design parameters, the frequency and NF of amplifier will also deviate from their desired values. To set frequency as well as NF at particular values and then increasing the gain of amplifier is a sufficiently complex problem. This is a constrained optimization problem, which can be very difficult when design has many control factors.

Design of Experiments (DOE) is an area of study in industrial engineering, in which the experiments are planned for designing of a product efficiently and reliably. Based on these experiments, the effects of various factors on design are calculated. Based on these calculations, initial design is modified to achieve better quality. Noise conditions are also applied to make design insensitive to uncontrollable factors. For this the same experiments are repeated at different noise conditions e.g. temperature or supply voltage in the case of circuits. Orthogonal arrays are used here for fractional factorial DOE. For more details, refer [5]-[8].

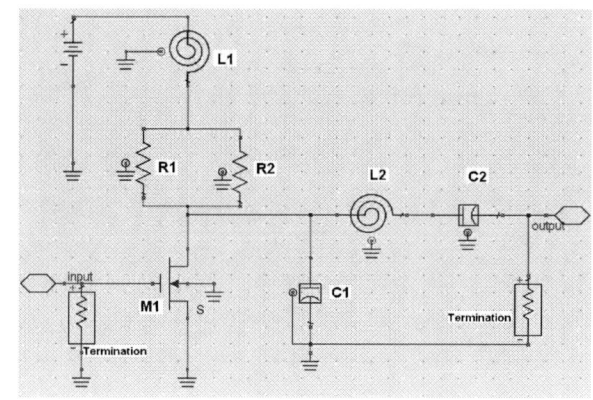

Fig. 1.   Circuit of 5 GHz Amplifier

There are three types of approaches, used for optimization, using Taguchi methods and ANOVA

1) Smaller the better (minimizing)
2) Larger the better (maximizing)
3) Nominal the best (tuning at a certain point)

First two approaches, minimizing or maximizing an objective function, are same as found in other optimization techniques.

The "Nominal-the-best" can't be achieved in all the optimization techniques. In this paper, since the target is to achieve maximum gain with the desired frequency and NF; the gain will be taken as 'larger the better' along with the frequency and NF both as 'nominal the best'. For larger-the-better approach the function $\eta = -10 log_{10}(\sum \frac{1}{\eta_i^2})$ is maximized and for nominal-the-best approach $\eta' = 10\ log_{10}(\frac{\mu^2}{\sigma^2})$ is maximized. Where μ is the overall mean and σ is the variance. The equation for nominal-the-best means that the overall mean should be put on target value with minimum variance and thus it sets the output at target value.

III. EXPERIMENTS

The output parameter for the analysis is gain of amplifier. The factors affecting this output, are chosen by the design. There are three levels of variation for all the 13 design factors. The components used in amplifier are all real components of 180 nm technology. Since we can not vary the width of transistor directly, we vary Finger No. instead, thus width is varied indirectly. Finger number is an integer which is multiplied by the unit width of transistor ($5\mu m$ in UMC 180 nm library), for the same reason it is unitless for the transistor M1 in table I. The 13 design factors are -

1) Width of finger of Transistor M1 ($w1$)
2) Width of Inductor L1 ($w2$)
3) Width of Inductor L2 ($w3$)
4) Width of Capacitor C1 ($w4$)
5) Width of Capacitor C2 ($w5$)
6) Width of Resistor R1 ($w6$)
7) Width of Resistor R2 ($w7$)
8) Length of Capacitor C1 ($l1$)
9) Length of Capacitor C2 ($l2$)
10) Length of Resistor R1 ($l3$)
11) Length of Resistor R2 ($l4$)
12) Diameter of finger of Inductor L1 ($d1$)
13) Diameter of finger of Inductor L2 ($d2$)

Table I shows the values of the 13 design factors at all levels. This is a 3 level DOE problem with 13 design factors so the orthogonal array used is $L_{27}(3^{13})$ which can accommodate 13 factors with 3 levels each. The number of experiments performed were 27. For applying noise condition, circuit was simulated at 5 different temperatures (readers should not be confused with noise conditions of experiments and noise of circuit ! Noise in term of experiments are unpredictable variations which cause the deviations in output).

IV. ANALYSIS

The output response corresponding to the experiments are shown in table II. There are three types of outputs of our interest for each experiment - gain, NF and frequency. For different experiments there are variations in outputs due to temperature as the noise factor. For each experiment row, the '*single summary statistic*' or objective function is calculated as shown in last column of table II. Using the objective function for each row of experiment, ANOVA is performed to obtain

TABLE I
DESIGN PARAMETERS AND THEIR LEVELS

Parameters	Level 1	Level 2	Level 3
$w1$	10	12	14
$w2, w3$	10 μm	12 μm	14 μm
$w4, w5$	18 μm	20 μm	22 μm
$w6, w7$	5 μm	6 μm	7 μm
$l1, l2$	27 μm	30 μm	33 μm
$l3, l4$	9 μm	10 μm	11 μm
$d1, d2$	165 μm	180 μm	195 μm

TABLE II
SIMULATION RESULTS OF EXPERIMENTS FROM L27

		Temperature				
		15^0 C	25^0 C	35^0 C	45^0 C	55^0 C
Exp 1	Gain (dB)	7.377	7.219	7.062	6.906	6.751
	Freq (GHz)	5.982	5.984	5.986	5.988	5.991
	Noise Figure	2.297	2.369	2.443	2.518	2.595
Exp 2	Gain (dB)	7.943	7.786	7.629	7.474	7.319
	Freq (GHz)	5.332	5.334	5.336	5.338	5.34
	Noise Figure	2.209	2.279	2.35	2.422	2.495
Exp 3	Gain (dB)	8.388	8.231	8.074	7.918	7.764
	Freq (GHz)	4.808	4.81	4.812	4.813	4.815
	Noise Figure	2.15	2.218	2.287	2.356	2.428
	.	.	.	.	.	.
	.	.	.	.	.	.
.						
Exp 25	Gain (dB)	8.723	8.567	8.411	8.257	8.103
	Freq (GHz)	4.932	4.935	4.939	4.942	4.945
	Noise Figure	1.909	1.972	2.037	2.102	2.169
Exp 26	Gain (dB)	8.73	8.569	8.409	8.249	8.091
	Freq (GHz)	4.828	4.834	4.84	4.846	4.852
	Noise Figure	1.942	2.008	2.075	2.144	2.213
Exp 27	Gain (dB)	8.36	8.198	8.037	7.878	7.719
	Freq (GHz)	4.843	4.848	4.852	4.857	4.861
	Noise Figure	1.97	2.037	2.105	2.175	2.245

tables III, IV and V for gain, frequency and NF respectively. Based on the mean at different levels (1,2,3), the percentage effect of each design factor on gain, NF and center frequency can be calculated. Third column in table III and table IV shows the percentage contribution by each design factor on output responses.

A. Design Constraints

As discussed earlier, the components used for the design are from umc 180 nm library. These components have following physical limitations (as mentioned in previous section, transistor finger no. is taken as design parameter) :

$$5 \leq w1 \leq 21\mu m$$
$$6\mu m \leq w2, w3 \leq 20\mu m$$
$$10\mu m \leq w4, w5, l1, l2 \leq 70\mu m$$
$$2\mu m \leq w6, w7 \leq 10\mu m$$
$$2\mu m \leq l3, l4 \leq 100\mu m$$
$$126\mu m \leq d1, d2 \leq 238\mu m$$

And there are some constraints on ratio of widths and lenghts of capacitors and resistors.

TABLE III
ANOVA TABLE FOR GAIN

Control Factor	DoF	Factor Effect (%)	F (before pooling)	Empty or pooled	F After pooling	Error Bar (dB)
$w1$	2	24	11939	no	81	± 0.09
$w2$	2	0	135	pooled	-	± 0.09
$w3$	2	0	45	pooled	-	± 0.09
$w4$	2	17	8324	no	56	± 0.09
$w5$	2	7	3347	no	23	± 0.09
$w6$	2	5	2390	no	16	± 0.09
$w7$	2	4	1789	no	12	± 0.09
$l1$	2	15	7430	no	50	± 0.09
$l2$	2	10	5027	no	34	± 0.09
$l3$	2	4	2112	no	14	± 0.09
$l4$	2	2.29	1146	no	8	± 0.09
$d1$	2	1	264	pooled	-	± 0.09
$d2$	2	12	6052	no	41	± 0.09

TABLE IV
ANOVA TABLE FOR FREQUENCY

Control Factor	DoF	Factor Effect (%)	F (before pooling)	Empty or pooled	F After pooling	Error Bar (dB)
$w1$	2	74	37151	no	206	± 0.59
$w2$	2	0	115	pooled	-	± 0.59
$w3$	2	0	119	pooled	-	± 0.59
$w4$	2	6	3197	no	18	± 0.59
$w5$	2	5	2412	no	13	± 0.59
$w6$	2	0	165	pooled	-	± 0.59
$w7$	2	0	28	pooled	-	± 0.59
$l1$	2	7	3578	no	20	± 0.59
$l2$	2	4	2218	no	12	± 0.59
$l3$	2	0	32	pooled	-	± 0.59
$l4$	2	0.14	69	pooled	-	± 0.59
$d1$	2	1	326	pooled	-	± 0.59
$d2$	2	1	589	pooled	-	± 0.59

TABLE V
ANOVA TABLE FOR NF

Control Factor	DoF	Factor Effect (%)	F (before pooling)	Empty or pooled	F After pooling	Error Bar (dB)
$w1$	2	71	35462	no	120	± 0.05
$w2$	2	0	1	pooled	-	± 0.05
$w3$	2	0	0	pooled	-	± 0.05
$w4$	2	1	737	pooled	-	± 0.05
$w5$	2	1	350	pooled	-	± 0.05
$w6$	2	9	4660	no	16	± 0.05
$w7$	2	8	4028	no	14	± 0.05
$l1$	2	1	535	pooled	-	± 0.05
$l2$	2	0	155	pooled	-	± 0.05
$l3$	2	4	1863	no	6	± 0.05
$l4$	2	3.23	1617	no	5	± 0.05
$d1$	2	1	556	pooled	-	± 0.05
$d2$	2	0	36	pooled	-	± 0.05

$$1 \leq (w6/l3) \leq 10 \; ; \; 1 \leq (w7/l4) \leq 10$$
$$1 \leq (l1/w4) \leq 6 \; ; \; 1 \leq (l2/w5) \leq 6$$

B. Best Settings

Initial design is having all the design factors at level 2, resulting the gain of 8 dB, center frequency of 5.008 GHz and NF of 2.096. The above constraints (in previous section) define the maximum limit of variation in case of monotonous increase or decrease at all the levels of design factors e.g. S/N ratio of gain (fig.2) is monotonously decreasing with increasing the levels (or increasing the dimensions) of $w4$ and $l1$. Since the S/N ratio should be as large as possible, the dimensions of $w4$ and $l1$ will be varied to the minimum possible values.

From analysis of variance (ANOVA), we have % effect of each design factor on all output responses as shown in tables III, IV and V. This information can be used for adjustment of outputs. The objective is to adjust the design factors in such a way that the gain should be as high as possible, center frequenct deviation should as low as possible and the same for NF. There are different design factors dominated for different output responses e.g. % effects of $d2$ on NF is 0% while the same on frequency is 4% so $d2$ can be used to adjust frequency without affecting NF. The adjustment of all the design factors will be done in three steps, taking into account the initial design.

Step 1 - S/N ratio of gain will be increased as much as possible by adjusting the design factors in the direction where S/N ratio of gain is increasing. This adjustment will affect frequency and NF by a huge amount. The design obtained after this step was [21,12,12,10,70,8,8,10,70,8,8,228,228](all in μm, except $w1$ which is finger number), for $w1,w2,...d1,d2$ respectively providing gain of 13.223 dB, center frequency of 2.458 GHz and NF (at center frequency) of 1.343.

Step 2 - In second step, frequency will be adjusted since it has been shifted hugely after first step. From tables III and IV, $w1$ has 24 % effect on gain while 74 % effect on frequency. But these % effects are in reverse directions thus will affect each other. There are two S/N ratios plotted for frequency as well as NF because these output responses are subjected to 'nominal-the-best' approach (fig.3 and 4). One S/N ratio which is plotted in blue color, indicates the factor effect on output response and its direction. While the other S/N ratio, plotted in yellow color, indicates the sensitivity of design factors on noise conditions (which is temperature change here). But in this analysis, we are interested in adjusting the factors for desired outputs. Thus, $w1$ will be adjusted to compensate the frequency, but for high gain, it should be as large as possible. So, it will be adjusted to level 3. This adjusts frequency to 5.027 GHz, gain to 11.835 dB and NF to 1.978. Similarly, we can see frequency is significantly affected by $w4$, $w5$, $l1$ and $l2$ also. But $l2$ has 0% effect on NF if so frequency will be adjusted by $l2$. The design obtained after this step was [12,12,12,10,70,8,8,10,50,8,8,228,228] giving gain of 12.176 dB, center frequency of 5.008 GHz and NF (at center frequency) of 1.953.

Step 3 - Once frequency is adjusted to the desired value,

NF can be adjusted without affecting frequency by resistor parameters. From ANOVA tables, it is clear that $w6, w7$, $l1$ and $l2$ have a rarge effect on NF and about 0% effect on frerquency. Thus NF will be adjusted according to these factors, taking into consideration the constraint given by the design library of components that ratio of width to length of resistor should not be lesser than 1 and greater than 10 (as mentioned earlier also). The design obtained after this step was [12,12,12,10,70,4.95,5,10,16.2,10,10,228,228] providing gain of 10.429 dB, center frequency of 5.008 GHz and NF of 2.096. This is the desired output.

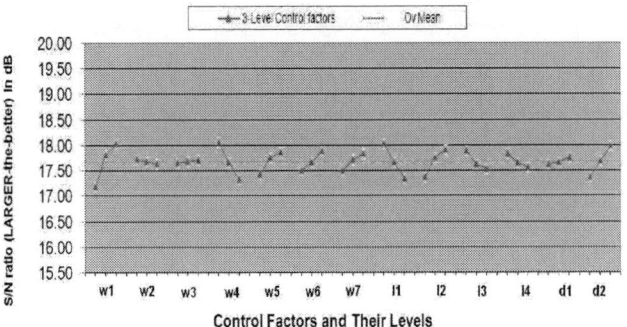

Fig. 2. Factor Effects for Gain (in dB) for larger-the-better

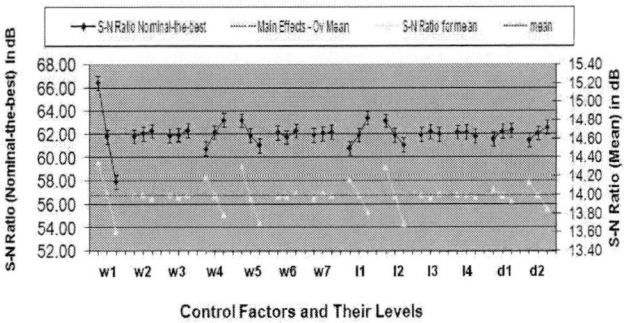

Fig. 3. Factor Effects for Frequency (in dB) for nominal-the-best

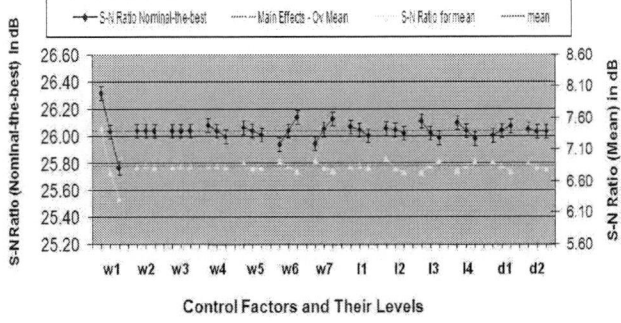

Fig. 4. Factor Effects for NF (in dB) for nominal-the-best

TABLE VI
COMPARISON OF INITIAL DESIGN AND OPTIMIZED DESIGN

Design	Gain	Frequency (GHz)	Noise Figure	Band Width (GHz)	Power (mW)
Initial	8.00	5.008	2.096	5.252	18.0721
Optimized	10.429	5.008	2.096	6.002	17.3304

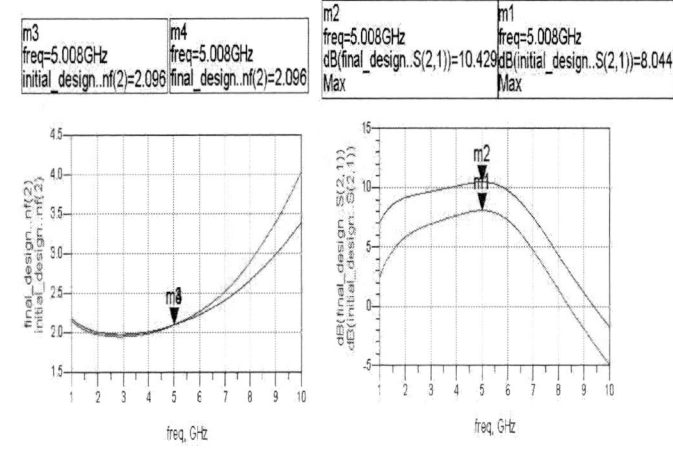

Fig. 5. Comparison of initial and final design

V. RESULTS AND CONCLUSION

A multi-constrained optimization problem for 5 GHz RF Amplifier has been solved using ANOVA. There is 2.4 dB improvement in gain without affecting other outputs which are center frequency of 5.008 GHZ and noise figure of 2.096, just by adjusting the values of design factors (components). Consequently, band width is increased by 750 MHz and power consumption is reduced by 0.6 mW. Larger-the-better for gain of amplifier alongwith nominal-the-best for center frequency as well as for NF, was used. A step by step method is presented which is useful for optimization of a particular response of RF circuits without affecting the other responses significantly.

REFERENCES

[1] D.M. Colleran et al, "Optimization of phase-locked loop circuits via geometric programming"; IEEE CICC 2003, pp. 377 - 380, Sept. 2003.
[2] F.Antonini, G. Antonini,"Genetic optimization of charge-pump PLL parameters"; The 45th MWSCAS 2002.
[3] Jai Narayan Tripathi, "Designing, Optimization and Modeling of Analog/RF Circuits by Design of Experiments", Ph.D. Forum, 19th IFIP/IEEE VLSI-SoC, pp.457 - 460, Oct. 2011, Hong Kong, China.
[4] Y.Deval et al,"Toward analog circuit synthesis: a global methodology based upon design of experiments", 13th Symposium on Integrated Circuits and Systems Design, pp. 295-300, 2000.
[5] M. S. Phadke; "Quality Engineering Using Robust Design", Pearson Education, 2008.
[6] J N Tripathi et al "Designing Asymmetric 2.4 GHz RF Oscillator for Improving Signal Integrity by Design of Experiments", APCCAS 2010, Dec 2010, Malaysia.
[7] J N Tripathi et al; "Designing Asymmetric 2.2 GHz RF Oscillator by Design of Experiments Using Taguchi Methods", 5th ECCSC'10, Nov. 2010, Belgrade, Serbia.
[8] J N Tripathi, J Mukherjee and P R Apte, "Optimizing Phase Noise of 2 GHz RF Oscillator with Minimum Frequency Deviation", 54th MWSCAS 2011, Seoul, South Korea.

A 5mA 2.4GHz 2-point modulator with QVCO for Zigbee tranceiver in 0.18-μm CMOS

Dan Lei Yan, Bin Zhao, M.Kumarasamy Raja and Yuan Xiaojun

Institute of Microelectronics , A*STAR

Singapore. 117685

yandl@ime.a-star.edu.sg

Abstract— **A low power 2.4GHz 2-point modulator with quadrature voltage control oscillator (QVCO) is demonstrated for Zigbee transceiver application using 1P6M 0.18-μm RF CMOS process. Special design techniques are proposed to achieve accurate frequency deviation Δf, in the presence of varactor mismatches and process variations. Measured error vector magnitude (EVM) of the transmitter is 6.5% even without auto calibration for the error in Δf and the spectrum meets 802.15.4 mask with margin. The implemented transmitter modulator covers 2405MHz to 2480MHz band with 16 channels each 5MHz wide as allocated by IEEE 801.15.4. The synthesizer also provides IQ output when operated in the receiver mode with a measured phase noise of -109dBc/Hz at 1MHz offset over the entire tuning range. The active die area is 1mm x 1mm. The chip operates over a wide range of supply voltage from 1.6 V to 2.0V and temperature from -40°C to +85°C. The chip draws 5 mA current from a +1.8V supply at +25°C.**

Index Terms—**CMOS, Zigbee, 2 point modulation, Transmitter, Fractional-N Synthesizer, Transceiver.**

I. INTRODUCTION

The 2.4GHz PHY of the IEEE 802.15.4 (Zigbee) standard attracts a lot of focus from the wireless industry because the globally available 2.4 GHz ISM band with 80MHz bandwidth promotes tremendous market opportunities [1] and enables connectivity with battery powered transceiver as the network node. Direct conversion/modulation architecture is a candidate which features low power operation, reduced component count, improved performance, high level of integration, and low cost. Zigbee transceivers based on 2-point modulation exhibit the lowest dc power and the highest circuit integration.

A synthesizer is one of the critical blocks in design of the transceiver to meet low power requirement for the transmitter and the receiver. In a Zigbee transceiver design, 2-point modulation with fractional-N synthesizer that directly drives the power amplifier is used for transmitter and the synthesizer with QVCO that is used to drive down conversion mixer for the receiver. Special design techniques are needed in the design to achieve accurate frequency deviation Δf, in the presence of varactor mismatches and process variations.

The ZigBee modulation is Offset Quadrature Phase Shift Keying (O-QPSK) with half-sine pulse shaping which is equivalent to Minimum Shift Keying (MSK) [1][3].

Conventional I/Q modulator plus up-converter type of transmitter will consume more power because it needs up-converting I/Q mixers driven by a pair of baseband DACs+LPFs to generate half-sine pulse shaped baseband signals[1]. For low power design, 2-point direct FSK modulation with frequency deviation (Δf) of ±500 KHz (frequency change is 1MHz) shall be a better choice [3-7]. In [6], auto calibration was required to achieve accuracy of frequency deviation Δf in the presence of mismatches in the extremely smaller sized varactors used for modulation and process variation. This increases the die area and power consumption. We propose a low power, fast settling direct modulation fractional-N synthesizer circuit using a pair of normal size varactors in series parallel combination with MIM capacitors to get low modulation sensitivity. As the normal size varactor has much less process variations, there is no need auto calibration scheme to meet Zigbee specification as in [7].

The paper is organized as follows. In Section II the chip architecture is explained in detail followed by key building blocks design challenges and circuit topologies described in Section III. In Section IV the circuit implementation and layout are briefly explained. Finally, the measured results are discussed in Section V and summarized in Section VI.

II. CIRCUIT ARCHITECTURE

The proposed architecture of 2-point modulation transmitter is shown in Fig. 1. The circuit includes a 32MHz crystal oscillator, a low power direct FSK modulation control circuit, a 250Kbps to 2Mcps digital spreading block with FSK coding, and a fractional-N synthesizer with external loop filter. The synthesizer uses 16MHz signal as reference and

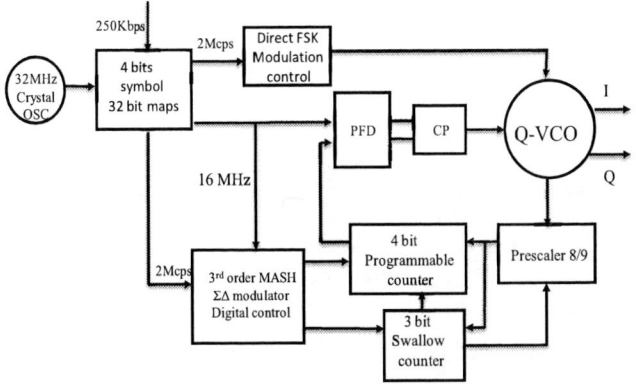

Fig. 1. Architecture of the two point modulation transmitter with QVCO

clock to delta-Sigma modulator. The output frequency of synthesizer can be obtained using the relation:

$$f_{SYNTH} = (8 \times P + S + \frac{F}{2^{21}}) \times f_{ref} \qquad (1)$$

Where, P and S are the values of the programmable counter and swallow counter respectively. F is the value of 21 bits 3rd -order $\Sigma\Delta$ modulator's register input.

III. KEY BUILDING BLOCKS

A. Quadrature Voltage-Controlled Oscillator (QVCO)

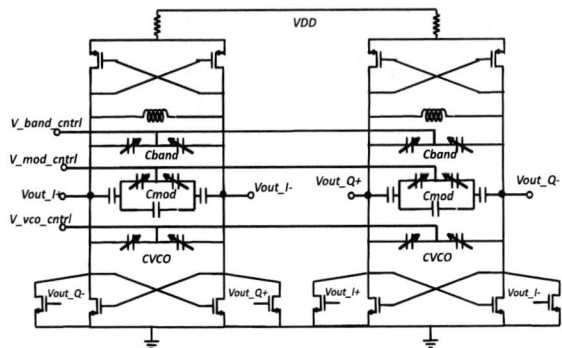

Fig. 2. Schematic diagram of QVCO.

The schematic of QVCO is shown in Fig. 2. In order to get low phase-noise and low reference spurs, the tuning sensitivity has been reduced. This is done by using 2 bands for the QVCO. In this design, the tuning sensitivity of VCO is smaller than 100MHz/V. The VCO has 2-bands and the band can be selected through digital controls V_band_cntrl. Accumulation-mode MOS varactors Cvco are used for frequency tuning. The phase-noise of VCO is more critical for the overall phase-noise of the synthesizer. For the direct VCO modulation to generate ±500KHz FSK signal, tuning sensitivity of about 3MHz/V is needed. Which demands very small modulation varactors (only few fF capacitance). Most of prior work uses very small size pair NMOS or PN diode varactor[5-7]. But very small size varactor are sensitive to mismatch and process variation, this variation will cause gain

mismatch of 2 point modulation [6].To meet the specified <35% EVM in IEEE 802.15.4, the gain mismatch should be maintained within ±10% to leave enough design margin[4], simulation show process and mismatch variation should be less than about 20%. By monte-carlo simulation a pair 3 fF NMOS varactor have 26% variationas shown in Fig.4.(a). This will cause gain mismatch is out of 10% margin. Hence, auto gain calibration circuit is employed [6]. In this design, a normal big size varactors pairs are used to form a equivalent very small capacitance, but have smaller process and mismatch variation.

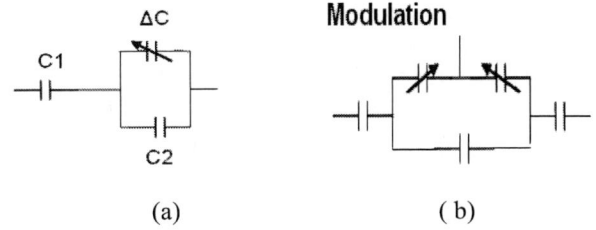

Fig. 3. Schematic diagram of equivalent varactor for FSK modulation.

The modulation varactors is shown in Fig.3.(a). Where ΔC is varactor capacitance change. So the equivalent capacitance change is

$$C_{\Delta mod} = \frac{C_1(C_2+\Delta C)}{C_1+C_2+\Delta C} - \frac{C_1(C_2)}{C_1+C_2} = \frac{C_1^2 \Delta C}{(C_1+C_2+\Delta C)(C_1+C_2)}$$

If $C_1+C_2 >> \Delta C$, then

$$C_{\Delta mod} = \frac{C_1^2 \Delta C}{(C_1+C_2)^2} \qquad (2)$$

Where C_1 and C_2 are MIM capacitors. Hence, the equivalent capacitance change is equal to change in varactor capacitance divided by the square of $C_1 / (C_1 + C_2)$. If C_2 is 10 times than C_1, the capacitance change is only less than 1/100 times. The circuit is very easy to implement for cross couple VCO as shown in Fig.3.(b).

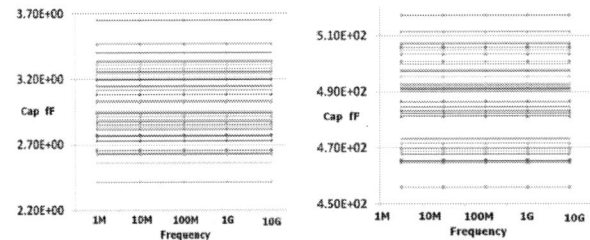

Fig. 4. Variation of process and mismatch for (a) very small NMOS and (b) normal size varactor.

The process and mismatch variation of equivalent varactor is only depend on normal size varactor. From monte-carlo simulation, the normal size varactor have 6% variation only

978-1-61284-863-1/11 $26.00 © 2011 IEEE

as shown in Fig.4.(b)., simulation show the gain mismatch should be within 4%. Hence, one time manual calibration is enough to compensate for mismatch. `

B. Direct FSK modulation control circuit

Direct FSK modulation control is used to control the modulation varactors and make VCO frequency shift between ±500KHz at 2Mbps chip rate. The control signal is square wave with very fast setting time at rising and falling edge. Most of 2 point modulator use DAC to generate this control signal[4-6], in order to generate fast setting control signal, a higher power DAC are needed. In this design, an ultra low power direct modulation circuit is used. Since the Δf is fixed value of ±500KHz, V_{high} and V_{low} are also fixed. The direct modulation circuit is design as shown in Fig.5.

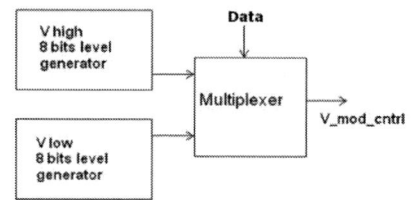

Fig. 5.   Schematic diagram of direct FSK modulation control

Where V_{high} and V_{low} are generated by 8 bits R-ladder, the 2Mbps modulation data is used to control Multiplexer switch, the direct modulation control signal (V_mod_cntrl) is switch between V_{high} and V_{low} base on the input modulation chips and with very fast setting time. Because V_{high} and V_{low} are static, total current of modulation circuit are only 60uA.V_{high} and V_{low} could be set by one time calibration.

C. Phase Frequency Detector (PFD), Charge-pump

A dead-zone free classic phase frequency detector (PFD) is employed for phase and frequency comparison to ensure wide lock range. The current of charge-pump is set to 150uA from low power design perspective.

D. Prescaler, Counters and 3rd–order MASH Σ-Δ modulator

A true-single-phase clock (TSPC) prescaler 8/9 is used for the synthesizer [9]. The prescaler consists of a divide-by-4/5 synchronous counter and a divide-by-2 asynchronous counter. In combination with 5-bit programmable counter and 4-bit swallow counter, the two counters are controlled by 3rd-order 21-bit multi-stage noise-shaping (MASH) sigma-delta modulator. For the 2 point modulation, fractional value related with ±500KHz was added fractional input and synchronized with direct modulation control signal was as shown in Fig.1. The signal injected to directly modulate the VCO is one time manual calibration to alleviate the mismatch effect. The direct modulate the VCO is subject to the high-pass shaping; while the signal applied to the fractional divider is low-pass filtered. These two signal paths are combined and produce an output which is free from the synthesizer loop bandwidth limitation [7].

IV.   CIRCUIT IMPLEMENTATION AND LAYOUT

The reference spur is mainly due to the substrate and power supply reference noise coupling. The reference noise is generated from the digital circuit of synthesizer [10], which includes the reference counter, PFD, dividers, MASH Σ-Δ modulator. In order to reduce the noise coupling from the substrate, the analog and digital circuit are separated wide apart. The inductor used in VCO core occupies large area and more sensitive to the substrate noise coupling. Hence, a guard ring is added for VCO core circuit separately. Guard-ring that connect to substrate ground is added to surround each digital circuit in order to reduce the suppress coupling of reference spur noise to synthesizer output. The spur noise from the power supply was suppressed by separating the power supplies of digital and analog circuits.

V.   MEASURED RESULTS

The 2 point modulator with QVCO for Zigbee transceiver is implemented in standard 1P6M 0.18-μm RF CMOS process with 2-μm top metal and MIM capacitor options. The microphotograph of the test chip is shown in Fig.6. The active area occupied by the circuitry is 1.0mm × 1.0mm.

Fig.6.   Chip microphotograph (active die size: 1.0.mm x 1.0mm).

Fig.7. shows the measured phase-noise plot at 2405MHz with the loop filter bandwidth of 45 kHz. The phase-noise at 1MHz offset is -109dBc/Hz, Measurement results show that synthesizer has constant phase noise performance over the entire 2405 MHz to 2480MHz band. Fig.8. shows the modulation spectrum and PSD MASK of Zigbee transmitter with 2Mcps Pseudo random pulse data modulation signal. Fig.9. shows the measured 6.5% EVM of transmitter signal by using VSA with 2.45GHz Zigbee option..

978-1-61284-863-1/11 $26.00 © 2011 IEEE

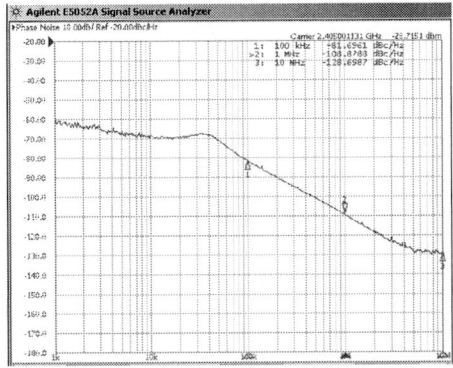

Fig. 7.   Measured phase-noise.

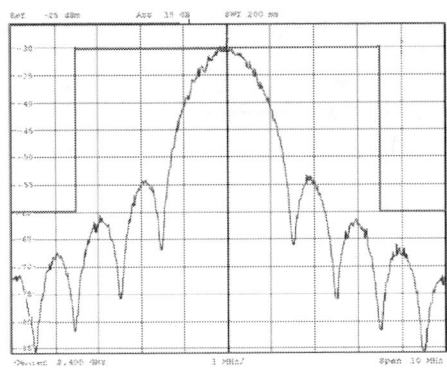

Fig.8.   Measured spectrum and PSD MASK of transmitter for PRBS data at 2 Mbps .

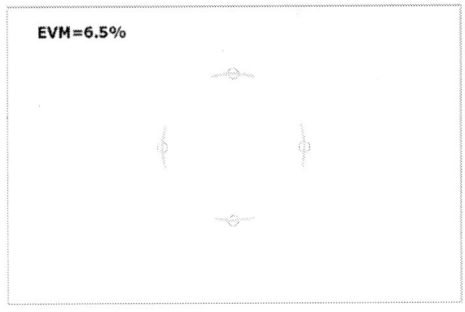

Fig. 9.   Measured 6.5% EVM of transmitter.

VI.   CONCLUSION

Design, implementation and measurement results of 2 point modulator for Zigbee transmitter have been presented. The chip consumes 5mA from a single +1.8V power supply achieving -109dBc/HZ at 1MHz offset over the entire band from 2405MHz to 2480MHz band. Measured result shows transmitter signal EVM is 6.5%, and Spectrum of modulation signal meet the PSD mask.

ACKNOWLEDGMENT

The authors would like to thank HealthSTATS International Pte Ltd, Singapore for their support, and thanks the assistance of the staff of Integrated Circuits and Systems Laboratory at the Institute of Microelectronics, A*Star, Singapore, in the successful realization of the presented IC.

REFERENCES

[1]   IEEE Standard, 802.15.4TM-2006.

[2]   M.Kumarasamy Raja, Xuesong Chen, Yan Dan Lei, Zhao Bin, Ben Choi Yeung and Yuan Xiaojun., " A 18mW Tx 22mW Rx Tranceiver for 2.45GHz IEEE 802.15.4 WPAN in 0.18- μm CMOS ", IEEE A-SSCC, *Dig. Tech. Paper,* Nov. 2010.

[3]   J. Notor, A. Caviglia, G.levy, "CMOS RFIC Architectures for IEEE 802.15.4 Networks," White Paper, Cadence Design Systems, 2003.

[4]   Y.S. Eo et al, "A Fully Integrated 2.4GHz Low IF CMOS Transceiver for 802.15.4 Zigbee Applications," in Proc. IEEE A-SSCC, Nov. 2007, pp. 164-167.

[5]   S. Beyer et al, "A 2.4GHz Direct Modulated 0.18μm CMOS IEEE 802.15.4 Compliant Transmitter for ZigbeeTM," in Proc. IEEE CICC, Sept.   2006, pp. 121-124..

[6]   Rui Yu, Theng-Tee Yeo, "A 5.5mA 2.4-GHz Two-Point Modulation Zigbee Transmitter with Modulation Gain Calibration," in Proc. IEEE CICC, Sept. 2009, pp. 21-24.

[7]   C.L. Ti, T.H. Lin, "A 2.4-GHz 18-mW Two-Point Delta-Sigma Modulation Transmitter for IEEE 802.15.4," in Proc. IEEE VLSI-DAT Symp., Apr. 2007. pp. 1-4.

[8]   D. Muer, and M. Borremans, "A 2GHz phase noise integrated LC VCO set with flicker noise up conversion minimization," *IEEE J. Solid-State Circuits, vol.35,* no. 7, pp. 1034-1038, Jul. 2000.

[9]   J. Yuan, and C. Svensson, "New single-clock CMOS latches and flip-flops with improved speed and power savings," *IEEE J. Solid-State Circuits,* vol. 32, no. 1, pp. 62-67, Feb. 1997.

[10]   S. Pellerano, S. Levantino, C. Samori, and A.L. Lacaita, "A dual-band frequency synthesizer for 802.11a/b/g with fractional-spur averaging technique," *ISSCC Dig. Tech. Papers,* pp. 104-106, vol. 1, Feb. 2005

978-1-61284-863-1/11 $26.00 © 2011 IEEE

A 2GHz Tx LO Generation Circuit with Active PPF and 3/2 Divider in 65nm CMOS

Markus Törmänen[1], Andreas Axholt[1], Jonas Lindstrand[1], and Henrik Sjöland[1,2]

[1]Electrical and Information Technology
Lund University
Lund, Sweden

[2]Ericsson Research
Ericsson AB
Lund, Sweden

Abstract— This paper demonstrates a 2GHz Tx LO generation circuit with 8% tuning range featuring a VCO, an active polyphase filter (PPF), and a 3/2 divider with 50% output signal duty cycle, implemented in a standard 65nm LP CMOS process. The circuit measures a phase noise of –158dBc/Hz at 20 MHz offset frequency, with a power consumption of 29.8mW. It occupies a chip area of 0.53 mm^2, including pads. The spurious tone at 1GHz, caused by mismatch in the active polyphase filter and frequency divider, is at -50dBc.

Index Terms — CMOS integrated circuits, Frequency dividers, Polyphase filters, Voltage controlled oscillators.

I. INTRODUCTION

The transmitter LO signal generation must be robust against power amplifier (PA) pulling of the voltage controlled oscillator (VCO), especially if both blocks should be implemented on the same chip. A solution is of course to take different actions to increase the isolation, which increases the cost, but also to have the VCO operate at a different, preferably non-integer, frequency than the transmitter. Having the VCO operate at non-integer multiple of the LO frequency, *e.g.* 1.5 times, followed by a frequency divider, also has some issues as the output signal is often required to have a 50% duty cycle.

Previously demonstrated 3/2 dividers with 50% duty cycle are based on regenerative dividers (mixers with feedback) [1][2], however, they tend to have poor suppression of spurious tones. An injection locked divider is demonstrated in [3], but they are often sensitive to PVT variations. The implemented 3/2 divider utilizes standard logic cells, however, quadrature input signals are required to produce a 50% duty cycle output signal.

The quadrature LO signal generation can be performed in a number of ways, *e.g.* quadrature oscillator (QVCO) or active/passive polyphase filter (PPF). The PPF typically occupies less area than a QVCO, but the passive PPF attenuates the signal, whereas the active PPF consumes DC power [4][5]. The QVCO can also suffer from reduced frequency tuning range and phase noise performance compared to a VCO followed by a PPF.

This research was sponsored by the VINNOVA Industrial Excellence Center – System Design on Silicon.

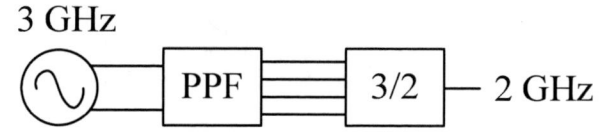

Figure 1. Block schematic.

II. CIRCUIT DESIGN

The block schematic of the implemented circuit is shown in Fig. 1. It consists of a voltage controlled oscillator (VCO), an active polyphase filter (PPF), and a 3/2 divider. The VCO provides a differential signal to the PPF, which then converts it to differential quadrature signal components for the divider. The divider produces a 2GHz single-ended output signal from the 3GHz quadrature signal input.

A. Voltage Controlled Oscillator (VCO)

The schematic of the VCO is shown in Fig. 2. It consists of a cross-coupled pair (M_1) that realizes the negative resistance to compensate for the resonator losses. Also, the source inductor L_s, FET current source (M_5), and capacitor C_s, form a filtered current source [6], which improves the VCO phase noise performance. It is used to control the output voltage swing, since the best phase noise performance is typically obtained when the amplitude in the resonance tank is on the border of the voltage limited region.

Large CMOS varactors, with their step-like C-V characteristics, have problems with AM to PM noise conversion [7]. To reduce this effect the implemented varactor consists of a smaller continuously tuned part (M_2) combined with a larger digitally controlled part (M_3 and $M_4=2M_3$) [6].

B. Active Polyphase Filter (PPF)

The schematic of the active PPF is shown in Fig. 3. To prevent the input resistance of the first stage of the PPF from loading the VCO, and thus degrading the phase noise

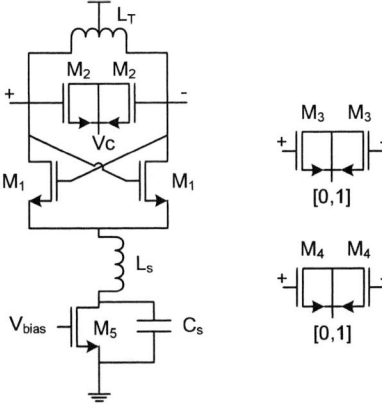

Figure 2. Circuit schematic of the VCO.

performance, buffer stages are used for isolation. Three stages are used and the chosen configuration gives best quadrature phase accuracy over the bandwidth, which is critical for suppressing spurious tones in the divider. The stages are designed to have progressively higher load impedances by dimensioning the transconductance amplifiers according to (1).

$$\tau = \frac{1}{f} = R \cdot C = \frac{C}{g_m} \qquad (1)$$

The different stages were sized to not degrade the VCO phase noise according to $\tau = 1/f$; $f_1 = 1.3f_0$, $f_2 = f_0$, $f_3 = 0.7f_0$, with f_0 equal to 3GHz. A MIM capacitor of 250fF is used in all stages for improved matching. Inverters are used to buffer the PPF output signals to the divider input.

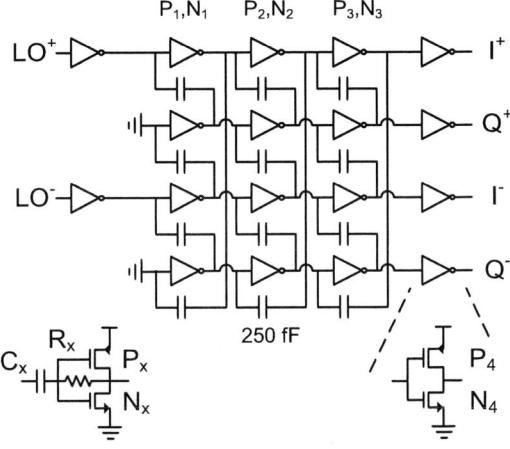

Figure 3. Circuit schematic of the active PPF.

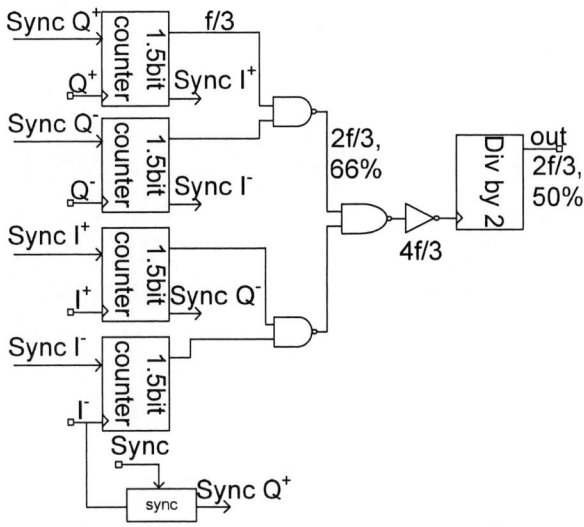

Figure 4. Block schematic of the 3/2 divider.

C. 3/2 Divider

A 3/2 divider can be implemented using just an inverter, a multiplexer, and a divide by 3 circuit, however, the duty cycle cannot be 50% using that approach. The presented 3/2 divider is shown in Fig. 4. It has a quadrature input, supplied by the active poly phase filter, and generates a 50 % duty cycle output. The quadrature signals drive the clock inputs of the four 1.5 bit counters. After the counters the frequency is converted to f/3. The signals from the counters are logically combined in two NAND-gates (Q+ NAND Q-, I+ NAND I-), generating a pulse train with the frequency f/1.5, however, the duty cycle is not 50 %. The two signals from the NAND-gates are fed to a NOR gate, generating 2*f/1.5. The signal is then divided by 2 to reach the desired division number, 3/2, with 50% duty cycle.

A well-defined start-up of the 1.5 bit counters is necessary for correct operation. The timing requirements become tough if a common synchronization signal, latched by one of the inputs, is used, since the next rising edge of the quadrature signals occurs just a quarter of a period later. By propagating a synchronization signal though the counters, starting with counter I-, followed by Q+, I+, and finally Q-, the timing requirement is relaxed. The timing margin is relaxed by a factor of 3, to ¾ of a period, which is sufficient to cope with gate delays and PVT.

The rise and fall time is balanced in the logic gates of the signal path; in *e.g.* in a NAND-gate two cascaded p-type transistors were implemented in parallel to the conventional two p-type transistors, ensuring similar rise and fall times with larger transistor dimensions. The phase noise also benefits from increasing the size of the transistors. However, the blocks determining only the startup behavior are kept small since they do not contribute to the phase noise. For measurement purposes an inverter is also implemented as a buffer at the divider output.

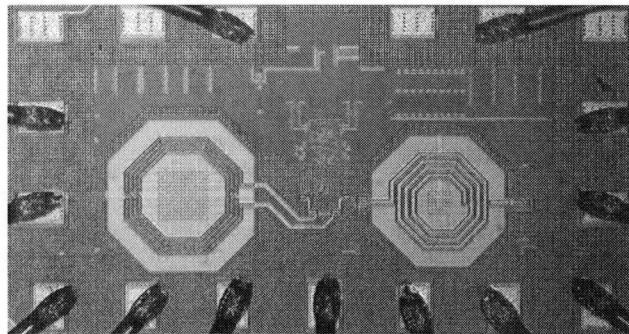

Figure 5. Die photograph, 0.53mm².

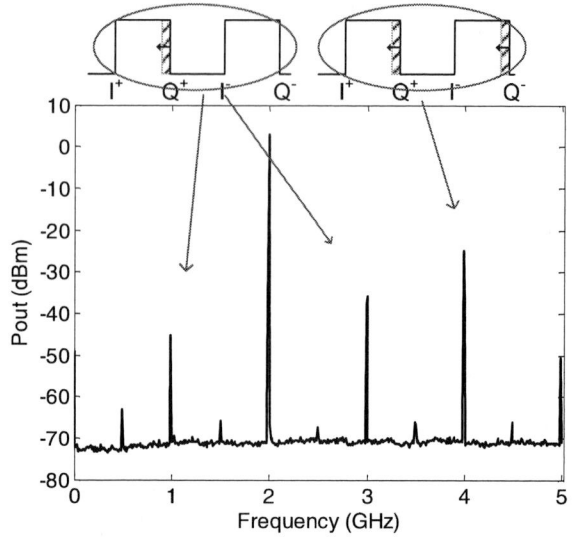

Figure 7. Measured output spectrum.

III. LAYOUT

The die photo of the implemented circuit is shown in Fig. 5, where the total area including pads measures 0.53mm². The layout was carefully drawn to reduce mismatch between the branches, where the layout of the PPF and divider is critical to obtain high quadrature accuracy and thus to reduce the spurs. Transistors critical for matching have dummy fingers, non minimum channel length is used in the PPF stages, and large transistor dimensions are used in the divider not to degrade the phase noise. The chip was mounted on a PCB to which the pads were wire bonded.

IV. EXPERIMENTAL RESULTS

A Rohde and Schwarz spectrum analyzer (FSEB) and a Europtest phase noise measurement system (PN9000) was used to measure the frequency tuning characteristic and phase noise, respectively, of the implemented chip.

The chip consumes a total DC power of 29.8mW, where the divider (8.7mA) and PPF (7.5mA) operate from a 1.2V supply. The VCO consumes 8mA from a 1.3V supply.

The measured frequency tuning characteristic is shown in Fig. 6. There are four different states for the 2-bit digitally controlled part of the varactor. The effective tuning range measures 8%.

A measured frequency spectrum of the divider output is shown in Fig. 7, where the fundamental 2GHz tone has an output power of 3dBm. There are three, in particular, interesting spuriouses in the measured spectrum, located at 1GHz, 3 GHz, and 4 GHz. The second harmonic at 4 GHz is at -30dBc, which corresponds to an average duty-cycle close to 50% (49 or 51%). The transitions of the output signal correspond to alternating transitions of the quadrature input signal, thus a small phase modulation is expected. The sideband at 1 GHz is just -50dBc indicating a small cycle to cycle variation. The 3GHz sideband is also a measure of this variation, however, the VCO leakage is also added to this

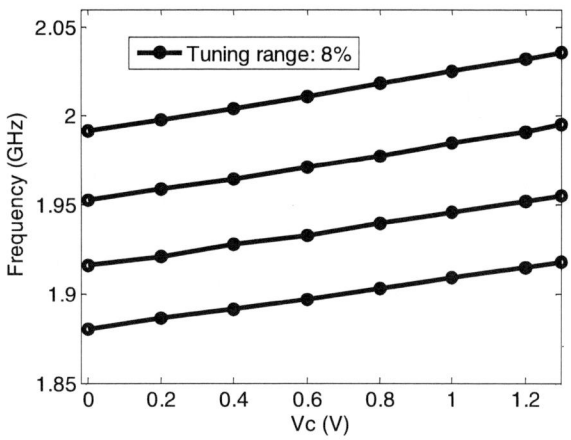

Figure 6. Measured frequency tuning characteristic.

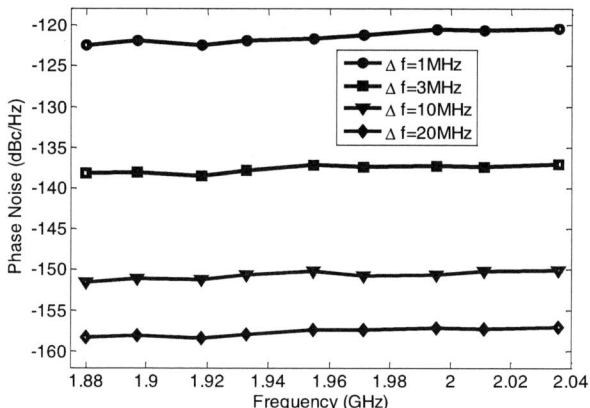

Figure 8. Measured phase noise at different offsets vs. frequency.

978-1-61284-863-1/11 $26.00 © 2011 IEEE   210

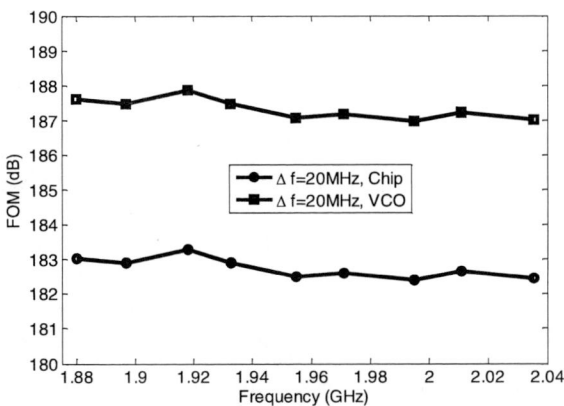

Figure 9. Measured phase noise figure of merit.

$$10 \cdot \log_{10}\left(\left(\frac{f_0}{\Delta f}\right)^2 \cdot \frac{1}{10^{L(\Delta f)/10} \cdot P(mW)}\right) \qquad (2)$$

sideband. Based on these measurement results the quadrature phase error is below 2 degrees and the differential phase error is less than 0.4 degrees.

The measured phase noise at different offset frequencies is shown in Fig. 8. At 20MHz offset the circuit achieves a phase noise of -158dBc/Hz. The phase noise figure of merit (FOM), (2), at this offset frequency is plotted in Fig. 9. Also shown in Fig. 9 is the de-embedded FOM of the VCO.

V. COMPARISON

A fair performance comparison to other 3/2 dividers is difficult, since spur levels, phase noise performance, and area occupied are often not reported. Some papers have embedded the divide-by-3/2 circuit in a phase locked loop [8], PLL, complicating a phase noise comparison. The most common 3/2 frequency division technique reported is based on the regenerative Miller divider [9]. In [2] the divider is based on a single-sideband balanced mixer, SSBM, driven by a quadrature VCO, and the reported spur at $f_{vco}/3$ is -36dBc. In [1] the SSBM is driven by a VCO and polyphase filter, where the measured spur at $f_{vco}/3$ is -20dBc. These numbers are 14 and 30dB, respectively, higher than in this work.

The output phase noise in [2] at 1 MHz and 10 MHz offset is -120dBc/Hz and -140dBc/Hz, respectively. At far-out frequency offsets, the phase noise is dominated by the divider

noise. The total current consumption is 34mA, of which 10mA is consumed by the divider from a 1.2V supply in a 0.13μm CMOS process.

The proposed divider consumes 8.7mA from a 1.2V supply, and occupies an area of 0.2mm x 0.07mm. The total phase noise, including VCO and active poly phase filter, measures -123dBc/Hz and -152dBc/Hz at 1MHz and 10MHz offset frequencies, respectively.

I. CONCLUSION

A Tx LO generation circuit featuring VCO, active polyphase filter, and a 3/2 divider has been demonstrated. At 29.8mW power consumption, the circuit produces a 2GHz output signal with 8% tuning range and a phase noise of -158dBc/Hz at 20MHz offset frequency, corresponding to a FOM of 183dB. Furthermore, the de-embedded FOM of the VCO is 188dB. The spur at 1GHz is at -50dBc, corresponding to a combined quadrature phase error below 2 degrees in the active polyphase filter and frequency divider.

REFERENCES

[1] H. Shin and B. Won, "A 4.5 to 9.2-GHz Wideband Semidynamic Frequency Divide-by-1.5 in GaInP/GaAs HBT," *IEEE Microwave and Wireless Components Letters*, vol. 17, no. 1, pp. 73-75, 2007.

[2] D. Guermandi, P. Tortori, E. Franchi, and A. Gnudi, "A 0.83-2.5-GHz continuously tunable quadrature VCO," *IEEE Journal of Solid-State Circuits (JSSC)*, vol. 40, no. 12, pp. 2620-2627, 2005.

[3] S. Hara, K. Okada, and A. Matsuzawa, "10MHz to 7GHz quadrature signal generation using a divide-by-4/3, -3/2, -5/3, -2, -5/2, -3, -4, and -5 injection-locked frequency divider," *IEEE Symposium on VLSI Circuits (VLSIC)*, pp. 51-52, 2010.

[4] F. Tillman and H. Sjöland, "A Polyphase Filter based on CMOS Inverters," *Analog Integrated Circuits and Signal Processing*, vol. 50, no. 1, pp. 7-12, 2007.

[5] J. Wernehag and H. Sjöland, "Analysis of a high frequency and wide bandwidth active polyphase filter based on CMOS inverters," *Analog Integrated Circuits and Signal Processing*, vol. 59, no. 3, pp. 243-255, 2009.

[6] E. Hegazi, H. Sjöland, and A. Abidi, "A Filtering Technique to Lower LC Oscillator Phase Noise," *IEEE Journal of Solid-State Circuits (JSSC)*, vol. 36, no. 12, pp. 1921–1930, Dec. 2001.

[7] E. Hegazi and A. Abidi, "Varactor Characteristics, Oscillator Tuning Curves, and AM-FM Conversion," *IEEE Journal of Solid-State Circuits (JSSC)*, vol. 38, no. 6, pp. 1033–1039, June 2003.

[8] Jongsik Kim et al, "A CMOS Direct Conversion Transmitter With Integrated In-Band Harmonic Suppression for IEEE 802.22 Cognitive Radop Applications," *IEEE Custom Integrated Circuits Conference (CICC)*, pp. 603–606, 2008.

[9] Miller R.L, "Fractional-Frequency Generators Utilizing Regenerative Modulation", *Proceedings of the IRE*, vol. 27, no. 7, pp. 446-457, 1939.

978-1-61284-863-1/11 $26.00 © 2011 IEEE

RF Receiver System for Cognitive Radio Application

Yuanzhong Xue
Nanyang Technological University
Singapore
xyzpirlo1819@hotmail.com

Yuanjin Zheng
Nanyang Technological University
Singapore
YJZHENG@ntu.edu.sg

YEO Kiat Seng
Nanyang Technological University
Singapore
eksyeo@ntu.edu.sg

Abstract— **In order to design RF receiver system, different topologies has been researched in the current industry. A novel RF receiver system with ability of spectrum sensing and demodulation had been successfully designed to meet the requirement for the project. It was designed using MATLAB and simulated based on Simulink. The RF receiver, recover the carrier signal as desired, was able to detect the frequency of input signal and demodulate the received signal. Recommendations for future work include looking into the spectrum sensing method for a more simply scheme and adding a feedback voltage to control VCO for killing the phase error etc.**

Keywords-RF receiver; cognitive radio ; spectrum sensing; demodulation; MATLAB

I. INTRODUCTION

With the increasing demand of wireless data transfer and rapidly emerging of multiple communication standards at crowd spectrum, spectrum scarcity at frequencies is becoming a more and more serious problem. However, despite the fixed spectrum assignment policy that characterizes current Wireless networks, the biggest part of the available spectrum is not well exploited and even that 15% to 85% of the spectrum availability is under used [1]. Thus, a new wireless communication system using spectrum more efficiently than in the past is intensely urged to create in wireless technology. Cognitive Radio (CR) technology has been proposed as a promising solution for improving the efficiency of spectrum usage by adopting a dynamic spectrum resource management concept [2]. The CR system is promising not only because it may improve the efficiency of spectrum usage, but also because it promises improved connectivity and self adaptability of channel environment [3]. Consequently, CR have the potential to revolutionize how devices perform wireless networking.

In this paper, we describe a novel RF receiver architecture that combines spectrum sensing and demodulation for Cognitive Radio purposes.

II. PROPOSED SYSTEM

A. System Design

The most straight-forward way to sense the spectrum is to add a high-speed ADC. But the method is not practical since it will cover the spectrum ranges from several hundred MHz to several GHz. An ultra-high sample rate (>10GHz) power consumption will be required for this ADC. To reduce the requirement of ADC, it is common to implement analog circuit to down convert the signal for further processing. However, the

analog circuit operates in RF frequency and therefore the real circuit implementation become difficulties and is not accurately controlled. Besides these short comes, the circuits have to implement an oscillator to generate a same carrier signal to coherent demodulate the input signal rather than use a recovery carrier signal.

As spectrum sensing need to detect the exact carrier frequency and coherent demodulation need to use the carrier signal, a carrier recovery component should be equipped to generate such a line spectrum. It is easy to use the carrier recovery signal to achieve the aim of spectrum sensing and demodulation.

A Novel RF receiver system which based on this design idea is proposed. Fig. 1 shows the proposed architecture for the cognitive radio system. It employs analog PLL to lock the carrier signal, and then to detect current user and/or interference and then switch circuits settings to other available channels. This sensing system can be easily realized in analog circuits, and also can achieve fast detection, low power consumption, and the low hardware complexity. The suggested sensing techniques potentially can detect a wide spread of wireless standards such as IS-95, WCDMA, EDGE, GSM, Wi-Fi, Wi-MAX, Zigbee, Bluetooth, digital TV (ATSC, DVB), etc.

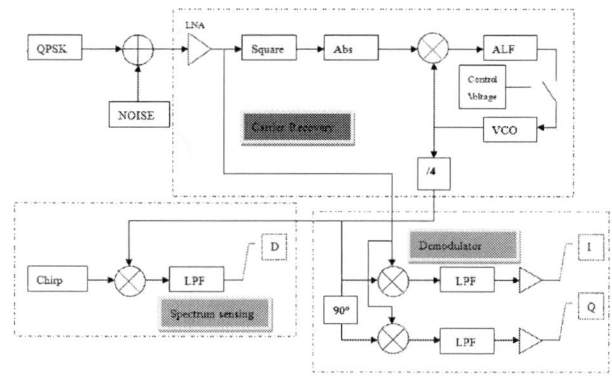

Figure 1:Block diagram of proposed receiver

The RF signals are induced by the antenna and then feed to a low noise amplifier (LNA) at the first stage of the receiver. The weak RF signal is amplified by the first LNA. The Square and Abs blocks after the LNA produce the spectral line at $4fc$. This 4 times frequency multiplication of carrier signal will be tracked by a followed Phase-Locked Loop (PLL). For the PLL part, a two-point modulate method is employed. The control voltage set up a start-up voltage to provide a fixed frequency

978-1-61284-863-1/11 $26.00 © 2011 IEEE

which can be changed among different spectrum segments. Then the switch is close with Active Loop Filter (ALF) to lock the signal precisely. The desired carrier at frequency f_c with almost the same phase of the received signal is then recovered by a divide-by-4 device. A chirp signal generated by a linear voltage controlled VCO sweep a wide range frequency to detect the carrier recovery signal. This carrier recovery signal is also used for coherent demodulating the received RF signals.

B. System Structure

1) Carrier Recovery: The PSK signals have no spectral line at carrier frequency. Therefore a nonlinear device is needed in the carrier recovery circuit to generate such a line spectrum. In [4], a type of carrier synchronizers, the *M*th power loop is explained.

Fig. 2 is the *M*th power loop for carrier recovery for M-ary PSK. For QPSK (or OQPSK, DEQPSK), $M = 4$, it is a quadrupling loop. It is the Mth power device that produces the spectral line at $M f_c$. The phase lock loop consisting of the phase detector, the LPF, and the VCO, tracks and locks onto the frequency and phase of the $M f_c$ component. The divide-by-*M* device divides the frequency of this component to produce the desired carrier at frequency f_c and with almost the same phase of the received signal. Before locking, there is a phase difference in the received signal relative to the VCO output signal. We denote the phase of the received signal as θ and the phase of the VCO output as $M\hat{\theta}$.

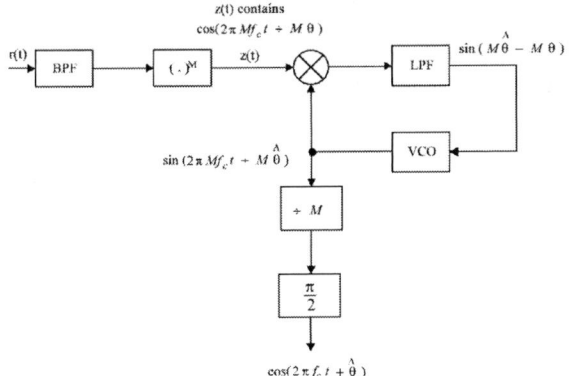

Figure 2: *M*th power synchronizer for carrier recovery

A convenient orthogonal realization of a QPSK waveform, $s(t)$, is achieved by amplitude modulating the in-phase and quadrature data streams onto the cosine and sine functions of a carrier wave, as follows:

$$s(t) = d_I(t)\cos(\omega_0 t + \theta) + d_Q(t)\cos(\omega_0 t + \theta) \quad (1)$$

$$d_I(t) = d_0, d_2, d_4, \dots (even\ bits)$$

$$d_Q(t) = d_1, d_3, d_5, \dots (odd\ bits)$$

For QPSK signal, using (1), noticing that $d_I^2(t) = d_I^2(t) = 1$, we have :

$$z(t) = [s(t) + n(t)]^4$$
$$= \{[d_I(t)\cos(\omega_0 t + \theta) + d_Q(t)\cos(\omega_0 t + \theta) + n(t)]^2\}^2 \quad (2)$$
$$= [1 - d_I(t)d_Q(t)\sin(4\pi f_c t + 2\theta)]^2 + noise\ terms$$
$$= 1 - 2d_I(t)d_Q(t)\sin(4\pi f_c t + 2\theta) + \frac{1}{2} - \frac{1}{2}\cos(8\pi f_c t + 4\theta) + noise\ terms$$

The last signal term contains a spectral line at $4fc$ which is locked onto by the PLL. A divide-by-four device is used to derive the carrier frequency. Note that the $4\pi f_c t + 2\theta$ term resulting from squaring operation cannot produce a line spectrum since $d_I(t)d_Q(t)$ has a zero mean value. Therefore fourth power operation is needed for QPSK (and OQPSK).

2) Spectrum Sesing Detector: For a CR system, the ability to accurately sense the spectrum usage status over a wide frequency range plays an important role. So far, many researchers have proposed various spectrum sensing methods which can be mainly categorized into two groups, i.e., feature detection and energy detection.

Energy detection measures a total power within a certain frequency band and compares it with a threshold level to determine the presence of a signal. This method provides faster sensing time and simpler implementation than those of feature detection. Since there is a trade-off between the detection time and the detection bandwidth or the sensing threshold, it is possible to find a compromise between sensing time and accuracy. Given the wide variation in signal bandwidth and formats that must be reliably sensed, it is preferable to have flexibility in selecting detection bandwidth, just like spectrum analyzer that can adjust its resolution bandwidth depending on the frequency span and sweep time [5].

For these reasons, our receiver has been proposed as a type of energy detection method. Differentiate from the conventional energy detect method that measuring a total power within a certain frequency band, the proposed receiver employ a Low Pass Filter to detect the DC component of the multiple signal produced by a chirp signal and recovery carrier signal. Once the chirp signal has the same frequency with the recovery carrier signal, the DC component will be detected. Based on the time of DC component appearance, it is convenient to find out the frequency of chirp signal which is equate to the frequency of the recovery carrier signal.

3) Coherent Demodulator: As described at previous section, the received RF signal is a QPSK waveform. The RF signals can be coherent demodulated by mixing with the carrier recovery signal and then filtering with a low pass filter. Proposed system incorporates quadrature mixing. So the data stream $d_I(t)$ and $d_Q(t)$ are demodulated in I and Q channels separately. Assume the carrier signal can be recovery without phase error; the carrier recovery signal is described as follows:

$$c_1(t) = \cos(\omega_0 t + \theta) \quad (3)$$

This signal is mixed with the received RF signal directly in I-channel, using (3), we have:

$$s(t) \times c_1(t) = [d_I(t)\cos(\omega_0 t + \theta) + d_Q(t)\cos(\omega_0 t + \theta)] \times \cos(\omega_0 t + \theta)$$
$$= \frac{1}{2}[d_I(t) + d_I(t)\cos(2\omega_0 t + 2\theta) + d_Q(t)\cos(2\omega_0 t + 2\theta)] \quad (4)$$

The carrier recovery signal through a $\pi/2$ phase shift form a sine waveform:

$$c_Q(t) = sin(\omega_0 t + \theta) \qquad (5)$$

Equation (5) is used in Q-channel similarly:

$$s(t) \times c_Q(t) = [d_I(t)cos(\omega_0 t + \theta) + d_Q(t)cos(\omega_0 t + \theta)] \times sin(\omega_0 t + \theta)$$
$$= \frac{1}{2}[d_Q(t) + d_Q(t)cos(2\omega_0 t + 2\theta) + d_I(t)cos(2\omega_0 t + 2\theta)] \qquad (6)$$

After a low pass filter ,the data stream $d_I(t)$ and $d_Q(t)$ are demodulated separately in I/Q channel.

III. KEY BUILDING BLOCKS

In practical design, improvement has been made in *M*th power loop and loop filter of PLL. This improvement lower hardware complexity and power consumption, furthermore, decrease sensing-time.

A. Mth Power Loop

The 4th power loop for carrier recovery for QPSK calls for 4th power device which is highly complex hardware realization. Two series Square function blocks achieving 4th power reduce the hardware complexity. To further lower the complexity, Absolute value function block series with Square block are implemented. The signal after the square function block is:

$$z_S(t) = [s(t) + n(t)]^2$$
$$= [d_I(t)cos(\omega_0 t + \theta) + d_Q(t)cos(\omega_0 t + \theta) + n(t)]^2 \quad (7)$$
$$= [1 - d_I(t)d_Q(t)sin(4\pi f_c t + 2\theta)] + noise\ terms$$

As a high pass filter followed by the Square block, the DC component is eliminated. So the signal pass the Absolute value function block could be written as follows:

$$z_{SA}(t) = |d_I(t)d_Q(t)sin(4\pi f_c t + 2\theta)| + noise\ terms$$
$$= |sin(4\pi f_c t + 2\theta)| + noise\ terms \qquad (8)$$

The $z_{SA}(t)$ produces a spectral line at $4 f_c$ as the same as 4th power device.

B. Loop Filter of PLL

A PLL, as shown in Fig. 3, is a control system, where phase is the variable of interest. The circuit is called a phase-locked loop because the feedback operation in the loop automatically adjusts the phase of the output signal $V_0(f_0, \phi_0)$ to follow the phase of the reference signal $V_i(f_i, \phi_i)$. Thus, both the phase and the frequency of the oscillator are "locked" to the phase and the frequency of the input signal.

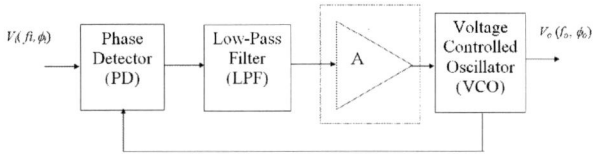

Figure 3: Phase-locked loop

Since we are concerned with the frequency at which ϕ_0 can vary and still be followed reasonably closely by ϕ_i, the bandwidth of a PLL will be discussed as follows. The PLL bandwidth is determined by the gain K_d of the Phase Detector (PD), the high-frequency gain K_h of the loop filter, and the gain K_o of the VCO, as follows:

$$\omega_{3dB} = K_d K_h K_o = K \qquad (9)$$

Since the PD and the VCO designs are usually less flexible, the design of the loop filter is the principle tool in determining the bandwidth.

We also saw from the phase error (in the frequency domain) which is given by (10) as:

$$\theta_e(s) = \theta_i(s) - \theta_o(s) = \frac{s\theta_i(s)}{s + K_d K_o F(s)} \qquad (10)$$

These errors are readily evaluated by means of the final value theorem of Laplace transforms, which states:

$$\lim_{t \to \infty} y(t) = \lim_{s \to 0} sY(s) \qquad (11)$$

Application of the final value theorem to the phase-error equation yields:

$$\lim_{t \to \infty} \theta_e(t) = \lim_{s \to 0} \frac{s^2 \theta_i(s)}{s + K_d K_o F(s)} \qquad (12)$$

In [6], steady-state error cause is classified into two situations: a step change of input phase and a step change of frequency.

For the first situation, steady-state error resulting from a step change of input phase of magnitude $\Delta\theta$ is considered. The Laplace transform of the input is therefore $\theta_i(s) = \Delta\theta/s$, which may be substituted into (12) to give:

$$\lim_{t \to \infty} \theta_e(t) = \lim_{s \to 0} \frac{s\Delta\theta}{s + K_d K_o F(s)} = 0 \qquad (13)$$

In other words, the loop will eventually track out any change of input phase; there is no steady-state error resulting from a step change of phase.

For the other situation, steady-state error resulting from a step change of frequency of magnitude $\Delta\omega$ is considered. Input phase is a ramp, $\theta_i(t) = \Delta\omega t$, so $\theta_i(s) = \Delta\omega/s^2$. Substitution of this value of θ_i into (12) result in:

$$\lim_{t \to \infty} \theta_e(t) = \lim_{s \to 0} \frac{\Delta\omega}{s + K_d K_o F(s)} = \frac{\Delta\omega}{K_d K_o F(0)} \qquad (14)$$

A large DC gain $F(0)$ of the loop filter is needed to keep the θ_e small. Fortunately, active loop filters can achieve $F(0)$ essentially infinite.

The design of an active loop filter [7] begins with an amplifier with gain K_h to modify the bandwidth of the PLL. In practice, most PLL designs call for $K_h < 1$. The second part of the loop filter design is to realize infinite gain at DC by an integrator. By summing the output of the amplifier with the output of the integrator, we get the complete control voltage.

IV. SIMULATION RESULTS

All the simulation of proposed system is done based on MATLAB. MATLAB is a high-level mathematical tool available on many types of computers. It has been designed with signal processing and system analysis in mind.

A. Proposed System Overview

The proposed RF receiver system model is built in Simulink. In this model, two Uniform Random Number blocks is adopted to generate data stream $d_I(t)$ and $d_Q(t)$. This baseband signal is modulated to carrier wave at 922MHz. The sum of the two modulated band pass signal forms a QPSK signal for testing. To close to reality, Gaussian noise is added to the QPSK signal. The received signal has been amplified by a LNA for further processing, including carrier recovery, spectrum sensing and demodulation.

B. Simulation Results

1) Carrier Recovery Signal: After the input signal has been locked by the PLL, the carrier signal has been recovered as desired. The carrier recovery signal has the same frequency with input signal as shown in Fig. 4.

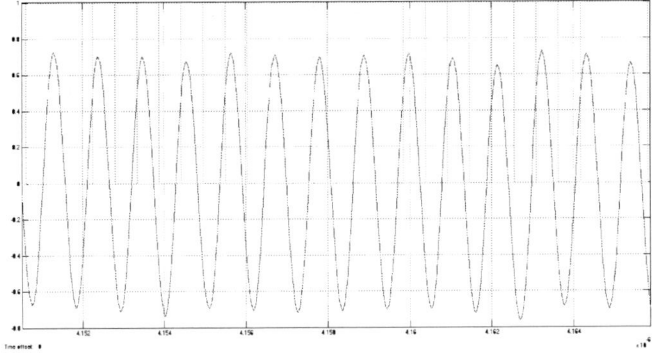

Figure 4: Comparison of Input signal and carrier recovery signal

2) Spectrum Sensing Signal: The chirp signal sweeps a wide range frequency from 200MHz to 1200MHz, once the chirp signal has the same frequency with carrier recovery signal, their multiplied result will pass the low pass filter, as shown in Fig. 5.

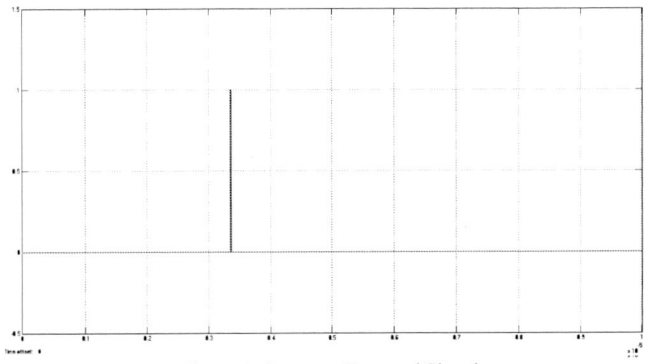

Figure 5: Spectrum Detected Signal

3)Demodulated Signals: The data stream $d_I(t)$ and $d_Q(t)$ are demodulated separately in I/Q channel. The demodulated signals in I/Q channel are plotted in 1th and 3rd axes in Fig. 6, with original data stream $d_I(t)$ and $d_Q(t)$ in 2nd and 4th axes. Comparison between the demodulated signal and original data stream indicates that the demodulator operates well.

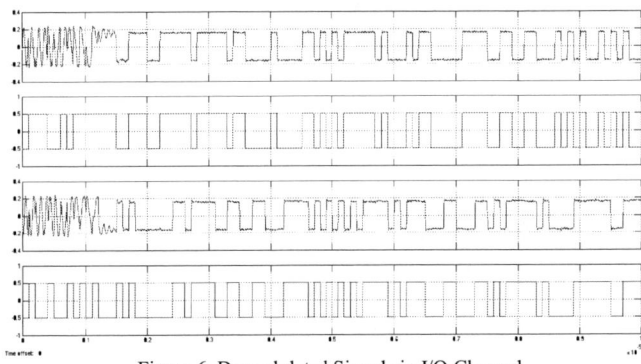

Figure 6: Demodulated Signals in I/Q Channel

V. CONCLUSION

In this paper, a novel RF receiver system had been designed, and successfully been simulated in MATLAB. In order to efficiently achieve spectrum sensing and demodulation, carrier recovery technique is adopted. In order to reduce phase error, PLL is modified by using active loop filter. Preliminary experimental results confirm the overall expected performance and receiver behavior.

REFERENCES

[1] Cabric, D.; Chen, M.S.W.; Sobel, D.A.; Jing Yang; Brodersen, R.W.; , "Future wireless systems: UWB, 60GHz, and cognitive radios," *Custom Integrated Circuits Conference, 2005. Proceedings of the IEEE 2005* , vol., no., pp.793-796, 21-21 Sept. 2005.

[2] Mitola, J., III; , "Cognitive radio for flexible mobile multimedia communications," *Mobile Multimedia Communications, 1999. (MoMuC '99) 1999 IEEE International Workshop on* , pp.3-10, 1999

[3] Kyutae Lim; Laskar, J.; , "Emerging opportunities of RF IC/system for future cognitive radio wireless communications," *Radio and Wireless Symposium, 2008 IEEE* , vol., no., pp.703-706, 22-24 Jan. 2008

[4] Fuqin Xiong, "Digital Modulation Techniques", pp. 192-193

[5] Jongmin Park et al., "A Fully Integrated UHF-Band CMOS Receiver With Multi-Resolution Spectrum Sensing (MRSS) Functionality for IEEE 802.22 Cognitive Radio Applications," *Solid-State Circuits, IEEE Journal of* , vol.44, no.1, pp.258-268, Jan. 2009

[6] Floyd M. Gardner, "Phaselock Techniques", pp. 43-45

[7] Dan H.Wolaver, "Phase-Locked Loop Circuit Design", pp. 25-33

A Divide-by-Two Injection-Locked Frequency Divider with 13-GHz Locking Range in 0.18-μm CMOS Technology

Xiang Yi, Chirn Chye Boon, Jia Fu Lin, Manh Anh Do, Kiat Seng Yeo, Wei Meng Lim

VIRTUS, School of Electrical Electronic Engineering
Nanyang Technological University, Singapore 639798
Email: e090036@e.ntu.edu.sg

Abstract—**In this paper, a new divide-by-two RC-oscillator-based injection-locked frequency divider is proposed. We present a symmetrical injection circuit, with only differential inputs, to realize multi-phase injection and hence the locking range is improved. Our proposed frequency divider can output quadrature signals which are useful for modern transceiver. Post-layout simulation results in a 0.18-μm CMOS Technology show that, the divider can be locked from 2 GHz to 15 GHz, draws a current less than 4 mA. The core area is only 30 μm by 30 μm.**

Index Terms—**Frequency divider, injection locked, ring oscillator, multi-phase injection, quadrature.**

I. INTRODUCTION

Under the influence of ever-increasing demand for higher data rate communication, the required operation frequency of phase-locked loops (PLLs) keeps getting higher. In high frequency PLL or frequency synthesizer, frequency dividers are critical components [1], [2]. The challenges of high frequency divider design are wide locking range, low power and small area. True-single-phase-clock (TSPC) dividers are known for their low power consumption and wide locking range. However, TSPC divider suffers from frequency limitation due to RC delay. For example, the maximum operation frequency of TSPC divider is less than 6 GHz, in a standard 0.18-μm CMOS technology [3], [4]. Current-mode logic (CML) static frequency divider is also widely used in high speed application for its simple design and robustness. However, the power consumption increases rapidly with its operation frequency. Recently, injection-locked frequency divider (ILFD) has attracted much attention for its high frequency and low power [5]–[15]. But they still face some problems such as narrow locking range and multi-phase inputs.

Another issue is the phase of inputs and outputs in a divider. It is well-known that, the quadrature LO signals are necessary for a typical transceiver. There are many approaches to generate quadrature signals. The first one is to use the quadrature voltage-controlled oscillator (VCO), but the power consumption will double and phase noise is generally worse compared with the conventional LC VCO. Secondly, we can employ poly-phase filter to generate quadrature signals. However, its insertion loss is high, so additional buffers are needed to compensate the loss. The last choice is to adopt the divider with quadrature outputs. Quadrature outputs are

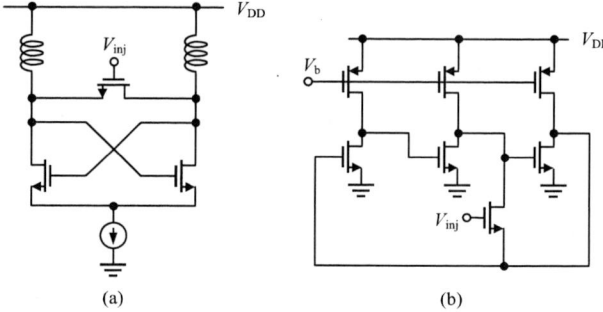

Fig. 1. Traditional ILFD (a) LC oscillator [10] and (b) based on RC ring oscillator [7].

common in CML divider, but not common in TSPC divider or traditional ILFD.

In this paper, we present an ILFD with a new architecture. The ILFD implements a division ratio of two, with differential inputs and quadrature outputs. Furthermore, the proposed topology can realize multi-phase injection naturally, which greatly improves the locking range. Section II presents the proposed ILFD structure after introduction of traditional ILFD, and analyzes its multi-phase injection concept. Post-layout simulation results will be discussed in section III. The paper is concluded in Section IV.

II. CIRCUIT ARCHITECTURE AND ANALYSIS

A. Traditional ILFDs

Generally, an ILFD is an oscillator synchronized by a reference signal at a frequency close to an integer multiple k of its free running frequency f_0. There are two types of ILFDs: one is based on LC oscillator, and the other is based on RC ring oscillator, as shown in Fig. 1. LC-oscillator-based ILFD can operate at high frequency, but its locking range is very narrow due to high-Q LC tank [6], mandating fine and frequent calibrations in PLL. In addition, The area-hungry LC tank also limits its utilization in low cost application. Thus, unless in very high frequency, such as millimeter wave application, LC-oscillator-based ILFD is not practical.

RC-oscillator-based ILFD has merits of wide locking range, compact area and low power. Both [8] and [9] have demonstrated that multi-phase injection technique can improve the

978-1-61284-863-1/11 $26.00 © 2011 IEEE

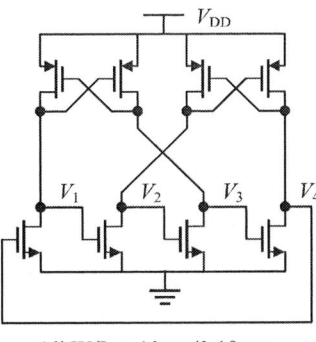

All W/L = 10μm/0.18μm

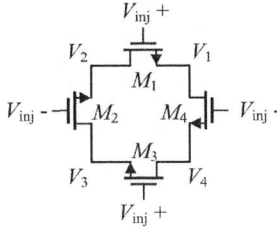

All W/L = 5μm/0.18μm

Fig. 2.   Schematic of proposed ILFD.

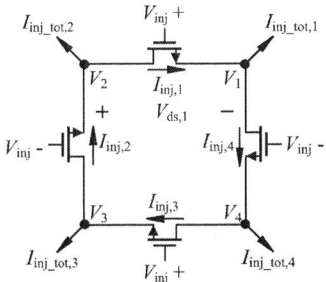

Fig. 3.   Schametic of injection circuit for multi-phase injection analysis.

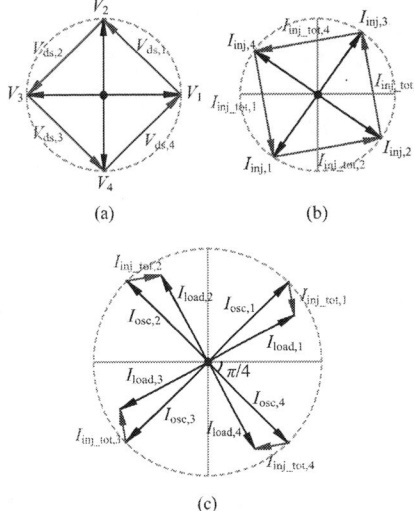

Fig. 4.   Phase relationship of (a) voltage, (b) current and (c) multi-phase injection.

locking range of ILFD. However, their proposed topologies need specific multi-phase inputs, which is not easy to obtain from the conventional LC VCO. Furthermore, the locking range will become worse if the phase difference of inputs departs from the optimum value [8]. These defects prevent the application of multi-phase injection technology in ILFD.

B. Architecture of Proposed ILFD

Fig. 2 shows the schematic of proposed design. Two-stage differential RC ring oscillator, with cross-coupled pMOS load in each stage, is used to get high free running frequency and quadrature outputs. Four nMOS transistors are connected back-to-back between the drain and source of adjacent transistors to form a loop. The four connection points, V_1, V_2, V_3, and V_4, are connected to the oscillation nodes in ring oscillator. All the four nodes are loaded by identical CMOS inverter buffers. $V_{inj}+$ is injected to the gates of M_1 and M_3, while $V_{inj}-$ is connected with the gates of M_2 and M_4. The size of all transistors in ring oscillator, including both nMOS and pMOS transistors, is the same (10 μm/0.18 μm). The size of all four injection transistors is 5 μm/0.18 μm. It seems that large injection transistors will result in larger injection currents and, consequently, larger locking range. However, the too high average conductance of the injection transistors will damp the oscillation of ring oscillator, making locking range smaller [11].

C. Analysis of Proposed ILFD

To explain our proposed design, let us start with the conventional ring-oscillator-based ILFD. Assuming that the gain of each stage is sufficiently large, only the phase condition of Barkhausen criteria needs to be taken into account. Without

injection, the ILFD operates at free running frequency f_0, and the load at each stage provides an identical phase shift to satisfy the phase condition. When the ILFD achieves locked, the phase shift provided by the load will change. Meanwhile, the frequency of ILFD will shift to a new value, that is $f_{inj}/2$ in divide-by-2 case. An extra phase shift must be generated by injection current to compensate the change of phase shift provided by the load, so as to meet the phase condition again. For conventional single-phase injection, the locking range of the ILFD is narrow since the total extra phase shift around the loop is generated by only one injection current. For multi-phase injection, the total extra phase shift can be provided by multiple injection currents. The locking range of this ILFD will be widened if the phase of the injections progress with the ring oscillator's intrinsic delays. However, as mentioned previously, the requirement of specific multi-phase inputs makes this technique impractical in low power application.

In our proposed ring-oscillator-based ILFD, this issue can be avoided. In fact, since the multi-phase always exists in ring oscillator inherently, the multi-phase injection can be generated in the symmetrical injection circuit by using differential inputs. For the qualitative analysis, we treat the injection

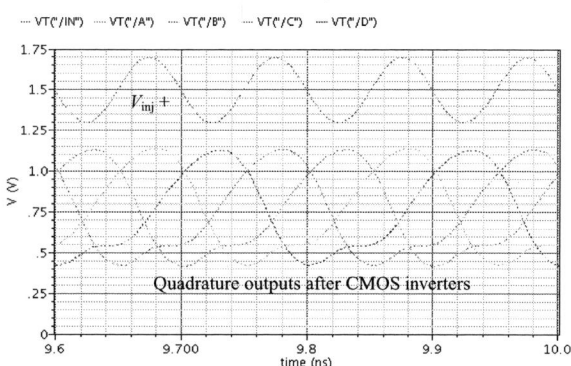

Fig. 5. Waveforms of proposed ILFD when reference input peak-to-peak voltage is 0.4 V and frequency is 10 GHz.

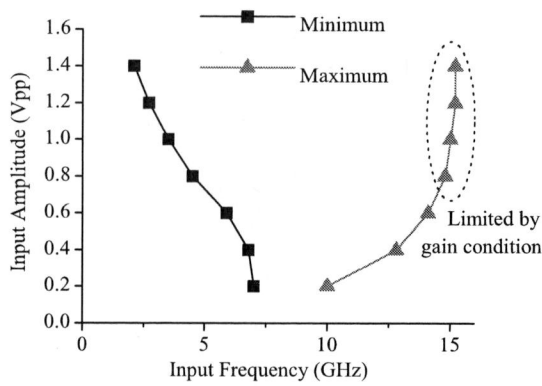

Fig. 6. Simulated input sensitivity of proposed ILFD at $V_{\mathrm{cm}} = 1.5$ V.

transistor as a mixer, as shown in Fig. 3. The injection current of injection transistor M_i is $I_{\mathrm{inj},i}$ ($i = 1, 2, 3, 4$). The total current injected into the node V_i of the ring oscillator is $I_{\mathrm{inj\_tot},i+1} = I_{\mathrm{inj},i+1} - I_{\mathrm{inj},i}$.[1] Without loss of generality, we assume the current $I_{\mathrm{inj},i}$ flows from drain to source. The drain-source voltage of M_i is $V_{\mathrm{ds},i} = V_{i+1} - V_i$. After injection-locked, the node voltage $V_i = \cos(\omega t + \varphi_i)$, so the drain-source voltage of M_i will be

$$V_{\mathrm{ds},i} = \sqrt{2}\cos\left(\omega t + \varphi_i + \frac{3\pi}{4}\right), \quad \varphi_i = \frac{(i-1)\pi}{2}. \quad (1)$$

The voltage phasor relationship is shown in Fig. 4 (a). We can assume the differential injection voltages are

$$V_{\mathrm{inj}}+ = V_{\mathrm{cm}} + A_{\mathrm{inj}}\cos(\omega_{\mathrm{inj}}t + \varphi_{\mathrm{inj}}), \text{ and} \quad (2)$$

$$V_{\mathrm{inj}}- = V_{\mathrm{cm}} + A_{\mathrm{inj}}\cos(\omega_{\mathrm{inj}}t + \varphi_{\mathrm{inj}} + \pi), \quad (3)$$

where ω_{inj} is equal to 2ω for divide-by-2 case, and V_{cm}, which can be tuned externally, is the common-mode voltage of differential input signals. After mixing, the injection current generated by M_i can be described as

$$I_{\mathrm{inj},i} = \sum_{m=0}^{\infty}\sum_{n=0}^{\infty} a_{mn}\cos(m\omega_{\mathrm{inj}}t + m\varphi_{\mathrm{inj}})$$
$$\cdot \cos\left(n\omega t + n\varphi_i + \frac{3n\pi}{4}\right), i = 1, 3 \quad (4)$$

$$I_{\mathrm{inj},i} = \sum_{m=0}^{\infty}\sum_{n=0}^{\infty} a_{mn}\cos(m\omega_{\mathrm{inj}}t + m\varphi_{\mathrm{inj}} + m\pi)$$
$$\cdot \cos\left(n\omega t + n\varphi_i + \frac{3n\pi}{4}\right), i = 2, 4 \quad (5)$$

where a_{mn} is an intermodulation transconductance coefficient of the mixer. Here we pay more attention to the fundamental current because other harmonics will be suppressed by the intrinsic low-pass filter in the loop of the ring oscillator, that is, $|m\omega_{\mathrm{inj}} \pm n\omega| = \omega$. For simplicity, only the terms with $m = 1$, and $n = 1$, are taken into consideration. Therefore, the

[1] Note that $i + 1 = 5$ is equal to $i + 1 = 1$ for expressing convenient.

injection currents can be simplified as

$$I_{\mathrm{inj},1} = \frac{a_{11}}{2}\cos\left(\omega t + \varphi_{\mathrm{inj}} - \frac{3\pi}{4}\right) \quad (6)$$

$$I_{\mathrm{inj},2} = \frac{a_{11}}{2}\cos\left(\omega t + \varphi_{\mathrm{inj}} - \frac{\pi}{4}\right) \quad (7)$$

$$I_{\mathrm{inj},3} = \frac{a_{11}}{2}\cos\left(\omega t + \varphi_{\mathrm{inj}} + \frac{\pi}{4}\right) \quad (8)$$

$$I_{\mathrm{inj},4} = \frac{a_{11}}{2}\cos\left(\omega t + \varphi_{\mathrm{inj}} + \frac{3\pi}{4}\right) \quad (9)$$

Four equations above show that the phases of $I_{\mathrm{inj},i}$ are in quadrature. It can be seen that, the phases of four total injection current $I_{\mathrm{inj\_tot},i}$ are also in quadrature, as depicted in Fig. 4 (b). In other words, the injection can be considered as multiphase injection, as depicted in Fig. 4 (c).

III. POST-LAYOUT SIMULATION RESULTS

Our proposed ILFD has been designed in GlobalFoundries 0.18-μm CMOS technology. This work has already been sent for fabrication in May 2011, and the expected die delivery date is in September 2011. Self-resonance frequency of this ILFD f_0 is at about 4 GHz while supply voltage is 1.8 V. As mentioned previously, the common-mode voltage of differential inputs V_{cm} can be tuned externally. To realize a large modulation of conductance of injection transistors at the synchronization rate, we chose $V_{\mathrm{cm}} = 1.5$ V as the optimized value [11]. Fig. 5 shows the waveforms of proposed ILFD when reference input peak-to-peak voltage is 0.4 V and frequency is 10 GHz.

A large locking range is achieved in our ILFD with divide-by-2 operation. Fig. 6 shows the simulated input sensitivity of the proposed ILFD at $V_{\mathrm{cm}} = 1.5$ V. The locking range is from 2 GHz to 15 GHz, or 153% without tuning, when input peak-to-peak voltage is 1.4 V. One would expect that large injection voltages can result in larger locking range. However, the high average conductance of the injection transistors will cause gain condition of Barkhausen criteria failed in high frequency. Therefore, the maximum operation frequency is limited by gain condition.

978-1-61284-863-1/11 $26.00 © 2011 IEEE

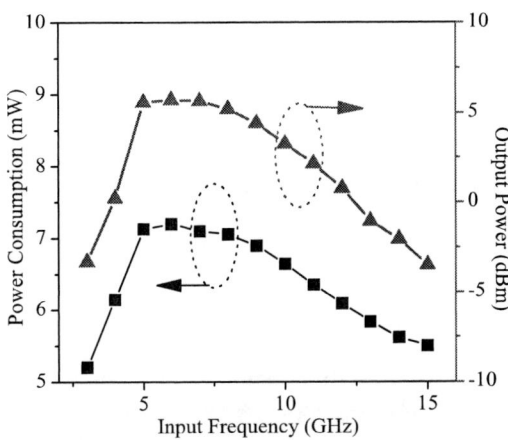

Fig. 7. Power consumption and output power of proposed ILFD when reference input peak-to-peak voltage is 1 V.

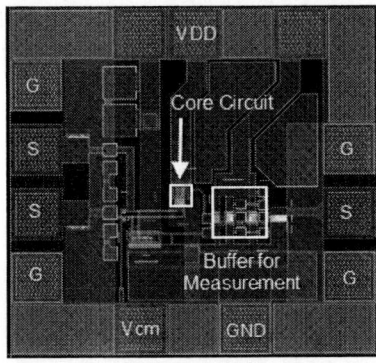

Fig. 8. Layout of proposed ILFD.

Fig. 7 shows the power consumption, as well as output power of ILFD. Both power consumption and output power have similar curves because the ILFD operates in an almost class-A fashion [11]. The power dissipation is lower than 7.2 mW in any case. The output signals are obtained from CMOS inverters, and one of them is loaded by an AC-coupled open-drain 48 μm/0.18 μm nMOS transistor for measurement.

Its core circuit occupies a small area of only 30 μm by 30 μm, as shown in Fig. 8. The performance of our ILFDs are compared with other works reported in similar CMOS technology, as summarized in Table I. Our proposed ILFD has best performance compared with dividers reported in 0.18-μm and 0.13-μm CMOS technology.

IV. CONCLUSION

A new divide-by-2 ILFD, with differential inputs and quadrature outputs, is proposed in 0.18-μm CMOS technology. This paper has demonstrated that our proposed topology can realize multi-phase injection by using symmetric injection circuits and only differential inputs. Post-layout simulation results show that our design can achieve a reasonable performance with very small area.

TABLE I
COMPARISON WITH OTHER WORKS IN SIMILAR CMOS TECHNOLOGY

Ref.	Tech. (nm)	V_{DD} (V)	Power (mW)	Area (μm^2)	Locking Range (GHz)	Locking Range (%)
[8]	180	1.8	24	3000	13–25*	63*
[12]	130	1.2	3.6	1944	11–15	31
[13]	180	1.8	17.6	N/A	1.95–5.5*	95*
[14]	130	2	10.4	N/A	4–6	40
[15]	180	1.8	6.8	6700	2.3–4.3	63
This work	180	1.8	7.2	900	2–15	153

* With tuning.

ACKNOWLEDGMENT

The authors would like to thank GlobalFoundries for chip fabrication.

REFERENCES

[1] C. C. Boon, M. A. Do, K. S. Yeo, and J. G. Ma, "Fully integrated CMOS fractional-N frequency divider for wide-band mobile applications with spurs reduction," *IEEE Trans. Circuits Syst. I: Reg. Papers*, vol. 52, no. 6, pp. 1042–1048, Jun. 2005.

[2] M. V. Krishna, J. Xie, W. M. Lim, M. A. Do, K. S. Yeo, and C. C. Boon, "A low power fully programmable 1MHz resolution 2.4GHz CMOS PLL frequency synthesizer," in *Biomedical Circuits and Systems Conference, IEEE*, Nov. 2007, pp. 187–190.

[3] M. V. Krishna, M. A. Do, K. S. Yeo, C. C. Boon, and W. M. Lim, "Design and analysis of ultra low power true single phase clock CMOS 2/3 prescaler," *IEEE Trans. Circuits Syst. I: Reg. Papers*, vol. 57, no. 1, pp. 72–82, Jan. 2010.

[4] M. V. Krishna, M. A. Do, C. C. Boon, and K. S. Yeo, "A low-power single-phase clock multiband flexible divider," *IEEE Trans. Very Large Scale Integr. (VLSI) Syst.*, to be published, 2011.

[5] H. R. Rategh and T. H. Lee, "Superharmonic injection-locked frequency dividers," *IEEE J. Solid-State Circuits*, vol. 34, no. 6, pp. 813–821, Jun. 1999.

[6] B. Razavi, "A study of injection locking and pulling in oscillators," *IEEE J. Solid-State Circuits*, vol. 39, no. 9, pp. 1415–1424, Sep. 2004.

[7] K. Yamamoto and M. Fujishima, "A 44-μW 4.3-GHz injection-locked frequency divider with 2.3-GHz locking range," *IEEE J. Solid-State Circuits*, vol. 40, no. 3, pp. 671–677, Mar. 2005.

[8] J.-C. Chien and L.-H. Lu, "Analysis and design of wideband injection-locked ring oscillators with multiple-input injection," *IEEE J. Solid-State Circuits*, vol. 42, no. 9, pp. 1906–1915, Sep. 2007.

[9] A. Mirzaei, M. E. Heidari, R. Bagheri, and A. A. Abidi, "Multi-phase injection widens lock range of ring-oscillator-based frequency dividers," *IEEE J. Solid-State Circuits*, vol. 43, no. 3, pp. 656–671, Mar. 2008.

[10] M. Tiebout, "A CMOS direct injection-locked oscillator topology as high-frequency low-power frequency divider," *IEEE J. Solid-State Circuits*, vol. 39, no. 7, pp. 1170–1174, Jul. 2004.

[11] S. Dal Toso, A. Bevilacqua, M. Tiebout, N. Da Dalt, A. Gerosa, and A. Neviani, "An integrated divide-by-two direct injection-locking frequency divider for bands S through K_u," *IEEE Trans. Microw. Theory Tech.*, vol. 58, no. 7, pp. 1686–1695, Jul. 2010.

[12] A. Bonfanti, A. Tedesco, C. Samori, and A. L. Lacaita, "A 15-GHz broad-band ÷2 frequency divider in 0.13-μm CMOS for quadrature generation," *IEEE Microw. Wireless Compon. Lett.*, vol. 15, no. 11, pp. 724 – 726, Nov. 2005.

[13] Y.-H. Chuang, S.-H. Lee, S.-L. Jang, J.-J. Chao, and M.-H. Juang, "A ring-oscillator-based wide locking range frequency divider," *IEEE Microw. Wireless Compon. Lett.*, vol. 16, no. 8, pp. 470–472, Aug. 2006.

[14] F. H. Huang, D. M. Lin, H. P. Wang, W. Y. Chiu, and Y. J. Chan, "20 GHz CMOS injection-locked frequency divider with variable division ratio," in *Radio Frequency integrated Circuits (RFIC) Symposium, IEEE*, Jun. 2005, pp. 469–472.

[15] M. Acar, D. Leenaerts, and B. Nauta, "A wide-band CMOS injection-locked frequency divider," in *Radio Frequency integrated Circuits (RFIC) Symposium, IEEE*, Jun. 2004, pp. 211–214.

978-1-61284-863-1/11 $26.00 © 2011 IEEE

Wideband RF Frontend Design for Flexible Radio Receiver

Fahad Qazi, Quoc-Tai Duong, and Jerzy J. Dąbrowski

Department of Electrical Engineering, Linköping University, Sweden

{qazi⏐tai⏐jdab}@isy.liu.se

Abstract – A wideband RF frontend for flexible radio applications is presented. A target is the performance adequate for multistandard 3G/ 4G systems operating in frequency range 0.8 – 6 GHz. Because of relaxed requirements on band select filters, more demands are placed on linearity while the necessary noise performance is assured. We discuss architecture with Low Noise Transconductance Amplifier (LNTA) and a passive mixer terminated by a low impedance load. One variant of it is a Miller integrator where its input impedance is upconverted by the mixer and provides partial attenuation for blockers at RF. We also investigate a variant with a capacitive load followed by a high impedance transconductance amplifier (TCA). In either variant there is a tradeoff between the frontend NF and IP3, and also between NF and the blocker attenuation. Optimization of the 65 nm CMOS frontend is attained by careful sizing of the mixer switches. The frontend simulation results show NF < 3 dB, IIP3 > +4 dBm and blocker attenuation > 20 dB. Finally, we also discuss the demands placed on the A/D converter resolution in terms of the frontend and the receiver overall NF.

Keywords: wideband receiver; multistandard receiver; current-mode frontend; impedance upconversion; current-mode mixer.

I. INTRODUCTION

The contemporary new generations, 3G and 4G, of communications systems place challenging demands on the RF frontends in terms of frequency range, linearity, noise, and power consumption. Addressing a variety of standards with bands occupying the range from hundreds of MHz up to several GHz, the multistandard transceivers require flexible filters that are still a challenge. As a consequence, a selectivity worse than of a band-dedicated radio must be assumed and hence, more demands should be placed on the receiver dynamic range (DR) and linearity. Another consequence is the frontend gain limitation imposed by blockers (strong interference). A possible blocker of -10 to 0 dBm (200 to 632 mV$_{pp}$) can only experience a small voltage gain under the available supply voltage in the submicron CMOS technology. One possible solution to this problem is a current mode frontend where LNA is a transconductance amplifier (LNTA) followed by a passive mixer which offers a possibly low impedance load [1,2]. Since current rather than voltage represents a received signal in this case, the mixer design is facilitated [3]. Additionally, the low voltage gain of LNTA is useful in terms of the frontend linearity [4].

However, for a capacitive load that provides embedded RF selectivity achieved by impedance upconversion the effective transconductance gain for a signal is less than for a blocker and also a lower IP3 is attained compared to a low impedance load. Additionally, parasitics of the passive mixer tend to affect the frontend performance so optimization of the transistor switches is necessary.

In this paper, we present an LNTA-based receiver frontend suitable for wideband applications. In Section II we present a current mode frontend architecture including LNTA [4] followed by a passive quadrature mixer with 25% clock duty cycle and a low resistive load provided by a transimpedance amplifier (TIA). The circuit is optimized by sizing the mixer switches. In Section III a capacitive load is considered which offers embedded RF selectivity and hence, the requirements on the receiver dynamic range are relaxed. The noise folding effect can be neglected in this case due to a low noise bandwidth. Two variants, one based on TIA with Miller capacitance and the other based on TCA are investigated. In Section IV we discuss the demands placed on the A/D converter in terms of the attained frontend performance and the receiver overall requirements. Conclusion is formulated in the last section.

II. CURRENT-MODE FRONTEND

The requirements for linearity and NF of a wideband frontend are usually tough. A systematic performance specification study for common communication standards can be found e.g. in [5,6].

The current-mode frontend composed of LNTA and a quadrature passive balanced mixer followed by a low-noise transimpedance amplifier (TIA) is shown in Fig.1. In the present discussion a resistive feedback TIA is assumed (i.e. Z_{BB} is real). With a low loading impedance the LNTA, designed in 65 nm CMOS, can achieve the best performance in terms of gain and linearity [4]. Its NF is as low as 1.5 dB and IIP3 > 10 dBm over 0.8 – 6 GHz. The passive mixer can provide good linearity and low noise while $1/f$ noise is largely avoided [7]. In fact, the size of mixer switches largely decides the frontend performance. For small-size switches the LNTA loading impedance is substantially elevated, so the LNTA linearity and the effective transconductance are lower. On the

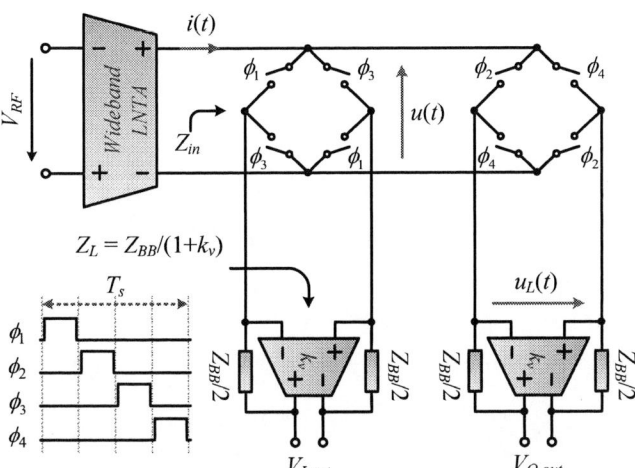

Figure 1. Architecture of current mode frontend.

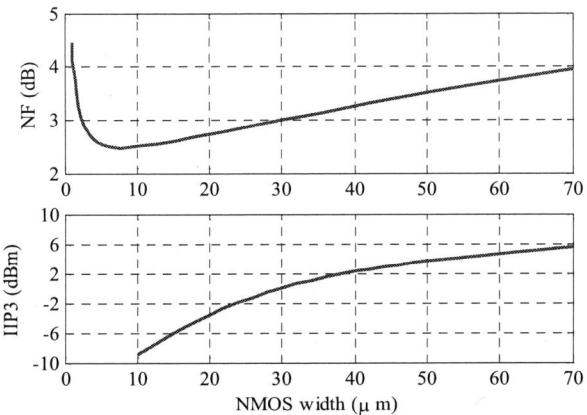

Figure 2. IIP3 and NF of the frontend at 2 GHz vs. size of mixer switches.

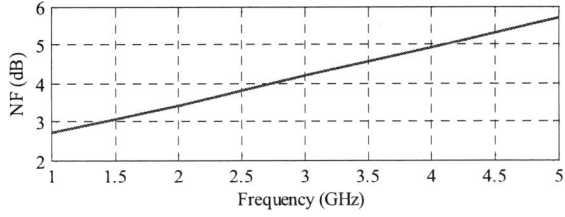

Figure 3. NF of the current mode frontend vs. local oscillator frequency.

other hand, for large-size switches with increased parasitic capacitances the mixer NF is deteriorated. As shown in Fig. 2 the frontend achieves a minimum NF for switches sized below 10μm. However, a much better linearity is obtained for larger switch sizes. To keep the NF below 4dB while maximizing IIP3, all switch sizes ≤ 70 μm are chosen. The simulations are carried out with Spectre RF® where the TIA input referred noise of 1.8nV/√Hz is used that corresponds to NF_{TIA} = 7 dB. The simulated NF of the front end with switch sizes of 50 μm over 1... 5 GHz local oscillator frequency is shown in Fig. 3.

III. EMBEDDED SELECTIVITY

A. Embedded selectivity.

Using the model in Fig. 1 while assuming that Z_{BB} is frequency dependent, we show the effect of impedance upconversion resulting in a tunable narrowband selectivity achieved at RF. The relation between the mixer input/output current can be expressed by a square waveform $w(t)$ with amplitudes equal ±1. $w(t)$ is defined according to the clocking scheme.

$$i_L(t) = w(t)i(t) \qquad (1)$$

Using Fourier series expansion for $w(t)$ and converting to the frequency domain the corresponding voltage appears as

$$U_L(\omega) = \frac{Z_L(\omega)}{2} \sum_{k=-\infty}^{\infty} a_k I(\omega - k\omega_s) \qquad (2)$$

where ω_s is the angle frequency of the local oscillator. The RF voltage $u(t)$ at the input of the mixer can also be expressed as $w(t)u_L(t)$ assuming a complementary mixer (I/Q) is connected in parallel so that the LNTA output is never open (when $w(t)$ = 0, for the complementary mixer $w^*(t) \neq 0$). Resistances of the switches are omitted for simplicity. Then using (2) we have

$$U(\omega) = \frac{1}{4} \sum_{i=-\infty}^{\infty} \sum_{k=-\infty}^{\infty} a_i a_k I(\omega - k\omega_s - i\omega_s) Z_L(\omega - i\omega_s) \quad (3)$$

Here, the components at the fundamental frequency are of interest, for which $\omega - k\omega_s - i\omega_s = \omega$. Hence, we find

$$U(\omega) = \frac{I(\omega)}{4} \sum_{i=-\infty}^{\infty} a_i^2 \left(Z_L(\omega - i\omega_s) + Z_L(\omega + i\omega_s) \right) + \ldots\ldots (4)$$

If a contribution of the complementary mixer is added it can be proved that $U(\omega)$ is doubled. Specifically, for the capacitive load $Z_L(\omega) = 1/(j\omega C)$ and when $\omega \approx \omega_s$ such as for zero-IF receiver, the first component in (4) largely dominates. For non-overlapping 25% duty cycle clock we find

$$Z_{in}(\omega) \cong \frac{4}{j\pi^2 |\omega - \omega_s| C}, \quad (\omega \approx \omega_s) \qquad (5)$$

that is compliant with a result reported in [2]. This upconverted impedance (in fact, Miller capacitance) is controlled by the LO frequency and can serve attenuation of interference at RF if the resistance of mixer switches is small. The simulated frequency characteristic of LNTA when connected to a passive mixer with a Miller capacitive load is demonstrated in Fig. 4. A TIA based on ideal amplifier with 40 dB gain is used for this simulation. The obtained attenuation of blockers is 15... 20 dB, so the corresponding requirements e.g. on the mixer IIP2 would be diminished by 30... 40 dB, respectively. The blocker maximum attenuation for local oscillator frequencies 1...5 GHz is shown in Fig. 5.

Linear distortions of the signal can be corrected digitally at baseband in this case. The capacitor value is set by channel bandwidth requirements of a particular standard. The receiver selectivity is also affected by the size of mixer switches as

978-1-61284-863-1/11 $26.00 © 2011 IEEE

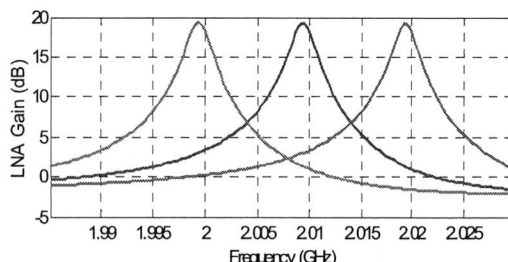

Figure 4. Frequency characteristic of LNTA with upconverted capacitive impedance at different local oscillator frequencies.

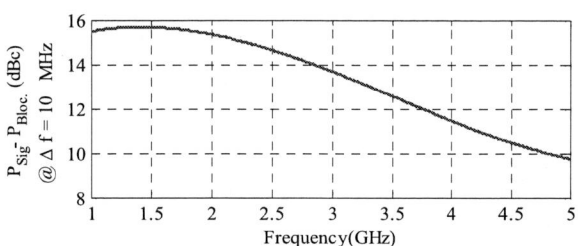

Figure 5. Blocker attenuation at 10 MHz offset vs. local oscillator frequency.

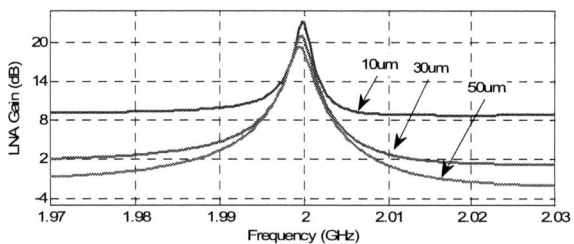

Figure 6. LNTA gain vs. size of mixer switches.

shown in Fig. 6. Compared to 10 μm a switch size of 70 μm improves blocker attenuation by up-to 4dB.

Similar results are achieved when the mixer is terminated by a high impedance TCA with a capacitance equal to the Miller capacitance in front of it [10], as discussed above.

While 50 μm wide switches are used with 40 dB TIA gain and a feedback capacitance of 3 pF, IIP3 = 4.4 dBm is achieved for two-tone test at 2 GHz with 10 MHz spacing. The IIP3 measured before the mixer is 5.5 dBm. Compared to the LNTA with equivalent nonselective load (IIP3$_{max}$ > 12 dBm) [4], here the observed lower IIP3 can be attributed to much different gain values achieved by the two-tone blocker and the corresponding IM3 product at the local oscillator frequency. In this case we have

$$IIP_3 = P_{Bl} + \frac{(P_{Bl} + G_{Bl}) - (P_{IM3}^{(in)} + G_{Sig})}{2} \qquad (6)$$

where IIP3 is reduced by $\Delta G = (G_{sig} - G_{Bl})/2$ as compared to a nonselective frontend . We note that the reduced IIP3 value is sufficient to keep the IM3 product at the same level as achieved by the nonselective frontend with IIP3 larger by ΔG (in this case $\Delta G \approx 7$ dB). The two-tone test of the selective frontend at 4 GHz gives IIP3 value only by 0.5 dB lower.

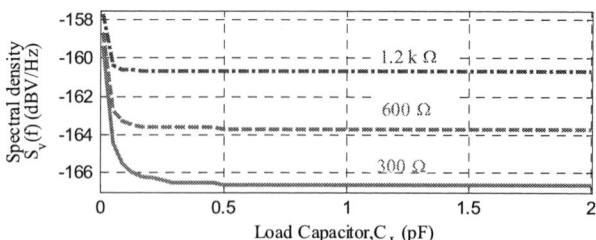

Figure 7. Noise PSD of mixer vs. loading cap. for different source resistances of LNTA.

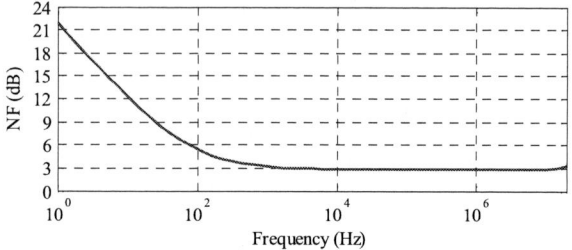

Figure 8. NF of the frontend with mixer capacitive load.

The achieved selectivity allows to limit the IM3 contribution of the mixer and it helps to keep the receiver NF low according to the large gain at the carrier frequency. Note that with missing selectivity the wanted signal gain is limited by the blocker power.

B. Noise performance

The output thermal noise of a passive mixer with a capacitive load (S/H) can be characterized as a combination a track and held components, where the latter (kT/C noise) originates from the folding phenomena [8]. The noise folding effect is subject to the equivalent noise bandwidth 1/(4RC), which according to the LNTA output resistance and a relatively large loading capacitance appears less than the sampling frequency so the folding does not occur in this case. Hence, at low frequencies the mixer output noise power spectral density can be estimated from

$$S_v(f) \cong \left(m + (1-m)^2\right)4kTR \qquad (7)$$

where m denotes the clock duty factor (25% in this application). The simulation results for a passive balanced mixer for different LNTA output resistances and loading capacitances are shown in Fig. 7 where the resistance of switches is 50 Ω. As seen, the kT/C noise only contributes for small capacitance values that are not of interest here. NF of the frontend with mixer capacitive load of 23pF, switch size of 50 μm and oscillator frequency of 2 GHz is shown in Fig. 8. In this case, a high LNA voltage gain keeps the frontend NF at 3dB except for low frequencies affected by 1/f noise. A 1.8nV/√Hz input referred noise of TCA is assumed in this simulation.

IV. ADC VS FRONTEND PERFORMANCE

In a flexible receiver with limited blocker filtering the ADC has to maintain SNR required by the demodulator, when the maximum blocker and a minimum wanted signal power S are present at the ADC input simultaneously. The required ADC dynamic range follows

$$DR = (P_B - R_B)_{\max} - (S + G_P - SNR_{\min}) + \Delta_{DR} \qquad (8)$$

where P_B is the blocker power at the receiver input, R_B is the corresponding rejection provided by the receiver filter (embedded RF and/or IF/BB select), and G_P is the possible processing gain typical of spectrum spreading systems. Here, $(S + G_P - SNR_{\min})$ stands for the maximum allowed noise and distortion level, also receiver input-referred. A reserve Δ_{DR} is used for possible variations in receiver gain and signal magnitude, such as peak to average ratio ($PAPR$). For a zero-IF receiver this reserve should also cover the maximum DC offset. Based on (8) with $\Delta_{DR} = 6$ dB the required DR values for common communication standards with in-band blocking vary between 68 and 93 dB minus the blocker attenuation R_B.

The minimum SNR requirement at the ADC output can be expressed as

$$SNR_{\min} - G_P = (S + G_{FE}) - 10\log\frac{N_{n+d,FE} + N_{n+d,ADC}}{1\text{mW}} \qquad (9)$$

where $N_{n+d,FE}$ is the frontend output noise including in-band distortions, $N_{n+d,ADC}$ is ADC noise and distortion power calculated over the signal bandwidth, and G_{FE} is the frontend gain. Assuming $N_{n+d,ADC} = \alpha N_{n+d,FE}$ with $\alpha < 1$, we find the ADC contribution to the receiver $SNR_{\min}$ as $10\log(1+\alpha)$. Also the receiver noise figure would be affected in a similar way

$$NF_{FE} + 10\log(1+\alpha) = SNR_{Ref} - (SNR_{\min} - G_P) \qquad (10)$$

where NF_{FE} is the frontend noise figure. As an example, consider the UMTS standard with $S = -114$ dBm over 4 MHz bandwidth and $G_P = 25$ dB, $P_B = -15$ dBm @ 25MHz offset, $SNR_{\min} = 8$ dB, and $\Delta_{DR} = 9$ dB. We find $SNR_{Ref} = -114dBm + 174dBm/Hz - 10\log(4{\times}10^6)Hz = -6dB$ and assuming the total $NF_{FE} = 9$ dB (including losses of the RF filter/duplexer) from (10) we calculate $\alpha_{\max} = 0.58$, which corresponds to 2 dB contribution by ADC. Clearly, with a smaller α value a reserve in SNR can be achieved.

In order to limit the receiver NF according to (10) the equivalent number of ADC bits (ENOB) can be estimated from

$$ENOB = \frac{DR - 1.76 + 10\log(1+1/\alpha)}{6.02} \qquad (11)$$

From (8) we find $DR = 91dB - R_B$ and from (11) $ENOB > 14.7 - R_B/6.02$, so a possible embedded blocker attenuation of 12 dB gives ENOB > 12.7. To meet this number, preferably an oversampling ADC can be used that also allows avoiding aliasing effects when blockers are only partly attenuated.

Finally, assuming 0 dBm ADC range with Δ_{DR} reserve the frontend gain adequate to the blocking condition is found as $G_{FE} = R_B + 6$ dB. When the blocker power drops by ΔP_B the automatic gain control mechanism (AGC) can increase the receiver gain by the same amount.

V. CONCLUSION

A design of a wideband RF frontend consisting of a high performance LNTA [4] and a balanced passive quadrature mixer followed by TIA is presented. The effect of mixer's switch size on NF, linearity, and selectivity is highlighted. Design tradeoffs among those quantities are discussed. Two types of TIA feedback are investigated, resistive and capacitive. The latter provides tunable embedded selectivity at RF for a passive mixer driven by a non-overlapping four-phase clock. Although the IP3 is reduced in this way, the effective IM3 distortion is lower compared to a nonselective case. The embedded selectivity relaxes the DR requirements for the succeeding ADC block by attenuating the blockers at RF up to 20 dB and also the frontend NF is improved. The requirements for the ADC block in a multi-standard receiver are formulated in terms of the attained frontend performance. This performance is sufficient according to the specifications of common communications standards [5,6].

REFERENCES

[1] R. B. Staszewski, et al.,"All-digital TX frequency synthesizer and discrete-time receiver for Bluetooth radio in 130-nm CMOS," *J. of Solid-State Circuits*, vol. 39, no. 12, 2004.

[2] Z. Ru, et al. "Digitally enhanced software-defined radio receiver robust to out-of- band interference", *J. of Solid-State Circuits*, vol. 44, no. 12, Dec. 2009.

[3] R. Bagheri et al., "An 800-MHz–6-GHz Software-Defined Wireless Receiver in 90-nm CMOS", *J. of Solid-State Circuits*, vol. 41, no. 12, Dec. 2006.

[4] Q-T. Duong, J. Dąbrowski, "Low Noise Transconductance Amplifier Design for Continuous-Time ΣΔ Wideband Frontend," Proc. ECCTD'11.

[5] M. Brandolini, et al., "Toward Multistandard Mobile Terminals—Fully Integrated Receivers Requirements and Architectures," *Transactions on Microwave Theory and Techniques*, vol. 53, no. 3, March 2005.

[6] Q. Gu, "RF system design of transceivers for wireless communications," Springer, 2005.

[7] E. Sacchi, et al., "A 15 mW, 70 kHz 1/f corner direct conversion CMOS receiver," Proc. CICC'03.

[8] R. Gregorian, G. Temes, "Analog MOS Integrated Circuits", *Wiley*, 1986

[9] S. Karvonen, et al., "A low noise quadrature subsampling mixer," Proc. ISCAS'01, vol. 4, 2001.

[10] J. Borremans et al., "A 40nm CMOS highly linear 0.4-to-6 GHz receiver resilient to 0dBm out-of-band blockers," Proc. ISSCC'11, pp.62-64, 2011.

Ultra-Low-Power Wireless Implantable Blood Flow Sensing Microsystem for Vascular Graft Applications

Rui-Feng Xue, Jia Hao Cheong, Hyouk-Kyu Cha, Xin Liu, Peng Li, Huey Jen Lim, Li Shiah Lim, Ming-Yuan Cheng, Cairan He, Woo-Tae Park, and Minkyu Je

Institute of Microelectronics, A*STAR (Agency for Science, Technology and Research)
11 Science Park Road, Singapore Science Park II, Singapore 117685
Email: xuerf@ime.a-star.edu.sg

Abstract—**Flow rate monitoring provides an indication for early intervention of vascular graft degradation or failure used in lower limb bypasses and renal haemodialysis. This paper presents an inductively powered implantable blood flow sensing microsystem with bidirectional telemetry capability, which fully integrates the silicon nanowire (SiNW) sensor with tunable giant piezoresistivity, the ultra-low-power ASIC and the high-efficiency transcutaneous coupling coils. Operating at 13.56 MHz carrier frequency, the micro-fabricated coils transfer the power and command forward and backscatter the processed sensor readout information to an external device. The ASIC fabricated in 0.18 μm CMOS process occupies an active area of 1.5×1.78 mm² and consumes 21.6 μW totally. The SiNW diaphragm-based sensor provides the gauge factor higher than 300 with tuning voltage below 0.5 V. The proposed solution has demonstrated the 0.176 mmHg/√Hz sensing resolution with small device dimension and the lowest power consumption to the authors' knowledge.**

Keywords-blood flow; implantable biomedical devices; inductive coupling; MEMS; silicon nanowires; piezoresistive sensor; rectifiers; regulators; SAR ADC; wireless telemetry.

I. INTRODUCTION

Prosthetic grafts are frequently used in vascular surgery in the context of bypass surgery for lower limb ischemia or as a conduit for haemodialysis in renal failure. At least 20–30% of the existing renal haemodialysis population has a prosthetic vascular graft in-situ. In addition, thousands of lower limb bypasses are performed all over the world yearly, of which at least 20% require the use of prosthetic grafts. In these settings, graft failure can result in deleterious outcomes for the patients i.e. worsening ischemia, inability to undergo haemodialysis. Insufficient blood flow rates in these grafts are predictive of subsequent graft thrombosis and failure. Underlying this is the presence of stenoses in the graft or downstream from the graft. Variations in flow rates can localize the position of significant stenosis that may result in graft thrombosis. Flow rate monitoring provides an indication for early intervention to prevent graft failure.

Various modalities of monitoring graft flow rate including ultrasound, computed tomography (CT) scan and formal angiogram are researched to detect graft failure vis-a-vis decreasing flow rates. However, these methods are not entirely risk-free for patients and come with some procedural morbidity and monitoring of flow rates are done at regular intervals (i.e.

once in 4–6 months). There are a few commercially available intravascular devices for flow rate detection but they are invasive in nature causing trauma to the patient for every measurement. Therefore an implantable sensing microsystem with wireless data telemetry and powering capability is highly desirable to provide convenient monitoring of blood flow in vascular prosthetic grafts.

In this paper, an ultra-low-power implant microsystem, embedded within the vascular graft and inductively powered, is proposed to sense the differential pressure as an indication of stenosis build-up and transmit the information wirelessly to an external hand-held device. The blood flow directly contacts and causes the mechanical deformation to the MEMS sensor which is translated into electrical signal through piezoresistive sensing element. The piezoresistive sensing element is silicon nanowire (NW) based and its conductance is electrically tunable. The sensor is sensitive enough while covering the required blood pressure range, but also small enough to not affect the blood flow or cause any blood clotting. The proposed flow sensing microsystem demonstrates better resolution with small dimension and the lowest power consumption compared to previous works. The overall microsystem fully integrating the coupling coil, the ASIC and the piezoresistive sensor is described in Section II. The ASIC building blocks are elaborated in Section III. The system characterization is discussed in Section IV with conclusions given in Section V.

II. WIRELESS IMPLANTABLE SENSING MICROSYSTEM

The overall system architecture is shown in Fig. 1, consisting of the sensor, the ASIC and the coil. The piezoresistive sensor produces resistance changes in correspondence to the blood flow rate. The ASIC converts the

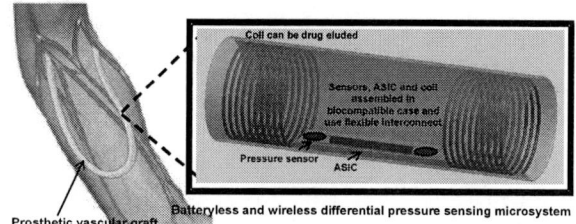

Fig. 1. Wireless implantable blood flow sensing microsystem.

This work was supported by the Science and Engineering Research Council of A*STAR (Agency for Science, Technology and Research), Singapore. The grant number for the project is 092 148 0069.

resistance changes into a form of voltage and communicates this acquired information through a passive wireless telemetry after processing. The coil serves two purposes – (a) through an inductive coupling with a primary coil placed outside the patient's body, delivers power and commands to the ASIC and backscatters the digitized sensor data; (b) acts as an anchor to hold the sensor in position when blood flows. The communication distance of the complete blood flow monitoring microsystem mainly depends on how efficiently the RF energy is coupled from the external device to the implanted coil and how efficiently the coupled RF power is converted to useful dc power apart from how low the building blocks consume power.

A. High-Efficiency Transcutaneous Inductive Coupling

Transcutaneous wireless powering can obviate the need for replacement or recharging of batteries, the risk of infection and the possibility of signal distortion induced by piercing wirings, as well as the post-implant trauma of patients. Inductive coupling is the bottle neck of the wireless powering link efficiency so far, the design of which is strictly limited by the available implant size, the power requirement and the living tissue safety considerations [1].

Within the design constraints, 13.56 MHz is chosen as the carrier frequency. The implant coil is wrapped around the implant PTFE graft with a diameter of 6 mm and the coil material is nitinol with biocompatible coating. The external coil made of copper surrounds the implant coil with a diameter of 10 cm [2]. The quality factor and the coupling coefficient of the two coils should be optimized to achieve the maximum power transfer efficiency so that the inductive coupling can provide enough energy for the microsystem implanted into 5~50 mm thick tissue with coil misalignment in practice. The coupling coefficient also affects the backscattering directly.

B. Ultra-Low-Power ASIC

The ASIC mainly consists of a rectifier and a regulator, a clock extractor, an ASK demodulator and a backscatter modulator, a sensor interface and a SAR ADC, as well as digital block [3]. The rectifier and on-chip LDO regulator are designed with high efficiency to provide power supply for other blocks. The demodulator and backscatter modulator

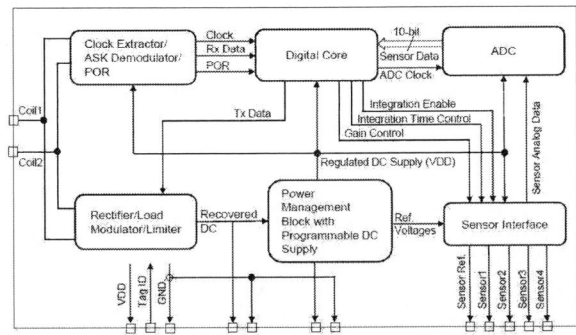

Fig. 2. ASIC architecture.

realize bidirectional wireless telemetry. The clock required by digital block and ADC is extracted from the power carrier. The sensor interface circuit operates with the piezoresistive sensors and translates the resistance change to a voltage signal that can be processed by the following stages and digitized by the ADC. The whole architecture is shown in Fig. 2 and will be elaborated in the next section.

C. NWFET based piezoresistive sensor

Giant piezoresistance in silicon nanowires (NWs) can be modulated by an external electric field perpendicular to the current flow within the NW. Mechanical stress applied on the NW will cause an increase of the charge carriers' concentration. The electrical biasing and the application of mechanical stress have been combined to create a stress-gated field-effect transistor (FET) exhibiting a low voltage tunable giant piezoresistivity [4]. To overcome the potential risk of sensitivity degradation of the cantilever sensor due to endothelialization, a novel NWFET diaphragm-based sensing structure is designed, shown in Fig. 3.

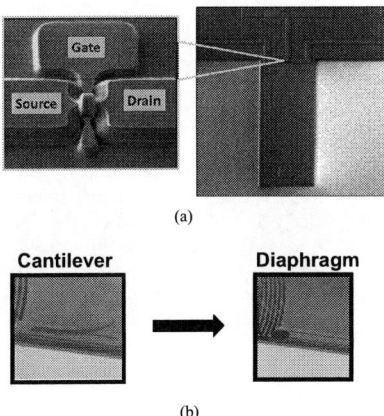

Fig. 3. (a) Integrating NWFET. (b) Diaphragm-based sensing structure.

The sensor is fabricated using nitride based membrane with piezoresistive single crystal silicon NW as sensing element. The piezoresistive sensing element is embedded at the edge of a 200 µm silicon nitride circular diaphragm. The presented solution provides 0.176 mmHg/√Hz resolution for 2 µm thick diaphragm. The gauge factor is higher than 300 with tuning voltage less than 0.5V. The design allows robust measurement with the gate surrounding sensitive NW and low noise level with negative bias.

III. ASIC BUILDING BLOCKS

The most important specification for the ASIC is how efficiently the power conversion and management perform as well as how low power the building blocks of the ASIC consume to function.

A. RF Front-End

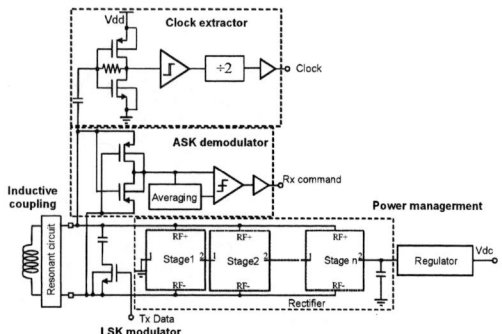

Fig. 4. Schematic of the ASIC RF front-end.

The front-end schematic is shown in Fig. 4. For converting AC energy to DC energy, an eight-stage differential-drive rectifier is used. The rectifier core has a cross-coupled bridge configuration. A differential-drive active gate bias mechanism enables to achieve both low ON-resistance and small reverse leakage of diode-connected MOS transistors at the same time, resulting in high power conversion efficiency. Unit stages are serially stacked along the DC path and connected in parallel to the input RF terminals. By using this multi-stage configuration, appropriate DC output voltage is obtained at the optimal operating point where the PCE is maximized. The unregulated DC voltage from the rectifier is regulated by the low-power LDO voltage regulators. The power management block also generates the desired reference voltages for the SAR ADC and sensor interface blocks. The clock extractor consists of an AC-coupled input amplifier and a Schmitt trigger. The clock is divided by two, buffered and fed to the digital core as its reference clock. The envelope of the received ASK-modulated signal is compared with the average value of the envelope to obtain the command from the external device. The load shift keying (LSK) modulator backscatters the acquired data to external device and provides the close-loop power control.

B. Digital Baseband

The state machine of the implemented digital baseband and controller is shown in Fig. 5. The state machine control block controls the states transition of the ASIC.

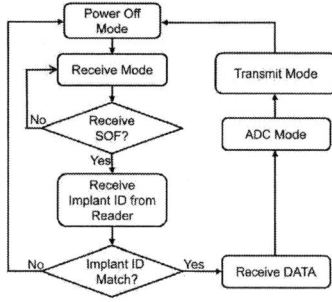

Fig. 5. Digital baseband state machine.

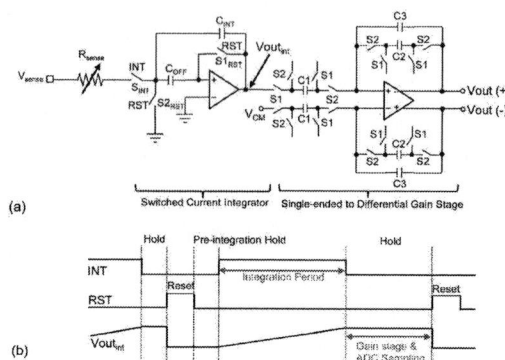

Fig. 6. (a) Schematic diagram of the sensor interface circuit. (b) Timing diagram and the output voltage waveform of the switched current integrator.

C. Low Noise Sensor Interface

Nanowire-based piezoresistive sensors are used to sense the blood flow. The change in resistance, ΔR, can be in the range of $\pm 10\%$ to $\pm 30\%$. The sensor interface circuit converts ΔR into analog voltage. Fig. 6 shows the schematic diagram of the sensor interface circuit and its timing diagram. The single ended output voltage from the integrator is amplified and converted to a differential signal by the gain stage. The gain stage consists of a fully differential folded-cascade op-amp with a switched-capacitor common-mode feedback (SC-CMFB), a switched-capacitor (SC) feedback, and a nonoverlapping clock generator.

D. SAR ADC

A 400-nW successive approximation analog-to-digital converter (SAR ADC) is presented and implemented in the ASIC, the architecture of which is shown in Fig. 7 [5]. It consists of a capacitor array, a switching array, a time-domain comparator and a switching logic. A novel common-mode resetting tri-level switching scheme is applied to the SAR ADC. Utilizing the proposed tri-level switching scheme, an N-

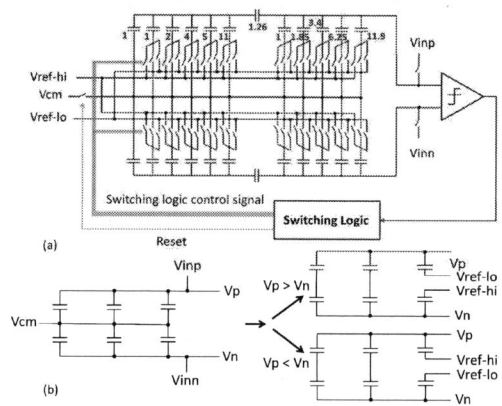

Fig. 7. (a) SAR ADC Architecture. (b) Common-mode resetting tri-level switching scheme.

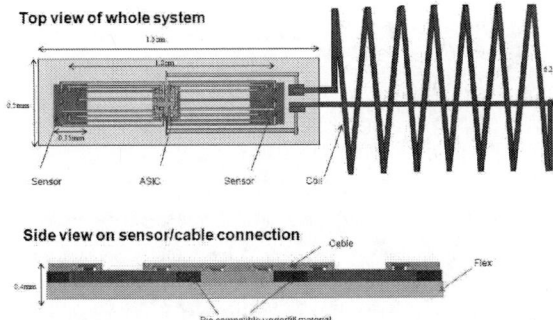

Fig. 8. Integration of the sensor, ASIC and coil within the prosthetic graft.

bit SAR ADC with M redundant bits needs N+M capacitors, and it takes only N+M cycles to complete the conversion. A fully differential structure and complementary switches are to reduce the effect that top-plate sampling may be subject to charge injection. In an integrated system, an additional voltage level would mean an additional buffer needed in the system to hold the voltage. Such low impedance buffer will significantly increase the overall power consumption of the whole system. In order to avoid using a third voltage level, the reference voltage can be designed in such a way that V_{ref-lo} is 0V, and V_{ref-hi} is set to V_{ref}, which is selected to be equal to the input common mode voltage (V_{in_common}).

IV. MICROSYSTEM IN-VITRO CHARACTERIZATION

The diagram of system integration including the silicon NW diaphragm-based sensors, the CMOS ASIC and the micro fabricated coupling coils is shown in Fig. 8. The ASIC, fabricated in 0.18μm CMOS process shown in Fig. 9, occupies a total active area of 1.5×1.78mm^2 and consumes a total power of 21.6 μW.

In-vitro characterization of the whole system has been conducted. The external reader module is interfaced with LabVIEW for sending RF energy and the commands to implant flow sensing microsystem at a distance of 20mm. The energy coupled at the implant coil is converted to DC supply to power the ASIC, and the command from the reader is demodulated for sensing configuration. The clock is extracted from the

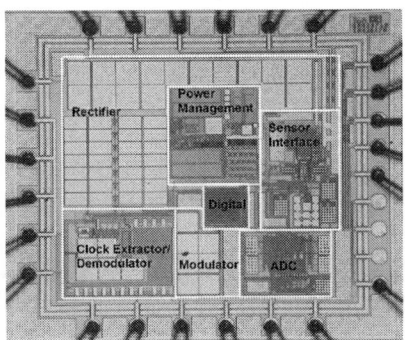

Fig. 9. Microphotograph of the ASIC die.

incoming carrier for the 10-bit ADC to convert analog sensor information to digital and fed to the load modulator for backscattering. Table I summarizes the measured results of the complete blood flow rate sensing microsystem.

TABLE I.   BENCHMARK OF IMPLANTABLE PRESSURE SENSING MICROSYSTEM

Device	Sensing Mechanism	Resolution (mmHg)	Power Consumption	Sensor Size (mm³)	Application
Ziaie (2001)	Capacitive	0.5	0.12 mW	12.5	Tonometry method External to vessel
DeHennis (2006)	Capacitive	3	0.34 mW	0.5	Measure with stent
Fassbender (2008)	Capacitive	1.5	N. A.	0.864	Measure with catheter
Cong (2009)	Capacitive	0.1	300 μW	0.08	External to the vessel
Pressurewire™ (2010)	Piezoresistive	2	N. A.	0.02	Measure with guidewire
This work	Piezoresistive	0.176	21.6 μW	0.064*	Measure stenosis-induced graft failure

*Die area used for sensor and essential bonding pads.

V. CONCLUSION

A wireless implantable blood flow sensing microsystem has been designed, implemented and verified in-vitro for early intervention of prosthetic vascular graft failure. The ultra-low-power CMOS ASIC, and the nanowire-based sensor with low-voltage tunable giant piezoresistivity, as well as the micro-fabricated high-efficiency coupling coil make the complete microsystem. Compared to previous works, our solution provides better resolution with smaller dimension and the lowest power.

ACKNOWLEDGMENT

The authors would like to thank the colleagues at IME for technical discussion, IC layout, PCB fabrication and measurement support. The authors also appreciate the valuable comments of the reviewers.

REFERENCES

[1] R.-F. Xue, P. B. Khannur, K. Kang, et al., "The RF front-end of a blood flow sensor for vascular graft applications," *Proceedings of the second APSIPA annual Summit and Conference*, pp. 761-764, Biopolis, Singapore, 14-17 Dec. 2010.

[2] K. Kang, P. B. Khannur, R.-F. Xue, et al., "The RF front-end of a blood flow sensor for vascular graft applications," *Proceedings of the second APSIPA annual Summit and Conference*, pp. 741-744, Biopolis, Singapore, 14-17 Dec. 2010.

[3] P. B. Khannur, K. L. Chan, J. H. Cheong, et al., "A 21.6 μW indutively powered implantable IC for blood flow measurement," *IEEE ASSCC Dig. Tech. Papers*, pp. 249-252, Beijing, China, Nov. 2010.

[4] P. Neuzil, C. C. Wong, and J. Reboud, "Electrically controlled giant piezoresistance in silicon nanowires," *Nanoletters*, vol. 10, pp. 1248-1252, Oct. 2010.

[5] J. H. Cheong, K. L. Chan, P. B. Khannur, et al., "A 400-nW 19.5-fJ/conversion-step 8-ENOB 80-kS/s SAR ADC in 0.18-μm CMOS," *IEEE Trans. Circuits Syst. II*, accepted.

APPENDIX

A. Flow Sensing System Configuration

The overall flow sensing system configuration is depicted in Fig. 10.

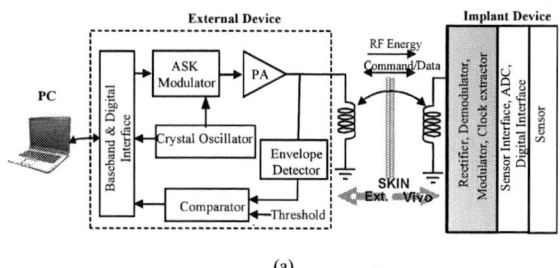

(a)

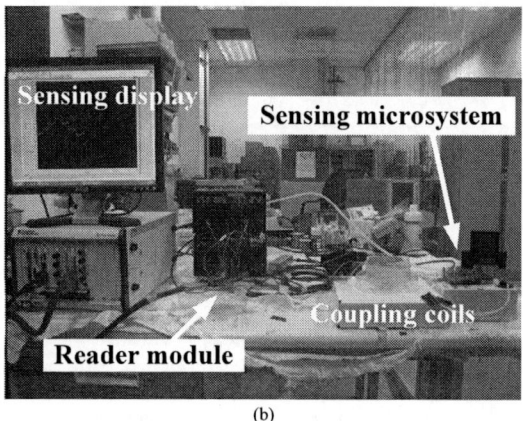

(b)

Fig. 10. (a) Simplified block diagram and (b) photograph of the overall measurement setup for the flood flow sensing system.

The RF power inductively coupled from the external reader is converted to DC supply to power the ASIC. The clock is extracted from the incoming carrier from the external device and the carrier frequency f_c is 13.56 MHz. The sampling clock for the ADC is 106 kHz which is f_c/128. The external device configures the implant ASIC by sending the modulated command. The command from the reader is demodulated and the power-on-reset signal is generated to reset the digital baseband. After selecting the sensor to be read and setting the parameters such as gain, integration time, etc., the ADC clock is generated. The sensors based on nanowires provide changes in resistance which is proportional to the flow rate. The sensor interface converts the resistance to the voltage. This analog voltage is in turn converted to the digital data by the 10-bit ADC and converted to the serial bit stream. The digital data is coded to a desired format in the digital baseband and sent to the external device by backscattering the incoming RF carrier through a load modulation.

The state machine control block controls the states transition of the ASIC. The measured waveforms of the demodulated command, power-on-reset, ADC clock and sensor backscattering data are shown in Fig. 11.

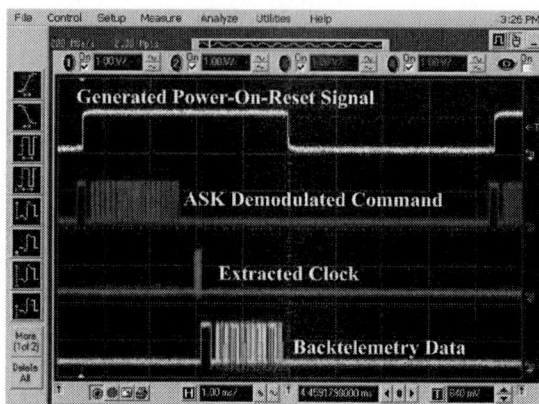

Fig. 11. Measured waveforms of the power-on-reset, ASK demodulated command, extracted clock and backtelemetry data.

B. The Proposed 400-nW SAR ADC

A tri-level switching scheme with common mode reset, redundant algorithm and a time-domain comparator is proposed and implemented in our SAR ADC to achieve ultra-low power consumption. It takes only $N+M$ cycles for an N-bit SAR ADC with M redundant bits to complete the conversion applying the proposed tri-level switching scheme. The redundant algorithm mitigates the offset error caused by the level mismatch of tri-level switching scheme, whereas the tri-level switching scheme simplifies the switching logic of the redundant algorithm. The proposed SAR ADC achieves a signal-to-noise-and-distortion ratio (SNDR) of 50 dB, equivalent to 8-bit effective number of bits (ENOB), at 80-kS/s conversion rate. The figure of merit (FOM) is 19.5 fJ/conversion-step, which is the best FOM by SAR ADC under 0.18 μm CMOS process to the author's knowledge.

TABLE II. BENCHMARK OF THE PROPOSED SAR ADC

Parameter	ISSCC 08'	ISSCC 10'	VLSI 09'	JSSC 07'	This work
CMOS Process	65 nm	65 nm	0.18 μm	0.18 μm	0.18 μm
Sampling Speed	1 MS/s	100 MS/s	100 KS/s	200 KS/s	80 KS/s
Resolution	10 bit	10 bit	10 bit	8 bit	10 bit
DNL	0.5LSB	0.58LSB	NA	0.53LSB	0.7LSB
INL	2.2LSB	0.69LSB	NA	0.9LSB	1.5LSB
ENOB	8.9 bit	9 bit	8.7 bit	7.44 bit	8 bit
Power Consumption	1.9 μW	1.13 μW	1.3 μW	2.47 μW	0.4 μW
FOM	4.4 fJ/Step	15.5 fJ/Step	31 fJ/Step	65 fJ/Step	19.5 fJ/Step

Table II shows the comparison with the previously published SAR ADCs.

C. The Proposed SiNW Diaphragm-based Sensor

A silicon nanowire (SiNW) diaphragm-based pressure sensor with low-voltage tunable giant piezoresistive (gauge factor higher than 300 with tuning voltage < 0.5 V) is designed,

978-1-61284-863-1/11 $26.00 © 2011 IEEE   228

fabricated, and characterized for the use in the flow sensing microsystem, shown in Fig. 12.

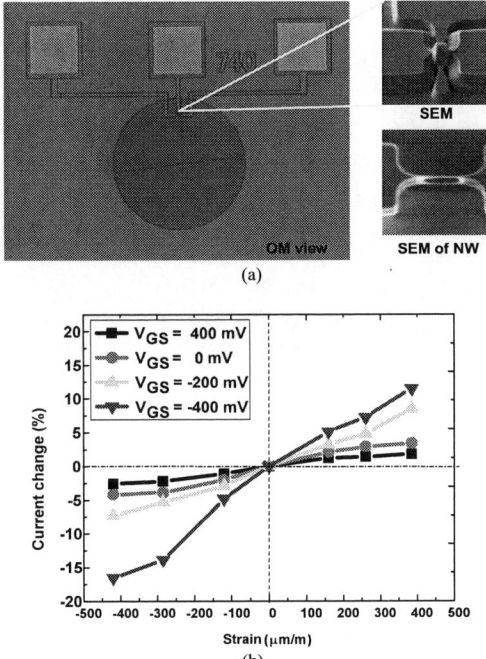

Fig. 12. (a) Fabricated circular SiNW pressure sensor. (b) Measured gauge factor tune-ability.

Circular membrane with a diameter of 200 μm is chosen for ease of backside deep reactive-ion etching (DRIE) release and 2-μm thick Nitride/oxide membrane is to minimize initial deflection. Characterization of the pressure sensor has been done using a pressure regulator and compressed air. Pressure sensor signal is measured using Agilent 4156 parameter analyzer. The measured resolution is 0.176 mmHg/√Hz.

D. The Presented Sensing Solution

Different modalities have been devoted to monitoring graft flow rate thus far. Disadvantages of these existing modalities include the need for significant amounts of procedural time (ultrasound, angiogram), the use of nephrotoxic contrast (CT scan, angiogram) and invasiveness (angiogram). More importantly, these procedures are not entirely risk-free and come with some procedural morbidities, and the monitoring is done at regular intervals (i.e. Once in 4-6 months).

Our solution is implantable direct (intra-graft) methodology for convenient monitoring of blood flow. With the sensing microsystem embedded graft, the failing graft can be detected at its earlier stage thus implementing early intervention strategies. This will reduce the number of graft surveillance scans (CTs, angiograms, etc.) per patient from a routine process to one done only when there are abnormalities detected in the flow rates.

TABLE III
MEASURED PERFORMANCES OF THE PRESENTED MICROSYSTEM

Parameter	Measured Result
Power carrier frequency	13.56 MHz
Communication protocol	Modified and simplified from ISO 14443 RFID standard
Forward command data demodulation	ASK (programmable modulation depth from 10% to 90% in steps of 10%)
Backtelemetry	LSK (programmable modulation depth from 10% to 70% in four steps)
ASIC	Active area: 1.5×1.78 mm^2 Power consumption: 21.6 μW (Rectifier: 5 μW; regulator: 12.8 μW; digital core: 2μW; ADC: 0.4μW; sensor interface: 1.4μW)
Power conversion and management efficiency	Rectifier: 66% Regulator: 56%
SAR ADC	Resolution: 10 bits ENOB: 8.6 bits @ 5 kHz input 7.4 bits @ 25 kHz input INL /DNL: ±1.5 LSB / ±0.7 LSB FOM: 19.5 fJ/step
MEMS sensor	SiNW diaphragm based sensor Size: 0.064 mm^3 Sensing mechanism: piezoresistive Resolution: 0.176 mmHg/√Hz
Implant coil	Helix diameter: 6.2 mm Wire material: Nitinol
External reader	Output power: 200 mW (maximum) Interrogating distance: 20 mm (nominal) Coil diameter: 10 cm
The safety guidelines for specific absorption rate (SAR)	2 W / kg for any 10 g of tissue

Table III summarizes the complete performance metrics and Fig. 13 demonstrates an example of the measured data. The system benchmark is listed in Table I.

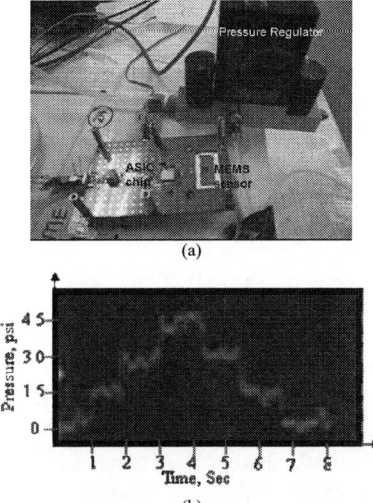

Fig. 13. (a) Evaluation board of the ASIC, the sensor and the coil. (b) Measured pressure data with the presented flow sensing microsystem.

An 80-dB SNR 4th-Order Discrete-Time Sigma-Delta Modulator

Chan-Keun Kwon, Chan-Hui Jeong
Department of Nano-Semiconductor Engineering
Korea University
Seoul, Korea
{ckkwon, kjch5286}@asic.korea.ac.kr

Young-Jae Min, Young-Mok Jung, Soo-Won Kim
Department of Electrical Engineering
Korea University
Seoul, Korea
{yjmin, ymin21, ksw}@asic.korea.ac.kr

Abstract—**This paper presents a 4th-order single-bit DT ΔΣ modulator for audio applications. To achieve higher signal-to-noise ratio (SNR), the compensation capacitors in the input of a comparator are exploited to reduce the metastability of a comparator. The proposed modulator has been fabricated in a quantizer level 0.18-μm CMOS process. Measurements show a SNR of 80.27 dB and a signal-to-noise and distortion ratio (SNDR) of 77.58 dB over a 40-kHz signal bandwidth with a sampling clock frequency of 10.24 MHz. A total power consumption including a clock generator is 340 μW with a 1.8-V supply voltage.**

Index Terms—**Analog-to-digital converter, discrete-time ΔΣ modulator, compensation capacitor, comparator-metastability**

I. INTRODUCTION

In modern audio codec applications, the demands for high-resolution and low-power analog-to-digital converters (ADCs) have increased. Among various ADCs, the delta-sigma (ΔΣ) ADC with the noise-shaping property is widely used for low-power and high-resolution data converters [1].

There are two types of ΔΣ modulators, which are the discrete-time (DT) ΔΣ modulator and the continuous-time (CT) ΔΣ modulator. Since the CT ΔΣ modulators have the inherent anti-alias filter and require the lower unit-gain bandwidth of operational amplifier, they consume less power than DT ΔΣ modulators [2]. However, CT ΔΣ modulators have several challenges. The CT ΔΣ modulators that suffer from process variations, require the additional circuitry to compensate a RC time constant variation [3]. In addition, both of clock jitter and excess loop delay lower the resolution and degrade the stability. The DT ΔΣ modulators, by contrast, have several advantages of switched-capacitor circuits such as coefficient matching, scalability and low sensitivity to clock jitter [4][5].

The order of loop filter, the oversampling ratio (OSR), and the quantizer level have to be carefully considered to design a high-resolution and low-power ΔΣ modulator. The in-band noise power in the single-loop ΔΣ modulator can be decreased with high loop filter order, but it is prone to instability. The alternative approach, which reduces the in-band noise power without the stability problem is the multistage noise-shaping (MASH) structure. While the MASH structure can guarantee the stability issue by cascading the single-loop low-order

modulators, the additional digital cancellation circuit is required. In addition, the performance of the ΔΣ modulator is limited by the noise leakage caused by the mismatches between the analog and digital noise transfer functions (NTFs). OSR determines the speed requirement of analog circuits. Consequently, the increase of the ratio between the clock frequency and signal bandwidth requires the higher sampling circuit and the higher performance integrator, which results in an increase of the power dissipation. The ΔΣ modulators using the multi-bit quantizer has become a popular topology because of its ability to reduce quantization noise, which relaxes the analog circuitry requirements and reduces the power dissipation. However, the matching among comparators causes the performance degradations [6]. Besides the above considerations, the low-noise sampling circuits, the low-distortion amplifiers, the linear capacitors, and the accurate digital-to-analog converters (DACs) are also required for high performances.

In this paper, a single-loop fourth-order single-bit DT ΔΣ modulator with compensation capacitors in the input of the comparator is proposed. The proposed ΔΣ modulator has been designed and fabricated in a 0.18-μm CMOS technology.

II. DESIGN CONSIDERATIONS

Figure 1 shows the block diagram of the proposed DT ΔΣ modulator. The proposed architecture is the single-loop fourth-order single-bit ΔΣ modulator with feedback and feedforward path. The single-loop structure is selected to avoid the mismatch problem between analog and digital NTF, instead of the MASH structure. Since the feedforward path allows the swing range of operational transconductance amplifiers (OTAs) to be lowered [7], the power consumption of OTAs can be reduced.

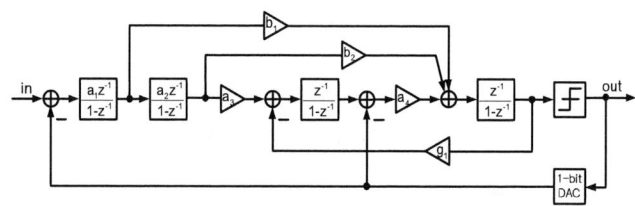

Figure 1. Block diagram of the 4th-order ΔΣ modulator

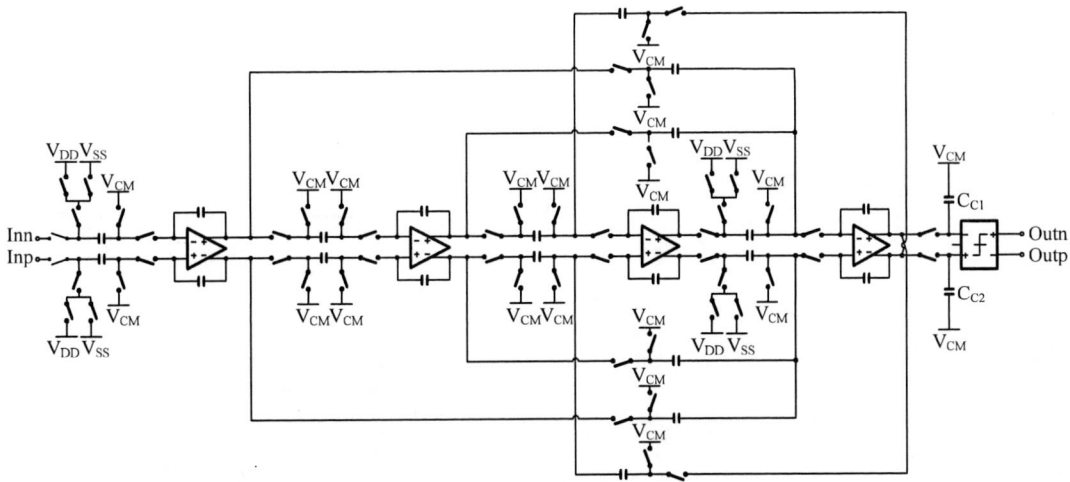

Figure 2. Schematic of proposed 4th-order ΔΣ modulator.

The NTF is determined by the gain of the integrators, feedforward, and local feedback paths. The stability is a critical problem in the design of the ΔΣ modulator with the high order loop filter. In order to solve the stability problem and obtain the higher signal-to-noise ratio (SNR) in this 4th-order ΔΣ modulator, the gain is determined through iterative behavior simulation in MATLAB [8]. The local feedback path in the loop filter makes a noise notch in the signal bandwidth by placing a zero of the NTF into the signal bandwidth from DC [9]. The lower total in-band noise power can be achieved by properly adjusting the position of the noise notch, which can be moved by changing the local feedback gain, g_I in Fig. 1.

By using the multi-bit quantizer, a SNR can be increased [10]. The linearity of the DAC is critical in the multi-bit ΔΣ modulator. If the number of level in the quantizer is increased, the design of the DAC becomes more challengeable. Additional circuitry, such as the dynamic element matching (DEM), is required to compensate for the linearity of the DAC. And the OSR must be carefully selected. Generally, the ΔΣ modulator with large OSR can achieve high SNR. However, if the OSR is increased with the same signal bandwidth, which causes the increase of sampling frequency, the OTAs may consume more power, because the unit-gain bandwidth of OTAs is proportional to sampling frequency. The SNR of ΔΣ modulator is given by [11]

$$SNR = \frac{\frac{\Delta^2}{8}}{P_N} = \frac{2^{2N}3(2L+1)M^{2L+1}}{2\pi^{2L}}, \quad (1)$$

where Δ, N, M, and L denote the input range of the ΔΣ modulator, the number of bits in the quantizer, the OSR, and the order of the ΔΣ modulator, respectively. In this design, an OSR of 128 is selected, and instead of a multi-bit quantizer, a single-bit quantizer that has an inherent linearity is used to reduce the circuit design complexity and additional power consumption.

The dynamic range (DR) that is one of the ADC specifications is proportional to the SNR. The comparator of the quantizer, which has the problem of metastability, that causes the performance of the DR to degrade. The maximum DR is given by

$$DR \geq 11.5 + 0.5\log_2 OSR \text{ bits}, \ f_s / f_T \ \leq 5\%$$
$$DR \approx 8.5 + 0.5\log_2 OSR + \log_2 \frac{f_s / f_T}{5} \text{ bits}, \ f_s / f_T \ \geq 6\% \quad (2)$$

where f_S denotes the clock frequency, and f_T the cutoff frequency of technology. In (2), it is shown that the *DR* is affected by the metastability of the comparator [6]. In order to reduce the performance degradation by the metastability, the compensation capacitors, C_{C1} and C_{C2} in the input of the regenerative comparator are exploited as shown in Fig. 2.

III. CIRCUIT DESCRIPTION

The overall schematic of the proposed single-bit 4th-order ΔΣ modulator is shown in Fig. 2. This modulator is designed with fully differential switched-capacitor circuits to reduce the common mode noise, such as the supply noise. The gain of the integrator, feedforward, and local feedback paths have been determined by behavior simulations.

A. Single-Bit Quantizer

In order to obtain higher SNR, the ΔΣ modulator needs not only high-performance OTAs but also a well-designed quantizer. In this work, the single-bit quantizer is implemented with a comparator.

The small-size capacitors, C_{C1} and C_{C2} in the input of the comparator, are added to reduce the metastability of the comparator as shown in Fig. 3. One side of the capacitor is connected to the common mode voltage *Vcm*, and the other side is connected to the input of the comparator. By adding compensation capacitors, the input of the comparator is stabilized and the degradation of the SNR by the metastability is reduced. Figure 4 shows the simulation results using HSPICE. The noise floor of the modulator with compensation capacitors is decreased, compared to one of the comparators without the compensation capacitors. The peak SNR is increased from 83.34 dB to 87.65 dB.

978-1-61284-863-1/11 $26.00 © 2011 IEEE

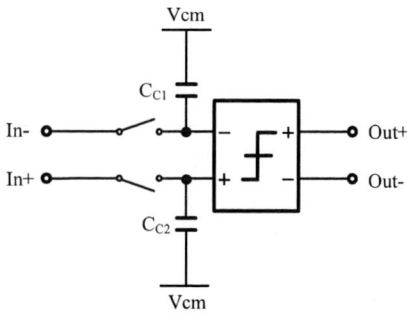

Figure 3. Proposed comparator with additional compensation capacitors.

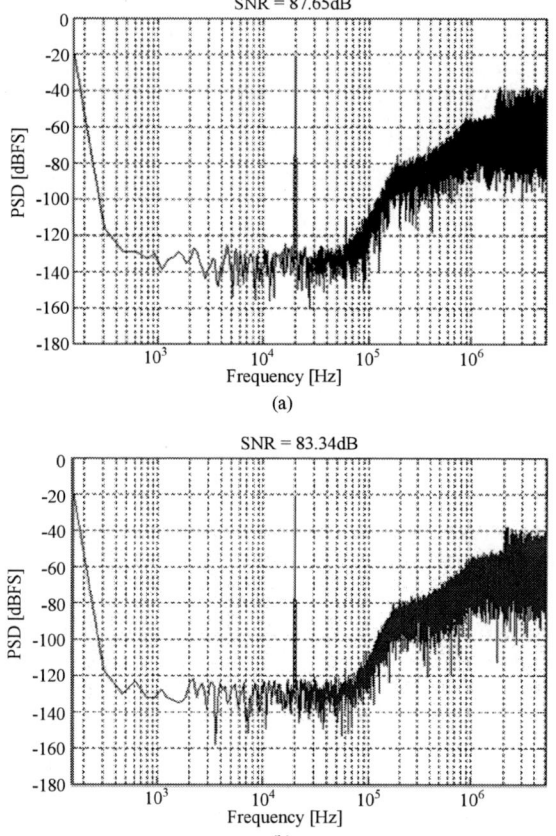

Figure 4. 4th-order DT ΔΣ modulator (a) with and (b) without compensation capacitors.

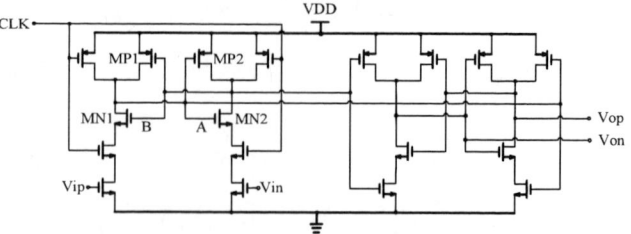

Figure 5. Schematic of comparator.

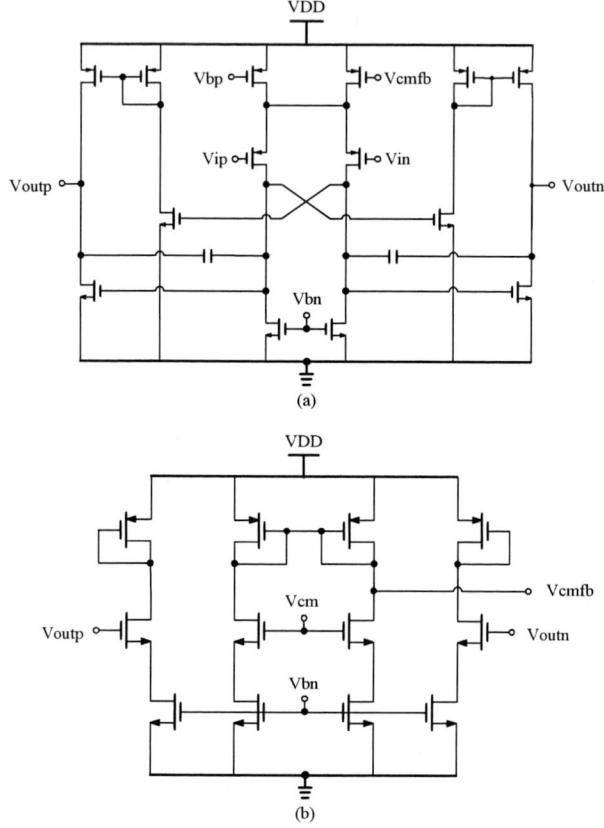

Figure 6. Schematic of (a) OTA and (b) common mode feedback

Figure 5 shows a schematic of the comparator, which consists of two stages to increase the gain. When *CLK* is low, the nodes *A* and *B* are charged to *VDD*, then all of the NMOS transistors in the second stage turn on, and all of the PMOS transistors turn off. Both outputs become high. When *CLK* is high, the nodes *A* and *B* are discharged by the input values. If *Vip* is larger than *Vin*, the node *B* is discharged faster than the node *A*, which affects the PMOS transistors *MP1* and *MP2*, and NMOS transistors *MN1* and *MN2*. This causes the regenerative process to increase the speed of the comparator decision. By adding the second stages, the larger gain and the higher speed can be obtained.

B. Operational Transconductance Amplifier (OTA)

The integrator that determines the noise and linearity of the ΔΣ modulator is the key building block. In this work, each integrator is implemented with the identical OTA, which consists of two stages with the common mode feedback (CMFB) circuit. The schematics of the OTA and CMFB circuit are shown in Fig 6. The CMFB circuit stabilizes both output levels of the first stage and makes it equal to the output level of the OTA and *Vcm* (*VDD*/2). Although the NMOS pair input stage can obtain a higher DC gain, the input stage is implemented with the PMOS pair, which has lower input-referred noises than the NMOS pair. The second stage is composed of the class AB stage to minimize the power consumption. The miller compensation is used to compensate the two-stage OTA. The designed OTA has a 66-dB DC gain, 57.3-degree phase margin, and 89.9-MHz unit-gain bandwidth with a capacitive loading of 5 pF.

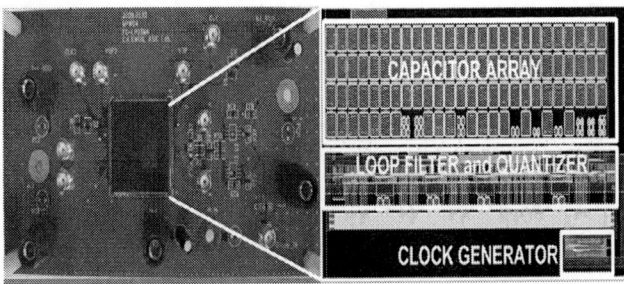

Figure 7. Test board and chip layout.

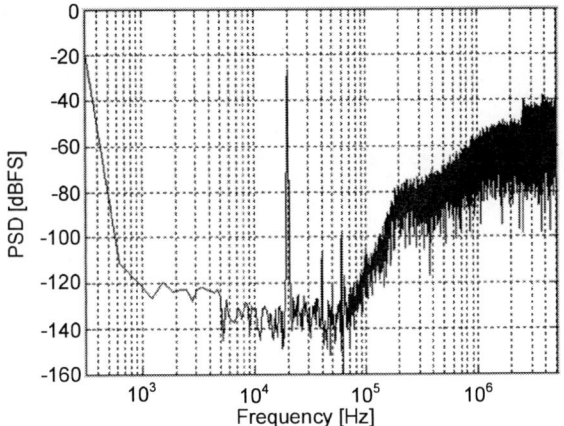

Figure 8. Measured output spectrum (input signal frequency = 20 kHz).

TABLE I.   SUMMARY OF RESULTS

Specification	Measurement results	Simulation results	
		With capacitor	*W/O capacitor*
SupplyVoltage	1.8V	1.8V	1.8V
Bandwidth	40kHz	40kHz	40kHz
Sampling Clock	10.24MHz	10.24MHz	10.24MHz
Peak SNR	80.27dB	87.65dB	83.24dB
Peak SNDR	77.58dB	87.62dB	83.28dB
Die Area w/o Pad	0.645mm²	-	-
Power Consumption	340μW	-	-
Process	1-poly 6-metal 0.18-μm CMOS		

IV.   EXPERIMENTAL RESULTS

The proposed $\Delta\Sigma$ modulator has been fabricated in a 0.18-μm single-poly six-metal CMOS process with the active area of 0.645 mm² (1.5 mm x 0.43 mm) including a clock generator. The test board and chip layout of the modulator are shown in Fig. 7. The prototype chips have been tested with a 20-kHz differential input signal under a 10.24-MHz sampling clock frequency.

Figure 8 shows a 32,768-points FFT plot of the output of the modulator. The measured peak SNR and SNDR over a 40-kHz signal bandwidth are 80.27 dB and 77.58 dB, respectively. Second and third harmonic distortions become large as the amplitude of the input signal is increased. These harmonic distortions could be much improved by using the bootstrap sampling switches, which could increase SNDR. The power consumption of 340μW, including a clock generator is measured. In this design, each integrator uses the identical OTA. If we replace the rest of the OTAs with scale-downed OTAs, except for the one in the first integrator, lower power consumption can be achieved [9]. Table I summarizes the simulated and measured performance of the modulator.

V.   CONCLUSION

In this paper, a 4th-order single-bit DT $\Delta\Sigma$ modulator has been presented. The SNR is increased by adding the compensation capacitors in the input of the comparator, which reduces the metastability of the comparator. This $\Delta\Sigma$ modulator has been fabricated in a 0.18-μm single-poly six-metal CMOS technology. The peak SNR of 80.27 dB and the peak SNDR of 77.58 dB with a 40-kHz bandwidth are measured. The total power consumption is 340 μW, including a clock generator with a 1.8-V supply voltage.

ACKNOWLEDGMENT

This work was supported by the National Research Foundation of Korea (NRF) grant funded by the Korea government (MEST) (No.K20902001448-10E0100-03010) and Seoul R&BD Program (10920).

REFERENCES

[1] S.R. Norsworthy, R. Schreier, and G.C. Temes, *Delta-Sigma Data Converters: Theory, Design, and Simulation*, IEEE Press, 1996.

[2] S. Yan, and E. Sánchez-Sinencio, "A Continuous-Time $\Sigma\Delta$ Modulator With 88-dB Dynamic Range and 1.1-MHz Signal Bandwidth," *IEEE J. Solid-State Circuits*, vol. 39, no. 1, pp. 75-86, Jan. 2004.

[3] B. Xia, S. Yan, and E. Sánchez-Sinencio, "An RC Time Constant Auto-Tuning Structure for High Linearity Continuous $\Sigma\Delta$ Modulator and Active Filters," *IEEE Trans. Circuits Syst. I, Reg. Papers*, vol. 51, no. 11, pp. 2179-2188, Nov. 2004.

[4] M.-Y. Choi, S.-N. Lee, S.-B. You, H.-J. Park, J.-W. Kim, H.-S. Lee, "A 101-dB SNR Hybrid Delta-Sigma Audio ADC using Post Integration Time Control," *in Proc. IEEE CICC*, Sept. 2008, pp. 89-92.

[5] J. Chen, Y.P. Xu, "A 94dB SFDR 78dB DR 2.2MHz BW Multi-bit Delta-Sigma Modulator with Noise Shaping DAC," *in Proc. IEEE CICC*, Sept. 2007, pp. 69-72.

[6] J.A. Cherry, W.M. Snelgrove, "Clock Jitter and Quantizer Metastability in Continuous-Time Delta–Sigma Modulators," *IEEE Trans. Circuits Syst. II, Analog and Digital Signal Processing*, vol. 46, no. 6, pp. 661-676, June 1999.

[7] L. Yao, M. Steyaert, W. Sansen, *Low-power Low-voltage Sigma-Delta Modulators in Nanometer CMOS*, Springer, 2006.

[8] The Mathworks, Inc., Category: Control Systems, File: SD Toolbox [Online].Available:http://www.mathworks.com/matlabcentral/fileexchange

[9] J. Roh, S. Byun, Y. Choi, H. Roh, Y.-G. Kim, J.-K. Kwon, "A 0.9-V 60-uW 1-Bit Fourth-Order Delta-Sigma Modulator With 83-dB Dynamic Range," *IEEE J. Solid-State Circuits*, vol. 43, no. 2, pp. 361-370, Feb. 2008.

[10] V. Colonna, G. Gandolfi, F. Stefani, A. Baschirotto, "A 10.7-MHz Self-Calibrated Switched-Capacitor-Based Multibit Second-Order Bandpass $\Sigma\Delta$ Modulator With On-Chip Switched Buffer," *IEEE J. Solid-State Circuits*, vol. 39, no. 8, pp. 1341-1346, Aug. 2004.

[11] P. Malcovati, S. Brigati, F. Francesconi, F. Maloberti, P. Cusinato, A. Baschirotto, "Behavioral Modeling of Switched-Capacitor Sigma–Delta Modulators," *IEEE Trans. Circuits Syst. I, Fundam. Theory Appl.*, vol. 50, no. 3, pp. 352-364, Mar. 2003.

978-1-61284-863-1/11 $26.00 © 2011 IEEE

A CMOS Single Stage Fully Differential Folded Cascode Amplifier Employing Gain Boosting Technique

S.A Enche Ab Rahim, I. M. Azmi
Advance Physical Technologies Lab
TM Research & Development Sdn. Bhd
Cyberjaya, Selangor, Malaysia
amalina@tmrnd.com.my

Abstract— **In this paper, a single stage fully differential folded cascode amplifier using gain boosting technique is presented. The amplifier was initially designed for 12 bits 22 MSPS pipelined analog-to-digital converter (ADC). However, the amplifier alone could not provide a sufficient DC gain in order to meet the ADC requirement. In order to obtain higher DC gain, a gain boosting technique was employed for this op-amp. The amplifier was implemented in 0.13μm CMOS process technology. The simulation results show that this amplifier has a DC gain of 95dB, a unity gain bandwidth of 414 MHz and a phase margin of 82 degrees.**

Keywords-fully differential folded cascode amplifier, gain boosting techniques

I. INTRODUCTION

In high performance CMOS analog circuits such as pipelined analog-to-digital converters, high accuracy and fast settling operational amplifiers are required. However, these requirements contradict each other as high accuracy requires high DC gain of the op-amps, whereas fast settling behavior requires a high unity gain frequency [1].

In [2], the performances of various op-amp topologies were compared. For a high speed op-amp, the single stage folded-cascode topology is the adequate choice as it offers large unity gain and it provides large output swing. However, this amplifier has a limitation in DC gain. To overcome the problem, a gain boosting technique is employed. This technique increases the DC gain of the op amp without sacrificing the output swing.

This paper presents the design of a single stage fully differential folded cascode amplifier employing the gain boosting technique. Fully differential topology is chosen because it can reduce the problems associated with noise coupling and it doubles the signal swing.

In this work, the amplifier is implemented in 0.13μm CMOS process technology. The organization of this paper is as follows: section II describes the gain boosting technique, section III presents the circuit implemented in 0.13μm CMOS process technology, and section IV discusses the simulation results. Finally, the conclusion is drawn in section V.

II. GAIN BOOSTING TECHNIQUE AND FREQUENCY ANALYSIS

Fig. 1 illustrates the gain boosted amplifier in cascode stage. T1 and T2, also known as input device and cascode device respectively, are the transistors from the main amplifier.

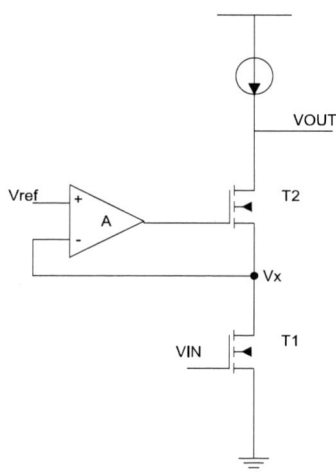

Figure 1.   Gain-boosted amplifier in cascode stage [1]

The idea of this technique is to increase the cascoding effect of T2 by adding the gain-boosted amplifier. Cascoding is a well-known means to enhance the DC gain of an amplifier without deteriorating the high-frequency performance. By increasing the cascoding effect, the DC gain of the amplifier will increased.

The gain-boosted amplifier, with a gain of A, provides a negative feedback loop to T2 and sets the drain voltage of T2, furthermore it drives the gate of T2 until Vx is equal to Vref. As a result, Vx is less affected by the voltage variation of the drain in T2 as it is regulated by the gain-boosted amplifier. Therefore the new expression of output impedance is given in (1):

978-1-61284-863-1/11 $26.00 © 2011 IEEE

$$R_{out} = r_{o1} + r_{o2} + \left[g_{m2}(A+1) + g_{mb2}\right] \times r_{o1} r_{o2} \quad (1)$$
$$R_{out} \approx A g_{m2} r_{o1} r_{o2}$$

It can be seen that the output impedance has been increased by a multiplication factor of A. Therefore the overall gain of the op amp increases, as shown in (2):

$$A_v = g_{m1} R_{out} = g_{m1} g_{m2} A r_{o1} r_{o2} \quad (2)$$

Considering the stability of the amplifier, the stability problem may occur if the gain-boosted amplifier is too fast, due to the fact that it forms a close-loop with T2. There will be a second pole for the whole circuit, which is the pole at the source of T2. Therefore, for stability reasons, the unity gain frequency of the gain-boosted amplifier has to be lower than the second pole of the main op amp [1].

On the other hand, the presence of the gain boosted amplifier introduces a pole-zero doublet, which can degrade the settling behavior of the op amp. This is because, a pole-zero doublet introduces a slow settling component to the amplifier. In order to remove this slow settling component, the unity-gain frequency of the gain-boosted amplifier has to be higher than the -3dB bandwidth of the main op amp [1].

Combining both conditions, the safe range for the unity gain frequency of the gain boosted amplifier is:

$$f_{-3dBmain\_opamp} < f_{u\_gainboosted} < f_{2ndpole\_main\_opamp} \quad (3)$$

III. CIRCUIT IMPLEMENTATION

Fig. 2 illustrates the fully differential folded cascode amplifier, which is the main op amp, designed for the ADC.

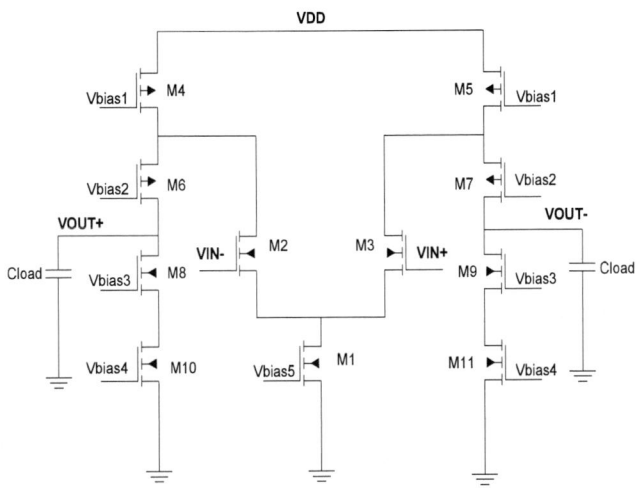

Figure 2. Fully differential folded cascode amplifier

The op amp is a fully differential folded cascode amplifier with NMOS input transistor. These input transistors were dimensioned in such a way to obtain a unity gain frequency higher than 350MHz. The relation between the unity gain frequency and the input transistors is given by (4):

$$g_m = C_{load} \times \frac{f_u}{2\pi} \quad (4)$$

where g_m is the transconductance of the input transistor, C_{load} is the load capacitance at the output node, and f_u, is the unity gain frequency of the amplifier.

However, this amplifier could not provide sufficient DC gain required by the ADC circuit. The performance of this amplifier is presented in section IV. Due to this reason, the gain boosting technique is employed. The implementation of gain-boosted amplifier is shown in Fig. 3.

The simplest implementation for the gain-boosted amplifier is a single MOS transistor [3]-[5]. In this work, the gain-boosted amplifiers, also presented in [6] are used in the gain-boosted stage. They are single MOS transistors amplifiers with current-source load. As the main op amp already consumes current, the gain-boosted amplifiers were designed with minimal current as possible in order to not add much current to the total current consumption.

The circuit was implemented in 0.13μm CMOS process technology. The power supply applied is 3.0V and the load capacitance of the main amplifier is set at 2pF.

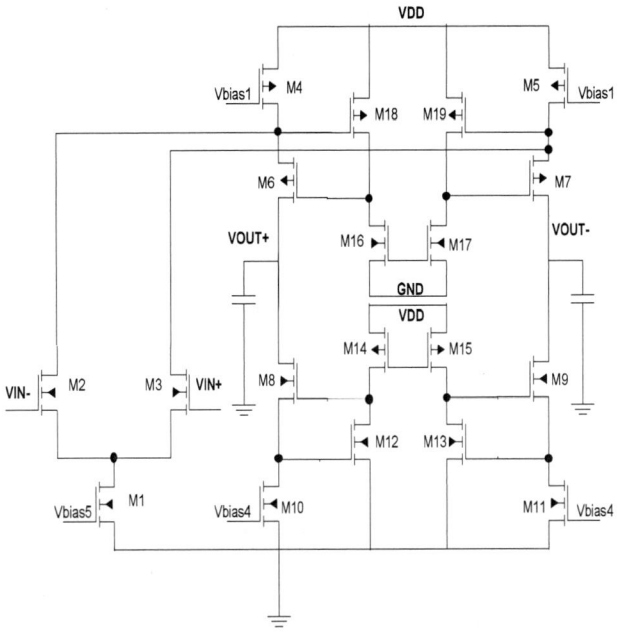

Figure 3. Fully differential folded cascode amplifier with gain-boosted stage

IV. SIMULATION RESULTS

The small signal performance of the main op amp without using the gain boosting technique is shown in Fig. 4. The DC gain is 60.8dB with a unity gain frequency of 383.8 MHz. The first order role-off of this op amp is at 350 KHz with a second pole situated at nearby 1.82 GHz. The phase margin of the amplifier is 81 degrees.

The performances of the op amp, with and without (dashed plot) gain-boosted amplifier are illustrated in Fig. 5 for comparisons purposes. It can be seen that the gain boosting technique increases the DC gain of the op amp, from 60.8dB to 94.9dB. The unity gain frequency also has been increased from 383.3 MHz to 414.0 MHz. The addition of the gain-boosted amplifier does not degrade the phase response of the main amplifier where the phase margin is 82.3 degrees.

In terms of power consumption, the op amp without the gain-boosted amplifier consumes a total current of 3mA. The added gain-boosted amplifiers increase the total current consumption to 3.67mA. Thus the addition of the gain-boosted amplifiers represents 18% increment of the overall total power consumption.

In order to see the effect of the gain-boosted amplifier on the op-amp, the performance of N-Booster circuit and P-booster (dashed plot) circuit are also plotted in the Bode diagram, as shown in Fig. 6. The unity gain frequency of N-Booster is 1.45 GHz while for P-Booster, it is 1.18 GHz.

The unity gain frequencies of both amplifiers are smaller than the second pole of the main op amp. At the same time, they are superior to the first order roll-off of the op amp. Thus the designed op amp with gain-boosted amplifier fulfils the requirements given by (3).

The settling behavior of the op amp is simulated in a closed-loop configuration of sample and hold amplifier. The transient simulation in Fig. 7 shows that the op amp settles to the final value of 1% accuracy of 1.4V step within 6.17ns. Table 1 summarizes the performance of the op amp with the gain-boosting technique, as well as the performance from other published research work for comparisons purposes with this work.

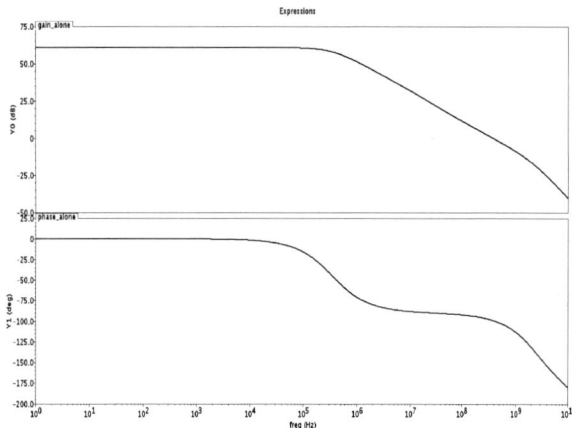

Figure 4.   Frequency response of fully differential folded cascode amplifier

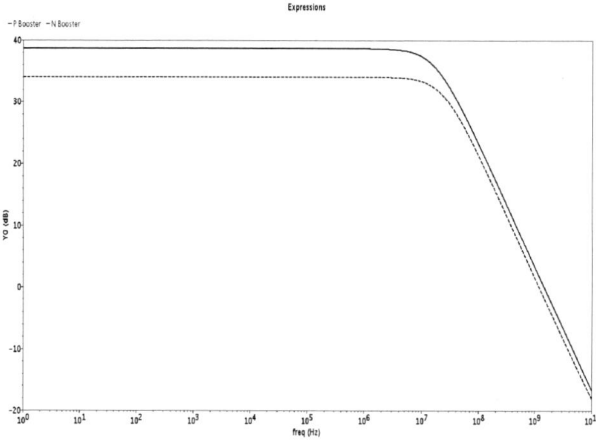

Figure 6.   Frequency response of gain boosted amplifiers

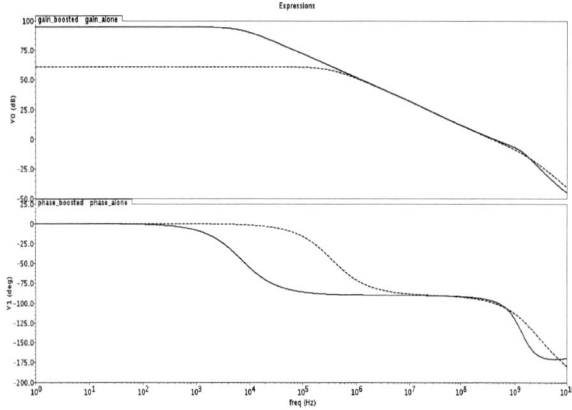

Figure 5.   Frequency response of fully differential folded cascode amplifier (with and without gain boosting technique)

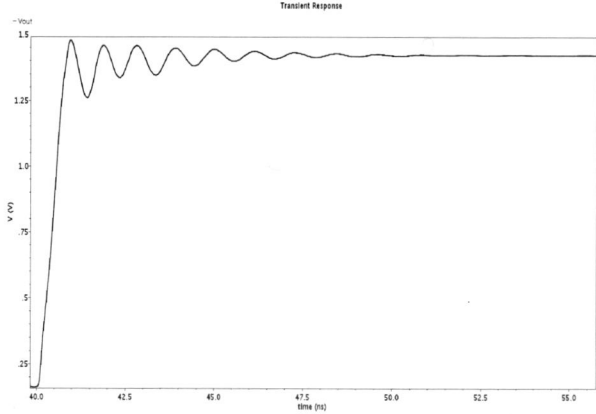

Figure 7.   Settling behavior of the amplifier with gain boosting technique

TABLE I.   THE SPECIFICATIONS OF THE OP AMPS

	This work	*Reference [6]*	*Reference [7]*
DC Gain, (dB)	94.9	95.0	112.0
Unity GBW, (MHz)	414.0	412.0	1150.0
Phase Margin, (deg)	82.3	75.0	69.0
Settling Time, (ns)	6.17 (1% accuracy)	7.5 (5% accuracy)	2.44 (unknown)
Load Capacitor, (pF)	2.0	1.9	3.0
Supply Voltage, (V)	3.0	3.0	3.0
Power, (mW)	11.0	12.8	19.5

V.   CONCLUSION

A single stage fully differential folded cascode amplifier was designed. In order to increase the DC gain of the amplifier while having a high unity gain frequency, the gain-boosting technique was used. As a result, the DC gain has increase approximately 34 dB, while the unity gain frequency has been improved to 414 MHz. The settling time of the op amp is 9.6ns. Finally, the impact of additional gain-boosted amplifier is minimal to the total power consumption.

ACKNOWLEDGMENT

The research is supported by TMRND Sdn Bhd. The authors would like to thank Telekom Malaysia for sponsoring the work through research project RDTC/080699.

REFERENCES

[1] Klaas Bult and Govert J.G. M. Geelen, "A Fast Settling CMOS Op Amp for SC Circuits with 90-dB DC Gain", IEEE Journal of Solid-State Circuits, Vol. 25, No 6, December 1990.

[2] Behzad Razavi, "Design of Analog CMOS Integrated Circuits", McGraw-Hill 2001, pg 313.

[3] B.J. Hosticka, "Improvement of the gain of MOS amplifiers," IEEE J. Solid-State Circuits, vol. SC-14, no-6, pp 1111-1114, Dec 1979.

[4] E. Sackinger and W. Guggenbuhl, "A high-swing, high-impedance MOS cascode circuit," IEEE J. Solid-State Circuits, vol. 25, no 1, pp 289-298, Feb 1990.

[5] H.C. Yang and D.J. Allstot, "An active-feedback cascode current source," IEEE Trans. Circuits Syst., vol. CAS-37, no 5, pp. 644-646, May 1990.

[6] Rohana Musa, Yuzman Yusoff, Tan Kong Yew, Mohd Rais Ahmad, "Design of Single-Stage Folded-Cascode Gain Boost Amplifier for 100mW 10-bit 50MS/s Pipelined Analog to Digital Converter", IEEE International Conference on Semiconductor Electronics, ICSE, 2006.

[7] Feng Wenxiao, Lu Tiejun, Wong Zongmin, "Analysis and Design of Fully Differential Gain-Boosted Op-amp for 14bit 100 MS/s Pipelined Analog-to-Digital Converter", Fifth International Joint Conference on INC, IMS and IDC, 2009.

A Novel Ripple Controlled Modulation for High Efficiency DC-DC Converters

Zhuochao Sun and Liter Siek
VIRTUS-IC Design Centre of Excellence
School of Electrical and Electronic Engineering, Nanyang Technological University
50 Nanyang Avenue, Singapore 639798
Email: sunz0008@e.ntu.edu.sg; elsiek@ntu.edu.sg

Abstract—To achieve high conversion efficiency, traditional DC-DC converters usually employ the hybrid-mode control, which is a combination of pulse-width modulation (PWM), pulse-frequency modulation (PFM) and pulse-skipping modulation (PSM) controllers. However, the hybrid-mode control does not maximize the efficiency because it only concerns about the load change of the converter under light-load condition, without taking the full range of load variation and the source voltage variation into consideration. In addition, a hybrid-mode controller also includes complex mode-switching circuit, which brings in more design difficulties. In this paper, a new ripple controlled modulation (RCM) is proposed. It aims to maximize the conversion efficiency with simple control. This is done by using a ripple controlling loop to optimize the switching frequency according to the converter output voltage ripple under all source and load conditions.

Keywords-Ripple Controlled Modulation (RCM); dual-edge ramp; high efficiency DC-DC converter

I. INTRODUCTION

Nowadays, portable devices have become part and parcel of our daily lives. Owing to the advance of modern microelectronic technology, these small devices are able to perform very complicated functions by integrating more circuit blocks into a single chip. As a result, the total power consumption increases. Thus, when enjoying these additional functions, we suffer from shorter operating time due to the limitation of the battery. Therefore, the conversion efficiency of the DC-DC converter inside those devices is of great concern.

Switching-mode converters are used in many portable devices due to their higher efficiency, as compared to the low dropout (LDO) regulators. The losses in a switching-mode converter can be generally categorized into several types [1], which is listed in Table I. As the static losses are much smaller than the others, the efficiency of a switching-mode converter is usually expressed as

$$\eta = \frac{P_{load}}{P_{load} + P_{cond} + P_{sw}}, \qquad (1)$$

where P_{load} is the power delivered to the load, P_{cond} is the conduction loss, and P_{sw} is the switching loss. Therefore, to

TABLE I. GENERAL LOSSES IN A SWITCHING-MODE CONVERTER

Switching (Frequency-dependent) Losses MOSFET output capacitance MOSFET gate capacitance Diode capacitance Diode stored minority charge Inductor and transformer core loss Snubber loss Gate driver losses
Conduction (Load-dependent) Losses MOSFET on-resistance Diode forward voltage drop Inductor winding resistance Capacitor equivalent series resistance
Static Losses Controller standby current MOSFET, diode and capacitor leakage currents

improve the conversion efficiency, both conduction loss and switching loss should be minimized.

The conduction loss is usually dominant at heavy load condition. Most high efficiency converters use a synchronous switch to replace the diode, thus the reduction on conduction loss is normally achieved by increasing W/L ratio of the power transistor to reduce its on-resistance. However, larger transistor size will result in higher switching loss due to the increase in parasitic capacitance. Because of this conflict, compact layout methods are used to maximize the W/L ratio within a fixed transistor area [2].

At light load, the switching loss becomes more significant. Because it is a frequency-dependent loss, the pulse-frequency modulation (PFM) is preferred due to its low switching frequency, as compared to the pulse-width modulation (PWM). Therefore, hybrid-mode control is commonly used for converters with wide load range, which is illustrated in Fig. 1. In recent literature, pulse-skipping modulation (PSM) is added in between of PWM and PFM to smoothen out the transition of the efficiency curve at medium load range [3]. Thus, by combining the PWM, PSM and PFM control techniques, the hybrid-mode converter can choose the most suitable modulation type which delivers higher efficiency at that particular load current.

This work is sponsored by Nanyang Technological University, Singapore; under the NTU Research Scholarship.

978-1-61284-863-1/11 $26.00 © 2011 IEEE

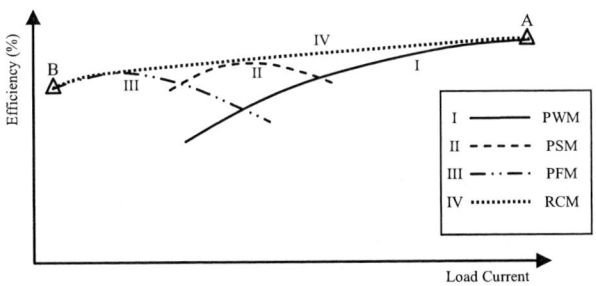

Figure 1. Conversion efficiency versus load current for different control techniques

Although the hybrid-mode control improves the light load efficiency, there are some drawbacks. Firstly, when it switches to PSM mode, the load regulation is usually sacrificed with ripple exceeding the tolerance. Secondly, the conversion efficiency in hybrid mode is only maximized at several points, so it can be further improved for the other conditions. Taking the boost converter as an example, its efficiency in PWM mode is only maximized at the minimum source voltage with the maximum load current (point A in Fig. 1). This operation point gives the largest output voltage ripple, which is designed to be equal to the maximum tolerance. When the source or load changes, this ripple reduces, and the conversion efficiency is no longer optimum. Similarly for the traditional fixed on-time PFM control, the optimum efficiency occurs at the maximum source voltage with the minimum load current (point B in Fig. 1), and can be improved for the rest of the efficiency curve. Lastly, hybrid-mode control requires more than one modulator with complex mode-switching circuits, complicating the design process.

At any source and load, a switching mode DC-DC converter can achieve better efficiency if it reduces the switching loss by lowering its frequency. Thus, for a well regulated converter to achieve the maximum efficiency, its switching frequency must be reduced to the lowest possible level at which its output voltage ripple will be equal to the maximum tolerance. Therefore in this work, a new ripple controlled modulation (RCM) is proposed. In order to maximizing the efficiency, the RCM control adjusts the switching frequency and fixes the output voltage ripple at a predefined tolerance.

The control method for the proposed modulation is discussed in Section II based on the boost converter topology. Section III shows the Matlab simulation results by comparing the boost converter in PWM, PFM and RCM modes. Conclusions are given in Section IV.

II. PROPOSED RIPPLE CONTROLLED MODULATION

The proposed modulation is applicable to all switching converter types, including buck, boost, and buck-boost converters, to maintain high efficiency under varying source and load conditions. As shown in Fig. 2, a boost converter with RCM control is discussed in this section.

Working Principle

There are two control loops in this architecture. The first one is the conventional voltage regulation loop through the compensation network EA1. It regulates the converter output voltage by varying the duty ratio of the power switches. This loop is equivalent to a PWM controller. To achieve high bandwidth, a type III compensator is preferred at EA1 [4].

The second loop is the ripple controlling loop through the compensation network EA2. It optimizes the switching frequency according to the output voltage ripple, and thus maximizes the conversion efficiency by reducing the switching loss.

The output voltage ripple of the boost converter shown in Fig. 2 can be expressed as

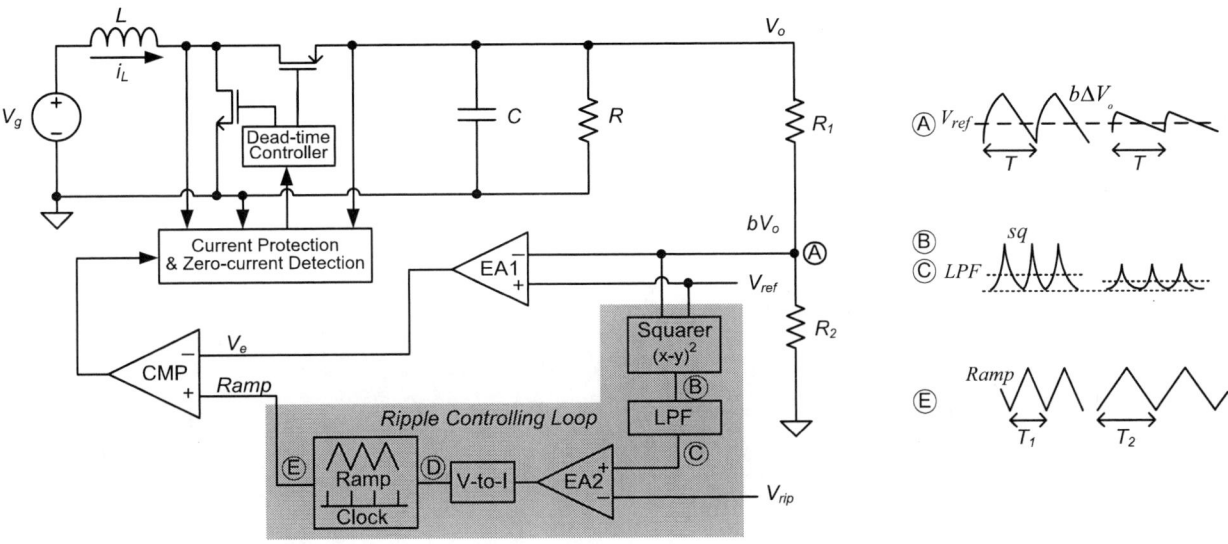

Figure 2. A boost converter with the proposed RCM control

978-1-61284-863-1/11 $26.00 © 2011 IEEE

$$\Delta V_o = \frac{V_o}{RC} DT = \frac{V_o - V_g}{RCf}, \qquad (2)$$

where V_g and V_o are the source and load voltages, R is the load impedance, C is the filtering capacitor, D is the duty ratio, T is the duty cycle, and f is the switching frequency. Equation (2) shows that, when either the source voltage V_g increases or the load decreases (i.e. R increases), the output voltage ripple will reduce. A ripple much smaller than the tolerance implies that the converter works under an unnecessarily high frequency and thus suffers from extra switching loss. To reduce this loss, the following process happens in the circuit of Fig. 2.

- At node A: The ripple of the output feedback voltage reduces.

- At node B: The squarer circuit calculates the square of the voltage difference between the feedback voltage $b \cdot V_o$ and the reference voltage V_{ref} , and its output voltage also reduces.

- At node C: The DC output of the low-pass filter (LPF) reduces.

- At node D: After going through EA2 and V-to-I converter, the voltage drop of the LPF output causes a current reduction at node D.

- At node E: The switching cycle time increases due to the reduction of the charging/discharging current in the ramp generator.

This process keeps reducing the switching frequency until the output voltage ripple increases to the predefined tolerance. After that, if the source and load conditions do not change any more, the switching frequency will be maintained, and the converter is actually controlled under PWM, either in continuous conduction mode (CCM) or discontinuous conduction mode (DCM).

Circuits of the Ripple Controlling Loop

A voltage squarer and a LPF are used in the first stage of the ripple controlling loop. Its output is equivalent to the magnitude of the output voltage ripple. Because this ripple is not synchronous with the duty control signal and its wave pattern is dependent on the actual component values in the converter, therefore, it is hard to sample the peak ripple. Hence in this work, the voltage squarer is used together with the LPF to calculate the mean square value of the ripple. The voltage squarer can use the design in [5] together with a pre-amplifier stage.

A type I compensator is preferred to be used in EA2 with a very low frequency dominant pole. So the slow adjustment in switching frequency along the ripple controlling loop will not affect the overall stability of the converter circuit. Its structure is shown in Fig. 3. The voltage V_{rip} is a predefined value, fixed to confine the output voltage ripple.

The output voltage of EA2 is sampled at the end of each cycle and converted into a current, which serves as the charging/discharging current of the ramp generator. A dual-edge ramp is preferred, rather than its single-edge counterpart [6]. As illustrated in Fig. 4(a) for the single-edge ramp, when switching frequency reduces, the average inductor current increases, which will in turn lead to a small fluctuation in the output voltage. However, for the dual-edge ramp as illustrated in Fig. 4(b), it does not have this problem as the inductor current always returns to the average level at the end of each switching cycle.

III. SIMULATION RESULTS

A boost converter with the proposed RCM control described in Section II was simulated in Matlab. The source voltage V_g was swept from 1.8V to 3.4V, and the output voltage V_o was regulated at 3.6V with a maximum load of 0.75A. The inductor value used is 2.5µH, and capacitor value used is 18µF with a 3mΩ ESR (equivalent series resistor).

Fig. 5 shows the simulation results of the boost converter with different modulation schemes. Fig. 5(a) shows an 800kHz PWM control under heavy load condition, whereas in Fig. 5(b) using PFM control under light load condition. In both simulations, the output voltage ripple varies when source or load changes, which implies the conversion efficiency is not maximized. The proposed RCM control in Fig. 5(c) achieves a

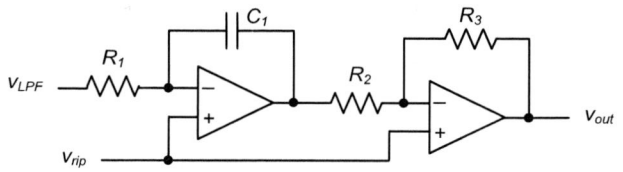

Figure 3.   Compensator EA2 in ripple controlling loop

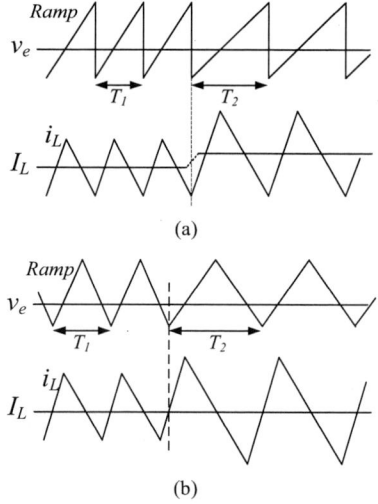

Figure 4.   Inductor waveforms when switching frequency reduces
(a) with single-edge ramp, and (b) with dual-edge ramp

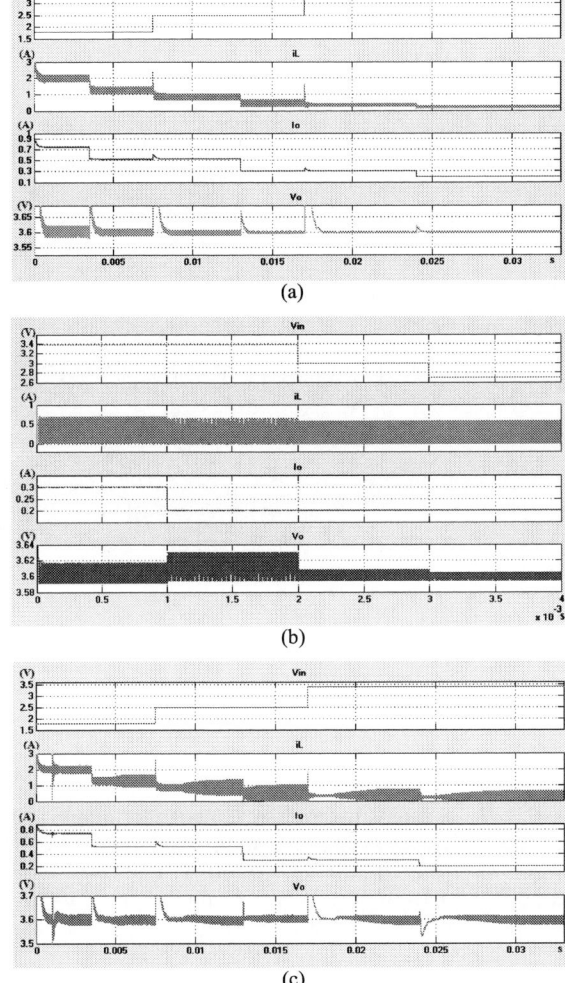

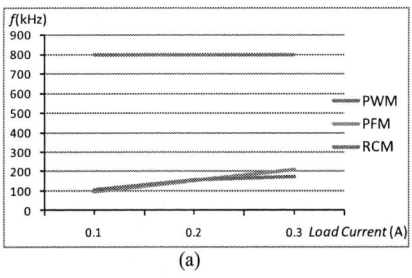

(a)

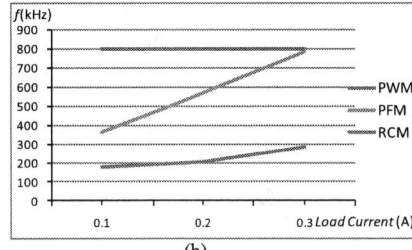

(b)

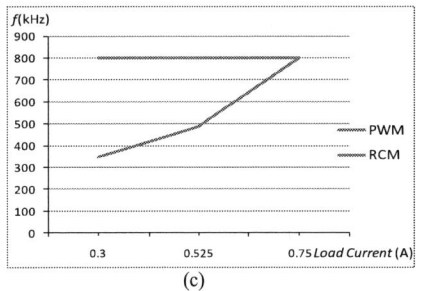

(c)

Figure 6. Switching frequency versus load at different source voltages: (a) light load with V_g=3.4V, (b) light load with V_g=2.7V, and (c) heavy load with V_g=1.8V

Figure 5. Simulation waveforms of a boost converter under varying source and load conditions (a) with PWM control, (b) with fixed on-time PFM control, and (c) with proposed RCM control

fixed ripple, predefined to be 30mV in this work, under all conditions.

The efficiency improvement of the proposed control can be reflected by the reduction of the switching frequency. At high source voltage with light load (Fig. 6(a)), the proposed RCM control has achieved roughly the same performance as the PFM control; while at reduced source voltage (Fig. 6(b)), the RCM control has achieved up to 64% further reduction of switching frequency as compared to the PFM control. The proposed RCM control also works in CCM with heavy load as in Fig. 6(c), which is not a suitable condition for conventional PFM working in extremely high frequency.

IV. CONCLUSION

There is a tradeoff between output voltage ripple and conversion efficiency in switching mode DC-DC converters. When achieving better regulation with smaller ripple by increasing the switching frequency, the switching loss increases as well, thus the efficiency drops; and vice versa. To

find the optimal balance, a new RCM control has been proposed to adjust the switching frequency and maintain the output voltage ripple at a fixed tolerance under all source and load conditions. With the implementation, the efficiency of the converter can be maximized without violating its regulating requirement, which improves the efficiency compared to that using the complicated hybrid-mode controller, and yet is able to achieve with a single RCM modulator.

REFERENCES

[1] R. Erickson and D. Maksimovic, "High efficiency DC-DC converters for battery-operated systems with energy management," Worldwide Wireless Communications, Annual Reviews on Telecommunications, 1995.

[2] A. Yoo, M. Chang, O. Trescases, and N. Wai Tung, "High Performance Low-Voltage Power MOSFETs with Hybrid Waffle Layout Structure in a 0.25μm Standard CMOS Process," in 20th International Symposium onPower Semiconductor Devices and IC's, 2008 (ISPSD '08), pp. 95-98.

[3] H. Hong-Wei, C. Ke-Horng, and K. Sy-Yen, "Dithering Skip Modulation, Width and Dead Time Controllers in Highly Efficient DC-DC Converters for System-On-Chip Applications," IEEE Journal of Solid-State Circuits, vol. 42, pp. 2451-2465, 2007.

[4] K. Wing-Hung, "Switching Converters: Stability and Compensation," Tutorial Notes of ISIC09, December 2009.

[5] S.-I. Liu and D.-J. Wei, "Analogue squarer and multiplier based on MOS square-law characteristic," ELECTRONICS LETTERS, vol. 32 No. 6, March 1996.

[6] Z. Sun and L. Siek, "A novel power line communication controller designed for point-of-load dc-dc converters," 2010 IEEE International Conference on Power and Energy (PECon), pp. 667-671.

Gm-enhanced Differential Colpitts VCO

Xue-Fei Xiao[1,2], Wang Ling Goh[2], Minkyu Je[1], Jae-Hong Chang[1]

[1]*Instite of Microelectronics, A*STAR (Agency for Science, Technology, and Research), 117685, Singapore*
[2]*Nanyang Technological University, Singapore*

Abstract—**This paper presents a novel *gm*-enhanced differential Colpitts voltage-control oscillator (VCO) that requires a lower start-up current. The central frequency is 1.8-GHz and the tuning range is about 15%. The tank inductors were designed with a quality factor of 9. The proposed *gm*-enhanced differential Colpitts VCO was compared with conventional cross-coupled VCO as well as differential Colpitts VCO, and demonstrated 3.5 dBc/Hz and 1.9 dBc/Hz better phase noise performances over the two. Comparison of the three VCOs was conducted at 2.5-V supply voltage and 3-mA biasing current, using 0.18-μm standard CMOS process. The proposed VCO showed a phase noise of -128.9 dBc/Hz at 1-MHz offset.**

Keywords—*voltage-controlled oscillator (VCO); Colpitts; gm-enhanced; phase noise; differential; cross-coupled;start-up current*

I. INTRODUCTION

The Voltage-controlled oscillator (VCO) is one of the most important functional blocks in a frequency synthesizer where it plays a vital role in determining the performance of a system. Due to the explosive growth in wireless communication in the last decade, growingly stringent requirements have been placed on the spectral purity of the VCOs. As a result, VCOs have been continuously studied and improved [1], [2] to attain lower phase noise, lower power consumption, wider tuning range, and so on.

Till now, many structures have been used to improve the performance of the VCOs. The cross-coupled LC VCOs have been a preferred choice compared with other topologies due to their relaxed start-up condition, ease of implementation, differential operation and relatively good phase noise [3], [4]. A typical cross-coupled LC VCO schematic is shown in Fig. 1. However, some barriers such as AM to FM conversion mechanism [5], [6] in active device have to be overcome before it can realise the desired superior performance. Another topology, the Colpitts VCOs [7], [8], can achieve lower phase noise because of its finer cyclostationary noise properties. Then again, the single-ended Colpitts VCOs are rarely adopted in integrated circuits because of their higher start-up requirements and single-end nature, until the differential Colpitts VCOs were reported.

Based on the differential Colpitts topology [9], a novel *Gm*-enhanced differential Colpitts VCO is presented in this paper. From experiments on 0.18-μm standard CMOS process, we achieved a VCO operating at 1.8 GHz with a 2.5 V supply voltage. The tank inductors have quality factors of 9 and the phase noise at 1 MHz offset is -128.9 dBc/Hz.

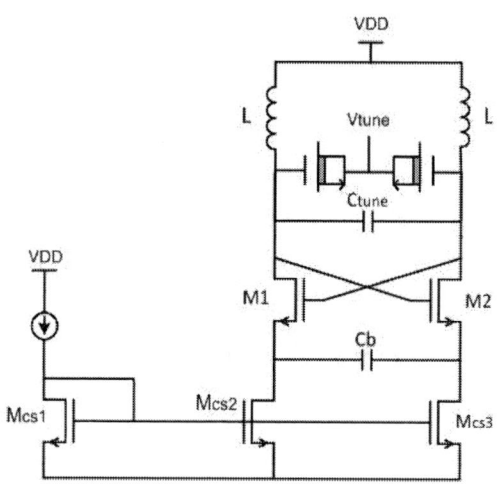

Fig. 1 Schematic of a typical cross-coupled LC VCO.

The rest of the paper is organized as follows. Section II reviews the *gm*-boosted differential Colpitts topology and the proposed circuit structure is explained in details in Section III. In Section IV, the simulation results of the *gm*-boosted differential Colpitts VCO, differential Colpitts VCO and conventional cross-coupled VCO are compared and discussed. Finally, Section V presents the conclusion of this paper.

II. DIFFERENTIAL COLPITTS VCO

Colpitts VCO has been exhibiting brilliant phase noise property. Most of the noise is injected into the tank during the energy transfer process. In the conventional cross-coupled VCO, the energy is delivered into the tank at the maximum impulse sensitive point of oscillation. This property significantly deteriorates the phase noise performance of the conventional cross-coupled VCO. In Colpitts VCO, the energy is instead delivered to the LC tank at the minimum impulse sensitive point [10] of oscillation. Thus, a better phase noise can be expected in Colpitts VCO as compared to the cross-coupled LC VCO. However, conventional Colpitts VCO suffers from the many limitations mentioned in the introduction section, especially the single-end topology. As a result, a differential Colpitts VCO is not only expected but also necessary.

Fig. 2 shows the schematic of a differential Colpitts VCO [9]. A differential output can be achieved by coupling two identical single-end Colpitts oscillators. The source-to-ground capacitors of the two oscillators can be replaced by one capacitor between their sources, which is C_2 in the Fig. 2.

978-1-61284-863-1/11 $26.00 © 2011 IEEE

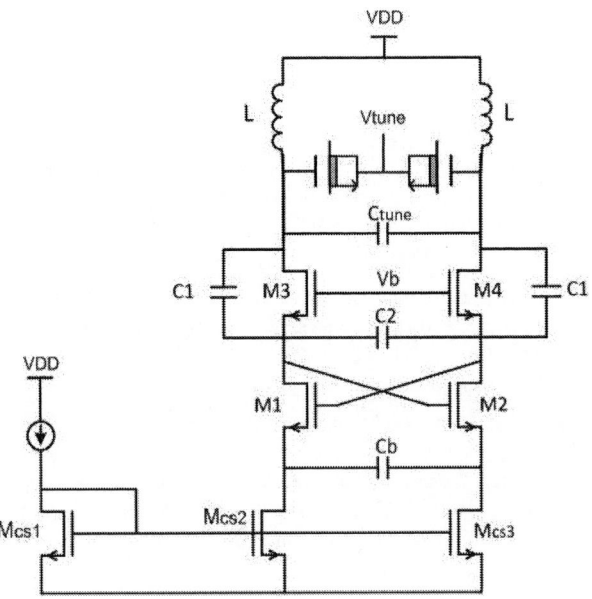

Fig. 2 A differential Colpitts VCO schematic.

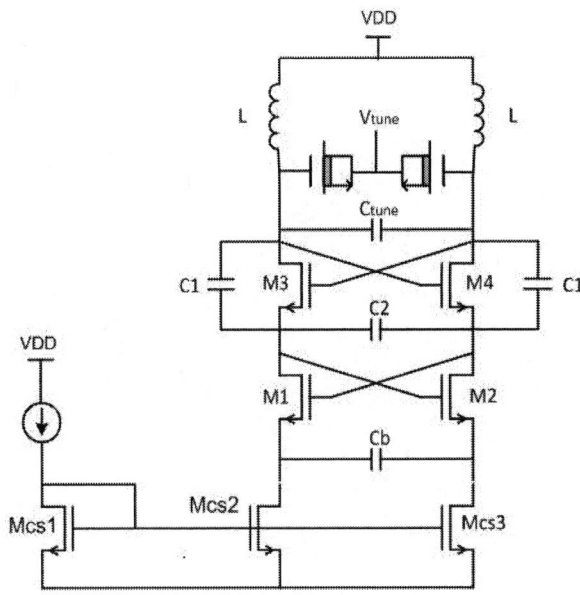

Fig.3 Schematic of the proposed *gm*-enhanced differential Colpitts VCO.

When the differential Colpitts VCO oscillates, the current flows in one transistor (i.e. M_3) switches to the other transistor (i.e. M_4) every half oscillation period. The switch has to occur in a synchronized manner and can be achieved by a pair of cross-coupled NMOS, denoted as M_1 and M_2 in Fig. 2. The fast current switching between M_1 and M_2 can suppress noise contributions from the active devices during the zero-crossings of the tank voltage and hence improving on the phase noise characteristics. Meanwhile, the tail cross-coupled NMOS pair can also provide a negative resistance to improve the small-signal loop gain, relaxing the start-up condition accordingly.

III. *GM*-ENHANCED DIFFERENTIAL COLPITTS VCO

Based on the differential Colpitts VCO structure, a *gm*-enhanced differential Colpitts VCO is proposed in this section and Fig. 3 depicts the schematic. Transistors Mcs_1, Mcs_2, Mcs_3 collectively construct the current source, C_b is used to block the extra noise from the current source, M_1-M_2 realize the tail current switching mechanism mentioned in Section II, and M_3-M_4 are the *gm*-enhanced transistors. According to [11], coupling signal from the source to gate terminal of a transistor can enhance the effective transconductance, to relax the start-up requirement of the Colpitts oscillator. In addition, due to the in-phase relationship between the source and drain voltage, the gate can be alternatively connected to the drain of the transistor. As seen in Fig. 3, instead of directly connecting the gates of M_3 and M_4 to the biasing voltage, the gate of M_3 is joined to the drain of M_4 and the gate of M_4 to the drain of M_3. Such connection allows M_3 and M_4 to be self-biasing, requiring no biasing resistor. In addition, the cross-coupled structure of M_3 and M_4 can further increase the transconductance of the VCO as well as the tank amplitude and overall small signal loop gain. As a result, a better phase noise performance and a lower start-up current can be expected.

The small-signal admittance looking into the drain of M_3 of the proposed VCO is

$$Y_{in} = \frac{(sC_1 - g_{m3})(2sC_2 - g_{m1})}{2sC_2 - g_{m1} + sC_1 + g_{m3}} \quad (1)$$

where, g_{m3} is the transconductance of M_3, and g_{m1} is the transconductance of M_1. The real part of Y_{in} is

$$R_e[Y_{in}]_A = -\frac{(2g_{m3}C_2 + g_{m1}C_1)(C_1 + 2C_2)\omega^2 - g_{m3}g_{m1}(g_{m3} - g_{m1})}{\omega^2(C_1 + 2C_2)^2 + (g_{m3} - g_{m1})^2} \quad (2)$$

Assuming $g_{m3} = g_{m1}$, $R_e[Y_{in}]_A = -g_{m1} = -g_{m3}$, the negative small-signal conductance of the differential Colpitts VCO of Fig. 2 can be expressed as follows:

$$R_e[Y_{in}]_B = -\frac{\omega^2(C_1^2 g_{m1} + 2C_1 C_2 g_{m3})}{\omega^2(C_1 + 2C_2)^2 + (g_{m3} - g_{m1})^2} \quad (3)$$

Letting $g_{m3} = g_{m1}$,

$$R_e[Y_{in}]_B = -\frac{g_m C_1}{C_1 + 2C_2} \quad (4)$$

By comparing (2) and (3), $\left| R_e[Y_{in}]_A \right| - \left| R_e[Y_{in}]_B \right|$ can be calculated.

978-1-61284-863-1/11 $26.00 © 2011 IEEE   243

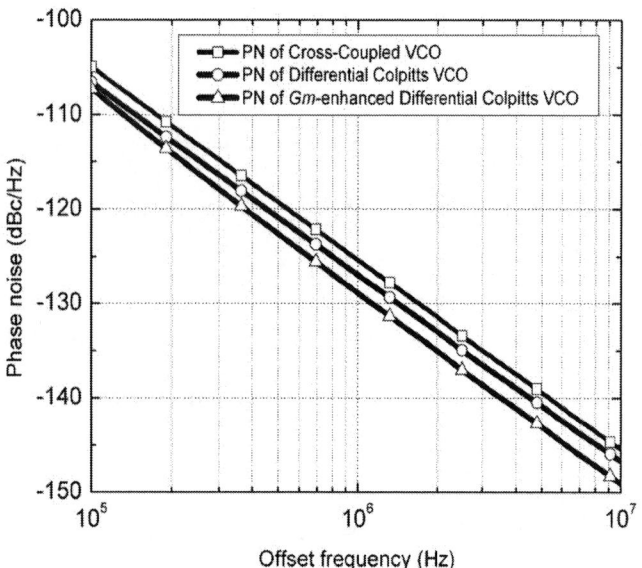

Fig. 4 Phase noise at 1MHz offset of three VCOs, with biasing current setting at 3 mA.

$$\left|R_e[Y_{in}]_A\right| - \left|R_e[Y_{in}]_B\right| =$$

$$\frac{\omega^2[4g_{m3}C_2^2 + g_{m1}C_1C_2] - g_{m3}g_{m1}(g_{m3} - g_{m1})}{\omega^2(C_1 + 2C_2)^2 + (g_{m3} - g_{m1})^2} \quad (5)$$

Since $\omega^2[4g_{m3}C_2^2 + g_{m1}C_1C_2] \gg g_{m3}g_{m1}(g_{m3} - g_{m1})$,
$\left|R_e[Y_{in}]_A\right| - \left|R_e[Y_{in}]_B\right| > 0$.

Assuming $g_{m3} = g_{m1}$,

$$\frac{R_e[Y_{in}]_A}{R_e[Y_{in}]_B} = 1 + 2\frac{C_2}{C_1} \quad (6)$$

If $C_2 = C_1$, $\dfrac{R_e[Y_{in}]_A}{R_e[Y_{in}]_B} = 3$.

From the above calculations, it is obvious that the negative conductance generated in the proposed *gm*-enhanced differential Colpitts VCO is augmented when compared with the differential Colpitts VCO. If $g_{m3} = g_{m1}$ and $C_2 = C_1$, the negative conductance increased by three times. As a result, the biasing current required to ensure a reliable start-up is reduced. Also, the tank amplitude is increased and a better phase noise performance can be expected.

IV. SIMULATION RESULT

To compare the performances of conventional cross-coupled VCO, differential Colpitts VCO and the proposed *gm*-enhanced differential Colpitts VCO, circuits in Fig. 1, Fig. 2 and Fig. 3 are simulated using Cadence SPECTRE with 0.18-μm standard CMOS process. The two 2.5 n tank inductors have quality factors of 9. The supply voltage is 2.5 V. The tanks of

three VCOs are constructed with the same varactors and inductors to make the results comparable. Only C_{tune} is changed in each VCO to tune the central frequency to 1.8 GHz. The channel length of the NNOS varactors is optimized to maximize the quality factor while maintaining a good tuning range.

We first bias the three VCOs using the same biasing current i.e. 3 mA, to compare the phase noise performances. The phase noise characteristics of three VCOs are presented in Fig. 4.

Table I shows the tank amplitude of the three VCOs discussed versus their respective biasing currents. As can be seen, the proposed *gm*-enhanced differential Colpitts VCO shows amplitudes that are larger than the other two VCOs when the biasing current is varied from 1 mA to 5 mA.

The phase noises of the three VCOs versus the respective biasing current are tabulated Table II. After changing the biasing current from 1 mA to 5 mA, the proposed *gm*-enhanced differential Colpitts VCO is seen to be able to yield a better phase noise performance over the other two VCOs. As seen in Table II, the conventional cross-coupled VCO has a phase noise of -125.3 dBc/Hz at 1 MHz offset while the differential Colpitts VCO attained -126.9 dBc/Hz at 1 MHz offset, and hence a better performance than the conventional one. The proposed *gm*-enhanced differential Colpitts VCO however has the best phase noise performance among the three, achieving a phase noise of -128.9 dBc/Hz at 1 MHz offset.

To have a better insight of the proposed *gm*-enhanced differential Colpitts VCO, the three oscillators are operated at1.8 GHz, and varying with biasing current only. The phase noise at 1 MHz offset is given in Fig. 5. The tank amplitude

TABLE I
TANK AMPLITUDE (mV)

VCO Type	Biasing Current				
	1mA	2mA	3mA	4mA	5mA
Conventional cross-coupled	162.2	338.7	504.4	655.1	801.0
Differential Colpitts	NA	283.0	479.3	659.6	835.2
Gm-enhanced differential Colpitts	160.5	373.1	566.3	733.0	884.4

TABLE II
PHASE NOISE (dBc/Hz)

VCO Type	Biasing Current				
	1mA	2mA	3mA	4mA	5mA
Conventional cross-coupled	-116.9	-122.5	-125.3	-126.3	-127.1
Differential Colpitts	NA	-122.7	-126.9	129.3	-130.9
Gm-enhanced differential Colpitts	-119.0	-125.7	-128.9	-130.5	-131.6

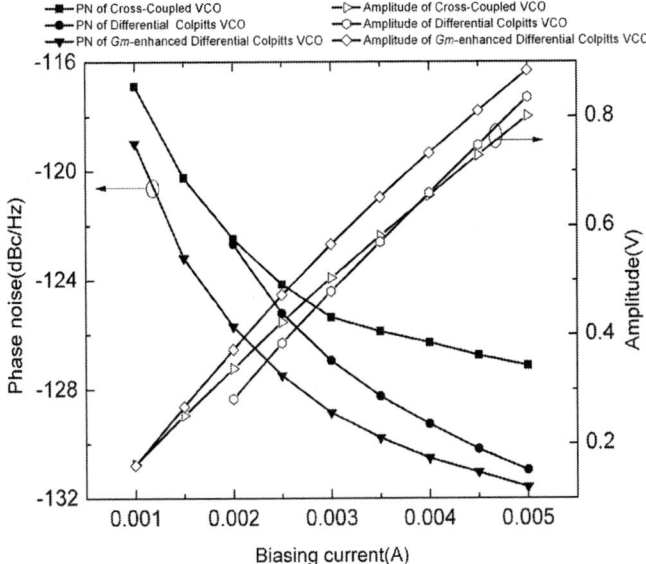

Fig. 5 Phase noise and tank amplitude versus biasing current.

changing with biasing current is also drawn in the same graph.

In Fig. 5, the tank amplitude of *gm*-enhanced differential Colpitts VCO is larger than the other two VCOs. Due to this increased tank amplitude, a better phase noise performance can be achieved. This has been verified by the simulation results. In addition, we also observed that the differential Colpitts VCO call for a 2 mA biasing current to start-up, while the *gm*-enhanced differential Colpitts VCO requires only 1 mA, which is comparable to the conventional cross-coupled VCO. As explained above, this improvement is attributed to the two cross-coupled pairs that help to increase the transconductance of the VCO, hence relaxing the start-up condition.

V. Conclusions

This paper presents a novel *gm*-enhanced differential Colpitts VCO. The proposed VCO reduces the start-up biasing current and increases the tank amplitude. The performance of the proposed *gm*-enhanced differential Colpitts VCO is demonstrated by simulations and has been compared with both the conventional cross-coupled VCO and differential Colpitts VCO.

When biased at 3 mA, the phase noise of the proposed *gm*-enhanced differential Colpitts VCO is -128.9 dBc/Hz at 1 MHz offset, which is 3.5 dBc/Hz better than that of the conventional cross-coupled VCO and 1.9 dBc/Hz better than that of differential Colpitts VCO.

Acknowledgment

The authors would like to thank the assistance provided by staff of the Institute of Microelectronics, Agency for Science Technology and Research (A*STAR), Singapore.

References

[1]. N. Troedsson and H. Sjoland, "An ultra low voltage 2.4 GHz CMOS VCO," in Proc. Radio and Wireless Conf. (RAWCON) 2002, Aug 2002, pp. 205-208

[2]. J. Rael and A. Abidi, "Physical process of phase noise in differential LC oscillators," in *Proc. IEEE Custom Integrated Circuits Conf. (CICC)*, 2000, pp. 569-572.

[3]. A. Kral, F. Behbahani, and A. A. Abidi, "RF-CMOS oscillators with switched tuning," *in Proc. Custom Integrated Circuit Conf. (CICC)*, May 1998, pp.555-558.

[4]. A. Tasic, W. A. Serdijn, and J.R.Long, "Adaptivity of voltage-controlled oscillators--theory and design," *IEEE Trans. Circuits Syst. I.*vol.52, no.5, pp.894-901, May 2005.

[5]. S. Levantino, C. Samori, A. Zanchi, and A. L. Lacaita, "AM-to-PM conversion in varactor-tuned oscillators," *IEEE Trans. Circuits Syst. II*, vol. 49, no. 7, pp. 509–513, Jul 2002.

[6]. B. Soltanian and P. Kinget, "AM-FM conversion by the active devices in MOS LC-VCOs and its effect on the optimal amplitude," in *Radio Frequency Integrated Circuits Symp. (RFIC) Dig.*, Jun 2006.

[7]. Viessmann. A., Damitz. F., Franke. R., Tempel. R.,"Comparison of a fully integrated differential voltage controlled Colpitts oscillator to the cross-coupled oscillator topology," SMI Circuits in RF Systems, 2006 Topical Meeting on 18-20, pp. 4, Jan 2006.

[8]. A. Shibutani, S. Moro, and S. Mori, "Transient response of Colpitts-VCO and its effect on performance of PLL system," *IEEE Trans. Circuits Syst. I.*vol.45, no.7, pp.717-725, Jul 1998.

[9]. R.Aparicio and Λ, Hajimiri, "A noise-shifting differential Colpitts VCO," *IEEE J. Solid-State Circuits,* vol. 37, no.12, pp.1728-1736, Dec. 2002.

[10]. A.Hajimiri and T.H.Lee, "A general theory ofphase noise in electrical oscillators," *IEEE J. Solid-State Circuits*, vol. 33, no.2, pp.179-194, Feb 1998.

[11]. X.Li, S.Shekhar, and D.J.Allstot, "Gm-boosted common-gate LNA and differential Colpitts VCO/QVCO in 0.18-μm CMOS," *IEEE J. Solid-State Circuits*, vol. 40, no.12, pp.2609-2619, Dec 2005.

32kHz MEMS-Based Oscillator for Implantable Medical Devices

Jae-Hong Chang, ShengXi Diao, Raja Muthusamy Kumarasamy, Minkyu Je

Integrated Circuits and System Laboratory
Institute of Microelectronics, A*STAR (Agency for Science, Technology and Research)
Singapore
changjh@ime.a-star.edu.sg

Abstract—**A MEMS based high stability 32kHz oscillator for real time clock (RTC) for medical device application using 0.13μm CMOS technology is designed and simulated. This oscillator achieves a power consumption of less than 1μA with 10pF load capacitor, and a functional temperature drift of ± 80ppm with temperature range from -20°C to 60°C**

Keywords-MEMS; 32kHz; Oscillator; RTC; CMOS; Cystal oscillator; Low power; RTC; Temperature compensation

I. INTRODUCTION

Due to high Q and inherently temperature stable properties, quartz crystal oscillators are popular and important reference clock sources in consumer, commercial, industrial, and military products for many years. The tremendous growth of implantable medical devices and portable devices markets in the past few years accelerates the demand of smaller size and low power consumption. The size of the packaged quartz crystal resonators was about 10mm^3 in 2003. However, as quartz crystal size decreased further, the rate of size reduction is saturated. This saturation is mainly due to process difficulties as well as packaging/assembly limits [2]. Silicon MEMS has shown promise for replacing quartz and other high Q resonators technologies such as ceramic, BAW, and SAW resonator. The major advantage of silicon MEMS resonators is that they are fabricated with batch processing technologies used in standard semiconductor industry for the sake of low cost [3]. Another advantage of Si MEMS resonator is the possibility for integration into standard CMOS process and 3D packaging. This further reduces the coat and size of electronic systems. However, the main drawback of Si MEMS resonator is a high temperature coefficient of frequency(TCF) in the range of -30ppm/°C dominated by the variation of the elastic properties of silicon related to doping level [5].

In this paper, an oscillator using 0.13μm CMOS process with a MEMS 32kHz Si resonator is described. This ASIC has been designed to compensate the MEMS oscillator's TCF using bias voltage generator for applying an electric field induced frequency changing. Making additional electrodes next to the spring of MEMS resonator provide frequency tuning to cover the inherently MEMS frequency drift over temperature. To cover the full range of electrostatic force over operating temperature, high voltage generation using charge pumping circuit, low power PTAT [6], and ring oscillator are also implemented. The whole power consumption for this ASIC is less than 1μA at 1.2V under 10pF load.

II. SILICON MEMS RESONATOR

The silicon MEMS resonator is designed and fabricated based on a 30μm SOI MEMS process with chip vacuum level packaging. A comb resonator is patterned and etched on a 30μm thick SOI wafer after highly doped to reduce the TCF as shown in Figure 1. The structure consists of a resonator mass, supported 4 sets of beams, and inter-digitated comb shaped electrostatic transducers with 2μm gap spacing between static and moving fingers. The inter-digitated comb-drive transducer is designed so that the motional resistance(Rs) of the resonator is 463kΩ with the polarization voltage (Vp) of 2.5V.

The quality factor (Q) of the resonator has been simulated over a range of vacuum from 1mTorr to 100mTorr. The quality factor of the resonator is 40,000 with the vacuum better than 35mTorr. To ensure the resonators work properly, vacuum packaging is required.

To compensate the TCF of MEMS resonator, the additional electrodes next to spring can change the frequency using electrostatic force. If bias voltage is applied to these tuning electrodes the mechanical spring stiffness is changed, which affected the resonance frequency of MEMS resonator. The simulated tuning gain is 500ppm/V.

$$f_o = \frac{1}{2\pi}\sqrt{\frac{k}{m}} = \frac{1}{2\pi}\sqrt{\frac{k_m + k_e}{m}} = \frac{1}{2\pi}\sqrt{\frac{k_m - \frac{2\varepsilon_0 WL}{d^3}V_P^2}{m}} \quad (1)$$

From Eq. (1), k is the stiffness and m is the mass of comb-driver. Here $k=k_o+k_e$, k_o is the intrinsic stiffness and k_e is the negative stiffness generated from electrostatic force (Vp).

The simulated MEMS resonator shows the Q is 40,000 and the frequency is 32.768kHz. The frequency tuning range is over 1500ppm. The design parameters of the resonator are listed in Table I.

The equivalent model for the resonator is shown in Figure 2. The model values used in ASIC are shown in the Table II and

978-1-61284-863-1/11 $26.00 © 2011 IEEE

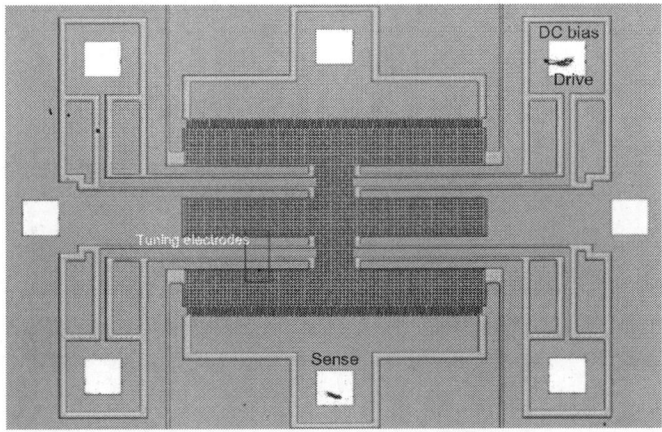

Figure 1.   SEM Photograph of a 32kHz MEMS Resonator

TABLE I.   DESIGN PARAMETER OF 32KHZ MEMS RESONATOR

Parameter	Number
Finger length	24μm
Finger width	4μm
Thickness of SOI	30μm
DC bias voltage	2.5V
Designed Motional Resistance(at 2.5V)	463.25kΩ
Designed Motional frequency	32.768kHz
Typical Resonator Q	40000
Tempereature Coefficeint	15ppm/°C

have been extracted from simulation results using Eq.(2). To simulate frequency accuracy over temperature, the ASIC uses a MEMS TCF (-15ppm/°C) and electric filed induced stiffness control to change resonance frequency (500ppm/V) is used.

$$f_o = \frac{1}{2\pi}\sqrt{\frac{1}{L_s C_s}}$$

$$R_s = \frac{\sqrt{km}}{Q\eta^2} = \frac{2\pi f_o m}{Q\eta^2} \qquad L_s = \frac{m}{\eta^2}$$

$$C_s = \frac{\eta^2}{k} \qquad \eta = V_P \cdot \left(\frac{\partial C}{\partial x}\right) = V_P \cdot \left(\frac{2N\varepsilon_o\varepsilon_r h}{d_f}\right)$$

$$\tag{2}$$

,where k is effective spring constant, m is total effective mass of the resonator, Q is quality factor at certain pressure level, h is the thickness, d_f is gap of comb driver, Vp is the bias voltage, and N is number of finger.

III.   SYSTEM LEVEL DESCRIPTION

The block diagram for the whole system is shown in Figure 3. The main VDD is 1.2V for 0.13μm CMOS process. The ASIC consists of a 32kHz oscillator core for compensating loss of MEMS resonator, a voltage boosting circuit using charge pump and ring oscillator, and temperature compensation circuit. Tri-state buffer is followed by oscillator core and provide a square wave digital output clock at the OUTPUT pin. The logic

high voltage is determined by the value of VDD. The output buffer can be turned off by setting the OUTE pin low.

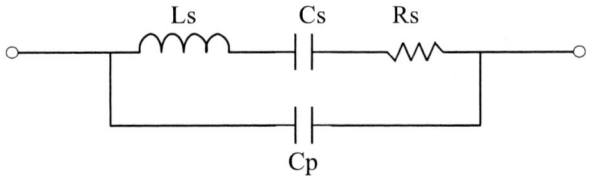

Figure 2.   MEMS Resonator model

TABLE II.   MODEL PARAMETER FOR THE MEMS RESONATOR

Model parameter	
Fs[kHz]	32.768
Ls[H]	92422
Cs[fF]	0.27
Rs[kΩ]	463
Cp[pF]	0.4
Q	40000
Vp[V]	2.5

The lower part of the diagram shows the voltage boosting circuit. The role of voltage boosting circuit is reduction the motional resistance (Rs) of the comb drive resonator. This resistance is inversely proportional to the square of the applied voltage Vp = 2.5V. The motional resistance is related to the current consumption of oscillator core, because higher motional resistance requires higher transconductance of FET. The voltage boosting circuit raises the voltage to a level where the motional resistance is below 450kΩ. This level provides sufficient margin on the device transconductance to reliably operate the oscillator circuit over operating temperature. Another role of voltage boosting circuit is to provide a temperature compensated bias voltage. This bias voltage (VB) generates electrostatic force to compensate the MEMS resonator's negative TCF. To cover the full range of operating temperature, required voltage is from 0 to 3.3V. The voltage boosting circuit consists of a ring oscillator and charge pump. The ring oscillator has been designed to operate under 1.2kHz to minimize power consumption and not to affect whole system of the startup time. Simulated settling time of charge pump is below 1ms and maximum output voltage is 3.5V.

The temperature compensation circuit consists of low power and small area PTAT circuit is designed to sense the temperature variation of MEMS resonator and bias voltage generator using above mentioned voltage boosting circuit. To minimize the power consumption, subthreshold CMOS PTAT generation circuit is used [6]. Compared to conventional PTAT using parasitic vertical BJT, subthreshold CMOS has small chip area, because it doesn't require hundreds of MΩ resistor. The area of conventional PTAT is 2.4mm^2 and subthreshold CMOS PTAT is 0.04mm^2.

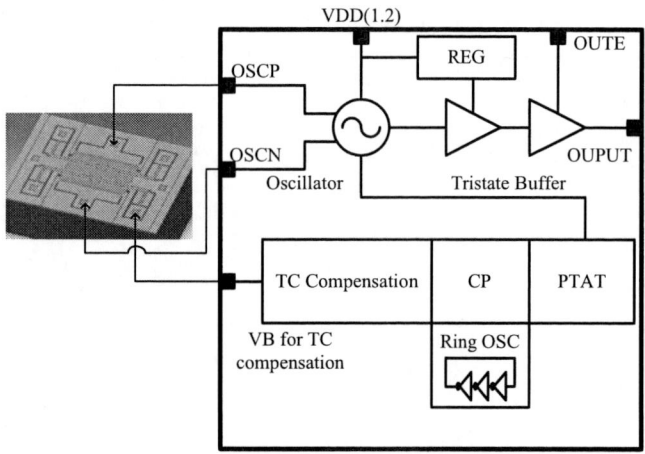

Figure 3. System Diagram of 32kHz oscillator system

Detailed temperature compensation method is described in Figure 4. To compensate MEMS resonator's negative TCF (-15ppm/°C, temperature compensation circuit generates corresponding bias voltage using 500ppm/V.

IV. CIRCUIT DESCRIPTIONS

The circuit for the oscillator core and low power PTAT bias generation block are shown in Figure 5. The simplicity and high stability of three point oscillators are widely used in crystal oscillator, such as Pierce [1]. The main drawback of this type of oscillator is to obtain a stable operation, large loading capacitor are required increasing the oscillator power consumption [5].

Differential circuits provide many advantages in integrated system such as high CMRR and suppress the excessive noise due to interference from the substrate of the oscillator circuit and from the bonding wires coupled between the oscillator circuit and resonator. As a result, differential cross coupled type of negative transconductance generation is popular in LC VCO. However this is not suitable for a MEMS resonator having high DC impedance since the circuit will merely latch in one of its two stable states. Differential crystal oscillator circuits [5] typically utilize a current source to provide the bias current to drive the differential oscillator circuit. However, this type of oscillator, the flicker noise induced by the current source contributes significantly to overall circuit's phase noise (jitter). To remove the current source, using voltage biasing with high pass filtering is presented. Cross-coupled transistor pair MN0 and MN1 forms the gain stage providing the negative differential transconductance $-2/g_m$ similarly to LC VCO. In order to cope with the high DC impedance of the resonator, C1, C2, MN4, MN5 forms the high pass filter to reduce the positive FB gain at DC. To get high resistor value not to degrade resonator's Q, under threshold region MOSFETs (MN2~MN5) is implemented using PTAT bias. MN2 and MN3 provide the bias voltage of MN0/MN1. For typical oscillator circuits, it is normal for negative resistance of oscillator circuit to be designed with a 5x margin for the loss of resonator.

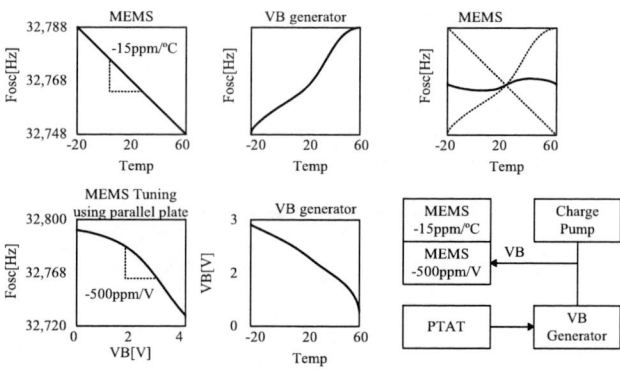

Figure 4. Tempereature compensation method

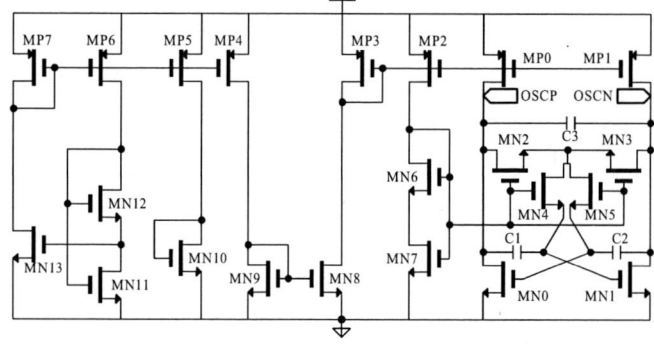

Figure 5. Schematic of oscillator core and low power PTAT bias

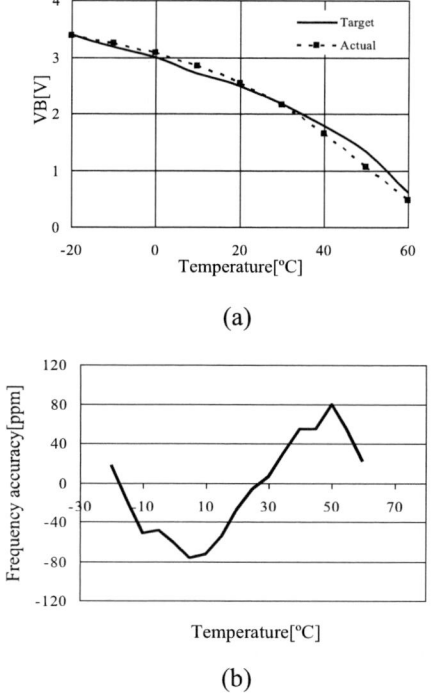

(a)

(b)

Figure 6. (a)VB generator (b) Accuracy of the output frequency after compensation using a temperature sensor and bias generator(-20~60°C)

978-1-61284-863-1/11 $26.00 © 2011 IEEE   248

To generate bias current for oscillator core and temperature compensation circuit, PTAT current generation is formed using MN10~MN13 and MP5~MP7 [6].

The I-V characteristics of the nMOS operating in subthreshold region is given by

$$I_D = S\mu V_T^2 \exp\left(\frac{V_{GS} - V_{th}}{mV_T}\right)\left[1 - \exp\left(-\frac{V_{DS}}{V_T}\right)\right] \quad (3)$$

$$I_D(T) = \alpha\mu T^2$$

From Eq. (3), the current is proportional to the square of temperature. In moderate temperature range and inverse relationship of mobility (μ) with temperature, this can be approximately modified as linear function. Using this current, bias voltage (VB) is generated as shown in Figure 6(a).

V. RESULTS

The ASIC was designed in a 0.13μm CMOS process. The simulated results of circuit performance are summarized in Table III. To reduce the startup time and guarantee the initial oscillation, the current of oscillator core is changed. After startup, the current of oscillator core is reduced to get better jitter performance and to get low power consumption. The simulated RMS jitter is under 60ns all PVT conditions.

Accuracy of output frequency after compensation using a temperature sensor and bias generation is +/-80ppm from -20°C to 60°C, as shown in Figure 6 (b).

Current consumption of temperature compensation circuit is 80nA and PTAT is below 8nA. The startup time of charge pumping circuit is 1ms and total startup time is under 1sec.

TABLE III.   SUMMARY OF SIMULATED RESULTS

Parameters	Simulation
Resonant Frequency	32.768 kHz
Biasing Voltage (Power supply)	1.2V
Current Consumption (Enabled)	0.9uA(1.2uA)*
Current Consumption (no load)	0.5uA(0.8uA)*
Operating Temperature	-20°C to 60°C
Functional Temperature Drift	+/-80ppm
Duty Cycle	50.40%
Rise/Fall Time (5pf to 10pf load)	<300n
Periodic RMS jitter	<60n
Die Size (Including Package)	1.5X1.5mm²

*At startup condition

VI. CONCLUSION

A 32kHz oscillator circuit using a Si MEMS resonator for medical device has been designed and simulated. This oscillator has shown performance similar to previously commercially used quartz oscillators. The presented differential oscillator structure allows minimizing the power consumption (<1μA) by reducing the load capacitance. The oscillator temperature stability from -20 ~ 60 °C is +/-80ppm using electrostatic force to change the stuffiness of spring which related resonance frequency.

ACKNOWLEDGMENT

The authors would like to thank Xu Jinghui, Liu Youhe, Lynn Khine, and Julius Tsai for performing the fabrication presented in this paper.

REFERENCES

[1] Kenneth R. Cioffi and Wan-Thai Hsu, "32kHz MEMS-Based Oscillator for Low-Power Application," Frequency Control Symposium and Expositon, 2005, pp.551-558

[2] Wan-Thai Hsu, "Resonator Miniaturization for Oscillator," Frequency Control Symposium, 2008, pp.392-395

[3] Clark T.-C. Nguten, Roger T. Howe, "An Integrated CMOS Micromechanical Resonator High-Q Oscillator," IEEE J. Solid-State Circuits, vol. 34, pp. 440-455, April 1999.

[4] Dvaid Ruffieux, "A High-Stability, Ultra-Low-Power Differential Oscillator Circuit for Demanding Radio Applications," Proc. ESSCIRC 2002, pp. 85-88.

[5] David Ruffieux, Francois Krummenacher, Aurelie Pezous, and Guido Spinola-Durante, "Silicon Resonator Based 3.2μW Real Time Clock with ±10ppm Frequency Accuracy," IEEE J. Solid-State Circuits, vol. 45, pp. 224-234, January 2010.

[6] Luca Magnelli, Felice Crupi, Pasquale Corsonello, Calogero Pace, and Giuseppe Iannaccone, "A 2.6nW, 0.45V Temperature-Compensated Subthreshold CMOS Voltage Reference," IEEE J. Solid-State Circuits, vol. 46, pp. 465-474, January 2011.

Pseudo Differential Operational Transconductance Amplifier using Common Mode Feed Forward and HD3 Feed Forward

Yen-Shuo Chang[1], Hong-Chong Wu[2], Miin-Shyue Shiau[2], Don-Gey Liu[2], Heng-shou Hsu[2]

[1] Graduate Institute of Electronic Engineering, Feng Chia University, Taichung, Taiwan 40724, R.O.C.
[2] Department of Electronic Engineering, Feng Chia University, Taichung, Taiwan 40724, R.O.C.
utadamatt@hotmail.com, hwu@fcu.edu.tw, msshiau@fcu.edu.tw, dgliu@fcu.edu.tw, hshsu@fcu.edu.tw

Abstract - In this study a low supply voltage CMOS pseudo differential operational transconductance amplifier (OTA) with CMFF and HD3-FF to improve CMRR and THD is proposed. Common mode rejection ratio (CMRR) is improved by using the common mode feed forward (CMFF) technology and the THD is simulated result of improved by using a HD3 feed forward (HD3-FF). The proposed OTA shown the CMRR is 68.6dB and the THD is lower than -66dB under the input voltage range of 0.1V with 1.2V supply voltage. The chip has been manufactured in a 0.18 μm-CMOS technology and the chip area is 0.69x0.38 mm².

Index Terms—pseudo differential, CMRR, CMFF, OTA

I. INTRODUCTION

Voltage to current converter, it is also called the operational transconductance amplifier (OTA). The OTA is an amplifier that with infinite input and output impedance. It is one of the most important building blocks in analog and mix-mode circuits, including transconductance-C Filter (GM-C filter)[1], front-end of the SI-ADC (Switched-Current Analog to Digital Converter)[2]. Thus, the OTA is required to own well linearity that means the OTA must be maintained a constant *Gm* in the OTA input range.

$$Gm = \frac{I_{out}}{V_{in}}$$

In recently research the OTA is designed for low supply voltage, low power consumption and high linearity. For high linearity requirement, the fully differential (FD) structure is a good choice for OTA design. FD structure has better dynamic range than the single-ended structure and has better common-mode noise rejection. The tail current source can inhibit the common-mode gain but the tail current source consumes a large voltage drop. So that the FD structure is not the best choice for low supply voltage design.

Pseudo differential (PD) structure eliminates the tail current source. So we can design the OTA with PD structure for lower supply voltage. In this study, we use the common mode feed forward (CMFF) [2] technology to solve the CMRR problem, happened in the PD structure. In recently CMOS process, the short-channel effects influence the performance of linearity. We use a third harmonic distortion feed forward (HD3-FF)

technology to suppress the third harmonic distortion and to get higher linearity. Section II will describe the OTA structure and the study of some circuit design problems and considerations. The proposed OTA design of our study will be presented in Section III. In section IV, show the simulations results and compared with the other works. Finally, some conclusions are presented in Section V.

II. OPERATIONAL TRANSCONDUCTANCE AMPLIFIER

A. Traditional fully differential structure

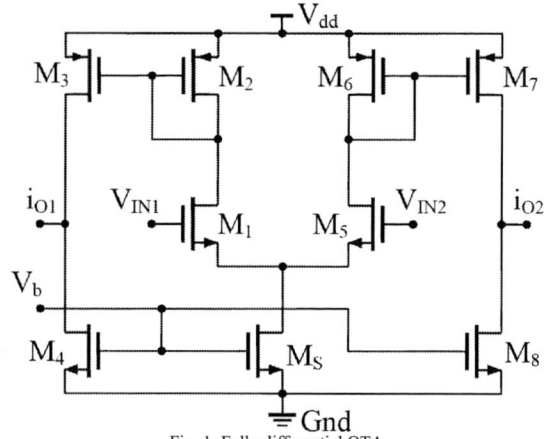

Fig. 1. Fully differential OTA

The fully differential (FD) OTA structure shown in Fig.1 has advantage of high dynamic range and high common mode noise rejection. The disadvantage of FD structure is there is an additional common mode feedback (CMFB) circuit. The CMFB circuit can fix the output common-mode dc level, but the FD structure will consume much more power. The FD structure usually uses the tail current source; it consumes a voltage drop described in Eq. 1. So the FD structure is not suitable for low supply voltage design.

$$V_{dd} = V_{SG2} + V_{OV1} + V_{OV(MS)} \qquad (1)$$

978-1-61284-863-1/11 $26.00 © 2011 IEEE

B. Pseudo differential structure

The pseudo differential (PD) OTA [3] shown in Fig. 2 is designed by removing the M_S in FD structure shown in Fig. 1 to have low supply voltage and more signal swing ranges. The common mode gain (A_{CM}) is equal to the differential mode gain (A_D) in PD structure, CMRR = A_D / A_{CM} = 1. This is a serious drawback in PD structure.

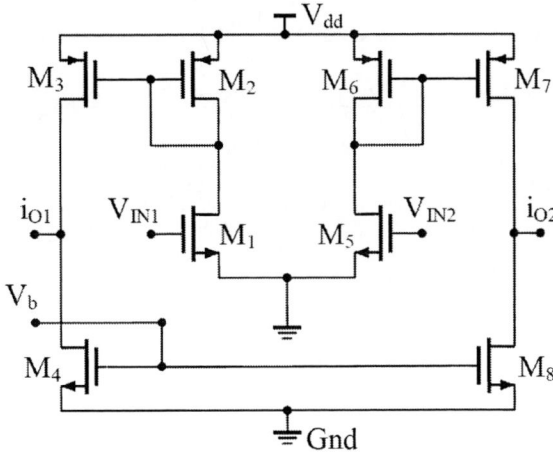

Fig. 2 Pseudo differential OTA

C. Common Mode Feed Forward

The conceptual structure of CMFF is shown in Fig. 3 it use to improve the drawback of CMRR. [5] The common mode circuit uses the same differential transconductance copy the common mode signal and reversed them for the output nodes. Then the the common mode signals will be canceled at the output nodes, the performance of CMRR will has significant improvement.

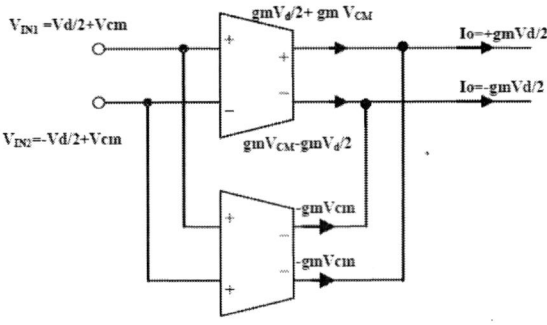

Fig. 3 Common Mode Feed Forward conceptual structure

D. Third order Harmonic Distortion Feed Forward

The short channel effect problem becomes important in advanced CMOS process. In short channel devices, the mobility is affected by the longitudinal and transversal electric fields. The drain current can be approximated as shown in Eq. 2

$$I_{D,sat} = \frac{u_n C_{ox}(\frac{W}{L})(V_{GS} - V_t)^2}{2[1 + \theta(V_{GS} - V_t)]}(1 + \lambda V_{DS}) \qquad (2)$$

W and L are the width and length of the device respectively. C_{ox} is the oxide capacitance per unit channel area. μ_n is the low-field mobility; θ is the mobility reduction coefficient. λ is the output impedance constant. V_t is the threshold voltage. The transistors operate in saturation region with short channel effect; the third order harmonic distortion term can be shown as the Eq. 3 $V_{(cm)}$ is the common mode signal.

$$HD_3 \cong \frac{\theta \cdot V_{(cm)}^2}{16V_{OV}(1 + \theta V_{OV})^2(2 + \theta V_{OV})} \qquad (3)$$

We use the feed forward concept to improve the THD. The third order harmonic distortion feed forward (HD3-FF) conceptual structure [6] is shown in Fig. 4 with the the same differential transconductance will produce a third order harmonic distortion term and forward to the output nodes.

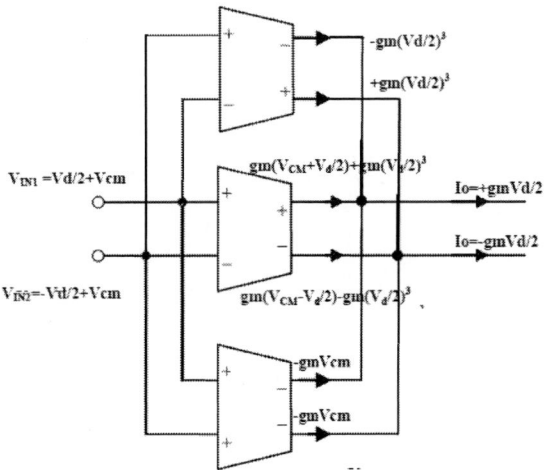

Fig. 4 Third order Harmonic Distortion Feed Forward conceptual structure

III. PROPOSED OTA

The pseudo differential OTA, shown in Fig. 2 can operate in low supply voltage but the output impedance is not large enough. The proposed OTA with wide swing cascade structure [4] shown in Fig. 5 has high output impedance, as shown in Eq. 4 and the operate voltage is shown in Eq. 5.

$$R_{out} = (A \cdot ro), A = gm \cdot ro \qquad (4)$$

$$\begin{aligned} V_{DD}1 &= V_{SG3} + V_{DS1} \\ V_{DD}2 &= V_{SD4} + V_{SD5} + V_{SD6} + V_{SD7} \end{aligned} \qquad (5)$$

The linearity of proposed OTA, show as Eq. 6~Eq. 7

978-1-61284-863-1/11 $26.00 © 2011 IEEE

$$I_{D3} = I_{D6} = gm \cdot (v_{cm} + \frac{1}{2} v_d) \qquad (6)$$

$$I_{D3} = I_{D6} = \beta \cdot I_{D7} = \beta \cdot I_{D11}$$
$$= \beta \cdot I_{D15} = \beta \cdot I_{D19} = \beta \cdot gm \cdot (v_{cm} + \frac{1}{2} v_d) \qquad (7)$$

β is the aspect ratio of M3, M7 and M15 aspect ratio. Vx is the differential mode signal virtual ac ground, so the common mode signal current we can get $I_{D7} = I_{D11} = \beta \cdot gm \cdot v_{cm}$, show the Eq. 8 at the output nodes. The common mode signal is canceled at the output nodes.

At the VO1 node: $I_{D15} - I_{D18} = \frac{1}{2} \cdot \beta \cdot gm \cdot vd$

$$(8)$$

At the VO2 node: $I_{D19} - I_{D12} = -\frac{1}{2} \cdot \beta \cdot gm \cdot vd$

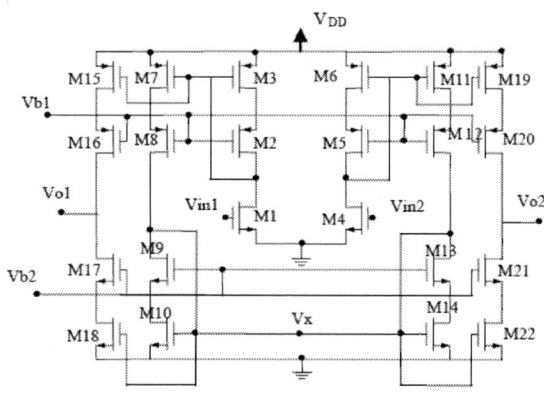

Fig. 5 Proposed OTA with wide swing cascade and CMFF

The third order harmonic distortion term cancel by adding the third order harmonic distortion feed forward (HD3-FF) circuit in proposed OTA. The proposed OTA will include the CMFF and HD3-FF in shown in Fig. 6.

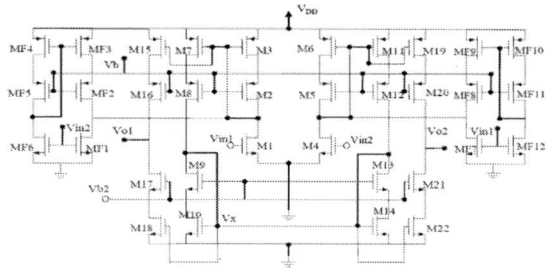

Fig. 6 Proposed OTA with CMFF and HD3-FF

Transistors MF1 to MF12 are third order harmonic distortion feed forward circuit and the currents at output nodes are $I_{MF9}-I_{MF10}$, $I_{MF7}-I_{MF12}$. The idea target of our design is to

generate the third order harmonic distortion term only. But it is very difficult to approach our target in real design. From the Eq. 9 we know the circuit not only generate the distortion term but also generates the main signal.

$$I_{D,3hd} = \frac{1}{16} \{ [\frac{\theta \cdot \beta_{MF1}}{(1+\theta V_{OV,MF1})^4}] v_{id}^3 - [\frac{\theta \cdot \beta_{MF6}}{(1+\theta V_{OV,MF6})^4}] v_{id}^3 \} \qquad (9)$$

$$I_{D,1st} = \frac{1}{2} \{ \frac{\beta_{MF1} V_{OV,MF1} - [(\theta \beta_{MF1} V_{OV,MF1}^2)/(2\theta V_{OV,MF1} + 2)]}{1+\theta V_{OV,MF1}} \} v_{id}$$
$$- \frac{1}{2} \{ \frac{\beta_{MF6} V_{OV,MF6} - [(\theta \beta_{MF6} V_{OV,MF6}^2)/(2\theta V_{OV,MF6} + 2)]}{1+\theta V_{OV,MF6}} \} v_{id}$$

IV. SIMULATION RESULT

The simulated open loop gain and phase of proposed OTA are shown in Fug. 7. Open loop gain is 64.8dB and phase margin is 68°. The simulation CMRR shown in Fig. 8 is equal to 68.6dB with bandwidth 12K Hz. In the bandwidth > 100K Hz the CMRR also maintain 50dB. Fig. 9 show the FFT graph. When the input signal range is 0.1-V the HD3 is -66.6dB. Table I. presents the comparisons of our work with other researches. That shows we have good CMRR and low HD3.

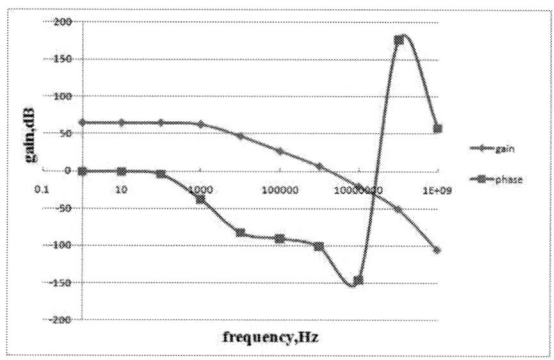

Fig. 7 Simulated frequency response with open loop gain and phase

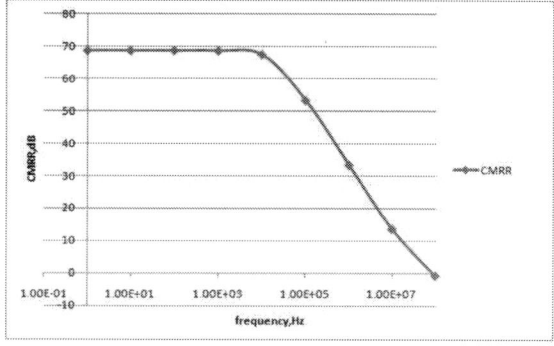

Fig.9 Simulated frequency response with CMRR

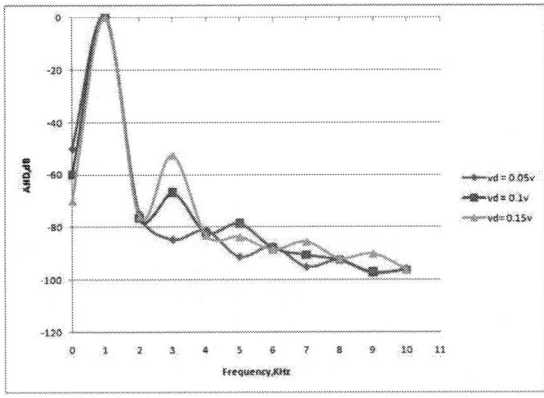

Fig. 10 Simulated HD3 with different input signal range

TABLE I. COMPARED TABLE

	This work	[7]	[9]	[6]
CMOS tech (um)	0.18	0.5	0.35	0.18
Power Supply	1.2	1.8	2	1
Unity Gain Freq	18 MHz	562 MHz	200 KHz	50 MHz
Open loop gain	64.8	73	30	30
Phase margin	68°	59°	90°	90°
CMRR (dB)	68.6	19	54	-----
A_{HD3} (dB)	-66.6	-----	-----	-56
Power	75.8 uW	11 mW	80 uW	1.8 mW
C_L	5pf	1.8pf	-----	2pf
$R_O(\Omega)$	11.8 M	-----	-----	-----

V. CONCLUSIONS

The proposed OTA use the CMFF to get higher CMRR and use the HD3-FF to lower THD. The circuit has been verified that it can operate with lower supply voltage at 1.2v and has high linearity. Power consumption is 75.8 uW.

ACKNOWLEDGMENT

This research was supported in part by a grant of making chips from National Chip Implementation Center (CIC).

REFERENCES

[1] Mingdeng Chen, Silva-Martinez, J., Rokhsaz, S., and Robinson, M.," A 2-Vpp 80-200-MHz fourth-order continuous-time linear phase filter with automatic frequency tuning," IEEE Journal of Solid-State Circuits, Volume: 38, Issue: 10, pp. 1745-1749, Oct. 2003.

[2] D.L. Shen, and W.T. Lee, " A negative resistance compensated switching current sampled-and-hold circuit," Circuits and Systems and TAISA Conference, 2009. NEWCAS-TAISA '09. Joint IEEE North-East Workshop on, July 2009.

[3] T.Y. Lo, and C.C. Hung, "A high speed and high linearity OTA in 1-V power supply voltage," Circuits and Systems, 2006. ISCAS 2006. Proceedings. 2006 IEEE International Symposium on, pp. 4, Sept. 2006.

[4] Mohieldin, A.N., Sanchez-Sinencio, E. and Silva-Martinez, J., "Nonlinear effects in pseudo differential OTAs with CMFB," Circuits and Systems II: Analog and Digital Signal Processing, IEEE Transactions on Volume 50, Issue 10, Oct. 2003 pp:762 – 770

[5] A.N. Mohieldin, Sanchez-Sinencio, E., Silva-Martinez, J., "A low–voltage fully balanced OTA with common mode feed forward and inherent common mode feedback detector," Solid-State Circuits Conference, 2002. ESSCIRC 2002. Proceedings of the 28th European, pp. 191, Sept. 2002.

[6] C.L. Chien, C.C. Hung, C.W. Chen," A pseudo-differential OTA with linearity improving by HD3 feedforward," Solid-State Circuits Conference 2009. A-SSCC 2009. IEEE Asian, pp. 237, Nov. 2009.

[7] Shankar, A., Silva-Martinez, J.and Sanchez-Sinencio, E., "A low voltage operational transconductance amplifier using common mode feedforward for high frequency switched capacitor circuits," Circuits and Systems, 2001. ISCAS 2001. The 2001 IEEE International Symposium on Volume 1, 6-9 May 2001 pp:643 - 646

[8] Ferreira, L.H.C., Pimenta, T.C. and Moreno, R.L "An Ultra-Low-Voltage Ultra-Low-Power CMOS Miller OTA With Rail-to-Rail Input/Output Swing," Circuits and Systems II: Express Briefs, IEEE Transactions on Volume 54, Issue 10, Oct. 2007 pp:843 - 847

[9] Stockstad, T. and Yoshizawa, H "A 0.9-V 0.5-μA rail-to-rail CMOS operational amplifier," Solid-State Circuits, IEEE Journal of Volume 37, Issue 3, March 2002 pp:286 - 292

High Gain, High Speed OTA for S/H Circuit in 14-b 100-MS/s Pipeline ADC

Shuai CHEN, Lenian HE, Lu ZHANG

Institute of VLSI Design, Zhejiang University
Hangzhou 310027, P.R. China
Email:helenian@vlsi.zju.edu.cn

Abstract—**In this paper, a high gain, high speed operational trans-conductance amplifier (OTA) is designed and implemented in TSMC 0.18μm mixed signal process. Gain-boost technique and other optimal designs are used to meet the requirements. Simulation result shows that the OTA can attain 120dB DC-gain, a unity gain bandwidth of 1.12GHz and a settling time of 4ns to 0.001% accuracy with 4pF load and 3.3V voltage supply. An S/H circuit is also designed using this OTA, with 81.2dB SNDR and 100.9dB SFDR for 10MHz input. The S/H circuit is used in 14bit 100MHz Pipeline ADC.**

Keywords-OTA; gain-boost; S/H; Pipeline ADC

I. INTRODUCTION

As the bridge between the analog and digital parts in modern communication systems, high speed and high resolution ADCs are urgently required than ever before. For example, in 3G cellular standard like UMTS, an ADC with a sampling rate in the range of 80-150MS/s, an SNR of 72-75 dB and an SFDR of 85-90dB is required [1]. Among various architectures of ADCs, a Pipeline ADC which is suitable for high-speed, high-resolution demands becomes the research hotspot in recent years, commonly achieves 100MS/s with 14bit resolution.

In Pipeline ADC, a sample and hold circuit (S/H) is usually designed at the front-end, which should reach to a small error in a short time, therefore it requires the core part of the S/H, the op-amp must be designed with high gain, high bandwidth and fast settling time. However, with the development of deep sub-micron technology, intrinsic gain of a MOSEFT keeps decreasing, which makes designing the high gain OTA more difficult [2]. In this paper, an OTA is designed by using TSMC 0.18μm mixed signal model. It achieves 120dB and 1.12GHz GBW, and 4ns settling time to 0.001% of the final value. As an application of this OTA, an S/H circuit is designed with 81.2dB SNDR and 100.9dB SFDR for 10MHz input. This circuit is used in a 14bit 100MS/s Pipeline ADC for RF communication systems.

The paper is organized as follows. Design of an OTA based on gain-boost technique is presented in Section II. Structure of the S/H circuit will be introduced in Section III. Section IV shows the layout design. Simulation results are summarized in Section V. The conclusion follows in Section VI.

II. OTA DESIGN

A. Specifications and Basic Structure

Since the OTA is used in front-end S/H circuit for Pipeline ADC, its specifications should be defined to satisfy certain requirements. Transfer function of the S/H circuit shown in (1) can determine the required op-amp specifications, such as the OTA gain and bandwidth.

$$V_{OUT} = (1 - e^{-\frac{t}{\tau}}) \cdot (1 + \frac{1}{A\beta})^{-1} \cdot V_{IN} \qquad (1)$$

where β is the feedback factor, A is the open loop DC-gain and τ is a time constant. In (1), an error term mainly caused by the finite open loop gain is introduced, which should be no more than half of the Least Significant Bit (LSB), and therefore the required OTA DC-gain can be obtained to meet the accuracy requirement of the S/H circuit. Meanwhile, within the settling time t, the output settles to N-bit accuracy, which determines the minimum unity gain bandwidth required. According to [3], DC-gain and GBW can be calculated in (2) and (3), respectively.

$$A > 2^{N+2} \qquad (2)$$

where N is the resolution of the ADC.

$$GBW > \frac{3 \ln 2 \cdot N \cdot f_T}{2\pi\beta} \qquad (3)$$

where f_T is the clock frequency.

In this design, N = 14, so the minimum DC-gain is 96.3dB, and for calculating GBW, slew rate constrains is also considered besides (3), thus the required bandwidth should be no less than 800MHz [3].

To achieve such high gain, OTA is designed with gain enhancement circuit. The principle is to increase the output impedance by additional amplifiers. Fig. 1 shows the OTA schematic. The folded cascode architecture is chosen to get a higher output swing than the telescope architecture does. PMOS transistors M_0 and M_1 are used as input differential pair to reduce the parasitic capacitance, and to get a high non-dominant pole, which will improve the phase margin. The gain-boost amplifiers A_P and A_N in Fig.1 are also using folded cascode topology, providing over 60dB DC-gain.

978-1-61284-863-1/11 $26.00 © 2011 IEEE

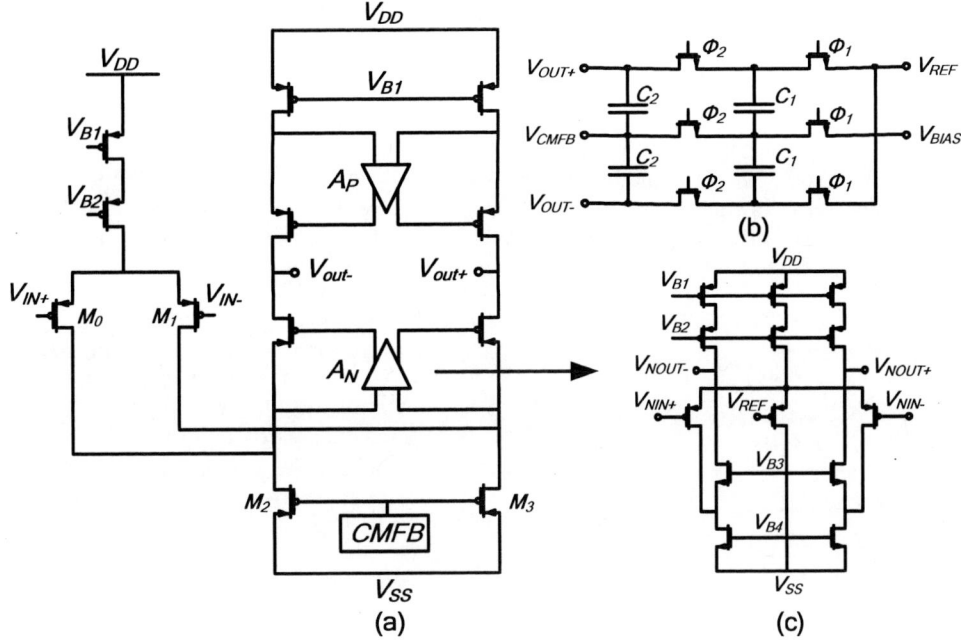

Figure 1. OTA schematic.(a) Main folded cascode Op-Amp. (b)Switch-capacitor common-mode feedback circuit. (c) Gain-boost amplifier

B. Settling Behavior and Design Constrains

According to [4], a pole-zero doublet is introduced at a frequency below the unity gain frequency of the OTA, which brings in a slow settling component. As a result, the doublet will persist a longer settling time if the gain boosting op-amps are not carefully designed. One way to reduce this impact is to push the pole-zero doublet above the unity gain frequency of the main OTA, but it will of course consume more power and may cause the gain boosting amplifier loop unstable. Another way is more practical, which is illustrated in [5], the basic concept is to make this 'slow' settling component fast enough, as in (4)

$$\frac{1}{\omega_{pz}} < \frac{1}{\beta\omega_{unity}} \qquad (4)$$

where the time constant of the doublet, $1/\omega_{pz}$, is smaller than the main time constant, $1/\beta\omega_{unity}$ with the closed loop feedback factor β, so that the settling time will not be affected by the doublet. What's more, for stability reasons, the unity-gain frequency of the gain-boost amplifier (ω_{uga}), where also lies the pole-zero doublet, should be smaller than the second pole (ω_{p2}) of the main OTA. Equation (5) shows the final design constrain [5].

$$\beta\omega_{unity} < \omega_{uga} < \omega_{p2} \qquad (5)$$

C. SC-CMFB Circuits

Switch-Capacitor Common-Mode Feedback Circuit (SC-CMFB) is essential in a fully differential amplifier such as the main OTA in Fig. 1. This structure has two benefits: no static

power consumption and no constrain to the output swing. The basic model is shown in Fig. 1(b), where V_{BIAS} is the desired biasing voltage for the gate voltage of M_2 and M_3, and V_{REF} is the desired common mode output voltage. The circuit works in this way: in phase Φ_1, C_1 get the different voltage between V_{REF} and V_{BIAS}, in phase Φ_2, the electric charge collected on C_1 in the previous phase redistributes between C_1 and C_2, and after several phases, it reach to a stable state that V_{CMFB} will be equal to V_{BIAS}, and the common mode output voltage will be set to V_{REF}, both these two conclusions can be easily derived from (6).

$$V_{CMFB} = \frac{(V_{OUT+} + V_{OUT-})}{2} - (V_{REF} - V_{BIAS}) \qquad (6)$$

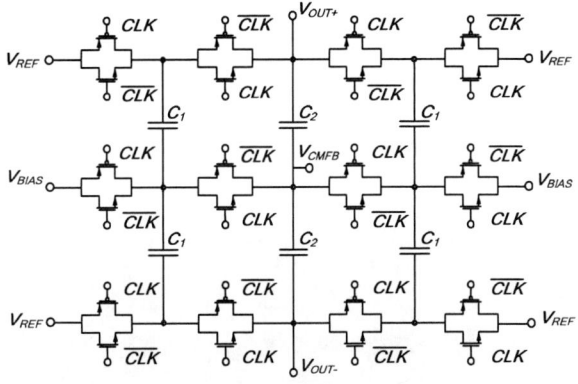

Figure 2. Dual phase switch-capacitor common-mode feedback circuit

However SC-CMFB circuit can bring in a severe problem, i.e., the clock jitter, mainly caused by charge injection and clock feed-through. To reduce this effect, an improved version of SC-CMFB circuit used in this design is shown in Fig. 2. Compared with Fig. 1(b), it has two advantages: first the transmission switch replaces the simple MOS switch so as to reduce the impact by the non-ideal factors; second an additional set of capacitors C_1 and an extra set of switches are used. Thus, during every clock phase, the total capacitance in the differential loop will be the same and the clock jitter will be reduced [6].

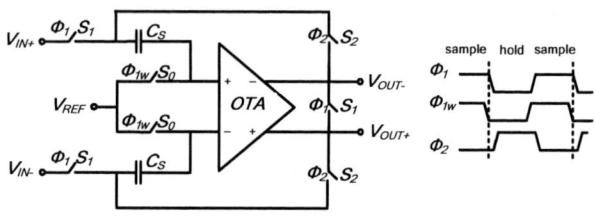

Figure 3.   Flip-around S/C circuit

III.   S/H CIRCUIT

As an implementation of the proposed OTA, S/H circuit is also designed. The S/H circuit using flip-around architecture is shown in Fig.3. The flip-around architecture is widely used for advantages in power, die area and noise [7]. Fig. 3 also demonstrates the clock diagram of the sample and hold operation. One particular thing should be noticed in this clock diagram is that at the end of sampling phase, S_0 (Φ_{1w}) closes slightly before S_1 (Φ_1), so the charge injection and clock feed-through in S_1 will not affect the charge on the sampling capacitor, simply because there is no other DC path. Moreover bootstrapping technique is used to stable MOSEFT switches' 'turn on' resistances.

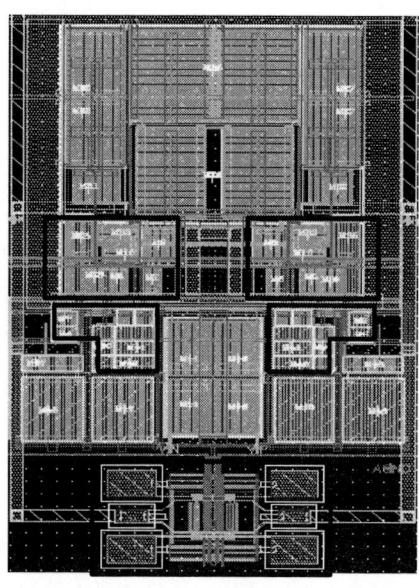

Figure 4.   Layout design

IV.   LAYOUT DESIGN

Layout design of the OTA is shown in Fig. 4. Main OTA can be easily distinguished, and for symmetry considerations, gain-boosting amplifiers, A_N and A_P are divided into two identical parts respectively, placing on both sides of the main OTA, as shown in four frames of the figure. SC-CMFB is placed on the bottom of the figure and it adopts symmetry structure as well to minimize the mismatch. The area of the OTA is 130μm × 215μm.

V.   SIMULATION RESULT

The proposed OTA has been designed and fabricated in TSMC 0.18μm mixed signal process. Fig. 5 shows the simulation results of open loop DC-gain and phase margin. Settling behavior of the OTA is shown in Fig. 6. Table I summarizes the simulated performances. Table II gives the comparison between this design and previous works. A 1024 point FFT test is carried out to test the S/H circuit based on this OTA, the input sinusoidal signal is 9.86328125MHz with 0.9Vpp. Post simulation results in Fig 7 show that S/H circuit achieves 81.2dB SNDR, 100.9dB SFDR. Experimental results will be shown in the oral presentation session of the meeting.

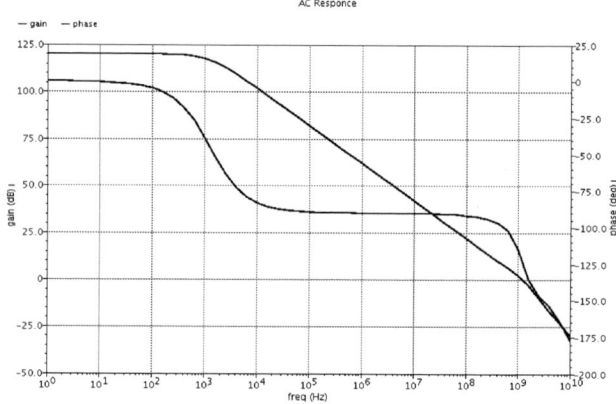

Figure 5.   AC response

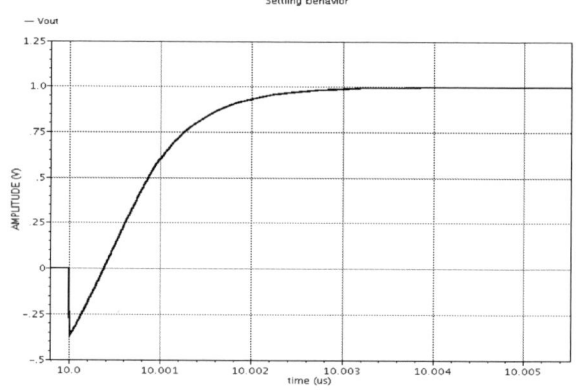

Figure 6.   Settling behavior

TABLE I.　SUMMARY OF THE PROPOSED OTA

DC-gain	120dB
GBW	1.12GHz
Load Cap.	4pF
Phase Margin	60°
Slew Rate	1100V/µs
0.001% Settling Time	4ns
Output Swing	2V
Supply Voltage	3.3V
Power Consumption	60mW

TABLE II.　SUMMARY OF COMPARISION

Spec.	This work	Ahmadi [2006]	Ma Lei [2010]
Gain	120dB	80dB	105dB
GBW	1.12GHz	660MHz	937MHz
Process	0.18µm	0.18µm	0.35µm

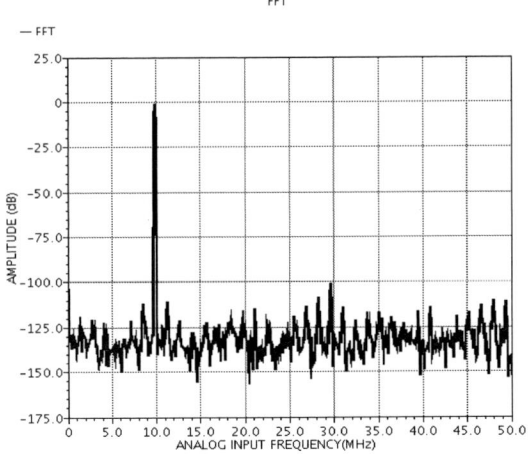

Figure 7.　1024-point FFT result for S/C cuicuit with 10MHz input

VI. CONCLUSION

This paper presents a high gain, high speed OTA with gain boosting technique and other optimal designs. It achieves 120dB DC-gain, 1.12GHz unity gain bandwidth and a settling time of 4 ns to 0.001% accuracy with 4pF load capacitor. Table II shows that it achieves the best performance at DC-gain and GBW, which are crucial in high speed high accuracy design, but it will of course increase the consumption as a compromise. An S/H circuit using this OTA is also designed with 81.2 dB SNDR and 100.9dB SFDR which is enough for 14bit 100MS/s Pipeline ADC.

REFERENCES

[1] Hans Van de Vel, et al. "A 1.2-V 250-mW 14-b 100MS/s digitally calibrated pipeline ADC in 90-nm CMOS," IEEE J. Solid-State Circuits, vol.44, no. 4, pp. 1047-1059, Apr. 2009.

[2] Anne-Johan, et al. "Analog circuits in ultra-deep-submicron CMOS," IEEE J. Solid-State Circuits, vol. 40, no. 1, pp. 132-143, Jan. 2005.

[3] Lauri Sumanen, "Pipeline analog-to-digital converters for wideband wireless communications," Thesis, Helsinki University of Technology, 2002.

[4] B. Y. Kamath, et al. "Relationship between frequency and settling time of operational amplifiers," IEEE J. Solid-State Circuits, vol. SC-9, pp. 347-352, Dec. 1974.

[5] Klass Bult, et al. "A fast-settling CMOS op amp for SC circuits with 90-dB gain," IEEE J. Solid-State Circuits, vol. 25, no. 6, pp. 1379-1384, Dec. 1990.

[6] Ojas Choksi, et al. "Analysis of switched-capacitor common-mode feedback circuit," IEEE Trans. Circuits Syst.□, vol. 50, no. 12, pp. 906-917, Dec. 2003.

[7] Wenhua (Will) Yang, et al. "A 3-V 340-mW 14-b 75Msample/s CMOS ADC with 85-dB SFDR at Nyquist input," IEEE J. Solid-State Circuits, vol. 36, no. 12, pp. 1931-1936, Dec. 2001.

[8] Mohammad Mahdi Ahmadi, et al. "A new modeling and optimization of gain-boosted cascode amplifier for high-speed and low-voltage applications," IEEE Trans. Circuits Ssyt. □, vol. 53, no. 3, pp. 169-173, March 2006.

[9] Ma Lei, et al. "A folded cascode OTA using current-mode gain-boost amplifier," ICSE2010 Proc., pp. 92-95, June 2010.

A 10 Gb/s 4-PAM transceiver with Adaptive Pre-Emphasis

Sungmin Yoo, Daeho Yun, Bongsub Song, Jinwook Burm[1], Jinil Chung[2], and Jun Hyun Chun[2]

Dept. of Electronic Engineering, Sogang University
Seoul, 121-742, Korea
[2]Hynix Semiconductor
Icheon, Gyeonggi- Do, 467-701, Korea
[1]burm@sogang.ac.kr

Abstract— **A 10 Gb/s 4-level pulse amplitude modulation (PAM) transceiver was implemented using a 0.13 μm CMOS process. The implemented 4-PAM transmitter employs current mode logics (CMLs) for high-speed operations. The proposed 4-PAM transceiver achieves a channel efficiency of 2 bit/symbol with adaptive pre-emphasis. The pre-emphasis was designed to be proportional to each 4-level's amplitude. The measured maximum data-rate was 10 Gb/s over 0.7-m cable and 3-cm printed circuit board (PCB) traces. The transmitter and receiver consume 245 mW and 69 mW, respectively. The measured bit-error rate (BER) was less than 10^{-12} at 10 Gb/s data rate.**

Index Terms— **Pre-emphasis 4-PAM transceiver CML**

I. Introduction.

A high speed serial-link is widely used more than low-speed parallel-links for high capacity data transmissions. The number of connection wires and power dissipation can be reduced with a high speed serial-link. However, for the cost effective serial links, bandwidth limits of cables and internal circuits set the maximum available transmission rate [1]. To overcome the bandwidth limit of data transmissions, pulse amplitude modulation (PAM) signalling schemes are employed for high-speed serial-links. N-PAM signalling reduces not only the effective symbol-rate by a factor of $\log_2(N)$ compared to a conventional binary signalling but also the required clock frequency. However, in the high speed serial-link, typically the data rate is up to a few Gb/s. In this case, the PCB traces or cable is not simple wire connections, but transmission lines. A transmission line acts like a low pass filter with a bandwidth limit. Therefore, the high frequency components are attenuated, resulting in a limited transmission data rate and distance [2]. Hence, the transmitter needs a pre-emphasis to circumvent attenuation of high frequency components. However, most reported papers on pre-emphasis [3, 4] have not considered N-PAM's different level transitions requiring different pre-emphasis current, *i.e.*, most reported pre-emphasis techniques use the same amount of current for different level transitions. In such a case, the attenuation of N-PAM's each transition level by transmission line will be different. Large transition has more high frequency

components so that they will be attenuated more. As a result, the receiver may receive erroneous data. To overcome the above problems, we propose a 10 Gb/s 4-PAM transceiver with adaptive pre-emphasis in the transmitter.

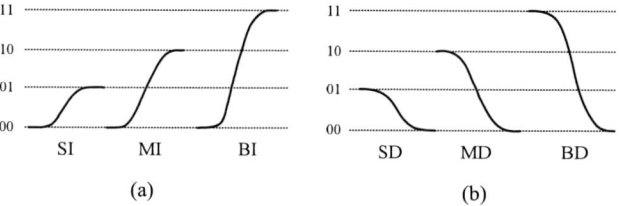

(a)   (b)

Fig. 1 6-case of 4-PAM transition : (a) 3-case of increase transition case and (b) 3-case of decrease transition case

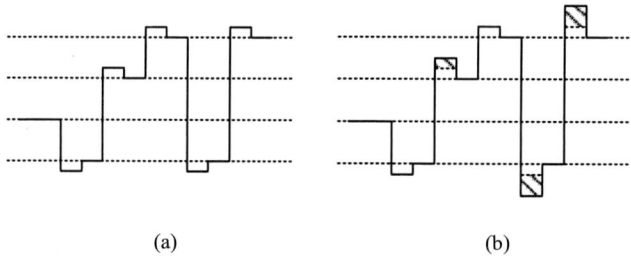

(a)   (b)

Fig. 2 4-PAM pre-emphasis output : (a) with conventional technique and (b) with proposed technique

II. Transmitter

As shown in Fig. 1 (a) and (b), 4-PAM transmitter has six kinds of transition case and be called small increase/ decrease (SI/ SD), medium increase/ decrease (MI/ MD), and big increase/ decrease (BI/ BD). And each six kinds of transition have a different slope. From the comparison of SI and BI, it is clear that BI has steep slope, *i.e.*, BI has more high frequency components than SI. Because a transmission line acts like a low-pass filter, high frequency components will be attenuated more, which means that more currents are needed in BI. However, as shown in Fig. 2 (a) conventional 4-PAM pre-emphasis output, the amount of pre-emphasis currents is the same. With the conventional 4-PAM pre-emphasis technique, BI and BD

978-1-61284-863-1/11 $26.00 © 2011 IEEE   258

is more attenuated than SI and SD. Therefore, as shown in Fig 2. (b), the proposed pre-emphasis was designed to be proportional to each 4-level's amplitude.

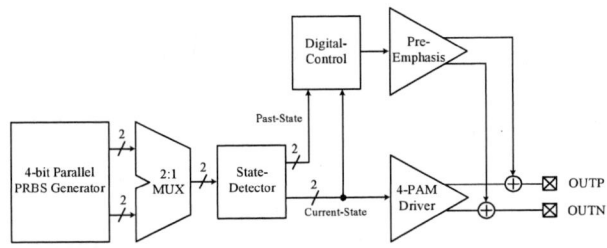

Fig. 3 Block diagram of proposed transmitter

The proposed transmitter shown in Fig. 3 consist of 4- bit parallel Pseudo Random Bit Sequence (PRBS) generator, 2-to-1 multiplexer, state detector, digital control, 4-PAM driver, and pre-emphasis. PRBS is integrated in the transmitter for Built In Self Test (BIST). 2-to-1 multiplexer makes data rate twice faster. State detector makes current state and past state. Digital control in Fig. 3 compares current state and past state, then makes six kinds of control signal (SI, MI, BI, SD, MD and BD). These control signal turn on pre-emphasis circuit and pre-emphasis current is summed with 4-PAM main driver.

A. PRBS Generator and 2-to-1 Multiplexer

PRBS is integrated in the transmitter [5]. 4 bits of pseudo random bit data from flip flop output is transferred in half clock period. And 2-to-1 multiplexor makes data rate twice faster. 2-to-1 multiplexer is used to prevent PRBS from being bottleneck of operation.

B. State Detector

Fig. 4 shows the block diagram of State detector [6]. For pre-emphasis of a signal, the current state and the past state are needed. The state detector makes these states. The first flip-flop makes a current state A and B. The second flip-flop makes a past state C and D. Current state and past state have a one clock difference. And to prevent undesired sampled data from flip-flop, the flip flop's clock is controlled by external clock.

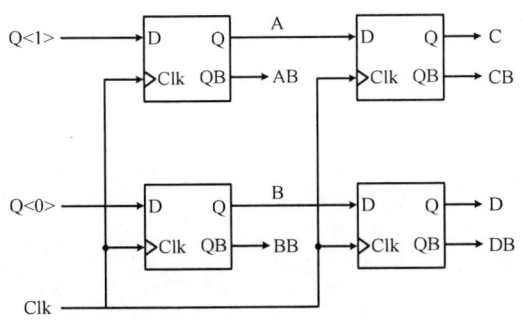

Fig. 4 State Detector

C. Digital Control

Table 1 shows the digital control signal according to current state and past state [6]. Transition between current state and past state makes a digital control signal. As shown in Fig. 1 (a) and (b), 4-PAM transmitter has six cases of transitions. To differentiate pre-emphasis currents between six cases, digital control signals (SI, MI, BI, SD, MD and BD) are needed. Digital control circuit makes these six kinds of control signals. Digital control circuit is implemented according to Table 1.

TABLE 1
CODING TABLE FOR DIGITAL CONTROL.

Current State		Past State		Digital Control Signal					
A	B	C	D	SI	MI	BI	SD	MD	BD
0	0	0	0	L	L	L	L	L	L
0	0	0	1	L	L	L	H	L	L
0	0	1.	0	L	L	L	L	H	L
0	0	1	1	L	L	L	L	L	H
0	1	0	0	H	L	L	L	L	L
0	1	0	1	L	L	L	L	L	L
0	1	1	0	L	L	L	H	L	L
0	1	1	1	L	L	L	L	H	L
1	0	0	0	L	H	L	L	L	L
1	0	0	1	H	L	L	L	L	L
1	0	1	0	L	L	L	L	L	L
1	0	1	1	L	L	L	H	L	L
1	1	0	0	L	L	H	L	L	L
1	1	0	1	L	H	L	L	L	L
1	1	1	0	H	L	L	L	L	L
1	1	1	1	L	L	L	L	L	L

D. Pre-Emphasis

Fig. 5 shows the proposed pre-emphasis circuit. To operate with CML level's input, the architecture of the pre-emphasis circuit is identical to that of the conventional PAM main driver. Each current sink transistor is designed to produce more current when bigger transitions occur. Decrease and increase switching transistors are connected to Fig. 6's OUTP branch and OUTN branches, respectively. For example, if the increase input signal SI is high, I_PE is flow through OUTN branch, then the OUTN node's voltage decreases. As a result, the differential output voltage, i.e., OUTP-OUTN, increases when SI is high. Likewise, if decrease or increase signal is high, the differential output voltage decreases or increases, respectively.

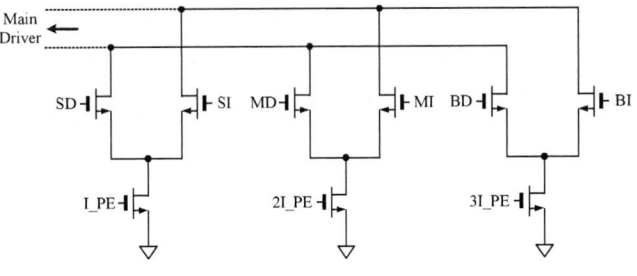

Fig.5 The Proposed pre-emphasis circuit

E. Main Driver with Enhanced Maximum Voltage Swing

Fig. 6 shows the 4-PAM main driver. To increase the output voltage swing range, the enhanced maximum voltage swing method is adopted [7] so that the chances of decoding the erroneous data in the receiver will be reduced. As shown in Fig. 6, thick gate transistor V_G is located below output matching resistor R_T. Thick gate transistors reduce high voltage stress for input gate transistors (A0, A0B, A1 and A1B). The 4-PAM driver, the mismatches of current sources from each branch can make inaccurate output voltage levels. For reducing mismatches of the current sources, a common-centroid layout technique for symmetric device placements is used.

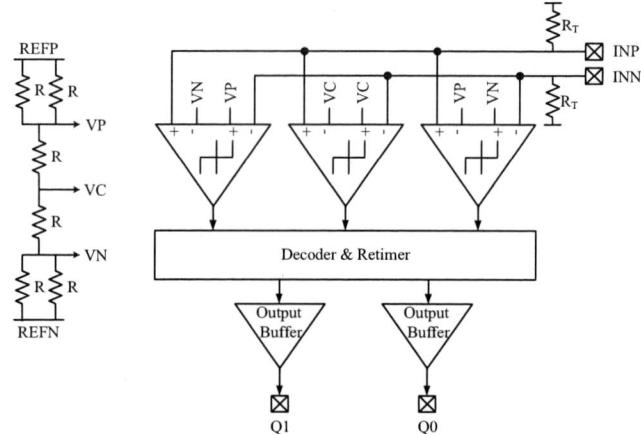

Fig. 7 The receiver's block diagram

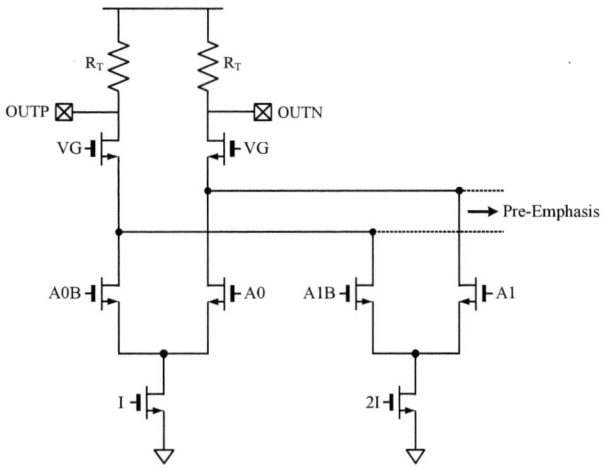

Fig. 6 Main driver with Enhanced maximum voltage swing

III. RECEIVERS

Fig. 7 shows a designed top level view of the receiver block diagram. The receiver is consists of a comparator blocks, decoder, re-timer and an output buffer. In a 4-PAM serial link transceiver, both the serialization in the transmitter and the de-serialization in the receiver occur [8].

The differential input signal is provided at the output of the transmitter with two wires. At first, input signal is compared with three reference voltages using three comparators, producing a thermometer code. The reference voltage made using a resistor string. Using the three reference values, the comparator outputs are a thermometer code [9].

To decode the thermometer code into the binary bits, the decoder is used. The decoder outputs are the pseudorandom bit sequence (PRBS) which is output of PRBS generator in the transmitter.

The last block of receiver is the output buffer. The 2-bit binary digital signals from the decoder output go to output buffer so that the data could be measured by oscilloscope. The output transistor size is optimized to have a 50-Ω output load resistance when the transistors are on for maximum power delivery and for minimum reflection at the output nodes.

IV. EXPERIMENTAL RESULTS

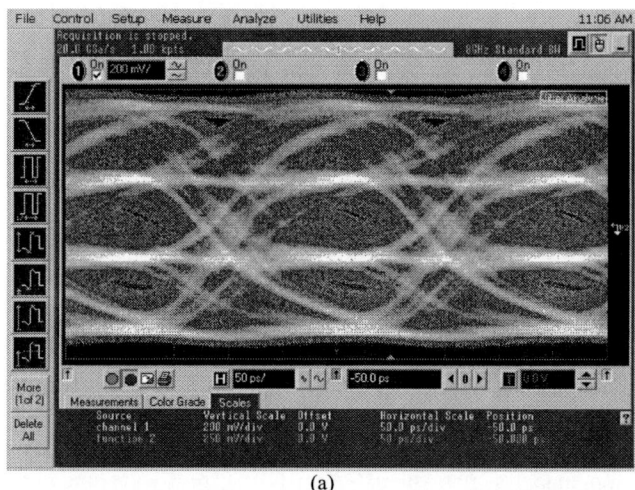

(a)

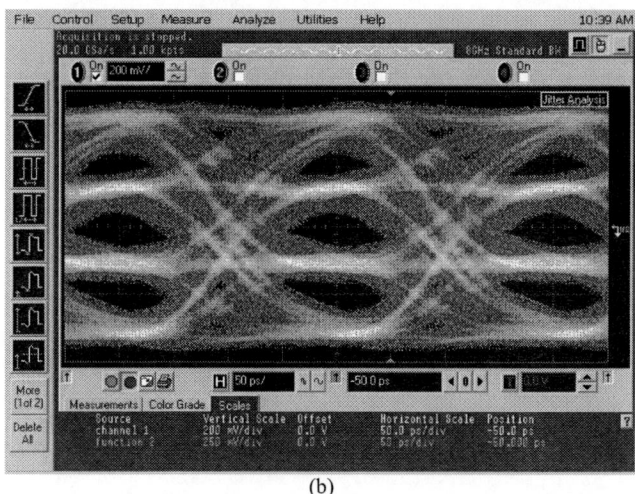

(b)

Fig. 8 The measured transmitter's eye-diagram at 10 Gb/s over 0.7-m cable and 3-cm PCB traces : (a) without pre-emphasis, and (b) with pre-emphasis

978-1-61284-863-1/11 $26.00 © 2011 IEEE

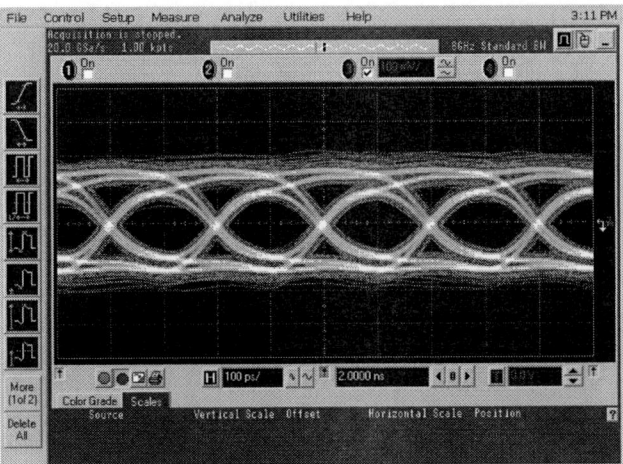

Fig. 9 The measured receiver's eye-diagram at 10 Gb/s over 0.7-m cable and 3-cm PCB traces

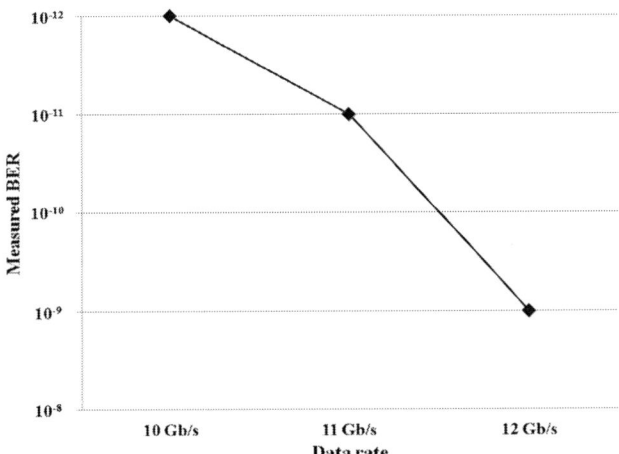

Fig. 10 The measured BER

The proposed 4-PAM transceiver with adaptive pre-emphasis for 2 bit/symbol was fabricated on in a 0.13 μm CMOS process. It was measured in a chip-on-board (COB) configuration. The transmitter and receiver consume 245 mW and 69 mW, respectively. The CMLs are used in the 4-PAM transmitter for high-speed operations at the cost of large power consumption.

Fig. 8 shows the measured eye-diagrams of the proposed 4-PAM transmitter with adaptive pre-emphasis over 0.7-m cable and 3-cm printed circuit board (PCB) traces at 10 Gb/s. Fig. 8 (a) and (b) show the measured eye-diagram without pre-emphasis and with pre-emphasis, respectively. The improvement in both the eye-opening width and eye-opening height are evident. The minimum eye opening of the pre-emphasis-mode is 80 ps wide and 200 mV high. The maximum output voltage swing was 1.8 V_{P-P}. Fig. 9 shows the measured eye-diagram of the receiver. The maximum data rate was 10 Gb/s. As shown in Fig. 10, we found that the bit-error rate is less than 10^{-12} at 10 Gb/s data rate measured with 2^7-1 pseudorandom bit sequence.

V. CONCLUSION

The proposed 4-PAM transceiver achieved 10 Gb/s data rate by using adaptive pre-emphasis method. The proposed 4-PAM transceiver with adaptive pre-emphasis fabricated on in a 0.13 μm CMOS process achieved 10 Gb/s data rate. The transmitter and receiver consume 245 mW and 69 mW, respectively. The measured bit-error rate is less than 10^{-12} at 10 Gb/s data rate.

ACKNOWLEDGMENT

This research was supported by Hynix Semiconductor. This research was also supported by the MKE(The Ministry of Knowledge Economy), Korea, under the ITRC(Information Technology Research Center) support program supervised by the NIPA(National IT Industry Promotion Agency) (NIPA-2011-(C1090-1111-0006)).

REFERENCES

[1] T. Granberg, *Handbook of Digital Techniques for High-Speed Design*, NJ: Prentice Hall PTR, 2004.

[2] D.J. Foley and M.P. Flynn, "A Low-Power 8-PAM Serial Transceiver in 0.5-μm Digital CMOS," IEEE J. Solid-State Circuits, vol. 37, no. 3, pp.310–316, Mar. 2002.

[3] F. R. Ramin, *A CMOS 4-PAM multi-Gbps serial link*, Ph.D. Dissertation, Stanford University, CA, 2000.

[4] R. Farjad-Rad, C.K. Yang, M.A. Horowitz, and T.H. Lee, "A 0.4-μm CMOS 10-Gbit/s 4-PAM pre-emphasis serial link transmitter," IEEE J.Solid-State Circuits, vol. 34, no.5, pp.580–585, May. 1999.

[5] E. Laskin et. al, "A 60 mW per Lane, 4 x 23-Gb/s 2^7-1 PRBS Generator," IEEE J. Solid-State Circuits, vol. 41, no. 10, pp. 2198-2208, Oct. 2006.

[6] Z.M. Lin and K.Y. Chang, "Level selection based pre-emphasis for PAM transmitter," Electronics letters, vol 42, no 7, pp.399-400, Mar. 2006.

[7] K. Farzan et. al, "A CMOS 10-gb/s power-efficient 4-PAM transmitter," IEEE J. Solid-State Circuits, vol. 39, no. 3, pp.529-532, Mar. 2004.

[8] J. Jeong, J. Lee, and J. Burm, "A 0.18 μm CMOS multi-Gb/s 10-PAM transmitter," Proc. International Symposium on Integrated Circuits, pp.85-88, 2009.

[9] J. Lee, J. Jeong, and J. Burm, "A multi Gbps 10-PAM receiver in 0.18um CMOS technology," Proc. International Symposium on Integrated Circuits, pp.89-92, 2009.

978-1-61284-863-1/11 $26.00 © 2011 IEEE

Processing *N*-ary Trees in Hardware Circuits

Valery Sklyarov
DETI/IEETA/HIPEAC,
University of Aveiro,
Aveiro, Portugal
skl@ua.pt

Iouliia Skliarova
DETI/IEETA,
University of Aveiro,
Aveiro, Portugal
iouliia@ua.pt

Dmitri Mihhailov
Computer Department,
University of Technology,
Tallinn, Estonia
d.mihhailov@ttu.ee

Alexander Sudnitson
Computer Department,
University of Technology,
Tallinn, Estonia
alsu@cc.ttu.ee

Abstract—The paper demonstrates that *N*-ary trees (*N*>2) can efficiently be used to model and process data in hardware. It is done through: 1) representation of data by *N*-ary trees; 2) compact coding of *N*-ary trees in memory; 3) common methods for data processing based on the model of a hierarchical finite state machine (HFSM). The proposed techniques have the following advantages: 1) similarity of processing *N*-ary trees with different characteristics such as the size of data *M*, the value *N*, and the depth *d* of trees; 2) fixed number of processing steps from the root to leaves for the given depth *d*; 3) the ease of reconfiguration (customization) of HFSM for different values of *N*, *d*, and *M*; 4) potential parallel processing of nodes' children. The results of experiments confirm effectiveness of the proposed techniques and their applicability for solving practical problems.

Keywords-N-ary tree; hierarchical finite state machine; graph and tree search strategies; special-purpose hardware

I.   INTRODUCTION

A tree is a connected graph that does not contain cycles [1]. An *N*-ary tree (*N*≥2) has up to *N* children for nodes and it can be seen as an effective model to support data sorting [1], to manage priorities in queues [2], to solve problems of combinatorial search [3], etc. An important advantage of *N*-ary trees compared to other alternative models is an opportunity of rapid adaptation to eventual modifications in input data. This is possible because address-based manipulations over tree nodes are simple and fast [1].

A number of recent research works in this area are targeted to the potential of advanced hardware accelerators. Notable results have been achieved through applying parallelism, pipelining, non-sequential circuits and other techniques and building specialized blocks in hardware. A special attention has been paid to such competitive implementation platforms as: FPGAs, graphical processing units (GPU) and multi-core CPU. The use of FPGA permits design constraints of CPU and GPU with predefined architectures to be eliminated [4]. Besides, parallelism in FPGA circuits can be implemented easier.

This paper suggests:

- Representation in memory and processing *N*-ary trees capable to be efficiently implemented in low-cost and widely available FPGAs;
- A computational model based on a hierarchical finite state machine (HFSM) that can be easily configured for processing trees with different characteristics such as the value *N*, the size of data *M* and the depth of the tree *d*.

The remainder of this paper is organized in five sections. Section II suggests different types of modeling with *N*-ary trees. Section III is dedicated to representation of trees in memory. Section IV suggests HFSM-based processing of *N*-ary trees in hardware. Section V describes implementations and experiments. The conclusion is given in Section VI.

II.   MODELING OF COMPUTATIONS BY *N*-ARY TREES

Many computations use *N*-ary trees (*N*>2) as a basic model. For example, an FPGA-based Boolean Constraint Propagation (BCP) accelerator [3] assists in solving the SAT problem. An *N*-ary tree in [3] permits to find out clauses of a given Boolean formula containing the selected variables. In [2] trees are used for fast re-arranging of newly received instructions according to their priorities, which is essential for priority managements. *N*-ary trees are also widely applicable to data sort [4,5] and to a number of other practical problems.

Let us agree to consider such *N*-ary trees (*N*>2) for which *N* is a power of 2, *i.e.* 4,8,16, *etc.* Let $G=\log_2 N$, and, thus, *G* is 2,3,4, etc. Binary input data of size *M* bits (*M*>*G*) can be partitioned into $N=2^G$ blocks (represented by sub-trees) and the blocks are created on the basis of *G*-bit groups within *M*-bit words much like it is done in [3] for clause index walk. Thus, the root of the tree enables the first *G* bits of *M*-bit data items to be analyzed. The remaining *M-G* bits of *M*-bit data items are distributed between the children of the root and each child deals with *M-G* bits of data items (*i.e.* the initial complexity of the problem is reduced). Any child of the root can be seen as a local root of sub-tree modeling similar type of processing. Thus, reduction of the problem complexity can be applied incrementally.

Let us consider an example with $G=3$, $N=2^G=8$ and for the following data: 111100010, 010110000, 111111010, 010111101, 111100111, 010111000, 010100011, 111111111, 010100110, 010100001, 111111100, 010110101, 010111110, 111111101, 111100100 (*M*=9). The given above input *M*-bit data are divided into *M*/*G* groups from left to right. For our example any data item (*e.g.* 111100010) is composed of three 3-bit groups and they are 111 100 010 for the first item. These three 3-bit groups will be associated with levels of *N*-ary tree (*8*-ary tree for our example) and every non-leaf tree node might have up to *N*=8 children (*i.e.* the tree will be *M*/*G* deep). If we apply the method described in [4] the rightmost group can be processed differently and, thus, the tree will be (*M-G*)/*G* deep. The depth *d* of the tree is equal to the number of edges and, consequently, to the number of steps needed to execute forward traversal from the root to a leaf (one step for every level of the tree). As soon as a new data item is received, G groups are sequentially analyzed to allocate the existing group on the tree or to create a new one. The following steps will be executed for each data item:

1) Take the most significant (the leftmost) group and allocate a new child node or use the existing child node of the tree;

2) Execute forward propagation step to the node taken in point 1);

3) Apply recursively the steps 1) and 2) to the remaining groups (from left to right).

The first data item is 111100010. The first segment 111 has to be associated with a new node of the tree connected with the root by the rightmost edge ($N=8$ and 111 is the biggest value from potential G-bit values: 000, 001,…, 111). In point 2) the control flow is passed to the newly created node considered as a local root of the relevant sub-tree. Then the same steps have to be executed for the child node and for the second G-bit (3-bit) code 100. Similarly, the entire tree for the presented above set of data items can easily be built and it is shown in Fig. 1.

As you can see from Fig. 1 the depth $(M-G)/G$ is 2 and two steps are needed to reach any leaf from the root. Thus, any newly arrived data item can be allocated on the tree during just two steps and the remaining nodes of the tree are not changed. Such a tree can be built in software by analyzing G-bit groups of incoming data items from left to right and traversing through the existing nodes or allocating new nodes until the last G-bit group permitting a leaf to be created. The number of such steps is $(M-G)/G$ (taking into account the method [4]). If we model this process in software then memory cells for nodes can be allocated dynamically. Repeated values can be taken into account by counters associated with the leaves. Each node might have up to $N=8$ children but many of them do not exist and the maximum number of actual children in Fig. 1 is 3. $N=8$ because each G-bit (3-bit) group (used for nodes) has 8 potential codes (000, 001,…,111), which might be associated with children.

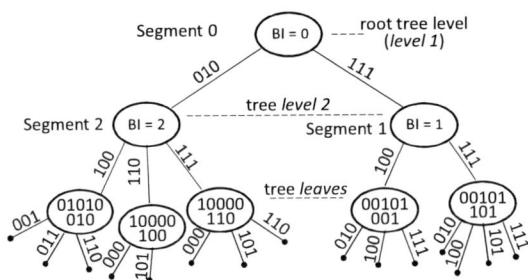

Figure 1. *N*-ary tree (*N*=8) that is built for data from section II

In fact, such a tree contains ordered data that can be extracted executing recursive C function [4]. Constructing the tree can also be done by a similar iterative function.

Given a non-leaf node, the address of the node in the tree is called the *base index* (*BI*). *BI* for the root is always 0. *BIs* for remaining (not associated with leaves) children are assigned sequentially in order of their creation. As soon as the first data item 111100010 is allocated on the tree a new non-leaf node

with $BI = (BI$ of the root $= 0) + 1 = 1$ is created and its *BI* is an increment of *BI* assigned to the root. As soon as the second data item 010110000 is allocated on the tree a new non-leaf node with $BI = 1 + 1 = 2$ is created and its *BI* is an increment of the previous *BI* with the biggest value.

III. REPRESENTATION OF *N*-ARY TREES IN MEMORY

N-ary trees need to be represented in memory, which can be based on the methods proposed in [3,4]. Any tree node (beginning from the root) with its children is coded in a segment that is composed of 2^G sequential memory addresses (eight memory addresses for our example). In the previous section we explained that segments are ordered sequentially in accordance with the following rules:

- *BI* of the root segment is assigned all zeros;
- *BI* of any new segment is assigned an increment of code of the previous segment (with the biggest value);
- Non expanded nodes are marked with *no-match* tag, which is a code with all zeros.

To locate the i_{th} child, the address can be calculated by adding i to the *BI* of the node. If a node does not have any child then the tree is not expanded and just a *no-match tag* (0) can be stored in the memory. All nodes (not containing *no-match tags*) are called *working nodes*. There is no cell in memory for the root node and the first memory address (all zeros) is just considered to be the *BI* for finding children of the root. Other *BIs* are assigned sequentially for newly processed *G*-bits groups taking into account that every tree node necessitates a space in memory for *N* potential children.

Let us agree to code up to *N* values for the rightmost group associated with leaves by *N*-bit code in such a way [4] that in the code $Bit_i=1$ if and only if there exists a value associated with the considered leaf node that is equal to i_2 (*i.e.* binary code of i), else $Bit_i=0$. For example, such code for the left leaf in Fig. 1 is 01010010. Indeed, there are *N*=8 bits in the code. There are three values (001,011,110) associated with the leaf. They have the following decimal representations: 1, 3, 6. Thus, bits 1, 3 and 6 (the leftmost bit is 0) of the code have to be set to 1 and all the remaining bits have to be assigned 0. Such codes are written inside the nodes for the leaves in Fig. 1.

For optimization purposes two types of memory are used. The first type represents all tree nodes, except leaves. The second type represents leaves (see Fig. 2).

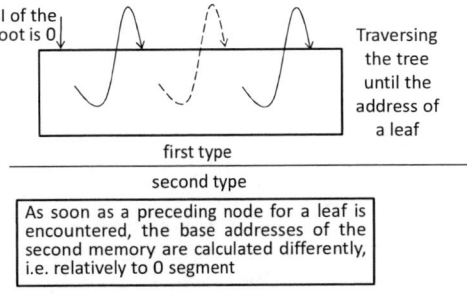

Figure 2. Representing trees by two types of memories

Addresses for the first type and for the second type of memory are chosen *independently*. As soon as the first leaf is created it is assigned *relatively* to the address 0 of memory of the second type (*i.e.* the first segment of the second memory is 0 and it is not equal to *the biggest previous BI+1*). Addresses of other *segments* of the second memory are calculated incrementally and relatively to the initial segment address 0. Thus, memories of each type can be of different size, which permits hardware resources to be optimized and the size of the rightmost group not equal to *G* to be considered.

Fig. 3 shows representation in memory of the tree in Fig. 1 assuming that two types of memories are not separated.

	address	BI		address	data		address	data
	0 (000)	00		8 (1000)	00000000		16 (10000)	00000000
	1 (001)	00		9 (1001)	00000000		17 (10001)	00000000
	2 (*010*)	10 (*01*)		10 (1010)	00000000		18 (10010)	00000000
	3 (011)	00		11 (1011)	00000000		19 (10011)	00000000
	4 (100)	00		12 (*1100*)	00101001		20 (*10100*)	01010010
	5 (101)	00		13 (1101)	00000000		21 (10101)	00000000
	6 (110)	00		14 (1110)	00000000		22 (*10110*)	10000100
	7 (*111*)	01 (*00**)		15 (*1111*)	00101101		23 (*10111*)	10000110

Segment 0 (00) – first type; Segment 1 (0) – second type; Segment 2 (1) – second type

* if there if a problem with 0 address it can be incremented in column *BI* in the first memory and decremented when we calculate addresses of the second memory

Figure 3. Representation of the tree from Fig. 1 in memory

IV. PROCESSING *N*-ARY TREES IN HARDWARE

There are two basic modules that create and traverse *N*-ary trees. The first module constructs the three, *i.e.* receives data items (such as that are given in section II) and builds the tree *i.e.* fills in memory in a way shown in Fig. 3. The second module traverses the tree in such a way that permits to visit all nodes in the required order. Suppose, we need to sort data allocated on the tree. In this case the following recursive procedure can be executed:

1. Begin from the root;
2. Find the address of the left most working node, which has not been processed yet. If all nodes have already been processed then finish the procedure;
3. If the address points to memory of the second type then record the result of sorting, else recursively execute points 2 and 3 for a newly selected node.

Since the number of steps from the root to the memory of the second type is known in advance, both recursive and iterative procedures can be applied easily.

Let us consider our example (see Fig. 1 and 3). The first working node has *BI*=2. Such node can easily be found in the first segment using either sorting networks or similar combinational circuits [4]. Note that only for working nodes the value in *BI* column of Fig. 3 is not equal to 0. The value *BI* at the address 2 is 10 and points to the second segment of memory of the second type. There are 3 non-empty leaves in the second segment: 20 – 01010010; 22 – 10000100, and 23 – 10000110. They are decoded sequentially. Thus we produce the first (initial) part of the sorted sequence: 010100001, 010100011, 010100110, 010110000, 010110101, 010111000, 010111101, 010111110. The second working node from the root has *BI*=01 pointing to the first segment of memory of the second type. There are two non-empty leaves in the second segment: 12 – 00101001, and 15 – 00101101. Similarly, we

produce the second and the last part of the sorted sequence: 111100010, 111100100, 111100111, 111111010, 111111100, 111111101, 111111111. We assumed that segments in Fig. 3 are allocated sequentially within the same memory. If memories of different types *are separated* (see Fig. 2) then underlined *BI* in the first memory have to be used and consequently the segments of the second memory begin from 0 (instead of 8) and 8 (instead of 16). Thus, the first/the second memory can be organized as 8 2-bit words/16 8-bit words. The total size of memories is 144 bit (taking into account that the first address of memory of the second type is 0 instead of 8 and, thus, *BI* at the address 7 in the first memory would contain 00 instead of 01 and at the address 2 - 01 instead of 10 – see underlined text in Fig. 3).

Synthesis and implementation in hardware were done using the model of a HFSM [6] and the following steps:

1. Modeling in general-purpose language (we used C language for such purposes);
2. Mapping C functions (like considered in [4]) to synthesizable VHDL templates [6] for HFSM;
3. Synthesis and implementation of circuits from the customized VHDL template using commercial tools such as Xilinx ISE;
4. Uploading the generated bitstream and testing the resulting circuits in FPGA.

For example, if *N*=4 C function sort considered in [4] can be converted to state transition diagram for HFSM that is shown in Fig. 4. Transitions between the states are presented in a simplified form and *l*, *lm*, *rm*, *r* are pointers to child nodes for the selected node from left to right, *l*/*lm*/*rm*/*r* with inversion denote an absence (indicated by *no match* tag) of *left*/*left-middle*/*right-middle*/*right* child nodes for the selected node. A curve with the condition *level=depth* indicates that the condition *level=depth* has to be valid for all relevant edges going from the state *a0* (*level*=1 for the root, *level*=2 for children of the root, *etc.*). Transitions take place where conditions written near output edges are satisfied. For example, transition from *a1* to *a4* takes place when for the considered tree node there are no *left middle* and *right middle* children and there is a *right child*. Transitions from *a0* to *a5-a8* take place when the condition *"level is not equal to depth"* plus conditions written near input edges for the states *a5-a8* are satisfied.

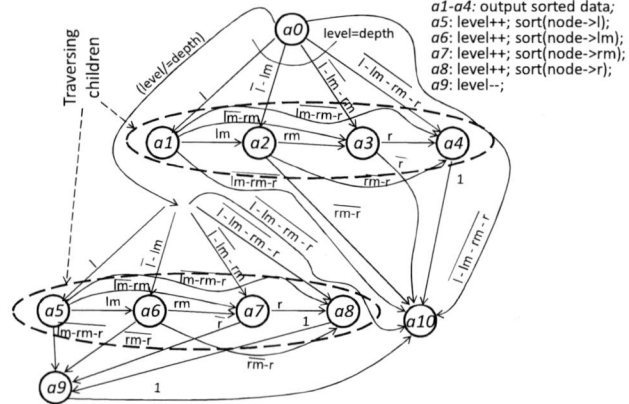

Figure 4. State transition diagram for the function sort

Note that the number of states enclosed by dashed ellipses is equal to *N*. Thus, if *N*=8 (see Fig. 1) then each ellipse will contain 8 states allowing children of each node to be traversed from left to right.

VHDL template for HFSM is given in [6]. All transitions from Fig. 4 have to be used to customize the template. Optimization technique from [6] enables the number of states to be minimized. Instead of a register in a conventional FSM, the HFSM contains stack memory. Recursive calls are implemented through an increment of a stack pointer and recording the initial state (*a0*) for the called function at the top level of the stack. The arguments (sub-roots within *N*-ary tree) are supplied through indicating *BI*s for the memory in a dedicated register. Returns from recursive modules are done in the state *a10* by decrementing the stack pointer.

As you can see any change in values *M*, *N* and *d* can be easily introduced. Indeed, just the number of states (such as *a1-a4* and *a5-a8* in Fig. 4) is changed and customizing the template [6] is exactly the same. Thus, the proposed technique enables *N*-ary trees with different characteristics to be implemented through simple configuration of the HFSM.

V. EXPERIMENTS AND RESULTS

Synthesis and implementation of the circuits from specification in VHDL were done in Xilinx ISE for FPGA Spartan3E-1200E-FG320 (NEXYS-2 prototyping board of Digilent). Data were processed either based on their static or dynamic representation. In the last case new data items are allocated on the tree as soon as they are produced by a random-number generator and removed when required.

N-ary trees (*N*=4 and *N*=8) were successfully tested for different problems. We found that in a single FPGA Spartan3E-1200E-FG320 any set of 18-bit and some sets of 19- and 20-bits data items can be processed. The size of memory (due to representing data in a form shown in Fig. 2) is reduced comparably with [4]. Thus, using *N*-ary trees enables fast hardware accelerators to be developed on the basis of low-cost widely available FPGAs, which is essential for data processing in real-time embedded systems.

We found that the developed circuits are faster than implementations in general-purpose software and in embedded to FPGA Power PC (the latter two cases were tested in HP EliteBook 2730p and PPC405 embedded to FPGA Virtex-4 FX12). Some results of comparison can be found in [7]. Performance is comparable with known results obtained for significantly more advanced FPGAs.

You can see that there is no data dependency between tree branches. Thus, the algorithm permits individual sub-trees with any desired level of parallelism to be processed. We found that parallel processing of trees is faster than sequential, but hardware resources are also increased.

The results of experiments have shown that the main restriction that limits the number of data items is the available embedded block RAMs on the FPGA microchip. The algorithms themselves are easily scalable.

The main advantage of the proposed technique for FPGAs is the ease of processing *N*-ary trees with different values of *N*, *M*, and *d*. Indeed, the considered HFSMs are easily customizable through generic VHDL statements [6]. Thus, a number of alternative and competitive implementations can be explored.

The following conclusion can be drawn:

- The considered *N*-ary trees permit to represent and process data for different computational problems, such as priority management [2], BCP [3], and data sort [4];
- Allocating new items does not reallocate previously accommodated items, which is essential for problems requiring fast respond to frequently changing input data;
- The number of steps from the root to leaves is fixed and is equal to the depth $d=(M-G)/G$, which enables children of the nodes to be traversed in parallel.

VI. CONCLUSION

An *N*-ary tree is an effective model for data processing in FPGAs: it can be compactly coded in memory and processed using the proposed technique. The main contributions of the paper are: 1) method of compact representation of data in memory common for different computational problems; 2) the use of HFSM as an effective model for processing *N*-ary trees in hardware.

ACKNOWLEDGMENT

This research was supported by the European Union through the European Regional Development Fund, FEDER through the Operational Program Competitiveness Factors - COMPETE and by National Funds through FCT - Foundation for Science and Technology in the context of project FCOMP-01-0124-FEDER-022682 (FCT reference PEst-C/EEI/UI0127/2011).

REFERENCES

[1] F.M. Carrano, *Data Abstraction and Problem Solving with C++*, Addison Wesley, 2005.

[2] V. Sklyarov and I. Skliarova, "Modeling, Design, and Implementation of a Priority Buffer for Embedded Systems", *Proc. 7th Asian Control Conference* – ASCC'2009, Hong Kong, August 2009, pp. 9-14.

[3] J.D. Davis, Z. Tan, F. Yu, and L. Zhang, "A practical reconfigurable hardware accelerator for Boolean satisfiability solvers," *Proc. 45th ACM/IEEE Design Automation Conf.*, Anaheim, CA, USA, June 2008, pp. 780-785.

[4] V. Sklyarov, I. Skliarova, D. Mihhailov, and A. Sudnitson, "Implementation in FPGA of Address-based Data Sorting", *Proceedings 21st Int. Conf. on Field Programmable Logic and Applications - FPL 2011*, Crete, Greece, September 2011, pp. 405-410.

[5] R. Mueller, *Data Stream Processing on Embedded Devices*, Ph.D. thesis, ETH, Zurich, 2010.

[6] V. Sklyarov, "Synthesis of Circuits and Systems from Hierarchical and Parallel Specifications", *Proc. 12th Biennial Baltic Electronics Conference,* invited paper, Tallinn, October 2010, pp. 37-48.

[7] V. Sklyarov, I. Skliarova, R. Oliveira, D. Mihhailov, and A. Sudnitson, "Processing Tree-like Data Structures in Different Computing Platforms", *Proc. Int. Conf. on Informatics and Computer Applications - ICICA' 2011*, Dubai, UAE, March 2011, pp. 112-116.

978-1-61284-863-1/11 $26.00 © 2011 IEEE

Simulation Environment for Visual Prototyping of Circuits and Systems

Iouliia Skliarova
DETI/IEETA, University of Aveiro,
Aveiro, Portugal
iouliia@ua.pt

Valery Sklyarov
DETI/IEETA, University of Aveiro,
Aveiro, Portugal
skl@ua.pt

Abstract— **The paper presents results in the following two areas: the visual graphical verification of hardware systems and the synthesis of digital circuits from modular, hierarchical, recursive, and parallel specifications. Within these areas a simulation multimedia environment has been developed and used for verification of the proposed methods that are based on new structural models. The applicability of the environment and the methods is demonstrated through examples.**

Keywords-visual prototyping; finite state machine; FPGA; simulation environment

I. INTRODUCTION

Nowadays FPGAs are used in a wide variety of practical contexts but prototyping, hardware/software co-design and co-simulation are still the application areas where FPGAs are most effective and promising. Many development efforts are dedicated to making progress in these directions. For example, paper [1] attempts to leverage the merits of software simulation and hardware (FPGA-based) emulation to combine flexibility and performance in multi-core research. Paper [2] compares software-only simulation with hybrid (software + FPGA-based hardware) co-simulation and reports an average 39x speed-up for the latter. Paper [3] presents four FPGA-based prototypes for network-on-chip and makes accurate evaluations of the various design trade-offs. A superscalar central processing unit (modeled in C language) communicating with a reconfigurable unit that is controlled through a bus is studied in [4] and the whole system demonstrates a significant speedup compared to pure software implementation. The methodology considered in [5] partitions a simulator (that models embedded and other systems) into an easily modifiable and extended predictive model that calculates some metrics, and a functional model that is considered to be an effective computing system. It is shown that implementing at least the predictive model in an FPGA improves performance significantly. All these results demonstrate the importance and benefits of the areas indicated above and the significance of the respective research efforts.

This paper focuses on a visual simulation and integration of relevant tools with the system implemented within the same hardware (FPGA). This technique simplifies the prototyping, evaluation and comparison of different design ideas by showing the results in the form of interactive images (dynamic visual items representing corresponding physical objects) on a monitor screen. The environment is capable of addressing real world problems and is also easy to understand and use. The capability of this technique is demonstrated through different examples.

Basic functional components of the system can be constructed by applying alternative models, architectures and design methods. We base the design approach on advanced models of finite state machines (FSM) with two primary objectives: 1) to verify the capabilities of the FPGA-based visual environment and simulation; 2) to evaluate the applicability and effectiveness of advanced FSM models that enable modular, hierarchical, recursive, and parallel algorithms to be implemented and mapped to hardware. Conceptually this approach provides tools for the evaluation and comparison of alternative design techniques and models, such as those proposed in [6-13] with the primary objectives being to support run-time modifiability [6], modularity [7], hierarchy [7,8], recursive calls [7-13], and parallelism [13].

II. EXAMPLES

Prototyping is a very important step for the design of circuits and systems. Nowadays FPGAs are widely used for such purposes. Typically an FPGA-based prototyping board includes many peripheral components, such as push buttons, LEDs, LCD displays, etc. Besides, the majority of available boards provide an interface with a VGA monitor and a keyboard/mouse. Visual prototyping permits many peripheral components of the board to be removed and replaced with the relevant visual objects on the attached monitor screen as it is shown in Fig. 1 [14]. Pre-designed hardware libraries enable the designer to choose such components that are needed for particular application. The developed circuits interact inside the FPGA with the components and the latter communicate with a monitor and a keyboard/mouse. As a result an interface with peripheral devices is replaced by interaction with images (such as that shown in Fig. 1) on the monitor controlled by a

978-1-61284-863-1/11 $26.00 © 2011 IEEE

single input device, such as a keyboard or a mouse. One of FPGA distinctive features is an opportunity for distant configuration/interaction [15]. Thus, the technique can also be used remotely applying any available wireless interface (see examples in [15,16]. Besides, an interaction can be provided through the Internet [14].

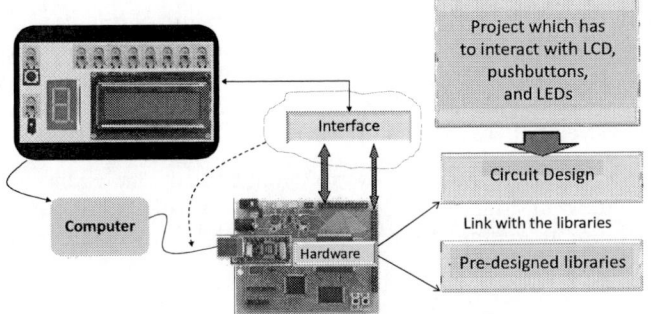

Figure 1. Visual representation of perhipheral components

Another example of visual prototyping is given in [16]. Let us consider a system that consists of two sub-systems executing functions of a garage control and a car control. Basic functionality of the system is implemented in hardware circuits interacting with a visual simulator that generates a static image of the garage and dynamic images of gates and cars and executes commands from the control sub-system (*e.g.* open/close doors, drive cars). Visual object interaction allows the capacity of the garage to be estimated for different parameters supplied through the keyboard.

The last example that uses similar prototyping is taken from [17] and it permits visual model of an urban traffic system to be created in such a way that is demonstrated in Fig. 2. Simulated part of city is displayed on a monitor screen (Fig. 2, a). Then a number of cars can be introduced together with traffic lights (see example in Fig. 2, b) and the relevant parameters (switching time of traffic lights, car speed, etc.).

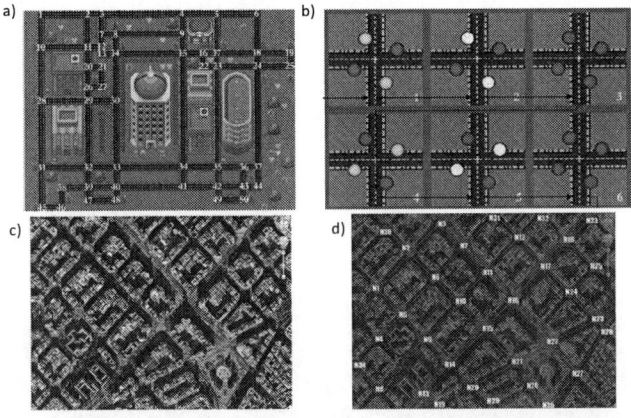

Figure 2. Visual models of an urban traffic system

Working with such model enables us to discover potential problems and to study how well urban traffic system is realized. Pictures (like shown in Fig. 2, a) can be taken from maps (such as that are available through the Internet – see Fig. 2, c), which can be converted to the appropriate model (see Fig. 2, d), which further can be evaluated visually.

III. GENERAL MODEL OF BASIC CONTROL UNITS

We suggest to simulate functionality of systems (similar to the considered in the previous section) using hierarchical specifications (hierarchical graph-schemes – HGS) [7,8] and models such as hierarchical [7,8] and parallel [14] FSMs (HFSMs and PHFSMs).

Fig. 3 presents a top-level structural FSM model that can be used in each sub-system. The left-hand part depicts a conventional FSM composed of a combinational circuit (CC) and a register (Rg). The remainder part shows the proposed communication mechanisms that are established through synchronization semaphores S_{out} and capabilities of advanced FSMs providing support for hierarchy [7,8] and parallelism [13]. The FSM performs relatively simple actions and interacts with a datapath (DP) composed of a set of registers. Many operations, like *count* and *shift*, are executed directly in FSM processes (see the gray rectangle on the outputs of the CC) and the proper synchronization is provided in a way very similar to [18]. As soon as it is necessary to activate new modules the relevant semaphores (flip-flops) are set by the signals **A** in the circuit S_{out}. As a result the output signals **C** of S_{out} activate the necessary modules running in parallel with the FSM shown in Fig. 3. Acknowledgements of some selected important states of the modules are provided through the signals **B** in Fig. 3 allowing the behavior of all active components to be synchronized. It is important that the structure shown in Fig. 3 enables us to verify the states of modules during each clock cycle. Thus, on the one hand, fast communication mechanisms between cooperating FSMs are provided (through S_{out}) and, on the other hand, such important properties as modularity, hierarchy, recursive module invocations and parallelism are also supplied. Parallelism is established in two ways. Firstly, it is directly supported by the model in Fig. 3, and, secondly, it is achieved when a macro-instruction includes more than one macro-operation [13].

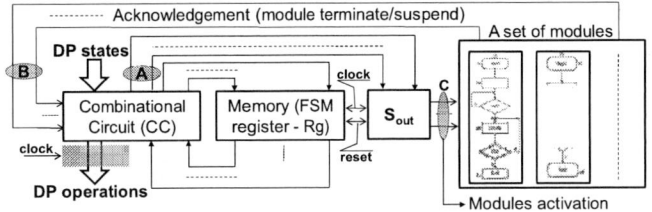

Figure 3. Top-level model

As an example let us demonstrate the description of a priority buffer that can be used in applications [16,17]. Suppose a sub-system (from [16,17]) has to be controlled by incoming instructions. The input instruction transfer rate is not the same as the instruction processing speed in the sub-system so it is necessary to use an input buffer and the instructions have to be processed not in the order of their arrival. Each instruction has an additional field pointing to the priority. A priority buffer (PB) is a device storing the incoming flow of events and allowing outputs that are selectively extracted from the incoming data to be supplied in accordance with their priorities. The PB considered implements the following functionality and it is organized in such a way that: 1) each new data item is pushed to the buffer with an extra field

978-1-61284-863-1/11 $26.00 © 2011 IEEE

indicating its priority; 2) at any time the buffer has to output the data item with the highest priority; 3) the buffer has to be able to remove all data items that are not longer required. The considered PB possesses the following distinctive features: 1) run-time data sorting in a single buffer memory; 2) dynamic memory allocation and de-allocation.

The proposed design technique includes: 1) description of the PB operations by parallel recursive algorithms applying methods [10,13]; 2) implementing the algorithms in hardware using a parallel hierarchical finite state machine [13].

The working functionality of the PB is based on the incremental construction and processing of a binary tree considered in [10]. In order to build the tree for a given set of data, we have to find the appropriate place for each new (incoming) data item in the current tree. In order to extract a data item, we can apply a special technique, which, in general, depends on selection rules and it can be based on steps that are exactly the same for each node. Thus, recursive technique might be helpful. The proposed PB is composed of 3 primary blocks, which enable the buffer: 1) building the tree [10] from incoming data; 2) extracting a data item by applying selection rules; 3) rebuilding the tree (removing the nodes that are no longer required).

Fig. 4 depicts simplified modules for the basic algorithm of the PB composed of the following 7 entities: Z_0 – the top level module; Z_1 – gets input data items and activates Z_4 for each of them (adding a new data item to the tree); Z_2 – extracts a data item from the buffer (with the aid of the module Z_6) using the selected priority rule and activates Z_3 when possible; Z_3 – provides synchronization with other modules and removes unneeded tree nodes (activating Z_5) if they have been already extracted by Z_6; Z_4 – adds a new data item to the tree; Z_5 – removes unneeded tree nodes; Z_6 – finds and supplies the data item with the highest priority.

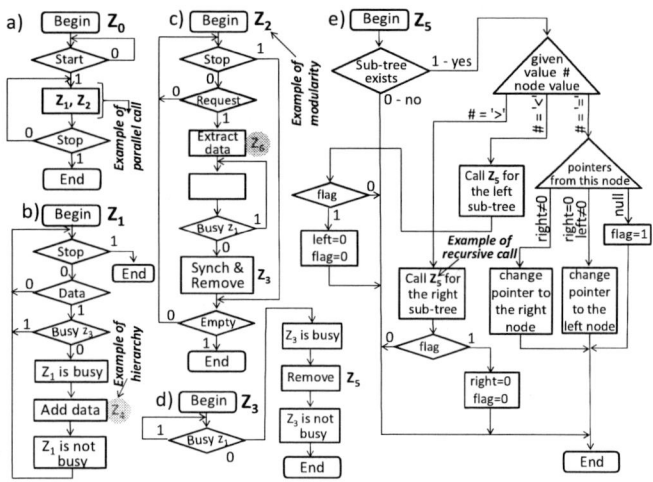

Figure 4. Modules describing the functionality of the PB

For simplicity Fig. 4 depicts just the modules needed for the primary functionality of the PB and the interface operations are hidden. Modularity as well as hierarchical, recursive and parallel module calls are explicitly indicated. All the buffer data items are kept in a dual port memory. Access to a shared buffer memory is properly synchronized using

semaphores z_1 and z_3. Indeed, the modules Z_4 and Z_5 alter the memory contents and they cannot be executed at the same time. That is why synchronization is done in the modules Z_1 and Z_3 (see Fig. 4). The modules that construct the binary tree (Z_4) and extract nodes from the tree (Z_6) are not shown in Fig. 4 (they are indicated by gray circles). This is because the module Z_4 is exactly the same as module z^m_1 in Fig. 3 in [11] and the module Z_4 is a simplified version of module z^m_2 shown in the same Fig. 3 [11]. Note that both these modules are recursive. The modules Z_0-Z_6 in Fig. 4 are called hierarchical graph-schemes (HGS) [8] that can formally be converted to PHFSM model [8,13].

Synthesis of the hardware circuits that implement the modules in Fig. 4 is done from specifications in VHDL incorporating templates proposed in [10,13]. Any template is a customizable VHDL code (such as that is sketched in Fig. 5) for a parallel hierarchical FSM - PHFSM (see [13] for additional details). The code shown in rounded rectangles describes reusable parametric stacks and a skeleton of the CC. The latter has to be customized through describing transitions between the HGS nodes as well as between different modules. All the rules are given in [10,13] with examples. Since the code for the CC indicates just module entrances, any of the modules can easily be replaced through trivial changes and each module can be tested, validated and debugged autonomously. Additional benefits are achieved by applying a technique that allows verification using simplified functionality (indeed, some modules can easily be blocked), the establishment of virtual links (e.g. module invocations can be re-assigned during run-time), etc.

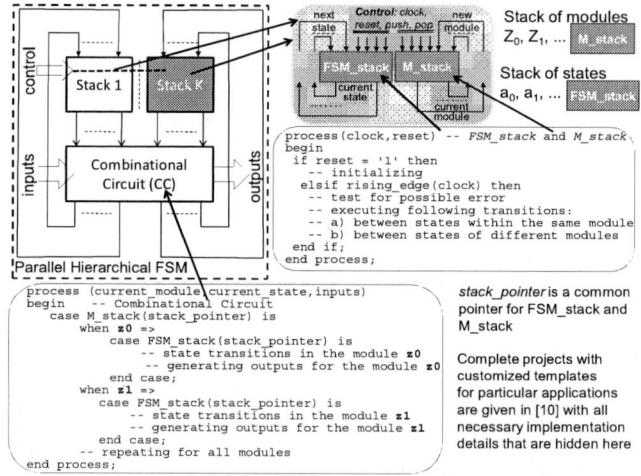

Figure 5. Architecture and VHDL templates for PHFSM

The stack control signals are *clock*, *reset*, *push* and *pop* and the latter two signals are generated in the VHDL process shown at the bottom of Fig. 5. After customization of the templates according to the HGSs [10,13] the proper code becomes a synthesizable VHDL specification and it can be processed in commercial CAD systems, such as Xilinx ISE.

IV.   IMPLEMENTATION AND EXPERIMENTS

Complete systems (briefly characterized in section II) are designed, verified, implemented and tested in NEXYS2 and Celoxica RC10 prototyping boards with a VGA monitor and a

978-1-61284-863-1/11 $26.00 © 2011 IEEE

keyboard connected. Functionality can be adjusted by defining such parameters as the number of cars in [16,17], car speeds, etc. All circuits for [16,17] occupy almost entire FPGA. Many resources are needed to support graphics and their minimization was not a target of the paper. Hardware circuits for a parallel hierarchical finite state machine implementing the functionality of Fig. 4, occupy only 276 Spartan-3 FPGA slices (the depth of the stacks is set 10). The PB (see Fig. 4) was used in the system [16] to rearrange dynamically association of arriving cars with the most preferable slots.

The following technique has been applied to allocate and free memory dynamically. PB storage was implemented in a memory block with a fixed number of cells. An auxiliary register contains the index of a memory cell that stores the root node of the tree Each cell is expanded with a one bit flag field – F, indicating if the cell is occupied (F=1) or not (F=0). The tree is constructed sequentially in such a way that for any new incoming data item, the first cell from the beginning for which F=0 is selected. As soon as a node is removed, the appropriate flag F is reset to 0 indicating that the cell can be used once again for new data. Thus, the cells are occupied and emptied during run time and this allows dynamic memory allocation and de-allocation.

The results of experiments and analysis of alternative implementations can be summarized as follows:

- Modularity, hierarchy, recursive and parallel module calls can be combined within the same synthesizable specification;
- Potential errors can easily be detected because of inherent simplicity of the modules;
- Thanks to modularity, the number of states in each module can be reduced significantly;
- The proposed structural model (Fig. 3) favorably combines fast reaction to potential input changes with the advantages of hierarchical reusable specifications;
- The visual tools that have been developed are very effective for experiments with alternative design techniques in an environment close to real world problems.

V. CONCLUSION

The paper presents: a simulation environment enabling alternative techniques to be verified and compared visually in a form that is very similar to their physical implementations; the structural model (Fig. 3) efficiently combining communicating and hierarchical state machines and allowing modularity, hierarchy, parallelism, and recursive calls to be mapped to hardware; an example illustrating specification and synthesis of a reusable priority buffer that implements modularity, hierarchy, parallelism, and recursive calls in hardware. All the developed circuits have been tested in FPGA-based prototyping boards.

ACKNOWLEDGMENT

This research was supported by FEDER through the Operational Program Competitiveness Factors - COMPETE and by National Funds through FCT - Foundation for Science

and Technology in the context of project FCOMP-01-0124-FEDER-022682 (FCT reference PEst-C/EEI/UI0127/2011).

REFERENCES

[1] T. Suh, H.H. S. Lee, S.L. Lu, and J. Shen, "Initial Observations of Hardware/Software Co-Simulation using FPGA in Architecture Research", Proc. 2nd Workshop on Architecture Research using FPGA Platforms, Austin, 2006, [Online]. Available: http://www.cag.csail.mit.edu/warfp2006/submissions/suh-git.pdf.

[2] E.S. Chung, E. Nurvitadhi, J.C. Hoe, B. Falsafi, and K. Mai, "A Complexity-Effective Architecture for Accelerating Full-System Multiprocessor Simulations Using FPGAs", Proc. FCCM'2008, pp. 77-86.

[3] U.Y. Ogras, R. Marculescu, H.G. Lee, P. Choudhary, D. Marculescu, M. Kaufman, and P. Nelson, "NoC Prototyping Using FPGAs: Challenges and Promising Results in NoC Prototyping Using FPGAs", IEEE Micro Special Issue on Interconnects for Multi-Core Chips, vol. 27 , issue 5, Sept. 2007, pp. 86-95.

[4] K.N. Vikram and V. Vasudevan, "Hardware-software co-simulation of bus-based reconfigurable systems", Microprocessors and Microsystems, vol. 29, issue 4, May, 2005, pp. 133-144.

[5] D. Chiou, D. Sunwoo, J. Kim, et al., "The FAST Methodology for High-Speed SoC/Computer Simulation", Proc. IEEE/ACM Int. Conf. on CAD, San Jose, 2007, pp. 295-302.

[6] M. Koster and J. Teich, "(Self-)reconfigurable Finite State Machines: Theory and Implementation", Proc. DATE'2002, Paris, 2002, pp. 559-566.

[7] V. Sklyarov, Synthesis of Finite State Machines Based on Matrix LSI, Minsk, Science and Techniques, 1984, 271 p.

[8] V. Sklyarov, "Hierarchical Finite-State Machines and Their Use for Digital Control", IEEE Trans. on VLSI Syst., 1999, vol. 7, no. 2, pp. 222-228.

[9] T. Maruyama, M. Takagi, and T. Hoshino, "Hardware Implementation Techniques for Recursive Calls and Loop", Proc. 9th Int. Workshop on Field-Programmable Logic and Applications, 1999, pp. 450-455.

[10] V. Sklyarov, "FPGA-based implementation of recursive algorithms", Microprocessors and Microsystems. Special Issue on FPGAs: Applications and Designs, vol. 28/5-6, 2004, pp. 197-211.

[11] V. Sklyarov, I. Skliarova, and B. Pimentel, "FPGA-based implementation and comparison of recursive and iterative algorithms", Proc. FPL'2005, Tampere, Finland, 2005, pp. 235-240.

[12] S. Ninos and A. Dollas, "Modeling Recursion Data Structures for FPGA-based Implementations", Proc. FPL'08, Heidelberg, Germany, pp. 11-16, 2008.

[13] V. Sklyarov and I. Skliarova, "Design and Implementation of Parallel Hierarchical Finite State Machines", Proc. 2nd Int. Conf. on Communications and Electronics – HUT-ICCE'2008, Hoi An, Vietnam, June 2008, pp. 33-38.

[14] V. Sklyarov, I. Skliarova, B. Pimentel, and M. Almeida, "Multimedia Tools and Architectures for Hardware/Software Co-Simulation of Reconfigurable Systems", Proc 21st Int. Conf. on VLSI Design, Hyderabad, India, January 2008, pp. 85-90.

[15] M. Almeida, B. Pimentel, V. Sklyarov, and I. Skliarova, "Design Tools for Rapid Prototyping of Embedded Controllers", Proc. 3rd Int. Conf. on Autonomous Robots and Agents - ICARA'2006, Palmerston North, New Zealand, December 2006, pp. 683-688.

[16] V. Sklyarov, I. Skliarova, and A. Neves, "Modeling and Implementation of Automatic System for Garage Control", Proc. ICROS-SICE Int. Joint Conf. – ICCAS-SICE'2009, Fukuoka, Japan, August 2009, pp. 4295-4300.

[17] S.T. Soldado, FPGA Urban Traffic Control Simulation and Evaluation Platform, M.Sc. thesis, University of Aveiro, 2009, available at: http://www.ieeta.pt/~skl/Research/From2008/MSc/Sergio.pdf.

[18] P.P. Chu, FPGA prototyping by VHDL examples, Jonh Willey & sons, Inc, 2008.

A Compact Millimeter-Wave CMOS Bandpass Filter Using a Dual-Mode Ring Resonator

Sha Luo
Deparment of Electrical and Computer Engineering,
National University of Singapore,
Singapore
eleluos@nus.edu.sg

Aaron V. Do
Analog Design Department,
Marvell Asia Pte Ltd,
Singapore
doaaron82@gmail.com

Chirn Chye Boon, Lei Zhu, and Manh Anh Do
School of Electrical and Electronic Engineering,
Nanyang Technological University,
Singapore

Abstract— A stub-loaded dual-mode ring resonator is proposed to design a millimeter-wave bandpass filter using 0.18-μm CMOS Technology. By increasing the length of the open-circuited stub at the inner corner of the ring resonator, the even-mode resonant frequency is moved to a lower frequency to separate it from the odd-mode resonant frequency. Therefore, the center frequency of the passband has been shifted to a lower frequency to achieve a reduced-size filter design. At the same time, an additional transmission zero is brought in at the upper stopband that can be adjusted by the length of the stub as well. Finally, a 60 GHz bandpass filter is fabricated and characterized. The filter has achieved an ultra-compact size of 0.092 x 0.56 mm^2. The measured results show good passband performance with two visible transmission poles and three transmission zeros thereby verifying the design principle.

Index Terms — *stub-loaded, dual-mode, ring resonator, millimeter-wave, bandpass filter.*

I. INTRODUCTION

Recently, design of millimeter-wave CMOS-based filters has become very popular. Several millimeter-wave bandpass filters have been explored in 0.18-μm CMOS Technology using thin film microstrip (TFMS) structures [1]-[5]. By using TFMS structures, conventional microstrip theory is still appropriate to analyze the design structures [6], [7]. In [1], two millimeter-wave bandpass filters were designed using simple sinuous-shaped TFMS lines. The measured insertion losses are around 3.7 dB for the 60 GHz filter and 2.7 dB for the 65 GHz filter. However, the fractional bandwidths for both filters are more than 50%, which is too wide for wireless personal area network (WPAN). Two second-order classical Chebyshev filters using parallel-coupled lines were designed to operate at 60 and 77 GHz, respectively, which both have 10% 3-dB fractional bandwidths in [2]. The two filters suffer from high insertion loss, which is 9.3 dB in both passbands. Dual-mode ring resonators have been successfully implemented in millimeter-wave narrow-band bandpass filter

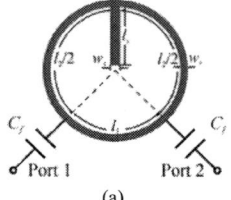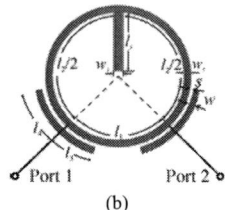

Fig. 1. Schematic of the proposed ring resonator with two distinct excitation structures. (a) By lumped capacitors. (b) By parallel-coupled lines

designs with low insertion loss, high filtering selectivity and compact size [3]-[5]. In [3], the ring filter was measured to have its center frequency at 64 GHz and a minimum in-band insertion loss of 4.9 dB. Its 3-dB fractional bandwidth is about 18.75%. The chip size is 1.148 x 1.49 mm^2. The ring filter reported in [4] operates at 70 GHz with a minimum in-band insertion loss of 3.6 dB. Its 3-dB fractional bandwith is around 25.71%. By folding the ring structure, the chip size is reduced to 0.65 x 0.67 mm^2. However, in both designs, additional matching networks with high impedance lines and pads were included, which enlarged the overall size and are also not practical when integrating with other devices on chip. Recently, a stepped-impedance dual-mode ring resonator was used to design a bandpass filter [5]. The stepped-impedance structure was formed using a pedestal, which was constructed from multiple vertically stacked finite metal planes connected to the ground plane by via holes. The designed filter achieves a passband with return loss better than 16 dB from 59 to 71 GHz. The minimum in-band insertion loss was measured to be 3.1 dB. The filter has a compact chip area of 0.30 x 0.24 mm^2. For all the filters reported in [3]-[5], one transmission zero appears at the each side of the desired passband.

In this paper, a stub loaded dual-mode ring resonator is proposed to design a 60 GHz narrow-band bandpass filter using a TFMS structure in 0.18 μm CMOS technology. Besides the transmission zero at the each side of the desired passband, the

978-1-61284-863-1/11 $26.00 © 2011 IEEE

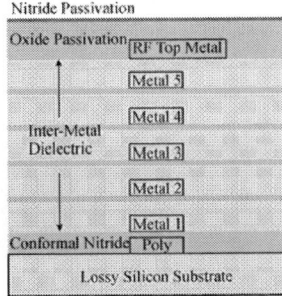

Fig. 2. Layer configuration of the TFMS structure in 0.18 μm CMOS

main novelty in this design is to bring a third transmission zero at the upper stopband of the filter and at the same time achieve an even more compact size by loading an open-circuited stub at the inner corner of the ring resonator. Finally, a prototype filter is designed and fabricated to verify design theory.

II. PRINCIPLE OF THE PROPOSED RING RESONATOR

The proposed dual-mode ring resonators with lumped capacitors (C_f) and parallel-coupled lines at the two excited ports are shown in Fig. 1(a) and (b), respectively. w_r is the width, and l_1 and l_2 are the lengths of the two paths of the ring. The open-circuited stub is loaded at the inner corner of the ring resonator with a width of w_s and a length of l_s. The two pairs of parallel-coupled lines have a width of w and a spacing of s. They are coupled to the ring resonator with lengths of l_3 and l_4, respectively. The layer configuration of a 0.18 μm CMOS technology with six-metal interconnects on a silicon substrate is shown in Fig. 2. The bottom metal layer (Metal 1) is used as a ground plane to minimize the electric field leaking into the silicon region; the top metal layer is used as a conductor plane to keep a maximum distance from the ground plane for high quality factor.

The equivalent circuit model of the ring resonator in Fig. 1(a) is developed as displayed in Fig. 3(a). Its even- and odd-mode circuits are illustrated in Fig. 3(b) and (c), respectively. Z_r and Z_s represent the characteristic impedances of the ring resonator and the open-circuit stub. θ_1 and θ_2 are the electrical lengths of the two paths in the ring; θ_s is the electrical length of the stub. The intrinsic even- and odd-mode resonant frequencies can be derived as algebraic equations as follows, respectively:

$$Z_r \cot(\frac{\theta_1 + \theta_2}{2}) + 2Z_s \cot \theta_s = 0 \qquad (1)$$

$$Z_r \tan(\frac{\theta_1 + \theta_2}{2}) = 0 . \qquad (2)$$

From (1) and (2), we observe that the emergence of the stub only influences the even-mode resonant frequency. By increasing the length of the stub, the fundamental even- and odd-mode resonant frequencies can be separated and make up the desired passband. As demonstrated in Fig. 3(d), the passband is stimulated with one or two poles when l_s increases from 200 to 400 μm, and the center

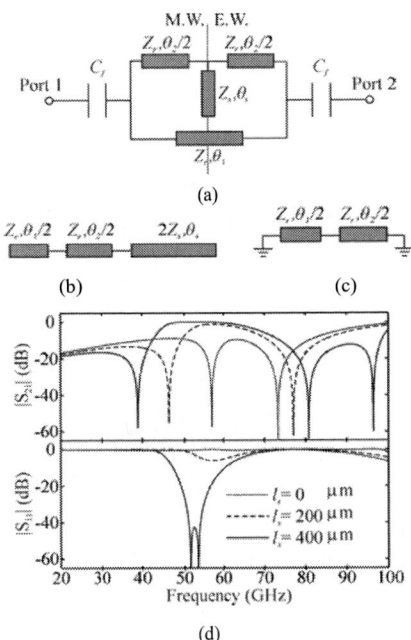

Fig. 3. Proposed ring resonator in Fig. 1(a). (a) The whole equivalent circuit with perfectly M.W. and E. W. at the diagonal line. (b) Even-mode circuit. (c) Odd-mode circuit. (d) Frequency responses of S_{21}-magnitudes for various stub lengths (l_s). (C_f=0.05 pf, w_r= w_s =1.5 μm, l_1=408 μm, and l_2=1880 μm)

frequency of the passband moves to a lower frequency. Meanwhile, two transmission zeros are observed at both sides of the passband. Moreover, there is one more transmission zero appearing at 96 GHz when the stub length is 400 μm.

The equivalent circuit of the proposed ring resonator in Fig. 1(b) is shown in Fig. 4(a). θ_3 and θ_4 are the lower and upper coupling sections, respectively. $Z_c = -2jZ_{0e}Z_{0o} / [(Z_{0e}+Z_{0o}) \tan(\theta_3 +\theta_4)]$, $N = (Z_{0e}+Z_{0o}) / (Z_{0e}-Z_{0o})$ and $Z = (Z_{0e}+Z_{0o})/2$, where Z_{0e} and Z_{0o} represent the even- and odd- mode characteristic impedances of the parallel-coupled lines, respectively. Fig. 4(b) shows the influences of the open-circuited stub on the transmission zeros and poles. For l_s = 400 μm, the third transmission zero appears at 97 GHz. As l_s increases from 420 to 440 μm, this zero moves from 94 to 91 GHz. With increasing stub length, only the even-mode resonant frequency moves to a lower frequency and the first two transmission zeros slightly shift away from the passband. The open-circuited stub brings in another transmission zero, which can be properly positioned to further improve the rejection in the upper stopband.

III. RESULTS AND DISCUSSION

Based on the above analysis, a dual-mode bandpass filter was designed on a 0.18 μm CMOS multi-layered structure. The folding technique was implemented to further reduce the overall size of the filter. Fig. 5(a) shows the physical layout of the designed filter. The microphotograph of the chip is shown in Fig.

978-1-61284-863-1/11 $26.00 © 2011 IEEE

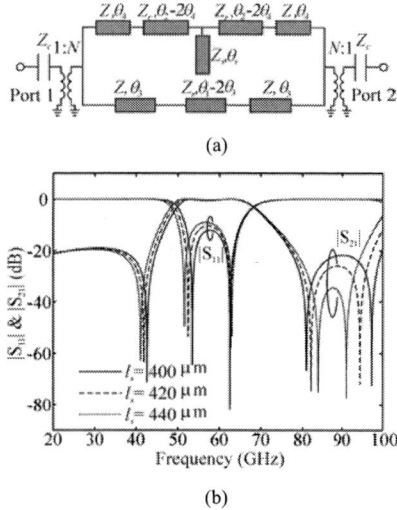

(a)

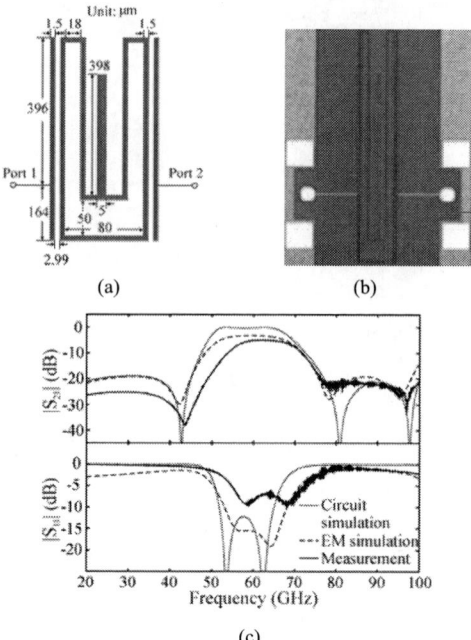

(b)

Fig. 4. Proposed ring resonator in Fig. 1 (b). (a) Equivalent circuit model. (b) Frequency responses of S-magnitudes under various stub lengths (l_s) ($w_r = w_s$ =1.5 μm, l_1=408 μm, l_2=1880 μm, s =2.99 μm, w =1.5 μm, l_3=164 μm, and l_4=396 μm).

Fig. 5. Fabricated filter. (a) Physical layout. (b) Microphotograph. (c) Simulated and measured S-magnitudes

5(b). To our best knowledge, this filter is the smallest reported CMOS 60 GHz filter, which occupies an area of 0.092x 0.56 mm^2. The filter was characterized on wafer after calibration with respect to the reference planes of the RF probes. The measurement data was further de-embedded to eliminate pad and feeding line influence, and other EM parasitic effects [8].

The simulated results from both circuit model and full-wave EM simulator [9], and the measured results after de-embedding are plotted together in Fig. 5(c). A reasonably good agreement between the three sets of results in the frequency range from 20 to 100 GHz is observed. The measured center frequency is at 62 GHz with a minimum in-band insertion loss of 4.9 dB. This insertion loss is mainly contributed by the conductor loss and the dielectric loss. The 3-dB passband is around 24 % with a frequency range from 55 to 70 GHz. The measured return loss is better than 7 dB in the passband. The three transmission zeros occur at 44, 77 and 97 GHz as expected. The rejection in the lower stopband is higher than 25 dB from DC to 47 GHz; the rejection in the upper stopband is greater than 21 dB from 76 to 100 GHz.

IV. CONCLUSION

In this work, a dual-mode stub-loaded ring resonator has been successfully implemented to design a 60 GHz bandpass filter using 0.18 μm CMOS technology. Conventional microstrip theory has been applied to develop the equivalent circuit model of the proposed ring resonator and to predict the filter performances. Finally, a 60 GHz bandpass filter has been fabricated and measured. The filter has achieved an ultra-compact size of 0.092 x 0.56 mm^2. Predicted results were verified experimentally, showing two transmission poles in the desired passband and three zeros in stopband.

[1] S. Sun, J. Shi, L. Zhu, S. C. Rustagi, and K. Mouthaan, "Millimeter-wave bandpass filters by standard 0.18-μm CMOS technology," *IEEE Electron Device Lett.*, vol. 28, No. 3, pp.220-222, Mar. 2007.

[2] L. Nan, K. Mouthaan, Y. -Z. Xiong, J. Shi, S. C. Rustagi, and B. -L. Ooi, "Design of 60- and 77-GHz narrow-bandpass filters in CMOS technology," *IEEE Trans. Circuit Syst. II: Express Brief,* vol. 55, no. 8, pp. 738–742, Aug. 2008.

[3] C. -Y. Hsu, C. -Y. Chen and H. -R. Chuang, "A 60 GHz millimeter-wave bandpass filter using 0.18-μm CMOS technology," *IEEE Electron Device Lett.*, vol.29, no.3, pp.246-248, Mar. 2008.

[4] C. -Y. Hsu, C. -Y. Chen and H. -R. Chuang, "70 GHz folded loop dual-mode bandpass filter fabricated using 0.18 μm standard CMOS technology," *IEEE Microw. Wireless Compon. Lett.*, vol. 18, no. 9, pp. 587-589, Sep. 2008.

[5] S. -C. Chang, Y. -M. Chen, S. -F. Chang, Y. -H. Jeng, C. -L. Wei, C. -H. Huang, and C. -P. Jeng, "Compact millimeter-wave CMOS bandpass filters using grounded pedestal stepped-impedance technique," *IEEE Trans. Microw. Tehory Tech.*, vol. 58, no. 12, pp. 3850-3858, Dec. 2010.

[6] G. E. Ponchak and A. N. Downey, "Characterization of thin film microstrip lines on polymide," *IEEE Trans. Compon., Packag., Manuf. Technol. B*, vol. 21, no. 2, pp. 171-176, May 1998.

[7] G. Prigent, E. Rius, F. L. Pennec, S. L. Magure, C. Quendo, G. Six, and H. Happy, "Design of narrow DBR planar filters in Si-BCB technology for millimeter-wave applications," ," *IEEE Trans. Microw. Tehory Tech.*, vol. 52, no. 3, pp. 1045-1050, Mar. 2004.

[8] E. P. Vandamme, D. M. M. -P, and C. V. Dinther, "Improved three-step de-embedding method to accurately account for the influence of pad parasitics in silicon on-wafer RF test-structures," *IEEE Electron Device Lett.*, vol.48, no.4, pp.737-742, Apr. 2001.

[9] *CST Microwave Studio, Computer Simulation Technology*, Wellesley Hills, MA 02481.

978-1-61284-863-1/11 $26.00 © 2011 IEEE

Transfer Function Analysis for Model Topology Determination of On-Chip Transmission Lines

Huang Wang
Key Laboratory of RF Circuits and Systems of Ministry of Education
Hangzhou Dianzi University
Hangzhou, China
School of Information Science and Technology
East China Normal University
Shanghai, China

Lingling Sun, Jun Liu, Jincai Wen, Zhiping Yu
Key Laboratory of RF Circuits and Systems of Ministry of Education
Hangzhou Dianzi University
Hangzhou, China
sunll@hdu.edu.cn

Abstract—**A novel method to determine the topology of the equivalent circuit model of on-chip transmission line by transfer function analysis is presented. The fitting capacities of the equivalent circuits with different number of 1-π segments are clearly evaluated through the calculation of their transfer functions. A broadband macromodel is proposed by rational approximation of measured network parameters. By comparing the poles and zeros of the transfer functions of the equivalent circuit model and macromodel, a reliable and efficient criteria for choosing model topology for transmission lines at various frequencies is developed. It is found that the number of poles and zeros provided by certain topology is a constant, while complex poles are responsible for the broadband fitting capacity of the model. Measured S-parameters of a coplanar waveguide (CPW) transmission line up to 50 GHz have verified that the proposed method is quite helpful in finding the simplest topology with sufficient accuracy. This transfer function analysis based method for model topology determination is also applicable to other passive devices, such as inductors and transformers.**

Keywords-transmission line; transfer function; passive devices; circuit topology; rational approximation

I. Introduction

The recent decade has witnessed the tremendous advancement of the radio frequency (RF) and even millimeter-wave (MMW) performance of CMOS, which potentializes the applications of silicon-based RF and MMW integrated circuits (ICs) [1]. Highly integration and extremely high operating frequencies make on-chip transmission lines widely used in matching networks, baluns, and VCOs [2], [3]. At millimeter wave frequency all interconnects are transmission lines to some extent. Therefore, fast and accurate compact model of on-chip transmission lines are desired in the emerging parasitic-aware RF and MMW circuit design tools.

Transmission lines are generally characterized by electromagnetic (EM) simulation [4] and measurement based models [5]–[7] which is directly obtained from the measurements of the on-chip test structures. Compared with the EM-based modeling approach, the measurement based modeling approach has more silicon-verified accuracy and

higher computational efficiency [5]. As transmission line can be electrically defined by the capacitance and shunt conductance (dielectric loss) of the dielectric materials per unit length [6], it can be fully characterized by cascading equivalent circuit segment for each unit length. These physics-based circuit models have advantage in length scalability but limitation in bandwidth [8]. However, the wave length based criteria for choosing model topology is an empirical method which lacks objectivity and universality for various transmission lines, model topologies and operating frequencies [7].

Macromodel (or black-box model) concerns the behavior of the device at the input-output ports, which is suitable for broadband modeling and can be synthesized by rational approximation to frequency-domain network parameters [8]. This is the main reason that this paper constructs a macromodel to determine the model topology.

Transfer function is a mathematical representation of the relation between the input and output of a continuous, linear, and time-invariant system in frequency domain. In electronic field, transfer function provides insight into the essential nature of the electric network. Thus, the equivalent circuit models of transmission lines can be analyzed by their transfer functions.

This paper is organized as follows: the transfer function analysis for on-chip transmission line models is presented in Section II, followed by experimental verification and discussions in Section III. Finally, the paper is concluded in Section IV.

II. Transfer Function Analysis for On-Chip Transmission Line Models

A. Transfer function analysis theory

Transfer function represents the input-output relation of a continuous, linear, and time-invariant electric network in frequency-domain. Poles and zeros of the transfer function are closely related to the topology and components of the network, and thus reveal the features of the network (e.g. bandwidth,

---

This work was supported by the 973 Program of China under Grant 2010CB327403 and the National Natural Science Foundation of China under Grant 60906015.

978-1-61284-863-1/11 $26.00 © 2011 IEEE

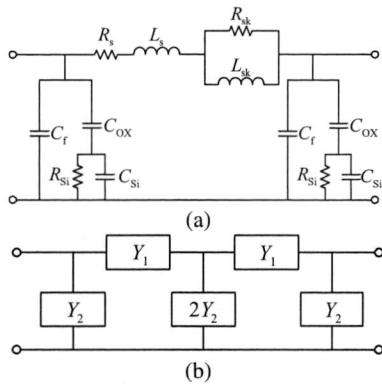

(a)

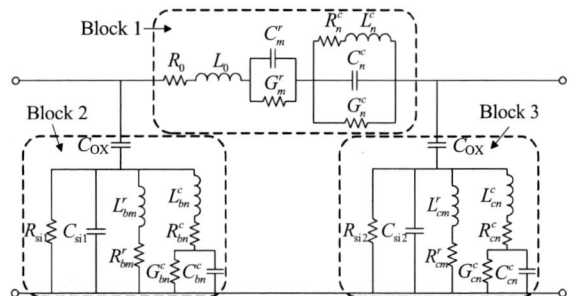

(b)

Figure 1. Equivalent circuit model of transmission line. (a) 1- segment of equivalent circuit model in [5]. (b) Schematic block diagram of 2- model.

Figure 2. Synthesized equivalent circuit of transmission line from measured S-parameters by VF.

gain and phase shift). The order of the transfer function or the number of its poles is determined by the complexity of the network (i.e. the number of nodes and energy-storage components such as inductors and capacitors). Obviously, transfer function is a useful tool for electric networks investigation, which is valid for the analysis of equivalent circuit models of passive devices.

B. Equivalent Circuit Model

The proposed method is applicable to various equivalent circuit models of transmission lines. Here, we just take the model in [5] for example as shown in Figure 1(a). In order to simplify the model construction, the series blocks of each π segment in the 2-π model depicted in Figure 1(b) are forced to be identical and so are the shunt blocks. This optimization is physically reasonable due to the symmetrical structure of the transmission line. Thus, the impedances of the blocks in Figure 1(b) are expressed as

$$
\left.
\begin{aligned}
Y_1 &= \frac{1}{R_s + sL_s + (sL_{sk}R_{sk})/(sL_{sk} + R_{sk})} \\
Y_2 &= sC_f + \frac{1}{1/sC_{ox} + R_{Si}/(1 + sC_{Si}R_{Si})}
\end{aligned}
\right\}. \tag{1}
$$

where $s = \quad + j$ is the variable of complex frequency domain. The Y-parameters of this circuit network are

$$
\left.
\begin{aligned}
Y_{22} &= \frac{Y_1^2 + 4Y_1Y_2 + 2Y_2^2}{2Y_1 + 2Y_2} \\
Y_{21} &= -\frac{Y_1^3 + 2Y_1^2Y_2}{2Y_1^2 + 6Y_1Y_2 + 4Y_2^2}
\end{aligned}
\right\}. \tag{2}
$$

Then the transfer function is obtained by

$$
H(s) = -\frac{Y_{21}}{Y_{22}} \tag{3}
$$

The above procedure has been implemented in a MATLAB program. Both the symbolic and numerical form (by substituting every element value of the circuit) of transfer functions are obtained simply by running this program. This transfer functions calculation method can be applied in a same manner to N-π models (N=1, 2, 3, …).

C. Macromodel by Rational Approximation

The popular macro-modeling method Vector fitting (VF) [9] is employed to obtains transfer function from measured frequency-domain responses with stability and passivity ensured [10]. A efficient method for model parameter extraction is proposed. The entire extraction procedure has been implemented by a MATLAB program. The final values of all parameters are obtained simply by running this program and no further optimization is needed. As redundant poles may cause ill-conditioned problem and high fitting error after passivity enforcement [10], the lowest order are employed in the approximation to achieve a sufficient accuracy, for instance, the RMS error is within 5%.

The fitting result of the impedance $Z(s)$ (Block 1 in Figure 2) by VF is expanded as follows

$$
\begin{aligned}
Z(s) &= d + se + \sum_{i=1}^{m} \frac{c_i^r}{s - a_i^r} + \sum_{i=1}^{n} \left(\frac{c_i^c}{s - a_i^c} + \frac{c_i^c}{s - \overline{a}_i^c} \right) \\
&= d + se + \sum_{i=1}^{m} \frac{c_i^r}{s - a_i^r} + \sum_{i=1}^{n} \frac{\lambda_i s + \gamma_i}{s^2 + \alpha_i s + \beta_i}
\end{aligned} \tag{4}
$$

where, d and e are real coefficients and the first two terms of (4) are optional in rational approximation, while α, β, λ, and γ are coefficients of the polynomials, a_i and c_i are and poles residues. The superscripts r and c stand for real and complex poles. The numbers of real poles and complex poles are m and n, respectively. In order to synthesize a lumped equivalent circuit by $Z(s)$, equation (4) is expressed by R, L, C, and G as

$$
\begin{aligned}
Z(s) &= R_0 + sL_0 + \sum_{i=0}^{m} \frac{1/C_i^r}{s + G_i^r/C_i^r} \\
&+ \sum_{i=0}^{n} \frac{(1/C_i^c)s + R_i^c/C_i^c L_i^c}{s^2 + (R_i^c/L_i^c + G_i^c/C_i^c)s + (1 + G_i^c R_i^c)/(C_i^c L_i^c)}.
\end{aligned} \tag{5}
$$

The synthesized parameter values are extracted by equaling (4) and (5) as the follows

$$
\left.
\begin{aligned}
R_0 &= d \\
L_0 &= e
\end{aligned}
\right\} \tag{6}
$$

978-1-61284-863-1/11 $26.00 © 2011 IEEE

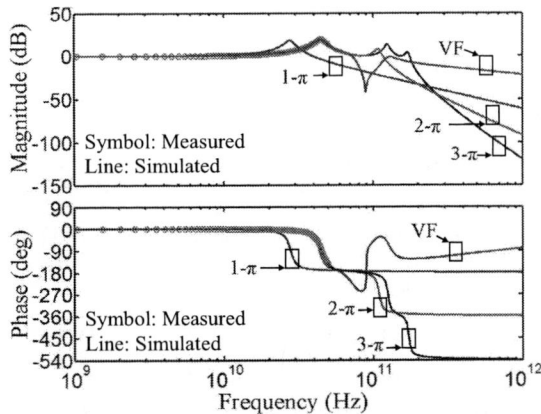

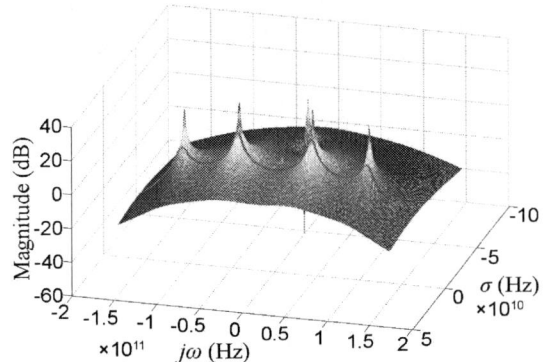

Figure 3. Bode diagram for comparison of the measured and simulated transfer functions by 1-π, 2-π, 3-π, and VF models of transmission line.

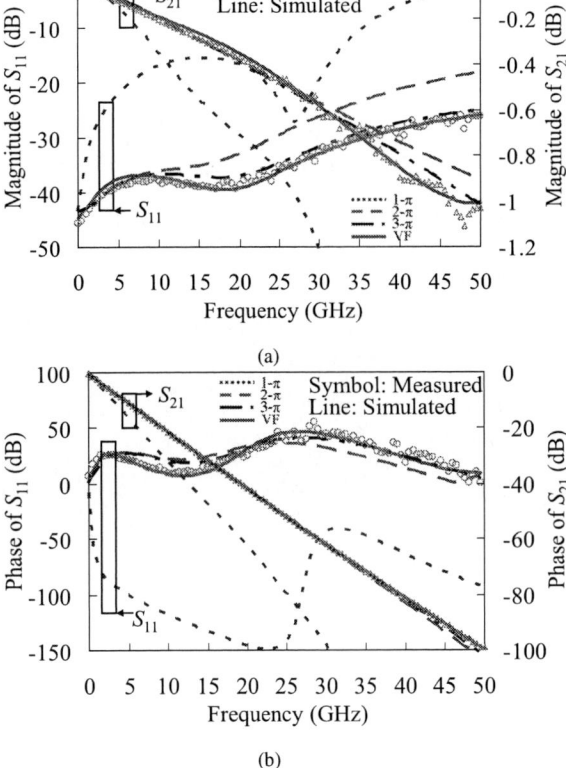

Figure 4. Surface of magnitude of transfer function for 2-π over complex frequency plane and the positive frequency half of the curve ($\omega > 0$, $\sigma = 0$) on the surface is the magnitude of transfer function in frequency plane.

$$\left. \begin{aligned} C_i^r &= 1/c_i^r \\ G_i^r &= -a_i^r/c_i^r \end{aligned} \right\} \quad (i = 1, 2, \cdots, m) \qquad (7)$$

$$\left. \begin{aligned} C_i^c &= 1/\lambda_i, \ G_i^c = C_i^c \left(\alpha_i - C_i^c \gamma_i \right) \\ R_i^c &= 1/\left(\beta_i/\gamma_i - G_i^c \right), \ L_i^c = R_i^c / \left(C_i^c \gamma_i \right) \end{aligned} \right\} \quad (i = 1, 2, \cdots, n). \quad (8)$$

All the circuit elements in Figure 2 can be simply extracted from impedance $Z(s)$ and admittance $Y(s)$ (Block 2 and Block 3 in Figure 2). This parameter extraction procedure has also been realized in the MATLAB environment. The transfer function is obtained by the procedure in Part B.

III. EXPERIMENTAL VERIFICATION AND DISCUSSIONS

For verification of the proposed method, test structure of a CPW is designed and fabricated employing a standard 90 nm RF CMOS technology. The width of the signal line, ground line, and the gap between them are all 10 μm. The lengths of the CPW is 800 μm after de-embedding. Two-port S-parameters up to 50 GHz were measured by GSG probes using an Agilent E8363B Network Analyzer and a CASCADE Summit probe station.

The bode diagram depicted in Figure 3 is the comparison of the measured and simulated transfer functions by 1-, 2-, 3-, and VF models of transmission line. The magnitude of the

Figure 5. Comparison of measured and simulated frequency responses of S-parameters of transmission line by 1-π, 2-π, 3-π, and VF. (a) Magnitude of S_{11} and S_{21}. (b) Phase of S_{11} and S_{21}.

TABLE I. POLES AND ZEROS ANALYSIS FOR TRANSFER FUNCTIONS OF TRANSMISSION LINE MODELS (c DENOTES COMPLEX CONJUGATE POLE OR ZERO PAIR) AND RMS ERRORS (%) BETWEEN MEASURED AND SIMULATED S-PARAMETERS

		1-π	2-π	3-π	VF
Zeros		2 (c 0)	4 (c 0)	16 (c 6)	8 (c 3)
Poles		4 (c 1)	8 (c 3)	22 (c 8)	10 (c 3)
Pairs of ECPs		1	2	3	3
mag	S_{11}	60.2	18.3	3.78	2.37
	S_{21}	56.3	9.22	6.60	5.62
ph	S_{11}	138	54.2	15.6	8.26
	S_{21}	36.6	7.81	5.34	6.41

measured transfer function (obtained by (3) from measured S-parameters) has one peak. Only 1- model can not trace this peak well. The maximum numbers of peaks provided by 1-, 2-, and 3- are 1, 2, 3, respectively, with the maximum phase shift -180, -360, and -540. The reason for this phenomena is that the peaks of the transfer function are formed by effective complex poles (ECPs) which are not cancelled by zeros, as observed in Figure 4, and each pair of ECPs causes a -180° phase shift, which means the model that provides more complex poles has better broadband fitting capacity.

Table I lists the numbers of poles and zeros provided by the four models, as well as the RMS errors between measured and simulated S-parameters depicted in Figure 5. The pairs of

ECPs of 1- , 2- , and 3- models are identical with the magnitude peaks of their transfer functions. Note that the numbers of poles and zeros increase very fast when cascading more segments. VF model has the highest accuracy but not the most poles and zeros, whose numbers are just between those of 2- and 3- models. Therefore, it is inferred that 3-model is adequate to model this CPW with sufficient accuracy up to 50 GHz, and more cascaded segments will not improve the fitting accuracy significantly. This is corroborated by the changing of RMS errors with more cascaded segments and the accuracy of 3- model is so close to that of VF model or even more accurate (the phase of S_{21}, 5.34<6.41). Thus, it is demonstrated that the proposed transfer function analysis based method for model topologies determination of on-chip transmission lines is reliable and efficient.

IV. CONCLUSION

An novel method to determine the topology of the equivalent circuit model of on-chip transmission line has been presented using transfer function analysis. A broadband macromodel of transmission line is constructed by rational approximation. By comparing the poles and zeros of the transfer functions of the equivalent circuit model and macromodel, a reliable criteria for choosing model topology for transmission lines at various frequencies is developed. The relationship between transfer function and circuit topology has been analyzed, which provides insight into the frequency response of transmission lines. Measured S-parameters of a CPW up to 50 GHz have verified that the proposed method is quite helpful in searching for the simplest topology with sufficient accuracy. Meanwhile, this method is also applicable to other passive devices, such as inductors and transformers [11], hence, may be useful for RFIC designs.

REFERENCES

[1] J.-W. Huang, C.-S. Wang, C.-K. Wang, and S.-H. Yeh, "Vertical-ground-plane transmission lines for miniaturized silicon-based MMICs," *IEEE RFIC Symp. Dig.*, pp. 563–566, Jun. 2007.

[2] H. Shigematsu, T. Hirose, and F. B. Rodwell, "Millimeter-wave CMOS circuit design," *IEEE Trans. Microwave Theory Tech.*, vol. 53, no. 2, pp. 472–477, Feb. 2005.

[3] H. Veenstra and M. G. M. Notten, "60GHz VCOs with transmission line resonator in a 0.25μm SiGe BiCMOS technology," *IEEE RFIC Symp. Dig.*, pp. 119–122, Jun. 2008.

[4] D. A. White and M. Stowell, "Full-wave simulation of electromagnetic coupling effects in RF and mixed-signal IC's using a time-domain finite-element method," *IEEE Trans. Microw. Theory Tech.*, vol. 52, no. 5, pp. 1404–1413, May 2004.

[5] X. Shi, J. Ma, B. H. Ong, K. S. Yeo, M. A. Do, and E. Li, "Equivalent circuit model for on-wafer CMOS RFICs," in *Proc. IEEE Radio and Wireless Conf.*, Sep. 2004, pp. 95–98.

[6] A. Sayag, D. Ritter, and D. Goren, "Compact modeling and comparative analysis of silicon-chip slow-wave transmission lines with slotted bottom metal ground planes," *IEEE Trans. Microw. Theory Tech.*, vol. 57, No. 4, pp. 840–887, Apr. 2009.

[7] X. Shi, J. Ma, S. Member, K. S. Yeo, M. A. Do, and E. Li, "Equivalent circuit model of on-wafer CMOS interconnects for RFICs," *IEEE Trans. Very Large Scale Integr. (VLSI) Syst.*, vol. 13, no. 9, pp. 1060–1071, Sep. 2005.

[8] S.-H. Min and M. Swaminathan, "Construction of broadband passive macromodels from frequency data for simulation of distributed interconnect networks," *IEEE Trans. Electromagn. Compat.*, vol. 46, no. 4, pp. 544–558, Nov. 2004.

[9] B. Gustavsen and A. Semlyen, "Rational approximation of frequency domain responses by vector fitting," *IEEE Trans. Power Del.*, vol. 14, no. 3, pp. 1052–1061, Jul. 1999.

[10] B. Gustavsen and A. Semlyen, "Enforcing passivity for admittance matrices approximated by rational functions," *IEEE Trans. Power Syst.*, vol. 16, no. 1, pp. 97–104, Feb. 2001.

[11] H. Wang, L. Sun, J. Liu, H. Zou, Z. Yu, and J. Gao, "Transfer function analysis and broadband scalable model for on-chip spiral inductors," *IEEE Trans. Microw. Theory Tech.*, vol. 59, no. 7, pp. 1696–1708, Jul. 2011.

Design Optimization of MOS Operational Amplifiers using Finite Difference Sensitivity*

Binbin Weng and Guoyong Shi

School of Microelectronics, Shanghai Jiao Tong University

Shanghai 200240, China

e-mail: {wengbinbin,shiguoyong}@ic.sjtu.edu.cn

Abstract—MOS analog circuit sizing is considered a highly complex task that requires experience and skills. Many methods proposed in the literature have arguable merits and limitations; none of them has become a widely recognized method being adopted in the design practice. This paper attempts to use a simply computable metric called the *finite difference sensitivity* computed mainly in the ac domain for the purpose of device sizing. Multiple design goals are formulated as a weighted optimization objective function and a gradient search is developed for optimizing the objective function. All constraints are subsumed in the objective function in the form of penalty functions. Experimental results show that such a simple formulation of circuit optimization is capable of finding satisfactory suboptimal sizing results which can be used for subsequent manual tuning or layout reference. The automated sizing procedure is compared to manual sizing and is demonstrated that the auto-sizing scheme has a better capability in balancing the multiple design objectives.

Index Terms—analog design automation, finite-difference sensitivity, MOS operational amplifier, optimization, sizing.

I. INTRODUCTION

The area of analog design automation has been evolving very slowly due to a number of self-evident reasons. The art of analog design has reached its maturity in the form of manual design quite early and this art of practice has been inherited from one generation to another without much revolutionary change. Despite the constantly advancing technology, the art of simulation-based analog design practice largely remains unchanged. Although the analog design practice is gradually becoming a standardized procedure, it still takes a beginner many years to master the essence of the design art. The lack of design automation aids in analog design has inspired many CAD researchers to develop some easy to use tools with interactive functionalities. It is now widely recognized that design automation tools must co-exist with the experienced designers in the analog design world.

Early works that centered around analog design synthesis are [1]–[12], among others. Typical methods used in these works are: 1) Knowledge-based optimization, which requires the construction of knowledge database incorporating skilled designers' experience. 2) Simulation-based optimization, which requires extensive simulations (transient, ac, noise, etc.) 3) Analytical equation-based optimization, which relies

*This research was supported by the National Natural Science Foundation of China (Grant No. 60876089).

on a symbolic simulator developed in the early days. The optimization methods employed are typically simulated annealing (SA) or geometric programming (GP). Using SA for analog optimization [7] is considered too computationally expensive because the analog search spaces are typically continuous and multidimensional. The once popular GP method [11], [12] also encounters difficulty when dealing with complicated MOS device models and design constraints that are not always convertible to posynomials.

Although those optimization methods proposed in the early literature with quite complexity can generate synthesized circuits with certain degree of optimality, it is demonstrated in this paper that very simple formulation of design optimization with a simple search strategy can achieve good optimization as well. This paper proposes to use a simply computable quantity called the *finite-difference sensitivity*, which is available just by using a standard circuit simulator like HSPICE. Although symbolically computed ac small-signal sensitivity can be useful for analog circuit optimization [13]–[15], it turns out that the dc large-signal sensitivity is also crucial for sensitivity-based optimization search. When the large-signal dc sensitivity is not available by a symbolic simulator, the *finite-difference sensitivity* can be used as a substitute for the purpose of optimization.

Computation techniques for ac response sensitivity are reviewed in section II. A formulation of design optimization that is tractable by the *finite difference* sensitivity is presented in section III. Experimental results on automatically sizing two operational amplifier (op-amp) circuits then are reported in section IV, where an auto-sizing result is compared to that of manual sizing. Conclusion is made in section V.

II. COMPUTATION OF AC RESPONSE SENSITIVITY

In analog design practice, sensitivity is a useful metric that helps the designer judge how the circuit performance depends on the small change of one or more circuit parameters. The device sizes in one MOS analog circuit are among the important circuit parameters to be tuned for optimal circuit performance. Hence, it would be of great importance to compute the sensitivity of a circuit performance metric with respect to (w.r.t) the MOS device sizes and other element values.

The standard sensitivity definition is given as

$$\text{Sens}(h, p) = \frac{\partial \ln h(p)}{\partial \ln p}, \tag{1}$$

where $h(p)$ is a performance metric as a function of the parameter p.

Performance metrics are typically defined in both the time-domain and the frequency-domain, the computation of their sensitivities would require computation methods with different computational complexities. An analog design cycle would normally need a large amount of repetitive computations of the performance metric sensitivities. Hence, the efficiency of sensitivity computation becomes an issue. In general, a time-domain sensitivity requires more computation than a frequency-domain sensitivity. Since a large portion of design performance metrics are specified in the frequency-domain, using the ac small-signal sensitivity for the purpose of auto-sizing can save lots of computation.

Let $H(s)$ be the frequency response function of a network. We would like to compute the sensitivity of $H(s)$ w.r.t. any MOSFET size W_k. Since $H(s)$ results from small-signal analysis, we must first run an operating point analysis to get the small-signal MOSFET parameter values, say, p_i^k, $i = 1, \cdots, m$, where the superscript k indicates the kth MOSFET and m is the number of small-signal model parameters for each MOSFET. If one would like to compute the sensitivity of $H(s)$ w.r.t. the kth MOSFET size W_k, then the following formula can be used

$$\text{Sens}(H(s), W_k) = \sum_{i=1}^{m} \text{Sens}(H(s), p_i^k) \, \text{Sens}(p_i^k, W_k), \tag{2}$$

where $\text{Sens}(H(s), p_i^k)$ is the sensitivity of $H(s)$ w.r.t. the small-signal parameter p_i^k of the kth MOSFET and $\text{Sens}(p_i^k, W_k)$ is the sensitivity of p_i^k w.r.t. the size W_k.

It seems that formula (2) is well-defined, but actually something is missing in the strict sense. One should be aware that the change of any device size would in general alter the circuit biasing condition, which in turn changes the small-signal parameter values of other nonlinear devices in the same circuit. Only all MOS devices in a circuit are in their saturation regions can we approximately use formula (2) for sensitivity computation [14], [15].

In the scenario of automatic sizing, it is restrictive to assume that all MOS devices are in saturation at the beginning. In that sense, the dc sensitivity with respect to the device sizes must be considered. However, not all circuit simulators implement the computation of dc sensitivity w.r.t. user specified device parameter such as W_k. To circumvent the computation difficulty, we propose to use the finite difference sensitivity defined by

$$\text{Sens}(H(s, w), w) = \frac{H(s, w + \Delta w) - H(s, w)}{\Delta w} \cdot \frac{w}{H(s, w)}, \tag{3}$$

where Δw is a small variation of the selected circuit parameter.

III. SIZING BY FINITE DIFFERENCE SENSITIVITY

Automatic MOSFET sizing can be formulated as a constrained optimization problem with the design specifications and device size limits as the constraints. Multiple design specs must be considered for an analog circuit. For op-amp design, typical design metrics are dc gain, phase margin (PM), unity-gain frequency, slew rate, chip area, and static power, etc., with alterable priorities. Some of the design metrics can be obtained directly from the frequency response $H(s)$, while some are available indirectly from SPICE simulation, such as a stage supply current that is used to estimate the slew rate.

The following design spec variables are considered in this work: dc-gain (g), unity-gain frequency (f), phase-margin (ϕ), static power (p), chip area (a), and slew rate (sr). Others also can be added with appropriate conversion. We require that the following inequalities must be satisfied by the spec variables:

$$g > G, \; f > F, \; \phi > \Phi, \; p < P, \; a < A, \; sr > SR, \tag{4}$$

where the upper-case letters are the designer given performance (lower or upper) bounds.

The auto-sizing objective function is then defined as follows

$$J(\mathcal{W}) := \alpha_g(g - G)^2 + \alpha_f(f - F)^2 + \alpha_\phi(\phi - \Phi)^2 + \\ \alpha_p(p - P)^2 + \alpha_a(a - A)^2 + \alpha_{sr}(SR - sr)^2, \tag{5}$$

where $\mathcal{W}$ is the vector containing all MOS device sizes and other compensation element values, and the coefficients α_g etc. are defined by

$$\alpha_g = \frac{(G - \hat{g})^+}{G}, \quad \alpha_f = \frac{(F - \hat{f})^+}{F}, \quad \alpha_\phi = \frac{(\Phi - \hat{\phi})^+}{\Phi},$$
$$\alpha_p = \frac{(\hat{p} - P)^+}{P}, \quad \alpha_a = \frac{(\hat{a} - A)^+}{A}, \quad \alpha_{sr} = \frac{(SR - \hat{sr})^+}{SR}.$$

Here we have used the notation $x^+ = \max\{x, 0\}$. The variables with cap ($\hat{g}$, etc.) are those just solved in the previous iteration step. The meaning of coefficients is obvious; when the spec constraint is not satisfied, the corresponding coefficient is nonzero, otherwise, it is set to zero. The optimization goal is to minimize the objective function $J(\mathcal{W})$.

Since the MOSFET sizes must be constrained as well, we set $W_k \in (W_{min}, W_{max})$ for all k. Define a function

$$g^2(W) := [(W_{min} - W)^+]^2 + [(W - W_{max})^+]^2. \tag{6}$$

We add the penalty term $\sum_{i=1}^{m} M_i g^2(W_i)$ to the objective function $J(\mathcal{W})$ to take into the consideration of the size constraints, where $M_i > 0$ are the penalty parameters properly chosen. Then the modified objective function becomes

$$\tilde{J}(\mathcal{W}) := J(\mathcal{W}) + \sum_{i=1}^{M} M_i g^2(W_i), \tag{7}$$

where M is total number of MOSFETs in the circuit.

The power is defined by $p := \left(\sum_{i=1}^{B} I_i \right) \times V_{dd}$, where I_i's are the stage currents supplying one path from V_{dd} to V_{ss}. The area is defined by $a := \left(\sum_{i=1}^{M} W_i \right) \times L$, where W_i's are the MOSFET widths and L is the effective channel length. The

978-1-61284-863-1/11 $26.00 © 2011 IEEE

slew rate is approximated by the standard formula $sr = I/Cc$, where I is a quiescent stage current depending on the circuit configuration.

Minimization of the objective function $\widetilde{J}(\mathcal{W})$ defined in (7) is carried out by a gradient search, which uses the finite difference gradient vector

$$\nabla \widetilde{J}(\mathcal{W}) := \left[\frac{\partial \widetilde{J}(\mathcal{W})}{\partial W_i} \right] \approx \left[\frac{\widetilde{J}(W_i + \Delta W_i) - \widetilde{J}(W_i)}{\Delta W_i} \right], \quad (8)$$

which is a length-P vector (called *size vector*) if P parameters are considered in the optimization. The gradient search direction is the negative of the gradient vector, i.e.,

$$\mathcal{W}^{(k+1)} = \mathcal{W}^{(k)} - \lambda_k \nabla \widetilde{J}(\mathcal{W}^{(k)}), \quad (9)$$

where k indicates the iteration step and $\lambda_k > 0$ is the kth search step size.

Along the gradient direction, we use Armijo-Goldstein one-dimensional (1-D) search criterion [16], which is described by the following steps. Let (a, b) be the 1-D search interval and ρ be the parameter in the Armijo-Goldstein criterion. Let $\widetilde{J}_k := \widetilde{J}(\mathcal{W}^{(k)})$. Let $d_k := \nabla \widetilde{J}(\mathcal{W}^{(k)})$ be the gradient vector.

Step 1. Choose $\lambda_0 > 0$, $\rho \in (0, 0.5]$, $a = 0$, $b = 2\lambda_0$, and the initial $\mathcal{W}^{(0)}$. Calculate the initial objective function and its gradient d_0.

Step 2. Update the size vector. Calculate the new objective function.

Step 3. If $\widetilde{J}_{k+1} > \widetilde{J}_k - \rho\lambda\|d_k\|^2$, set $b = \lambda$, $\lambda = (a+b)/2$. Go to Step 4.
Else If $\widetilde{J}_{k+1} \leq \widetilde{J}_k - (1-\rho)\lambda\|d_k\|^2$, set $a = \lambda$, $\lambda = (a+b)/2$. Go to Step 4.
Else go to step 5.

Step 4. If $|a - b| < 10^{-8}$, go to Step 5. Else go to Step 2.

Step 5. Quit 1-D search.

IV. EXPERIMENTAL RESULTS

The two op-amp circuits given in Figs. 1 and 2 are used as the test circuits. Simulation uses a $0.18\mu m$ process technology. The supply voltage is $V_{dd} = 1.8V$.

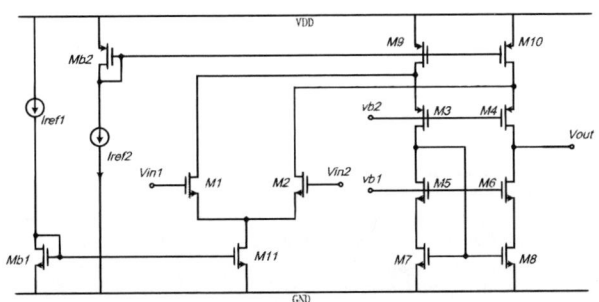

Fig. 1.   One-stage cascode MOS op-amp (test circuit 1).

The first circuit is a one-stage cascode op-amp. Its performance specification is given in the second row of Table I and the initial device sizes are given in the second column of Table II. The automatic sizing procedure starts from the initial device

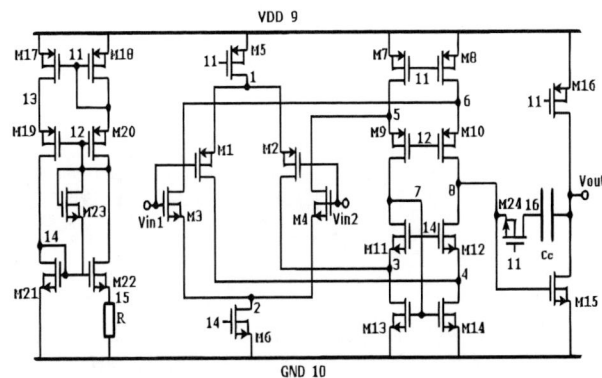

Fig. 2.   Two-stage cascode MOS op-amp (test circuit 2).

TABLE II
AUTOMATIC SIZING RESULT FOR TEST CIRCUIT 1.

Design parameters	Initial Value	Final Value
$(W/L)_{1,2}$	$45\mu/0.5\mu$	$32.6\mu/0.5\mu$
$(W/L)_{3,4}$	$45\mu/0.5\mu$	$27.9\mu/0.5\mu$
$(W/L)_{5,6}$	$45\mu/0.5\mu$	$16.8\mu/0.5\mu$
$(W/L)_{7,8}$	$45\mu/0.5\mu$	$3.12\mu/0.5\mu$
$(W/L)_{b1}$	$1.5\mu/0.5\mu$	*unchanged*
$(W/L)_{b2}$	$1.5\mu/0.5\mu$	*unchanged*
$(W/L)_{9,10}$	$40\mu/0.5\mu$	*unchanged*
$(W/L)_{11}$	$60\mu/0.5\mu$	*unchanged*
C_L	$20pF$	*unchanged*

sizes and ends after 12 iterations with all spec metrics satisfied. The generated sizes are listed in the third column of Table II. Note that the sizes of some biasing MOS devices are pre-chosen and kept unchanged during optimization. Also the two biasing currents I_{ref1} and I_{ref2} are kept unchanged at $10\mu A$.

The auto-sizing results of the second circuit in Fig. 2 are given in Tables III and IV. After 13 iterations, the optimized device sizes satisfying the performance spec are listed in the third column of Table IV with the circuit performance listed in the forth row of Table III. For comparison this circuit is manually sized as well. We admit that the manual sizing results are very much dependent on the designer's experience and skill. In this experiment it is assumed that the designer is inexperienced and uses the tutorial design procedure introduced in [17].

The performances corresponding to the three manual sizing results are listed in Table III as well for the purpose of comparison. We observe that manual sizing can only achieve part of the specs in each attempt but not all, while the automated sizing procedure is capable of simultaneously achieving the multiple design goals.

V. CONCLUSION

A simple optimization procedure is formulated in this work for analog sizing. The search algorithm uses the finite difference sensitivity as the gradient and the Armijo-Goldenstein criterion as the 1-D search. It is demonstrated that the multi-

TABLE I
PERFORMANCE SPEC AND AUTO-SIZING RESULT FOR TEST CIRCUIT 1.

Spec	DC Gain (dB)	f_{GBW} (Hz)	PM (°)	Power (mW)	Area (μm^2)	Slew Rate ($V/\mu s$)
Target	> 70	> 200M	> 60	< 1.5	< 40	> 20
Initial	53.5	185.4M	64.34	1.244	88.5	28.2
Final	69.88	204M	72.46	1.14	40.33	29.5

TABLE III
PERFORMANCE SPEC AND RESULTS BY AUTO-SIZING AND MANUAL SIZING (TEST CIRCUIT 2).

Spec	DC Gain (dB)	f_{GBW} (Hz)	PM (°)	Power (mW)	Area (μm^2)	Slew Rate ($V/\mu s$)
Target	> 100	> 100M	> 60	< 2	< 500	> 10
Auto (initial)	61.76	5.1M	99.38	0.7931	477.51	16.857
Auto (final)	**103.6**	**107.3M**	**60.91**	**1.524**	**461.6**	**26.58**
Manual attempt 1	79.12	78.0M	107	0.7744	615.5	10.12
Manual attempt 2	127.3	98.6M	45.3	1.671	704.2	17.48
Manual attempt 3	103.6	101.3M	60.91	1.924	648.15	16.58

TABLE IV
AUTOMATIC SIZING RESULT FOR TEST CIRCUIT 2.

Design parameters	Initial value	Final value
$(W/L)_{1,2}$	$60\mu/0.5\mu$	*unchanged*
$(W/L)_{3,4}$	$20\mu/0.5\mu$	*unchanged*
$(W/L)_5$	$24\mu/0.5\mu$	*unchanged*
$(W/L)_6$	$4\mu/0.5\mu$	*unchanged*
$(W/L)_{7,8}$	$45\mu/0.5\mu$	$49.85\mu/0.5\mu$
$(W/L)_{9,10}$	$45\mu/0.5\mu$	$45.22\mu/0.5\mu$
$(W/L)_{11,12}$	$45\mu/0.5\mu$	$47.46\mu/0.5\mu$
$(W/L)_{13,14}$	$45\mu/0.5\mu$	$12.91\mu/0.5\mu$
$(W/L)_{15}$	$126\mu/0.5\mu$	$167.8\mu/0.5\mu$
$(W/L)_{16}$	$129\mu/0.5\mu$	*unchanged*
$(W/L)_{17,18}$	$59\mu/0.5\mu$	*unchanged*
$(W/L)_{19,20}$	$59\mu/0.5\mu$	*unchanged*
$(W/L)_{21}$	$10\mu/0.5\mu$	*unchanged*
$(W/L)_{22}$	$40\mu/0.5\mu$	*unchanged*
$(W/L)_{23}$	$10\mu/0.5\mu$	*unchanged*
$(W/L)_{24}$	$2\mu/0.5\mu$	*unchanged*
$Rref$	$1.24k$	*unchanged*
C_C	$2pF$	*unchanged*
C_L	$10pF$	*unchanged*

objective optimization with penalty terms is effective for MOS circuit device sizing.

REFERENCES

[1] M. G. R. Degrauwe, O. Nys, and E. Dijkstra, *et al.*, "IDAC: An interactive design tool for analog CMOS circuits," *IEEE J. of Solid State Circuits*, vol. SC-22, pp. 1106–1115, Dec. 1987.

[2] W. Nye, D. C. Riley, , A. L. Sangiovanni-Vincentelli, and A. L. Tits, "DELIGHT.SPICE: An optimization-based system for the design of integrated circuits," *IEEE Trans. on Computer-Aided design*, vol. 7, no. 4, pp. 501–519, April 1988.

[3] M. G. R. Degrauwe, B. L. Goffart, and C. Meixenberger, *et al.*, "Towards an analog system design environment," *IEEE J. of Solid State Circuits*, vol. 24, no. 3, pp. 659–671, June 1989.

[4] F. El-Turky and E. E. Perry, "BLADES: An artifical intelligence approach to analog circuit design," *IEEE Trans. Computer-Aided Design*, vol. 8, pp. 680–692, June 1989.

[5] R. Harjani, R. A. Rutenbar, and L. R. Carley, "OASYS: A framework for analog circuit synthesis," *IEEE Trans. Computer-Aided Design*, vol. 8, pp. 1247–1265, Dec 1989.

[6] G. E. Gielen, H. Walscharts, and W. Sansen, "ISAAC: a symbolic simulator for analog integrated circuits," *IEEE J. Solid-State Circuit*, vol. 24, no. 6, pp. 1587–1596, December 1989.

[7] ——, "Analog circuit design optimization based on symbolic simulation and simulated annealing," *IEEE J. Solid-State Circuit*, vol. 25, no. 3, pp. 707–713, June 1990.

[8] H. Y. Koh, C. H. Séquin, and P. R. Gray, "OPASYN: A compiler for CMOS operational amplifiers," *IEEE Trans. on Computer-Aided Design*, vol. 9, no. 2, pp. 113–125, Feb. 1990.

[9] F. Medeiro, F. V. Fernandez, R. Dominquez-Castro, and A. Rodriguez-Vazquez, "A statistical optimization based approach for automated sizing of analog cell," in *Proc. IEEE/ACM International Conference on Computer-Aided Design*, 1994, pp. 594–597.

[10] E. S. Ochotta, R. Rutenbar, and L. R. Carley, "Synthesis of high-performance analog circuits in ASTRX/OBLX," *IEEE Trans. on Computer-Aided Design*, vol. 15, pp. 273–294, March 1996.

[11] M. Hershenson, S. Boyd, and T. Lee, "Optimal design of a CMOS op-amp via geometric programing," *IEEE Trans. on Computer-Aided Design of Circuits and Systems*, vol. 20, no. 1, pp. 1–21, Jan. 2001.

[12] P. Mandal and V. Visvanathan, "CMOS op-amp sizing using a geometric programing formulation," *IEEE Trans. on Computer-Aided Design of Circuits and Systems*, vol. 20, no. 1, pp. 22–38, Jan. 2001.

[13] G. Shi and X. Meng, "Variational analog integrated circuit design by symbolic sensitivity analysis," in *Proc. International Symposium on Circuits and Systems (ISCAS)*, Taiwan, China, May 2009, pp. 3002–3005.

[14] D. Ma, G. Shi, and A. Lee, "A design platform for analog device size sensitivity analysis and visualization," in *Proc. Asia Pacific Conference on Circuits and Systems (APCCAS)*, Malaysia, Dec. 2010, pp. 48–51.

[15] X. Li, H. Xu, G. Shi, and A. Tai, "Hierarchical symbolic sensitivity computation with applications to large amplifier circuit design," in *Proc. International Conference on Circuits and Systems (ISCAS)*, Rio de Janeiro, Brazil, May 2011, pp. 2733–2736.

[16] P. Wolfe, "Convergence conditions for ascent methods," *SIAM Review*, vol. 11, no. 2, pp. 226–235, 1969.

[17] G. Palmisano, G. Palumbo, and S. Pennisi, "Design procedure for two-stage CMOS transconductance operational amplifiers: A tutorial," *Analog Integrated Circuits and Signal Processing*, vol. 27, pp. 179–189, 2001.

Automated Synthesis Design Flow of Power Converter Circuits Aimed at SOC Applications

Hsin-Yu Luo, Hsiu-Wen Li*, Long-Ching Yeh and Chien-Nan Jimmy Liu.

Department of Electrical Engineering, National Central University, Taiwan, ROC

{985201030, 945401024}@cc.ncu.edu.tw and {lcyeh916, jimmy}@ee.ncu.edu.tw *the corresponding author

Abstract—**In this paper, we propose a power converter synthesis design flow aimed at SOC applications. A buck DC-DC converter and a low dropout (LDO) linear regulator, both with controllers are studied. We apply both the knowledge-based and the simulation-based methods in the proposed flow and they lead to an accurate result when it is compared with the design specification. Demonstration cases validate our work.**

I- INTRODUCTION

Recent development of SOC integrates the digital and the analog/RF circuits, as well as the power processing circuit on a chip. The computer aid design tools for SOC have provided a complete and successfully support for the digital circuit design already. However, designers still need to adjust the analog and the RF circuit designs manually. Moreover, on chip power processing circuits to power different blocks of a SOC chip become a trend and their designs become very complicated [1]. Therefore, since the electronic industry is concerned about time to market, this may require the EDA tool to assist those designs.

TABLE I shows the comparisons of linear regulators and switching converters [2]. The buck converter can be applied to portable electronic products with high efficiency, like LCD-TVs, internet communication chips and cell phones. However, the inductor in the buck converter occupies a large area and it defeats the purpose of a SOC design. To reduce the size of switching converters, one can integrate the switches and the controller circuit in the SOC but leave the inductor and capacitors for off-chip connections. In the case of LDO, its efficiency is dominated by the input voltage and the output voltage difference, and thus the low drop out voltage is necessary.

To combine the advantages of both the switching converter and the LDO's, one can connect a switching converter followed by a LDO to improve the fast transient response, such as a CPU load under a demand of intensive calculation [3]. An

LDO alone can also be used for sensitive circuits, like audio amplifiers, analog and RF circuits. Because the buck converter and the LDO are often used as the power conversion units in SOC design, we select these two circuits as the test vehicle of automatic synthesis flows in this paper.

II- BUCK CONVERTER SYNTHESIS FLOW

Fig. 1 shows the buck converter circuit and Fig. 2 shows the buck converter synthesis flow. We use the closed form equations to calculate the design parameters. We also use the closed form equations to do the small signal analysis where the equations are the linearized equations unique to the switching power converter topology. Finally, we add HSpice to simulate the entire circuit to verify the design. As shown in Fig. 2, it starts with the input of specs listed as follows: the supply voltage (V_i), the output voltage (V_o), the max load current ($I_{o,max}$), the ripple current ($I_{L,pp}$), the ripple voltage ($V_{o,pp}$), the switching frequency (f_s) and the target efficiency. Next, steady state analysis is carried out to find the corresponding parameters of the buck converter in steady state operation, such as: the duty cycle (D), the inductance (L), the capacitance (C), and the width of the MOSs (W_p & W_n) as switches. The small signal analysis is then carried out. The control circuit with a PID controller is also synthesized at this time and its small signal model is derived, the net-list is then generated before the final simulation of the entire circuit with HSpice.

Among these parameters, the current ripple and voltage ripple represent slight variations of inductor current and output capacitor voltage state variables during switching respectively [4], their functions are given by equations (1) and (2).

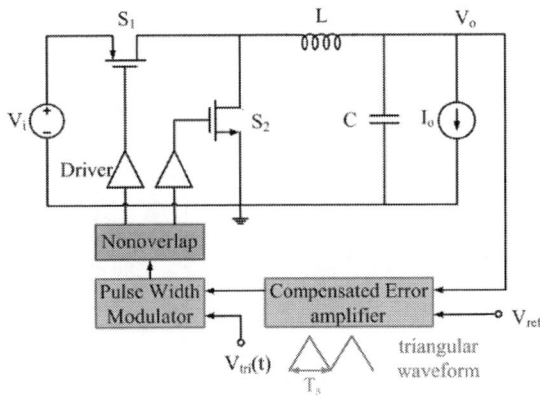

Figure 1. Architecture of buck converter

TABLE I.   COMPARISONS OF DC-DC CONVERTERS

Parameters	Linear Regulator	Switching Converter
Efficiency	Low	High
Power Rating	Medium	High
Size(PCB Real Estate)	Compact	Large
Cost and Complexity	Low	High
Noise	Low	High

978-1-61284-863-1/11 $26.00 © 2011 IEEE

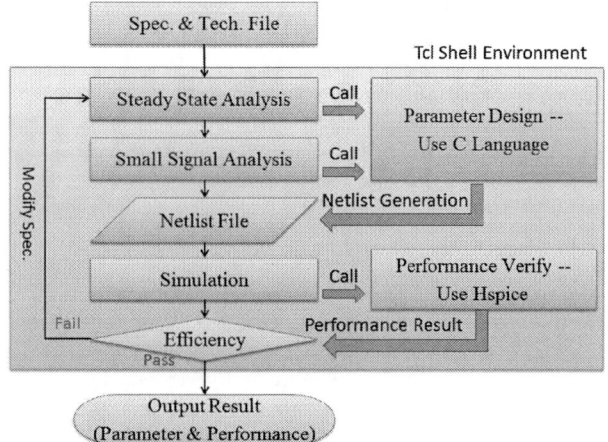

Figure 2.   The buck converter synthesis flow

$$\Delta I_L = i_L(DT_s) - i_L(0) = \frac{V_o(1-D)T_s}{L} \qquad (1)$$

$$\Delta V = \frac{\Delta Q}{C} = \frac{1}{C}\frac{1}{2}\frac{T_s}{2}\frac{\Delta I_L}{2} = \frac{\Delta I_L}{8Cf_s} \qquad (2)$$

The required component values of the inductor and the capacitor are given by equations (3) (4). The converter efficiency is given by equation (5).

$$L \geq \frac{V_o(1-D)T_s}{\Delta I_{L,pp}f_s} \qquad (3)$$

$$C \geq \frac{\Delta I_L}{8\Delta V_{o,pp}f_s} \qquad (4)$$

$$\eta = \frac{P_o}{P_o + P_{loss}} \qquad (5)$$

After the selection of the inductor and capacitor components from the vendors according to their spec., we include their series resistance for the efficiency calculation. Furthermore, the on-resistance of the MOS switch is also calculated. The equivalent circuit of buck converter of Fig. 3 is used for the efficiency calculation. r_p and r_n are the MOSs' on-resistance, r_{dcr} is the DC resistance of the inductor, and r_{esr} is the equivalent series resistance of the capacitor C.

The power loss P_{loss} is given by equation (6). P_{s1}, P_{s2}, P_{ind} and P_{cap} are the power loss of S_1, S_2, L and C respectively.

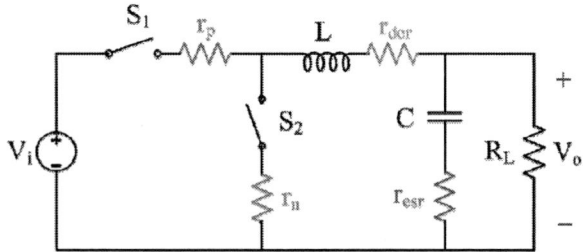

Figure 3.   The equivalent circuit of buck converters

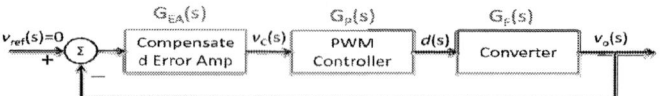

Figure 4.   Feedback control system of buck converter

$$P_{loss} = DP_{s1} + (1-D)P_{s2} + P_{ind} + P_{cap} \qquad (6)$$

The whole circuit analysis with the controller [5] is shown in Fig. 4.

The open loop gain ($G_{OL}(s)$) is given by equation (7).

$$G_{OL}(s) = G_F(s)G_P(s)G_{EA}(s) \qquad (7)$$

The Buck converter's transfer function $G_F(s)$ is given by equation (8) which consists of L and C mainly.

$$G_F(s) = \frac{\frac{1}{LC}}{s^2 + \frac{1}{LC}} \qquad (8)$$

The transfer function of the PWM modulator $G_P(s)$ is given by equation (9) where d(s) is the duty cycle, and $v_c(s)$ is the error amplifier's output voltage, both in the s-domain.

$$G_P(s) = \frac{d(s)}{v_c(s)} \qquad (9)$$

$G_{EA}(s)$ is the PID compensator transfer function.

After we obtain all the coefficients needed and the small signal analysis using equations (1) to (9), the net-list of the design will be generated and the next step is to simulate the entire buck converter using HSpice. In an unusual situation, one might need to improve the specs. This is because the CMOS process technology used in this work might have some limitations like the allowable size and width selection of the MOSs and their corresponding switch on-resistance.

III- LDO SYNTHESIS FLOW

Fig. 5 shows the circuit of LDO and Fig. 6 is the synthesis flow for LDO design. The small signal analysis is done with an equivalent circuit as shown in Fig. 7. Firstly, we require the user to key-in the process information and design specs, such as the process technology, the supply voltage (V_o), the output voltage (V_o), and the max load current ($I_{L,max}$). The synthesis flow generates the parameters of the sample resistance (R_{s1}, R_{s2}), the width of the power MOS (MP), the output capacitor (Co), and the compensation series resistance (R_{csr}).

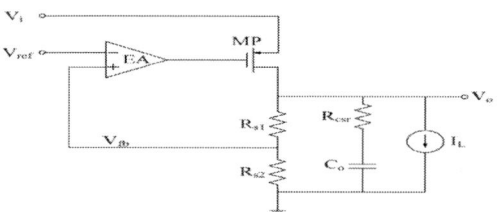

Figure 5.   Architecture of LDO

978-1-61284-863-1/11 $26.00 © 2011 IEEE   282

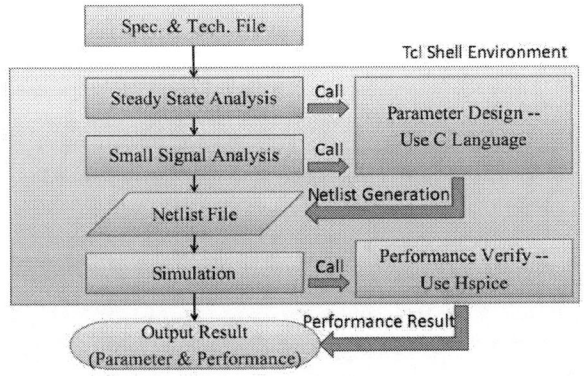

Figure 6. The LDO synthesis flow

For LDO, we usually focus on the line regulation and the load regulation. Line regulation is given by equation (10). It is the ability to maintain a stable output voltage (V_o) under perturbation of input voltage (V_i). The load regulation is given by equation (11). It is the ability to maintain a stable output voltage (V_o) under load current I_L perturbation.

$$\text{Line regulation} = \frac{\Delta V_o}{\Delta V_i} \qquad (10)$$

$$\text{Load regulation} = \frac{\Delta V_o}{\Delta I_L} \qquad (11)$$

The controller of the LDO uses a feedback loop by comparing a reference voltage (V_{ref}) with the feedback voltage (V_{fb}), which is the partial voltage of V_o given by equation (12).

$$V_{fb} = \frac{R_{s2}}{R_{s1} + R_{s2}} V_o \qquad (12)$$

As for the efficiency, it is given by equation (13). The quiescent current is given by equation (14). I_{qea} is the quiescent current of error amplifier (EA). I_{qs} is the quiescent current of the sample resistance.

$$\text{efficiency} = \frac{I_L V_o}{(I_L + I_q)V_i} \times 100\% \qquad (13)$$

$$I_q = I_{qea} + I_{qs} \qquad (14)$$

In Fig. 7 of the small signal model of LDOs [6], g_{ea} is the transconductance of the EA, R_p is the output resistance of EA, g_{mp} is the transconductance of power MOS (MP), C_p is the

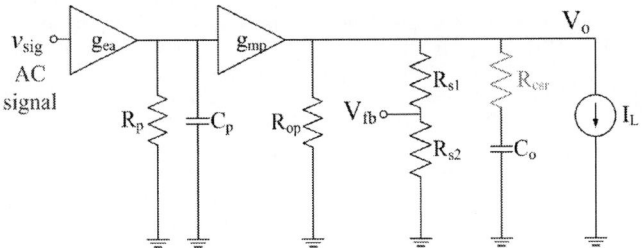

Figure 7. The equivalent small signal model of LDOs[6].

gate-source capacitor of MP, and R_{op} is the on-resistance of MP. The loop gain of the LDO consists of two poles (P_{EA} and P_{MP}) and one zero (Z_{out}) which are given by equations (15) (16) and (17).

$$P_{EA} = \frac{1}{2\pi R_p C_p} \qquad (15)$$

$$P_{MP} = \frac{1}{2\pi R_{op} C_o} \qquad (16)$$

$$Z_{out} = \frac{1}{2\pi R_{csr} C_o} \qquad (17)$$

In an unusual case, if the system is unstable, we can change the value of Z_{out} by fixing the value of R_{csr} and add a resistance (R_{add}) in series with it as given by equation (18).

$$R_{csr} = R_{esr} + R_{add} \qquad (18)$$

IV- AUTOMATION SELECTION FLOW

This paper does not only automate the design of power converters but it also optimizes the design by selecting the most suitable architecture according to the specifications and the foundry process corners.

At first, the user keys-in the input voltage (V_i), the output voltage (V_o), the maximum output current ($I_{L,max}$), and the efficiency (η) as before, then the tool generates the steady-state design parameters and it carries out the small signal analysis for both the buck converter and the LDO respectively with different process corners. The flow automatically selects a suitable architecture according to the highest efficiency that is set by the user. Third, it provides the proposed architecture. However, if there is a determined architecture, the flow still can complete the design after the user selects it directly.

V- DESIGN EXAMPLES

We use a few demo cases to verify the correctness of our work. TABLE II shows two demo cases and the results of the buck converter with the process technology of TSMC 0.18μm 3.3volt. TABLE III shows the output design parameters of two test cases. TABLE IV shows different corner case results of buck converter when MOSFET at FF (fast-fast or best) and SS (slow-slow, or worst). TABLE V shows the performance of the buck converter with the inductor tolerance of ±10%. Fig. 8~9 shows the simulated output voltage ripples of case 1 and 2. They are within the specs.

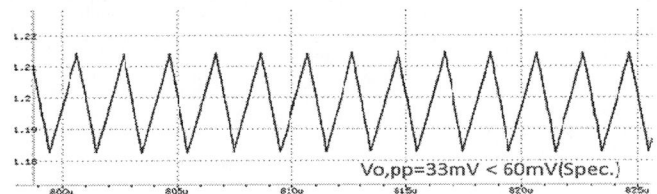

Figure 8. Output voltage ripple of case1, buck converter

978-1-61284-863-1/11 $26.00 © 2011 IEEE

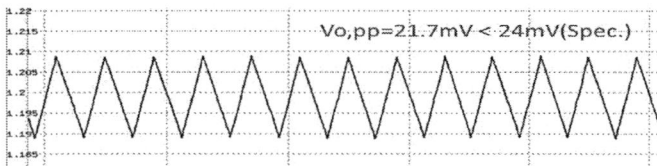

Figure 9. Output voltage ripple of case2, buck converter

TABLE II. DEMO CASES OF BUCK CONVERTER

	Case1		Case2	
	Spec	*Simulation*	*Spec*	*Simulation*
V_i(V)	2.4		3	
V_o(V)	1.2	1.20	1.2	1.20
$I_{L,max}$(mA)	300	299.6	600	599.3
P_o(mW)	360	359.0	720	718.3
$V_{o,pp}$(mV)	$\leq$60	33.3	$\leq$24	21.2
η(%)	$\geq$80	80.6	$\geq$90	90.7

TABLE III. DESIGN COEFFICIENTS OF CASES

Item	Case1	Case2
Duty cycle	0.5	0.4
R_L(Ω)	4	2
L(μH) / r_{dcr} (Ω)	22 / 0.11	27 / 0.12
C (μF) / r_{esr} (Ω)	10 / 0.76	33 / 0.44
Power PMOS	W=10u L=0.35u M=596	W=10u L=0.35u M=5784
Power NMOS	W=10u L=0.35u M=298	W=10u L=0.35u M=2892
r_p (Ω)	1.187	0.081
r_n (Ω)	0.528	0.043
Area(μm^2)	77900	237000

TABLE IV. DIFFERENT CORNER CASE OF BUCK CONVERTER

	Case1		Case2	
	Best	*Worst*	*Best*	*Worst*
V_i(V)	2.4		3	
V_o(V)	1.20	1.20	1.20	1.20
$I_{L,max}$(mA)	299.6	299.6	599.3	599.3
P_o(mW)	359.1	359.0	718.3	718.2
$V_{o,pp}$(mV)	33.6	32.0	21.7	21.2
η(%)	82.8	76.9	91.0	90.7

TABLE V. PERFORMANCE OF DIFFERENT INDUCTORS

	Case1		Case2	
	L+10%	*L-10%*	*L+10%*	*L-10%*
V_i(V)	2.4		3	
V_o(V)	1.20	1.20	1.20	1.20
$I_{L,max}$(mA)	299.6	299.6	599.3	599.3
P_o(mW)	359	359	718.2	718.3
$V_{o,pp}$(mV)	29.7	36.2	19.6	23.5
η(%)	80.6	80.6	90.2	90.6

TABLE VI. DEMO CASES AND RESULTS OF LDO

	Case1		Case2	
	Spec	*Simulation*	*Spec*	*Simulation*
V_i(V)	2$\pm$1%		2.5$\pm$5%	
V_o(V)	1.2	1.20	1.2	1.20
$I_{L,max}$(mA)	200	200.0	300	300.6
P_o(mW)	240	240.0	360	361
η(%)	59.9	59.9	47.9	47.9
I_q(μA)	400	246.8	600	520.9
Line Regulation(mV/V) (Vin$\pm$1%,$\pm$5%@ $I_{L,max}$)	$\leq$0.1	0.032	$\leq$0.5	0.157
Load Regulation(mV/mA) (0 to $I_{L,max}$@ nominal V_i)	$\leq$1	0.60	$\leq$1.5	0.624

TABLE VII. DESIGN COEFFICIENTS OF TWO CASES

Item	Case1	Case2
Rs1 (kΩ)	1.5	1
Rs2 (kΩ)	4.5	3
Power PMOS	W=10u L=0.35u M=509	W=10u L=0.35u M=398

TABLE VIII. DIFFERENT CORNER CASE OF LDO

	Case1		Case2	
	Best	*Worst*	*Best*	*Worst*
V_i(V)	2$\pm$1%		2.5$\pm$5%	
V_o(V)	1.20	1.20	1.20	1.20
$I_{L,max}$(mA)	200.0	200.0	300.0	300.0
P_o(mW)	240.0	240.0	359.9	360.0
η(%)	59.8	59.90	47.9	47.9
I_q(μA)	574.2	324.1	677.5	424.8
Line Regulation(mV/V) (Vin$\pm$1%,$\pm$5%@ $I_{L,max}$)	0.032	0.033	0.157	0.158
Load Regulation(mV/mA) (0 to $I_{L,max}$@ nominal V_i)	0.604	0.745	0.696	0.843

For the LDO, we also use two demo cases and results are shown in TABLE VI. TABLE VII shows the output design parameters of two test cases. Simulation results show the line to output and load to output regulations. TABLE VIII shows synthesis results of LDO with MOSFET Spice parameters at FF (best) and SS (worst) cases. They are within the specs.

VI- CONCLUSION

Automated synthesis design flow of power converter blocks aimed at SOC application is proposed. The design flow consists of the equation-based and the simulation-based methods. A DC-DC buck converter and a LDO designs are demonstrated with the proposed flow. We include the parasitic resistance of both the passive and active devices in the design to analyze the power conversion efficiency. This work also includes the controller designs to regulator the output voltage of the converters. Demonstration cases show that the designs meet the specs. However, we need a real silicon verification to verify our work which is the limitation of this paper. In the future, we will focus on the optimization of synthesis flow and include more power converter architectures and more controller options to generate on chip power converter designs for multiple-voltage SOC applications.

REFERENCE

[1] T. Hattori, etal., "A Power Management Scheme Controlling 20 PowerDomains for a Single-Chip Mobile Processor," InternationalSolid-State Circuit Conference, 2006.

[2] Biranchinath Sahu, "Integrated, dynamically adaptive supplies for linear RF power amplifiers in portableapplications," Publish for PhD degree, Georgia Institute of Technology, Nov. 2004.

[3] JULIANA GJANCI, "On-Chip Voltage Regulation for Power Management in System-on-Chip," For the degree of Master of Science in Electrical and Computer Engineering In the Graduate College of the University of Illinois at Chicago, 2008

[4] R.W. Erickson and D. Maksimovic, "Fundermental of Power Electronics," 2nd edition, Kluwer Academic Publishers, 2004

[5] Ned Mohan, Tore M. Undeland, William P. Robbins, "Power Electronics: Converters, Applications and Design," 3rd edition, John Wiley & Sons, Inc. 2003

[6] Gupta, V.; Rincon-Mora, G.A.; Raha, P., "Analysis and Design of Monolithic, High PSR, Linear Regulators for SoC Applications," IEEE International SOC Conference, 2004.

A Resource Binding Technique for TSV Number Minimization in High-Level Synthesis of 3D ICs

Wei-Kai Cheng[*)] and Yi-Chun Yen
Department of Information and Computer Engineering
Chung Yuan Christian University
Chung Li, Taiwan, R. O. C.
[*)] Email: wkcheng@cycu.edu.tw

Abstract—**Three dimensional integrated circuits allow multiple devices to be stacked on multiple layers. Therefore, utilize the area of each layer efficiently and minimize the number of through silicon vias (TSVs) are crucial to the 3D IC design. In this paper, we propose an integer linear programming (ILP) model to perform simultaneous resource binding and layer assignment in high-level synthesis of 3D ICs. Our objective is to minimize the number of TSVs under both the layer number constraint and the footprint area constraint. By means of duplicating hardware resources properly during resource binding and layer assignment, our approach has obvious improvement in reducing the number of TSVs while not increasing the footprint area. Experimental results show that our methodology is effective indeed for this 3D ICs synthesis problem.**

Keywords-High-Level Synthesis; Three Dimensional Integrated Circuits; Integer Linear Programming; Through Silicon Vias; Layer Assignment.

I. INTRODUCTION

Technology scaling has made the design of system in a single chip becomes feasible. However, the increasing of system complexity leads to long wire connection between circuit modules. In addition, different types of circuits will require different semiconductor processes. Integrating all these semiconductor processes for a SOC design will be very expensive. In recent years, many research efforts have concentrated on the integration strategies at the system level to resolve this problem. Among these strategies, partitioning the system into multiple dies and stacking these dies vertically provide a promising solution.

Three dimensional integrated circuits integrate the system by vertically stacking multiple dies together [8][11]. By this technique, not only the problem of manufacture integration can be resolved, long wiring delay in traditional two dimensional integrated circuits could also be enhanced by system partition, and connect modules in different layers by short vertically interconnects named through silicon vias [9][10]. Besides to reduce long wiring delay, power consumption in the connection wires could also be reduced..

However, the size of a TSV is much larger than the size of the silicon via. In addition, TSVs become obstacles for placement and routing during physical synthesis [1][2]. Furthermore, TSVs have a negative impact on the die yield of semiconductor manufacture. Therefore, minimizing the number of TSVs will be an important issue in high-level synthesis of 3D ICs.

There have been a lot of researches for TSV number minimization in the high-level synthesis. Under the constraint of floorplaning criteria, Mukherjee and Vemuri [3][4] Proposed an integer linear programming formulation for simultaneous scheduling, binding, and layer assignment, and their objective function was composed of both the number of TSVs and the critical path length. Krishnan and Katkoori [5] proposed a framework to integrate the resource binding and floorplaning problems together in the 3D ICs structure. Lee et al. [6][7] also proposed an integer linear programming model for high-level synthesis of 3D ICs, and guarantee to get the optimal solution for the number of TSVs under the resources and footprint area constrains.

However, all the previous binding and layer assignment approaches minimized the number of TSVs under the number of resources obtained from traditional high-level synthesis of 2D circuit design, and this constraint may not be best for the 3D IC structure. In this paper, we propose a resource duplication approach to overcome this resources usage problem in high-level synthesis of 3D ICs. By modifying and extension of the integer linear programming model proposed by Lee et al. [6], our model can fully utilize the footprint area of 3D ICs structure to further reduce the number of TSVs without increasing the footprint area.

The rest of this paper is organized as follows. Section II introduces our motivation and the concept of resource duplication. In section III, we propose our integer linear programming model for resource binding in high-level synthesis of 3D ICs. In Section IV, we show the experimental results for our methodology in TSV number minimization. Finally, we draw the concluding remarks in Section V.

II. MOTIVATION

Traditionally, high-level synthesis of two dimensional integrated circuits design aims at reducing the need of hardware resources. However, the least number of resources used may not lead to the best result in the high-level synthesis of 3D ICs. For the hal example shown in Fig. 1, there are totally six multiplication operations, two addition operations, two subtraction operations, and one comparison operation. These operations are scheduled into four control steps, and the least number of resources required for the synthesis of 2D ICs will be two multipliers, one adder, one subtracter, and one comparator. Suppose the area of a multiplier is 12, the areas of both adder and subtracter are all 4, and the area of a comparator is 2. Our problem is to bind each operation to a resource, and assign each resource to a layer in the 3D ICs structure such that

978-1-61284-863-1/11 $26.00 © 2011 IEEE

the number of TSVs is minimized, under both the footprint area constraint and the layer number constraint. In this example, we suppose the footprint area constraint is 16, and the layer number constraint is 3.

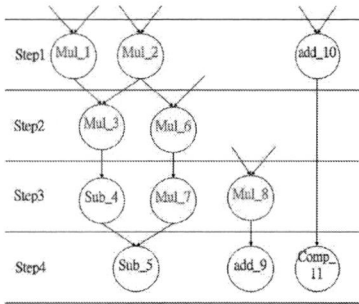

Figure 1: Scheduled DFG of the HAL example.

Fig. 2 illustrates the motivation of our resource duplication technique. Fig. 2(a) shows the optimal result of resource binding and layer assignment when the resources used are the same as the lower bound required for the scheduling result of 2D ICs as shown in Fig. 1. In this case, operations $\{O_2, O_6, O_8\}$ were bound to multiplier M1, operations $\{O_1, O_3, O_7\}$ were bound to multiplier M2, operations $\{O_4, O_5\}$ were bound to subtracter S1, operations $\{O_9, O_{10}\}$ were bound to Adder A1, and operation $\{O_{11}\}$ was bound to comparator C1. Because of the footprint area constraint, a multiplier could be assigned to a layer together with at most one non-multiplier function unit. Therefore, M2 and S1 were assigned to Layer1, M1 and A1 were assigned to Layer2, and C1 was assigned to Layer3. There exists a TSV in the direction from M1 to M2 because of the data dependences between operations $O_2 \rightarrow O_3$ and $O_6 \rightarrow O_7$ inter Layer1 and Layer2. Occurrence of this TSV was inevitable because operations O_1 and O_2 were scheduled to the same control step such that they must be bound to different multipliers, and there existed data dependences between both $O_1 \rightarrow O_3$ and $O_2 \rightarrow O_3$. Another TSV exists in the direction from A1 to C1 because of data dependence between operations $O_{10} \rightarrow O_{11}$ inter Layer2 and Layer3. Although we can assign both A1 and C1 to Layer3 to eliminate the TSV between them, it will introduce another TSV in the direction from M1 to A1 because of data dependence between operations $O_8 \rightarrow O_9$ inter Layer2 and Layer3. Therefore, the minimal number of TSVs required is 2 when the resources used were constrained to the lower bound of the scheduled result for 2D ICs.

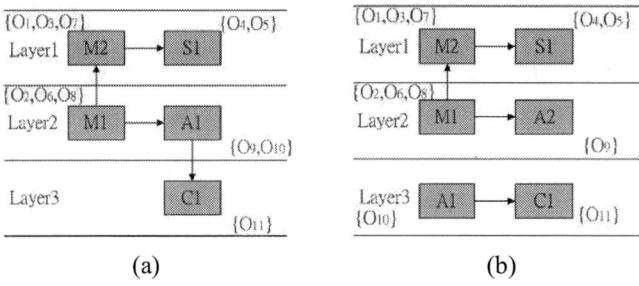

Figure 2: (a) Binding and layer assignment with no duplication of resources.
(b) Binding and layer assignment with duplication of resources to further minimize the TSV number.

However, the TSV between A1 and C1 in Fig. 2(a) could actually be eliminated without introducing another TSV if the resource duplication technique could be applied. In the Layer3 of Fig. 2(a), there still exists area space to be utilized because of that only the comparator C1 was assigned to it. If we use additional one adder as shown in Fig. 2(b), in which operation $\{O_9\}$ was bound to adder A2 and assigned to Layer2, operation $\{O_{10}\}$ was bound to adder A1 and assigned to Layer3, other resource binding and layer assignment were the same with that in Fig. 2(a). Because the data dependence between operations $O_8 \rightarrow O_9$ was intra Layer2, and the data dependence between operations $O_{10} \rightarrow O_{11}$ was intra Layer3, the second TSV in Fig. 2(a) was no longer necessary as shown in Fig. 2(b), and the number of TSVs required could be further reduced to 1. In this case, we successfully reduce the number of TSVs without increasing either the footprint area or the layer number, and hence no extra cost was introduced.

Therefore, if the use of resources was not constrained to the lower bound obtained from the high-level synthesis of two dimensional integrated circuits, there will be much more opportunities to further reduce the number of TSVs in the high-level synthesis of 3D ICs by the application of resource duplication technique. In this paper, we will propose an integer linear programming methodology to select and add resources automatically in order to get the minimal number of TSVs for resource binding and layer assignment in synthesis of 3D ICs.

III. RESOURCE BINDING AND LAYER ASSIGNMENT

In this section, we propose an ILP methodology to select and add resources for simultaneous resource duplication, resource binding and layer assignment in high-level synthesis of 3D ICs. Given a scheduled DFG, the layer number constraint, the footprint area constraint, and the upper bound on each type of resources available, our object is to minimize the number of TSVs. Detailed notations and ILP formulations are described as below.

A. Notations

- N: An integer constant to denote the total number of operations.

- TS: An integer constant to denote the total number of control steps.

- $v_{i,k}$: A binary variable that denotes the resource binding. If operation i is bound to resource k, $v_{i,k} = 1$; otherwise, $v_{i,k} = 0$;

- k_x: Denotes the resource x.

- Rmax: An integer constant to denote the total number of resources.

- via_{k1k2} : An integer variable that denotes the number of TSVs between the resources k_1 and k_2.

- D_{k1k2} : A binary variable that denotes the communication relation of resources. If the output of resource k_1 is the input of the resource k_2, $D_{k1,k2} = 1$; otherwise, $D_{k1,k2} = 0$.

- $r_{k,l}$: A binary variable that denotes the layer assignment of resource. If the resource k is assigned to layer l, $r_{k,l} = 1$; otherwise, $r_{k,l} = 0$.

- A_k: An integer constant to denote the area of resource k.

978-1-61284-863-1/11 $26.00 © 2011 IEEE   286

- A_{max}: An integer constant to denote the footprint area constraint.

- N_{layer}: An integer constant to denote the layer number constraint.

- $OP_{i1,i2}$: A binary constant that denotes the communication relation of operations. If the output of operation j_1 is the input of operation j_2, $op_{j1,j2} = 1$; otherwise, $op_{j1,j2} = 0$.

- OF_{i1}: Denotes the type of resource that bound to operation i_1.

- OS_{i1}: An integer constant to denote the control step that operation i_1 is scheduled to.

B. ILP Formulations for TSV Number Minimization

We modified the object function and constraints from the previous work of Lee et al. [6]. Some of the constraints including (1),(2),(3),(4),(5),(6) are the same with that of [3][6], and others were designed by us for the purpose of simultaneous resource duplication, resource binding and layer assignment.

- Object function

  Our object function is to minimize the total number of TSVs and described as below.

$$\text{Min:} \quad \sum_{k1=1}^{Rmax} \sum_{k2=1}^{Rmax} via_{k1,k2}$$

- TSV number constraint

  Suppose that resources k_1 and k_2 are assigned to layers l_1 and l_2, respectively. If $D_{k1,k2} = 1$, the number of TSVs between resources k_1 and k_2 is equal to the absolute value of $(l_1 - l_2)$; otherwise, no constraint on the number of TSVs between them. Thus, for each pair of resources k_1 and k_2, we have the following two constraints on the number of TSVs between them.

$$\sum_{l=1}^{N_{layer}} l * r_{k1,l} - \sum_{l=1}^{N_{layer}} l * r_{k2,l}$$
$$\leq via_{k1,k2} + (1 - D_{k1,k2}) * N_{layer} \quad (1)$$

$$\sum_{l=1}^{N_{layer}} l * r_{k2,l} - \sum_{l=1}^{N_{layer}} l * r_{k1,l}$$
$$\leq via_{k1,k2} + (1 - D_{k1,k2}) * N_{layer} \quad (2)$$

$$0 \leq via_{k1,k2} \quad (3)$$

- Uniqueness constraint

  Each operation must be bound to one and only one resource. Thus, we have the constraint described as below for each operation.

$$\sum_{k=1}^{Rmax} v_{i,k} = 1 \quad \forall v_i \in V \quad (4)$$

- Resource usage constraint

  For each control step, a resource can only be bound to at most one operation. Thus, we could have the following constraint for each resource in every control step.

$$\sum_{i=1}^{N} v_{i,k} \leq 1, \quad \forall \ k = 1 \dots Rmax, J = 1 \dots TS, OS_i = J \quad (5)$$

- Footprint area constraint

  The total area of resources at the same layer must be less than or equal to the footprint area constraint. Thus, we have the following constraint for each layer.

$$\sum_{k=1}^{Rmax} A_k r_{k,l} \leq A_{max} \ , \quad \forall l = 1 \dots N_{layer} \quad (6)$$

- Layer assignment for resources

  Because the total number of resources is larger than the lower bound of resources required, a resource must be assigned to at most one layer. Thus, we have the layer assignment constraint for each resource.

$$\sum_{l=1}^{N_{layer}} r_{k,l} \leq 1, \quad \forall k \in 1 \dots Rmax \quad (7)$$

- Correctness constraint

  If a resource was bound by at least one operation in the DFG, it must be assigned to one and only one layer; otherwise, no layer assignment is necessary for the resource. Thus, we have the following two constraints to link the relation of resource binding and layer assignment.

$$\sum_{l=1}^{N_{layer}} r_{k,l} \geq v_{i,k} \ , \quad v_i \in V, \forall k \in 1 \dots Rmax \quad (8)$$

$$\sum_{i=1}^{N} v_{i,k} \geq \sum_{l=1}^{N_{layer}} r_{k,l} \ , \quad \forall k \in Rmax \quad (9)$$

- Resource connect constraint

  For each pair of operations (j_1, j_2) that the output of operation j_1 is the input of operation j_2, if j_1 is bound to resource k_1 and j_2 is bound to resource k_2, then there is a communication link between resources k_1 and k_2. Thus, we could have the resource connect constraint described as below.

$$D_{k1,k2} \geq v_{i1,k1} + v_{i2,k2} - 1 \ \forall \ k_1 \in Rmax, k_2 \in Rmax,$$
$$OF_{i1} \in k_1 \ , OF_{i2} \in k_2 \ , i_1 \ 、 i_2 \in OP_{i1,i2} = 1 \quad (10)$$

- Layer usage constraint

  For each layer, we enforce that at least one resource is assigned to it. Therefore, we have the layer usage constraint described as below.

$$\sum_{k=1}^{Rmax} r_{k,l} \geq 1 \quad \forall l \in 1 \dots N_{layer} \quad (11)$$

IV. Experimental Results

We use Extended LINGO Release 11.0 as the ILP solver. The platform is Windows-7 x64 running on i5-650 dual-cores CPU with 4GB RAMs. Four benchmarks are used to test the performance of our approach. Table 1 lists the characteristics of the four benchmark circuits. The columns "Steps" and "Total Operations" list the number of control steps and total operations, respectively. The column "Operation Type" lists the number for each type of operations. The column "Resource Lower Bound" gives the least number of resources required for each type of operations. We assume all functional units are 8-bit designs, and implemented by the Synopsys DesignWare targeted to the TSMC 0.18um process. The area of each design is (*, +, -, <, mux) = (1057, 278, 305, 117, 108).

Table 1: Characteristics of the benchmarks

Benchmark	Steps	Total Operations	Operation Type (*, +, -, <, mux)	Resource Lower Bound (*,+,-,<.mux)
HAL	4	11	{6,2,2,1,0}	{2,1,1,1,0}
BF	10	29	{11,12,6,0,0,}	{2,2,1,0,0}
AR	10	28	{16,12,0,0,0}	{2,2,0,0,0}
G2	10	24	{9,9,0,3,3}	{2,2,0,1,1}

Table 2 shows the experimental results of our resource duplication approach for high-level synthesis of 3D ICs. The column "Layer" gives the layer number constraint. The column "Footprint Area" gives the footprint area constraint. The "No Resource Duplication" column lists the optimal result of TSV number minimization when each type of resources used are exactly equal to the lower bound as listed in Table 1. The column "Resource Duplication" lists the number of TSVs and resources used after resource binding and layer assignment. Experimental results show that our approach can effectively reduce the TSV number for all the benchmarks.

Table 2: Comparisons on the effect of resource duplication in term of the TSV number

Benchmark	Layer	Footprint Area	No Resource Duplication		Resource Duplication	
			Resources	TSV	Resources Clone	TSV
HAL	3	1340	{2,1,1,1,0}	2	{2,2,1,1,0}	1
	4	1060	{2,1,1,1,0}	3	{2,2,1,1,0}	3
BF	2	1670	{2,2,1,0,0}	3	{2,2,2,0,0}	2
	3	1340	{2,2,1,0,0}	6	{2,3,2,0,0}	5
AR	3	1340	{2,2,0,0,0}	5	{3,2,0,0,0}	3
	4	1060	{2,2,0,0,0}	8	{2,3,0,0,0}	6
G2	3	1980	{2,2,0,1,1}	4	{2,3,0,1,1}	3
	4	1340	{2,2,0,1,1}	7	{2,3,0,1,1}	6

V. Conclusions

In this paper, we present a resource duplication approach to minimize the number of TSVs in high-level synthesis of 3D ICs. We develop new ILP methodology to formally draw up this optimization problem under the footprint area constraint and the layer number constraint. Different from previous researches to use the lower bound of resources for resource binding and layer assignment, our ILP formulations can automatically select and add the binding resources to further reduce the TSV number. Experimental results show that our approach is effectively indeed for this 3D ICs synthesis problem.

VI. Acknowledge

This work was supported in part by the National Science Council of Taiwan, R.O.C., under grant number NSC 99-2218-E-033-007.

References

[1] T.-Y. Chiang, S. J. Souri, C. O. Chui, and K. C. Saraswat, "Thermal Analysis of Heterogeneous 3-D Ics with Various Integration Scenarios", IEEE International Electron Devices Meeting, pp. 681-684, 2001.

[2] D. H. Kim, S. Mukhopadhyay, and S. K. Lim, "Through-Silicon-Via Aware Interconnect Prediction and Optimization for 3D Stacked Ics", Proc. of ACM/IEEE International Workshop on System Level Interconnect Prediction, pp. 85-92, 2009.

[3] M. Mukherjee and R. Vemuri, "Simultaneous Scheduling, Binding and Layer Assignment for Synthesis of Vertically Integrated 3D Systems", Proc. of IEEE/ACM International Conference on Computer Design, pp. 222-227, 2004.

[4] M. Mukherjee and R. Vemuri, "On Physical-Aware Synthesis of Vertically Integrated 3D Systems, " Proc. of International Conference on VLSI Design, pp. 647-652, 2005.

[5] V. Krishnan and S. Katkoori, "A 3D-Layout Aware Binding Algorithm for High-Level Synthesis of Three-Dimensional Integrated Circuits", Proc. of IEEE International Symposium on Quality Electronic Design, pp. 885-892, 2007.

[6] C.-H. Lee, T.-Y. Huang, C.-H. Cheng, and S.-H. Huang, "A Post-Processing Approach to Minimize TSV Number for High-Level Synthesis of 3D ICs", Proc. of IEEE International Symposium on Computer, Communication, Control and Automation, pp. 434-437, 2010.

[7] C.-H. Lee, S.-H. Huang, and C. H. Cheng, "High-Level Synthesis of 3D IC Designs for TSV Number Minimization", Proc. of Synthesis And System Integration of Mixed Information technologies, pp. 260-265, 2010.

[8] S. Das, A. Chandrakasan, and R. Reif, "Three-Dimensional Integrated Circuits: Performance, Design Methodology, and CAD Tools", Proc. of IEEE Computer Society Annual Symposium on VLSI, pp. 13-18, 2003.

[9] J. A. Davis, R. Venkatesan, A. Kaloyeros, M. Beylansky, S. J. Souri et al., "Interconnect Limits on Gigascale Integration in the 21st Century", Proc. of the IEEE, vol. 89, no. 3, pp. 305-324, March 2001.

[10] J. W. Joyner, R. Venkatesan, P. Zarkesh-Ha, J. A. Davis, and J. D. Meindl, "Impact of Three-Dimensional Architectures on Interconnects in Gigascale Integration", IEEE Transactions on VLSI Systems, vol. 9, no. 6, pp. 922-928, Dec. 2001.

[11] B. Black, M. M. Annavaram, E. Brekelbaum, J. DeVale, L. Jiang et al., "Die Stacking (3D) Microarchitecture", Proc. of International Symposium on Microarchitecture, pp. 469-479, 2006.

Low Power Digital Type ADC

Richard Wee Tar Ng[1] and Liter Siek[2]

email: [1]e080049@ntu.edu.sg, [2]elsiek@ntu.edu.sg

VIRTUS-IC Design Centre of Excellence,

50, Nanyang Avenue, Singapore 639798, Nanyang Technological University

Abstract – **Shrinking complementary metal-oxides-semiconductor (CMOS) technology has caused analog type analog-to-digital converters (ADCs) to face mounting challenges from reduced signal-to-noise ratio (SNR), lower intrinsic gain in CMOS devices, increased device leakage, larger transistor mismatches, and lower quality passives in lower transistor geometries. Digital functions instead have benefited from this technology scaling into smaller scales. Due to increased challenges of analog integration into these smaller scales, there is a drive to push these analog functions into the digital domain. Various digital type ADCs have been proposed, but such ADCs faced challenges from having large silicon area, and high power consumptions as compared to their analog counterparts. In this paper, we proposed to use Braun's D-ADC [1] as the main approach to a low power digital type ADC. The approach to achieve low power is through gate count and frequency operation reduction. The proposed low power design is implemented and tested on field programmable gate arrays (FPGA), with a bandwidth of 20 kHz and achieving a SNR of -58 dB.**

I. INTRODUCTION

As CMOS technology rapidly progresses into deep-sub micron scales, the average digital design space for digital designers has increased exponentially and cheaply, following Moores Law. The technologies of digital integrated circuits have advanced tremendously in speed and size over the past two decades. This has lead to possibilities of highly sophisticated digital signal processing (DSP) systems. These systems operate on a wide variety of digital data from fields of speech, communication, audio, video, medical, power and so on.

And one of the main successes to the use of such sophisticated DSP systems is the advancement in the designs of ADCs. These converters have played an important role in digitalizing our physical world which is analog in nature. Due to the need to digitalized large number signal types, much ADCs architecture has evolved [2]. In most applications, both analog and digital functions co-exist on the same integrated circuit (IC) chip. It has been a preferred or mandatory requirement that the analog functions are implemented in the same digital CMOS technology so as to avoid additional processing steps, and thus avoiding addition cost of production to the IC chip. However, the analog integration has gained progress on a much slower scale than its digital counterpart. While digital functions have benefited from technology scaling into deep-sub micron scales, this has been counter effective on analog

functions. Due to increased challenges of analog integration in deep-sub micron scales, there is a drive to push more and more analog functions into the digital domain. This includes the ADC [3]-[9].

However, there are many challenges to achieve a fully digital ADC. In general, some form of analog elements is still necessary to be in a full digital ADC in order to receive signal from the physical world, which is analog in nature. This is inevitable. One of the main challenge is to keep the amount of analog elements low, and the other is that the performance of the analog elements that are used have to be unaffected by silicon deviation or low power supplies. Gate count is also high in the digital ADCs as the translated analog signals needs some form of signal processing to further process it. Frequency operation of the signal processing can be high also. These factors inevitably lead to high power consumption in the digital ADCs.

In this paper, we proposed to adopt C. Braun's D-ADC as the main approach to a low power digital type ADC. The approach to reduce power consumption in this digital type ADC is by digital gate count reduction, and lowering of the operating frequency of these digital gates in the digital processing unit. Proposal for power consumption reduction in the analog front end (AFE) is also discussed.

The paper is organized as follows. In section II, we introduced the principle of operation of the D-ADC. In section III, we present the proposed low power approach for the digital functions in the D-ADC. In section IV, we present the experimental results that shows the power estimation from the low power approach, followed by the implementation and testing of the proposed design on FPGA. Last, but not least, low power on the AFE is discussed in section V.

II. PRINCIPLE OF OPERATION

The D-ADC architecture is originally proposed by Christoph Braun. The design is based on a linear pulse modulation with constant impulse amplitude. The analog input signal is sampled with a sine wave carrier which is generated by a digital read-only memory (ROM). The crossing of these two sine waves leads to the generation of impulses at specific time locations, which in the idealized case will have a frequency modulated function that gives a relationship between the input frequency and the carrier frequency. The overall architecture consists of mostly digital CMOS gates and simple resistor-capacitor (RC) network. The D-ADC AFE sampling architecture is shown in Fig. 1. The 1st order frequency

978-1-61284-863-1/11 $26.00 © 2011 IEEE

spectrum of the time location of the impulses is shown in Fig. 2. fc is the sine carrier, and fa is the sampled analog input signal. The 1st order of the frequency spectrum of the impulses is typically a band-pass spectrum with a bandwidth of 2 times the highest input signal that is being sampled.

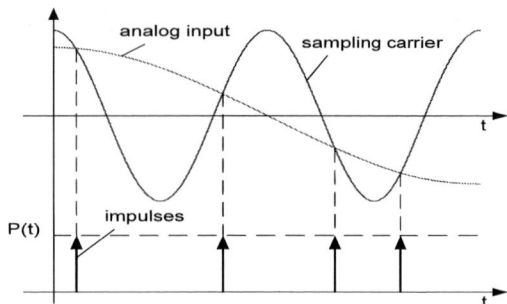

Fig. 1: Analog Front End Sampling Architecture

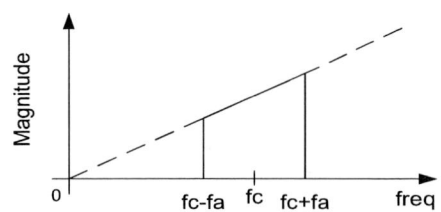

Fig. 2: 1st Order Frequency Spectrum of the Time Location of the Impulses

The time stamps that locate the impulses are recorded, and these values are then passed on to a digital band-pass FIR filter for filtering, followed by baseband down-conversion through decimation, convolution and digital low-pass filtering in order to recover the original analog input frequency. The quantized data processing sequence is illustrated in Fig. 3. It involves extracting the 1st order band-pass type frequency spectrum using a digital band-pass filter, followed by a baseband down-conversion of the extracted spectrum down by digital mixing with the carrier frequency. A digital low-pass filter is used to remove secondary images formed by the digital mixing. This results in the input analog frequency component being correctly extracted.

III. Low Power for Digital Functions

D-ADC has a good non-linearity performance of +/- 0.024% at full scale of 1.8 V. However, this good performance comes at an expense of high gate count of more than 100,000 gates, not including the analog front end components, and having a power consumption of 7 mW for a differential channel ADC operation [1]. In order to achieve a low power D-ADC, we proposed 3 approaches for the digital functions:

(A) Reduction of Digital Processing Unit for Baseband Recovery

In this sub-section, we reviewed a simplified approach to baseband recovery [10]. A band-pass signal band confined in the region of:

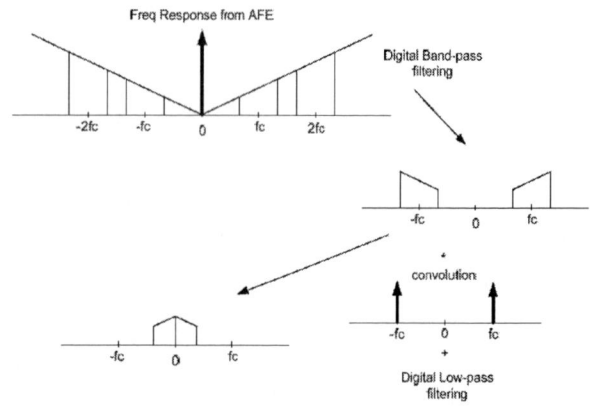

Fig. 3: D-ADC Digital Signal Processing Basics

$$\frac{m\pi}{D} < \omega < \frac{(m+1)\pi}{D} \qquad (1)$$

where D is the decimation factor, can be down-converted to baseband of:

$$0 < \omega < \frac{\pi}{D} \qquad (2)$$

purely by decimation. In order to remove the carrier frequency from the signal spectrum under conventional approach, convolution of the quantized signal data with the carrier signal using multipliers in digital mixers is needed. However in this case, since the sampling frequency is 4 times the frequency of the carrier sine signal which is always given as repetitions of 0, 1, 0, -1, there is no need for multipliers in the digital mixer. Only an inversion of the Most Significant Bit (MSB) is needed. Following this, by applying a decimation of 2 of the spectrum, all the signals are aliased into one another, giving only the desired signal fa in the baseband region. Thus by applying the simplified approach to baseband recovery, the need for multiplication and digital low pass filter resources for the removal of the carrier sine frequency and baseband signal recovery can be avoided by choosing an appropriate carrier sine frequency for the D-ADC.

(B) Reduction of Coefficients for Digital Filters

In order to accurately track the time locations of the impulses in the AFE, a high frequency sampling rate greater than 1 GHz is necessary in order to reach a ADC resolution of 16-bits. This will in turn dictate the need for high sampling rate in the digital band-pass filter to extract 1st order band-pass spectrum. Using a Finite Impulse Response (FIR) band-pass filter, with a combination of interpolation and decimation to reach the desired sampling output rate, the number of coefficients can reach more than 2 million if the desired sampling output rate is low. In order to limit the number of coefficients, we should avoid interpolation. Taking desired sampling output rate as 44.1 kHz for audio applications: For a 1 GHz sampling rate in the AFE, the number of coefficients required in the FIR filter is approximate 900,000 with an

978-1-61284-863-1/11 $26.00 © 2011 IEEE

optimal fix-point width of 22-bits in order to reach a -90 dB quantization noise floor. This takes up 2.5 Mbytes of ROM space if this band-pass filter is to be implemented on a silicon chip. Thus there is a need to further reduce the number of coefficients in order to reduce the silicon footprint taken up by the FIR filter.

The proposal to achieve a lower silicon area for the coefficients is to interpolate the points of the coefficients of the filter from a smaller set of points. Since the FIR-type filter coefficients is symmetrical in its centre, only half of the coefficients need to be in the silicon. By using point interpolation illustrated in Fig. 4, the filter coefficients footprint can be further reduced. The black dots shown in Fig. 4 represents the coefficients recorded and that is used to calculate the actual filter response of the filter.

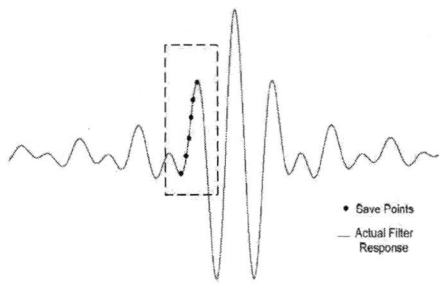

Fig. 4: Point Interpolation for a Typical FIR Filter Response

(C) Optimal Frequency Operation of digital logic

In order to ensure that digital computation rate is optimal, the computational rate in the digital filter should be kept at a value not exceeding twice the highest frequency component contained in the sampled domain. For audio applications, the optimal computation rate is 44.1 kHz. The optimal computational rate of the digital FIR band-pass filter in combination with decimation function can be achieved by using polyphase decomposition technique [11]. Assuming that the desired output sampling rate is 44.1 kHz, and the AFE sampling rate greater than 1 GHz, the AFE sampling rate is chosen to be 1.08486 GHz. The decimation rate is then calculated to be 24,600, and needed number of coefficients is 885,600. The block diagram of the designed polyphase digital band-pass filter is illustrated in Fig. 5.

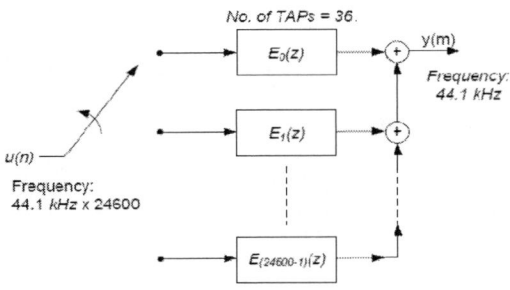

Fig. 5: Block Diagram of Designed Poly-phase Digital Band-pass Filter

IV. EXPERIMENTAL RESULTS

The D-ADC being an almost full digital design is implemented in VHSIC Hardware Description Language (VHDL), and adopts the proposed low power approaches introduced in section III. In order to do a power estimation of the D-ADC design, gate count estimation is required. The digital part of the design is synthesized with a 0.18 μm technology node and the digital gate count estimation for a single channel is given in Table I . Digital power estimation for a single channel is given in Table II. The total power consumption from the digital part is an approximate 0.053 mW.

TABLE I: Table for Digital Gate Count Estimation for a Single Channel

Parameters	Size	Area (mm^2)
ROM	64 kbits	0.055
SRAM	1.6 kbits	0.02
Digital Part	16343 gates	0.1343

TABLE II: Table for Power Estimation for a Single Channel

Parameters	Switching Power (mW/MHz)	Freq (MHz)	Power (mW)
ROM	0.05375	0.0441	0.00237
SRAM	0.00400	0.0441	0.0001764
Digital Part	1.14750	0.0441	0.0506

Test Setup

The proposed low power D-ADC is implemented and tested on a Xilinx Virtex 4 ML402 FPGA platform. This allows a pre-silicon evaluation of the D-ADC design on real hardware. A single channel is realized. The digital portion of the proposed D-ADC is synthesized and implemented on the FPGA platform, and the analog portion is represented on a veroboard using passive resistors and capacitors. The block diagram of the design setup is illustrated in Fig. 6.

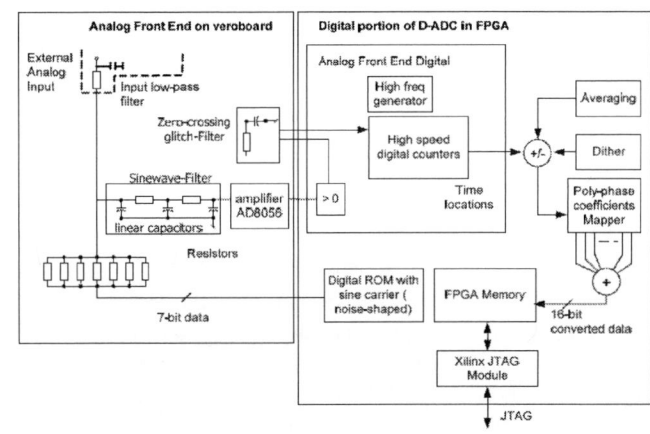

Fig. 6: Design Setup on FPGA and Veroboard

The digital sine wave filter is achieved using a 2nd order low-pass resistor-capacitor (RC) filter. The smoothed digital sine wave is directed into the AD8056 amplifier which acts as a high-input impedance buffer using a voltage follower configuration. This prevents the input pad of the FPGA from distorting the smoothed digital sine wave as it draws current from the setup. The input pad of the FPGA also acts as a zero-crossing detector which detects the zero-crossing point at the mid-point voltage level. Glitches occur at the zero-crossing point due to random noise on the smoothed digital sine wave, thus there is a need to implement a 1st order low pass RC glitch filter to remove these glitches. This setup in turn averaged out the random noise on the smoothed digital sine wave. The sampling rate of the zero-crossings used is 300 MHz. This is the maximum frequency that the clock routing resources can be supported by the Virtex 4 FPGA device family. The output sampling rate of the converted data is 44.1 kHz. The converted data is dumped into a FPGA FIFO memory which is 8192 in depth, sufficient for performing a Fast Fourier Transform (FFT) to analyze the amplitude and frequencies present in the converted data. These contents are extractable by using Joint Test Action Group (JTAG) protocol using a Personal Computer (PC). A photo of the FPGA setup is shown in Fig. 7.

Fig. 7: Photo of FPGA Setup

Test Results

A single tone sine wave with a frequency of 6.559 kHz is used to do a dynamic test on the D-ADC design. The sine wave amplitude is a factor of 0.8 with respect of the digitally generated sampling sine wave. The test is conducted for the voltage level type of 3.3 V. The achieved dynamc SNR is an approximate -58 dB, shown in Fig. 8.

V. LOW POWER CHALLENGES FOR ANALOG FUNCTION: A DISCUSSION

2 components of the AFE design consume the most power. They are the high frequency counters and then the RC network. The counters consume high power due to its high frequency switching. The RC network draws a worst case of 0.126 mA of current through its 7 resistors in parallel considering a supply voltage of 1.8 V. There is not yet an immediate solution how to overcome these 2 challenges. One possible method could be to use transistors in place of the bulky and power consuming resistors. With regards to the high frequency switching, a better approach to sample the time locations with a low frequency counter without impacting the SNR is required.

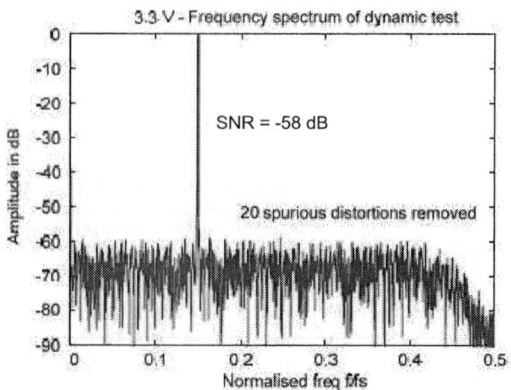

Fig. 8: Signal Spectrum of Single Tone Test

VI. CONCLUSION

In this paper, we presented a feasible approach to low power digital type ADCs. The low power digital type ADC uses D-ADC with a proposed low power approach which reduces the amount of digital logic and silicon area in the digital processing unit. This involves using a simplified approach to baseband recovery, and reduction of the cofficients in the FIR band-pass filter by interpolation. The low power approach for D-ADC is implemented and tested on FPGA achieving a SNR of -58 dB. The low power challenges for the analog function remains to be a topic for future research work.

REFERENCES

[1] C. Braun and B.Engl, "A 1.8v digital a/d converter in 0.18 um cmos," *Symposium On VLSI Circuits Digest of Technical Papers*, pp. 144 – 145, 2002.

[2] H. Lee and C.Sodini, "Analog-to-digital converters: Digitizing the analog world," *Proceedings of IEEE*, vol. 96, no. 2, February 2008.

[3] T. M. T. Watanabe and Y. Makino, "An all-digital analog-to- digital converter with 12-v/lsb using moving-average filtering," *IEEE Journal of Solid-state Circuits*, vol. 38, pp. 120 – 125, January 2003.

[4] L. C. J. Mainardi, A. Adao and A. Susin, "A comparison of totally digital adcs for socs," *IEEE International Symposium on Circuits and Systems*, vol. 1, pp. 641 – 644, 2004.

[5] J. Ortega and C. Janer, "Analog to digital and digital to analog conversion based on stochastic logic," *IEEE International Conference on Industrial Electronics, Control and Instrumentation*, vol. 2, pp. 995 – 999, 1995.

[6] S. Toral and Quero, "Stochastic a/d sigma-delta converter on fpga," *IEEE 42nd Midwest Symposium on Circuits and Systems*, vol. 1, pp. 35 – 38, 1999.

[7] J.-t. X. Fu-yuan Wang, Yong-liang Li and J. Chicharo, "Implementation of a quasi-digital adc on pld," *IEEE Conference on Solid-State and Integrated Circuit Technology*, pp. 1791 – 1793, October 2006.

[8] M. M.-N. H. Farkhani and M. Sachdev, "A fully digital adc using a new delay element with enhanced linearity," *IEEE International Symposium on Circuits and Systems*, pp. 2406 – 2409, May 2008.

[9] H. C. Hor and L. Siek, "K-locked-loop and its application in time mode adc," *12th International Symposium on Integrated Circuits*, pp. 101 – 104, December 2009.

[10] R. W. T. Ng and L. Siek, "A simplified approach for baseband recovery in sdr architectures," *IEEE Asia Pacific Conference on Circuits and Systems*, pp. 1059 – 1062, December 2010.

[11] G. B. M.G. Bellanger and M. Coudreuse, "Digital filtering by polyphase network: Application to sample rate alteration and filter banks," *IEEE Transaction on Acoustic Speech Signal Process*, vol. 24, pp. 109 – 114, April 1976.

Ultra Low Power SOC for Portable Health Monitoring Platforms

Richard Wee Tar Ng[1], Attard Laurent[2] and Sie Boo Chiang[3]

email: [1]richard.ng@infineon.com, [2]laurent.attard@infineon.com,
[3]sieboo.chiang@infineon.com

Infineon Technologies Asia Pacific,
8, Kallang Sector, Singapore 349282

Abstract – **The use of portable health monitoring devices has re-shaped the conventional "face-to-face" medical diagnostic mode for the patients. Most of these portable monitoring devices required low-energy, low-duty cycle hardware which help to detect various medical conditions of a patient, and transmit these information wired or wirelessly back to the medical personnel. The current offerings of such devices, consisting of using a standard micro-computer interfacing with high-end analog standard products and wireless modules are power consuming. Such a platform is not optimum in power usage and cost. In this paper, we present a System-On-Chip (SOC) silicon chip that has power aware capabilities and completes the need of a portable wireless health platform for monitoring various medical conditions of a patient. The power measurement results during idle and low power operation are shown, and has demonstrated that a health monitoring platform running on our SOC can have a standby time of 145,000 hours.**

I. INTRODUCTION

Patients with varying health conditions required different forms of monitoring ranging from short term to long term monitoring. The health conditions for monitoring typically includes basic vital signs such as heart beating rate, body temperature, blood pressure, blood glucose level and so on. While some health conditions required non-stop continuous monitoring, others required low-duty cycle of one time monitoring every few minutes or weeks [1]. These long-term health monitoring devices are often advised to be highly portable, energy misery, low-maintenance, and can be invasive or non-invasive type in nature. Such features will allow these devices to be carried closely to the patient and can be easily managed by the patient on a long term basis. The collected health data is then sent back to the medical personnel for proper diagnosis of the patient's conditions. This form of wireless diagnosis mode allows a patient to live his normal life routine without the need to visit the medical personnel "face-to-face" all the time, re-shaping the conventional health diagnostic mode [2].

Many portable health monitoring platforms have been proposed [1][3]. These platforms typically involve using a standard microcomputer interfacing with high-end analog standard product as sensors and wired or wireless modules such

as RS232, Universal Serial Bus (USB), ethernet and Radio Frequency (RF) serving as communication interfaces. Display driver is needed for platform-to-human communication, and Power Management Unit (PMU) is a must for long term, low duty monitoring platforms. Such a platform generally required interfacing multiple silicon chips, and this result in high cost, and high power usage due to Inputs and Outputs (IOs) communication between the silicon chips. Proper power saving modes are difficult to achieve due to multiple number of chips integrated on a hardware platform. Fig. 1 shows the typical system architecture of a portable health monitoring platform used for blood glucose measurement.

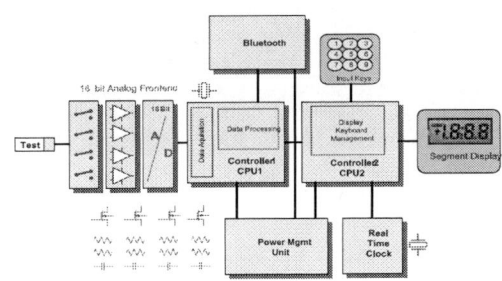

Fig. 1: Typical System Architecture of a Portable Health Monitoring Platform

In this paper, we proposed a complete integration of all the requirements of the portable health monitoring platform into a single silicon chip with power aware capabilities. Such a value proposition allows a power aware portable health monitoring platform to be realized with a single SOC. Various low power capabilities are designed into our SOC, and offers good control of the power consumption of the platform by software. Results from experimental measurements show that our SOC can entered into a ultra low power mode on our evaluation platform drawing a low current of 20 μA, providing a standby time of 145,000 hours.

The paper is organized as follows. In section II, we provide an overview of the chip architecture of our proposed SOC solution. The power aware design concept for various modules in our SOC is presented in section III. In section IV, the low-power verification strategies to realize our SOC is discussed.

978-1-61284-863-1/11 $26.00 © 2011 IEEE

Lastly, the power measurements in idle and low power mode are presented in section V.

II. ARCHITECTURE OVERVIEW

The internal architecture of our SOC solution is shown in Fig. 2. All its digital and analog modules are integrated and realised in 0.13 μm technology node. The use of 0.13 μm technology node helps to limit leakage power in the chip during power off mode or clock stop mode. Its digital modules consist of a high performance 32-bit Advanced RISC Machines (ARM) processor as a main control unit, a subsystem with Interrupt Controller Unit (ICU), and a Direct Memory Access (DMA) controller. An extensive variety of digital interfacing options includes Univeral Asynchronous Receiver Transmitter (UART), Serial Peripheral Interface (SPI), Inter-Integrated Circuit (I²C), General Purpose Inputs and Outputs (GPIOs) and USB for wired communications and bluetooth for wireless communications. A fully digital class D output stage provides an audio output functionality. In addition, a display controller with hardware graphic acceleration is available for driving matrix Liquid Crystal Displays (LCDs), and Organic Light-Emitting Diodes (OLEDs) displays. Its analog modules consist of an embedded PMU which generates all voltage sources for the platform from a single voltage source such as a 3.0 V battery. The PMU handles all the different power savings and wake up scenarios as well as battery charging and fuel gauging. Sigma Delta type Analog-to-Digital Converters (ADCs) and Digital-to-Analog Converters (DACs) with resolutions up to 16-bits are available to fulfil various medical applications. DCDC controllers are integrated into the SOC to work with external power components to generate various voltage sources for the platform. Such a SOC proposition completes the platform shown in Fig. 1 with a single silicon chip.

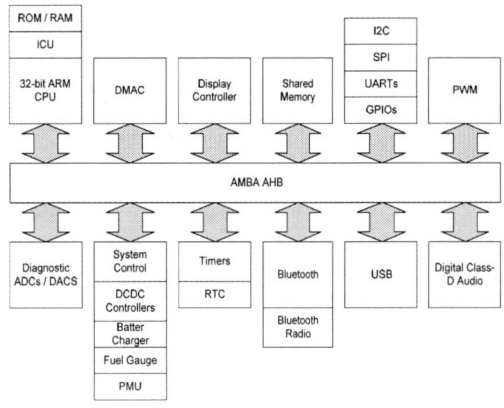

Fig. 2: Internal Architecture of our SOC

III. POWER AWARE DESIGN CONCEPT

For a portable health monitoring platform, power consumption is crucial, as it determines the number of standby hours,

the number of tests that the platform could provide for the patient. Power aware design concept is incorporated into the design of our SOC to reduce power consumption. Chip power aware design is achieved with the following 4 methods:

(A) Clock Control Network

The availability of an extensive clock control network on the SOC system level allows dynamic power reduction by clock frequency reduction or gating when the whole system or part of system enters into low activity or idle mode, where low duty cycle operations are required. This is fully controlled by software. The clock control network consists of a centralized clock generation unit (CGU) and clock control units (CCUs) at module levels. The structure of the complete clock control involving the CGU and CCU is shown in Fig. 3. The CGU provides an extensively variety of programmable clock values down to 32.768 kHz or to gate the clock to each individual modules. The CCU forms the interfacing block between the Advanced Microcontroller Bus Architecture (AMBA) Advanced High-performance Bus (AHB) bus to the specific module. It allows an additional programmable divider to the module and to gate the clock to the configuration registers of the module when write to configuration registers are idling.

(B) Multi-Voltage Domains and Voltage Scaling Application

Our SOC implement multi-voltages domain where modules are grouped according to their functions, and powered by dedicated LDOs to each functional voltage domain. According to the speed and performance of the modules in the blocks required, one may reduce the supply voltage to the modules when the modules are in a slower or idling mode. Since power consumption is proportional to square of the supply voltage, reduction in supply voltage could be one of the most effective ways for dynamic power reduction in the SOC operation. The control unit could realize a software-based power management control to adjust the supply voltage and the clock frequency operation dynamically according to a pre-determined voltage-to-frequency table, depending on the workload running on the SOC.

In general, all digital modules are placed into a single voltage domain. Modules, that are constantly "ON", such as

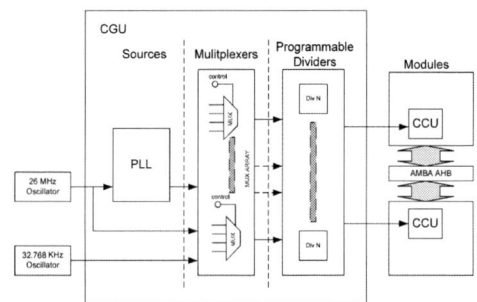

Fig. 3: Clock Control Network

978-1-61284-863-1/11 $26.00 © 2011 IEEE

the PMU, and Real-Time Clock (RTC) are placed together into another voltage domain, so that all other functions can be power gated when not in use. Fig. 4 shows a general planning of the voltage domains grouping in our SOC. It also shows that other voltage domains includes dedicated voltage domains for DCDC controllers, battery monitoring circuitries, charger circuitries, RF circuitries, high precision voltage references and analog power supply.

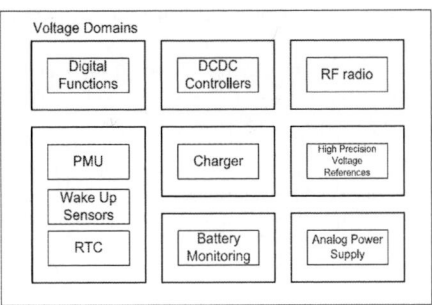

Fig. 4: Multi-Voltage Domains Grouping

(C) Deep Sleep Modes

Various modules are powered by dedicated LDOs, allowing deep sleep mode where modules not in use can be power gated by shutting off the specific LDOs. Modules with such deep sleep mode are isolated by special isolation cells and level shifters, not affecting other modules function when they are power gated. The SOC provides an almost complete power gating of the whole chip while keeping only the Real-Time Clock (RTC) module alive. This has allowed the SOC to achieve a low current consumption of 20 μA driven from a 3.0 V battery source in sleep mode measured on silicon, and be waken up by a pre-programmed clock alarm, or other external triggered events.

However, there are verification challenges posed by such power gating design. This is due to the absence of power gating modeling on Register Transfer Level (RTL) level design, thus not being verified by digital simulations. Transistor level verification of the power gating design is limited due to the complexity and size of the SOC. In our SOC development, we overcome this verification challenge by enhancing a conventional low power verification flow for digital designs extended to mixed-signal SOC using Spyglass from Atrenta. This enhanced low power verification flow is presented in section IV.

(D) Power "Smart" Features

Power "smart" features are designed into modules such as voltage references, DCDC controllers, and bluetooth. Low power with lower precision voltage references are used to reference voltage for chip operation that need to be continuously powered, especially during sleep mode. High precision voltage references can be selectively activated and used for analog circuitries when requiring higher accuracy. The DCDC

controller has 2 operation modes: Pulse Frequency Modulation (PFM) and Pulse Width Modulation (PWM). PFM mode runs with better power savings, but supports only lower loading. On the other hand, PWM mode runs with higher power and supports higher loading. Thus PFM mode is mainly used for booting up, and low power operations that uses less than 100 mA of current, while PWM mode is used for high power and performance operations. This, in turn, helps to limit the power consumption of the DCDC controllers during low power operation. Sleep mode is also designed into the DCDC modules to allow a sleep current drawn of less than 1 μA. The bluetooth version 2.1 protocol supports sleep modes such as hold, sniff and park. Bluetooth sleep modes allow the connected devices to go into sleep mode, and wakes up periodically to communicate with the master. The maximum time period that can be assigned can be approximately 40 seconds. Our SOC on-chip bluetooth module has the capability to clock or power gate all functions leaving only a low power timer to run and to trigger wake up during these sleep mode periods. Our bluetooth module with power "smart" feature draws as low as 10 μA of current during such sleep modes. This helps to save precious battery power on long term operations using bluetooth communication.

IV. ENHANCED LOW POWER VERIFICATION FLOW

A systemic and predictive way for low power checks for SOC has been established in-house using Atrenta Spyglass Predictive Analyzer. The tool has the capabilities to analyze designs written in two major Hardware Description Languages (HDL): VHDL and Verilog or a mixture of both. Spyglass is used in different stages in the implementation flow to do a static check against a set of rules. A set of low power checks is established in-house to ensure that the best chance of a first time silicon success for a low power SOC. The low power checks typically verify the design against a set of low power rules which includes verifying the correctness of the use of level-shifters at voltage domain crossings. This static rule check process flow is illustrated in Fig. 5. A different set of rules is applied at each development stage to check for correctness of the design. Unfortunately, such checks are only offered for designs following a digital implementation flow environment and limited to checking only digital designs. Low power checks on complex mixed signal chips still rely heavily on transistor level simulation which is inefficient and time consuming.

The enhanced low power verification flow is proposed to allow static rule checking on complex mixed signal chips such as our SOC. One key criteria for using the proposed flow is that a power and ground Verilog netlist must be generated for the mixed signal module to be checked. This is achieved in a 2-step approach explained below, and is applied to the design after the layout development stage in Fig. 5.

Step 1: Mixed Signal Module Level Power Check

Mixed signal modules such as PMU is designed and layout using Cadence Framework. Using a self-developed script, the Verilog netlist with power and ground information is generated

---

978-1-61284-863-1/11 $26.00 © 2011 IEEE   295

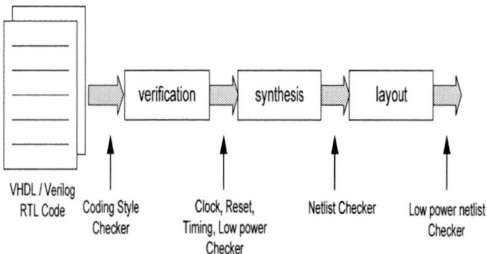

Fig. 5: Static Rule Check Process Flow

from the layout database of the PMU. This Verilog netlist is then analyzed by Spyglass against the low power rules. This ensures the correctness of the power net connection, and also the power intent functionalities of the module, performed in a systemic and predictive way introduced by Spyglass. With this approach done, we proceed to step 2.

Step 2: Chip Level Power Check

The Verilog netlist from step 1 is then emerged together with the chip level Verilog netlist with power and ground information, and re-analyzed by Spyglass against the low power rules. With this 2-step approach, our SOC, together with the mixed signal modules, is completely verified against the low power rules.

V. POST SILICON MEASUREMENTS

The SOC was fabricated in 0.13 μm CMOS process. Fig. 6 shows the micrograph of the SOC with the die area at approximately 42 mm^2.

Fig. 6: Micrograph of our SOC

An extensive number of functional and power measurements were conducted on this SOC, but this paper has been limited to present only the Idle and power modes.

Idle and Power Down Modes

On booting up of the SOC, the chip runs with following minimum functionalities: The DCDC controllers run with low power PFM modes generating the required pad IOs and core voltages. The PMU runs with low precision voltage reference for its analog circuitries. The main control subsystem unit runs with a low clock frequency of 5 MHz to operate the installed boot code. The bluetooth module is disabled. If there is no application code to be executed, the SOC is said to be in "IDLE" mode, and there is an option in the software control to allow the main control subsystem to enter into "Wait-For-Interrupt" (WFI) mode where the main control subsystem clocks are stopped until an interrupt request comes in. The SOC can also entered into "Ultra Low Power" (ULP) mode by software control by only keeping the 32.768 kHz clock oscillator, RTC module and wake-up logic powered on.

Table I shows the measured current consumption of the SOC in the mentioned modes with power drawn from a 3.0 V battery source.

TABLE I: Measured Current Consumption

Power Modes	Current Drawn (mA)
IDLE	16
WFI	12
ULP	0.020

Experimental Conclusion

This SOC has achieved an almost first time success with most of its functionalities working as expected. Our enhanced low power verification flow has contributed much to this success.

In ULP mode, the power consumption is a low 20 μA. In general, a 1.5 V AA alkaline batteries can provide up to a drain rate of 2900 mAh. A health monitoring platform can be set up by using 2 AA batteries to power up our SOC providing a 3.0 V battery source. In ULP mode, this will translate to a standby lifespan of 145,000 hours for a platform using our SOC.

CONCLUSION

In this paper, we presented a ultra low power SOC solution for portable health monitoring platforms. It integrates all the requirements needed by a portable health monitoring platform, and introduces power aware capabilities to the SOC with the use of power aware design concept. A hardware evaluation platform using our SOC has demonstrated its ultra low power sleep mode of a low 20 μA.

REFERENCES

[1] A. B. Dolgov and R. Zane, "Low-power wireless medical sensor platform," *28th Annual International Conference of the IEEE Engineering in Medicine and Biology Society*, p. 2067, Aug 2006.
[2] K. Wu and X. Wu, "A wireless mobile monitoring system for home healthcare and community medical services," *International Conference on Bioinformatics and Biomedical Engineering*, pp. 1190 – 1193, July 2007.
[3] L. Hui and L. W. Ding, "Low-power and portable design of bioelectrical impedance measurement system," *WASE International Conference on Information Engineering*, vol. 3, p. 38, Sep 2010.

978-1-61284-863-1/11 $26.00 © 2011 IEEE

A Scalable Strategy for Runtime Resource Management on NoC based Manycore Systems

Xiongfei Liao[†] and Thambipillai Srikanthan[‡]

[†]Institute of Microelectronics, A*STAR (Agency for Science, Technology and Research), Singapore
[‡]School of Computer Engineering, Nanyang Technology University, Singapore
Email: liaox@ime.a-star.edu.sg

Abstract—Centralized resource management and communication contentions among applications limit the scalability of the existing resource management strategies on NoC-based manycore systems. A scalable hierarchical strategy is proposed in this paper to overcome the above limitation.

I. INTRODUCTION

"Manycore" processors with hundreds/thousands of small but energy-efficient CPU cores will become the mainstream [1] [2]. Some of them have identical tiles connected by Network-on-Chip (NoC) [3] with 2D mesh and are of interest to researchers [4] [5] [6]. They are called "embedded manycore NoCs" here and an example is Tilera's TILE-Gx100 [7].

Embedded manycore NoCs can execute several applications concurrently. Each application can have multiple tasks running in parallel, using multiple CPU cores. As these NoCs will be used in devices such as smart phones and PDAs, users can start/stop applications that lead to dynamic system configurations in term of the use of CPU cores. Such dynamic configurations are extremely difficult to model off-line. Hence, run-time techniques are indispensable.

Several run-time strategies [4] [6] have been proposed for event-driven resource management. When an application *App* enters the system, suitable resources, i.e., CPU cores (a core is a basic unit to execute tasks) are identified and allocated to tasks of *App* (*allocation process*). When *App* finishes, the resources *App* occupies are reclaimed by the system (*deallocation process*). These strategies designate a core as the global manager (*GM*) that is responsible for resource allocation and deallocation when events occur. Moreover, to fully utilize resources, *GM* identifies and allocates resources in forms of irregular regions to applications. Fig. 2 shows how applications in Fig. 1 (described in Application Communication Graphs, i.e., ACGs [10]) are mapped to a 5×5 embedded NoC.

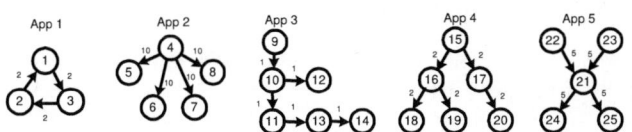

Fig. 1. The application characteristics of five applications

However, the above strategies have difficulties when applied to embedded manycore NoCs. The centralized resource management

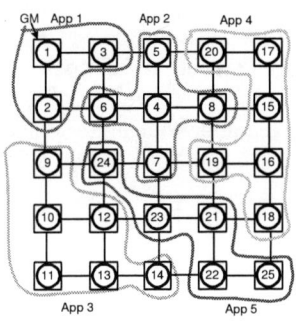

Fig. 2. Mapping five applications in Fig. 1 to a 5×5 NoC

agement by *GM* could make *GM* the performance bottleneck of the system and impede the scalability as the number of cores of embedded manycore NoCs increases. As irregular regions are identified and allocated to applications, the applications suffer both *internal* and *external communication contentions* caused by messages from tasks belonging to the "same" or "different" applications contending for links, respectively.

To overcome above limitation of scalability, we propose a scalable hierarchical strategy for runtime resource management on embedded manycore NoCs. Our strategy uses submeshes in resource organization and accommodates applications with resources in forms of submeshes. A *submesh* is defined as a rectangular area consisting of multiple cores which helps avoid external communication contentions. Our strategy handles resource management in a hierarchical way. *(1)* A scalable scheme is adopted by the *GM* to manage submeshes: a submesh is identified and allocated to an incoming application and the submesh is reclaimed after the application finishes. *(2)* A CPU core on a submesh is chosen to be the *local manager (LM)* which manages the resources within the submesh. *(3)* In managing resources of a submesh, the *GM* communicates with its *LM* and the *LM* communicates other cores within the submesh. So, the communication costs including latency and energy consumption can be reduced compared to those under the centralized resource management of the previous strategies.

II. SUBMESHES FOR RESOURCE MANAGEMENT

As embedded manycore NoCs are similar to large-scale parallel and distributed systems where submesh is used in resource management, submesh is introduced into our strategy.

As cores of embedded manycore NoCs are small and energy efficient [1], it is necessary that multiple cores cooperate to offer reasonable performance. So, submesh is a good choice for organizing resources. It reduces the complexity in resource management and keeps communication cost low.

When tasks of an application are run on a separate submesh, this leads to the below benefits. *(1)* It has been shown in [6] that tasks of the same application don't suffer from external communication contents when tasks run on a convex region. Submesh is such a convex region. *(2)* Submeshes help achieve performance isolation and Quality of Service (QoS) of applications on embedded manycore NoCs [11].

Based on the "utilization wall" [18], only fraction of the on-chip transistors can be used at full-speed at one time. Thus, *internal fragmentation*, i.e., existence of idle CPU cores within submeshes, is allowed under our strategy. The idle cores within submeshes are put to sleep to save power.

III. THE PROPOSED HIERARCHICAL STRATEGY

A. Overview

Under the previous strategies in [4] [6], the resource management is carried out by the GM alone, which could make the GM become the performance bottleneck of the system when the number of the CPU cores increases. In our proposed strategy, after submesh is introduced, the resource management is carried out by the GM and local managers in a hierarchical way. The GM manages resources for applications in forms of submeshes. The resource management within submeshes is off-loaded from the GM to the local managers so as to reduce the GM's chance of becoming the performance bottleneck.

For an example application App, the processes that it goes through under our strategy are elaborated as follows. (1) When ACG of App is given, the off-line preprocessing of App decides the submesh size for App and how to map tasks of App onto the submesh. (2) When App enters the system, the GM allocates a submesh S to App based on the required submesh size. The CPU core at the top-left corner of S is chosen as the local manager of S, denoted as LM_S. (3) After S is allocated to App, the LM_S maps the tasks of App onto the cores on S. (4) When a task of App completes, the core running this task sends a message to the LM_S to notify the task's completion. When all tasks of App complete, the LM_S sends a message to the GM to start a deallocation process.

The components of our strategy consist of preprocessing of applications, submesh management, and hierarchical resource management, which are discussed in following sections.

B. Off-line Preprocessing of Applications

For an application App whose characteristics are given, the off-line preprocessing aims to decide: *(1)* dimensions of the submeshes that can accommodate App; *(2)* how to map App's tasks to the corresponding submesh when a dimension is chosen. As dimension has direct influence on the mapping of tasks, these two factors are considered together. A method

for off-line preprocessing is presented here and more methods can be investigated in future.

As applications run on separate submeshes, external communication contentions among applications are avoided. When App runs on a submesh S, the communication contentions within S are caused by App's own tasks. Hence, to achieve optimal submesh dimension and mapping of tasks, the goal is to minimize the communication energy consumption by App.

Assume that App is described in ACG_{App}, using the bit energy metric in [15], the total communication energy consumption of App per time unit can be calculated using the following equation:

$$E_{App} = \sum_{\forall e_{ij} \ \in \ E \ in \ App} w(e_{ij}) \times E_{bit}(e_{ij})$$

where $w(e_{ij})$ is the communication rate of an edge in ACG_{App} (in bits per time unit), and $E_{bit}(e_{ij})$ stands for the energy consumption to send one bit between the cores where vertices v_i and v_j are allocated to (in Joules per bit).

Assume that App has n tasks. We search for a submesh $S_{w \times h}$ whose number of cores is equal or larger than n and the difference between width (w) and height (h) is minimal. Such a submesh has enough resources to run App and the average number of communication hops is minimized [16]. After $S_{w \times h}$ is decided, App's tasks are mapped to cores of the submesh where a task is allocated to a core. All possible mappings are checked to find one that makes E_{App} minimal. An exhaustive search for submeshes and their mappings is done off-line, which is rational as n usually is small. Commonly, applications from the E3S benchmarks [14] have less than a dozen tasks. When n is large, other algorithms can be explored.

For the five applications whose application characteristics are shown in Fig. 1, after the preprocessing, the dimensions of submeshes and mapping of tasks are shown in Fig. 3.

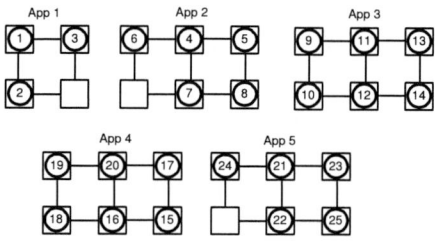

Fig. 3. Mapping tasks of applications to submeshes

C. Submesh Management

As submesh-based processor allocation schemes have been applied to large-scale parallel and distributed systems, many algorithms have been proposed in the literature. The free-list submesh allocation scheme proposed in [12] is adopted by our strategy as it is better than other schemes in terms of scalability and algorithmic complexity. A submesh consisting of non-occupied cores is a *free submesh*. The time complexity of the scheme is linear to the number of free submeshes.

978-1-61284-863-1/11 $26.00 © 2011 IEEE

The adopted scheme maintains an unordered list, named *LIST*, of possibly overlapped free *maximal submeshes* (defined in [12]). When none of cores is occupied, the list contains only a free submesh, i.e., the whole mesh.

For allocation process, the scheme selects the first free submesh that has at least the same size as the request, and when the selected submesh is larger than the request, the part actually allocated is one that has the largest number of busy neighbors and boundary cores [12]. For deallocation process, submeshes released by applications are marked as free again. After successful allocation and deallocation of applications, the scheme dynamically splits or combines submeshes to make sure that only maximal submeshes are kept in the list.

D. Hierarchical Resource Management

The hierarchical resource management has the below benefits. A local manager is responsible for handling the messages from the cores inside its submesh. So, the other cores inside the submesh don't exchange messages directly with the GM. GM only communicates with local managers and its communication is decreased. So, the overall energy consumed in communication is reduced as most communications are within the submeshes. Moreover, the local manager can implement application level optimizations within its submesh.

Local managers are designed in a way to separate the time for managing their submeshes and for carrying out computation tasks so as not to impact their computation efficiency.

1) Hierarchical Resource Allocation: For an application *App* after preprocessing, the GM first identifies a suitable submesh for it. After a submesh is allocated to *App*, the core at the left-top corners of the submesh is chosen as "local manager" as it has the minimal distance to the GM among cores within the submesh. This local manager maps tasks of *App* onto the other cores of its submesh. To support the hierarchical allocation, related data structures are designed.

2) Hierarchical Resource Deallocation: When a task of *App* finishes, it sends a control message to its local manager to notify its completion. When all cores inside of a submesh have finished the processing of their tasks, the local manager notifies the global manager that the application has been completed and the submesh is ready for deallocation.

E. An Illustrative Example

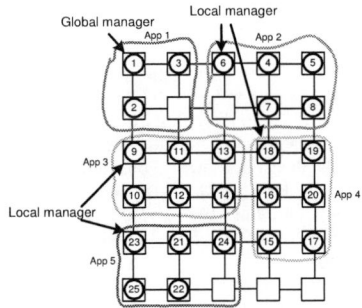

Fig. 4. A hierarchical mapping of five applications to a 5 × 6 NoC

Fig. 4 shows an example of a hierarchical resource management. The GM decides the 5 submeshes for applications App_i, $1 \leq i \leq 5$. The local manager of the submesh for App_2 maps tasks 4, 5, 6, 7, and 8, to the cores of its submesh.

In Fig. 2, the GM exchanges control messages with the other 24 cores during resource management. However, in Fig. 4, only the local managers exchange messages with the GM. The number of cores that the GM communicates with is greatly decreased, from 24 to 4. The disadvantage shown in Fig. 4 is that some resources are allocated but not used.

IV. EVALUATION OF THE PROPOSED STRATEGY

A. Experimental Setup

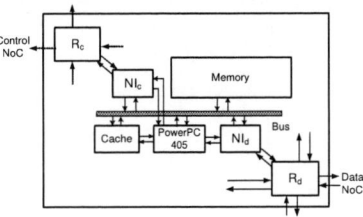

Fig. 5. The microarchitecture of a tile of embedded NoC

1) The Simulator: A cycle-accurate simulator for an embedded manycore NoC with a 5 × 6 mesh is generated from the simulation framework [8] [9]. The microarchitecture of its tile is shown in Fig. 5. Two instructions "send" and "recv" are added to enable communication using data NoC. Similar to the data NoC, two instructions are added to enable the CPU core to access the NI of the control NoC (NI_c). The CPU core can read (send) control messages from (to) NI_c.

2) The Bit Energy Metric Model: The bit energy metric model proposed in [15] is adopted in above simulator to evaluate the communication energy consumption in experiments where E_{link} is set to $4.49 \times 10^{-13} (J/bit)$ and $E_{R_{bit}}$ contains the energy consumes by the routing engine (10^{-13}J/packet), arbiter request (1.155×10^{-12} J/packet), switch fabric (2.84×10^{-13} J/bit), and buffer reading and writing (1.056×10^{-12} J/bit and 2.831×10^{-12} J/bit, respectively).

B. Experimental Results

Two experiments have been done to compare the strategy in [6] and our strategy. The GM supports different resource management strategies in these experiments. The algorithms of these strategies are implemented in C language and are compiled to be executed on the PowerPC 405 CPU core. We focus on the comparisons of *1)* the time spent on allocation process by the GM and *2)* the communication energy consumed during the resource management.

Five applications are chosen from the embedded system benchmark suite (E3S) [14] as they have been used in experiments in [6]. They are partitioned off-line using the method in [10]. Their application characteristics and preprocessing results are shown in Fig. 1 and Fig. 3, respectively.

These applications are assumed to invoke the events, which are shown in Table I, when they enter the system and leave

978-1-61284-863-1/11 $26.00 © 2011 IEEE

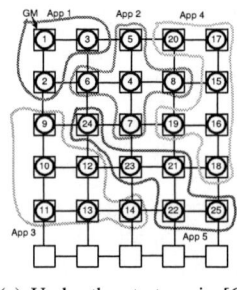 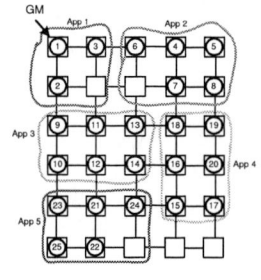

(a) Under the strategy in [6]   (b) Under our strategy

Fig. 6.   System Configurations at Time 4

the system, in both experiments. In each experiment, the corresponding resource management strategy is invoked in handling these events. Accordingly, the system configuration changes after each event occurs.

TABLE I
EVENTS IN EXPERIMENT

Event	Start Time	Incoming application	End Time
1	0	App_1	5
2	1	App_2	6
3	2	App_3	12
4	3	App_4	13
5	4	App_5	9

1) Time of Executing Allocation Algorithms: The cycles used to execute the algorithms during simulations under above events are recorded in the Table II. At time 4, after resources have been allocated to the application App_5, two different system configurations following the two different strategies are shown in Fig. 6. Note that the bottom row of Fig. 6 (a) is not used in the allocation process.

TABLE II
EXECUTION CYCLES OF ALLOCATION ALGORITHMS BY THE GM

Event	1	2	3	4	5
Algorithm in [6]	122,776	892,368	1,092,368	1,322,875	1,262,536
Our Algorithm	121,890	281,337	132,047	301,752	298,266

We can see from Table II that the execution of allocation algorithm in [6] by the GM generally takes more time than our proposed allocation algorithm. The reason is as follows. The allocation algorithm in [6] implements functionalities including identifying resources for an application and mapping tasks of the application to the identified resources. However, our allocation algorithm only identifies a suitable submesh.

2) Communication Energy Consumption: We compare these two configurations in terms of communication energy consumption as follows. Communication energy consumption is calculated using the aforementioned bit energy metric.

For configuration in Fig. 6 (a), $7.92 \times 10^{-10} J$ energy is used for transmitting messages. However, for the configuration in Fig. 6 (b), the energy is $2.93 \times 10^{-10} J$. The reduction ratio is 63%. The reason is that our strategy reduces the number of cores, which exchange messages with the GM, from 24 down to 4. Most cores only exchange the messages with their local managers, leading to shorter communication distances. Therefore, the communication energy is greatly reduced.

3) Internal Fragmentation: The cost for saved time in allocation process and saved energy in transmitting control messages is that some cores allocated in certain submeshes are not used for executing tasks. They can be considered as "internal fragmentation". For submeshes allocated to the applications App_1, App_2 and App_5, the "internal fragmentation ratio" are 25%, 16.7% and 16.7%, respectively.

V. RELATED WORK

OS has been proposed to optimize communication resource usage with the NoC support and ensure the QoS requests of applications [17]. Runtime strategies [4] [6] have been proposed for resource management on embedded multicore NoCs. Submesh-based processor allocation schemes such as [12] [13] have been used in parallel and distributed systems for performance optimization and resource management.

VI. CONCLUSION

A submesh-based scalable hierarchical strategy for runtime resource management on embedded manycore NoCs has been presented. Experiments show that our strategy reduces the time spent on centralized allocation process and saves communication energy consumed in resource management.

REFERENCES

[1] K. Asanovic et al., The Landscape of Parallel Computing Research: A View from Berkeley. TR UCB/EECS-2006-183, U. C. Berkeley, 2006.
[2] S. Borkar. Thousand Core Chips: A Technology Perspective. In *DAC '07: Proc. of the 44th Design Automation Conference*, 2007.
[3] L. Benini et al., Networks on Chips: A New Paradigm for Component-Based MPSoC Design. In IEEE Computer, pp. 70-78., Jan. 2002.
[4] C. Chou et al., Energy- and Performance-aware Incremental Mapping for Networks on Chip with Multiple Voltage Levels. *CAD of Integr. Circuits and Syst., IEEE Trans. on*, 27(10):1866 –1879, Oct. 2008.
[5] S. V. Tota et al., A Case Study for NoC-based Homogeneous MPSoC Architectures. *IEEE Trans. VLSI Syst.*, 17(3):384–388, 2009.
[6] C. Chou et al., Run-time Task Allocation Considering User Behavior in Embedded Multiprocessor Networks-on-Chip. IEEE Trans. on CAD of Integr. Circuits and Syst., 29, 2010.
[7] Tilera. TILE Processors, http://www.tilera.com, 2009.
[8] X. Liao et al., A Modular Simulator Framework for Network-on-Chip Based Manycore Chips Using UNISIM. Transactions on High-Performance Embedded Architectures and Compilers, Vol.4 (4), 2009.
[9] X. Liao et al., A UNISIM Based Simulator Framework for NoC Based Manycore Systems. http://sourceforge.net/projects/nocsim-unisim.
[10] M. Pastrnak et al., Parallel Implementation of Arbitrary-Shaped MPEG-4 Decoder for Multiprocessor Systems. Proc. Visual Comm. and Image Processing, 2006.
[11] M. Azimi et al., Integration Challenges and Tradeoffs for Tera-Scale Architectures. *Intel Technology Journal, Volume 11, Issue 3*, 2007.
[12] I. Ababneh. An Efficient Free-List Submesh Allocation Scheme for Two-Dimensional Mesh-Connected Multicomputers. *Journal of Systems and Software*, 79(8):1168–1179, 2006.
[13] S. Bani-Mohammad et al., An Efficient Non-Contiguous Processor Allocation Strategy for 2D Mesh Connected Multicomputers. *Information Sciences*, 177(14), pages 2867-2883, 2007.
[14] R. Dick, Embedded System Synthesis Benchmarks Suites (E3S)[online], available http://ziyang.eecs.umich.edu/~dickrp/e3s, 2010.
[15] T. Ye et al., Analysis of Power Consumption on Switch Fabrics in Network Routers. In Proc. of DAC'02, 2002.
[16] M. Bender et al., Communication-Aware Processor Allocation for Supercomputers: Finding Point Sets of Small Average Distance. *Algorithmica*, 50(2):279–298, 2008.
[17] Vincent Nollet et al., Operating-System Controlled Network on Chip. In *DAC '04: Proc. of the 41st Design Automation Conference*, 2004.
[18] G. Venkatesh et al., Conservation Cores: Reducing The Energy of Mature Computations. In *Proc. of ASPLOS'10*, 2010.

Hybrid Non-preemptive/Cooperative Multitasking on NoC Based Manycore Systems

Xiongfei Liao[†], Thambipillai Srikanthan[‡] and Xiao He[‡]

[†]Institute of Microelectronics, A*STAR (Agency for Science, Technology and Research), Singapore

[‡]School of Computer Engineering, Nanyang Technology University, Singapore

Email: liaox@ime.a-star.edu.sg

Abstract—**Existing resource management strategies for embedded NoCs have limitations due to their adopted non-preemptive multitasking approach. This paper proposes a hybrid multitasking technique in order to overcome the above limitations.**

I. INTRODUCTION

"Manycore" processors with hundreds/thousands of small but energy-efficient CPU cores will become the mainstream [1] [2]. Some of them have identical tiles connected by Network-on-Chip (NoC) [3] with 2D mesh and are of interest to researchers [4] [5] [6]. They are called "embedded manycore NoCs" here and an example is Tilera's TILE-Gx100 [7].

Embedded manycore NoCs can execute several applications concurrently. Each application can have multiple tasks running in parallel, using multiple CPU cores. As these NoCs will be used in devices such as smart phones and PDAs, users can start/stop applications that lead to dynamic system configurations in term of the use of CPU cores. Such dynamic configurations are extremely difficult to model off-line. Hence, run-time techniques are indispensable.

Several run-time strategies [4] [6] have been proposed for resource management by providing suitable resources, i.e., CPU cores (a core is a basic unit to execute tasks), to tasks of an application when it enters the system. Resources are reclaimed after the application finishes. In addition, these strategies adopt *the non-preemptive multitasking*. Under this multitasking, after resources are allocated to an application, tasks of the application, which have been allocated with CPU cores, can not be interrupted until these tasks finish.

Though the non-preemptive multitasking is simple, it has its limitations. Fig. 1 illustrates a scenario of failure in launching a new application. There are 5 applications, App_i, running on the areas S_i respectively (these areas are assumed to be rectangular for simplicity), where $i = 1, 2, 3, 4, 5$. Now, the user tries to start an application J, which requests an area of 3×3. Due to the non-preemptive multitasking, no resource can be offered to J before some application completes. In another scenario, after App_i, $i = 1, 2, 3, 4$, complete and the resources they used are released, but App_5 is running, J still cannot get started because a 3×3 area cannot be found for J. Though there are idle CPU cores, J has to wait until App_5 is completed. This leads to low resource usage efficiency.

To the best of our knowledge, Tessellation manycore OS [8] is a preemptive multitasking approach that has been proposed

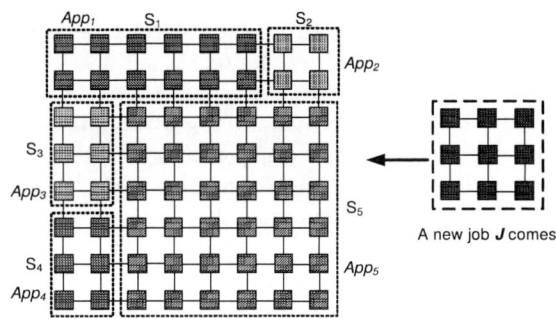

Fig. 1. Concurrent applications on a 64-tile embedded manycore NoC

to improve the resource usage on manycore systems in above scenarios. However, it is still at its early research stage.

In this paper, we propose a hybrid multitasking technique to overcome the limitations of the non-preemptive multitasking. We propose to introduce *cooperative multitasking* to embedded manycore NoCs. A *cooperative application*, which supports cooperative multitasking, can cooperate with the OS at run-time, after resource allocation is completed, for *1)* giving up some of its computing resources to an incoming application to get started; *2)* obtaining extra resources so as to achieve higher performance if possible. For embedded manycore NoCs, it is not required for all applications to be cooperative. The non-preemptive and cooperative applications can co-exist and we further propose a *hybrid non-preemptive/cooperative multitasking* to enable interactions between applications so as to reduce the chance of failure in launching applications and improve the resource usage efficiency.

II. COOPERATIVE MULTITASKING

An alternative for the non-preemptive multitasking is *preemptive multitasking*. On single-CPU systems supporting this multitasking, the CPU is shared by processes based on time slices. When the running of a process is interrupted after its time slice expires, various temporary values must be stored in memory until the process resumes its execution. Thus, this multitasking is resource costly because of the large amount of memory needed. It is also time consuming due to lots of accesses to memory. Preemptive multitasking for manycore systems is even more time and resource demanding. Tessellation OS is trying to support preemptive multitasking.

978-1-61284-863-1/11 $26.00 © 2011 IEEE   301

Another alternative is the cooperative multitasking, which has been applied on single-CPU systems where applications are designed with interruption points. When an interruption point of a process is reached, the OS is permitted to switch to another process. To save memory for storing temporary values, designers set interruption points where few values needed to be stored. However, on single-CPU systems under this multitasking, each application must be designed as co-operative. Otherwise, because the OS has no way to preempt operation, once an application starts, if it does not freely give up control, it will continue to run until it terminates. This leads to no multitasking at all. As the efforts of designing all applications as cooperative are paramount, cooperative multitasking became obsolete in single-CPU systems.

Observing that a parallel application has multiple CPU cores for running its tasks on embedded manycore NoCs, we propose to introduce a cooperative multitasking for multi-/many-core systems where an application manages its computing resources in terms of CPU cores rather than time slices. Further, an application can be designed with one or more interruption points. When an interruption point is met during its execution, the application may give up or take in CPU cores after negotiating with the OS. After that, the application can continue with an adjusted amount of resources and possibly changed performance. Before meeting next interruption point, tasks of the application run on CPU cores in the non-preemptive style.

A cooperative application can be designed in a way that the number of CPU cores it uses has a *1)* low bound (LB) for the application to maintain the minimal performance for correct execution and *2)* a high bound (HB) desired for maximal performance. Moreover, a cooperative application can run with n cores ($LB \leq n \leq HB$) for variable performance.

Different from the cooperative multitasking on single-CPU systems, it is not mandatory for all applications on embedded manycore NoCs to be cooperative. The proposed cooperative multitasking can be realized if at least one application cooperates with the OS. Other applications that are not cooperative run in the non-preemptive style.

III. THE HYBRID MULTITASKING

A. Overview

As discussed above, it is possible that non-preemptive and cooperative applications coexist on an embedded manycore NoC. Hence, we propose a novel hybrid multitasking approach which combines features of non-preemptive multitasking and cooperative multitasking and aims at improving the flexibility of resource management at runtime. This hybrid multitasking is named "hybrid non-preemptive/cooperative multitasking".

The resource management on embedded manycore NoCs under the hybrid multitasking has the following features: *1)* when there are enough resources for applications, the OS allocates desired resources to them such that they are executed for maximal performance; *2)* when a new application comes but available resources are insufficient for it, the OS tries to find resources for it via negotiations between the OS and the running cooperative applications; *3)* when any application

releases its resources upon completion, the OS communicates with running cooperative applications such that some of them may gain extra resources to achieve higher performance.

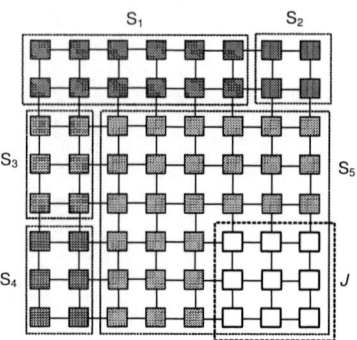

Fig. 2. An example of hybrid non-preemptive/cooperative multi-tasking

With the proposed hybrid multitasking, for the example in Fig. 1, if App_5 is designed as cooperative, a 3×3 area can be cut from the area S_5 to get J started and the resulted system configuration is shown in Fig. 2.

The inseparable components of the proposed hybrid multitasking consist of the following:

1) A negotiation mechanism between OS and applications for cooperative multitasking.
2) Architectural supports on embedded manycore NoCs for realizing the hybrid multitasking.
3) A method for designing cooperative application or parallelizing applications to support cooperative multitasking.

These components are discussed in the ensuing subsections.

B. The Negotiation Mechanism

The negotiation between the OS and applications is realized by exchanging *control messages*. The OS initializes negotiations when resource management events occur. The OS broadcasts cooperation requests to applications. Non-cooperative applications ignore requests and cooperative applications reply with POSITIVE or NEGATIVE messages. Further the OS can confirm or cancel a cooperation. After its cooperation is completed, an cooperative application notices the OS.

Each message has an ID that is used to match responses to requests. The brief information of control messages is listed in Table I. Other information of messages such as amount of resources is omitted due to limited space.

TABLE I
CONTROL MESSAGES USED IN NEGOTIATION

Message	Source	Destination	Meaning
REQUEST	OS	Application	Request for cooperation
NEGATIVE	Application	OS	Deny to cooperate
POSITIVE	Application	OS	Agree to cooperate
CONFIRM	OS	Application	Cooperation is confirmed
CANCEL	OS	Application	Request is cancelled
COMPLETE	Application	OS	Cooperation is completed

A cooperative application doesn't concede all of its resources for new applications and must have at least LB CPU cores for its execution. The number of conceded CPU cores from an application depends on the available resources. On the other hand, the OS guarantees that an application doesn't have more than HB CPU cores.

C. The Architectural Supports

Like the strategies in [4] [6], besides an NoC for transferring data, a separate NoC is added to transfer the control messages. In previous strategies, the control NoC is implemented as a broadcast tree and support the unidirectional communication. In our proposal, a bidirectional control NoC is adopted.

To avoid deadlocks, virtual channels are adopted in the control NoC. The number of virtual channels for a physical channel is configured to be at least 2. To avoid deadlocks, messages from the OS to applications are always and only sent via the virtual channel 0. The messages from applications to the OS can only be sent through other virtual channels.

The network interface (NI) of the control NoC is designed to enable interactions between the OS and cooperative applications. The NI has a buffer for storing "REQUEST" messages before an application handles them. The NI removes a "REQUEST" message from the buffer if a "CANCEL" message is received before the request is handled. This situation happens when another application has responded to the request. The NI of a non-cooperative application responds to "REQUEST" messages with "NEGATIVE" messages.

D. The Method of Designing Cooperative Applications

A cooperative application has the following features: *1)* it is a parallel application with tasks executed on multiple cores; *2)* interruption point(s) must be designed where the application checks if there is any resource request; *3)* the application can adjust the number of CPU cores in use based on request; *4)* the application can continue execution to achieve correct results with the adjusted number of CPU cores.

For an application with data parallelism, we present the below method for designing it for cooperative multitasking. During processing of multiple data units, a data unit is divided into smaller parts and these parts are computed by several CPU cores in parallel. After all parts of a unit are processed, these partial results are combined to get the final result. An interruption point can be placed before processing the following unit where the cooperative application can check resource requests and change the amount of its resources. After resource adjustment, the volume of partial data for each core can be adjusted accordingly. The parts of data should be approximate such that the workloads of cores can be balanced to achieve high performance.

In processing data, the cores of a cooperative application are categorized into a master and slaves, following the *master and slave* pattern for parallel programming. The master divides a unit of data into approximate parts and distributes them to other cores which act as slaves. Slaves compute their partial data respectively and send partial results to the master. The master combines partial results into a final result. Before proceeding to the next unit, the master reaches an interruption point where it checks and handles resource request if there is any. Then, the master proceeds to the next unit if there is any.

IV. EVALUATION OF THE HYBRID MULTITASKING

To evaluate the hybrid multitasking, an MPEG-2 encoder is parallelized into a cooperative application whose behaviour is studied under resource cooperation requests at runtime.

A. Cooperative Application Example: MPEG-2 Encoder

An MPEG-2 encoder [9] is chosen to be parallelized into a cooperative application. The hierarchy of layers, which allows random access, in an MPEG-2 stream is: *sequence, group of frames, frame, slice, macro-block* and *block*. As shown in Fig. 3, the data parallelism is exploited at the granularity of slice and a frame is split into slices of equal size at master which are distributed to be processed by slaves. The partial results are merged into a whole result at output slave. An interruption point is set at the source code executed by the master where after a frame is processed and before the next frame is handled.

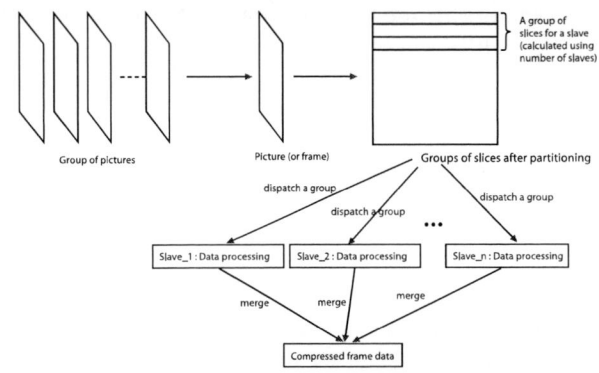

Fig. 3. Data parallelism at slice level

B. Experimental Setup

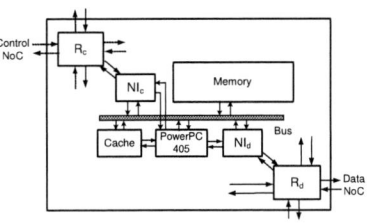

Fig. 4. The microarchitecture of a tile

1) The Simulator: A cycle-accurate simulator for an embedded manycore NoC with an 8×8 mesh is generated from the simulation framework [10] [11]. The microarchitecture of its tile is shown in Fig. 4. Similar to the data NoC, two instructions are added to enable the CPU core to access the NI of the control NoC (NI_c). These instructions are only executed when the interruption point is met. The CPU core can read (send) control messages from (to) NI_c.

2) The Video Sequence: We use the beginning 30 frames of the Foreman video sequence [12] in experiments. The size of a frame is 352×288 pixels. As the height of a slice is chosen to be 16 pixels, each frame has 18 slices.

978-1-61284-863-1/11 $26.00 © 2011 IEEE

3) Simulated Scenarios: Assumptions are made as below. Initially the parallelized MPEG-2 encoder runs on 20 cores (1 master, 1 output slave and 18 processing slaves). For simplicity, among all running applications only the encoder supports cooperative multi-tasking. Other applications come to (depart from) the system during the encoding of 30 frames, leading to resource management events. Under the requests from the OS, the encoder cooperates on resource management.

TABLE II
EVENTS IN EXPERIMENTS

Experiment	New application	Start Time*	Resource Request	End Time*
1	J_1	frame 4	3 cores	frame 16
1	J_2	frame 4	9 cores	frame 24
2	J_1	frame 4	3 cores	frame 24
2	J_2	frame 4	9 cores	> frame 30 **

*: The time is referred as the period during which a frame is under processing.
**: J_1 is still running when the processing of frame 30 is completed.

Two experiments are done to study the behaviour of the MPEG-2 encoder when resource management events occur. The information of events is shown in Table II. For example, the first row of the table indicates: during the processing of frame 4, application J_1 comes to the system, requesting 3 CPU cores; J_1 finishes when frame 16 is under processing.

We use speedups to describe the behaviour of the encoder. As the baseline configuration, one processing slave is used and clock cycles spent on executing each frame are recorded. As the advanced configuration, the encoder runs under hybrid multi-tasking and clock cycles used in processing each frame are also recorded. The speedups for individual frames are calculated using recorded cycles under two configurations.

C. Experimental Results

In Experiment 1, the execution of encoder undergoes stages shown in Fig. 5. *1)* The encoder starts with 18 processing slaves and continues encoding frames from 1 to 4. *2)* After frame 4 is processed, at the interruption point, the master finds two requests for conceding cores. The encoder concedes a total of 12 (= 3 + 9) cores for J_1 and J_2. *3)* So, for frames from 5 to 16, the encoder has 6 slaves for encoding. *4)* After frame 16 is processed, the master finds one request for returning cores. After being given back 3 cores, the encoder runs with 9 cores to process frames from 17 to 24. *5)* After frame 24 is processed, the master finds another request for returning cores. After being given back 9 cores, the encoder runs with 18 cores to process frames from 25 to 30. The overall speedup achieved in the experiment is 8.02.

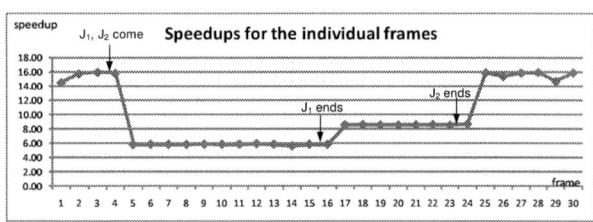

Fig. 5.   Speedups for the individual frames in Experiment 1

The stages of encoder in Experiment 2 are shown in Fig. 6 and the achieved overall speedup is 6.66.

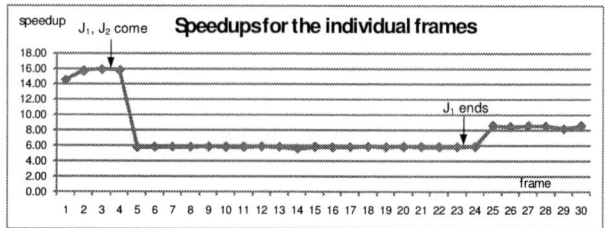

Fig. 6.   Speedups for the individual frames in Experiment 2

Experiments show that the parallelized encoder supports cooperative multi-tasking by dynamically adjusting its resources when resource management events occur, achieving variable performance.

V. RELATED WORK

OS has been proposed to optimize communication resource usage with the NoC support and ensure the quality-of-service requests of applications [13]. Runtime strategies [4] [6] have been proposed for task allocation on homogeneous embedded multicore NoCs. They support centralized and event-driven resource management, and non-preemptive multitasking. Many-core OSes such as Tessellation [8] take an aggressive approach of preemptive multitasking in resource management.

VI. CONCLUSION

We have presented a novel technique for hybrid multitasking to overcome the limitations resulted from the non-preemptive multitasking approach. Experiments with an MPEG-2 encoder supporting the cooperative multitasking demonstrate the feasibility of the proposed hybrid multitasking.

REFERENCES

[1] K. Asanovic et al., The Landscape of Parallel Computing Research: A View from Berkeley. TR UCB/EECS-2006-183, U. C. Berkeley, 2006.
[2] S. Borkar. Thousand Core Chips: A Technology Perspective. In *DAC '07: Proc. of the 44th Design Automation Conference*, 2007.
[3] L. Benini et al., Networks on Chips: A New Paradigm for Component-Based MPSoC Design. In IEEE Computer, pp. 70-78., Jan. 2002.
[4] C. Chou et al., Energy- and Performance-aware Incremental Mapping for Networks on Chip with Multiple Voltage Levels. *CAD of Integr. Circuits and Syst., IEEE Trans. on*, 27(10):1866 –1879, Oct. 2008.
[5] S. V. Tota et al., A Case Study for NoC-based Homogeneous MPSoC Architectures. *IEEE Trans. VLSI Syst.*, 17(3):384–388, 2009.
[6] C. Chou et al., Run-time Task Allocation Considering User Behavior in Embedded Multiprocessor Networks-on-Chip. IEEE Trans. on CAD of Integr. Circuits and Syst., 29, 2010.
[7] Tilera. Tilera, http://www.tilera.com, 2009.
[8] R. Liu et al., Tessellation: Space-Time Partitioning in a Manycore Client OS. *Proc. of HotPar'09*, 2009.
[9] M. Li et al., The ALPbench Benchmark Suite for Complex Multimedia Applications. In *Workload Characterization Symposium, 2005. Proc. of the IEEE International*, pages 34 – 45, 6-8 2005.
[10] X. Liao et al., A Modular Simulator Framework for Network-on-Chip Based Manycore Chips Using UNISIM. Transactions on High-Performance Embedded Architectures and Compilers, Vol.4 (4), 2009.
[11] X. Liao et al., A UNISIM Based Simulator Framework for NoC Based Manycore Systems. http://sourceforge.net/projects/nocsim-unisim.
[12] ASU. Forman Video Sequences. http://trace.eas.asu.edu/yuv/index.html.
[13] Vincent Nollet et al., Operating-System Controlled Network on Chip. In *DAC '04: Proc. of the 41st Design Automation Conference*, 2004.

978-1-61284-863-1/11 $26.00 © 2011 IEEE

An Autonomous Vehicle Using a Multi-Thread and Event-Driven Processor

Touta Hayashi
Graduate School of Computer and Information Sciences
Hosei University
3-7-2, Kajino-cho, Koganei-shi, Japan
touta.hayashi.4a@stu.hosei.ac.jp

Kenji Ohmori
Computer and Information Sciences
Hosei University
3-7-2, Kajino-cho, Koganei-shi, Japan
ohmori@hosei.ac.jp

Abstract—**The conventional microcomputers often used in autonomous vehicles suffer from the disadvantage of having long complex codes containing unavoidable bugs. This paper describes how to resolve the complexity of both software and hardware development of an embedded system using a new XMOS processor that can perform concurrent processes. In recent years, more and more software controllers have been installed in many parts of a vehicle. Conventional microcomputers based on sequential execution are unsuitable for concurrent processes in a real-time system. So-called spaghetti codes with their insufficient interruption handling bring about serious problems. In contrast, an event-driven, multi-thread XMOS processor can accommodate simple and user-friendly codes using highly abstract modeling. A secure embedded system that takes advantage of event-driven, multi-thread processors has been developed for a radio-controlled car with some sensors and simple codes on an XMOSX K-1 board. Successful results have been obtained with modest efforts.**

Index Terms—**Concurrency, embedded system, event-driven, multi-thread, XMOS**

I. Introduction

Autonomous vehicles are used worldwide as public transportation in cities, rescue machines in disaster-affected areas, or sweepers in living rooms. These vehicles work in a stand-alone mode with preprogrammed patterns. The basic mechanism of an autonomous vehicle is quite simple, involving (1) getting information from sensors, (2) analyzing the information to recognize problems in the environment, and (3) carrying out appropriate tasks to solve the problems. Regardless of the simplicity, conventional microprocessors need quite complicated software and hardware organization because of poor performance, the use of low-level program languages, and an inappropriate design method that gives priority to hardware.

In the development of software systems for personal computers, object-oriented programming is popular because it offers efficient and secure software development, an advantage over traditional embedded systems. Embedded systems are featured as event-driven concurrent processes. In the 1970s, serious problems in concurrent processes were pointed out and many ideas to conquer them have been proposed. In Hoare's famous paper "Communicating Sequential Processes" [1] he introduced process algebra for concurrent systems. In accordance with process algebra, the programming language

Occum [2] and the concurrent processor Transputer [3] have been developed. These technologies have never been utilized in the marketplace since the device technology at that time was not mature enough to develop such concurrent processors. The basic ideas of Transputer and Occum, and of their successor XMOS [4], are suitable for use in the marketplace. In recent years, some applications for business [5] and vehicle control [6] have been announced.

II. XMOS

XMOS, a new programmable semiconductor from XMOS, Ltd., in the United Kingdom, has been built on the basis of the concept of communicating sequential and event-driven processes. XMOS uses a new concept, called software-defined silicon, in which hardware mechanisms are realized by software programs using emulation technology. This concept is based on the following idea: There are no differences between signals from hardware circuits and signals from software emulation if they are delivered on time. XMOS technology features a compact, event-driven, multi-threaded processor, called an XCore, and a high-speed low-latency switch, called an XLink. An XCore is a general-purpose processor and is tightly coupled to the outside world through a set of event-driven input–output ports. One or multiple XCores are embedded in an XMOS chip. Each XCore can run eight threads concurrently. When an XMOS chip is constructed with multiple XCores, the XMOS chip can run many (eight times the number of XCores) threads. Because it is possible to connect multiple XMOS chips, a massive parallel system can be achieved by XMOS chips. Threads can communicate with each other by message passing. To guarantee low-latency communication between threads located at different XCores, special communication components, called XLinks, are provided for each XCore. Programs on an XCore are written in the XC or the C language and are compiled with an extended GNU Compiler Collection. XC enhances parallelism and message passing in C. An XMOS Development Environment (XDE) is a supporting tool, based on an eclipse IDE, that includes a source code editor, compiler, debugger, device manager, and simulator. A timing simulator for serious real-time management is also supplied with an XDE. Since XCore's thread execution is deterministic and the

978-1-61284-863-1/11 $26.00 © 2011 IEEE

Fig. 1. XMOS car.

TABLE I
SENSOR LIST.

	sensor	min (cm)	max (cm)	number
Short range	GP2Y0A21YK	10	80	6
Long range	GP2Y0A710K	100	550	1

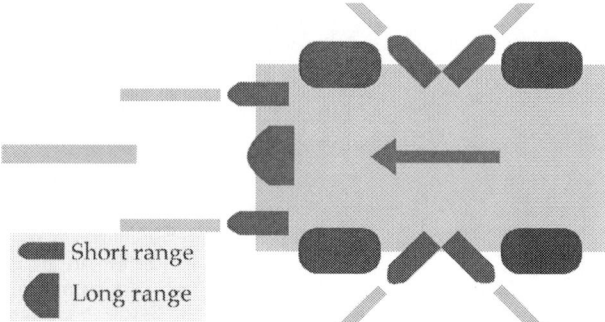

Fig. 2. Sensors.

time to execute a sequence of instructions can be accurately predicted, the behavior of processes can be simulated precisely.

III. XMOS CAR

An XMOS car has been developed by remodeling a radio-controlled (RC) car using an XMOS XK-1 development board and some sensors. The RC car is a hobby model controlled remotely with a radio-frequency link. A series of throttle positions and rudder angles that a driver generates by operating a remote controller are delivered to an on-board device on the car through radio waves. The RC car runs according to the control operator's operation. The remodeling principle is not so complicated. It is realized by replacing the car's radio-frequency receiver with an XMOS XK-1 development board and by having the XMOS board act exactly like the removed radio-frequency receiver. If the signals generated by the XMOS board are identical to the ones that would be generated by the removed radio-receiver, the XMOS car works perfectly as if nothing has been changed from the original analog circuits of the RC car.

By using an RC car chassis, the XMOS car (Fig. 1) be made to run at full speed with minimal effort and less knowledge of machine engineering. If only speed is of concern, a handmade chassis with a powerful drivetrain is a reasonable choice. When a vehicle is running at high speed, it is difficult for a handmade chassis to run straight since speed depends sensitively on power and weight. A crude car will meander even at the neutral steering position. A computerized controller might be adopted to stop meandering as has been done by Toyota for real cars. Because this kind of controller is too complicated and belongs to future work, it is not implemented in the study presented in this paper. In this study, an RC car chassis equipped with a well-designed suspension and steering system is installed to enable a prompt study of the usefulness of event-driven, multi-thread processors.

Even more useful are the RC cars sold as kits in hobby shops. These have components that are modularized and easily

replaceable. Because these RC cars are just hobbyist models, specification sheets, circuit diagrams, and guarantees are not available. Although these shortcomings may be disadvantageous for mass production, component replaceability, which is important for research, outweighs these disadvantages. In this study, an old Tamiya electric radio-controlled car is used. It emulates a Honda S-MX and uses an M-01M chassis. M-01M is the first chassis in Tamiya's mini-chassis series. Hence it is a prototype model of front-wheel-drive chassis; its design has not been polished and there is considerable margin for payload extra devices, unlike the more recent chassis.

An XMOS XK-1 development board with a single-core XS-1 processor was installed on the RC car. The XK-1 board has the smallest footprint of all XMOS development boards, has less capability than others with four-core XS-1 processors, and is scalable by means of powerful XLinks. The XK-1 board is powerful enough and its small size is important for installing it on a vehicle.

A Futaba motor driver unit for the RC car was also installed because the original radio receiver integrates a motor driver unit inside, which prevents it from being accessed from the XK-1 board. A general motor integrated circuit is used instead of a commercial motor driver unit. Since a modern chip using FET elements is hard to obtain and an old-style chip using bipolar elements is less efficient and produces heat, alternatively, FET elements to compose circuits are installed in the system (although this poses some risk of fire caused by a circuit design error). Using homemade FET elements for RC cars is a good choice for completing a study in a short time and at low cost. Their light-weight package and highly efficient FET elements are sufficient for the purpose of winning an RC car race.

For sensors, Sharp optical ranging sensors operated by infrared rays are used, as shown in Table I. Six short-range sensors and one long-range sensor (Fig. 2) are installed on the board. Sensors are placed so that three pairs of short-

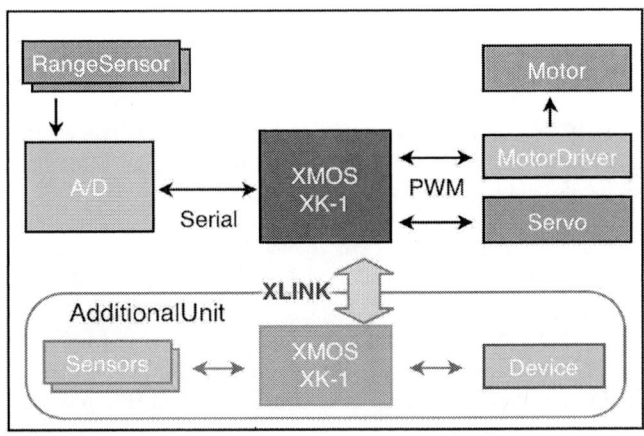

Fig. 3.   Hardware block diagram.

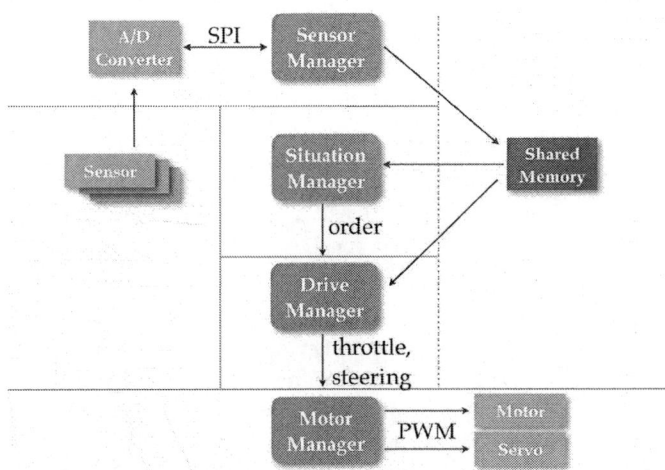

Fig. 4.   Software block diagram.

range sensors can face every direction to obtain the angle of an obstacle. The long-range sensor is provided to find a path on which the car can run. The data obtained by the sensors are transported via an A/D converter. A Serial Peripheral Interface (SPI) of four-wire serial buses connects the A/D converter to the XMOS XK-1 board. Four pins of the SPI—designated to clock, upstream, downstream, and enable signals—are connected to the XMOS's universal port. The hardware block diagram is shown in Fig. 3.

IV. XMOS CAR SOFTWARE

A. Modeling

The software system (Fig. 4) has been modeled with four functions: a sensor manager, a situation manager, a drive manager, and a motor manager. Four functions, each of which is implemented as a thread, are a reasonable limiting number to run all threads in an XCore at full speed. The sensor manager interfaces with the SPI to obtain data from the A/D converter. Data are written out to the shared memory. The situation manager decides what to do with environmental information stored in the shared memory and makes a decision for subsequent behavior. The drive manager controls the direction and speed of the car by analyzing sensor information stored in the shared memory and the decision made by the situation manager. The motor manager outputs pulse width modulation (PWM) waveforms to motor drivers and servomotors to control the accelerator and the steering.

Each manager as a thread runs in parallel by communicating with another thread using message passing and the shared memory. The use of shared memory is not favored in XMOS technology. The XC language does not allow sharing memory, which is only available in the C language. Sharing memory is only valid when two threads run on the same core. It never scales on multi-cores, despite scalability provided by message passing. However, shared memory is a valuable way to simplify the model.

B. Threading

The ordinary system uses interruptions or a thread priority for real-time execution. An interruption occurs by an event and a trigger to run specified codes. An event itself is a simple concept, but it is hard to model and design at an abstract level. Furthermore, interruption multiplexing is necessary in a large-scale system and makes the program more complex since timing problems, which are difficult to resolve, arise.

The thread priority is a flexible, high-level descriptive method; however, its behavior is difficult to predict. The thread with the highest priority runs first, the next higher thread can run while the highest one is idling, and the third and lower priority threads can run when all the higher priority threads are idling. If a thread with a higher priority takes a long time, then a thread with a lower priority may never be executed. To solve this problem, some real-time operating systems provide timeout count attributes for threads. Of course, this may increase complexity in designing and implementing programs.

In an XCore, threads share CPU time by time-slicing. There is no priority to be managed. All threads on the same core are evenly executed by hardware scheduling. The program does not need to care about scheduling because there is nothing to be configured. It is necessary only to describe parallel execution of threads using a *par* statement of the XC language.

C. Message passing

In XMOS programming, synchronized message passing is normally used; when thread A wants to talk with thread B, thread A sends a message and waits until thread B receives it. The behavior of threads is described by a formal language or process algebra. Process algebra is a valuable approach in real-time or parallel systems for eliminating dead-locks or reducing bugs. Process algebra requires extra understanding of the basic concepts for using it in the right way. Therefore, asynchronous message passing with a buffer is used in this paper. An asynchronous design with the following simple rule is much more flexible and clear: Using message channels

978-1-61284-863-1/11 $26.00 © 2011 IEEE

with streaming attributes, a thread cannot be blocked when a message is thrown to an empty buffer.

The system has been designed so that message passing is carried out in a pipeline-streaming way. The situation manager sends messages to the drive manager. The drive manager then sends a message to the motor manager. For guaranteeing an empty buffer, the two managers must perform message passing in such a way that the receiving manager accesses the channel more frequently than the sending manager. This is accomplished by providing the two managers with different speeds such that the receiver manager is provided with more frequent access than the driver manager. According to this rule, the motor manager is provided with the highest speed and the situation manager with the lowest speed. The drive manager has a middle speed. In the conventional system, these speeds are supplied by thread priority, but an XMOS chip provides it by harmonization of software, which gives a flexible design.

D. I/O

Unlike other processors based on conventional architecture such as PIC or SoC, an XCore does not have peripheral circuits such as a PWM modulator, an A/D converter, or an SPI interface. Only a virtual I/O, a named port, is available. Everything has to be implemented by software. Of course, peripheral circuits can be used as external chips, but communication has to be provided by software. A virtual port encapsulates fussy hardware problems. Signals are inputted or outputted to a buffer on a port and sent or received with a related clock signal or an inner timer. Clock signals for timing synchronization are obtained from an inner frequency divider of the CPU clock or the signal from another port. The program is thus liberated from timing problems. The program has to output data at an output timing or request an input of data at a specific timing.

For SPI communication, a conventional sequential protocol is used to pull down a clock signal, to send or receive data bits, to wait a half clock, to pull up a clock, to wait the remaining half clock, and to return to the starting point. Waiting half-clock time is simply described with a *when* syntax of XC. A *when* syntax waits for an event that occurs at the given port and satisfies the given condition. An inner timer has to be set at the port where synchronization is carried out. A *timeafter* function causes the timer to generate a trigger when the specified time has passed. Although a clock-synchronized port has to be used, clock signals are created by software. To improve code readability, XC codes are wrapped as a macro. XC supports defined macros that are exactly the same as in C. The call function in C of the macro *wait*(*timer*, *basetime*, *duration*) is described in XC as the statement *timer when timeafter* (*basetime* + *duration*) :> *time*, which means to wait until the current time becomes after *basetime* + *timeduration*, and then to update *basetime* by the current value of timer.

When a PWM wave is outputted, timed output is used. Executing a timed output with the specified time blocks a thread until the given time passes. The thread sleeps and awakes automatically by an event supported by the hardware mechanism. Only by setting a low value for the pulse time and a high value for the rest duration repeatedly can PWM waves be generated. By implementing functions by software, algorithms used in the system are be easily updated without any hardware modification. For example, PWM waves are currently used for motor control and the plan is to use a phase looked loop in future, which requires only adding a feedback sensor and reprogrammed codes.

E. Intelligence

The system control is divided into higher and lower parts. The higher part is realized by the situation manager and the lower one by the drive manager. The separation is carried out in a way similar to that of the human brain and spinal marrow. The higher part controls over a wide span, and the lower one does over a narrow span or the current time. In other words, the higher part controls abstract matters and the lower part does specific tasks. The higher controller works with a state transition model. A state transition is triggered by a conditional event that occurs when the car deviates from the expected range. Each state gives a decision of movement. Moving forward slowly, moving along the right wall, following ahead, and stopping immediately are examples of such decisions. The higher part will pass a decision to the lower one as an order. The lower part analyzes the decision given by the higher part in the current environment. The decision may be wrong and not suitable for the current situation. In this case, the lower part has priority over the higher part, much like spinal reflexes. The lower part may ignore an order to save the car from a crash. Collaboration of two controllers makes the program simple and helps developers reduce development cost and time.

V. Conclusion

This paper has described how an autonomous vehicle has been achieved with XMOS, a multi-thread, event-driven processor. The paper has also shown how new values are added using an event-driven, multi-thread processor by replacing the old controller of existing hardware. Ubiquitous computing with sophisticated embedded devices in the coming world will require appending a controller chip to the existing device for rapid transformation. XMOS abstract models represented by event-driven processes enable software engineers to develop hardware-oriented embedded systems. Thus, XMOS removes a barrier and encourages people who have never dealt with embedded systems to engage in hardware development.

References

[1] C.A.R. Hoare, Communicating sequential processes, *Commun. ACM*, 21(8): 667–677, 1978.

[2] INMOS, *Occam Programming Manual*, Prentice-Hall, 1984.

[3] A. Kent and J.G. Williams, *Encyclopedia of Computer Science and Technology*, Dekker, 1998.

[4] D. May, Communicating process architecture for multicores, *30th Communicating Process Architectures Conference*, 21-32, 2007.

[5] K. Ohmori and T.L. Kunii, Designing and modeling cyberworlds using the incrementally modular abstraction hierarchy based on homotopy theory, *Visual Comput.*, 26(5): 297-309, 2010.

[6] G. Martins, A. Moses, M. Rutherford, and K. Valavanis, Enabling intelligent unmanned vehicles through XMOS technology, *J. Defense Model. Simul.* (in press).

978-1-61284-863-1/11 $26.00 © 2011 IEEE

A Dynamic Comparator with Analog Offset Calibration for Biomedical SAR ADC Applications

M.M.J Herath and P.K. Chan

School of Electrical and Electronic Engineering, Nanyang Technological University, Singapore
(Email: jaya0028@e.ntu.edu.sg, epkchan@ntu.edu.sg)

Abstract— A new ultra-low voltage and power digital comparator using analog offset calibration technique is presented in this paper. The comparator's input transistors are working in subthreshold region for the entire input range and is used as a zero crossing detector for the analog offset calibration technique. Realized using GLOBALFOUNDRIES 0.18μm CMOS process technology, at the voltage supply of 0.7V and 1MHz clock frequency, the standalone comparator dissipates 153nW but increasing to 494nW with calibration function. For voltage supply down to 0.4V and 20 kHz clock, the single comparator dissipates only 8.054nW. The calibration circuit is capable of reducing a 20mV of input-referred dc offset down to 655.3μV, making it suitable for 10-bit biomedical SAR ADC applications.

I. INTRODUCTION

Reducing power consumption of the comparator benefits low-power Successive Approximation (SAR) ADC design and prolongs the device battery life in wearable biomedical electronics. Comparators can be mainly classified in to two groups. They are analog comparators (comparators with a pre-amplifier) and dynamic comparators (comparators triggered by a clock). Ignoring leakage current factor, the dynamic comparators are most often preferred than the analog counterparts due to their almost zero static power dissipation. However, the dynamic comparators suffer from larger offset which limits the desired resolution of the ADC.

Resistive divider comparators belong to one group of dynamic comparator architectures where the comparator's input transistors are operating in the triode region. Fig. 1 shows typical resistive divider architecture [1]. However, it has several limitations that restrict them in using in ultra-low power applications. Firstly, it cannot operate with very low supply. The main reason is that of M2 and M9 cross couple transistors being literally off during metastable state and hence it requires some time to turn on under low supply voltages. Secondly, there is no mechanism to remove memory due to the previous decision at the drain nodes of M1&M2 and M9&M10 transistors when the comparator is not making a decision. Thirdly, the input transistors M1 (Vin+) and M10 (Vin-) are operating in the triode region which limits the gain. In addition, the current flow is not only dependent on the input signals, but also on drain source voltages (V_{DS}). More importantly, the triode operation consumes large amount of current that is not suitable for ultra-low power applications. In final remark, the kick-back noise due to the switching activity is a common problem in many dynamic comparators. This limits the sensitivity of the comparator by disturbing the input signals.

This paper presents a modified version of the resistive divider comparator architecture to achieve ultra-low power in the range

of nano watts with the proposed analog offset calibration technique. As a result, the comparator can be used as a zero crossing detector in a SAR ADC. Fig. 2 shows a typical zero crossing detector where the positive input terminal is connected to the ground (GND) and the negative input terminal to the output of a DAC (V_{DAC}). Zero crossing detector's maximum possible input range of V_{DAC} is from -500mV to 500mV with reference to ground. This is dedicated to the SAR ADC using the offset-biased comparator [1].

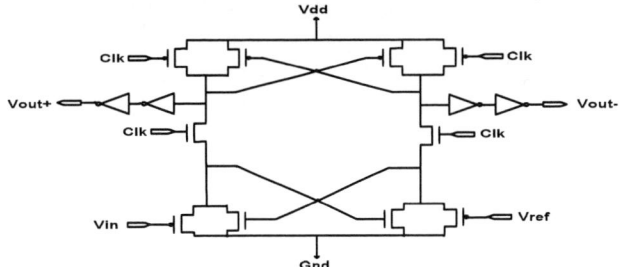

Fig. 1 Typical resistive divider comparator

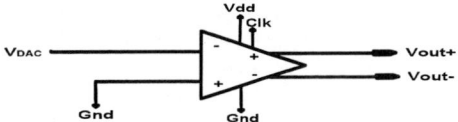

Fig.2 Typical Zero Crossing Detector

Section II presents the proposed concept of the comparator incorporating the offset calibration circuit. Section III describes the realization of the proposed design. Section IV provides the results and discussions. This is then followed by the concluding remarks in Section V.

II. PROPOSED CONCEPT

The proposed dynamic comparator with reduced power technique is shown in Fig. 3. There are several modifications in this comparator with respect to the conventional one.

In order to increases the input transistors V_{th}, a negative voltage charge pump is used [2]. Thus, the transistors can operate in sub-threshold region for the maximum possible input range of -500mV to 500mV. Negatively increasing the bulk potential increases the depletion region, leading to the increase of V_{th} [3] which is given by

$$V_{th} = V_{th0} + \gamma(\sqrt{\phi_s - V_{bs}} - \sqrt{\phi_s}) \quad (1)$$

where ϕ_s and γ are the surface potential at the threshold and body effect coefficient respectively, V_{th0} is the threshold

978-1-61284-863-1/11 $26.00 © 2011 IEEE

voltage of the device when the body bias is zero and V_{bs} is the body to source potential. When the gate-source voltage (V_{GS}) is less than the V_{th}, the input transistors will work in the sub-threshold region. This reduces power consumption significantly. Changing the NMOS cross-coupled configuration (M5 and M6 transistors) helps reduce the supply voltage and kick back noise, avoiding memory effect due to previous decision and increase the gain at the trip point (M1 to M6 transit through sub-threshold region). Operating in the sub-threshold ($V_{GS}<V_{TH}$) with drain-source voltage (V_{DS}) greater than thermal voltage (V_T), it makes the transistors to be independent of V_{DS} and operate like MOSFET in saturation but with larger gain [4].

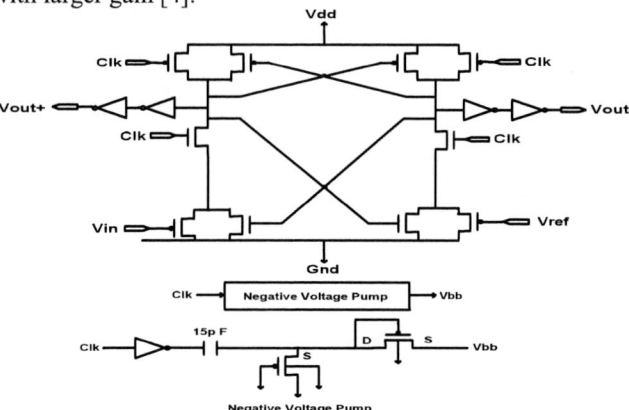

Fig. 3 Proposed Dynamic Comparator with Negative Charge Pump

The zero crossing detector with the offset calibration is shown in Fig. 4. Offset is calibrated at the input terminals with respective 50pF capacitor. A ramp input signal (+1mV/µS) is generated by either system or ramp circuit. During the calibration phase, the ramp input signal creates an analog test voltage and it is stored in a capacitor. The analog calibration is controlled digitally by the output whereas the comparator is in open loop configuration. It is important to have capacitors (50pF) to retain charges as calibration is done once in each conversion cycle. Calibration time depends on the magnitude of the offset. If the offset is large, the system takes longer time to calibrate the offset which will affect the overall speed of the ADC. Nevertheless, it is adequate for majority of biomedical applications in view of the usual low bandwidth of bio-signals.

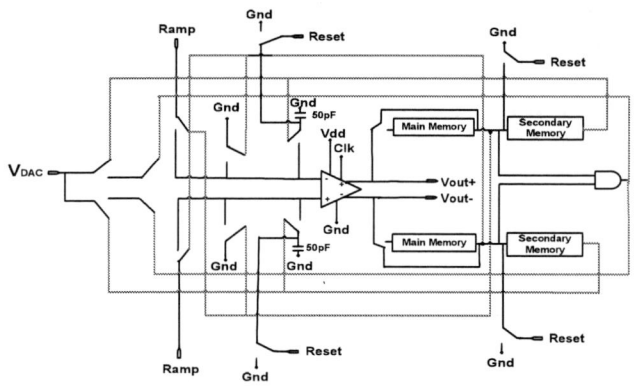

Fig. 4 Block diagram of the zero crossing detector with the offset calibration circuit schematic

III. REALIZATION OF PROPOSED DESIGN

A. Dynamic Comparator

Increasing the V_{TH} using a negative voltage pump can reduce currents flowing through input transistors. When the currents through input transistors are low, the kick back noise associated with C_{GS} and C_{DS} is minimized [5]. Since the NMOS cross-coupled transistors are on before the clock goes high, it requires less time to response when compared to the traditional configuration. Due to the reason, the circuit can operate at low supply voltage of 0.4V using 0.18 µm CMOS. In the meta-stable point, both the output nodes V_{out+} and V_{out-} are charged up to Vdd by applying the clock signal to the PMOS transistors. However, the previous decision memory at the drain node of the input transistors still affects in the conventional comparator configuration. Those parasitic charges may create an offset in the next comparison cycle. In the modified configuration, the NMOS cross-coupled transistors are on during the non-comparison phase (at the meta-stable point). This removes the memory (charges) at the drain nodes of the input transistors. Moreover, large portion of kick back noise comes from the switching activity from the output will be grounded immediately by transistors M_5 and M_6 during the transition. Fig. 5 shows the exemplary nodal voltages of the comparator at the trip point. In conventional design with the large supply voltage, the power consumption is high as transistors M1-M6 transit through saturation region to triode region. However, with low supply voltage, the transistors M1-M6 transit between the sub-threshold region and the triode region. Compared to the transistors operating in triode region, it gives a relatively higher transconductance to amplify the signal variation at the drain nodes of input transistors with low current consumption.

Consider M5 and M6, at the trip point $V_{GSM5}=V_{GSM6}=340mV$ and $V_{thM5}=V_{thM6}=420mV$. $V_{GSM5}=V_{GSM6} < V_{thM5}=V_{thM6}$ (M5 and M6 are in weak inversion), $V_{DSM5}=V_{DSM6}=212mV \gg V_T$ (26mV at 300K)

Consider M3 and M4, at the trip point $V_{GSM3}=V_{GSM4}=488mV$ where $V_{clk}=700mV$ and $V_{thM3}=V_{thM4}=520mV$ (after increasing the V_{th}). $V_{GSM3}=V_{GSM4} < V_{thM3}=V_{thM4}$ (M3 and M4 are in weak inversion). $V_{DSM3}=V_{DSM3}=128mV > V_T$. Similarly, M1 and M2 are also in subthreshold region with $V_{DSM1}=V_{DSM2}>V_T$.

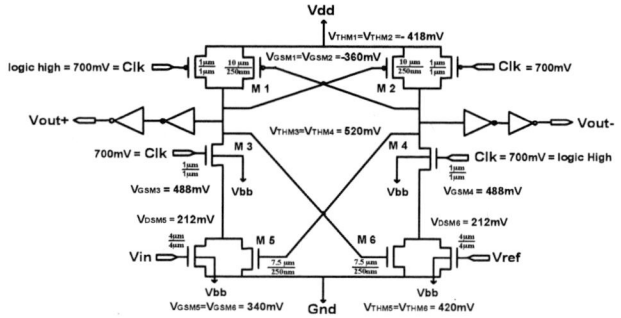

Fig. 5 Exemplary nodal voltages at the trip point in 0.18µm CMOS weak-inversion based comparator design

This confirms that the nodal voltages to satisfy the sub-threshold conduction with a larger gain. Moreover, at the trip point the transistors M1-M6 that operate in the subthreshold region have transconductance of 6µA/V.

978-1-61284-863-1/11 $26.00 © 2011 IEEE

B. Offset Calibration Circuit

Fig. 6 shows the transistor level design of the offset calibration circuit with the comparator. The main memories are used to initialize and control the calibration phase and the comparison phase. The secondary memories are used to store the information of the first comparison cycle as it is important to locate the offset affected terminal. Hence, the input signal can be applied to the desired terminal (offset unaffected terminal) immediately after the calibration phase. V_{out+} and V_{out-} are taken from a NAND based latch at the outputs. This is used to keep the previous output decision when the comparator is not making a decision (when the clock is logic "0").

Each output terminal consists of one main memory and one serially-connected secondary memory. The main memory, which control several switches, consists of two inverters and one control switch. Firstly, during the calibration phase, the main memory at positive output terminal controls the ramp input switch at the negative input terminal whereas the negative output terminal's main memory controls positive input terminal's ramp input switch. Secondly, after the first comparison, one of the main memory circuits is immediately disabled while the other main memory controls the calibration. Disabling one of the main memories automatically connects the V_{DAC} to the comparator before the calibration finishes. To avoid the possibility, an AND gate is used where inputs are connected to both the main memories. Using AND gate, it is possible to initialize the comparison phase. AND gate controls a TG at each input terminal. During the calibration AND gate outputs a logic "0". The output of the AND gate goes high by receiving a logic "1" from both the main memories; which indicates calibration is finished. The secondary memories are controlled by the main memories through the transmission gates (TGs). They are arranged in a way such that the content of the secondary memory will not change after the first clock cycle. After the first clock cycle, the secondary memory selects the input terminal so that V_{DAC} can be applied. It also waits for the main memories to initialize the comparison phase using the AND gate. The operation procedure of the calibration circuit is explained as follows:

Initially, both the input terminals are grounded. In the first comparison cycle, the intrinsic offset of comparator makes the decision and the logic information is transferred to the main memory element and secondary memory element. This turns on one of the ramp input switches (initial ground connection becomes disconnected) while the other ramp input is not connected to the circuit (ground connection still in enable mode). This initializes the calibration phase. During the calibration, the offset is stored in a capacitor with the help of the ramp signal while the comparator makes decision at every clock pulse. Calibration progresses until the initial output logic changes to the opposite logic. The calibration phase is terminated.

The actual comparison starts immediately after the calibration as it is controlled automatically by the memory units. V_{DAC} is injected only to the offset unaffected input terminal via two TGs in series connection. One TG is controlled by the main

memory whereas the other TG controlled by the secondary memory. The external "Reset" signal can be used to reset the memory.

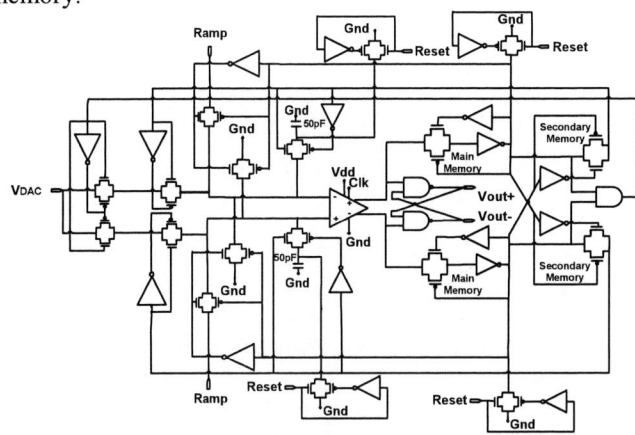

Fig. 6 Transistor level design of offset calibration circuit with the dynamic comparator

IV   RESULTS AND DISCUSSIONS

The standalone comparator can discriminate smallest voltage of 0.5mV difference; which can support 11-bit resolution. Besides, the input signal frequency is important when considering the comparator's resolution. If the input signal variation is very fast, the comparator cannot response immediately. Therefore, the rate of change of input voltage must satisfy with the comparator's response time. The response time of the comparator is 70.03nS. In this simulation, the applied input frequency is 167 kHz. Most of the bio-signals have much lower frequencies than that of the applied input. Fig. 7 shows the output transient responses of the dynamic comparator when the applied inputs (V_{DAC} and V_{ref}) and their difference is minimum. Note that ($V_{DAC(max)}$ - V_{ref})=(V_{ref} - $V_{DAC(min)}$)= 0.5mV.

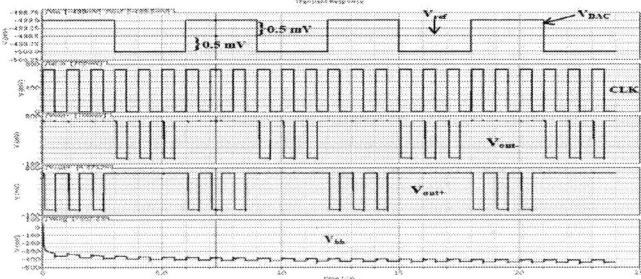

Fig. 7 Transient Responses of the Comparator

Transient time -12µs, CLK=> duty cycle =500ns, period= 1000ns, delay =10ns,t_{rise}=t_{fall}=10ps,($V_{DAC(min)}$=-500mV,$V_{DAC(max)}$=-499mV)➔ (Pulse),period= 6µs, duty cycle=3µs, delay=10ns, t_{rise}= t_{fall}=10ps, V_{ref}=-499.5mV.

TABLE 1(a) shows the simulated values of power consumption with respect to different supply voltages and clock frequencies. The maximum power consumption is 153.335nW at the maximum applied voltage (500mV). TABLE 1(b) shows simulated performance of the comparator circuit that is used for the offset calibration using a 0.7V supply voltage and 1 MHz clock frequency.

TABLE 1(a) Power @ different Vdd & clock freq.

Supply	Clock (Hz)	Power Consumption
0.7V	1 M	153 nW
0.7V	50 k	85.5 nW
0.6V	500 k	90.7 nW
0.6V	50 k	59.44 nW
0.4V	50 k	9.11 nW
0.4V	20 k	8.054 nW

TABLE 1(b) Performance of comparator

Power Supply	0.7V
Clock Frequency	1MHz
Response Time	70.03nS
Power Consumption	153.35nW
Input Range	-500 to 500 mV
Output range	0-700mV

To observe the input-referred offset calibration performance, an external DC source of +20mV is intentionally added to the drain terminal of the V_{in} input transistor of the comparator. The simulation results of the calibration circuit were observed by applying triangular inputs [6]. The ramp input stores 70.77mV of voltage in the capacitor at the positive terminal to balance the offset. Fig. 8(a) shows the output curves of calibration circuit when V_{DAC} varies between -200mV and +200mV. Fig. 8(b) shows the output curves of the calibration when V_{DAC} varies between -2mV and +2mV. Fig. 8(b) is the zoomed curve which illustrates the offset after calibration. It is shown that the comparator yields correct output results after the calibration. Refer to Fig 8(b), the comparator gives an valid output when any input difference (V_{DAC} -V_{ref}) is above 655.53μV during the comparison phase. Hence, it can meet 10 bit resolution. Moreover, the calibration time needed to counteract +20mV DC offset is 42μS. TABLE 2 summarizes the performance of the proposed calibration circuit. TABLE 3 gives the comparison with the reported works, demonstrating ultra-low power consumption in the proposed work.

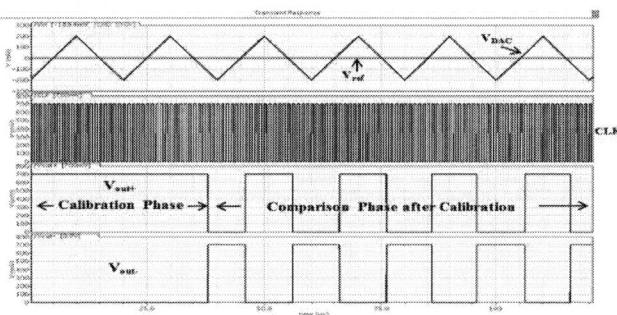

Fig. 8 (a) Output curves of calibration circuit (large V_{DAC} variation)
V_{DAC}= -200mV to 200mV to -200mV (input variation of 400mV) triangular step. Period= 12μs and transient analysis time =160 μs, V_{ref} = GND (0V)

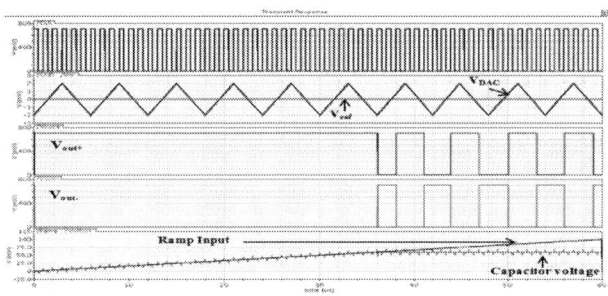

Fig. 8 (b) Output curves of the calibration circuit (small V_{DAC} variation)
V_{DAC}= -2V to 2mV to -mV (input variation of 4mV) triangular step. Period= 12μs and transient analysis time =160 μs, V_{ref}= GND (0V)

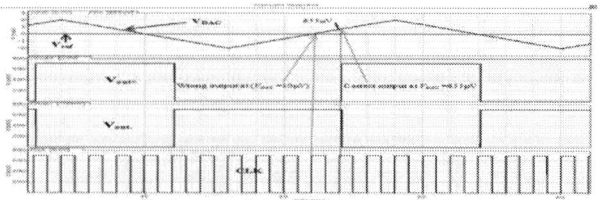

Fig.8 Zoomed Output Curves of Fig. 8(b)

TABLE 2 Performance of the proposed calibration circuit

Parameter (When 20mV DC source at the input to create an offset)	Calibration Circuit Performance
Offset Voltage	Reduced to 655.3μV
Power Consumption	494nW
Calibration Time	42μS
Resolution	10bits

IV. CONCLUSION

A modified resistive divider based dynamic comparator is proposed. The comparator works well for a maximum input range from -500mV to 500mV while consuming only 153nW of power at a 0.7V supply voltage and 1MHz clock. A new analog offset calibration circuit is proposed on the basis of the comparator working as a zero crossing detector. The overall circuit consumes ultra-low power, suggesting its attraction for realizing a 10-bit SAR ADC for biomedical applications. The comparison has also shown that the proposed comparator has achieved appreciable performance with respect to the previously-published works.

TABLE 3 Performance comparison with other reported works

Parameter	Ref. [7] [CMOS 0.18μm]	Ref. [4] [CMOS 0.13μm]	This Work [CMOS 0.18μm]
Type	Dynamic Comparator	Analog Comparator	Dynamic Comparator
Power Supply	0.8V	0.4V	0.7V
Power Consumption	640nW	213nW	153.35nW
Input Range	1V	0.4V	-500mV-500mV

REFERENCES

[1] Y. Susanti, P.K. Chan and V.K.S. Ong "An Ultra Low-Power Successive Approximation ADC Using an Offset-Biased Auto-Zero Comparator", Proc. IEEE Asia Pacific Conference on Circuits and Systems (APCCAS), pp. 284-287, Dec. 2008.

[2] R. Jacob Baker. *CMOS Circuit Design, Layout, and Simulation: Chapter 18.* Wiley-IEEE Press, 2nd edition, Nov. 2007.

[3] Y. Cheng and C. Hu, *MOSFET MODELING & BSIM3 USER'S GUIDE: Chapter 3, Kluwer Academic Publishers,* 2002

[4] K. Abdelhalim, L. MacEachern and S. Mahmoud, "A Nanowatt Successive Approximation ADC with Offset Correction for Implantable Sensor Applications", *Proc. IEEE International Symposium on Circuits and Systems* (ISCAS), pp. 2351-2354, May 2007.

[5] J. Chen, S. Kurachi, S. Shen; H. Liu, T. Yoshimasu and J. S. Yong, "A low-kickback-noise latched comparator for high-speed flash analog-to-digital converters", *IEEE International Symposium on Communications and Information Technology,* vol. 1, pp. 259- 262, Oct. 2005.

[6] A. Graupner. (2006, Oct 1). *A Methodology for the Offset-Simulation of Comparators* (Version 1) [Online]. Available: http:// www.designers-guide.org.

[7] S.B. Kobenge and H. Yang, "A 250KS/s, 0.8V ultra low power successive approximation register ADC using a dynamic rail-to-rail Comparator", *IEICE Electron. Express,* vol. 7, no. 4, pp. 261-267, April. 2010.

Automated Wafer Defect Map Generation for Process Yield Improvement

Cher Ming Tan
School of EEE, Nanyang Technological University
Singapore
ecmtan@ntu.edu.sg

Kheng Tuan Lau
School of EEE, Nanyang Technological University
Singapore

Abstract—**Spatial Signature Analysis (SSA) is used to detect a reoccurring failure signature in today wafer fabrication. In order for SSA to be effective, it must correlate the signature to a wafer defect maps library. However, classifying the signatures for the library is time consuming and tedious. The Manual Visual Inspection (MVI) of several failure bins in a wafer map for multiple lots can lead to fatigue for the operator and resulted in inaccurate representation of the failure signature. Hence, an automated wafer map extraction process is proposed here to replace the MVI while ensuring accuracy of the failure signature library. Clustering tool namely Density-Based Spatial Clustering of Applications with Noise (DBSCAN) is utilized to extract the wafer spatial signature while ignoring the outliners. The appropriate size for the clustered signature is investigated and its performance is compared to the MVI signature. The analysis shows that for 3 selected failure modes, 20% occurrence rate clustered pattern provide similar performance to a 50% MVI signature. The proposed technique leads to a significant reduction in the time required for extracting current and new signatures, allowing faster yield response and improvement.**

Keywords- SSA, Library, DBSCAN, Wafer Defect Map, Failure Bins, Clustering

I. INTRODUCTION

In wafer fabrication, a wafer-map is created during wafer electrical test. This map shows the characteristic patterns or signatures of dies consist of a particular failure bin which could give insight into a manufacturing process that is out of specification. Conventionally, optical imaging and spatial signature analysis (SSA) are used to compare the wafer map with that in a library consists of reference failure signatures of interest for matching features. This SSA is an automated procedure developed in 1998 to address the issue of intelligent data reduction while providing feedback on current manufacturing processes [1]. It is to eliminate the need of a human operator checking a huge amount of wafer data for any discrepancies.

To ensure the effectiveness of SSA, generation of wafer map for a given failure bin on a wafer and the establishment of signature library are essential. Currently, the manual process of visually inspecting the wafers in the lot for signature is tedious and when fatigue set in for the operator, wafers with contributing signature might be missed. This will result in a

less accurate representation of the reference pattern in the library.

In this work, we develop a method that can extract the signature from a wafer with little human intervention, and the collection of these signatures over many wafers can also help to establish the signature library. The method employed is built using Density-Based Spatial Clustering of Applications with Noise (DBSCAN).

II. DESCAN METHOD

DBSCAN is a density based clustering method proposed by Ester et al in 1996 [2]. It is based on the distribution condition of the data point as a parameter to group and connect points. Therefore within each cluster there will have a higher concentration of points as compared to that outside of the cluster. DBSCAN has two input parameters Eplison "Eps" and "Minpts". "Eps" is a parameter that determines the maximum radius of the neighbourhood for which a point can be considered to be part of the cluster. "Minpts" sets the criterion for which there must be a minimum number of points in an Eps-neighbourhood of that point before it can be considered. The DBSCAN algorithm is as follows [3]

Step 1: Select a random point x
Step 2: Compare all points from x with the criteria of *Eps* and *Minpts*.
Step 3: If x is a core point, a cluster is formed.
Step 4: If x is a border point and no points are density-reachable based on *Eps*, the process move to the next point in the data set.
Step 5: Continue the process until all of the points have been processed.

III. SIGNATURE LIBRARY WORKFLOW

The data used in this work are obtained from Wafer Sorting (WS) test from a commercial wafer fab. The WS Bins are categorized into the main failure modes of Hard Bins 'HB' and Soft bins 'SB'. These bins consist of die coordinates and failure modes will be used for clustering. RapidMiner is chosen among other due to its simple interface and its acceptance to MS Excel spreadsheet which will be constructed during the building of the signature library as will be

978-1-61284-863-1/11 $26.00 © 2011 IEEE

discussed later. This clustering technique is able to capture the entire signature while ignoring the outliners on the wafer. To setup the signature library, the workflow to extract the failure signature from the database is shown in Figure 1. Two Excel VBA macros are written to sort the required failure Bins, and display the resultant signature. The first macro imports the raw data from the database of defects in a wafer, extracts the user defined failure Bins and exports the failure Bins coordinates of individual wafer into wafer file in Excel format. DBSCAN clustering is then applied to the wafer file.

In order to extract the failure pattern without outliners, the two parameters of DBSCAN are evaluated. The parameters Minpts are varied from one to five and Epsilon from two to five. The results showed that by increasing Epsilon, the enlarged radius allowed the outliners which are distant from the signature to be cluster together. When Epsilon is reduced to 2, the outliners are excluded from the main cluster. As the wafer Bin data are in Cartesian coordinate, the distance for each dice is a value of one. This indicates that for a dice to be part of the cluster, the maximum distance should only be two die away. This allowed DBSCAN to generate the arbitrary shape while excluding the outliners. Each dice/point on the wafer is corresponding to four other die. Therefore the minimum number of Minpts should be 1 to allow for linking of points to form the cluster in all direction. Once the conditions for the DBSCAN parameters (Epsilon and Minpts) have been set, the operator will load and run the wafer files in RapidMiner. The program will generate a list of cluster numbers correspond to each coordinate.

The second macro automatically imports the clustered result. It will search each wafers coordinates for the highest occurrences of a particular cluster number. These coordinates with high counts will constitute into the major cluster list and the rest as minor cluster list. Subsequently, the macro will generate the major and minor cluster patterns based on the occurrence rate. This occurrence rate is generated by the number of occurrence of a given Bin failure at a given die coordinate in the lot. Thus the coordinate with an occurrence rate that does not meet the pre-defined rate will be filtered out. This pre-defined occurrence rate will be discussed in the following section. The remaining coordinates will formed the resultant pattern. It will be the desired defect signature that will contribute to the signature library as showed in Figure 2. Hence with minimum intervention, the user will be able to establish a SSA library and transfer this pattern for SSA processing to examine the wafer defect map, thus producing an alert for investigation with suggested root causes.

IV. VERIFICATION OF THE DBSCAN METHOD

In order to verify the accuracy of the DBSCAN method, the resultant wafer defect signature obtained will be compared to Manual Visual Inspection (MVI) signature. The MVI signature is obtained from the database through KLA AceXP, by inspecting each wafer in the lots for reoccurring signature. These wafers with the selected signatures are stacked in AceXP

software to reveal the failure occurrence rate on each dice. The outliners or die that are detached from the main signature are ignored. Each dice on a wafer will exhibit an occurrence rate. Hence the stacked wafer is separated into different groups based on different pre-defined rate ranging from 20% to 70% occurrence rate. These patterns are designed in AceXP software and are feed into the pattern recognition node. The node will generate a score for each wafer which is a measure of it similarity to the template pattern based on KLA AceXP proprietary calculation.

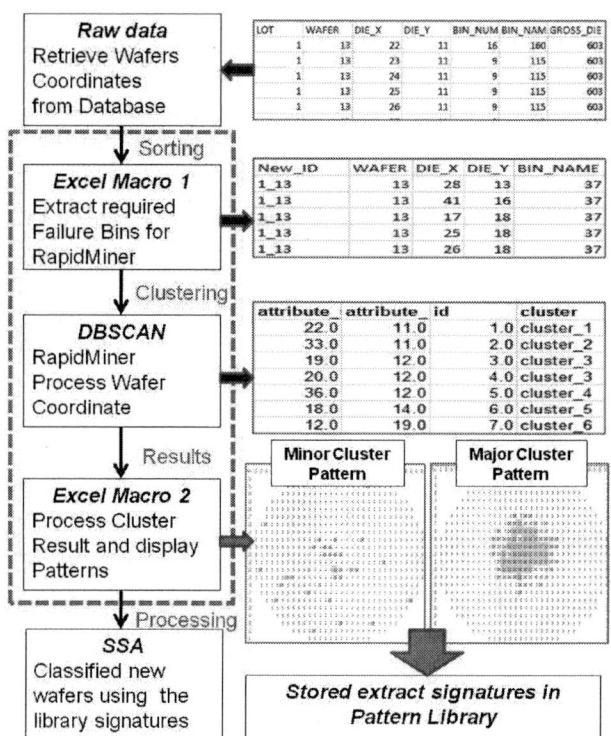

FIGURE 1: DBSCAN CLUSTER LIBRARY WORKFLOW CONSISTING 3 COMPONENTS; SORTING OF REQUIRED FAILURE BINS, DBSCAN CLUSTERING AND DISPLAYING LOT FAILURE PATTERN.

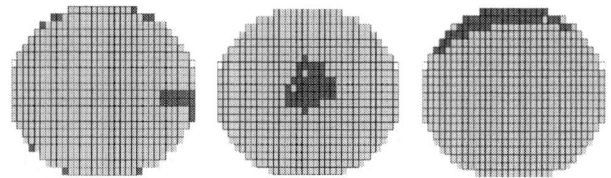

FIGURE 2: WAFER DEFECT SIGNATURE FOR FAILURE MODES OF HB15, SB37 AND SB115 RESPECTIVELY. THE OCCURRENCE RATE IS SET AT 20%.

Next, there is a need to select an appropriate MVI pattern weightage for comparison with the cluster pattern. A MVI pattern with high occurrence rate will generate very high scores for the wafers due to low die count in the pattern size. Whereas MVI pattern with a low occurrence rate will generate very low scores due to it high die counts. The choice of the former is not feasible as it does not reflect the comprehensiveness of the failure signature. Hence it is more appropriate to select 50% MVI occurrence rate as it is a balance between pattern size and score. Similarly, different stacked cluster pattern using DBSCAN with occurence rate ranging from 10% to 30% will also obtained. For occurrence rate above 30%, the number of defect dies is very low and will be ignored.

In order to establish the equivalent weightage of the cluster and MVI methods, the wafer signature scores obtained from AceXP software will be examined. The wafers score for 10%, 20%, 30% clusters pattern and 50% MVI pattern are plotted on a normal distribution chart. This normal probability density function of the data is obtained using Maximum likelihood method. The results in Figure 3 showed that out of the 3 different cluster pattern, 20% cluster (Black Line) and 50% MVI (Pink Line) scores distribution are the closest matched in terms of its mean and sigma. The observation also showed that 10% cluster mean is always lower than the 50% MVI and 30% cluster mean is higher than 50% MVI. These charts indicate that 20% cluster is the most likely representation of 50% MVI. These results are consistent for HB15, SB37 and SB115 failure signatures.

To be more quantitative, the difference in score for each wafer between 20% cluster and 50% MVI are plotted in Figure 4. The average differences in score for HB 15, SB 37 and SB 115 are 4.8, 2.3 and 5.3 respectively which are the smallest as compared to 10% / 30% - 50% MVI. The comparison of the average difference in score across all 3 patterns for 10%, 20% and 30% cluster are as followed 18.3, 6.4 and 9.4 respectively and one can see that 20% cluster demonstrate the closest correlation to 50% MVI.

V.   RESULTS

Multiple lots verification is done for the 3 Bin patterns. Wafer signatures from 13 lots of wafers are processed by 20% cluster and 50% MVI pattern in AceXP recognition node, and the score results are shown in Figure 5. The result are analyzed for scores greater than 40% as wafers' signature with score below 40% does not resembled the reference signature.

From Figure 5, we see that both the DBSCAN / 20% Cluster and MVI methods managed to capture 6 wafers of similar signature for failure mode SB115, 2 of it are showed in Figure 6. Also, the MVI pattern captures 2 additional signatures as compared to 1 additional signature from the DBSCAN cluster method. It is observed that the additional wafers captured by MVI pattern of SB 115 in Figure 5 shows less die counts as compared to the template pattern in Figure 2. Whereas the wafer that is missed by 50% MVI method but is captured by DBSCAN method bears a closer representation to the 20% Cluster \ 50% MVI template pattern.

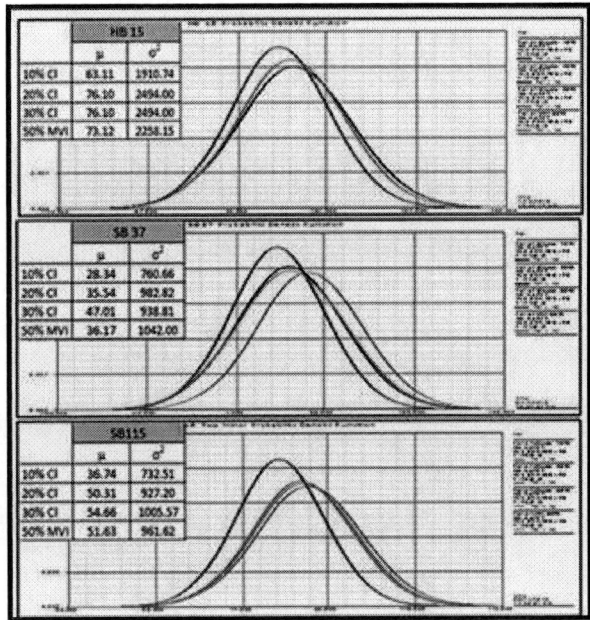

FIGURE 3: PDF DISTRIBUTION FOR 10%, 20%30% CLUSTER PATTERN AND 50% MVI PATTERN: TOP – HB15, CENTRE -SB37, BOTTOM – SB115

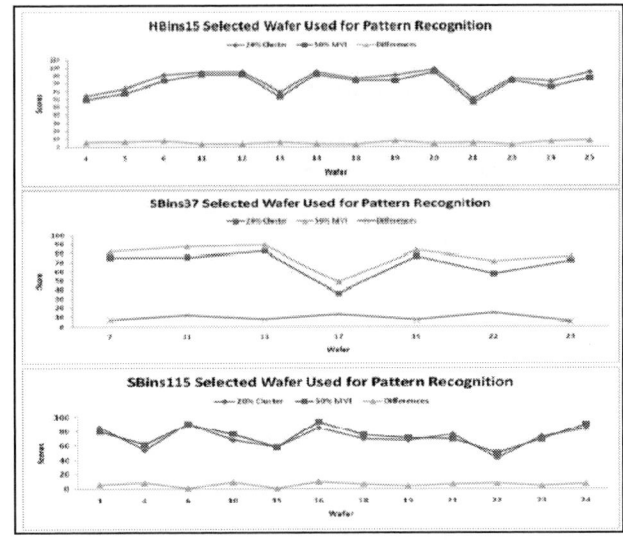

FIGURE 4: PATTERN RECOGNITION SCORE FOR 20% CLUSTER AND 50% MVI PATTERN: TOP- HB15, CENTRE-SB37, BOTTOM- SB 115

Figure 5 shows that for SB37, both 50% MVI and 20% cluster patterns managed to capture the same wafer signatures from the 13 lots. In addition, 20% cluster method was able to detect another similar signature wafer that was missed by 50% MVI. Lastly, the results for HB 15 showed the wafers captured by both methods are similar but with a slight variation in the score. The wafers' scores for these 2 methods have an average difference of 6.2%, which is in line with the differences that were observed in Figure 4. In summary, the validations for DBSCAN clustering method have demonstrated that 20% cluster die occurrence is the desired pattern per lot for the signature library.

VI.   CONCLUSION

In this work, we developed a technique of using DBSCAN to extract the wafer signature. A comprehensive evaluation was done to determine the correct cluster pattern weightage for a lot. The analysis showed that out of the three occurrence patterns, 20% cluster provides the closest match when compared to the 50% MVI pattern.

Although the workflow is semi automatic due to the combination of Microsoft VBA macros and RapidMiner program, this process is designed to replace the tedious MVI process of building the pattern library. The user can expect to obtain accurate and consistent results from this unsupervised clustering technique. Also, a significant reduction in the time required for extracting current and new signatures.

In conclusion, the technique allows corrective actions to be taken sooner, and ultimately resulting in faster yield recovery and improvements.

REFERENCES

[1] T. P. Karnowskia, K.W. Tobin, S.S. Gleason, and F. Lakhani, "The Application of Spatial Signature Analysis to Electrical Test Data: Validation Study," Proceeding of SPIE, v. 3677, pt 1-2, p.530-541, 1999.

[2] M. Ester, H-.P. Kriegel, J. Sander, and X. Xu "A density-based algorithm for discovering clusters in large spatial databases with noise," Proceeding of the Second International Conference on Knowledge Discovery and Data Mining (KDD-96) p. 226–231, 1996.

[3] P.N. Tan, M. Steinbach and V. Kumar, "Introduction to Data Mining" Addison-Wesley, 2005.

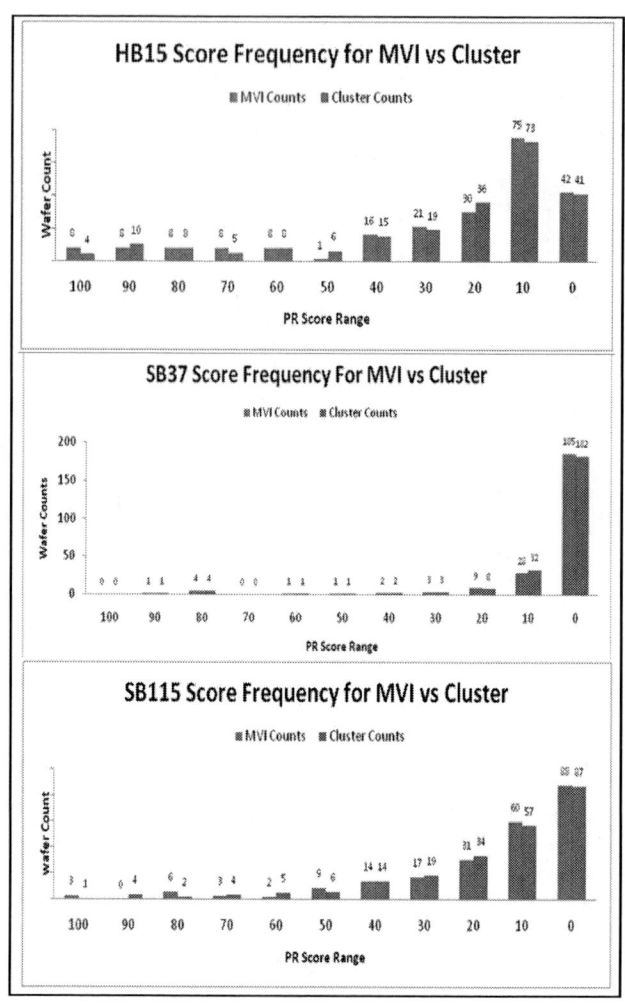

FIGURE 5 50% MVI AND 20% CLUSTER SCORE FREQUENCY FOR 13 LOTS: TOP-HB15, CENTRE-SB37 AND BOTTOM-SB115

SB115			SB37		
Both	50% MVI	20% Cluster	Both	50% MVI	20% Cluster

FIGURE 6 WAFERS WITH SIMILAR SIGNATURE CAPTURE BY 50% MVI AND 20% CLUSTER PATTERN DURING PROCESSING OF 13 LOTS. LEFT: SB115 AND RIGHT: SB37

Delay Defect Diagnosis Methodology using Path Delay Measurements

Eun Jung Jang, Jaeyong Chung, and Jacob A. Abraham

Computer Engineering Research Center
The University of Texas at Austin
{ejang,chung,jaa}@cerc.utexas.edu

Abstract—With aggressive device scaling, timing failures have become more prevalent due to manufacturing defects and process variations. When timing failure occurs, it is important to take corrective actions immediately. Therefore, an efficient and fast diagnosis method is essential.

In this paper, we propose a new diagnostic method using timing information. Our method approximately estimates all the segment delays of measured paths in a design using inequality-constrained least squares methods. Then, the proposed method ranks the possible locations of delay defects based on the difference between estimated segment delays and the expected values of segment delays. The method works well for multiple delay defects as well as single delay defects. Experiment results show that our method yields good diagnostic resolution. With the proposed method, the average first hit rank (FHR), was within 7 for single delay defect and within 8 for multiple delay defects.

I. INTRODUCTION

As IC technology scales, designs are more susceptible to manufacturing defects. There are basically two types of defects. One makes the design malfunction producing undesired logic values. The other prevents the design from meeting timing specifications by introducing unexpected delays on the signal paths. The latter is referred to as *delay defects*.

In the nanometer era, many chips do not meet performance goals after fabrications due to process parameter variations. Diagnosis is performed to locate the physical sources of chip failures. For yield enhancements, we need to identify root causes of performance degradation so that we can take corrective actions based on the diagnosis results for the next silicon stepping immediately.

Diagnostic methods usually build a list of potential sources of failure, which are referred to as *candidates* or *suspects*. These candidates will be examined for root cause analysis. Therefore, it is more efficient if the total number of reported candidates is small. The number of candidates determines the quality of a diagnosis method which is measured by *diagnosis resolution*. Diagnosis resolution is defined as the ratio of the number of actual defects to the total number of reported candidates.

For better diagnosis resolution, a high degree of observability is required. If we can observe all the segment delays, we can identify which segments are defective. To improve observability, DFT techniques such as scan chains can be implemented with a design. However, most designs do not provide such level of observability even with scan insertion. Instead of measuring individual segment delays, we can think of deconvolving segment delays from path delay measurements.

Extracting individual segment delays from measured path delays will make delay defect diagnosis much easier. It will uncover the locations that need to be fixed. If we obtain the segment delays on silicon, we can compare the segment delays on silicon with the expected values. Based on the mismatches of the two, we can prioritize the locations that need to be looked into. However, working backwards to get individual segment delays from measured path delays is a difficult problem because most practical designs have a limited number of observable paths [8]. Thus, our goal is to solve all the segment delays from smaller number of path measurements.

In this paper, we propose a delay defect diagnosis method that approximately estimates all the segment delays of the observed paths using inequality-constrained least squares methods. Here, we assume that the random component of the mismatch between the expected segment delays and the actual delays of non-defective segments on manufactured chips is smaller than the delay defect size. Based on this assumption, we can now find a unique solution that minimizes the Euclidean distance (L2-norm) of the difference of segment delays and expectation value of segment delays. Unlike traditional diagnosis methods, our method enables collective analysis of silicon timing information. Also, the proposed method does not require any assumption on fault model, number of faults or size of faults.

II. PREVIOUS WORK

Delay defect diagnosis based on backtrace algorithms has been studied. The authors of [6] have presented an effect-cause diagnosis method for single gate delay faults. With a six-valued simulation, they can take static hazards into consideration. However, they assumed only a single delay fault in a circuit to simplify the algorithm. They assumed the possible suspects could be identified in the intersection of the common sensitized paths from failed outputs. In [7], a path delay fault diagnosis approach is presented using a backtracing method. First, robust path delay fault simulation on a circuit is performed to obtain a set of paths that pass robust test. Then, for each test pattern that causes circuit failure, they perform backtracing to determine a set of paths (a set of suspects) using 5-valued algebra. Among those paths, those that have been robustly tested, and hence fault free, are removed from the set of suspects. These backtrace algorithms usually lack the ability to consider fault masking effects and multiple path sensitization.

In [11], the authors proposed a forward approach based on the six-valued simulation to rectify the problem of backtrace algorithms. First, they execute fault-free simulation for each vector pair to check the arrival times of fault-free cases. Then, all the nodes with transitions during fault-free simulation will be considered as candidates. For each candidate, a delay symbol representing the delay fault will be injected and forward-propagated to outputs. This step gives information on which outputs will be affected by the injected delay fault. Then, the similarity between the faulty output responses of circuit under diagnosis and the syndromes of the faulty chip is calculated. Both the syndrome of the faulty chip and the symbolic delay propagation results are considered to calculate the similarity. Finally, based on the calculated similarity, they rank the candidates. The candidates with higher similarity are ranked higher. The accuracy of the forward approach can be higher because they consider the fault masking effect and multiple-path sensitization. However, multiple faults were not considered in this approach.

In deep submicron technology, assuming multiple point faults is more realistic. In [4], the authors applied a critical tracing method to multiple delay defects. From each faulty output, they perform

978-1-61284-863-1/11 $26.00 © 2011 IEEE

backtracing to find a set of suspects using 6-valued algebra. If there exists an intersection between the sets of suspects, the intersection will be referred to as prime suspects, which can cause all the observed failures. Otherwise, they assume that there are multiple point defects, which cause the output failure separately. After generating a list of suspects, they rank the suspects based on the expected delay defect sizes of the suspects. They use static timing information to estimate upper and lower bounds of delay defect sizes. In order to improve diagnosis resolution further, they proposed adaptive test pattern generation in [5]. To reduce the size of suspects, they derive the adjacency tests or the two-pattern test which has the minimum number of input transitions among the test patterns that can reproduce the failure. The less the number of input transitions, the smaller the size of suspects, because there are less number of sensitized paths.

Delay defect diagnosis methods using statistical timing models have been studied [10], [9]. These methods enable estimation of the delays that are difficult to predict with process variations in the current technology. They model the delays and the defects as random variables, and then build a probabilistic error dictionary. After observing the faulty responses, authors compare them with the data in the dictionary and rank the suspects based on the probabilities that sensitized paths do not meet timing specification when suspects are defective. In [3], authors proposed a diagnosis method for small delay defects caused by process variations. Their method produces good diagnostic results. However, their method assumes that the probability density functions of each delay segments are known, which may not be easily obtainable in reality.

In [12], the authors proposed a delay defect diagnosis method using timing information. They obtain a list of suspects from the traditional diagnosis method which does not require timing information. They propagate possible lower and upper bounds of defect sizes in a forward direction and update them using their timing-based simulator. Then, suspects will be ranked based on the size of the defect. However, the resolution can be improved further.

Chen, et al., proposed a diagnosis framework using linear programing in [1], [2]. They take test data directly to get information on failed and passed paths. First, they build a set of linear equations for observed paths during the test because each path delay can be expressed as the linear summation of segments on the path. Then the linear equations are fed to a linear programing solver. The linear programing solver tries to approximately estimate all the segment delays while maximizing the summation of delays of segments on failed paths. Based on the linear programing results, they rank the suspects. Their method provides good diagnostic resolution, but it requires run-time which increases linearly with the number of failing paths. In case there are many failed paths, the run-time can be high.

III. PROPOSED METHOD

A path delay is the linear summation of segment delays on it. Assume that $p_1, p_2, ..., p_M$ are M sensitizable paths in the circuit with N segment delay variables. Each path p_j can be represented as below.

$$p_j = \sum_{i=1}^{N} c_{ij} \cdot d_i, \quad j = 1, 2, ..., M$$

where

$$c_{ij} = \begin{cases} 1 & \text{if } d_i \text{ belongs to the path } p_j; \\ 0 & \text{otherwise.} \end{cases}$$

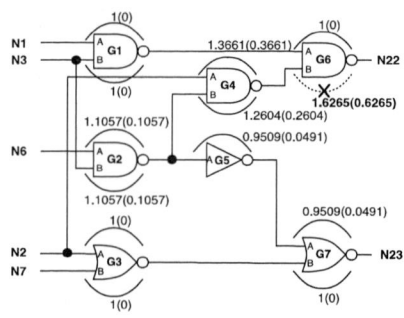

Fig. 1.   Example of estimated segment delays and size of faults - c17

TABLE I
PATH DELAY MEASUREMENTS OF FIG. 1

Path	linear sum. of segments	measured	expected
path 1	G1/A + G6/A	2	2
path 2	G4/A + G6/B	3	2
path 3	G1/B + G6/A	2	2
path 4	G2/B + G4/B + G6/B	4	3
path 5	G2/A + G4/B + G6/B	4	3
path 6	G3/A + G7/B	2	2
path 7	G2/B + G5/A + G7/A	3	3
path 8	G2/A + G5/A + G7/A	3	3
path 9	G3/B + G7/B	2	2

The equations can be represented with matrices. We define the coefficient matrix C, and the path measurement vector $\mathbf{p}$.

$$C \cdot \mathbf{x} = \begin{pmatrix} c_{11} & c_{12} & \cdots & c_{1N} \\ c_{21} & \cdots & \cdots & c_{2N} \\ \cdot & \cdot & \cdot & \cdot \\ \cdot & \cdot & \cdot & \cdot \\ \cdot & \cdot & \cdot & \cdot \\ c_{M1} & \cdots & \cdots & c_{MN} \end{pmatrix} \begin{pmatrix} d_1 \\ d_2 \\ \cdot \\ \cdot \\ \cdot \\ d_N \end{pmatrix} = \begin{pmatrix} p_1 \\ p_2 \\ \cdot \\ \cdot \\ \cdot \\ p_M \end{pmatrix} = \mathbf{p}$$

We want to obtain a solution of $\mathbf{x}$ here. If the coefficient matrix is full rank, the solution is unique: $\mathbf{x} = C^{-1} \cdot \mathbf{p}$. However, C is always rank deficient with a practical design.

There can be numerous solutions that satisfies $C \cdot \mathbf{x} = \mathbf{p}$. To find an optimal solution among all possible solutions, we can use nominal timing information of each segment delay, which is the expected value of each segment delay. Now, our problem can be formulated as

$$min \, || \, \mathbf{x} - \mathbf{x_0} \, ||_2$$

subject to

$$\mathbf{p} - t_{measure} \leq C \cdot \mathbf{x} \leq \mathbf{p} + t_{measure},$$

where $\mathbf{x_0}$ is the expected values of segment delays which can be obtained from the timing model, and $t_{measure}$ is the error comes from the resolution of path delay measurement. This method is referred to as the *least squares with inequality constrained method* (LSI). We can find a unique and optimal solution from an underdetermined system using this method.

A. A Numerical Example

We will give an example of our method using circuit c17 from the ISCAS-85 benchmark circuits. Figure 1 shows circuit c17. In this example, all the segments in the design are characterized to have expected delays of unit delay. To show how the proposed method works, we assumed that the expected delay of G6 from input B to output is 1 but the actual delay is 2 due to the delay defect. Table I shows how each path of c17 is composed of and the delays of the corresponding paths. There are nine observable paths in total when we do not differentiate the rising and the falling transitions for simplicity. We can see that paths 2, 4, and 5 have mismatches between the measured delay and the expected delay due to the defect of G6/B. The segments on paths 2, 4, and 5, which are G2/A, G2/B, G4/A, G4/B, and G6/B, will share these mismatches. Because G6/B is on all three paths, the estimated delay size of fault of G6/B will be the largest. The segment delays obtained using the proposed method are denoted in Figure 1. The numbers in parentheses are the differences between the expected delays and the delays obtained using the proposed method. As we can see, the delay of G6 from input B to output is 1.6265 while the expected value was 1. Thus, it has the largest estimated delay size of fault, 0.6265. Based on the estimated size of fault, the rank (high to low) will be determined: G6/B - G4/A - G4/B - G2/A, G2/B - G5/A, G7/A. As we can see, G6/B is ranked first so that it will be examined first during the failure analysis. Also, we can see that all the segments that were on paths 2, 4, and 5 are ranked higher than other segments. With this example, we can see that the proposed method successfully finds the location of the delay defect.

B. Limitation of the Proposed Method

If defective segments appear on the measured paths infrequently and the delay size of the fault is small, it can lower FHR. For example, if there is only one measured path that has a defective segment on it, all the segments on the path will be the suspects. The proposed method tries to allocate the additional delay due to the defect to all the suspects as far as the suspects meet other inequality constraints. Thus, it will be hard to find the defective segment among all segments on the failing path when the delay size of the fault is small. However, this limitation exists for other traditional diagnosis methods as well (i.e., finding the intersection of the cones of failing outputs). Another limitation of the proposed method is that the proposed method does not provide a good diagnostic resolution in case the size of delay fault is smaller than $t_{resolution}$ even if there are multiple failing paths. However, the proposed method provides not only the list of suspects and the rank, but also the approximate delay size of the fault. In case the estimated delay fault size is smaller than the measurement error, we can take alternative diagnosis methods.

IV. EXPERIMENTS

A. Experiment Setup

In this work, we used ISCAS-85 benchmark circuits to show the proposed method works well for different circuits. All benchmark circuits were mapped to IBM 45nm technology library. The experiments were run on a system with Intel(R) Xeon(R) X5670, 2.93GHz. First, the gate-level designs were fed into a static timing analysis (STA) tool. The STA tool extracted paths that would be tested. Then, the information of extracted paths was fed into an automatic test pattern generation (ATPG) tool. The ATPG tool determined whether each path was testable or not. For the testable paths, it generated test patterns that sensitize those paths. Among the testable paths, linearly independent paths were selected since the rest of the paths do not help estimation of segment delays.

To emulate the timing behavior of a manufactured chip better, we assumed delay of each segment could have up to ±10% random deviation from its model value. Then, we sampled segment delays from the timing model library. There can also be a systematic shift from the timing model, which can be shared by all segments in the design. However, we only considered random variations since the proposed method works well when the common mode noise exists.

We will give an example for c17 of ISCAS-85 circuit when there is a global deviation of segment delays from the timing model. Figure 2 shows the results obtained with our method when we assumed a systematic shift of actual segment delays from the timing model. Each

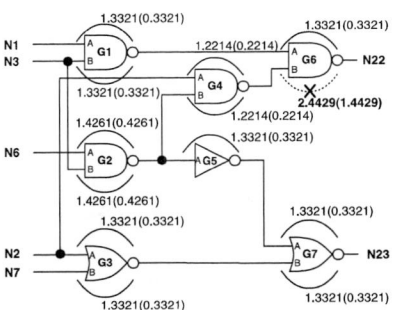

Fig. 2. Example of estimated segment delays and errors - c17, common mode noise exists

TABLE II
FHR FOR SINGLE DELAY DEFECT FOR DIFFERENT SIZE OF FAULTS

Circuit	Size of Faults (FO4)				# Paths	CPU time (s)
	3	5	7	10		
c432	1.74	1.13	1.02	1.01	303	4.809
c499	1.26	1.05	1.11	1.01	262	7.336
c880	2.59	1.10	1.01	1.00	317	9.582
c1355	1.35	1.02	1.01	1.00	178	6.565
c1908	3.21	1.02	1.01	1.01	450	14.670
c2670	2.00	1.07	1.00	1.01	663	8.125
c3540	6.53	1.09	1.03	1.02	583	23.515
c5315	1.78	1.01	1.00	1.02	514	6.480
c6288	4.98	1.27	1.11	1.07	526	22.753
c7552	3.02	1.08	1.02	1.00	940	75.484

segment is characterized to have a delay of 1, but the actual delay is 1.5 due to the shift except for G6/B. The actual delay of G6 from pin B to output is 2.5 due to the shift and the delay defect of size of 1. Therefore, the measured delays of paths 1~9 are 3, 4, 3, 5.5, 5.5, 3, 4.5, 4.5, and 3, respectively. In Figure 2, we can see that the proposed method correctly locates the most problematic segment, G6/B. In this case, all the segments have significant differences between the expected value and the estimated delay. Therefore, the results suggest the global shift of segment delays from the timing model. With this example, we can see that the proposed method successfully finds the location even when common mode noise exists.

We assumed both single delay defects and multiple point delay defects. Also, we tried different sizes of delay faults from 3 times of FO4 inverter delay to 10 times of FO4 inverter delay. For single delay defects, we randomly chose one segment and injected delay fault on the segment. For the same circuit and the same size of delay fault, we injected delay fault 100 times making 100 test cases. For multiple delay defects, we injected two and three faults on different segments. 100 defective test cases were sampled for multiple delay defects experiments as well as the single delay defects.

To obtain path delays, we used the results that the STA tool outputs since we do not have measurements. Also, to build matrix $\mathbf{x_0}$, the expectation values of segment delays were reported by the STA. $t_{resolution}$ was set as the delay of FO4 inverter.

B. Experimental Results

1) Single Delay Defect: Our method ranks the probable defective segments so that the most probable defective segment is reported first. Table II shows the average FHR of each circuit. The results in Table II were obtained by averaging 100 randomly injected test cases for each circuit as mentioned previously. The second row of columns 2~5 shows the size of delay faults. The unit of fault size is the delay of one FO4 inverter. The sixth column represents the number of paths that were used for the calculation, and the seventh column shows the

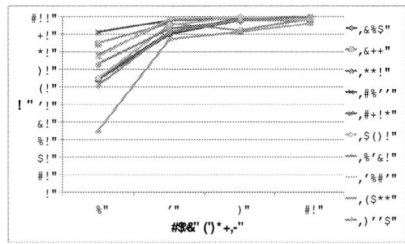

Fig. 3. Percentage that the top ranked suspect is the faulty segment, single fault exists

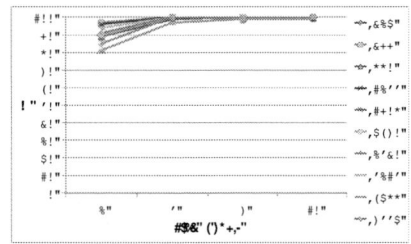

Fig. 4. Percentage that the faulty segment is found among top 3 ranked suspects, single fault exists

CPU time consumed. FHR represents the position of the first true fault [12]. The lower FHR is, the easier to locate the segment whose delay does not match the timing model. An FHR of 1 means the segment that has the highest rank is the faulty segment. Experiment results show that FHR ranges from 1 to 6.53 for various circuits and sizes of faults. We can see that FHR becomes better when the delay fault size is larger.

Figure 3 shows the percentage that the first candidate using the proposed method is the faulty segment. For every circuit and the various sizes of faults, the percentage was 35% and above. Figure 4 shows the percentage that the faulty segment is found among the top 3 ranked suspects. For any test case, the percentage of the fault was found among the top 3 ranked candidates was 81% and higher.

One thing to note is that there can be indistinguishable segments [1]. For example, if some segments are connected in series, there is no way to observe their individual delays since we can only observe path delay values. Indistinguishable segments were grouped together to enhance diagnostic resolutions.

2) Multiple Delay Defect: We executed separate experiments to show the proposed method works well for multiple delay defects as well. We injected two or three faults on each circuit at the same time. Table III shows the FHR when two and three faults are present in each circuit. As in the single defect experiments, results were obtained from 100 randomly fault-injected test cases and averaged out. To use

FHR as the measure of diagnosis resolution for multiple faults, we averaged out the rank of multiple faults. However, adding the ranks of multiple faults as is and averaging them out is not appropriate. For example, if there are two faults and one is ranked top (1) and the other is ranked second, FHR will be 1.5 instead of 1. To adjust FHR for multiple faults, we subtracted the rank of previously found faulty segment. Thus, the new FHR of the previous example will be $\{1+(2-1)\}/2 = 1$, which reflects the true rank of faulty segments correctly. Experiment results show that FHR ranges from 1 to 5.10 when we injected 2 faults, and from 1 to 7.15 when there are 3 faults injected. Diagnostic resolution of our method is slightly lower than the methods proposed in [1] and [2]. However, our proposed method does not require longer CPU time when the number of failing paths increases, while their method requires CPU time that increases linearly with the number of failing paths.

V. CONCLUSIONS

Deconvolving segment delays from path delay measurements will make delay defect diagnosis much easier. A new method that approximately estimates all the segment delays based on the path delay measurement has been presented in this paper. With least squares with inequality-constrained methods, we estimated all the segment delays in an underdetermined system. The proposed method also gives rank of probable segments that need to be inspected during failure analysis. Rank information will reduce the search space, and thus will shorten the time required for direct probing.

REFERENCES

[1] Y. Chen, M. Kuo, and J. Liou. Diagnosis framework for locating failed segments of path delay faults. In *Test Conference, 2005. ITC 2005. IEEE International*, page 8, 2005.

[2] Y. Chen and J. Liou. Diagnosis framework for locating failed segments of path delay faults. *Very Large Scale Integration (VLSI) Systems, IEEE Transactions on*, 16(6):755–765, 2008.

[3] Y. Chen and J. Liou. A non-intrusive and accurate inspection method for segment delay variabilities. In *2009 Asian Test Symposium*, pages 343–348. IEEE, 2009.

[4] J. Dastidar and N. Touba. A systematic approach for diagnosing multiple delay faults. In *Defect and Fault Tolerance in VLSI Systems, 1998. Proceedings., 1998 IEEE International Symposium on*, pages 211–216. IEEE, 2002.

[5] J. Ghosh-Dastidar and N. Touba. Adaptive techniques for improving delay fault diagnosis. In *VLSI Test Symposium, 1999. Proceedings. 17th IEEE*, pages 168–172. IEEE, 2002.

[6] P. Girard, C. Landrault, and S. Pravossoudovitch. A novel approach to delay-fault diagnosis. In *Design Automation Conference, 1992. Proceedings., 29th ACM/IEEE*, pages 357–360. IEEE, 2002.

[7] Y. Hsu and S. Gupta. A new path-oriented effect-cause methodology to diagnose delay failures. In *Test Conference, 1998. Proceedings., International*, pages 758–767. IEEE, 2002.

[8] E. Jang, A. Gattiker, S. Nassif, and J. Abraham. Efficient and product-representative timing model validation. In *VLSI Test Symposium (VTS), 2011 IEEE 29th*, pages 90–95. IEEE.

[9] A. Krstic, L. Wang, K. Cheng, J. Liou, and M. Abadir. Delay defect diagnosis based upon a statistical timing model-the first step. In *Computers and Digital Techniques, IEE Proceedings-*, volume 150, page 346. IET, 2003.

[10] A. Krstic, L. Wang, K. Cheng, J. Liou, and T. Mak. Enhancing diagnosis resolution for delay defects based upon statistical timing and statistical fault models. In *Proceedings of the 40th annual Design Automation Conference*, page 673. ACM, 2003.

[11] H. Wang, S. Huang, and J. Huang. Gate-delay fault diagnosis using the inject-and-evaluate paradigm. In *Defect and Fault Tolerance in VLSI Systems, 2002. DFT 2002. Proceedings. 17th IEEE International Symposium on*, pages 117–125. IEEE, 2003.

[12] Z. Wang, M. Marek-Sadowska, K. Tsai, and J. Rajski. Delay-fault diagnosis using timing information. *Computer-Aided Design of Integrated Circuits and Systems, IEEE Transactions on*, 24(9):1315–1325, 2005.

TABLE III
FHR FOR MUTIPLE DELAY DEFECTS (2 AND 3 FAULTS ON A CIRCUIT)
FOR DIFFERENT SIZE OF FAULTS

Circuit	2 faults injected				3 faults injected			
	3	5	7	10	3	5	7	10
c432	3.25	1.20	1.02	1.07	3.81	1.26	1.10	1.04
c499	1.28	1.32	1.13	1.07	1.45	1.21	1.09	1.10
c880	2.19	1.21	1.04	1.01	3.18	1.31	1.07	1.04
c1355	1.32	1.32	1.03	1.01	1.20	1.16	1.05	1.00
c1908	2.18	1.17	1.03	1.00	2.45	1.12	1.05	1.04
c2670	3.18	1.14	1.02	1.05	2.93	1.19	1.05	1.01
c3540	5.10	1.25	1.10	1.05	7.15	1.43	1.16	1.10
c5315	2.06	1.13	1.03	1.01	3.46	1.21	1.04	1.05
c6288	4.78	1.27	1.15	1.06	5.20	1.44	1.15	1.11
c7552	2.39	1.15	1.02	1.04	4.61	1.17	1.07	1.06

978-1-61284-863-1/11 $26.00 © 2011 IEEE

A Compact Model of AlGaN/GaN on Silicon Schottky Diode and Its Application

Yihu Li[1,2], Lei Wang[2], S. Arulkumaran[3], Yong-Zhong Xiong[2], Geok Ing Ng[1], Wang Ling Goh[1], Shane Todd[2], Patrick Lo[2]

[1]Nanyang Technological University, Singapore

[2]Instiute of Microelectronics, A*STAR (Agency for Science, Technology, and Research), 117685, Singapore

[3] Temasek Laboratories@NTU, Nanyang Technological University, Singapore

Abstract—The I-V characteristics and S-parameters of AlGaN/GaN on silicon Schottky diodes are investigated. A compact circuit model is constructed for RF application. Simulated results are compared with measured results using AlGaN/GaN on Silicon diodes as RF switches. The simulated and measured results illustrated good agreement within a wide frequency range. The output power from the switch is also measured by sweeping the input power. No suppression is observed when the highest input power equals to 14 dBm at 10 GHz for small device, which proved that the AlGaN/GaN on Silicon devices offer good potential for RF power applications.

Keywords— AlGaN/GaN on Silicon, RF switch, Schottky diode, circuit model, S-parameters

I. Introduction

These years, hetero-junction devices are popular in microwave circuit design due to their high frequency and high power handling capacity. Gallium Nitride (GaN) based electronics have become a promising solution for high temperature, high power and high frequency applications due to its wide direct band gap [1]-[10]. But the traditional GaN growth on Sapphire or SiC process is expensive and only available on small substrates. On the other hand, GaN on silicon offers potential solution for low-cost GaN devices on large substrates.

Schottky diodes based on GaN on silicon process appear to be an optimal candidate for the RF switch application and as a result, a circuit model of GaN on silicon Schottky diode for RF switch application is desirable in order to perform schematic level simulation. In this paper, the power handling performance of n-GaN on silicon Schottky diodes is characterized with DC bias sweep from -5 V to 1.5 V. I-V characteristics and S-parameters are measured. A compact model of the GaN on Silicon Schottky diode RF switch is then extracted from the data obtained. A comparison between the measurement and simulation results is also investigated.

II. Device Structure and Measurement Setup

The Aluminium Gallium Nitride/Gallium Nitride on silicon (GaN on silicon) devices were fabricated in-house at NTU [11]-[13]. The length of the Schottky contact was fixed at 0.25 μm and the widths were varied to control the sizing. Two finger contacts were engaged. The measurement setup for RF

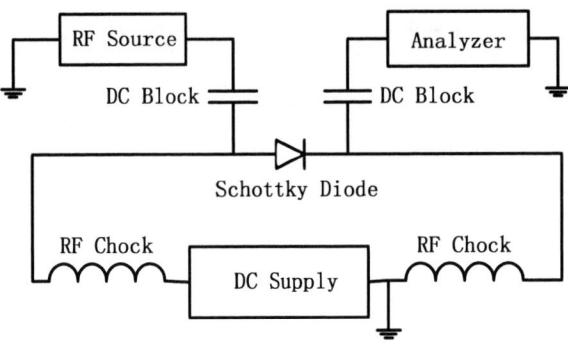

Fig.1 RF power measurement setup.

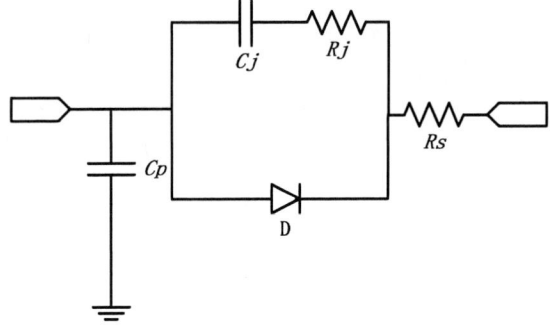

Fig. 2 Proposed model of Schottky diode RF switch.

power switching is given in Fig. 1. The RF signal generator used is the Agilent Technologies SMF100A. The output signal was measured using Agilent Technologies E4448A spectral analyser. DC bias had been added but isolated from the RF source via DC bias tee.

III. Measurements and Modelling

Schottky diodes with contact widths of 10 μm and 25 μm were measured and modelled. The proposed circuit model of a Schottky diode RF switch is shown in Fig. 2. D is an ideal thermionic emission diode, Rs is the series resistance of the contact, Cp models the parasitic capacitor to the substrate, Cj represents the junction capacitor of the contact and Rj is connected in series with Cj to resist any high frequency leakage.

978-1-61284-863-1/11 $26.00 © 2011 IEEE

A. Extraction of Is, n and Rs

For a Schottky diode, the current transportation mechanism is dominated by thermionic emission. The equation can be expressed by the following equation,

$$I = I_S \exp(\frac{qV_A}{nKT}) \quad (1)$$

where *Is* represents the sub threshold current of the Schottky diode; V_A stands for the applied voltage across the contact; *n* denotes the ideality factor of the Schottky contact; and finally, *K, T* and *q* retain their usual meanings. In order to determine

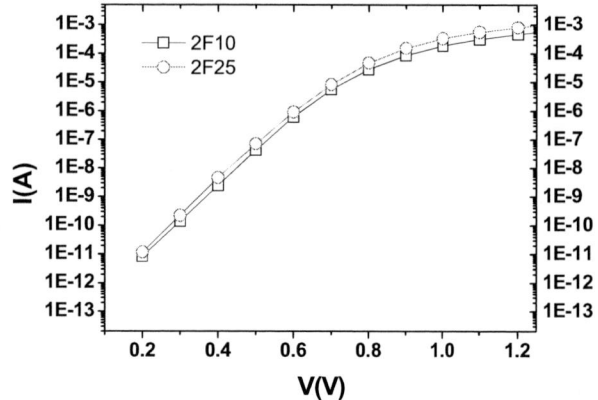

Fig. 3 *Is* and *n* extractions from I-V curve.

the values of *Is* and *n*, the I-V characteristics are plotted in Fig. 3. The square sampled-line denotes the two-finger 10-μm contact whereas the circle sampled-line represents the two-finger 25-μm one. The slope of the linear range on the I-V curve, from 0.2 V to 0.7 V, is selected to extract the ideal factor, *n*. The reason for using this voltage range is to avoid the tunnelling current transportation at low voltage and also the effect of series resistance at higher voltage. By extrapolating the slope of the curve such that it intersects with the *y*-axis, the value of the sub threshold current, I_s can be found.

The extraction of the series resistance *Rs* depends on the I-V characteristics at higher voltage. A plot of I-V curve from 1.0 V to 2.0 V is presented in Fig. 4. The current increases linearly with the biasing voltage after 1.5 V, the gradient of the slope then indicates the reciprocal of the series resistance. It can be seen that Schottky diode with 10-μm contact width

TABLE I
PARAMETER VALUES FROM I-V

Contact Width (μm)	Is (fA)	n	Rs (Ω)
10	30.5	1.37	62.9
25	37.3	1.34	27.4
50	45.5	1.41	17.2
150	66.7	1.27	8.4

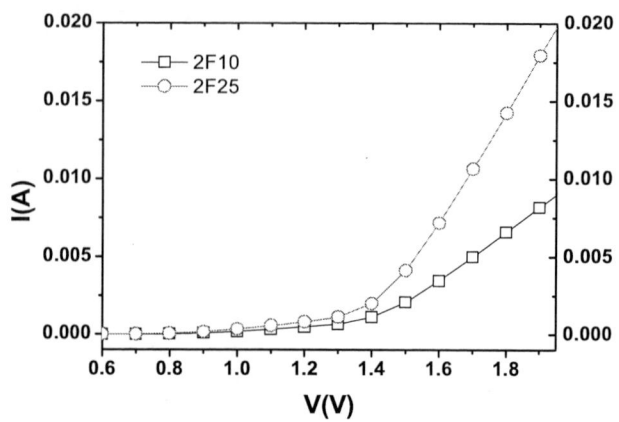

Fig. 4 *Rs* extraction from I-V curve.

has higher series resistance, which is expected since it has smaller contact area. The values of the parameters extracted from the I-V characteristics are tabulated in Table I. The parameters of even larger size contacts are also included in the table. A low series resistance and a small deality factor are found.

B. Extraction of Cj and Cp

Cj and *Cp* are important parameters of the performance of the RF switch. *Cj* affects the RF isolation when the switch is in "OFF" state and together with *Rs, Cp* defines the insertion loss when the switch is in the "ON" state. The values of these two capacitors can be extracted from the Y-parameters under low frequency where the equations deployed are illustrated below.

$$C_p = \frac{1}{\omega} \text{Im}(Y_{11} + Y_{12})_{LF} \quad (2)$$

$$C_j = \frac{1}{\omega} \text{Im}(Y_{12})_{LF} \quad (3)$$

The Y-parameters are derived from the S-parameters measured through the Agilent E8364B Network Analyser after calibration. The values extracted serve as the references

TABLE II
PARAMTER VALUES FROM S-PARAMETER

Contact Width	Cj (pF)	Cp (pF)	Rj (Ω)	Ir (pA)
10 μm	0.12	0.03	30	11.3
25 μm	0.18	0.1	11.4	22.7

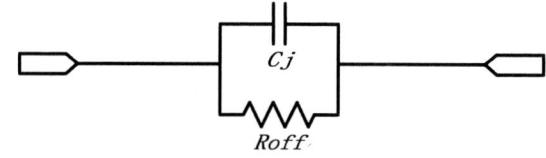

Fig. 5 Reversed bias circuit model for the RF switch.

values. The final values are obtained by optimization using the Advanced Design System (ADS). Rj is added to limit the high frequency leakage when the switch is in the "OFF" state. This also makes the model applicable in a wider frequency range. The value of Rj will not affect the I-V behaviour and the insertion loss during "ON" state since it is opened by Cj and shorted by D under the above two conditions. The optimized values of the parameters are indicated in Table II. Note that Ir is the reverse bias leakage current.

C. Reversed bias modelling

When the Schottky diode is under reversed bias, the switch is in the "OFF" state, and so the RF isolation is dominated by the junction capacitance Cj since large Rs is found. The variation of Cj can be used to model the switch. Under forwards bias, the diode is on. The change in value of Cj does not have a significant effect since it is almost shorted by the diode. Hence, the circuit model can be simplified to that shown in Fig. 5. Here, R_{off} represents the off-state resistance of the diode. It can be traded as an equivalent resistor of the model in Fig .2 under reverse bias so that one needs not to adjust the model circuit during simulation. Cj is modified to a voltage controlled capacitor from the junction capacitor derived in Part B of this section, i.e. equation (3). The relationship between the value of Cj and the applied reversed biasing voltage is given by the following formula.

$$C_j = \frac{C_{j0}}{\sqrt[\gamma]{1 - \dfrac{V_R}{V_i} - \dfrac{KT}{q}}} \qquad (4)$$

Cj_0 represents the capacitance under zero bias; V_R denotes the value of reversed voltage applied; V_i is the built-in voltage of the Schottky contact; and K, T and q have their usual meanings. Lastly, γ is an arbitrary number which is set to 1 here.

D. Measured and simulated results

To verify the performance of the GaN on silicon Schottky diode RF switch and the accuracy of the circuit model, two devices with contact widths of 10 µm and 25 µm were characterised. The measured and simulated results are presented in Fig. 6. The RF frequencies are of 5 GHz and 10 GHz, respectively, with an input power of 0 dBm for both cases. The biasing voltage was swept from -5 V to 1.5 V, to smoothly control the switch from the "OFF" to "ON" state. For both devices, a relatively smooth attenuation was observed when the reversed bias was applied. In contrast, an abrupt increased in the output power was noted at the on-voltage of the Schottky diode. For the 10-µm contact, an insertion loss of 4.56 dB was observed at a forward biasing voltage of 1.5 V at 5 GHz and a 16 dB isolation with a reverse bias of 5 V. When the frequency was raised to 10 GHz, the insertion loss and isolation dropped to 4.6 dB and 11.9 dB, respectively.

The performance of the 25-µm switch is slightly poorer than the 10-µm case. At 5 GHz, although the insertion loss is

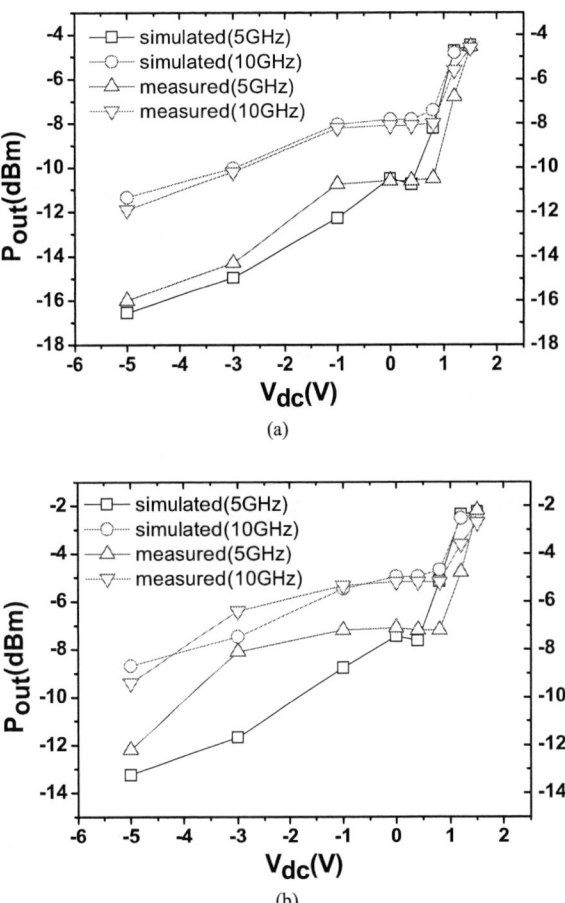

Fig. 6 Measured and simulated results at 5 GHz and 10 GHz for contact width of: (a) 10 µm and (b) 25 µm.

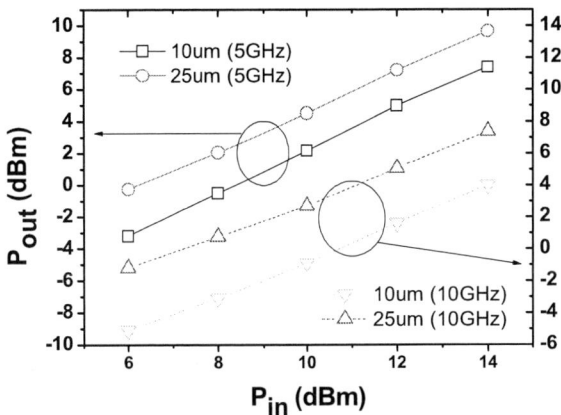

Fig. 7 Measured output power with increasing input power.

only 2.2 dB, the isolation is as low as 12.2 dB due to the large size of the contact. The large contact area makes a big junction capacitor, permitting RF leakages through the junction at even the "OFF" state, thereby degrading the isolation. At 10 GHz, insertion loss and isolation 2.7 dB and 9.4 dB are achieved, respectively. Fortunately, the two GaN on silicon devices are able to maintain good linearity as the input power increased. In Fig. 7, the input RF power increases from 6 dBm to 14 dBm. Even for the 10-μm contact, no output suppression can be seen which verfiy the capability of the GaN on silicon devices for power applications.

The circuit model is simulated in ADS, and a good agreement between the simulated and measured results is obtained in Fig. 6. Comparing with the measured results, the model performs well when the barrier is under forward bias. In the region of reverse bias, some discretion exists between the simulated and measured results, especially for the case of the 25-μm contact. For larger device size, the parasitic effect is more pronounced, which makes the behaviour of the device hard to predict. One possible solution is to adjust the value of γ to modify the value of the voltage controlled capacitor. It can also be observed that the proposed circuit model offers a better match to the measured results at 10 GHz than at 5 GHz. This is attributed to Rj. One may therefore choose a suitable value of Rj to derive a relatively wide band model.

IV. CONCLUSIONS

In this paper, the performance of GaN on silicon Schottky diodes for used as RF power switches has been measured. A good linearity is observed from the output power plot versus the input power. A compact circuit model for the RF switch has been constructed with the value of component parameters extracted from I-V characteristics and S-parameter. The simulated results of the model are able to achieve good matching with the measured results across a wide frequency range.

ACKNOWLEDGMENT

The authors would like to thank the assistance provided by staff of the Institute of Microelectronics, Agency for Science Technology and Research (A*STAR), Singapore. The authors would also like to express gratitude to the Temasek Laboratories@NTU for supporting the fabrication of the GaN on silicon devices. This work is part of the project entitled, "Device Characterization and Process-Design-Kit (PDK)

Establishment", funded by A*STAR, Singapore. Grant Number: 102 169 0131

REFERENCES

[1] G. Hellings, J. Join, A. Lorenz, P. Malinowski and R. Mertens, "ALGaN schottky diodes for detector application in UV Wavelength range", *IEEE Trans. Electron Devices*, vol. 56, no. 11, Nov, 2009.

[2] L. Liu, J. L. Hesler, H, Xu, A. W. Lichtenberger and R. M. WeikleII, "A broadband quasi-optical Tera-Hertz detector utilizing a zero bias schottky diode", *IEEE Microwave Wireless Components Lett*, vol. 20, no. 9, Sep, 2009.

[3] J. H. Oh, S. W Moon, D. S. Kang and S. D. Kim, "High-performance 94-GHx single balanced mixer using disk-shaped GaAs schottky diode", *IEEE Electron Device Lett*, vol 30. No. 3, Mar, 2009.

[4] S. Aslam, R. E. Vest, D. Franz, F. Yan, and Y. Zhao, "Large area GaN, Schottky photodiode with low leakage current", *Electron. Lett.*, vol. 40, no. 17, pp. 1080–1082, Aug. 2004.

[5] E. Monroy, T. Palacios, O. Hainaut, F. Omnes, and J.-F. Hochedez, "Assessment of GaN metal–semiconductor photodiodes for high-energy ultraviolet photodetection", *Appl. Phys. Lett.*, vol. 80, no. 17, pp. 3198–3200, Apr. 2002.

[6] M.J. Shin & R.J. Trew, "GaN MESFETs for high-power and high temperature microwave applications", *Electron. Lett.*, vol. 31, no. 6, Mar. 1995.

[7] Y. Ando, Y. Okamoto, H. Miyamoto, N. Hayama, T. Nakayama, K. Kasahara, and M. Kuzuhara, "A 110-W AlGaN/GaN heterojunction FET on thinned sapphire substrate", in *IEDM Tech. Dig.*, 2001, pp. 381–384.

[8] K. Kasahara, H. Miyamoto, Y. Ando, Y. Okamoto, T. Nakayama, and Kuzuhara, "Ka-band 2.3 Wpower AlGaN/GaN heterojunction FET", in *IEDM Tech. Dig.*, 2002, pp. 693–696.

[9] W. Lu, J. Yang, A. Khan, and I. Adesida, "AlGaNGaN HEMTs on SiC with over 100 GHz fT and low microwave noise", *IEEE Trans. Electron Devices*, vol. 48, no. 3, pp. 581–585, Mar. 2001.

[10] Y. Hirose, Y. Ikeda, M. Ishii, T. Murata, K. Inoue, T. Tanaka, H. Ishikawa, T. Egawa, and T. Jimbo, "Low noise and low distortion performances of an AlGaN/GaN HFET", *IEICE Trans. Electron.*, vol. E86-C, pp. 2058–2064, 2003.

[11] Z. H. Liu, G. I. Ng, S. Arulkumaran, Y. K. T. Maung, K. L. Teo, S. C. Foo, and. V. Sahmuganathan, "Improved 2-D electron gas transport characteristics in AlGaN/GaN MIS HEMT with atomic layer-deposited Al_2O_3 as gate insulator", *Appl. Phys. Lett.*, 95, 223501 (2009).

[12] Z. H. Liu, S. Arulkumaran, and G. I. Ng, "Temperature dependence of ohmic contact characteristics in AlGaN/GaN HEMT from -50 to 200C", *Appl. Phys. Lett.*, 94, 142105 (2009).

[13] S. Arulkumaran, S. L. Selvaraj, T. Egawa, and G. I. Ng, "Sheet carrier density enhancement by Si_3N_4 passivation on non-polar a-plane (1120) sapphire grown AlGaN/GaN heterostructures", *Appl. Phys. Lett.*, 92, 092116 (2008).

978-1-61284-863-1/11 $26.00 © 2011 IEEE

Prototyping A Bidirectional Processor Design Based on Reversible Principles

Dilip Vasudevan and Michel Schellekens
CEOL*, Department of Computer science
University College Cork
Cork, Ireland
dv2@cs.ucc.ie and michel.schellekens@cs.ucc.ie

Nasim Zeinolabedini and Emanuel Popovici
Department of Electrical and Electronic Engineering
University College Cork
Cork, Ireland
nasim@ue.ucc.ie and e.popovici@ucc.ie

Abstract— **This work is motivated towards building a complete bidirectional architecture based on the reversible principles and to study the feasibility of implementation and bottlenecks encountered during this design. A new bidirectional architecture with each core built using infamous single cycle MIPS architecture is proposed. The bidirectional design facilitates the reversible principle of forward and reverse direction of execution of the architecture. The main goal of this work is to implement and empirically study the outcomes of the reversible system built. The architecture is implemented using commercially available 90 nm technology node. The processor operates at 200 MHz with medium mapping and optimization effort and the performance can be improved with further optimization.**

Keywords-component; Reversible Computing; MIPS; Bidirectional Processor Design

I. INTRODUCTION

Principles of reversibility and reversible computing had been researched for more than several decades with one of the goals of reduction in power through information lossless computing. Several methods and techniques were proposed to design the reversible computing system from circuit level to software language design.

Recent literatures on reversible computing promise low energy dissipation due to the information lossless computing. These literatures are backed by the facts from Landauer's and Bennett's work [1] , [2] , which are more at thermodynamics point of view. So bridging the gap between thermodynamics of reversible computing and the actual silicon level implementation of reversible computing system is still an Achille's heel. This work does not promise low energy/power achievement through reversible architecture design. Rather this work aims at implementing an architecture based on reversible principles to practically realize the bidirectional execution of the program at microarchitecture level.

First architecture based on reversible principle was introduced in [4] namely Pendulum. This architecture was designed based on MIPS R2000 micro-architecture. Complete system on charge recovery logic (CRL) based design was designed and implemented. Following same charge recovery logic based designed several 8 bit and 16 microcontroller designs were also reported in the literature [6]. In [11], an abstract model of the reversible architecture was proposed by adding a dir, pc and branch registers to the conventional architecture. A

reversible language construct for this abstract architecture was also proposed in that work. Other reversible languages proposed were Janus [10], R Language (language for Pendulum [4]) etc. On the hardware design part, several adder and ALU designs were proposed. In [7], a reversible CLA adder design was proposed based on control gate based pass transistor designs [5, 9]. Recently in [8], a reversible ALU design was proposed which is more suitable for quantum reversible computing. Apart from ALU and adder designs, circuits for applications in digital signal processing (DSP) is also implemented as reversible designs. Once example is the reversible discrete integer transform implementation in [7]. With the increased interest in the reversible computing system design, designing a new reversible architecture with the current state of the art silicon technology is the point of interest of this work. With the multicore paradigm shift and parallel programming development in the processor design, experimenting the reversible multicore architectures will be a good step forward to find the applicability of reversible computing.

This work is motivated towards developing a dual-core architecture based on reversible principles. Main contributions of this work are 1) two bidirectional designs namely DEM1 and DEM2 which is designed based on the infamous MIPS architecture, 2) Implementation of the layout of the DEM2 architecture in 90 nm technology and 3) future steps to improve the proposed architecture. Rest of the paper is organized as below starting from the background of the MIPS architecture. Background on the MIPS single cycle architecture and ISA is provided in section 2. Microarchitecture of the proposed bidirectional architecture is described in section 3. Implementation results of the architecture are analyzed in the section 4. Lessons learned from the design of the proposed architecture are discussed in section 5. In section 6, the possible next steps to improve the architecture are discussed by introducing adder and SRAM designs. The paper is concluded in section 7.Background

A. MIPS Architecture

In typical single cycle MIPS architecture a 32x32 bit register forms the core of the system with a data memory and instruction memory forming the main two memory blocks. The datapath consists of a 32 bit ALU with basic operation of ADD/SUB, OR, AND, EXOR and SLT operations. A single

---

* CEOL – Centre for Efficiency Oriented Languages
This work was funded by the Science Foundation Ireland under Grant number 07/IN.1/I977.

978-1-61284-863-1/11 $26.00 © 2011 IEEE

cycle comprises of Fetch, Decode, Execute, Memory and Write-back operations. The control path of this architecture is only a combinational block compared to a FSM in pipelined and multicycle versions. The Instruction Set Architecture (ISA) of the MIPS architecture is mainly classified in to R, I and J type instructions. Table 1 shows the instruction set types of the MIPS architecture.

II. PROPOSED BIDIRECTIONAL ARCHITECTURE

With the goal of designing a reversible architecture and as a system designer, the way the system can be made reversible can be viewed in four different perspectives namely 1) circuit/device level reversibility, 2) architecture/logic level reversibility, 3) instruction set level reversibility and 4) high level language level reversibility. Theoretically, each and every view has the same goal of implementing the reversible Turing tape. Engineering the above mentioned four levels will lead to the reversible architecture design. With this view, a bidirectional architecture based on single cycle MIPS micro-architecture was chosen. To start with, out of the two cores, first core was used for forward direction and the second core was used for reverse direction of execution of the processor. The direction of operation is controlled by a direction port which will update the Program Counter (PC) based on the value of the direction signal. When designing this architecture, the main goals are to preserve MIPS architecture and to realize the R, J and I instruction with the dual-core reversible architecture that will execute these instruction in both forward and reverse direction. It might look like merely taking two copies of the MIPS single-cycle datapath and control path will realize the goal. But it is not that trivial. When designing the forward and reverse direction cores, the datapath and control path have to be updated to accommodate the direction of execution with the program counter to read the correct instruction to decode and execute. Thus the control signals of the forward core and reverse core was different. Arithmetic instructions pose no problem during the forward and reverse execution as the addition and its inverse subtraction can be chosen in the ALU. The main concern will be dealing with the branch and jump instructions. Since the architecture is bidirectional, the jump and branching from forward and reverse core has to be taken care of. With this in mind, the branch and jump hardware of the conventional MIPS are extended to accommodate the reversible bidirectional architecture.

A. DEM1 Architecture

The block diagram of this new bidirectional architecture named DEM1 (Bidirectional rEversible Mips)1 architecture is shown in Fig.1. The blue and red lines and blocks shows the new blocks incorporated for handling jump and branch in reverse and forward core. Since both cores are going to read the same program, the instruction memory is shared between the two cores. This will be updated in the DEM2 design. The reversible execution of the processor is controlled by the 8 "R_controlname" port and the forward execution are controlled by the 8 "controlnames" port (shown as blue). These 16 ports "controlnames" and "R_controlnames" are the output of the

main decoder and ALU decoder of the control path. Since the architecture is a single cycle MIPS the control path is a combinational circuit. It should be noted that the data memory for the forward and reverse cores are separate and hence the forward or reverse execution of a program won't disturb the forward or reverse data. The direction port plays the main role in the selection of forward or reverse operation.

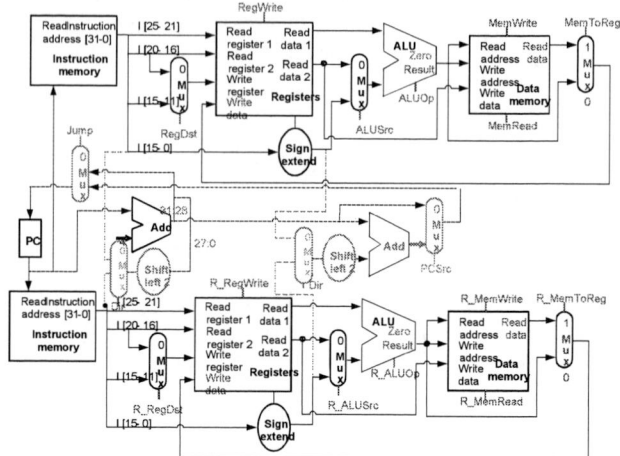

Fig. 1. Bidirectional Reversible MIPS Architecture –DEM1

The direction port is not shown in the figure as it forms the input of the main decoder of the control path as shown in the Fig 2 of the DEM2 design. The operation of the processor is controlled by the PC and direction signals. To run a program in forward direction, the system is clocked keeping the direction signal at '0'. Once the forward execution is completed the direction signal is set to '1' and the system is clocked to run the program in reverse direction. Based on the direction signal value, the control path selects "controlname" ports or "R_controlname" ports to control the respective the core. The selection is implemented by multiplexing the output of the decoder between the "controlnames" and "R_controlnames" controlled by the direction pin. Since the forward and reverse core use different control ports, the decoder takes extra overhead (almost double the output pins of single core). This can be optimized further to use same control ports for both the forward and reverse core operation. This is done in the DEM2 design. Detailed control path and data path of the DEM2 design will be discussed in the next subsection

B. DEM2 Architecture

The microarchitecture of the DEM2 design is briefed in this subsection. As noted in the previous subsection, the main decoder of the control path of DEM1 needs double the usual number of output ports and extra multiplexer to control the forward and reverse cores. This will lead to the higher over head. To overcome this, main decoder design is updated in the DEM2.

Data path

The datapath consists of the 32 x 32 register file, a 32 bit ALU, instruction memory and data memory. The instruction memory is shared by both the cores and hence only one copy

978-1-61284-863-1/11 $26.00 © 2011 IEEE   326

of the instruction memory is used in the DEM2 design. The output of the instruction memory feeds the register file of each core.

Control path

The control path of the DEM2 architecture is shown in Fig.2. The input of the decoder has 13 signals composed of 6 bit funct, 6bit opcode and a single bit di port. The output is composed of 10 signals composed of MemtoReg, MemWrite, Branch, ALUcontrol (3bit) , ALUSrc, RegDst, RegWrite and Jump signals. Compared to DEM1 decoder which used 13:20 I/O, this decoder used 13:10 I/O. The dir pin controls the ALU decoder to decode the Add signal to Adder during forward direction and Add signal to subtract during reverse direction. This avoids the usage of forward and reverse "controlnames" as in DEM1. Since Memory, Branch, and Jump instructions operate in the same manner for both the forward and reverse instructions. The ALU and main decoder encodings of the control path of DEM2 are shown in the Table 2 and Table 3 and 4 respectively. Main instructions namely lw, sw, addi, subi, jmp and bne are shown. The role of the direction signal in decoding the control signals is shown by the shaded cells. Table 3 shows the encoding for the forward direction and the table 4 shows the encoding for the reverse direction. It should be noted that the "dir" signal is 0 for forward and 1 for reverse execution. By carefully noticing the encoding for the addi and subi pair and the lw and sw pair, it should be evident that the decoder decodes the subi signals for addi in reverse direction and viceversa. Same condition applies for the lw and sw pair. This enables the forward and reverse execution of the architecture without changing the ISA of the original MIPS.

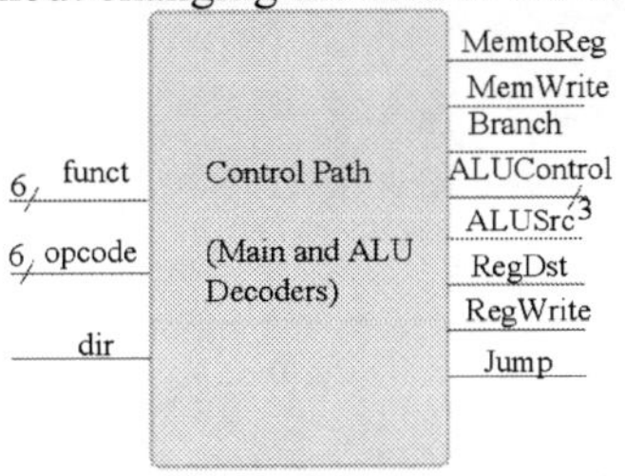

Fig.2 Control Path Decoder(DEM2)

The block diagram of the DEM2 architecture is shown in Fig 3. The blocks in red and blue shows the extended blocks incorporated for bidirectional execution. The control path of the architecture is not shown in the figure for clarity. It is quite intuitive from the encoding tables in Table III. It should be noted in the Fig.3 that the branching hardware between the "zero" and the pcSrc signals are not shown for clarity. In the complete design, the zero signal will drive a 2 input AND gate along with the "branch" signal from the main decoder. The output of this AND gate is connected to the PCSrc signal which enables the correct execution of the branching operation.

TABLE I. MIPS Instruction Format

Format	Operations	Description
R Type	lw, sw	register and memory
I Type	addi	Arithmetic, immediate
J Type	jmp	Jump

TABLE II. ALU Control Encoding

ALUOp	Funct	ALUControl
00	X	010(add)
X1	X	110(sub)
1X	100000(add)	010
1X	100010(sub)	110

The layout of the design DEM2 is shown in Fig.4 and Fig.5. With the layout mapping effort set to medium, the total area of the design is 166112 μm^2. The combinational area was 70152 μm^2 and the sequential area was 95960 μm^2. The designs consisted of totally 4985 standard cells with 5049 ports. The layout is done with the standard cell 90nm library. The area and performance can be further improved by designing the layout with the custom design libraries.

TABLE III Main Decoder Encoding (Forward and Reverse)

Op	dir	Op Code	Reg Wr	Reg Dst	ALU Src	Br	Mem Wr	Mem toRg	ALU Op	Jmp
Rtype	0	000000	1	1	0	0	0	0	10	0
	0	100011	1	0	1	0	0	1	00	0
	1	100011	0	X	1	0	1	X	00	0
	0	101011	0	X	1	0	1	X	00	0
	1	101011	1	0	1	0	1	X	00	0
Beq	0	000100	0	X	0	1	0	X	01	0
addi	0	001000	1	0	1	0	0	0	00	0
(Rev)	1	001000	1	0	1	0	0	0	01	0
J	0	000010	0	X	X	X	0	X	XX	1

Table IV shows the comparison of the DEM1 and DEM2 processor based on their core execution. The column1 gives the design name, the column 2 gives the number of actual hardware cores present in the architecture and the column 3 gives the number of possible reverse and forward execution cores in the architecture. It should be noted that with the same number of hardware cores (totally 2), the DEM2 can be run as four possible cores(both can operate in reverse and forward direction) whereas the DEM1 design can run only 2 possible cores (one for forward and another for reverse).

III. CONCLSION

A new bidirectional/reversible processor design was proposed in this work. Two architectures namely DEM1 and DEM2 were presented. Bidirectional functionality of the architecture proposed enables the execution of the program both in forward and reverse direction. The DEM2 processor was implemented in 90nm technology with 200 MHz performance and the area of the chip layout is 166112um². By carefully, creating the program to handle the jump and branch executions the whole program can be run in forward and reverse direction in the processor by using the direction signal. Replication of the core and reversing the control logic with the control of the direction signal have resulted in efficiently implementing a reversible architecture with a conventional MIPS microarchitecture.

978-1-61284-863-1/11 $26.00 © 2011 IEEE

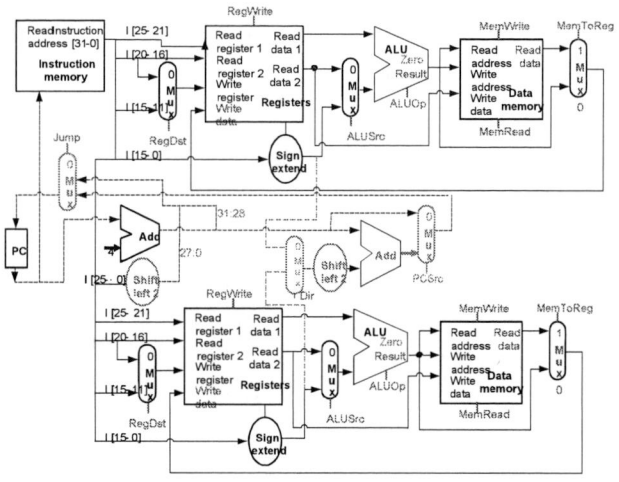

Fig. 3 Bidirectional MIPS Architecture- DEM2

TABLE IV Number of cores in DEM1 and DEM2 Processors

Design	Number of Cores	No of Fw/Rw Cores
DEM1	2	2
DEM2	2	4

TABLE V Design Features

Features	DEM2 Processor
Technology	90nm
Frequency	200 MHz
Number of Cores	2
Area	166112 μm^2

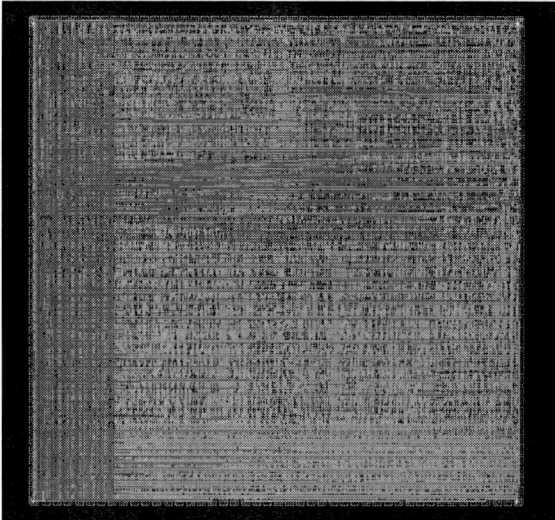

Fig.4 DEM2 Processor - Layout – 90 nm Technology

At this stage, the limitation of the processor to run the whole program forward and then to run the program in the reverse direction is being studied. As a next step, new SRAM and ALU designs will be incorporated to improve the reversibility of the architecture further. Also the ISA of the architecture will be extended with the new SRAM and ALU added.

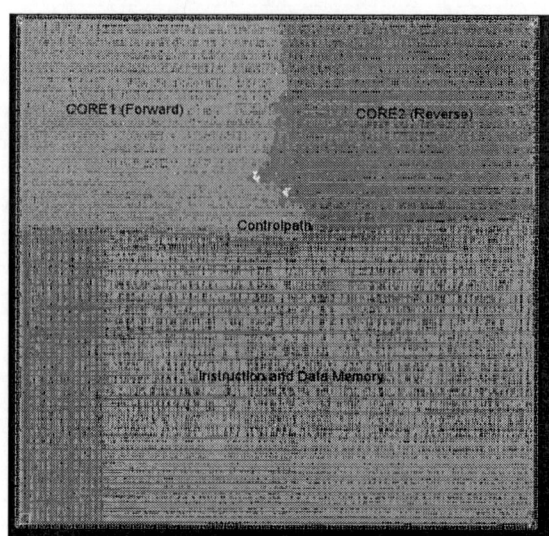

Fig.5 DEM2 Processor – (Hierarchical view) Core 1 (Orange) and Core 2 (Red) are placed at the top. Data and Instruction memory (Green) are placed at bottom.

REFERENCES

[1] Bennett, C. H, Logical Reversibility of Computation, "IBM J. Res. Dev. 6 (1973), 525-532.

[2] Landauer, R, "Irreversibility and Heat Generation in the Computing Process, "IBM J. 5 (1961), 183-191.

[3] T. Toffoli, "Reversible computing," in Automata, Languages and Programming, W. de Bakker and J. van Leeuwen, Eds. Springer, 1980, p.632, technical Memo MIT/LCS/TM-151, MIT Lab. for Comput. Sci.

[4] Carlin Vieri, Pendulum: A Reversible Computer Architecture, M.S Thesis, MIT, 1995.

[5] Desoete, B.; De Vos, A. A reversible carry-look-ahead adder using control gates Integration, the VLSI Journal._ Elsevier Science B.V.. Vol. 33. 2002. pp. 89-104

[6] Kim, S. and Chae, S. 2005. Complexity reduction in an nRERL microprocessor. In Proceedings of the 2005 international Symposium on Low Power Electronics and Design (San Diego, CA, USA, August 08 - 10, 2005). ISLPED '05. ACM, New York, NY, 180-185.

[7] De Vos, A.; Burignat, S. and Thomsen, M. K. Reversible Implementation of a Discrete Integer Linear Transform. In Multiple-Valued Logic and Soft Computing, 2011.

[8] Thomsen, M. K.; Glück, R. and Axelsen, H. B. Reversible arithmetic logic unit for quantum arithmetic. In Journal of Physics A: Mathematical and Theoretical, 43 (38): 382002, 2010.

[9] De Vos, A. Reversible Computer Hardware. In Electronic Notes in Theoretical Computer Science, 253 (6): 17-22, 2010.

[10] Yokoyama, T. Reversible Computation and Reversible Programming Languages. In Electronic Notes in Theoretical Computer Science, 253 (6): 71-81, 2010.

[11] Axelsen H B, Gl¨uck R and Yokoyama T 2007 Reversible machine code and its abstract processor architecture Computer Science—Theory and Applications. Second International Symposium on Computer Science inRussia (Lecture Notes in Computer Science vol 4649) (Berlin: Springer) pp 56–69.

978-1-61284-863-1/11 $26.00 © 2011 IEEE

Efficient Pipelined VLSI Architectre with Dual Scanning Method for 2-D Lifting-Based Discrete Wavelet Transform

Anand Darji

Electrical Engineering Department
Indian Institute of Technology Bombay
Powai, India
anand@ee.iitb.ac.in

S.N.Merchant,A.N.Chandorkar

Electrical Engineering Department
Indian Institute of Technology Bombay
Powai, India
mechant@ee.iitb.ac.in, anc@ee.iitb.ac.in

Abstract—**In this paper, we describe a high speed, memory efficient, very low power and dual memory scan based pipelined VLSI architecture for 2-D Discrete Wavelet Transform (DWT) based on Legall 5/3 filter. Proposed architecture consists of two 1-D pipelined architectures along with transpose unit (TU). Architecture consumes two inputs per clock cycle and produces two outputs per cycle. Moreover dual scan technique is employ to enhance throughput with 100% hardware utilization efficiency without significant increase in power. This architecture uses 2N on chip buffer and five transpose register to process single level 2-D DWT of image size of NxN. RTL (Register Transfer Level) is written using VHDL and netlist is compiled using Synopsys Design Vision using UMC 180 nm MMRF technology cell library. After formal verification netlist is imported to cadence Soc encounter for GDS-II file generation for (Application Specific Integrated Circuit) ASIC. Simulation results show positive slack with 200 Mhz frequency. Core area of proposed architecture is only 0.73 mm^2 with low power consumption such as 13.38 mw.**

Keywords- ASIC ,DWT, Dyanamic Power, JPEG 2000, RTL, VHDL

I. INTRODUCTION

The DWT is used in many image processing applications such as compression, bioinformatics, texture discrimination [1]. There are lots of disadvantages of (Discrete Cosine Transform) DCT in comparison to DWT This template, modified such as blocking artifact, less resolution capability. DWT understands Human Visual System (HVS) better so that it has been accept in JPEG 2000 standard and adopted as the transform coder in MPEG-4 still texture coding. DWT has many useful properties like symmetrical transform, integer-to-integer transform and in-place computation. The conventional implementation using filter bank approach for 2-D DWT demands very high computational power, and most of the applications demands real-time processing with low power consumption. High speed yet low power implementation of 2-D DWT to meet the timing requirement of real-time and low power applications is therefore, considered as challenging task.

As convolution based approach of 2-D DWT demands more silicon area and power, Swelden *et al.* [2] have suggested lifting based scheme for biorthogonal filters, in which liner filter is factorized into few lifting steps. Lifting based scheme

utilized far less numbers of adders and multiplier compared to convolution based approach.

The 2-D DWT usually implemented as 1-D DWT as leave cell along with TU and storage place of $O(N^2)$. Line based architectures in [3]-[5] to reduce the size of transportation buffer. Liao *et al.*[5] have proposed two DWT architectures with recursive and dual scan methods for multi-level and single-level 2-D DWT, respectively. Xiong *et al.*[4] have suggested improved method to reduce the size of on-chip buffers. Wu et al. have [6] proposed pipelined architecture with modified lifting scheme to reduce critical path to only one multiplier delay. Y. Lai *et al.*[7] have proposed pipelined architecture using 1-D core based on dual scan method. This paper, we have proposed 5/3 lifting filter based 2-D DWT using two pipelined 1-D as sub cell and its VLSI implementation. We have used decimation property of DWT for interleaving to enhance throughput.

The rest of this paper is organized as follows. In section II, the proposed high speed and low power 2-D DWT structure is presented. ASIC implementation results and performance comparison is presented in section III and concluding remark are presented in section IV.

II. PROPOSED ARCHITECTURE

Proposed 2-D DWT architecture is as depicted Fig.1 has been used for ASIC implementation. Image or video frame is read from dual port ROM/RAM with dual scan method and given to Row Processing Unit (RPU) to calculate 1-D DWT. These 1-D coefficients are given to TU. In dual scan from each row two pixels are scanned per clock and given to RPU then in next clock two pixels from second row is scanned. This process is repeated till the end. The TU manages the column wise input to Column Processing Unit (CPU) and calculates 2-D DWT coefficients. This architecture is scalable and can be used for multilevel DWT.

The proposed architecture shown in Fig. 2 takes two inputs and gives two outputs per cycle. Data1 and Data2 are the odd and even input samples given to hardware in single clock for 100 % hardware utilization. This architecture is very simple design as compared to other architectures suggested in [5] and [8] which have complex control path to achieve 100%

hardware utilization efficiency. Simple control path helps in power minimization by introducing less switching. Usually 2-D DWT architecture has to wait for one complete row to be processed to start column processing and has requirement of line buffer with size N for image size NxN. Proposed architecture has a property to process two rows at alternate clocks which gives required 1-D coefficients to start column processing simultaneously to get 2-D DWT coefficients. This dual scanning not only saves line buffer but reduces latency as well. Transpose Unit (TU) is responsible for sequencing coefficient available from RPU to CPU. Architecture of CPU is same as RPU but has 2N line buffers. Here, in order to reduce power we use direct mapped divide by two and four instead of shifter or multiplier. Further, designed TU utilizes only five registers compare to normal requirement of 1.5N as transposing buffers [9] and uses two multiplexers which work on half clock rate. This design reduces number of transitions to reduce power. Moreover design of TU is independent of size on input image size. Proposed architecture has very low memory requirement and produces 2-D coefficients at a latency of only three cycles. This way lot of parallelism is introduced to save clocks as shown in Table I for data sequencing. This architecture has critical path of four adder delay can be reduced to only two adder delay by inserting pipeline registers.

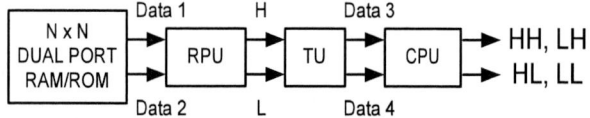

Figure 1. Pipelined 2-D DWT Architecture

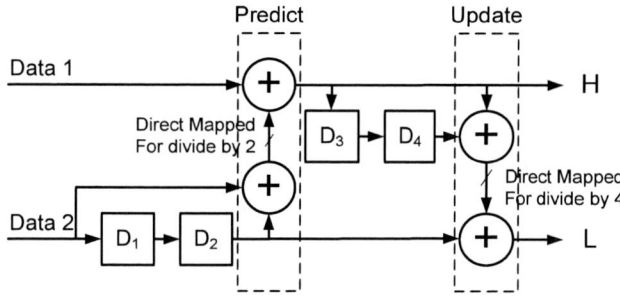

Figure 2. Row Processing Unit (RPU)

III. ASIC IMPLEMENTATION AND PERFORMACE ANALYSIS

The proposed architecture is implemented using UMC 180 nm technology standard cell library with clock frequency of a 200 MHz. The RTL design is synthesized using Synopsys Design Vision using standard cell library to calculate estimated power and area. Gate level netlist obtained after synthesis is exited using test vectors for verification and resultant waveforms are shown in Fig.3. The power estimated by Design Vision is based on the voltage, capacitance and time values provided as a standard.

Table I DATA SEQUENCE OF RPU AND CPU

Clk	Input	1D DWT Output	2D DWT Output
1	$X_{1,1}$; $X_{1,2}$		
2	$X_{2,1}$; $X_{2,2}$	$L_{1,1}$; $H_{1,2}$	
3	$X_{1,3}$; $X_{1,4}$	$L_{2,1}$; $H_{2,2}$	
4	$X_{2,3}$; $X_{2,4}$	$L_{1,3}$; $H_{1,4}$	$LL_{1,1}$; $LH_{1,2}$
5	$X_{1,5}$; $X_{1,6}$	$L_{2,3}$; $H_{2,4}$	$HL_{2,1}$; $HH_{2,2}$
6	$X_{2,5}$; $X_{2,6}$	$L_{1,5}$; $H_{1,6}$	$LL_{1,3}$; $LH_{1,4}$
7	$X_{1,7}$; $X_{1,8}$	$L_{2,5}$; $H_{2,6}$	$HL_{2,3}$; $HH_{2,4}$
..	....	....	....

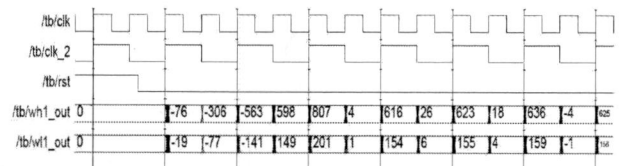

Figure 3 Post Synthesis Simulation Results of Gate level netlist

Then Design Vision synthesized netlist is imported in Cadence SOC encounter for ASIC implementation. SOC encounter is responsible for place and route the imported netlist as per user constraints in terms of area and speed and produces circuit level netlist. Detail report from SOC encounter is described in Table II. Post layout circuit power is calculated using Synopsys prime power for 2-D DWT operation on 256x256 image tile. We can see low Post routed netlist is simulated with same input vectors to see the effect of wire load the waveforms for the same is shown in Fig. 4. It is clearly visible from Fig. 3 and Fig. 4 that post synthesis is matches with post route simulation for same input vectors .

Table II CHIP LAYOUT REPORTS OF PROPOSED ARCHITECTURE USING SoC ENCOUNTER

Parameter	Value
Standard Cells	28459
Total area of Standard Cells	0.7386 mm^2
Total area of Core	0.7388 mm^2
Core Density	99.976 %
Core Density	99.976 %
Total Wire Length	0.3611 um
Cell with Maximum Capacitance	4.8244 pF

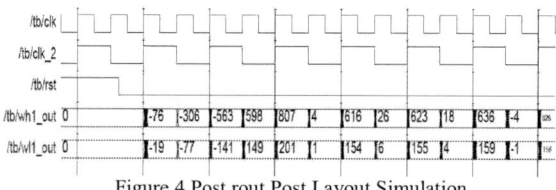

Figure 4 Post rout Post Layout Simulation

978-1-61284-863-1/11 $26.00 © 2011 IEEE

Table III COMPARISONS OF CHIP LEVEL IMPLEMENTATION

Parameter	Liao et al.[5]	Lai et al. [7]	Proposed
Specification	2-D 9/7 Filter (RA)	2-D 5/3 & 9/7 filter	2-D 5/3 filter
Technology	TSMC 180nm CMOS	TSMC 180nm CMOS	UMC 180nm CMOS
Core Area	2.25 mm^2	0.704 mm^2	0.7388 mm^2
Frequency	50 MHz	100 MHz	200 MHz
Power	---	102.6 mW	13.38 mW

IV. CONCLUSION

Our design is compared with other chip level implementation as shown in Table III. The core size of our design is slightly more than in [7] but very much less than [5] which is recursive architecture (RA) and the power is reduced drastically. The increase in core area is negligible in view of the power reduction achieved. The reduction in power is due to the optimized design using pipeline, dual scanning and less transpose buffer compare to other familiar architecture for same throughput rate. Both designs use 256x256 image tile for design evaluation.

REFERENCES

[1] S.G.Mallat," A theory for multiresolution signal decomposition: the wavelet representation,"IEEE Trans. Pattern Analysis and Machine Intelligence, vol. 11, no. 7, pp. 674-693, July 1989.

[2] Daubechies I. and W. Sweldens, "FactoringWavelet Transforms into Lifting Schemes," The Journal of Fourier Analysis and Applications, Vol. 4, No. 1, pp. 247–269,1998.

[3] P.C. Wu and L.G.Chen," An efficeint architecture for two-dimensional discrete wavelet transform", IEEE Trans. Circuits and Systems for Video technology, vol.11, no.4, pp.536-545, Apr. 2001

[4] Chengyi Xiong, Jinwen Tian, and Jian Liu, "Efficient Architectures for Two-Dimensional Discrete Wavelet Transform Using Lifting Scheme," IEEE Transaction on Image Processing , Vol. 16, No. 3, March 2007.

[5] Hongyu Liao, Mrinal Kr. Mandal," Efficient Architecture for 1-D and 2-D Lifting Based Wavelet Transform," IEEE Transaction on signal Processing, Vol. 52, N0. 5, May 2004.

[6] Bing-Fei Wu and C-F. Lin, "A High-Performance and Memory-Efficient Pipeline Architecture for the 5/3 and 9/7 Discrete Wavelet Transform of JPEG2000 Codec," IEEE Transaction on circuits and systems for Video Technology, Vol. 15, No. 12, December 2005

[7] Yeong-Kang Lai,L-F Chen and Y.-C. Shih, "A High-Performance and Memory-Efficient VLSI Architecture with Parallel Scanning Method for 2-D Lifting-Based Discrete Wavelet Transform," IEEE Transactions on Consumer Electronics, Vol. 55, No. 2, May 2009.

[8] M. Ferretti and D. Rizzo, "A parallel architecture for the 2-D discrete wavelet transform with integer lifting scheme," J. VLSI Signal Processing, vol. 28, pp. 165–185, July 2001.

[9] P-C.Tseng,C-T.Huang, and L-G.Chen," Generic RAM-based architecture fro two-dimnetional discrete wavelet transform with line-based method,"IEEE Trans. Circuit Syst.Video Technology,vol. 15,no.7,July 2005.

A Fast-locking Clock and Data Recovery Circuit with A Lock Detector Loop

Chih-Lin Chen, *Student Member, IEEE,*
and Chua-Chin Wang[†], *Senior Member, IEEE*

Chun-Ying Juan

Department of Electrical Engineering
National Sun Yat-Sen University
Kaohsiung, Taiwan 80424
Email: ccwang@ee.nsysu.edu.tw

Metal Industries Research & Development Centre (MIRDC),
Taipei 106, Taiwan.
Email: chunying@mail.mirdc.org.tw

Abstract—**This work presents a PLL-based (phase-locked loop) clock and data recovery (CDR) circuit with a lock detector loop for fast locking and low jitter. We use an adjustable charge pump to change the charge current according to the state of the lock detector loop, which is determined by seven clocks with equal phase difference. An experimental prototype was implemented using a typical 0.18 μm CMOS process. The post-layout-extracted simulation results reveal that the worst case jitter of the recovery clock is less than 199.66 ps (peak-to-peak) and the settling time is less than 4 μs at all PVT (Process, voltage, and temperature) corners.**

Index Terms—**fast-locking, phase shift, CDR, and lock detector loop.**

I. INTRODUCTION

Recently, FlexRay [1] is considered as a total solution to be integrated with different in-car communication specifications, i.e., CAN and MOST. FlexRay is mainly aimed at safety and reliability. In next-generation FlexRay specification, the data rate might be drastically increased for adding more audio/video equipments in a vehicle, e.g., mobile TV receiver, GPS, video player, video game, and so on. Therefore, clock and data recovery (CDR) circuit will be required in the future FlexRay systems.

Two major architectures for CDR designs are PLL-based CDR and phase interpolation CDR. The former is close to a PLL architecture, including phase/frequency detector (PFD), voltage controlled oscillator (VCO), charge pump (CP), and low pass filter (LPF). However, the disadvantages include: the clock frequency of VCO can not surpass ±50% of the center frequency [2], the settling time is long depending on bandwidth and jitter of the system, and the CDR circuit might be locked in a wrong frequency called harmonic lock. Traditionally, FlexRay systems need a high frequency (8 times data rate) phase-locked loop (PLL) circuit to achieve over-sampling function and data recovery. For example, if FlexRay system operates at a high date rate (e.g., 100 Mbps), a 800 Mbps PLL is required for over-sampling. A reliable 800 Mbps PLL is not easy to be designed and integrated in system-on-

†: Prof. C.-C. Wang is the contact author.

chip (SOC) such that it is not reliable to be used in a in-vehicle network for the sake of safety.

Phase interpolation CDR utilizes the output clock of VCO with different phases to sample data, and then the sampled results are transferred to recover data. Notably, each bit of the data only needs 3 sampled bits to recover. Meanwhile, the jitter performance depends on equalization of the clock phase shift generated by VCO. Notably, it is also very hard to have equal phase shift between two adjacent clocks.

CDR is mainly used to generate clock, synchronize received data, and reduce jitter. In a receiver design, receivers need a CDR circuit to synchronize data with the clock, because incoming data are usually asynchronous with respect to the system clock. If the incoming data are coupled with noise, the receiver with CDR should reject noise to reduce jitter. In prior reports, CDR designs usually had a trade-off between settling time and clock jitter for different applications. If a CDR circuit operates in a short settling time, it will have a poor jitter performance. By contrast, a low-jitter CDR circuit must spend longer time to lock the incoming data. To achieve short settling time and low jitter, a lock detector was reported to adjust loop bandwidth in a receiver system [3]- [7]. PLL designs [3] use a digital frequency difference detector (DFDD) [4] to adjust resistors of LPF to change system bandwidth for short settling time. In prior CDR designs [5]- [6], a lock detector was proposed to change system bandwidth for fast locking. Another kind of CDR design [7] utilized a lock detector to detect the transition with respect to a reference clock. If the clock transition occurs before or after the reference clock, a counter is counted up. Otherwise, the counter is counted down. The counter then determines the CDR circuit to operate in a frequency detecting loop or a phase detecting loop.

This work proposes a 100 Mbps CDR circuit with short locking time and low jitter. The proposed CDR circuit is used to recover data bits given a 100 Mbps data rate for FlexRay specifications. In Section II, we introduce the architecture of PLL-based CDR with a lock detector loop and show the data flow diagram. In Section III, we demonstrate the simulation results of CDR circuit by MATLAB and HSPICE. We compare our CDR circuit with the prior works in a comparison table

978-1-61284-863-1/11 $26.00 © 2011 IEEE

as well. A brief conclusion is given in Section IV.

II. ARCHITECTURE

The proposed CDR circuit includes a phase detector (PD), a frequency detector (FD), two charge pumps for PD and FD (CP_PD and CP_FD), a second-order low-pass filter (R1, C1, and C2), a voltage controlled oscillator (VCO), a Divider, and a lock detector loop, as shown in Fig. 1. The function of each block is given in the following text.

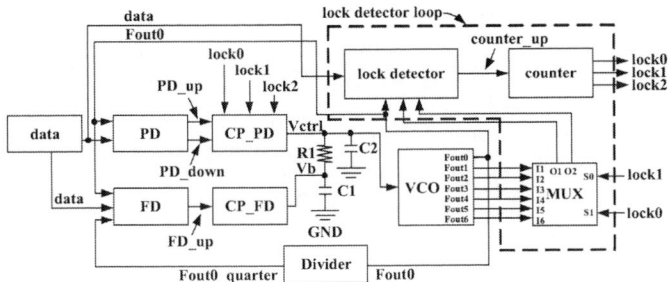

Fig. 1. The proposed CDR circuit with a lock detector loop

A. Phase detector (PD)

The Alexander binary phase detector in [8] is used as our PD. If the data transition occurs after the Fout0's falling edge, the PD_up outputs logic '1' to increase clock frequency of VCO, where Fout0 is the main base output clock. Otherwise, if the data transition occurs before the Fout0's falling edge, the PD_up outputs logic '0' to decrease clock frequency of VCO. Because the PD does not have a linear frequency response, the PD causes a constant jitter in the CDR system.

B. Frequency detector (FD)

The FD in [9] is used to detect whether the data stream have two rising edges during Fout0_quarter is logic '0' or logic '1', where the frequency of Fout0_quarter is a quarter of the output clock frequency of VCO that is generated by the Divider. If the FD detects two rising edges, FD_up outputs logic '1'. In other words, the clock frequency of VCO is less than 2 times of data rate such that the clock frequency of VCO must be increased.

C. Charge pump for PD and FD (CP_PD and CP_FD)

Fig. 2 shows the schematic of the charge pump for PD (CP_PD). R_{PD} is used to generate a bias current, I_{RPD}. Notably, I_P is determined by I_{RPD}, lock0, lock1, and lock2. The maximum charge current, I_P, is 8 times of I_{RPD}. If PD_up is logic '1', I_P flows into Vctrl from VDD. If PD_down is logic '1', I_P flows into GND from Vctrl. The architecture of the charge pump for FD (CP_FD) is identical to that for PD.

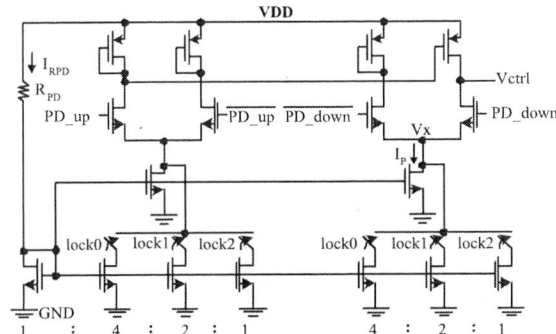

Fig. 2. The schematic of CP_PD

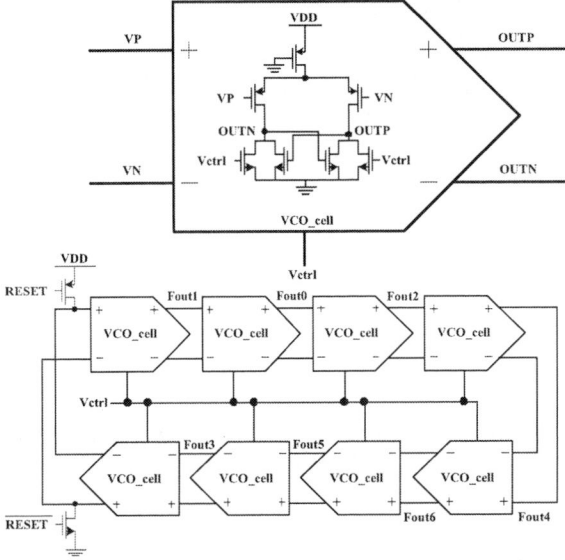

Fig. 3. The architecture of VCO and the schematic of VCO cell

D. Voltage controlled oscillator (VCO)

Fig. 3 shows the VCO block diagram, where the schematic of the VCO_cell is included. A differential 8-stage voltage controlled ring oscillator is used to generate clocks with equal phase shift. When RESET is activated, the VCO starts oscillating. Vctrl is used to adjust the clock frequency of VCO. VCO_cells accumulate different phase delays, respectively, to generate a bank of clocks with seven different phases, i.e., Fout0 to Fout6. Fout0 is the main clock of VCO to be synchronized with the data. Fout1, Fout3, and Fout5 lead Fout0 in phase, while Fout2, Fout4, and Fout6 lag in phase.

E. Lock detector loop

The lock detector loop is composed of a lock detector, a MUX, and a counter. Three D-flip-flops (DFFs) and an XNOR are used to implement the lock detector, as shown in Fig. 4. The lock detector is used to detect if the positive and negative edges of incoming data are synchronized with Fout0. Referring to Fig. 1, lock0 and lock1 are used to select different clocks fed into the lock detector by the MUX, which are described as follows.

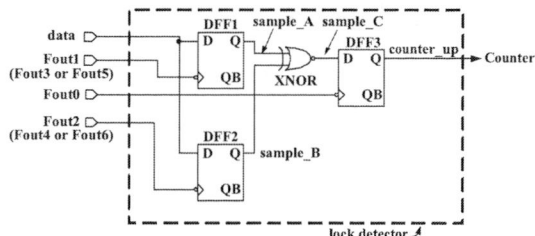

Fig. 4. The architecture of lock detector

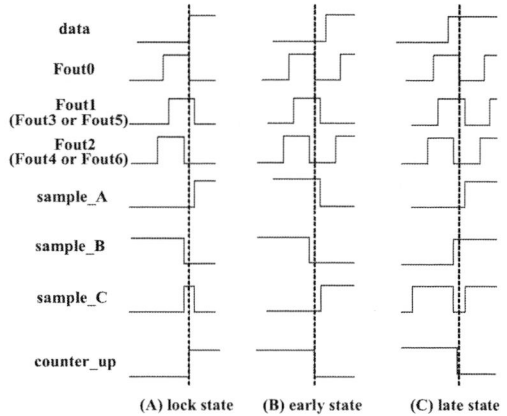

Fig. 5. Three different scenarios in the lock detector loop

- If {lock0, lock1}='00', MUX selects Fout1 and Fout2 fed into the lock detector.
- If {lock0, lock1}='10', MUX selects Fout3 and Fout4 fed into the lock detector.
- If {lock0, lock1}='11', MUX selects Fout5 and Fout6 fed into the lock detector.

Fig. 5 shows the three scenarios in the lock detector loop, which are described as follows. DFF1 uses three clocks, Fout1, Fout3, and Fout5, to sample data depending on lock0 and lock1. If the data transition occurs before Fout1, sample_A is set to logic '1'. Otherwise, if the data transition occurs after Fout1, sample_A is set to logic '0'. DFF2 uses three clocks, Fout2, Fout4, and Fout6, to sample data also depending on lock0 and lock1. If the data transition occurs after Fout2, sample_B is set to logic '1'. Otherwise, if the data transition occurs before Fout2, sample_B is set to logic '0'. If CDR circuit is the lock state, the data transition occurs after Fout2 and before Fout1. In other words, sample_A and sample_B are switched to logic '0', as shown in Fig. 5(A). By contrast, if CDR circuit is not locked, sample_A and sample_B are either switched to logic '0', as shown in Fig. 5(B) or Fig. 5(C). Sample_A and sample_B are coupled to the inputs of XNOR to decide which scenario occurs. DFF3 is used to sample the output of XNOR, sample_C, by Fout0. The sampled result is "counter_up", which is used to trigger the counter or reset it.

The phase shift of each clock is shown in Fig. 6. Because the CP_PD initially operates with a large current for short settling time, the clock jitter is large. In other words, the phase shift of any two adjacent clocks is large. To enhance the reliability

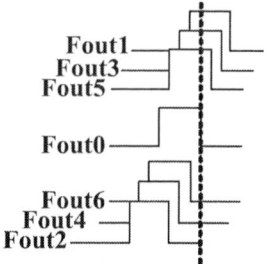

Fig. 6. The clocks with equal phase shift

TABLE II
THERE PARAMETERS IN
EQN. (1)

Parameter	value
I_{RPD}	36 uA
VCO gain	88 MHz
R1	4.5 KΩ
C1	100 pF
C2	3 pF

of the lock detector, the MUX selects Fout1 and Fout2 for the lock detector in the beginning, because the phase difference between Fout1 and Fout2 is maximum. Therefore, the phase shift between Fout1 and Fout2 can cover variation of the clock jitter given a wrong detection. If the CDR circuit continues to be in the lock state, the counter continues to count up. If the counter runs up to a pre-defined value, lock0 is switched to logic '1' to ensure that CDR circuit is locked. Therefore, the current of CP_PD is decreased to reduce the clock jitter. Then, the MUX selects Fout3 and Fout4 for the lock detector. If the counter runs up to a pre-defined value, lock1 is switched to logic '1' to reduce more current of CP_PD. Finally, the MUX selects Fout5 and Fout6 for the lock detector. Similarly, if the counter runs up to a pre-defined value, lock2 is switched to logic '1' to further reduce more current in CP_PD. In short, depending on lock0, lock1, and lock2, the current of CP_PD is reduced step by step such that the clock jitter behaves the same.

III. SIMULATION RESULTS

For the sake of stability, we need to avoid CDR system from oscillating such that the phase margin of the the CDR system must be larger than 0°. Eqn. (1) is the transfer function of our CDR shown in Fig. 1, which is simulated using MATLAB to derive all of the system parameters given in Table II. By Eqn. (1), the phase margin is 71° and the bandwidth is 13.7 MHz.

Frequency response =

$$I_{RPD} \times VCO \text{ gain} \times \frac{1}{s} \times \frac{R1 + \frac{1}{s \times C1}}{s \times C2 \times (R1 + \frac{1}{s \times C1} + \frac{1}{s \times C2})} \quad (1)$$

The proposed design is implemented using a typical 0.18 μm CMOS process to justify the performance. Notably, all of the process corners: [-40°C, +125°C] and [SS, TT, FF] models are simulated. Fig. 7 shows the transient response between Vctrl and the lock detector. The voltage amplitude of Vctrl is

TABLE I
COMPARISON BETWEEN THE PROPOSED DESIGN AND PRIOR WORKS

	Year	Process	Frequency	Clock jitter	Settling time	Power	Supply voltage	Core area	FOM
		(μm)	(Mbps)	(pk-pk)	(bits/time)	(mW)		(μmm^2)	$\frac{\text{Mbps}\times\mu\text{m}\times\text{V}^2}{\text{bits}\times\text{mW}\times\text{ps}\times\text{mm}^2}$
Ours	2011	0.18	100	199.66 ps	400 bits/4μs	6.649	1.8 V	0.16	686.42
[7]	2009	0.18	5120/6400	2.12 ps(rms)	N/A	136	1.8 V	0.8	N/A
[11]	2008	0.18	662-3125	62.2 ps	>331000 bits / >500 μs	60	1.8 V	0.1326	118.8
[5]	2006	0.18	2500	88 ps	4862 bits/3.89 μs	N/A	1.8 V	0.133	N/A
[10]	2006	0.18	155.52-3125	467 ps	>15552 bits/ >100μs	95	1.8 V	0.88	3
[12]	2006	0.35	200-2000	120 ps	>60000 bits/ >30 μs	170	3.3 V	0.4	15.56

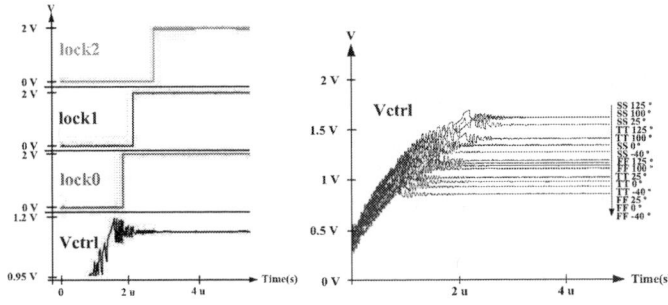

Fig. 7. Simulation waveform of the lock detector   Fig. 8. Simulation waveform of Vctrl

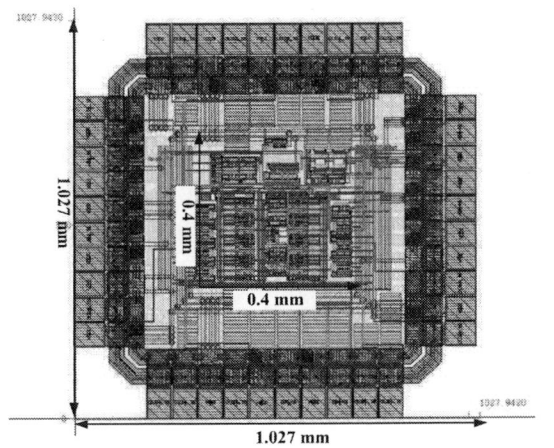

Fig. 9. Layout of the proposed design

decreased from 80 mV to 0.3 mV. Fig. 8 shows the simulation waveforms of Vctrl at different process corners. Fig. 9 shows the layout of the proposed design. The chip area is 1.027 × 1.027 mm^2, and the core area is 0.4 × 0.4 mm^2. Table I shows the comparison between the proposed design and several prior works. Besides attaining the least power dissipation and the shortest settling time, our design also shows the best FOM.

IV. CONCLUSION

The proposed CDR circuit resolves the difficulty of traditional CDR designs, because it does not need to make a trade-off between settling time and jitter. When CDR circuits include a lock detector loop, they can use a larger current in charge pumps to shorten settling time. When CDR circuits are locked, they decrease the current in the charge pumps to reduce jitter.

Therefore, our proposed design has a good performance in both settling time and jitter.

ACKNOWLEDGEMENT

This investigation is partially supported by National Science Council under grant NSC99-2221-E-110-082-MY3, NSC99-2220-E-110-001. It is also partially supported by Metal Industries Research Development Centre (MIRDC) and Ministry of Economic Affairs, Taiwan, under grant 100-EC-17-A-01-1010. The authors would like to express their deepest gratefulness to Chip Implementation Center of National Applied Research Laboratories, Taiwan, for their thoughtful chip fabrication service.

REFERENCES

[1] FlexRay Communications System - Protocol Specification V2.1 (http://www.flexray.com), 2005.

[2] R. J. Baker, H. W. Li, and D. E. Boyce, CMOS Circuit Design, Layout, and Simulation, 2nd ed. New Jersey: John Wiley & Sons, Inc., 1997.

[3] Y. Tang, M. Ismail, and S. Bibyk, "A new fast-settling gearshift adaptive PLL to extend loop bandwidth enhancement in frequency synthesizers," in Proc. IEEE International Symposium on Circuits and Systems, May 2002, vol. 4, pp. 787-790.

[4] I. Hwang, S. Lee, and S. Kim, "A digitally controlled phase loop with fast locking scheme for clock synthesis application," in Proc. IEEE International Solid-State Circuits Conference Digest Technical Papers, Feb. 2000, pp. 168-169.

[5] J.-K Woo, H. Lee, W.-Y. Shin, H. Song, D.-K. Jeong, and S. Kim, "A fast-locking CDR circuit with an autonomously reconfigurable charge pump and loop filter," in Proc. IEEE Asian Solid-State Circuits Conference, Nov. 2006, pp. 411-414.

[6] J.-K Woo, D.-K. Jeong, and S. Kim, "Fast-locking CDR circuit with autonomously reconfigurable mechanism," Electronics Letters, vol. 43, no. 11, pp. 624-626, May 2007.

[7] F.-T. Chen and J.-M. Wu, "An extended phase detector 2.56/3.2Gb/s clock and data recovery design with digitally assisted lock detector," in Proc. IEEE International Symposium on Circuits and Systems, May 2009, pp. 1831-1834.

[8] J. D. H. Alexander, "Clock recovery from random binary signals," Electronics Letters, vol. 11, no. 22, pp. 541-542, Oct. 1975.

[9] D. Dalton, K. Chai, E. Evans, M. Ferriss, D. Hitchcox, P. Murray, S. Selvanayagam, P. Shepherd, and L. DeVito, "A 12.5-Mb/s to 2.7-G/s continuous-rate CDR with automatic frequency acquisition and data-rate readback," IEEE Journal Solid-State Circuits, vol. 40, no. 12, pp. 2713-2725, Dec. 2005

[10] R.-J. Yang, K.-H. Chao, S.-C. Hwu, C.-K. Liang, and S.-I. Liu, "A 155.52 Mbps-3.125 Gbps continuous-rate clock and data recovery circuit," IEEE Journal Solid-State Circuits, vol. 41, no. 6, pp. 1380-1390, Jun. 2006.

[11] S.-H. Lin and S.-I. Liu, "Full-rate bang bang phase/frequency detectors for unilateral continuous-rate CDRs," IEEE Transactions on Circuits and Systems II, vol. 55, no. 12, pp. 1214-1218, Dec. 2008.

[12] R.-J. Yang, K.-H. Chao, and S.-I. Liu, "A 200-Mbps 2-Gbps continuous-rate clock-and-recovery circuit," IEEE Transactions on Circuits and Systems I, vol. 53, No. 4, pp. 842-847, Apr. 2006.

A 5-bit 500-MS/s Time-Domain Flash ADC in 0.18-μm CMOS

Young-Jae Min, Ammar Abdullah, Hoon-Ki Kim, and Soo-Won Kim

Department of Electrical Engineering
Korea University
Seoul, Korea
yjmin@asic.korea.ac.kr

Abstract— **A 5-bit 500-MS/s time-domain flash ADC is presented. The proposed ADC consists of a reference resistor ladder, two voltage-to-time converter arrays, a time-domain comparator array and a digital encoder without sample-and-hold. In order to achieve low-power consumption with high conversion-speed and to enhance design reusability in terms of a highly digital implementation with more regular mask patterns, the time-domain comparison is devised in the flash ADC. The prototype has been implemented and fabricated in a standard 0.18μm CMOS technology and occupies 0.132mm^2 without pads. The measured SNDR and SFDR up to the Nyquist frequency are 26.6 and 35.1 dB, respectively. And the peak DNL and INL are measured as 0.43 LSB and 0.58 LSB, respectively. The prototype consumes 8mW with a 1.8-V supply voltage.**

Keywords-component; analog-to-digital converter; time-domain comparison; flash converter; voltage-to-time converter

I. INTRODUCTION

As improving the speed of the data rate in digital wireless communication systems and emerging communication technologies such as ultra-wideband (UWB) communication [1], the demand for high-speed analog-to-digital converters (ADCs) continues to increase. In addition, state-of-the-art disk drive systems also demand such ADCs to achieve higher transfer data rates [2].

Since the ADC power consumption takes a significant portion of the overall power consumption in the whole analog front end of wireless communications and portable devices, the low-power AD conversion is the key constraint. A lot of efforts have been devoted to reduce the power consumption of high-speed low-resolution ADCs. A recent attractive way to realize low-power high-speed ADCs is exploiting parallelism to increase the speed of the simple energy-efficient ADCs such as time-interleaved successive approximation register (SAR) ADCs [3], [4]. And the architecture of binary-search ADCs between flash and SAR ADCs has been presented [5]. However, the use of the binary-search limits the latency and conversion speed. Flash ADCs are still one of the most preferred architectures for high-speed and low-to-medium resolution ADCs. Flash ADCs require 2^b comparators, where b is resolution in bits, which results high-power consumption. Interpolating [6] and folding [7] ADC architectures have been presented to reduce power consumption by reduction of the

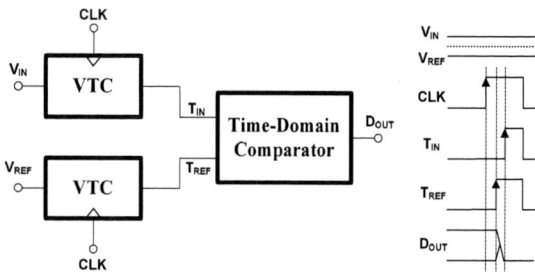

Fig. 1. Operation principle of the time-domain ADC.

number of comparators with preamplifier and latch, but the fundamental problem in terms of power consumption by comparators in flash ADCs remains [8]. For low-power consumption and design reusability with more regular mask pattern, an SAR ADC employing a time-domain comparator [9] and inverter-based flash ADC [10] have been proposed.

In this paper, a highly digital 5-bit 500-MS/s flash ADC with time-domain comparison to achieve low-power consumption and to enhance design reusability is presented. This paper is organized as follows. In Section II, the operation principle of the time-domain ADC and proposed flash ADC architecture are presented. Section III describes details of the implemented circuits. And Section IV shows experimental results of our fabricated circuit. Finally conclusions are drawn in Section V.

II. TIME-DOMAIN FLASH ADC ARCHITECTURE

Comprising the time-domain comparator in a SAR ADC that operates in the time domain, instead of the voltage or current domain has been proposed [9]. Figure 1 shows the operation principle of the time-domain ADC. The input voltage V_{IN} and reference voltage V_{REF} are converted to corresponding time-domain signals T_{IN} and T_{REF} by voltage-to-time converters (VTCs). The time-domain comparator checks whether T_{IN} arrives earlier than T_{REF} or not and then the digital output signal D_{OUT}, which corresponds to the magnitude of the input voltage V_{IN} over V_{REF}, is generated. As mentioned, using the inverter threshold in flash ADC also has been proposed [10]. However, the nonlinearities in the circuit, which is primarily caused by

978-1-61284-863-1/11 $26.00 © 2011 IEEE

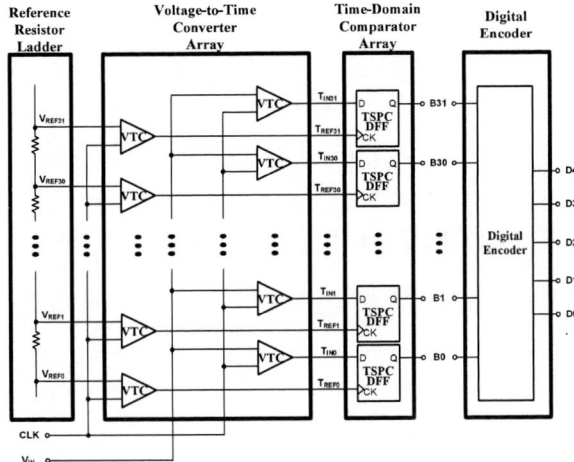

Fig. 2. Block diagram of the proposed time-domain flash ADC.

process and environmental variations, can degrade performances in terms of robustness for the design of highly digital ADC circuits.

The architecture of the proposed flash ADC is shown in Fig. 2. The proposed 5-bit single-ended time-domain ADC without sample-and-hold (S&H) consists of a reference resistor ladder, two VTC arrays, a time-domain comparator array and a digital encoder. A VTC array consists of 32 VTCs and a time-domain comparator array consists of 32 true single phase clock (TSPC) D flip-flops (DFFs). In this ADC, no S&H was used because of the high speed of the converter and the corresponding extremely hard demands on the performance of the S&H, which can lead to the serious degradation of the ADC performances.

The input voltage and 32 reference voltages are converted to time-domain digital signals with corresponding delays by two VTC arrays. 32 TSPC DFFs of the time-domain comparator array compare the delay differences. If the time-domain input signal T_{INi} arrives later than the time-domain reference signal T_{REFi} when the input voltage is lower than the reference voltage, the output of the comparator goes to low as shown in Fig. 1. The digital encoder converts 32 outputs of the comparator array to 5-bit binary outputs.

III. CIRCUIT DESCRIPTION

A. Voltage-to-Time Converter

There are three popular techniques of shunt capacitor, current starved and variable resistor techniques available for the design of the voltage controlled delay element [11] as a VTC. In this work, considering the controllable delay range and noise rejection [12], the current starved technique is utilized in the VTC. Figure 3 shows the schematic of the proposed VTC. At the instant of the high-to-low transition on the current starved inverter when transistor M2 turns on, the capacitor C_L at its output node starts to discharge. The

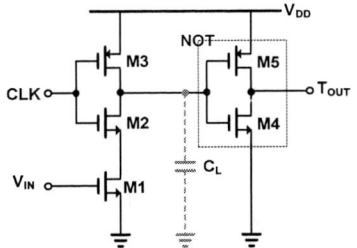

Fig. 3. Schematic of the voltage-to-time converter.

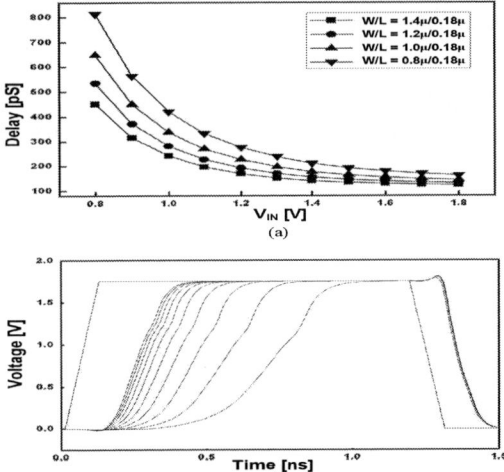

Fig. 4. Simulation results of the voltage-to-time converter. (a) Rising delay range versus W/L ratio. (b) Rising delay versus V_{IN}.

discharge current is controlled by transistor M1 acting as a current source. Therefore, the rising delay t_d together with the delay of the second inverter composing of transistors M4-M5 can be derived as [11]

$$t_d = \tau \ln \frac{1 + \lambda_1 V_{DD}}{1 + \lambda_1 \frac{V_{DD}}{2}}, \qquad (1)$$

where

$$\tau = \frac{C_L}{\frac{\mu_n C_{OX}}{2} \left(\frac{W}{L}\right)_{M1} (V_{IN} - V_T)},$$

μ_n, V_T, and C_L are the mobility of electrons, threshold voltage of transistor M1 and the overall capacitance at output node of the current starved inverter. From (1), the range of the rising delay t_d can be controlled by the W/L ratio of transistor M1, and the delay t_d is determined by V_{IN}. Figure 4 depicts the simulation results of rising delay range for different W/L ratios of transistor M1 and delayed output of a VTC for variable voltages of V_{IN}.

978-1-61284-863-1/11 $26.00 © 2011 IEEE   337

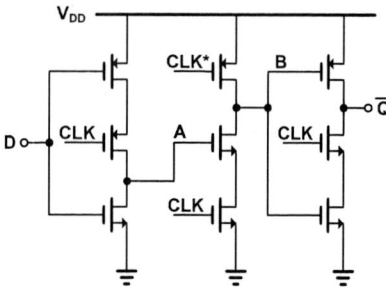

Fig. 5. Schematic of the time-domain comparator with TSPC-DFF.

B. Time-Domain Comparator with TSPC DFF

Since the small time difference can be applied to the time-domain comparator in this ADC, it requires DFFs with the high-speed operation. TSPC DFF offers the superior advantage of using a single clock. Thus, higher frequency operation can be achieved. The disadvantage is the slight increase in the number of transistors [13]. Figure 5 shows the schematic of the implemented TSPC DFF. When *CLK* remains high to suppress extra charge and discharge at node B and D keeps low, the middle-stage gate functions as an inverter and generates glitches on node B, which incurs unwanted power consumption [14]. Thus, the delayed *CLK** is applied as shown in Fig. 5.

C. Other Building Blocks

The reference resistor ladder without linear functions such as averaging subdivides the converter reference voltage in a set of 32 reference voltages, which are compared in parallel with the input signal. To suppress the kick-back noise, shunt MOS capacitors are added in the outputs of the reference resistor ladder. The digital encoder converts the thermometer code generated by 32 time-domain comparators to the binary code every clock cycle.

IV. EXPERIMENTAL RESULTS

The prototype of the proposed time-domain flash ADC has been fabricated in a standard 0.18μm single-poly four-metal (1P4M) CMOS technology and occupied 1100μm × 130μm without pads. The die photo and layout of the proposed ADC core are shown in Fig. 6. The chip has mounted on a standard PCB and directly wire-bonded for testing. The output digital stream is down-sampled by a factor of 4 to allow acquisition by the logic analyzer. The on-chip clock divider generates the divided by 4 clock to avoid sampling uncertainty. The testing was only conducted at a supply voltage of 1.8V.

Figure 7 shows the measured static characteristics such as differential nonlinearity (DNL) and integral nonlinearity (INL) using the histogram method. The peak DNL and INL are measured as 0.43 LSB and 0.58 LSB, respectively. Also, the dynamic performance is shown in Fig. 8 by using a Fast-Fourier Transform (FFT) spectrum analysis. The signal-to-noise-and-distortion ratio (SNDR) and spurious free dynamic range (SFDR) with a 239MHz input signal are achieved 26.6

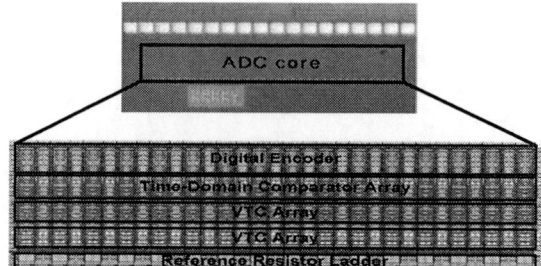

Fig. 6. Die photo and layout of the proposed ADC core.

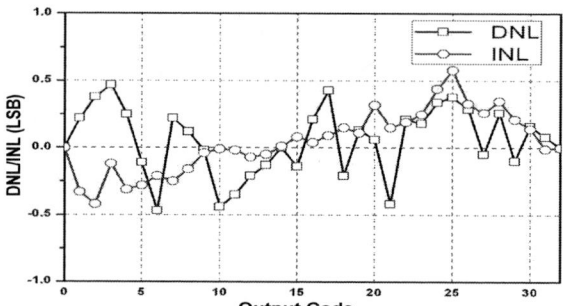

Fig. 7. Measured DNL and INL.

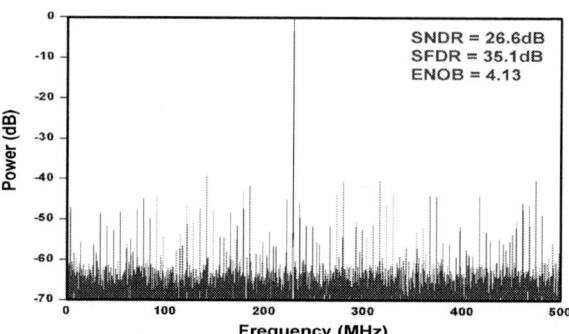

Fig. 8. Measured FFT spectrum with a 239MHz input frequency.

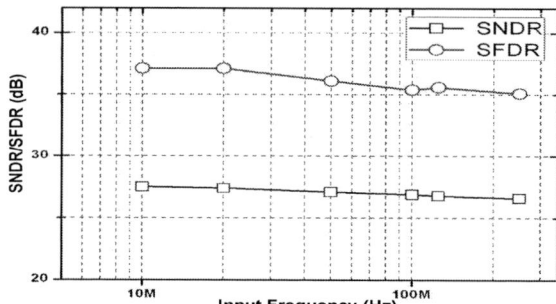

Fig. 9. Measured SNDR and SFDR vs. input frequency.

TABLE I
PERFORMANCE SUMMARY AND COMPARISONS OF PUBLISHED ADC.

	This work	[3]	[4]	[5]	[6]	[10]
Technology	CMOS 0.18μm	CMOS 90nm	CMOS 65nm	CMOS 65nm	CMOS 0.13μm	CMOS 90nm
Architecture	Time-domain Flash	Time-interleaved SAR	Time-interleaved SAR	SAR + Flash	Flash	Inverter-based Flash
Sampling Rate	500MS/s	600MS/s	250M, 500MS/s	800MS/s	600M, 1.2GS/s	300M, 600MS/s
Resolution	5 bits	6 bits	5 bits	5 bits	6 bits	5 bits
ENOB	4.13	5.1	4.10, 4.04	4.40	5.6, 5.7	4.45, 4.08
Peak DNL/INL (LSB)	0.43/0.58	N/A	0.26/0.16	0.56/0.62	0.4/0.6	0.4/0.54, 0.31/0.87
Supply Voltage	1.8V	1 ~ 1.2V	1.2V	1V	1.5V	0.8, 1V
Power Consumption	8mW	10mW	1.8, 5.9mW	1.97mW	90m, 160mW	3.2m, 6.7mW
Area	0.132mm^2	N/A	0.91mm^2	0.018mm^2	0.12mm^2	0.11mm^2
FOM	0.91 pJ/conv	0.5 pJ/conv	0.44, 0.75 pJ/conv	0.116 pJ/conv	1.5, 2.2 pJ/conv	0.49, 0.66 pJ/conv

and 35.1dB. Figure 9 summarizes the measured SNDR and SFDR as a function of the input frequency when using a 500MHz sampling frequency.

The performance summary of the proposed flash ADC is given in Table I. And performance comparisons with published ADCs are also listed. The power consumption by 8mW at a 1.8V supply voltage and 500MHz sampling frequency is measured. Therefore, the proposed time-domain flash ADC achieves a figure-of-merit (FOM) with effective resolution bandwidth (ERBW) of

$$FOM = \frac{POWER}{2^{ENOB} \cdot 2 \cdot ERBW} = 0.91 \, J/conv. \quad (2)$$

V. CONCLUSION

In this paper, a 5-bit 500-MS/s time-domain flash ADC is presented. To achieve low-power consumption with high conversion-speed and to enhance design reusability in terms of highly digital implementation with more regular mask patterns, the time-domain comparison is devised in the flash ADC. The prototype, which has been fabricated in a standard 0.18μm CMOS technology, achieves a FOM of 0.91pJ/conv. Although no low-power digital circuit technique has been comprised, further low-power operation can be easily achieved by voltage scaling or reduction techniques of leakage power.

ACKNOWLEDGMENT

This work was supported by the National Research Foundation of Korea (NRF) grant funded by the Korea government (MEST) (No.K20902001448-10E0100-03010) and Seoul R&BD Program (10920).

REFERENCES

[1] D. D. Wentzloff, R.l Blazquez, F. S. Lee, B. P. Ginsburg, J. Powell and A. P. Chandrakasen, "System design considerations for ultra-wideband communication," *IEEE Commun. Mag.*, vol. 43, no. 8, pp. 114-121, Aug. 2005.

[2] E. F. Haratsch and Z. A. Keirn, "Digital signal processing in read channels," *in Proc. IEEE Custom Integrated Circuits Conf. (CICC)*, pp. 683-690, Sept. 2005.

[3] D. Draxelmayr, "A 6 b 600 MHz 10 mW ADC array in digital 90 nm CMOS," *IEEE Int. Solid-State Circuits Conf. (ISSCC) Dig. Tech. Papers*, pp. 264-265, Feb. 2004.

[4] B. P. Ginsburg and A. P. Chandrakasan, "500-MS/s 5-bit ADC in 65-nm CMOS with split capacitor array," *IEEE J. Solid-State Circuits*, vol. 42, no. 4, pp. 739-747, Apr. 2007.

[5] Y. Lin, S.Jyh Chang, Y.Liu, C.Liu and G. Huang, "A 5b 800MS/s 2mW asynchronous binary-search ADC in 65nm CMOS," *IEEE Int. Solid-State Circuits Conf. (ISSCC) Dig. Tech. Papers*, pp. 80-81, Feb. 2009.

[6] C. Sandner, M. Clara, A. Santner, T. Hartig and F. Kuttner, "A 6-bit 1.2-GS/s low-power flash-ADC in 0.13-μm digital CMOS," *IEEE J. Solid-State Circuits*, vol. 40, no. 7, pp. 1499-1505, Jul. 2005.

[7] B. Verbruggen, J. Craninckx, M. Kuijk, P. Wambacq and G. Van der Plas, "A 2.2 mW 1.75 GS/s 5 bit folding flash ADC in 90 nm digital CMOS," *IEEE J. Solid-State Circuits*, vol. 44, no. 3, pp. 874-882, Mar. 2009.

[8] B. P. Ginsburg and A. P. Chandrakasan, "Dual time-interleaved successive approximation register ADCs for an ultra-wideband receiver," *IEEE J. Solid-State Circuits*, vol. 42, no. 2, pp. 247-257, Feb. 2007.

[9] A. Agnes, E. Bonizzoni, P. Malcovati and F. Maloberti, "A 9.4-ENOB 1V 3.8μW 100kS/s SAR ADC with time-domain comparator," *IEEE Int. Solid-State Circuits Conf. (ISSCC) Dig. Tech. Papers*, pp. 246-247, Feb. 2008.

[10] J. E. Proesel and L. T. Pileggi, "A 0.6-to-1V inverter-based 5-bit flash ADC in 90nm digital CMOS," *in Proc. IEEE Custom Integrated Circuits Conf. (CICC)*, pp. 153-156, Sept. 2008.

[11] M. Maymandi-Nejad and M. Sachdev, "A digitally programmable delay element: design and analysis," *IEEE Trans. Very Large Scale Integration (VLSI) Syst.*, vol. 11, no. 5, pp. 871-878, Oct. 2003.

[12] M. G. Johnson and E. L. Hudson, "A variable delay line PLL for CPU-coprocessor synchronization," *IEEE J. Solid-State Circuits*, vol. 23, no. 5, pp. 1218-1223, Oct. 1988.

[13] J. M. Rabaey, A. Chandrakasan, and B. Nikolic, *Digital Integrated Circuits, A Design Perspective*, Second Edition, Prentice Hall, Upper Saddle River, NJ, 2003.

[14] K. Y. Kim, W. K. Lee, H. K. Kim and S. W. Kim, "Low-power programmable divider for multi-standard frequency synthesizers using reset and modulus signal generator," *in Proc. IEEE Asian Solid-State Circuits Conf. (ASSCC)*, pp. 77-80, Nov. 2008.

Parallel Background Calibration with Signal-Shifted Correlation for Pipelined ADC

Kexu Sun, Xuan Wang, Lenian He
Institute of VLSI Design, Zhejiang University
Hangzhou 310027, P. R. China
Email: sunkx@vlsi.zju.edu.cn

Abstract—**Correlation-based background calibration methods have been used to correct capacitor mismatch and finite opamp open-loop gain errors of pipelined analog-to-digital converter (ADC). However, the correlation takes long time to converge. A novel parallel background calibration for a 14-bit 100Msps ADC with signal-shifted correction is proposed to overcome the above constraint by three means. First, a modified 1.5-bit stage is proposed in order to allow the injection of a large pseudo-random dither without missing code. Second, before correlating the signal, it is divided into 18 sub-ranges via some additional comparators and shifted for the purpose that the error in correlation converges fast. Finally, the front pipeline stages are calibrated simultaneously rather than stage by stage to reduce calibration tracking time constants. In the proposed background calibration, the capacitor mismatch and gain errors in the modified pipeline stage are measured and calibrated as one error. With calibration, the simulations show a signal-to-noise-and-distortion-ratio performance of 77.1 dB and a spurious-free dynamic range performance of 98.2 dB.**

Index Terms —**Background calibration, capacitor mismatch and gain calibration, digital calibration, pipelined analog-to-digital converter, signal-shifted correlation**

I. INTRODUCTION

The correlation-based background calibration has been used in pipelined for measuring and correcting capacitor mismatch and finite opamp gain error [1]-[5]. This method modulates the signal in MDAC with a pseudo-random noise (PN) sequence in the analog domain and then demodulates it in the digital domain to derive the measurement and cancel code errors in conversion. The correlation based background calibration has two limitations. One is the fact that the correlation converges slowly, due to a large uncorrelated signal, resulting in long measurement time. The other is the injected PN magnitude constraint, so that the signal plus PN injection would not saturate the next pipeline stage.

To inject PN dither into the signal path, several background calibration schemes split the sample capacitors into a certain number of fragments [2][3]. However, those schemes increase the number of mismatch capacitor that need to measure or unnecessarily introduce extra error due to mismatch between split capacitors. Furthermore, in the reported background calibration scheme [1][3], PN dithers are injected only in a certain range of input full-scale range, which means a large

proportion of samples outside of the calibration range are not used for digital calibration. Moreover, the calibration proceeds from rear stages to the front repeatedly in paper [2][3], thus for one calibration cycle, it takes the summation of the correlation time of each calibration stage. Here we proposed a parallel background calibration with signal-shifted correlation to improve the drawback mentioned above.

This paper is organized as follows. Section II describes the proposed modified 1.5-bit stage, error sources and its digital calibration concept. In Section III, a parallel digital background calibration scheme based on signal-shifted correlation is proposed. Section IV provides the simulation results for verification. Discussion and conclusion are given in Section V.

II. ERROR SOURCE AND DIGITAL CALIBRATION

A. Modified 1.5-bit/Stage Pipelined ADC

Figure 1 shows a modified tri-level 1.5-bit pipeline stage in two phases, where "00,01,10" is the sub-ADC output. In ideal, V_{refT} and V_{refB} equal to $+V_{ref}$ and $-V_{ref}$ respectively. During the amplification phase, the sample input is multiplied by a gain of 2. When the sub-ADC output is "00", both C_1 and C_2 are connected to V_{refB}; When the sub-ADC output is "01", both C_1 and C_2 are connected to 0;When the sub-ADC output is "10", both C_1 and C_2 are connected to V_{refT}. The residue plot of V_{res} is shown in Figure 1(c), where the comparator thresholds of the sub-ADC are set to $\pm V_{ref}/2$.

As for the typical 1.5-bit pipeline stage with RSD (Redundant Sign Digit) technique proposed by Ginetti and Jespers in 1990, the error resulting from comparator threshold offset can be canceled by introducing an redundant bit and setting the comparator thresholds at $\pm V_{ref}/4$, so that certain amount of output range margin is left when input signal is at the threshold point. Unlike the typical 1.5-bit pipeline stage, which confines the output within the range of $[-V_{ref}, +V_{ref}]$ not to exceed the next stage input range, the modified 1.5-bit pipeline stage cancels its comparator threshold offset error, by extending the next stage input range to $[-1.5V_{ref}, +1.5V_{ref}]$. Even if the output of a 1.5-bit pipeline stage exceed the range of $[-V_{ref}, +V_{ref}]$, as long as it is still in the range of $[-1.5V_{ref}, +1.5V_{ref}]$, the error due to comparator threshold offset can be canceled. The main purpose of modifying the typical 1.5-bit pipeline stage is to inject a large PN dither to signal path, so as to

978-1-61284-863-1/11 $26.00 © 2011 IEEE

reduce measurement time during a background calibration cycle [3].

Since the output of the modified 1.5-bit pipeline stage may exceed the range $[-V_{ref}, +V_{ref}]$, two additional comparators with threshold set at $\pm V_{ref}$ should be added in the last flash ADC stage, so that input range of flash ADC can be extended and the output is with an redundant bit. A 2.5-bit flash ADC with one redundant bit is shown in Figure 2(a). As for an N-bit pipelined ADC consisting of modified 1.5-bit stages and 2.5-bit flash ADC, the reconstruction of code is presented in Figure 2(b). As the input range of 1.5-bit stage extends to $[-1.5V_{ref}, +1.5V_{ref}]$, the input signal range can be judged by two bit $\{S_1, S_0\}$, as shown in Figure 2(b). When $\{S_1, S_0\}$ equals to 00, it implies input signal is within the range $[-V_{ref}, +V_{ref}]$; When $\{S_1, S_0\}$ equals to 01 it implies input signal is larger than $+V_{ref}$; When $\{S_1, S_0\}$ equals to 11, it implies input signal is less than $-V_{ref}$. And $\{D_N, D_{N-1}, \ldots, D_2, D_1\}$ represents the analog-to-digital conversion result.

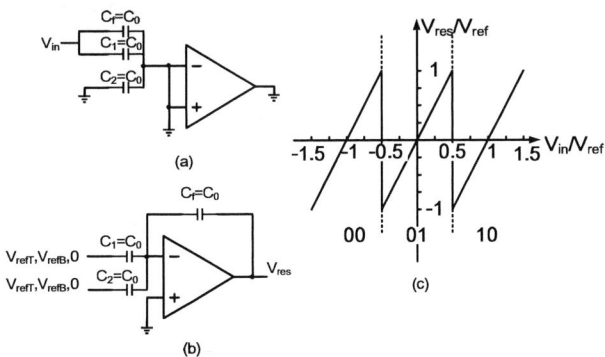

Figure 1. Modified Tri-level MDAC. (a) Sampling phase. (b) Amplification phase. (c) Residue plot.

B. Nonlinearity Error of the 1.5-bit Stage

The pipeline stage model with nonideal factors is presented in Figure 3. ε_{subADC} represents sub-ADC comparator threshold offset, which can be corrected in this modified 1.5-bit stage as discussed above; ε_{DAC0}, ε_{DAC1} and ε_{DAC2} represent the error caused by the mismatch between C_f, C_1 and C_2, which can be calibrated by the proposed digital calibration in this paper; coefficients b_1 and b_3 represent the closed-loop gain and third-order nonlinearity coefficients, where they can be expressed as [6]

$$b_1 = \frac{a_1}{1 + fa_1} \tag{1}$$

$$b_3 = \frac{a_3}{(1 + fa_1)^4} \tag{2}$$

where coefficient a_1, a_3 and f represent the open-loop amplifier gain, third-order distortion and the closed-loop feedback factor, respectively. In this paper, for a 14-bit pipelined ADC, it is assumed that the open-loop amplifier gain a_1 is sufficiently large and third-order closed-loop nonlinearity coefficient b_3 is negligibly small. The error resulting from finite open-loop opamp gain a_1 can be calibrated by the proposed digital calibration in this paper. As the opamp offset V_{os} only results in

an overall offset to the pipeline ADC, which has no impact on the performance, thus the opamp offset is not calibrated.

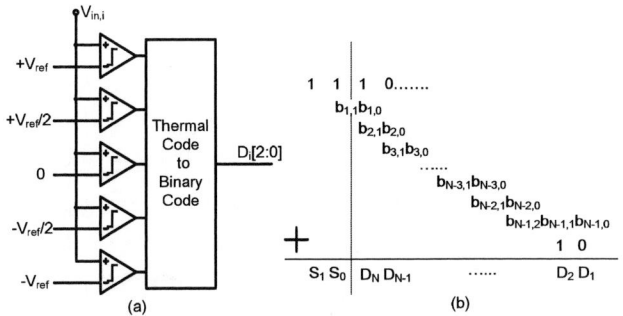

Figure 2. (a) A 2.5-bit flash ADC. (b) The algorithm of code reconstruction.

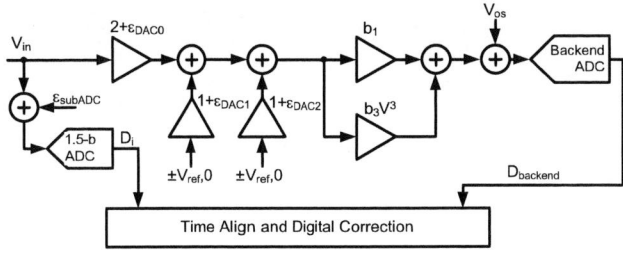

Figure 3. Model of a pipelined ADC with nonideal factors.

C. Digital Calibration Concept

According structure of MDAC shown in Figure 1, the transfer function of practical 1.5-bit MDAC is

When $V_{in} < -0.5V_{ref}$, $V_{res} = GV_{in} - K_1 V_{refB} - K_2 V_{refB}$ (3)

When $-0.5V_{ref} \leq V_{in} \leq +0.5V_{ref}$, $V_{res} = GV_{in}$ (4)

When $V_{in} > +0.5V_{ref}$, $V_{res} = GV_{in} - K_1 V_{refT} - K_2 V_{refT}$ (5)

where V_{refB} and V_{refT} equal to $-V_{ref}$ and $+V_{ref}$, respectively in ideal, and coefficients G, K_1 and K_2 are

$$G = \frac{C_f + C_1}{\dfrac{C_1 + C_2 + C_f}{A} + C_f} \tag{6}$$

$$K_1 = \frac{C_1}{\dfrac{C_1 + C_2 + C_f}{A} + C_f} \tag{7}$$

$$K_2 = \frac{C_2}{\dfrac{C_1 + C_2 + C_f}{A} + C_f} \tag{8}$$

in which A is the finite opamp open-loop gain. The digital calibration concept is to measure the terms $K_1 V_{refB}$, $K_2 V_{refB}$, $K_1 V_{refT}$ and $K_2 V_{refT}$, respectively. While these terms are subtracted in the MDAC, the digital value of them are added back instead of the ideal digital value of V_{refT} or V_{refB}. Despite the error still exists in the closed-loop gain slope coefficient G, however it does not deteriorate the performance of pipelined

978-1-61284-863-1/11 $26.00 © 2011 IEEE

ADC, for the ADC transfer function is linear, thus it is not necessary to correct this error.

III. PROPOSED DIGITAL BACKGROUND CALIBRATION

A. Capacitor Mismatch and Gain Error Measurement

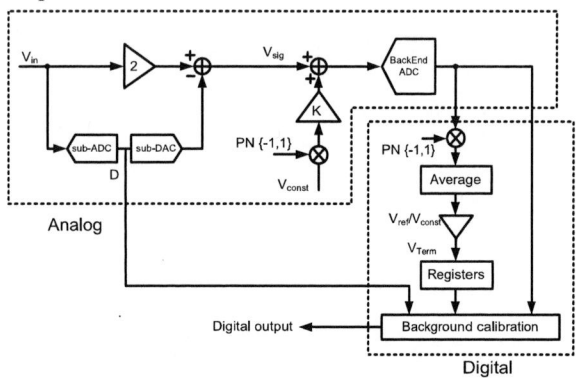

Figure 4. Background term "K" measurement

In order to estimate the terms K_1V_{refB}, K_2V_{refB}, K_1V_{refT} and K_2V_{refT}, a correlation-based background calibration is utilized in this work. Figure 4 shows these term measurement scheme by PN dithering, where terms K_1 and K_2 are regarded as the same variable K for simplicity, but in reality, they are measured separately. To measure the term K, a PN-modulated calibration signal is added to the signal path, V_{sig}. The PN is a zero-mean sequence of 1 and -1. At the same time, the calibration signal is modulated by a constant V_{const}. V_{Term} is the estimation of term K and can be inducted from the block diagram as follows:

$$V_{Term} = E\left(\frac{V_{sig}}{V_{const}}V_{ref}PN\right) + KV_{ref}PN^2 \qquad (9)$$

where V_{sig} is uncorrelated to PN. Since $PN^2=1$, the first term in eq.(9) could be regarded as the measurement error, and after averaging a large number of samples, it can converge to a negligibly small value. It is obvious that the ratio of V_{sig}/V_{const} should be as small as possible so that the measurement error can converge fast. After the accumulator averages a large number of samples, the value of V_{Term} is obtained and stored in registers for digital calibration.

B. Signal-Shifted Correlation

In order to shorten the averaging time and improve the measurement accuracy, the signal-to-dither ratio V_{sig}/V_{const} should be as small as possible. There are two approaches to achieve this goal. One way is to inject a large PN dither into the signal path to largen the value of V_{const}. As for the modified 1.5-bit pipeline stage proposed in this paper, the signal plus dither is in effect a large fixed-magnitude dither of $0.5V_{ref}$, as shown in Figure 5. The PN dithering injection scheme is presented in Figure 6. Another way is to minimize the variation range of V_{sig}, thus a signal-shifted correlation scheme is proposed in this paper as shown in Figure 5. V_{sig} belongs to the range $[-0.5V_{ref}, +0.5V_{ref}]$. The signal-shifted algorithm is as follows:

As for $V_{in} \in [-V_{ref}+0.125nV_{ref}, -0.875V_{ref}+0.125nV_{ref}]$,

$$V_{sig}+DV_{ref}-5V_{ref}/8-nV_{ref}/4 \in [-V_{ref}/8, +V_{ref}/8] \qquad (10)$$

$$V_{sig}=2V_{in}+2.5V_{ref}-DV_{ref} \qquad (11)$$

where $n=0, 1, 2\ldots15$ and $D=1,2,3,4$. The equivalent fixed-magnitude PN dithering is shown in Figure 5(b). It is obvious that after shifting the signal, the variation range of V_{sig} reduces from $[-0.5V_{ref}, +0.5V_{ref}]$ to $[-0.125V_{ref}, +0.125V_{ref}]$, resulting in the reduction of measurement time. When input signal is outside the range $[-V_{ref}, +V_{ref}]$, no dithering is added and errors in MDAC are not calibrated, because probability of this case is small and the digital calibration will not become complex.

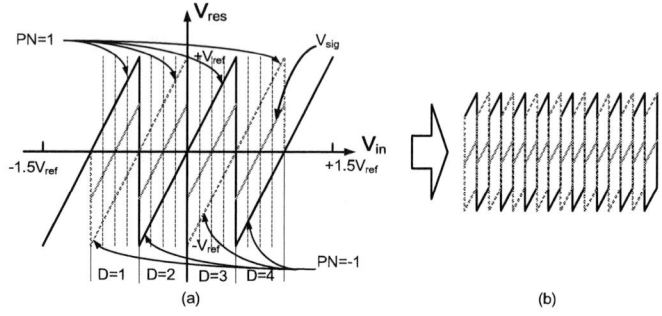

Figure 5. Dithering for signal-shifted correlation. (a)18 sub-ranges. (b) Equivalent fixed-magnitude PN dithering.

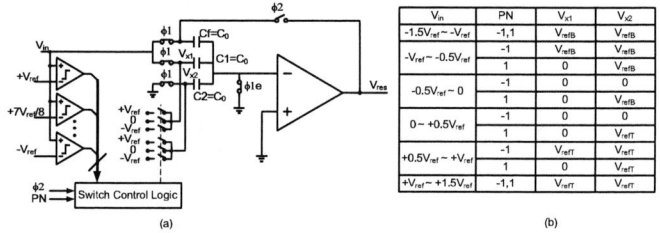

Figure 6. PN dithering injection scheme. (a)modified residue of an MDAC stage. (b)adding dithering rules.

C. Parallel background calibration

Parallel background calibration scheme is proposed as another way to reduce the background calibration tracking time, which is to measure the parameter of calibration stages simultaneously. Unlike the background calibration in paper[3], which proceeds with the rear stage and works its way toward the front repeatedly, the correction parameters in all calibration stages are continuously measured and updated in a concurrent as a pipeline rather than sequential fashion. Upon startup of the system, however, calibration parameters are measured and stored from the rear stage to the front stage, as a 6 calibration stages startup process illustrated in Figure 7, where 1st, 2nd…and 7th represents 1st calibration cycle, 2nd calibration cycle, … and 7th calibration cycle. As for the calibration cycle, the stage i-1 begins to measure correction parameters only after the correction parameters of stage i have been determined.

The highlight of parallel background calibration is that the time for one calibration cycle is equivalent to the time for calibrating one stage and no longer dependent on the number of calibration stages.

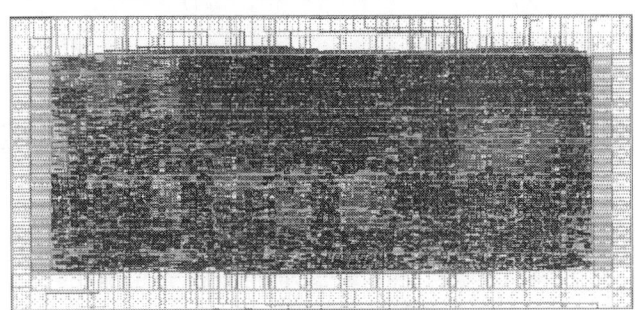

Figure 7. Startup process of the 6 calibration stages system.

IV. SIMULATION RESULTS

The architecture of a 14-bit pipelined ADC is illustrated in Figure 8. It has one S/H, 14 1.5-bit stages and a 2.5bit flash ADC. Two extra bits are added to diminish the digital calibration quantization noise, and only the first six stages are calibrated. The ADC was simulated by Matlab Simulink and Modelsim co-simulation, and the analog part of pipelined ADC is simulated in the Simulink environment while the digital calibration part is simulated as a verilog design implemented in modelsim environment.

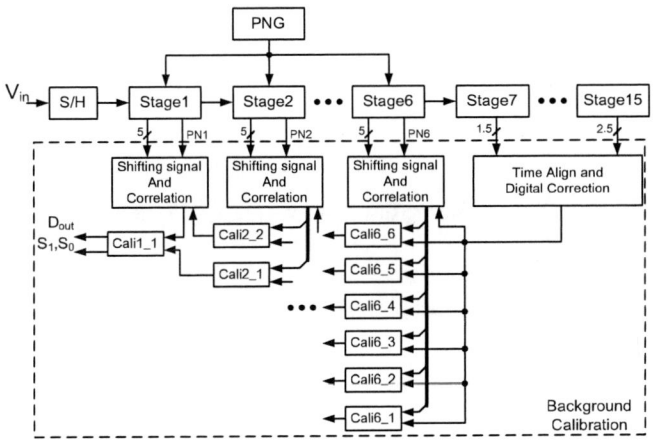

Figure 8. Architecture of a 14-bit pipelined ADC

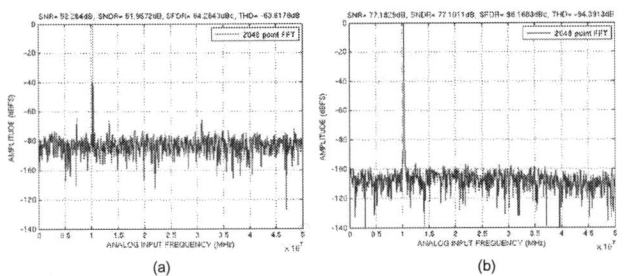

Figure 9. Sine-wave spectral response.

In the ADC model, and the first and second stage utilize 4pF capacitance, a capacitor mismatch within 1%. At the same time, thermal noise is simulated into the ADC model. When the correlation takes 2^{22} samples for each DAC path, the sine-wave spectral responses are illustrated in Figure 9. Before calibration,

the ENOB is 8.3bit and SFDR is 64.3dB due to capacitor mismatch and opamp gain error. After calibration, the ENOB is increased to 12.5bit and SFDR is 98.2dB, proving the efficiency of digital calibration. The number of averaging samples for one calibration cycle is 4×2^{22} and it needs to average $6 \times 4 \times 2^{22}$ samples for digital calibration startup. In order to be integrated into a 14-bit 100Msamples/s pipelined ADC, the calibration engine is implemented in the TSMC-0.18um 1P6M Digital process. The area is 1.4mm². Figure10 shows the layout of the calibration circuit.

Figure 10. Layout of the calibration circuit.

V. CONCLUSION

A digital background calibration scheme is proposed, for the purpose of reducing calibration cycle. This technique allows the injection of a large dithering when the input signal is within the range of $[-V_{ref}, +V_{ref}]$. Moreover, by dividing and shifting the signal, the signal-to-dither ratio is further reduced. At the same time, the front stages are calibrated simultaneously as a pipeline. Benefiting from the measure s taken above, the tracking and convergence rates are fast. Simulation results show that the proposed calibration technique is capable of correcting capacitor mismatch and opamp gain errors of the pipelined ADC.

REFERENCES

[1] J. P. Keane, P. J. Hurst, and S. H. Lewis, "Background interstage gain calibration technique for pipelined ADCs," IEEE Trans. Circuits Syst. I, Fundam. Theory Appl., vol. 52, no. 1, pp. 32–43, Jan. 2005.

[2] H.-C. Liu, Z.-M. Lee, and J.-T. Wu,"A 15-b 40-MS/s CMOS pipelined analog-to-digital converter with digital background calibration," IEEE J. Solid-State Circuits, vol. 40, no. 5, pp. 1047-1056, May 2005.

[3] Y. S. Shu, B. S. Song, "A 15-bit Linear 20-MS/s CMOS Pipelined ADC Digitally Calibrated With Signal-Dependent Dithering", IEEE J. Solid-State Circuits, vol. 43, pp. 342–350, Feb 2008.

[4] J. Ming and S. H. Lewis, "An 8-bit 80-Msample/s pipelined analog-to-digital converter with background calibration," IEEE, J. Solid-State Circuits, vol. 39, no. 10, pp. 1489–1497.

[5] I. Galton, "Digital cancellation of D/A converter noise in pipelined A/D converters," IEEE Trans. Circiuts Syst. II, Analog Digit. Signal Process., vol. 47. no. 3, pp. 185-196, Mar. 2000.

[6] W. Sansen,"Distortion in elementary transistor circuits," IEEE Trans. Circiuts Syst. II, Analog Digit. Signal Process., vol. 46, no. 3, pp. 315-325, Mar. 1999.

A 3 bit 36 GS/s Flash ADC in 65 nm Low Power CMOS Technology

Damir Ferenci, Simon Mauch, Markus Grözing, Felix Lang, Manfred Berroth

Institute of Electrical and Optical Communications Engineering, University of Stuttgart
Pfaffenwaldring 47, 70569 Stuttgart, Germany

damir.ferenci@int.uni-stuttgart.de

Abstract—**A 36 GS/s 3 bit flash ADC with a large analog input bandwidth is realized in a 65 nm CMOS technology. By employing a fourfold parallelization a high sampling rate is achieved, while a large input bandwidth is maintained. The measured effective resolution is about 2 bit up to 20 GHz input signal frequency at a sampling rate of 36 GS/s. The power consumption of the ADC core is 2.6 W, the core area is 0.16 mm². The ADC is intended for the cost-effective integration with an equalizer circuit on a single CMOS chip.**

I. INTRODUCTION

The dispersion in fiber optical systems limits the maximum transmission distance in 10 to 40 Gbit/s fiber-optic systems. A cost effective solution for this problem is the electronic dispersion compensation by means of a digital equalization circuit. This solution requires a high speed analog-to-digital converter (ADC) with a minimum sampling rate that is equal to the data transmission rate. For on-off keying transmission systems a relatively low nominal resolution of about 3 bit is sufficient [1], [2].

The advances in CMOS technology allow the design of analog circuits in the range of multiple Gigahertz, which was traditionally dominated by bipolar technologies. Due to the high integration in CMOS technology and the fast transistors expensive multi-chip solutions can be replaced by a cost-efficient single chip CMOS solution.

The fastest ADC available today is a 8 bit 56 GS/s CMOS ADC [3]. Although the performance of this ADC is higher, the chip area is more than 100 times larger than the chip area of the 3 bit ADC presented here.

The simulation results of the ADC were presented in [4] and the first measurement results at 12.8 GS/s and 18 GS/s were presented in [5]. The layout and the circuit are also covered in detail in [4], [5]. In this paper the linearity measurement and the SNDR measurement results up to 36 GS/s are presented.

The micrograph of the chip is depicted in Fig. 1. With a core area of only 0.16 mm², this ADC is the smallest compared to any state-of-the-art ADCs with sampling rates in this range. This allows for a cost-effective integration of multiple converters on a single chip. In the above described application of a 40 Gbit/s equalizer the ADC can be integrated together with an equalizer on a single chip in a state-of-the-art CMOS technology to realize a cost-effective data transmission system [6].

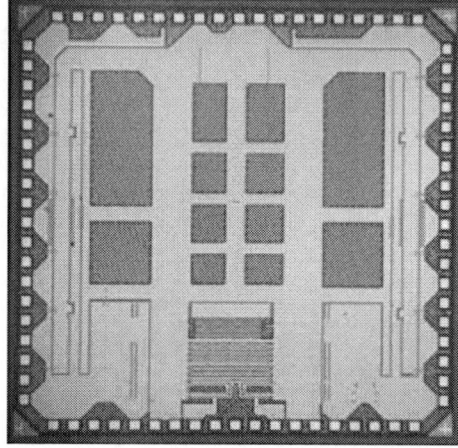

Fig. 1. Micrograph of the ADC (2.3 x 2.3 mm²).

II. ADC ARCHITECTURE

The basic concept of the ADC is a fourfold parallelized flash ADC which is fed by a four phase clock divider (Fig. 2). The analog input is demultiplexed by the sample and hold circuits. Following the sample and hold circuits quantizers and deciders are performing the quantization of the sampled input signal. Subsequently the thermometer code is encoded into the binary code. Finally all four channels are synchronized by a three step synchronizer to allow for a synchronous data output of all four channels. The detailed description of the building blocks is covered in [4], [5].

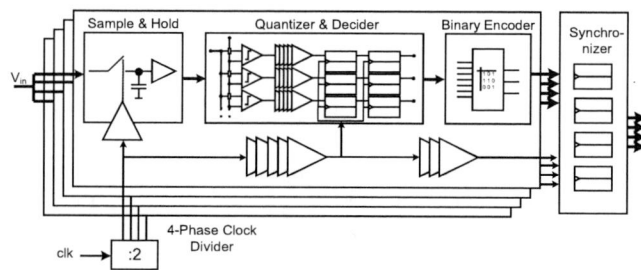

Fig. 2. Block diagram of the 3 bit ADC.

978-1-61284-863-1/11 $26.00 © 2011 IEEE

III. MEASUREMENT RESULTS

The ADC chip is mounted on Taconic RF-60A substrate for the measurements. The chip is placed in the center of the substrate and bonded to microstrip transmission lines. The layout has a star-like appearance with SMP connectors that are arranged in about 6 cm distance around the chip.

In the first measurement the DC characteristics are measured and the integral non-linearity (INL) and the differential non-linearity (DNL) is calculated. In the second measurement the signal to noise and distortion ratio (SNDR) is measured up to a total sampling rate of 36 GS/s.

A. DC Characteristics

The measurement setup for the linearity measurement is shown in Fig. 3. A Parameter Analyzer is used to generated the static analog input voltages. The ADC output data is stored in a Virtex4 FPGA and read out by a PC where the data is evaluated.

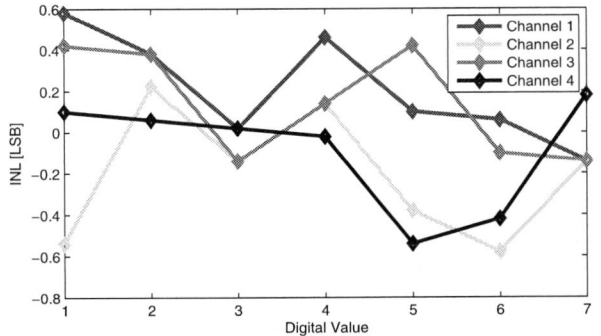

Fig. 3.   Measurement setup for the INL and DNL measurement.

Fig. 4 shows the INL plot. The INL of channel 3 is 0.3 LSB, the INL of the other channels is 0.4 LSB. The maximum mismatch between the channels is 1.2 LSB.

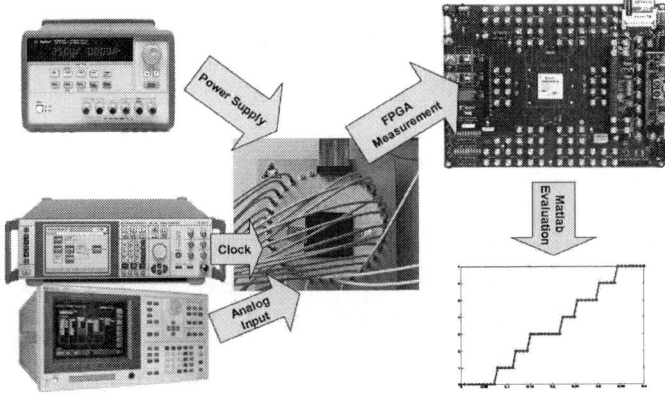

Fig. 4.   The INL of the four ADC channels.

The DNL plots are depicted in Fig. 5. The DNL of the channels 1 to 4 is 0.4 LSB, 0.8 LSB, 0.6 LSB and 0.6 LSB.

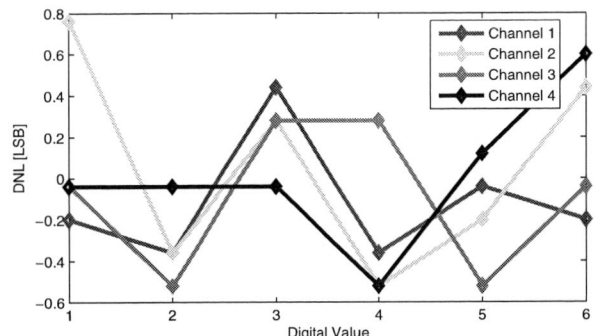

Fig. 5.   The DNL of the four ADC channels.

B. SNDR Measurement

For the SNDR measurement the setup in Fig. 3 is slightly modified. The input signal is generated by a sinusoid signal generator and balanced with a 180°-hybrid. The amplitude is kept constant at the maximum input voltage range of the ADC over the measured frequency range. The losses of the hybrid, the RF cables and the RF substrate are compensated. The digital output data is either stored in the FPGA or captured by a four channel 40 GS/s real-time scope. The measurement data is transferred to a PC where a discrete Fourier transform (DFT) is computed by means of Matlab to calculate the SNDR.

The real-time scope provides only four measurement channels. Thus only one of the four ADC channels can be measured simultaneously. The advantage of the FPGA measurement is that the twelve output channels of the four sub-ADCs can be measured simultaneously. The drawback of the current FPGA design is that the data transmission speed is limited to 3.2 Gbit/s per channel. This limits the sampling rate of the ADC to 12.8 GS/s. The FPGA measurement environment is explained in detail in [7].

1) FPGA Measurement Results: In Fig. 6 the single channel measurement results for a combined sampling rate of 12.8 GS/s are shown. Due to mismatch the results of the individual channels differ by up to 3 dB. All channels have a SNDR above 15 dB up to 15 GHz.

In Fig. 7 the SNDR of the ADC is shown after multiplexing the digital output data by means of Matlab and subsequent calculation of the SNDR. The resulting SNDR is about 15 dB up to 15 GHz input signal frequency. This corresponds to a resolution of at least 2 bit.

2) Real-Time Scope Measurement Results: Figure 8 shows the single channel measurement result for a single channel sampling rate of 6.4 GS/s. The increased sampling frequency doesn't decrease the SNDR compared to the FPGA measurement results at 3.2 GS/s. A further increasing of the sampling rate to 9 GS/s per channel (cf. Fig. 9) reduces the SNDR by about 2 dB between 12 GHz and 16 GHz input signal frequency in comparison to the measurement at the lower sampling rate.

Due to a limited number of scope channels only one ADC channel and the most significant bit (MSB) of the adjacent

978-1-61284-863-1/11 $26.00 © 2011 IEEE   345

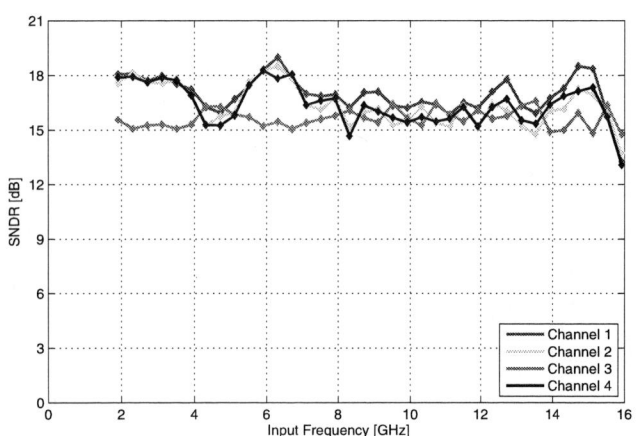

Fig. 6. Signal to noise ratio versus the input signal frequency of the four ADC channels, each channel is operating at 3.2 GS/s.

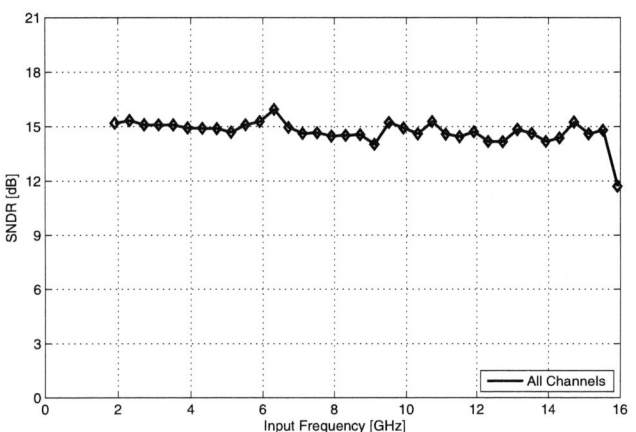

Fig. 7. SNDR of the multiplexed output at 12.8 GS/s versus the input signal frequency of the ADC.

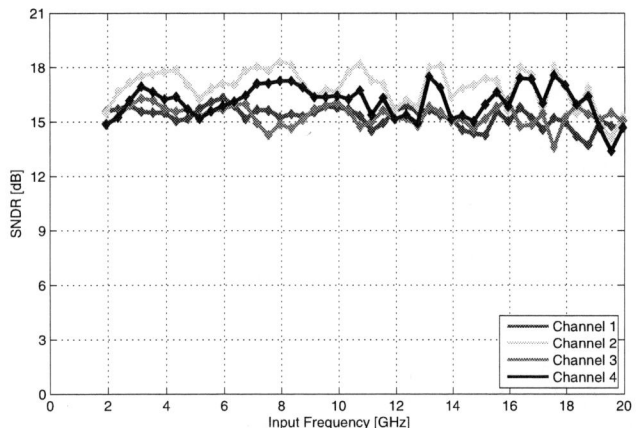

Fig. 8. Signal to noise ratio versus the input signal frequency of the four ADC channels, each channel is operating at 6.4 GS/s.

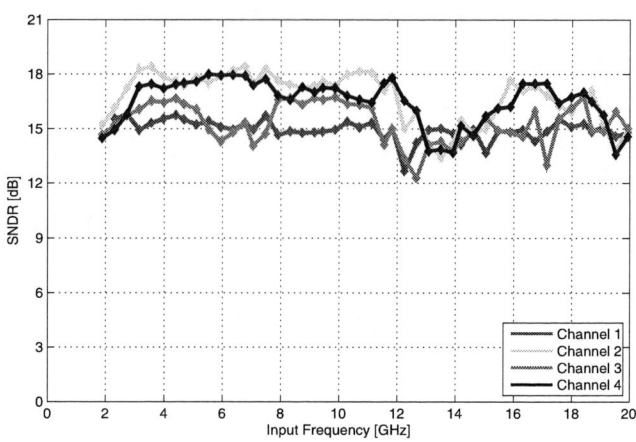

Fig. 9. Signal to noise ratio versus the input signal frequency of the four ADC channels, each channel is operating at 9 GS/s.

ADC channel can be measured simultaneously. To calculate the SNDR of the time interleaved ADC (TIADC) from the single channel measurements, it is necessary to take offset errors, gain errors and timing mismatches into account [8].

The calculation of the offset error is performed in the frequency domain by calculating the DFT for each ADC channel. The resulting frequency spectrum gives also the DC signal power for each channel. The offset mismatch is now calculated as the deviation of each channels DC component from the cumulative mean of all DC components. This way, the offset error can be incorporated into the SNDR calculation in the frequency domain.

According to the offset calculation, the gain error is calculated as the deviation of each channels signal amplitude from the mean value of all channels amplitudes. This way, the power at the frequency bin of the DFT associated with the input signal can be separated into signal power and power resulting from the gain error.

The timing mismatch calculation is more sophisticated, as it has to be estimated from the measurement of the MSB of the adjacent ADC channel. The calculation of the timing mismatch is performed in the time domain. The input phase of a sinusoid input signal is varied until the mean error between the input signal and the measured signal is minimized. This is done with the measured channel and with the MSB of the adjacent channel. The limited resolution of the ADC leads to a limited accuracy of this estimation.

The difference in the estimated input signal phases allows to calculate the timing deviation between the channels, which in turn leads to a worst-case estimation of the associated error power that adds to the whole error power when calculating the SNDR of the TIADC. In Fig. 10 the mean timing mismatch at different ADC input signal frequencies is shown for all channels. The worst case timing deviation is 0.1 Ts, the mean timing deviation is about 0.05 Ts, where Ts is the sampling period of the TIADC.

By taking the estimated offset errors, gain errors and timing missmatch into account, the SNDR of the TIADC in Fig. 11

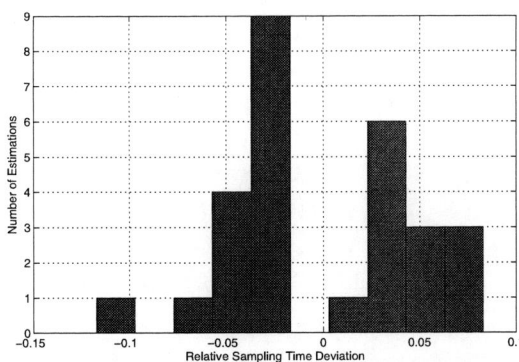

Fig. 10. Distribution of the estimated sampling time deviation, normalized with the sampling time Ts of the TIADC.

is calculated. The topmost curve is the SNDR without timing mismatch. With a timing deviation of 0.05 Ts the resolution is reduced marginal compared to an ADC without timing deviation. The lowermost curve is a worst case estimation with a timing deviation of 0.1 Ts, this reduces the SNDR by up to 5 dB for very high input signal frequencies.

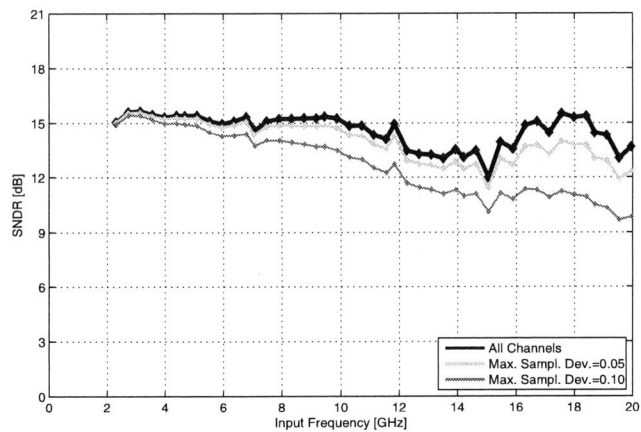

Fig. 11. The SNDR versus the input signal frequency of the TIADC at 36 GS/s, without timing deviation and with a timing deviation of 0.05 Ts and 0.1 Ts. Offset and gain errors are included in all curves.

The algorithms and the complete procedure is verified in Matlab with a modeled TIADC which exhibits offset errors, gain errors and timing mismatch.

IV. COMPARISON BETWEEN STATE-OF-THE-ART ADCS

Table I shows some representative state-of-the-art ADCs. The converters [9] and [10] are intended for the same application as the ADC in this work. Converter [3] is the ADC with the highest resolution and sampling rate available.

When comparing this work with the SiGe converters [9], [10] it is obvious that CMOS has the same performance with a lower power consumption. When the power saving principles that we used in [11] are applied to this design, the power consumption of this design can be reduced by a factor of 10 to about 250 mW. Another advantage of CMOS is the ability

TABLE I
HIGH SPEED STATE-OF-THE-ART ADCS

Vendor	[9]	[10]	[3]	This work
Technology	SiGe	SiGe BiCMOS	CMOS	CMOS
Publication	2004	2010	2010	
Sampling rate	40 GS/s	40 GS/s	56 GS/s	36 GS/s
Nom. Resolution	3	4	8	3
ENOB@15 GHz	2 bit	3 bit	6 bit	2 bit
Core Power	3.8 W	5.5 W	2 W	2.6 W
Core Size	1.8 mm^2	1.4 mm^2	16 mm^2	0.16 mm^2

to integrate the ADC and a digital signal processer on the same chip. This is a large cost benefit compared to the multi chip solution. Also the core area is about 10 times smaller than in SiGe Technology [9], [10].

Comparing this work with the converter in [3] is difficult, because the applications are different. The advantage of the three bit ADC for the application in a digital equalizer is the small core size, the lower price of the IP due to a lower complexity and a potentially lower power consumption after rescaling of the ADC components.

For the application together with a digital equalizer the presented 3 bit ADC is the best choice concerning the chip area, the power consumption, the performance and the overall system costs.

V. CONCLUSION

The presented 3 bit 36 GS/s CMOS ADC achieves an effective resolution of 2 bit, while occupying a very small core area of 0.16 mm². This is the smallest chip area ever reported for an ADC with a sampling rate of 36 GS/s, and a hundred times smaller than the currently fastest CMOS ADC [3]. This allows for the cost-effective integration with a digital equalizer for high speed fiber-optic systems on a single CMOS chip [6].

REFERENCES

[1] J. Lee, "A 5-b 10-GSamples/s A/D Converter for 10-Gb/s Optical Receivers," *IEEE Journal Solid-State Circuits*, vol. 39, no. 10, pp. 1671–1679, Oct. 2004.
[2] H. Tagami, "A 3-bit soft-decision IC for powerful forward error correction in 10-Gb/s optical communication Systems," *IEEE Journal Solid-State Circuits*, vol. 40, no. 8, pp. 1695–1705, Apr. 2005.
[3] I. Dedic, "56Gs/s ADC Enabling 100GbE," *OFC*, 2010.
[4] D. Ferenci, M. Grözing, and M. Berroth, "Design of a 3 Bit 20 GS/s ADC in 65 nm CMOS," *PRIME*, July 2009.
[5] D. Ferenci, M. Grözing, F. Lang, and M. Berroth, "A 3 bit 20 GS/s Flash ADC in 65 nm Low Power CMOS Technology," *EuMIC*, Oct. 2010.
[6] T. Veigel, M. Grözing, M. Berroth, and F. Buchali, "Design of a Viterbi Equalizer Circuit for Data Rates up to 43 Gb/s," *ESSCIRC Fringe*, Sept. 2009.
[7] D. Ferenci and M. Berroth, "A 100 Gigabit Measurement System with State of the Art FPGA Technology for Characterization of High Speed ADCs and DACs," *PRIME*, July 2010.
[8] C. Vogel, "The impact of combined channel mismatch effects in time-interleaved ADCs," *IEEE Trans. Instrum. Meas.*, vol. 54, no. 1, pp. 415–427, Feb. 2005.
[9] W. Cheng, "A 3b 40GS/s ADC-DAC in 0.12um SiGe," *ISSCC*, 2004.
[10] M. Chu, "A 40 Gs/s Time Interleaved ADC Using SiGe BiCMOS Technology," *IEEE Journal of Solid-State Circuits*, vol. 45, Feb. 2010.
[11] F. Lang, T. Alpert, D. Ferenci, M. Grözing, and M. Berroth, "Design of a 25 GS/s 6-bit Flash-ADC in 90 nm CMOS technology," *ESSCIRC Fringe*, Sept. 2009.

Adaptive Spread Spectrum Clock Tracking for Interpolator-based Clock and Data Recovery

Chuan-Thim Khor*, Alan Chai
IC Design Engineering
Altera Corporation
Penang, Malaysia
*email: ctkhor@altera.com

Abstract— **A 3Gb/s half-rate interpolator-based clock and data recovery (i-CDR) with asynchronous spread spectrum clock (SSC) tracking capability is presented. The i-CDR comprises a proportional path to track phase error and an integral path to compensate frequency offset. The integral path consists of ppm detector and ppm decoder with a combiner acts as the integration point between the two paths. The ppm decoder exploits averaging algorithm to facilitate wide range ppm tracking for triangular modulation profiles. The i-CDR, implemented in TSMC 60nm LP 1P11M CMOS technology occupies an active area of 390μm x 800μm and consumes 48.8mA from a 1.2V power supply when operated at 3Gb/s. Silicon measurements substantiate that the i-CDR has ppm tolerance up to +/-5000ppm.**

Keywords-spread spectrum clock (SSC), interpolator-based clock and data recovery (i-CDR), ppm

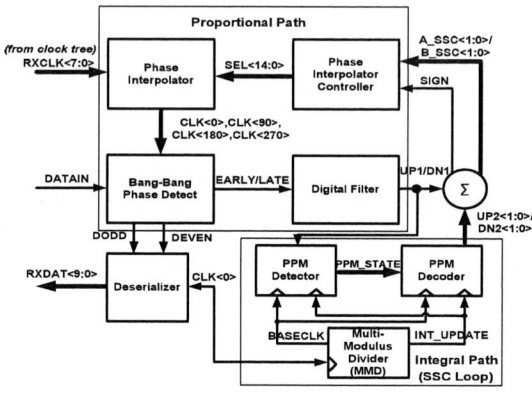

Figure 1.   Block diagram of proposed i-CDR

I.   INTRODUCTION

With the rapid proliferation of high speed electronic devices, the amount of electromagnetic emissions becomes more severe. This is highly undesirable especially for high-end portable appliances because it stands as a potential source of electromagnetic interference (EMI) to nearby wireless systems. To ameliorate this objectionable radiation, spread-spectrum clock (SSC) modulation is utilized in some synchronous digital systems to widen the clock signal's harmonics while reducing the peak amplitude of the spectral[1]. It is indispensable to have the receiver-end equipped with the capability to track the modulated data accurately. An example of SSC applications is Serial AT-Attachments (SATA) which employs triangular down-spreading profile of 5000ppm with a modulation frequency from 30 to 33kHz[2].

This paper is organized as follows. Section II briefs the proportional path. The formulas on ppm tracking capability of the interpolator-based clock and data recovery (i-CDR) are given in Section III. Detailed implementation of the SSC loop is elaborated in Section IV. Experimental results are reported in Section V, while conclusions are summarized in Section VI.

II.   PROPORTIONAL PATH OF I-CDR

As die size becoming a core consideration in the competitive integrated circuit marketplace, digitization is one of the attractive yet feasible solutions. Compared to its analog-based counterpart (a-CDR) which requires dedicated clock multiplication unit (CMU), interpolator-based CDR enables phase-locked loop (PLL) sharing between multiple channels.

This will contribute to substantial area and power savings[3]. Apart from this, the highly digitized architecture enables the number of external reference clock pins to be minimized. The top-level block diagram of the interpolator-based CDR is shown in Figure 1. The proposed half-rate i-CDR is composed of: (a) proportional path which keeps track the phase error between the incoming data and the recovered clock (b) integral path which detects the frequency offset between DATAIN and recovered clock and injects corresponding compensation for frequency tracking (Section IV devoted to comprehensive review on the integral path).

The proportional path includes an Alexander bang-bang phase detector (BBPD) which extracts the lead-lag information between the incoming data and the recovered clock by utilizing three consecutive data samplings taken from the quadrature clock edges[4]. The lead-lag information are combined to produce EARLY/LATE. The digital filter is required to minimize the probability of false-compensation due to jittery DATAIN and/or CLK<270:0> as well as meta-stability of sampling flip-flops. After a predefined CLK<0> cycles, the digital filter will update UP1 and DN1 based on majority voting from samplings within that particular window. The summation block takes the outputs of both digital filter and SSC loop and generates magnitude and sign on shift indications to the phase interpolator controller. Upon receiving the updated A_SSC<1:0>, B_SSC<1:0> and SIGN, the finite state machine of the controller will update the sampling phase by changing the current magnitude of the mixers' legs and/or the interpolated clocks (quadrant change).

978-1-61284-863-1/11 $26.00 © 2011 IEEE

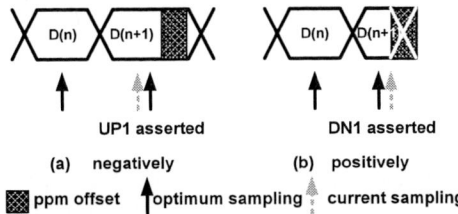

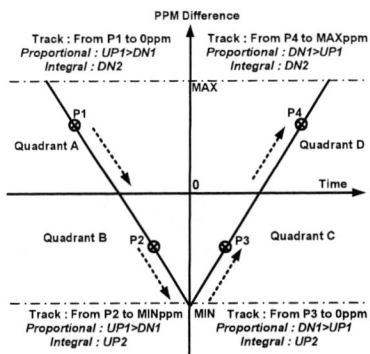

Figure 2. Proportional path's response to cases when ppm increases (a) negatively (b) positively

III. PPM TRACKING CALCULATION

The proportional path's response to ppm changes is demonstrated in Figure 2. As the DATAIN ppm difference increases negatively, the data period will be larger compared to the fixed sampling clock period. To maintain optimum sampling, a slower clock phase should be deployed. Hence, the proportional path will issue UP1 as shown in Figure 2(a). By the same token, the opposite happens when ppm difference increases positively as illustrated in Figure 2(b).The concept from Figure 2 is extended to explain the interaction between the proportional and integral paths for a complete SSC cycle (center-spread triangular modulation profile). In principle, the integral path is responsible for the coarse frequency offset compensation with any over- or under-compensated residuals covered by the finer proportional path. The responses of the two paths can be classified into 4 quadrants as illustrated in Figure 3.

Quadrant A : the positive ppm indicates that the data period is smaller than clock period. The system needs to use a faster phase to achieve optimum sampling, as a result the integral path issues DN2. However, from P1 to 0ppm, data period is increasing, demanding more UP1 assertions (unavoidably there will be DN1 assertions to recover the quantization error from SSC loop compensation).

Quadrant B : the increasing negative ppm trend manifests continuously expanding data period compared to the fixed clock period. Hence, both paths operate in the same direction thru UP2 and more UP1 (than DN1) issuances.

Quadrant C : the negative ppm reveals that the data period is larger than clock period. To sustain optimum sampling, a slower clock phase should be opted, therefore UP2 is activated. From P3 to 0ppm, data period is dwindling, hence more DN1 (than UP1) are delivered.

Quadrant D : the increasing positive ppm imparts continuously diminishing data period relative to the fixed clock period. Therefore, both paths operate towards the same direction with DN2 and more DN1 (than UP1) assertions.

The ppm tolerance is calculated based on equation (1) :

$$\text{ppm tolerance} = \left(\frac{1}{P*L}\right)*1e6 \qquad (1)$$

where   P = total phase step of interpolator (32 per UI)
    L = correction rate

Exploiting the concept from equation (1), the accumulated ppm difference is computed using equation (2):

Figure 3. Proportional and integral paths' interaction for SSC modulation

$$\text{accumulated ppm} = \left(\frac{C}{P*L_2}\right)*1e6 \qquad (2)$$

where   C = net count of UP1-DN1 in L_2
    P = total phase step of interpolator (32 per UI)
    L_2 = integral path update rate (1200UI)

IV. INTEGRAL PATH

The SSC loop includes ppm detector, ppm decoder and multi-modulus divider (MMD). The ppm detector serves the purpose of determining the frequency offset between the incoming data and the recovered clock based on the net count of UP1-DN1 for each window of integral path update cycle, INT_UPDATE, (L_2 in equation (2)). The frequency offset information is encoded into a multi-bit signal, labeled PPM_STATE. Upon receiving the adjusted ppm state, the ppm decoder will output the most optimum compensation pattern (thru UP2<1:0> and DN2<1:0>) for the current ppm difference. The MMD provides the required clocks for proper operation. To avoid detrimental interaction among the two loops which possibly leads to system instability, the integral path is operated at a much slower pace than the proportional path (integral loop update rate is 1200UI versus proportional loop's 32UI). To guarantee loop convergence, the proportional loop's update rate is designed to be integer multiple of the SSC loop's pattern shifting rate (32UI versus 8UI).

A. PPM Detector

The ppm detector adopts 2 binary bidirectional counters: mod-10 and mod-23 (cascaded mod-8 and mod-3) counters as depicted in Figure 4. In order to simplify the counter design and decoding system for supporting ±5500ppm, the ppm step resolution is set to be 250ppm. From equation (2), this is approximately tantamount to net count of 10 for UP1-DN1, hence mod-10 counter. To accommodate the complete ppm range from 0ppm to 5500ppm (the additional 500ppm is meant for static ppm coverage) with a step size of 250ppm, it necessitates a counter with 23 states. For every net count of 10 from mod-10 counter, the mod-10 overflow logic will be activated to increment or decrement the mod-23 counter. The direction generator resided as part of the mod-23 counter is meant for providing the integral path's compensation direction

978-1-61284-863-1/11 $26.00 © 2011 IEEE   349

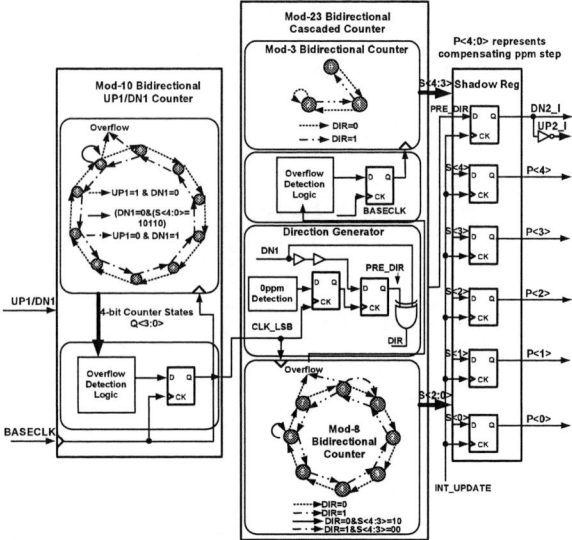

Figure 4. Block diagram of ppm detector

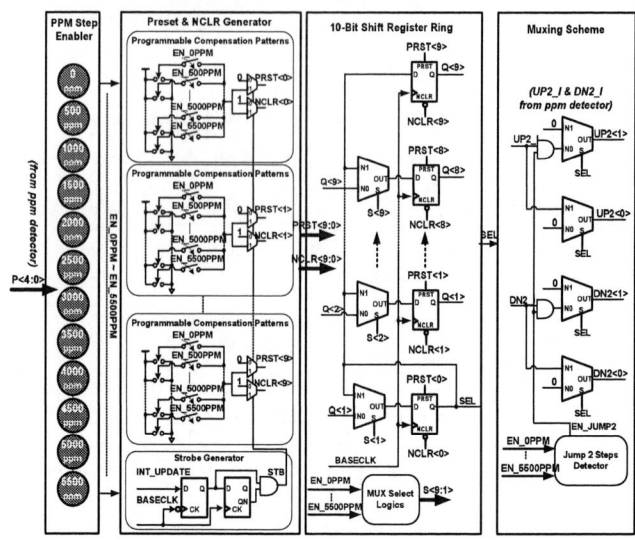

Figure 5. Block diagram of ppm detcoder

(PRE_DIR) as described in Figure 3 apart from generating the counter's increment/decrement direction, DIR. At every rising edge of INT_UPDATE, the output of the mod-23 counter, S<4:0> is transferred to ppm decoder as P<4:0> together with PRE_DIR (as UP2_I and DN2_I). The shadow registers function to store the active ppm step as P<4:0> while S<4:0> keeps updating (based on the ppm deviation) during each INT_UPDATE cycle.

B. PPM Decoder

The ppm decoder shown in Figure 5 involves a ppm step enabler which decodes the ppm state from ppm detector. The preset and NCLR generator is a "sea-of-switches" with hard-coded (look-up table is amendable as well for programmability) logic levels (1 or 0). The embedded logic levels represent the compensation pattern associated with each ppm step. One set of these patterns will be transferred to PRST<9:0> and NCLR<9:0> (dedicated to the active ppm step) after every INT_UPDATE active transition. The strobe generator is employed to provide a platform for loading in the activated set of pattern to PRST<9:0> and NCLR<9:0>. The STB signal needs to be released after the shift registers preset or cleared (controlled by the pulse width of the STB signal) to facilitate the shifting process. The PRST<9:0> and NCLR<9:0> control the string of pattern shifted out from the register ring as SEL. The unique pattern is repeated until a new set of pattern arrived (indicating new ppm step) after subsequent INT_UPDATE active transitions. Considering not all the ppm steps occupy the same amount of registers to cover their respective unique compensation patterns, multiplexing approach is manipulated between flip-flops to cater for the different needs on register consumption. With this concept, a common structure is reusable for the whole ppm range leading to substantial area and power savings. The output of the ring is deployed as the select signal of the pairs of multiplexers for

UP2<1:0> and DN2<1:0>. UP2<1:0> and DN2<1:0> are the binary representations of the number of phase interpolator step movement contributed by the integral path. For low ppm range (monitored by the jump-2-steps detector), UP2<1> and DN2<1> are fixed to 0 whereas UP2<0> and DN2<0> are a function of SEL, UP2_I and DN2_I. For high ppm range, there are instances where 2 phase steps shift is required to track fast data drift, hence, UP2<1> and DN2<1> can be logic 1. Table I tabularizes the ppm amount (based on equation(1)) and rate of corrections from both proportional and integral paths for the supported ppm steps. For the proposed i-CDR, proportional path's compensation rate is fixed to 32UI, indicating a ppm tolerance of 977ppm (based on equation (1)). A few important insights to highlight are described below :

a) for each ppm step, majority of the ppm difference will be absorbed by the integral path with residuals (over and under-compensations) covered by the proportional path.

b) to ensure convergence, the correction rate of both paths needs to be an integer multiples of a base which in this design is 8UI (BASECLK).

c) for some cases, correction rate of 8N (N=integer≥1) is not sufficient as the integral path SSC residual is out of proportional path's tolerance. Hence, averaging concept is applied to widen the horizon of ppm tracking. For example, at

3000ppm, without deploying averaging, the closest integer correction rate is 8UI (ppm correction of 3906ppm), resulting over-compensation of 906ppm. Conversely, by manipulating averaging, we can achieve correction rate of 10UI as illustrated in Figure 6(a). This yields 3125ppm which is much closer to the target of 3000ppm. This is realized by masking off (delivering 0) one time in every five corrections.

d) for high ppm range (>4000ppm), even updating at the maximum rate of 8UI is not adequate to compensate optimally. Under worst case scenario, the data drift might be

TABLE I. PPM COMPENSATION OF PROPORTIONAL & INTEGRAL PATHS

PPM	Loop	Correction	Correction Rate (UI)
0	Int	0	0
	Prop	0	32
500	Int	488	64
	Prop	12	32
1000	Int	977	32
	Prop	23	32
1500	Int	1563	20
	Prop	-63	32
2000	Int	1953	16
	Prop	47	32
2500	Int	2604	12
	Prop	-104	32
3000	Int	3125	10
	Prop	-125	32
3500	Int	3472	9
	Prop	28	32
4000	Int	3906	8
	Prop	94	32
4500	Int	4464	7
	Prop	36	32
5000	Int	5208	6
	Prop	-208	32
5500	Int	5208	6
	Prop	292	32

Notes : Int=Integral; Prop=Proportional

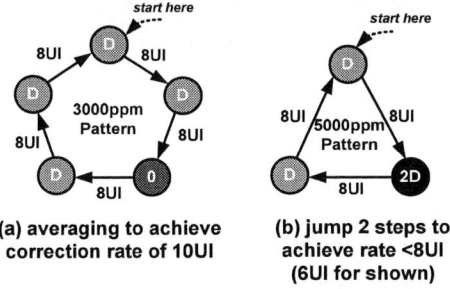

(a) averaging to achieve correction rate of 10UI

(b) jump 2 steps to achieve rate <8UI (6UI for shown)

Figure 6. Averaging algorithm for fractional correction from 8UI base

too far away leading to lose of lock. For instance, at 5000ppm, if the integral path compensates at 8UI, the compensated ppm is only 3906ppm, under-compensates by 1094ppm. Incorporating the capability to contribute 2 steps from the SSC loop as demonstrated in Figure 6(b) is instrumental to remain intact in tracking fast data drift. For this particular example, there are 4 corrections in 24UI, which thru fractional effect results in correction rate of 6UI. At this rate, the integral path is capable of compensating 5208ppm.

V. MEASUREMENT RESULTS

The proposed i-CDR has been fabricated in TSMC 60nm Low-Power (LP) 1P11M CMOS technology. The i-CDR occupies an active area of 390μm x 800μm and draws 48.8mA from a 1.2V power supply when operated at 3Gb/s.

Figure 7 shows the silicon measurements of the proposed i-CDR in tracking +/-2500ppm center-spread triangular SSC modulation at 3Gb/s. The SSC modulated data shows 20dB EMI reduction. Figure 8 confirms that the proposed design is capable of tracking both center-spread and down-spread SSC modulations of 5000ppm at 3Gb/s. This feature is of significant importance for multi-protocols support for example SATA and DisplayPort require down-spread SSC whereas V-by-1 needs center-spread triangular SSC tracking.

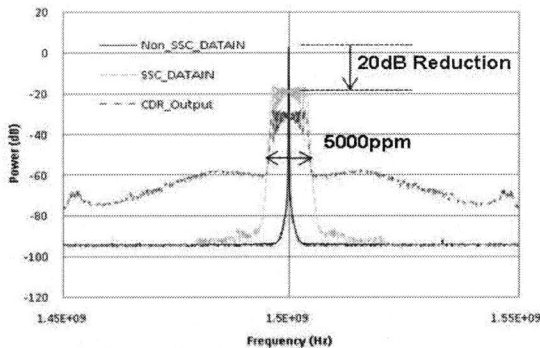

Figure 7. Power spectrum of i-CDR output tracking for center-spread SSC modulated at +/-2500ppm

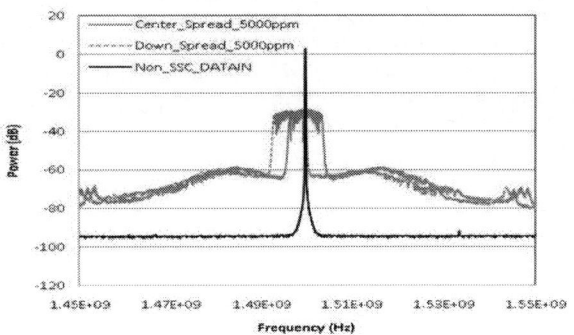

Figure 8. Power spectrum of i-CDR output tracking for center-and down-spread SSC modulated at +/-2500ppm and 0~-5000ppm, respectively

VI. CONCLUSIONS

This paper has introduced an interpolator-based clock and data recovery with a new asynchronous spread spectrum clock tracking topology. The proposed integral path includes a counter-based ppm detector and a fractional-based ppm decoder. The concept of averaging effect in the context of ppm compensation algorithm is presented which enables significant saving in terms of area and power. Experimental results validate that the proposed design is capable of tracking both center- and down-spread SSC modulated data with ppm difference from -5000ppm to +2500ppm at data-rate of 3Gb/s.

REFERENCES

[1] Matsumoto Yasushi, Ishigami Shinobu, Gotoh Kaoru, "Effects of Spread Spectrum Clocking on Measured Noise Spectra," Journal of the National Institute of Information and Communications Technology, vol.53 No.1, pp. 101-115, 2006.

[2] Ming-ta Hsieh and Gerald E. Sobelman, "Clock and Data Recovery with Adaptive Loop Gain for Spread Spectrum SerDes Applications," IEEE International Symposium on Circuits and Systems (ISCAS), vol.5, pp. 4883 – 4886, 2005.

[3] Armin Tajalli, Paul Muller, Mojtaba Atarodi, and Yusuf Leblebici, "A Multichannel 3.5mW/Gbps/Channel Gated Oscillator Based CDR in a 0.18μm Digital CMOS Technology," Solid-State Circuits Conference, 2005. ESSCIRC 2005. Proceedings of the 31st European, pp. 193 – 196,2005.

[4] Behzad Razavi, "Challenges in the Design of High-Speed Clock and Data Recovery Circuits," IEEE Communications Magazine, 2002.

Temperature Behavior Mismatch of Halo Implanted Short Channel Transistors and its Influence on PUF Circuits

Maximilian Hofer, Christoph Böhm, Wolfgang Pribyl

Institute for Electronics

Graz University of Technology

Email: {maximilian.hofer,christoph.boehm,wolfgang.pribyl}@tugraz.at

Abstract—So-called physical unclonable functions (PUFs) are circuits that generate IDs or keys from manufacturing variations. Since the estimation of the error rates of PUFs should already be done during the design phase, Monte Carlo circuit simulations should provide realistic results. Unfortunately, it turns out that the temperature behavior is not covered well in the Monte Carlo parameters if halo implanted transistors are used. In this paper we analyze the influence of halo and substrate doping concentration variations on the temperature behavior of minimum size transistors and suggest an adaptation of the Monte Carlo parameters with respect to this problem. Since temperature depending behavior of transistors is a general issue, the proposed approach is not restricted to PUF circuits.

I. INTRODUCTION

Secrecy of information plays an important role in today's communication. Here, generation and storage of key material are basic and problematic steps. Access by an attacker has to be avoided to guarantee secrecy. For some years now, so-called physical unclonable functions (PUFs) are one approach to this problem. PUFs use manufacturing variations of devices to generate a function that provides numbers [1]. The numbers are used for different identification and security applications. These include authentication of devices where a challenge is sent to a PUF which responds corresponding to its inherent properties. Another example of an application is key material generation. Here, the output of a PUF is used to en- and decrypt data. Since PUFs utilize properties which cannot be controlled during manufacturing, the output of a PUF also cannot be controlled. This makes them inherently unclonable and thus qualifies PUFs to be used for key generation.

The first PUF was introduced by Pappu in 2002 [2]. He used an optical medium containing class spheres to produce a speckle on a detector as a function of a laser beam. Since the distribution of the glass spheres cannot be controlled during production, the output is unclonable and will differ from the output of other PUFs of that kind. Since manufacturing costs are quite high, recent research [3]–[8] focuses on so-called *silicon* PUFs. This kind of PUF mostly compares properties of transistors to generate the output. For example threshold voltages (v_{th}) of transistors are compared. The ratio defines the output. Since the threshold voltage depends on the doping concentration and the doping concentration varies between the transistors, v_{th} is qualified for PUF purposes.

An important property of PUFs is that they provide the same output within a defined region of operation every time they are called. So the output must not depend e.g. on supply voltage or temperature within the range. This is especially important if the PUF is used to provide a key. Here, the error rate must be zero. Since this cannot be guaranteed, error correction codes (ECC) [9] or similar approaches [10] have to be used. So it is important to get an idea of the resulting error rate already during the design of the PUF circuits to be able to choose an adequate ECC. Usually, this is done using Monte Carlo simulation, simulating the *local* mismatch of the transistors' parameters. In our case, the local mismatch is simulated changing the threshold voltage v_{th} and the mobility μ.

Unfortunately, it turns out that the simulation results and the measurement results of the final PUF chip do not match. The error rate over the temperature of the real case exceeds the simulation results.

The remainder of the paper is on the analysis of this problem and the solution we propose to fix it. The first section describes the problem in more detail. Then, the results of TCAD simulations are shown. Finally, an adaption of the Monte Carlo BSIM model parameters is suggested to include the temperature behavior. The final section concludes the paper.

II. TEMPERATURE DEPENDENCE OF PUF OUTPUTS

As already mentioned above, PUFs mostly produce results by comparing properties of transistors. To make the analysis clearer, in the following text only two transistors are used to define the output. This is done without loosing generality. One pair of transistors is called a PUF-cell and provides one bit of information. The whole PUF consists of a number of those cells depending on the number of required output bits. Fig. 1 shows such a cell.

If I_1 exceeds I_2, a '0' appears at the output, if I_1 is smaller then I_2, a '1' appears. If the operation point of a PUF is not changed, the ratio between the transistors will not change. That means, that only noise can produce errors at the output. Fig. 2a and Fig. 2b show a schematic drawing of two I_d vs. V_{gs} curves at different temperatures. In this example, the mismatch between the transistors is small and a crossing

978-1-61284-863-1/11 $26.00 © 2011 IEEE

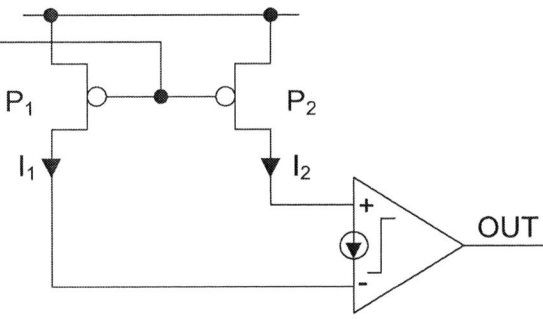

Fig. 1. Circuit which describes the functionality of a two transistor PUF using an ideal current comparator.

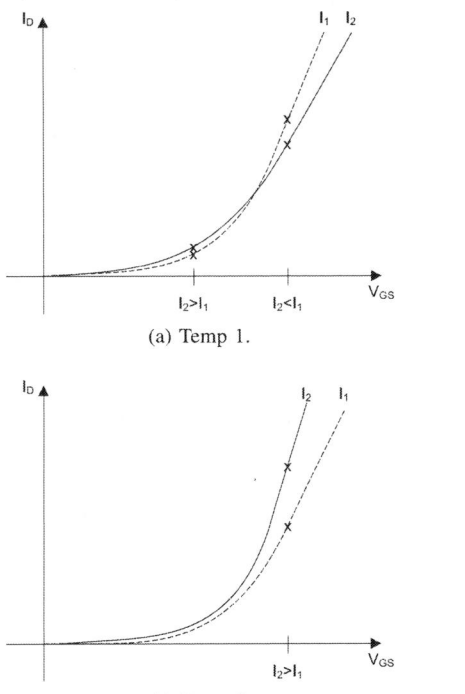

(a) Temp 1.

(b) Temp 2.

Fig. 2. I_d/V_{gs} curves of two different transistors at two temperatures. I_1 and I_2 are compared at different operation points.

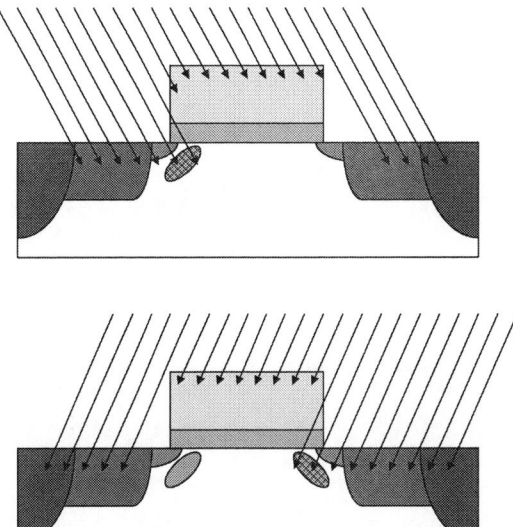

Fig. 3. Procedure of halo implantation [11].

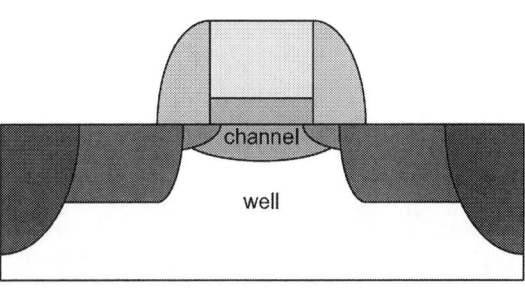

Fig. 4. In the short channel case, the halo implants touch/overlap each other and a more homogeneous channel doping concentration appears.

of the two curves happens at some point at temperature 1. Thus, an error occurs if V_{gs} is altered around the crossing. As already said in the introduction Monte Carlo simulation results and measurement results do not fit when looking at the temperature behavior. The error rate of the simulation is too small which is not acceptable when relying on the results during the design of the error correction. In our case, the error rate increased from 3% up to 6%. Looking at these result it turns out that the temperature behavior of the involved transistors which is shown in Fig. 2b is not included correctly in the simulation. Here, due to a mismatch in the temperature coefficients an error may occur when moving from temperature 1 to temperature 2. The mismatch in the temperature behavior is analysed using TCAD. The results are shown in the following section.

III. ANALYSIS OF A HALO IMPLANTED 65NM TRANSISTOR

Since in PUF design minimal size transistors are used to maximize the local mismatch between the transistors, the analyzed transistor is a 65nm minimum size transistor with halo implant. The halo implants are positioned at both sides of the channel region and are implanted on top of the substrate implants as shown in Fig. 3.

For the minimal size transistors, the halo regions overlap which results in a special case where the transistor can be divided vertically into two regions: The *channel region* and the *well region* as shown in Fig. 4. The channel region is defined by the doping concentration of both, the substrate doping (DC_{sub}) and the halo doping (DC_{hal}). The well region is defined by the substrate doping alone. The nominal halo doping concentration is chosen to be $0.5E18/cm^3$, the nominal substrate doping is chosen to be $1.0E18/cm^3$. This ends up in the following nominal doping concentration of the

978-1-61284-863-1/11 $26.00 © 2011 IEEE   353

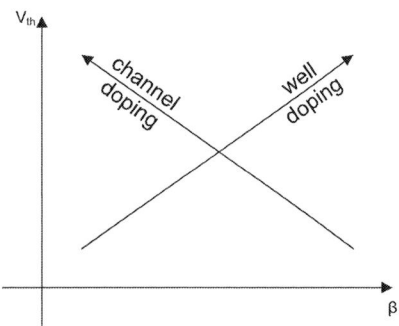

Fig. 5. Effect of doping concentration variation on v_{th} and β of halo implanted short channel transistors.

TABLE I
RELATIONS OF DOPING CONCENTRATION AND TRANSISTOR PARAMETERS.

	$DC_{ch} \pm 1\%$	$DC_w \pm 1\%$
v_{th}	$\pm 7.14\%$	$\pm 0.20\%$
β	$\mp 0.25\%$	$\pm 0.07\%$
dV_{th}/dT	$\pm 0.09\%$	$\pm 0.06\%$
$d\beta/dT$	$\mp 0.34\%$	$\pm 0.05\%$

channel (DC_{ch}) and the well (DC_w) region:

$$
\begin{aligned}
DC_{ch} &= DC_{sub} + DC_{hal} \\
&= 0.5E18/cm^3 + 1.0E18/cm^3 \\
&= 1.5E18/cm^3 \\
DC_w &= DC_{sub} \\
&= 0.5E18/cm^3
\end{aligned}
$$

By altering the concentration of the two types of doping, the effect of mismatch can be simulated. To do so, DC_{sub} and DC_{hal} were varied by up to $\pm 20\%$. The schematic view of the resulting correlation between doping concentration and v_{th} and the slope of the I_d/V_{gs} curve β are shown in Fig. 5.

This relation is important since it shows that the same threshold voltage can be realized by different combinations of well doping and channel doping concentrations which at the same time result in different β. Furthermore it turns out that different doping concentration combinations leading to the same v_{th} lead to different temperature behaviors of the transistors. Namely the temperature coefficients of v_{th} and μ may change. The relations for the analyzed transistor which could be extracted from the TCAD simulations are shown in Table I. The influence of the doping concentration on the temperature behavior especially on the v_{th} is rather small. But if the difference between the v_{th} of the two compared transistors is very small, a mismatch of dV_{th}/dT or $d\beta/dT$ can lead to different behavior of the PUF-cell at different temperatures. The doping concentrations of a simulation example are shown in Table II. In the example, the threshold voltage difference between the two transistors is chosen to be minimal. For the simulated transistor this can be done by altering DC_{sub} by a factor that is minus two times the altering factor of DC_{hal} ($factDC_{sub} = -2 \cdot factDC_{hal}$). The resulting change of

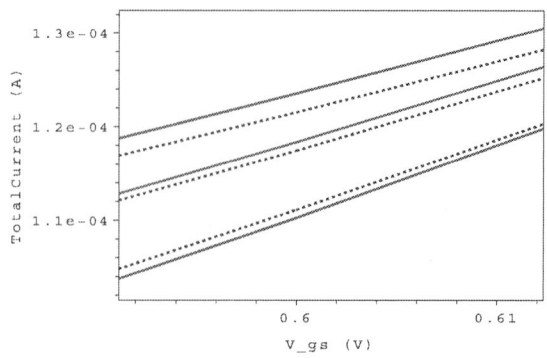

Fig. 6. TCAD result: Current output of two different transistors at three different temperatures (bottom-up: -50C, 27C, 120C). At -50C, the sign of the ratio between the currents changes which would lead to an erroneous PUF-cell output.

TABLE II
DOPING CONCENTRATION OF EXAMPLE TRANSISTORS.

Doping Concentration	Transistor 1	Transistor 2
$DC_{hal}/1/cm^3$	$1.2E18$	$0.9E18$
$DC_{sub}/1/cm^3$	$0.4E18$	$0.6E18$
$DC_{ch}/1/cm^3$	$1.6E18$	$1.5E18$
$DC_w/1/cm^3$	$0.4E18$	$0.6E18$

DC_w and DC_{ch} lead to a high difference in the temperature coefficients of β and v_{th}. The crucial part of the outcome of the simulation is shown in Fig. 6. Here, the currents through the two transistors are shown at different temperatures. At -50°C, the sign of the current ratio changes compared to 27°C and 120°C which would result in an error of a PUF circuit.

Even if the example looks somewhat constructed it happens in real circuits. This is due to the Gaussian distribution of the mismatch of v_{th} and mobility μ. Thus, the mean v_{th} difference between two transistor ends up to be zero. The same is true for μ. So the behavior over temperature often is defined by the mismatches between the temperature coefficients. This kind of mismatch is currently not included in the Monte Carlo simulation parameters which leads to wrong temperature behavior estimations of the analyzed circuits.

IV. MODEL PARAMETER ADAPTATION

To be able to estimate the error rate more realistically, the BSIM4 [12] model parameters have to be adapted. Up to now, only the parameters U0 and VTH0 are varied during Monte Carlo local mismatch simulations using Gaussian distributions:

$$VTH0_{MC} = \mathcal{N}(VTH0_{NOM}, \sigma_{VTH0}) \qquad (1)$$
$$U0_{MC} = \mathcal{N}(U0_{NOM}, \sigma_{U0}), \qquad (2)$$

where $VTH0_{NOM}$ and $U0_{NOM}$ are the mean values and σ_{VTH0} and σ_{U0} are the variances of the Gaussian distributions. The values depend on the technology and the size of the particular transistor. In the BSIM4 model, the temperature dependence is included mainly by the following formulas:

$$V_{th}(T) = V_{th}(VTH0_{MC}, TNOM) \tag{3}$$
$$+ KT1 \cdot \left(\frac{T}{TNOM} - 1\right)$$

$$U0(T) = U0_{MC}(TNOM) \cdot \left(\frac{T}{TNOM}\right)^{UTE}, \tag{4}$$

where in $V_{th}(T)$, the influences of $KT1L$ and $KT2$ have been omitted. To also include the temperature coefficient mismatch, the temperature coefficient for the threshold voltage KT1 and the mobility temperature exponent UTE have to be varied as well. As shown above, all the parameters needed for the Monte Carlo simulation correlate with both, the halo and the substrate doping concentration. Thus, it would be helpful to extract the correlation parameters and include them into the simulation.

For the purpose of this work it turned out that two more uncorrelated Gaussian distributions for KT1 and UTE were sufficient to model the mismatch behavior:

$$KT1_{MC} = \mathcal{N}(KT1_{NOM}, \sigma_{KT1}) \tag{5}$$
$$UTE_{MC} = \mathcal{N}(UTE_{NOM}, \sigma_{UTE}), \tag{6}$$

where the parameters $KT1_{NOM}$, σ_{KT1}, UTE_{NOM}, and σ_{UTE} depend on the technology. Thus these parameters have to be extracted by analyzing the transistor mismatch behavior over the required temperature range. Using the additional Monte Carlo parameters, the simulation results match the measurement results of the PUF test chip. This shows the usefulness of the suggested approach.

V. CONCLUSION

We presented the temperature behavior mismatch of halo implanted minimum size transistors. This kind of mismatch leads to unexpected high error rates in so-called physical unclonable functions. We showed that this behavior is based on the varying doping concentration of both, the halo implants and the substrate. To be able to see this effect in Monte Carlo circuit simulation, an approach was suggested that uses Gaussian distributions to vary the BSIM4 model parameters KT1 and UTE. With the adaptation of the Monte Carlo parameters well-matching simulation results could be produced.

To use this approach with arbitrary channel length transistors, temperature behavior mismatch should be analyzed as well for long channel devices. This could provide a more realistic outcome of Monte Carlo simulations for any kind of circuit.

ACKNOWLEDGMENT

This work has been partially funded by the Österreichische Forschungsförderungsgesellschaft mbH (FFG) within the project PUCKMAES.

REFERENCES

[1] R. Maes and I. Verbauwhede, "Physically unclonable functions: A study on the state of the art and future research directions," in *Towards Hardware-Intrinsic Security*, ser. Information Security and Cryptography, D. Basin, U. Maurer, A.-R. Sadeghi, and D. Naccache, Eds. Springer Berlin Heidelberg, 2010, pp. 3–37. [Online]. Available: http://dx.doi.org/10.1007/978-3-642-14452-3_1

[2] R. Pappu, R. Recht, J. Taylor, and N. Gershenfeld, "Physical one-way functions," *SCIENCE*, vol. 297, no. 5589, pp. 2026–2030, SEP 20 2002.

[3] K. Lofstrom, W. Daasch, and D. Taylor, "Ic identification circuit using device mismatch," *Solid-State Circuits Conference, 2000. Digest of Technical Papers. ISSCC. 2000 IEEE International*, pp. 372–373, 2000.

[4] B. Gassend, D. Clarke, M. van Dijk, and S. Devadas, "Silicon physical random functions," pp. 148–160, 2002.

[5] D. Lim, J. Lee, B. Gassend, G. Suh, M. van Dijk, and S. Devadas, "Extracting secret keys from integrated circuits," *Very Large Scale Integration (VLSI) Systems, IEEE Transactions on*, vol. 13, no. 10, pp. 1200–1205, Oct. 2005.

[6] Y. Su, J. Holleman, and B. Otis, "A digital 1.6 pj/bit chip identification circuit using process variations," *Solid-State Circuits, IEEE Journal of*, vol. 43, no. 1, pp. 69–77, Jan. 2008.

[7] J. Guajardo, S. Kumar, G.-J. Schrijen, and P. Tuyls, "Fpga intrinsic pufs and their use for ip protection," in *Cryptographic Hardware and Embedded Systems - CHES 2007*, ser. Lecture Notes in Computer Science, P. Paillier and I. Verbauwhede, Eds. Springer Berlin / Heidelberg, 2007, vol. 4727, pp. 63–80. [Online]. Available: http://dx.doi.org/10.1007/978-3-540-74735-2_5

[8] D. E. Holcomb, W. P. Burleson, and K. Fu, "Initial SRAM state as a fingerprint and source of true random numbers for RFID tags," in *Proceedings of the Conference on RFID Security*, July 2007. [Online]. Available: http://www.cs.umass.edu/~kevinfu/papers/holcomb-FERNS-RFIDSec07.pdf

[9] M. Hofer and C. Boehm, "Error correction coding for physical unclonable functions," in *Austrochip, Workshop on Microelectronics*, 2010.

[10] ——, "An alternative to error correction for sram-like pufs," in *CHES - Workshop on Cryptographic Hardware and Embedded Systems*, 2010, pp. 335–350.

[11] J. D. Cheek, "Angled halo implant tailoring using implant mask," US Patent 6 372 587, 2002.

[12] *BSIM4.6.4 MOSFET Model - Users Manual*, Department of Electrical Engineering and Computer Sciences, Univ. California, Berkeley, 2009. [Online]. Available: http://www-device.eecs.berkeley.edu/~bsim/

Low-Power Wireless Receivers for Healthcare Applications

Alper CABUK[1], Yuan GAO[1], Shengxi DIAO[1], Yuanjin ZHENG[1,2], Minkyu JE[1], and Chun Huat HENG[3]

[1]Institute of Microelectronics-A*STAR (Agency for Science, Technology, and Research)
[2]School of Electrical and Electronic Engineering, Nanyang Technological University
[3]Department of Electrical and Computer Engineering, National University of Singapore
Singapore

Abstract—**This paper discusses the major challenges that the research community is facing when designing circuits for low-power wireless receiver systems targeting healthcare applications. Possible solutions are offered to some of these challenges. Particular attention is given to *in-vivo* applications such as ingestible endoscopy capsules and medical implants.**

Keywords-healthcare, low-power, sensors, capsule, in-vivo

I. INTRODUCTION

The recent research activity for ultra low-power, low-voltage circuits and systems has generated some interesting outputs targeting body area networks (BAN) that could monitor/transmit vital signals to a healthcare provider, and major research groups are showcasing their miniaturized medical devices (MMD) that can not only monitor health-related problems, but also be part of the treatment procedures [1]-[3] (Fig. 1). Because in most of the cases, the end-users of these healthcare products are patients who need extra care without being subjected to additional physical/emotional burden, power consumption, form factor and interference specifications for the circuits that make up such systems are very stringent and these pose significant challenges to the designers. In this paper, we will review some of these challenges related to receiver design at the system and the circuit level along with brief discussions of possible solutions for medical implants and capsule endoscopy.

II. RECEIVER ARCHITECTURES FOR LOW-POWER IMPLEMENTATIONS

A. General Requirements

In general, monitoring an individual's health status via a wireless link happens in three ways: 1) planting a medical implant into the body that stays functional throughout the life span of the individual, 2) deploying multiple sensor nodes 'within and outside' the body that can remain there for a very long time and can transmit data as-and-when required, 3) passing a -preferably bidirectional- transceiver unit through an individual's digestive system to collect data and execute certain commands/actions as it travels in the gastro-intestinal (GI) tract.

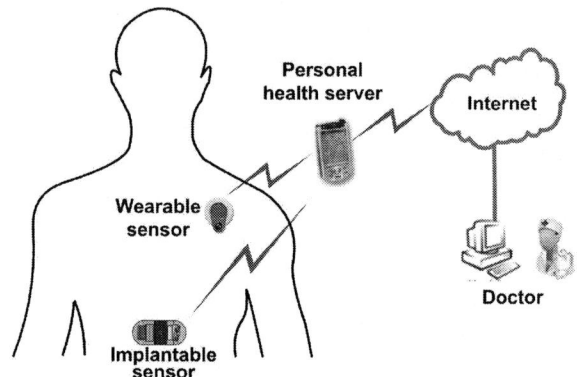

Figure 1. Wireless monitoring for healthcare.

Low power consumption, small form factor, and interference immunity are the major requirements a designer should look into when designing a system for all of the above applications:

A *pacemaker* powered by a lithium iodine battery is expected to work for 5-10 years without interruption until the next scheduled battery change. It should provide robust operation under interference that can come from a mobile phone, magnetic resonance imaging (MRI) units and other strong magnetic field generators.

A *bionic pancreas* developed by the Imperial College of London aims to electrochemically sense the glucose levels in the body and adjust the blood sugar level using an external pump [4]. Although the pump outside the body can be powered by external battery, the sensing implant needs to be very prudent with the available power for long-term operation. Moreover, the sensing unit should be made as small as possible so as not to interfere with the daily activities of the patient.

An *ingestible capsule* that is used to make the endoscopy process less disturbing for a patient needs to stay functional for as long as it stays in the GI tract [5]. This being the ideal, either the stringent requirements or the preferred mode of operation (real-time video image or delayed image recovery) usually dictates the design specifications for the wireless transceiver part of the capsule.

978-1-61284-863-1/11 $26.00 © 2011 IEEE

B. Lowering the Power Consumption

In order for a button-cell battery with 1.5-V 150-mAh capacity to sustain the operation of a miniaturized medical device for more than 2 years in the body, the average power consumption of the circuits should be below 10 μW, excluding the leakage currents. Given that the power consumption of a typical receiver is in the range of milliwatts, maintaining its operation even for a week is not possible unless duty cycling is employed. A cycling rate of 0.1% may result in average power consumption 1000 times less than that of a continuous mode operation. Ultra-low power wake-up receivers are necessary to sniff the transmitter's wake-up call signal without draining the battery in such implementations [6].

The selection of operation frequency will be another major issue: Currently, there are three main frequency bands that are allocated for Medical Implant Communication Service (MICS) and Industrial, Scientific, and Medical (ISM) applications: 400-MHz, 900-MHz, and 2.4-GHz bands. Each has its own merits and drawbacks. The 400-MHz band offers reduced power consumption in the expense of small channel bandwidth and increased circuit/antenna size; the 2.4-GHz band can deliver somewhat the opposite; and the 900-MHz band looks like the perfect region for operation, avoiding inefficient antennas of the lower end and the dramatic tissue losses of the upper end. However, the interference from the mobile telecommunication standards such as GSM at 900-MHz band may prove too difficult to eliminate, especially when one does not have much power budget to spare for improvements in linearity performance of the receiver. Trade-offs will have to be made as there is not a *one-size-fits-all* solution that can satisfy all the requirements.

The modulation scheme for the entire system should be selected carefully: While a 2 frames/s rate may be sufficient for a capsule endoscope that records images of the GI tract for reviewing them at a later time [5], the need for a more accurate diagnosis dictates bidirectional operation and transmission of much higher data rates. In order to handle high data traffic, the circuit blocks of a receiver should have sufficient bandwidth. Wideband operation, on the other hand, limits the choice of receiver architectures that can be adopted as the power consumption of building blocks can easily add up to impractical values if complex architectures were to be used .

The Amplitude Shift Keying (ASK) modulation scheme can help minimize the power consumption of a transceiver. In ASK modulation, a 'logic 1' is represented by a short pulse and a 'logic 0' is represented by no pulse transmission. This can be realized using a fast-turn-on voltage-controlled oscillator (VCO) instead of a power hungry full-fledged "transmitter + phase locked-loop (PLL)" circuit. The receiver at the other end of the transmission will naturally have to demodulate the ASK-based signal. Demodulation can be either coherent or non-coherent, where the former is the choice for good sensitivity and the latter is the preferred method for minimized die size and power consumption.

One such receiver is described in [7] with excellent discussion on link budget, antenna constraints, and digitally-controlled oscillator (DCO) design (Fig. 2). It is based on super-regenerative receiver (SRR) architecture.

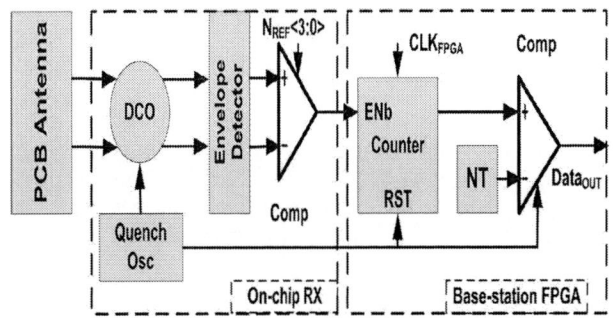

Figure 2.   ASK Receiver for sensor applications.

SRR has been used mostly in low-cost applications such as remote controlled toys, garage door openers, etc. Its potential for ultra low power consumption has increased its popularity for the medical implant applications in recent years. Coupled to a DCO that is also part of the transmitter, the SRR is comprised of a quench oscillator, a fully-differential envelope detector, a comparator with programmable offset, a digital counter, and a digital comparator. The counter is reset and it starts counting at the beginning of each quench cycle. Receiving a 'logic 1' is equivalent to receiving a signal with an oscillation frequency that is equal to the SRR's operating frequency. When such a signal is passed to the envelope detector, it is transformed into a frequency-doubled time varying signal. Once the peak amplitude exceeds the offset voltage of the comparator, it triggers the comparator to change its output state and hold the counter value. Obviously, the time it takes for a 'logic 1' to trigger the comparator is less than that for a 'logic 0'. The decision as to whether a *zero* or *one* is received is made based on the output of this digital counter. Fabricated in a 90-nm CMOS technology, the receiver consumes 400 μW from 1.2-V supply, delivering a data rate of 120 kbps with a receiver sensitivity of -93 dBm and energy efficiency of 3.3 nJ/b.

Although its non-coherent demodulation may be rather straight forward, ASK modulation has its own drawbacks such as low noise immunity and lower bandwidth efficiency when compared with phase-shift keying (PSK) and frequency-shift keying (FSK) modulation schemes. Furthermore, the use of ASK demodulation in the receiver -theoretically- necessitates an ASK transmitter, particularly in the case where bidirectional communication is required. However, these transmitters are susceptible to interferers and require highly linear power amplifiers (PA) because they deal with amplitude-modulated signals [8]. An FSK transmitter that can work with an efficient nonlinear PA may be more favorable for low-power applications. The drawback with this solution is that the FSK receiver implementation will be more complicated than that of an ASK case. Combining the merits of ASK and FSK under one roof is possible if one can transmit using an FSK transmitter and perform the demodulation in ASK at the receiver end. A recent work reported in [9] shows that this can be realized using injection-locked frequency dividers (ILFD). Unlike the case in ASK where an RF gain stage is directly connected to the envelope detector, the receiver in [9] inserts an injection-locked oscillator (ILO) based ILFD between the LNA and the envelope detector so that 1) the noise generated by the envelope detector is prevented from degrading the

978-1-61284-863-1/11 $26.00 © 2011 IEEE   357

sensitivity of the RF front-end, 2) the gain and the power consumption specifications for the LNA block is relaxed, 3) frequency-to-amplitude conversion of the input signal is performed (Fig. 3). The end result is a *hybrid receiver* with an impressive power consumption figure of 420 µW from a 0.7-V supply voltage. Implemented in 0.18-µm CMOS technology for the 900-MHz ISM band, the receiver sensitivity can go as low as -73 dBm and the data rate can reach up to 5 Mbps, resulting in an energy efficiency of 84 pJ/b.

Another demodulation scheme that is getting popular due to its ability to handle very high data rates in ultra low-power receiver implementations is the binary phase shift keying (BPSK).

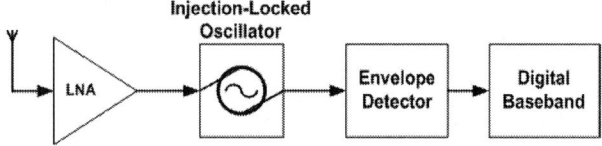

Figure 3.   Block diagram of ILFD-based FSK receiver in [9].

Coherent detection of BPSK modulated signals requires synchronization of the received signal with the local oscillator reference. In general, a *Costas Loop* is employed for this purpose [10]; however, in MMD applications, the power consumption and the complexity make Costas Loop somewhat impractical. Non-coherent, non-linear demodulation methods have been proposed to replace it. Reported in [11], one such BPSK receiver relies on two ILOs to perform the demodulation directly from the received RF signal. The *hybrid receiver* concept similar to the one described in [9] is used here; the BPSK signal is converted to an ASK signal to simplify the demodulation process and eliminate the need for carrier synchronization. Additional power savings is achieved by sacrificing the receiver sensitivity.

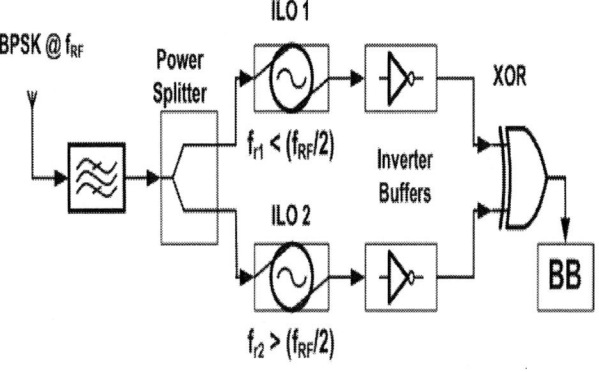

Figure 4.   Ultra low-power BPSK receiver with two injection-locked RC oscillators, reported in [11].

The phase-modulated BPSK signal is passed through a simple bandpass filter and a power divider and fed to two injection-locked RC oscillators. The oscillator-inverter pair pushes the signal to rail-to-rail and the digital information is

reproduced using an XOR gate (Fig. 4). In order to ensure correct demodulation in this implementation, the initial RF signal amplitude must exceed the injection locking level of the oscillators. This leads to poor receiver sensitivity in the range of -35 dBm at maximum data rate. Furthermore, the use of ILO makes the receiver susceptible to interference as the ILOs can lock onto a strong interference instead of the desired signal if the interference is in the vicinity of its free running frequency [12]. Nevertheless, this architecture also offers good insight and demonstrates the great potentials of the injection locking concept in designing ultra low-power receivers. Implemented in a 90-nm CMOS technology, operating at 300 MHz, the receiver achieves a power consumption figure of 120 µW from a 1-V supply while delivering a data rate of 1 Mbps and an energy/bit efficiency of 0.12 nJ/bit.

C.   Improving Linearity

Low-power receivers for healthcare applications reported earlier may differ in terms of architecture, modulation scheme, and of course performance; but they seem to have at least one thing in common: the linearity was not given much consideration. Lowering the power consumption and making the devices as small as possible were the first two priorities, and they still are. On the other hand, the amount of resources poured into the biomedical electronics research is constantly increasing, and this results in new medical devices, implants, and capsules making their ways into the healthcare industry. As commercial products become more available and affordable, the interference of these medical devices with each other and with the other communication apparatus operated on different mobile communication standards such as GSM, Bluetooth, and UWB is set to introduce its own set of problems.

Based on its dielectric properties and different types of tissues it is made of, human body can attenuate the transmitted signals as much as 60 dB as the distance between the transmitter and the receiver antennas increases [13], [14]. Attenuation characteristics depend heavily on the operation frequency as well. But we should not solely rely on the body itself as the interference suppressing agent. Multiple implants with different functionalities may exist in the same body in close proximity with each other [15]. The interaction of these implants needs further study. Until we get a more comprehensive picture, we can adopt a flexible design approach in order to account for strong interferences.

The burden lies with the receiver units that operate in and on the body rather than the transmitters as the output power of transmitters has already been strictly controlled by the government agencies that oversee the usage of frequency spectrum in a particular country or region. If size permits, a channel-select filter with low insertion loss and tight bandwidth may be placed between the antenna and the LNA so as to further suppress the out-of-band interference.

At the IF portion, the variable-gain amplifier (VGA) block in the receiver chain is the critical block that can either *sink or lift* the linearity performance of the overall system. To this end, VGAs with high IP3 values are essential. The RSSI circuits can be incorporated into VGAs so that the gain can be continuously

978-1-61284-863-1/11 $26.00 © 2011 IEEE

adjusted and interference can be minimized. RSSI signal can be utilized by a variable gain LNA circuit as well.

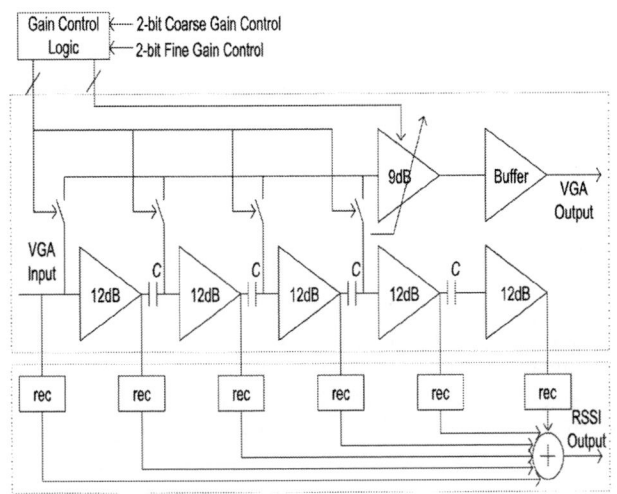

Figure 5. VGA circuit merged with the RSSI in [16].

One such combination is reported in [16]. Five stages of 12-dB fixed-gain amplifiers provide the coarse tuning for the gain variation and 60-dB dynamic range for the RSSI. A VGA with 9-dB total gain (3-dB steps) also ensures that 0-45 dB gain variation can be achieved as per the requirements of that particular system (Fig. 5). Thanks to the closed-loop amplifier topology, the extrapolated IIP3 is more than 18 dBm.

TABLE I.   COMPARISON OF RECEIVERS WITH VARIOUS MODULATION SCHEMES

Parameters	[7]	[9]	[11]
Frequency (MHz)	391-415	902-928	300
Modulation	ASK	FSK+ASK	BPSK
Data rate (Mbps)	0.12	5	1
Sensitivity (dBm)	-93	-73	-34
Power consumption (μW)	400	420	120
Energy per bit (RX) (nJ/b)	3.3	0.084	0.12
Supply voltage (V)	1.2	0.7	1
Area (mm²)	1 x 2	1.1 x 1.5	0.5 x 0.5
Technology	90nm CMOS	0.18μm CMOS	90nm CMOS

III. CONCLUSION

The efforts to design versatile, low-power and small form-factor receivers for healthcare applications are discussed in this paper. Major requirements and challenges related to the μ-power receivers are also highlighted .With the introduction of new circuit topologies or by exploiting the existing techniques -both at circuit and system level- that were once

considered unattractive, significant advancements have been made towards this goal (Table I). While the ultra-low power consumption and miniaturization maintain their priority in the *specs sheets*, a closer look into interference performance may bring about the successful integration of multiple biosensor applications.

ACKNOWLEDGMENT

This work was supported by Science and Engineering Research Council (SERC) of A*STAR (Agency for Science, Technology, and Research) - Singapore, under SERC MedTech Research Program Grant Number: 082 140 0033.

REFERENCES

[1] I. Karhonen, J. Parkka, and M. van Gils, "Health monitoring in the home of the future," *IEEE Eng. Med. Biol. Mag.*, vol. 22, pp. 66-73, May 2003.

[2] E. Y. Chow, A. L. Chelebowski, S. Chakraborty, W. J. Chappell, and P. P. Irazouqui, "Fully wireless implantable cardiovascular pressure monitor integrated with a medical stent," *IEEE Trans. Biomed. Eng.*, vol. 57, no. 6, pp. 1487-1496, Jun. 2010.

[3] Y. Gao, et al., "Low-power ultrawideband wireless telemetry transceiver for medical sensor applications," *IEEE Trans. Biomedical Eng.*, vol. 58, no. 3, pp. 768-772, March 2011.

[4] M. F. El Sharkawi, P. Georgiou and C. Toumazou, "A silicon pancreatic islet for the treatment of diabetes," in IEEE Proc. Intl. Symp. Circuits and Systems (ISCAS), pp. 3136-3139, May 2010.

[5] G. Meron, "The development of swallowable video capsule (M2A)," *Gastrointestinal Endoscopy*, vol. 52, no. 6, pp. 817-819, Dec. 2000.

[6] I. Demirkol, C. Ersoy, and E. Onur, "Wake-up receivers for wireless sensor networks: benefits and challenges," *IEEE Wireless Communications*, vol. 16, no. 4, pp. 88-96, April 2009.

[7] J. L. Bohorquez, A. P. Chandrakasan, and J. L. Dawson, "A 300-μW CMOS MSK transmitter and 400-μW OOK super-regeneretive receiver for medical implant communications," *IEEE J. Solid-State Circuits.*, vol. 44, no. 4, pp. 1248-1259, April 2009.

[8] B. Otis, Y. H. Chee, and J. Rabaey, "A 400 μW-RX, 1.6mW-TX super-regenerative transceiver for wireless sensor networks," in IEEE Int. Solid-State Circuits Conf. Dig. Tech. Papers, 2005, pp. 396-397, 606.

[9] J. Bae, L. Yan, and H-J. Yoo, "A low energy injection-locked FSK transceiver with frequency-to-amplitude conversion for body sensor applications," *IEEE J. Solid-State Circuits*, vol. 46, no. 4, pp. 928-937, April 2011.

[10] J. Costas, "Synchronous communications," *IRE Trans. Communications Systems*, vol. 5, no. 1, pp. 99-105, Jan. 1957.

[11] H. Yan, et al., "A 120-μW fully-integrated BPSK receiver in 90nm CMOS," in IEEE Proc. Radio Freq. Integrated Circuits Symposium (RFIC), pp. 277-290, May 2010.

[12] B. Razavi, "A study of injection locking and pulling in oscillators," *IEEE J. Solid-State Circuits*, vol. 39, no. 9, pp. 1415-1424, Sept. 2004.

[13] C. Gabriel, "Compilation of dielectric properties of body tissues at RF and microwave frequencies," in Final Technical Report, June 1996 (AL/OE-TR-1996-0037): Airforce Materiel Command, Brooks Air Force Base, Texas U.S.A.

[14] C. C. Johnson and A. W. Guy, "Nonionizing electromagnetic wave effects in biological materials and systems," in Proc. of IEEE, vol. 60, no. 6, pp. 692-718, June 1972.

[15] N. Cho, J. Bae, and H-J. Yoo, " A 10.8 mW body channel communication/MICS dual-band transceiver for a unified body sensor network controller," *IEEE J. Solid-State Circuits*, vol. 44, no. 12, pp. 3459-3468, Dec. 2009.

[16] T. H. Teo, M. A. Arasu, W. G. Yeoh, and M. Itoh, "A 90nm CMOS variable-gain amplifier and RSSI design for wideband wireless network application," in Proc. European Solid-State Circuits Conference (ESSCIRC), pp. 86-89, Sept. 2006.

978-1-61284-863-1/11 $26.00 © 2011 IEEE

Versatile MIMO Voltage-Mode OTA-C Universal Biquadratic Filter

Montree Kumngern

Department of Telecommunications Engineering, Faculty of Engineering,
King Mongkut's Institute of Technology Ladkrabang,
Bangkok 10520, Thailand
E-mail: kkmontre@kmitl.ac.th

Abstract–**This paper presents a new versatile voltage-mode universal filter with multiple-input multiple-output (MIMO) based on single-ended operational transconductance amplifiers (OTAs). The proposed circuit can realize a two-input, four-output biquad filter and a three-input, single-output biquad filter without changing the circuit topology. The natural frequency and the quality factor can be set orthogonally by adjusting the circuit components. The natural frequency can also be controlled electronically by adjusting the bias currents. The simulation results are performed to confirm the presented theory.**

Keywords–*versatile filter, voltage-mode circuit, multiple-output multiple-output, operational transconductance amplifier*

I. INTRODUCTION

Operational transconductance amplifiers (OTAs) have exhibited some advantages in the circuit design. The OTA provides an electronic tunability, a wide tunable range and powerful ability to generate various circuits. Moreover, OTA based circuits require no resistors and, therefore, are suitable for integrated circuit implementation [1].

A universal filter is the circuit that can simultaneously realize various filter functions, i.e. low-pass (LP), band-pass (BP), high-pass (HP), band-stop (BS) and all-pass (AP), which plays an important role in the fields of electronic measurement, communication, automatic control and neural networks. Several voltage-mode biquadratic filters based on OTAs have been reported [2]-[20]. Considering the number of input and output ports, these filters can be classified into three categories: (i) a single-input, multiple-output (SIMO) type [2]-[8], (ii) a multiple-input, single-output (MISO) type [9]-[14] and (iii) a multiple-input, multiple-output (MIMO) type [15]-[20]. Generally, the SIMO filters can simultaneously realize three basic filter functions, i.e., low-pass (LP), band-pass (BP), and high-pass (HP), at a time without altering the connection way of the circuits and without input signal matching. However, for the realizations of all-pass (AP) and band-stop (BS) functions, additional adder and subtractor circuits are usually required. The MISO filter can realize multifunction outputs by altering the way in which the input signals are connected. On the other hand, in comparison with the SIMO filter, the MISO and MIMO configurations provide a variety of circuit characteristics with different input voltage and usually do not require any parameter matching conditions. In addition, MISO and MIMO filters may lead to a reduction in the number of active elements used. Moreover, to realize a larger variety of filter functions such as inverting and/or non-inverting-type functions, the MISO and MIMO configurations seem to be more suitable than the SIMO configuration.

As the reported voltage-mode OTA-based MISO and MIMO filtering circuits, circuit [9] contains minimum elements and enjoys very low sensitivities, but it cannot offer orthogonal tuning capability of the characteristic parameters ω_o and Q. The circuits [10], [16], [17] involve only active components, and enjoy orthogonal tuning capability of parameters ω_o and Q and low sensitivities, but it use two kinds of active components (OTA and op-amp). The circuit [11] contains two OTAs, one current conveyor and two capacitors, and enjoys very low sensitivities but still no orthogonal tuning capability of the characteristic parameters (ω_o and Q) and use two kinds of active components (OTA and DDCC). The circuit in [12] operates in mixed-mode filter but employs several OTAs.

In this paper, a new versatile multiple-input multiple-output voltage-mode universal filter using six single-ended OTAs and two capacitors is presented. The filter can provide both MIMO biquad filter and MISO biquad filter into a single topology. Also, the proposed filter can be set an independent electronic control of the natural frequency and the quality factor. The circuit employs only one kind of active component which is ideal for integrated circuit implementation. All the incremental parameter sensitivities are low. PSPICE simulation results are performed to confirm the theoretical analysis.

II. PROPOSED CIRCUIT

The circuit symbol of the OTA is shown in Fig. 1. The characteristic of the ideal OTA can be described by

$$I_o = g_m(V_1 - V_2) \qquad (1)$$

where I_o is the output current, g_m is the transconductance gain, V_1 and V_2 denote non-inverting and inverting input voltage, respectively.

Fig. 2 shows the addition/subtraction circuit using two OTAs. Referring [21]-[22], this circuit may be called a "pool circuit". Assume two OTAs in Fig. 2 are identical, the voltage output V_o can be expressed as

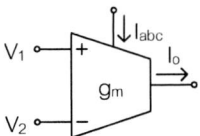

Figure 1. Circuit symbol of OTA.

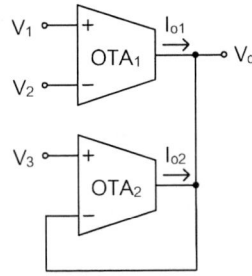

Figure 2. Addition/subtraction circuit using two OTAs.

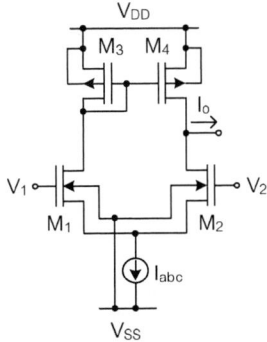

Figure 3. CMOS implementation of the simple OTA.

$$V_o = V_1 - V_2 + V_3 \tag{2}$$

Fig. 3 shows the CMOS implementation of simple OTA. It employs only four MOS transistors and one current source. Assume four MOS transistors operating in saturation regions, the transconductance gain (g_m) can be expressed by

$$g_m = \sqrt{\mu C_{ox}(W/L)I_{abc}} \tag{3}$$

where I_{abc} is the biasing current, μ is the carrier mobility, C_{ox} is the gate oxide capacitance per unit area, W and L are the channel width and length, respectively.

Using the OTA in Fig. 2 and the addition/subtraction in Fig. 3, the proposed filter can be shown in Fig. 3. If $V_{in1}=V_{in2}=V_{in}$ and $V_{in3}=0$, the transfer function is realized as

$$\frac{V_{o1}}{V_{in}} = -\frac{s^2 C_1 C_2}{s^2 C_1 C_2 + sC_2 g_{m1} + g_{m1}g_{m2}} \tag{4}$$

$$\frac{V_{o2}}{V_{in}} = \frac{sC_2 g_{m1}}{s^2 C_1 C_2 + sC_2 g_{m1} + g_{m1}g_{m2}} \tag{5}$$

$$\frac{V_{o3}}{V_{in}} = \frac{s^2 C_1 C_2 + g_{m1}g_{m2}}{s^2 C_1 C_2 + sC_2 g_{m1} + g_{m1}g_{m2}} \tag{6}$$

$$\frac{V_{o4}}{V_{in}} = \frac{g_{m1}g_{m2}}{s^2 C_1 C_2 + sC_2 g_{m1} + g_{m1}g_{m2}} \tag{7}$$

Thus, the circuit realizes a HP signal at V_{o1}, a BP signal at V_{o2} and a BS signal at V_{o3} and a LP signal at V_{o4}. In this case, the filter bases grounded capacitors.

In addition, if V_{in1}, V_{in2} and V_{in3} are three input signals and $V_{o3}=V_{out}$, the transfer function can be expressed as

$$V_{out} = \frac{s^2 C_1 C_2 V_{in1} + sC_2 g_{m1} V_{in3} + g_{m1}g_{m2} V_{in2}}{s^2 C_1 C_2 + sC_2 g_{m1} + g_{m1}g_{m2}} \tag{8}$$

It is clearly seen from equation (8) that:

(1) The LP response can be obtained when $V_{in1}=V_{in3}=0$ and $V_{in2}=V_{in}$.
(2) The BP response can be obtained when $V_{in1}=V_{in2}=0$ and $V_{in3}=-V_{in}$.
(3) The HP response can be obtained when $V_{in2}=V_{in3}=0$ and $V_{in1}=V_{in}$.
(4) The BS response can be obtained when $V_{in3}=0$ and $V_{in1}=V_{in2}=V_{in}$.
(5) The AP response can be obtained when $V_{in1}=V_{in2}=-V_{in3}=V_{in}$.

Thus, the proposed filter can realize all the standard types of the biquadratic filtering function without component-matching condition requirements. The natural frequency (ω_o) and the quality factor (Q) of the proposed circuit can be expressed as

$$\omega_o = \sqrt{\frac{g_{m1}g_{m2}}{C_1 C_2}} \tag{9}$$

$$Q = \sqrt{\frac{g_{m2}C_1}{g_{m1}C_2}} \tag{10}$$

Letting $g_{m1}=g_{m2}=g_m$, the circuit parameters are simplified to

$$\omega_o = g_m \sqrt{\frac{1}{C_1 C_2}} \tag{11}$$

$$Q = \sqrt{\frac{C_1}{C_2}} \tag{12}$$

Thus, the parameter Q can be set by C_1 and C_2 while the parameter ω_o can be controlled by g_m without disturbing Q. Therefore, the biquadratic filter has orthogonal tuning capability for the circuit parameters Q and ω_o. The parameter ω_o can be tuned by the transconductance gain through adjusting the bias currents of the OTAs, hence the name "electronically tunable filter".

978-1-61284-863-1/11 $26.00 © 2011 IEEE

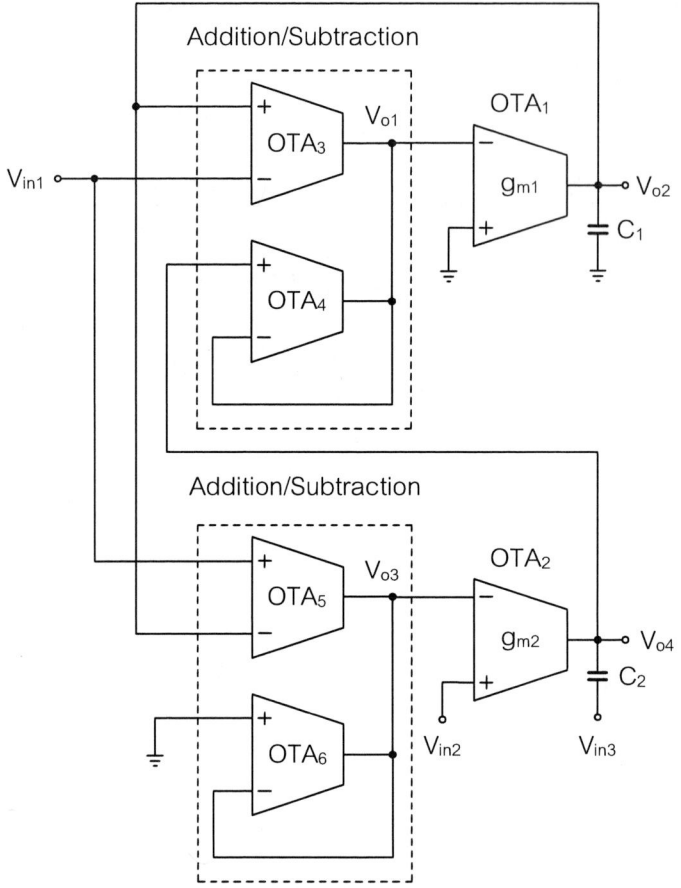

Figure 4. Proposed voltage-mode MIMO universal filter using OTAs.

III. SIMULATION RESULTS

The proposed universal biquadratic filter is verified through PSPICE simulation using the circuit in Fig. 4 with 0.35 μm CMOS process from TSMC. The OTA in Fig. 2 and the addition/subtraction in Fig. 3 are used. The aspect ratio of transistors are W/L = 10 μm / 1 μm for nMOS devices and W/L = 5μm / 1 μm for pMOS devices [23]. The biasing currents for OTA_3 to OTA_6 are chosen as 20 μA. The power supplies are selected as $V_{DD} = -V_{SS} = 1.65$ V.

As an example design, the capacitors $C_1 = C_2 = 15$ pF and the biasing currents $I_{abc1} = I_{abc2} = 50$ μA ($g_m = 181.97$ μS) are given. This setting has been designed to obtain the LP, BP, HP, BS and AP filter responses with $f_o \cong 1.93$ MHz and Q=1. The simulated responses of the LP, HP, BP and BS of the proposed filter are shown in Fig. 5 ($V_{in1}=V_{in2}=V_{in}$ and $V_{in3}=0$). In this figure, the pole frequency of 1.92 MHz is obtained.

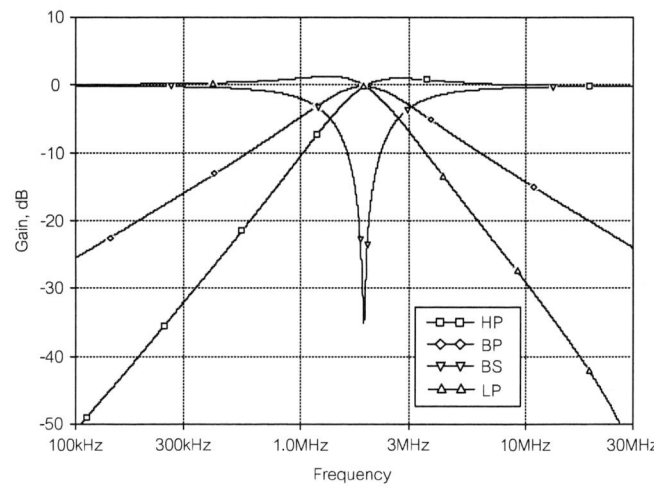

Figure 5. Simulated HP, BP, BS and LP responses.

978-1-61284-863-1/11 $26.00 © 2011 IEEE

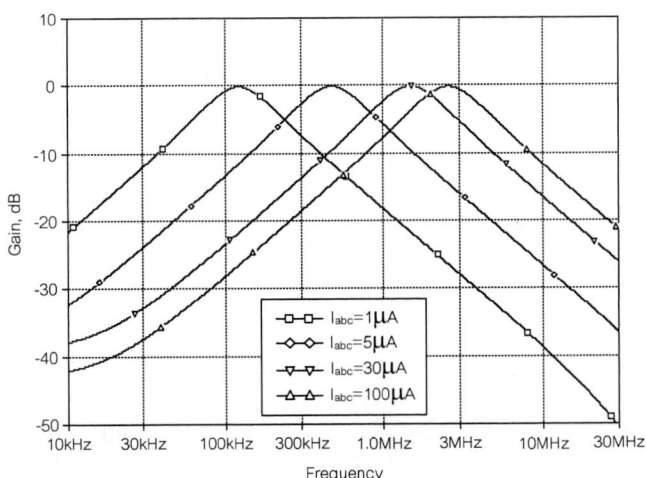

Figure 6. Simulated frequency responses of BP filter when I_{abc} is varied.

Fig. 6 shows the simulated a BP filter response when the biasing currents I_{abc} (i.e., $I_{abc} = I_{abc1} = I_{abc2}$) were simultaneously adjusted for the values 1, 5, 30 and 100 µA, respectively, while keeping $C_1 = C_2 = 15$ pF. This result is confirmed by equation (11).

IV. CONCLUSIONS

In this paper, a new versatile voltage-mode multiple-input and multiple-output universal filter using six simple OTAs and two capacitors was proposed. The proposed filter can realize a two-input four-output biquad filter and a three-input single-output biquad filter without changing the circuit topology. The new circuit offers several advantages, such as no need to component-matching conditions, electronically tunable of parameter ω_o and low active and passive sensitivities performance.

REFERENCES

[1] E. Sanchez-Sinencio, R. L. Geiger, and H. Nevarez-Lozano, "Generation of continuous-time two integrator loop OTA filter structure," IEEE Transactions on Circuits and Systems, vol. CAS-35, pp. 936–949, 1988.

[2] P. V. A. Mohan, "Generation of OTA-C filter structures from active RC filter structures," IEEE Transactions on Circuits and Systems, vol. 37, pp. 656–660, 1990.

[3] C. M. Chang, "New multifunction OTA-C biquads," IEEE Transactions on Circuits and Systems–II, vol. 46, pp. 820-824, 1999.

[4] T. Tsukutani, M. Higashimura, N. Takahashi, Y. Sumi, and Y. Fukui, "Novel voltage-mode biquad without external passive element," International Journal of Electronics, vol. 88, pp. 13–22, 2001.

[5] J.-W. Horng, "Voltage-mode universal biquadratic filter with one input and five outputs using OTAs," International Journal of Electronics, vol. 89, pp. 729–737, 2002.

[6] C.-M. Chang, "Analytical synthesis of the digitally programmable voltage-mode OTA-C universal biquad," IEEE Transactions on Circuits and Systems-II, vol. 53, pp. 607–611, 2006.

[7] W.-T. Lee and Y.-Z. Liao, "New voltage-mode high-pass, band-pass, and low-pass filter using DDCC and OTAs," International Journal of Electronics and Communications, vol. 62, pp. 701–704, 2008.

[8] M. Kumngern and K. Dejhan, "Voltage-mode low-pass, high-pass, band-pass biquad filter using simple CMOS OTAs," in Proceedings of IEEE International Instrumentation and Measurement Technology Conference 2009, (I2MTC 2009), Singapore, 2009, pp. 924–927.

[9] I. A. Khan, M. T. Ahmed, and N. Minhaj, "A simple realization scheme for OTA-C universal biquadratic filter," International Journal of Electronics, vol. 72, pp. 419–429, 1992.

[10] T. Tsukutani, Y. Sumi, Y. Kinugasa, M. Higashimura, and Y. Fukui, "Versatile voltage-mode active-only biquad circuits with loss-less and lossy integrators," International Journal of Electronics, vol. 91, pp. 525–536, 2004.

[11] J.-W. Horng, "High input impedance voltage-mode universal biquadratic filter using two OTAs and one CCII," International Journal of Electronics, vol. 90, pp. 183–191, 2003.

[12] M. T. Abuelma'atti and A. Bentrcia, "A novel mixed-mode OTA-C universal filter," International Journal of Electronics, vol. 92, pp. 375–383, 2005.

[13] M. Kumngern, B. Knobnob, and K. Dejhan, "Electronically tunable high-input impedance voltage-mode universal biquadratic filter based on simple CMOS OTAs," International Journal of Electronics and Communications, vol. 64, pp. 934–939, 2010.

[14] C.-N. Lee "Multiple-mode OTA-C universal biquad filters," Circuits, Systems and Signal Processing, vol. 29, pp. 263–274, 2010.

[15] J. Wu and C.-Y. Xie, "New multifunction active filter using OTAs," International Journal of Electronics, vol. 74, pp. 235–239, 1993.

[16] T. Tsukutani, M. Higashimura, Y. Sumi, and Y. Fukui, "Voltage-mode active-only biquad," International Journal of Electronics, vol. 87, pp. 1435–1442, 2000.

[17] T. Tsukutani, M. Higashimura, N. Takahashi, Y. Sumi, and Y. Fukui, "Novel voltage-mode biquad using only active devices," International Journal of Electronics, vol. 88, pp. 339–346, 2001.

[18] J.-W. Horng, "Voltage-mode universal biquadratic filter using two OTAs," Active and Passive Electronic Components, vol. 27, pp. 85–89, 2004.

[19] M. Kumngern and K. Dejhan, "Voltage-mode multifunction biquadratic filter based on simple CMOS OTAs," in Proceedings of 5th International Colloquium on Signal Processing and its Applications 2009 (CSPA 2009), Malaysia, 2009, pp. 317–322.

[20] M. Kumngern and K. Dejhan, "Electronically tunable voltage-mode universal filter with three-input single-output," in Proceedings of International Conference on Electronics Devices, Systems & Applications 2010 (ICEDSA 2010), Kuala Lumpur, Malaysia, 2010, pp. 317–322.

[21] R. R. Torrance, T. R. Viswanathan, and J. V. Hanson, "CMOS voltage to current transducers," IEEE Transactions on Circuits and Systems, vol. CAS-32, pp. 1097–1104, 1985.

[22] S.-I. Liu and C.-C. Chang, "CMOS analog divider and four-quadrant multiplier using pool circuits," IEEE Journal of Solid-State Circuits, vol. 30, pp. 1025–1029, 1995.

[23] S.-H. Tu, C.-M. Chang, J. N. Ross, and M. N. S. Swamy, "Analytical synthesis of current-mode high-order single-ended-input OTA and equal-capacitor elliptic filter structures with the minimum number of components" IEEE Transactions on Circuits and Systems–I, vol. 54, pp. 2195–2210, 2007.

New Current-Mode First-Order Allpass Filter Using a Single CCCDTA

Montree Kumngern

Department of Telecommunications Engineering, Faculty of Engineering,
King Mongkut's Institute of Technology Ladkrabang,
Bangkok 10520, Thailand
E-mail: kkmontre@kmitl.ac.th

Abstract—This paper presents a new current-mode first-order allpass filter employing only one current-controlled current differencing transconductance amplifier, one grounded capacitor and one grounded resistor. The use of grounded capacitor makes the circuit more suitable for integrated circuit implementation. The circuit also possesses high output impedance which enables easy cascading in the current-mode circuits. Simulation results are given to verify the presented theory.

Keywords—first-order allpass filter; current-mode circuit; current-controlled current differencing transconductance amplifier

I. INTRODUCTION

First-order allpass filters are widely useful in many applications, for example, they can be employed as quadrature oscillator [1], multiphase oscillator [2], and high-Q band-pass filter [3]. Therefore, several techniques for realize of the first-order allpass filters have been proposed in the literature technique; see, for example [4]-[18]. Note that the allpass filters in [4]-[10] are the voltage-mode circuits and the one in [11]-[18] are the current-mode circuits. The current-mode filter can easily be cascaded to the next stage without additional buffer circuits, if it has the property of high impedance outputs [19]. On the other hand, it is attractive for monolithic integrated circuit implementation if the filters employ grounded capacitors [20]. In the proposed current-mode first-order allpass filters, circuits [11]-[18] enjoy a variety of first-order allpass characteristics. However, the reported filters suffer from one or more of the following disadvantages: (i) the realizations use floating capacitor and/or floating resistor [11]-[15], (ii) the circuits do not provide high-output impedance [11]-[13], [15], (iii) the circuits suffer from an excessive active component [11]-[17], (iv) the gain can not variable [11]-[18].

Recently, a new current-mode active element, which is called a current differencing transconductance amplifier (CDTA) has been proposed [21]. It is a synthesis of the well-known advantages of the CDBA and OTA. Our survey found that several CDTA-based first-order allpass filters have been reported [22]-[24]. However, they still suffer from excessive active or passive components.

In this paper, a new current-mode first-order all-pass filter using current-controlled differencing transconductance amplifier (CCCDTA) is presented. The proposed circuit employs one CCCDCC, one grounded capacitor and one

grounded resistor. The use of grounded capacitor and grounded resistor makes the proposed circuit suitable for integrated circuit implementation. The pole frequency and the current gain of the filter can be electronically controlled by the bias currents. The inverting and non-inverting first-order allpass filters can be achieved into a single topology. The proposed circuit possesses high output impedance, which enables easy cascading. PSPICE simulation results verifying theoretical analyses are also included.

II. CIRCUIT DESCRIPTION

The equivalent circuit and the electrical symbol of the CCCDTA are shown in Fig. 1. The port relation of the CCCDTA can be characterized by the following matrix equation:

$$
\begin{pmatrix} V_p \\ V_n \\ I_z \\ I_x \end{pmatrix} = \begin{pmatrix} R_p & 0 & 0 & 0 \\ 0 & R_n & 0 & 0 \\ 1 & -1 & 0 & 0 \\ 0 & 0 & \pm g_m & 0 \end{pmatrix} \begin{pmatrix} I_p \\ I_n \\ V_x \\ V_z \end{pmatrix} \tag{1}
$$

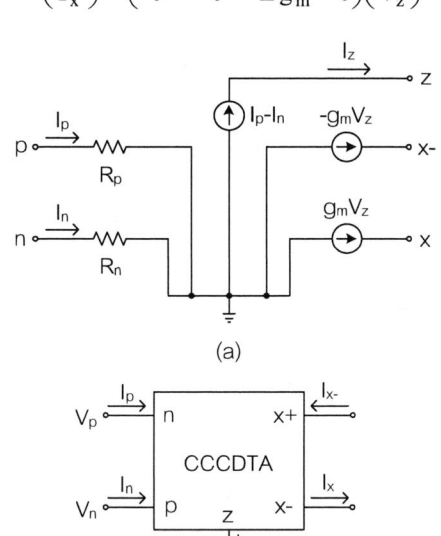

(a)

(b)

Figure 1. CCCDTA: (a) equivalent circuit, (b) electrical symbol.

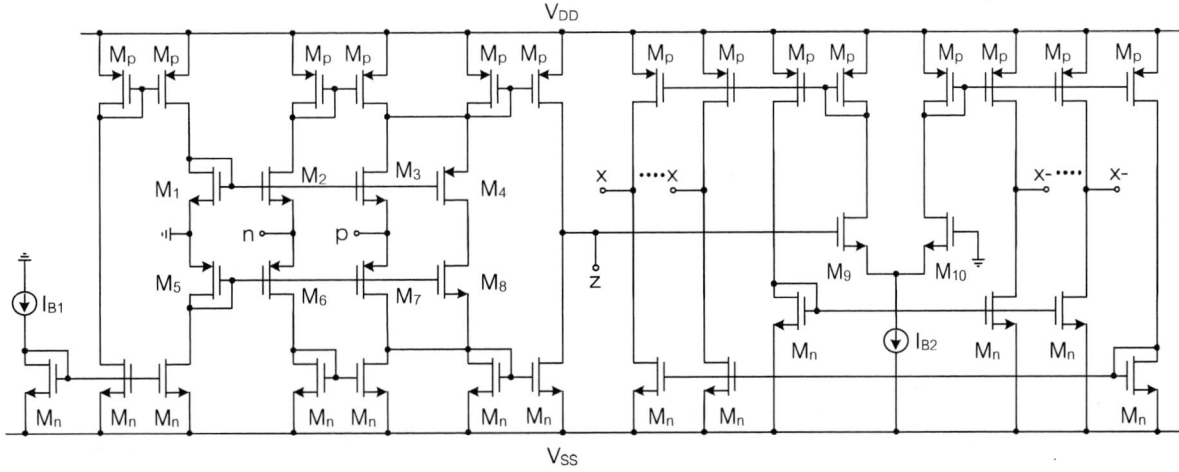

Figure 2.   Possible CMOS implementation of the CCCDTA.

It has four terminals where the terminals x and z possesses a high output impedance and the terminals p and n possesses resistances R_p and R_n, respectively. Fig. 2 shows the CMOS implementation of the CCCDTA. Assume that transistors M_1 to M_8 are matched and operated in saturation region, the p and n terminal resistances (R_p and R_n) can be approximated as

$$R_p \cong R_n \cong \frac{1}{\sqrt{8\mu C_{ox}\left(W/L\right)I_{B1}}} \qquad (2)$$

where μ is the carrier mobility, C_{ox} is the gate oxide capacitance per unit area, and W and L are the channel width and channel length of MOS transistor, respectively. From equation (2), the input parasitic resistances R_p and R_n can be controlled by varying the biasing current I_{B1}. This property makes it different from conventional CDTA. Again assume transistors M_9 and M_{10} are operated in saturation regions and matched. The transconductance gain (g_m) can be expressed by

$$g_m = \sqrt{\mu C_{ox}\left(W/L\right)I_{B2}} \ . \qquad (3)$$

Also the transconductance gain can be controlled by adjusting the biasing current I_{B2}.

Fig. 3 shows the proposed current-mode allpass filter. It can be seen that the circuit uses only one CCCDTA, one grounded capacitor and one grounded resistor. The current transfer function can be expressed as

$$\frac{I_{o1}}{I_{in}} = -\frac{I_{o2}}{I_{in}} = \left(R_1 g_m\right)\left(\frac{1-sCR_p}{1+sCR_p}\right). \qquad (4)$$

The pole frequency (ω_C) of the filter is;

$$\omega_C = \frac{1}{CR_p} \ . \qquad (5)$$

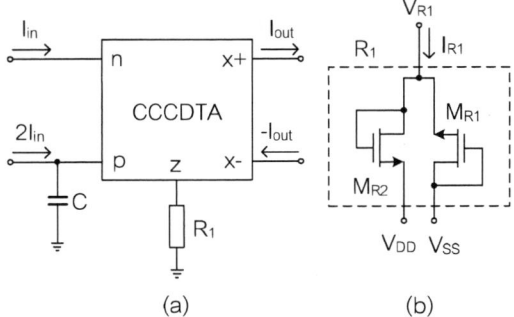

Figure 3.   (a) Proposed current-mode allpass filter, (b) MOS resistor.

It is evident from equation (4) that the circuit in Fig. 3 can realize the current-mode first-order allpass transfer function. By the use of multiple-output CCCDTA that provides both plus-type and minus-type output currents, hence the proposed first-order allpass filter yields the phase-shifter of 0° to -180° as I_{o1} and 180° to 0° as I_{o2} without changing the circuit configuration. The current gain of filter can also be controlled by g_m through adjusting the bias current I_{B2}. Because the x terminal impedance of CCCDTA is high, the output terminals of proposed first-order allpass filter can be directly connected to the next stage without additional buffer circuit. Furthermore, the use of only grounded capacitor and grounded resistor in the design is beneficial from the point of view of integrated circuit realization [20]. The passive sensitivities of Fig. 3 to C and R_p are no more than unity in magnitude.

The resistor R_1 can be realized by two MOS transistors M_{R1} and M_{R2} [25]. Assume that transistors M_{R1} and M_{R2} have the same characteristics remaining in the saturation region, the resistance value can be expressed as [25]

$$R_1 = \frac{1}{2K(V_{DD} - V_{TH})} \qquad (6)$$

978-1-61284-863-1/11 $26.00 © 2011 IEEE

where $K=\mu C_{ox}(W/L)$ is the transconductance parameter, V_{TH} is the threshold voltage, V_{DD} is the supply voltage ($V_{DD}=|-V_{SS}|$).

III. NON-IDEAL EFFECT

Taking the non-idealities of the CCCDTA into account, the relationship of the voltages and currents can be rewritten as:

$$
\begin{pmatrix} V_p \\ V_n \\ I_z \\ I_x \end{pmatrix} = \begin{pmatrix} R_p & 0 & 0 & 0 \\ 0 & R_n & 0 & 0 \\ \beta_p & -\beta_n & 0 & 0 \\ 0 & 0 & \pm\gamma g_m & 0 \end{pmatrix} \begin{pmatrix} I_p \\ I_n \\ V_z \\ V_x \end{pmatrix} \tag{7}
$$

where $\beta_p=1-\varepsilon_{pi}$ and ε_{pi} ($|\varepsilon_{pi}|\ll1$) denotes the current tracking error from p terminal to z terminal, $\beta_n=1-\varepsilon_{ni}$ and ε_{ni} ($|\varepsilon_{ni}|\ll1$) denotes the current tracking error from n terminal to z terminal and $\gamma=1-\varepsilon_v$ and ε_v ($\varepsilon_v\ll1$) denotes the transconductane error from z terminal to x terminal of CCCDTA. Re-analysis yields the characteristic equation of Fig. 3 becomes

$$
\frac{I_{out}}{I_{in}} = \pm\gamma R_1 g_m \left(\frac{2\beta_p - \beta_n\left(1+sCR_p\right)}{1+sCR_p} \right). \tag{8}
$$

The current tracking error will be deviated the equation (8) from ideal first-order allpass characteristic and also effect to the pole frequency of the filter.

IV. APPLICATION EXAMPLE

As an application of the proposed dual-input first-order all-pass filter, a current-mode quadrature oscillator configuration is constructed. It is well-known that the quadrature oscillator can be implemented in a cascade of two first-order all-pass filters (phase lead and phase lag) and feedback [2]. By using two first-order allpass filters in Fig. 3, the quadrature oscillator is shown in Fig. 4. Letting $C_1=C_2=C$, $R_1=R_2=R$, $R_{p1}=R_{p2}=R_p$ and $g_{m1}=g_{m2}=g_m$, where R_{p1} and R_{p2} are the resistances at p terminals of CCCDTA$_1$ and CCCDTA$_1$, respectively, g_{m1} and g_{m2} are the transconductances of CCCDTA$_1$ and CCCDTA$_2$, respectively, the characteristic equation of Fig. 4 can be expressed by

$$
-\left(Rg_m\right)^2\left(\frac{1-sR_pC}{1+sR_pC}\right)^2 = 1. \tag{9}
$$

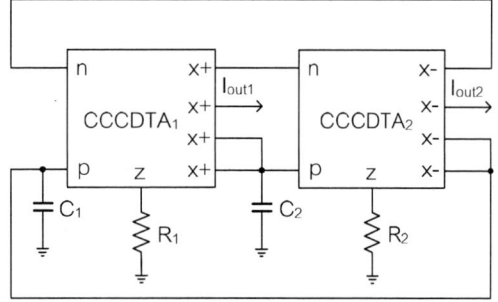

Figure 4. CCCDTA allpass filter-based quadrature oscillator.

It implies that the oscillation condition of Fig. 4 can be controlled electronically by g_m through adjusting the biasing current I_{B2} and the oscillation frequency can be given as

$$
f_o = \frac{1}{2\pi CR_p}. \tag{11}
$$

Also the oscillation frequency can be controlled electronically by R_p through adjusting the biasing current I_{B1}. Therefore it is confirmed that the proposed allpass filter can be used to realize the quadrature oscillator well.

V. SIMULATION RESULTS

To verify the theoretical prediction of the proposed circuit, the circuit in Fig. 3 was simulated using PSPICE program. The CCCDTA in Fig. 2 was simulated using the TSMC 0.25 μm CMOS technology. The transistor aspect ratios of the transistors were set to be same as [26]. The supply voltages used are ±1.5V. PSPICE simulations have been verified that when current I_{B1} increasing from 1 to 200 μA, achieved resistances R_p and R_n are decreasing from 13.58 kΩ to 1.52 kΩ.

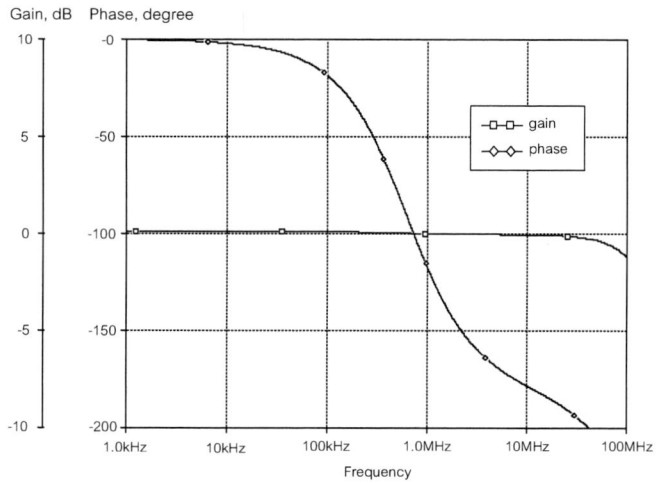

Figure 5. Gain and phase responses of proposed allpass filter.

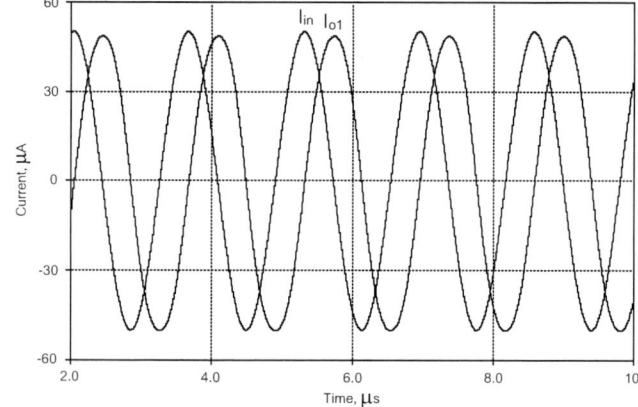

Figure 6. Input and output waveforms for the circuit at 612.6 kHz.

978-1-61284-863-1/11 $26.00 © 2011 IEEE

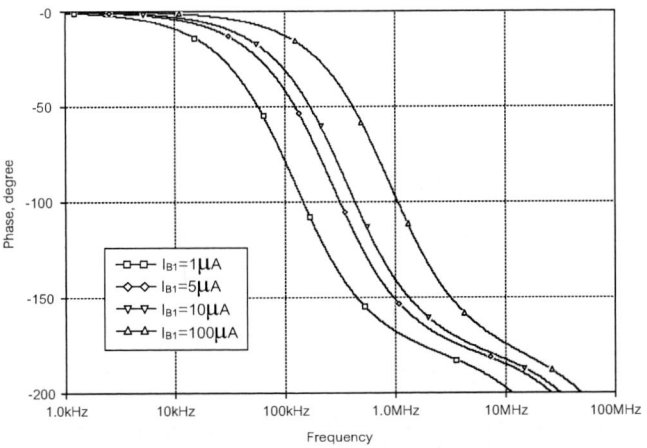

Figure 7. Pole-frequency tuning with bias current I_{B1}.

For example design, the proposed allpass filter in Fig. 3 was designed with $C_1 = 100$ pF, $I_{B1} = 35$ μA ($R_p = 2.5$ kΩ), $I_{B2} = 47$ μA ($g_m = 305.4$ μA/V) and the aspect ratio of transistors M_{R1} and M_{R2} are W/L = 1.4 μm / 1 μm ($R_1 = 3.33$ kΩ). This setting has been designed to obtain the allpass filter response with $f_o = 636.6$ kHz and $R_1 g_m \approx 1$. The simulation results for the magnitude and phase responses of I_{o1} are shown in Fig. 5. In this figure, the pole frequency of 612.62 kHz is obtained. The pole frequency is 612.62 kHz instead of 636.6 kHz owing to the effect described in Section III. According to equation (8), this error would be caused by voltage tracking errors of CCCDTA. To confirm Fig. 5, the circuit is inputted with a sinusoidal signal of 612.6 kHz, the input and -90° phase shifted output (I_{o1}) are shown in Fig. 6. Fig. 7 shows the variability of the pole frequency of Fig. 3 with the bias current I_{B1}. This result is confirmed by equation (5).

VI. CONCLUSIONS

In this paper, a new current-mode first-order all-pass filter employing one CCCDTA, one grounded capacitor and one grounded resistors is presented. The filter configuration is very suitable for implementation in both bipolar and CMOS technologies. The proposed filter has following advantages: (i) it provides both inverting and non-inverting types of first-order allpass responses, (ii) it possesses high-output impedance, (iii) it uses only grounded passive component, (iv) it provides an electronic control and (v) the current gain can be controlled. The theoretical results are verified using PSPICE simulation program. In addition, the quadrature sinusoidal oscillator using CCCDTA-based all-pass filter is also introduced.

REFERENCES

[1] M. T. Ahmed, I. A. Khan, and N. Minhaj, "On transconductance-C quadrature oscillators," International Journal of Electronics, vol. 83, pp. 201-207, 1997.

[2] S. J. G. Gift, "The application of all-pass filters in the design of multiphase sinusoidal systems," Microelectronics Journal, vol. 31, pp. 9-13, 2000.

[3] D. T. Comer, D. J. Comer, and J. R. Gonzales, "A high frequency integrable band pass filter configuration," IEEE Transactions on Circuits and Systems–II: Analog and Digital Signal Processing, vol. 44, pp. 856-861, 1997.

[4] O. Cicekoglu, H. Kuntman, and S. Berk, "All-pass using a single current conveyor," International Journal of Electronics, vol. 86, pp. 947-955, 1999.

[5] I. A. Khan and S. Maheshwari, "Simple first-order section using a single CCII," International Journal of Electronics, vol. 87, pp. 303-306, 2000.

[6] S. Maheshwari, "High input impedance VM-APSs with grounded passive elements," IET Circuits and System, vol. 1, pp. 72–78, 2007.

[7] N. Pandey and S. K. Paul, "All-pass filters based on CCII- and CCCII-," International Journal of Electronics, vol. 91, pp. 485-489, 2004.

[8] C. Cakir, U. Cam, and O. Cicekoglu, "Novel allpass filter configuration employing single OTRA," IEEE Transactions on Circuits and Systems–II: Analog Digital Signal Processing, vol. 52, pp. 122-125, 2005.

[9] S. Minaei and O. Cicekoglu, "A resistorless realization of the first-order all-pass filter," International Journal of Electronics, vol. 93, pp. 177-183, 2006.

[10] M. A. Ibrahim, S. Minaei, and H. Kuntman, "DVCC based differential-mode all-pass and notch filters with high CMRR," International Journal of Electronics, vol. 93, pp. 231-240, 2006.

[11] A. M. Soliman, "Generation of current conveyor based all-pass filter from opamp based circuits," IEEE Transactions on Circuits and Systems–II: Analog Digital Signal Processing, vol. 44, pp. 324–330, 1997.

[12] O. Cicekoglu, H. Kuntman, and S. Berk, "All-pass using a single current conveyor," International Journal of Electronics, vol. 86, pp. 947–955, 1999.

[13] I. A. Khan and S. Maheshwari, "Simple first-order section using a single CCII," International Journal of Electronics, vol. 87, pp. 303–306, 2000.

[14] S. Maheshwari and I. A. Khan, "Novel first order all-pass sections using a single CCIII," International Journal of Electronics, vol. 88, pp. 773–778, 2001.

[15] S. Maheshwari, "New voltage and current-mode APS using current controlled conveyor," International Journal of Electronics, vol. 91, pp. 735–743, 2004.

[16] N. Pandey and S. K. Paul, "All-pass filters based on CCII- and CCCII-," International Journal of Electronics, vol. 91, pp. 485–489, 2004.

[17] M. Kumngern, P. Sampattavanich, P. Prommee, and K. Dejhan, "A capacitor-grounded current-tunable current mode all-pass network," in Proceedings of 2004 The IEEE Region 10 Conference (TENCON 2004), Chiang-mai, Thailand, 2004, pp. 384-386.

[18] N. A. Shah, M. F. Rather, and S. Z. Iqbal, "SITO electronically tunable high output impedance current-mode universal filter," Analog Integrated Circuits and Signal Processing, vol. 47, pp. 335-338, 2006.

[19] U. Torteanchai and M. Kumngern, "Current-tunable current-mode all-pass section using DDCC," in Proceedings of 2011 International Conference on Electronics Devices, Systems & Applications (ICEDSA 2011), Kuala Lumpur, Malaysia, April 25-27, 2011

[20] M. Bhusan and R. W. Newcomb, "Grounding of capacitors in integrated circuits," Electronics Letters, vol. 3, pp. 148-149, 1967.

[21] D. Biolek, "CDTA building block for current-mode analog signal processing," in Proceedings of the ECCTD'03, vol. III, pp. 397-400, 2003.

[22] N. A. Shah, M. Quadri, and S. Z. Iqbal, "CDTA based transimpedance type first-order all-pass filter," WSEAS Transactions on Electronics, vol. 5, pp. 260-264, 2008.

[23] W. Tanjaroen and W. Tangsrirat, "Resistorless current-mode first-order allpass filter using CDTAs," in Proceedings of 5th International Conference on Electrical Engineering/Electronics, Computer, Telecommunications and Information Technology 2009, Thailand, 2009, pp. 721-724.

[24] W. Tangsrirat, T. Pukkalanun, and W. Surakampontorn, "Resistorless realization of current-mode first-order allpass filter using current differencing transconductance amplifiers," Microelectronics Journal, vol. 41, pp. 178-183, 2010.

[25] Z. Wang, "2-MOSFET transistor with extremely low distortion for output reaching supply voltage," Electronics Letters, 26, 951-952, 1990.

[26] M. Kumngern, "Current-mode multiphase sinusoidal oscillator using current-controlled current differencing transconductance amplifiers," in Proceedings of 2010 IEEE Asia Pacific Conference on Circuits and Systems (APCCAS 2010), Kuala Lumpur, Malaysia, 2010, pp. 728-731.

978-1-61284-863-1/11 $26.00 © 2011 IEEE

A 15nV/√Hz Noise 0.2µV Offset Chopper Conditioning Amplifier for Monolithic Infrared Sensing Systems

Juanda, Wei Shu, Joseph Chang and Wenfeng Yu
School of Electrical and Electronic Engineering
Nanyang Technological University, Singapore

Abstract – **Conventional chopper conditioning amplifiers with a differential architecture are largely inappropriate for the monolithic infrared sensing systems with the requirements of single-ended architecture, low power, and yet low noise and low offset. In this paper, a novel chopper amplifier is proposed to satisfy all the aforesaid requirements. This is achieved by means of a novel chopper demodulator that not only incurs no hardware overhead but also inherently suppresses noise and offset. On the basis of computer simulations, the proposed chopper amplifier features very low input referred noise (15nV/√Hz), very low input referred offset (0.2µV), high Signal-to-Noise Ratio (~87dB) and high Common Mode Rejection Ratio (98dB), and dissipates low quiescent current (57.6µA). When compared to reported differential chopper amplifiers, the proposed amplifier depicts very competitive Figure-Of-Merit.**

I. INTRODUCTION

Infrared (IR) sensing systems are widely employed in various commercial and military applications including photography, video & audio systems [1,2], security devices, night-vision, etc. In a general IR sensing system, an IR sensor senses the IR radiation and outputs a photovoltage signal, and Read-Out Integrated Circuit (ROIC) subsequently conditions the photovoltage signal and provides amplification.

Conventionally, the IR sensor and ROIC are separate entities and are interconnected by either wire bonding [3] or bump bonding [4]. A monolithically-integrated IR sensing system (IR sensor-cum-ROIC) is however more desirable because of the potential higher robustness, lower power, smaller form-factor and lower cost. This integration heretofore is not practical, in part because the IR sensor that allows for monolithic integration generally has relatively low sensitivity to IR illumination (its photovoltage signal can be < 1µV) and high noise, resulting in photovoltage outputs with very poor Signal-to-Noise Ratio (SNR) [5]. Moreover, in view of the monolithic structure where the IR sensor is fabricated directly on top of the ROIC [5], the heat dissipated by the ROIC may inadvertently degrade the performance of the IR sensor (as the ROIC may be an erroneous IR source) and further deteriorate the SNR.

To mitigate these problems, the ROIC for IR sensing usually embodies a low-power conditioning amplifier with low noise and low offset [6]. The low noise attribute is particularly critical at very low frequencies where the photovoltage signals are often present. That is because flicker noise is significant at low frequencies and it may further compromise the SNR. The low offset attribute is also important because it is usually difficult to extract/differentiate the photovoltage signal from the offset.

Autozero and chopper techniques are widely adopted for low-noise and low-offset conditioning amplifiers [6,7]. The autozero technique serves to extract photovoltage signals and to cancel (a first order) flicker noise and offset by means of two-phase double sampling [6]. However, the wideband noise therein is increased due to aliasing. The chopper technique is essentially a modulation processing, where the photovoltage signal is differentiated from flicker noise and offset by modulating the latter to higher chopping frequency and subsequently suppressing them via a low-pass filter [6,7]. Compared to autozero, the chopper technique generally features lower noise characteristics and higher SNR, and does not have the aliasing issue. In view of these, the chopper technique is adopted in this paper.

Despite the attributes of low noise, low offset and higher SNR, a few issues arise when the chopper technique is applied to the monolithic IR sensing system with single-ended architecture. The first problem is that the modulator and demodulator are inherently differential in conventional chopper amplifiers [6,8,9]. Hence, the additional differential-to-single-ended conversion circuits are required, incurring hardware overhead. The second problem is that the chopper amplifier with differential architecture dissipates more power (than its single-ended counterpart), potentially resulting in (higher) erroneous IR radiation. The third problem is that the attributes of low noise low offset and high SNR may be compromised by the single-ended operation (compared to differential). Put simply, there is a real need to design a single-ended low-power chopper amplifier without compromising noise, offset and SNR, thereby featuring equivalent Figure-Of-Merit (FOM) to its differential counterpart.

In this paper, we propose a novel chopper conditioning amplifier embodying a single-ended architecture particularly applicable to a monolithically-integrated IR sensing system. The single-ended architecture is achieved by means of a novel chopper demodulator – it not only incurs no hardware overhead (hence low quiescent current, 57.6µA), but also provides inherent suppression to noise and offset. In addition to the single-ended operation, the simulations (based on 0.35µm CMOS) show that the proposed chopper amplifier achieves very low input referred noise (15nV/√Hz), very low input referred offset (0.2µV), high SNR (~87dB) and high Common Mode Rejection Ratio (CMRR) (98dB). Put simply, the proposed novel chopper amplifier features the low-power single-ended operation, yet without compromising low-noise low-offset and high SNR attributes.

978-1-61284-863-1/11 $26.00 © 2011 IEEE

II. DESIGN OF A CHOPPER AMPLIFIER

In this section, we will first briefly review the chopper technique, and thereafter propose a novel chopper amplifier design for monolithically-integrated IR sensing systems.

A. Review of Chopper Technique

Figure 1 depicts the block diagram of a conventional differential chopper amplifier, where the chopper modulator and demodulator are designed at the input and output of the amplifier respectively. The operating principle of the chopper technique is as follows. The chopper modulator modulates the differential input signals from at a low frequency (close to DC) to a higher chopping frequency, and the chopping frequency is determined by the biphasic non-overlapping control signals Φ_1 and Φ_2. The amplifier amplifies both the modulated differential input signals (v_{IN+} and v_{IN-}) and its combined input-referred noise and offset ($V_N + V_{OS}$). The chopper demodulator subsequently demodulates the amplified input signal back to its original low frequency but modulates the amplified noise and offset to the chopping frequency – hence, the respective frequencies of the amplified input signal and the amplified $V_N + V_{OS}$ are effectively separated. Finally, the low-pass filter recovers the amplified input signal and attenuates the noise and offset.

For completeness, note that although the chopper technique can significantly reduce the noise and offset from the amplifier, it inevitably introduces an additional residual offset [7,9-11]. This residual offset is largely due to the charge-injection and clock-feedthrough of the chopper modulator.

As depicted in Figure 1, the chopper technique is inherently a differential architecture, and it undesirably requires additional hardware to accommodate single-ended operation. We will now describe the proposed single-ended low-noise low-offset chopper amplifier without hardware overhead.

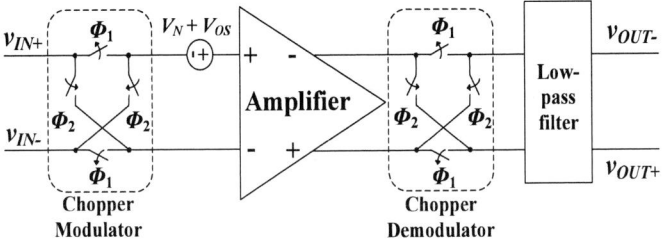

Figure 1 Block diagram of a conventional differential chopper amplifier

B. Proposed Single-ended Chopper Amplifier

Consider first the block diagram of the IR sensing system in Figure 2, where the photoresistive IR sensor is monolithically integrated. The IR sensor is biased by a current source, and the photovoltage output of the IR sensor is amplified (×200) by the proposed chopper amplifier. The amplified analog signal is finally converted to digital by the ensuing Analog-to-Digital Convertor (ADC).

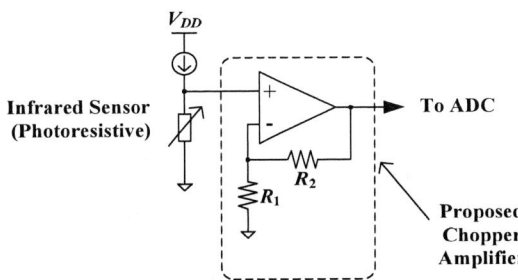

Figure 2 Block diagram of the monolithically-integrated IR sensing system

Figure 3 depicts the schematic of the proposed chopper amplifier whose component values are tabulated in Table 1. At this outset, note that the proposed design is surprisingly simple, yet featuring nearly equivalent FOM of a more complex conventional differential design. The proposed design embodies a two-stage amplifier with a source follower as the driving stage. The first stage (M_1-M_7) is the input stage. PMOS is employed for the input transistors M_1 and M_2 because PMOS has lower flicker noise than NMOS. The second stage (M_8 and M_9) is the gain stage. The output stage is the load driving stage consisting of M_{10} and M_{11}. The chopper modulator (MC_1-MC_4) is at the input of the amplifier. Note that the chopper modulator herein is the conventional architecture and it is appropriate for the differential input stage of amplifiers.

The novel chopper demodulator (MD_1-MD_4) is proposed to achieve the desired single-ended architecture without hardware overhead. The operating principle of the proposed chopper demodulator is briefly explained as follows. When MD_1-MD_4 alternately switch, M_3-cum-M_5 and M_4-cum-M_6 behave as current mirror and current reference respectively in one phase, and they alternate in functionality in the next phase. In this fashion, only the useful signal from the high-impedance current mirror is passed to the second stage. Unlike the conventional chopper demodulator that is differential and located at the output of the amplifier, the proposed demodulator is single-ended and integrated into the first stage of the amplifier.

The proposed chopper single-ended (see Figure 2) demodulator offers three primary advantages over its conventional differential counterpart:

(i) Unlike its conventional differential counterpart, the noise and offset at high frequencies are now suppressed. This is because the additional low-pass filtering is achieved when the proposed chopper demodulator is integrated into the amplifier due to the low frequency Miller-compensated pole.

(ii) Single-ended demodulation is achieved without hardware overhead. In conventional chopper amplifiers, additional differential-to-single-ended hardware is required for conversion to single-ended operation.

(iii) Following (ii), the lower power attribute arises from the hardware-simplified single-ended architecture.

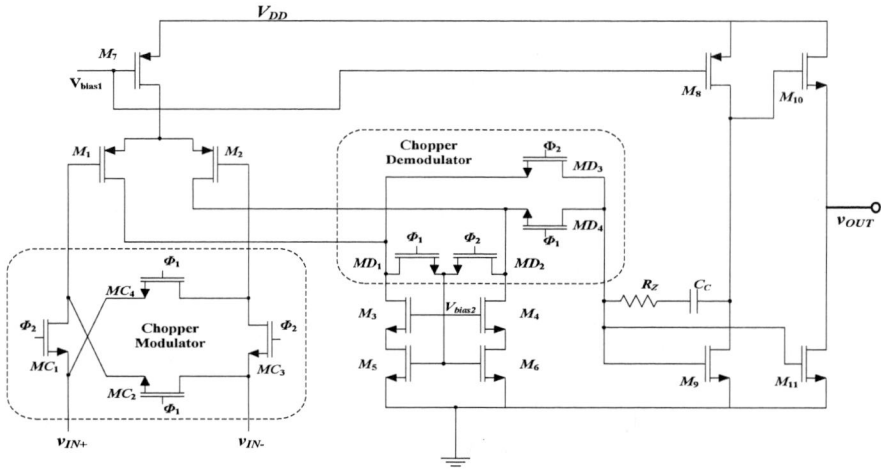

Figure 3 Schematic diagram of the proposed single-ended chopper amplifier

Table 1 Component values of the proposed single-ended chopper amplifier

Component	Value	Component	Value
M_1-M_2	200/1	M_9	4/5
M_3-M_4	13/1	M_{10}	12/3
M_5-M_6	26/10	M_{11}	12/3
M_7	27/1	MC_1-MC_4	2/0.35
M_8	5/1	MD_1-MD_4	0.4/0.35
R_z	20kΩ	C_c	1.4pF

As depicted in Figure 3, the chopper operation is performed at the first stage of the amplifier, and hence the noise and offset therein are largely reduced. On the other hand, as the remaining stages operate as a conventional amplifier without chopping, the noise and offset introduced by these stages are not suppressed. Nonetheless, the latter noise and offset are usually insignificant compared to the former because of the high gain provided by the first stage. In short, the optimization of the proposed chopper amplifier is largely confined to the first stage therein.

C. Design Optimization

To optimize the proposed chopper amplifier, particularly its low noise and low offset attributes, several important design parameters need to be considered.

(i)   Width of $M_{1,2}-W_{1,2}$ and length of $M_{5,6}-L_{5,6}$

On the basis of well-established modeling [12] for thermal noise and flicker noise, thermal noise and flicker noise can be generally optimized by increasing $g_{m1,2}$ (transconductance of $M_{1,2}$) and decreasing $g_{m5,6}$ (transconductance of $M_{5,6}$). Specifically, $W_{1,2}$ and $L_{5,6}$ are designed large to reduce both thermal noise and flicker noise. Further, large $W_{1,2}$ also reduces the offset of the amplifier.

(ii)   Chopping frequency

The chopping frequency needs to be optimized to obtain a good compromise between noise and offset. The specific trade-off is that while a higher chopping

frequency reduces noise (particularly flicker noise), it conversely increases the offset that is induced by charge injection and clock-feedthrough at the chopper modulator. In general, the chopping frequency is designed to be higher than the noise corner frequency [9,13]. The chopping frequency is judiciously chosen to be 5kHz in view of the ~200Hz noise corner frequency.

(iii)   Size of chopper modulator

The size of the chopper modulator switches (MC_1-MC_4) requires optimization to obtain a good compromise between charge injection and thermal noise. The specific trade-off is that while a larger switch size reduces thermal noise, it conversely increases charge injection. The size of MC_1-MC_4 is judiciously chosen to be 2μm/0.35μm. For completeness, the minimum size is adopted for the chopper demodulator switches (MD_1-MD_4) as their size has negligible impact on noise and offset.

(iv)   Bias current

The bias current of M_1-M_2 needs to be optimized to obtain a good compromise between flicker noise and thermal noise. The specific trade-off is that a higher bias current reduces thermal noise at the cost of increased flicker noise. The bias current of M_1-M_2 is judiciously selected to be 13.5 μA.

III. SIMULATION RESULTS

The proposed chopper amplifier is designed using AMS0.35μm CMOS process, and is simulated in Cadence. The characteristics of the proposed design are tabulated in Table 2. Of particular interest, the proposed design achieves very low input offset, 0.2μV, very low noise, 15nV/√Hz, and dissipates 57.6 μA quiescent current.

For an overall FOM, we employ a composite FOM based on offset, noise and quiescent current: composite FOM = 1000 / (offset(μV)×noise(nV/√Hz) ×quiescent current(μA)).

978-1-61284-863-1/11 $26.00 © 2011 IEEE   370

On the basis of this composite FOM, the proposed chopper amplifier is >3× better than all reported designs; this magnitude of improvement is nevertheless probably lesser for practical measurement.

Table 2 Characteristics of the proposed chopper amplifier

Parameter	This Work	[6]	[8]	[9]
Supply voltage (V)	3.3	5	1.8	5
Quiescent current (µA)	57.6	260	13	200
Bandwidth (Hz)	100	500	-	4.5k
Input offset (µV)	0.2	1.5	1.3	0.1
Input noise (nV/√Hz)	15	15	95	27
PSRR (dB)	106	70	135	-
CMRR (dB)	98	70	125	140
Chopping frequency (Hz)	5k	5k	50k	16&2k
Composite FOM	5.79	0.17	0.62	1.85

Figure 4 depicts the output noise spectrums with chopping and without chopping. It is apparent that the output noise without chopping is high (59µV/√Hz@1Hz) at low frequencies where flicker noise is dominant. The noise gradually decreases as the frequency increases. On the other hand, the noise with chopping is significantly lower (~3µV/√Hz@1Hz) at low frequencies where flicker noise is reduced by chopping. In other words, a much higher SNR at low frequencies is achieved by means of chopping. Both noises are comparable at higher frequencies where thermal noise is dominant. As expected, a spike at 5kHz chopping frequency is present in the output noise with chopping, and this inconsequential spike constitutes the energy of the modulated offset and noise.

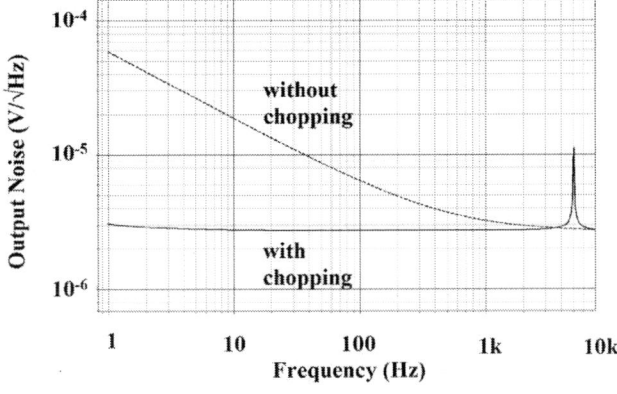

Figure 4 Output noise spectrums with chopping and without chopping

Figure 5 depicts the CMRR of the proposed design over frequencies. The high CMRR (~98dB) is achieved over a relatively high bandwidth (up to ~10kHz). High CMRR is imperative for low noise and low offset. As expected, CMRR decreases as the frequency increases.

In summary, the proposed single-ended chopper amplifier features very low noise and offset, yet low power dissipation and very simple hardware.

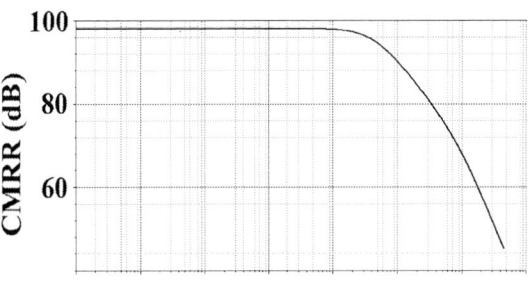

Figure 5 CMRR of the proposed chopper amplifier

IV. CONCLUSIONS

A low-power low-noise and low-offset chopper amplifier based on a novel chopper demodulator has been proposed for a single-ended monolithic IR sensing system. The computer simulations have shown that the proposed chopper amplifier features very low input referred noise (15nV/√Hz), very low input referred offset (0.2µV), high SNR (~87dB) and high CMRR (~98dB), dissipates low quiescent current (57.6µA), and yet embodies very simple hardware. Those attributes, including its composite FOM, have been shown to be very competitive when compared to reported differential chopper amplifiers.

REFERENCES

[1] Wei Shu and Joseph S. Chang, "Power Supply Noise in Analog Audio Class D Amplifiers", *IEEE Transactions on Circuits and Systems I*, Regular Paper, vol. 56, iss. 1, January 2009, pp. 84-96.

[2] Wenfeng Yu, Wei Shu, and Joseph S. Chang, "A Low THD Analog Class D Amplifier based on Self-Oscillating Modulation with Complete Feedback Network", *IEEE International Symposium on Circuits and Systems*, May 2009.

[3] M. Blomberg, O. Rusanen, K. Keranen and A. Lehto, "A Silicon Microsystem - Miniaturised Infrared Spectrometer", *Solid State Sensors and Actuators*, 1997.

[4] F. Serra-Graells, B. Misischi, E. Casanueva, C. Méndez and L. Téres, "A 60ns 500×12 0.35µm CMOS Low-Power Scanning Read-Out IC for Cryogenic Infra-Red Sensors", *IEEE ISCAS*, 2005.

[5] X. C. Sun, *et al.*, "Multispectral pixel performance using a one-dimensional photonic crystal design", *Applied Physics Letters* 89, 2006.

[6] C. Menolfi and Q. Huang, "A Low-Noise CMOS Instrumentation Amplifier for Thermoelectric Infrared Detectors", *IEEE JSSC*, vol. 32, no. 7, 1997.

[7] C. C. Enz and G. C. Temes. "Circuit techniques for reducing the effects of op-amp imperfections: autozeroing, correlated double sampling, and chopper stabilization" *Proc. IEEE*, vol. 84, no. 11, pp. 1584-1614, Nov. 1996.

[8] Y. Kusuda, "Auto Correction Feedback for Ripple Suppression", *IEEE JSSC*, vol.45, no.8, pp. 1436-1445, 2010.

[9] A. Bakker, K. Thiele and J. H. Huijsing, "A CMOS Nested-Chopper Instrumentation Amplifier with 100-nV Offset", *IEEE JSSC,* vol.35, no.12, pp. 1877-1883, Dec. 2000.

[10] C. C. Enz, E. A. Vittoz and F. Krummenacher, "A CMOS Chopper Amplifier", *IEEE JSSC*, vol. sc-22, no. 3, pp. 335-342, June. 1987.

[11] J. F. Witte, K. A. A. Makinwa, and J. H. Huijsing, "A CMOS Chopper Offset-Stabilized Opamp", *IEEE JSSC*, vol. 42, no. 7, July 2007.

[12] D.A. Johns and K. Martin, "Analog Integrated Circuit Design ", John Wiley & Sons, Inc., 1997.

[13] Q. Huang and C. Menolfi, "A 200nV offset 6.5nV/√Hz Noise PSD 5.6kHz Chopper Instrumentation Amplifier in 1µm Digital CMOS", *ISSCC Digest of Technical Papers*, pp. 362-363, Feb. 2001.

A CMOS Circuit Design of a Loss of Signal and the Application in Optical Receivers

Feiyan Qin, Guoqing Xu and Huiyun Li

Center for Automotive Electronics
Shenzhen Institutes of Advanced Technology, Chinese Academy of Sciences
The Chinese University of Hong Kong
Shenzhen, China
E-mail: {fy.qin, hy.li}@siat.ac.cn

Abstract—**This paper presents a new gigabit optical receiver structure with a circuit of loss of signal (LOS). The LOS is placed between the transimpedance amplifier (TIA) and the limiting amplifier (LA) of the optical receiver, so as to mute the data output signal when the input voltage falls below the expected threshold. The optical receiver provides a conversion gain up to 79.01dB Ω and -3dB bandwidth of 3.85GHz from a dynamic range of 40µApp - 200µApp input currents. When the input current amplitude is below 20µApp, the limiting amplifier is turned off by the LOS. The simulation results also show that the LOS circuit achieves 74.49% power saving when the input signal of the optical receiver is below 20µApp.**

Keywords-Loss of signal (LOS); transimpedance amplifier; saving power; opitcal receivers; CMOS

I. INTRODUCTION

Optical communications become more and more attractive for the merits of wider bandwidth, lower loss, lower crosstalk and lower electromagnetic interference, compared to the traditional electric connection communications.

Signal detection or so-called loss of signal (LOS) circuits, are indispensable parts in optical communications, which are used to monitor the status of communication links [1] and to switch off the circuit modules afterwards along communication links to save power and cool the circuits, if the input signal is out of the dynamic ranges of optical receiver circuits [2].

In the past, several LOS circuits have been reported, for example, the logarithmic form [3][4]. However, although the logarithmic form enjoys a wide dynamic range, the precision of circuits is related to the gain of the limiting amplifier (LA) and it requires the LA a high gain (for example, 80dB). In this work, we present a LOS featuring a sensitive amplifier, a peak-detection circuit, and a comparator. The LOS is placed between the transimpedance amplifier (TIA) and the LA of the optical receiver, so as to mute the data output signal when the input voltage falls below the expected threshold.

This paper is organized as follows. Section II presents the architecture of the proposed LOS with its simulation results. The general architecture of the proposed optical receiver is presented in Section III. Section IV demonstrates the experiment and simulation results of the proposed optical receiver. Section V concludes this work with a brief summary.

II. THE PROPOSED LOS

A. Circuit Description: Proposed LOS

Figure 1 shows the schematic diagram of the proposed LOS.

The source follower (M1 and R1) is used to isolate the impacts of other circuits to the LOS. M2, R2, M3 and R3 act as

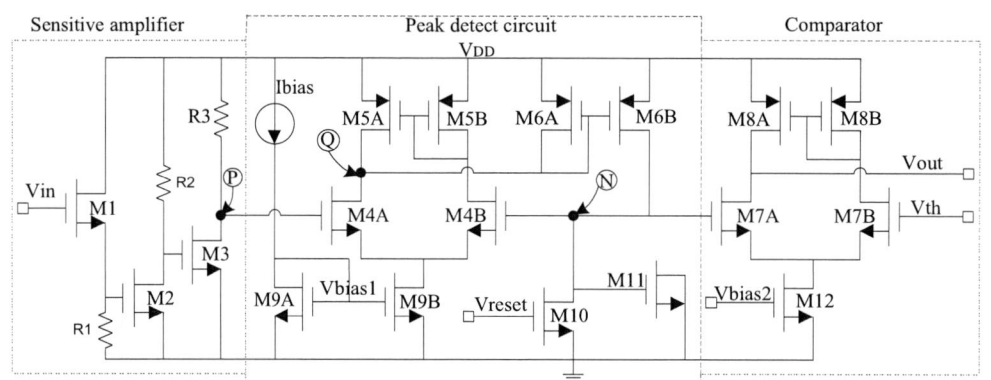

Figure 1. Schematic diagram of the proposed LOS.

The research funding from National S&T Major Project with the Grant No. 2009ZX02038 and from Shenzhen Basic Research Fund with the Grant No. JC200903160412A are greatly acknowledged.

978-1-61284-863-1/11 $26.00 © 2011 IEEE

an amplifier and a DC level shift.

This peak-detection circuit follows the idea from paper [5]. However, different from [5], the peak-detection circuit in this paper has no source follower output buffer, but exports directly from the charging capacitor instead. Because V_N meets the requirement of latter comparator, the output buffer of the peak-detection circuit is dismissed to save area. M5A, M5B, M4A, M4B and M9B constitute a differential pair with active current mirror. M6A, used as a diode, and M6B constitute a rectify mirror. M10 is a reset switch. And M11 is a charging capacitance. When $V_P = V_N$ and $g_{m,4A} = g$ is destined,

$$V_Q = 2V_P g\left(\frac{2r_{o,4A} + \dfrac{1}{g_{m,5B}} \| r_{o,5B}}{2} \| \frac{1}{g_{m,6A}} \| r_{o,6A} \| g_{m,5A}\right) .$$

If $V_{peak,th} = V_{Q,V_P=V_N}$ is destined, the following can be concluded. When $V_P > V_N$, $0 < V_Q < V_{peak,th}$, the more current through M4A than M4B, produced by M6A, can be mirrored to M6B. Then the current charging the M11 and V_N becomes higher than before. When $V_P < V_N$, $V_{peak,th} < V_Q < V_{DD}$ and M6A and M6B are switched off. Then M10 is resetted, providing a leakage path for the charging voltage $V_{G,M11}$. Because the output of the peak-detection circuit (V_N) is hoped to be compared with V_{th}, V_N should hold a long time. So, $I_{d,M10}$ is destined to a current, on the order of a few μApp.

In the comparator V_{th} is a predetermined threshold voltage. If $V_{out,th} = V_{N,V_N=V_{th}}$ is destined, we can draw that when $V_N > V_{th}$, $0 < V_{out} < V_{out,th}$ and when $V_N < V_{th}$, $V_{out,th} < V_{out} < V_{DD}$.

B. Simulation Results

Cadence Virtuoso simulations with the SMIC 0.18μm CMOS technology were conducted to analyze the performance of the proposed LOS. The transient simulation results show that when $V_{in} < 3.6mV$, the output of the proposed LOS is at the high voltage level of about 1.61V; meanwhile, when $V_{in} > 6.9mV$, the output of the proposed LOS is at the low voltage level of about 1.05V.

III. GIGABIT OPTICAL RECEIVER WITH THE PROPOSED LOS AND TIA

The proposed LOS was applied to realize a gigabit optical receiver. Figure 2 illustrates the functional block diagram of the proposed optical receiver, which consists of a TIA, a LA, a LOS and a bias.

A. Transimpedance Amplifier

The TIA is a key block in an optical receiver. Figure 3 shows the schematic diagram of the proposed TIA. The pseudo-differential configuration is adopted to isolate the power and the substrate noises. Various configurations have

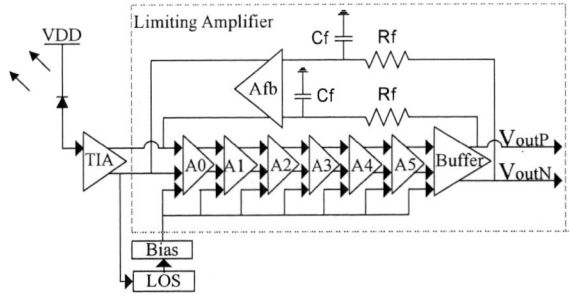

Figure 2. Functional block of diagram of the proposed optical receiver.

been discussed in [6], [7], [8] and [9]. And among these configurations, the regulated cascade (RGC) configuration, with better noise performance, isolating the large input capacitance from TIA effectively and relaxing the tradeoff between gain and bandwidth [7], is explored as the input configuration in this paper. The output node of RGC is a high impedance node. A voltage-voltage feedback amplifier adopting an inverter is used as gain stage for the advantage of the use of n- and p-channels, providing a higher gain [6] and low input impedance. M5 and R5 act as a DC level shift.

According to the small signal analysis, based on paper [7], the transfer function of the proposed TIA in Figure 3 can be derived as following:

$$T_{TIA}(s) \approx \frac{-R_f' g_{m5} R_5 (g_{m3} + g_{m4})(r_{o3} \| r_{o4})}{[1 + s\dfrac{C_{tot}}{g_{m1}(1 + g_{m2}R_2)}][1 + s\dfrac{C_{gs1} + C_{gd2}}{g_{m1}}]}$$

$$\times \frac{1}{1 + sR_f' C_f'}$$

(1)

Where $C_{tot} \approx C_{pd} + C_{ESD} + C_{pad} + C_{gs2}$ (2)

$$R_f' \approx R_1 \| \frac{R_f}{1 + (g_{m3} + g_{m4})(r_{o3} \| r_{o4})}$$

and $C_f' \approx C_{gd1} + C_{gs3} + C_{gs4} + (C_{gd3} + C_{gd4}) \times [1 + (g_{m3} + g_{m4}) \cdot (r_{o3} \| r_{o4})]$

In (2), C_{tot}, including photodiode capacitance C_{pd},

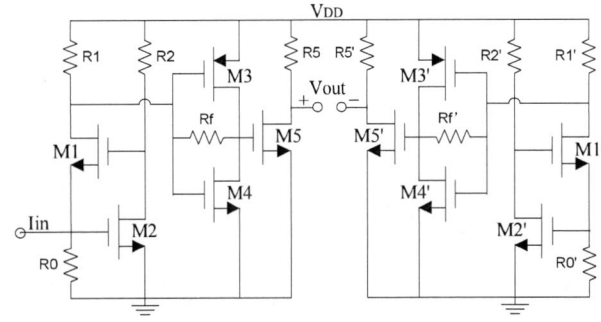

Figure .3. Schematic diagram of the proposed TIA.

978-1-61284-863-1/11 $26.00 © 2011 IEEE

electrostatic discharge (ESD) protection diode capacitance C_{ESD}, pad parasitic capacitance C_{pad} and C_{gs2}, is the input capacitance of the TIA. It can be seen from (1) that the third pole is the dominant pole and if $1+(g_{m3}+g_{m4})(r_{o3}\parallel r_{o4})\gg 1$, the DC transimpedance gain is nearly $R_f g_{m5} R_5$, because R_f cannot be too large for the tradeoff between the gain and the bandwidth. So, $(g_{m3}+g_{m4})>40mS$ and $(r_{o3}\parallel r_{o4})>100\Omega$.

Based on paper [7], the -3dB bandwidth of the TIA can also be derived from as following:

$$\omega_{-3dB,TIA}\approx\frac{1+\dfrac{R_f}{R_1}\times\dfrac{1}{1+g_{m3}+g_{m4}}}{2\pi R_f(\dfrac{C_{gd1}+C_{gs3}+C_{gs4}}{1+g_{m3}+g_{m4}}+C_{gd3}+C_{gd4})}. \qquad (3)$$

It can be inferred from (3) that M3 and M4 are expected to have small parasitic capacitance. So, there is a tradeoff to make between the large g_{m3}, g_{m4} and small parasitic capacitance of M3 and M4.

B. The New Structure of Optical Receivers with the LOS and TIA

Figure 2 is the functional block diagram of the proposed optical receiver, in which the bias circuit provides steady DC bias voltages for the LA and the LOS separately. In Figure 2, the output of the LOS controls the work condition of the LA through controlling the bias voltages for the LA. The proposed optical receiver in Figure 2 is different from the normal one, where the LOS locates between the last gain stage of the LA and the buffer [2] [3] [10]. In Figure 2, the proposed LOS locates between the TIA and the LA. Because the DC current consumption of the LA is far larger than that of the LOS, the optical receiver in Figure 2 can save more power consumption than in configuration of [2], [3] and [10] when the LOS switches off the latter circuits.

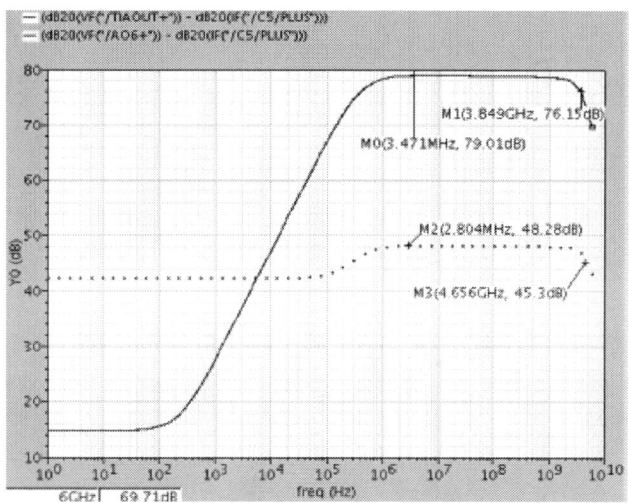

Figure 4. Simulated frequency response of the proposed optical receiver and the proposed TIA.

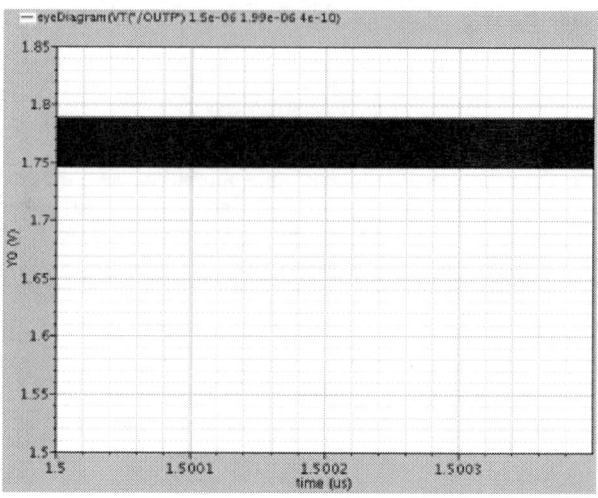

(a)   20μApp

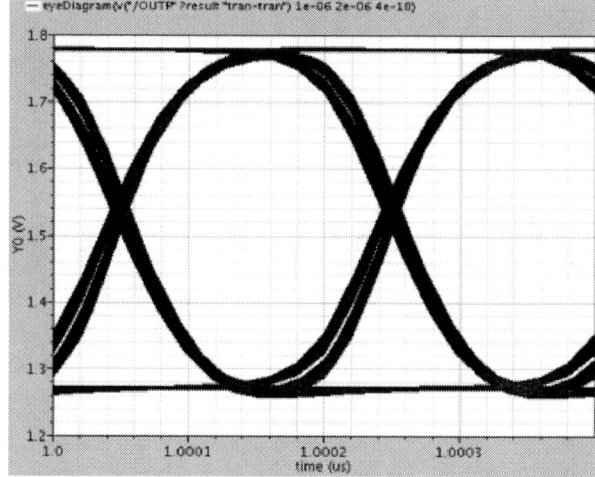

(b)   40μApp

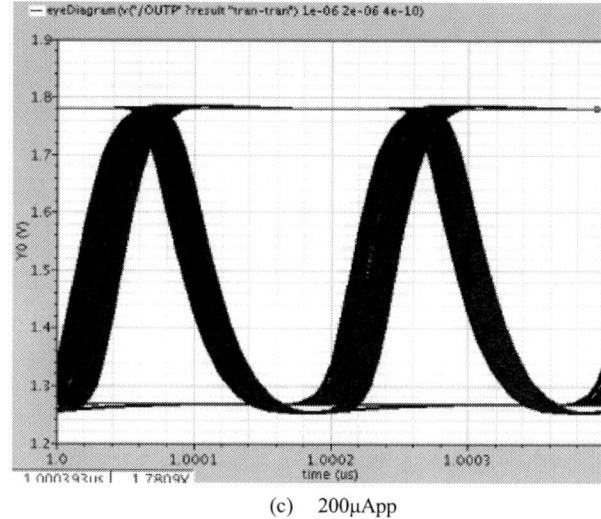

(c)   200μApp

Figure 5. 5Gb/s eye-diagrams with PRBS in the following input currents. (a) 20μApp. (b) 40μApp. (c) 200μApp.

978-1-61284-863-1/11 $26.00 © 2011 IEEE   374

IV. Experimental Results

Cadence Virtuoso and the same CMOS technology were used to analyze the proposed optical receiver. PRBS (Pseudo-Random Binary Sequence), with the rise time of 10% and the fall time of 10%, were used as the input signals in simulations.

Figure 4 illustrates the simulated frequency response of the proposed optical receiver, which provides a conversion gain of 79.01dBΩ with the bandwidth of 3.85GHz. The proposed TIA provides a conversion gain of 48.28dBΩ and a bandwidth of 4.67GHz.

Figure 5 shows the eye-diagrams of the proposed optical receiver when the input current is 20μApp, 40μApp and 200μApp separately. From Figure 5, it can be concluded that when the input currents $I_{in} \leq 20 \mu A_{pp}$, the outputs of the LA is nearly the power supply, which means that the proposed LOS shuts down the LA. When the input currents are in the ranges of $40 \mu A_{pp} \leq I_{in} \leq 200 \mu A_{pp}$, the optical receiver has an output voltage of 540mV. The work conditions of the proposed optical receiver are not concluded when the input currents of the proposed optical receiver are about $20 \mu A_{pp} < I_{in} < 40 \mu A_{pp}$ in Figure 5. This period is the buffer region of the comparator in the proposed LOS. Figure 6 shows the experiment result of the proposed optical receiver. It can be seen that the measurement results is nearly half of the simulation results, which results from the input impedance of test instruments parallel with the output impedance of the optical receiver.

Finally, the DC measurements reveal that the power dissipation of the TIA (containing bias), the LA, and the LOS are separately 39.57mW, 158.37mW, and 14.66mW. It can be inferred the LOS can save as much as 74.49% power consumption when the input signal of TIA $I_{in} < 20 \mu A_{pp}$ in this proposed optical receiver.

V. Conclusion

A gigabit optical receiver with a circuit of loss of signal (LOS) was designed with the SMIC 0.18μm CMOS technology. The LOS is placed between the transimpedance amplifier (TIA) and the limiting amplifier (LA) of the optical receiver, so as to mute the data output signal when the input voltage falls below the expected threshold. Cadence Virtuoso simulation and measured results confirm that when the input current is between 40μApp and 200μApp, the proposed optical receiver provides a conversion gain of 79.01dBΩ with bandwidth of

3.85GHz, among which 48.28dBΩ is provided by the TIA with the bandwidth of 4.67GHz and it is also capable of delivering 540mV differential voltage swings to 50Ω output loads directly with an output signal of half. Otherwise, when the input current is below 20μApp, the proposed LOS shuts down the LA, and saves 74.49% power consumption. The simulation and measured results also indicate the LOS to be a potential solution for applications of gigabit optical links.

Acknowledgment

The authors want to thank the team members: Fanquan Mu, Shuaifeng Liu and Qi An.

References

[1] Robert G. Meyer, "Low-power monolithic RF peak detector analysis," IEEE Journal of Solid-State Circuits, vol. 30, Jan. 1995, pp. 65–67, doi:10.1109/4.350192.

[2] Jae J. Chang, M. Abrams and Young Kim, "Four-channel Parallel 3.125Gbit/s/ch/s Fiber Optic Receiver/transmitter Chip Signal Detection Circuit," Electronics Letters, vol. 38, Nov. 2002, pp. 1556–1558, doi:10.1049/el20021083.

[3] W. B. Chen, H. J. Zou, M. Z. Deng, C. Y. Li, and L. F. Deng, "A limiting amplifier with LOS indication for Gigabit Ethernet", IEEE International Conference Electron Devices and Solid-State Circuits, 2008, Hong Kong.

[4] Po-Chiun Huang, Yi-Huei Chen, and Chorng-Kuang Wang, "A 2-V 10.7-MHz CMOS limiting amplifier/RSSI", IEEE Journal of Solid-State Circuits, vol. 35, No. 10, pp.1474-1479, Oct. 2000.

[5] P.Y. Chang and H.P. Chou, "A High Precision Peak Detect Sample and Hold Circuit," Nuclear Science Symposium Conference Record, Oct. 2006, pp. 329-331, doi:10.1109/NSSMIC.2006.356168.

[6] Karl Schrodinger, Jaro Stimma, and Manfred Mauthe, "A fully integrated CMOS receiver front-end for optic Gigabit Ethernet", IEEE Journal of Solid-State Circuits, vol.37, No.7, pp.874-880, Jul. 2002.

[7] Sung Min Park, Hoi-Jun Yoo, "1.25-Gb/s regulated cascode CMOS transimpedance amplifier for Gigabit Ethernet applications", IEEE Journal of Solid-State Circuits, vol. 39, no. 1, pp. 112-121, Jan. 2004.

[8] Mark Ingels, Geert Van der Plas, Jan Crols, and Michel Steyaert, "A CMOS 18THz Ω 240Mb/s transimpedance amplifier and 155Mb/s LED-Driver for low cost optical fiber links", IEEE Journal of Solid-State Circuits, vol.29, Dec. 1994, pp.1552-1558.

[9] Beehnam Analui and Ali Hajimiri, "Bandwidth Enhancement for Transimpedance Amplifiers", IEEE Journal of Solid-State Circuits, vol.39, No.8, pp.1263-1270, Aug. 2004.

[10] Rentaro Yoshikoshi, "Loss of Signal Detection Circuit for Light Receiver," United States Patent, US 6819880 B2, Nov. 16, 2004.

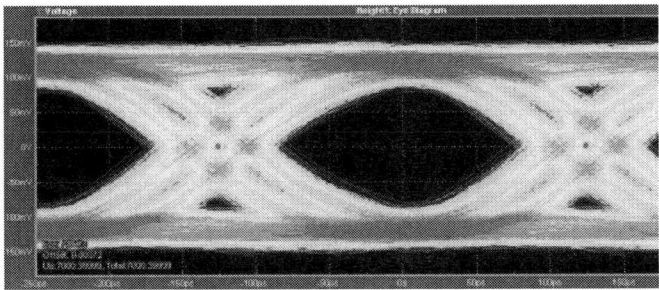

Figure 6. 5Gb/s eye-diagrams in measurement result.

A Power-Efficient Integrated Input/Output Completion Detection Circuit for Asynchronous-Logic Quasi-Delay-Insensitive Pre-Charged Half-Buffer

Weng-Geng Ho*, Kwen-Siong Chong, Bah-Hwee Gwee, Joseph. S. Chang, Ming-Fatt Yee

Nanyang Technological University
Nanyang Avenue, Singapore 639798
*howe0031@ntu.edu.sg

Abstract—**We propose a power- and area-efficient completion detection circuit for improved asynchronous-logic (async) quasi-delay-insensitive (QDI) Pre-Charged Half-Buffer (PCHB) handshake communications. These improved attributes can be achieved by integrating the (separate) input and output completion detection circuits (within the async QDI PCHB circuit) into an integrated input/output completion detection circuit in the transistor level, hereby reducing leakage current paths (leakage power dissipation) and circuit overheads. Moreover, by integrating a reset signal into the integrated input/output completion detection circuit, a stable output state can be achieved during the initialization stage. Based on the simulations on 4×4-bit pipeline multipliers (@1V, 65nm CMOS process), we show that the multiplier with our proposed approach is 37% lower power dissipation (@400MHz input rate), yet 39% lower energy dissipation (per-operation), 35% lower energy-delay product and 21% lesser number of transistors (compared to a conventional design with separate input and output completion detection circuits). These improved results are achieved with an insignificant cost of 5% longer delay (slower speed).**

I. INTRODUCTION

As the process technology continues to scale downwards (to 10nm and even smaller) [1], VLSI system design has become increasingly challenging to accommodate higher design specification, especially with lower power dissipation, higher speed and smaller IC chip area. Therefore, the circuit overhead (in terms of number of circuit components) is one of the critical issues for circuit designer, while obeying strictly with a number of pre-stated design rules and constraints in order to ensure a robust and yet optimized operation in digital systems. Besides, timing information must be specified clearly and efficiently to reduce the propagation delay of data flowing from inputs to outputs, hence to maximize the overall operation speed, especially in clock-driving (synchronous) circuits. As an alternative to synchronous approach, a clock-less (asynchronous) circuit is able to eliminate global clocking infrastructure, by implementing the request-acknowledge handshake protocol [2] in between the pipeline structure.

Asynchronous-logic (async) circuit [3] has demonstrated potential benefits, including improvements in high speed, low-power and reduced electromagnetic interference (EMI), in many aspects of system design. However, async design implementation has faced substantial obstacles such as lack of CAD tools, and shortage of tools for testing and test vector generation [2]. Nonetheless, due to good design efforts, various async designs have been successfully implemented and

realized, including Caltech's CAM (first asynchronous microprocessor in the world, 1989), Japanese's TITAC (in 1994), France's ASPRO-216 (in 1998), Caltech's Lutonium 8051 (in 2003), Intel's 8051 and Tiempo's TAM (both are widely commercialized nowadays) etc [4]-[6].

Indeed, there is a number of realization approaches for async design, and quasi-delay-insensitive (QDI) approach [7] is shown to be the most potential direction for async field due to its virtually "delay insensitive" (over bundled-data/matched-delay [8]-[9] which needs timing assumption) and practical realization (over delay-insensitive and speed-independent [7] which are unrealistic for design and implementation). The design methodology proposed by Caltech focused on ease of using QDI templates, yet achieving good performance through fine-grain pipelining and parallelism [10]. Some examples of QDI pipeline templates are Caltech's Weak-Conditioned Half-Buffer (WCHB), Pre-Charged Half-Buffer (PCHB), and Pre-Charged Full-Buffer (PCFB) templates [11].

Of particular interest, the PCHB realization approach [11] has been reported as a truly QDI approach and works on assumption that the wire fork between the input completion detection and functional block is isochronic [3]. The PCHB realization approach [11] is basically an integrated gate-level 4-phase pipeline dual-rail circuit realization where a dual-rail logic, two handshake signals including an acknowledge-in (*Rack*) and an acknowledge-out (*Lack*), and an internal enable signal (*En*) are embedded into a microcell. This methodology involves incorporating *En* signal into a DCVSL circuit, and including an input completion detection (ICD) for validating the inputs availability and an output completion detection (OCD) for validating the outputs availability, and hence generating a complete acknowledge signal, *Lack*. Due to the *En* signal and ICD, the NMOS stack in the pull-down network is configured in weakly-indicating [2]. However, the ICD and OCD circuits, which contain several standard gate cells (OR gates, NAND gates and C-Muller elements), contribute a large circuit overheads (occupies ~74% transistor count in the overall circuit), resulting in undesirable high power dissipation. Moreover due to the initial condition of *En* and *Rack* (= logic '1'), before inputs data arrive (valid), the outputs state is floating (due to no direct path to the power supply rails).

In this paper, we propose an integrated input/output completion detection circuit for an improved async QDI PCHB approach which features low power and small area advantages over the reported PCHB approach. These improved attributes can be achieved by implementing completion detection in transistor-level (instead of gate-level in the reported approach),

978-1-61284-863-1/11 $26.00 © 2011 IEEE

and integrating both ICD and OCD into a single CMOS logic circuit (instead of using several gate cells for ICD and OCD circuits respectively in the reported approach), therein reducing leakage current paths and circuit overheads (number of transistors). Apart from that, by integrating reset logic at output of the completion detection circuit, a stable dual-rail outputs state (for initialization purpose) can be achieved, for robust circuit operation. To depict the efficacy of our proposed design, we compare our improved PCHB approach against the reported PCHB approach by means of a 4×4-bit pipeline multiplier (@65nm CMOS process). Based on the computer simulations, our proposed PCHB multiplier dissipates 37% lower power (@400MHz input rate), consumes 39% lower energy per-operation, features 35% better energy-delay product and features 21% lesser number of transistors. Nonetheless our design features 5% longer delay – an insignificant cost as compared to the huge advantages in the other attributes.

The paper is organized as follows. Section II reviews the reported PCHB approach, Section III describes our improved PCHB approach, and their comparison is presented in Section IV. Finally conclusions are drawn in Section V.

II.   REVIEW: REPORTED PCHB APPROACH

This section reviews the reported PCHB approach [11] and serves as a preamble to our improved PCHB in Section III.

Fig. 1 depicts the schematic of the async QDI dual-rail 2-input PCHB template [11], which is a DCVSL circuit comprising a pull-up network and a pull-down network (both are controlled by *Rack* and *En* signals), an pull-down NMOS stack, two inverters and two weak feedback inverters (keepers) at both sides of the outputs. The pull-up network pre-charges the outputs while the pull-down network performs evaluation, the pull-down NMOS stack evaluates the logic function from the dual-rail inputs (*A.T*, *A.F*, *B.T* and *B.F*) to the intermediate outputs (*S.T* and *S.F*) and the inverters generate the desired dual-rail outputs (*Q.T* and *Q.F*). The weak feedback inverters maintain the states and serves as implicit latches.

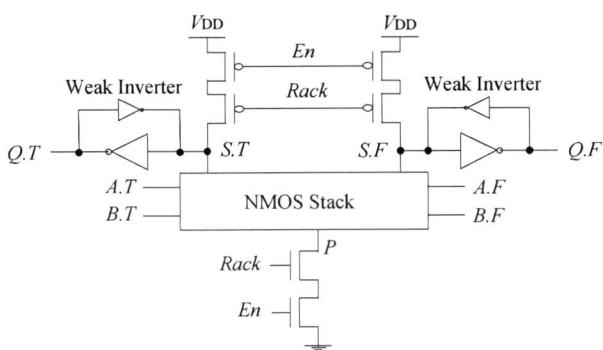

Fig. 1: Reported 2-input PCHB Template [11]

Fig. 2 depicts the 2-input PCHB completion detection circuit [11], which comprising an input completion detection (ICD), an output completion detection (OCD), an inverted C-Muller element and an inverters chain. ICD validates the inputs (*A.T*, *A.F*, *B.T* and *B.F*) availability and OCD validates the immediate outputs (*S.T* and *S.F*) availability. The inverted C-Muller element validates availability of the outputs from both ICD and OCD, to generate *Lack* and *En* signals through the inverters chain, which serves as buffer in between.

Fig. 3 illustrates the pipeline structure of reported PCHB approach. The functional block represents the async QDI dual-

rail PCHB template (Fig. 1), which integrates with the PCHB completion detection circuit (comprising an ICD, an OCD, an inverted C-Muller element and an inverters chain, as shown in Fig. 2) to form a pipeline structure.

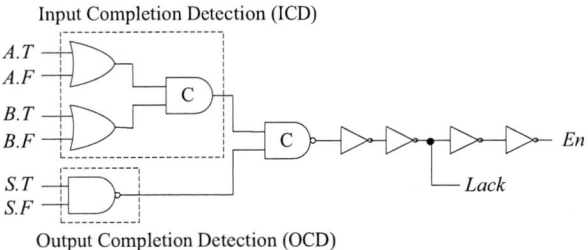

Fig. 2: Reported 2-input PCHB Completion Detection Circuit [11]

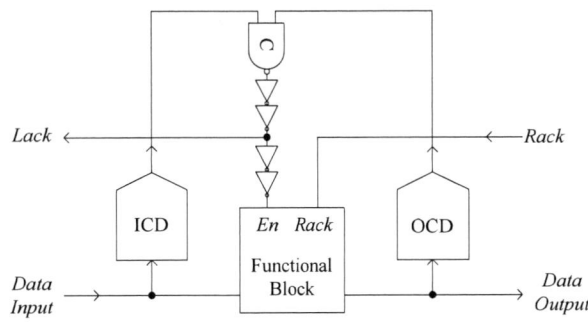

Fig. 3: Pipeline Structure of Reported PCHB Approach [11]

The operation of a PCHB circuit is as follows. Initially, *Data Input* and *Data Output* are logic '0' (empty); the handshake signals (*Lack* and *Rack*), enable signal (*En*) and intermediate outputs (*S.T/S.F*) are logic '1'. When *Data Input* arrives (valid), the pull-down network evaluates the logic function, asserting *S.T/S.F* and *Data Output* (valid). The validities of *Input Data* and *S.T/S.F* through ICD/OCD assert *Lack* and *En* to logic '0' (acknowledge completion of evaluation). Once the succeeding stage has completed evaluation (*Rack* to logic '0'), the circuit resets (pre-charges). When *Data Input* and *Data Output* are both logic '0' (empty), ICD/OCD re-asserts *Lack* and *En* to logic '1'. Noted that once the succeeding stage has completed the reset operation (*Rack* to logic '1'), a new computation can be initiated.

In the staged pipeline, *Data Output* should be initialized at logic '0' for a proper operation. However due to the initial condition of *Lack* and *En* signals (= logic '1'), *S.T/S.F* and hence *Data Output* are floating (due to no direct path to the power supply rails), resulting in malfunction and high short-circuit current, hence high power dissipation. Between, implementation of the PCHB completion detection circuit is neither robust nor power-efficient (due to large circuit overheads).

III.   PROPOSED INTEGRATED INPUT/OUTPUT COMPLETION DETECTION CIRCUIT

We propose an integrated input/output completion detection circuit for an improved async QDI PCHB approach with emphases on the lower power dissipation and more robust operation compared to the reported PCHB approach [11]. Our improved PCHB approach adopts the same dual-rail functional template (Fig. 1), but with improved implementation of the transistor-level completion detection circuit, as shown in Fig. 4.

978-1-61284-863-1/11 $26.00 © 2011 IEEE   377

The proposed 2-input integrated input/output completion detection circuit is basically a single CMOS logic circuit, which simultaneously accepts the input signals (*A.T*, *A.F*, *B.T* and *B.F*) and the output signals (*Q.T* and *Q.F*) to generate *Lack* and *En*, through a series of transistors and a inverters chain (as opposed to the separate gate-level ICD and OCD in the reported approach). The top-right and bottom-left transistor stacks serve as completion detection for the data (inputs and outputs) from empty to valid state, and the top-left and bottom-right transistor stacks do the converse (for the data from valid to empty state).

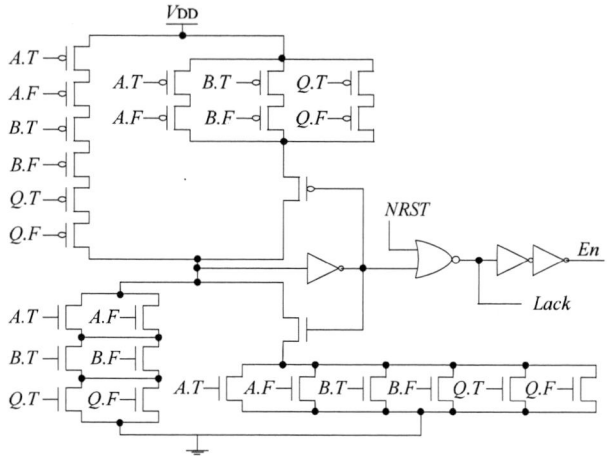

Fig. 4: Proposed 2-input Integrated Input/output Completion Detection Circuit

The operation for the proposed integrated input/output completion detection circuit is as follows. Initially all data signals is logic '0' (empty), *Lack* and *En* are logic '1'. The top pull-up network (including the top-left and top-right transistor stacks, feedback PMOS transistor) is 'on'. The bottom pull-down network (including the bottom-left and bottom-right transistor stacks, feedback NMOS transistor) is 'off'. When the partial data starts to arrive (valid), the top-left stack is 'off' immediately. The bottom-right stack (controlled by the feedback NMOS transistor) remains 'off'. When all data arrives (valid), the top-right stack is 'off' and the bottom-left stack is 'on', changing *Lack* and *En* to logic '0'. Now the feedback PMOS transistor is 'off' and the feedback NMOS transistor is 'on'. The top-right stack (controlled by the feedback PMOS transistor) is now 'off'. When the partial data starts to become 'empty', the bottom-left stack is 'off' immediately. When all data become 'empty', the top-left stack is 'on' and the bottom-right stacks is now 'off', changing *Lack* and *En* to logic '1'. Now the feedback PMOS transistor is 'on' and the feedback NMOS transistor is 'off', turning the top (-left and -right) pull-up network 'on' and the bottom (-left and -right) pull-down network 'off'. A new cycle can be initiated.

Besides, our proposed integrated input/output completion detection circuits integrates *NRST* reset signal, which NORs with the output from completion detection CMOS logic circuit. Initially, *NRST* is logic '1', generating *Lack* and *En* signals to logic '0', initializes the intermediate outputs (*S.T* and *S.F* in PCHB functional template, Fig. 1) at logic '1' and hence the dual-rail outputs (*Q.T* and *Q.F*) at logic '0'. After initialization and before the first data inputs arrives, *NRST* is changed to logic '0', generating *Lack* and *En* signals to logic '1', making the PCHB functional template ready for evaluation. Since *NRST* is maintained at logic '0' after that, *Lack* and *En* signals depend solely on the output from the completion detection circuit, in the continuing cycles.

Fig. 5 illustrates the pipeline structure of improved PCHB approach. It combines ICD and OCD into a single block (as opposed to the separate ICD and OCD in the reported PCHB pipeline structure, Fig. 3), and includes a *NRST* reset signal to generate the *Lack* and *En* signals, for the dual-rail outputs state initialization purpose.

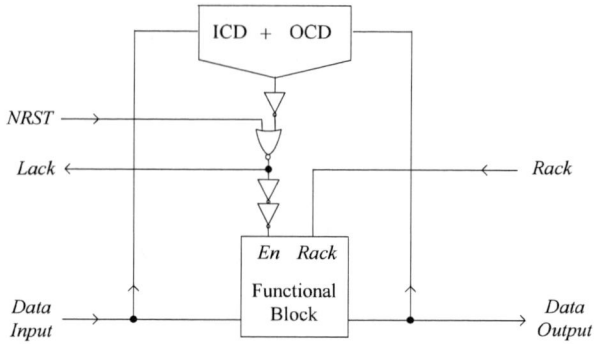

Fig. 5: Pipeline Structure of Improved PCHB approach

IV. SIMULATION RESULTS

To illustrate the power advantage of our proposed PCHB approach, we compare our proposed PCHB against the reported PCHB by means of a 4×4-bit pipeline multiplier. Both the reported and proposed 4×4-bit multipliers have the same architecture as in Fig. 6 to Fig. 8, but in different circuit template implantation (different completion detection circuit structure).

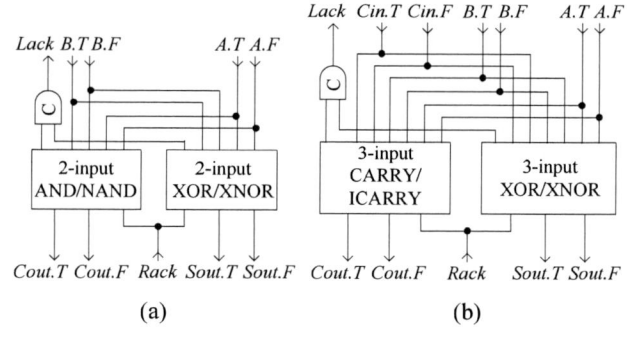

Fig. 6: Block Diagram: (a) 1-bit Half Adder (HA) and (b) 1-bit Full Adder (FA)

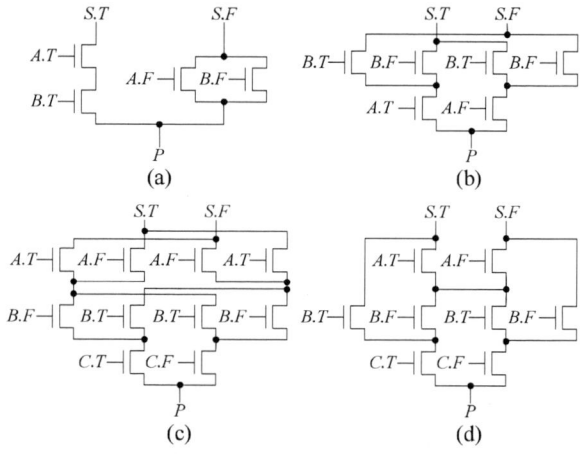

Fig. 7: NMOS Transistor Stacks: (a) 2-input AND/NAND, (b) 2-input XOR/XNOR, (c) 3-input XOR/XNOR, and (d) 3-input CARRY/ICARRY

978-1-61284-863-1/11 $26.00 © 2011 IEEE

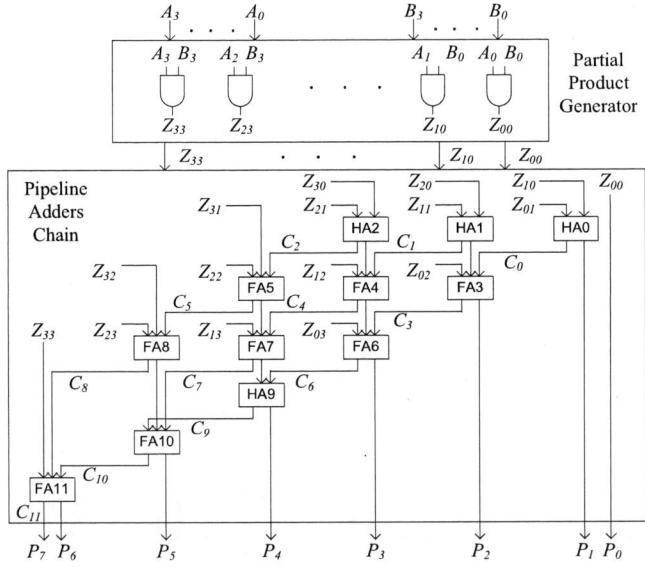

Fig. 8: 4×4-bit Multiplier Architecture

Fig. 6 depicts the block diagram of a 1-bit full adder (HA) and a 1-bit half adder (FA). Fig. 7 depicts the schematic for NMOS transistor stacks which perform the cell functionalities for 2-input AND/NAND, 2-input XOR/XNOR, 3-input XOR/XNOR and 3-input CARRY/ICARRY. Fig. 8 depicts the 4×4-bit multiplier architecture.

For a fair comparison between our proposed 4×4-bit multiplier and the reported design, both of them are implemented on 65nm CMOS technology process and simulated at $V_{DD} = 1V$ and input switching rate = 400MHz using Cadence Spectra and Synopsys Nanosim simulators. The minimum-size transistor sizing is used for all transistors (except for the weak inverters). Table I tabulates the forward delay (from *Data Input+* to *Lack-*), backward delay (from *Data Input-* to *Lack+*), total delay (sum of forward and backward delay), power dissipation, energy dissipation per-operation, energy-delay product and transistor count. Positive percentage (in improvement column) indicates how much our proposed design has outperformed the reported design and vice-versa.

TABLE I. COMPARISON BETWEEN PROPOSED DESIGN AND REPORTED PCHB 4×4-BIT MULTIPLIER (V_{DD} = 1.0V, 400MHz, 65NM CMOS)

4×4-bit Multiplier	Reported PCHB [11]	Proposed PCHB	Improvement (%)
Forward Delay (ns)	1.04	1.09	-5%
Backward Delay (ns)	0.44	0.47	-5%
Total Delay (ns)	1.48	1.56	-5%
Power (μW)	347	217	+37%
Energy/operation (pJ)	2.36	1.45	+39%
Energy-Delay (10^{-21} Js)	3.49	2.26	+35%
Transistor Count	3488	2752	+21%

From Table I, we remark that our proposed 4×4-bit multiplier features slightly longer delays of 5% in both forward delay and backward delay, resulting in 5% longer overall delay (cost). However, our proposed 4×4-bit multiplier dissipates 37% lower power and 39% lower energy per-operation respectively. Despite of 5% longer delay, our proposed design achieves a 35% improvement in terms of the energy-delay

product. Finally, our proposed design has 21% lesser transistor count.

Based on our analysis, the longer delay is attributed to the long series of the transistors between the power rails and output port in the integrated input/output completion detection circuit, and the using of $Q.T/Q.F$ (instead of $S.T/S.F$) for the inputs of OCD. View differently, since ICD and OCD are integrated into a single CMOS logic circuit, it reduces leakage current paths and circuit overheads (number of transistors), which in turn potentially lower the leakage power dissipation and overall transistor count. Besides, the reset logic in the completion detection circuit has contributed greatly to stabilize the initial output states, thereby removing the floating nodes, and lowering short-circuit power dissipation.

V. CONCLUSIONS

We have proposed a transistor-level integrated input/output completion detection circuit with improved power and area attributes in the asynchronous-logic QDI PCHB approach (as compared to the reported PCHB approach). We obtained these attributes by reducing leakage current paths and circuit overheads (number of transistors) in the completion detection circuit. The improved PCHB template features simpler completion detection circuit, which integrates the reset logic for initialization purpose. We have shown that by means of 4×4-bit pipeline multiplier (@1V, 400MHz, 65nm CMOS) simulations, our proposed PCHB design outperforms the reported PCHB design, by ~37% lower power dissipation, ~39% lower energy dissipation per-operation, ~35% better energy-delay product and ~21% lesser transistor count. The drawback of our design have shown an insignificant ~5% longer delay (slower speed).

REFERENCES

[1] International Technological Roadmap for Semiconductors

[2] J. Sparsø, and S. Furber, *Principle of Asynchronous Circuit Design: A System Perspective*. Norwell, MA: Kluwer Academic, 2001.

[3] P. A. Beerel, R. O. Ozdag, and M. Ferretti, *A Designer's Guide to Asynchronous VLSI*. Cambridge University Press, 2010.

[4] A. J. Martin, M. Nsytrom and C. G. Wong, "Three generations of asynchronous microprocessors," *IEEE Design & Test of Computers*, vol. 20, no. 6, pp. 9-17, 2003.

[5] T. Nanya, Y. Ueno, H. Kagotani, M. Kuwako, and A. Takamura, "TITAC: design of a quasi-delay-insensitive microprocessor," *IEEE Design & Test of Computers*, vol. 11, no. 2, pp. 50-63, Feb. 1994.

[6] M. Renaudin, P. Vivet, and F. Robin, "ASPRO-216: a standard-cell QDI 16-bit RISC asynchronous microprocessor," in Proc. *Symp. Advanced Research on Asynchronous Circuit Syst.*, pp.22-31, 1998.

[7] A. J. Martin, and M. Nsytrom, "Asynchronous techniques for system on-chip designs," *IEEE Proc.* vol. 94, no. 6, pp. 1089–1120, Jun. 2006.

[8] K.-S. Chong, B.-H. Gwee, and J. S. Chang, "Energy-efficient synchronous-logic and asynchronous-logic FFT/IFFT processors," *IEEE J. of Solid-State Circuits*, vol. 42, no. 9, pp. 2034–2045, Sep 2007.

[9] B.-H. Gwee, J. S. Chang, Y. Shi, C.-C. Chua, and K.-S. Chong, "A low-voltage micropower asynchronous multiplier with shift-add multiplication approach," *IEEE Trans. Circuits Syst. I: Regular Papers*, vol. 56, no. 7, pp. 1349–1359, Jul, 2009.

[10] R. O. Ozdag and P. A. Beerel, "High-speed QDI asychronous pipelines," in Proc. *Eighth International Symposium on Async Circuits and Systems*, pp.13-22, 2002.

[11] A. J. Martin *et al.*, "The design of an asynchronous MIPS R3000 microprocessor," in Proc. *Conf. Advance Research in VLSI*, pp. 164–18, 1997.

A Low-Cost and High-Throughput Architecture for H.264/AVC Integer Transform by Using Four Computation Streams

Yuan-Ho Chen[1], Tsin-Yuan Chang[2], and Chih-Wen Lu[1]

[1]Department of Engineering and System Science,
[2]Department of Electrical Engineering,
National Tsing Hua University, Hsinchu 30013, Taiwan, R.O.C.
Email: yhchen@larc.ee.nthu.edu.tw

Abstract—In this paper, a four paths H.264/AVC integer transform, which employs four computation paths to achieve a high throughput rate and is implemented by a using single one-dimensional (1-D) DCT core with one transpose memory (TMEM) to reduce the area cost, is proposed. The proposed 1-D integer transform can calculate first-dimensional (1st-D) and second-dimensional (2nd-D) transformations simultaneously in four parallel streams. The two-dimensional (2-D) integer transform utilizes a single 1-D transform core and one TMEM. Therefore, a high throughput rate and a low area cost are achieved in the proposed 2-D transform core. To evaluate the circuit performance of the proposed integer transform, the transform core is implemented in a TSMC 0.18-μm CMOS process. The proposed transform core can achieve a high throughput rate of 1 G-pels/s with only 17.7 K gate area.

Index Terms—Integer transform, H.264/AVC, four paths, DA-based, Simultaneous operation.

I. INTRODUCTION

The H.264/AVC is widely used standard in high definition video compression [1]-[3]. To eliminate the mismatching of the reconstructed image in encoder and decoder, ITU-T and ISO presented the integer transform applying in H.264/AVC. Therefore, the integer transform for H.264/AVC can avoid the mismatch problem.

Recently, many researcher aim to implement the H.264/AVC integer transforms [4]-[11]. Two methods are widely used to design the two-dimensional (2-D) transform core: direct 2-D [4]-[6] and row-column decomposition [7]-[11] methods. The Direct 2-D methods are expanded the 2-D transform as the matrix form and derived to a hardware sharing format. In the row-column decomposition method, the work in [7] derives a matrix to share the hardware resources by using two one-dimensional (1-D) core with a transposed memory (TMEM), which is shown in Fig. 1(a). Similar [7], many architectures use the structure in Fig. 1(a) to implement the integer transform -[9]. According to the cost consideration, the multiplexer 2-D structure shown as Fig. 1(b) is induced in [10]-[11]. The multiplexer controls the 1-D core computing first-dimensional (1st-D) or second-dimensional (2nd-D) operation. Thus, the throughput rate will be dropped off compared with the original speed.

(a) Parallel structure 2-D integer transform.

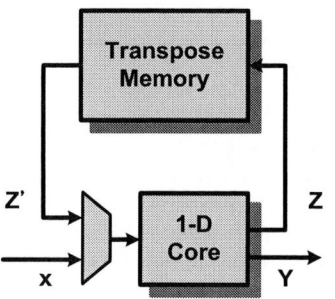

(b) Multiplexer structure 2-D integer transform.

Fig. 1. Structure of 2-D integer transform.

In this paper, a four path integer transform is proposed to achieve a high throughput rate and low area cost designs. The proposed transform core utilizes a single 1-D core with a TMEM in order to reduce the circuit area. The 1-D core can calculate 1st-D and 2nd-D transforms simultaneously during two-clock-cycle period in four parallel computation streams. For this reason, the MP-DCT can four fold throughput rate compared with the operated frequency. The implementation results show that the proposed integer transform core has a high hardware efficiency supporting H.264/AVC $8 \times 8 / 4 \times 4$ transforms.

This paper is organized as follows. In Section II, the mathematical derivation of the integer transform is given. The proposed 2-D integer transform architecture is introduced in Section III. Comparisons and discussions are presented in Section IV, and conclusions are drawn in Section V.

978-1-61284-863-1/11 $26.00 © 2011 IEEE

II. MATHEMATICAL DERIVATION OF H.264/AVC INTEGER TRANSFORM

A. 8 × 8 Integer Transform

The 2-D H.264/AVC 8×8 integer transform is defined as [1]:

$$\mathbf{Y}_8 = \mathbf{C}_8 \mathbf{x}_8 \mathbf{C}_8^T$$

where the 1-D 8-point integer transform can be expressed as follows:

$$\mathbf{Z}_8 = \mathbf{C}_8 \mathbf{x}_8 \qquad (1)$$

and

$$\mathbf{C}_8 = \begin{bmatrix} c_4 & c_4 & c_4 & c_4 & c_4 & c_4 & c_4 & c_4 \\ c_1 & c_3 & c_5 & c_7 & -c_7 & -c_5 & -c_3 & -c_1 \\ c_2 & c_6 & -c_6 & -c_2 & -c_2 & -c_6 & c_6 & c_2 \\ c_3 & -c_7 & -c_1 & -c_5 & c_5 & c_1 & c_7 & -c_3 \\ c_4 & c_4 & c_4 & c_4 & c_4 & c_4 & c_4 & c_4 \\ c_5 & -c_1 & c_7 & c_3 & -c_3 & -c_7 & c_1 & -c_5 \\ c_6 & -c_2 & c_2 & -c_6 & -c_6 & c_2 & -c_2 & c_6 \\ c_7 & -c_5 & c_3 & -c_1 & c_1 & -c_3 & c_5 & -c_7 \end{bmatrix}$$

$$\mathbf{x}_8 = \begin{bmatrix} x_0 & x_1 & x_2 & x_3 & x_4 & x_5 & x_6 & x_7 \end{bmatrix}^T$$

$c_1 = 12, c_2 = 8, c_3 = 10, c_4 = 8, c_5 = 6, c_6 = 4,$ and $c_7 = 3$. According to the symmetry feature of the coefficients, the 8-point integer transform in Eqn. (1) can be divided into even and odd parts : $\mathbf{Z}_e$ and $\mathbf{Z}_o$ as listed in Eqns. (2) and (3), respectively.

$$\mathbf{Z}_e = \begin{bmatrix} Z(0) \\ Z(2) \\ Z(4) \\ Z(6) \end{bmatrix} = \begin{bmatrix} c_4 & c_4 & c_4 & c_4 \\ c_2 & c_6 & -c_6 & -c_2 \\ c_4 & -c_4 & -c_4 & c_4 \\ c_6 & -c_2 & c_2 & -c_6 \end{bmatrix} \begin{bmatrix} a_0 \\ a_1 \\ a_2 \\ a_3 \end{bmatrix}$$
$$= \mathbf{C}_e \cdot \mathbf{a} \qquad (2)$$

$$\mathbf{Z}_o = \begin{bmatrix} Z(1) \\ Z(3) \\ Z(5) \\ Z(7) \end{bmatrix} = \begin{bmatrix} c_1 & c_3 & c_5 & c_7 \\ c_3 & -c_7 & -c_1 & -c_5 \\ c_5 & -c_1 & c_7 & c_3 \\ c_7 & -c_5 & c_3 & -c_1 \end{bmatrix} \begin{bmatrix} b_0 \\ b_1 \\ b_2 \\ b_3 \end{bmatrix}$$
$$= \mathbf{C}_o \cdot \mathbf{b} \qquad (3)$$

where

$$\mathbf{a} = \begin{bmatrix} a_0 \\ a_1 \\ a_2 \\ a_3 \end{bmatrix} = \begin{bmatrix} x_0 + x_7 \\ x_1 + x_6 \\ x_2 + x_5 \\ x_3 + x_4 \end{bmatrix}$$

$$\mathbf{b} = \begin{bmatrix} b_0 \\ b_1 \\ b_2 \\ b_3 \end{bmatrix} = \begin{bmatrix} x_0 - x_7 \\ x_1 - x_6 \\ x_2 - x_5 \\ x_3 - x_4 \end{bmatrix}. \qquad (4)$$

The even part $\mathbf{Z}_e$ can be further decomposed into even and odd parts:

$$\mathbf{Z}_{ee} = \begin{bmatrix} Z(0) \\ Z(4) \end{bmatrix} = \begin{bmatrix} c_4 & c_4 \\ c_4 & -c_4 \end{bmatrix} \begin{bmatrix} A_0 \\ A_1 \end{bmatrix}$$
$$= \mathbf{C}_{ee} \cdot \mathbf{A} \qquad (5)$$

$$\mathbf{Z}_{eo} = \begin{bmatrix} Z(2) \\ Z(6) \end{bmatrix} = \begin{bmatrix} c_2 & c_6 \\ c_6 & -c_2 \end{bmatrix} \begin{bmatrix} B_0 \\ B_1 \end{bmatrix}$$
$$= \mathbf{C}_{eo} \cdot \mathbf{B} \qquad (6)$$

where

$$\mathbf{A} = \begin{bmatrix} A_0 \\ A_1 \end{bmatrix} = \begin{bmatrix} a_0 + a_3 \\ a_1 + a_2 \end{bmatrix}$$

$$\mathbf{B} = \begin{bmatrix} B_0 \\ B_1 \end{bmatrix} = \begin{bmatrix} a_0 - a_3 \\ a_1 - a_2 \end{bmatrix}. \qquad (7)$$

B. 4 × 4 Integer/Hadamard Transform

The 2-D H.264/AVC 4×4 integer and Hadamard transform is defined as [1]:

$$\mathbf{Y}_4 = \mathbf{C}_4 \mathbf{x}_4 \mathbf{C}_4^T \qquad (8)$$

where

$$\mathbf{C}_4 = \begin{bmatrix} c_4 & c_4 & c_4 & c_4 \\ c_2 & c_6 & -c_6 & -c_2 \\ c_4 & -c_4 & -c_4 & c_4 \\ c_6 & -c_2 & c_2 & -c_6 \end{bmatrix} \qquad (9)$$

and

$$\begin{cases} c_2 = 2, c_4 = 1, c_6 = 1 & \text{as Interger Tranform} \\ c_2 = 2, c_4 = 1, c_6 = 1 & \text{as Hadamard Tranform} \end{cases}.$$

Similar to the even part of 8-point in (2), the 1-D 4-point integer/Hadamard transform can be derived as the format of the (5) and (6). For this reason, the hardware resources can be shared in H.264/AVC 8-point and 4-point integer transforms.

III. PROPOSED FOUR-PATH TRANSFORM CORE

The proposed 2-D integer transform core consists of one 1-D transform core and one TMEM to achieve low-cost design. The 1-D transform core includes a 4P-MBF2, a DAE, a DAO, eight ECATs, and a Post-Reorder modules. The architecture of the proposed 2-D integer transform is illustrated in Fig 2, and each module will be described in detail in the following.

A. 4P-MBF2

The proposed 4P-MBF2 employs four MBF2 modules [12], and each MBF2 has two one-word registers, one adder, and one subtracter. Due to the four computation paths in four MBF2s, the proposed 4P-MBF2 can calculate data in four fold rate compared with the operation frequency of the transform core. Furthermore, the 4P-MBF2 can compute 8-point 1st-D and 2nd-D transformation data in each two-clock-cycle period. For this reason, the 1st-D and 2nd-D transform will be changed in each cycle, and the proposed 4P-MBF2 can execute four 1st-D and 2nd-D transformation simultaneously.

B. DAE and DAO

The DAE is implemented (2) by using adder-based DA architecture. According to the DA sharing strategy, six adders are used in the DAE module. Similarly, the DAO utilizes 11 adders to implement the function of the odd part of the 8-point integer transform (3) and 4-point interger/Hadamard transform (9). Therefore, the throughput rate of 4-pint DCT computation can achieve the same as that of 8-point DCT computation.

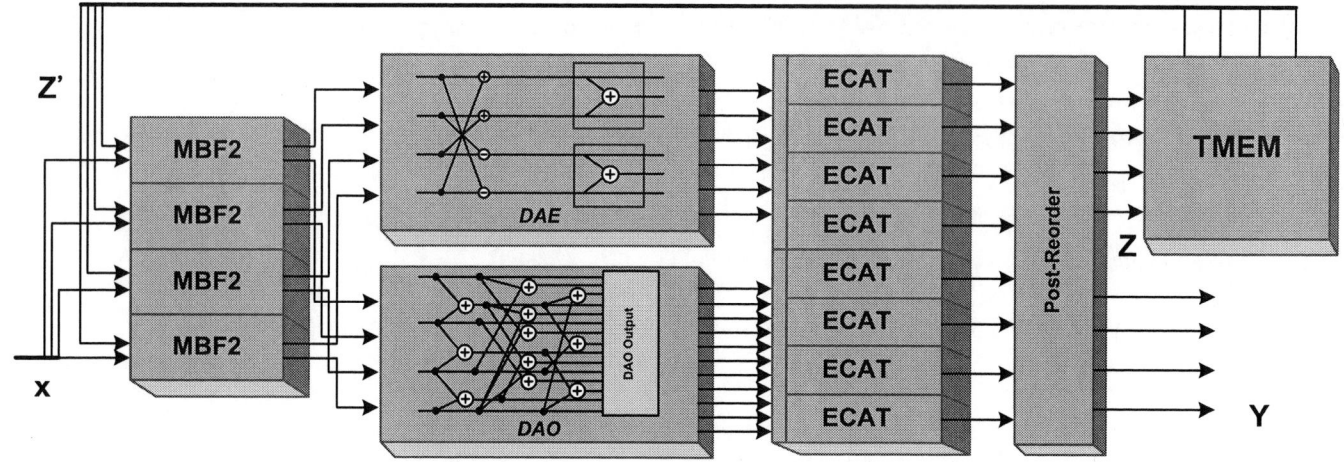

Fig. 2. Architecture of the proposed 2-D 4P-DCT.

C. ECATs

Eight ECATs are followed the DAE and DAO modules. Each ECAT is designed by using error-compensated method to aim the goal of high-accuracy, low-cost, and high-throughput rate.

D. Post-Reorder

Due to the two-clock-cycle period in 1st-D and 2nd-D transformation, the 1st-D $\mathbf{Z}_e$ and 2nd-D $\mathbf{Z}_o$ are obtained in the first cycle, and the 1st-D $\mathbf{Z}_o$ and 2nd-D $\mathbf{Z}_e$ are obtained in the next cycle. Therefore, the Post-Reorder module can permutates the output data from ECATs in normal order.

E. TMEM

TMEM transposes the 1st-D output data $\mathbf{Z}$ to 2nd-D input data $\mathbf{Z}'$. There are 64 12-bit registers with transposed control signal in the TMEM.

In conclude, the proposed 2-D integer transform uses one 1-D transform core and one TMEM to reduce the area cost and achieve a high-throughput rate based on four parallel computation paths.

IV. DISCUSSION AND COMPARISONS

A. Comparison With Other 2-D Integer Transform Works

Table I compares the proposed 2-D integer transform core with the existing works. In [5] and [6], the direct 2-D method is presented to implement the 2-D transform core. Lee *et al.* use permutation matrices to implement the 2-D integer transform core [5]. However, the hardware cost is large due to the multiplier-based structure. The high-performance core is presented by expanding 4×4 transform as an eight pixels per cycle in [6]. However, the 2-D core only supports 4×4 transform in H.264/AVC standard. The distributed arithmetic (DA)-based row-column decomposition methods are introduced in [8] and [9]. Due to the two transform cores are used in the 2-D core, the circuit area will consume large hardware cost. A low cost multiplexer 2-D integer transform is implemented by

using a single 1-D core with a TMEM. Thus, a high hardware efficiency is achieved in [10]. The hardware efficiency is defined as throughput rate divided by circuit gate counts.

$$\text{Hardware Efficiency} \left(10^3 \text{ pels/sec-gate}\right) = \frac{\text{Throughput Rate}}{\text{Gate Counts}} \quad (10)$$

The proposed 2-D integer transform has a high hardware efficiency supporting 8×8 and 4×4 integer transform in the comparison Table I. Therefore, the proposed 2-D integer transform achieves a high throughput rate with low cost design.

B. Chip Implementation

In order to verify the proposed architecture, the proposed 2-D integer transform is implemented in a 1.8 V TSMC 0.18-μm 1P6M CMOS process by using a Synopsys Design Compiler to synthesize the RTL code and a Candence SoC Encounter for placement and routing (P&R). The implementation results show that 2-D transform core has a high throughput rate of 1 G pixels per sec when operated at 250 MHz $(250 \times 4 = 1000)$. The core layout and simulated characteristics of the proposed 2-D transform core are shown in Fig. 3 and Table II, respectively. The Test Module (TM) in the layout block will test the proposed 2-D transform core. In test mode, the external data can input and output in the TM serially. The TM feeds the input data to the proposed transform core in parallel, and captures the transform paralleled output data in the function mode.

V. CONCLUSION

In this paper, a low-cost and high-throughput 2-D integer transform core for H.264/AVC is proposed. By executing four computation streams, the throughput rate of the proposed transform core can be improve to four fold of the clock frequency. Furthermore, the 2-D transform core employs only one 1-D transform core and one TMEM to reduce the area cost. For this reason, the proposed 2-D integer transform can achieve 1 G-pels/s throughput rate with only 17.7 K gate area as implemented in a TSMC 0.18-μm process. In summary,

TABLE I
A COMPARISON OF DIFFERENT 2-D INTEGER TRANSFORM ARCHITECTURES WITH THE PROPOSED ARCHITECTURE

		Lee *et al.* [5]	Chen *et al.* [6]	Huang *et al.* [8]	Chen *et al.* [9]	Chao *et al.* [10]	**Proposed**
Method		Direct 2-D	Direct 2-D	Two 1-D DCTs + TMEM	Two 1-D DCTs + TMEM	Single 1-D DCT + TMEM	**Single 1-D DCT + TMEM**
Gate Counts		36.6 K	6.5 K	39.8 K	15.2 K (1-D)*	18.5 K	**17.7 K**
Technology		0.13 μm	0.18 μm	0.18 μm	0.13 μm	0.18 μm	**0.18 μm**
Operation Freq.		103 MHz	100 MHz	50 MHz	200 MHz	125 MHz	**250 MHz**
Processing Rate	8×8	32/15	$\times$	8	1	64/16	**4**
(pels/cycles)	4×4	48/13	8	8	1	16/2	**4**
Throughput Rate	8×8	218-M	$\times$	400-M	200-M	500-M	**1-G**
(pels/sec)	4×4	380-M	800-M	400-M	200-M	1-G	**1-G**
Support Transform	8×8	○	$\times$	○	○	○	○
	4×4	○	○	○	○	○	○
Hardware Efficiency	8×8	5.96K	$\times$	10.1K	13.2K	27.3K	**56.5K**
(10^3 pels/sec-gate)	4×4	10.4K	123.4K	10.1K	13.2K	54.1K	**56.5K**
Power		N/A	24.2mW	39mW	N/A	N/A	**54 mW**

* The gate counts result is estimated for 1-D core.

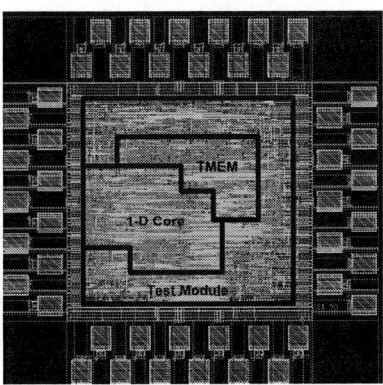

Fig. 3.   Core layout of the proposed 2-D integer transform core.

TABLE II
CHIP CHARACTERISTICS OF THE PROPOSED 2-D INTEGER TRANSFORM CORE

Process Technology	0.18-μm CMOS, 1P6M.
Supply Voltage	1.8 V
Max. Clock Frequency	250 MHz
Core Area	685 $\times$ 677 μm^2
Gate Counts	17.7 K
Power Consumption	53.9 mW for 8×8 Integer
	52.8 mW for 4×4 Integer
	53.4 mW for 4×4 Hadamard
Throughput Rate	1 G-pels/s

the proposed architecture is suitable for high-throughput and low-cost applications in VLSI designs.

ACKNOWLEDGMENT

The authors would like to thank the National Chip Implementation Center (CIC), Taiwan, for providing the circuit design automation tools.

This work was supported in part by the National Science Council under project number NSC 100-2221-E-007-992.

REFERENCES

[1] I. E. G. Richardson, *H.264 and MPEG-4 Video Compression: Video Coding for Next-generation Multimedia.* England: John Wiley & Sons, 2003.

[2] T. Wiegand, G. J. Sullivan, G. Bjontegaard, and A. Luthra, "Overview of the H.264/AVC video coding standard," *IEEE Trans. Circuits Syst. Video Technol.*, vol. 13, no. 7, pp. 560–576, Jul. 2003.

[3] H. Schwarz, D. Marpe, and T. Wiegand, "Overview of the scalable video coding extension of the H.264/AVC standard," *IEEE Trans. Circuits Syst. Video Technol.*, vol. 17, no. 9, pp. 1103–1120, Sep. 2007.

[4] Z. Y. Cheng, C. H. Chen, B. D. Liu, and J. F. Yang, "High throughput 2-D transform architectures for H.264 advanced video coders," in *Proc. IEEE Asia-Pacific Conf. Circuits Syst.*, 2004, pp. 1141–1144.

[5] S. Lee and K. Cho, "Design of high-performance transform and quantization circuit for unified video codec," in *Proc. IEEE Asia-Pacific Conf. Circuits Syst.*, Nov. 2008, pp. 1450–1453.

[6] K. H. Chen, J. I. Guo, and J. S. Wang, "A high-performance direct 2-D transform coding IP design for MPEG-4AVC/H.264," *IEEE Trans. Circuits Syst. Video Technol.*, vol. 16, no. 4, pp. 472–483, Apr. 2006.

[7] G. A. Su and C. P. Fan, "Low-cost hardware-sharing architecture of fast 1-D inverse transforms for H.264/AVC and AVS applications," *IEEE Trans. Circuits Syst. II*, vol. 55, no. 12, pp. 1249–1253, Dec. 2008.

[8] C. Y. Huang, L. F. Chen, and Y. K. Lai, "A high-speed 2-D transform architecture with unique kernel for multi-standard video applications," in *Proc. IEEE Int. Symp. Circuits Syst.*, 2008, pp. 21–24.

[9] J. W. Chen, K. H. Chen, J. S. Wang, and J. I. Guo, "A performance-aware IP core design for multi-mode transform coding using scalable-DA algorithm," in *Proc. IEEE Int. Symp. Circuits Syst.*, 2006, pp. 1904–1907.

[10] Y. C. Chao, H. H. Tsai, Y. H. Lin, J. F. Yang, and B. D. Liu, "A novel design for computation of all transforms in H.264/AVC decoders," in *Proc. IEEE Int. Conf. Multimedia Expo*, Jul. 2007, pp. 1914–1917.

[11] G. Pastuszak, "Transforms and quantization in the high-throughput H.264/AVC encoder based on advanced mode selection," in *Proc. IEEE Comput. SoC. Annu. Symp. VLSI.*, Apr. 2008, pp. 203–208.

[12] Y. H. Chen and T. Y. Chang, "A high performance video transform engine by using space-time scheduling strategy," *IEEE Trans. VLSI Syst.*, 2011, to be published.

Design of Support Vector Machine Circuit for Real-time Classification

Soojin Kim , Seonyoung Lee*, Kyoungwon Min* and Kyeongsoon Cho
Department of Electronics and Information Engineering
Hankuk University of Foreign Studies
*Convergent SoC Research Center
Korea Electronics Technology Institute
{ksjsky9888, kscho}@hufs.ac.kr, *{drleesy, minkw}@keti.re.kr

Abstract—**This paper describes the design of support vector machine circuit for real-time classification. By unifying the algorithms and architectures of linear and non-linear SVM classifications, the proposed circuit can support both linear and non-linear classifications. The circuit size is minimized by sharing most of the resources required in the computation for both classification types. The sliding window size of 64x64 or 48x96 with 18 window strides is applied to the proposed circuit for efficient classification. Since the proposed circuit can process up to 31 640x480 image frames per second, it is sufficient to be used for real-time classification. The synthesized circuit using 65nm standard cell library consists of 654,435 gates and its maximum operating frequency is 178MHz.**

Keywords-support vector machine; unified; real-time; pattern recognition; classification

I. INTRODUCTION

In general pattern recognition, the centers of gravity in two groups are used to determine the optimal hyper-plane for binary classification. The main disadvantage of this method is that the possibility of misclassification for new data is high. Therefore, Vladimir Vapnik and AT&T Bell laboratory team proposed support vector machine (SVM)[1] to improve the accuracy of classification.

Instead of considering the centers of gravity in two groups, SVM algorithm focuses on the support vectors which are on the boundaries of each group as shown in Fig. 1. The unique feature of SVM is that it determines the optimal hyper-plane by setting some limitations. The distance between two boundaries must be the maximum and there should not be any data between the boundaries for accurate classification. By using the designated optimal hyper-plane, SVM can provide great performance on the pattern recognition and classification.

SVM has been employed for several applications such as ultra wide band (UWB) channel equalization, channel estimation in orthogonal frequency division multiplexing (OFDM) systems, voice activity detection, and target recognition[2]. Since SVM is considered to be the state-of-the-art tool for binary classifications, various algorithms and architectures of SVM circuit have been proposed[2-9].

SVM algorithm is comprised of two steps: SVM learning and SVM classification. Support vectors which make the maximum margin between two groups are found and the optimal hyper-plane is determined in SVM learning procedure. It is important that the optimal hyper-plane is determined to provide the best performance with the minimum errors for efficient classification. By using the designated optimal hyper-plane, newly incoming data is classified into one of two groups in SVM classification procedure. SVM classification should be performed in real time on newly obtained data for efficient classification[3][4]. Therefore, it is important to increase the operating speed of SVM classification, in which large amount of computation is required.

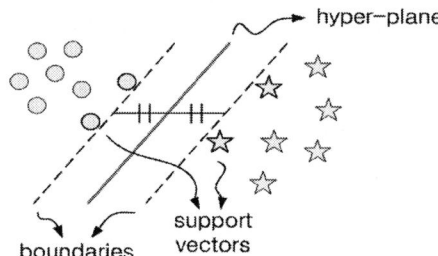

Figure 1.   SVM classification using optimal hyper-plane

This paper proposes a unified SVM circuit for real-time classification. The proposed circuit can support both linear and non-linear classifications by unifying the algorithms and architectures for both types. To minimize the proposed circuit size, most of the resources required for linear and non-linear SVM classifications are shared. By adopting parallel architecture to accelerate the operating speed, the proposed SVM circuit processes up to 31 640x480 image frames per second. The 48x96 or 64x64 sliding windows with 18 window strides are applied for efficient classification.

II. SVM CLASSIFICATION

In binary SVM classification, an object is classified into one of two groups by the optimal hyper-plane which is determined in advance during SVM learning process.

978-1-61284-863-1/11 $26.00 © 2011 IEEE

A. Linear SVM

Linear SVM is used when the two groups can be classified linearly as shown in Fig. 1. The optimal hyper-plane used for linear SVM classification is determined when d(x) in (1) is zero. In this equation, X represents support vector, Y represents feature of the object, i represents the dimension for data, and b represents bias.

By applying newly incoming data to (1), new data can be classified into one of two groups according to the value of d(x), which means the distance between optimal hyper-plane and the data. For example, new data is classified into class A if d(x) is lower than zero. Otherwise, new data is classified into class B.

$$d(x) = X_i^T Y_i + b \qquad (1)$$

B. Non-linear SVM

The optimal hyper-plane for non-linear SVM classification is determined when d(x) in (2) is zero, and the newly incoming data is classified into one of two groups according to the value of d(x). Non-linear SVM is used when it is difficult to classify the new data in linear way. However, the data can be classified linearly by mapping the present dimensions to the higher dimensions. Although the classification is able to be performed linearly by making the dimensions higher, the amount of required computation becomes enormous burden. To resolve this problem, kernel trick is adopted in non-linear SVM as shown in (2). In this equation, svnum represents the number of support vector data, α represents Lagrange multiplier, and y is the parameter used in Lagrange function. Lagrange function is used to determine the optimal-hyper plane in SVM learning procedure.

$$d(x) = \sum_{N=1}^{svnum} \alpha y K(X_i, Y_i) + b \qquad (2)$$

Kernel trick is used when vector data are required in inner product operations. By using the kernel trick, the effect for computation in high dimension is achieved in non-linear SVM. The general kernel functions in (2) for non-linear SVM classification are polynomial kernel, radial basis function (RBF) kernel, and sigmoid kernel. Equation (3) is RBF kernel that is typically used for non-linear SVM classification. σ in (3) is the width of Gaussian window.

$$K(X_i, Y_i) = e^{-\|X_i - Y_i\|^2 / 2\sigma^2} \qquad (3)$$

III. PROPOSED SVM CIRCUIT

The proposed linear and non-linear SVM circuit for real-time classification uses support vector data and feature data with 3,780 dimensions. Histograms of oriented gradients (HOG)[10] are used for extracting the features of the object in the proposed SVM circuit. A 48x96 or 64x64 sliding window

and 18 window strides are applied to each image frame for efficient classification.

Fig. 2 shows the proposed unified SVM circuit for real-time classification. It operates in linear SVM mode when 'kernel_type' is 0 and in non-linear SVM mode when 'kernel_type' is 1. The proposed circuit is based on the parallel with pipeline architecture and it processes 112 dimensions for one data in one clock cycle. Since the proposed SVM circuit uses input data with 3,780 dimensions, 34 clock cycles are required for one pair of support vector and HOG feature input data. 'BUF #1' in Fig. 2 is used to accumulate the results of 'Unified Inner Product Calculator' circuit for 34 clock cycles.

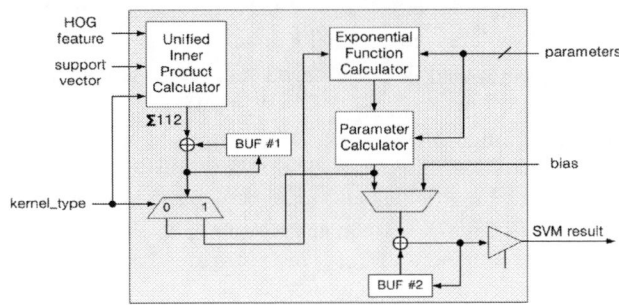

Figure 2. Proposed unified SVM circuit

While the operations for kernel function and multiplications with parameters are not included in linear SVM, they are required in non-linear SVM as shown in (2). In proposed SVM circuit, 'Exponential Function Calculator' and 'Parameter Calculator' circuits are responsible for the operations in (2) and (3). The results of operations for 3,780 dimensions are not necessary to be accumulated in linear SVM. In non-linear SVM, however, the results of operations for 3,780 dimensions should be accumulated for the number of svnum as shown in (2). 'BUF #2' in Fig. 2 is used for this accumulation in non-linear SVM.

A. Unified inner product calculator circuit

As shown in (1) and (3), inner product operations for vector data are necessary for both linear and non-linear SVM classifications. Inner product equation (4) is used in linear SVM, and the Euclidean distance equation (5) is used in non-linear SVM. To share the required resources in proposed circuit, we rearranged (5) into (6) for non-linear SVM classification.

$$X_i^T Y_i = X_i \bullet Y_i = X_0 \bullet Y_0 + X_1 \bullet Y_1 + \ldots\ldots + X_N \bullet Y_N \qquad (4)$$

$$\| X_i - Y_i \|^2 = X_i \bullet X_i - 2X_i \bullet Y_i + Y_i \bullet Y_i \qquad (5)$$

$$\| X_i - Y_i \|^2 = X_i \bullet X_i + Y_i(Y_i - 2X_i) \qquad (6)$$

Fig. 3 shows the proposed 'Unified Inner Product Calculator' circuit using parallel architecture with two-stage

978-1-61284-863-1/11 $26.00 © 2011 IEEE

pipeline. As shown in the figure, the proposed 'Unified Inner Product Calculator' circuit shares multipliers and adders to support both linear and non-linear SVM classifications. The adders in the first pipeline stage are only used for non-linear SVM classification.

By adopting parallel architecture with two-stage pipeline, the proposed circuit can process large amount of operations quickly. The proposed circuit processes 112 dimensions out of 3,780 for one data in one clock cycle to accelerate the operating speed. Since three multipliers and two adders are required in (5), 336 (112x3) multipliers and 224 (112x2) adders are necessary to process 112 dimensions in one clock cycle. Because support vectors are already determined in SVM learning procedure, the values for inner product of ($X_i \bullet X_i$) can be pre-computed. We rearranged (5) into (6) to exclude the inner product operation for ($X_i \bullet X_i$) and applied it to the proposed 'Unified Inner Product Calculator' circuit. Since the equation requires only one multiplier and adder except ($X_i \bullet X_i$), 112 multipliers and adders are used in the proposed circuit. Furthermore, most of the resources required for both linear and non-linear SVM classifications are shared as shown in Fig. 3 to decrease the circuit size. In this way, the circuit size is significantly reduced. The computation results of 'Unified Inner Product Calculator' circuit are accumulated for 34 clock cycles.

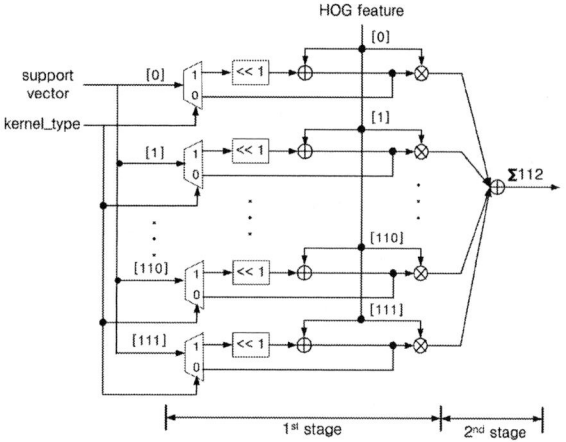

Figure 3.   Proposed unified inner product calculator circuit

B. Exponential function calculator circuit

The result of the Euclidean distance operation is divided with $2\sigma^2$, and then it is used to calculate the kernel function in (3). Table-driven algorithm was proposed in [11] and we adopted it to proposed 'Exponential Function' circuit for efficient calculation. Table-driven algorithm accelerates the operation speed for exponential function in which fixed-point numbers are required. Since the table-driven algorithm can increase not only the operation speed but also the accuracy, it has been widely used in fast operation for fixed-point numbers.

Fig. 4 shows the proposed 'Exponential Function' circuit. The operation of exponential function for one data requires four clock cycles, and the result of proposed 'Exponential

Function' circuit is used in 'Parameter Calculator' circuit in non-linear SVM mode.

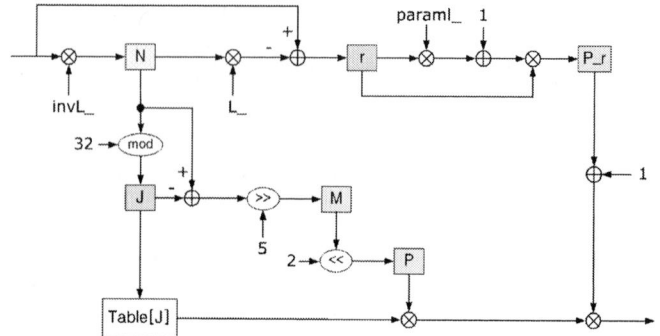

Figure 4.   Proposed exponential function calculator circuit

C. Parameter calculator circuit

Fig. 5 shows the proposed 'Parameter Calculator' circuit, and it operates only when 'kernel_type' is 1, i.e. non-linear SVM mode. 'Parameter Calculator' circuit is used to multiply the kernel function, α, and y in (2). The result of kernel function can be obtained after the operations for inner product and exponential function, so the multiplication with kernel function should be performed afterward. However, α and y can be multiplied with each other before obtaining the result of kernel function. They are multiplied in advance to increase the operating speed in the proposed circuit. Although three multipliers are required for the operations, we used only one multiplier as shown in Fig. 5. The pre-computed value for R_2 in the register is used in the final multiplication after the kernel function operation is finished.

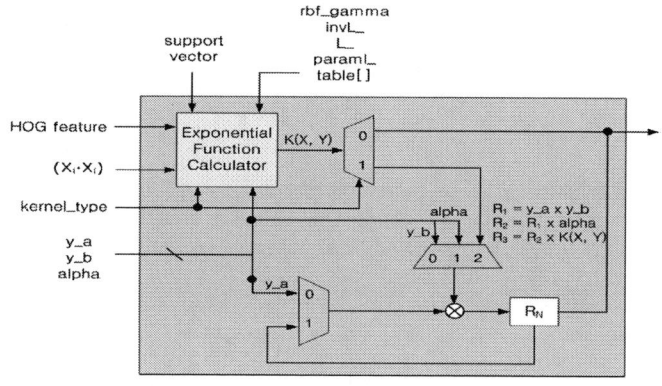

Figure 5.   Proposed parameter calculator circuit

IV.   EXPERIMETAL RESULTS

We described the proposed unified SVM circuit for real-time classification using Verilog hardware description language (HDL) and synthesized the gate-level circuits using 65nm standard cell library. Fig. 6 shows the timing diagram of the proposed SVM circuit when the circuit is processing one sliding window. As shown in the figure, 7,249 clock cycles are required for the non-linear SVM classification in which 213 input data with 3,780 dimensions per sliding window are required to be processed. Therefore, the proposed unified SVM circuit is sufficient to process linear SVM classification, in which only 37 clock cycles per sliding window are required.

Table I shows the synthesized results and performance of the proposed SVM circuit. The synthesized circuit consists of 654,435 gates and its maximum operating frequency is 178MHz.

792 sliding windows per one 640x480 image frame are required to be processed since 48x96 or 64x64 sliding window size with 18 window strides is applied. We considered that the sliding window moves to both of horizontal and vertical directions. Since it requires 7,249 clock cycles to process one sliding window, the proposed unified SVM circuit can process up to 31 640x480 image frames per second, and newly incoming data can be classified in real-time.

TABLE I.   SYNTEHSIS RESULTS

Image size	640x480	# of clock cycles /sliding window	7,249 cycles
Sliding window size	64x64 48x96	Maximum delay	5.62 ns
Window strides	18	Maximum operating frequency	178 MHz
# of sliding windows/frame	792	Gate count	645,435

V.   CONCLUSIONS

Because of the great performance on the pattern recognition, SVM has been widely used for binary classification. This paper proposed unified SVM circuit for real-time classification. The proposed circuit supports both linear and non-linear SVM classifications by unifying their algorithms and architectures. The unification in the proposed SVM circuit is achieved by sharing most of circuit resources such as adders and multipliers to minimize the circuit size. Parallel architecture with two-stage pipeline is adopted to accelerate the processing speed to handle large amount of data for real-time classification. Two kinds of sliding window sizes are applied to the proposed SVM circuit for efficient classification. The proposed SVM circuit for real-time classification can process up to 31 640x480 image frames per second when 65nm standard cell library is used.

Therefore, the proposed unified SVM circuit can be used for real-time classification.

ACKNOWLEDGMENT

This work was supported by Ministry of Knowledge Economy (MKE) and IDEC Platform center (IPC) and the grant from the industrial technology development program (KI002162) of the Ministry of Knowledge Economy (MKE) of Korea.

REFERENCES

[1] Vapnik, V. N., *Statistical Learning Theory*, John Wiley & Sons, 1998.

[2] Gomes Filho, J., Raffo, M., Strum, M., and Wan Jiang Chau, " A General-purpose Dynamically Reconfigurable SVM," 2010 VI southern Programmable Logic Conference, pp. 107-112, Mar. 2010.

[3] Papadonikolakis, M. and Bouganis, C., "A Novel FPGA-based SVM Classifier," Proc. of IEEE International Conference on Filed-programmable Technology, pp. 283-286, Dec. 2010.

[4] Kyrkou, C., and Theocharides, T., "Scope: Towards a Systoloc Array for SVM Object Detection," IEEE Embedded Systems Letters, vol. 1, no. 2, pp. 46-49, Aug. 2009.

[5] Demir, B. and Erturk, S., "Improving SVM Classification Accuracy using a Hierarchical Approach for Hyperspectral Images," Proc. of IEEE International Conference of Image Processing, pp. 2849-2852, Nov. 2009.

[6] Hsu, C. F., Mong-Kai Ku, and Li-Yen Liu, "Support Vector Machine FPGA Implementation for Video Shot Boundary Detection Application," Proc. of IEEE International SoC Conference, pp. 239-242, Sep. 2009.

[7] Ruiz-Llata, M., Guarnizo, G., and Yebenes-Calvino. M., "FPGA Implementation of a Support Vector Machine for Classification and Regression," Proc. of International Joint Conference on Neural Networks, pp. 1-5, Jul. 2010.

[8] Irick, K., DeBole, M., Narayanan, V., and Gayasen, A., "A Hardware Efficient Support Vector Machine Architecture for FPGA," Proc. of International Symposium on Filed-programmable Custom Computing Machines, pp. 304-305, Apr. 2008.

[9] Manikandan, J. and Venkataramani, B., "Design of a Modified One-against-all SVM Classifier," IEEE International Conference on Systems, Man and Cybernetics, pp. 1869-1874, Oct. 2009.

[10] Dalal, N. and Triggs, B., "Histograms of Oriented Gradients for Human Detection," IEEE Computer Society Conference on Computer Vision and Pattern Recognition, vol. 1, pp. 886-893, Jun. 2005.

[11] Tang, P. T. P., "Table-driven Implementation of the Exponential Function in IEEE Floating-point Arithmetic," ACM Trans. on Mathematical Software, vol. 16, no. 2, pp. 144-157, Jun. 1989.

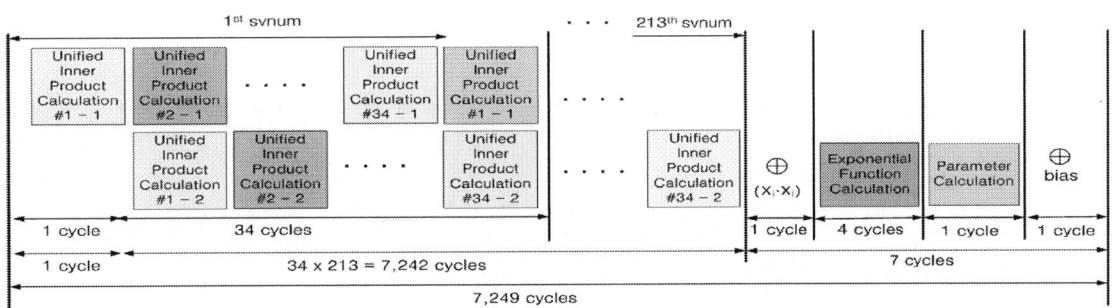

Figure 6.   Timing diagram of proposed SVM circuit

An All-Digital DLL with Dual-Loop Control for Multiphase Clock Generator

Yu-Lung Lo, Pei-Yuan Chou, Hsiang-Hui Cheng, Shu-Fen Tsai*, and Wei-Bin Yang**

Department of Electronic Engineering
National Kaohsiung Normal University
Kaohsiung, Taiwan, R.O.C.
Email: yllo@nknu.edu.tw

*Department of Microelectronics Engineering
National Kaohsiung Marine University
Kaohsiung, Taiwan, R.O.C.
Email: 981544110@mail.nkmu.edu.tw

**Department of Electrical Engineering
Tamkang University
Tamsui, Taipei, Taiwan, R.O.C.
Email: robin@ee.tku.edu.tw

Abstract—This paper describes a low power and low area multiphase digital DLL. The architecture of the proposed DLL uses coarse tune loop and fine tune loop to reduce the static phase error and accomplish faster locking time. The DLL was designed using a 0.35 μm standard CMOS process with a 3.3V supply voltage. Simulation results show that the proposed DLL can generate four-phase clock signals ranging from 320 MHz to 500 MHz within a single cycle. At 500 MHz, the peak-to-peak jitter is 6 ps and the total power consumption is 28.3 mW. Moreover, the proposed DLL's locking time is less than 24 clock cycles and the core area is 0.17 mm².

Keywords-Digital DLL, Dual-Loop Control, Multiphase

I. INTRODUCTION

In high-integration density very large scale integration (VLSI) system and high speed memory system, clock signal is usually the most important part of the operation of the whole circuit. However, the clock signal generated from external oscillator is sent into the integrated circuit or the chip through bond wires, thus introducing the clock skew. On the situation of asynchronization, the circuit can't read the data in time and thus can't operate correctly. Therefore, the phase-locked loop (PLL) and delay-locked loop (DLL) have been widely adopted to eliminate clock signal skew and jitter. Compared with the PLL, the DLL is often used because of design simplicity, higher loop stability, small area and better jitter performance [1]. Generally, DLL can be classified into two categories, digital DLL [2]–[8] and analog DLL [9], [10]. Digital DLL is preferred in modern design because of small chip area, locking time, power dissipation compares with analog DLL. Moreover, the digital DLL is less process variation-sensitive, and can be migrated easily to future process technologies. Fig. 1 shows the simplified block diagram of the conventional digital DLL. It is composed of a digital-controlled delay line, a phase detector, and a control unit. The phase detector compares the input and output clock signals. The control unit adjusts the digital-controlled delay line to adjust the delay time of the output signal according to the output of phase detector. If the output clock (Out_clk) leads the input reference clock (Ref_clk), the delay time will be increased, and the phase detector detects the Out_clk signal and the Ref_clk

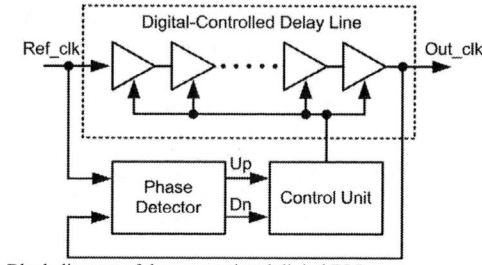

Fig. 1. Block diagram of the conventional digital DLL.

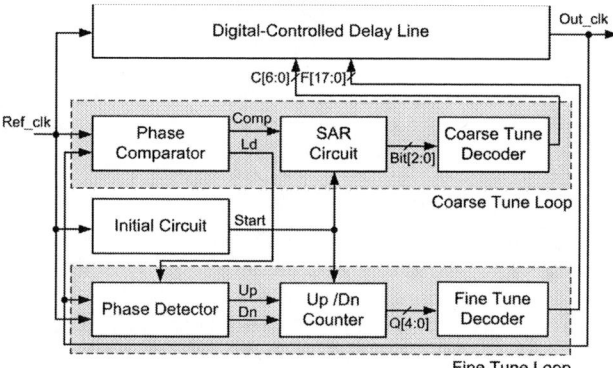

Fig. 2. The architecture of the proposed DLL.

signal repeatedly until the two signals are synchronized. Conventional DLLs may suffer from the problem of false lock and harmonic lock. Therefore, the proposed digital DLL uses the dual-loop architecture with initial circuit to overcome the limited range problem and to improve the locking time.

II. ARCHITECTURE AND OPERATING PRINCIPLE OF PROPOSED DLL

The architecture of the proposed DLL is shown in Fig. 2. It is composed of the digital-controlled delay line (DCDL), the coarse tune loop (CT Loop), the fine tune loop (FT Loop), and the initial circuit (IC). The CT loop comprises a phase

978-1-61284-863-1/11 $26.00 © 2011 IEEE

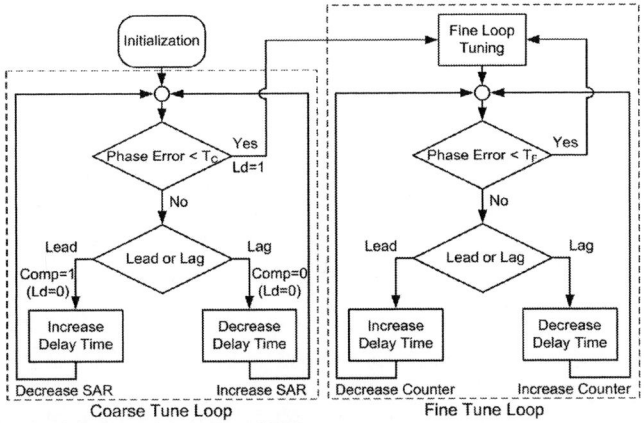

Fig. 3. Flow chart of the proposed DLL.

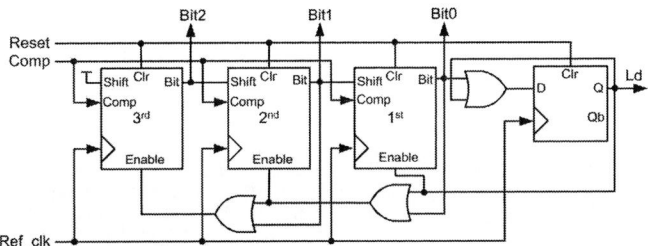

Fig. 4. The block diagram of 3-bits SAR.

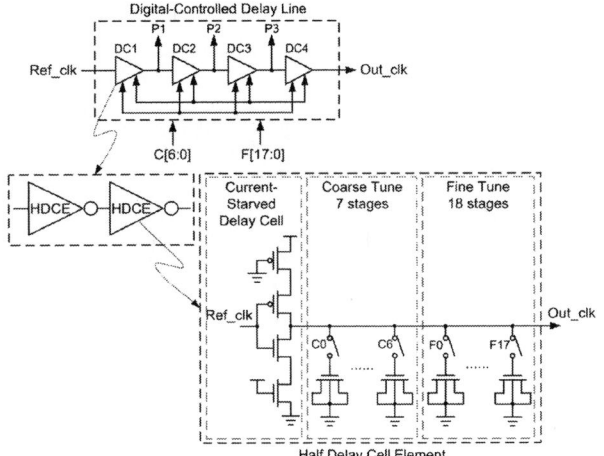

Fig. 5. The structure of digital-controlled delay line.

comparator (PC), a successive approximation register (SAR) circuit, and a coarse tune decoder (CT Decoder). The FT loop comprises a phase detector (PD), an up/dn counter, and a fine tune decoder (FT Decoder). The initial circuit sets the initial value of the SAR and up/dn counter to medium value. The DCDL, which is made up of four identical delay cell, can generate equally spaced four-phase clocks in just one cycle.

Fig. 3 shows the flow chart of the proposed DLL. After initialization, the Ref_clk passes through the delay line and becomes Out_clk. The delay time of the DCDL is initially set to medium value and the CT loop is activated. The CT loop is designed to reduce locking time and the phase error between Ref_clk and Out_clk in a greater step. After the PC compared the relationship between Ref_clk and Out_clk, the PC generates a signal to control the operation of the SAR. And the SAR's code makes the CT decoder generate different digital code. The delay time of DCDL is changed by the digital code, thus the phase error between Ref_clk and Out_clk will be reduced until the Out_clk is located within the PC's lock window (T_C). After the CT loop completes locking, the lock detective (Ld) signal is set to high, to enable the FT loop. Consequently, the phase error is reduced in a more precise scale by the FT loop. The PD detects the difference between Out_clk and Ref_clk signals, and then the counter generates the signal to control the FT decoder's digital code to change the DCDL's delay time until the circuit being locked. Under this condition, the phase error between the Out_clk and the Ref_clk is less than the dead zone (T_F) of the PD. Therefore, using dual-loop architecture can reduce locking time, and also provide a smaller phase error in the proposed DLL.

III. CIRCUIT DESCRIPTION

A. Coarse Tune Loop

To design an appropriate DLL, an important parameter to evaluate the performance is the locking time. In a case of n-bit register-controlled DLL, the longest locking time will be 2^n input clock periods. In a case of n-bit CDLL (Counter-Based DLL), the longest locking time will be 2^n-1. In this DLL, the CT loop uses the PC to determine the relationship between Ref_clk and Out_clk. The successive approximation register (SAR) and CT decoder are used to offer the faster locking time. Fig. 4 shows the block diagram of SAR [11]. The SAR uses a binary searching to raise the lock speed extremely. The longest locking time of an n-bit SAR is n cycles. The initial value of the control word is set to "100". If the Out_clk leads the Ref_clk, the Comp signal is high and the Ld signal is low. The most significant bit (MSB) of the SAR's output consequently changes to low. If not, both the Comp and the Ld signals are low, and the MSB of the SAR's output consequently remains high. This process is repeated for each subsequent bit until the least significant bit (LSB) is determined. Then the Ld signal is set to high, indicating that the CT loop has completed locking. In other words, when a 3-bit binary-weighted delay line is used, the CT loop can lock within 3 input clock periods, faster than conventional register-controlled DLLs and counter-controlled DLLs [11].

B. Fine Tune Loop

A dead lock problem could occur in the conventional SAR DLL after the searching procedure is completed because of the open-loop characteristic. To reduce the phase error more precisely, the FT loop is applied. After the CT loop is locked, the Ld signal is at a high logic level and the FT loop is activated. To achieve the goal, the proposed phase detector with the control logic to enable the FT loop when the CT loop is locked. In this

978-1-61284-863-1/11 $26.00 © 2011 IEEE   389

Fig. 6. The layout of the proposed DLL.

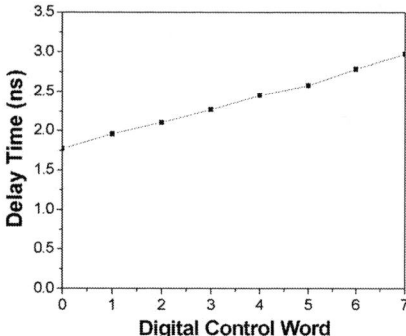

Fig. 7. The relationship diagram of coarse tune line delay time versus digital code.

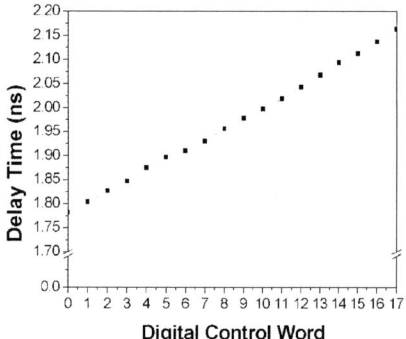

Fig. 8. The relationship diagram of fine tune line delay time versus digital code.

work, the phase detector is performed using a structure similar to the one presented in [12]. Using a PD with small dead zone, the phase error can be obviously smaller when the DLL is locked. That is, the maximal static phase error between the Out_clk and Ref_clk will less than the dead zone of the PD. In addition, the up/dn counter is initially set medium value "01001" to reduce the locking time. If the Out_clk is not located within the dead zone of the PD, one of the following two conditions may be the case. When the Out_clk leads the Ref_clk, the control value will decrease, and the delay time will be increased. In contrast, when the Out_clk lags Ref_clk, the control value will increase, and the delay time will be decreased. Once the Out_clk is located in the dead zone of the PD, the whole DLL is locked. Furthermore, the closed-loop is adopted to accommodate the process, voltage, temperature, and loading (PVTL) variations in the FT loop.

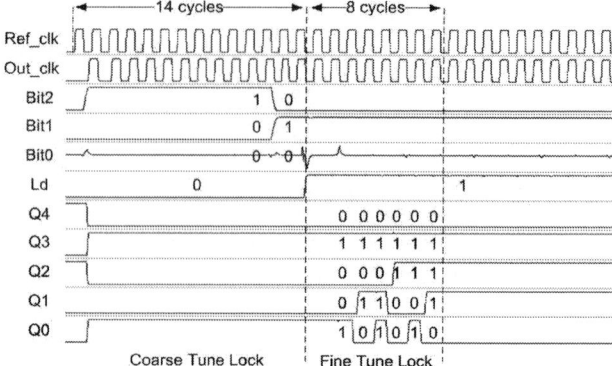

Fig. 9. The simulation waveform at 500MHz.

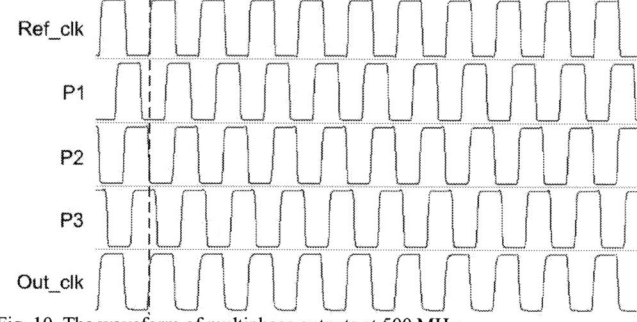

Fig. 10. The waveform of multiphase outputs at 500 MHz.

C. Digital -Controlled Delay Line

In the proposed DLL, the digital-controlled delay line with CT control code (C [6:0]) and FT control code (F [17:0]) is shown in Fig. 5. It is composed of the basic half delay cell element (HDCE). Two HDCEs form a delay cell (DC) and the DCDL is made up of four DCs. Each HDCE contains one current-starved inverter and 7 switches (C0~C6) to form CT delay line, and 18 switches (F0~F17) to form FT delay line. By controlling these switches, the MOS capacitors can be changed to adjust the delay time of the DCDL. In the CT delay line, the output of the SAR will control the switches of 7 capacitors whose size is larger. Moreover, the output of the up/dn counter will control the switches of 18 capacitors whose size is smaller. The conventional DLL may suffer from the problem of harmonic lock or false lock. To overcome these problems, the delay time (T_{DCDL}) of the DCDL has a minimal and a maximal boundary ($T_{DCDL.min}$ and $T_{DCDL.max}$). Consequently, to operate correctly, the period (T_{CLK}) of the reference clock should satisfy following expression [13]

$$Max(T_{DCDL.min}, 2/3 \times T_{DCDL.max}) < T_{CLK} < Min(T_{DCDL.max}, 2 \times T_{DCDL.min}) \quad (1)$$

IV. THE SIMULATION RESULTS

The simulation results are based on 0.35 μm CMOS process with a 3.3 V power supply voltage. Fig. 6 shows the layout of the proposed DLL, the chip area without input and output buffers is

TABLE I. COMPARISON WITH OTHER WORKS

Performance Parameter	REF [7]	REF [8]	**This work**
Technology	0.18 μm	0.35 μm	**0.35 μm**
Supply	1.8 V	3.3 V	**3.3 V**
Phase Output	4	7	**4**
Operating Frequency	510-1100 MHz	20-85 MHz	**320-500 MHz**
Jitter (pk-pk)	20.4 ps @ 800 MHz	310 ps @ 32 MHz	**6 ps @ 500 MHz**
Lock Time	< 80 cycles	< 124 cycles	**< 24 cycles**
Power	12 mW	75.1 mW	**28.3 mW @ 500 MHz**
Area	0.0161 mm²	1.9 mm²	**0.17 mm²**

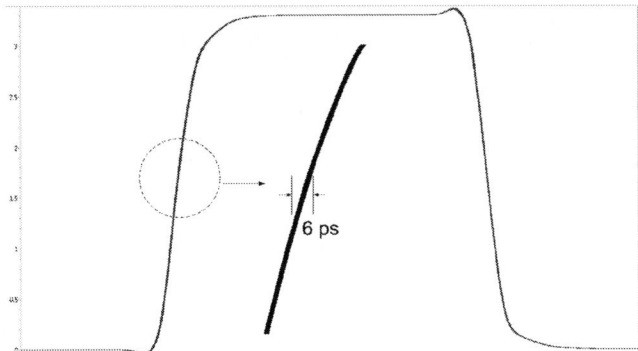

Fig. 11. The jitter histogram of the DLL at 500MHz.

0.23×0.73 mm². The delay time of the CT line versus digital code is shown in Fig. 7 and the delay time of the FT line versus digital code is shown in Fig. 8. According to Fig. 7 and Fig. 8, it has a linear relationship between delay time and digital code. The delay resolution of the coarse tune is about 210 ps and about 20 ps for the fine tune. Fig. 9 shows the simulation waveform of the DLL when the operating frequency is 500 MHz. As shown in Fig. 9, the output of the SAR is Bit2-Bit0 and the output of the up/dn counter is Q4-Q0. When the PC output signal (Ld) is at a high logic level, the CT loop is completed, then the FT loop is activated. After 22 clock cycles, the DLL is locked. Fig. 10 shows the waveforms of input reference clock (Ref_clk), phase1-phase3 (P1-P3), and the output clock (Out_clk) in the locked state at 500 MHz. Fig. 11 shows the jitter histogram of the DLL output clock at 500 MHz and the peak-to-peak jitter is 6 ps. Table I gives the performance summary of the proposed DLL. The comparison between this work and previously published all-digital multiphase DLLs are also listed in table I. The results show that the proposed DLL has faster locking time than [7], [8]. Furthermore, the area and power consumption are smaller than [8].

V. CONCLUSION

In this paper, an all-digital DLL for multiphase clock generator is designed in standard 0.35 μm CMOS process. The proposed DLL uses dual-loop architecture to offer a faster locking time and reduce the static phase error. According to the simulation results, the proposed DLL can operate correctly with an input clock frequency ranging from 320 MHz to 500 MHz. Moreover, the DLL can generate equally spaced four-phase clocks within a single cycle. The peak-to-peak jitter is 6 ps and the power consumption is 28.3 mW at 500 MHz. If more advanced technologies can be used to design this DLL, the main performances such as operating frequency range, power consumption, locking time, static phase error, and jitter could be further improved.

ACKNOWLEDGMENT

The authors thank the National Chip Implementation Center for fabricating the chip and the National Science Council for supporting this work.

REFERENCES

[1] M. J. E. Lee, W. J. Dally, T. Greer, H. T. Ng, R. Farjad-Rad, J. Poulton, and R. Senthinathan, "Jitter Transfer Characteristics of Delay-Locked Loops– Theories and Design Techniques," *IEEE J. Solid-State Circuits*, vol. 38, no. 4, pp. 614-621, Apr. 2003.

[2] R. J. Yang, S.I. Liu, "A 2.5 GHz All-Digital Delay-Locked Loop in 0.13 μm CMOS Technology," *IEEE J. Solid-State Circuits*, vol. 42, no. 11, pp. 2338-2347, Nov. 2007.

[3] S. K. Kao, B. J. Chen, and S. I. Liu, "A 62.5–625-MHz Anti-Reset All-Digital Delay-Locked Loop," *IEEE Trans. Circuits and Systems- II: Express Briefs*, vol. 54, no. 7, pp. 566-570, July 2007.

[4] Bo Ye, X. Han, and Min Luo, "A Fast-lock Digital Delay-Locked Loop Controller," *IEEE Int. Conf. ASIC*, pp. 809-812, Oct. 2009.

[5] J. S. Wang, Y. M. Wang, C. H. Chen, and Y. C. Liu, "An Ultra-Low-Power Fast-Lock-In Small-Jitter All-Digital DLL," *in Dig. Tech. Papers IEEE Int. Solid-State Circuit Conf.*, pp. 422–423,607, Feb. 2005.

[6] H. H. Chang and S. I. Liu, "A Wide-Range and Fast-Locking All-Digital Cycle-Controlled Delay-Locked Loop," *IEEE J. Solid-State Circuits*, vol. 40, no.3, pp. 661–670, Mar. 2005.

[7] K. I. Oh, L. S. Kim, K. I. Park, Y. H. Jun, and K. Kim, "Low-Jitter Multi-Phase Digital DLL with Closest Edge Selection Scheme for DDR Memory Interface," *Electron. Lett.*, vol. 44, no. 19, pp. 1121-1123, Sep. 2008.

[8] C. C. Chung and C. Y. Lee, "A New DLL-Based Approach for All-Digital Multiphase Clock Generation," *IEEE J. Solid-State Circuits*, vol. 39, no. 3, pp. 469-475, Mar. 2004.

[9] L. K. Soh, M. S. Sulaiman, and Z. Yusoff, "Fast-Lock Dual Charge Pump Analog DLL using Improved Phase Frequency Detector," *in Proc. International Symposium on VLSI Design, Automation and Test (VLSI-DAT)*, pp. 1-5, Jun. 2007.

[10] Ghaffare, Abrishamifar, "A Wide-Range Delay-Locked Loop with a New Lock-Detect Circuit," *IEEE Int. Conf. on Electron., Circuits and Syst.*, pp. 1168-1171, Dec. 2006.

[11] G. K. Dehng and S. I. Liu, "Clock-Deskew Buffer Using a SAR-Controlled Delay-Locked Loop," *IEEE J. Solid-State Circuits*, vol. 35, no. 8,pp. 1128-1136, Aug. 2000.

[12] Y. Moon, J. Choi, K. Lee, D. K. Jeong, M. K. Kim, "An All-Analog Multiphase Delay-Locked Loop Using a Replica Delay Line for Wide-Range Operation and Low-Jitter Performance," *IEEE J. Solid-State Circuits*, vol. 35, no. 3, pp. 377–384, Mar. 2000.

[13] H. H. Chang, J. W. Lin, C. Y. Yang, and S. I. Liu, "A Wide-Range Delay-Locked Loop With a Fixed Latency of One Clock Cycle," *IEEE J. Solid-State Circuits*, vol. 37, no. 8, pp. 1021-1027, Aug. 2002.

Design and Fabrication of Configurable Digital Controller Interface for Micro Mirror Projector ASIC

Jianwen Luo[*], Peng Li[*], Chin Yann Pang[*], Pradeep Kumar Gopalakrishnan, Tal Langer[†], Minkyu Je[*]

[*]Institute of Microelectronics, A*STAR (Agency for Science, Technology and Research), Singapore

E-mail: luojw@ime.a-star.edu.sg Tel: +65-6770 5517

[†]Maradin Ltd., 2 HaCarmel St. Yokneam 20692, Israel

Email: tal.langer@maradin.co.il

Abstract—This paper presents the design and implementation of digital interface for micro mirror projection ASIC for MEMS scanner control and sensing. The design includes a microprocessor for sensor configuration, MEMS control registers access and video interface for providing scanning and synchronizing signals. It has been fabricated on 0.18μm high voltage CMOS process, and silicon proves the full functionality of the design concept.

I. INTRODUCTION

Micro mirror display [1] has become increasingly popular recently due to the advancement of MEMS technology and the exploration of new applications such as personal mobile devices, health care apparatus and gaming console for micro display system. With the need to interface and control the MEMS projector ASIC [2], a digital interface has been designed and targeted for this device with configurability for MEMS control, calibration and sensing requirements. In this paper, we have designed a digital system targeting for micro mirror projection ASIC with video synchronizer.

II. DESIGN ARCHITECTURE

A. System Block

The main role of MEMS projector ASIC is to control and sense micro mirror MEMS position both in X and Y directions and to synchronize with video bit stream in order to project a video image on a 2D screen. In this work, the MEMS is resonant in X axis at 10 kHz resonance frequency after power on, which makes mirror to scan at X direction. By providing saw tooth signal to drive MEMS movement in Y direction for each video frame, the micro mirror scanner is able to project the image in the 2D screen. In this design, SVGA [3] resolution has been achieved. The digital controller interface for micro mirror ASIC is to provide the controllability for MEMS position adjustment as well as synchronizing scanning signals with video bit stream timing.

The digital interface consists of an 8051 core, memory system, SPI/I2C for serial communication, GPIOs, video synchronizer for synchronizing video signal with MEMS movement and eye safety alarm generation. The internal memory bus is mapped to 8051 memory address. The control registers for alarm, PLL clock phase control and other analog blocks are accessible by 8051 through internal serial bus

controller. Figure 1 shows the components residing in the digital controller interface.

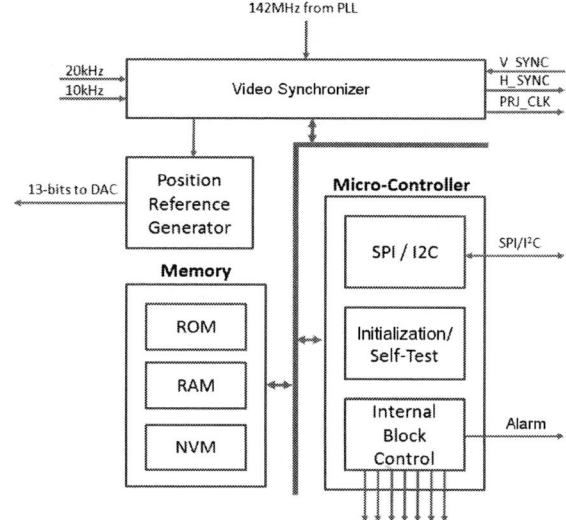

Figure 1 Digital block diagram

B. Memories

The memories mapped to the 8051 core address consists of 8K Bytes ROM, 8K Bytes program SRAM, 256 Bytes internal memory and 32 Bytes Non Volatile Memory (NVM).

The 8KB ROM contains the 8051 boot up code to initialize 8051 primitive routine and commands after power on. The ROM module used is TSL18RO160, 0.18 micron, 1.8 volt, generated by Synchronous via ROM Compiler from Synopsys [4].

The 8KB SRAM provides storage for external program download and execution after chip has been fabricated. The memory module used is TSL18RS160, 0.18 micron, 1.8Volt, High Density Synchronous RAM Compiler from Synopsys.

The 256 Bytes SRAM is the internal memory of the 8051 core which is to be used as temporary data storage and stacks. The memory module used is TSL18RS160.

The 32 Bytes NVM consists of 4 64-bit NVM Y-flash [5] modules. Each 64-bit Y-flash NVM shares the same program, erase and read control signals and is enabled by address mapping from address decoder block. The address decoder is part of Glue Logic design which is to be addressed in Section C.

Figure 2 shows the address map for the memories and I/O used in the 8051 core in the digital interface design.

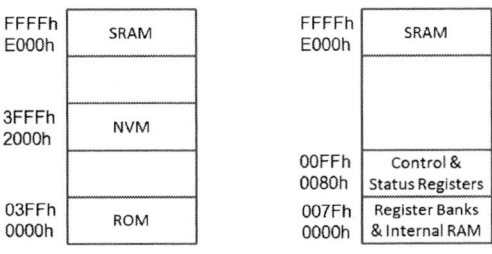

Program Memory Address Map   Data Memory / IO Address Map

Figure 2 Memory and I/O address Mapping

C. Glue Logics

The Glue Logics comprise the following functional blocks:
- Chip select generation logic
- Video synchronizer
- Position reference generator
- Alarm generation
- Internal serial bus for analog control registers

Each functional block is described in the following section:

1) Chip select generation logic

The Mask ROM, SRAM and NVM are allocated with addresses as shown in the Memory map in Section B. They share the same memory bus in the 8051 architecture. The chip select logic generates the chip select signals for these memory modules based on the memory map address to enable the access of the each memory.

2) Video synchronizer

One of the key functions for digital interface is to synchronize the micro mirror projection position with the video stream timing. Due to process variation, the actual projector resonance frequency may vary. The video synchronizer is taking VSYNC from video stream and generates HSYNC and projection clock based on the 10 kHz MEMS sensor resonance frequency.

3) Position reference generator

The purpose of position reference generator is to generate 13-bit saw tooth reference signal to DAC block to produce the row scanning signal to MEMS projector. Between each HSYNC pulse generated by video synchronizer, the 13-bit position reference data is generated and converted to the reference voltage which drives the motor inside the MEMS. A clock frequency of 20 kHz is output to DAC module, where the 13-bit reference data has been sampled on the falling edge of this clock and converter to the row position for the MEMS row scanning control.

Figure 3 illustrates the timing between VYSNC, HYSNC and 13-bit reference data. There are two fields (odd & even) between two VYSNC pulses, where each field has an approximate duration of 16.67msec. There are 300 reference steps within each field. For the SVGA resolution, there are 600 HSYNC pulses between two VSYNCs. Each VSYNC appears at the beginning of each frame and each HSYNC appears at the beginning of each line. Between each HSYNC, the 13-bit reference data is generated to the DAC module. Here, the variation of resonance frequency is taken care of by control the programmable delay registers.

4) Alarm generation

Due to the light source of the micro mirror projector is laser, eye safety concern needs to be taken account during design phase. This is to prevent too much laser light absorbed by human eyes if the projector happens to project the laser light against someone's eyes. During projection stage, alarm signal is to be generated to shut down the laser module if the absence of MEMS resonance has been monitored.

5) Internal serial bus

In order to control the parameters for calibration of analog/digital blocks and MEMS sensor, internal serial communication bus has been adopted. The internal serial bus is connected in the daisy-chain scheme with each slave register resides inside individual blocks as shown in Figure 4.

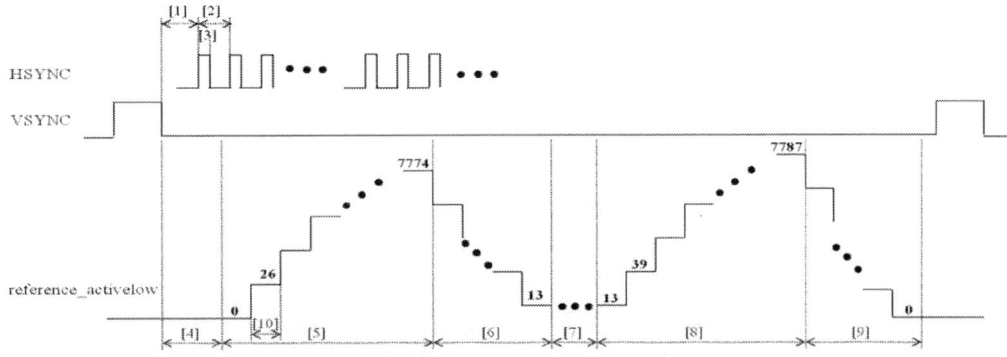

[1] The programmable steps (142M) between VSYNC to HSYNC (spi_input_H): [0,1023]=[0.007us,7.211us]
[2] The period of HSYNC: 50us
[3] The high level in one period of HSYNC: 3.2us
[4] The programmable steps (20k) between VSYNC to the 1st reference step (spi_input_V): [0,25]=[0,1.25ms]
[5] [8] The 300 steps: add 26 every step
[6][9] The programmable retrace steps (20k) (spi_input_R): [1,20]=[0.05ms,1ms]
[7] The programmable blanking steps (20k) (spi_input_B): [0,15]=[0,0.75ms]
[10] The time of one step: 50us

Figure 3 Video timing and reference data generation

978-1-61284-863-1/11 $26.00 © 2011 IEEE   393

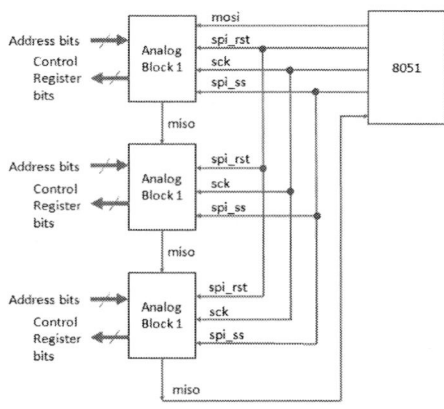

Figure 4 Internal serial bus configuration

Each block contains one or more slave registers with each to be assigned with a unique slave address. The 8051 core accesses each slave register with a 15-bit serial command, with first 4-bit to be slave address followed by the 2-bit register address, the 7th bit is to control the data direction - write or read, followed by the 8-bit of data to be written or read from the targeting slave registers. The serial command and data are MSB first.

D. I/O Multiplexing

In this work, both SPI and I2C interface are supported. In order to reduce the pin out count, the I/O mux scheme has been proposed which allows the SPI and I2C interface to share the same pins. Figure 5 shows the SPI/I2C multiplexing.

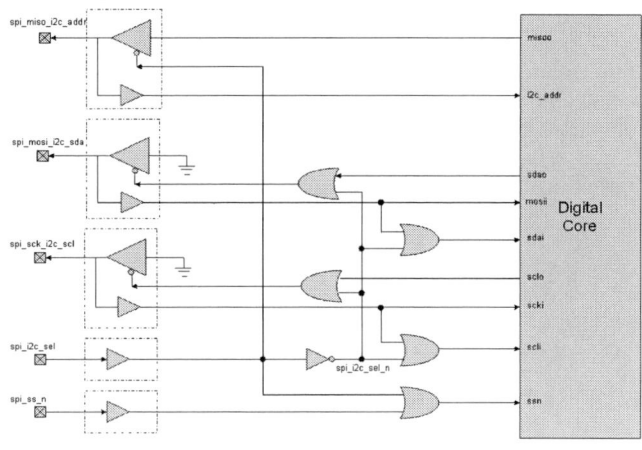

Figure 5 SPI/I2C multiplexing

E. Hardware/Software Co-design

The main functionality of the digital control block is realized through the execution of firmware running on the 8051 core.

The following are the major functions performed by the firmware:

- *Boot up – Initialization of ASIC registers/memory*
- *Decoding commands from external controller to support*
  - a. *Read / write registers*
  - b. *Read / write SRAM*
  - c. *Program/ erase / read NVM*
  - d. *Download and execute a program to internal SRAM from the external controller*

The firmware is mask programmed into the embedded 8K ROM during fabrication stage. It enables execution of primitive commands such as fetching, decoding and executing instructions for write/read slave registers and memories and downloading customized program into the on-chip 8K SRAM to perform sub-routine function extension and debugging purpose.

III. FABRICATION

The presented digital interface is implemented and fabricated in 0.18μm high voltage process, with standard cell gate count of 50k and total transistor size of 500k. The system clock runs at 142 MHz, and the total power consumption is 1.1mw at 1.8V core voltage under operational condition. The die size is 1.9 mm^2 of which over 70% is taken by memories and NVM, the rest is standard cells and routing resources. The test chip is packaged in the QFP52 package.

Figure 6 shows the microphotograph of the digital interface test chip.

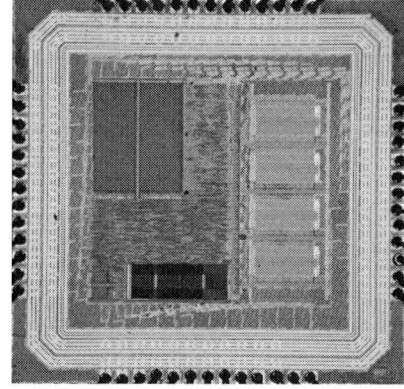

Figure 6 Digital interface test chip microphotograph

IV. MEASUREMENT RESULTS

The design has been tested and the idea has been proved working. Figure 7 shows the measured result of the I2C interface signals with write command to register 0x3F with data 0xAA. The similar test has been verified for SPI interface as well. The communication is setup through the serial interface with on-chip 8051. Figure 8 shows the programming signals for NVM which have met the specification requirements of Y-flash. The program, read and

978-1-61284-863-1/11 $26.00 © 2011 IEEE

erase operations of NVM have also been verified. Figure 9 is the alarm generation when the 10k resonance stops. Figure 10 shows the video synchronizer and position reference data generated for MEMS control with respect to the HSYNC timing.

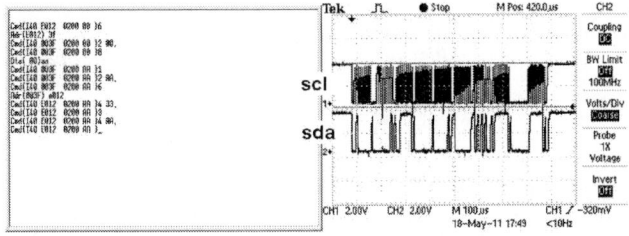

Figure 7 SPI/I2C serial interface communication

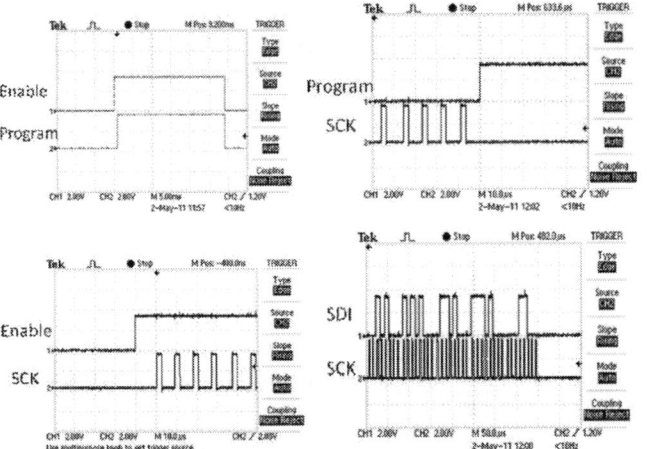

Figure 8 NVM program signals

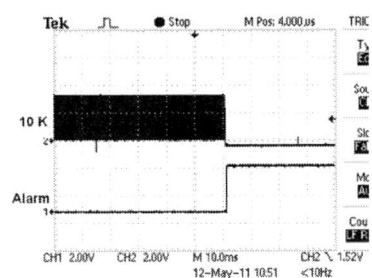

Figure 9 Alarm generation

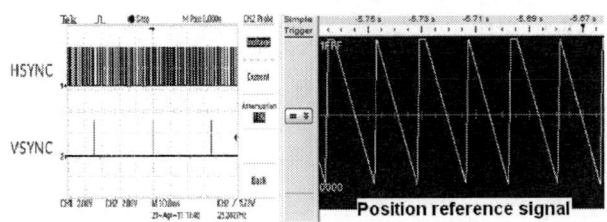

Figure 10 Video synchronize signal & position reference data

V. CONCLUSIONS

A digital interface design for micro mirror projector ASIC has been presented with embedded 8051 for MEMS sensor calibration and video timing synchronization. This is part of the design blocks to be fully integrated with the SoC projector ASIC. The embedded 8051, on-chip NVM modules and SRAM provide a wide range of configurability for sensor calibration and versatile function expansion. The design has been silicon validated and ready for the SoC integration.

ACKNOWLEDGMENT

The authors would like to thank for Maradin Ltd. [6] for their invaluable advice and assistance of staff from Integrated Circuits and Systems laboratory from Institute of Microelectronics, A*Star for their support for this project.

REFERENCES

[1] C. Liao, J. Tsai, "The Evolution of MEMS Displays", IEEE Transactions on Industrial Electronics, 2009
[2] Duy-Dong Pham, Ravinder Pal Singh, Dan-Lei Yan, Kei-Tee Tiew, Minkyu Je, "Position Sensing and Electrostatic Actuation Circuits for 2-D Scanning MEMS Micromirror," IEEE Defence, Science & Research (DSR 2011)
[3] Super VGA standard, http://www.vesa.org/
[4] Synopsys, http://www.synopsys.com/
[5] Y-flash, Tower Semiconductor Ltd, http://www.towersemi.com/
[6] Maradin Ltd., http://www.maradin.co.il/

A Low Power JPEG Image Compression IC for Wireless Ingestible Endoscopy

Wei-Da Toh, Bin Zhao, Yuan Gao, Yuanjin Zheng, Minkyu Je and Chun-Huat Heng[1]

Institute of Microelectronics, A*STAR (Agency for Science, Technology and Research), Singapore
[1] Department of Electrical and Computer Engineering, National University of Singapore, Singapore
tohwd@ime.a-star.edu.sg, zhaobin@ime.a-star.edu.sg

Abstract—**In this paper, a low power baseline JPEG image compressor architecture, to be interfaced directly with an image sensor, has been presented. It compresses the raster YCbCr 4:2:2 image data from the image sensor into JPEG format. The compressed image data are being packed into byte format and DMA can be used to retrieve the JPEG image. The JPEG image compression chip is designed using 0.18-μm CMOS technology. It consumes 1.69mW when compressing at 7.49 (VGA) frames per second.**

Keywords-Discrete Cosine Transform (DCT); Direct Memory Access (DMA); Huffman Coding; JPEG; Image Compression;

I. INTRODUCTION

Imaging devices, such as consumer cameras and medical imaging devices usually capture high resolution images at high frame rate. This results in large amount of data processing and leads to high storage and communication costs. Therefore image compression is commonly used to reduce the image redundancy.

There are two types of image compression methods, namely lossless and lossy. For lossless method, the exact original image can be reconstructed after compression and decompression. On the other hand, for lossy method, the reconstructed image only gives closed approximation to the original image. Standard lossless image compression formats [1] such as bitmap (BMP), graphics interchange format (GIF), portable network graphics (PNG) and tagged image file format (TIFF) have a compression ratio of around 3:1 or less. Lossy image compression formats, such as joint photographic experts group (JPEG) [1-2] and JPEG2000 [3], have a scalable compression ratio which can trade off with image quality. As higher compression ratio is used, more pixelated and image artefacts will be observed for image with lossy compression. JPEG and JPEG2000 employ discrete cosine transform (DCT) [4-7] and wavelet transform respectively. JPEG algorithm is less computational intensive. Furthermore, there is slight deterioration in image quality at compression ratio less than 20:1 [8]. Hence, JPEG compression is generally preferred when simplicity and computational cost are important consideration.

In this paper, a full chip solution with a simple, low power image compressor, targeting for wireless ingestible capsule

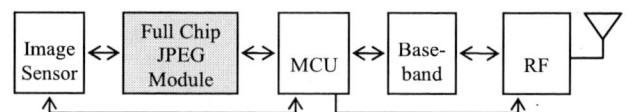

Fig. 1. Typical wireless ingestible capsule system architecture

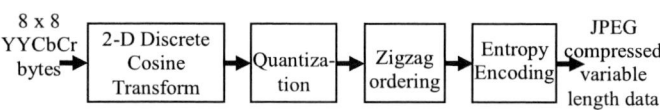

Fig. 2. JPEG compression algorithm block diagram

applications [9] is proposed. It employs compression ratio higher than 5:1 while maintaining the quality of image. By exploiting the quality factor and compression ratio, some techniques are proposed to lower the computation complexity, which helps reduce the latency, power and area compared to conventional implementation. The typical wireless ingestible capsule system architecture is shown in Fig. 1. The microcontroller unit (MCU) helps initializing the image sensor as well as changing the parameter of RF transceiver. In addition, it dictates the switching of baseband from transmit mode to receive mode and vice versa. During transmit mode, the image sensor does raster scanning of the captured image and sends the image data to the JPEG module. The proposed full chip JPEG module will receive, store and arrange the raster scan image data prior performing JPEG compression. The compressed image data will then be sent to the baseband via MCU for coding and framing before actual transmission through RF transceiver. During receive mode, upon receiving a packet, the baseband will unpack and decode the data before sending it to the MCU for further decision making. This includes re-initializing the image sensor, changing the RF transceiver parameters or switching transmit/receive mode.

This paper is divided into four main sections. Section II describes the overview of the JPEG compression algorithm. Section III presents the hardware implementation. The implementation results and performance of this work are presented in section IV. Last but not least, the conclusions are shown in section V.

II. JPEG COMPRESSION ALGORITHM OVERVIEW

The JPEG compression algorithm consists of four main

978-1-61284-863-1/11 $26.00 © 2011 IEEE

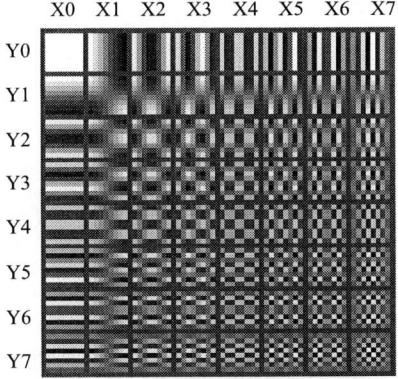

Fig. 3. 2-D DCT frequencies diagram

Fig. 4. Zigzag arrangement diagram

steps namely, 2-dimentional (2-D) DCT, quantization, zigzag ordering, and entropy encoding as shown in Fig. 2.

The 2-D discrete cosine transform process transforms 8 rows × 8 columns of each component in the colour space separately into spectral sub-bands as illustrated in Fig 3. The zero-frequency component is known as the DC coefficients of the 2-D DCT and the rest of the 63 components are the AC coefficients of the 2-D DCT. The DC coefficient is at the top left corner (X0, Y0) whereas the highest frequency AC component is at the bottom right corner (X7, Y7). The horizontal and vertical spatial frequencies increase with increasing horizontal and vertical index value respectively.

The quantization process performs division on the 2-D DCT coefficients. In general, the high frequencies AC coefficients for natural images (not computer graphics) are of small value. In addition, higher compression ratio will lead to smaller quality factor and thus poorer image quality. This will result in the use of larger divisor values for quantization process. Considering these facts, the high frequencies AC coefficients will likely to become zero after quantization process given its initial smaller value and larger divisor value employed.

The zigzag ordering arranges the 2-D DCT coefficients from the lowest spatial frequency to the highest spatial frequency as shown in Fig. 4. Due to the high likelihood that the quantized high frequency AC coefficients will have zero value, this process will group most of the zero AC coefficients together.

The standard JPEG compression uses Huffman encoder as the entropy encoder to encode the quantized 2-D DCT coefficients. Huffman coding is a lossless coding scheme. Through the coefficients reduction in the quantization operation, the Huffman encoder will result in lesser amount of coded data. The DC coefficient will go through differential

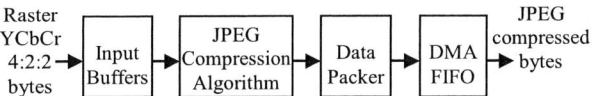

Fig. 5. Full chip JPEG compression system architecture

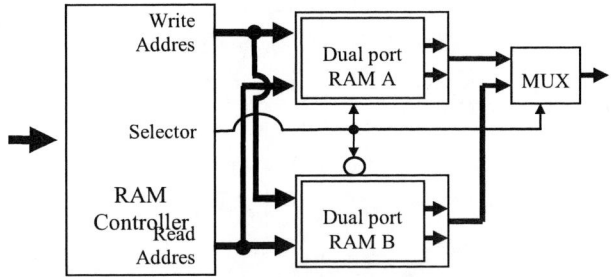

Fig. 6. Input buffers block diagram

pulse code modulation (DPCM) before Huffman encoding. DPCM is a modulation technique taking the difference between the current quantized DC coefficient and the previous quantized DC coefficient. Due to the zigzag ordering which groups all the zero AC coefficients, the AC coefficients will be coded with zero run length when there is one or more zeros between two non-zero zigzag arranged quantized AC coefficients. The JPEG compression of the 8×8 image is completed after Huffman encoding.

III. DIGITAL IMPLEMENTATION

The full chip JPEG compression architecture is shown in Fig. 5. It consists of four main modules namely, the input buffers, a JPEG compression algorithm module, a data packer and a DMA first-in-first-out (FIFO). In this work, we assume the image sensor output is raster scan of YCbCr 4:2:2 images with resolution of 640×480. YCbCr is the colour space where Y is the luminance component, and Cb and Cr are the chrominance blue and chrominance red components. 4:2:2 is the chroma subsampling scheme where every pixel consists of one luminance component and one chrominance component. The full chip JPEG compression is designed to operate at the same frequency as the image sensor pixel clock. The JPEG quality factor is fixed at 90 after considering the trade-off between compression ratio and image quality.

A. Input Buffers

The input buffers module is used to stores 16 rows of the raster YCbCr 4:2:2 image data. The input buffers module will send the required image data to the JPEG compression algorithm when 8 rows of image data are ready. While sending these 8 rows of image data to the JPEG compression algorithm, it is also receiving and storing the incoming image data from the image sensor simultaneously.

The main components of the input buffers are the random access memory (RAM) controller and two dual port RAMs, as shown in Fig. 6. Each dual port RAM can store 8 rows of image data and two are needed for storing and processing of the image in parallel. The RAM controller writes to the input buffers in raster format whereas reading will be done in 8 × 8.

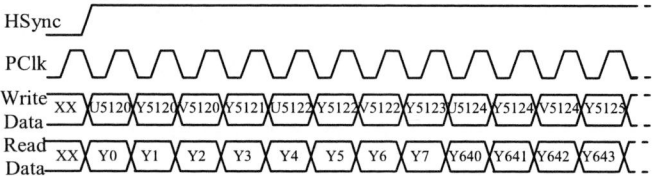

Fig. 7. Input buffers write and read waveforms

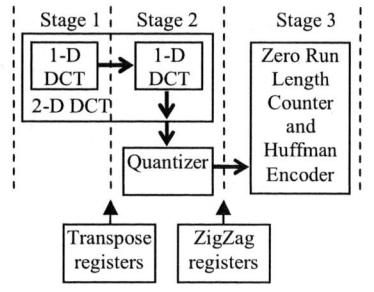

Fig. 8. JPEG compression algorithm block diagram

One of the RAM will be in read mode while the other will be in write mode controlled by the selector signal. The outputs of the RAMs are multiplexed and are controlled by the selector signal as well. The selected RAM outputs will be sent into the JPEG compression algorithm module. The write and read modes timing waveforms of the RAM are shown in Fig. 7. The HSync and PClk are the horizontal synchronization output signal and the pixel clock of the image sensor respectively.

B. JPEG Compression Algorithm

As mentioned in section 2, JPEG compression consists of four major steps. However, the implemented JPEG compression algorithm module consists of only three modules as shown in Fig. 8. It comprises of two 1-D DCTs, a quantizer and a Huffman encoder with zero run length counter. Different DCT algorithms had been proposed [4]-[5], simplified hardware implementations had been proposed [6]-[7] and successfully demonstrated in FPGA [7]. In [7], the 1-D DCT is being pipelined into 6 stages for ease of FPGA implementation. However, the simple and fast DCT algorithm proposed in [6] is adopted in this paper. The pipelining within 1-D DCT has been eliminated in this paper to improve the latency. In addition, full ASIC implementation allows power and timing optimization, making pipelining unnecessary.

The 2-D DCT operation is achieved by pipelining the two 1-D DCT modules as shown in Fig. 8. The first DCT module will perform the DCT operation on 8 columns of the image, row by row. The second DCT module will perform the DCT operation of the outputs of the first DCT on 8 rows of the 1-D DCT image, column by column. The two DCT modules use only 4 fractional constant multipliers. To simplify the calculations and save resources, fixed-length decimal multipliers are employed whereby only 8 bits are used to represent the fractional part of the constants. The 4 constant multipliers are implemented using adders only. The transpose registers which converts row operations to column operations also function as the desired pipelining registers.

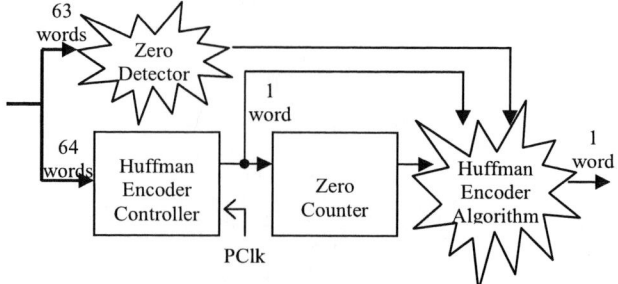

Fig. 9. Huffman encoder block diagram

Quantization operation is performed right after the second DCT operation without any pipelining. For parallelism, 8 sets of quantizers are needed for the incoming eight 2-D DCT data. Data width of 12 bits is chosen to represent the luminance and chrominance divisors. However, due to the chosen JPEG quality factor of 90, only 9 bits representation is needed. Within the quantizer, incoming data is multiplied with the 9-bit divisor. The quantized results are obtained after truncating the 12 least significant bits and stored in the zigzag registers. It should be pointed out that the zigzag ordering shown in Fig. 4 has been incorporated into quantizer to speed up the JPEG algorithm.

The zigzag arranged data is then sent for Huffman encoding as shown in Fig. 9. Static compression technique and standard JPEG Huffman codes [3] are employed here. The sixty four 2-D DCT coefficients are multiplexed through the Huffman encoder controller, one coefficient at a time. It is triggered by the same pixel clock. The DC coefficient is encoded in a manner described earlier through Huffman encoder algorithm. The zero counter will count the number of zero 2-D DCT AC coefficients between two non-zero 2-D DCT AC coefficients and input to the Huffman encoder algorithm for zero run length encoding as well as Huffman encoding. At the same time, the sixty-three 2-D DCT AC coefficients will also go through a zero detector for locating the index of the last non-zero 2-D DCT AC coefficient. This index will be sent to the Huffman encoder algorithm for end of block encoding.

C. Data Packer

The data packer is employed to organize the variable length output of the JPEG compression into 8-bit data format. The data packer consists of a FIFO and a control module. If the JPEG compressed data is less than 8 bits, the data will be stored in the data packers' registers and accumulate until there is at least 8 bits of data available before being written to the FIFO. Reading will be done on the next clock cycle to enable swift clearing of FIFO to minimize the required FIFO capacity. As the maximum JPEG compressed data length is 26 and the maximum unpacked data within data packers' registers is 7, the FIFO capacity is chosen to be 32 bits with the additional 1 bit stored in the data packers' register. Therefore, the write operation is performed in 32 bits whereas the read operation is done in 8 bits format at a maximum of 4 clock cycles for each memory location.

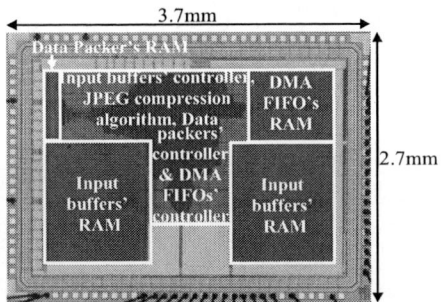

Fig. 10. Die photo of the full chip JPEG compression

Fig. 11. Original (left) and JPEG compressed (right) Lenna images

D. DMA FIFO

The DMA FIFO is employed to interface with DMA. It consists of a controller and a dual port RAM. The data packer output will be sent to the DMA FIFOs' controller and will be stored into the RAM. The controller is responsible for generating the required control or FIFO status signals for DMA interface such as "FIFO empty", "FIFO filled" and etc. The write and read operations are performed on different ports of the RAM to enable direct access the RAM by the device.

IV. MEASUREMENT RESULTS

The full chip JPEG compression is implemented using Verilog hardware description language (HDL) and fabricated in 0.18 μm CMOS technology. The chip occupies a die area of 3.7 mm × 2.7 mm as shown in Fig. 10. The chip is tested together with Pixelplus image sensor, and a FPGA is employed for DMA reading. The Pixelplus image sensor does raster scanning of 716 × 516 pixels with configurable frame rate up to 30fps (frames per second), which has an equivalent pixel clock frequency of 24 MHz. For the testing, the Pixelplus image sensor is configured to output YCbCr 4:2:2 images at the targeted 7.49 fps. The full JPEG compression chip operates under 1.3 V and consumes only 1.69 mW at 7.49 fps, when running at pixel clock frequency of 6 MHz. To evaluate PSNR and compression ratio, we first convert Lenna image to YCbCr 4:2:2 format prior sending it into the JPEG compression chip. The conversion is achieved with Verilog coding and FPGA to mimic the image sensor function. The original Lenna image and the JPEG compressed Lenna images are illustrated in Fig. 11 without any significant loss of image details or quality. We have also employed open-source JPEG compression software to verify the full chip JPEG compression algorithm. Additional tests on other images were also carried out and the JPEG compression algorithm produces JPEG compressed images with typical peak signal-to-noise ratio (PSNR) of 38 dB with typical compression ratio of 6:1.

Table 1 Full chip JPEG compression performance

System clock	< 24 MHz
Typical compression ratio	≈ 6:1
Typical PSNR	≈ 38 dB
Area	3700 μm × 2700 μm
Operating voltage	1.3 V
Power consumption	1.69 mW @ 7.49 fps
Technology	0.18 μm CMOS

The performance of the JPEG compression ASIC is summarized in Table 1.

V. CONCLUSION

A full chip JPEG compression for a wireless ingestible capsule application is presented in this paper. By exploiting the quality factor and compression ratio, some techniques have been proposed to simplify the implementation to achieve smaller power and area. Implemented in 0.18-μm CMOS process, the JPEG compression chip employs interleaving baseline JPEG image compression method with quality factor of 90 and achieved more than the targeted compression ratio of 5:1. The proposed techniques do not degrade the image quality much from the measured result. It can be employed in applications which require direct interface with image sensor and DMA.

ACKNOWLEDGMENT

This work was supported by Science and Engineering Research Council of A*STAR (Agency for Science, Technology and Research), Singapore, under Science and Engineering Research Council (SERC) MedTech Research Program grant number 082 140 0033.

REFERENCES

[1] J. Miano. *Compressed Image File Formats – JPEG, PNG, GIF, XBM, BMP*, Addison Wesley Longman Inc, USA, 1999.

[2] International Organization for Standardization, "Information technology — Digital compression and coding of continuous-tone still images," ISO/IEC 10918, 1994.

[3] International Organization for Standardization, "Information technology — JPEG 2000 Image Coding System," ISO/IEC 15444, Oct. 2004.

[4] C.W. Kok, "Fast algorithm for computing discrete cosine transform," *IEEE Trans. on Signal Processing*, vol. 45, no. 3, pp. 757-760, Mar. 1997.

[5] N. I. Cho and S. U. Lee, "Fast algorithm and implementation of 2-D discrete cosine transform," *IEEE Trans. on Circuits and Systems*, vol. 38, no. 3, pp. 297-305, Mar. 1991.

[6] M. Kovac and N. Ranganathan, "JAGUAR: A fully pipelined VLSI architecture for JPEG image compression standard," *Proceedings of the IEEE*, vol. 83, no. 2, pp. 247-258, Feb. 1995.

[7] L. Agostini and S. Bampi, "Pipelined fast 2-D DCT architecture for JPEG image compression," *Proceedings of the 14th Annual Symposium on Integrated Circuits and Systems Design*, pp. 226-231, 2001.

[8] F. Ebrahimi, M. Chamik, S. Winkler, "JPEG vs. JPEG2000: An objective comparison of image encoding quality," *Proc. of SPIE Applications of Digital Image Processing*, vol. 5558, pp. 300-308, Aug. 2004.

[9] Y. Gao, Y. Zheng, S. Diao, W.-D. Toh, C.-W. Ang, M. Je and C.-H. Heng, "Low power ultra-wideband wireless telemetry transceiver for medical sensor applications," *IEEE Trans. Biomed. Eng.*, vol. 58, no. 3, pp. 768-772, Mar. 2011.

Design and Implementation of a Bio Sensor Array (BSA) for Cancer Cell Detection

Lim Lay Keng, Antoine Jalabert and Roshan Weerasekera
Institute of Microelectronics
A*STAR (Agency for Science, Technology and Research)
Singapore

Abstract—**Advances in bio-medical technologies have led to a great attraction in the design of bio-sensor array for rapid, qualitative and quantitative detection. In this paper we present the CMOS implementation of a high throughput bio-sensor array that consists of 96 x 96 electrodes. Each individual electrode is addressable in the system. The sensor array adopted the impedance sensing detection method. It is a single cell-based biosensor, which is targeted to achieve both detection and identification of cell type.**

Keywords: Array sensor, Cell Detection, Cell-Based, Impedance

I. INTRODUCTION

Cancer is a leading cause of death worldwide and was projected to overtake cardiovascular disease as the leading cause of global mortality by 2010. In Singapore, it is the second commonest cause of mortality. Thus, an effective cancer cell detector or indicator during the prognostic, diagnostic and therapy stage is essential. There are two clinical ways of tackling the cancer epidemic:

a) Early diagnosis, and
b) Effective and personalized therapy.

Either one of the above requires a technology that allows the physician to asses, a) the present condition of the tumor and b) tumor load. This information helps the physician in assessing the present state of the patient and the response of the patient to therapeutic intervention.

Currently, such tools are non-existent. Specifically, for patients undergoing therapy, the proof of disease regression is lack of metastasis. This leads to unnecessary deaths. Ideally, a patient could be monitored through a routine blood test in regular basis, every few days, to assess his response to therapy. This will allow the physician to intervene and change the course if necessary to achieve most effective treatment.

Circulating tumor cell (CTC) detection has emerged as a promising alternative in allowing assessment of the tumor through a simple blood draw. The term, liquid biopsy or real-time biopsy, alluded to its vast potential in cancer monitoring. Technological solutions for isolating and counting or molecularly profiling circulating tumor cells have been reported, however, none of these solutions have been able to meet the clinical needs. Subsequently, a grand challenge for CTC detection technologies was formulated and published in the Lab-on-a-chip journal in 2011. It proposed a multipoint addressable design with high throughput electro-chemical detection [1].

In this paper, we present the CMOS implementation of a Bio Sensor Array that consists of an addressable matrix of 9216 cell-based electrodes. This large array sensor could fulfill the requirement of high throughput detection of cells and their properties. Each cell-based electrode is of single-cell-sized 25 μm in diameter in order to achieve the sensitivity and precision of detection. It allows high accuracy detection and identification of the cell type [2]. The system has included an interface with automation mode for measurement to achieve a high time resolution.

The design of this BSA is targeted on high density array sensor and single cell-cased electrode sensor for both quantitative and qualitative measurement. The BSA system can be controlled by an FPGA in auto or manual mode. The area of sensor can also be controlled based on user requirement. This creates the flexibility during the measurement.

II. DESIGN CONCEPT OF BIO-SENSOR

A. State-of-Art of Bio-sensor system

Biosensor is a sensor that employs biological components such as cell, protein, etc as its sensing element to detect and/or quantify biochemical molecule [3]. Presently, there is an increase of interest on the development of array-based biosensor due to the demand for rapid, comprehensive and high throughput analyses. Several detection method of the array-based biosensor was introduced such as optic or fluorescence detection, electrical detection, mechanical detection, and etc. Fluorescence detection is the most commonly used methods for biosensor because of its high sensitivity. However, this method has undesired fluctuations due to emission from non-target materials, shielding by turbid solution, and require a label attached to the target [1]. Labeling cell will drastically vary or damage its properties.

Electrical impedance is an alternative method of label-free biosensors. It could provide measurement almost in real-time

978-1-61284-863-1/11 $26.00 © 2011 IEEE

basis. Besides, impedance is frequency dependent, thus the flow of detection could be simplified and processed in high speed but lower cost [4]. In conventional impedance sensor, a large cell population is often cultured over a large electrode. It is insufficient to detect cells with single cell precision. Thus, single cell-based analysis not only can achieve higher sensitivity but also contribute in the studies of cell nature and metabolism [5].

The BSA proposed in this paper adopted the single-cell based electrical impedance method of detection. The magnitude and phase of the cell impedance will be measured at a frequency range from 100 Hz to 100 KHz for cell analysis. When cells are immersed on the electrodes, the impedance changes based on several factors such as the frequency of measurement, cell coverage, and cell-electrode gap.

B. System Desgin

The BSA system involves 9216 cell-based electrodes that are arranged in a matrix form with 96 columns and 96 rows. Each electrode is individually addressable by the sensor address de-multiplexer. The cell-based working electrodes are post-processed and deposited on top of the chip. In the first phase of the BSA design, stimulation, measurement, and control or interface circuitries will be situated outside the CMOS chip. This allows the flexibility and focus on high-density electrodes integration. Figure 1 illustrates the design concept. It involves configuration for stimulation on the working electrodes and measurement on the common counter electrode. The size of common counter electrode is at least 2.875 mm that could cover all the working electrodes on the chip. It will be placed above the chip together with the micro-fluidic system.

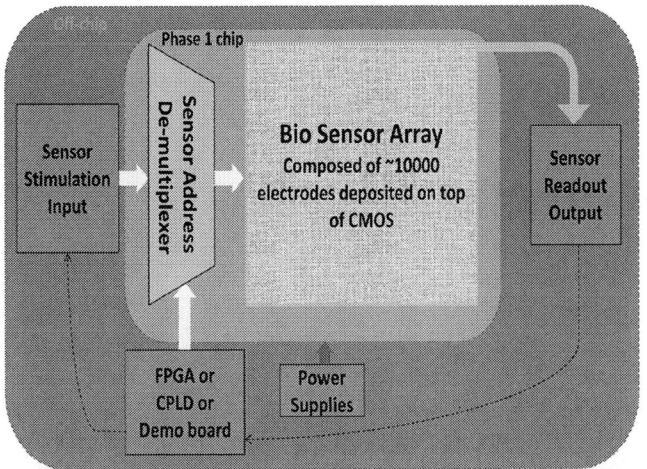

Figure 1 : Block diagram of the BioSensor Array system

III. ARCHITECTURE

The fabrication of first phase system consists of 9216 + 96 analog switches or transmission gates. Each transmission gate structure is placed underneath an electrode for shortest connection path and the individual electrode can be accessed by standard CMOS address decoding scheme controlling transmission-gate structures. The structure allows bi-directional stimulation and measurement, i.e. it is possible to stimulate the working electrode and to do the measurement on the common counter electrode, and vice-versa.

Figure 2 shows a detailed view of the proposed architecture. The full system will be modeled, designed and simulated using Cadence Analog and Mixed-Signal silicon design and verification environment. The decoders will be described in Verilog HDL while the array of transmission gates is schematic-based.

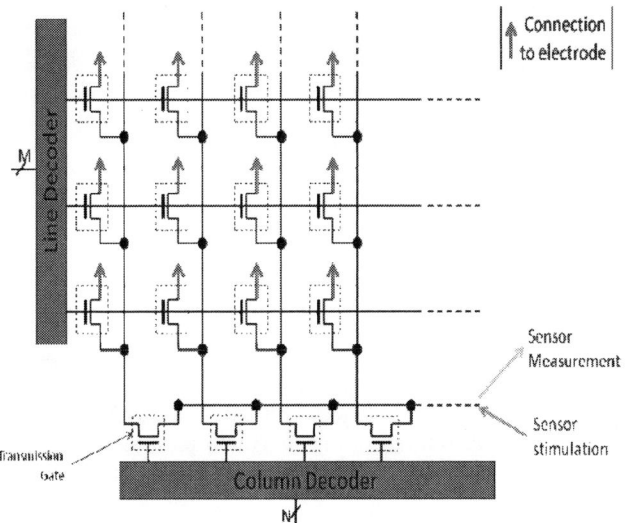

Figure 2 : First phase system architecture

A. Array sensor

The transmission gate was constructed using CMOS transistors in a 180 nm technology. The control input to the switch is a standard logic input. For the matrix of 96 x 96 transmission gates, the control signals are connected to the Line Decoder (M). On the other hand, an array of 96 transmission gates are controlled by a Column Decoder (N) for selecting the column of the matrix switches. Thus, the output of these 96 transmission gates are connected to the respective input of the matrix switches (See Figure 2).

The signal range of the transmission gate is up to 1.8 V. To achieve a smaller chip size, the CMOS gate parameter was set to (W/L)p=0.46/0.18(μm) and (W/L)n=0.23/0.18(μm). The simulated combined parallel resistance of the transmission gate has a maximum value of 17kΩ.

The physical design (layout) of the sensor architecture involved in an array of 17 μm × 17 μm Aluminum Copper (AlCu) pads (with a pitch of 30μm). Gold (Au) electrode with

---

Identify applicable sponsor/s here. If no sponsors, delete this text box. (sponsors)

978-1-61284-863-1/11 $26.00 © 2011 IEEE

a size of 25 μm × 25 μm will be post-processed on top of each AlCu pad. Figure 3 illustrates the top view of the die (note that the gold electrode is post-processed).

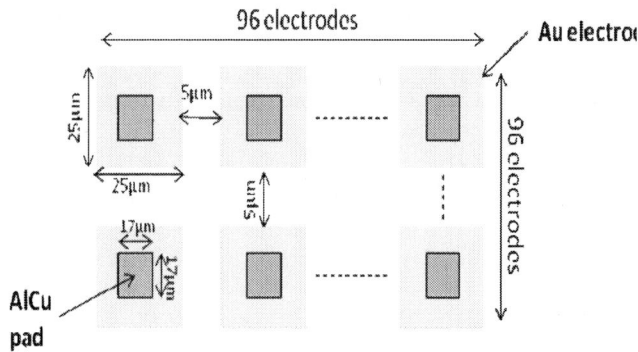

Figure 3 : Physical layout of array sensor

B. Sensor Address De-multiplexer

The sensor address de-multiplexer was formed by two 7-to-128 digital decoders: column decoder (N) & line decoder (M). This decoder block generates the control signals for selecting the 9216 analog switches. The column decoder controls an array of 96 switches from transmitting stimulation signal to the sensor. The line decoder scans the array and accesses each individual sensor at a time.

This module included dual-mode for this application: the automation mode allows selection of all sensors simultaneously and the manual mode is for selection of sensor one at a time depending on the 7-bits input. A buffer has been added at the output to ensure its driving capability to drive the load imposed by the electrode and the cell. The sensor is selected and activated according the control signal from the decoder outputs as shown Table 1.

Outputs		Results
Column Decoder<95:0>	Line Decoder<95:0>	Selected sensor
0	0	Column 0, Row 0
0	1	Column 0, Row 1
:	:	:
1	0	Column 1, Row 0
:	:	:
45	60	Column 45, Row 60
:	:	:
95	45	Column 95, Row 45

Table 1 : Control signals for selecting the electrode sensor array

C. FPGA

A commercial FPGA board (Digilent Spartan 3E Starter Board) was used in this BSA system. The FPGA is programmed to supply the required digital data to the DAC channel. It is also programmed as an IC controller. User can enter the commands to select the amplitude and frequency for the required stimulation signal. Besides, the board is to supply the input conditions to the decoder block based on user's selected command. Information of the commands will be displayed on the LCD on the board.

IV. MEASUREMENT

The integrated circuit will be fabricated by Global Foundry using a 0.18 μm technology node. The wafer will then be post-processed to implement the top structure including the gold electrodes. Figure 4 illustrates the formation of the working electrode (Au). The process begins with deposition of 1-2 μm of passivation layer, Figure 4(b), for protection from damaging the chip. It is composed of silicon dioxide (SiO2) and silicon nitride (Si3N4). Then, via regions are etched and the openings are filled with Copper (Cu) electroplating, Figure 4(c). Gold (Au) lift-off process is then employed to form the sensing/working electrodes, Figure 4(d). Finally, 2um thicknesses of passivation layers are deposited as the electrode separation gap, Figure 4(e). This allows for monitoring cultured cell network properties while limiting crosstalk due to neighboring cells.

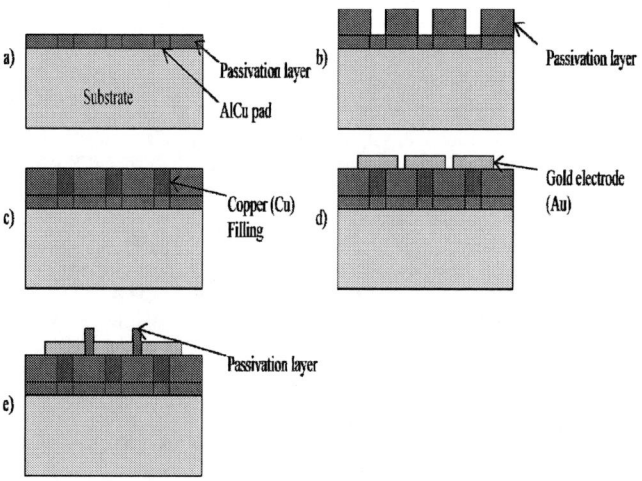

Figure 4 : Post-process flow for the electrode structure

The measurement setup is shown in Figure 5. The inputs to the BSA chip are the stimulation signal and the digital control inputs to the decoder, from the FPGA. The stimulation signal of approximately 100mV peak-to-peak will be applied to the electrode one at a time, depending on the command from the user. A micro-fluidic system, together with a common electrode, will be placed on top of the BSA chip. The CTC cell will be transported through system and adhered on the surface of working electrodes. The resulting current can be monitored and measured at the common counter electrode to provide the real and imaginary parts of the impedance of each electrode.

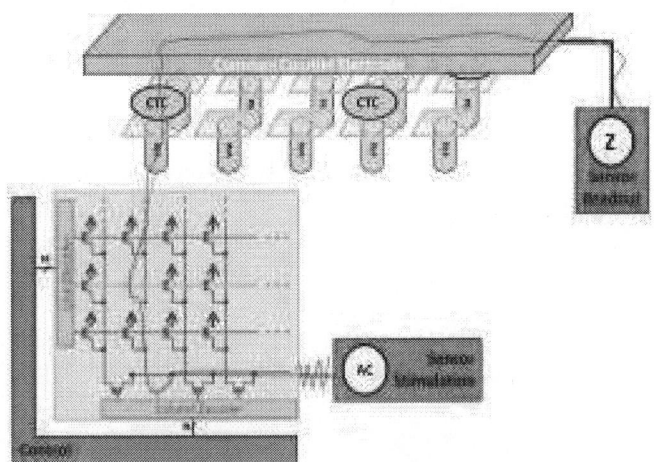

Figure 5 : Measurement setup

V. CONCLUSIONS

CMOS implementation of a high density BioSensor Array for CTC detection was presented for measurement analysis in this paper. The system will be tested to assure reliable high-throughput impedance detection.

Future works will focus on fully integrated the stimulation and readout into a CMOS IC as a fully integrated CMOS BSA offers the advantages of high accuracy, low cost, and small size for biomedical application. This will achieve a complete workflow, including the cell delivery, processing, and measurement, on a single chip.

REFERENCES

[1] Wataru S., Mashiro K., Taizo U., Hitoshi S., Tomokazu M. Kosuke I., "Addressable electrode array device with IDA electroeds for high-throughput detection," *Lab on a Chip*, vol. 10.1039/c01c00437e, p. 4, Dec 2010.

[2] David W.G., Nguyen D., Michael M.D. Huang X., "Simulation of Microelectrode Impedance Changes Due to Cell Growth," *IEEE Sensors Journal*, vol. 4, no. 5, pp. 576-583, Oct 2004.

[3] B.R. Eggins, *Chemical Sensors and Biosensors*.: ISBN, 2002.

[4] Nader P. Jonathon S.D., "Label-Free Impedance Biosensors: Opportunities and Challenges," *Electroanalysis*, vol. 19, no. 12, pp. 1239-1257, 2007.

[5] Fareid A., An C., Ryan B., Miqin Z., Jian X. Myo T., "Response Characteristics of Single-cell Impedance Sensors employed with surface-modified Microelectrodes," *Biosensors and Bioelectronics*, pp. 1363-1369, Jan 2010.

Closed loop wireless power transmission for implantable medical devices

Luis Andia, Rui-Feng Xue, Kuang-Wei Cheng, Minkyu Je
Institute of Microelectronics – A*STAR (Agency for Science, Technology & Research)
11 Science Park Road, Singapore Science Park II, Singapore 117685
monteslaa@ime.a-star.edu.sg

Abstract—This paper describes a closed loop wireless inductive power transfer system for an implantable brain machine interface. The proposed system is designed to ensure optimal power transfer by an off-body unit, battery powered, to an in-body implanted unit while guarantying a minimum transmitted power level for proper operation and a maximum level to avoid brain tissue damage due to implanted electronics overheating. The closed loop ensures a minimum power of 20mW delivered to the implanted unit with a power amplifier measured power added efficiency of 69%. As part of the closed loop a microcontroller interprets a backscattered pulse train signal modulating a PWM signal duty cycle in order to control power amplifier drain bias voltage by using a PWM controlled DC/DC converter.

Keywords-Biomedical power supplies; implantable biomedical devices; RF powering;transcutaneous power transfer, switching amplifiers.

I. INTRODUCTION

World population aging and implied healthcare rising costs forces research community to find new solutions to alleviate public expenses on this field. Recent improvements in low and ultra-low power design techniques in deep sub-micron CMOS technologies have facilitated market introduction of a big number of inexpensive sensors specially conceived to monitor various human physiological signals as well as to stimulate different human body tissues for medical and well-being purposes [1] [2]. These sensors must operate for a long period of time powered by batteries, most of the time small button cell ones. Off-body sensors operation relies on attached batteries which could be easily replaced. On the other hand, in-body sensors battery replacement implies a chirurgical intervention. This could be acceptable if battery lifetime lasts for a decade or longer, but is not acceptable for lower time periods due to high risks related with this kind of interventions [3] [4]. An alternative method to remotely power sensors is then required.

In the recent years neuroscientists and clinicians have begun to use implantable MEMS multi-electrode arrays to further explore brain function. These electrode structures are inserted into the cerebral cortex allowing, together with an electronic system, to record, process and stimulate the nearby nerve cells signals. Data from implanted multi-electrode array could be collected using bundles of fine wires that tether the array to a skull-mounted connector which is wired to an off-body amplification and recording system. This method implies not only high health risks of infections but also provides a path to couple noise to the wires conveying weak neural signals [5]. Different research teams, [5], [6], [7], have proposed fully implantable neural recording systems powered wirelessly. In all these systems a closed loop control is essential to adequate power level delivered to the implant as elevated temperatures produced by overheated electronics can easily kill the neurons one is trying to observe. This paper proposes a closed-loop wireless inductive power transmission subsystem intended for such fully-implantable systems. It is organized as follows: after a brief introduction a second section refers to the system design; there a highly efficient power amplifier stage is presented and its design steps are detailed, second section also deals with the inductive link as well as closed loop design. Then a third section resumes the simulated and measured performances of the system. A fourth one is dedicated to the conclusions.

II. SYSTEM DESIGN

Figure 1 depicts the block diagram of the proposed wireless power transmission subsystem. Output power level control closed loop signals are there represented and they will be further explained in section C.

As coils generally have better quality factor, Q, at higher frequencies, ω; free space inductive link loses could be reduced choosing a frequency in the range of some hundreds to thousands MHz. Nevertheless, for a human body implant, higher frequency makes the link much more sensitive to parasitics, increase tissue power absorption and require much more precise tuning of the power amplifier and its drivers. Considering these tradeoffs we have chosen an operation frequency of 13.56MHz, which is a free of access ISM band.

A. High Efficiency Power Amplifier

In a class E power amplifier [8] a capacitor is shunted with the transistor (switch) to control drain voltage charge/discharge time and then, minimize current and voltage overlapping during commutation, resulting in higher efficiency. Output network resonates at carrier frequency allowing only the fundamental component to flow on the load.

This work was supported by the Science and Engineering Research Council of A*STAR (Agency for Science, Technology and Research), Singapore under Neurodevices program. The grant number for the project is 102-171-0161.

978-1-61284-863-1/11 $26.00 © 2011 IEEE

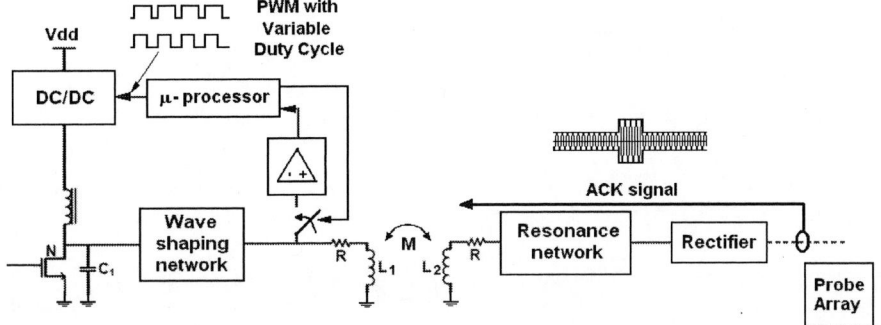

Figure 1. Wireless power transmission subsystem block diagram

For class E design, ideally an infinite DC feed inductor should be used for transistor biasing in order not to modify charge/discharge time of load network. In [9] analytical design equations with finite DC inductance are proposed, we have employed these equations to design the power amplifier.

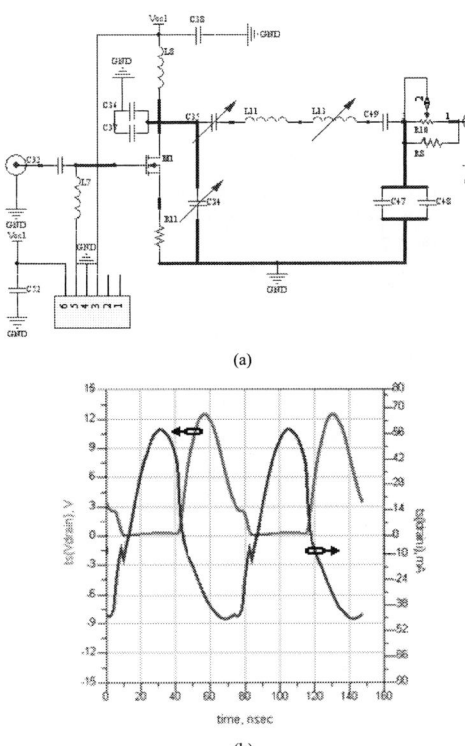

(a)

(b)

Figure 2. Class E power amplifier (a) schematic and (b) simulated drain waveforms

In order to validate the ideas here presented the power amplifier is implemented with off-the-shelf inexpensive discrete components. We have employed a N-channel MOSFET transistor (model 2N7002K) from Vishay Siliconix which has a low ON resistance, r_{sat} of 2Ω, high drain to source breakdown voltage, BV_{DS} of 60V, and fast switching speed of 25ns. Spice model of such transistor is available at manufacturer website [10] and has been used, together with passive components, inductors and capacitors, models from different vendors for simulation purposes. For tuning purposes, in order to compensate implementation parasitics, trimmer capacitors and tunable inductors have been included in the power amplifier circuit depicted on Figure 2(a). Agilent ADS has been used to perform harmonic balance and transient simulations of the circuit which drain voltage and current transient waveforms are illustrated on Figure 2(b).

B. Inductive Link

Inductive link coils had been designed on a commercially available PCB substrate, Rogers 4350B. Design had been made considering all human brain dielectric properties, as conductivity and permittivity. Unfortunately, to the best of our knowledge, brain tissue dielectric properties at 13.56MHz are not available in scientific literature; approximations have been made from available data [11] [12].

Using Ansoft HFSS software, PCB substrate specifications and brain tissue dielectric data, inductive link, external and implanted coils depicted in Figure 3, had been designed. For a nominal distance of 10 mm between coils and considering skin in between we have estimated a minimum coupling coefficient, k of 0.05. As a compromise between size and quality factor at 13.56 MHz transmitter coil and receiver coil of inductance values of 9.6 and 2.1µH, respectively, are implemented. External coil of Q factor 82, counts 14 concentric turns of 0.5mm width copper line, each turn separated 0.4mm from the next one. Implanted coil of Q factor 67, counts 18 concentric turns of 0.3mm width copper line, each turn separated 0.1mm from the next one. External coil PCB has a size of 620 x 25mm while implanted one has a size of 25 x 10mm.

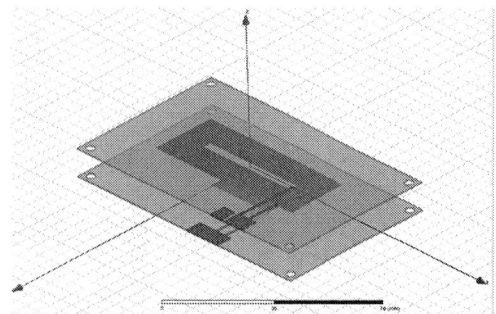

Figure 3. External and implanted coils

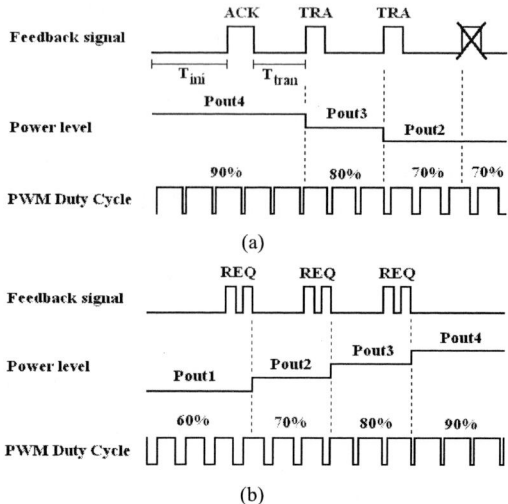

(a)

(b)

Figure 4. Closed loop signals to (a) decrease power level and (b) increase power level

C. Closed Loop

As we mention before, a closed loop for power delivery control is essential to prevent implanted system failure due to insufficient received power to guarantee regular operation or health risks due to implanted electronics overheating caused by excessive power dissipation. Power transmitter works in a time division multiple access basis; during an initial period, T_{ini} carefully establish to avoid health risks, the power amplifier is biased to deliver its maximum output power, after T_{ini}, an acknowledgement signal, ACK on Figure 1 and Figure 4(a), is send back by modulation of the implanted inductor impedance. ACK acknowledges power reception on implanted side. Once the external unit has detected ACK signal the microprocessor is programmed to wait a period of time T_{tran}, for another pulse signal TRA. After T_{tran}, if TRA is detected the power amplifier bias is set to reduce delivered power; this procedure is repeated until the optimum power level is received at implant side, as shown in Figure 4(a). Alternatively, if implanted system detected a near to insufficient power level a REQ pulse is send back to the external unit. REQ signal trigger an exception of the microcontroller translated in increased delivered power level, as in Figure 4(b). REQ and ACK pulses are sending back to the external unit whenever a power level adjustment is necessary.

Power level delivered to the implant is directly dependant on power amplifier output power which is controlled by its transistor drain bias voltage. The microprocessor represented in Figure 1, part of the closed loop, is programmed to deliver a PWM signal with duty cycle controlled by ACK and REQ pulses as detailed in Figure 4(a) and Figure 4(b). A DC/DC converter translated this PWM signal into a DC voltage level which magnitude is set by its duty cycle. DC/DC converter is set to deliver up to nine different DC voltages levels for duty cycles ranging from 10%, for the lower DC voltage, to 90% for the highest DC voltage level corresponding to highest output power level.

III. SIMULATIONS & MEASUREMENTS RESULTS

The whole wireless power transmission subsystem has been implemented using off-the-shelf inexpensive components, as shown in Figure 5.

Figure 5. Wireless transmitter power amplifier test board

Power amplifier simulations results have shown a power added efficiency, PAE of more than 80% for power levels of 27 to 65mW at the implant rectifier output. Simulated PAE is more than 60% for power levels comprised between 15 and 80mW, as shown in Figure 6. Inductive link k factor of 0.05 and rectifier efficiency of 90% have been considered for the simulation setup.

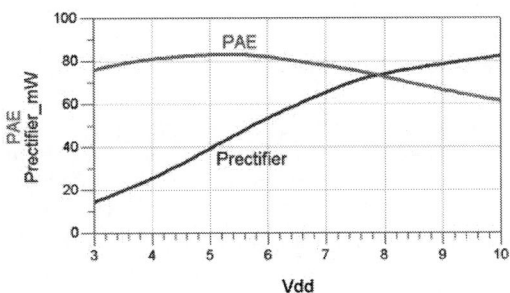

Figure 6. PAE and implant rectifier available power level as function of drain bias voltage

Measurements results of efficiency are close to simulated ones, PAE is 69% for an amplifier output power of 110mW which could be translated to around 20mW power available at implant rectifier output. Drain bias voltage is 6V and peak drain voltage is around 20V which is far below 60V BV_{DS} of transistor.

IV. CONCLUSIONS

We have designed, implemented and measured a high efficiency switch-mode power amplifier using off-the-shelf inexpensive discrete components. At 20mW implant power consumption we have measured an efficiency of 69% while transistors drain voltage swinging is far below its drain-source breakdown level. An inductive link composed of one external and one implanted coil has been designed to maximize power transmission capability at 13.56MHz while keeping a reduced shape factor. Furthermore, adding a closed loop to the system has permitted to ensure a minimum power delivered to the implant (enough to avoid failures due to insufficient power resources) as well as maximum power level.

ACKNOWLEDGMENT

Authors would like to thanks C. S. Yong and K. C. Lim from CDS/ICS laboratory for his precious help in the test board design and P. B. Khannur as well as K. W. Cheng and all members of BMIC/ICS laboratory for their support.

REFERENCES

[1] R. Bashirullah, "Wireless implants," IEEE Microwave Magazine, vol. 11, no. 7, pp. S14-S23, Dec. 2010

[2] M. Sivaprakasam, W. Liu, G. Wang, J. D. Weiland, and M. S. Humayun, "Architecture tradeoffs in high-density microstimulators for retinal prosthesis," IEEE Trans. Circuits Syst. I, vol. 52, no. 12, pp. 2629–2641, Dec. 2005.

[3] B.A. Walker, A.H. Khandoker and J. Black, "Low cost ECG monitor for developing countries," Intelligent Sensors, Int. Conf. on ISSNIP, pp. 195-199, Dec. 2009.

[4] L. S. Y. Wong, S. Hossain, A. Ta, J. Edvinsson, D. H. Rivas, and H. Nääs, "A very low-power CMOS mixed-signal IC for implantable pacemaker applications," IEEE J. Solid-State Circuits, vol. 39, no. 12, pp. 2446–2456, Dec. 2004.

[5] R. R. Harrison, P. T. Watkins, R. J. Kier, R. O. Lovejoy, D. J. Black, B. Greger, and F. Solzbacher, "A low-power integrated circuit for a wireless 100-electrode neural recording system," IEEE J. Solid-State Circuits, vol. 42, no. 1, pp. 123–133, Jan. 2007.

[6] M. Mojarradi, D. Binkley, B. Blalock, R. Andersen, N. Ulshoefer, T. Johnson, L. Del Castillo, "A miniaturized neuroprosthesis suitable for implantation into the brain," IEEE Trans. on Neural Systems and Rehabilitation Engineering, vol. 11, no. 1, pp.38-42, March 2003.

[7] G. Wang, W. Liu, M. Sivaprakasam, and G. A. Kendir, "Design and analysis of an adaptive transcutaneous power telemetry for biomedical implants," IEEE Trans. Circuits Syst. I, vol. 52, no. 10, pp. 2109–2117, Oct. 2005.

[8] N. Sokal and A. D. Sokal, "A class-E: A new class of high-efficiency tuned single-ended switching power amplifier," IEEE J. Solid State Circuits, vol. 10, no. 3, pp. 168–176, June 1975.

[9] M. Acar, A. J. Anema and B. Nauta, "Analytical design equations for class-E power amplifiers," IEEE Trans. on Circuits and Systems, December 2007, vol. 54, no. 2, pp. 2706-2717.

[10] http://www.vishay.com/mosfets/v-ds-gteq-31-v-lteq-80-v as in May 2011.

[11] G. Schmid, G. Neubauer and P.R. Mazal, "Dielectric properties of the human brain measured less than 10 hours post mortem," Proceedings of the General Assembly of the International Union of Radio Science, pp. 1759-1762, The Netherlands, August 2002.

[12] S. Gabriel, R. W. Lau, and C. Gabriel, "The dielectric properties of biological tissues: II. Measurements in the frequency range 10 Hz to 20 GHz," Phys. Med. Biol., vol. 41, no. 11, pp. 2251–2269, Nov. 1996.

High-voltage Pulser for Ultrasound Medical Imaging Applications

Dongning Zhao[1], Meng Tong Tan[1], Hyouk-Kyu Cha[1], Jinli Qu[1], Yan Mei[2], HaoYu[2], Arindam Basu[2], Minkyu Je[1]

1. Institute of Microelectronics, A*STAR (Agency for Science, Technology and Research)
2. Nanyang Technological University
Singapore
zhaod@ime.a-star.edu.sg, jemk@ime.a-star.edu.sg

Abstract— **This paper presents the design and implementation of a fully integrated high-voltage (HV) front-end transducer for MEMS ultrasonic applications. Each of the front-end transducers in the array includes a HV transmitting driver and a 50MHz capacitive micro-machined ultrasound transducer (CMUT). The design adapts low voltage integrated circuit (IC) design techniques to create integrated high voltage interfaces. Effective methods to implement high voltage interfaces to integrated CMOS/DMOS circuits are presented. Using a 0.18μm CMOS/DMOS process, simulation results of this interface IC show that can generated a 30V high voltage pulse from a 1.8V input triggering signal to drive a 50MHz MEMS ultrasound transducer with the largest 44pF capacitive load.**

Keywords-ultrasound; MEMS transducer; high voltage pulser

I. INTRODUCTION

In recent years, the capacitive micro-machined ultrasound transducer (CMUT) technology had become a popular alternative to the conventional piezoelectric transducer after the development of silicon micromachining techniques. CMUT's offer various advantages over piezoelectric transducers, including integration with readout electronic circuits, wider bandwidth which results in better axial resolution, increased flexibility in array design, and operating over a wider temperature range, offering high frequency/high-resolution capability to replace the bulky, discrete solutions.

Three-dimensional (3D) ultrasound imaging provides important clinical benefits beyond those of traditional two-dimensional (2D) ultrasound imaging. It also greatly increases the utility of analyzing images after the examination, potentially leading to less difficult and less expensive examinations. Currently three-dimensional ultrasound imaging system is substantially more complex than 2D imaging system. The system functional block diagram for one channel is shown in Fig 1. Multiple channels and transducers are required for high-performance system. As shown in Fig. 2, two major issues associated with 3D ultrasound imaging are: increased element count and the reduced signal strength of individual array elements due to limitations in physical dimensions. One solution to interfacing electronics with 2D transducer arrays is to combine the transducer array with the integrated circuit (IC). Implementing more of the system electronics with an IC can reduce the cost of 3D imaging systems. Each of the front-end

interface circuit designed for ultrasound system consists of a driver circuit, a protection circuit, and a readout amplifier. The advantage of IC pulser is that they can be provided to every element in the array without expensive external electronics or numerous cables.

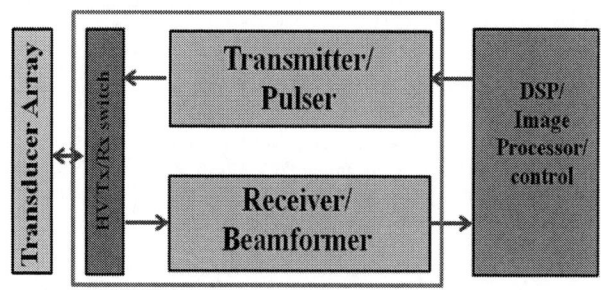

Figure 1. System diagram of ultrasound imaging system.

This paper presents a design methodology that integrates HV components with low-voltage (LV) components in order to achieve compact density, thus reducing layout and board area, thereby decreasing production costs and increasing packaging density for the same area. Several integrated HV integrated design techniques were reported [1-4] to create various high voltage interfaces. However, the resonance frequencies of those transducer elements are far less than 50MHz, and the equivalent capacitances are less than 5pF or not reported. In this design, the technique and process technology presented can be used for circuits operating at voltages higher than 30V in the future. The driver is capable of generating narrow HV pulses with the largest capacitive load and with shorter rise and fall times as compared to other works [2-5]. This fast transition time will mitigate the common problem of having a large short circuit current and thus results in reduced power consumption of the HV driver.

As shown in Fig. 2, the capacitive micro-machined ultrasonic transducer (CMUT) in our work consists of a suspended membrane built on a conductive silicon substrate. The CMUT is operated by means of electrostatic forces: an applied DC bias voltage causes the membrane to deflect toward the substrate, while an AC pulse imposed on the device causes the membrane to vibrate, emitting acoustic power to the

978-1-61284-863-1/11 $26.00 © 2011 IEEE

surrounding medium. When used for reception, the incident acoustic field causes a change in the device capacitance, which will be detected by the sensing circuit for further digital signal processing to create medical imaging for evaluation. The front-end interface circuit designed for 3D ultrasound imaging contains a 2D array of circuits, each consisting of a driver circuit, a protection circuit, a readout amplifier, and a bonding pad for vertical connection to the same sized transducer array by flip-chip bonding or CMUT-on-CMOS.

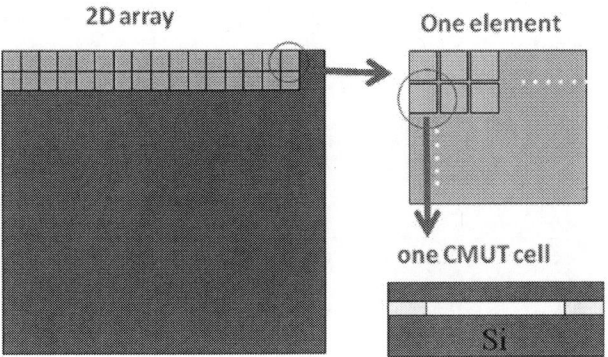

Figure 2.   Diagram of 2D CMUT and cross-section view of one CMUT device.

A common problem for virtually any ultrasound system is to protect the receiver from the high-voltage pulse. Several schemes have been developed to either isolate or minimize the voltage present at the input of the receiver. One common approach is an expander/limiter scheme [6-7]. The expander is a diode bridge placed in series with the transmitter circuit that isolates it from the MEMS element and the receiver circuit during the receiver mode. The limiter, on the other hand, is a diode bridge placed in parallel with the receiver circuit and shut the transmitter current at the input of the receiver circuit during transmit. But the diode limiter must be used in conjunction with a series resistor or more sophisticated tuning circuits to assure that the transmit circuit and the MEMS element are not short circuited to ground. However, these termination approaches introduce signal loss that results in lower signal-to-noise ratio [6-7], and exhibit excessive ringing [8].

An alternative approach can be achieved by high-voltage MOSFET switches, which require drivers and level-shifters, instead of bridge diodes to isolate the receive circuit [8]. Though the series switches will still introduce an insertion loss due to the series resistance from the physical switch, this approach resolves the voltage-dependant mismatch and oscillations present in the diode-bridge limiter circuits and allow for wideband applications. This typically requires high-voltage power supplies.

In this paper, we present the design of the outlined front-end integrated pulser circuit for 2D CMUT arrays. The high-voltage CMOS front-end pulser and protection circuits, as well as the application of the non-overlapping clocks regarding to short-circuit current reduction during the output transitions are presented in Section II. The full functional verification of the circuit is presented in Section III.

II.   CIRCUIT DESIGN & IMPMENTATION

A. High Voltage Pulser

The primary objective of the driver circuit is to generate the pulse signal at the output without violating the safe operating requirement of the high-voltage MOSFETs. The double diffused metal oxide semiconductor (DMOS) transistor has become the primary choice for HV integrated circuits. While the drain-source voltages of the HV MOSFET can stand up to 30v, the source-gate voltage cannot exceed 6V. Implementing a complex circuit with only a high voltage supply is usually impractical. HV transistors are much larger than LV transistors, which results in higher parasitic capacitance and higher dynamic power consumption. With most of a circuit functionality implemented in LV circuits, a key circuit design issue is to develop effective LV-HV interfaces, i.e. a level-up shifter that transforms signals from the LV core power supply domain to the HV I/O power supply domain.

The high-voltage CMOS (HVCMOS) digital output interface shown in Fig. 3 is similar to a simple CMOS inverter, except for the gate connections. The two transistors, M1 and M2, are driven by two signals created from low-voltage logic (triggering). The gate of the HVNMOS transistor can be readily controlled by the low-voltage logic level 0 and V_{DD} (low), where V_{DD} is the low-voltage power supply. The gate of the HVPMOS transistor needs a level shift to operate between V_{DDH} and V_{DDH}-V_{DD}.

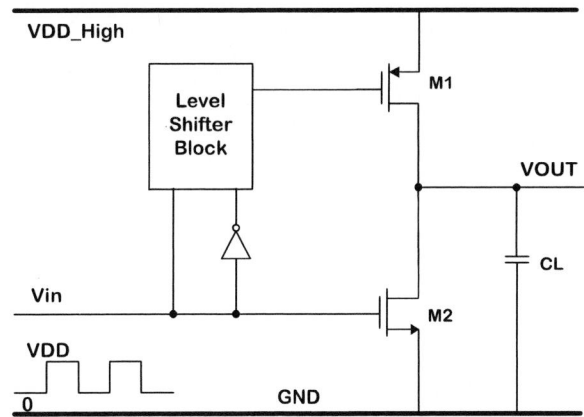

Figure 3.   Function diagram of pulser with level-shifter.

As shown in Fig. 4, a fully integrated drive amplifier is proposed. It creates ultrasound pulses when the ultrasonic transducer element is triggered. The drive amplifier is composed of a static level-up stage and a class-D switching output stage. Transistors, M10 and M11, are the high voltage transistors which constitute the output driver of the circuit. The remaining transistors (M3, M4, M5 and M6 have isolated bulk and M1 and M2 are the high voltage transistors) in the circuit do the level shifting of the signal that drives the gate of the transistor M10. The signals, Pulse1 and Pulse2, are the low-voltage signals which are complementary to each other.

978-1-61284-863-1/11 $26.00 © 2011 IEEE

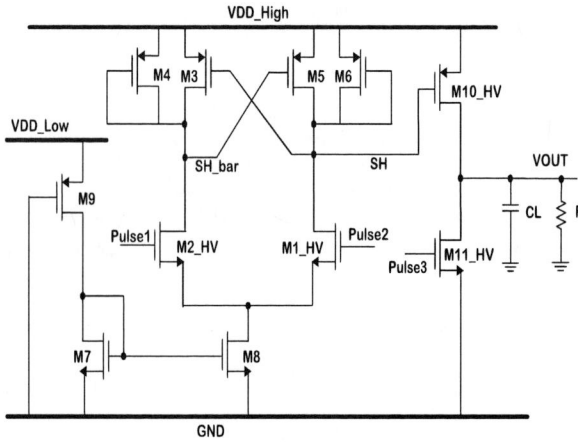

Figure 4. Circuit Schematic of the pulser.

During the operation, transistor M9 is a low-voltage PMOS transistor with fixed V_{GS} which will be mirrored to the transistor M10. The power supply of the drive amplifier connected to the output node of V_{DD_high} is 30V. The excitation HV pulse generated by the drive amplifier is triggered by the Pulse1, Pulse2 and Pulse3. The signal Pulse1 is used to control a unipolar HV pulse applied across the transducer element, Pulse2 with phase delay is used to completely turn off the transistor M10 during the pulse repetition time, and finally Pulse3 is used to discharge M11 by connecting it to ground. The triggering pulse duration is 20ns, which corresponds to the 50MHz resonating frequency of the transducer element. Notice that the width of M10 and M11 is determined by the output current stage. To avoid turning on M10 and M11 simultaneously, and to decrease the power consumption of the driver, the timing of Pulse1, Pulse2 and Pulse3 are adjusted carefully. Therefore, by changing the polarity of the triggering pulse signal, either transistor M10 or M11 turns on, connecting the output to the high-voltage source or to ground, generating high-voltage unipolar pulses at the output of the driver.

One of the main requirements of the ultrasound pulser is small output capacitance. The parasitic capacitances associated with the interconnect lines and the transistors M10, M11 at the output driver stage are of an order of magnitude larger than the transducer capacitances. Those large capacitances can significantly reduce the output signal amplitude and will seriously degrade the element sensitivity. Therefore, optimized physical sizes of M10 and M11 are preferred, so that the generated pulse can have sharp rise and fall times to better excite on the ultrasound transducer elements.

B. Non-Overlapping Clocks for Short-Circuit Current Protection

As mentioned earlier, it is important to prevent the PMOS and NMOS transistors of the 30V output stage from turning on at the same time during the signal transition of the input pulse. This is because in the ultrasound array, the short-circuit current of each pulser will create a huge current spike in the high voltage supply line. The PMOS and NMOS transistors can be prevented from turning on at the same time during signal

transition by using two non-overlapping clock signals to drive the PMOS and NMOS transistors.

Fig. 5 shows the simulated short-circuit currents of the Pulser with and without the non-overlapping clock. From the figure, it can be observed that the short-circuit leakage current of 270mA can be removed effectively by adding a delay time for the two control signal, Pulse 1 and Pulse 3, of the PMOS and NMOS transistors, respectively, during the signal transitions.

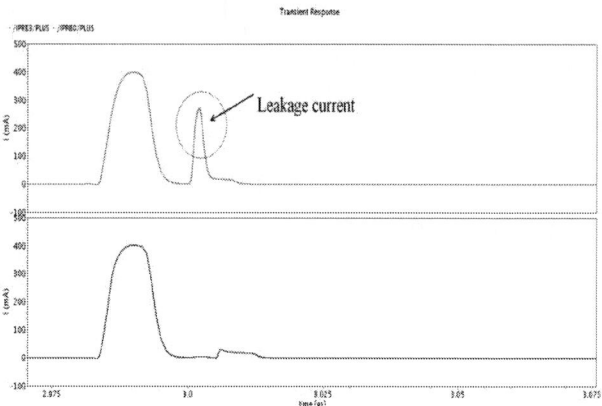

Figure 5. Short-circuit current of PMOS and NMOS transitors with and without non-overlapping clock singals.

C. Protection Scheme

The driver circuit generates high-voltage pulses (30V) on the node where the CMUT is connected. The receiving readout circuit was designed for low-voltage operation, making it feasible to fit inside the unit cell area. To prevent high-voltage pulses from damaging the input transistors of the readout amplifier, a protection circuit must be placed between the CMUT and the receiving readout circuit.

In this work, as shown in Fig. 6, a passive protection circuit, comprising an expander at driving side and HV switch at sensing side, is used in our protection scheme. A single HV NMOS transistor is used as a switch. During the transmission, the pulser generates a high-voltage pulse, the expander conducts, and the pulser is connected to the transducer. The switch will be at OFF state to protect the sensing circuit, while the drain terminal of the switch withstands the high-voltage pulse. When the pulse terminates, the expander diodes stop conduction, isolating the pusler from the transducer during the receiving phase. The NMOS transistor turns on with very small ON-resistance, compared to the transducer output resistance.

Another alternative protection scheme is to add an extra digital logic control to the circuit to turn off M9 and put M10 and M11 in Fig. 4 at high-impedance state. This will save the static power consumption when the circuit is not transmitting pulse, providing extra protection to the circuit. In this scheme, the diode bridge is optional to save the IC die area in the array.

978-1-61284-863-1/11 $26.00 © 2011 IEEE

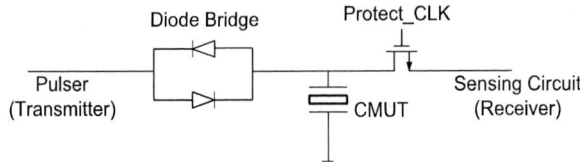

Figure 6.   Protection scheme between pulser and receiver.

The high voltage Pulser is fabricated using Global Foundry 0.18μm 30V technology process. This technology provides high-voltage-enabled MOSFETs (HV MOSFETs) operational up to 30V as well as low-voltage standard 5/6V MOSFETs. In order to avoid channel hot carrier effects and eventual avalanche breakdown, the method of extending the vertical length and doping concentration of the n-well at the drain region of a HV MOS transistor is used. As a result, HV MOSFETs occupy a wide area. The rest of the driver triggering circuits uses standard 0.18μm 5/6V GF process components to optimize the area.

III.   FUNCTION VERIFICATION

The functional verification of the overall design was verified by using Cadence Spectre simulator and the overall circuit transient simulation result is shown in Fig. 7. The nodes are observed from trigger input to the amplifier output. A 5V square wave with 20ns pulse width was used as the triggering signal for the driver circuit, which produces a 29V driving pulse at the output with protection scheme ~~is~~ implemented in the circuit. A second triggering pulse with 5ns delay was applied to the pull-down NMOS.

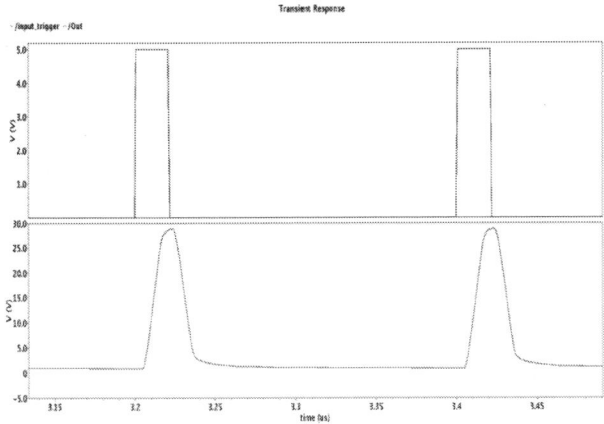

Figure 7.   Simulated output waveform (bottom) vs. input triggering signal (top)

IV.   CONCLUSION

We present the design and implementation of a fully integrated front-end transmitter dedicated to interface 2D CMUTs in ultrasonic applications, using 0.18μm CMOS/DMOS 30V HV fabrication technology from Global Foundries. New fully integrated HV drive amplifier has been described. The drive amplifier consists of a level-up stage and a Class-D switching output block. Simulation results of the fully integrated front-end transmitter validate its ability to meet the specifications of medical ultrasonic applications. The proposed circuit has been sent for fabrication and measurements will be done when the prototypes are ready.

REFERENCES

[1]   I. Ladabaum, X. C. Jin, H. T. Soh, A. Atalar, and B. T. Khuri-Yakub, "Surface micromachined capacitive ultrasonic transducers," IEEE Trans. Ultrason., Ferroelect., Freq. Contr., vol. 45, no. 3, pp. 678–690, May 1998.

[2]   I. O. Wygant, et al. "An integrated circuit with transmit beamforming flip-chip bonded to a 2-D CMUT array for 3-D ultrasound imaging,"IEEE Trans. Ultrason., Ferroelect. Freq. Contr. vol.56, no.10, pp. 2145-2156, 2009.

[3]   I. Cicek, A. Bozkurt and M. Karaman, "Design of a front-end integrated circuit for 3D acoustic imaging using 2D CMUT arrays," IEEE Trans. Ultrason., Ferroelect. Freq. Contr. vol.52, no.12, pp. 2235-2241, 2005.

[4]   I. O. Wygant, et al. " Integration of 2D CMUT arrays with front-end electronics for volumetric ultrasound imaging," IEEE Trans. Ultrason., Ferroelect. Freq. Contr. vol.55, no.2, pp.327-342, 2008.

[5]   R. Chebli and M. Sawan, "Fully integrated high-voltage front-end interface for ultrasonic sensing applications," IEEE Trans. Circuit and Systems-I: Reg. papers, vol. 54, no. 1, pp. 179-190, 2007.

[6]   J. K. Poulsen, "Low loss wideband protection circuit for high frequency ultrasound," in Proc. IEEE Ultrason. Symp., 1999, pp. 823–826.

[7]   G. R. Lockwood, J. W. Hunt, and F. S. Foster, "The design of protection circuitry for high-frequency ultrasound imaging systems," IEEE Trans. Ultrason., Ferroelect., Freq. Contr., vol. 38, no. 1, pp. 48–55, Jan. 1991.

[8]   N. C. Chaggares, R. K. Tang, and A. N. Sinclair, "Protection circuit and time resolution in high frequency ultrasonic NDE," in Proc. IEEE Ultrason. Symp., pp. 819–822, 1999.

ACKNOWLEDGEMENTS

This work is funded by the Biomedical Engineering Program (BEP), Agency for Science, Technology, and Research (A*STAR), Singapore.

Design of a Radiation Tolerant CMOS Image Sensor

Xinyuan Qian[1], Hang Yu[1], Bo Zhao[1], Shoushun Chen[1] and Kay Soon Low[2]

[1]VIRTUS IC Design Center of Excellence, [2]Satellite Research Center,
School of Electric and Electronic Engineering,
Nanyang Technological University, Singapore

Abstract—**This paper presents the design of a radiation tolerant CMOS image sensor for space applications. The pixel is based on a commercially available 4T pinned photodiode architecture and is designed using a number of radiation-tolerant physical layout techniques. In addition, a simple yet robust programmable column biasing current is proposed to deal with the dramatic temperature fluctuations. A prototype chip consisting 256×256 pixel array has been implemented using TSMC 0.18 CIS process.**

I. INTRODUCTION

CMOS image sensors (CIS) have overwhelmed their CCD counterparts in many applications for their predominant advantages of providing low power consumption at low voltage operation, high integration capability for SoC design, cost effective solution from its standardized fabrication process [1][2][3]. However, for space applications, the CIS has to be tolerant to space radiation and dramatic temperature fluctuation. Extensive reviews of radiation damage to microelectronic devices were well documented in the literature [4][5]. The impact of radiation has historically been categorized into two groups: one reflects the effects over a long period of time, termed as Total Ionizing Dose (TID) and the other is the immediate result of a single radiant charged particle, known as single event effects (SEE). The TID effects are cumulative, in which the absorption of radiation energy makes permanent changes in the device. All forms of radiation are capable of generating cumulative effects and the impact on device performance is determined by the integrated history of radiation exposure. The three major and consequential effects of total ionizing radiation on standard CMOS devices are shift of threshold voltages, leakage current in NMOS transistors, and n-channel inter-transistor (isolation field) leakage current.

In addition to sharing the various radiation-induced undesirable characteristics with other semiconductor devices, image sensors are inherently susceptible to pixel leakage current. Threshold shifts may result in an inversion region connecting n-channel sources and drains along the gate oxide to field oxide transition. This produces leakage current that can be very serious in the charge sensitive applications found in image sensors. Secondly, leakage current arises from the increased surface generation/recombination rate due to the formation of interface states. These are energy levels within the bandgap of the silicon, located at the silicon-oxide interface, so that they can communicate with the carriers in the silicon. Wherever interface states are in a depletion region, they result in electron-

hole pair generation, leading to dark current and leakage. The pinned 4T APS structure (Fig.1) is now widely used because of its capability of minimizing dark current generated by interface defects in photodiode region. Extensive efforts have been made to analyze and understand the radiation effects on 4T APS [6][7][8][9][10]. These studies have outlined suggestive guidelines for designing radiation tolerant CIS based on 4T pixel architecture in deep submicron technology.

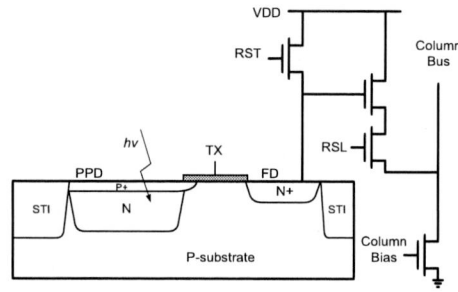

Fig. 1. Schematic and cross-sectional view of a typical 4T pixel architecture.

In this paper, we introduce the design of a CMOS Image sensor for space application. In order to achieve high stability, a variety of radiation-harden-by-design techniques are employed. Four pixel arrays with different configurations were fabricated. Due to page length constraint, we focus on one architecture and discuss the radiation-tolerant design considerations, namely: critical transistors in the pixel use enclosed layout transistor (ELT) and P+ guard ring; the N implant for the pinned photodiode is spaced from the STI; floating diffusion was carefully sized to deal with leakage current. In addition, we proposed a simple yet robust programmable column biasing current to deal with the dramatic temperature fluctuation. This paper is organized as follows. Section II discusses the pixel design and operation principle. Section III describes the VLSI implementation. Section IV concludes this paper.

II. SENSOR ARCHITECTURE

The schematic and cross-sectional view of a typical 4T pixel architecture is shown in Fig.1. It has four NMOS transistors: transfer gate (TX), reset transistor (RST), source follower (SF) and row select (RSL) and a pinned photodiode (PPD), in which a thin P+ pinning layer is utilized on the top of the photodiode

978-1-61284-863-1/11 $26.00 © 2011 IEEE

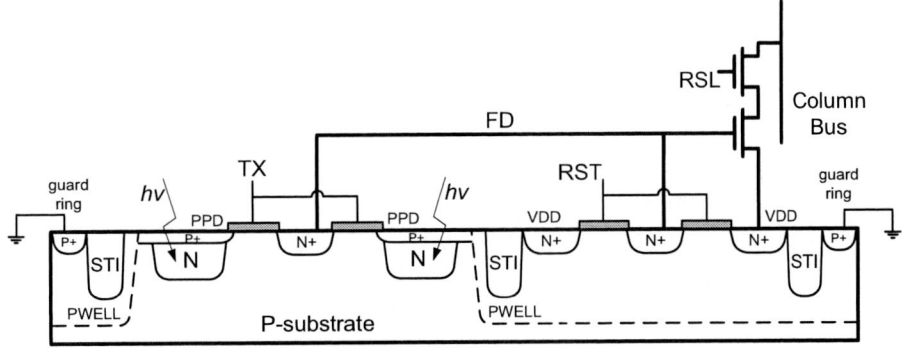

Fig. 2. Proposed radiation tolerant pixel architecture with its highlighted cross sectional view of the pinned photodiode and floating diffusion node.

to shield it from the Si-SiO2 surface defects and suppress the noise charges. The floating diffusion (FD) node is where the integrated charges in the PPD are transferred via TX and converted to the voltage signal. This is followed by a unity gain buffer (SF) and the voltage signal is conveyed to the output bus when the row is selected for readout.

The proposed radiation tolerant pixel architecture is illustrated in Fig.2. It has the same schematic as the typical 4T pixel architecture. However, extra design efforts are made to deal with dark current: critical transistors in the pixel are designed using enclosed layout transistor (ELT) and P+ guard ring; the N implant for the pinned photodiode is spaced from the STI in order to prevent the photodiode junction from interaction with the defective sidewalls and edge; floating diffusion is carefully sized to deal with leakage current. These approaches primarily highlights charge-sensitive regions of the PPD and FD node in the pixel and are explained in detail in the following sections.

A. Design of the Pinned Photodiode

The PPD is very susceptible to the ionizing radiation where the increased leakage current contributes to destroy the collected charges, and accordingly falsify the real signal. During the integration time, the PPD is open-circuited by the TX. It senses the incident light and accumulates the generated charges in the boundary of the depletion region. Fig.3 shows the cross-sectional view of the PPD in the proposed design. The photo collection region is composed of the N-implant/P-substrate junction. The pinning P+ layer has a much higher doping than the N implant so the depletion region extends only slightly into the pinning P+ layer. This efficiently isolate the photodiode from the surface defects. However, it has been found that another significant source of leakage stems from the defective sidewalls and the edges of shallow trench isolations (STIs) separating the photodiodes [11]. Protecting the STIs by P-Well structures has been proven to be effective against this type of leakage [8]. More spacing between the N-implant to STI results in higher immunity to radiation degradation, but trade off should be taken to balance the sensitivity and saturation level of the sensor (fill factor).

As shown in Fig.3, P-Well structure is used to protect

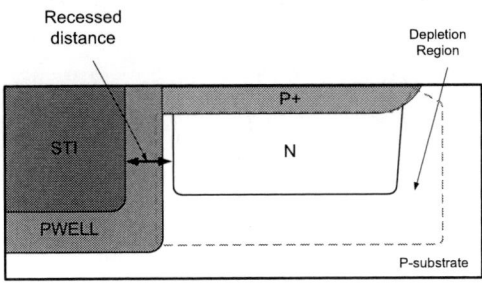

Fig. 3. Highlighted Cross-Sectional View of a Pinned Photodiode.

the PPD from STI and the space between the two was set to 0.2μm. Since the doping density of the P-Well is higher than the intrinsic P-substrate, the extension of the depletion region of the photodiode peripheral junction into the P-Well is curtailed and short compared to the intrinsic P-substrate, which additionally pushes the depletion region of the photodiode away from the STI sidewalls. It is worth to mention that the recessed STI, in fact, is realized by reducing the geometric size of the N implant of the PPD from the STI.

B. Design of the Floating Diffusion

One major effect of ionizing radiation is the increase of leakage current which arises from the inversion region connecting n-channel sources and drains along the gate oxide to field oxide transition. This produces edge leakage current that can be very serious in the charge sensitive applications found in image sensors. In our pixel, the floating diffusion (FD) node is where the integrated charges in the PPD are transferred via TX and converted to the voltage signal. Since the array is readout by sequentially scanning using column and row scanners. During readout, the FD node of the last pixels in a row are electrically floated as dynamic memories and suffer edge leakage from transistors RST and TX. Physical design techniques of enclosed layout transistor (ELT) and P+ guard ring[12] are proved to be very effective for significantly reducing leakage current in NMOS transistors. The source/drain diffusion is completely isolated by the gate and the edge leakage is significantly reduced. By employing the P+ guard ring, the inter-transistor leakage through the inverted

978-1-61284-863-1/11 $26.00 © 2011 IEEE

field oxide is substantially curtailed.

In our pixel, both transistors RST and TX are designed using ELT. As shown in the Fig. 2. The floating diffusion is composed of two enclosed geometry. In this configuration, the floating diffusion has no directly interaction with the defective STI, and the edge leakage current from these two NMOS transistors is considerably reduced. The leakage on the floating diffusion is only dominated by the junction leakage which does not increase in the presence of radiation. On the other hand, the adoption of ELT transistors and P+ guard rings results in silicon area penalty and leads to reduced fill-factor. ELT transistor is usually several times larger than traditional layout and should only be used for critical transistors. The other two transistors, SF and RSL, are therefore designed using non-ELT. This is consolidated by the fact that the biasing current in the source follower is usually larger at several orders of magnitude than the leakage current.

C. Programmable Source Follower Biasing Current

Circuits designed for space applications must be able to operate under wide temperature range. The biasing current in the source follower determines the settling time on the column bus and therefore affects the readout speed. Although a bandgap reference circuit can be used to generate a stable column biasing voltage, the basing transistor itself can be vulnerable to temperature fluctuation (variation of mobility and threshold voltage). Fig. 4 shows the simulated biasing current under different temperatures. The gate voltage is constantly biased using a bandgap-generated 550 mV. It shows that the current increases almost linearly with temperature. The biasing current is designed to be $1.3\mu A$ at room temperature, but it deviates to maximum over $\pm 30\%$ within the range of temperature from $-40°C$ to $110°C$.

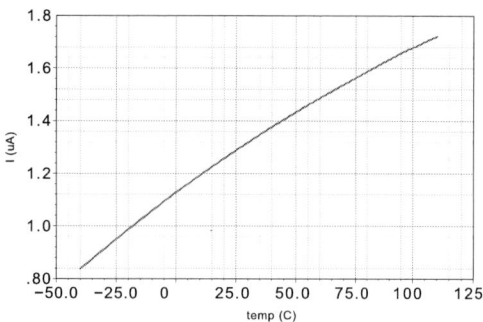

Fig. 4. Simulation result of the biasing current due to temperature change.

When switching from one pixel to another, the settling time on the column bus depends on the signal swing on the FD node. In the worst-case scenario, the FD nodes of two consecutive pixels can experience the maximum possible swing and thus need the longest settling time. Fig. 5 shows the simulation waveform on the column bus at the worst-case scenario with respect to different temperatures. The discharge requires around 350ns before the column bus attains the steady

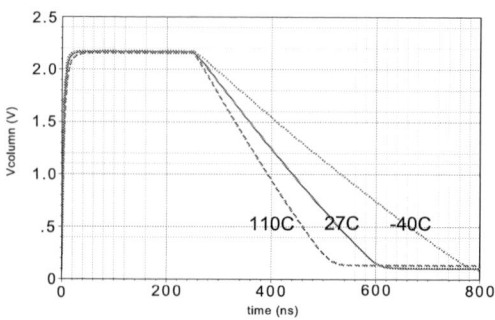

Fig. 5. Simulated waveform of the column bus showing the operation of charging and discharging under different temperature in a 256×256 pixel array. The waveform denotes the worst-case readout scenario of two consecutive pixels with maximal voltage difference. The charging operation shows no significant difference. However, discharging through the biasing transistor is strongly affected under different temperature

state at room temperature, but much longer at lower temperature ($-40°C$). The incomplete discharging within 350ns fails to attain the steady state on the column bus and violate the readout timing.

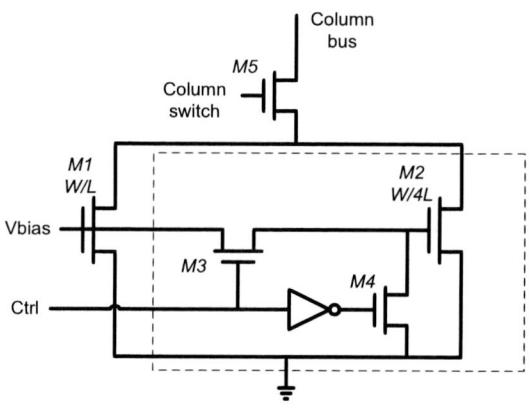

Fig. 6. Schematic of the proposed programmable column biasing circuit for temperature compensation. The circuit in the dotted-line box is enabled at low temperature when digital $Ctrl$ signal generated from the off-chip controller is high. $M3$ is turned on and $M4$ is turned off, so $M2$ is connected in parallel with $M1$ to supply an additional 25% biasing capability.

A programmable column biasing circuit is designed in order to compensate the biasing current variation due to temperature change. As illustrated in Fig. **??**, $M1$ is sized to provide nominal current for high temperature ($110°C$). While at low temperature, off-chip controller can produce a signal ($Ctrl$) which turns on transistor $M2$. The W/L ratio of $M2$ is a quarter of $M1$, the biasing current is thus scaled to 125%. This enlarged biasing current offers the opportunity to compensate the reduced biasing current due to temperature fall and therefore assure the discharge time within the timing requirements.

III. VLSI IMPLEMENTATION

A prototype chip including four arrays of 256×256 pixel was implemented using TSMC 0.18 μm CIS process (2-poly

6 metal layers). Fig. 6 (a) illustrates the layout of the chip as well as one pixel. In the pixel, the recess distance between the N implant and STI is set to 0.2μm and the P-Well structure is used in between for protection. The PPD is also geometrically designed to be as square as possible to reduce the junction perimeter to minimize the dark current. The RST transistor and the TX transistor, as discussed earlier, are designed to be ELT transistors. P+ guard ring is employed around the photodiode and the RST transistor. SF transistor and RSL transistor are designed using minimum size, and they are protected by the P+ guard ring as well to avoid inter-transistor leakage current. The pixel pitch of 6.5μm is approximately the size limitation in this compact pixel configuration for a fill factor of about 30%.

Fig. 7. (a) Layout of the proposed pixel architecture which has been embedded into a test chip. (b) Chip test platform which uses a FPGA system for chip controller and image readout. (c) Reproduced sample image

TABLE I
CHARACTERIZATION RESULTS OF THE IMAGE SENSOR

Technology	TSMC 0.18 μm CIS
Supply voltage	3.3V
Pixel pitch	6.5μm
Pixel array format	256×256
Voltage swing	1.28V
Dynamic range	60dB
Sensitivity	2.92V/lux·s
FPN	1.02%
Dark current	16.2mV/s

As illustrated in Fig.6 (b), the chip test platform is used to characterize the chip. Table I gives the summary of the preliminary characterization results and Fig.6 (c) shows a reproduced sample image under outdoor illumination conditions with about 2ms integration time and off-chip correlated-double sampling.

IV. CONCLUSION

In this work, we have presented a radiation tolerant 4T pixel architecture for space application. The design approaches of the pixel were described in detail and major design concerns lie within the design of the photodiode and the floating diffusion node against radiation-induced dark current. In order to adapt the sensor to the dramatic temperature change in space environment, a programmable column biasing current circuit was proposed. Circuit simulations demonstrate the successful operation of the sensor. A prototype chip including four arrays of 256×256 pixel was fabricated using the TSMC 0.18 μm CMOS image sensor process. We have included our preliminary characterization results of the chip.

V. ACKNOWLEDGEMENTS

This work was supported by Nanyang Assistant Professorship (M58040012) and ACRF Project (M52040132).

REFERENCES

[1] C. Shoushun, F. Boussaid, and A. Bermak, "Robust intermediate readout for deep submicron technology cmos image sensors," *IEEE Sensors Journal*, vol. 8, no. 3, pp. 286 –294, march 2008.

[2] S. Chen, W. Tang, and E. Culurciello, "A 64 × 64 pixels uwb wireless temporal-difference digital image sensor," in *Proceedings of 2010 IEEE International Symposium on Circuits and Systems (ISCAS)*, june 2010, pp. 1404 –1407.

[3] S. Chen, A. Bermak, and Y. Wang, "A cmos image sensor with on-chip image compression based on predictive boundary adaptation and memoryless qtd algorithm," *IEEE Transactions on Very Large Scale Integration (VLSI) Systems*, vol. 19, no. 4, pp. 538 –547, april 2011.

[4] H. Hughes and J. Benedetto, "Radiation effects and hardening of mos technology: devices and circuits," *IEEE Trans. Nuclear Science*, vol. 50, no. 3, pp. 500 – 521, 2003.

[5] D. Mavis and D. Alexander, "Employing radiation hardness by design techniques with commercial integrated circuit processes," in *AIAA/IEEE 16th Digital Avionics Systems Conference, 1997*, vol. 1, Oct. 1997, pp. 2.1 –15–22 vol.1.

[6] V. Goiffon, M. Estribeau, and P. Magnan, "Overview of ionizing radiation effects in image sensors fabricated in a deep-submicrometer cmos imaging technology," *IEEE Trans. Electron Devices*, vol. 56, no. 11, pp. 2594 –2601, 2009.

[7] V. Goiffon, P. Magnan, O. Saint-pe, F. Bernard, and G. Rolland, "Total dose evaluation of deep submicron cmos imaging technology through elementary device and pixel array behavior analysis," *IEEE Trans. Nuclear Science*, vol. 55, no. 6, pp. 3494 –3501, 2008.

[8] P. Rao, X. Wang, and A. Theuwissen, "Degradation of spectral response and dark current of cmos image sensors in deep-submicron technology due to gamma-irradiation," in *37th European Solid State Device Research Conference (ESSDERC)*, 2007, pp. 370 –373.

[9] J. Tan, B. Buettgen, and A. Theuwissen, "Radiation effects on cmos image sensors due to x-rays," in *2010 8th International Conference on Advanced Semiconductor Devices Microsystems (ASDAM)*, 2010, pp. 279 –282.

[10] M. Innocent, "A radiation tolerant 4t pixel for space applications," in *International Image Sensor Workshop, Bergen, Norway, 2009.*, 2009.

[11] H. I. Kwon, I. M. Kang, B.-G. Park, J. D. Lee, and S. S. Park, "The analysis of dark signals in the cmos aps imagers from the characterization of test structures," *IEEE Trans. Electron Devices*, vol. 51, no. 2, pp. 178 – 184, 2004.

[12] E. El-Sayed, "Design of radiation hard cmos aps image sensors for space applications," in *17th National Radio Science Conference, 2000*.

A Review of CMOS Multimodal Neuromonitoring Sensors and Systems

Wai Pan Chan
Institute of Microelectronics,
A*STAR (Agency for Science, Technology & Research)
11 Science Park Road
Singapore 117685
chanwp@ime.a-star.edu.sg

Minkyu Je
Institute of Microelectronics,
A*STAR (Agency for Science, Technology & Research)
11 Science Park Road
Singapore 117685
jemk@ime.a-star.edu.sg

Abstract—**A review of recent CMOS techniques applicable for multimodal neuro-monitoring including intracranial pressure, partial pressure of brain tissue oxygen, and brain temperature measurement is presented here. It summarizes how multimodal neuro-monitoring sensing can be interfaced with modern CMOS technologies. Various system architectures for interfacing generic sensors are evaluated.**

Keywords-Multmodal neuro-monitoring; Intracranial pressure (ICP); oxygen sensing; temperaure sensing.

I. INTRODUCTION

Traumatic brain injury (TBI), which is an injury to the brain due to an external force, happens mostly from transportation accidents, physical assaults, or sport injuries. In severe cases, TBI leads to fatalities, long term sensory damages, or post-traumatic amnesia, an impaired memory of events that happened before or after TBI [1]. In addition to the damage done at the moment of injury, TBI can be worsened by a secondary brain injury, which arises from an increase in the intracranial pressure (ICP) caused by an excess of cerebrospinal fluid going into the brain. If the ICP increases significantly, it can restrict the blood flow to the brain resulting tissue hypoxia and ischemia [2]. Therefore, the key management to a secondary brain injury is to have an accurate and reliable multimodal neuro-monitoring system which can measure both ICP and partial pressure of brain tissue oxygen in real time. Besides, measuring the brain temperature is also important so as to compensate the temperature sensitivity associated with the pressure sensor, oxygen sensor, and the associated electronics. Although pH or glucose concentration [3] were proposed to add extra information for the management of TBI, only will pressure sensing, oxygen sensing, and temperature sensing be discussed here as they are more adopted by hospitals toward TBI treatments.

II. SENSORS REALIZATION IN SILICON PLATFORMS

A. Pressure sensor

ICP monitoring has been commercially offered by Codman's ICP catheters for brain pressure monitoring. The key technology associated with the catheter is a silicon chip with diffused piezoresistive strain gauges. Both sides of the strain gauge are connected by resistive wires, with one side of the gauge is terminated by a movable diaphragm which is in direct contact to the brain tissue. When the diaphragm moves in respond to a change in the brain pressure, one group of the resistive wires stretches, so the electrical resistance increases.

With recent advances in Microelectromechanical Systems (MEMS) processes, capacitive pressure sensors are adopted by many designers who optimize the sensors for human body pressure sensing and accelerometer applications [4]-[5].

Note that the sensing capacitance derived from those capacitive sensors is generally very small (in the order of several tens of femto-Farad). The relative change in the capacitance associated with the sensors will also be very small. Realizing that the resolution for ICP monitoring generally confines to 1 *mmHg*, the required sensitivity of the sensor can be determined from Equation 1:

$$\frac{\frac{V_{ref}}{2^N}}{Sensitivity} < 1 \, mmHg \qquad (1)$$

where V_{ref} is the reference voltage for a N-bit ADC converter. Typical sensitivity found in capacitive CMOS pressure sensors range from 200 *μV/mmHg* to 1 *μV/mmHg* [5], so a 10-bit ADC with a reference voltage of 1 *Volt* will need a minimum sensitivity of 976 *μV/mmHg* to achieve 1 *mmHg* resolution. Besides, the useful range for ICP monitoring is generally less than 50 *mmHg* [6], so signal amplification is practically used to extend the dynamic range of the sensor's output. Note also that the sensor instrument needs to be calibrated prior measurements to maintain an inaccuracy of $\pm 2 \, mmHg$ [7].

B. Oxygen sensor

Oxygen measurements have also been commercially offered based on an optical oximetry (Neurotrend catheters) or an electrochemical detection (Licox catheters) [8]. An optical oximetry needs a light source to generate the stimuli, and it needs a photo-detector to do the optical sensing. On the other hand, an electrochemical based oxygen sensor is also popular in academic and industrial research. An electrochemical oxygen sensor consists of 3 electrodes, collectively known as Clark Cell [9]. The electrodes can be set up by using off-chip electrodes such as platinum or gold.

The correct operation of the Clark cell will enforce a reduction of oxygen to take place at the working electrode, only when the potential of the working electrode is lower than the reference electrode by approximately from 0.6 to 0.8 *Volt* [10]-[11]. In addition, the use of the counter electrode is to prevent polarizing the reference electrode by forcing the potential of the counter electrode to be the same as the reference electrode. Sosna reported that the sensitivity of the oxygen sensor can be obtained at a minimum of few hundreds of *pA/ppm*, whose value depends on the size of the working electrode [12]. For a disc-type platinum electrode, Sosna

978-1-61284-863-1/11 $26.00 © 2011 IEEE

modeled the limiting current i_{lim} generated by the reduction of the dissolved oxygen as:

$$i_{lim} = 4n_{app}FD_{O_2}ac \qquad (2)$$

where n_{app} is the number of electrons transferred during an oxygen reduction, F is the Faraday constant, D_{O2} is the dissolved oxygen coefficient, a is the radius of the working electrode, and c is the dissolved oxygen concentration. Equation 2 reveals that the sensitivity of the Clark Cell is directly proportional to the size of the working electrode.

When the sensor sensitivity is not high enough to achieve a specified resolution, signal amplification will be done at the sensor front-end. A typical resolution requirement to monitor oxygen level in the brain tissue is 0.07 *ppm* (1 *mmHg*) [8]. For an N-bit ADC, the minimum sensitivity can be determined from Equation 4:

$$\frac{\frac{V_{ref}}{2^N}}{Sensitivity} < 0.07ppm \qquad (3)$$

where V_{ref} is the reference voltage for the ADC converter. Note that the output of the Clark Cell oxygen sensor is a current, and a trans-impedance amplifier will usually be used to convert the current into a voltage before A to D conversion. Note also that to achieve an inaccuracy of ± 1 *mmHg* for oxygen measurements, the system also needs calibrations prior usage.

C. Temperature sensor

Temperature sensing in CMOS platform is perhaps the most developed area in academic research and business environments as compared to the other multimodal neuro-monitoring parameters. Unfortunately, most commercial CMOS temperature sensor chips only provide an inaccuracy of ± 1 ^oC, which is not accurate enough to be used in TBI investigations. One exception of the commercial temperature sensor chip is the high quality temperature sensor from Analog Device (ADT7420) which achieves an inaccuracy of ± 0.2 ^oC from -10 ^oC to 85 ^oC. Generally speaking, the difficulties to make a high accurate temperature measurement instrument can be attributed to the process variations during manufacturing, and a high grade temperature sensor often requires expensive calibrations and a circuit trimming to correct the manufacturing error.

On the contrary, efforts from academic research produce much better result in temperature sensing prototypes, and inaccuracy of few tenths of Celsius degree error and even an inaccuracy of ± 0.1 ^oC were reported from literature [13]-16].

Temperature sensors in CMOS platforms exist in passive or active devices. For instance, a thermopile, which can generate a voltage due to a temperature gradient over a junction of two materials, is available in CMOS technology by connecting a metal interconnect layer (Al) to a p-substrate; thermopile is regarded as a passive temperature sensor. On the other hand, the gate-source voltage of a MOSFET, or the base emitter voltage of a parasitic bipolar transistor is an active temperature sensor. These active devices are generally chosen for temperature sensing as they are a typical element used in circuit building blocks.

Both the gate-source voltage of a MOSFET and the base-emitter voltage of a bipolar transistor exhibit a temperature sensitivity of approximately -2 $mV/^oC$, and bipolar transistors are more preferred to be used as a temperature sensing device. Unfortunately, both types of devices always have mismatch.

The threshold voltage of a MOSFET can be adjusted by ion-implantation [17], but the process never yields devices with identical threshold voltages. The value of the mismatch is more significant in devices with different in size, location, batch-to-batch, or lot-to-lot samples. Similarly, the temperature sensitivity of the base-emitter voltage of bipolar transistors will never be consistent as there are mismatch over the saturation current parameter and the current gain parameter [18]. In order to achieve an inaccuracy of ± 0.2 ^oC toward temperature sensing, a voltage error cannot be made larger than ± 0.4 *mV*. Thus, circuit calibrations and trimmings are always needed to improve the accuracy of CMOS temperature sensors.

III. SYSTEM ARCHITECTURES REVIEW

A general representation of system architectures for interfacing the aforementioned sensor is constructed according to whether the system processes signals in (1) continuous time or discrete time, and in (2) continuous value or discrete value. So, there are 4 signal processing environments in total: (A) Analog signal processing, (B) Mixed signal processing, (C) Frequency domain processing, and (D) Synchronous digital processing. Fig. 1 illustrates all the possible signal processing environments for the sensors interfacing.

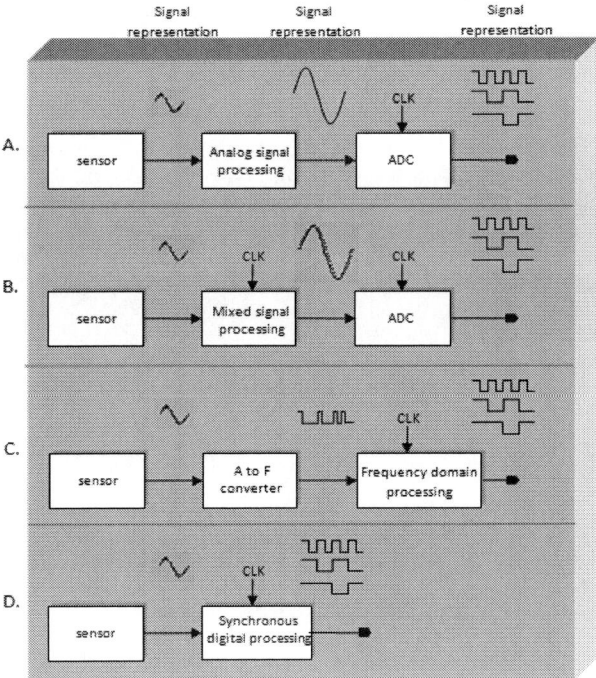

Fig. 1: Classification of various signal processing environments. (A) an analog signal processing which processes signals in continuous times and continuous values, (B) a mixed signal processing which processes signals in discrete times and continuous values, (C) a frequency domain processing which processes signals in continuous times and discrete values, and (D) a synchronous digital processing which processes signals in discrete times and discrete values.

A. Analog signal processing

This is a system where signals are processed in continuous times and in continuous values. Ordinary arithmetic operations (addition, integration, etc) can be used directly to manipulate the signals in linear ways. In applications where sensors have a very low sensitivity in a way that the instrumentation noise is stronger than the sensors' output, a continuous time chopper amplifier can be used as the analog front-end amplifier. A chopper amplifier has two attributes: (1) a signal gain, and (2) a chopped flicker noise. The principle of chopper stabilization is that the signal is first modulated and amplified at AC, where the flicker noise is not significant over that frequency band. After amplification, the amplified signal is demodulated down to DC. The flicker noise of the amplifier will be up-converted to the chopping frequency only once, and it will not be demodulated back at the output. The expression of the low frequency input noise can be approximated as Equation 4 [19].

$$S_N \approx S_{n0} \left(1 + \frac{17 f_k}{2\pi^2 f_{chop}} \right) \qquad (4)$$

where S_{n0} is the amplifier's thermal noise, f_k is the corner frequency of the flicker noise, and f_{chop} is the chopping frequency. A good discussion of chopper amplifier can be referred to [19].

For instances, Pertijs [13] reported a temperature sensor with an inaccuracy of $\pm 0.1\ ^oC$, Chai [20] reported a 118-dB dynamic range capacitance to voltage converter used in accelerometer, and Wei [21] reported a high sensitive amplifier for oxygen probe. They all made use of a chopper amplifier to achieve a low noise system for interfacing the sensors with very low sensitivities. Nonetheless, the demodulated signal has strong spikes that have to be low pass filtered before the signal can be further processed. Therefore, an extra circuitry and power are needed to post-process the analog chopped signal.

B. Mixed signal processing

This system processes signals in continuous values but in discrete times. Switched–capacitor or switched-current circuits belong to this class of system. A good discussion of switched-capacitor techniques can be referred to [22]. The principle of this system is that during the first half of the sampling phase, signal, offset, and noise are sampled. In the second half signal processing phase, the signals are amplified whereas the offset and the low frequency flicker noise are zeroed. [22] describes how the low frequency noise will be reshaped in the baseband; the folded flicker noise is modeled by Equation 5:

$$S_{fold-1/f} = 2S_0 f_k T_s \left[1 + ln\left(\frac{2}{3} f_c T_s\right) \right] sinc^2(\pi f T_s) \quad (5)$$

where S_0 is the amplifier thermal noise, f_k is the corner frequency of the flicker noise, f_c is the cut-off frequency of the amplifier, and T_s is the sampling period. Note that the 1/f spectrum characteristic disappears in Equation 5.

Note that the sampling nature of the system also causes the high frequency thermal noise to be aliased back to the baseband. Therefore, the noise performance of switched-capacitor amplifiers is generally not superior than as in chopper amplifiers, which have no high frequency noise folded back to the baseband. However, the sampling nature of this system makes the subsequent A to D conversion more straight forward, and mixed signal processing environment is commonly found in the $\Sigma\Delta$ ADCs.

For instances, Cong reported an implantable blood pressure monitoring system which only needs 18 μA for running the whole system; the analog front-end to the pressure sensor is realized by a capacitance to voltage converter [23]. Sebastiano reported a temperature sensor with an inaccuracy of $\pm 0.2\ ^oC$ in 10 μW operation [16]. They all utilized switch capacitor techniques either at the sensor front end or at the A to D data conversion. The overall performance, when evaluated along

with the power dissipation, could be better than by using pure analog techniques such as chopper stabilization amplifiers.

C. Frequency domain processing

This is a system where signals are processed in continuous time, discrete value manner. The signal of interest is not a voltage nor a current but a frequency. In other words, information is not represented by amplitude, but it is encoded in the instantaneous phase of the signal. Nonetheless, a voltage (or current) to frequency converter is needed at the frontend, preferably interfaced closely to the sensors. The advantage of this signal representation will be a more immune to the amplitude noise along the signal path. However, the amplitude to frequency conversion is not always linear, and the phase noise associated with the frequency modulated signal cannot be eliminated or reduced by any means. Therefore, this class of system architecture will not be a suitable choice in applications where high precision and high accurate measurements are important.

For instances, Huang reported an integrated pressure sensor with a capacitance to frequency conversion [25], and Park reported a ring oscillator based temperature sensor for RFID tags [26]. In particular, the temperature sensor system report by Park achieved only 0.4 ^oC resolution, which is not precise enough in TBI neuro-monitoring measurements. However, the system is still an attractive solution to RFID applications as it only consumes 95 nW of power.

D. Synchronous digital processing

This class of systems process signals in discrete times and discrete values. This system architecture is generally not applicable to directly incorporate the sensors as there is no well defined signal conversion from the external analog stimuli to the digital processing medium. In fact, no suitable example is found in this particular case.

A qualitative evaluation of the various system architectures is illustrated in Table 1. Note that the evaluation of the synchronous digital processing is not applicable here. In general, a low noise system can be designed under pure analog techniques, but it requires a higher power to achieve the low noise performance. On the other hand, the frequency domain processing can be optimized with power, but it is generally noisier than the other systems.

Table 1: Qualitative evaluation of system architectures.

System	Power	Noise
A. Analog signal processing	Highest	Lowest
B. Mixed signal processing	⇑	⇓
C. Frequency domain processing	Lowest	Highest
D. Synchronous digital processing	N.A.	N.A.

IV. Conclusion

An overview of the multimodal neuro-monitoring comprising pressure, oxygen, and temperature sensing has been presented. Possible system processing environments have been identified and evaluated. For applications found in TBI patients where high accuracy and high precision pressure and oxygen measurement is the prime concern, an analog signal processing or a mixed signal processing architecture will be the right choice to be used as the sensor interface.

ACKNOWLEDGMENT

The work was supported by the Science and Engineering Research Council of A*STAR (Agency for Science, Technology and Research), Singapore, under the grant number: 102 148 0002.

REFERENCES

[1] http://www.ninds.nih.gov/disorders/tbi/detail_tbi.htm

[2] L. Littlejohns, M. Bader, and K. March, "Brain Tissue Oxygen Monitoring in Sever Brain Injury, I" *CrticalCareNurse*, **vol**. 23, pp. 17-25, 2003.

[3] C. Dai, and M. Liu, "Complementary Metal-Oxide Semiconductor Microelectromechanical Pressure Sensor Integrated with Circuits on Chip," *Applied Physica, Japanese Journal of*, **vol**. 46, pp. 843-848, 2007.

[4] C. Li, C. Ahn, L. Shutter, and R. Narayan, "Toward real-time continuous brain glucose and oxygen monitoring with a smart catheter," *Biosensors and Bioelectronics*, **vol**. 25, pp. 173-178, 2009.

[5] C. Sun, C. Wang, M. Tsai, H. Hsie, and W. Fang, "Monolithic integrateion of capacitive sensors using a double-side CMOS MEMS post process," *J. Micromech. Microeng.* **vol**. 19, 2009.

[6] B. North, *Head Injury*. Published by Chapman & Hall, 1997.

[7] J. Zhong, et al., "Advances in ICP monitoring techniques," *Neurological Research*, **vol**. 25, pp. 339-350, 2003.

[8] B. Hoelper, "Brain oxygen monitoring: in-vitro accuracy, long-term drift and respond-time of Licox- and Neurotrend sensors," *Acta Neurochir*, **vol**. 147, pp. 767-774, 2005.

[9] L. Clark, "Monitor and Control of Blood and Tissue Oxygen Tensions," *American Society for Artifical Internal Organs, Trans. of*, **vol**. 2, pp. 41-48, 1956.

[10] C. Wu, H. Luk, Y. Lin, and C. Yuan, "A Clark-type oxygen chip for *in situ* estimation of the respiratory activity of adhering cells," *Talanta*, **vol**. 81, pp. 228-234, 2010.

[11] H. Suzuki, T. Hirakawa, I. Watanabe, amd Y. Kikuchi, "Determination of blood pO_2 using a micromachined Clark-type oxygen electrode," *Analytic Chimica Acta*, **vol**. 431, pp. 249-259, 2001.

[12] M. Sonsa, et al., "Development of a reliable microelectrode dissolved oxygen sensor," *Sens and Actuators **B***, **vol**. 123, pp. 344-351, 2007.

[13] M. Pertijs, K. Makinwa, J. Huijsing, "A CMOS Temperature Sensor with a 3σ Inaccuracy of $\pm0.1°C$ from $-55°C$ to $120°C$," *IEEE International Solid-State Circuits Conference*, Feb. 2005, pp. 238–239.

[14] F. Sebastiano, et al., "A 1.2V 10µW NPN-Based Temperature Sensor in 65nm CMOS with an Inaccuracy of $\pm0.2°C$ (3σ) from $-70°C$ to $125°C$," *IEEE ISSCC*, Feb. 2010, pp. 312–313.

[15] Caspar P. L. van Vroon, Dan d'Aquino, Kofi A.A. Makinwa, "A Thermal-Diffusivity-Based Temperature Sensor with an Untrimmed Inaccuracy of $\pm0.2°C$ (3σ) from $-55°C$ to $125°C$," *IEEE ISSCC*, Feb. 2010, pp. 314–315.

[16] F. Sebastiano, et al., "A 1.2V 10µW NPN-Based Temperature Sensor in 65nm CMOS with an Inaccuracy of $\pm0.2°C$ (3σ) from $-70°C$ to $125°C$," *IEEE J. Solid-State Circuits, vol. 45, no. 12*, pp. 2591–2601, 2010.

[17] B. Streetman, and S. Banerjee, *Solid State Electronic Devices*. Prentice Hall, 2000.

[18] M. Pertijs, and J. Huijsing, *Precision Temperature Sensors in CMOS Technology*, Springer, 2006.

[19] C. Enz, E. Vittoz, and F. Krummenacher, "A CMOS Chopper Amplifier," *Solid-State Circuits, Journal of*, **vol**. 22, pp. 335-342, 1987.

[20] K. Chai, et al., "118-dB Dynamic range, Continuous-Time, Opened-Loop Capacitance to Voltage Converter Readout for Capacitive MEMS Accerometer," *Asian Solid-State Circuits, IEEE Conf.*, 2010.

[21] S. Wei, and H. Lin, "CMOS chopper amplifier for chemical sensor," *Instrumentation and Measurement, IEEE Trans.*, **vol**. 41, pp. 77-80, 1992.

[22] C. Enz, and G. Temes, "Circuit techniques for Reducing the Effects of Op-Amp Imperfections: Autozeroing, Correlated Double Samping, and Chopper Stabilization," *Proceedings of IEEE*, vol. 84, pp. 1584-1614, 1996.

[23] P. Cong, W. Ko, and D. Young, "Low Noise µWatt Interface Circuits for Wireless Implantable Real-Time Digital Blood pressure Monitoring," IEEE CICC, pp. 523-526, 2008.

[24] M. Hovin, A. Olsen, T. Lande, and C. Toumazou, "Delta-Sigma Modulators using Frequency Modulated Intermediate Values," *IEEE J. Solid-State Circuits*, **vol**. 32, pp. 13-22, 1997.

[25] X. Huang, J. Huang, M. Qin, and Q. Huang, "A Fully Integrated Capacitive Pressure Sensor with High Sensitivity," *IEEE Sensors Conference*, pp. 1052-1055, 2007.

[26] S. Park, C. Min, and S. Cho, "A 95nW Ring Oscillator-based Temperaure Sensor for RFID Tags in 0.13um CMOS," *ISCAS*, pp. 1153-1156, 2009.

On-Chip RF Energy Harvesting Circuit for Image Sensor

Jun Wu Zhang, Xiang Yu Zhang, Zhuang Liang Chen, Kye Yak See, Cher Ming Tan, Shou Shun Chen

Division of Circuit & system, School of EEE
Nanyang Technological University
Singapore
Zhan0291@e.ntu.edu.sg

Abstract— An on-chip RF energy harvesting circuit is integrated into an image sensor. The RF energy harvesting circuit is designed to function at 900 MHz. A DC voltage of 1.8V is achievable from the energy harvesting circuit output with -3 dBm input RF power. By charging up a 2.2 mF capacitor to 1.867 V, it has been demonstrated that the low-power image sensor functions well with the ability to obtain more than 1 frame of the image.

Keywords- Image sensor, RF energy harvesting, RFID, autonomous sensor, Charge pump.

I. INTRODUCTION

Autonomous sensing and monitoring platform is an autonomous device that incorporating sensor, micro-controller, signal processor, power management circuitry, and wireless transceiver, etc [1]. This kind of sensing technology has gained increasing interests because it can be deployed without considering the wiring for power and signal transfer. It has promising applications such as Wireless Sensor Network and Structural Health Monitoring [2][3][4].

Battery remains to be a major limiting factor for the developing of truly autonomous deploy-and-forget sensing platform. If the sensing system could be powered by harvesting energy wirelessly from a low-power transmitter, it can almost require no attention once deployed. Prototypes of wirelessly powered sensing platforms for different applications have been reported in [1][6].

In this paper, a prototype of 64 by 64 pixels temporal difference image sensor with on-chip RF energy harvesting circuit is presented. The paper will be organized as follows. The design process and considerations for the RF energy harvesting circuit will be briefly described in Section II. Section III presents the measured I-V curve and the input impedance of the AC-DC voltage multiplier. The characterization result of the whole image sensor will be discussed in Section IV and finally Section V concludes this paper.

II. RF ENERGY HARVESTING CIRCUIT

The design of the far-field wireless RF energy harvesting circuit is illustrated in Fig. 1. The RF power received by the off-chip antenna is rectified and the DC power is stored in a storage capacitor. A matching circuit is designed for maximum power transfer between antenna output and rectifier circuit input. A voltage regulator is employed to provide stable DC voltage for the image sensor circuitry.

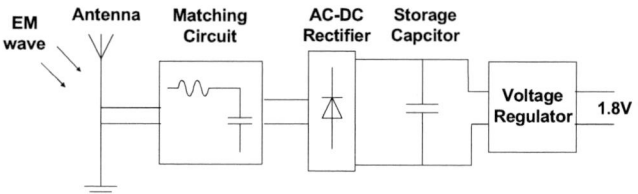

Fig. 1   Block diagram of RF energy harvetsing system

Due to the relatively large power consumption of the image sensor comparing to that of RFID (a few mini-watt vs. micro-watts), the image sensor must be designed to operate in the sleep-active cycle in order to function properly when integrated with the RF energy harvesting module. The image sensor remains in sleep mode until the voltage at the storage capacitor reaches a specific DC voltage. When the DC voltage reaches the required value, the image sensor wakes up and takes a snap of a scene. The image is then processed and stored. When the DC voltage drops below the required operating voltage level, the image sensor goes into the sleep mode again.

III. AC-DC VOLTAGE RECTIFIER

A multi-stage diode-capacitor voltage rectifier is adopted. The circuit is fabricated using UMC 0.18 μm CMOS process with the diode implemented using the diode-connected PMOS transistor. The schematic of the circuit is shown in Fig. 2. The circuit is not only capable of rectifying the AC voltage but also capable of boosting the output DC voltage to a higher level. The DC output voltage of this diode-capacitor rectifier can be expressed as follows:

$$V_{OUT} = 2N(V_{in} - \Delta V)) \qquad (1)$$

Where Vin is the amplitude of the input voltage, N is the number of stages of the rectifier, and ΔV is the voltage drop across the PMOS transistor when the diode-connected transistor is turned on. Equation (1) shows that reducing ΔV will increase V_OUT. Thus a large W/L ratio is desirable but other factors like parasitic resistance, capacitance and leakage as well as the area occupied by the circuit needs to be considered during the design. A systematic analysis of designing the circuit on CMOS is presented in [7].

978-1-61284-863-1/11 $26.00 © 2011 IEEE

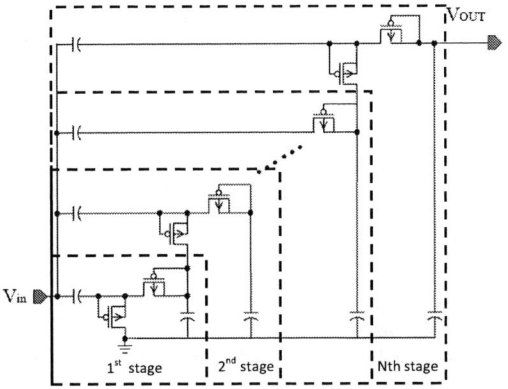

Fig. 2   PMOS implentation of AC-DC rectifier

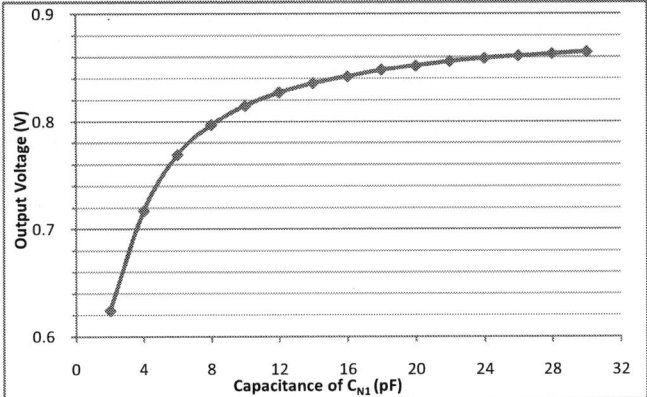

Fig. 3 Capacitor size versus output DC voltage of a single stage voltage rectifier. (note: the vertical axis represents the open-circuit DC voltage at the output)

Transistor with large W/L ratio leads to a significant leakage current when the transistor is reverse biased and it also introduces larger parasitic capacitance. The size of the transistor is chosen carefully by trading-off between the lower turn-on voltage and leakage. The body of PMOS is tied to the source to reduce the leakage current from the body.

The size of the capacitor for each stage also affects the performance of the whole circuit [7]. The capacitor cannot be too small otherwise so that the effect of the parasitic capacitance becomes significant and degrades the performance of the circuit. On the other hand, it cannot be too large due to area constraints. The value of the capacitor is chosen such that it is large enough to suppress the ripple at the output of each stage so that it can be nearly constant. To determine the optimal value of the capacitor, a simulation was carried out on a single stage rectifier. The input voltage is fixed at 0.6 V, the open-circuit DC voltage at the output was observed with varying value of the capacitor and the result is plotted in Fig. 3.

Fig. 3 shows that the output DC voltage increases with the size of the capacitor. But when the capacitor is larger than 10 pF, further increase in the capacitor value contributes little to further increase in output voltage. Thus the optimal capacitor size for our design was chosen to be 10 pF.

Once the sizes of transistor and capacitor are determined, the design of the each stage of the circuit is fixed. Multiple

stages are stacked up in order to reach the required output DC voltage. More stages leads to a higher DC voltage with the input RF power kept constant. However, the increase in number of stages also increases the parasitic resistance and capacitance. These parasitic components degrade the Q factor of the input impedance and causes lower input voltage [8]. The number of stages is finalized to be six and the AC-DC voltage multiplier is integrated with the image sensor on the same die as shown in Fig. 4.

IV.   OUTPUT CAHRACTERIZATION

Measurements are done to characterize the DC output with varying input received power. An RF signal generator at 900 MHz is used as a signal source and a source meter configured as current sink was connected at the output of the circuit as a load. The current drawn by the source meter is adjusted and the voltage across the source meter is measured. The input impedance at different input power can be measured with a Vector Network Analyzer. The power received at the input of the circuit is calculated based on the methodology presented in [9].

Fig. 5 shows the measured I-V curve. The various color zones represent different input power level in dBm. It can be seen that to charge up a capacitor to 1.8 V, a minimum input power of -3 dBm is needed.

Fig. 4 IC photo of the image sensor with AC-DC rectifier

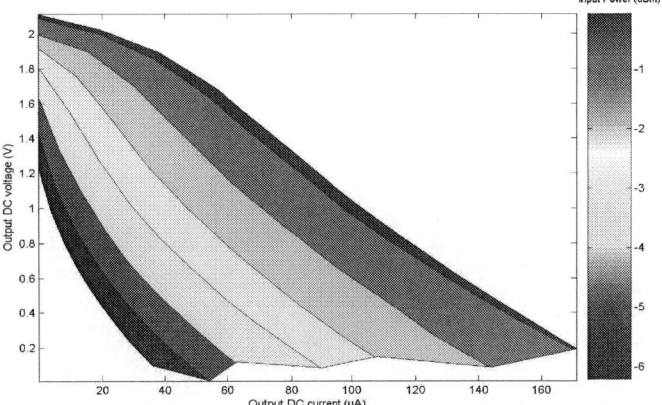

Fig. 5 Output DC I-V curve of the voltage mulitplier at 900 MHz.

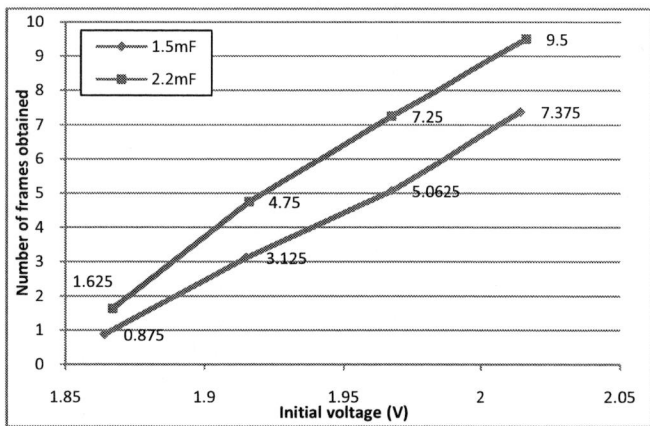

Fig. 6 No. of frames versus the initial voltage on the storage capacitor. (note: the value of the storage capacitor is 2.2mF)

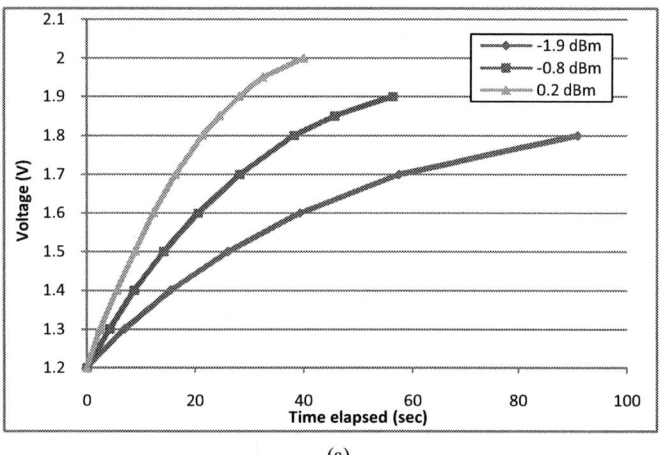

(a)

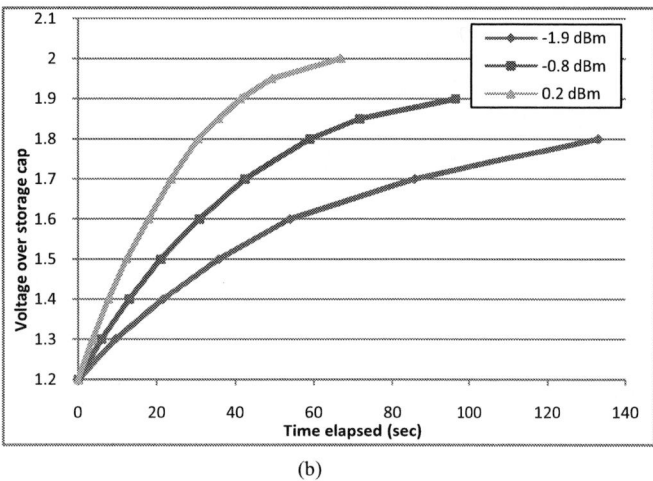

(b)

Fig. 7 (a) Voltage over 2.2mF storage capacitor versus time elapsed. (b) Voltage over 1.5mF storage capacitor versus time elapsed. The legend shows the calculated input power.

V.   MEASUREMENT RESULTS

The performance of the complete circuit is characterized with different input power, size of capacitor and amount of energy stored. Firstly, the storage capacitor is charged up to a

certain voltage level and the image sensor is turned on. The energy stored in the capacitor will be consumed by the image sensor and the voltage across the storage capacitor decays. The number of frames captured by the image sensor during the process is recorded.

Fig. 6 presents the measured results. The horizontal axis represents the DC voltage over the storage capacitor right before the image sensor is turned on. The vertical axis shows the number of frames recorded. Since only one pixel is accessed at one clock cycle, and what has been recorded is actually the total number of pixels accessed during the process. No. of frames are obtained by divide the number by the resolution (64x64), which explains the reason why the number of frames being fractional.

As presented in Fig 6, either higher voltage or larger capacitor can provide more energy for the operation of the image sensor circuit as more energy being stored. The measurement result shown in Fig. 7(a) illustrates that it takes longer time to charge up larger capacitor to the same voltage level. Time was recorded from the instance when the voltage over the capacitor exceeds 1.2V until voltage over the capacitor stabilizes. The reason why 1.2V is chosen other than 0V is that the voltage over the capacitor will not drop to 0V during the operation as the circuits will cease to function when the supply voltage is below certain level. Throughout the measurement, the voltage is found out to be around 1.2V for the image sensor.

The maximum voltage over the capacitor is determined by the available input power and the performance of the RF-DC circuit. And the power consumption of the image sensor to obtain the required data determines the size of the capacitor. The above two factors together determines the charging time needed.

Although the RF signal generator serves as a good emulator as the terminal of the receiving antenna, still the device was attached to a receiving antenna to evaluate the performance of the prototype. The receiving antenna used was a planar monopole antenna fabricated on FR4 substrate. A log-periodic antenna was used as the transmitting antenna. The distance between the transmitter and the receiving antenna was set to be 1 meter due to the space constraint in the lab. The setup was surrounded by computers and equipment with people passing by from time to time. Thus the transmission-path loss is expected to be higher than that of the ideal free space.

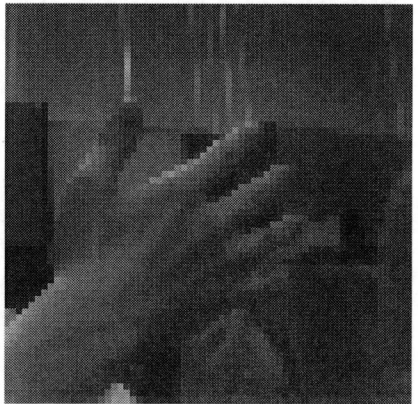

Fig. 8 Image obtained by the 64x64 image sensor prototype

As a result, at the transmitting power of 29.5 dBm EIRP (23.4 dBm transmitting power and 6.1 dBi gain) the 2.2mF storage capacitor was charged up to 1.85V from 1.2V in 123 seconds and 1 frame of image was captured. As the maximum transmission power is capped at 36dBm EIRP, the maximum operating distance of the device is expected to be larger. Fig. 8 shows one of the sample image captured by the image sensor.

VI. CONCLUSION

A prototype of an autonomous image sensor powered by harvesting RF energy from a transmitter is presented. The energy harvesting circuit is integrated with the image sensor on the same die. The harvesting circuit operates at 900 MHz and a DC voltage of 1.8V is achievable with -3dBm input power. The performance of the sensor has been evaluated against the size of charge storage capacitor and the voltage over the storage capacitor. The circuit was attached to a receiving antenna and the measurement was done in the lab. The power from was 29.5 dBm EIRP (23.4 dBm transmitting power and 6.1 dBi transmitting antenna gain). With the receiving antenna 1 meter away from the transmitter, the circuit was able to charge the 2.2mF storage capacitor to 1.85V from 1.2V in 123 seconds and with this amount of energy, 1 frame of image was captured.

The performance of this prototype was limited by the internal leakage and the efficiency of the AC-DC voltage multiplier circuit. Further work is in progress to design the AC-DC voltage multiplier so as to optimize its conversion efficiency as well as reducing the leakage so that the required input power level can be reduced. A compact multi-antenna-multi-rectifier array will also be explored in order to capture more power so that the operating distance can be significantly increased for the next prototype.

ACKNOWLEDGMENT

This work was supported by start-up grant of Assistant Professorship (M58040012) from NTU and AcRF research grant (M52040132) from Ministry of Education, Singapore. The authors would like to thank Mr. Tang Ding for his assistance in measurement.

REFERENCES

[1] A. Sample, D. Yeager, P. Powledge, A. Mamishev, and J. Smith, "Design of an rfid-based battery-free programmable sensing platform," IEEE Transactions on Instrumentation and Measurement, vol. 57, no. 11, pp. 2608-2615, 2008.

[2] I. F. Akyildiz, W. Su, Y. Sankarasubramaniam, and E. Cayirci, "Wireless sensor networks: a survey," Computer Networks, vol. 38, no. 4, pp. 393-422, 2002.

[3] J. Lynch, and K. Loh, "A summary review of wireless sensors and sensor networks for structural health monitoring," Shock and Vibration Digest, vol. 38, no. 2, pp. 91-130, 2006.

[4] Shoushun Chen, Akselrod P., and Culurciello E."A biologically inspired system for human posture recognition"Biomedical Circuits and Systems Conference, 2009. BioCAS 2009. IEEE, Beijing, 2009.

[5] Thad Starner, and Joseph A. Paradiso, "Human Generated Power for Mobile Electronics," Low Power Electronics Design, vol. 1, pp. 30332-30280, 2004.

[6] Paing T, Morroni J, Dolgov A, Shin J, Brannan Jand Zane R et al.: "Wirelessly-powered wireless sensor platform"Proc. IEEE 37th Eur. Microw. Conf, Munich. Germany: 2007.

[7] J. Yi, W. H. Ki, and C. Y. Tsui, "Analysis and Design Strategy of UHF Micro-Power CMOS Rectifiers for Micro-Sensor and RFID Applications," Circuits and Systems I: Regular Papers, IEEE Transactions on, vol. 54 , no. 1, pp. 153- 166, 2007.

[8] T. Le, K. Mayaram, and T. Fiez, "Efficient far-field radio frequency energy harvesting for passively powered sensor networks," Solid-State Circuits, IEEE Journal of, vol. 43, no. 5, pp. 1287-1302, 2008.

[9] L. Mayer, and A. Scholtz, "Sensitivity and impedance measurements on UHF RFID transponder chips," 2nd Int. EURASIP RFID Technol. Workshop, Budapest, Hungary, 2008.

Scalable Modeling Based on Fill Ratio for Planar Spiral Inductors

Lin Zhong, Lingling Sun, Jun Liu, Huang Wang,

Key Laboratory for RF Circuits and Systems of Ministry of Education, Hangzhou Dianzi University,
Hangzhou 310037, China
E-mail: sunll@hdu.edu.cn

Abstract—**This paper presents a scalable model based on the fill ratio with an enhanced 1-π topology for on-chip spiral inductor. Usually, scalable modeling requires so many parameters such as the width, the turn spacing, number of turns and inner radius. To reduce the number of variables and coefficients in the scalable rules, this work utilizes fill ratio to represent geometry properties of an inductor and all the equations for the elements in the topology have simple form. The proposed scalable model is further verified by a set of spiral inductors fabricated by using 0.18-μm 1P6M RF CMOS technology.**

Keywords-on-chip spiral inductor; scalable model; fill ratio; enhanced 1-π.

I. INTRODUCTION

In recent years, with the rapid development of wireless communication technology, the application frequency of integrated circuits (ICs) increases continuously. Thus, in addition to active components with excellent radio frequency (RF) performance, high-quality passive components such as spiral inductors, resistors, varactors and capacitors are inevitable [1-3]. As important passive devices, on-chip spiral inductors are widely used in RF circuits, such as power amplifiers mixers, low-noise amplifiers and voltage controlled oscillators. Therefore, an accurate scalable inductor library is essential for the rapid and successful design of RFIC.

So far, many papers on the spiral inductors have been reported concerning the equivalent circuit modeling and parameter extraction methods [4-11]. There is little work about scalable models of inductors. In this paper, an enhanced 1-π scalable model based on *fill ratio* (ρ) is established. For a given shape, an inductor is usually specified by the number of turns n, the turn width w, the turn spacing s, and inner radius R. Above-mentioned layout parameters can be attributed to *fill ratio*, defined as $\rho=(d_{out}-d_{in})/(d_{out}+d_{in})$ [12], d_{out} is the outer diameter, and d_{in} is the inner diameter. The ratio ρ characterizes the hollowness of the inductor: the larger ρ is, the fuller the inductor is ($d_{out}>>d_{in}$) and the smaller ρ is, the hollower the inductor is ($d_{out}\approx d_{in}$). The top view and cross sectional view of a planar spiral inductor is shown in Figure. 1.

Scalable modeling procedures are composed of circuit topology choice, parameters extraction for various dimensions of inductors, scalable fitting of each parameter as a function of physical design parameters [13].

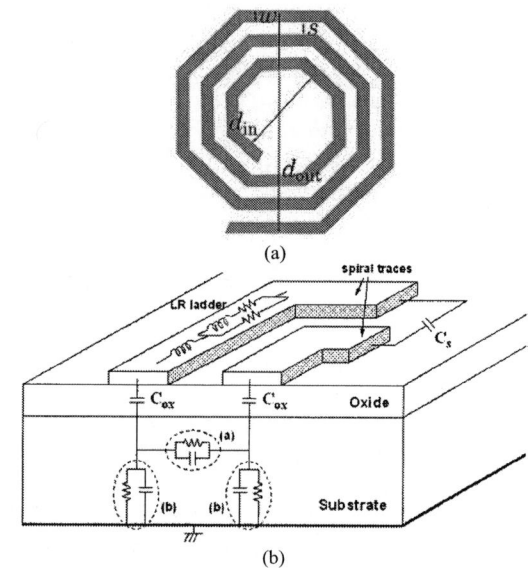

Figure. 1. (a) Top view of an octagonal inductor.
(b) Cross-sectional view of an on-chip spiral inductor.

In this paper, the equivalent circuit and parameters extraction flow will be introduced in Section II, followed by a set of scalable formulas and verification results in Section III. Finally, conclusions are drawn in Section IV.

II. EQUIVALENT CIRCUIT AND PARAMENTERS EXTRACTION

The equivalent circuit models for on-chip spiral inductors can be generally categorized into three types: T-model [9], 2-π model [6]–[8] and 1-π model [4]–[5]. The T-model and 2-π model are widely used to achieve the wideband modeling accuracy, but their parameters extraction and scalable fitting are difficult due to the complexity of the equivalent circuit and the increase of circuit elements. However, in the conventional 9-element 1-π model the distributed effects and the higher order loss effects are not properly considered, so the enhanced 1-π model is adopt in this paper. This topology was first proposed in [5] and its equivalent circuit is shown in Figure. 2.

This work was supported by 973 Program of China under Grant 2010CB327403

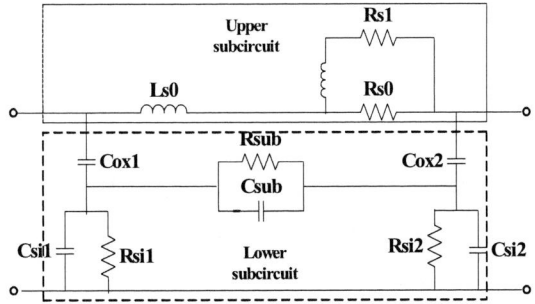

Figure. 2. Enhanced 1-π circuit model

TABLE I. GEOMETRY PARAMETERS OF THE INDUCTORS

DUT#	d_{in} (μm)	w(μm)	s(μm)	n
D1~D5	60	10	2	2.5,3.5,4.5,5.5,6.5
D6~D10	120	10	2	2.5,3.5,4.5,5.5,6.5
D11~D15	180	10	2	2.5,3.5,4.5,5.5,6.5

L_{s0} and R_{s0} are used to model series inductance and resistance. The series branch L_{s1} and R_{s1} in parallel with R_{s0} is used to capture wire skin and proximity effects at high frequencies. $C_{ox1, 2}$ represent the oxide capacitance between the inductor and substrate. $C_{si1, 2}$ and $R_{si1, 2}$ represents the substrate capacitance and resistance to ground, respectively. The parallel networks R_{sub} and C_{sub} are utilized to model the lateral substrate coupling. The initial values of the equivalent circuit parameters can be calculated from measured S-parameters according to the analytical method as presented in [10]. A set of reasonable values in accordance with physical meanings will be obtained through optimization.

III. SCALABLE MODEL AND VERIFICATION RESULTS

In this section, the scalable equations are constructed for each parameter. To construct the scalable model and verify the accuracy, a set of octagonal spiral inductors with various dimensions which were fabricated on 0.18-μm 1P6M RFCMOS technology are used and their geometry parameters are outlined in Table I.

A. L_{s0}

Mohan presented several formulas for planar spiral inductors, which were concerned with *fill ratio* [12]. In this work, a simple modification is made to Mohan's formula and an expression that is valid for planar spiral inductors is obtained as

$$L_{s0} = a_0 \mu_0 \frac{N^{a_1} d_{avg}}{1 + a_2 \rho} \qquad (1)$$

Where μ_0 is the permeability of vacuum, ρ is the *fill ratio* defined previously. The average diameter d_{avg} is calculated as

$$d_{avg} = \frac{d_{out} + d_{in}}{2} \qquad (2)$$

The coefficients a_0, a_1 and a_2 should be optimized in order to get good fit between the extracted values and scalable

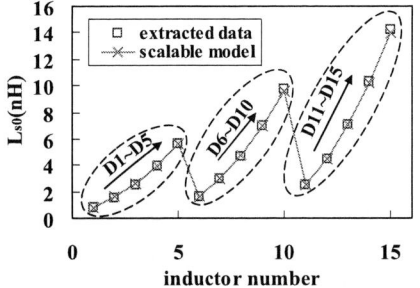

Figure. 3. Comparison of extracted and calculated L_{s0}

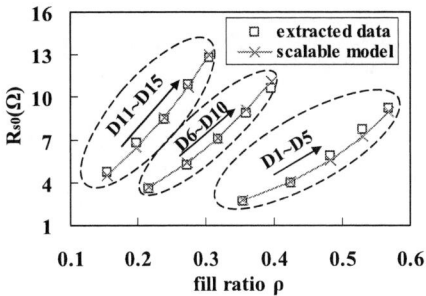

Figure. 4. Comparison of extracted and calculated R_{s0}

functions. The values of L_{s0} extracted and calculated from the scalable equations (1)-(2) are depicted in Figure. 3.

B. R_{s0}

To represent R_{s0}, an analytical method is employed in this paper. From the extracted values of R_{s0} shown in Figure. 4, it is observed that for the inductors with the same inner radius, they are nearly linear with ρ. So the expression of R_{s0} is based on *fill ratio*. Furthermore, the resistance is also related to the inner diameter and the number of turns from the picture, so a monomial in the variables d_{in} and n is added. Hence,

$$R_{s0} = b_0 \rho^{b_1} N^{b_2} d_{in}^{b_3} \qquad (3)$$

C. L_{s1} and R_{s1}

In many papers, R_{s1} are in proportion to R_{s0}, but a fixed proportional coefficient is not applicable to all the inductors with different geometric sizes, so a same form equation with R_{s0} is introduced to express R_{s1}. The extracted values of L_{s1} from a series of inductors indicate that L_{s1} has a consistent tendency as R_{s1}, as shown in Figure. 5(a)-(b). Therefore, the scalable formula of L_{s1} is similar to that of R_{s1}.

$$R_{s1} = d_0 \rho^{d_1} N^{d_2} d_{in}^{d_3} \qquad (4)$$

$$L_{s1} = c_0 \rho^{c_1} N^{c_2} d_{in}^{c_3} \qquad (5)$$

D. C_{ox1} and C_{ox2}

The extracted data of C_{ox1} also display approximate linear relationship with ρ, as show in Figure. 5(c). According to [13], the oxide capacitance is also proportional to the area of the metal lines and influenced by the number of turns [13]. Therefore, $C_{ox1, 2}$ can be characterized by

978-1-61284-863-1/11 $26.00 © 2011 IEEE   425

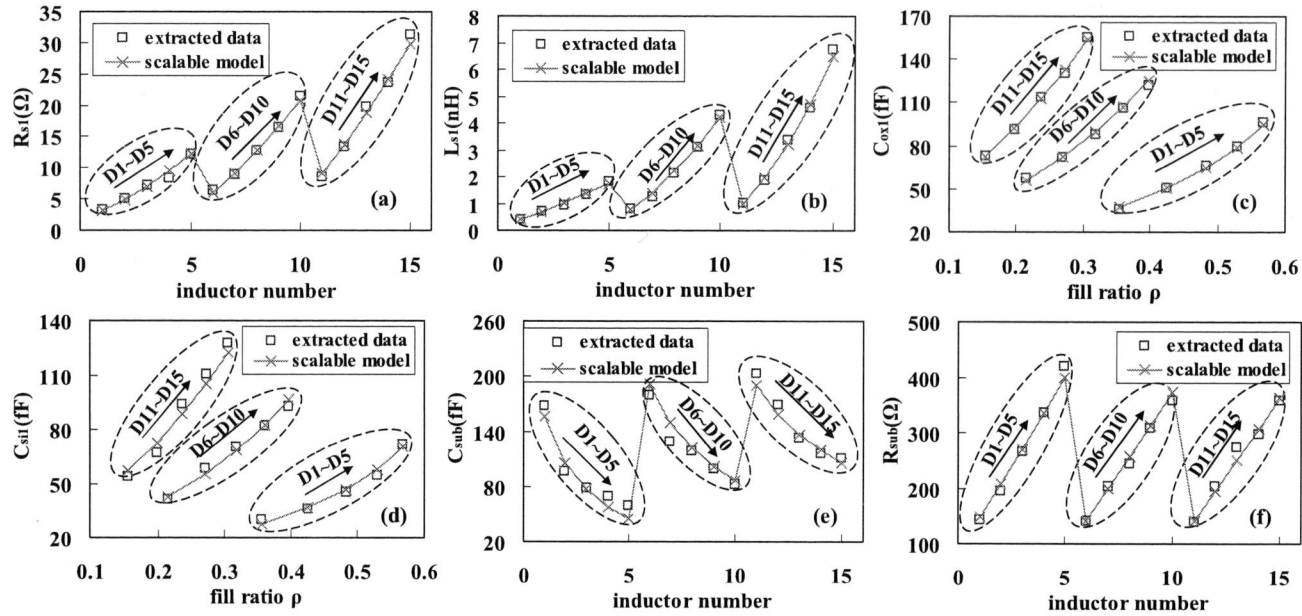

Figure. 5. Salabilities of the parameters extracted in the equivalent circuit

$$C_{ox1} = e_0 \rho^{e_1} N^{e_2} (lw)^{e_3} \qquad (6)$$

$$C_{ox2} = e_4 \rho^{e_5} N^{e_6} (lw)^{e_7} \qquad (7)$$

where, l is the length of metal lines.

E. $C_{si1}, C_{si2}, R_{si1}, R_{si2}$

From the extract data, as shown in Figure. 5, it is observed that C_{si1} and C_{si2} have similar behavior as C_{ox1} and C_{ox2}, so they can be modeled as follows

$$C_{si1} = f_0 \rho^{f_1} N^{f_2} (lw)^{f_3} \qquad (8)$$

$$C_{si2} = f_4 \rho^{f_5} N^{f_6} (lw)^{f_7} \qquad (9)$$

$R_{si1, 2}$ have inverse relationship with $C_{si1, 2}$ according to [7], hence, their expressions are

$$R_{si1} = \frac{f_8}{C_{si1}} \qquad (10)$$

$$R_{si2} = \frac{f_9}{C_{si2}} \qquad (11)$$

F. C_{sub} and R_{sub}

Both of the parameters are determined based on their physical characteristics and trends, therefore

$$C_{sub} = g_0 \rho^{g_1} N^{g_2} d_{in}^{g_3} \qquad (12)$$

$$R_{sub} = g_4 \rho^{g_5} N^{g_6} d_{in}^{g_7} \qquad (13)$$

The values of C_{sub} and R_{sub} extracted and calculated from the (9)-(10) are depicted in Figure. 5(e)-(f).

Some of the model coefficients in enhanced 1-π scalable rules are listed in Table II. The values of these parameters are determined through curve fitting and global optimization.

TABLE II. MODEL COEFFICIENTS FOR THE SCALABLE MODEL

element	coefficient	value
L_{s0}	a_0, a_1, a_2	2.267, 1.834, 2.098
R_{s0}	b_0, b_1, b_2, b_3	1.990e-3, -1.243, 1.938, -0.4248
L_{s1}	c_0, c_1, c_2, c_3	2.641, 1.777, 0.6975, 2.206
R_{s1}	d_0, d_1, d_2, d_3	4.677, -0.8924, 1.873, 0.3146
C_{ox1}	e_0, e_1, e_2, e_3	0.3672, 0.7039, -1.503, 1.488
C_{si1}	f_0, f_1, f_2, f_3	0.7007, 0.643, -1.521, 1.543
R_{si1}	f_8	16.73e-12
C_{sub}	g_0, g_1, g_2, g_3	4.918, 3.191, -2.877, 2.586
R_{sub}	g_4, g_5, g_6, g_7	4.431, -0.244, 1.166, -0.2241

For verification of the proposed model, simulated data of the inductors are compared with measured data. The two-port S-parameters are measured with the frequency ranging from dc to 40 GHz using an Agilent PNA E8363B network analyzer and Cascade Microtech coplanar GSG probes. The measured data are de-embedded with "open" and "thru" patterns

One-port characteristics of the inductors are derived using the relation of inductance

$$L = imag(1/Y_{11})/\omega \qquad (14)$$

and the quality factor is defined as

$$Q = -imag(Y_{11})/real(Y_{11}) \qquad (15)$$

In this paper, six inductors are chosen to verify the accuracy of the scalable models. The fitting results are shown in Figure. 6 and Figure. 7.

The root-mean-square (RMS) errors of the S-parameters, inductance (L) and quality factor (Q) between the measured and simulated results are listed in Table III. Definition of the RMS errors is as follows

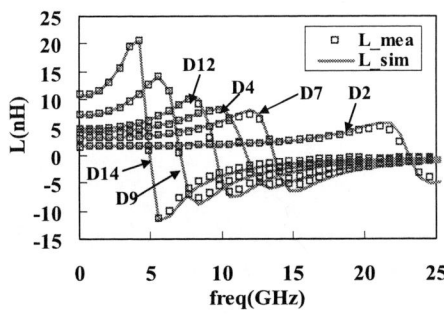

Figure. 6.   Comparison between the measured and the simulated L

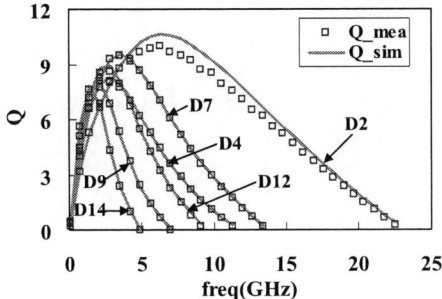

Figure. 7.   Comparison between the measured and the simulated Q

$$RMS\_error = 100 \cdot \sqrt{\left[\frac{1}{n} \sum_1^n \frac{\left(X_{mea} - X_{sim} \right)^2}{\left(\sum_1^n X_{mea}^2 / n \right)} \right]} \qquad (16)$$

where, n is the total number of data point. The RMS error calculation is executed within the frequency range from dc to the self reasonably frequency (SRF) of devices.

TABLE III.   LIST OF RMS ERRORS(%) BETWEEN THE MEASUREMENT AND SIMULATION FOR INDUCTANCE(L), QUALITY FACTOR(Q) AND S-PARAMETER

DUT#	L	Q	Re(S11)	Im(S11)	Re(S12)	Im(S12)
D1	1.904	5.828	6.407	4.346	5.342	3.573
D2	2.648	3.523	3.284	2.795	5.728	4.501
D3	2.388	1.763	1.772	2.649	4.675	2.939
D4	1.150	2.312	0.947	0.932	3.155	2.825
D5	0.918	3.911	0.705	1.807	3.046	2.716
D6	2.135	2.790	3.504	3.328	5.137	3.846
D7	2.051	3.164	1.550	1.769	5.085	2.189
D8	2.272	2.138	1.263	1.572	4.806	1.932
D9	1.142	1.174	0.851	1.726	2.230	2.590
D10	1.151	3.154	0.454	1.838	1.637	2.643
D11	0.809	5.589	2.450	2.914	4.790	3.080
D12	1.704	3.372	1.291	2.258	5.558	2.789
D13	1.643	2.369	1.025	0.935	4.330	3.211
D14	1.072	2.417	0.974	2.069	2.352	2.242
D15	2.041	3.665	3.372	1.604	1.896	2.136

From Figure.6 and Figure.7, it is found that the measured curves of L and Q are fitted well with modeled curves, which shows that high accuracy is achieved by the proposed models. The result is also proved by the RMS errors listed in Table III. The average RMS error of L is below 3% and for most device, the RMS errors of Q are below 5%.

IV.   CONCLUSION

The scalable model based on the *fill ratio* with an enhanced 1-π topology has been proposed. From the values of ρ of the inductors used in this experiment, it is observed that each inductor has a unique *fill ratio*. Therefore, it is reasonable to represent the features of an inductor with an equation including ρ. Simple and unite scalable rules are utilized to construct the scalable model. To demonstrate the validity of the proposed modeling technique, a series of inductors were fabricated and modeled as a scalable library. The results show that excellent agreement between measured and simulated data has been achieved.

REFERENCES

[1] Herbert S. Bennett, Ralf Brederlow, Julio C. Costa, et al., "Device and Technology Evolution for Si-based RF Integrated Circuits," *IEEE Trans. Electron. Devices*, vol. 52, no. 7, pp. 1235–1258, July. 2005.

[2] M. Je, I. Kwon, H. Shin, and K. Lee, "MOSFET modeling and parameter extraction for RF-ICs," in *CMOS RF Modeling Characterization and Applications*. Singapore: World Sci., 2002, pp.67–120.

[3] Yuhua Cheng, "An Overview of Device Behavior and Modeling of CMOS Technology for RFIC Design," *Electron Devices for Microwave and Optoelectronic Applications*, pp. 109–114, 2003.

[4] C. P. Yue and S. S. Wong, "Physical modeling of spiral inductors on silicon," *IEEE Trans. Electron. Devices*, vol. 47, no. 3, pp. 560–568, Mar. 2000.

[5] J. Gil and H. Shin, "A simple wide-band on-chip inductor model for silicon-based RF ICs," *IEEE Trans. Microw. Theory Tech.*, vol. 51, no.9, pp. 2023–2028, Sep. 2003.

[6] Fengyi Huang, Jingxue Lu, Nan Jiang, et al., "Frequency -Independent Asymmetric Double-π Equivalent Circuit for On-Chip Spiral Inductors: Physics-Based Modeling and Parameter Extraction," *IEEE J. Solid - State Circuits*, pp. 2272–2283, Oct. 2006.

[7] W. Gao and Z. Yu, "Scalable Compact Circuit Model and Synthesis for RF CMOS Spiral Inductors," *IEEE Trans. Microw. Theory Tech.*, vol. 54, no. 3, pp. 1055–1064, Mar. 2006.

[8] C. Wang, H. Liao, C. Li, R. Huang, W. Wong, X. Zhang, and Y Wang, "A Wideband Predictive 'Double-π' Equivalent-Circuit Model for On-Chip Spiral Inductors," *IEEE Trans. Electron Devices*, vol. 56, no. 4, pp. 609–619, Apr. 2009.

[9] T. S. Horng, J. K. Jau, C. H. Huang, and F. Y. Han, "Synthesis of a Super Broadband Model for On-Chip Spiral Inductors," *IEEE RFIC Symp. Dig.*, pp. 453–456, Jun. 2004.

[10] H. H. Chen, H. W. Zhang, S. J. Chung, J. T. Kuo, and T. C. Wu, "Accurate Systematic Model-Parameter Extraction for On-Chip Spiral Inductors," *IEEE Trans. Electron Devices*, vol. 55, no. 11, pp. 3267–3273, Nov. 2008.

[11] M. Kang, J. Gil, and H. Shin, "A simple parameter extraction method of spiral on-chip inductors," *IEEE Trans. Electron. Devices*, vol. 52, no. 9, pp. 1976–1981, Sep. 2005.

[12] S. S. Mohan, M. del Mar Hershenson, S. P. Boyd, and T.H. Lee, "Simple Accurate Expressions for Planar Spiral Inductances," *IEEE J Solid-State Circuits*, vol.34, pp.1419-1424, Oct. 1999.

[13] Young-Ghyu Ahn, Seong-Kyun Kim, Jung-Hoon Chun, Byung-Sung Kim, "Efficient Scalable Modeling of Double-π Equivalent Circuit for On-Chip Spiral Inductors," *IEEE Trans. Microw. Theory Tech.*, vol. 57, no. 10, pp. 2289-2300, Oct. 2009.

[14] Angelo Scuderi, Tonio Biondi, Egidio Ragonese, and Giuseppe Palmisano, "A Lumped Scalable Model for Silicon Integrated Spiral Inductors," *IEEE Trans. Circuits and Systems*, vol. 51, no. 6, pp. 1203-1209, Jun. 2004

Model of On-Chip VGP-CPW with P+ Implant in CMOS Process

Jincai Wen, Jia Lou, Lingling Sun*

Key Laboratory for RF Circuits and Systems of Ministry of Education,
Hangzhou Dianzi University
Hangzhou, 310018, China
E-mail: sunll@hdu.edu.cn

Abstract—**This paper presents a vertical-ground-plane coplanar waveguide (VGP-CPW) line with P+ implant, and compared the characteristic parameters of transmission line with the general VGP-CPW structures. The ground plane connected to the substrate makes the proposed CPW line have higher isolation and lower characteristic impedance. The physical model of the VGP-CPW with P+ implant is presented as well as the general VGP-CPW line. Two different structures of transmission lines are fabricated in 65 nm 1P8M RF CMOS process and models are verified by the measured *S*-parameters up to 50 GHz.**

Keywords-CMOS; VGP-CPW; P+ implant; model

I. INTRODUCTION

As the frequency goes up, on-chip transmission lines (Tline) become more and more popular and have been the basic components in millimeter-wave (MMW) and terahertz circuits. Coplanar waveguide (CPW) structures are commonly used in high-speed and high-frequency circuits, and some researches have been reported in recent years. Some CPWs with various widths of signal lines and different substrate resistivities were compared and low-loss coplanar lines were used in the design of distributed amplifiers and oscillators [1]. Various slow wave coplanar waveguides (SW-CPW) have been studied and applied for MMW circuit design, due to their advantages in reducing chip size and good isolation from the lossy substrate[2~5].

A VGP-CPW coplanar waveguide structure is proposed in reference [6], which provides several advantages, such as compact meandering, wide range of characteristic impedance and high isolation between nearby lines, which make it suitable to realize compact silicon-based monolithic microwave integrated circuits (MMICs). This paper presents a novel VGP-CPW structure with P+ implant, and its characteristics and physical model are compared with the general VGP-CPW structures.

This paper is organized as follows. The configurations and characteristics of the two Tlines are presented in Section II, followed by the compact physical models and the parameter extraction method. After that, the measurement results and experimental verifications will be discussed in Section IV. Finally, the paper is concluded in Section V.

Figure 1. Cross-sections of VGP-CPW struchures. (a) general pattern (b) with P+ implant

II. VGP-CPW STRUCTURES WITH P+ IMPLANT

A. VGP-CPW Structure with P+ Implant

Figure 1(a) shows the general VGP-CPW structure (named "General"), which is comprised of one signal line surrounded by two vertical ground planes formed by stacked-metal-via layers. The VGP-CPW structure with P+ implant (named "PPI") is shown in Figure 1(b), and the vertical ground planes are connected to the substrate with contact via layer and P+/P-junction (generally, the mask layer are CT, AA and SP, etc.).

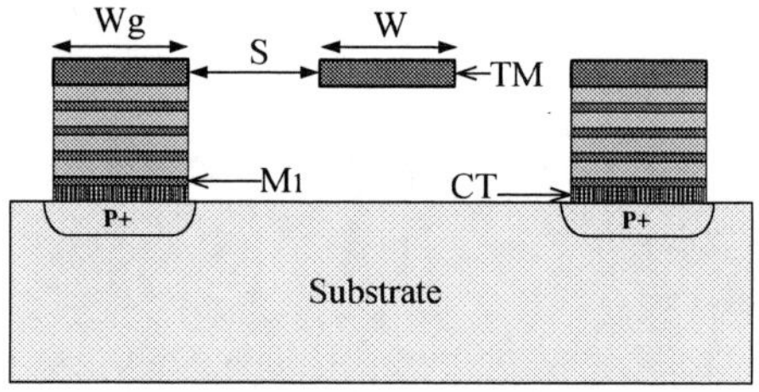

Since the ground planes are biased in substrate, the VGP-CPW has better isolation between nearby lines compared with the general VGP-CPW structures. For example, the coupling effect of the two "PPI" line can be decreased nearly 5 dB per millimeter at 40GHz differ from the two "General" line in 65nm RF CMOS process thru EM simulation, and the layout parameters W=S=Wg=10 μm, the inner ground planes of the two adjacent VGP-CPW lines are incorporated as the common ground line.

This work was supported by the National Natural Science Foundation of China under Grant (60906015) and the Natural Science Foundation of Zhejiang Province under Grant (Y1090877).

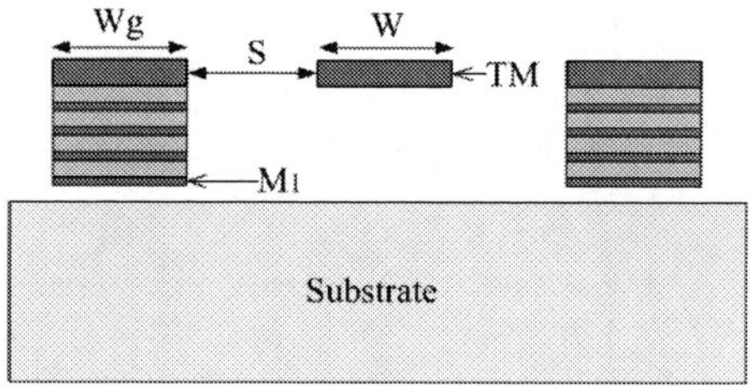

978-1-61284-863-1/11 $26.00 © 2011 IEEE

B. Measured Performances of "General" and "PPI" VGP-CPW Structures

Two Tline structures were fabricated in 65 nm 1P8M RF CMOS process. This process contains eight metal layers with the top metal layer thickness of 3.4 μm. Both of VGP-CPWs use the top metal layer (M8) as the signal lines with the width of 10 μm. The width of all the layers from M8 to M1 as ground line is 10 μm, and the space of 10 μm. Two different lengths (600 μm and 100 μm) are used for de-embedding [7].

The frequency dependences of the characteristic impedance Z0, attenuation constant LOSS, and R/L/G/C parameters are shown in Figure 2. The "PPI" Tline has approximately equal inductance and relatively larger capacitance per unit length compare with the "General" Tline, which result in larger range of characteristic impedance of the "PPI" Tline. The attenuation constants of "General" and "PPI" Tlines are almost equal in the low frequency band (less than 20GHz), while as the frequency increases, the "PPI" Tline exhibits larger loss. This is because the conductor loss plays a dominant role in the low frequency band, and since the two structures have the same signal line, they have the similar propagation loss. However, the substrate loss will dominates in the high frequency band and due to the side-wall connects into the substrate, the "PPI" Tline will suffer much more substrate loss in the high frequency band.

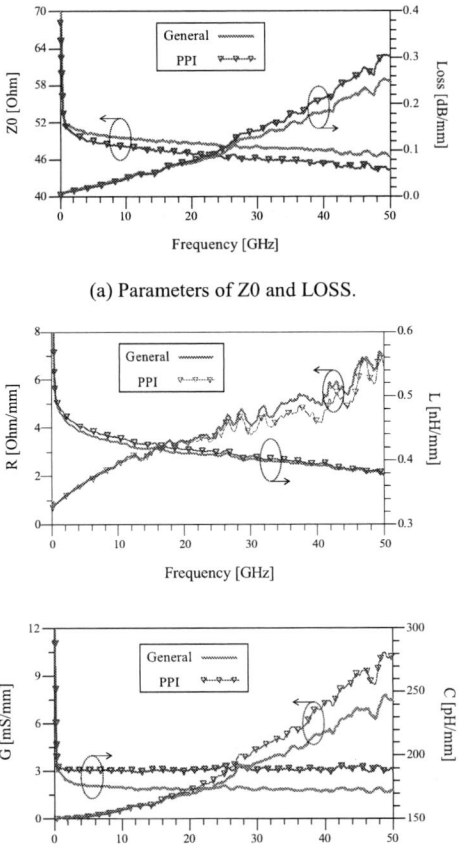

(a) Parameters of Z0 and LOSS.

(b) Parameters of R/L/G/C.

Figure 2.   Measurd performances of the VGP-CPW structures

III.   COMPACT PHYSICAL MODEL AND PARAMETERS EXTRACTION

A. Compact Physical Model

According to layout structures and physical mechanism of the VGP-CPW with P+ implant, the physical model is presented in Figure 3, as well as the general VGP-CPW.

Figure 3(a) shows the cross-section description of the two structures. Since the two Tlines have the same structure between signal line and ground as well as substrate, the same elements are adopted here. Ct represents the capacitance due to the field coupled from the top surface of signal line to the top surface of ground line in the air. Csg represents the capacitance between the signal line and the perpendicular surface of the ground sidewalls [11]. Csox is the capacitance between the signal line and substrate, and a parallel RC network (Rsub and Csub) is used to describe the electric coupling loss between the signal line and the ground line through the conductive substrate.

As shown in Figure 3(a), we use a capacitor Cgox to model the electrical coupling from the bottom metal of the ground plane to the substrate for the "General" Tline. Meanwhile, for the "PPI" Tline, the ground plane is connected to the substrate to form an ohmic-contact modeled by a resistor. The P+/P-junction between the P+ implant and the substrate is modeled by a parallel resistor and capacitor network. Since the width and length are relatively large, the resistance is very small and the capacitance is very large. Therefore, the ohmic-contact resistor, the junction resistor and capacitor are all negligible.

The complete physical model is shown in Figure 3(b). An additional RL ladder is included to capture the skin effect of metal conductor at high frequencies along the signal propagation direction. C1 equal the sum of Ct and Csg for the Tlines. Cox is the series capacitance of Csox and Cgox for the general VGP-CPW line, while Cox is just Csox for the VGP-CPW line with P+ implant.

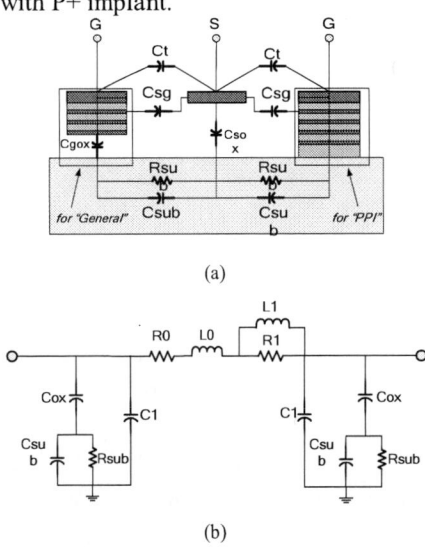

(a)

(b)

Figure 3.   Illustration of physical models of VGP-CPW lines. (a) the cross-section description of the two structures (b) the complete physical models

978-1-61284-863-1/11 $26.00 © 2011 IEEE

B. Parameter Extraction

Both of the two models have the same series elements including R0, L0, R1 and L1, which can be extracted directly from the measured data [9] by,

$$\begin{cases} R0 = R_{LF}, \\ L0 = L_{HF}, \\ R1 = R_{HF} - R0, \\ L1 = L_{LF} - L0. \end{cases} \quad (1)$$

where, R_{LF} and R_{HF} are the resistances at the lowest frequency and highest frequency over test range, respectively. L_{LF} and L_{HF} are the inductances at the lowest frequency and highest frequency over test range, respectively.

An empirical formula has been proposed to calculate Csox [11].

$$Cox = \varepsilon_0 \times \varepsilon_n \times \left\{ \frac{W}{H} - \frac{T}{2H} + \frac{2\pi}{\ln\left[1 + \frac{2H}{T}\left(1 + \sqrt{1 + T/H}\right)\right]} \right\} \quad (2)$$

where, W, T and l are the width, thickness and length of the signal line, respectively. H and ε_n is the height and effective dielectric constant between the bottom of the line and the upper surface of the substrate, respectively. ε_n can be calculated by

$$\varepsilon_n = \frac{H}{\sum\limits_{i=0}^{n} \dfrac{H_i}{\varepsilon_i}} \quad (3)$$

where, H_i and ε_i is the height and dielectric constant of the ith dielectric layer, respectively.

The capacitors Cgox and Csg in Figure 3 can also be obtained by (2). However, different process parameters and layout parameters need to be substitutes [11].

The capacitor Ct, represents the capacitance due to the electrical field in the air in Figure 3(a) and Figure 3(b), can be calculated according to [11] as follows,

$$C_t = 2\varepsilon_0 \frac{K(k)}{K(k')} \quad (4)$$

where K denotes the first kind complete elliptic integral, and k obtained by

$$k = \frac{W}{W+2S} \sqrt{\frac{1 - (\dfrac{W+2S}{W+2S+2W_g})^2}{1 - (\dfrac{W}{W+2S+2W_g})^2}} \quad (5)$$

where, W, S and Wg are the width, space and ground line width of the signal line, respectively. k' equals to $\sqrt{1 - k^2}$.

Generally, Rsub and Csub can also be extracted from the measured data, and for the relationships between the two parameters please refer to [9].

IV. VERIFICATION

The proposed models of VGP-CPW Tlines were verified based on-chip measured S-parameters up to 50 GHz. The extracted and simulated results of the two models parameters are compared with the corresponding measured ones and the curves of the "General" and "PPI" VGP-CPWs are present is Figure 4 and Figure 5, respectively.

From Figure 4 and Figure 5, it can be observed that the modeled results agree well with the measurement results, which indicates that the proposed physics-based equivalent circuit model is capable of describing the behavior of the general VGP-CPW Tline and the VGP-CPW with P+ implant over a large range of frequency up to 50 GHz.

The extracted parameters of the two Tlines are shown in Table I.

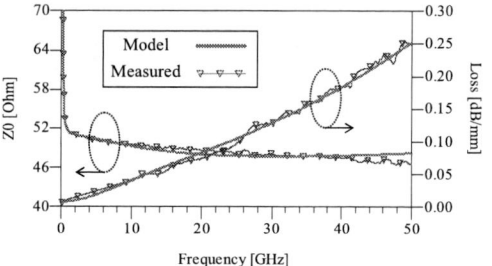

(a) Parameters of Z0 and LOSS.

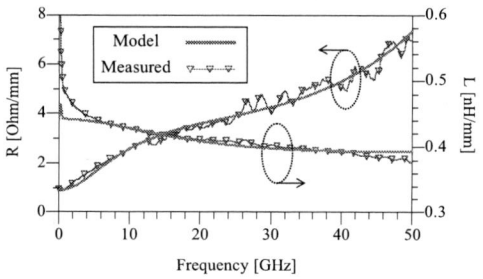

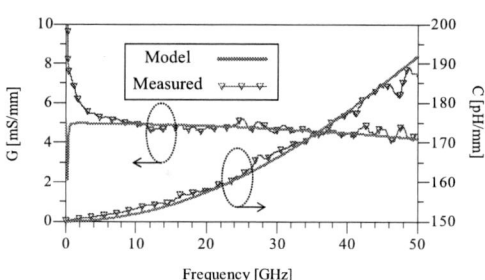

(b) Parameters of R/L/G/C.

Figure 4. The comparison of modeled and measured data of the general VGP-CPW structure

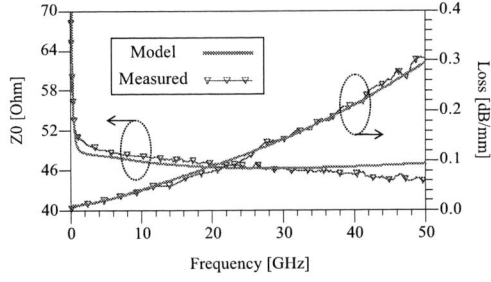

(a) Parameters of Z0 and LOSS.

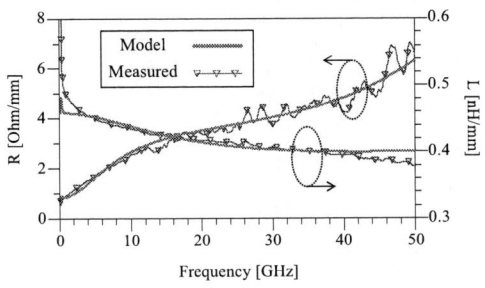

(b) Parameters of R/L/G/C.

Figure 5. The comparison of modeled and measured data of the VGP-CPW structure with P+ implant

TABLE I.   VALUE OF THE EXTRACTION PARAMETERS

Parameters	Unit	"General"	"PPI"
R0	mΩ	386.1	368.8
L0	pH	196	200.3
R1	Ω	2.07	2.12
L1	pH	35.37	26.31
C1	fH	11.49	14.63
Cox	fH	76.68	81.66
Csub	fH	18.92	
Rsub	Ω	46.49	

V.   CONCLUSION

In summary, this paper presents the experimental result and model characteristics of VGP-CPW with P+ implant in CMOS substrate, which has been detailedly compared with the general VGP-CPW structure. The two different structures of Tlines are fabricated in 65 nm 1P8M RF CMOS process. The proposed physical model is verified by the measured S-parameters up to 50 GHz.

ACKNOWLEDGMENT

The authors wish to thank Dr. Hui Hong for his help in CMOS technology, and would like to acknowledge the support and assistance of the staff of the Institute of Microelectronic CAD.

REFERENCES

[1] Bendik Kleveland, Carlos H. Diaz, Dieter Vook, Liam Madden, Thomas H. Lee, and S. Simon Wong, "Exploiting CMOS Reverse Interconnect Scaling in Multigigahertz Amplifier and Oscillator Design," IEEE Journal of Solid-State Circuits, Vol. 36, No. 10, pp. 1480-1488, October 2001.

[2] A.-L. Franc, D. Kaddour, E. Pistono, N. Corrao, J.-M. Fournier, and P. Ferrari, "Miniaturized high performance shielded CPWtransmission lines from RF to mm-waves," Solid State Device Research Conference 2009 (ESSDERC'2009), pp.129-133, 2009.

[3] Hsiu-Ying Cho, Tzu-Jin Yeh, Sally Liu, and Chung-Yu Wu, High-Performance Slow-Wave Transmission Lines With Optimized Slot-Type Floating Shields, IEEE Transactions on Electron Devices, Vol. 56, No. 8, pp.1705-1711, August 2009.

[4] Avraham Sayag, Dan Ritter, and David Goren, "Compact Modeling and Comparative Analysis of Silicon-Chip Slow-Wave Transmission Lines With Slotted Bottom Metal Ground Planes," IEEE Transactions on Microwave Theory and Techniques, Vol. 57, No. 4, pp. 840-847, April 2009.

[5] T. S. D. Cheung and J. R. Long, "Shielded passive devices for silicon-based monolithic microwave and millimeterwave integrated circuits," IEEE Journal of Solid-State Circuits, vol.41, no. 5, pp. 1183-1200, May 2006.

[6] Juin-Wei Huang, Chao-Shiun Wang, Chorng-Kuang, "Vertical-Ground-Plane Transmission Lines for Miniaturized Silicon-Based MMICs," Radio Frequency Intergrated Circuits Symposium. 2004. pp. 563-566.

[7] Alain M. Mangan, Sorin P. Voinigescu, Ming-Ta Yang, and Mihai Tazlauanu, "De-Embedding Transmission Line Measurements for Accurate Modeling of IC Designs," IEEE Trans. on Electron Devices, vol. 53, no. 2, pp. 235–241, Feb. 2006.

[8] William R.Eisenstadt, Yungseon Eo, "S-Parameter-based IC Interconnect Transmission Line Characterization," IEEE Transactions On Components, Hybrids, And Manufacturing Technology, 1992. vol.15, no. 4, pp. 483-490.

[9] Wen Jincai, Lou Jia, Sun Lingling, Zhang Nan. Comparison and Model of On-chip transmission lines with and without metal grounding in CMOS Process. 2011 China-Japan Joint Microwave Conference, pp. 419-422, 2011.

[10] N.v.d. Meijs, J.T. Fokkema, "VLSI Circuit Reconstruction," Mask Topology, Integration, vol. 2, 1984, pp. 85-119.

[11] Javad Yavand Hasani, Mahmoud Kamarei, and Fabien Ndagijimana, "Sub-nH Inductor Modeling and Design in 90-nm CMOS Technology for Millimeter-Wave Applications," IEEE Transactions on Circuits and Systems—II: Express Briefs, Vol. 55, No. 6, June 2008, pp. 517-521.

978-1-61284-863-1/11 $26.00 © 2011 IEEE

A Novel Accurate dB-Linear Control Circuit Topology for Variable Gain Amplifiers in BiCMOS Technology

[1]Zhenghao Lu, [2]C. H. Hu
[1]Soochow University, Suzhou, China
[2]Marvell Semiconductor, Santa Clara, CA 95054, USA
luzhh@suda.edu.cn

[3]X. P. Yu, [3]W. M. Lim, [3]Y. Liu, [3]K. S. Yeo
[3]School of Electrical and Electronics Engineering
Nanyang Technological University, Singapore
xpyu@pmail.ntu.edu.sg

Abstract—We present in this paper a novel accurate decibel linear control circuit topology for variable gain amplifiers. According to our knowledge, the proposed linear-in-dB control circuit features the simplest control mechanism and topology for current summing or current steering VGAs to date. Based on the proposed circuit topology, a wideband variable gain amplifier with accurate linear-in-dB gain variation is implemented in Tower Jazz 0.18μm SiGe BiCMOS technology. The simulation shows the designed wideband VGA operates up to 5.2GHz with a linear-in-decibel gain ranging from -50dB to +20dB. The VGA possesses nearly constant bandwidth over the gain variation range and dissipates about 6mA current from a single 1.8V supply. The simulated noise figure at maximum gain is less than 6.5dB when matched to 50Ω and the chip layout area is less than 0.3mm² excluding pads.

Keywords-Variable Gain Amplifier; dB-Linear; wideband; VGA; BiCMOS

I. INTRODUCTION

Variable gain amplifiers are widely used in various applications where dynamic range needs to be improved or automatic gain control function needs to be included [1-4]. For example, in RF transmitters, a variable gain amplifier is needed to adjust the output power to minimize interference problems [2]. In many RF receiver applications, the system has to amplify the received signal under circumstances that the wanted signal itself has a very wide variation range or many unwanted signals with almost the same power existing in the same frequency band. Thus, the receiver needs to provide fine gain adjustment with very wide tuning range to maximize the receiver dynamic range [3] [4].

Among all the performance characteristics of RF VGAs, two important factors need to be taken care of, one is the gain variation control range and the other is the bandwidth. To maximize the gain control range, a linear-in-dB gain control characteristic is preferred [4] [5], which means the gain control signal changes linearly while the gain changes linearly in decibel. There has been a large variety of linear-in-decibel VGA circuit topology reported. Generally speaking, various dB-linear gain control methodologies can be categorized into two major groups. One is simply using Programmable Gain Amplifiers with digitally controlled gain steps or digitally

switchable gain stages [6] [7]. The other is using continuous analog gain control signal to control a variable resistance or transconductance [8-10]. The continuous analog gain control mechanism is preferred in many situations not only due to its relative simplicity but also because in many situations such as the OFDM-based applications, an accurate continuous gain control without glitches is necessary [3] [8].

In the continuous control situation, CMOS implementations usually employ pseudo-exponential and Taylor series approximations [5] [11] to realize dB-linear gain control, which may lead to complicated control circuitries and limited accuracy and control range of dB-linear characteristics.

In Bipolar or BiCMOS implementations, the exponential I-V characteristic of the bipolar transistors can be used to realize accurate and much simpler linear-in-dB gain control [1-2] [4] [9-10]. Although by using bipolar transistors it is much easier to realize the exponential V-I conversion, designers still need some correction circuitry to achieve a wide dB-linear control range for VGAs [2] [4]. In this paper, we present a novel wide range dB-linear control circuit in BiCMOS technology, which features much simpler gain control circuit topology than previous reported ones and possess the merits of wide control range, accurate dB-linear characteristic, low power and very wide bandwidth.

II. CIRCUIT DESIGN AND ANALYSIS

A. Current Steering Variable Gain Cell

Bipolar or BiCMOS VGAs have been studied for many years. Figure 1 shows a conventional current steering bipolar gain control cell which is advantageous in terms of high frequency, low noise, low distortion etc. [1] [4]. According to Figure 1, the input signal V_{RFi} is fed to the input transistor Q_2 in the common emitter manner, the amplified collector current I_2 is steered between Q_3 and Q_4 by the voltage difference between the biasing voltage V_{Bias} and the control voltage $V_{Control}$. In this case, $Q_3 = Q_4$, the biasing voltage is usually set to be constant, the control voltage is variable and Q_4 is the output transistor. When the control voltage is high, most of I_2 is steered to the output transistor Q_4 and the gain cell is in

---

This work is partially supported by National Natural Science Foundation of China under Grant Number 60906017.

978-1-61284-863-1/11 $26.00 © 2011 IEEE

the high gain mode. When the control voltage is low, most of I_2 is steered to the Q_3 and shunted to the supply voltage line. Therefore, the gain cell is in the low gain mode. If we define $V_c = V_{Bias} - V_{Control}$, we have

$$\frac{I_3}{I_4} = \exp\left(\frac{V_c}{V_T}\right) \quad (1)$$

where I_3 and I_4 are collector currents of Q_3 and Q_4 respectively and $V_T = kT/q$ is the thermal voltage. The transconductance gain of this gain cell is therefore given by

$$\frac{I_{out}}{V_{RFi}} = g_{m1}\frac{I_4}{I_2} = g_{m1}\frac{I_4}{I_3 + I_4} = g_{m1}\frac{1}{1 + \exp\left(\dfrac{V_c}{V_T}\right)} \quad (2)$$

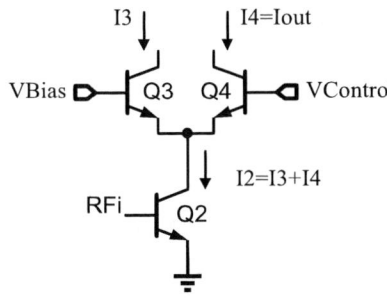

Figure 1.   Current Steering Variable Gain Cell without dB-linear Control

According to (2), one can find that when $V_c = V_{Bias} - V_{Control} \gg 0$ so that $\exp(V_c/V_T) \gg 1$, which means the gain cell is in the low gain mode, (2) can be simplified to

$$\frac{I_{out}}{V_{RFi}} = g_{m2}\frac{1}{1 + \exp\left(\dfrac{V_c}{V_T}\right)} \approx g_{m2}\exp\left(-\frac{V_c}{V_T}\right) \quad (3)$$

Based on (3), as $-V_c$ increases linearly, the output current increases exponentially, the characteristic is fairly linear-in-dB. However, as $-V_c$ increases and approaches 0, the preceding approximation is no longer valid. There will be significant deviation from the ideal dB-linear characteristic when the gain is high. Without any correction or compensation, the dB-linear control range will be significantly limited by such kind of deviation. Therefore, it is desirable to design a dB-linear control circuit to enforce the current relationship as give by (4) at any biasing conditions which is not limited to low gain condition approximation as given by (3).

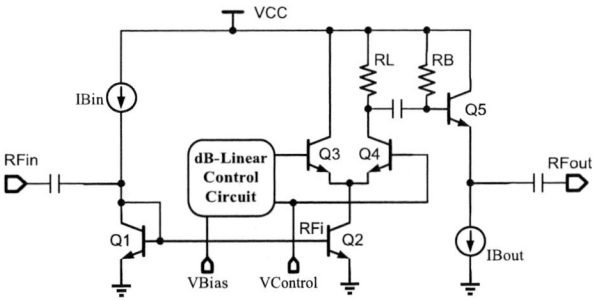

Figure 2.   Schematic of the proposed VGA with dB-linear Control Circuit

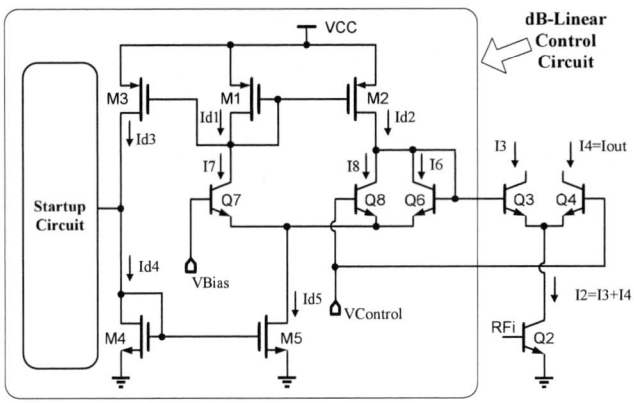

Figure 3.   Schematic of the Proposed Novel dB-linear Gain Control Circuit

$$\frac{I_4}{I_2} = \exp\left(-\frac{V_c}{V_T}\right) \quad (4)$$

B.   VGA Design with Proposed dB-Linear Control

Before we proceed to the detailed description on our proposed dB-linear circuit design, the complete VGA circuit is shown by Figure 2 in the first place. The 50Ω input impedance matching is accomplished by proper biasing of the transistor Q_1 as described by

$$r_{in} = \frac{1}{g_{m1}} = \frac{I_{Bin}}{V_T} = 50\Omega \quad (5)$$

According to (5), one can design the biasing current I_{Bin} to be a PTAT biasing current to design the circuit with temperature indepedant characteristic. Q_2, Q_3 and Q_4 form the current steering variable gain cell. The input signal is fed to transistor Q_2 in the common emitter manner, the amplified collector current I_2 is steered between Q_3 and Q_4 by the voltage difference between the biasing voltage V_{Bias} and the control voltage $V_{Control}$ as described previously. The maximum gain is about $g_{m2}R_L$ when I_2 is 100% steered into Q_4. Q_5 is the emitter follower output transistor for output matching purpose. To

978-1-61284-863-1/11 $26.00 © 2011 IEEE   433

achieve wideband accurate linear-in-dB gain control, a dB-linear control circuit is designed in parallel with the current steering pair Q_3 and Q_4 which will be described in detail in the following.

C. The Proposed Self-Biased dB-Linear Control Circuit

The dB-linear or linear-in-dB control circuit suitable for wide range gain control in current steering VGAs has also been reported in some well known publications such as [2] [4]. In [2], multiple references and operational amplifiers are employed in the dB-linear control circuit. In [4], complicated voltage to current and current to voltage conversion circuits are used. In this paper, we propose a very simple yet very effective dB-linear control circuit shown in Figure 3.

To simplify the description, bipolar transistors Q_3, Q_4, Q_6, Q_7 and Q_8 are assumed to be identical. We define $V_c = V_{Bias} - V_{Control}$ and according to the exponential V-I characteristics of bipolar transistor, we have

$$I_8 = I_7 \exp\left(-\frac{V_c}{V_T}\right) \tag{6}$$

Furthermore, if we design the PMOS transistors M_1 to be identical to M_2, the current mirror forces $I_{d1} = I_{d2}$ which leads to

$$I_7 = I_8 + I_6 \tag{7}$$

$$\frac{I_8}{I_6} = \frac{\exp\left(-\dfrac{V_c}{V_T}\right)}{1 - \exp\left(-\dfrac{V_c}{V_T}\right)} \tag{8}$$

Because the bases of Q_3/Q_4 and Q_6/Q_8 are connected in together and due to the exponential V-I characteristics of bipolar transistors, we can write

$$\frac{I_4}{I_3} = \frac{I_8}{I_6} = \frac{\exp\left(-\dfrac{V_c}{V_T}\right)}{1 - \exp\left(-\dfrac{V_c}{V_T}\right)} \tag{9}$$

Since the relationship exists that $I_2 = I_3 + I_4$ and based on (9) we have

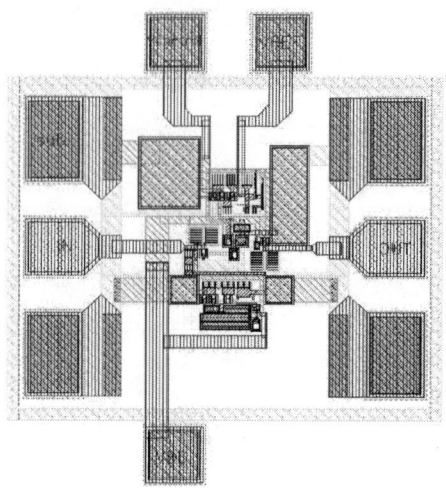

Figure 4.   Layout of the proposed VGA circuit

$$\frac{I_4}{I_2} = \frac{I_4}{I_3 + I_4} = \exp\left(-\frac{V_c}{V_T}\right) \tag{10}$$

The accurate exponential relationship between the input current I_2 and the output current I_4 which is exactly as expected as (4) is therefore achieved. According to (10), the output current increases linearly in decibel as the control voltage $V_{Control}$ increases linearly, which is expressed by

$$\ln(I_{out}) = \ln(I_2) + \frac{V_{Control} - V_{Bias}}{V_T} \tag{11}$$

The proposed dB-linear control circuit in Figure 3 is self-biased with current mirrors. A startup circuit is necessary to ensure the circuit works properly. The startup circuit is not shown in detail for simplicity.

III.   SIMULATION RESULTS AND DISCUSSIONS

The proposed VGA design with novel dB-linear control circuit is implemented in Tower Jazz 0.18μm SiGe BiCMOS technology. Figure 4 shows the layout of the proposed circuit. The proposed circuit performance is simulated with extracted parasitic in Cadence SpectreRF. The simulation testbench is set as follows: the supply voltage is VCC 1.8V, the reference voltage is VBias 1.3V, the control voltage is sweeping from 1.0V to 1.3V. Figure 5 shows the simulated gain S21 versus the sweeping control voltage. According to Figure 5 and (10), when $V_c = V_{Bias} - V_{Control} = 0$, I_2 is 100% steered into Q_4, the gain is highest at this control voltage 1.3V, which is about 21dB. As the control voltage decreases, the gain also decreases dB-linearly. The linear-in-dB gain variation range is approximately from -50dB to +20dB.

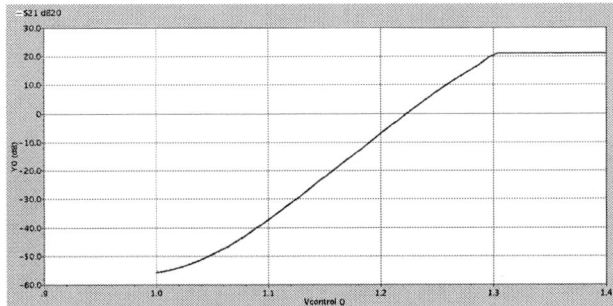

Figure 5. Simulated gain S21 versus the sweeping control voltage

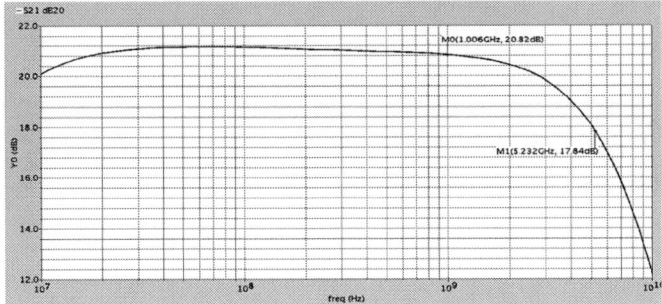

Figure 6. Simulated gain S21 versus frequency at 1.3V control voltage

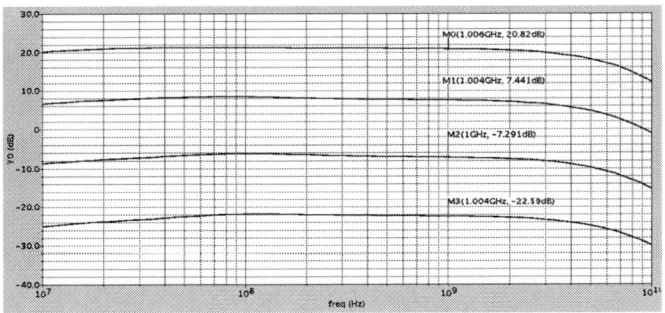

Figure 7. Simulated gain versus frequency at different control voltages

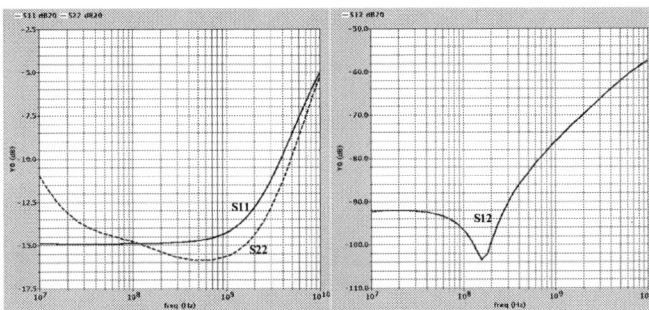

Figure 8. Simulated S11, S22 and S12 at maximum gain

Figure 6 shows the simulated S21 frequency response at 1.3V control voltage which means at the highest gain. The $-3dB$ bandwidth can reach up to 5.2GHz. Figure 7

shows the simulated gain-frequency response at different control voltages. From top to bottom, the control voltages are 1.3V, 1.25V, 1.2V and 1.15V respectively. As shown by figure 8, the bandwidth response at different gain level is nearly constant. Figure 8 shows the simulated S11, S22 and S12.

IV.   CONCLUSIONS

A wideband Variable Gain Amplifier with 5.2GHz bandwidth is designed. A novel dB-linear gain control circuit is proposed to extend the linear-in-dB gain variation range of the current steering gain cell to the extremity. The achieved dB-linear gain variation range is roughly from -50dB to +20dB. The simulated current consumption is 6mA and noise figure at maximum gain is less than 6.5dB.

ACKNOWLEDGMENT

The authors wish to acknowledge the MPW provided by Jazz Tower. This work is also partially supported by National Natural Science Foundation of China under Grant Number 60906017.

REFERENCES

[1] W. M. Sansen and R. G. Meyer, "An integrated wideband variablegain amplifier with maximum dynamic range," IEEE J . Solid-state Circuits, vol. SC-9, no. 4, pp. 159-166, Aug. 1974.

[2] Danielle Coffing, Eric Main, Mark Randol and Gina Szklarz, "A variable gain amplifier with 50-dB control range for 900-MHz applications," IEEE Journal of Solid State Circuits, Vol. 37, No. 9, pp. 1169-1175, September 2002.

[3] Hassan Elwan, Ahmet Tekin, and Kenneth Pedrotti, "A Differential-Ramp Based 65 dB-Linear VGA Technique in 65 nm CMOS," IEEE Journal of Solid State Circuits, Vol. 44, No. 9, pp. 2503-2514, September 2009.

[4] Shoji Otaka, Gaku Takemura, and Hiroshi Tanimoto, "A Low-Power Low-Noise Accurate Linear-in-dB Variable-Gain Amplifier with 500-MHz Bandwidth" IEEE Journal of Solid State Circuits, Vol. 42, No. 2, pp. 1942-1948, February 2007.

[5] Yuanjin Zheng, Jiangnan Yan, and Yong Ping Xu, "A CMOS VGA With DC Offset Cancellation for Direct-Conversion Receivers," IEEE Transactions on Circuits and Systems-I: Regular Papers, Vol. 56, No.1, pp. 103-113, January 2009.

[6] H. O. Elwan et al., "A buffer-based baseband analog front end for CMOS bluetooth receivers," IEEE Trans. Circuits Syst., vol. 49, no.8, pp. 545–554, Aug. 2002.

[7] Jianhong Xiao, Iuri Mehr, and Jose Silva-Martinez, "A High Dynamic Range CMOS Variable Gain Amplifier for Mobile DTV Tuner," IEEE J. Solid-State Circuits, vol. 40, no. 12, pp. 2536–2546, Dec. 2005.

[8] P. Antoine et al., "A direct-conversion receiver for DVB-H," IEEE J. Solid-State Circuits, vol. 42, no. 2, pp. 292–301, Feb. 2005.

[9] W. M. C. Sansen and R. G. Meyer, "Distortion in bipolar transistor variable-gain amplifiers," IEEE J. Solid-State Circuits, vol. SC-8, pp. 275–282, Aug. 1973.

[10] Robert G. Meyer, and William D. Mack, "A DC to 1-GHz Differential Monolithic Variable-Gain Amplifier," IEEE Journal of Solid State Circuits, Vol. 26, No. 11, pp. 1673-1680, November 1991.

[11] Q. H. Doung et al., "A 95-dB Linear Low-Power Variable Gain Amplifier," IEEE Transactions on Circuits and Systems-I: Regular Papers, Vol. 53, No.8, pp. 1648-1657, January 2006.

A 6-GHz dual-modulus prescaler using 180nm SiGe technology

C. Z. Nan[1], X. P. Yu[1], B. Y. Hu[1], Z. H. Lu[2], W. M. Lim[2], Y. Liu[2], K. S. Yeo[2], ChangHui Hu[3]

[1] Institute of VLSI Design, Zhejiang University, Hangzhou, P. R. China
[2] Nanyang Technological University, Singapore, 639798
[3] Marvell Semiconductor, Santa Clara, CA 95054, USA
Email: xpyu@vlsi.zju.edu.cn

Abstract—A 6-GHz divide-by-3/4 prescaler for fractional-N frequency synthesizers is presented in this paper. Optimization in the power consumption is carried out to ensure a high efficiency. Meanwhile, the critical cells, current-mode-logic (CML) building blocks, are optimized for high frequency and low power consumption. A prototype has been implemented in 180nm SiGe BiCMOS technology offered by TowerJazz. The silicon area for the divide-by-3/4 prescaler is only 40 μ m x 50 μ m. The maximum operating frequency can reach 6-GHz, with 7.92mW power consumption in 1.8V voltage supply.

Keywords- dual-modulus prescaler; power and speed optimization; SiGe BiCMOS

I. INTRODUCTION

With the development of the new generation communications system, the demands of multi-mode interoperability wireless transceiver solutions are stronger[1-3]. However, the requirements for fast switching and high operating frequencies make the design of frequency synthesizers a challenging task. Because of the limitation of step size by the reference frequency, it is often difficult to cover the multiple frequency bands using an integer frequency synthesizer with fine resolution. Conceptually, in order to achieve fine step size, one has to lower the reference frequency in an integer-N synthesizer design, which means a high division ratio for the PLL and thus high in-band phase noise [1]. On the contrary, fraction-N synthesizers allow the PLL to achieve a fine step size by taking advantage of the loop division ratio between integer numbers; while still maintain a high reference frequency. As a common part of this configuration, dual modulus prescalers are often considered as the most challenging part in the high speed frequency divider design because of its highest operating frequency. Therefore, a delicate design of dual modulus prescaler with power and speed optimization is highly desired in fraction-N frequency synthesizers.

In this paper, a 6-GHz divide-by-3/4 prescaler optimized for high frequency and low power dissipation is presented. In section II, three different modulus topologies of prescaler are compared for their power efficiency under unit of division ratio. The implementation of the proposed multi-modulus prescaler is presented in section III. At last, experiment results are given in section IV.

II. MULTI-MODULUS TOPOLOGIES OF PRESCALER

A conventional topology of divide-by-4/5 prescaler, which consists of three D flip-flops, is shown in Fig. 1(b). MC, as a modulus control signal, will decide the division ratio to be 4 or 5. To realize these two different division ratios, extra logic gates and D flip-flops are added to a conventional divide-by-2 frequency divider. As a result, the operating frequency is enormously reduced since an additional propagation delay is introduced [2]. Compared with the power consumption of a divide-by-2 unit, the prescaler mentioned above also consume a higher power, which takes the largest portion of the total power consumption in the divider chain.

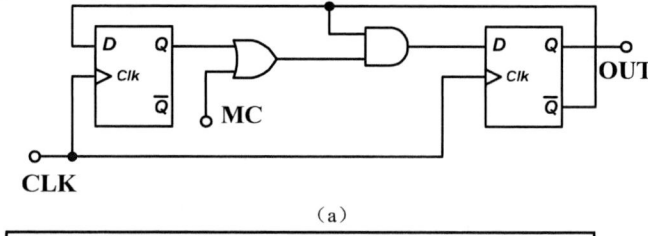

(a)

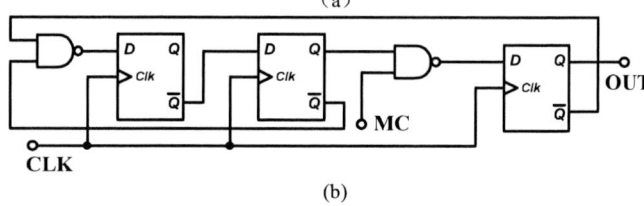

(b)

Fig. 1   (a) typical topology of divide-by-2/3 prescaler (b) typical topology of divide-by-4/5 prescaler

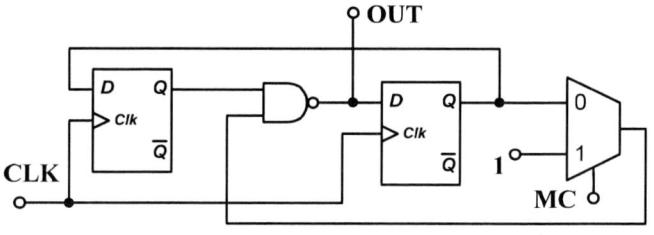

Fig. 2   divide-by-3/4 prescaler

The operation of the prescalers at the highest input frequency makes it the bottleneck of the PLL design. Theoretically, it is a consensus that the less components operate at the full speed, the higher frequency prescalers can achieve. Therefore, when referring to high speed applications, a typical divide-by-2/3 prescaler as shown in Fig. 1(a), is widely used to replace the divide-by-4/5 unit. Comparing the difference between the two topologies of prescaler, it is necessary to add extra logic gates and a D flip-flop when divide-by-4/5 unit is chosen. Unfortunately, A D flip-flops, especially realized by CML (current mode logic) architecture, often consume large power when enable.

The proposed divide-by-3/4 prescaler, shown in Fig. 2, provide a compromise of the dividing ratio and power consumption between the two dual-modulus prescaler. In practice, the ratio of modulus to power consumption of a prescaler has been introduced as a performance measure of a prescaler design, e.g, to demonstrate the power efficiency. From the perspective of power efficiency, shown in Fig. 3, it is obvious that modulus divide-by-3/4 is superior to the other two modulus mentioned above. Thus, divide-by-3/4 is chosen as the proposed topology.

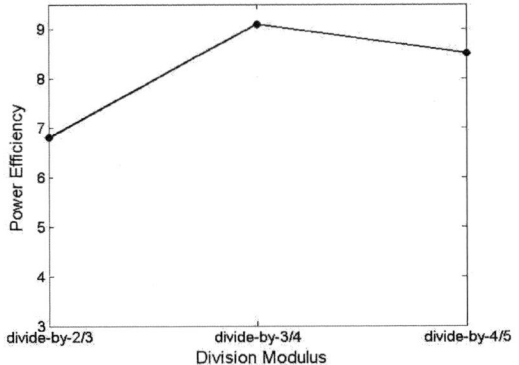

Fig. 3   power efficiency of differency modulus prescaler

III.   PROPOSED DESIGN

A.   D-latch

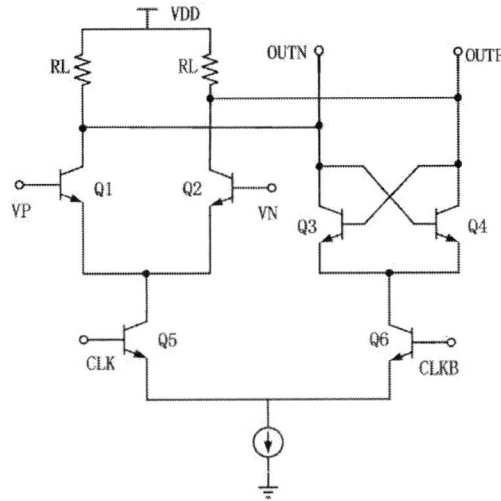

Fig. 4   schematic of CML D-latch

For high-speed and robust operation, current-mode-logic (CML) architecture shown in Fig. 4, is introduced to compose the D-flip flops[4-6], using in divide-by-3/4 prescaler. As the former study[7] revealed, D-latch often makes a large contribution to the propagation delay of whole design, due to its repeatability in a conventional topology. Therefore, it is necessary to optimize the performance of D-latch in power consumption and maximum operating frequency.

The D-latches always operate at two different modes periodically[8]. Conceptually, the operation mode of such architecture seems as follow: when the differential clock signal CLK is low, one of the latches is under sensing mode. The D-latch senses the input signal and flips it to the output. On the contrary, when the input signal CLK is high, the D-latch is under latching mode. It keeps the current state latched.

Basically, the speed of this type architecture heavily depends on how fast the sensing operation is [8]. Thus, the analysis of maximum operating frequency is based on the assumption of sensing mode. At the maximum operating frequency $f_{in,\,max}$, negative transconductance $gm_{bjt,\,max}$ takes charge of compensating the load C_L. Then a simplified term of the maximum operating frequency can be given as follow,

$$f_{in,\,max} = \frac{gm_{bjt,\,max}}{2\pi C_L} \tag{1}$$

$$gm_{bjt,\,max} = \frac{\delta I}{\delta V_{be}} = \frac{I_s}{V_T} e^{\frac{V_{be}}{V_T}} \approx \frac{I}{V_T} \tag{2}$$

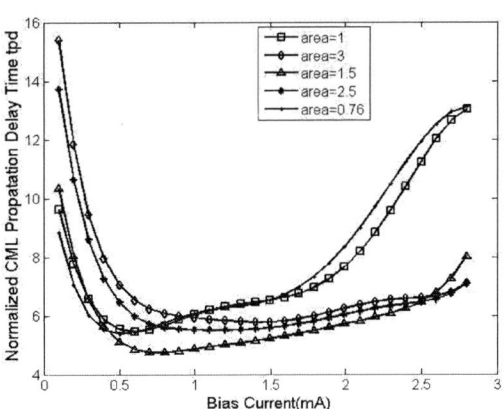

Fig. 5   Normalized CML Propatation Delay vs Bias Current at different transistor area

From the equation (1), (2) the operating frequency of the D-latch can be improved by increasing $gm_{bjt,\,max}$, which can be achieved either by increasing the bias current or the HBT's emitter area. Nevertheless, the increasing of HBT's emitter area, would also lead to the increasing of output loading, which would deteriorate the operating frequency[7]. As a result, a trade-off between the maximum operating frequencies, transistor emitter area and power consumption must be taken into consideration. Simulation of this trade-off with 180 nm SiGe technology offered by Jazz Tower, is presented as Fig. 5, where an area of 1, 2, 3 refers to 0.15 μ m x 1 μ m, 0.15 μ m x 2 μ m, and 0.15 μ m x 3 μ m, respectively. The delay behavior in Fig. 5 indicates that there is always an optimum bias current for a certain transistor area, to achieve the maximum switching speed.

B. High Frequency CML MUX

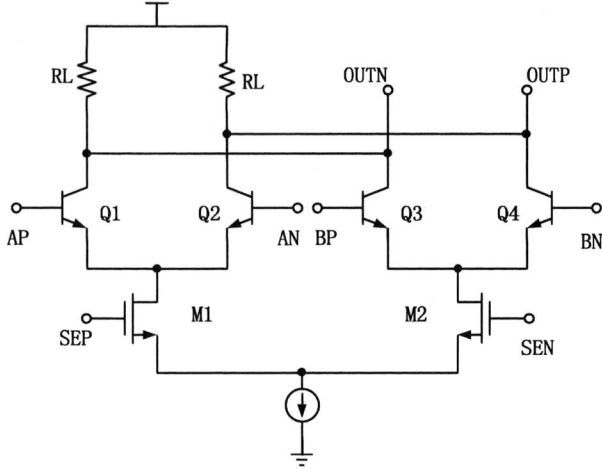

Fig. 6   schematic of CML MUX

Fig. 6 shows a conventional CML 2-1 MUX. Considering the GHz range application, current mode logic architecture has been chosen to take advantage of its high operating frequency and constant power consumption[6]. Due to the more ideal switch performance than bipolar transistor, a pair of NMOS transistors is assigned as the function of mode control. In the proposed circuit as shown in Fig. 3, the MUX chooses VDD as an output signal, when the signal SEP is logically high. Thus, the divide modulus is 3; On the other hand, when the signal SEP is logically low, the MUX steer the BP as an output signal. Consequently, the simplified topology comes to a typical divide-by-4 CML divider.

C. High Frequency CML NAND Gate

The propagation delay of the NAND gate in dual-modulus 3/4 prescaler is also critical[2]. In order to operate in GHz range, the NAND gate, shown in Fig. 7, is implemented in full differential CML architecture. A full switched differential voltage swing 800m $V_{pp,dif}$ is assigned to matching the input terminal of the D-flip flops.

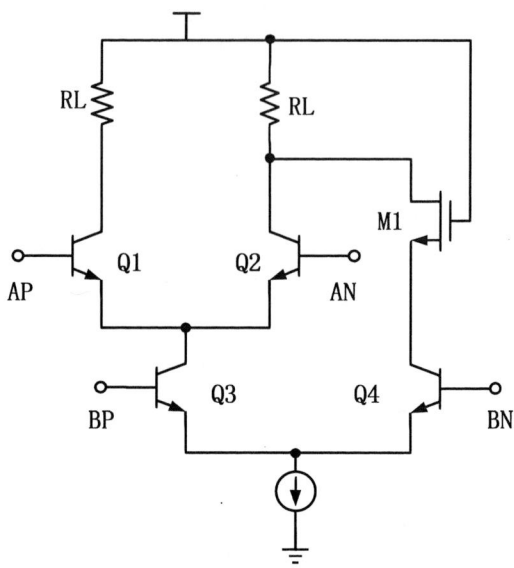

Fig. 7   schematic of CML NAND gate

IV.   SIMULATION AND EXPERIMENTAL RESULT

Following the CML optimization analysis, the transistor areas of the differential pair and cross-coupled pair in D-latch are designed to be 0.15 μ m x 1 μ m, 0.15 μ m x 1.2 μ m, respectively. This choice not only meets the relation of the optimized propagation delay and transistor area, bias current, but also takes the minimum transistor size of the technology (180nm SiGe process offered by TowerJazz) into consideration, which means less parasitic to increase speed. Fig. 8 shows the simulated waveform of the dual-modulus prescaler, which indicates that with a differential 800m $V_{pp,dif}$ input, the modulus of the divide-by-3/4 prescaler perform correctly. The prescaler was simulated using Cadence Spectre RF. When the MC signal is logically high, the ratio of division is 3. On the contrary, the ratio of division is 4, if the MC signal is logically low.

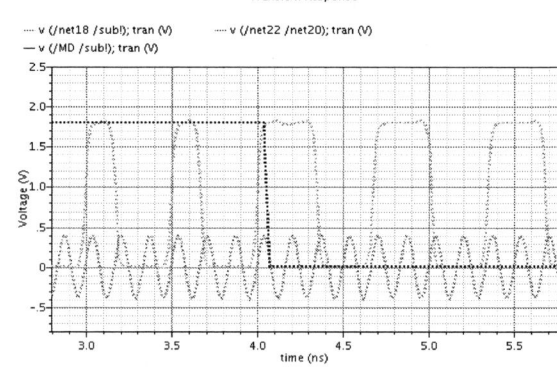

Fig. 8   Transient simulation of the proposed ciucuit

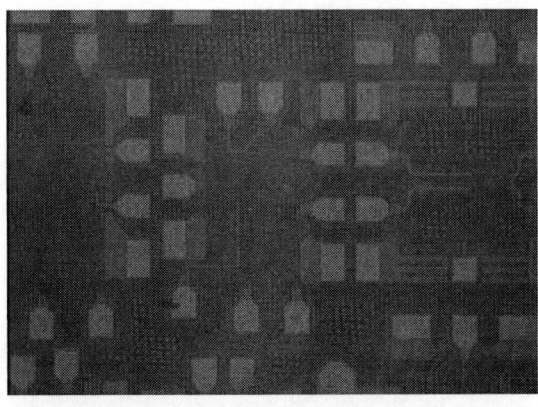

(a)

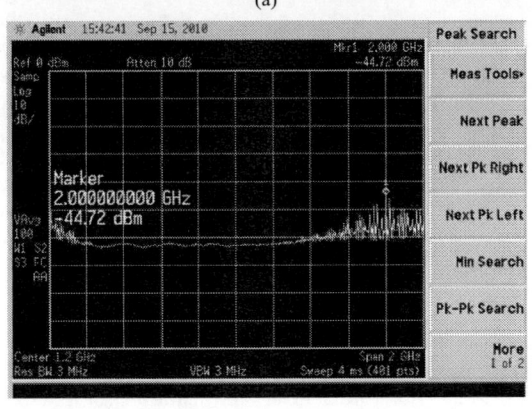

(b)

Fig. 9   (a) die photo (b) spectrum of the 6-GHz input under a ratio of 3:1

the proposed prescaler provide a remarkable improvement in performances in terms of low power and operating frequency. The high-speed and low-power nature of the prescaler is mainly attributed to the optimization of the CML architecture in transistor area and bias current, which lead to a small normalized propagation delay and small loading to the prescaler.

V.   CONCLUSION

A 6-GHz divide-by-3/4 prescaler has been demonstrated. The most critical cell, current-mode-logic D-latch, has been optimized for high frequency application and low power consumption. An analysis of the trade-off among the operating speed, transistor area, and bias current is presented. The optimization technique is verified by the measurement of a high performance dual-modulus prescaler implemented in an 180nm SiGe BiCMOS technology.

ACKNOWLEDGMENT

The authors wish to acknowledge the MPW provided by TowerJazz. This work is supported by National Natural Science Foundation of China under Grant Number 60906017 and   foundation for promote the education of RFICs in graduate studies, College of EE, Zhejiang University.

REFERENCES

[1] Ray M, Souder W, Ratcliff M, Dai F, Irwin J.D, "A 13GHz Low Power Multi-Modulus Divider Implemented in 0.13μm SiGe Technology," *IEEE Topical Meeting on Silicon Monolithic Integrated Circuits in RF Systems*, pp.1-4, Jan 2009

[2] Wenguan Li, Honglin Chen, Ruohe Yao, "A 5.5-GHz multi-modulus frequency divider in 0.35μm SiGe BiCMOS technology for delta-sigma fractional-N frequency synthesizers," *2010 International Conference on Microwave and Millimeter Wave Technology (ICMMT)*, pp.1937-1940, May 2010.

[3] Changhui Hu, Patrick Chiang, "A Fully-Integrated, 3-5GHz, 500Mbps, IR-UWB Transceiver with Pulse Injection-Locking for Receiver Phase Synchronization in 90nm CMOS", *IEEE J. Solid-State-Circuits*, May 2011.

[4] Chengzhi Li, Xiaofeng Qu, Kejie Lu, Wenquan Sui, "Design of a 18 GHZ divide-by-4 prescaler using 1μM GaAs HBT technology," *International Conference on Communications (MICC), 2009 IEEE 9th Malaysia*, pp.354-357, Dec 2009.

[5] Craninckx J, Steyaert M.S.J, "A 1.75-GHz/3-V Dual-Modulus Divide-by-128/129 Prescaler in 0.7-μM CMOS," *IEEE Journal of Solid-State Circuits*, Vol.31, No.7, pp.890-897, Jul 1996.

[6] Wei-Yu Tsai, Ching-Te Chiu, Jen-Ming Wu, Shuo-Hung Hsu, Yar-Sun Hsu, "A novel MUX-FF circuit for low power and high speed serial link interfaces," *Proceedings of 2010 IEEE International Symposium on Circuits and Systems (ISCAS)*, pp.4305-4308, Aug 2010.

[7] Joseph M. C. Wong, Vincent S. L. Cheung, and Howard C. Luong, "A 1-V 2.5-mW 5.2-GHz Frequency Divider in a 0.35-um CMOS Process". *IEEE Journal of Solid-State Circuits*, Vol.38, No.10, pp.1643-1648, Oct 2003

[8] Wong, M C, "A 1.8-V 2.4-GHz Monolithic CMOS Inductor-less Frequency Synthesizer for Bluetooth Application," MS thesis, University of Science and Technology, Hong Kong.

[9] Detratti M, Cabo J, Pascual J.P, Herrera A, "A 4.5 GHz 3-4 dual-modulus frequency divider IC in GaAs technology," 2005 European Microwave Conference, pp.4-7, Oct 2005

The divide-by-3/4 prescaler is followed by an output buffer used to convert the output as a rail to rail signal. The circuits are designed and fabricated in 180nm SiGe technology (sbc18h2) offered by TowerJazz. Fig. 9(a) shows the die photo of the proposed design. The silicon area of the core is about 40 u m x 50 u m. Measurement is carried out on-wafer using a cascade probe station while the circuit runs from a 1.8V power supply. The spectrum of 6-GHz input is shown in Fig. 9(b), exhibiting a correct division of 3, with the power consumption of 7.92mW. The power consumption in the output rail-to-rail buffers is not included in the above calculation.

TABLE I.   COMPARISON OF PERFORMANCES

reference	technology	Maximum operating frequency(GHz)	Supply voltage(V)	Supply current(mA)
[2]	0.35 μ m BiCMOS	5.5	3	6.8mA
[9]	GaAs pHEMT	4.5	3.5	75mA
This work	180nm BiCMOS	6	1.8	4.4mA

Table I shows comparison of the performances between the proposed circuit and other literatures. It can be found that

978-1-61284-863-1/11 $26.00 © 2011 IEEE

Self-demodulated Receiver at MM-wave Range Using SiGe Technology

X. P. Yu[1], B. Y Hu[1], X. L. Yan[1]

[1]Institute of VLSI Design, Zhejiang University, Hangzhou, P.R. China, 310007
xpyu@vlsi.zju.edu.cn

Z. H. Lu[2], W. M. Lim[2], Y. Liu [2], K. S. Yeo[2], C. H. Hu[3]

[2]School of Electrical and Electronic Engineering, Nanyang Technological University, Singapore, 638798
[3]Marvell Semiconductor, Santa Clara, CA 95054, USA
zhlu@pmail.ntu.edu.sg

Abstract—**In this paper, a new self-demodulated receiver at MM-wave range is proposed. In difference with the conventional local oscillator down-conversion, the OOK modulated data is recovered by mixing two paths of amplified RF signals. Thanks to the simplified architecture, the system is able to work at a significantly wider range with a lower power consumption and faster settling time. Implemented in TowerJazz 180nm SiGe technology, the proposed receiver is able to work from 50GHz-70GHz with a data rate higher than 1Gbps while consuming a current less than 15mA from a standard 1.8V supply voltage.**

Keywords- Low power;MMIC;receiver; SiGe; Self-demudulated

I. INTRODUCTION

The emerging high data-rate short-range personal-area network becomes increasingly promising under the great demand of ultra high speed wireless link. Millimeter-wave range frequency is an excellent candidate as the carrier for such applications. For example, IEEE 802.15.3c, which is defined at the frequency ranges from 57 to 64 GHz unlicensed spectrum, offer a data rate over 1Gbps. Moreover, the circuits at MM-wave usually have a smaller silicon area which makes it a potential solution for massive product. A substantial work has been dedicated to designing of RF transceivers at MM-wave range in recent years [1-3]. Several prototypes have been successfully developed to demonstrate the possibility of high data rate (>Gbps) within a short range. The feasibility of implementation of MM-wave range wireless link has been well-proven.

However, before the popularity and massive production of MM-wave wireless links in consumer electronics, two key issues should be settled: The system should be low-cost and of low power consumption. In recent decades, for its relatively low cost and easy of full-integration, CMOS integrated circuit has being celebrated as an overwhelming solution for communication circuits and system on chip. Recently, with the rapid progress transistor scaling down, the operation frequency at MM-wave range, which is conventionally achieved using IIIV materials, is no longer a technology difficulty to CMOS technology. For example, the cut off frequency of nowadays commercial deep sub-micro CMOS technology can go as high as 200 GHz. However, the deep sub-micro CMOS technology is currently very expensive for MM-wave integrated circuits. Unlike the digital circuits whose size can be reduced a lot in advanced technology, the RF/MM range circuits requires a lot of passive components which usually occupy large silicon areas. Meanwhile, the area or physical size of these components is determined by the value of inductance or capacitance. Hence, it does not benefit much from technology scaling down. This makes the deep sub-micro technology not cost-efficient for MMIC design. SiGe BiCMOS, on the other hand, offers high performances for high frequency applications and it has MOS technology for SOC solutions. More important, it is generally much cheaper than cutting edge CMOS technology. For example, 180nm SiGe technology (sbc18h2) by TowerJazz offers a f_T of 200GHz, the cost of fabrication is much lower compared with current nano-scale CMOS technology which offers similar cut-off frequency.

Another issue is power consumption. To operate at a frequency which is tens of that of conventional carrier frequency, the mm-wave circuits demand much higher power consumption. This limits their applications for battery-based portable devices. A straightforward way is to effectively reduce the power consumption of each block, e.g. LNA, VCO [3-5]. A more effective way is to optimize the system. For example, to reduce the power consumption, a direct down-conversion receiver which has less high frequency components can be used. Besides, simple modulation scheme like ASK or OOK is often used for their lower complexity and low requirement for the system linearity [6].

In this paper, a new self-demodulator is proposed to overcome above-mentioned difficulties. In this work, the receiver has a simple topology which is of high data-rata but low power consumption. The designs considerations in designing MM-wave receivers are described in Section II, followed by detail description of the proposed receiver in section III. Simulation results are shown in Section IV and conclusions are drawn in Section V.

II. DESIGN CONSIDERATIONS

The design considerations in a RF receiver include a lot of key aspects like sensitivity, linearity, power consumption, etc. They are determined by wireless communication standards. The emergence of MM-wave range communications within a short range brings new aspects in the design of RF receivers for such applications. Firstly, silicon based MM-wave range wireless link is mainly used in a short range; the requirement of sensitivity is therefore less stringent. The typical input signal power of the receiver for short range receiver can be as high as -40dBm, compared with <-100dBm in conventional long range communications. Secondly, because of the carrier frequency

978-1-61284-863-1/11 $26.00 © 2011 IEEE

becomes several tens of GHz, it is much easier to obtain a high transfer rata using very simple modulation schemes, e.g. ASK, OOK.

However, it is still necessary to overcome some technical difficulties in the implementation of receiver at MM-wave range. The implementation of building blocks become very challenging in silicon based technology. For instance, the voltage controlled oscillator which should have a wide tuning range but low phase noise is very hard to realize. The conventional way of digital-controlled capacitor array at low frequency becomes inefficient. The low quality factor of varactor at tens of GHz becomes very low; this makes the phase noise poor. A typical 60GHz VCO has a power consumption of 30mW [4]. Once a frequency synthesizer is used, a long settling time is foreseeable and large power consumption in divider chain is unavoidable. In [5], the power consumption of PLL reaches 43mW. To overcome this difficulty, in [9], the receiver with an injection locked oscillator is proposed, it has a wide but the injection locked oscillator still need additional settling time. And to increase the sensitivity of the system, an input buffer is still needed.

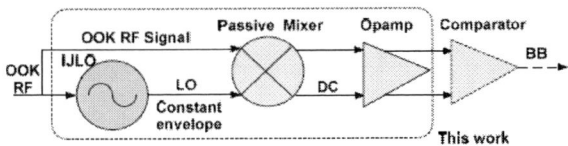

Figure 1.   The demodulator in [9]

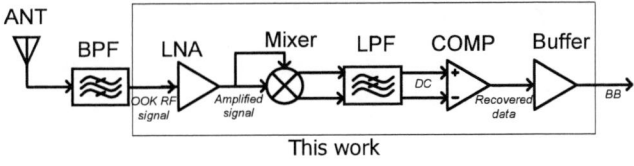

Figure 2.   Proposed Demodulator

III.   PROPOSED DESIGN

A.   Architecture

Figure 2 shows the architrave of proposed receiver. In this design, the oscillator is removed which a high gain amplifier is used. After passing this amplifier, the signal is strong enough to drive the mixer.   The proposed architecture has several advantages. Firstly of all, it is simple; the power consumption can be reduced significantly.   Without the oscillator or frequency synthesizer which requires settling and with a narrow locking rang, the data rate and bandwidth of the proposed receiver can be improved as well. Nevertheless, to achieve such a system, new requirements in the key building blocks emerge. The low noise amplifier should have a high gain with wide band-width. The mixer should have a good linearity and the IF amplifier must have a good noise performance and high gain.

B.   Low noise amplifier

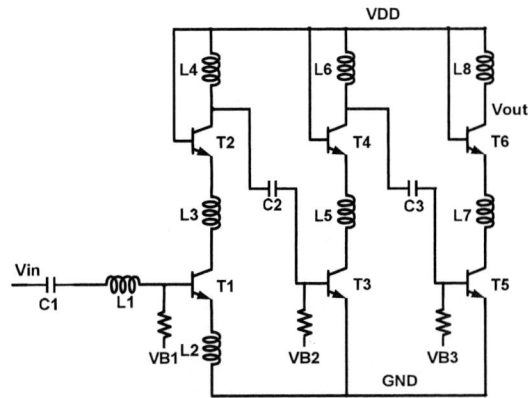

Figure 3.   Topology of the LNA

The low noise amplifier is the most challenging block in the proposed system. Due to the simple OOK modulation scheme and self-demodulation receiver architecture, the requirements of linearity and noise performance on the LNA are relaxed. Instead, significant RF path gain is needed to amplify the weak input signal to a proper level so that it can demodulate itself in a following mixer.

Considering the design requirements mentioned above, we proposed a three-stage single input single output LNA, the schematic of which is shown in Figure 3. Lumped inductors and MiM capacitors are adopted for frequency tuning and inter-stage matching separately. The first stage adopts the classical inductor-degeneration topology and is optimized for simultaneously input impedance and noise matching. The second stage is the same as the first one except for the emitter inductor. *L3, L5* and *L7* are inserted between transistor pairs *T1-T3, T2-T4* and *T5-T6* respectively. They compose *pi* networks together with the parasitic capacitances at the collector of *T1, T3, T5* and emitter of *T2, T4 and T6* respectively. These parasitic capacitances provide low impedance path to the noise current from *T2, T4* and *T6*, and steer part of the input signal to ground. By proper choosing the network's resonant frequency, both gain and noise performances can be greatly improved [10]. Figure 4 shows the simulated *S11, S21* and noise figure performances. The LNA has a gain over 18dB gain from 50GHz to 70GHz.

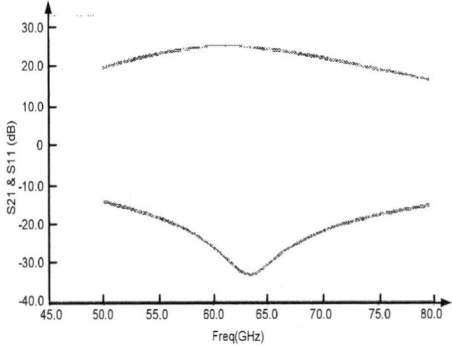

Figure 4.   Simulated Gain of the LNA

C. Mixer with IF amplifer

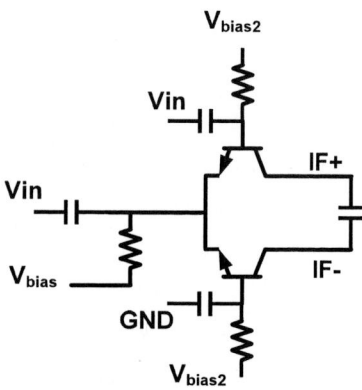

Figure 5. Topology of the mixer with filter

For better linearity, a double balanced passive mixer is used. Fig. 5 shows the topology of the mixer with loop filter. The mixer is followed by a high gain comparator. As the comparator operates a much lower frequency which is determined by the data rate of the receiver, it input stage of implemented in MOS transistors.

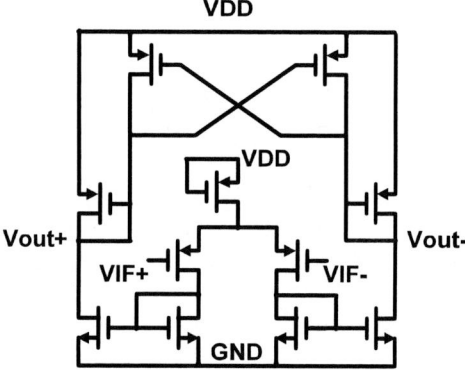

Figure 6. Topology of IF comparator

The sizing of the emitter length of transistors in mixer is targeted for better gain. The figure shows the voltage gain of the mixer plus the IF amplifier, which is simulated by periodical steady state plus periodical AC analysis and sweeping the emitter length from $1\,\mu$ m to $3\,\mu$ m. Based on the simulation, the peak gain of the mixer with comparator can be achieved around $1.2\,\mu$ m emitter length as shown in Fig. 7. Considering possible process variations and other non-idealities, it is more desirable and safer to choose the emitter length to be around $1.5\,\mu$ m according to the simulated conversion gain.

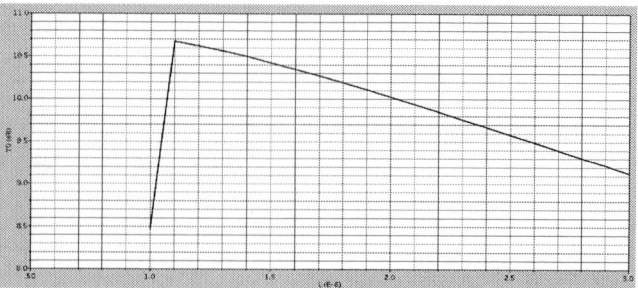

Figure 7. Conversion Gain vs. transistor size

D. Other circuits

To effectively transfer the power of RF signals, inter-stage matching circuits is necessary as well. A series peaking inductor is used at the input stage of mixer. By proper sizing of the input inductor, e.g. 100pH in this case, the good input matching of the mixer can be obtained. Simulated *S11* and *S21* of the mixer with series inductor matching are shown in Fig. 8. Within the range of 30-70GHz, S11 is less than -0.35 dB, while the S21 is roughly above 1 dB.

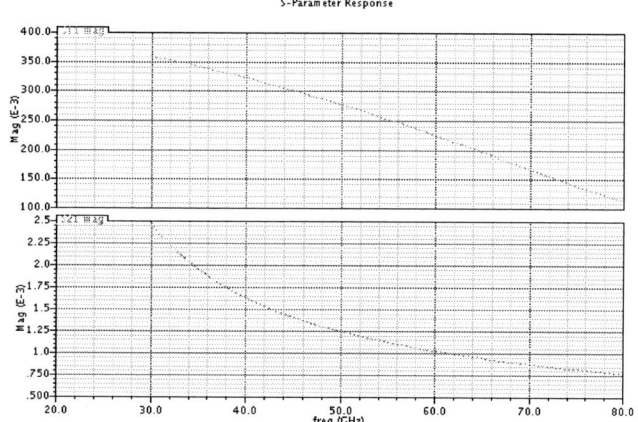

Figure 8. Input matching of the demodulator

IV. RESULTS AND DISCUSSION

Fig. 9 shows the receiver's die photo. The silicon area of the demodulator is bondpad-limited and the effective areas are 0.7 mm^2. Coplanar waveguide transmission (CPW) lines are used to connect the circuit core to the GSG. A 7-pin eye pass pads are used for DC biasing. The simulation is carried in Cadence SpectreRF with EM co-simulation (layout) with Agilent ADS-Momentum. The proposed system has a very wide operating range. In simulation, the carrier frequency can ranges from 50GHz to 70 GHz with an input signal of -40dBm.

978-1-61284-863-1/11 $26.00 © 2011 IEEE

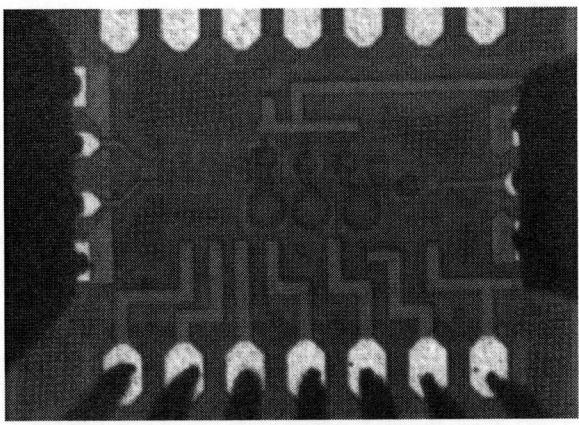

Figure 9. Layout of the proposed receiver

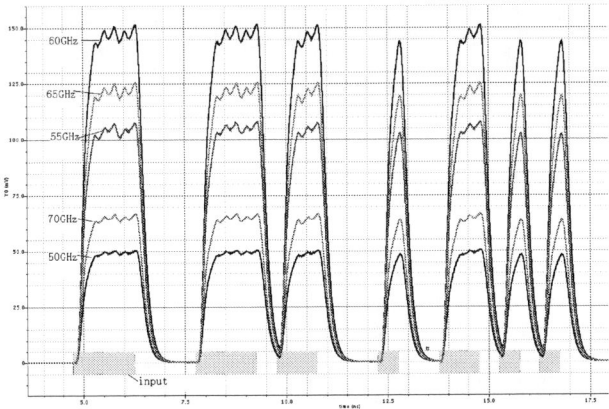

Figure 10. Output of the demodulator at different carrier frequency

The data rate can be as high as 2G bps in such frequency as carrier. Fig. 10 shows the input and output data of the proposed demodulator, the input signal has a data rate of 2 Gbps with 50GHz to 70GHz carrier, the demodulator can work properly. The input amplitude of signal is 5mV. Since the LNA is optimized at 60GHz, maximum gain of the receiver appears at 60GHz, while the signal power is minimized at 50 GHz input signal. After output buffer, the data can be correctly recovered as rail-to-rail logic. The typical current consumption is 10mA. The total current consumption can be as low as 8mA if the biasing current is further reduced. For test proposal, a power consuming buffer is added.

V. CONCLUSION

A 50 to 70 GHz self-demodulator is realized in TowerJazz 180 nm SiGe BiCMOS technology. The receiver is silicon-verified in a 180nm SiGe BiCMOS technology. By using a wide band input amplifier and self-demodulated topology, the receiver has a bandwidth of 20GHz and the conversion gain is about 15 dB. The DC power consumption of the core is about 15 mW from a 1.8-V supply. The theoretical maximum energy per bit (calculated by power consumption over the data rate) of the self-demodulator is 13.5 pJ/bit at 1 Gbps data rate. The

power consumption over data rate is comparable to ultra-low power receiver in CMOS technology[13].

TABLE I.　COMPARISON OF PERFORMANCES

	This work	[6]	[7]	[8]	[9]
technology	180nm SiGe	90nm CMOS	130nm SiGe	180nm SiGe	65nm CMOS
Carrier (GHz)	50-70	60	60	43	70-79
Modulation	OOK	OOK	QPSK	ASK	OOK
Area (mm2)	0.7	3.8	5.78	0.62	0.07
Date rate (Gbs)	2	1.8	2	5	N/A
Power (mW)	15	51	526	60	10.2

ACKNOWLEDGMENT

The authors wish to acknowledge the MPW provided by TowerJazz. This work is also supported by National Natural Science Foundation of China under Grant Number 61106034 and 60906017.

REFERENCES

[1]　S. E. Gunnarsson et al., "60 GHz Single-Chip Front-End MMICs and Systems for Multi-Gb/s Wireless Communication," IEEE Journal of Solid-State Circuits, vol. 42, no. 5, pp. 1143–1157, May 2007.

[2]　Behzad Razavi, "A 60-GHz CMOS Receiver Front-End," IEEE Journal of Solid-State Circuits, vol. 41, No. 1, pp. 17-22, Jan. 2006.

[3]　Ali M. Niknejad, Hossein Hashemi, mm-Wave Silicon Technology 60 GHz and Beyond, 2008 Springer Science and Business Media, ISBN 978-0-76558-7, pp. 47-49.

[4]　B. Floyd et al., "SiGe Bipolar Transceiver Circuits Operating at 60 GHz," IEEE Journal of Solid-State Circuits, vol. 40, no. 1, pp. 156–167, Jan. 2005.

[5]　Y. Yu et al., Integrated 60GHz RF Beamforming in CMOS, Analog Circuits and Signal Processing, Springer 2011.

[6]　T.Yao, etc, "Algorithmic design of CMOS LNAs and PAs for 60GHz radio ," IEEE J.Solid-State Circuits, vol. 42, no.5,pp. 1044–1057, May 2007.

[7]　Burak Catli and Mona M. Hella, "A 60 GHz CMOS Combined mm-wave VCO/Divider with 10-GHz Tuning Range," IEEE Custom Integrated Circuits Conference, San Jose, 2009.

[8]　Changhua Cap, Yanping Ding, Kenneth K. O, "A 50-GHz Phase-Locked Loop in 0.13-μm CMOS," IEEE Journal of Solid-State Circuits, vol. 42, pp. 1649-1656, Aug. 2007.

[9]　Kai Kang; Fujiang Lin; Duy-Dong Pham; Brinkhoff, J.; Chun-Huat Heng; Yong Xin Guo; Xiaojun Yuan; , "A 60-GHz OOK Receiver With an On-Chip Antenna in 90 nm CMOS," IEEE J. of Solid-State Circuits, vol.45, no.9, pp.1720-1731, Sept. 2010.

[10]　S. K. Reynolds, B. A. Floyd, U. R. Pfeiffer, T. Beukema, J. Grzyb, C. Haymes, B. Gaucher, and M. Soyuer, "A silicon 60-GHz receiver and transmitter chipset for broadband communications," IEEE J. Solid-State Circuits, vol. 41, no. 12, pp. 2820–2831, Dec. 2006.

[11]　W.-H. Chen, S. Joo, S. Sayilir, R. Willmot, T.-Y. Choi, D. Kim, J. Lu, D. Peroulis, and B. Jung, "A 6-Gb/s wireless inter-chip data link using 43-GHz transceivers and bond-wire antennas," IEEE J. Solid- State Circuits, vol. 44, no. 10, pp. 2711–2721, Oct. 2009.

[12]　L. Xia, P. Baltus, P. van Zeijl, D. Milosevic, A. van Roermund, "A 70 GHz 10.2 mW self-demodulator for OOK modulation in 65-nm CMOS", IEEE Custom Integrated Circuits Conference, San Jose, pp. 1-4, Jun. 2010.

[13]　Changhui Hu, R. Khanna, et. al, "A 90 nm-CMOS, 500 Mbps, 3–5 GHz Fully-Integrated IR-UWB Transceiver With Multipath Equalization Using Pulse Injection-Locking for Receiver Phase Synchronization", IEEE J. Solid-State Circuits, vol. 41, no. 12, pp. 1076-1088, May. 2011

Programmable Low-Dithering-Jitter Interpolator-based CDR

Lip-Kai Soh*, Wai-Tat Wong, Swee-Wah Lee, Chuan-Thim Khor
Altera Corporation
IC Engineering
*email: lksoh@altera.com

Abstract— **This paper presents a methodology to determine the optimum filter settings for interpolator-based CDR. The proposed methodology quantifies the relationship between filter settings, latency of the CDR loop and input data frequency offset. A programmable interpolator-based CDR is designed to verify the proposed methodology. The CDR is implemented in TSMC 60nm LP CMOS technology. Measurement results show correlation between calculated results versus simulation and characterization results.**

Keywords-Clock data recovery, dithering

I. INTRODUCTION

The development of the transceiver focuses heavily on the design of the clock and data recovery (CDR) circuit with small area to achieve lower product cost. One of the most commonly used CDR architecture in the area of low-cost transceiver is the interpolator-based CDR [1]–[3] as shown in Figure 1. The interpolator-based CDR (I-CDR) offers the advantages of a relatively simple implementation and is considerably smaller compared to the PLL-based CDR [4]-[6] due to the absence of the analog loop filter circuit and the usage of single PLL to generate the reference clock for multiple I-CDRs [7].

Conventional I-CDR uses negative feedback of the CDR loop to align data sampling clock to the middle of incoming data. The half-rate sampler samples even-bit and odd-bit data to determine the data position with respect to the clock. If the data is leading, the early signal will go high; and, if the data is lagging, the late signal will go high. Since the clock samples data at half rate using four clock phases, the subsequent circuit after the sampler can operate at a frequency that is half of the data rate. The digital filter is used to compensate for loop latency, and decrease loop dithering. This will be explained in more detail in Section 3. A PI controller is used to shift the phase of the phase interpolator output clocks until it fall at the middle of the data eye. Despite the simplicity of I-CDR architecture, this architecture has a problem of inherent dithering jitter. These problem is discussed in the next section.

This paper is organized as follows. The problem with the conventional I-CDR is described in Section 2. The proposed methodology to estimate the optimum filter setting is discussed in Section 3. The implementation of the I-CDR to test the proposed methodology is presented in Section 4. The simulation and measurement results are presented in Section 5 and the conclusion is given in Section 6.

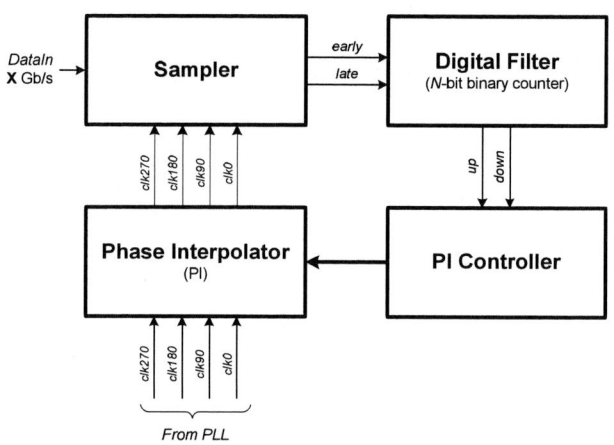

Figure 1. Conventional interpolator-based CDR architecture

II. PROBLEMS WITH CONVENTIONAL INTERPOLATOR-BASED CDR

The steady state of the I-CDR is determined by the feedback loop delay and the phase resolution of the phase interpolator. This oscillatory steady state manifests itself as recovered clock dithering jitter. This is explained in Figure 2. The vertical dotted line represents the middle of the data eye. The shaded regions at the boundary of the data are the uncertainty region due to input data sinusoidal jitter. The blue arrows show the movement of data sampling point caused by the phase shifting of phase interpolator clock, and the red arrows show the movement of data sampling point on every unit interval (UI) due to frequency deviation. The red arrow moves to the right by t_{PPM} for every UI because the frequency of the reference clock is slower than the data rate. Assuming that the phase step of the phase interpolator is $5t_{PPM}$ and the loop latency of the CDR is 3UI. This means that the I-CDR will only correct itself after 3 data samples. The movements of the sampling point are illustrated in Figure 2. If the green dot is the original sampling point of the data, the data would be shifted to the uncertainty region at the left boundary of the data eye after 5UI. This phenomenon is known as loop dithering and is inherent to all I-CDR architecture. Having the sampling point of the CDR at the uncertainty region

978-1-61284-863-1/11 $26.00 © 2011 IEEE

would decrease jitter tolerance of the CDR and increase bit error rate (BER) of the system.

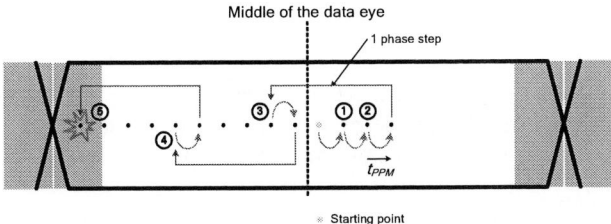

Figure 2. Dithering behavior of interpolator-based CDR without any filtering

To decrease the loop dithering of the I-CDR, a digital filter is included in the I-CDR to filter the phase detector outputs before propagating the signal to the PI controller. This causes the PI controller to delay its next decision until the results of its previous action have been propagated to the phase detector output. This reduces the inherent peripheral loop dither to one phase interpolation interval. However, having an incorrect digital filter setting will cause the I-CDR to dither by more than two phase steps. Insufficient filtering would increase loop dithering because the filter is unable to compensate for the feedback latency of the I-CDR. Over filtering will decrease the frequency offset tolerance of the I-CDR. To date guidelines and explanations on the optimum digital filter settings have yet to be found. And, the settings that are used in some literature seem to be arbitrary [8].

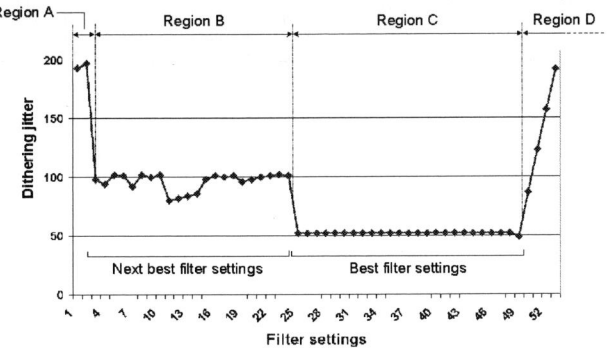

Figure 3. Dithering jitter characteristics of the I-CDR for different filter settings

III. DITHERING JITTER ANALYSIS

Figure 3 shows the dithering jitter characteristic of the I-CDR for different filter settings for feedback latency of 3UI, input data UI time drifting of x ps per UI, and phase interpolator step size of $50x$. The dithering jitter results are obtained through spreadsheet calculation. From the figure, we observe that the dithering of the I-CDR can be divided into 4 regions. Region C has the best dithering performance with loop dithering of one phase step and region B has the next best dithering performance with loop dithering of 2 phase steps. Region A shows higher

dithering jitter compared to Region B and C because the amount of filtering is insufficient to compensate for the loop latency. In region D, the large amount of filtering has cause the I-CDR loop to lose its capability to track the frequency deviation of the input data. The dithering jitter will continue to increase indefinitely if the amount of filtering is increased beyond region D.

A. Best Filter Settings

From the observation on Figure 3, we can derive the filter settings to achieve the lowest dithering (region C in Figure 3) as:

$$\frac{\Phi}{2t_{PPM}} \leq best\ filter\ settings \leq \frac{\Phi}{t_{PPM}} \qquad (1)$$

where Φ is the phase interpolator step size and t_{PPM} is the amount of drift in 1UI due to the difference between reference clock frequencies. Using the same drawing convention as Figure 2, Figure 4 shows the movement of the sampling point for the filter settings in region C. The red arrow shows the data transition that produces late outputs and the gray arrow shows the data transition that produces the early outputs from the phase detector. The amount of filtering is set to 5. Since the late outputs count hit the maximum filter count first, the phase of the phase interpolator is shifted to the left. This dithering behavior occurs for all I-CDR with best filter settings.

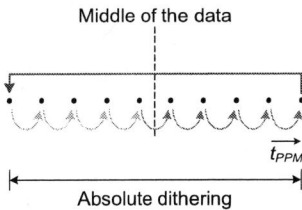

Figure 4. Dithering behavior of the I-CDR with best filter settings

To understand why the loop behaves this way, we can refer to equation (1). The best filter setting occurs when the amount of filtering is sufficient for the I-CDR to correct itself with bidirectional phase shift. With bidirectional phase shift, the loop behavior is more stable and therefore the dithering can be minimized. By using the best filter settings, the I-CDR can achieve the smallest absolute dithering. Also note that the best filter settings is independent of the loop latency.

B. Next Best Filter Settings

From observation on Figure 3, we can derive the filter settings to achieve the next lowest dithering (region B in Figure 3) as:

$$feedback\ latency \leq next\ best\ filter\ settings \leq \frac{\Phi}{2t_{PPM}} \qquad (2)$$

The next best filter settings is lower bounded by the amount of filtering required to compensate for the feedback latency and is upper bounded by the lowest best filter setting. Feedback latency refers to the sum of the latency from the phase detector, FSM and the phase interpolator delay. Dithering of the I-CDR loop is directly proportional to its feedback loop latency. In this region, although the sampling points dither between two phase steps, the dithering is unidirectional as illustrated in Figure 5. Due to the unidirectional dithering, the absolute dithering region of the I-CDR using the next best filter settings are larger compared to the I-CDR using the best filter settings.

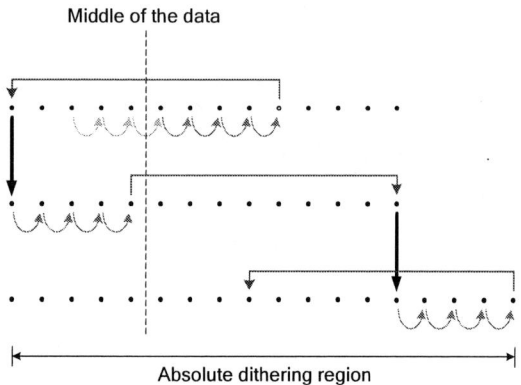

Figure 5.   Dithering behavior of the I-CDR with next best filter settings

The dithering performance of a I-CDR depends on its latency and frequency offset of the input data.

IV.   CIRCUIT IMPLEMENTATION

A I-CDR with programmable digital filter is implemented to verify the proposed methodology to determine the optimum filter settings. The individual building blocks of the I-CDR are described below.

A.   Half-Rate Sampler

The half rate sampler used in the proposed I-CDR is a half rate version of bang-bang phase detector [9]. Quadrature clocks generated by the phase interpolator are used to sample input data, and to determine if the data is leading or lagging the recovered clock. Clock with 0° phase shift (clk0), and clock with 180° phase shift (clk180) are used to sample the even and odd bit data respectively. Clock with 90° phase shift (clk90) and clock with 270° phase shift (clk270) are used to detect data transitions. When the loop is locked, clk0 and clk180 should be aligned to the middle of the data.

B.   Programmable Digital Filter

The digital filter circuit consists of programmable bi-directional binary counters to count the *early/late* output from the sampler. The counter can be programmed to the desired filter setting. The programmming of the digital filter is done using configuration shift registers (CSRs) [10].

C.   Phase Interpolator

In order to update the sample position, we use a phase interpolator that is controlled by a Gray coded up/down counter. Four reference clock phases generated by the PLL are fed into the phase interpolator (PI). The phase interpolator, shown in Figure 6, consists of a phase selection stage followed by signal conditioners and mixers. The decoder block is used to generate phase step control bits, W0-W14 for the PI.

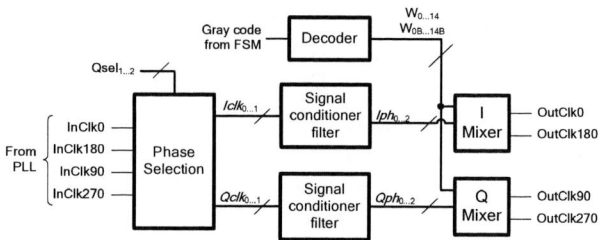

Figure 6.   Phase Interpolator

The phase selection block selects 2 sets of complementary clock phases that are 90° apart for I and Q clock paths, respectively. The clocks are, then, shipped to signal conditioner filters to convert CMOS clocks to triangular low-swing clock signals. The filter can be programmed to cater for clock frequencies range from 1.5GHz to 6.25GHz. Feeding clocks with triangular wave shape into mixers can improve differential non-linearity (DNL) of the PI. I-mixer and Q-mixer shown in Figure 6 are used to perform clock phase shifting by mixing 2 clocks that are 90° apart. The amount of phase shift that is generated by each mixer is controlled by W0..14.

D.   Other circuits

The PI controller consists of sequential logic and a Gray counter to control the steps of the phase interpolator.

V.   MEASUREMENT RESULTS

The dithering jitter performance of the proposed I-CDR is analyzed by simulating the I-CDR using different filter settings with different frequency deviation. Using formula (1) and (2), we estimate that the best filter settings for input data with frequency deviation of 100PPM is from 20 to 39 and the next best filter settings is from 8 to 20. These estimated values are compared with the value obtained from the simulation results.

Simulation results for the proposed I-CDR with different filter settings are shown in Figure 7. At 0PPM, the I-CDR dithers between 2 phase interpolator steps for all filter settings. This dithering behavior is caused by the metastability of the bang-bang phase detector. At 100PPM, the loop dithers between 2 phase interpolator steps when filter settings of 8 and 16 are used. These settings correspond to the next best filter settings. The loop dithers by only one step when filter setting of 32 is used and this correspond to the best filter settings. The

978-1-61284-863-1/11 $26.00 © 2011 IEEE   446

best filter and next best filter settings from the simulation result tally with the numbers that we estimated earlier. For filter settings of 64, the loop dithers by 6 steps because the loop is no longer able to track the frequency deviation due to the large loop latency. The same trend is observed when data with frequency deviation of 200PPM is used. Based on equation (1) and (2), higher PPM would require lower filter settings in order to achieve better dithering performance. As shown in Figure 7, for good dithering performance, small filter setting value needs to be used for high data rate deviation.

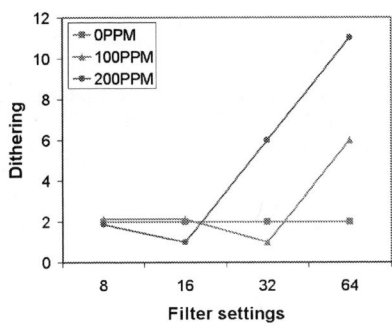

Figure 7. Loop dithering vs Filter settings

A I-CDR with programmable digital filter is implemented in TSMC 60nm CMOS LP process and the measurement results show correlation with the data collected from simulation. The data is collected using input data frequency offset of 100PPM. The characterized results are shown in Figure 8. The optimum digital filter setting is achieved at around the filter value of 32 which matches with our simulated and estimated value. Using this filter setting we see that the I-CDR produces the lowest amount of total jitter. The simulation and characterization results confirm the validity of equation (1) and (2).

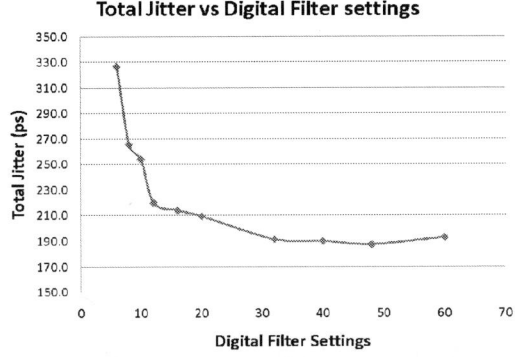

Figure 8. Loop dithering vs Filter settings

VI. CONCLUSION

A methodology to determine the optimum filter settings for interpolator-based CDR (I-CDR) is proposed and analyzed. By using the best and the next best filter

setting, minimum dithering jitter can be achieved. A I-CDR) with programmable digital filter is designed and fabricated using TSMC 60LP CMOS process. The measurement and simulation results show optimum dithering jitter is achieved using the proposed methodology.

REFERENCES

[1] J. Sonntag and R. Leonowich, "A monolithic CMOS 10 MHz DPLL for burst-mode data retiming", IEEE ISSCC Dig. Tech. Papers, 1990, pp. 194–195.

[2] T. Lee, K. Donnelly, J. Ho, J. Zerbe, M. Johnson, and T. Ishikawa, "A 2.5 V CMOS delay-locked loop for 18 Mbit, 500 Mbyte/s DRAM," IEEE J. Solid-State Circuits, vol. 29, no. 12, pp. 1491–1496, Dec. 1994.

[3] S. Sidiropoulos and M. Horowitz, "A semidigital dual delay-locked loop," IEEE J. Solid-State Circuits, vol. 32, no. 11, pp. 1683–1692, Nov. 1997.

[4] J. Savoj, B. Razavi, "A 10-Gb/s CMOS clock and data recovery circuit with a half-rate linear phase detector", IEEE Journal of Solid-State Circuits, May 2001, pp. 761-768

[5] J. Savoj and B. Razavi, "A 10-Gb/s CMOS Clock and Data Recovery Circuit With a Half-Rate Binary Phase/Frequency Detector," IEEE Journal of Solid-State Circuits, vol. 38, pp.13-21, Jan. 2003.

[6] M. Rau et al., "Clock/data recovery PLL using half-frequency clock," IEEE J. Solid-State Circuits, vol. 32, pp. 1156–1159, July 1997.

[7] R. Kreienkamp, U. Langmann, C. Zimmermann, T. Aoyama, H. Siedhoff, "A 10-gb/s CMOS clock and data recovery circuit with an analog phase interpolator", IEEE J. Solid-State Circuits, vol. 40, pp. 726-743, 2005

[8] Hanumolu P.K., Gu-Yeon Wei, Un-Ku Moon, "A Wide-Tracking Range Clock and Data Recovery Circuit", IEEE J. Solid-State Circuits, vol. 43, no. 2, pp. 425–439, Feb. 2008.

[9] J. D. H. Alexander, "Clock Recovery from Random Binary Data". Electronics Letters, 11:541-542, October 1975.

[10] U.S. Patent #7112992, Configuration Shift Register, issued September 2006.

A Power Efficient $\Sigma\Delta$ Modulator Based on CBSC IIR Filter in 0.18µm CMOS

Mehdi Taghizadeh[1a], Majid Zamani[2b], Payman Goodarzi[2] and Ammar Rahimi Kazerooni[1]

(1 – Department of Electrical Engineering, Kazerun Branch, Islamic Azad University, Kazerun, Iran)
E-mail (a) – m.taghizadeh@kau.ac.ir
(2-Department of Electrical Engineering, Science and Research Branch, Islamic Azad University, Tehran, Iran)
E-mail (b) –m.zamani@srbiau.ac.ir

Abstract— **In this paper a second-Order low distortion Sigma-Delta Modulator (SDM) with utilization of comparator-based switched-capacitor (CBSC)-based IIR filter, is explored. The advantages of this new structure are justified by the reductions of power and area. For this purpose IIR filter block can be made with single CBSC gain stage that has the same accuracy and performance of 2nd-order filter with two Op-amps. This design is intended to minimize the power consumption, and maximize dynamic performance. As shown in the simulation result, for a 20-KHz signal bandwidth, the modulator achieves a dynamic range of 70.2 dB and a peak signal-to-noise and distortion ratio (SNDR) of 68 dB with an over-sampling ratio of 64. In addition it consumes 198 µW from a 1.8-V power supply at 2.56MS/s.**

Keywords-Sigma Delta Modulator; IIR Filter; CBSC gain stage.

I. INTRODUCTION

Technology scaling, which is the main challenging factor in sub-micron processes, makes some issues on design of high performance op-amps which are one of the crucial analog building blocks in the Switch-Capacitor (SC) circuits using in Analog-to- Digital Converters (ADCs)[1] especially Sigma-Delta ADCs.

One of the important problems is decreasing the op-amp gain. It causes that, the precision in the feedback circuits are dramatically reduced, because in the traditional SC circuits an op-amp with high dc gain guarantees the accuracy of the charge transfer.

Recently, a Comparator-Based Switched-Capacitor (CBSC) technique was reported in [2], [7] to replace the op-amp with comparator and current sources which has same operation like as op-amp-based architecture. In CBSC technique, a comparator and switched current sources are used to sense the virtual ground condition, instead of forcing it with an op-amp. In this method by replacing op-amp with CBSC, finite gain and nonlinear slow-rate which is resulting from the non-ideal effect of op-amp, would be diminished. Also this method reduces power consumption of circuit dramatically.

In this paper a low-distortion second-order SDM which uses CBSC gain stage, is designed and simulated. To save power consumption and area, we use an IIR filter instead of two integrators in conventional second-order SDMs. This causes the IIR filter can be implemented with only single

CBSC gain stage [3]. After SDM architecture description in section II, the circuit level implementation is discussed in section III. Section IV covers the simulation results and section V concludes.

II. SIGMA DELTA MODULATOR ARCHITECTURE

In conventional SDM architectures, two integrators are required for implementing a second order modulator; where in its structure, two feedback paths from quantizer output is connected to the input of integrators.

But in this paper, the SDM architecture uses a low-distortion structure [10] where the input signal is not entered the loop filter and modulator just processes quantization noise. Therefore, due to lower swing in the outputs of integrators, loop filter function becomes more relax. A conventional second order low-distortion SDM is shown in Fig. 1.

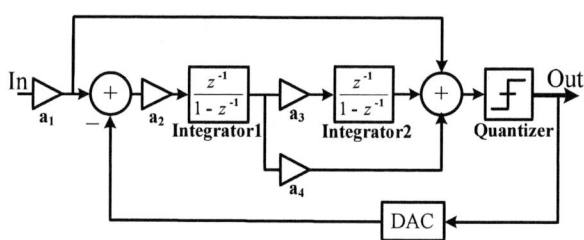

Fig. 1 Conventional second order low-distortion SDM.

Figure 2 shows the chosen structure for investigating second order Sigma-Delta Modulator using IIR Filter. We have used a 2nd-order IIR-filter instead of two integrators where the IIR filter is implemented via CBSC gain stage instead of op-amp.

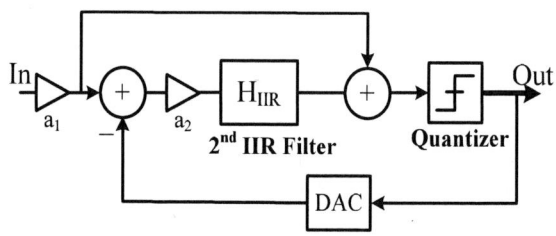

Fig. 2 Proposed second order Sigma-Delta Modulator.

978-1-61284-863-1/11 $26.00 © 2011 IEEE

This configuration has several properties: 1) because of implementing two integrators requires two CBSC gain-stages, by replacing them with 2nd-order IIR-filter, it can be implemented only with single CBSC gain-stage (we will discuss this method in section 3-A). Thus the total power consumption will be reduced.

2) It is possible to place the zeros of Noise Transfer Function (NTF) at any frequency and reach to desired noise-shaping condition, by using IIR filter.

3) Second order SDM structure with 2nd-order IIR filter reduces the need of feed-forward path towards a low-distortion integral's structures and thus decreases complexity of the modulator and number of coefficients.

If the 2nd-order IIR filter transfer function is written as:

$$H_{IIR}(z) = \frac{c_1 z^{-1} + c_2 z^{-2}}{1 - d_1 z^{-1} - d_2 z^{-2}} \quad (1)$$

So the NTF of SDM can be obtained easily as follows, to calculate the modulator coefficients.

$$NTF(z) = \frac{1 - d_1 z^{-1} - d_2 z^{-2}}{1 + (a_1 c_1 - d_1)z^{-1} - (a_1 c_2 - d_2)z^{-2}} \quad (2)$$

If the NTF function be as a second order FIR structure, then the coefficients should be chosen as $a_1 c_1 = d_1$ and $a_1 c_2 = d_2$. Selecting NTF Poles with zero values wouldn't decrease the stability margin [4]; also both non-delaying feed-forward path for input signal and main feedback path increase it. The values of d_1 and d_2 are determined so that the quantization noise power in the modulator output is minimized. Therefore, referred to [4], we obtain $d_1 = 1.998$ and $d_2 = -1$ for an Over Sampling Ratio (OSR) equal to 64.

III. CIRCUIT IMPLEMENTATION

The implementation of main building blocks of proposed SDM is performed as follow:

A. 2nd-Order IIR Filter Description

Recently, using IIR filter blocks instead of Integrators in Sigma-Delta Modulator structures was reported as a new method in designing of SDM loop filter [5], [6]. Equation (1), 2nd order IIR filter block, can be implemented similarly as in [6] with two stage, that in each stage we must use an OTA. With respect to [3], this filter can be implemented by means of single OTA too. This method reduces the area and complexity of the filter in circuit level. In this section, we use mentioned method for implementation of 2nd-order IIR filter block, but a CBSC gain-stage was used instead of OTA. For this purpose, equation (1) can be written as multiple of two terms:

$$H_{IIR}(z) = H_I(z).H_F(z) \quad (3)$$

That $H_I(z)$ and $H_F(z)$ are 2nd-order IIR and 1st-order FIR filters respectively. Fig. 3 shows the segregated diagram of the main IIR filter. The modified structure for realization of first

part, $H_I(z)$, which forms the feedback paths and denominator coefficients, was sketched in Fig. 4 as a single-ended circuit, including the timing diagram. It was designed by single CBSC-gain-stage, based on the work proposed in [2]. The p-path part of Fig. 4 is realized by capacitor C_{h1} that samples the output of the CBSC-gain stage in Φ_1, and transfers the relative charge to the output in Φ_2. The q-path part is realized with two paths capacitors, C_{h2}, having a delayed-sampling at feedback paths, sampling the CBSC-gain stage output alternately. When one capacitor samples the output in Φ_{s1}, the other one transfers $v_O(n-2)$ in Φ_{h2} phase to the integrating capacitor, C_I, and realizes the second order denominator.

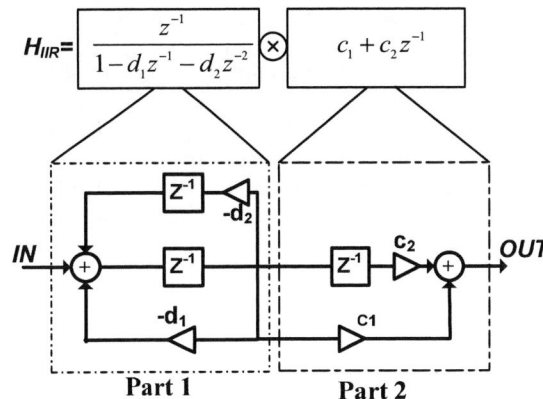

Fig. 3 segregated diagram of 2nd-order IIR filter transfer fuction.

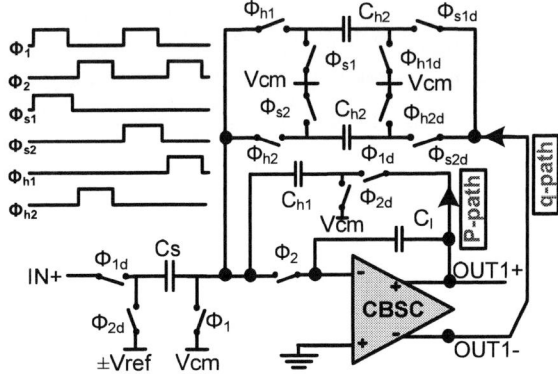

Fig. 4 The single-ended circuit realization of $H_I(z)$ function.

It can be easily shown that the structure introduced in Fig. 4, has the following transfer function:

$$H_I(z) = \frac{\dfrac{C_S}{C_I}z^{-1}}{1 - \dfrac{C_{h1} - C_I}{C_I}z^{-1} - \dfrac{C_{h2}}{C_I}z^{-2}} \quad (4)$$

By comparing above equation with (1) and circuit or quantization noise power, we can calculate the capacitors value.

Also the second part of Fig. 3, $H_F(z)$, can be mixed with the summation block in front of the quantizer.

978-1-61284-863-1/11 $26.00 © 2011 IEEE

B. CBSC Gain-Stage Description

The operation of a CBSC-based integrator with conventional CBSC circuit implemented in fully differential architecture [2] during its charge transfer phase and the corresponding timing diagram are depicted in Fig.5 (a) and (b), respectively. This architecture replaces the function of the op-amp with using threshold-detection comparator and a ramp generator instead of op-amp in gain stage (and now in integrator). The ramp generator in CBSC consists of two coarse current sources, two fine current sources, and a common-mode feedback (CMFB) circuit. In Φ_1 phase, the input is sampled by the input capacitors (C_{S1}, C_{f1} and C_{S2}, C_{f2}). The charge transfer phase is Φ_2 divided into four following sub-phases: a) preset phase for preset switches (P_1), b) coarse transfer phase (E_1) and c) fine transfer phase (E_2) d) settling phase (S). After sampling, the outputs are preset to V_{REFP} and V_{REFN}. Preset phase (P_1) sets the output nodes to preset levels away from the common voltage level (V_{CM}). After preset phase, coarse charge transfer phase is started. Ia current source charges the positive half circuit and discharges the negative half circuit. As a result, input nodes of the comparator (V_c and V_d) are charged in opposite directions to cross each other to make the first decision as seen in Fig. 5(b). After first detection ($V_c=V_d$), comparator and coarse current sources are forced to turn off. However, there is overshoot in the output due to the delay in the comparator and ramp rate ($Vovp$), which is given by:

$$Vovp = \frac{Ia.t_{dc}}{C_T} \qquad (5)$$

Here, Ia is the coarse current, t_{dc} denotes the delay of the threshold-detection comparator during coarse phase, and C_T is the total loading capacitance at the output of the gain stage, given by:

$$C_T = C_Y + C_n \qquad (6)$$

Where C_Y is the series combination of the input capacitors (C_S and C_f) and C_n is the equivalent capacitance of the next stage. To achieve more accurate output, the fine transfer phase is used. During this phase, fine current sources (I_b) are turned on to reduce the overshoot voltage and also achieve to the second detection as seen Fig. 5(b). Second detection leads to sampling switch is opened and final value sampled on C_T. The following equation shows the final overshoot analysis:

$$Vovf = \frac{Ib.t_{df}}{C_T} \qquad (7)$$

Where t_{df} denotes the delay of the threshold-detection comparator during the fine phase. The accuracy of the ADC is limited by final overshoot and it affects on SNDR directly. Sensitivity list of $Vovf$ consists of three parameters; t_d, C_T, and fine current (I_b) which they must be calculated as well as possible. The coarse current source circuit is composed of I_{ref} and wide-swing cascade current mirrors. Wide-swing cascade current mirrors are used to have an appropriate constant current in output nodes. To improve the balance of current in output branches, devices are designed in big sizes that do not have minimal gate length. In the fine phase, using of a series switch can improve the output resistance of single transistor instead of switch on gate. A good version of SC-CMFB circuit is used; This CMFB has the total capacitance load of $C_t = C_1 + C_2$ in $\Phi1$ and $\Phi2$. C_2 can be selected to achieve sufficient CM loop bandwidth (BW). C_1 can be designed 10 times bigger than C_2 in order to make settling time, leakage errors and less change injection. The CMFB circuit and current sources are shown in Fig. 6.

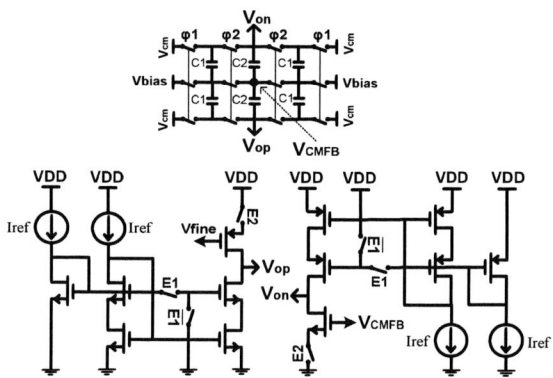

Fig. 6 Current sources with CMFB circuit.

C. Realization of Other Parts

The summing block, where located in front of quantizer, is implemented by a CBSC gain stage. The second part of IIR filter transfer function in Fig. 3, $H_F(z)$, has been mixed by summing block as shown in Fig. 7.

Also, one-bit quantizer is implemented by a comparator and DFF. With respect to the quantizer output digital code, a

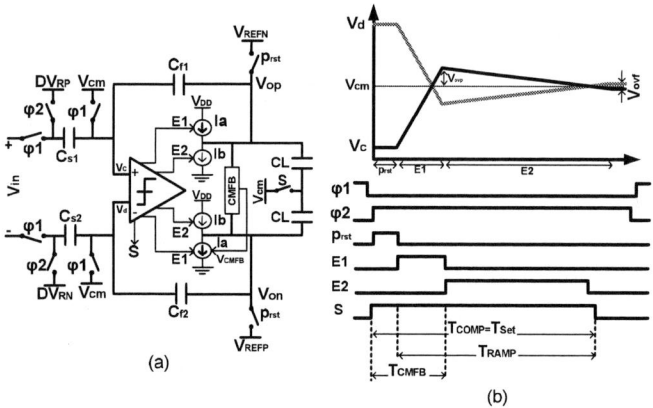

Fig. 5 (a) CBSC-based integrator (b) timing diagram.

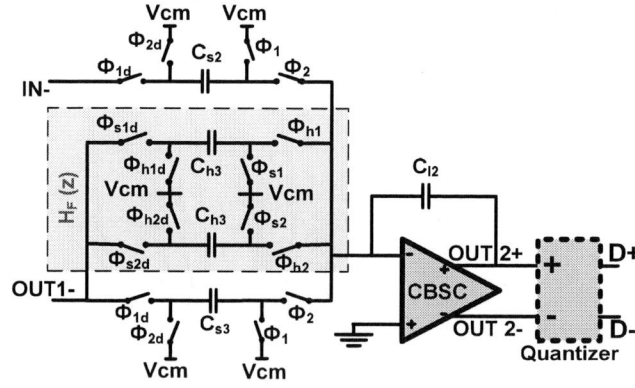

Fig. 7 Realization of summing block in front of quantizer.

978-1-61284-863-1/11 $26.00 © 2011 IEEE

reference voltage ($\pm V_{ref}$) is added to the modulator input in Φ_2.

IV. SIMULATION RESULTS

In this section, results obtained from system level simulation (MATLAB) and circuit level simulation using Hspice in 0.18 μm CMOS process are compared. Figure 8 shows the output power spectrum density (PSD) of the conventional and CBSC-based second order modulators with input signal amplitude -4 dBFS and input frequency fin=6.875KHz. Sampling frequency and oversampling ratio are fs=2.65MS/s and OSR=64 which yields a signal bandwidth (BW) of 20 KHz. This spectrum, which computed via a 4096-FFT point, confirms the performance of this technique through comparison with conventional opamp-based modulators. The maximum values for the SNDR are 74.8 dB and 75.2dB for the conventional and proposed structures in the SIMULINK respectively. Also the SNDR is 85 dB for the CBSC-based SDM in HSPICE. Fig. 9 shows the simulated SNDR versus the input signal amplitudes normalized by reference voltage (Vref). The dynamic range (DR) of CBSC-based modulator is 70.2 dB, which shows good agreement to the modulator simulated in MATLAB. The simulated power consumption is 198 μW from 1.8V power supply. Table I shows the performance comparison of state-of-the-art SDMs.

Table I: Performance comparison

	[8]	[9]	This work
Technology	0.18μm	0.13μm	0.18μm
Band width	20 KHz	62.5KHz	20 KHz
Sampling Frequency	2.56 MHz	8 MHz	2.56 MHz
SNDR (dB)	65.3	71.3	68
Power Consumption	420 uW	253 uW	198 uW

V. CONCLUSION

This paper describes a second order low distortion Sigma-Delta Modulator by using comparator-based switched-capacitor based IIR filter. Furthermore we have employed CBSC gain stage to implement IIR filter block which shows efficient architecture can be achieved for this application. In comparison with similar ones, the power dissipation and area is lower because of using CBSC integrators and single CBSC gain stage IIR filter. With OSR=64, the peak of SNDR is 68-dB, the bandwidth is about 20 KHz, and the power is 198 μW. Simulation results verify the usefulness of this implementation method and fabrication of the proposed architecture.

REFERENCES

[1] A.-J. Annema, B. Nauta, R. van Langevelde, and H. Tuinhout, "Analog circuits in ultra-deep-submicron CMOS," *IEEE J. Solid-State Circuits*, vol. 40, no. 1, pp. 132–143, Jan. 2005.

[2] T. Sepke, J. K. Fiorenza, C. G. Sodini, P. Holloway, and H.-S. Lee, "Comparator-based switched-capacitor circuits for scaled CMOS technologies," in ISSCC Digest of Technical Papers, pp. 220–221, 2006.

[3] A. Safarian, F. Sahandi, S. Atarodi, "A new low-power deltasigma modulator with the reduced number of op-amps for speech band applications", *ISCAS*, Bangkok, Thailand, pp. 1033-1036, May 2003.

[4] J. Marrkus and G. Temes, "An Efficient ΔΣ ADC Architecture for Low Oversampling Ratios," IEEE Trans. Circuits Syst. I, vol. 51, no. 1, January 2004.

[5] M. Yavari, O. Shoaei, "Low-Voltage Sigma-Delta Modulator Topologies for Broadband Communications Applications" IEICE Transactions on Electronics, vol. E87-C, no. 6, Jun 2004.

[6] R. Jiang, T. Fiez "A 14bit Delta-Sigma ADC with 8x OSR and 4MHz Conversion Bandwidth in a 0.18um CMOS process", *IEEE JSSC*, pp. 63-74, 2004.

[7] M.-C. Huang, and S.- I. Liu, "A 10MS/s to 100kS/s power-scalable fully-differential CBSC 10-bit pipelined ADC with adaptive biasing," IEEE Trans. Circuits and Syst. II: Express Briefs, vol. 57, pp. 11-15, Jan. 2010.

[8] M.-C. Huang, and S.- I. Liu, "FULLY DIFFERENTIAL COMPARATOR-BASED SWITCHED-CAPACITOR □□ MODULATOR", IEEE Trans. Circuits Syst. II: Express Briefs, Vol. 56, Issue 5, pp.369-373, May 2009.

[9] M. Momeni, D. Prelog, B. Horvat, M. Glesner, "Comparator-Based Switched-Capacitor Delta-Sigma Modulation", Contemporary Engineering Sciences, Vol. 1, No. 1, pp. 1 – 13, 2008.

[10] J. Silva, U. Moon, J. Steensgaard, and G. Temes, "Wideband low distortion delta-sigma ADC topology," *Electronics Letters*, Vol. 37, No. 12, pp. 737–738, June 2001.

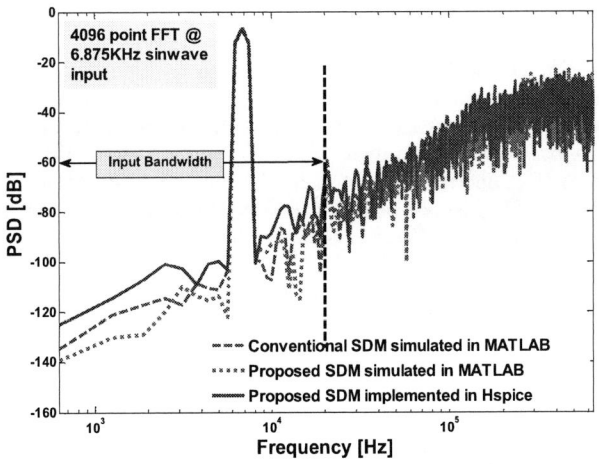

Fig. 8 Simulated output spectrum of the conventional and proposed 2nd order SDM in MATLAB and Hspice environment.

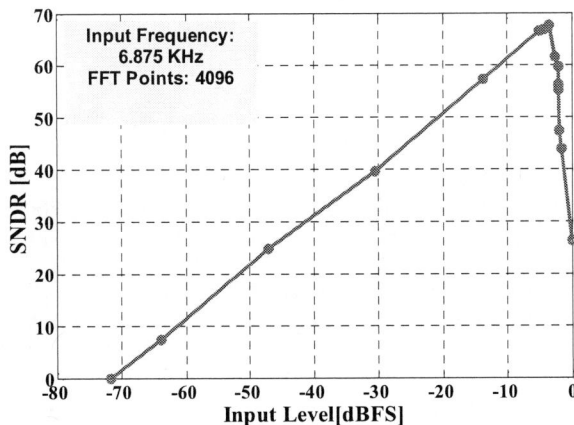

Fig. 9 SNDR versus input signal amplitude of the SDM

Low-Power Design Techniques with Process Tagging and Dynamic Power Management

Daniel Cooley, Yuwono Rahman, Jin Ruan, Xun Yu, Lei Chen, Jianyuan Deng
Silicon Laboratories International Pte., Ltd.
Singapore
Email: Daniel.Cooley@silabs.com

Abstract—**This paper proposes a power optimization strategy that takes into consideration the effects of operating frequency, process, voltage, and temperature variation (FPVTV) in both analog and digital circuits. Traditional designs set the biasing points to guarantee performance over all working conditions, resulting in wasted power in most instances. By collecting information about FPVTV, the designer can make intelligent decisions to reduce power consumption and extend battery life in mobile products. The proposed optimization strategy was implemented in silicon and successfully proven on an RF/mixed-signal chip with DSP, resulting in a power savings of 11.2% in typical conditions. The usage of such a strategy is discussed, as well as the production flow for gathering process corner information in high volume situations.**

Index Terms – dynamic power management, process tagging, process variation, low-power design

I. Introduction

As consumer electronic and medical devices continue to advance in feature set and battery life, chip suppliers are pressured to extend the power reductions that have fueled a two-decade boom in portable devices. Lowering power consumption without sacrificing performance requires design teams to develop new architectures that take advantage of advances in semiconductor processing and lithography. New designs are also incorporating information such as frequency, process, voltage, and temperature variation (FPVTV) into the biasing of digital, mixed-signal, and RF circuits [1]–[3]. Extensive searches to locate the optimal biasing for a given set of conditions have been made, but analysis has shown that increased optimization of biasing points provide diminishing returns [4]. In other words, the largest power savings come from simple ideas that prove challenging to implement.

In this work, we present a simple and flexible process tagging and biasing algorithm that reduced the typical power consumption in a fully-integrated RF/mixed-signal chip with DSP by 11.2%. More sophisticated use of the existing platform can further reduce power by another 5-10%. Like existing designs, voltage scaling according to FPVTV was used in the core digital circuits. However, unlike most designs, half of the power savings came from reductions in analog, RF, and mixed-signal circuit biasing points. To date, previous designs have considered power scaling in analog circuits according to

voltage, temperature, and frequency [3]. Other designs strive to reduce sensitivities to process variation [5]. This work goes one step further and incorporates process variation into the biasing algorithm of the analog, RF, and mixed-signal circuits to actually reduce the power consumption averaged over large quantities of chips.

After this introduction, Section II explains the basic concepts for dynamically biasing circuits according to FPVTV. Section III provides a simple yet flexible circuit for determining process variation. Section IV addresses system integration issues that arise while using FPVTV information to dynamically adjust power consumption during normal operation. Section V discusses the production flow considerations in a high-volume environment. Section VI describes experimental data captured in production and summarizes overall power savings. Section VII discusses additional concerns in implementing the ideas in this paper. Section VIII is dedicated to conclusions, and further experiments and data are presented in the appendix.

II. FPVT Biasing Background

Traditionally, the power consumption of a finished design is limited by the worst-case conditions that a design will experience over FPVTV. Designers must provide adequate margin over the lifetime of the chip to withstand the full FPVTV, resulting in wasted power when operating in typical conditions. Design teams have developed many methods to cope with FPVTV to minimize power consumption, and the techniques vary based on whether the circuits are digital, mixed-signal, or analog/RF in nature.

As standard CMOS process geometries have scaled from 1.0 μm in the early 1990's down to the 45 nm node available today, digital power consumption should grow 2.7X each two years without process modifications to threshold voltages and new design strategies to limit power growth. Starting in 1990, Macken et al. first suggested scaling the supply voltage of digital circuits in order to minimize current drawn from the battery over FPVTV [1]. Researchers over the next two decades extended this idea to include sophisticated modeling of critical path sensitivity to FPVTV, voltage adjustment during normal operation, advanced processor speed scheduling, immunity to supply ripple caused by voltage scaling, and systems with dynamic frequency and voltage scaling [2], [4].

978-1-61284-863-1/11 $26.00 © 2011 IEEE

These trends are expected to continue as new challenges appear in advanced digital processes at 40 nm and below.

Despite their common use in digital FPVTV compensation schemes, pure analog and RF circuits have not received the same attention. However, the idea has gained traction in recent years. Designers are now considering dynamic threshold voltage (DTMOS) control of critical transistors to reduce FPVTV effects in ultra-low power analog circuits. Additionally, one proposal optimizes power consumption by monitoring temperature and signal conditions to dynamically adjust biasing currents in RF circuits such as low-noise amplifiers [3]. As high-performance analog circuits continue to be integrated within standard digital CMOS processes, the trend in FPVTV compensation is expected to continue.

Mixed-signal circuits that bridge the analog-digital domains have experienced large power reductions with regard to FPVTV. A key objective in designing these circuits is to transition traditional analog functions into the digital domain to take advantage of existing FPVTV power saving techniques. New ADCs are continually executing more functions in the digital domain, going so far as using open-loop amplifiers with digital correction [6]. The analog portions of these data converters utilize the FPVTV compensation techniques listed above.

III. PROCESS IDENTIFICATION AND TAGGING

A. Methods

Many methods have been proposed to identify process variation for both critical transistors and broad chip-wide trends. Research has generally been split between device-level and functional tests. Device testing includes test structures that are used to extract information about transistor threshold voltage, oxide thickness, mobility, etc. [7]–[8]. These circuits can be operated in a number of regimes, including saturation, triode, and weak inversion. The device-level testing can be used for specific circuits or to gain a broad understanding of process variation across a single or multiple die. Beyond single-device testing, researchers have used circuits such as ring oscillators as a proxy for critical paths in digital circuits since the advent of FPVTV compensation schemes [1]. These structures provide a function-based way to identify FPVTV, rather than revealing more detailed device-level information.

B. Device Measurement and Storage

The disclosed circuitry minimizes complexity and maximizes flexibility. As shown in Fig. 1, an external current is driven onto the chip and passed over various transistor and resistor structures in succession during manufacture. At each step, the voltage is recorded, translated into a process corner, and then stored into non-volatile memory (NVM) within the chip. If NVM is not available, an alternative is shown in Fig. 2. Here, an on-chip MCU or digital state machine can control an integrated current source and ADC. The current source needs to be insensitive to FPVTV or can be calibrated during manufacture. The results can be stored in registers or RAM for later use.

Translation of the I-V characteristics of transistor and resistor structures require thorough comparisons and correlation of the simulation models, wafer acceptance test (WAT) data from the fabrication facility, and laboratory testing. Multiple device characteristics can be extracted, depending on the amount of current supplied to the test structures. For example, large currents yield information about I_{DSAT}, and small currents provide information about threshold voltage.

IV. SYSTEM INTEGRATION

Knowing the process variation of transistors and digital structures on the chip allows us to use that information to optimize the power consumption of the die. In addition to the process corner, it is useful to calculate what frequencies the digital core and various analog circuits –mixers, VCOs, ADCs, DACs, etc.– require, the supply voltage, and the junction temperature of the die itself. All of this information can be built into a comprehensive biasing algorithm. An on-chip MCU, external MCU, or digital state machine can implement such an algorithm, described in Fig. 3. First, the chip needs to boot with biasing points that are guaranteed to ensure full functionality and performance over FPVTV. Once this is complete, process information is either measured or retrieved from memory, followed by calculation of operating frequencies and measurement of temperature. Finally, this information is translated into biasing points (e.g., voltages and currents).

Each subcircuit will have a different dependency on and sensitivity to FPVTV and must therefore be biased independently. This can include both voltage and current scaling. In general, slower process corners and higher temperatures require higher biasing voltages and currents to maintain equivalent performance as typical corners and temperatures.

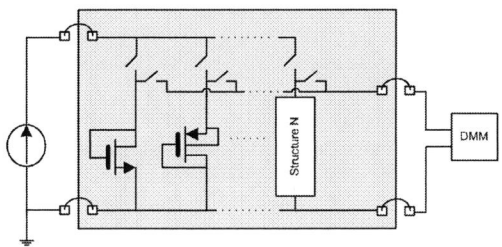

Figure 1. Process measurement structure using off-chip components during manufacture.

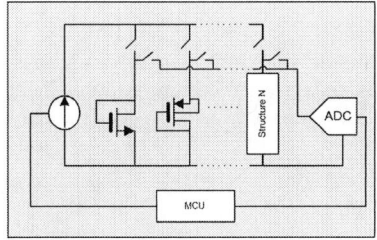

Figure 2. Fully-integrated process measurement structure.

978-1-61284-863-1/11 $26.00 © 2011 IEEE   453

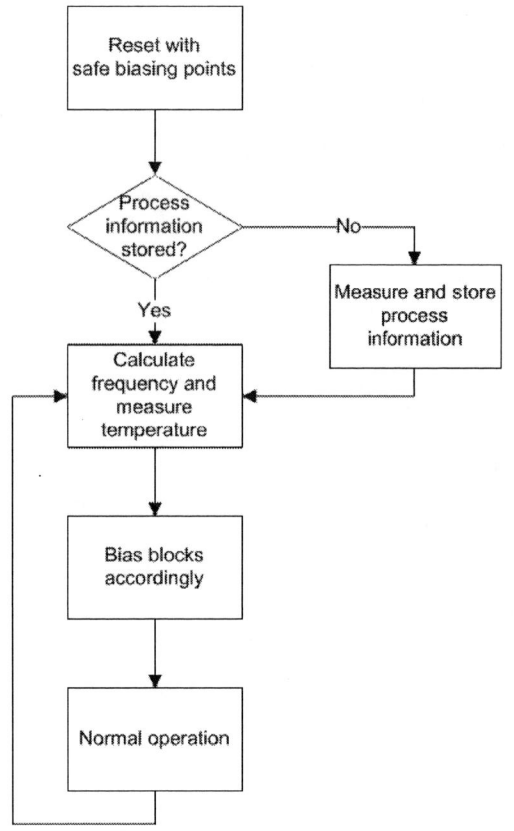

Figure 3. Flow diagram of system operation.

V. PRODUCTION FLOW

Process tagging is implemented on a previously established production flow for a similar product. The flow is demonstrated in Fig. 4 and starts with wafer fabrication, followed by wafer sort and probe, package assembly, and final test. In this flow, there are two options to implement process tagging: wafer probe and final test. Regardless of where process tagging is executed, temperature control is vital to producing consistent results. At wafer sort, the wafer under test sits in a chuck that can be set to a desired temperature. One prober, the TEL P12XL, has a ±2°C tolerance. Testing above room temperature is advantageous for distributing temperature evenly across the wafer. Since the wafer is in direct contact with the chuck, the junction temperatures are controlled easily.

At final test, a temperature sensor is placed near the device under test (DUT). During process measurement, the temperature is read by the tester and used to determine the final tagging result. However, controlling the temperature in test chamber better than ±5°C is difficult. The temperature sensor may not reflect the real temperature in the chamber or the silicon junction temperature inside of the package. This is due to the large volume of test chamber, which is opening and closing during testing. Thus, a stable temperature of either the chamber or DUT is difficult to achieve. In production, wafer probe is the preferred step to execute process tagging, with final test tagging as a backup. The process tagging results are then monitored in full production to ensure that the process is centered and the wafer probe is functioning properly. A systematic process shift can cause undesired changes in average power consumption of large numbers of die. In addition, chips whose process corner results fall outside of a defined window can be easily screened.

VI. MEASURED DATA

There are three stages of data collection during product development. Initial I-V data from all transistor corners are collected on a set of skewed samples to determine the boundaries for process tagging. The second set of data is taken at wafer sort, where the chips are measured and tagged according to boundaries that have been set. Subsequently, the wafer is sent to assembly house for packaging so that the process tagging results can be correlated in final test, bench measurement, and WAT comparison. An example wafer probe dataset for low-voltage (LV) transistors is shown in Table I.

TABLE I.   PROCESS TAGGING RESULTS FROM TWO WAFERS

LV Corner [a]	Wafer A (Typical)		Wafer B (Skewed)	
	Count	*%*	*Count*	*%*
TT	*20176*	*97.42%*	3339	16.28%
TF	192	0.93%	29	0.14%
TS	192	0.93%	137	0.67%
FT	150	0.72%	*12645*	*61.66%*
FF	0	0.00%	4358	21.25%
FS/SF/ST/SS	0	0.00%	0	0.00%
Total	20710	100.00%	20508	100.00%

a. Split naming is NMOS and then PMOS (e.g. FT = fast NMOS, typical PMOS).

Both wafer A and B were fabricated as typical material, and Table I shows that wafer A is mostly typical. Unexpectedly, wafer B is heavily skewed toward the faster corner. This observation was confirmed by WAT data from the fabrication facility, as well as lab testing. A map of wafer B's process tagging results is shown in Fig. 5.

When studied closely, many patterns emerge. First, the high-voltage transistors are all typical, as the fabrication facility has much better control over the larger geometries. Second, the FF die are clustered in the center, surrounded by FT die. The mask reticles are also visible as patterns of straight lines running the length of the wafer in the horizontal and vertical directions. Finally, note that even on this highly skewed wafer, there are still some slow chips. After viewing large amounts of wafers, it is reasonable to assume that every wafer will have fast, typical, and slow process variation in the low-voltage transistors.

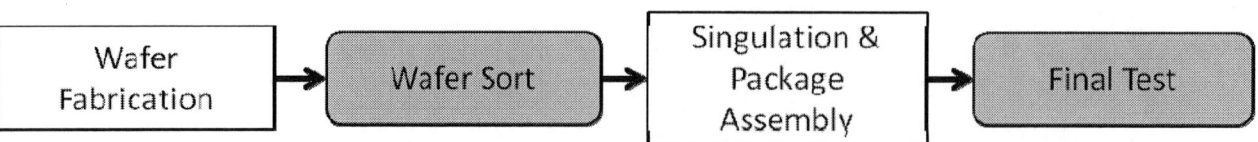

Figure 4. Production flow with process tagging capabilities at wafer sort and final test.

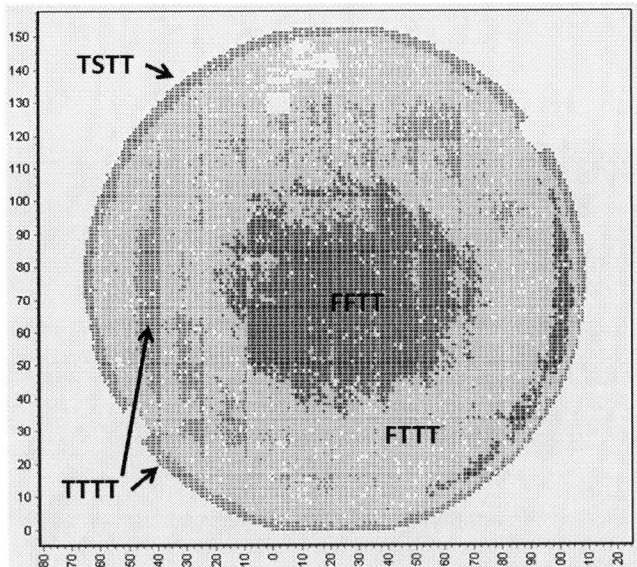

Figure 5. Wafer map of process tagging results on skewed wafer. The wafer is 300 mm in diameter and yields roughly 21000 chips. Process corners are defined as LV NMOS, LV PMOS, HV NMOS, and finally HV PMOS. Thus, the FTTT corner is fast for LV NMOS and typical for all other transistors.

The last set of production data involves long term process corner measurements, where a large amount of production wafers are tested and examined. Fig. 6 shows the process tagging results in a histogram of 2.42 million die. As expected, most of the die are centered in the typical region, but there are more chips shifted toward the faster corner when compared to the slow corner. Since the slow process corner generally requires the highest voltage and current biasing points, this set of data gives us confidence that most die will benefit from reductions in voltage and current while still maintaining performance.

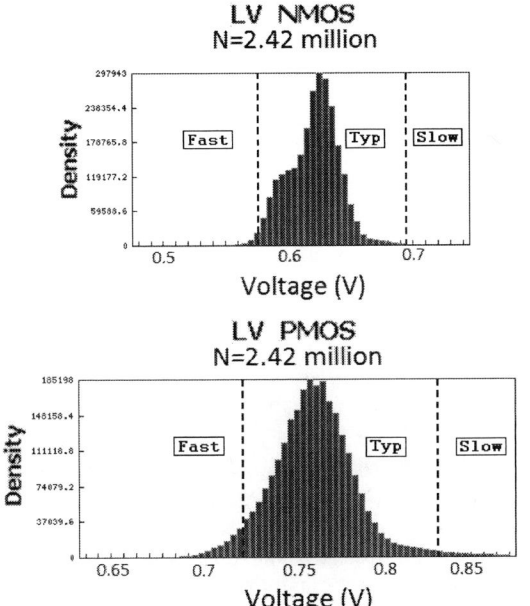

Figure 6. LV NMOS and PMOS process tagging results for 2.42 million die.

The final FPVTV compensation results are listed in Table II and show that typical and maximum power consumption can by reduced by 11.2% and 5.8%, respectively. The reductions in power consumption were achieved primarily through voltage scaling of the digital core and VCO partnered with current scaling in the LNA and ADC. Wafers yielded more than 98% for large volume.

TABLE II.   FPVTV RESULTS

Result	Chip Power Consumption (mA)	
	Typical	*Maximum*
Without FPVTV compensation	21.4	24.0
With FPVTV compensation	19.0	22.6
Power savings	2.4	2.6
	11.2%	5.8%

VII.   Additional Topics

Many additional effects must be considered but are not covered here. These include the effects of backgrinding the wafer, package-induced process shift, proximity to wafer edge, process variation and thermal gradients across a single die, and lifetime aging. The successful implementation of biasing algorithms depends on inclusion of these effects.

VIII.   Conclusion

A method for optimizing power consumption over FPVTV was presented, including measurement data. This algorithm was implemented in silicon and resulted in power savings of 11.2%, which can be extended 5-10% more on the existing platform.

References

[1] P. Macken, M. Degrauwe, M. Van Paemel, and H. Oguey, "A voltage reduction technique for digital systems," Proc. IEEE Int. Solid-State Circuits Conf., pp. 238–239, 1990.

[2] V. Gutnik, A.P. Chandrakasan, "Embedded power supply for low-power DSP," IEEE Trans. Very Large Scale Integration (VLSI) Systems, vol. 5, no. 4, pp. 425-435, Dec. 1997.

[3] D. Kawazoe, H. Sugawara, T. Ito, K. Okada, K. Masu, "A reconfigurable RF circuit architecture for dynamic power reduction," Proc. IEEE Region 10 TENCON, pp. 1–5, Nov. 2005.

[4] Lin Yuan; Gang Qu, "Analysis of energy reduction on dynamic voltage scaling-enabled systems," IEEE Trans. Computer-Aided Design of Integrated Circuits and Systems, vol. 24, no. 12, pp. 1827–1837, Dec. 2005.

[5] Z. Wang, H.S. Savci, J.D. Griggs, N.S. Dogan, E. Arvas, "Coping with process variations in ultra-low power CMOS analog integrated circuits," Proc. IEEE SoutheastCon, pp. 54–57, 2007.

[6] B. Murmann, "A/D converter trends: Power dissipation, scaling and digitally assisted architectures," Proc. IEEE Custom Integrated Circuits Conf., pp. 105–112, Sep. 2008.

[7] B. Datta, W. Burleson, "Calibration of on-chip thermal sensors using process monitoring circuits," Int. Symposium on Quality Electronic Design (ISQED), pp. 461–467, 2010.

[8] M. Meterelliyoz, P. Song, F. Stellari, J.P. Kulkarni, K. Roy, "Characterization of random process variations using ultralow-power, high-sensitivity, bias-free sub-threshold process sensor," IEEE Trans. Circuits and Systems I: Regular Papers, vol. 27, no. 8, pp. 1838–1847, Aug. 2010.

978-1-61284-863-1/11 $26.00 © 2011 IEEE

CMOS Based 16-Channel Neural/Muscular Stimulation System with Arbitrary Waveform and Active Charge Balancing Circuit

Lei Yao, *Member, IEEE*, Minkyu Je, *Member, IEEE*

Institute of Microelectronics,
A*STAR (Agency for Science, Technology and Research), Singapore
yaol@ime.a-star.edu.sg

Abstract—**Modern implantable neural/muscular stimulation system for neurological and physiological studies as well as for clinical treatment requires small, effective, safe and power efficient circuit system. CMOS technology is one of the best candidates to implement the circuit system to meet these requirements. In this paper, a 16-channel current mode stimulation circuit system with arbitrary waveform and active charge balancing circuit designed in CMOS 0.18µm 24V high voltage process is described. The design considerations are reviewed and discussed. The simulation results show that the proposed circuit system can successfully generate arbitrary waveforms and perform active charge balancing meeting the safety requirement.**

Keywords-Neural stimulation; muscle stimulation; active charge balancing; arbitrary waveform

I. INTRODUCTION AND BACKGROUND

Neural and muscular electrical stimulation has been used as a valid method for neurological and physiological studies as well as clinical treatment for decades [1-4] since the first fully electrical based stimulator was reported by O. H. Schmitt in early 1930s [5]. In recent years, driven by the increasing demands for implantable neural/muscular stimulators from biomedical scientific field, many research groups are developing different kinds of CMOS based micro electrical stimulation systems for deep brain stimulation (DBS) [6], retinal/cochlear prosthesis [7-8], functional electrical stimulation (FES) [9] and brain to brain interface [10]. These stimulation systems assist clinical doctors treat neurological disease/disorders such as Parkinson disease, neuralgia and psychological disorders, as well as to improve the daily life quality of disabled patients through recovering their physical functionalities such as visual/hearing sense or limb movement. A typical implantable stimulation system and its application are shown in Figure 1. Though the current existing stimulation systems achieve great success as mentioned above, there are several major issues left for further improvement and optimization.

A. Stimulation Mode

There are basically three modes for a stimulation system: current, voltage and charge mode. Voltage mode stimulation is rarely used in modern stimulation system because it is hard to control the total stimulation charge as the load impedance varies with time and stimulation polarity. It leads to serious safety issue in chronic applications. Charge mode stimulation can accurately control the stimulation charge. However, due to the high capacitance present in the load, large capacitors (~µF) are required [11] in charge mode to deliver sufficient charge into the load. Hence it is not suitable for implantable biomedical applications. Current mode stimulation is the most widely used. It has control on the stimulation charge with different load impedance. One of the shortcomings for current mode stimulation is that it requires high voltage to accommodate certain level of current when high load impedance is present, which can be solved using nowadays well developed high voltage CMOS process in semiconductor industry. In this paper, current mode stimulation is utilized due to the aforementioned advantages.

B. Stimulation Waveform

Stimulation waveform determines the number of stimulation parameters to be stored and affects the stimulation effect. Rectangular waveform is the most ofen used waveform in the stimulation systems mainly because of its simplicity. Other stimulation waveforms such as pulse

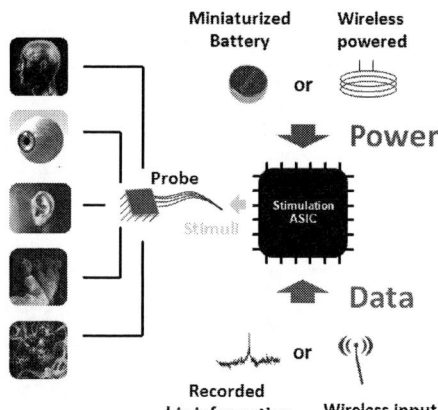

Figure 1. Typical implantable stimulation system and its applications

---

This work was supported by A*STAR SERC Grant Programme (SERC grant number: 1021520013).

trains [12] and non-rectangular waveforms [13] are also under investigation and some of them are found to have better stimulation effect than rectangular waveform. On the other hand, for different stimulation targets such as brain, peripheral nerve and muscle, there may be different optimized waveform which would reveal the unknown bioelectrical characteristics of the target. Therefore an arbitrary waveform stimulation system is of great importance for scientific research and stimulation system optimization.

C. Charge Balancing Strategy

Safety is an important issue in stimulation systems particularly for chronic biomedical applications. The safety problem in stimulation system is caused by the unbalanced charge injected by the stimulator into the tissue resulting in excess charge accumulation over time, which can lead to electrolysis with electrode dissolution and tissue destruction [14]. Biphasic stimulation is widely used to achieve better charge balancing performances [15]. However, when the stimulator is implemented using integrated circuit (IC), more than 1% to 5% mismatch in the biphasic waveform has to be taken into account due to the imperfections in IC fabrication process. Most of the current stimulators are using passive charge balancer of which the performance is unpredictable under variable load impedance. Active charge balancing is also applied in a few designs [8] which still need improvement to reduce their complexity.

D. Power Efficiency

Another important issue for neural/muscular stimulator is the power efficiency which is becoming increasingly important due to the limited power budget for the implantable circuit systems nowadays. As the tissue impedance is unpredictable for the stimulator, while maintaining the system functionality for the highest load impedance, the stimulator's power efficiency would drop dramatically when small load impedance is present. Ortmanns et al. [8] reduced this power efficiency drop through adapting the power supply voltage to the corresponding load impedance.

E. Number of Channels

The number of stimulation channels is determined according to different applications. Retinal prosthesis requires more than 1000 stimulation channels to increase the space resolution in visual recovery. On the other hand, a single channel can complete the therapy in DBS and pain management to induce or block the release of chemical transmitter. The peripheral nerve and muscle stimulation application has a moderate requirement for number of channels since the number of muscles to be simulated by the system is in tenth order, so 16 channels is chosen for our stimulation system as a prototype. To realize comprehensive access to the different pools of sensory and motor nerve

fibers for smoother control, more number of channels may be needed [16].

In this paper, we focus on the stimulation waveform and charge balancing strategy. A current mode 16-channel stimulation system with arbitrary waveform and active charge balancing circuit designed in 0.18μm CMOS 24V High Voltage process is described.

II. STIMULATION CIRCUIT SYSTEM DESIGN

A. System Architecture

The system architecture of the 16-channel stimulation system is shown in Figure 2. The main functionality of the system is to deliver controlled amount of charges into 16 different stimulation sites through constant current. The system consists of three main blocks: global digital control (GDC), 16 stimulation cells array and biasing circuitry. The global digital control is used to address the 16 cells and coordinate the stimulation pattern and timing among the 16 cells. Biasing circuitry is used to provide reference voltages and currents for each stimulation cell. Each cell consists of four main blocks: local digital control (LDC), digital to analog converter (DAC), high voltage output stage and charge balancing circuitry. The LDC receives the command from GDC and stores the parameters in the local registers to control the stimulation waveform. DAC is used to convert the digital controls from LDC into analog current. This current is then amplified and delivered into stimulation site through high voltage output stage. After each stimulation, charge balancing circuitry is enabled to clear the residual charges caused by the unbalanced stimulation waveform.

B. Digital Controller

The digital controller in our stimulation system mainly consists of two blocks: global digital controller (GDC) and local digital controller (LDC). The GDC consists of an addressing circuitry to locate each of the 16 stimulation cells and a global stimulation pattern control register to control the sequence of the stimulation for the 16 cells. Local digital control consists of four registers and one local finite state machine (FSM) to control the stimulation parameters such as waveform, amplitude and duration for each channel. The command structure and LDC FSM operation are shown in Figure 3.

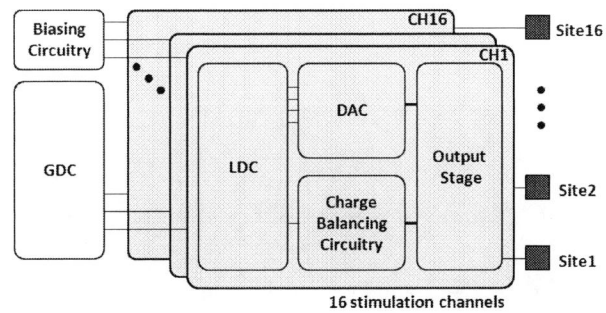

Figure 2. System architecture of 16-channel stimulation system

978-1-61284-863-1/11 $26.00 © 2011 IEEE   457

Command structure

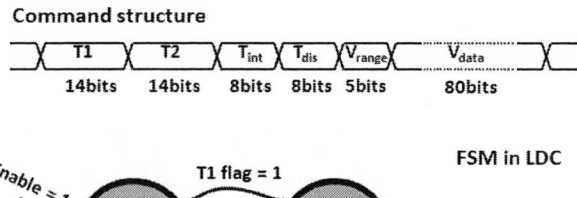

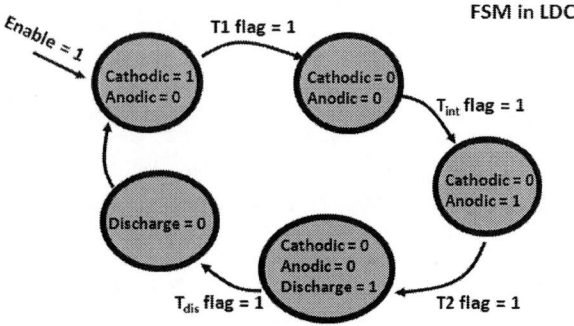

Figure 3. Command structure and FSM operation of the logical controller

C. DAC

The DAC used in the stimulation circuit is a 5 bit current steering DAC with range control. The circuit schematic is shown in Figure 4. I_{ref} is the reference current fixed as 100nA. r<0:4> defines the maximum value of the output current, it is fixed during the stimulation phase. b<0:4> controls the amplitude of the output current to form arbitrary current waveform, the output current is varying based on the pre-defined waveform parameter. The ratio of the transistor size is also shown in Figure 4. The resolution of the DAC is 5 bit, with LSB ranging from 100n A to 3.1µA, and the supply for DAC is 1.8V.

D. High Voltage Output Stage (HVOS)

The high voltage output stage further amplifies the output current from DAC and delivers this current into load through high output voltage output stage. It operates under high voltage supply *VDD_h* (as high as 24V). One main design consideration in HVOS is to keep the circuitry as simple as possible in order to 1) reduce the design complexity since only asymmetric high voltage transistor is available in the

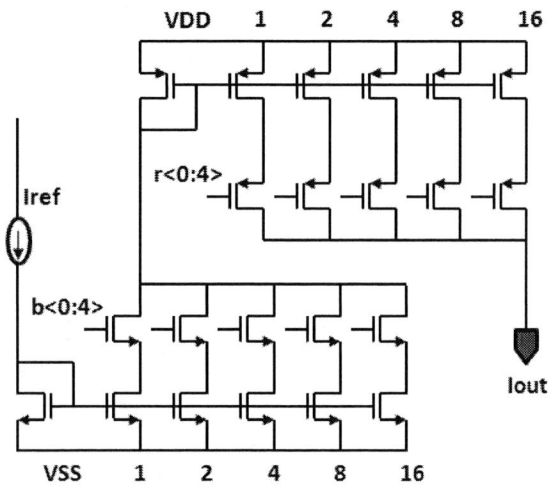

Figure 4. Circuit schematic of DAC

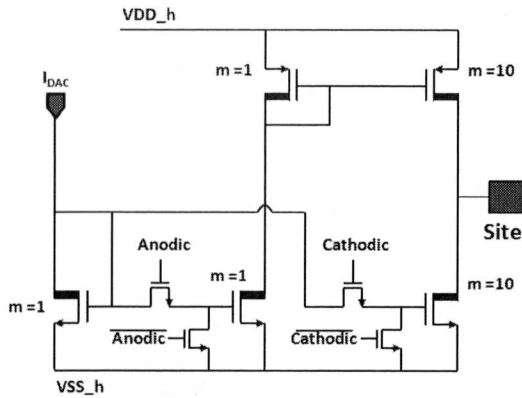

Figure 5. Circuit schematic of HVOS

process and 2) reduce the risk of circuit breakdown caused by parasitic components. The HVOS circuit schematic is shown in Figure 5. The drain of the high voltage transistor is illustrated with bolder line in the figure. *VDD_h* and *VSS_h* are high voltage supply and ground respectively. The digital signals *Anodic* and *Cathodic* are provided by LDC and control the polarity of the stimulation current. The input current I_{DAC} is amplified by 10 times through a current mirror and then is delivered to the stimulation site.

E. Charge Balancing Circuit

Passive charge balancing takes time and for our stimulation system it is not efficient due to the lack of high voltage switch in the chosen process. A residual voltage based current compensation charge balancing circuit which is similar to [8] is used in our stimulation system.

III. SIMULATION RESULT AND DISCUSSION

One 10kΩ resistor in serial with one 100nF capacitor is used as the load in the circuit simulation. The load is connected between the stimulation site of the high voltage output stage and the reference voltage (12V). The simulation result is shown in Figure 6 and Figure 7. As shown in **Error! Reference source not found.**, two different stimulation waveforms: sawtooth waveform and rectangular waveform

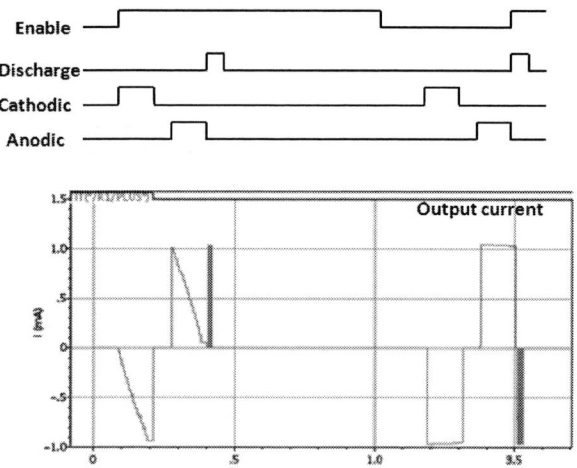

Figure 6. Control signals and the simulated stimulation current waveforms

978-1-61284-863-1/11 $26.00 © 2011 IEEE   458

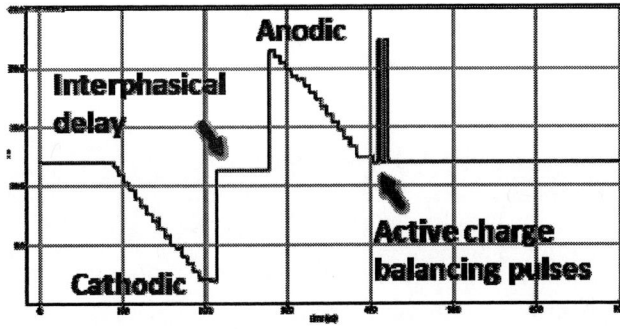

Figure 7. Close-up of the sawtooth waveform in figure 6

are generated. The digital control signals from LDC are also shown in the figure. The enable signal triggers the operation of the FSM in LDC, followed by the cathodic pulse, anodic pulse and discharge pulse. The width of these pulses is based on the parameters stored in the local registers in LDC. Figure 7 shows the close-up of sawtooth waveform which includes cathodic phase, anodic phase, interphasical delay between the two phases and the charge balancing phase. After the stimulation, the voltage on the stimulation site remains within +/- 50mV range comparing to the reference voltage.

IV. SUMMARY AND CONCLUSION

The current research trend in developing electrical stimulation system for neurological and physiological studies as well as clinical applications is to make the stimulation systems small, safe, effective and power efficient. CMOS technology is utilized to integrate the control electronics and stimulation waveform generation electronics on a single chip in millimeter scale. In this paper we reviewed the design considerations and the current state of art for CMOS based electrical stimulation circuit system. We presented one 16-channel current mode stimulation circuit system with arbitrary waveform and active charge balancing circuit designed in CMOS 0.18μm 24V high voltage process for neural/muscular stimulation applications. The simulation results show that the proposed circuit system can successfully generates arbitrary waveforms and eliminates the charge accumulation using active charge balancing method.

REFERENCES

[1] S. P. Hooker, S. F. Figoni, M. M. Rodgers, R. M. Glaser, T. Mathews, A. G. Suryaprasad and S. C. Gupta, "Physiologic effects of electrical stimulation leg cycle exercise training in spinal cord injured persons," *Arch. Phys. Med. Rehabil.*, vol. 73, no. 5, pp. 470–476, 1992.

[2] Glenn WW, Phelps ML, Elefteriades JA, Dentz B and Hogan JF, "Twenty years of experience in phrenic nerve stimulation to pace the diaphragm," Clin Electrophysiol 1986; 9:780.

[3] A. M. Kuncel and W. M. Grill, "Selection of stimulus parameters for deep brain stimulation," Clinical Neurophysiology 115 (2004) 2431-2441.

[4] T. R. Gheewala, R. D. Melen and R. L. White, " A CMOS implantable multielectrode auditory stimulator for the deaf," IEEE J. Solid-State Circuits, vol. 10, No. 6, December 1975.

[5] O. H. Schmitt and E. O. Schmitt, "A Universal Precision Stimulator," Science, vol.76, pp. 328-330, October 1932.

[6] J. Lee, H. G. Rhew, D. R. Kipke and M. P. Flynn, "A 64 channel programmable closed-loop neurostimulator with 8 channel neural amplifier and logarithmic ADC," IEEE J. Solid-State Circuits, vol. 45, No. 9, September 2010.

[7] K. Chen, Z. Yang, L. Hoang, J. Weiland, M. Humayun and W. Liu, "An integrated 256-channel epiretinal prosthesis," IEEE J. Solid-State Circuits, vol. 45, No. 9, September 2010.

[8] M. Ortmanns, A. Roche, M. Gehrke and H. J. Tiedtke, "A 232-channel epiretinal stimulator ASIC," IEEE J. Solid-State Circuits, vol. 42, No. 12, December 2007.

[9] S. Y. Lee and S. C. Lee, "An implantable wireless bidirectional communication microstimulator for neuromuscular stimulation," IEEE J. Solid-State Circuits, vol. 52, No. 12, December 2005.

[10] M. Azin, D. J. Guggenmos, S. Barbay, R. J. Nudo and P. Mohseni, "A battery-powered activity dependent intracortical mocrostimulation IC for brain-machine-brain interface," IEEE J. Solid-State Circuits, vol. 46, No. 4, April 2011.

[11] M. Ghovanloo, "Switched-capactior based implantable low-power wireless microstimulating systems," in Proc. IEEE Int. Symp. On Circuits and Systems, May 2006, pp. 2197–2200.

[12] P. Walter and K. Heimann, "Evoked cortical potentials after electrical stimulation of the inner retina in rabbits," Graefe's Arch. Clin. Exper. Ophthalmol., vol. 238, no. 4, pp. 315–318, Apr. 2000.

[13] P. R. Troyk, I. E. Brown, W. H. Moore, and G. E. Loeb, "Development of BION technology for functional electrical stimulation: bidirectional telemetry," in Proc. 31st IEEE/EMBS Conf., SEP. 2009, pp. 642–645.

[14] T. Jochum, T. Denison and P. Wolf, "Integrated circuit amplifiers for multi-electrode intracortical recording," J. Neural Eng., vol 6, 2009

[15] D. R. Merrill, M. Bikson, and J. G. Jefferys, "Electrical stimulation of excitable tissue: Design of efficacious and safe protocols," J. Neurosci. Meth., vol. 141, no. 2, pp. 171–198, Feb. 2005.

[16] B. K. Thurgood, D. J. Warren, N. M. Ledbetter, G. A. Clark and R. R. Harrison, "A Wireless integrated circuit for 100-channel charge-balanced neural stimulaton," IEEE Trans. Biomed. Circuits Syst., vol. 3, No. 6, December 2009.

Physical Design Exploration of 3DIC Wireless Transceiver using Through-Si-Vias

Mini Jayakrishnan[1], Xin Liu[1], Hong Yu Li[1], Jingjing Lan[2], Wang Ling Goh[2]

[1]Institute of Microelectronics (IME), A*STAR (Agency for Science Technology & Research),
11 Science Park Road, Singapore Science Park II, Singapore 117685
jayakrishnanm@ime.a-star.edu.sg, liux@ime.a-star.edu.sg, lihy@ime.a-star.edu.sg
[2]School of EEE, Nanyang Technology University, Singapore, 639798
E-mail: lanj0002@e.ntu.edu.sg, ewlgoh@ntu.edu.sg

Abstract- **3DIC's, the flagship for the "More than Moore Law" movement are already an integral part of the Semiconductor and Manufacturing industry and lot of investigations are going on within different sectors of the industry to efficiently fabricate 3DIC's. If you peep deep into the design process, everyone is trying to make it upward compatible with the 2D design for effective reusability. The methodology to design and verify the 3DIC involves a collaboration of different people from EDA, CAD, Physical design, Fabrication, TSV and Modeling departments. Different aspects need to be considered for the 3DIC apart from the miniaturization and the resulting space, cost & signal loss savings. The considerations include Power & clock Distribution, Thermal analysis, EMI, ESD protection scheme etc. This paper focuses on the physical design process steps carried out to develop a 3DIC Wireless Transceiver and it throws light into some of the process intricacies which are faced while carrying out the design and verification of such an IC stacking.**

I. INTRODUCTION

Physical Design of 3DIC [1] is very much similar to that of a 2D chip except that we should be able to build up the whole system in the vertical dimension rather than being confined to the horizontal dimension. From the manufacturing perspective the additional process which needs to be developed is for the Through Silicon Via (TSV) which can propagate connectivity through different dies rather than being confined to the intra die via. So the fabrication team has to explore various TSV processes and their ultimate aim is to achieve the minimum possible via dimensions so that 3DICs can take the advantage of the most modern 2D process nodes. The TSV manufacturing team sets standards like the TSV dimension, pitch and other spacing rules to be followed while doing the physical design. From the design perspective, the TSV need to be incorporated as a separate cell in the design kit with proper extraction rules for which TSV modeling and characterization is very essential. Apart from the TSV creation, there is a requirement for a shared environment for different technology files that represents different process nodes which can be ultimately integrated in the 3DIC. These steps ensure the use of TSV in existing 2D designs. The design flow can be similar to the 2D chip which starts with design exploration, floor planning, place & route, extraction and analysis. Thus 3DIC Design is collaboration between different sectors of the Semiconductor industry as shown in Fig. 1. Depending on the target application, the design can be low/high power, low/high data rate, analog/digital/mixed signal, memory/logic, RF/mixed signal etc and the third dimension should be able to take the advantage of all these combinations.

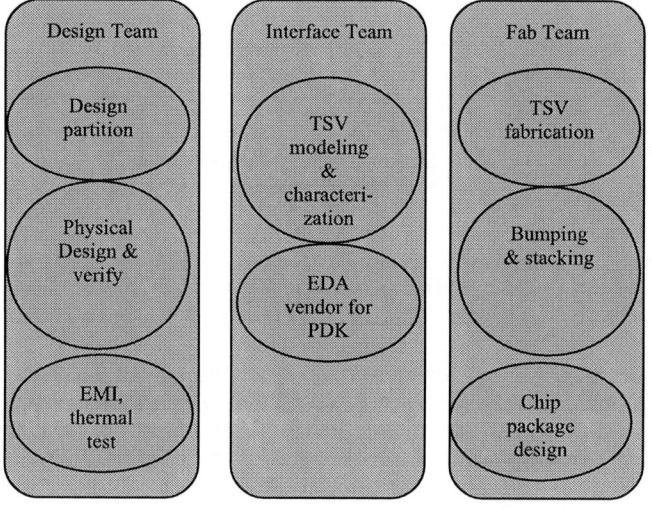

Fig. 1. Different players in the 3DIC industry.

II. DESIGN EXPLORATION

Aligned with the 2D design flow, design exploration is the first step essential for the 3DIC design. The purpose of this step is to analyze the design carefully and arrive at a conclusion whether 3D stacking will have an advantage on the cost/functionality/size of the circuit. The design need to be divided into different dies that can take full advantage of the 3D concept. The partition should be 3D aware in the sense that it should take into account the side effects that emerge out of vertical stacking like EMI between different dies, thermal effects of the stacking [2] [3] etc. The stack order depends on the number of TSVs that a particular die can afford because the lower dies should carry the TSVs related to input/output signals of the upper dies. Fig. 2 shows the design of the Wireless Transceiver which is to be stacked in the 3D domain. The design is logically separated into power management, radio frequency (RF), intermediate frequency (IF), baseband and signal conditioning (SC) units. The transceiver was designed to be of low power and so any possibility of thermal degradation in the stacking was negligible. From Fig. 2, it can

978-1-61284-863-1/11 $26.00 © 2011 IEEE

be observed that the possible area of EMI effects is in the RF block.

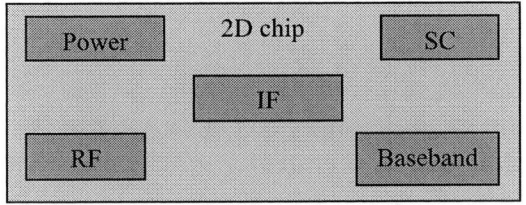

Fig. 2. Block diagram of the Wireless Transceiver.

III. FLOOR PLANNING

Apart from the 2D floor planning for the individual dies the additional factors which come into picture for the 3D chip is the area estimation for the TSVs and the associated floor planning should take into account the vertical stacking of dies. This includes the decision of how many dies to use in the stack up, the order of the stack up, the process technology node to be used for the dies, and the architecture of stacking itself. The stack up also need to consider the EMI effects between different dies used in the stack up, the thermal effects, the number of I/Os that are going to the PCB etc. In this process the die with minimum size core can be on the bottom side of the stack so that it can have more space to place and route the TSVs that need to go to the PCB on the bottom. Fig. 3 explains the stack up for the wireless transceiver where the sensor core is put at the bottom which has the minimum circuitry and the RF core at the top. Once the stack up is fixed, the TSV s can be inserted to satisfy the minimum dimensions. The other structure that comes with TSVs is the top and bottom bump pairs which acts as the interface between the different dies.

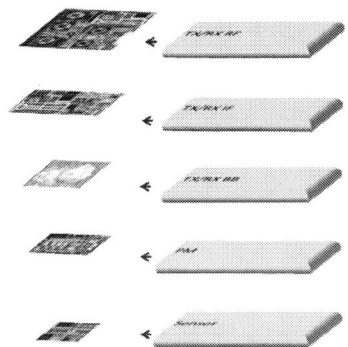

Fig. 3. The stack order for the Wireless Transceiver.

The whole stack needs to be interfaced with the PCB for which there are different options like Ball Grid Array (BGA) with and without a separate substrate or wire bonding. If the BGA is to be attached directly to the die bottom, then space need to be allocated for the BGA bumps also which can make the bottom die bigger than the other dies of the stack up.

Silicon interposers also come into picture at this point which can integrate different process node die stacks and finally interface it to the PCB [4] [5].

IV. PLACE & ROUTE

Before placing the new components such as TSVs and bumps in the actual design, it needs to be well characterized, modeled and placed as part of the Process Design Kit (PDK). Fig. 4 shows the cross section of the Die including TSVs, top and bottom bumps for the via last process. It can be treated as a separate component in the library which can be instantiated in the schematic and the layout. Its characteristics can be edited according to modeling. The place and route algorithms used for placing TSVs [6] should take into account the proper design rules for the TSVs and the associated bumps like diameter, pitch etc. The place and route algorithms should be able to reduce the overall routing length and optimize the power distribution, signal distribution, clock distribution, shielding, EMI effects etc. Ideally, the routing tool should recognize the signal lines which are too long so that the source and destination can be put in two dies which can be connected together using a TSV to reduce the routing length. There are some additional metal layers which come into picture other than the normal metal layers of the 2D chip. These layers are the Re Distribution Layers (RDL) which helps to route the signals from the top metal layer of the 2D die to the TSVs and from TSV top side to the micro bumps and bump pads. There should be synchronization between the process side and the circuit design side for the spacing and design rules. There may be high frequency differential signal lines for which the designer may prefer least possible RDL routing and absence of sharp bends and discontinuities.

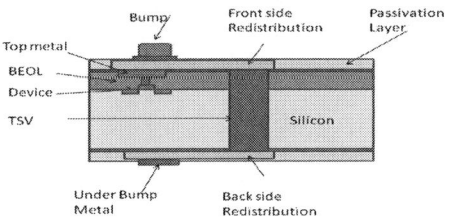

Fig. 4. Cross section of the die for via last TSV process.

The number of redistribution layers on the front side as well as the back side of the die is again determined by the TSV process. Proper ESD protection scheme is also required in the 3D stack especially since lot of post processing is required on the 3D chip stack. Fig. 5 shows the layout of the module of TSVs and bumps in the layout. In Fig. 6, an example of the 3D layout is presented. Here the original 2D layout design is placed in the core of the 3D layout, and the TSV and bumps modules are placed around the core design.

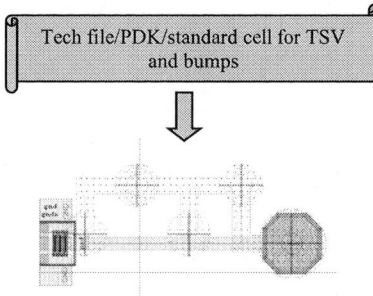

Fig. 5. Layout with TSVs and bumps routed in RDL.

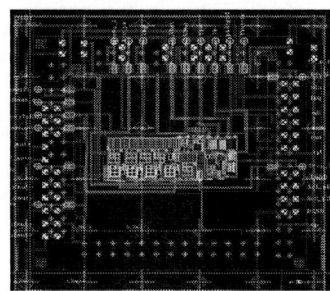

Fig. 6. Layout of one die including TSVs and bumps.

V. PHYSICAL VERIFICATION/EXTRACTION

One of the most important verification steps is the physical verification of the 3D chip which involves the Design Rule check (DRC) as well as the Layout Versus Schematic check (LVS). Fig. 7 shows the new rules which are required for the physical verification of the TSVs. For DRC check, the new layer geometries, dimensions and spacing for the additional TSV/bump structures need to be specified in the rule file of the EDA tool. Once the rule file is ready it needs to be integrated with the 2D design rules so that the DRC check for the complete 3D design can be performed. If the tape out is shared between different foundries for the core and the post processed layers, then proper metal filling process with some keep out zone from TSVs is required. For LVS check, LVS rules need to be added for the TSV and associated bump structures. There should be continuity between the 2D and 3D LVS rules so that the check can be performed effectively on the complete 3D stack up. Functional verification can also be performed through simulations both in 2D and 3D domain which takes the TSV effects into consideration. Similar to 2D extraction, 3D extraction should be able to extract parasitic for the TSVs and bump structures. For each TSV the signal frequency, the inter TSV coupling and TSV to substrate coupling effects should be taken into account. Other aspects like timing analysis, clock skew, and power distribution should also be extended to the 3D domain for timing and power critical designs so that they can take full advantage of the 3D space.

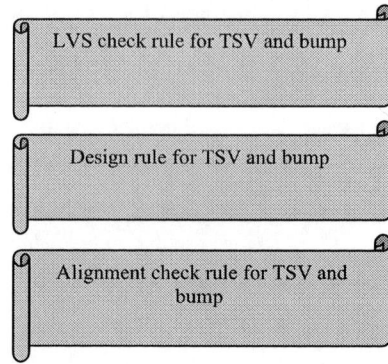

Fig. 7. The additional rule files required in the PDK for TSV physical verification process.

Apart from the physical verification for the TSVs and the core circuitry, alignment checks should also be done between different dies. Proper alignment marks should be inserted both on the die level as well as on the wafer level which ensures the correct stacking of the individual dies and thus the overall functioning of the IC stacking.

VI. 3D SIMULATION

The simulations in the 3DIC scenario will take into account the TSV parasitic which depend on factors like frequency, TSV to substrate as well as inter TSV coupling [8] [9]. The 3D simulation will give the effects of TSV on the performance of the design. If there is significant deviation from the required performance, then it can be rectified by making some changes in the 2D design. In addition to 3D simulation, some HFSS simulations may be required to analyze the high frequency signal EMI effects in the 3D domain because we need to consider the effects between the adjacent dies and the redistribution layers on the front side and back side of the die. For example, Fig. 8 shows the simulation results of receiver noise response of 2D and 3D , respectively, in which we take TSV and RDL resistance and capacitance into consideration for 3D simulation. Fig. 9 shows the 2D and 3D simulation results of receiver signal response. It can be observed that there are some difference between 2D and 3D performance. Furthermore, Fig. 10 provides the simulations results of the amplitude of power amplifier output corresponding to different TSV and RDL modeling with different capacitance. It can be observed that the output amplitude will be decreased with capacitance of TSV and RDL increasing.

978-1-61284-863-1/11 $26.00 © 2011 IEEE

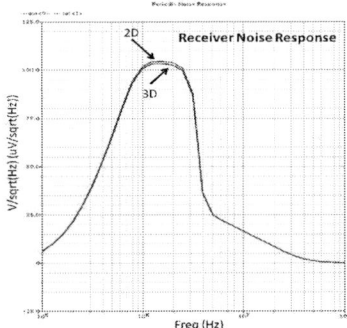

Fig. 8. Simulation results of RF receiver with and without TSV.

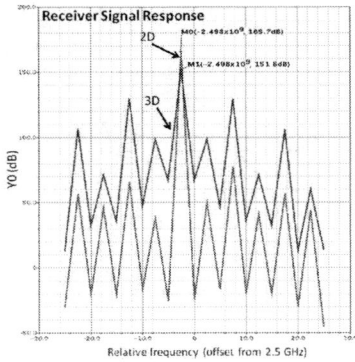

Fig. 9. Simulation results of RF receiver with and without TSV .

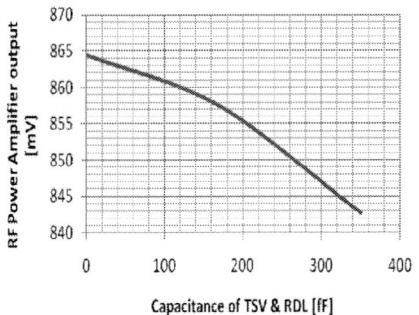

Fig. 10. Graph showing RF transmitter performance with TSV and RDL layer capacitance.

VII. PCB INTERFACE

The 3DIC stack can be mounted onto the PCB in different ways. If different process technology nodes are involved in the design, then a silicon interposer can be used to interface the IC stacks onto the PCB. A BGA type interface can also be used where a substrate comes in between the micro bumps of the IC stack and the large BGA balls that get connected to the bottom die of the IC stack instead of the substrate. There is another option of going without the BGA substrate where the BGA balls get directly attached to the bottom die. In Fig. 6, the bottom die layout of the 3DIC Transceiver is provided where the BGA pads are directly inserted as part of the die layout.

CONCLUSION

In this paper, we discussed that careful planning and feasibility analysis is important and required before we decide to go for the 3DIC design. The manufacturing cost and yield factors need to be carefully analyzed for the 3D design. It should be used for applications where there is absolute need of miniaturization and short signal interconnects without degrading the performance. Once the 3DIC standards are set by the manufacturing and EDA industry, it would be easy to use it as part of the daily design process. Finally, to take the real advantage of the miniaturization achieved through 3DIC, we should be able to integrate the passives as well as related components into the IC stack/interposer so that the whole system size can be reduced in proportion with the IC stack.

REFERENCES

[1] www.cadence.com, "3D ICs with TSVs—Design Challenges and Requirements", *Whitepaper*.

[2] Young-Joon Lee and Sung Kyu Lim, "Co-Optimization of Signal, Power, and Thermal Distribution Networks for 3D ICs", in *IEEE Electrical Design of Advanced Packaging and Systems Symposium*, 2008, pp. 163-166.

[3] Muhannad S. Bakir, Calvin King, Deepak Sekar, Hiren Thacker, Bing Dang, Gang Huang, Azad Naeemi, and James D. Meindl, "3D Heterogeneous Integrated Systems: Liquid Cooling, Power Delivery, and Implementation", *IEEE Custom Integrated Circuits Conference (CICC)*, 2008, pp. 663-670.

[4] Kouichi Kumagai, Yuko Yoneda, Hitoshi Izumino, Hiroko Shimojo, Masahiro Sunohara, Takashi Kurihara, Mitsutoshi Higashi, and Yoshihiro Mabuchi, "A Silicon Interposer BGA Package with Cu-Filled TSV and Multi-Layer Cu-Plating Interconnect", *IEEE Electronic Components and Technology Conference*, 2008, pp. 571-576.

[5] Navas Khan, Vempati Srinivasa Rao, Samule Lim, Ho Soon We, Vincent Lee, Zhang Xiao Wu, Yang Rui, Liao Ebin, Ranganathan, TC Chai, V.Kripesh and John Lau, "Development of 3D Silicon Module with TSV for System in Packaging", *Electronic Components and Technology Conference*, 2008, pp. 550-555.

[6] Dae Hyun Kim, Krit Athikulwongse, and Sung Kyu Lim, "A study of Through-Silicon-Via Impact on the 3D stacked IC Layout", *ICCAD*, 2009, pp. 674 – 680.

[7] Guruprasad Katti, Michele stucchi, Kristin de Meyer, and Wim Dehaene, "Electrical Modeling and Characterization of Through Silicon via for Three_Dimensional ICs", *IEEE Transactions on Electronic Devices*, 2010, pp. 256-261.

[8] Chuan Xu, Robert Suaya and Kaustav Banerjee, "Compact Modeling and Anaysis of Coupling Noise Induced by Through-Si-Vias in 3-D ICs", *IEDM* 2010, pp. 8.1.1 - 8.1.4.

[9] Michael B. Healy and Sung Kyu Lim, "A Study of Stacking Limit and Scaling in 3D ICs: An Interconnect Perspective", *Electronic Components and Technology Conference*, 2009, pp. 1213-1218.

978-1-61284-863-1/11 $26.00 © 2011 IEEE

A PLL with a VCO of Improved PVT Tolerance

Kok-Foong CHONG[1], Liter SIEK[2], and Benjamin LAU[1]

[1]Design Enablement, GLOBALFOUNDRIES Inc.
[2]VIRTUS-IC Design Centre of Excellence, School of Electrical & Electronic Engineering
Nanyang Technological University, 50 Nanyang Avenue, Singapore 639798
[1]chongkf@globalfoundries.com and [2]elsiek@ntu.edu.sg

Abstract—VCO has been the central block of a PLL used for on-chip clock generation. Due to the variation in process, voltage and temperature, the ratio of the VCO frequency at the fastest condition to the slowest condition could be a factor of 2~3. Also, significant variation of VCO gain is expected. For the same target VCO frequency, the loop bandwidth and stability could thus vary greatly due to the variation in the VCO gain. The designer needs to design the bandwidth and stability of the PLL so that it can function properly under all different PVT conditions. For example, the bandwidth is selected based on the worst-case condition, i.e. the fastest condition, trading off for less optimal bandwidths at other conditions. In this paper, the design of a compensation scheme to improve the PVT tolerance of a VCO is described. A four-stage differential ring oscillator with voltage regulation is employed to achieve the voltage tolerance. Depending on the process corner and temperature, the proposed compensation scheme works by changing the control current, thereby maintaining the desired frequency. The variation of frequency at 25°C is within ± 0.6% for the tuning voltage range from 0.4V to 1.5V comparing the fastest to slowest conditions. The variation of frequency at each control voltage point is within ±3.9% across a temperature range of -40°C to 125°C. A charge pump PLL is implemented with this VCO, demonstrating desired advantages as compared to a conventional VCO without compensation.

Keywords-PLL, VCO, PVT, process compensation, temperature compensation.

I. INTRODUCTION

Commonly, PLLs are used to multiply low-frequency clocks from off-chip to generate clocks that are of high frequency to be used on-chip. A voltage controlled oscillator (VCO) is the central building block of such PLLs. It is well-known that typical VCO frequencies could vary by a factor of 2~3 between its slowest and fastest conditions due the variations in process, voltage, and temperature [1]. Fig. 1 shows how the frequency of a conventional VCO varies with respect to the control voltage (V_{CTRL}), from the fastest FF corner to the slowest SS corner. With this variation, the upper bound of the usable frequency range is limited to that of the slowest running condition within the tuning voltage range, as indicated by the horizontal dotted line in Fig.1. The lower would normally be bound by the fastest running condition. It is clear that for the same control voltage, the fastest running frequency is more than 30% higher than the slowest one. The VCO gain (K_{VCO}), could vary by more than 60% for the same target frequency. Due to the sampling nature of a PLL, a common rule of thumb is to design the loop bandwidth (ω_{BW}) to be less than 1/10 of the reference frequency.

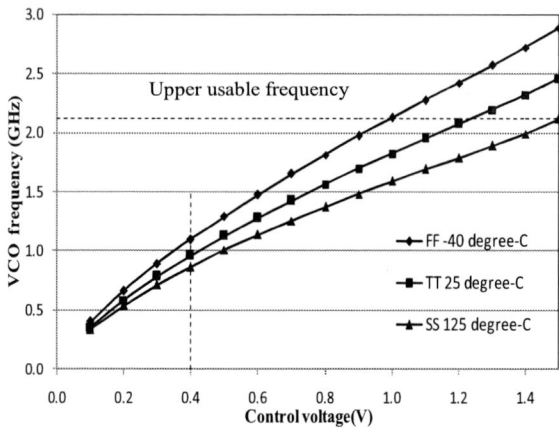

Figure 1. Variation in VCO frequency due to process and temperature

Loop bandwidth is proportional to K_{VCO} so the bandwidth selected must satisfy the highest K_{VCO} (FF corner) for a targeted frequency. The damping factor is also proportional to K_{VCO}, so the design has to cater for the lowest K_{VCO} (SS corner) to maintain a minimum damping factor. For a given charge pump current and divider ratio, the resistor and the capacitor values have to be selected based on the lowest K_{VCO}. Also, it is known that the output phase noise of a PLL associated with the VCO can be reduced by increasing the loop bandwidth. In contrast, a smaller bandwidth is desired in order to filter out the noise that comes from the reference clock. As such, there is an optimum loop bandwidth trading off between these two noise sources. With all these design constraints and the variation in K_{VCO}, a designer is left with some choices of design margins with more conservative design parameters that guarantees functionality and stability for all possible PVT variation, but at suboptimal performance. There is a class of complicated adaptive bandwidth PLL design [1] targeting PVT tolerance on loop bandwidth and damping factors by means of balancing the ratio among design parameters. Nevertheless, a VCO with less variation in K_{VCO} is still beneficial as it helps to design PLL with a narrower range of loop bandwidths as well as the damping factors, apart from the wider tuning range. Several attempts had been taken to reduce the variation in VCO frequency and VCO gain [4]-[7].

Relatively few papers focus solely on achieving PVT tolerant VCO for PLL. Paper like [3] focus on achieving a PVT tolerant VCO with a single frequency point but not for multiple

978-1-61284-863-1/11 $26.00 © 2011 IEEE

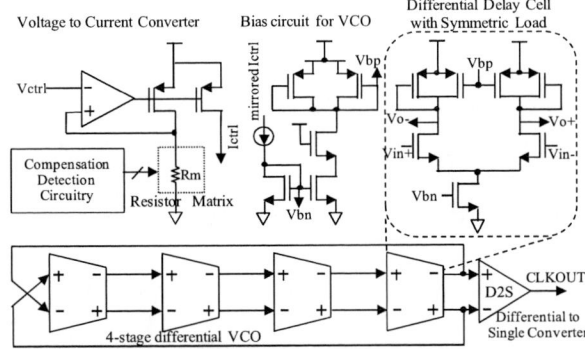

Figure 2. Block diagram of the compensated VCO

frequencies suitable for PLL application. [4] makes use of a separate VCO calibration block to perform calibration and then producing digital words to adjust the loading capacitors of the VCO which is powered by a regulated supply, thereby adjusting the VCO frequency. A faster 3-stage oscillator is used in the slow conditions and a slower 4-stage oscillator is used when in the fast condition in [5]. In [6], one VCO tracks the temperature and process variation and feed back a control voltage to control the second VCO to achieve lesser variation. A calibrated bias voltage is obtained by interpolating between two points and the resulting voltage is applied to tune the bias current of the differential delay cell [7]. This paper presents an on-chip compensation technique which adjusts the current going to the VCO depending on the process and temperature conditions detected. An implementation example to the design of a PLL is also presented.

II. THE COMPENSATION TECHNIQUE

A. Basic Compensation

Fig. 2 shows the block diagram of the compensated VCO. Not shown in the diagram is the regulated voltage supply of 1.2V that powers the VCO to provide voltage tolerance. A 4-stage differential VCO with the popular Maneatis delay cells [2] are used. The load is composed of a diode-connected PMOS biased in saturation state and the other equally sized PMOS biased in the triode region. The VCO dummy stages are required to provide symmetry and uniform loading that could help in reducing the common mode noise rejection and phase alignment. The upper swing of the VCO output is bounded by the V_{DD} and the lower swing is bounded by Vbp. The V_{CTRL} is converted into a control current (I_{CTRL}) by the voltage-to-current converter. A mirrored copy of the I_{CTRL} is used by the biasing circuit to set Vbp and Vbn. The control current is given by

$$I_{CTRL} = V_{CTRL} / R_m \qquad (1)$$

where R_m is the resistance at the switchable resistor matrix. The delay time at each state is very roughly given by

$$t_d = \frac{C_L(V_{DD} - V_{bp})}{I_{bn}} \qquad (2)$$

where I_{bn} is the bias current at the tail of the delay stage and is proportional to I_{CTRL}. The oscillation frequency is given by

$$f = 1/(2Nt_d) \qquad (3)$$

where N=4 is the number of delay stages and t_d the delay time of each stage. By changing the V_{CTRL} or the R_m, the I_{CTRL} is changed and hence the delay time of each stage and the resulting frequency is changed.

The compensation detection circuitry consists of a set of comparators as shown in Fig. 3(a). Basically, the threshold voltage (Vth) detected is compared against a predetermined set of reference voltage generated by a resistor divider. The input of the reference voltage (Vref) is taken from the bandgap voltage reference (1.2V). Digital control bits are generated and the resistor matrix is switched to a desired resistance value.

Fig. 3(b) shows the frequency of the VCO with and without the compensation. It can be seen that the compensated VCO frequency at the extreme operating condition is shifted closer to the typical case at 25°C. The upper usable frequency range has been extended from around 2.10GHz to 2.40GHz. Assuming 0.4V is the selected lower tuning voltage, the lower frequency range is extended from around 1.1GHz to 1GHz. Overall, the usable frequency range is extended by a factor of 1.4. The variation of K_{VCO} has also been significantly reduced.

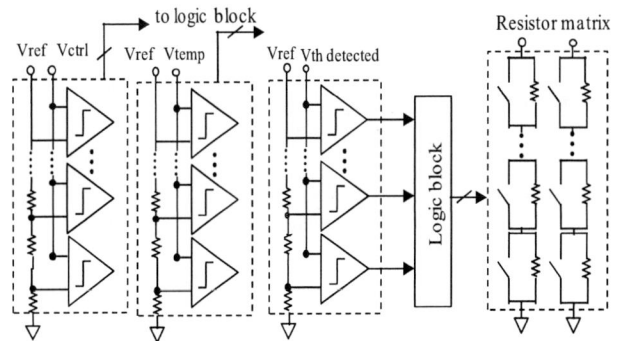

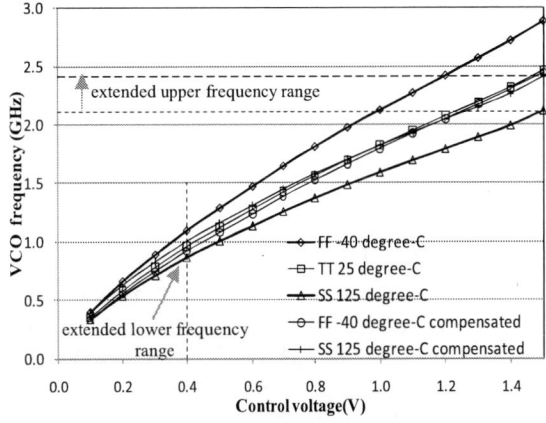

Figure 3. (a)Compensation detection circuitry block diagram. (b)Comparison of VCO frequency with and without compensation

978-1-61284-863-1/11 $26.00 © 2011 IEEE

VCO gain versus frequency

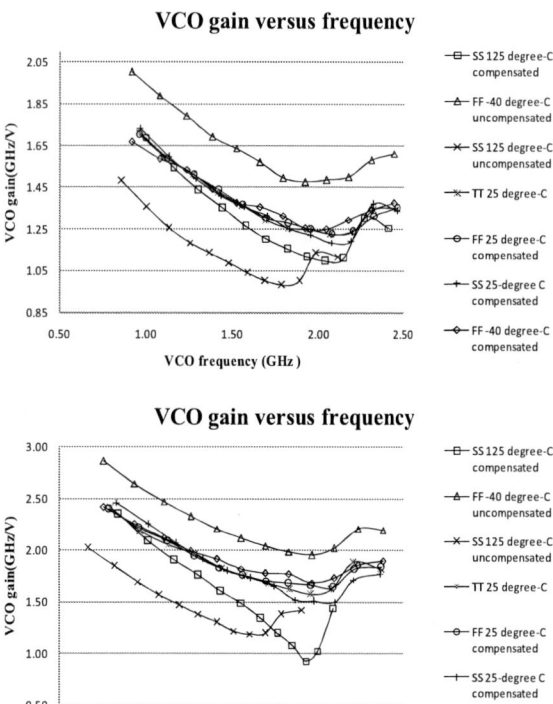

Figure 4. (a)Variation of VCO gain at higher Rm, centred at 7.6k ohm.(b)Variation of VCO gain at lower Rm, centred at 4.47k ohm.

TABLE I. COMPARISON BETWEEN DIFFERENT WORKS

Specification		[4]	[5]	[6]	[7]	This work
Supply voltage (V)		1.8	3.3	N/A	1.8	1.8
Temperature range (°C)		-25~125	0~100	0~100	0~100	-25~125
Technology (µm)		0.18	0.5	N/A	0.18	0.18
V_{CTRL} =0.50V	Slowest(MHz)	N/A	N/A	1924	40	1092
	Fastest(MHz)			1979	80	1177
	variation			2.9%	100.0%	7.8%
V_{CTRL} =0.80V	Slowest(MHz)	84	N/A	N/A	195	1534
	Fastest(MHz)	89			205	1606
	variation	6.0%			5.1%	4.7%
V_{CTRL} =1.00V	Slowest(MHz)	N/A	140	1998	N/A	1802
	Fastest(MHz)		180	2006		1857
	variation		28.6%	0.4%		3.1%
V_{CTRL} =1.25V	Slowest(MHz)	N/A	N/A	2031	N/A	2096
	Fastest(MHz)			2101		2156
	variation			3.4%		2.9%
V_{CTRL} =1.40V	Slowest(MHz)	510	275	N/A	575	2264
	Fastest(MHz)	526	360		675	2332
	variation	3.1%	30.9%		17.4%	3.0%

B. Complex Compensation

Fig. 4(a), (b) show that the variation for K_{VCO} greatly reduced. For the same frequency, the compensated K_{VCO} is about 10% at most and resulting in about 5% variation in bandwidth by proportion. In contrast, the uncompensated K_{VCO} show variation of as high as 50%. At 25°C, the K_{VCO} curves align well from process corner to corner. Though the basic compensation itself is good in reducing the VCO frequency variation for practical use, more complex compensation algorithm is required for higher requirement. In Fig. 4(b), for lower Rm, i.e. at higher frequency range and K_{VCO} at the extreme SS corner 125°C show deviation from the rest, especially at higher frequency. As shown in Fig.3. (a), the temperature and/or the V_{CTRL} can be combined to provide a more robust compensation. The VCO demonstrates different sensitivity to high and low temperature as well as high V_{CTRL}.

C. Comparison Between Different Works

Table I summaries the comparison of the proposed compensated VCO with prior arts [4]-[7]. [6] demonstrated that small frequency variation could only be reached within a narrow band of 2GHz. In [4]-[6], the circuits implemented incur the silicon area as they all require two VCOs. In [7], fewer components are used at the expense of performance, i.e. the variation in frequency is slightly higher. This work shows an overall smaller frequency variation across the control voltage range from 0.5V to 1.4V. The highest variation is at the lower V_{CTRL} of 0.5V.

III. PLL IMPLEMENTATION

A. Overview

The block diagram of the PLL implemented with the compensated VCO is shown in Fig. 5. The design is implemented in a 0.13µm CMOS technology. The reference frequency is injected from an external source. The operation of the VCO has been explained in Section II. All circuitry is powered by a single 1.8V power supply. A bandgap reference voltage is designed to provide a voltage (Vref) of 1.2V. Vref is used as the voltage reference for the compensation circuitry, the reference current generation and the voltage regulator to generate the 1.2V power supply. The regulator provides power supply that is used for the phase-frequency detector (PFD), VCO, differential to single converter (D2S), feedback divider and the bias circuitry for the VCO.

The current from charge pump across all PVT corners has been made relatively constant by mirroring the current which is derived from the bandgap-based current reference generator. The charge pump current can be adjusted from 20µA to 60µA to allow for adjustments of the loop bandwidth. The unity gain Op Amp is used for better performance [8].

The phase frequency detector (PFD) is designed in dynamic logic scheme. By inserting delay circuit in the feedback signal path, the PFD generates short coincidental pulses of UP and DN, even when the PLL is in lock, so that the dead zones are removed. The reset signal ensures a valid output logic state after power-on. The delay for reset signal is introduced by a chain of NAND gates and inverters. A divide-by-32 [Fig.5] Dlatch-based feedback divider has been designed with five stages of half-dividers. The half-divider consists mainly of a negative edge latch and a positive edge latch working in pair to achieve a divide-by-2. The divider has been simulated over the PVT corners and can work up to 5GHz with a good output duty cycle of 51.5±0.5%.

B. Simulation Results

Fig. 6 shows the transient response of the PLL simulated at 25°C. The V_{CTRL} is at 1.1V for the typical (TT) case, at 0.9V for the uncompensated FF case and at 1.4V for the uncompensated SS case. With compensation, both the FF and SS case is shifted closer to 1.1V. The operating range has thus been extended. The PLL is tuned to an optimum bandwidth by adjusting the charge pump current. As shown in Table III, the PLL with compensated VCO demonstrates improved jitter.

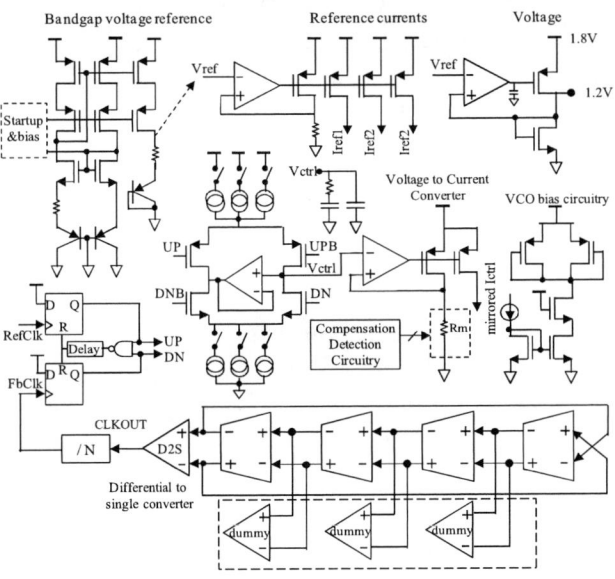

Figure 5.   Block diagram of the PLL

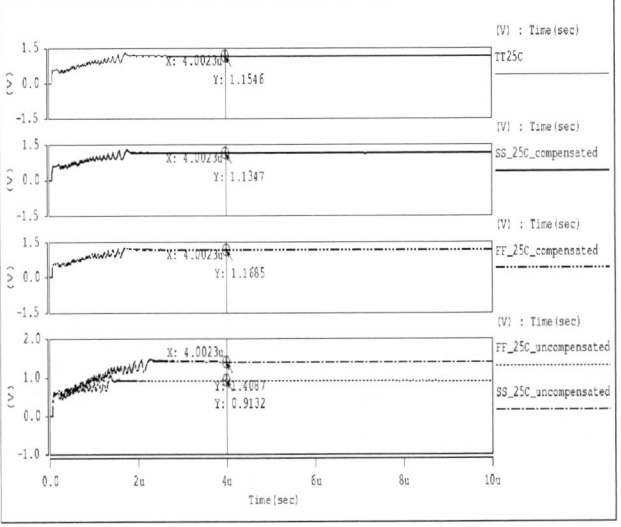

Figure 6.   Transient response of the PLL

The compensated VCO at the FF and SS corners is able to provide the K_{VCO} closer to the TT optimal operating point than the VCO without compensation. The lower jitter performance for the PLL with compensated VCO at the FF and SS corner is attributed to this unique characteristic.

TABLE II.   VALUES USED FOR THE SIMULATION

F_{REF}	I_{CP}	F_{VCO}	C_{RIPPLE}	C_{MAIN}	R_{ZERO}
80MHz	40µA	2.56GHz	3pF	58.35pF	8.66kΩ

TABLE III.   PERIODIC JITTER BASED ON 10K CYCLES

	TT	SS	FF	SS^a	FF^a
P2P Jitter	5.17ps	6.97ps	6.84ps	6.63ps	6.69ps

a. with compensated VCO.

IV.   CONCLUSION

A differential VCO is made with improved PVT tolerance by means of digitally assisted compensation. The variation of frequency at 25°C is within ± 0.6% for the V_{CTRL} range from 0.4V to 1.5V across the fastest to slowest conditions. The variation of frequency at each control voltage point is within -±3.9% across a temperature range of -40°C to 125°C. The usable frequency range is extended by a factor of 1.4. The variation of K_{VCO} for a targeted frequency has also been greatly reduced due to the reduction in frequency variation and hence the bandwidth shift, resulting in improved jitter performance over process variation. The overall feature of the proposed VCO helps to relax the design constraint. This approach could easily be ported over from process node to process node, different frequency range. Also, the relatively simple analysis makes it easy to adapt even for the inexperienced designers to the VCO or PLL design.

REFERENCES

[1] J. Kim, M. A. Horowitz, and G.Y. Wei, "Design of CMOS adaptive-bandwidth PLL/DLLs: a general approach," IEEE Transactions on Circuits and Systems—II: Analog and Digital Signal Processing, vol. 50, no. 11, pp. 860-869, Nov. 2003.

[2] J. G. Maneatis, "Low-jitter process-independent DLL and PLL based on self-biased techniques," IEEE J. Solid-State Circuits, vol. 31, no.11, pp. 1723–1732, Nov. 1996.

[3] K. Sundaresan and P. E. Allen, "Process and temperature compensation in a 7-MHz CMOS clock oscillator," IEEE J. Solid-State Circuits, vol. 41, no.2, pp. 433-442, Feb. 2006.

[4] J.B. Shin, S.H. Park, S.J. Bae, and H.J. Park, "A dual-loop CMOS PLL with the max-to-min frequency ratio larger than five guaranteed under PVT corners," International SoC Design Conference, pp. 313-316, Oct. 2004.

[5] J. Routama, K. Koli, and K. Halonen, "A novel ring-oscillator with a very small process and temperature variation," Proc. Int. Symp. Circuits Syst., Jun. 1998, vol. 1, pp. 181-184.

[6] H. Chen, E. Lee, and R. Geiger, "A 2 GHz VCO with process and temperature compensation," Proc. IEEE Int. Symp. Circuits Syst. (ISCAS), May 1999, pp. 569-572.

[7] S-J. Park, S. Woo, H. Ha, Y. Suh, H-J. Park, and J-Y. Sim, "A transistor-based background self-calibration for reducing PVT sensitivity with a design example of an adaptive bandwidth PLL," IEEE Asian Solid-State Circuits Conf., Nov. 2008, pp. 433-436.

[8] Y. Sun, P.Y.Song, and L.Siek, "Design of a high performance charge pump circuit for low voltage phase-locked loops," IEEE International Symposium on Integrated Circuits, Sep. 2007, pp.295-298.

The phase locked loop for MEMS horizontal scanning control of micro-laser projection ASIC

Dan Lei Yan[1], Luo Jian Wen[1], Li Peng[1], Ravinder Pal Singh[1], Duy-Dong Pham[1], Tal Langer[2] and Minkyu Je[1]

[1]Institute of Microelectronics, A*STAR Singapore. 117685

Email: yandl@ime.a-star.edu.sg

[2]Maradin Ltd. Yokneam, Israel.20692.

Email: tal.langer@maradin.co.il

Abstract— A phase locked loop (PLL) for MEMS horizontal scanning control of micro-laser projection ASIC is demonstrated for pico projectors application using high voltage 0.18-μm CMOS process. Special design techniques of phase frequency detector(PFD) and VCO control voltage with resetting are proposed to achieve resetting initial phase of horizontal scanning control, which make sure the PLL have stable performance when system power on. The PLL generates 142MHz signal with LVDS buffer also, which is used for generating timing signals providing a 1/8-pixel resolution for vertical control. The 142MHz of PLL has the measured phase noise of -112dBc/Hz at 1MHz offset and the integrated RMS jitter is 963ps from 100Hz to 10MHz. The active die area is 1mm x 1mm. The chip operates over a wide range of supply voltage from 1.6 V to 2.0V and temperature from -40°C to +85°C. The PLL chip core draws 6 mA current from a +1.8V supply at +25°C.

Keywords-component; Phase locked Loop, projector, MEMS, micromirror

I. INTRODUCTION

MEMS scanning mirror-based are drawing so much attention in Laser Projector application.. With pico projectors, one can project a full-size image onto any surface near at hand, whether it be the wall, your shirt, or a piece of paper. The projector control is ASIC based on a two dimensional(2-D) micro scanning mirror technology, which drives the MEMS mirror and generates horizontal and vertical the timing signals to control the laser module through the data manipulator. The MEMS scanner is based on a gimbaled mirror, with two uncoupled actuators – one for horizontal scan and the other for vertical scan. The horizontal scan is under phase closed loop control[1][2][3]. It's motion is created by running the horizontal axis at its resonant frequency (10KHz) based on an electrostatic resonating actuator driven by a comb structure. This paper describes the design and implementation of the phase locked loop for MEMS horizontal scanning control.

II. ARCHETECHURE

The proposed architecture of the phase locked loop (PLL) for MEMS horizontal scanning control is shown in Fig. 1 which includes a phase frequency detector with reset control, a 5 stage of differential ring oscillator (VCO), charge-pump , loop filter and a fixed 14200 counter and digital logic control circuit. At same time generate 142MHz with LVDS buffer output clock signal which is used for generating timing signals providing a 1/8-pixel resolution for vertical control.

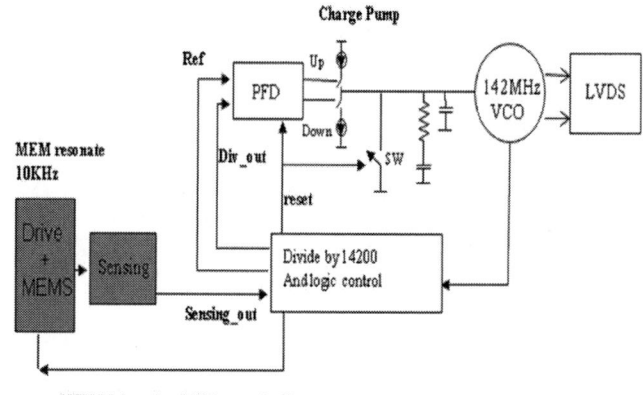

Fig.1. Block diagram of the phase locked loop for MEMS horizontal scanning control

After dividing the VCO output frequency by 14,200, the MEMS driven signal output is used to drive MEMS mirror, the sensing block is used to extract the capacitor phase change of MEMS and sensing out for PLL to maintain its horizontal resonance. Between MEMS driven signal and Sensing out have the phase frequency response as shown in Fig.2.[1] , It is obtained from the phase angle will shift for 90° at the 10KHz resonant point and have a sharp response due to high Q factor of MEMS . The MEMS mirror is driven by high voltage (80V max.) to achieve large angle.

978-1-61284-863-1/11 $26.00 © 2011 IEEE

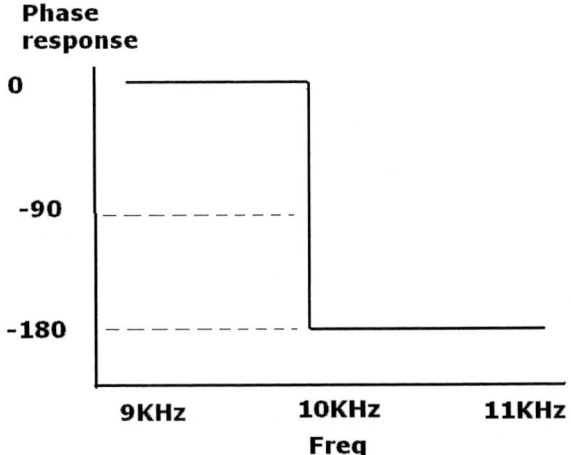

Fig.2. The phase frequency response between MEMS driven signal and Sensing out

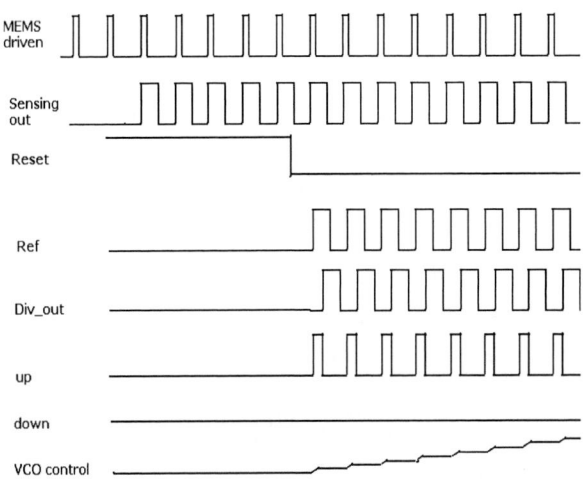

Fig.3. The phase relation for PLL signals

The phase locked loop use phase frequency detector (PFD) with charge pump(CP), which have unlimited lock range and easy implementation. The 5 stage ring oscillator structure VCO have had been used, which have the tuning range from 130 to 160MHz, in order to cover the MEMS resonate frequency after divide 14200. Div_out and MEMS driven signal are from divide 14200, but have differential phase, MEMS driven signal have 90° ahead than Div_out, when PLL is power on, MEMS driven signal will directly be output, Div_out will be hold until Ref have input. Due to PFD is sensitive for the frequency and phase initial condition. In order to make sure the PLL always have a stable initial condition after power on, some Special design techniques had been used. When power on, PFD is disable through reset signal, up and down control signal of CP are low. At same time, reset signal turn on the SW to pull the VCO control signal to LOW, that means MEMS driven signal is about 9.1KHz, due to HV drive and sense circuit need some time to get stable condition, so logic control detect sensing out signal, after 5 continuous period, the logic control will disable reset signal, make sure PLL is functional, then release sensing out signal to Ref. then release Div_out, make sure Div_out is phase delay than Ref.

Due to Ref is always ahead Div_out, so UP signal of charge pump is turn on, so charge will be pumped in to the loop filter, the VCO control signal will be pull to high. The frequency MEMS driven signal will be increased. At idea condition, if HV drive and sense block don't have phase delay, Ref should be 90° ahead than Div_out. When the frequency of MEMS driven signal increased to 10KHz, Sensing out signal will be have 90° phase delay, so Ref and Div_out have the same phase,at this frequency , the PLL is Locked at MEMS resonate frequency. The Fig.3 shows the detail of the above signals. MEMS_driven signal have 15% duty cycle based on MEMS specification. Sensing out have the same phase with the MEMS_driven signal and have 50% duty cycle.

The Fig.4 shows the VCO control voltage of simulation result for The phase locked for MEMS horizontal scanning control of micro-laser projection Resonator .

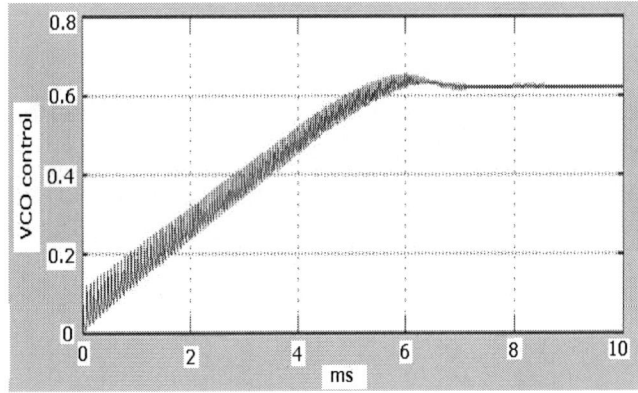

Fig.4. VCO control signal of MEMS horizontal scanning control.

III.   Key Building Blocks

A.   Phase frequency detector (PFD) with reset

A dead zone free PFD as shown in Fig.5[4] was used by insert a delay after AND gate. Compare with conventional PFD, a OR gate is added between the reset of D flip flop and delay output. When reset signal is low, the PFD will performance as the conventional PFD, when reset signal is high, the PFD is disable, up and down signals are at low level. That means the PFD was reset to initial condition, base on this condition, PLL have the correct initial response with Ref and Div_out input.

978-1-61284-863-1/11 $26.00 © 2011 IEEE   469

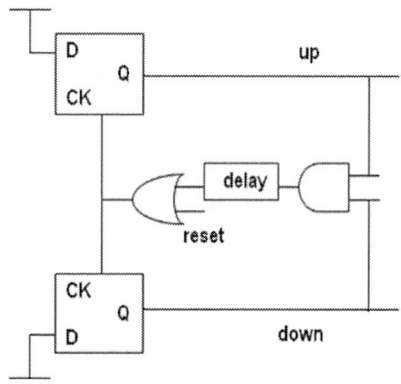

Fig.5. Phase frequency detector(PFD) with reset

B. Charge-pump

The charge pump schematic is shown in Fig.6[4] Charge injection is carefully minimized by implementing the switches of charge-pump with a minimal size. In parallel with the NMOS switches is PMOS switches are used. By optimizing the size of switch transistors, the injected charge of PMOS is made equal to the injected charge of NMOS achieving the overall zero charge injection.

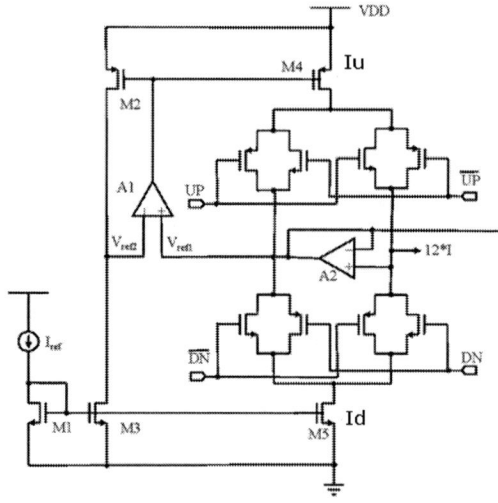

Fig.6. the schematic of charge pump

In order to compensate mismatches of Iu and Id dummy branch M2 and M3 is used. A1 is dynamic feedback loop, so the voltage of charge pump output , Vref1 and Vref2 have same voltage, the feedback loop will regulate Iu and make sure Iu to be equal to Id.

C Voltage Controlled oscillator (VCO)

For a given power budget, LC oscillators consume less power , but considerably chip area, For 140MHz LC oscillator, the inductor value could be bigger than 150nH, the

big area will be occupied and easy noise coupling from big size inductor. Due to their integrated nature, ring oscillators have become an essential building block in this design. For the ring oscillator, dominant noise sources in IC environment are common-mode signals in nature (e.g. power supply noise, substrate-coupled noise)[5].In order to remove the above noise and get low noise oscillator, fully differential design is a must as show in Fig.7.

The VCO circuit includes voltage to current bias control circuit and 5 stages Maneatis symmetric load delay cell of ring oscillator[5]. The current bias control circuit is the circuit(self-biased replica stage) to provide vpbias and vnbias which the two voltage are dependent, that granting a maximum output voltage swing . a single delay cell of the VCO is composed of a differential pair and two symmetric loads. Within a specific output swing, the $I - V$ curve of one load has odd symmetry with equal slopes on either side of an inflection point. When symmetric loads are used in an ideal differential stage, the impedance of the two loads varies with the swing but is always equal to each other. Any supply noise becomes common-mode at the cell output because of this impedance equality and is rejected by the common-mode rejection of the following delay cell. The bias circuit maintains the output swing to within the symmetric range of the loads and that varies with the VCO control voltage.

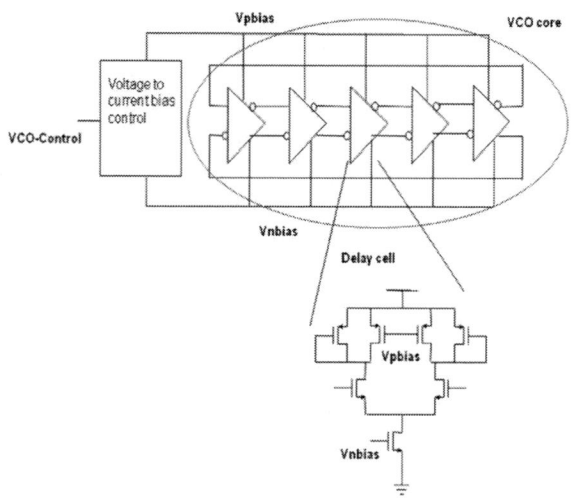

Fig.7. the schematic of Voltage Controlled Oscillator (VCO)

D LVDS, digital circuit

The 142MHz output clock signal which is used for generating timing signals providing a 1/8-pixel resolution for vertical control by FPGA. Normally single end digital buffer would generate more noise. In order to reduce noise of system, low voltage differential signaling(LVDS) buffer are used for 142MHz signal output.

The fixed 14200 counter , MEMS driven signal and Div_out are generated by digital circuit. In order to compensate the phase delay between MEMS_driven and sensing_out, the phase of Div_out and MEMS_driven are programmable

978-1-61284-863-1/11 $26.00 © 2011 IEEE   470

through the serial peripheral interface (SPI) control lines, namely data, clock and latch enable signals.

IV. CIRCUIT IMPLEMENTATION AND LAYOUT

The reference spur is mainly due to the substrate and power supply reference noise coupling. The reference noise is generated from the digital circuit of synthesizer [4], which includes the reference counter, PFD, dividers,. In order to reduce the noise coupling from the substrate, the analog and digital circuit are separated wide apart. VCO core occupies large area and more sensitive to the substrate noise coupling. Hence, a guard ring is added for VCO core circuit separately. Guard-ring that connect to substrate ground is added to surround each digital circuit in order to reduce the suppress coupling of reference spur noise to synthesizer output. The spur noise from the power supply was suppressed by separating the power supplies of digital and analog circuits.

V. MEASURED RESULTS

The phase locked loop for MEMS horizontal scanning control is implemented in high voltage 0.18-μm CMOS process. The microphotograph of the test chip is shown in Fig.8. The active area occupied by the circuitry is 1.0mm × 1.0mm.

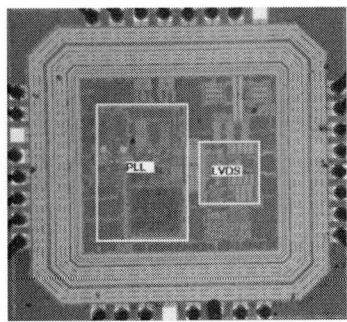

Fig.8. Chip microphotograph (active die size: 1.0.mm x 1.0mm).

Fig.9. shows the measured phase-noise plot at 142MHz with the loop filter 1KHz of PLL. The phase-noise at 1MHz offset is -112dBc/Hz, the integrated RMS jitter is 963ps from 100Hz to 10MHz. that meet the system specification of RMS jitter less than 2ns. Fig.10. shows the phase relation of Div_out and MEMS_driven signal, which is 90 ° differences as show in Fig.10.

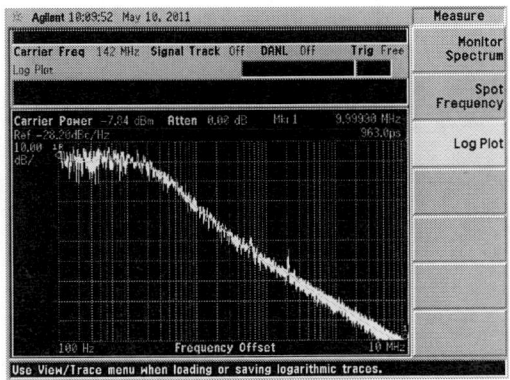

Fig. 9. Measured 142MHz phase-noise and jitter.

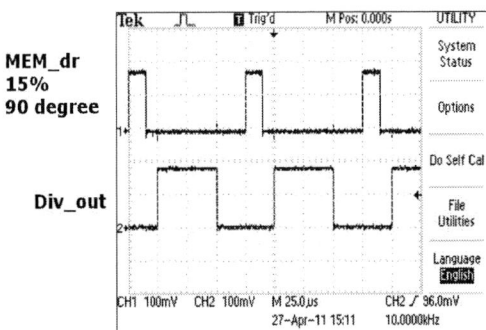

Fig.10. Phase relation of MEMS_driven and Div_out

Design, implementation and measurement results of The phase locked loop for MEMS horizontal scanning control have been presented. The circuits were fabricated in 0.18-μm high voltage CMOS process. closed-loop horizontal scanning were achieved. The PLL generates 142MHz signal with LVDS buffer also, which is used for generating timing signals providing a 1/8-pixel resolution for vertical control. The 142MHz have the measured phase noise of -112dBc/Hz at 1MHz offset and the integrated RMS jitter is 963ps from 100Hz to 10MHz The chip core consumes 6mA from a single +1.8V power supply. The measurement result show the PLL meet the design specification. The compact design is fully integrated onto the ASIC chip.

ACKNOWLEDGMENT

The authors would like to thank Maradin Ltd, for their support, and thanks the assistance of the staff of Integrated Circuits and Systems Laboratory at the Institute of Microelectronics, A*Star, Singapore, in the successful realization of the presented IC.

REFERENCES

[1] Chuanwei Wang; Hung-Hsiu Yu; Chingfu Tsou; Weileun Fang;, "Application of phase locking loop control for mems resonant devices," IEEE/LEOS Optical MEMS 2005 International Conference on Optical MEMS and Their Applications, pp. 165-166

[2] X. Sun, R. Horowitz, K. Komvopoulos, "Stability and Resolution Analysis of a Phase-Locked Loop Natural Frequency Tracking System for MEMS Fatigue Testing" Journal of Dynamic Systems, Measurement, and Control, pp.599–605. December,2002

[3] Duy-Dong Pham, Ravinder Pal Singh, Dan-Lei Yan, Kei-Tee Tiew, Minkyu Je1, "Position Sensing and Electrostatic Actuation Circuits for 2-D Scanning MEMS Micromirror," IEEE Defence, Science & Research (DSR 2011)

[4] Yan Dan Lei, "A low power CMOS 2.4-GHz monolithic integer-N synthesizer for wireless sensor ," Radio-Frequency Integration Technology: Integrated Circuits for Wideband Communication and Wireless Sensor Networks, 2005. Proceedings. 2005 IEEE International Workshop , pp. 219–222,

[5] Maneatis, J.G, "Low-Jitter Process-Independent DLL and PLL Based on Self-Biased Techniques," IEEE J. Solid-State Circuits, vol.31, no. 11, pp. 1723-1732, Novmber. 1996

A Flipped Voltage Follower Based Low-Dropout Regulator with Composite Power Transistor

S. S. Chong and P. K. Chan

School of Electrical and Electronic Engineering, Nanyang Technological University, Singapore
Email: chon0157@ntu.edu.sg, epkchan@ntu.edu.sg

Abstract—This paper presents an improved design for the low-dropout regulator (LDO) based on the Flipped Voltage Follower (FVF) structure. A FVF based LDO regulator with composite power transistor (CPT) is proposed. It reduces the minimum biasing current requirement of the power transistor in Single-Transistor-Control (STC) LDO regulator whilst enhancing the loop gain as well as bandwidth performance. The proposed LDO regulator has been validated using GLOBALFOUNDRIES 0.18-µm CMOS process. Using the FVF STC structure as the benchmark, the simulation results have shown that the proposed CPT LDO regulator has a better load transient and accuracy than that of the STC and Buffered Flipped Voltage Follower (BFVF) counterpart whilst maintaining low voltage operation.

I. INTRODUCTION

Recently, the emerging low voltage Integrated Circuit (IC) systems have been driven heavily by rapid development of the semiconductor technology. The advancement of the integration process technology allows the integration of microprocessors, memory, digital, analog and RF blocks onto one single chip. Therefore, the power management has an ever increasing role in electronic industry. A low-dropout (LDO) regulator is one of the important power sources in power management unit due to its low noise, low cost and fast transient characteristics. Low power consumption is always the key to prolong the battery life of portable devices. Many works [1]-[3], [5]-[6] have been reported in literature to improve the LDO performance such as low voltage, low quiescent current or fast transient response. Most of the works [1]-[3] are based on the conventional structure approach, in which the power transistor is driven by an error amplifier to form a regulating loop. Of particular interests, using the Flipped Voltage Follower (FVF) [4], the Single-Transistor-Control (STC) LDO regulator [5]-[6] provides another attractive solution for LDO regulator design on the basis of simplicity and good transient response. However, the price paid for the FVF LDO regulators are the tradeoffs in load regulation, minimum biasing current, minimum operation supply, accuracy and so forth. In this paper, an improved FVF LDO regulator, which makes use of a composite power transistor (CPT) as a replacement of single power transistor is proposed. It aims to improve the drawbacks as stated above.

Following the introduction, Section II reviews the FVF based STC LDO regulators and Buffer Flipped Voltage Follower

(BFVF). Section III describes and analyses the proposed FVF based LDO regulator with CPT. Section IV gives the simulated results and discussions. Conclusion is drawn in Section V.

II. REVIEW OF STC AND BFVF LDO REGULATORS

A. FVF Based STC LDO Regulator

The architecture of the FVF based STC LDO regulator [5] is shown in Fig. 1. The configuration of power transistor (M_P), control transistor (M_c) and biasing current source (I_N) form a typical FVF structure, where the output voltage (V_{out}) is always one $V_{SG(MC)}$ higher that the gate voltage (V_{CTRL}) of M_c. C_L is the off-chip capacitor with equivalent series resistance (ESR) of R_{esr}. I_L models the loading circuit. The control voltage V_{CTRL} can be generated based on the circuit [5] shown in Fig. 2. It is basically an amplifier in unity-gain configuration with a diode-connected transistor M_{C1}, which is biased by I_N, so as to generate the control voltage from the voltage reference. In STC LDO regulator, the dominant pole created by C_L is located at the output of the LDO regulator. Ideally, the non-dominant pole appears at the drain terminal of M_C is canceled by a zero which is generated by the ESR of the output capacitor to maintain stability.

The architecture of the STC LDO regulator is rather simple and is able to support low voltage operation environment. The required minimum supply voltage can be expressed as $V_{DSAT(I_N)} + V_{SG(M_P)}$ or $V_{out} + V_{dropout}$, whichever is larger. However, it still requires a minimum biasing current, I_N, in order to prevent M_C from being pushed into triode region under no

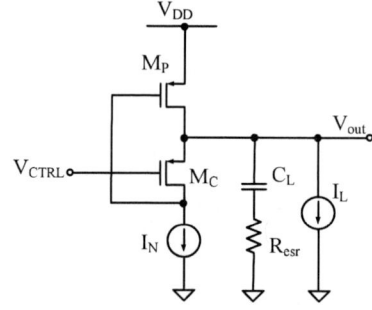

Fig. 1. Structure of FVF based STC LDO regulator.

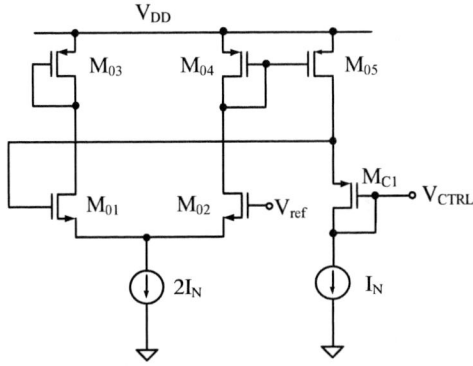

Fig. 2. Control voltage generator.

load condition. When the biasing current is too small, the potential at the gate of M_P has to increase so as to reduce the overdrive voltage. The action forces M_C to operate in triode region. As a result, the regulating loop is jeopardized whereas the output voltage is altered. This phenomenon is more obvious when the size of the power transistor is larger. The overdrive voltage can be increased by reducing the size of M_P. However, as the size of M_P decreases, I_N will be driven into triode region at heavy loading condition. It is because M_P requires a stronger overdrive voltage to supply current at heavy load when the size is smaller. Moreover, due to its simple folded circuit structure, the STC LDO regulator does not have a high loop gain. This influences directly the reported load regulation [5] as well as the accuracy.

B. BFVF LDO Regulator

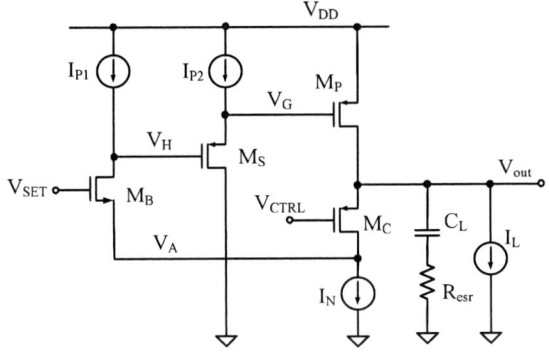

Fig. 3. Structure of BFVF LDO regulator.

In order to solve the stated problems, an improved FVF LDO regulator is reported in [6]. Figure 3 depicts the circuit architecture of BFVF LDO regulator. The transistor M_B is introduced to alleviate the minimum biasing requirement of STC LDO regulator. The potential of V_A is dictated by V_{SET}-$V_{GS(MB)}$. The gate-source voltage $V_{GS(MB)}$ can be set by the basing current I_{P1} as well as the size of M_B. With transistor M_B, the transistor M_C is ensured to work in saturation region even with a small biasing current I_N. In addition, the transistor M_B serves as a common gate amplifier providing extra gain in the feedback loop. As a result, the load regulation is enhanced.

On the other hand, the transistor M_S acts as a simple source follower or buffer stage which has a low output impedance of $1/g_{ms}$. The transconductance g_{ms} is governed by the biasing current I_{P2} and the size of M_S. With this low impedance node connected to the gate of M_P, the parasitic pole associated with the gate of M_P is shifted to higher frequencies. Consequently, the bandwidth and stability of the BFVF LDO regulator are enhanced. However, the BFVF LDO regulator requires higher voltage headroom which makes it not suitable for low voltage operating environment. The minimum supply voltage requirement of BFVF LDO regulator can be expressed as $V_{SET} - V_{GS(M_B)} + V_{DSAT(M_B)} + V_{SG(M_S)} + V_{SG(M_P)}$ or $V_{out} + V_{dropout}$, whichever is larger.

In addition, it also requires one extra biasing voltage for V_{SET}. This biasing voltage V_{SET} has to be designed carefully in order to prevent M_B from entering the triode region, especially at heavy load condition.

III. STRUCTURE OF PROPOSED LDO REGULATOR

A. Composite Power Transistor

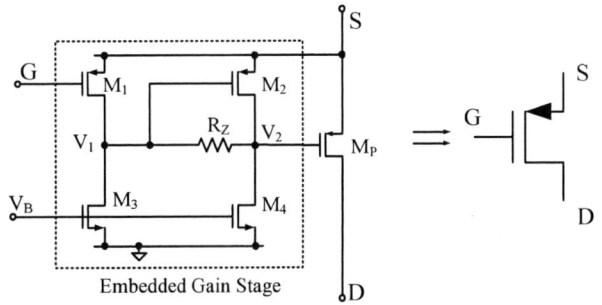

Fig. 4. Structure of composite power transistor.

The schematic of the CPT is shown in Fig. 4. The CPT has an open loop structure which is favorable to stability as well as bandwidth extension. The CPT can be viewed as a three terminals transistor with body tied to source terminal. The applications of CPT in LDO regulator have been reported in [7]-[8]. The embedded gain stage in the CPT has a frequency-dependent gain [7] which can be approximated as

$$A_v(s) = -\frac{g_{m1}R_z}{\left(1 + \frac{C_2}{g_{m2}}s\right)(1 + R_z C_1 s)} \tag{1}$$

where g_{mi} is defined as the transconductance of the respective device and C_i denotes the lumped parasitic capacitance. Due to the shunt feedback configuration, the output resistance of the embedded gain stage is approximately $1/g_{m2}$. Similarly, the pole associated with the parasitic gate capacitance of M_P is located at higher frequencies. Since the value of R_z and C_1 is not made large in this design, the overall transconductance of the CPT is approximately frequency-independent and can be expressed as

978-1-61284-863-1/11 $26.00 © 2011 IEEE   473

$$G_{mp} = g_{m1}R_z \times g_{mp} \qquad (2)$$

It can be seen that the gain and bandwidth of the embedded gain stage can be adjusted independently. This gives more design flexibility in CPT design.

B. Proposed LDO Regulator

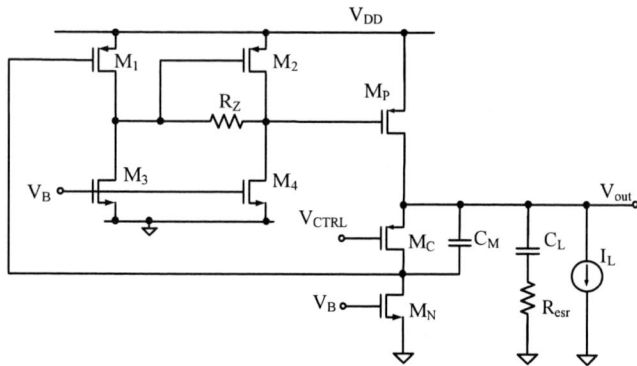

Fig. 5. Schematic of proposed FVF based CPT LDO regulator.

The schematic of the proposed FVF based CPT LDO regulator is shown in Fig. 5. As can be seen, the power transistor in STC LDO regulator is replaced by the CPT as discussed in previous section. The embedded gain stage has the same function as the buffer stage in BFVF LDO whilst providing some gain to enhance the current driving capability of the regulating transistor. The proposed CPT LDO regulator can be stabilized by a small on-chip capacitor, C_m. The design objective is to prevent transistor M_C from entering the triode region when the biasing current (I_N) is low and permit the low voltage operation of the LDO regulator. With the embedded gain stage, the voltage swing at the gate of M_1 is not as large as the voltage swing at the gate of M_P. Furthermore, the gate source voltage of M_1 can be set by the biasing current as well as the size of M_1. Therefore, neither M_C nor M_N will be pushed into the triode region at no load and full load condition. The minimum supply voltage can be obtained as $V_{DSAT(M_4)} + V_{SG(M_P)}$ or $V_{DSAT(M_N)} + V_{SG(M_1)}$ or $V_{out} + V_{dropout}$, whichever is larger.

The proposed CPT LDO regulator can be viewed as a two-stage amplifier compensated by a Miller capacitor shown in

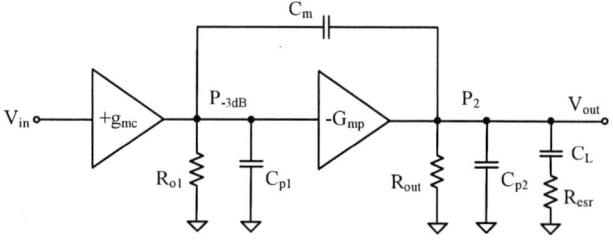

Fig. 6. Structure of proposed LDO regulator.

Fig. 6. It is noted that g_{mc} and G_{mp} are transconductance of transistor M_C and composite power transistor respectively. C_i and R_i denote the respective lumped output parasitic capacitance and output resistance of each node. Therefore, the transfer function can be derived as

$$A(s) = \frac{-g_{mc}G_{mp}R_{o1}R_{out}\left(1 - s\dfrac{C_m}{G_{mp}}\right)(1 + sC_LR_{esr})}{(1 + sC_mG_{mp}R_{o1}R_{out})\left(1 + s\dfrac{C_L}{G_{mp}}\right)} \qquad (3)$$

The derivation is based in the following assumptions: (i) $g_{mc}R_{o1}$ and $G_{mp}R_{out} \gg 1$, (ii) C_L and $C_m \gg C_{P1}$ and C_{P2}. From the transfer function, it can be observed that there are two left-hand-plane (LHP) poles, one LHP zero and one right-hand-plane (RHP) zero. The locations are shown in (4), (5), (6) and (7) respectively.

$$P_{-3dB} = -\frac{1}{C_mG_{mp}R_{o1}R_{out}} \qquad (4)$$

$$P_2 = -\frac{G_{mp}}{C_L} \qquad (5)$$

$$Z_{LHP} = -\frac{1}{C_LR_{esr}} \qquad (6)$$

$$Z_{RHP} = +\frac{G_{mp}}{C_m} \qquad (7)$$

When $I_L = 0$, despite of the weak g_{mp}, the Miller effect is still valid due to the gain provided by the embedded gain stage in the composite power transistor. The dominant pole P_{-3dB} at the drain terminal of M_C is located at very low frequency whereas P_2 at V_{out} and the RHP zero are well located beyond the unity-gain-frequency (UGF). The stability is ensured. When $I_L \neq 0$, the loop gain is boosted by g_{mp}, which is proportional to I_L. Due to the higher loop gain, P_2 appears within the UGF. Under this condition, the LHP zero is used to cancel P_2. The RHP zero is shifted to even higher frequencies due to large G_{mp}. As a result, a stable system is obtained.

IV. RESULTS AND DISCUSSIONS

The proposed CPT LDO regulator, STC LDO regulator and BFVF LDO regulator have been implemented and simulated with GLOBALFOUNDRIES 0.18-µm CMOS process. The sizes of the power transistor M_P of all the regulators are set equal to 1800µm/0.18µm. The LDO regulators are able to deliver 50mA load current. In order to have a fair comparison, the quiescent biasing current of all the LDO regulators are set to be equal at 17µA. For the proposed CPT and STC LDO regulators, the output voltage is regulated to 1V from a 1.2V supply voltage. On the other hand, due to meeting the headroom, the BFVF LDO regulator is powered by a supply voltage of 1.8V. It gives a regulated 1.6V at the output. The LDO regulators are stabilized by a 4.7µF output capacitor with

978-1-61284-863-1/11 $26.00 © 2011 IEEE

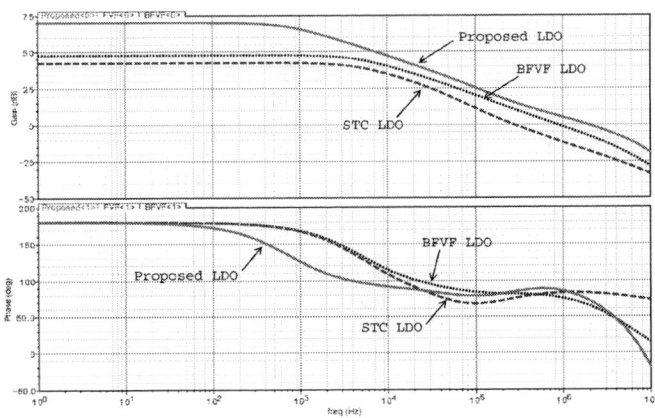

Fig. 7. Loop-gain frequency responses at I_L=50mA.

an ESR of 250mΩ. The Miller capacitor used in the proposed LDO regulator is only 1pF which can be integrated easily.

Figure 7 shows the loop-gain frequency response of the regulators at the identical load current of 50mA. The proposed CPT LDO regulator displays a higher loop gain and larger bandwidth than that of the STC and BFVF LDO regulators with similar phase margin. The UGF for STC, BFVF and CPT LDO regulator is 0.28MHz, 0.865MHz and 1.9MHz respectively. The proposed CPT LDO regulator displays a better efficiency in term of power per bandwidth. The loop-gain and bandwidth are the important parameters for fast transient and good regulation performance metrics. Furthermore, the stability of the proposed CPT LDO regulator is not compromised.

Figure 8 depicts the load transient responses of the regulators when the load current changes from 0 to 50mA and vice versa. It shows that the proposed CPT LDO regulator has the smallest overshoot and undershoots among the LDO regulators. The overshoot and undershoot is 1.8mV and 9.05mV respectively. The simulation result also indicates that the proposed CPT LDO regulator has a better load regulation

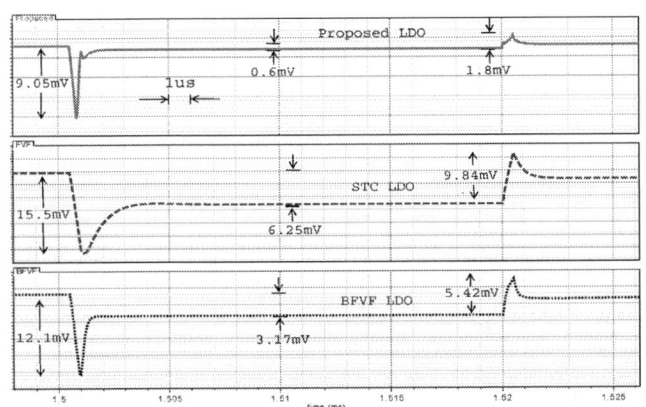

Fig. 8. Transient responses from 0 to 50mA and vice versa.

TABLE I
COMPARISON OF 3 FVF BASED LDO REGULATORS

	STC LDO	BFVF LDO	Proposed CPT LDO
V_{DD} (V)	1.2	1.8	1.2
V_{OUT} (V)	1	1.6	1
C_m (pF)	-	-	1
ΔV_{OUT} (mV)	15.5	12.1	9.05
Load Regulation (μV/mA)	125	63.4	12
Power Consumption (μW)	20.4	30.6	20.4
Bandwidth per Power (MHz/μW)	0.016	0.051	0.11

between 0 and 50mA load whilst without increasing the supply voltage to cater for adequate headroom. Table I summaries the performance of the LDO regulators. The technical merits of CPT LDO regulator are demonstrated.

V. CONCLUSION

A FVF based LDO regulator with CPT has been presented. It reduces the minimum biasing current requirement of STC LDO and allows a lower voltage operation than the BFVF LDO regulator. Under same design condition, the proposed CPT LDO regulator displays a higher loop gain, larger bandwidth and reduced overshoot/undershoot. This also greatly improves the load transient as well as regulation. Therefore, the proposed CPT LDO regulator offers an alternative design solution to the LDO regulators based on FVF structures.

REFERENCES

[1] G. A. Rincon-Mora and P. E. Allen, "A low-voltage, low quiescent current, low drop-out regulator," *IEEE J. of Solid-State Circuits*, vol. 33, no. 1, pp. 36-44, Jan. 1998.

[2] M. Al-Shyoukh, H. Lee and R. Perez, "A transient-enhanced low quiescent current low-dropout regulator with buffer impedance attenuation", *IEEE J. Solid-State Circuits*, vol. 42, no. 8, pp. 1732-1742, Aug. 2007.

[3] K. N. Leung and Y. S. Ng, "A CMOS low-dropout regulator with a momentarily current-boosting voltage buffer," *IEEE Trans. on Circuits and Systems I: Regular Paper*, vol. 57, no. 11, Nov. 2010.

[4] R. G. Carvajal, J. ramirez-Angulo, A. J. Lopez-Martin, A. Torralba, J. A. G. Galan, A. Carlosena and F. M. Chavero, "The flipped voltage followe: a useful cell for low–voltage low-power circuit design," *IEEE Trans. On Circuits and Systems I: Regular Paper*, vol. 52, pp/ 1276-1291, 2005.

[5] T. Y. Man, K. N. Leung, C. Y. Leung, P. K. T. Mok and M. Chan, "Development of single-transistor-control LDO based on flipped voltage follower for SoC," *IEEE Trans. on Circuits and Systems I: Regulator Paper*, vol. 55, pp. 1392-1401, 2008.

[6] H. Chen and K. N. Leung, "A fast-transient LDO based on buffered flipped voltage follower," *IEEE International Conference on Electron Devices and Solid-State Circuits*, pp. 1-4, Dec. 2010.

[7] Y. Tian and P. K. Chan, "Design of high-performance analog circuit using wideband g_m-enchanced MOS composite transistor," *IEICE Trans. on Electronics*, vol. E93-C, no. 7, pp. 1199-1208, Jul. 2010.

[8] S. S. Chong, H. KWANTONO and P. K. Chan, "A 4.7μA quiescent current, 450mA CMOS low-dropout regulator with fast transient response," *IEICE Trans. on Electronics*, vol. E94-C, no. 8, pp. 1271-1281, Aug. 2011.

978-1-61284-863-1/11 $26.00 © 2011 IEEE

20 MHz Accurate Peak Detector for FPW Allergy Biosensor With Digital Calibration

Tzung-Je Lee*, *Member, IEEE*
Department of Computer Science and
Information Engineering
Cheng Shiu University
Kaohsiung, Taiwan 83347
Email: tjlee168@gmail.com

Wei-Chih Hsiao, and Chua-Chin Wang, *Senior Member, IEEE*
Department of Electrical Engineering
National Sun Yat-Sen University
Kaohsiung, Taiwan 80424
Email: ccwang@ee.nsysu.edu.tw

Abstract—This paper proposes a 20 MHz peak detector utilizing digital calibration for a FPW allergy biosensor. The traditional peak detector operates at the frequency from several KHz to 1 MHz. By using the proposed digital calibration methodology, the operation frequency of the peak detector is simulated to be up to 20 MHz and the detected voltage error due to the delay time of the sample-and-hold circuit and the comparator is improved to be -0.8125%. Thus, the proposed design can be applied to the FPW biosensor, which requires the frequency up to 20 MHz. The proposed peak detector is implemented using TSMC 0.18 μm CMOS process. The post-layout simulation results show that the proposed peak detector possesses the power consumption of 12.69 mW given a input frequency of 20 MHz. Besides, the peaking time of the input signal is only 25 ns. The area of the proposed design is 1.05363 mm^2.

Keywords— FPW, biosensor, peak detection, peaking time

I. INTRODUCTION

Peak detector plays a critical role in lots of high accurate measurement systems, e.g., nuclear radiation spectrometer [1], electrostatic field sensor [2], X-ray image detection [3], and medical instrumentation [4], [5]. In the measurement systems, the miniaturization of detected voltage error is the most important factor for the peak detector.

Recently, FPW (Flexural Plate-Wave) allergy biosensor is developed to show the allergic level for helping the diagnosis of allergic diseases [6], [7]. The FPW allergy biosensor outputs a sinusoidal signal with the frequency of 20 MHz [6]. The amplitude of the 20 MHz sinusoidal signal varies according to the concentration of the IgE (Immunoglobulin E). By detecting the amplitude of the 20 MHz sinusoidal signal, the allergy level can be determined. In order to detect the amplitude of the 20 MHz sinusoidal signal output by the FPW allergy biosensor, it needs high operation frequency and short peaking time for the peak detector.

There were several topologies to implement the peak detector in the state-of-the-arts. Firstly, a comparator and a sample-and-hold circuit are used to detect the peak voltage [8]. Secondly, an OTA and a diode (or a source follower)

---

*Prof. Lee is with the Department of Computer Science and Information Engineering, Cheng Shiu University, Kaohsiung, Taiwan 83347. (e-mail: tjlee168@gmail.com). He is the contact author.

are utilized to detect the peak voltage of the input signal. The detected peak voltage is then stored in a capacitor [9]. A voltage error is induced due to the delay of the comparator and the OTA in the two topologies. To improve the accuracy of the peak detector, Dlugosz used current-mode latching mechanism and a flag generating circuit to detect and amplify the input signal [3]. Moreover, Lin presented an approach using dual ramp sampling to reduce the effect of the comparator delay [1]. However, in these modified topologies, the peak detector can only detect the pulse signal with pulse width of 1 us. Besides, it needs a long peaking time to generate the correct output signal in these prior works. The mentioned peak detectors are difficult to be applied to the FPW allergy biosensor, which requires high operation frequency and short peaking time.

Therefore, this paper proposes a novel peak detector for the FPW allergy biosensor. By using the proposed digital calibration methodology, the peak detector possesses a voltage error of -0.8125% with operation frequency of 20 MHz. Besides, the peaking time of the input signal is only 25 ns.

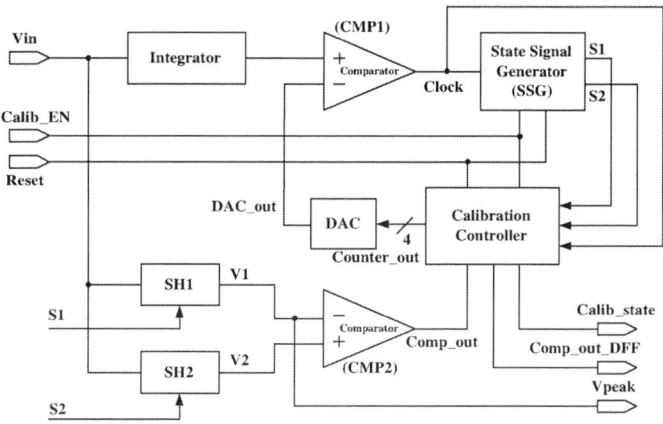

Fig. 1. Block diagram of the proposed peak detector using didital calibration.

II. DIGITAL CALIBRATION PEAK DETECTOR

Fig. 1 shows the block diagram of the proposed peak detector. The proposed peak proposed detector is composed of an

Integrator, a State Signal Generator (SSG), a Digital-to-Analog Converter (DAC), a Calibration Controller, two sample-and-hold circuits (SH1 and SH2), and two Comparators (CMP1 and CMP2).

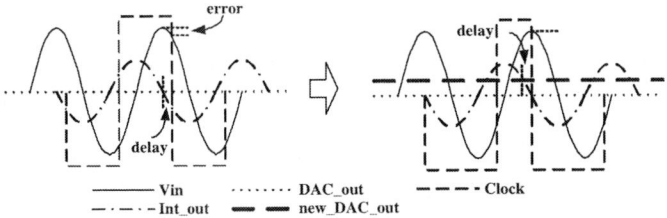

Fig. 2. The waveforms of the signals, Vin, Int_out, DAC_out, and new_DAC_out, for revealing the digital calibration of the proposed peak detector.

The input signal Vin is received by the Integrator. It generates a signal, Int_out, which leads Vin by $\pi/2$, as shown in Fig. 2. If the Integrator and the Comparator (CMP1) are ideal, the middle point of Int_out and the peak of Vin occur at the same time. By comparing Int_out and DAC_out which is biased at the voltage of the mid-point of Int_out, a negative edge of Clock at the peak of Vin could be generated If the negative edge of Clock is used to activate a sample-and-hold circuit, the peak of Vin could be sampled and held [2]. However, the Integrator and CMP1 are not ideal, the delay of the signal propagation path results in the offset at the negative edge of Clock. Thus, a voltage error at the peak voltage is caused.

In order the calibrate the voltage error due to the propagation delay, DAC_out is modified to be new_DAC_out. Because new_DAC_out > DAC_out, the negative edge of Clock can be shifted to the peak of Vin. Thus, the peak of Vin can be sampled correctly and the voltage error due to the propagation delay is calibrated. The detailed calibration process is described in subsection D.

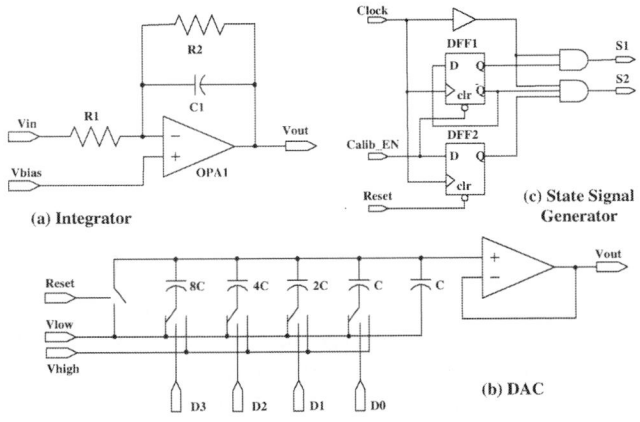

Fig. 3. Schematics of (a) the Integrator, (b) the DAC, and (c) the State Signal Generator (SSG) in the proposed design.

A. The Integrator

Fig. 3 (a) shows the schematic of the Integrator, which is composed of a OPA, a capacitor, C1, and two resistors, R1 and R2. By choosing the parameters R1 = 5 kΩ, R2 = 50 kΩ, and C1 = 5 pF, the DC gain of the Integrator is 20 dB (= 20 $\log(\frac{R2}{R1})$). Moreover, the output voltage can be expressed by Eqn. (1).

$$Vout = -\frac{1}{R1 \times C1} \int Vindt, \qquad (1)$$

where Vbias is biased at VDD/2 and is serving as a DC ground. Besides, R2 is 10 times larger than R1 so that it is ignored for simplicity.

B. The Digital-to-Analog Converter (DAC)

Referring to Fig. 3 (b), a charge-scaling DAC is employed. To achieve the required calibration mechanism, it requires only 4 bits resolution. Based on the post layout simulation with HSPICE, it shows that the detected voltage error is reduced to one fifth of that without any calibration.

C. The State Signal Generator (SSG)

Fig. 3 (c) reveals the schematic of the State Signal Generator (SSG). By using two D-type flip-flops, two AND gates, and one buffer, two state signals, S1 and S2, are generated, as shown in Fig. 4. Notably, the buffer is included to remove the unwanted glitch occurred at S1 and S2.

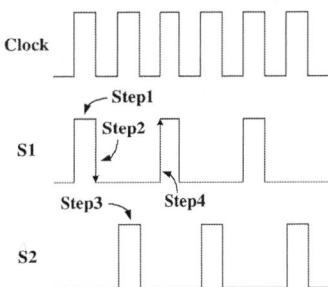

Fig. 4. Waveforms of the signals, Clock, S1, and S2, for revealing the steps of the proposed digital calibration.

D. The Calibration Controller

The schematic of the Calibration Controller is shown in Fig. 5. The Calibration Controller is composed of a Counter and the other logic circuits. The Calibration Controller receives the signals, S1, S2, Clock, and Comp_out, and output a 4-bit control signal Counter_out to control the DAC. When Calib_EN signal is received, the calibration process is activated and Calib_state is pulled high. The calibration process is divided into four steps, as shown in Fig. 4.

Step 1: The first high state of S1 signal is denoted as the step 1. At the step 1, Vin is sampled by SH1 and the corresponding first sampled peak voltage V1 is generated.

Step 2: Step 2 is at the negative edge of the state signal, S1. At the Step 2, the Calibration Controller activate the Counter

978-1-61284-863-1/11 $26.00 © 2011 IEEE   477

to accumulate its counted number. Thus, the output of the DAC, DAC_out, is pulled high by one LSB.

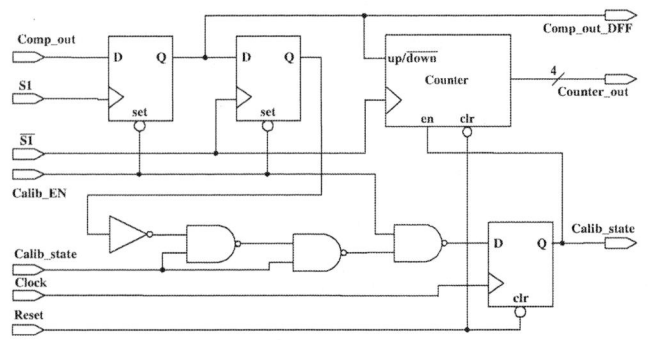

Fig. 5. Block diagram of the Calibration Controller in the proposed peak detector.

Step 3: Step 3 is defined by the following high state of the state signal, S2. At the Step 3, Vin is sampled by SH2 and the second sampled peak voltage, V2, is generated.

Step 4: Step 4 is denoted by the following positive edge of the state signal S1. At the Step 4, the voltages of V1 and V2 are compared by CMP2. The comparison results, Comp_out is read by the Calibration Controller. If V2 > V1, it means that increasing the voltage of DAC_out is predicted to reduce the detected voltage error of the peak voltage, Vpeak, due to the propagation delay. The four calibration steps, Step 1 to Step 4, are repeated to pull up the voltage level of DAC_out. Until V2 < V1, it indicates that increasing the voltage of DAC_out is failed to calibrate the detected voltage error. Thus, the counted number of the Counter is decreased by one at the next negative edge of S1. The best choice of the voltage level of DAC_out is obtained. Therefore, the calibration process is terminated and Calib_state is pulled low.

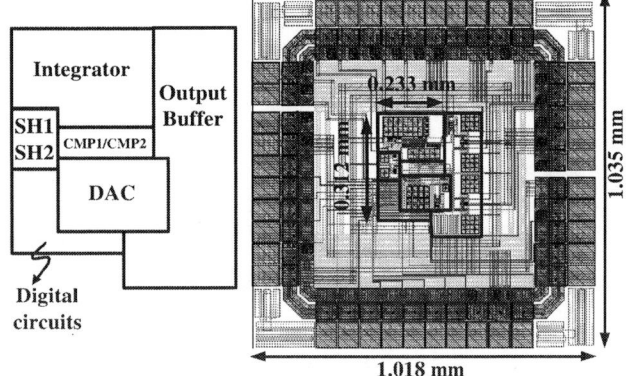

Fig. 6. The layout diagram of the proposed peak detector.

III. IMPLEMENTATION AND SIMULATION

The proposed peak detector is carried out using TSMC 0.18 μm CMOS process. The layout diagram is shown in Fig. 6. The chip area is 1.018×1.035 mm^2 and the core area

is 0.233×0.312 mm^2. Fig. 7 shows the simulated waveforms using HSPICE for the the input signal with the peak voltage of 1.2 V and the frequency of 20 MHz. Notably, the signals shown in Fig. 7 refer to the signals on the PADs. Because the digital PAD provided by the process vendor is in the configuration of an inverter, the signals are at the inversed state of the corresponding signals in the core. During the calibration process, DAC_out is accumulated gradually to calibrate the voltage error due to the propagation delay. Simultaneously, the detected peak voltage, Vpeak, is getting close to the peak of Vin. Fig. 8 shows the comparisons of the detected peak voltage with calibration and that without calibration. By using the digital calibration, the detected voltage error is improved from -6.85% to -0.8125% for Vin biased at 600 mV and 650 mV, respectively.

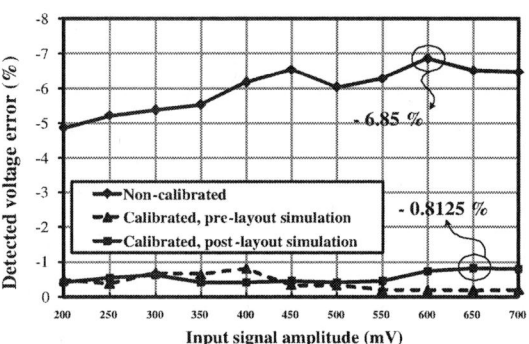

Fig. 8. The comparison of the detected voltage errors with digital calibration to that without digital calibration for the amplitude of Vin from 200 mV to 700 mV.

Table I reveals the specifications of the proposed peak detector compared to the prior works. The input frequency is 20 MHz and the input peak range is from 1.1 to 1.6 V. The calibrated voltage error is -0.8125% and the peaking time of the input signal is only 25 ns. A Figure of Merit (FOM) is provided to reveal the performance compared to the prior works. The proposed design possess the best performance by considering the detected accuracy, the peaking time and the input frequency, simultaneously.

IV. CONCLUSION

The paper proposes a novel peak detector using digital calibration. By using the digital calibration mechanism, the detected voltage error is improved from -6.85% to -0.8125%. Besides, the frequency of the input signal is up to 20 MHz, which meets the requirement of the FPW biosensor. Moreover, the peaking time is reduced only 25 ns. According to the post-layout simulation, it is evident that the proposed design possesses the outstanding performance on the operational frequency, the peaking time and the detected voltage error.

ACKNOWLEDGMENT

This research was partially supported National Science Council under grant NSC99-2923-E-110-002-MY2, NSC99-

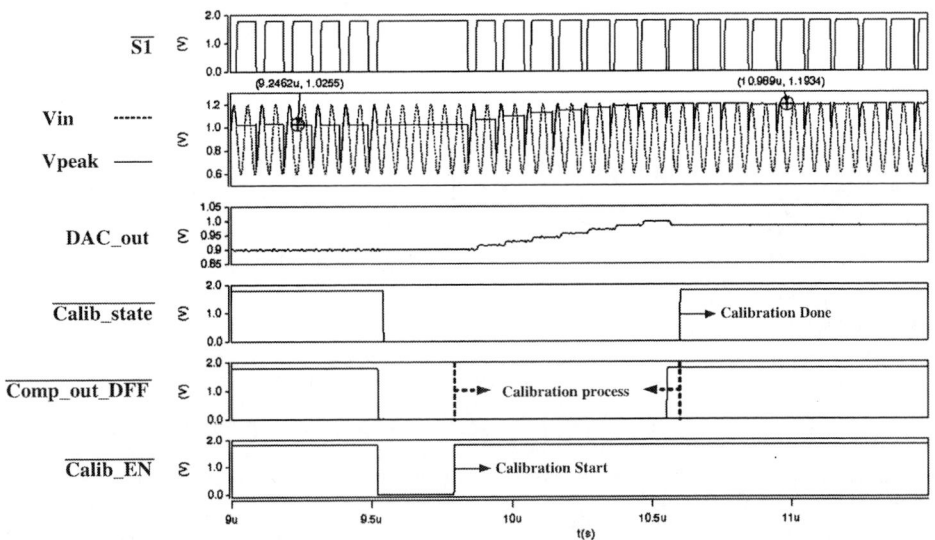

Fig. 7. Simulated waveforms for the amplitude and the frequency of Vin at 1.2 V and 20 MHz, respectively.

	This work	[1]	[2]	[5]	[3]	[4]
Process (μm)	0.18	0.18	0.35	0.35	N/A	0.5
Power supply (V)	1.8	1.8	3.3	1.2	1.0	4.0
Input frequency (MHz)	20	0.5†	0.004	0.005	0.5†	0.001
Input peak range (V)	1.1~1.6	0.6~1.6	N/A	50~100 (nA)	0.2~0.9	N/A
Voltage error (mV)	13	1	0.1	N/A	N/A	N/A
Voltage error (%)	-0.8125	0.1	N/A	N/A	0.7	2.6
Peaking time‡ (ns)	25	1000	250000	190	1000	3200000
Power dissipation (mW)	12.69	0.293	N/A	0.001	0.02¶	18
Chip size (mm^2)	1.05363	N/A	0.54	N/A	N/A	1.9856
Year	2011	2009	2007	2010	2007	2008
FOM§	0.98	0.005	N/A	N/A	0.0007	1.20×10^{-10}

Note: †The input frequency is obtained by $\dfrac{1}{2\times\text{Peaking time}}$.

‡ Peaking time refers to the time from the valley to the peak of the input signal.

¶The power dissipation is obtained by taking the multiplications of the energy and the input frequency.

§ FOM $= \dfrac{\text{Input frequency}}{(\text{Peaking time})(|\text{Voltage error}(\%)|)}$

TABLE I
COMPARISON WITH SEVERAL PRIOR WORKS

2221-E-110-081-MY3, NSC99-2220-E-110-001, NSC100-2218-E-230-001, and NSC100-2221-E-230-026. Besides, the authors would like to express their deepest gratefulness to CIC (Chip Implementation Center) of NARL (National Applied Research Laboratories), Taiwan, for their thoughtful chip fabrication service.

REFERENCES

[1] J. R. Lin, and H. P. Chou, "A peak detect and hold circuit using ramp sampling approach," *2009 IEEE Nuclear Science Symp. Conf. Record (NSS/MIC)*, pp. 309-312, 2009.

[2] G. Cui, H. Yang, and S. Xia, "CMOS digitalized peak detector for a MEMS-based electrostatic field sensor," *7th IEEE Conference on Nanotechnology, 2007 (IEEE-NANO 2007)*, pp. 131 - 134, 2007.

[3] R. Dlugosz, and K. Iniewski, "High-precision analogue peak detector for X-ray imaging applications," *Electronics Letters*, vol. 43 , no. 8, pp. 440-441, Apr. 2007.

[4] M. Chen, O. Boric-Lubecke, and V. M. Lubecke, "0.5-um CMOS implementation of analog heart-rate extraction with a robust peak detector,"

IEEE Transactions on Instrumentation and Measurement, vol. 57 , no. 4, pp. 690-698, 2008.

[5] C. Sawigun, W. Ngamkham, and W. A. Serdijn, "An ultra low-power peak-instant detector for a peak picking cochlear implant processor," *2010 IEEE Biomedical Circuits and Systems Conference (BioCAS2010)*, pp. 222-225, 2010.

[6] I.-Y. Huang, M.-C. Lee, Y.-W. Chang, and R.-S. Huang, "Development and characterization of FPW based allergy biosensor," *IEEE International Symposium on Industrial Electronics, 2007 (ISIE 2007)*, pp. 2736-2740, 2007.

[7] C.-C. Hsu, Y.-R. Lin, Y.-D. Tsai, Y.-C. Chen, and C.-C. Wang, "A frequency-shift readout system for FPW allergy biosensor," *IEEE International Conference on IC Design and Technology, 2011 (ICICDT11)*, CD-ROM version, N2, May 2011.

[8] G. De Geronimo, P. O'Connor, and A. Kandasamy, "Analog CMOS Peak Detect and Hold Circuits. Part 1. Analysis of the Classical Configuration, " *Nucl. Instrum. Methods Phys. Res. A*, vol. A484, pp.533543, May 2002.

[9] S.-B. Park, J. E. Wilson, and M. Ismail, "The CHIP - peak detectors for multistandard wireless receivers," it IEEE Circuits and Devices Magazine, vol. 22, no. 6, pp/ 6-9, Nov. 2006.

2.45 GHz ZigBee Receiver Frontend for HAN With Smart Meter

Tzung-Je Lee*, *Member, IEEE*
Department of Computer Science and
Information Engineering
Cheng Shiu University
Kaohsiung, Taiwan 83347
Email: tjlee168@gmail.com

Wayne Luo[†], Shang-Hsien Yang[†], Ming-Hung Shih[‡],
Ko-Chi Kuo[‡], and Chua-Chin Wang[†], *Senior Member, IEEE*
Department of Electrical Engineering
National Sun Yat-Sen University
Kaohsiung, Taiwan 80424
Email: ccwang@ee.nsysu.edu.tw

Abstract—**This paper proposes a 2.45 GHz ZigBee receiver frontend for the communication in a HAN (Home Area Network) with the smart meter. By employing a cascode LNA, the gain is simulated to be 17.376 dB at 2.45 GHz. Besides, by using the double-balanced Gilbert mixer with a current bleeding MOS transistor, the NF and the IIP3 of the mixer are only 5.074 dB and -7.234 dB, respectively. In order to reduce the phase noise of the receiver, a fractional-N frequency synthesizer with a complementary cross-coupled VCO and a multi-modulus divider are utilized. Moreover, a delta-sigma modulator is included for the noise shaping. The phase noise of the fractional-N frequency synthesizer is 137.7 dBc/Hz. The proposed circuit is carried out using the standard 0.18 μm CMOS process. The core area is 3.57 mm^2.**

Keywords— ZigBee, HAN, smart meter, frontend

I. INTRODUCTION

In order to reduce the carbon footprint, an organic building becomes a popular research field recently. The smart meter is one of the important device in the organic building. The smart meter can monitor the power consumption of each electrical device and collect the information of the consuming electricity to the remote database. With a proper management methodology, the smart meter can even shut down the unnecessary electrical device to avoid the waste of energy. For the purpose of the communication in a organic building, the smart meter must operate in a home area network (HAN). With the features of low-cost and low-power, the ZigBee transceiver is suitable for the wireless communication in the HAN, as shown in Fig. 1 [1].

Besides, 2.45 GHz ZigBee possesses the high transmission bit rate of 250 Kbps owing to the O-QPSK (offset-quadrature phase shift keying) modulation [1]. It is sufficient for conveying the security information and is suitable for the HAN. In order to maintain the transmission quality in the RF band, the

---

*Prof. T.-J. Lee is with the Department of Computer Science and Information Engineering, Cheng Shiu University, Kaohsiung, Taiwan 83347. (e-mail: tjlee168@gmail.com). He is the contact author.

[†]W. Luo, S.-H. Yang, and Prof. C.-C. Wang are with the Department of Electrical Engineering, National Sun Yat-Sen University, Kaohsiung, Taiwan 80424.

[‡]M.-H. Shih, and Prof. K.-C. Kuo are with the Department of Computer Science and Engineering, National Sun Yat-Sen University, Kaohsiung, Taiwan 80424.

paper proposes a 2.45 GHz ZigBee receiver frontend which employs a double-balanced Gilbert mixer and a fractional-N frequency synthesizer (FNFS). Based on the worst-case simulation results, the phase noise is -137.7 dBc/Hz.

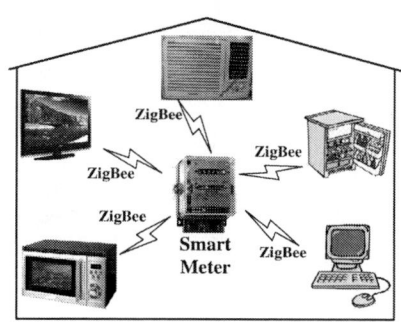

Fig. 1. The applications of ZigBee utilized in the HAN with the smart meter.

II. 2.45 GHZ ZIGBEE RECEIVER FRONTEND

The proposed ZigBee receiver frontend is composed of a RF circuit and a baseband circuit, as shown in Fig. 2. The RF circuit receives the differential RF signals from the antenna by the low-noise amplifier (LNA). The amplified differential RF signals are then downconverted to the baseband I/Q signals by the Mixer, which operates with a polyphase filter (PP Filter) and a fractional-N frequency synthesizer (FNFS). The differential baseband I/Q signals are firstly converted to the single-ended I/Q signals by the differential-to-single amplifiers (DtoS AMP). The single-ended signals are then filtered and amplified by the the multiple-feedback filter (MFB Filter) and the programmable gain amplifier (PGA).

A. The Low-Noise Amplifier (LNA)

Fig. 3 (a) shows the schematic of the LNA. The LNA is based on a single-ended common-source amplifier. The MOS transistor M2 serves as a cascode MOS, which can increase the gain of the LNA, avoid the Miller effect, and resist the reflected signal from the output. The input capacitor, C1, is a coupling capacitor to protect the antenna from the DC LNA bias. Besides, the source-inductor, Ls, and the gate-inductor,

---

978-1-61284-863-1/11 $26.00 © 2011 IEEE

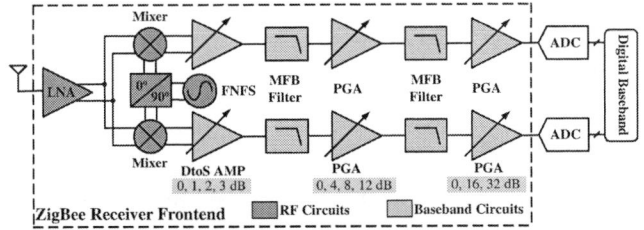

Fig. 2. Block diagram of the proposed ZigBee receiver.

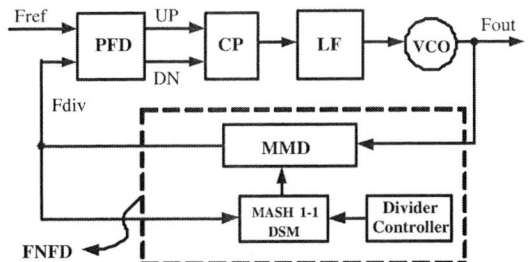

Fig. 4. Block diagram of the fractional-N frequency synthesizer.

Lg, the drain-inductor, Ld, and the output capacitor C2 are utilized for the impedance matching.

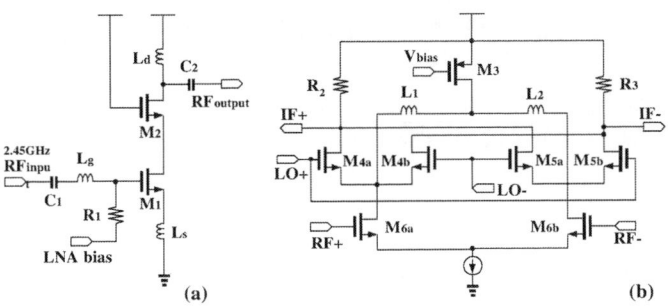

Fig. 3. Schematic of (a) the LNA and (b) the Mixer.

B. The Mixer

The Mixer is in the double-balanced Gilbert configuration, as shown in Fig. 3 (b). M6a and M6b provide the transconductance for the Mixer. By using the current bleeding MOS transistor, M3, and the resonate inductors, L1, and L2, the noise resulted from the bias current can be reduced such that a better noise figure (NF) can be resulted.

C. The Fractional-N Frequency Synthesizer (FNFS)

The LO is based on a fractional-N frequency synthesizer (FNFS). Referring to Fig. 4, the FNFS is composed of a phase-frequency detector (PFD), a charge pump (CP), a loop filter (LF), a voltage-controlled oscillator (VCO), and a fractional-N frequency divider (FNFD). The FNFD performs the integer division for the frequency of Fout. Until the loop is in the lock state, the MASH 1-1 DSM starts the non-integer frequency division for Fout. With the two step fractional-N frequency division, the phase noise and the locking time can be reduced. Besides, the unwanted unlock state is avoided.

1) The VCO: In order to reduce the phase noise, the VCO is in the complementary cross-coupled configuration, as shown in Fig. 5. By using the complementary configuration of M3, M4, M5, and M6, the VCO possesses a higher transconductance than that with only PMOS or NMOS configuration. Besides, the switching speed of the cross-coupled configuration is faster. Moreover, the rise time and the fall time of the VCO are symmetric, such that the phase noise can be reduced. Furthermore, the complementary cross-coupled MOS transistors, M3, M4, M5, and M6, provide the negative resistances, which

compensate the equivalent resistance of the LC tank. Thus, the power consumption can be reduced.

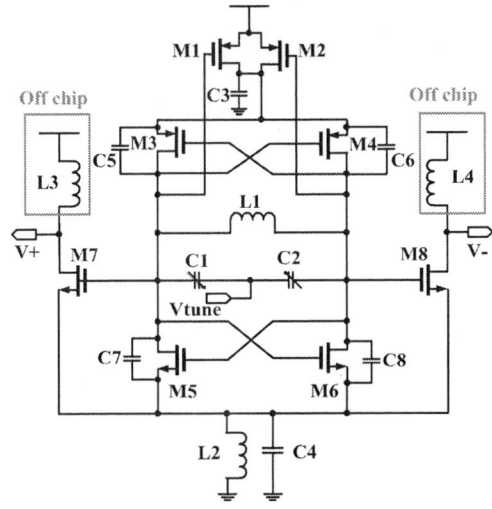

Fig. 5. Schematic of the VCO.

2) The Multi-Modulus Divider (MMD): Fig. 6 (a) shows the schematic of the multi-modulus divider (MMD), which is composed by foure NOR gates, five NAND gates and five 2/3 Dividers. The MMD provides a division ratio from 32 to 63. The division ratio is controlled by the control signals, B0 $\sim$ B4, which are generated by the MASH 1-1 delta-sigma modulator (DSM). Referring to Fig. 6 (b), the MASH 1-1 DSM is composed of five registers (REG), two 8bits CLA adders, and one Noise Cancellator. The MASH 1-1 DSM receives a control signal from the Divider Controller, and a reference clock signal, CK. The MASH 1-1 DSM provides the noise shaping for the MMD, thus, the phase noise can be reduced.

D. The Differential-to-Single-Ended Amplifier (DtoS AMP)

In order to reduce the physical cost, the baseband signal is converted to be single-ended by the differential-to-single-ended amplifier (DtoS AMP). Referring to Fig. 7, the DtoS AMP is composed of three operation amplifiers (OPA), 12 resistors, and 8 switches. Notably, because OP_{pos} and OP_{neg} are in the instrumentation amplifier configuration, the DtoS

978-1-61284-863-1/11 $26.00 © 2011 IEEE   481

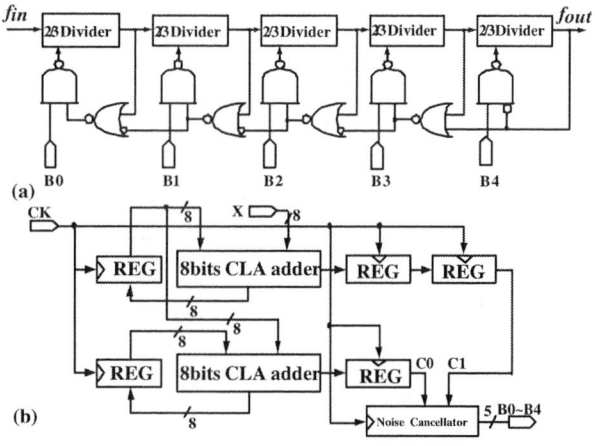

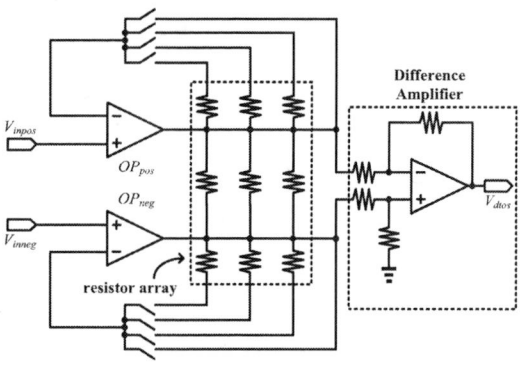

Fig. 6. Block diagram of (a) the multi-modulus divider (MMD) and (b) the MASH 1-1 DSM.

AMP possesses high CMRR. Thus, the DC offset resulted from the Mixer can be canceled. Besides, the DtoS AMP provides four selection for the gain of 0 dB, 1 dB, 2 dB, and 3 dB, by the 8 switches and the resistor array. Furthermore, the equivalent resistors of the turned-on switches can be ignored by connecting the switches to the negative input nodes of the OP_{pos} and OP_{neg} , which possess the infinite input resistance.

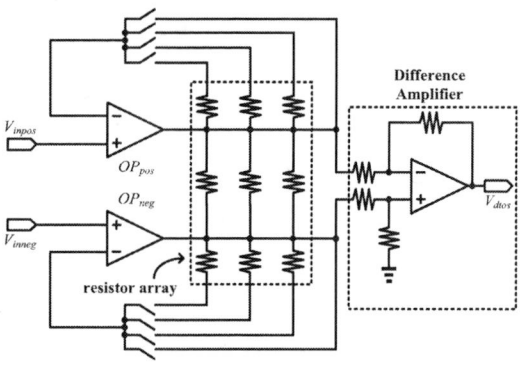

Fig. 7. Schematic of the differential-to-single-ended amplifier (DtoS AMP).

E. The Multiple Feedback (MFB) Filter

Fig. 8 (a) shows the schematic of the Multiple Feedback (MFB) Filter. Notably, the resistances should be determined by considering the loading effect. Besides, the ground symbol in Fig. 8 (a) refers to the AC ground, which are biased at the DC voltage of $1/2 \times$ VDD.

F. The Programmable-Gain Amplifier (PGA)

The programmable-gain amplifier is composed of an operational amplifier and the feedback resistor array, as shown in Fig. 8 (b). The feedback resistor array is composed of 6 resistors and four switches, which can provide 4 selection for the voltage gain. Referring to Fig. 2, the first PGA provide 4 selection for the gain of 0 dB, 4 dB, 8 dB, and 12 dB. The

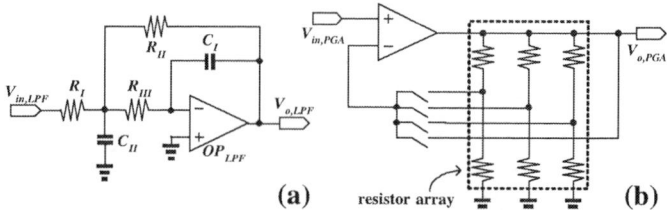

Fig. 8. Schematic of (a) the MFB filter and (b) the PGA.

second PGA provides the selectable gain of 0 dB, 16 dB, 32 dB. Thus, the gain of the baseband circuit is selectable from 0 dB to 47 dB with the gain step of 1 dB.

G. The Rail-to-Rail Operation Amplifier (OPA)

Fig. 9 show the schematic of the rail-to-rail operational amplifier, which is employed in the DtoS AMP, the MFB Filter, and the PGA. The rail-to-rail OPA is composed of the N-type and P-type input stages, the gain stage, and the output stage. By using the N-type and P-type input and the output stage, the rail-to-rail OPA can provide a full swing for the input and output signals. Besides, the cascode structure in the gain stage provides a large resistance at the output nodes of the gain stage. Thus, two MOS transistors, M_{M1} and M_{M2}, and 2 Miller capacitors C_{M1} and C_{M2} are employed for canceling the RHP zero. Therefore, the rail-to-rail OPA possesses the phase margin of 60°, the open loop gain of 100 dB, and the unity gain frequency of 30 MHz.

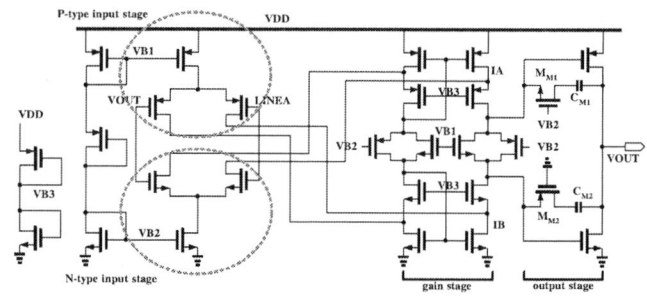

Fig. 9. Schematic of the rail-to-rail OTA.

III. IMPLEMENTATION AND SIMULATION

The proposed 2.45 GHz ZigBee receiver is carried out using TSMC 0.18 μm CMOS process. Fig. 10 shows the layout of the proposed design. The core area is 1.325 $\times$ 2.697 mm^2. Fig. 11 shows the gain of the LNA to be 17.376 dB with the worst-case of SS, 70°C. Fig. 12 shows that the gain of the baseband circuit is programmable from 0 dB to 47 dB with the gain step of 1 dB. Besides, the cutoff frequency is at 2 MHz determined by the MFB filter. Moreover, phase noise of the fractional-N frequency synthesizer is simulated to be -137.7 dBc/Hz. The specifications of the proposed design are shown in Table I. Moreover, the performances compared to the prior works are shown in Table II. A FOM (figure of merit) is given by considering the gain, IIP3, power consumption,

978-1-61284-863-1/11 $26.00 © 2011 IEEE

core area, and the phase noise. According to the FOM, the proposed design possesses the best performance.

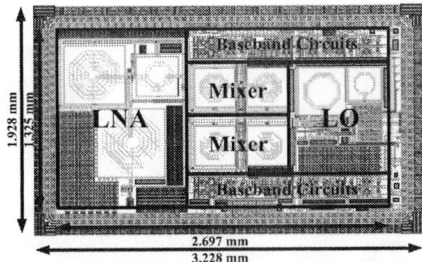

Fig. 10. Layout of the proposed ZigBee receiver.

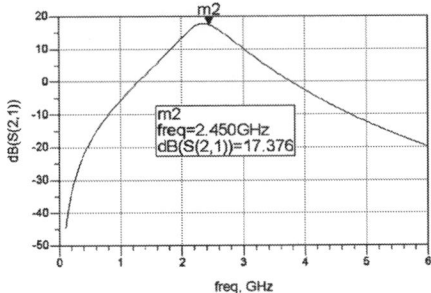

Fig. 11. Simulated waveform of the gain of the LNA at 2.45 GHz for the worst-case of SS, 70°C.

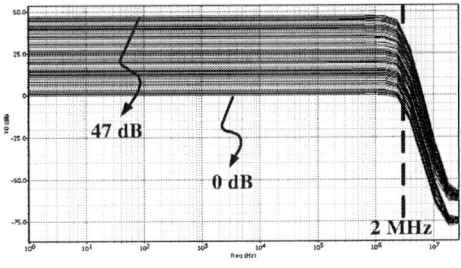

Fig. 12. Simulated waveform of the gain of the baseband circuit from 0 dB to 47 dB.

IV. CONCLUSION

A 2.45 GHz ZigBee receiver frontend is proposed for the HAN in this paper. The proposed ZigBee receiver employs the cascode LNA, the double-balanced Gilbert mixer, and the fractional-N frequency synthesizer to reduce the phase noise. The phase noise is -137.7 dBc/Hz at the frequency of 3 MHz.

ACKNOWLEDGMENT

This research was partially supported National Science Council under grant NSC99-2220-E-110-001, NSC99-2923-E-110-002-MY2, NSC99-2221-E-110-081-MY3, NSC100-2218-E-230-001, and NSC100-2221-E-230-026. Besides, the authors would like to express their deepest gratefulness to CIC (Chip Implementation Center) of NARL (National Applied

LNA	S$_{11}$ @2.45 GHz	-43.688 dB
	S$_{22}$ @2.45 GHz	-11.967 dB
	Gain @2.45 GHz	17.376 dB
	Noise Figure	2.712 dB
Mixer	Noise Figure	5.074 dB
	Gain	7.876 dB
	Out-of-channel IIP3	-7.234 dB
FNFS	Phase Noise @1 MHz	-127 dBc/Hz
	Phase Noise @3 MHz	-137.7 dBc/Hz
	Phase Noise @10 MHz	-148.2 dBc/Hz
	Lock time @ 2.45 GHz	< 20 μs
	VCO Gain	130 MHz/V
	Bandwidth	200 KHz
Baseband Circuits	Max. Gain	47 dB
	Min. Gain	0 dB
	Filter Bandwidth	2 MHz
Power	Whole chip	88 mW
	LNA	53 mW
	Mixer	5 mW
	FNFS	21 mW
	Baseband circuits	9 mW
Core area		3.57 mm^2

TABLE I
SPECIFICATIONS OF THE PROPOSED ZIGBEE RECEIVER.

	This work	[2]	[3]	[4]
Gain	> 73 dB	N/A	> 75 dB	> 67 dB
IF Bandwidth	2 MHz	2 MHz	2 MHz	2 MHz
Phase Noise (dBc/Hz)	-137.7 @3 MHz	N/A	-107.8 @3.5 MHz	-127 @3 MHz
IIP3	-7.234 dBm	-16 dBm	-12.5 dBm	-10.5 dBm
Core Area	3.57 mm^2	1.36 mm^2	0.35 mm^2	1.7 mm^2
VDD (V)	1.8	1.8 / -3.75	1.2	0.6
Power (mW)	88	30.8¶	4.8¶	22¶
Process	0.18μm	0.18μm	90 nm	90 nm
Publication	ISIC	ISSCC	ISSCC	JSSC
Year	2011	2006	2008	2010
FOM†	16.1	N/A	1.13	2.67

Note: ¶ The power includes the power consumption of the receiver and the analog circuits.

$$† \text{ FOM} = \frac{\text{Gain(in amplitude ratio)} \times \text{IIP3}}{\text{Power} \times \text{Core Area} \times \text{Phase Noise} \times 10^{13}}.$$

TABLE II
COMPARISON WITH SEVERAL PRIOR WORKS

Research Laboratories), Taiwan, for their thoughtful chip fabrication service.

REFERENCES

[1] C.-C. Wang, J.-M. Huang, L.-H. Lee, S.-H. Wang, and C.-P. Li, "A low-power 2.45 GHz ZigBee transceiver for wearable personal medical devices in WPAN," *2007 IEEE Inter. Conf. on Consumer Electronics (ICCE 2007)*, pp. 1-2, Jan. 2007.

[2] W. Kluge, F. Poegel, H. Roller, M. Lange, T. Ferchland, L. Dathe, D. Eggert, "A fully integrated 2.4GHz IEEE 802.15.4 compliant transceiver for ZigBee applications," in *Proc. IEEE Intl. Solid-State Circuit Conf., 2006*, pp. 1470-1479, Feb. 2006.

[3] A. Liscidini, M. Tedeschi, R. Castello, "A 2.4 GHz 3.6mW 0.35 mm^2 quadrature front-end RX for ZigBee and WPAN applications," in *Proc. IEEE Intl. Solid-State Circuit Conf., 2008*, pp. 370-371, Feb. 2008.

[4] A. Balankutty, S.-A. Yu, Y. Feng, and P. R. Kinget, "A 0.6-V zero-IF or low-IF receiver with integrated fractional-N synthesizer for 2.4-GHz ISM-band applications," *IEEE J. Solid-State Circuits*, vol. 45, no. 3, pp. 538-553, Mar. 2010.

High Frequency Tow-Thomas Tunable Filter using OTA based voltage op-amp

Walid Zemouri
Electrical and Electronics
Engineering Department
German University in Cairo,
Egypt
Walid.zemouri@student.guc.edu.eg

Eman A. Soliman
Electrical and Electronics
Engineering Department
German University in Cairo,
Egypt
eman.azab@guc.edu.eg

Soliman A. Mahmoud
Electrical and Computer Engineering
Department
Sharjah University, Sharjah,
UAE
solimanm@sharjah.ac.ae

Abstract—**A high frequency Tow-Thomas filter with tunable cutoff frequency is introduced in this paper. The filter is realized using operational trans-conductance amplifier (OTA) based voltage op-amp, resistor arrays and constant capacitors. The OTA used is digitally tuned to realize a voltage op-amp with tunable unity gain frequency from 3GHz to 35.48GHz. The Tow-Thomas filter's cutoff frequency is tuned using a digitally programmable resistor array. The filter is simulated two times, one for the band-pass response and another for the low-pass response with the filter's cutoff frequency tuned from 16.6MHz to 120.2MHz and from 52MHz to 501MHz respectively. The filter is realized using IBM 90nm CMOS technology model from MOSIS under 1V supply. The filter is simulated using LTSPICE.**

Index Terms— **Constant Capacitance Scaling, Feed-forward Compensation, Operational Trans-conductance Amplifier, Tow-Thomas filter, voltage op-amp**

I. INTRODUCTION

CONTINOUS-TIME filters with wide tunable bandwidth are needed in communication and storage systems such as magnetic storage (disc drives), optical storage (CD-ROM drives) and high speed local area network (LAN) transceivers [1-3]. For such systems the voltage op-amp active-RC structure is appealing as it provides excellent linearity and low noise levels. Unfortunately, this will not be the case at higher operating frequencies due to the op-amp's low unity gain frequency. This problem can be solved using Operational Tans-conductance Amplifier (OTA) based voltage op-amp. These voltage op-amps have larger unity gain frequency compared to conventional two stage voltage op-amp.

The conventional two stage op-amp compensation techniques were based on using Miller theorem. This method depends on adding a feedback capacitor across the gain stage of the op-amp. This method splits the poles of the op-amp to a dominant and a non-dominant pole. However, it adds a right half-plane zero that causes the op-amp phase margin to decrease. A nulling resistor or a buffer in series with the Miller capacitor can be added to eliminate this zero.

Another compensation technique was introduced in [4]. The authors in [4] used a feed-forward trans-conductance amplifier

stage along with the op-amp's second stage to improve the unity gain frequency of the op-amp without using any additional capacitor. This technique creates a left half plane zero. This zero can be designed to cancel out the dominant pole of the op-amp transfer function. The previously mentioned method is called No Capacitor Feed-Forward (NCFF) compensation technique [4].

In this paper, a fully differential Tow-Thomas bi-quad filter is realized using an OTA based voltage op-amp. The used op-amp unity gain frequency is large. This is due to the fact that the op-amp is realized with NCFF compensation technique. In addition; the parasitic capacitances of the OTA used to realize the op-amp are kept constant all over the tuning range due to the use of dummy circuits to fix them at the input/output nodes of the OTA [5-6]. The filter's cutoff frequency is varied using programmable resistor network. Moreover; the proposed filter's frequency magnitude response is constant all over the tuning range due to the constant parasitic capacitances of the OTA circuit used. This paper is organized as follows; Section II contains a detailed review of the fully differential Tow-Thomas filter circuit. Section III discusses the op-amp structure used to realize the proposed filter. In section IV the filter's simulation results are presented.

II. FULLY DIFFERENTIAL TOW-THOMAS FILTER

The Tow-Thomas filter was first presented in [7-8]. The filter consists of one lossy integrator and another lossless one. The filter fully differential version is shown in Fig.1. The filter has two responses: a band-pass and a low-pass. The filter band-pass and low-pass transfer functions are given by equation 1 and 2 respectively. The filter's cutoff frequency, quality factor, center frequency gain and DC gain are given by equations 3, 4, 5 and 6 respectively.

$$\frac{V_{BP}}{V_{in}} = \frac{S \dfrac{1}{R_4 C_1}}{S^2 + S \dfrac{1}{R_1 C_1} + \dfrac{1}{R_2 R_3 C_1 C_2}} \quad (1)$$

$$\frac{V_{LP}}{V_{in}} = \frac{\dfrac{1}{R_2 R_4 C_1 C_2}}{S^2 + S \dfrac{1}{R_1 C_1} + \dfrac{1}{R_2 R_3 C_1 C_2}} \quad (2)$$

978-1-61284-863-1/11 $26.00 © 2011 IEEE

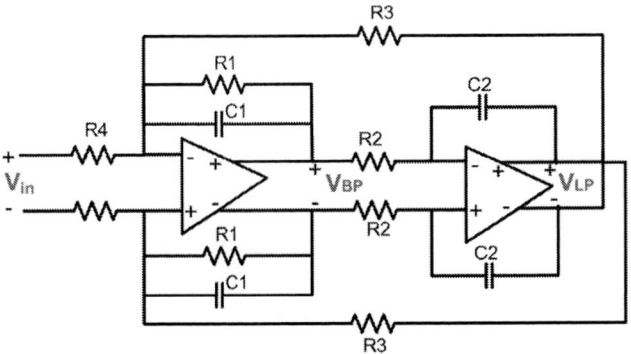

Figure 1 Fully Differential Tow-Thomas Filter

$$f_o = \frac{1}{2\pi\sqrt{R_2 R_3 C_1 C_2}} \quad (3)$$

$$Q = R_1 \sqrt{\frac{C_1}{C_2 R_2 R_3}} \quad (4)$$

$$\left.\frac{V_{BP}}{V_{in}}\right|_{S=j\omega_o} = \frac{R_1}{R_4} \quad (5)$$

$$\left.\frac{V_{LP}}{V_{in}}\right|_{S=0} = \frac{R_3}{R_4} \quad (6)$$

In order to tune the filter's cutoff frequency; a programmable resistor network is used [6]. The resistor network is shown in Fig.2. The programmable resistor network used is realized by connecting binary weighted resistor units in parallel. The resistor unit is realized as a series combination of a fixed poly-silicon resistor and a triode-operated MOST. The network has constant parasitic capacitance due to the presence of dummy circuit connected in parallel with the original network. This will fix the filter's magnitude response all over the tuning range.

III. FEED-FORWARD COMPENSATED OP-AMP

The two stage conventional op-amp consists of a differential amplifier and a voltage amplifier connected in cascade. The differential amplifier is used to generate an amplified differential voltage signal from the differential input voltage signal. The voltage amplifier is used to amplify the op-amp open loop gain.

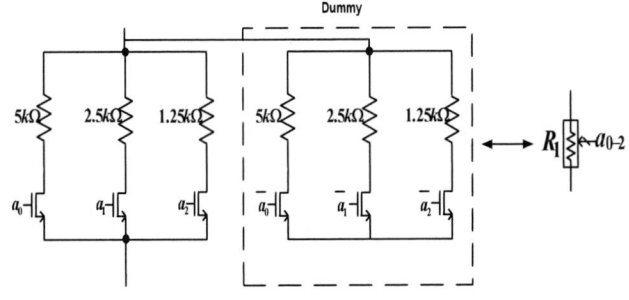

Figure 2 Resistor network [6]

Fig. 3 shows an OTA based voltage op-amp [6]. The op-amp consists of three OTA cells. The first one (G_{m1}) resembles the differential amplifier used in the conventional op-amp. The open loop gain amplification is realized using cascaded connection of another OTA (G_{m2}). For compensation, the NCFF compensation technique is used by adding another OTA (G_{m3}) is the feed-forward path between the input and output voltages.

Each OTA cell shown in Fig. 3 is based on a simple digitally controlled differential amplifier. The differential amplifier is shown in Fig. 4. M1 and M2 are the main differential pair; M5 and M6 are their biasing current sources. The main differential pair is controlled on/off by the bit 'a' which controls the switches Ms3-Ms4. M9 and M10 are used as active loads. They are switched on/off using Ms1-Ms2. To apply the concept of constant capacitance scaling; a dummy circuit is added formed with transistors M3 and M4. These transistors act as a dummy differential amplifier with M7-M8 their biasing current sources and Ms4-Ms5 are used as switches to turn the dummy pair on/off.

Constant capacitance scaling is achieved upon opposite operation between the main pair and the dummy amplifier; when one is turned on the other has to be turned off. The concept of constant capacitance scaling is used to maintain the frequency response shape of the OTA by multiplying all the parasitic poles and zeros by the same factor while programming the OTAs.

The OTA cells are digitally programmed by a three bit digital word (a_{0-2}). The OTA cell is composed of seven digitally controlled differential amplifiers connected in parallel [6] as shown in Fig.5; where a_0, a_1 and a_2 controls one, two and four differential amplifiers respectively.

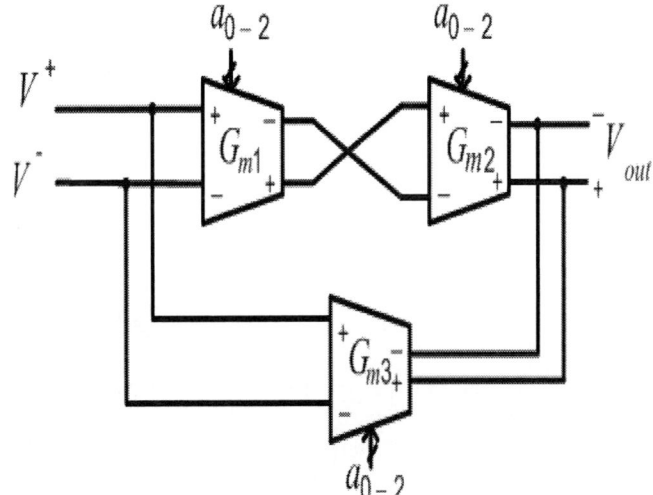

Figure 3 OTA based voltage op-amp block diagram

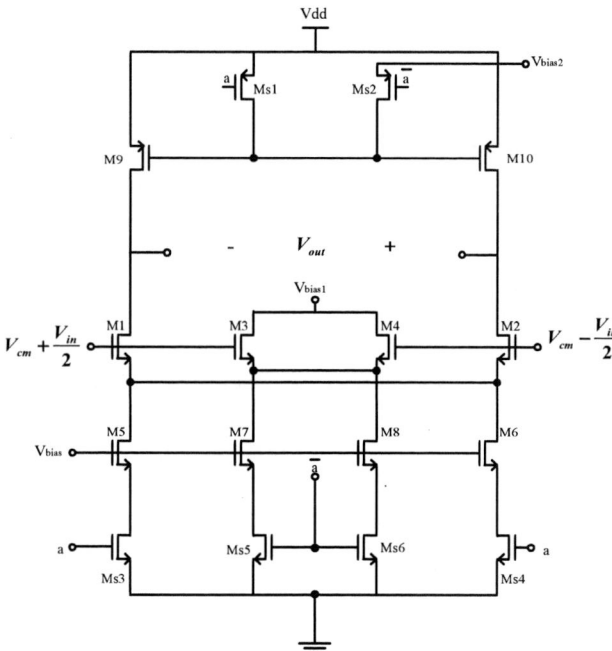

Figure 4 Digitally programmable Differential Amplifier Circuit Diagram

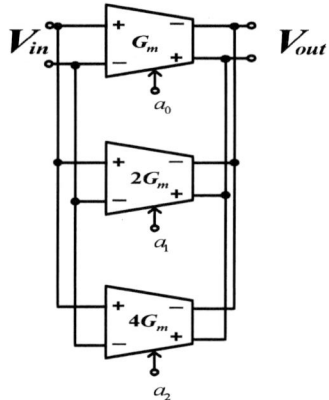

Figure 5 Three-bit digitally controlled OTA cell

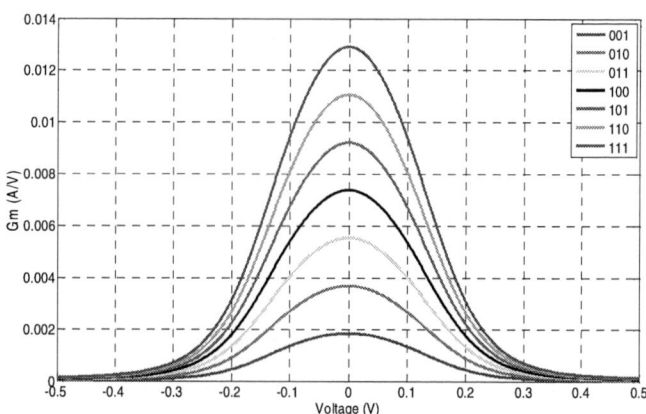

Figure 6 Trans-conductance gain of the OTA cell vs. Input voltage

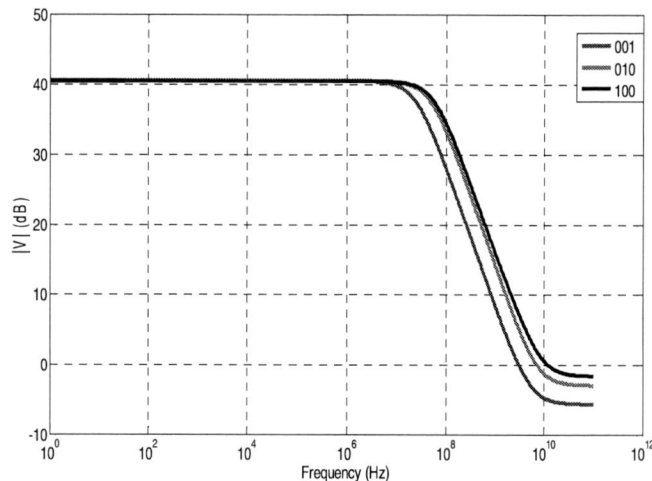

Figure 7 Magnitude Response of the Op-amp voltage gain

Table I
Op-amp Simulation results

'$a_2a_1a_0$'	Unity Gain Frequency	Phase Margin
001	3GHz	60°
010	7.3GHz	46°
011	9.02GHz	42°
100	12.1GHz	36°
101	14.45GHz	30°
110	20.9GHz	26°
111	35.48GHz	16°

IV. SIMULATION RESULTS

The basic circuits of the filter are simulated separately at first. All the circuits are realized using 90nm IBM CMOS technology model from MOSIS under 1V single ended supply voltage. The simulations are performed using LTSPICE.

The OTA cell DC simulation result is shown in Fig.6; where the trans-conductance gain versus the input voltage is shown for different control bit combinations. The trans-conductance gain varies from 2mA/V to 14mA/V.

The OTA based voltage op-amp simulation result is shown in Fig.7. The voltage gain magnitude response of the OTA based voltage op-amp is inspected under open circuit load condition. The unity gain frequency and the phase margin of the op-amp are given in Table I while varying the digital control bits '$a_2a_1a_0$'. The value of logic '1' and '0' are given by 1V and 0V respectively.

The resistor network is simulated as shown in Fig.8. The network is simulated while connecting its terminals in parallel with a voltage source and measuring the derivative of the current flowing through the network versus the applied voltage. The Tow-Thomas bi-quad filter is simulated. The band-pass response is shown in Fig.9 at equal R. The values of C_1 and C_2 are given by 1pF and 0.5pF respectively. The center frequency gain is unity; however due to variations in the resistor networks a slight increment in the gain is shown in the simulations.

978-1-61284-863-1/11 $26.00 © 2011 IEEE

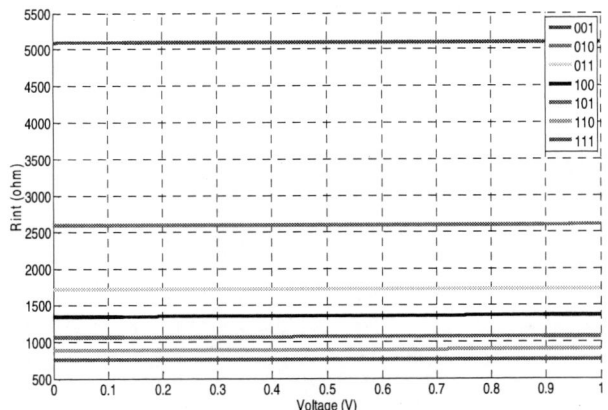

Figure 8 Resistor Network DC analysis

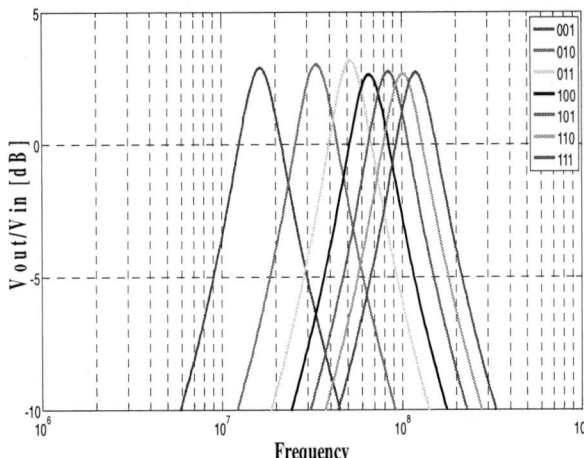

Figure 9 Magnitude Response of the band-pass filter

The Low-pass response is shown in Fig.10. The values of R are equal. The values of C_1 and C_2 are given by 1pF and 0.1pF respectively. A summary of the band-pass and low-pass simulation result is given in Table II and Table III respectively. The cutoff frequency 'f_o', input referred noise density ($\overline{V_{in}}$), third order harmonic distortion (HD3) and fifth order harmonic distortion (HD5) are given in the tables.

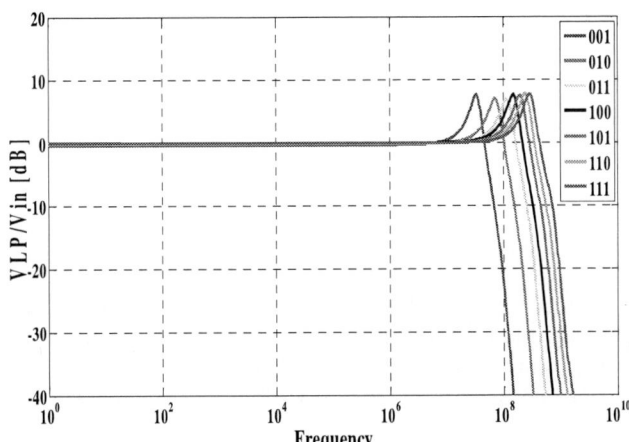

Figure 10 Magnitude response of the low-pass filter

Table II. Summary of the band-pass filter Simulation Results

'$a_2a_1a_0$'	Parameter			
	f_o [MHz]	$\overline{V_{in}}$ [nV/√Hz]	HD3 [dB]	HD5 [dB]
001	16.7	51	-38	-48.9
010	33.88	37	-39.7	-54.5
011	51.3	30.2	-39.5	-52.3
100	66	27.25	-41	-51.4
101	83.17	24.14	-41.3	-55.4
110	102.44	22	-41.5	-53.6
111	120.27	20.16	-41.24	-52.4

Table III. Summary of the low-pass filter Simulation Results

'$a_2a_1a_0$'	Parameter	
	f_o [MHz]	$\overline{V_{in}}$ [nV/√Hz]
001	52.49	50.35
010	115	40.3
011	188	35.5
100	245	31.54
101	323.6	30
110	402	27.5
111	501	28.18

V. CONCLUSION

A realization of a high frequency tunable Tow-Thomas filter is introduced. The filter DC power consumption is 12.8mW. The filter is realized using OTA based voltage op-amp, resistor array and constant capacitors. The op-amp introduced has large unity gain frequency. The filter is simulated twice to give a band-pass and low-pass responses with cutoff frequency tuning ranges from 16.7MHz to 120.27MHz and 52.49MHz to 501MHz respectively.

REFERENCES

[1] S.A. Mahmoud, "Low power Low-Pass Filter with Programmable Cutoff Frequency Based On a tunable Unity Gain Frequency Operational Amplifier," *Journal of Circuits, Systems and Computers*, vol. 19, no. 8, pp. 1-13, Dec. 2010.

[2] S. Pavan, "High Frequency Continuous-Time Filters in Digital CMOS Processes," Ph.D. Dissertation, Columbia University, New York, NY, May 1999.

[3] Y. Tsividis, "Integrated continuous-time filter design—An overview," *IEEE J. Solid-State Circuits*, vol. 29, no. 3, pp. 166–153, Mar. 1994.

[4] B. K. Thandri, and J. Silva-Martinez, "An overview of feed-forward design techniques for high-gain wideband operational transconductance amplifiers," *Microelectronics Journal*, vol. 37, pp. 1018-1029, Apr. 2006.

[5] M. O. Shaker, S. A. Mahmoud, and A. M. Soliman," New CMOS Fully-differential Transconductor and Application to Fully-Differential Gm-C Filters," *ETRI Journal*, vol. 28, no.2, pp.175-181, Apr. 2006.

[6] T. Laxminidhi, V. Prasadu, and S. Pavan "Widely Programmable High-Frequency Active RC Filters in CMOS Technology," *IEEE J. Solid-State Circuits*, vol. 56, no. 2, pp. 327-336, Feb. 2009.

[7] J. Tow, "Active RC Filters – A State Space Realization," *Proc. IEEE (Lett.)*, vol. 56, pp. 1137-1139, 1968.

[8] L. Thomas, "The Biquad: Part I – Some Practical Design Considerations," *IEEE Trans. Circuit Theory*, vol. CT-18, pp. 350-357, 1971.

Cascaded Third-order Tunable Low-Pass Filter using Low Voltage Low Power OTA

Sondos H. Ismail
Electrical and Electronics
Engineering Department
German University in Cairo,
Egypt
Sondos.ismail@student.guc.edu.eg

Eman A. Soliman
Electrical and Electronics Engineering
Department
German University in Cairo,
Egypt
eman.azab@guc.edu.eg

Soliman A. Mahmoud
Electrical and Computer
Engineering Department
Sharjah University, Sharjah,
UAE
solimanm@sharjah.ac.ae

Abstract—A low voltage low power tunable Operational Trans-conductance Amplifier (OTA) is presented in this paper. The OTA is tuned via a control voltage. The OTA power consumption is 0.15mW at 1V supply voltage. A digitally controlled Master/Slave tuning scheme is applied to generate the OTA control voltage. The trans-conductance gain of the OTA is varied from 9.3µA/V to 35.29µA/V. The proposed OTA is used to realize a cascaded third-order low-pass filter. The filter cutoff frequency is tuned from 4.75MHz to 12.79MHz. The circuit is realized using 90nm CMOS technology model from MOSIS under 1V single ended supply. Simulations using ADS are presented.

Index Terms—Master/Slave Tuning Scheme, Operational Trans-conductance Amplifier (OTA), OTA-C low-pass filter, Phase-Locked Loop (PLL), Voltage Controlled Oscillator (VCO).

I. INTRODUCTION

Multi-standard transceivers represent an important part of wireless communication systems such as Wi-Fi, WIMAX, and many other systems. Continuous time filters are essential building blocks in multi-standard transceivers since they are needed to filter out the undesired channels and interferences. Therefore, these filters need to have tunable cutoff frequency f_c to cover multiple standards [1-6].

Conventional voltage operational amplifier filters have been used in this field earlier. However; active op-amp filters do not work adequately at high frequencies. Many attempts have been made to overcome these drawbacks. The most successful approach is to use the operational trans-conductance amplifier (OTA) to realize tunable filters; since OTAs have larger bandwidth [7] and consumes less power compared to voltage op-amps. The filter's cutoff frequency can be tuned by changing the OTA's trans-conductance gain (G_m).

In this paper, a low voltage low power differential OTA circuit is proposed. The OTA is used to realize a tunable third-order low-pass filter based on cascaded technique. The OTA is tuned using a control voltage signal. This control voltage is generated using a Master/Slave tuning scheme [8] to adjust the G_m value and consequently the filter's cutoff frequency.

An OTA based voltage controlled oscillator (VCO) (master device) is locked to a reference frequency using a control voltage. This generated control voltage from the VCO is divided by a factor N using a trans-conductance division scheme and is fed to the proposed OTA circuit used to realize the third-order low-pass filter (slave device) to tune its cutoff frequency.

This paper is organized as follows; section II presents the proposed fully differential CMOS tunable OTA cell with its simulation results. In section III; the Master/Slave tuning scheme is discussed. Section IV; the proposed OTA is used to realize a tunable third-order low-pass filter using cascading topology. ADS simulation for the filter is included in section IV.

II. TUNABLE CMOS FULLY DIFFERENTIAL OTA

The proposed CMOS fully differential OTA cell is shown in Fig. 1. It is composed of two NMOS differential pairs (M_1-M_2, M_3-M_4), two level shifters (M_7-M_8) and current sources (M_5, M_6, M_9, M_{10} and M_{11}). The output currents are taken from the drains of M_1, M_2 and M_3, M_4.

The differential input (V_1-V_2) is applied to the gates of M_1, M_7, M_4 and M_8. The control voltage (V_c) is applied to the gates of M_5 and M_6. Assuming all transistors working in the saturation region and M_1-M_4 and M_5-M_8 are group matched; the relation between the differential output current and the differential input voltage is given by the following equation:

$$I_{out1} - I_{out2} = KV_C(V_1 - V_2) \qquad (1)$$

Where K is the trans-conductance parameter of M_1-M_4,

The OTA trans-conductance gain (G_m) is given by equation (2). The G_m is independent of the transistors threshold voltage; thus the circuit is insensitive to process variation.

$$G_m = KV_C \qquad (2)$$

The fully differential OTA is simulated using 90nm CMOS technology model under 1V single ended supply voltage. Fig.2 shows the trans-conductance gain results at different values of the control signal V_c with 0.04V step under 1kΩ load. The differential input voltage is varied from -0.1V to 0.1V. The trans-conductance gain ranges from 9.3µA/V to 36.29µA/V at V_c changing from 0.2V to 0.5V respectively.

978-1-61284-863-1/11 $26.00 © 2011 IEEE

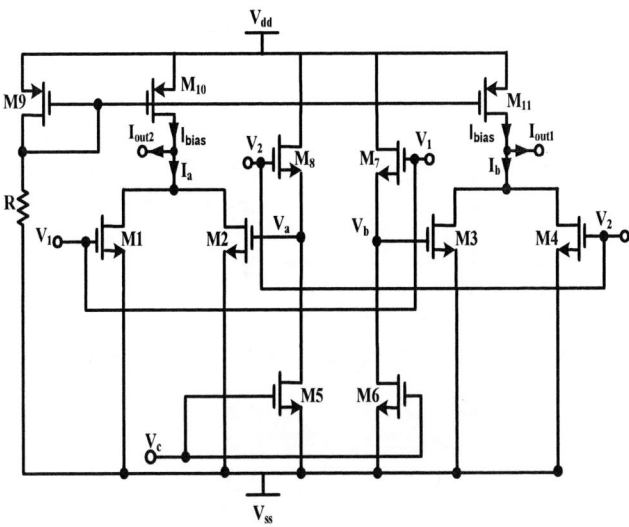

Figure 1 Proposed Fully Differential OTA Circuit Diagram

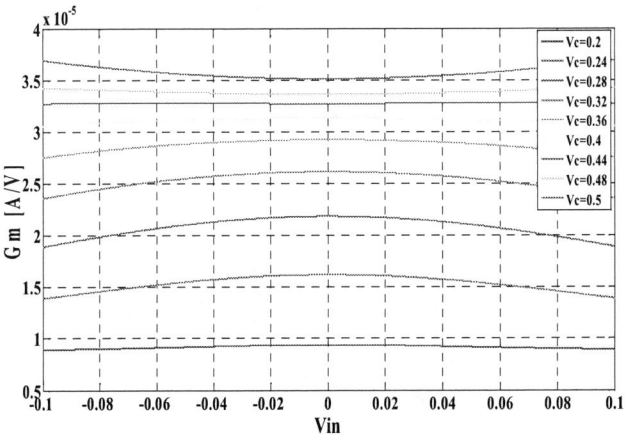

Figure 2 OTA DC Simulation Result

The G_m gain is almost constant over the input voltage range. In addition, the common mode voltage of the OTA is almost fixed at 0.5V; thus no need for a common mode feedback circuit to adjust its value. The frequency response of the OTA is shown in Fig.3. The magnitude response is given under resistive and capacitive load connected in parallel.

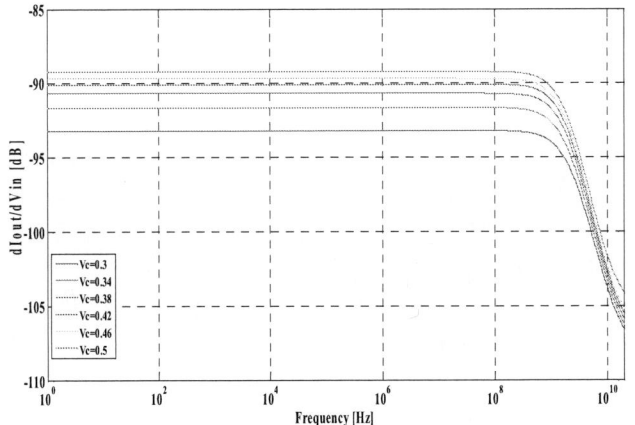

Figure 3 G_m Magnitude Response of the OTA

Table I contains the G_m values varying with the control voltage V_c. The third order harmonic distortion (HD3) of the OTA is measured at input voltage signal of amplitude 10 mV and 100 kHz frequency.

Table I. Summary of OTA Simulation Results

Control Voltage Vc	Parameter	
	Gm [μA/V]	HD3 [dB]
0.2 V	9.304	-54.713
0.24 V	15.88	-48.726
0.28 V	25.91	-67.614
0.32 V	31.26	-67.028
0.36 V	33.76	-61.673
0.4 V	34.64	-60.123
0.44 V	35	-59.566
0.48 V	35.5	-59.433
0.5 V	36.29	-59.268

III. MASTER/SLAVE TUNING SCHEME

The subject of automatic tuning has been widely developed over the past decades. The Master/Slave scheme is one of the most widely used automatic tuning techniques concerning continuous time filters .The Master/Slave tuning scheme's basic idea is to lock the master device which in our case is the OTA based VCO to a certain reference frequency. This process is done using a Phase Locked Loop (PLL) [9].The PLL generates a control voltage V_c to achieve this purpose. If this V_c is fed to the OTA cell's of the filter that is required to be tuned, this will make the filter's f_c -slave device- equal to the frequency of the VCO.

In order to make the filter's f_c tunable, different approaches for the Master/Slave tuning scheme were proposed; such as the use of programmable reference frequency for the PLL, programmable PLL or multiplying digital to analog converter (MDAC) between the VCO and the PLL. However, all these previous approaches came at the expense of complex hardware.

Fig. 4 shows a modified Master/Slave tuning scheme using a simple trans-conductance division scheme [8]. It is composed of two OTAs G_{m1} and G_{m2} and a control loop where G_{m1} and G_{m2} are matched to the OTA used to realize the VCO and the filter respectively. An arbitrary DC voltage V_2 is applied as the input of both G_{m2} and a simple divider with integer division N while the divider output (V_2/N) is fed to G_{m1}.The idea is to make the two output currents of the OTAs -G_{m1} and G_{m2}- equal using the control loop which is formed of two resistors and a simple differential amplifier.

This way, a division factor N will accurately sets the ratio between G_{m1} and G_{m2}. This will result in a relation between the filter's cutoff frequency and the VCO's frequency as given by the following equation:

$$f_{filter} = \frac{f_{vco}}{N} \qquad (3)$$

The division cell is simulated using ideal block models for the PLL and two of the proposed OTA to investigate the performance of the overall system. The PLL used is locked at 2.5MHz and the VCO's control voltage is equal to 0.5V.

978-1-61284-863-1/11 $26.00 © 2011 IEEE

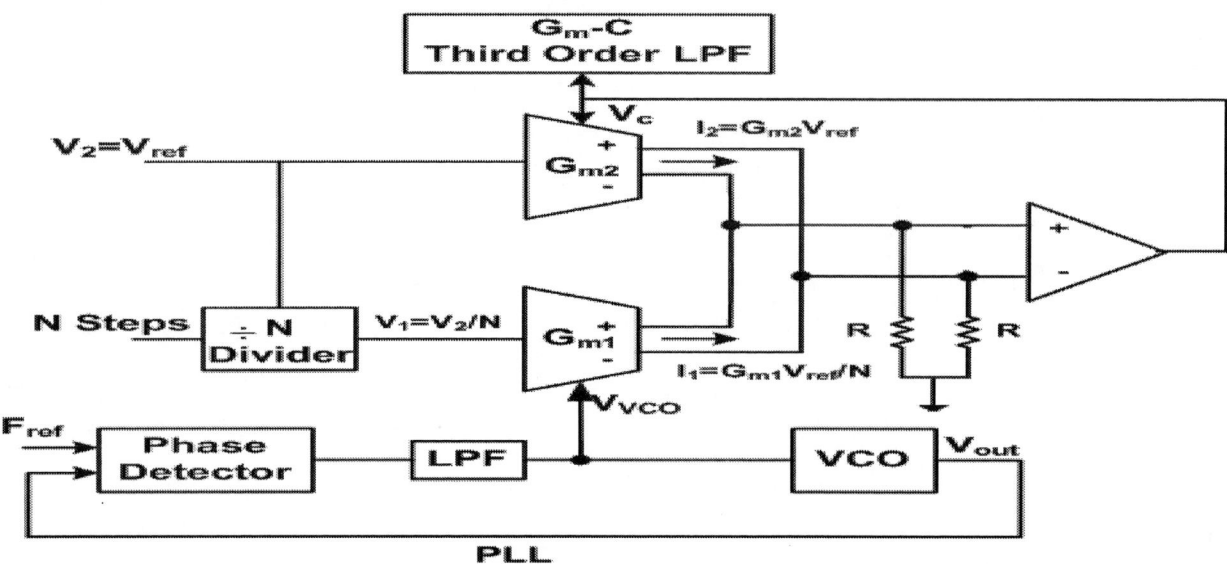

Figure 4 Trans-conductance Division Cell Block Diagram [8]

The OTA cell control voltage transient analysis simulation result is shown in Fig.5. The value of the control voltage V_c changes with the N value as discussed.

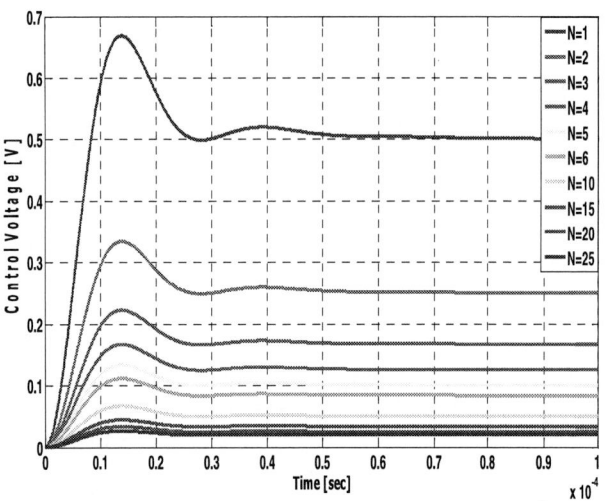

Figure 5 OTA's Control Voltage 'V$_c$' transient response

IV. THIRD ORDER TUNABLE LOW PASS FILTER

As an application for the proposed OTA a third-order tunable fully differential low-pass filter is presented. A cascaded third order low-pass filter is shown in Fig. 6. It consists of a lossy integrator cascaded with a Tow-Thomas second-order section [10]. The transfer function for the cascaded filter is given as follow:

$$\frac{V_{out}}{V_{in}} = \frac{\dfrac{G_m}{C}}{s + \dfrac{G_m}{C}} \cdot \frac{\dfrac{G_m^2}{C^2}}{s^2 + \dfrac{G_m}{C}s + \dfrac{G_m^2}{C^2}} \qquad (4)$$

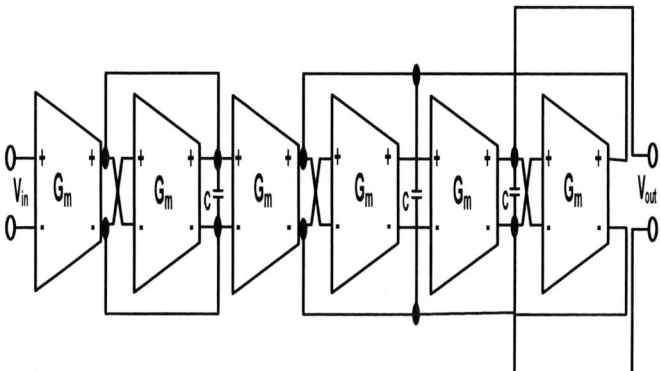

Figure 6 Fully Differential Cascaded Third-Order LPF

Simulations are done using ADS tool with 90nm CMOS transistor model. Fig. 7(a) and (b) shows the magnitude frequency response of the cascaded filter with the VCO's frequency equals to 15 MHz with V_c varying from 0.2V to 0.3V and from 0.34V to 0.5V respectively.

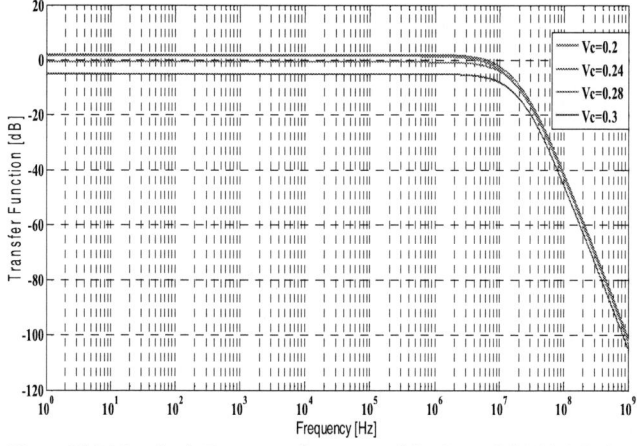

Figure 7(a) Magnitude Frequency Response of the Cascaded Third-Order LPF

978-1-61284-863-1/11 $26.00 © 2011 IEEE

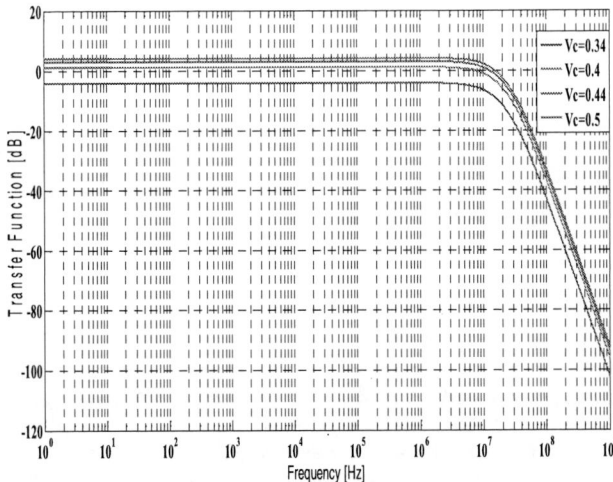

Figure. 7(b) Magnitude Frequency Response for the Cascaded Third-Order LPF

The control voltage is varied from 0.5V to 0.2V respectively with fourteen fractional steps. Table II shows a summary of the realized filter's specifications. Table III contains the filters' cutoff frequency tuning ranges with respect to the control voltage.

Table II
Tunable Third-order Low-Pass Filter Simulation Results

Parameter	Cascaded Filter
Technology	90 nm CMOS
Voltage Supply	1V
Power Consumption	1.16mW
Cut-off frequency Tuning Range	4.57MHz to 12.59MHz
In-Band IIP3	51dBm to 64dBm
Out-Of-Band IIP3	50dBm to 84dBm
In-Band IIP2	61dBm to 84dBm
Output Noise Power Spectral Density	28nV√Hz to 95nV√Hz
HD3 @ 10mVpp and f=0.1MHz	-41 dB to -61 dB
IM3@ 10mVpp and f=0.9MHz &1MHz	-51dB to -69dB

V. CONCLUSION

A low voltage low power tunable fully differential OTA is introduced. A tunable third-order low-pass filter using cascaded topology is presented. The filter's cutoff frequency is tuned using a Master/Slave tuning scheme through a trans-

conductance division scheme consisting of a PLL and a simple control loop. The filter is realized using 90nm CMOS technology model under 1V single ended voltage supply. The filter's cutoff frequency is varied from 4.76MHz to 12.95MHz.

Table III
Tunable Third-order Low-Pass Filter Cut-off Frequency

N-Value	Cascaded filter f_c	Control Voltage V_c
Step 1	12.59MHz	0.5V
Step 2	11.35MHz	0.44V
Step 3	10.25MHz	0.42V
Step 4	9.75MHz	0.4V
Step 5	9.3MHz	0.38V
Step 6	8.4MHz	0.36V
Step 7	8.01MHz	0.34V
Step 8	7.6MHz	0.32V
Step 9	7.2MHz	0.3V
Step 10	6.7MHz	0.28V
Step 11	6.2MHz	0.26V
Step 12	5.5MHz	0.24V
Step 13	5MHz	0.22V
Step 14	4.57MHz	0.2V

REFERENCES

[1] T.M. Hassan and S.A. Mahmoud, "New CMOS DVCC Realization and Applications to Instrumentation Amplifier and Active –RC filters," *International Journal of Electronics and Communications, AEU*, vol. 64, pp. 47-55, Jan. 2010

[2] H. Shin, and Y. Kim, "A CMOS Active-RC Low-Pass Filter With Simultaneously Tunable High- and Low-Cutoff Frequencies for IEEE 802.22 Applications," *Circuits and Systems II: Express Briefs, IEEE Transactions on* , vol.57, no.2, pp.85-89, Feb. 2010.

[3] M. A. Dawoud, S. A. Mahmoud, and A. K. El-Kafrawy, "Different Baseband Chain Architectures for Multi-standard Reconfigurable Receivers", *International Conference on Microelectronics (ICM.2009)*, pp.179-182, Dec. 2009

[4] S.A. Mahmoud, "A Gain / Filtering Interleaved Baseband Chain Architectures for Multi-standard Reconfigurable Receivers," *Journal of Circuits, Systems and Computers (JCSC)*, vol. 21, no. 1, Feb. 2012.

[5] T. El-Zomor, E.A. Soliman and S. A. Mahmoud, "Reconfigurable Baseband Chain for Software- Defined Radio Receivers", *International Conference on Microelectronics (ICM.2009)*, pp.270-273, Dec. 2009

[6] S. D'Amico, V. Giannini, and A. Baschirotto, "A 4th-order Active-*Gm*-RC reconfigurable (UMTS/WLAN) filter," *IEEE J. Solid-State Circuits*, vol. 41, no. 7, pp. 1630–1637, Jul. 2006.

[7] T.Delyiyamis, Y. Sun, and J. K. Fidler, "Continuous-Time Active Filter Design", *CRC Press LLC*, 1999.

[8] D. Chamla, A. Kaiser, A. Cathelin, and D. Belot "A Switchable-Order G$_m$-C Baseband Filter with Wide Digital Tuning for Configurable Radio Recievers," *IEEE Journal of Solid-State Circuits*, vol. 42, no. 7, Jul. 2007.

[9] Q. Carlos, B. Guillemo, and A. Inigo, *Design Methodology for RF CMOS Phase Locked Loops*, ARTEC House, ch.2, 2009.

[10] L. Acosta, M. Jiménez, R. G. Carvajal, A. J. Lopez-Martin, and J. Ramírez-Angulo, "Highly linear Tunable CMOS Gm-C Low-Pass Filter," *IEEE Trans. Circuits and Systems. I*, vol. 56, no. 10, Oct. 2009.

978-1-61284-863-1/11 $26.00 © 2011 IEEE

Modularized Development Platform for Hardware/Software Design

Kai-Chao Yang, Yu-Tsang Chang, Chien-Ming Wu, and Chun-Ming Huang
National Chip Implementation Center
Hsinchu, Taiwan
kcyang@cic.narl.org.tw

Abstract—**During the hardware or software developing process, the designer usually uses a particular hardware platform that meets the purpose of design goal. In school, since more and more hardware/software design courses are delivered, different kinds of platforms have to be prepared for students. This increases extra cost to buy and learn these platforms. In this article, we propose a modularized development platform for FPGA and embedded system design. The proposed development platform is modularly designed and divided to core modular boards and peripheral modular boards. Designers can combine different modules for various purposes to fit requirements in the lab. Thus more flexible hardware and software projects can be realized with lower cost due to reusable modules. We have built up several projects to demonstrate the functions of the proposed platforms. A pedestrian traffic signal controller, a digital photo frame, and a video surveillance system realized by the proposed platform are introduced in this article.**

Keywords- FPGA design; development platform; embedded system

I. INTRODUCTION

In recent years, cultivating software and hardware design manpower has become more and more important because of the increasing requirements of high-level electronic products in our life, such as smart phone and tablet PC. In the university, the number of software or hardware design courses also continuously increases to fulfill the manpower requirements in industry. Due to the extensive range of electronic applications, different kinds of hardware platforms are usually used for various purposes. To decide a platform used in the developing process, some suggesting systems have been proposed to integrate information and compare the difference between hardware platforms [1][2]. There is also a series of short courses proposed to teach design skills on different hardware platforms [3]. However, numerous different kinds of hardware platforms still confuse designers, and also bring out a problem, resource-wasting. Because the applicable area of a hardware platform is limited, the hardware platform used in one project may not be reusable in others. Thus, for different projects, different hardware platforms have to be prepared.

A solution for the above problem is Field programmable gate arrays (FPGAs), which provide the salient features of flexibility of design approach and hardware platform reuse. They are now considered as an appropriate solution to boost the performances of controllers, enabling the implementation of new control methods and/or designing concurrent architectures [4-6]. However, FPGA still cannot completely simplify the design environments. An obvious example can be observed in universities of Taiwan. In Taiwan, FPGA development platforms applied in universities are mostly from either Altera [7] or Xilinx [8]. Fig. 1 shows two evaluation boards usually used for FPGA design. From Fig. 1, we can see that each board has different functions and features. In many cases, some design projects might be favorable for Altera users over Xilinx ones, or vice versa. As a result, the design projects and curriculum materials might not be shared by different users. Furthermore, due to the limited functionalities in common for both development boards, the design projects become much less challenge.

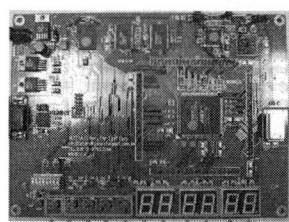

Figure 1. XILINX SPARTAN3 evaluation board (left) and ALTERA ACEX evaluation board (right).

In order to offer higher quality of design projects for software and hardware development, we have successfully developed a flexible development hardware platform, which consists of FPGA Core Modular boards and Peripheral Modular boards. Both boards have the same versatile peripherals with enhanced functionality. The key features of the architecture of the proposed platform are "flexibility" and "modularization," which enhances the expandability and configurability. Users can develop their own peripheral boards and stack them on top of the basic FPGA core modular board for curriculum development or for research use. Besides, different core modular boards and peripheral modular boards can be arbitrarily combined to fit the requirements in the lab, such that the designer can realize new projects without extra cost due to reusable modules. For example, a Xilinx user can select a Xilinx-based FPGA core modular board stacked by camera modular boards to design video surveillance applications.

978-1-61284-863-1/11 $26.00 © 2011 IEEE

This article is organized as follows. The design concept and implementation results of the proposed development platform are described in Section 2. Then, we describe some case studies to show how the proposed platform works in Section 3. Finally, conclusions are given in Section 4.

II. FLEXIBLE FPGA DEVELOPMENT PLATFORM

Our objective is a flexible development hardware platform, such that users can build up their own development environments according to requirements. Actually, a FPGA development platform can be viewed as the combination of the Core part and Peripheral part. During the developing process, the creativity of design projects are usually confined to limited functionalities of peripheral or unfamiliar FPGA core. Therefore, the main idea to achieve our objective is modularization of the core and peripheral, such that multiple modules can be combined to a development platform with different features.

The basic design concept is shown in Fig. 2(a). A FPGA core module board and a peripheral board are combined to a complete FPGA development platform. Different module boards are stacked together through the connectors. Due to the modularized design, the proposed platform can provide many configurations for different applications by stacking different modules. For example, two FPGA core modular boards can be stacked together as shown in Fig. 2(b). In addition, Fig. 2(c) illustrates the combination of one peripheral board and two FPGA core module boards.

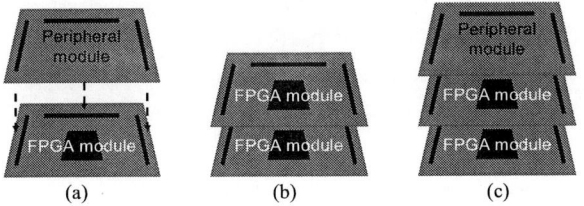

Figure 2. Combination of FPGA core board and peripheral board.

Based on the above design concept, we realized a FPGA core modular board, including the FPGA and its corresponding serial configuration device, LEDs, Buttons, GPIO, JTAG, and memory modules. The memory module must include the memory components such as SRAM, SDRAM, MRAM, DDR2, DDR3, NOR-flash, and NAND-flash. Besides, a peripheral modular board containing LCDM, Switches, Buttons, Keypad, 7-segment displays, Dot Matrix, RS232, VGA, PS/2, and Audio was also realized. Note that the peripheral modular board has to be stacked on top of the FPGA core module board as illustrated in Fig. 2(a).

Fig. 3(a) illustrates the implementation of the Xilinx FPGA core modular board, implemented by the Xilinx FPGA chip and the associated serial configuration device (XCF16P). The FPGA chip has 1.5M system gates, or approximately 30 K logic cells, 208K distributed RAM, 576 block RAM (each block contains 4K bits), and 487 user I/O pins. Similarly, Fig. 3(b) illustrates the circuit implementation of the Altera FGPA Core Module board which is realized with Altera EP2C35F672 FPGA chips and the serial configuration device (EPCS16).

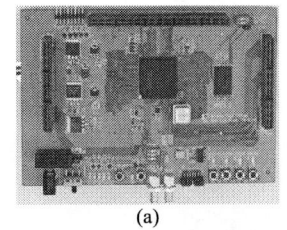 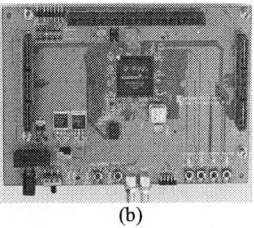

(a)   (b)

Figure 3. (a) Xilinx FPGA core modular board and (b) Altera FPGA core modular board.

Several peripheral module boards were also realized. Fig. 4 shows two different peripheral modular boards stacked on a FPGA core modular board. In Fig. 4(a), the top is a basic peripheral module containing switches, key pad, LEDs, buttons, Rotary switch, RS232 serial ports, PS/2, and GPIO. Fig. 4(b) shows a touch panel module containing some multimedia I/Os such as Audio jacks, VGA connector, touch panel, direction switches, and Ethernet connector. Besides, we also designed different memory modular boards, such as SRAM and SDRAM with different sizes. These memory modular boards can be plugged into the Small-Outline Dual In-Lin Memory Module (SODIMM) on the FPGA core modular board.

The stacked platform allows users comparing the implementation of their design with different FPGA cores in terms of cost and performance.

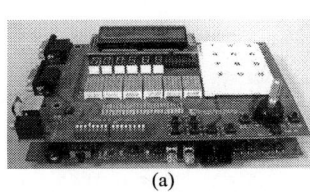 

(a)   (b)

Figure 4. Different peripheral modular boards stacked on the FPGA core module. (a) Basic peripheral module; (b) touch panel module.

III. CASE STUDY

A. Pedestrian Traffic Signal Controller

The design project aims to imitate the pedestrian traffic signal in Taiwan [9]. When pedestrians are allowed passing through the intersection, the remaining time and a walking green man are displayed on the traffic signal. Then the green man starts rushing in order to remind pedestrians hurry up while the remaining time is running out. When pedestrians are not allowed passing through the intersection, the remaining time and a still red man are shown on the traffic signal instead of the little green man.

In this project, we stack a basic peripheral modular board and an Altera FPGA core modular board. Fig. 5(a) shows the block diagram of the design project. The circuit takes three 1-bit input signals, reset, clk, and pause. As their names imply, reset and clk are the reset and clock signals, while pause is to temporally suspend or toggle the mode.

978-1-61284-863-1/11 $26.00 © 2011 IEEE

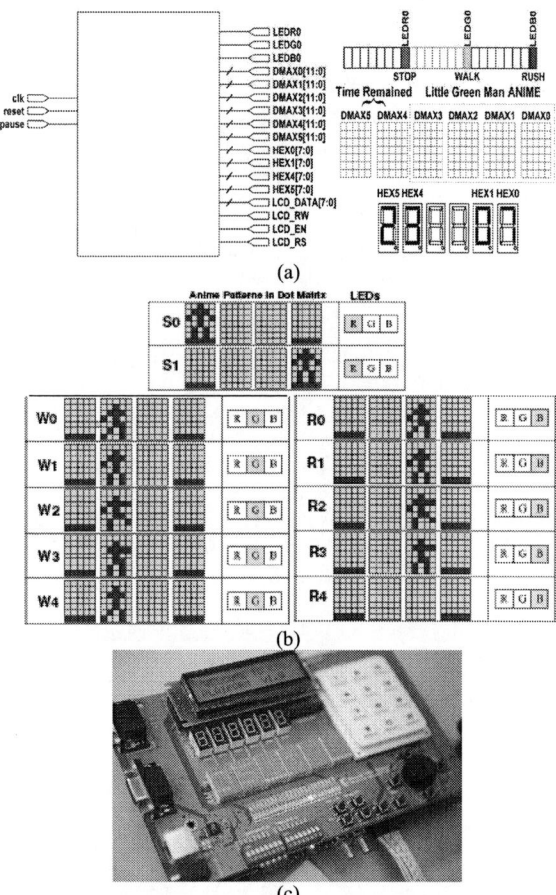

(a)

(b)

(c)

Figure 5. Predestrian traffic signal controller. (a) Block diagram; (b) anime patterns for the dotmatrix; (c) the traffic signal demo.

The controller has three modes: RUSH, WALK, and STOP, indicated by the LEDR0, LEDG0, and LEDB0, respectively. The six 5x7 dot matrix displays are labeled DMAX5~DMAX0, where DMAX3~DMAX0 show the animation of the Green Man in different modes, while DMAX5 and DMAX4 display the timing information (i.e., time remaining to change the mode). The six 7-segment displays are labeled by HEX5~HEX0, where HEX1 and HEX0 represent the time remaining at the STOP state. Note that one can use pause to temporally suspend the mode.

Basically, the three modes (STOP, WALK, RUSH) are repeatedly cycled in duration of 20 seconds, where the durations of STOP, WALK, and RUSH are 10, 7, and 3 seconds, respectively. During the STOP mode, LEDR0 is lighted up in *red*, Dot matrix plays the anime patters, as shown in Fig. 5(b), for STOP, and 7-segment display shows the timing information (i.e., time remained), as described above. Note that the number patterns in both 7-segment displays and dot matrix are illustrated in Fig. 5(b) and Fig. 5(b), respectively. Another two modes (WALK, RUSH) are operated in a similar way, except that LEDG0 (in green) and LEDB0 (in blue) are employed for WALK and RUSH modes, respectively. Fig. 5(c) shows the practical demonstration of this project.

B. Digital Photo Frame Album s

Fig. 6 presents the environment of the embedded system for digital photo frame application. The architecture is comprised of several layers.

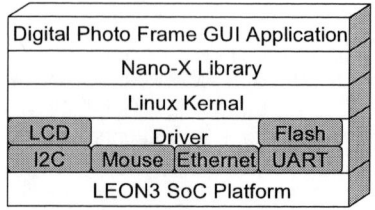

Figure 6. Architecture of the digital photo frame album.

The lowest layer is the hardware platform. We used a LCD modular board and a Xilinx FPGA core board to build up a LEON3 SoC platform. LEON3 is a synthesizable VHDL model of a 32-bit processor with the SPARC V8 architecture [10]. Adding a new peripheral to the LEON-based SoC platform requires not only to design hardware but also to build kernel driver, so that the new peripheral can be used by OS. Thus, in the first step, the interface between the peripheral and SoC system must be established. Then, based on Linux device model, the device driver must be designed.

The LEON3 SoC System was ported to Xilinx-based FPGA module. TABLE I shows the resource utilization of the porting, where the core uses approximately 55% of the available hardware logic resources (4-LUTs).

TABLE I.   RESOURCE USAGE FOR LEON3 SOC SYSTEM

	LEON3 SoC Implementation		
	Available	*Used*	*Utilization*
Number of slices	26624	6173	23%
Number of 4-LUTs	26624	14821	55%
Equivalent Gate Count	1123885		
JTAG Gate Count	15024		
Peak Memory Usage	611 MB		

Based on the LEON3 SoC system, the version 2.6.12 Linux and the root file system were installed in the system. The kernel provides the drivers for LCD, UART, Flash, I2C, Ethernet, and mouse.

On the above of the Linux kernel is the Nano-X (or microwindow) embedded GUI, which is an open source project allowing creating closed source drivers and applications. The Nano-X is comprised of three levels in structure. The lower level of the Nano-X contains the drivers for monitor, mouse/touchpad, and keyboard. The middle level includes the portable and device independent graphics engine. The top level provides the programmable API which is used to develop digital photo frame application.

Finally, the highest layer of the software architecture is the digital photo frame GUI application that calls the Nano-X library to display the photo pictures. The Album provides icons for the functions of auto-play, zoom-in, zoom-out, previous picture, and next picture. It also provides the function of playing music. Besides, through the NFS (Network File System) of LAN in the developed system, the user is allowed to view the photo pictures stored in the internet. Fig. 7 shows the demonstration of this project.

Figure 7. The digital photo frame album.

C. Video Surveillance

This project uses a core modular board, a LCD modular board, and a CMOS image sensor board. The surveillance video frames are continuously captured by the image sensor board which supports zoom in and zoom out functions. Then these frames are processed through the edge detection and motion detection algorithm implemented using FPGA on the core modular board. The LCD modular board can shows either the original frames or processed frames. If the motion activity is greater than an assigned threshold, an alarm will be announced on LCD. Fig. 8 (a) shows the modular boards used for video surveillance, and Fig. 8 (b) shows the demo of hardware edge detection algorithm for sensed video frames.

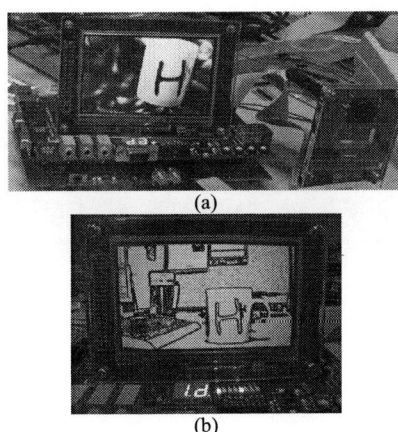

Figure 8. (a) Combination of the core modular board, LCD modular board, and CMOS image sensor board. (b) Edge detection demo.

IV. CONCLUSION

In this article, we propose a flexible hardware/software development platform. The key feature of the proposed platform is "modularization." We divide a development platform to several smaller modular boards, so that designers can combine different modular boards to their own platform. This design saves cost to buy and learn a new hardware platform for different purposes.

In the future, more modular boards are planned to be developed. The development process welcomes participants, including scholars, teachers, students, and engineers, to jointly create and share the educational resource, so the participants can widen and deepen their knowledge and design skills.

ACKNOWLEDGMENT

This work was supported in part by National Science Council of Taiwan. The authors thank Y.-S. Lin, C.-T. Kuo, H.-H. Luo, and C.-L. Wey.

REFERENCES

[1] K.-C. Yang, Y.-T. Chang, C.-M. Wu, and C.-M. Huang, "Suggesting Hardware Platform for Embedded System Education with Visualized Information," accepted by INNOVATIONS, 2011.

[2] K.-C. Yang, C.-T. Kuo, Y.-T. Chang, C.-M. Wu, J.-R. Chang, C.-M. Huang, and C.-L. Wey, "Case Study: A Universal Study Platform for Embedded Software Education," in International Conference on Engineering Education, pp. 1-8, 2010.

[3] K.-C. Yang, Y.-T. Chang, C.-M. Wu, and C.-M. Huang, "Application-Oriented Teaching of Embedded Systems," accepted by IEEE International Conference on Microelectronic Systems Education, 2011.

[4] E. Monmasson, and M.N. Cirstea, "FPGA Design Methodology for Industrial Control Systems—A Review," IEEE Transactions on Industrial Electronics, vol. 54, No. 4, pp. 1824-1842, 2007.

[5] J.J. Rodriguez-Andina, M.J. Moure, M.D. Valdes, "Features, Design Tools, and Application Domains of FPGAs," IEEE Trans. on Industrial Electronics, vol. 54, no. 4, pp. 1810-1823, 2007.

[6] M.N. Cirstea, A. Dinu, "A VHDL Holistic Modeling Approach and FPGA Implementation of a Digital Sensorless Induction Motor Control Scheme," IEEE Trans. on Industrial Electronics, vol. 54, no. 4, pp. 1853-1864, 2007.

[7] "Altera website," Available: http://www.altera.com

[8] "Xilinx website," Available: http://www.xilinx.com

[9] Taiwan traffic signal, Video clip: http://www.youtube.com/watch?v=GWaueEk1pE8

[10] LEON Specification, LEON, http://www.gaisler.com/cms/

978-1-61284-863-1/11 $26.00 © 2011 IEEE

An Improved Dynamic-Biased CMOS Operational Amplifier for Biomedical Circuit Applications

H. L. Tan, G. T. Ong and P. K. Chan
School of EEE, Nanyang Technological University, Singapore
tanh0122@e.ntu.edu.sg, ongg0009@e.ntu.edu.sg, epkchan@e.ntu.edu.sg

Abstract— **An improved dynamic-biased CMOS operational amplifier for biomedical circuits is presented in this paper. The proposed ultra-low power operational amplifier comprises a weak-inversion biased differential input stage, a pseudo class-AB output stage and a multi-phase master bias dynamic biasing circuit. Using GLOBALFOUNDRIES 0.18 µm CMOS process, the proposed amplifier, without dyamic biasing circuit, consumes a static supply current of 5.94 µA at a 1.8V supply. The simulated result shows a DC gain of 89.63dB and an unity gain bandwidth of 2.55 MHz at a capacitive load of 30.5 pF. When activating the proposed dynamic biasing circuit to the same amplifier in a reported sample-and-hold (S/H) circuit for biomedical application, the S/H circuit consumes 8.16 µW at a sampling frequency of 128 kHz. In response to 1-Vpp and 1-kHz sinusoidal input, the S/H circuit has achieved -76.67 dB of total harmonic distortion (equivalent to 12.73 bits of linearity-based ENOB) and 5.21×10-3 µA/MHz of Figure of Merit .**

I. INTRODUCTION

MOST of the medical devices processes biomedical signals to diagnose the state of underlying biologic and physiologic structures and dynamics. Portability is highly valued while in other cases an essential for some devices. As such, it is important to ensure every component of the devices consume as little power as possible in order to lengthen the devices' usage lifespan. Since operational amplifier (op-amp) is one of the main power consuming units in biomedical circuits, an ultra-low power op-amp is focused in this paper.

Refer to the earlier works [1]-[2], the dynamic biasing technique had been reported to reduce power consumption of amplifiers or op-amps effectively. Due to the periodical on-off the biasing network, a circuit will exhibit lower power dissipation with respect to that of static biasing technique.

Fig. 1 depicts a dynamic-biased two-stage op-amp [2]. Both the tail current source transistor and the output stage current source transistor are commonly biased by a diode-connected transistor on which the top pmos transistor serves as a shield so as to avoid over-voltage biasing to the mirror

transistor. When Clk 2 is high and Clk 1 is low, the capacitor Co is reset and the op-amp is isolated with biasing network. When Clk 2 is low and Clk 1 is high, the source of shield transistor, which is initially charged at V_{DD}, starts to decrease due to the flow of charging current to Co. As a result, the op-amp is powered during this time period. However, the drawback of this type of dynamic biasing technique is suitable for op-amp structures without employing cascode transistors. Due to the dynamic variation of the voltage with respect to time to the fixed gate bias cascode transistors, there exists risk of stressing the cascode transistors beyond operational limits. This leads to headroom issue which results in the degradation of circuit linearity in op-amp. Thus, this jeopardizes the performance of dynamic parameters of op-amp at the expense of reduced power consumption in switched-capacitor circuit applications.

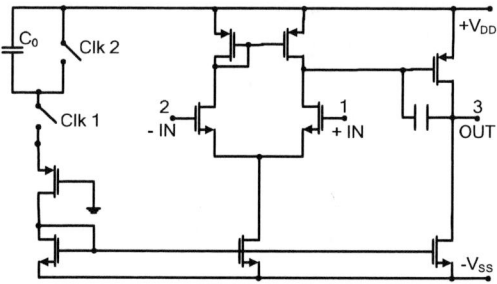

Fig. 1 Dynamic 2-stage CMOS op-amp using dynamic current sources

This motivates the investigation of an improved dynamic biasing scheme for the design of an ultra-low power op-amp with sustained dynamic performance metric when referencing from the benchmark circuit using static biasing technique. Its performance is demonstrated by employing the improved op-amp in the previously reported sample-and-hold (S/H) circuit.

Following the Introduction, Section II describes the proposed op-amp for use in the S/H circuit. Section III presents the results and discussions for the proposed design. The concluding remarks are given in Section IV.

978-1-61284-863-1/11 $26.00 © 2011 IEEE

II. LOW-POWER CMOS OP-AMPS

Fig. 2 shows a typical micropower weak-inversion based op-amp which is applied in a biomedical S/H circuit [3] shown in Fig. 3. It is a low-power CMOS two-stage op-amp that consists of the front-end cascode mirror differential gain stage (P1-P3, N1-N4) and the push-pull output stage (P6-P7, N6-N7). The cascode biasing network is formed by the devices P4 and N5. The wide-swing cascode mirror serves as an active load and provides the cascode frequency compensation for obtaining good power-bandwidth efficiency at low power. The push-pull stage supports rail-to-rail output swing while improving the load driving ability.

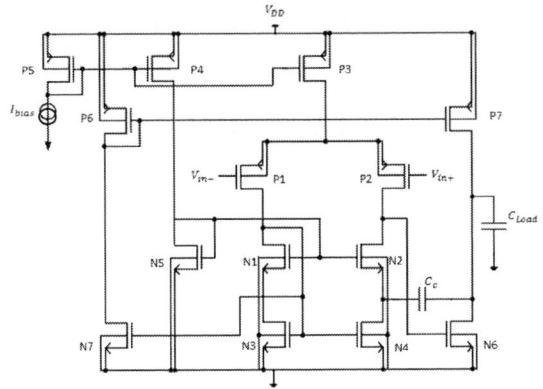

Fig. 2 A low-power cascode-compensated push-pull op-amp

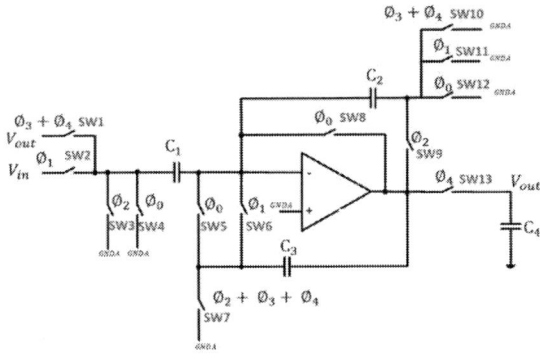

Fig. 3 Op-amp of Fig. 2 in the reported S/H circuit [3]

A. Ultra-Low Power Dynamic-Biased Class AB Op-Amp

In the reduced power design, Fig 4 depicts the dynamic biased class-AB topology. P-channel transistors are used as switches in the dynamic biasing circuitry. The tail current source P3 is added with a multiphase dynamic biasing network due to multiple phase operation of the S/H circuit as required in Fig. 3. The resistor Rc is added to produce a zero that ensures good stability of the operational amplifier even biased at lower biasing currents. Cascode compensation is still adopted for bandwidth efficiency [4]. In order to further save the power consumption contributed by P4 and N5 in Fig. 2, the push-pull output stage is replaced by a pseudo class-AB output stage [5]. It is explained as follows. Transistors P7 and N6 form a structure that is similar to the conventional Class A

structure. The capacitor, $C_{coupling}$, serves as the floating battery source for ac coupling. The pseudo-resistor transistor implemented using PMOS transistor P6, operates in the cut-off region and provides a very large resistance for passing direct current (DC) bias voltage, but blocking the alternating current (AC) signal from the capacitor, $C_{coupling}$, to the dc biasing circuit. When $C_{coupling}$, serves as the floating battery source for ac coupling for N6, this gives the push output stage with P7. An economical rail-to-rail Class AB output stage is realized.

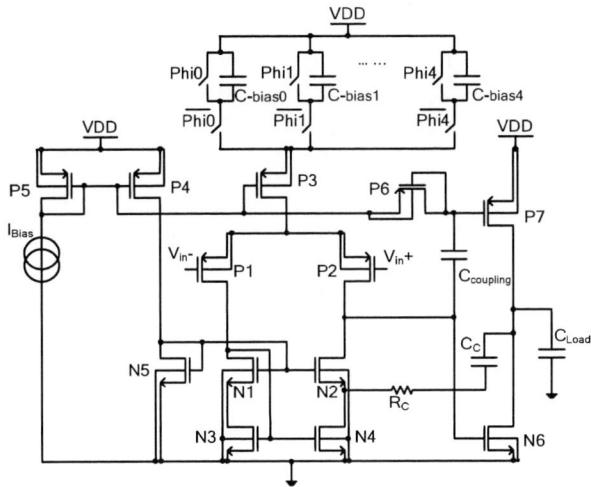

Fig. 4 Ultra-low power Class AB op-amp employing current source dynamic biasing (CSDB) circuit [2]

B. Proposed Ultra-Low Power Dynamic-Biased Class-AB Op-Amp

In order to optimize the circuit performance of the op-amp in the context of ultra-low power consumption and dynamic parameter performance metrics, a master bias dynamic biasing (MBDB) scheme is proposed in Fig. 5. As can be seen, both the current source P3 and cascode transistors, N1-N2, are simultaneously applied with dynamic bias through the central master bias. The MBDB network is mainly formed by the current source I_{Bias}, devices N5, N8-N9 and P4-P5 together with the associated switches and storage capacitors.

N-channel transistors are used as switches in the dynamic biasing circuitry. Initially, the gate voltage of N8 is fixed by the diode-connected transistor N9. Consider MBDB op-amp and taking phase 1 as an example. During dynamic operation, Phase 1_bar is used to reset C-bias1 and discharge the voltage across the capacitor to analog ground. During phase 1, C-bias1 is charged up, causing $V_{C-bias1}$ to increase. Hence V_{gs} of N8 will drop. As a result, the current flowing through branch P5/N8 reduces. This process is repeated for the rest of the other phases. Since the transistors P3 and P4 are current mirror transistors of P5, the biasing current to the differential input stage will be reduced. Due to the dynamic current mirror action to the devices P4 and N5, the cascode biasing

978-1-61284-863-1/11 $26.00 © 2011 IEEE

voltage becomes dynamic biased, keeping V_{gs1} and V_{gs2} more or less constant. This is contrasting to the conventional cascode bias using a fixed dc source. Consequently, the reduction in current flow to the branches P5/N8, P4/N5 and P3 will cause the op-amp power consumption to be reduced.

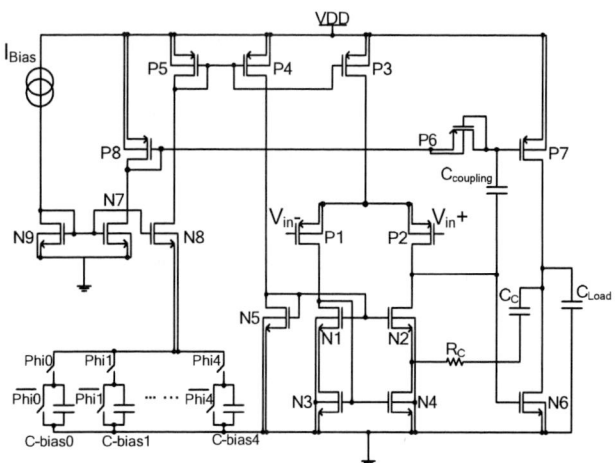

Fig. 5 Proposed ultra-low power Class AB op-amp employing master bias dynamic biasing (MBDB) circuit

III. RESULTS AND DISCUSSIONS

The op-amp and its respective S/H circuit is realized using GLOBALFOUNDRIES 0.18 μm CMOS process technology. It is simulated using Cadence Spectre and BSIM3 models. All the circuits are designed to work with a supply voltage of 1.8 V. For S/H circuits, the dynamic performance parameters are verified using an input of 1-Vpp and 1-kHz sinusoidal input at 128 kHz of sampling frequency.

A. Op-Amp Performance

Three op-amp performance metrics are compared in Table 1. Without applying the dynamic biasing techniques (all capacitor switches are on to reset capacitors, and the rest of connection switches are on so that all the biasing branches are on) in the CSDB and MBDB op-amp topologies, it can be seen that the two op-amps display similar ac performance parameters with respect to that of the benchmark op-amp [3] when driving identical capacitive loads.

Table I Performance comparison of three op-amps at static biasing condition

Parameter	Ref. [3]	CSDB	MBDB
Open-Loop Gain	93.64 dB	89.58 dB	89.63 dB
Phase Margin	59.36	50.82	49.19
Unity Gain Bandwidth	2.53 MHz	2.19 MHz	2.55 MHz
Capacitance Loading	30.5 pF	30.5 pF	30.5 pF
Total static supply current	6.69 μA	3.55 μA	4.94 μA

The two op-amps employing pseudo Class-AB output stage plus improved frequency compensation network are still stable while consuming lower static power. The MBDB op-amp topology suffers from higher power consumption with respect to CSDB op-amp. This can be attributed by two more biasing branches in the biasing network design.

B. Power Consumption of S/H Circuits

For the case of switched-capacitor circuit application, the respective dynamic biasing network for CSDB and MBDB op-amp is activated with the clock signals. When the op-amps are applied in the given S/H circuit topology, the respective mean power consumption of each S/H circuit is compared in Table II.

Table II Average power consumption in each S/H circuit

Parameter	Ref. [3]	CSDB	MBDB
Power Consumption	12.04 μW	4.63 μW	8.16 μW
Current Consumption	6.69 μA	2.57 μA	4.53 μA
Power Efficiency	0.39 μW/pF	0.15 μW/pF	0.27 μW/pF

It can be seen that both dynamic-biased S/Hs display lower power consumption when compared to the benchmark.

C. Total Harmonic Distortion of S/H circuits

The total harmonic distortion of a signal is defined as the ratio of the total power of the second and higher harmonic components to the power of the fundamental for that signal [6]. THD presented as a percentage value is as follows:

$$THD = \frac{\sqrt{V_{h2}^2 + V_{h3}^2 + V_{h4}^2 + \cdots}}{V_f} \times 100\%$$

where V_f and V_{hi} is the amplitude of the fundamental and ith harmonic component respectively. The spectral components of two dynamic-biased S/H circuits are compared in Discrete Fourier Transform (DFT) plots. The MBDB scheme displays lower magnitude in harmonic components, starting from the 3th harmonic. This suggests the technical merit of the proposed MBDB scheme with respect to the CSDB scheme.

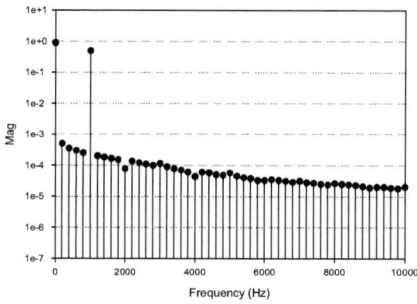

Fig. 6 CSDB DFT plot of V_{out} in response to 1-Vpp 1-kHz input frequency

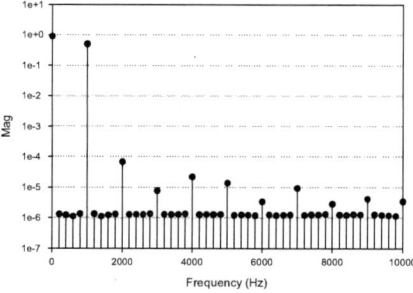

Fig. 7 MBDB DFT plot of V_{out} in response to 1-Vpp 1-kHz input frequency

978-1-61284-863-1/11 $26.00 © 2011 IEEE

In S/H context, THD reflects the held values as well as the tracking component of the waveform where a substantial source of non-linearity exists. Consider THD is translated into ENOB through the following formula,

$$ENOB_{THD} = \frac{log\left(\frac{100}{THD\%}\right)}{log2}$$

The higher the ENOB, the better the resolution is. Table III shows that the proposed MBDB scheme offers better THD performance metric than CSDB scheme whilst maintaining similar with that of [3].

Table III THD and ENOB of the S/H circuits using respective op-amp

Parameter	Ref. [3]	CSDB	MBDB
THD (%)	0.0159%	0.0333%	0.0147%
THD (dB)	-76.00	-69.55	-76.67
ENOB$_{THD}$	12.62 bits	11.55 bits	12.73 bits

D. Figure of Merit (FOM) of S/H Circuits

In order to normalize the power efficiency, the total current consumption is used instead of the total power consumption. It factors out the influence of the different supply voltages and gives a fair comparison in term of power efficiency. Hence, the FOM is defined as follows,

$$FOM = \frac{Total\ Current\ Consumption}{2^{ENOB_{THD} \times f_s}}$$

The lower the FOM, the better the power efficiency of a sample-and-hold circuit is.

Table IV FOM of the S/H circuits using respective op-amp

Parameter	[3]	CSDB	MBDB
FOM (µA/MHz)	8.30×10^{-3}	6.70×10^{-3}	5.21×10^{-3}

Although the CSDB scheme displays relatively lower power consumption, it suffers from higher linearity that degrades the FOM value. However, the proposed MBDB scheme offers the best FOM value. The suggests that MBDB scheme provides a better balance performance in terms of power consumption, THD, linearity based ENOB, leading to the

improved FOM performance metrics despite of having higher power consumption with respect of CSDB.

IV. CONCLUSION

The proposed master bias dynamic biasing op-amp offers reduced power consumption and sustains low-distortion characteristic with respect to the continuous-time counterpart. It achieves the best FOM value when compared to the same S/H circuits using op-amp with either static biasing technique or the current source dynamic biasing technique. The improved dynamic biasing op-amp will be useful for ultra-low power biomedical circuit applications.

REFERENCES

[1] R. Copeland, "Dynamic amplifier for M.O.S. technology", *Electronics Letters,* vol. 15, no. 10, pp. 301-302, Oct. 1979.

[2] B. J. Hosticka, "Dynamic CMOS amplifiers," *IEEE Journal of Solid-State Circuits,* vol. SC-15, no.5, pp. 887-894, May 1980.

[3] S.L. Mah, P.K. Chan and Shiv Kumar Mishra. "A precision low-power mismatch-compensated sample-and-hold circuit for biomedical applications", *Proc. IEEE Asia Pacific Conference on Circuits and Systems,* (APCCAS), pp. 192-195, Dec. 2010.

[4] P. R. Gray, P. J. Hurst, S. H. Lewis, and R. G. Meyer, "Analysis and design of analog integrated circuit", Fourth Edition, Chapter 9, John Wiley & Sons, Inc., 2001.

[5] J. Ramirez-Angulo, R. G. Carvajal, J. A. Galan and A. Lopez-Martin, "A free but efficient low-voltage class-AB two-stage operational amplifier," *IEEE Transactions on Circuits and Systems II: Express Briefs,* vol. 53, no. 7, pp. 568-571, July 2006.

[6] D. A. Johns and K. W. Martin, *Analog Integrated Circuit Design.* John Wiley & Sons, 1997.

[7] T.S. Lee, C.C. Lu and J.T. Zhan, "A 250MHz 11Bit 20mW CMOS low-hold-pedestal fully differential track-and-hold circuit", *International Symposium on VLSI Design, Automation and Test,* pp.1 - 4, 2006.

[8] A. Boni, A. Pierazzi and C. Morandi, "A 10-b 185-MS/s track-and-hold in 0.35-µm CMOS," *IEEE Journal of Solid-State Circuits,* vol. 36, pp. 195-203, 2001.

[9] A. Baschirotto, "A low-voltage sample-and-hold circuit in standard CMOS technology operating at 40 Ms/s", *IEEE Trans. on Circuits and Systems II: Analog and Digital Signal Processing,* vol. 48, pp. 394-399, Apr. 2001.

[10] H. H. Ou, B. D. Liu and S. J. Chang, "A 0.8-V 250-MSample/s double-sampled inverse-flip-around sample-and-hold circuit based on switched-opamp architecture", *IEICE Trans. on Electronics,* vol. E91-C, no. 9, pp. 1480-1487, Sept. 2008.

Table V Performance comparison with previous reported S/H circuits

Design	Boni [8]	Baschirotto [9]	Hsin-Hung [10] *	Sai-Lei [3] *	This work *
CMOS Technology	0.35 µm	0.5 µm	0.13 µm	0.18 µm	0.18 µm
Sampling Rate	185 MHz	40 MHz	250 MHz	128 kHz	128 kHz
Full-scale Input Range	1 Vpp	0.6 Vpp	0.8 Vpp	1 Vpp	1 Vpp
THD	-63dB @ 45 MHz	-50dB @ 2 MHz	-67.3dB @ 25 MHz	-75.96 dB @ 1 kHz	-76.67 dB @ 1 kHz
Supply Voltage	3.3 V	1.2 V	0.8 V	1.8 V	1.8 V
Power Consumption	70 mW	1.2 mW	3.5 mW	12.04 µW	8.16 µW
Total Current Consumption	21.2 mA	1.00 mA	4.38mA	6.69 µA	4.53 µA
ENOB$_{THD}$	10	8	11	12.62	12.73
FOM (µA/MHz)	1.12×10^{-1}	9.77×10^{-2}	8.54×10^{-3}	8.30×10^{-3}	5.21×10^{-3}

*simulation results

978-1-61284-863-1/11 $26.00 © 2011 IEEE

An Angina Diagnosing System using Fuzzy Clustering and Correlation in FPGA

Evaldo R F Cintra, Tales C Pimenta and Robson L Moreno
Universidade Federal de Itajuba
Itajuba, Brazil
e-mail tales@unifei.edu.br

Abstract—**In this paper, we present a signal processing method capable of detecting angina in electrocardiograms, that was implemented in FPGA. The adopted procedure is based on fuzzy clustering to reduce the amount of data sampling and correlation to compare with samples from a previously established database. By using the correlation method on the samples, it is possible to establish an initial indication of angina . The reduced number of samples of the clustering process turns the processing simpler and allows its hardware implementation in FPGA to validate it. According to the tests conducted, the method achieves 85% correct diagnoses.**

Keywords-cardiopathy; heart; correlation; clustering; electrocardiogram insert

I. Introduction

Due to the large number of death caused by heart diseases, researchers have been working on the search for solutions that can provide early detection of heart problems, and thus increase the chance of survival [1-3].

In order to diagnose a cardiopathy some factors are taken into account such as patient's age and physical activity. Nevertheless, the main analysis is based on the electrocardiogram. The waveform presented in Figure 1 represents a typical electrocardiogram signal. This signal is obtained by detecting and amplifying tiny electrical changes on the skin that are caused when the heart muscle "depolarizes" during each heartbeat. A typical electrocardiogram waveform is obtained in millivolts per second according to the PhysioNet database.

The diagnosis of a cardiopathy is made by assessing and detecting any amplitude or time variation during each interval.

Some studies [1–14] describe computer systems running signal processing techniques that evaluate the characteristics of the electrocardiograms to obtain a preliminary diagnosis of any disease.

This article intends to show a signal processing technique that uses fuzzy clustering to reduce the amount of data to be processed, and a correlation method used to identify angina on

The electrocardiogram. The smaller amount of data means also hardware simplification. It does allow FPGA implementation and provides diagnosis similar to software implementations [4-6], as it will be shown in Section 5.

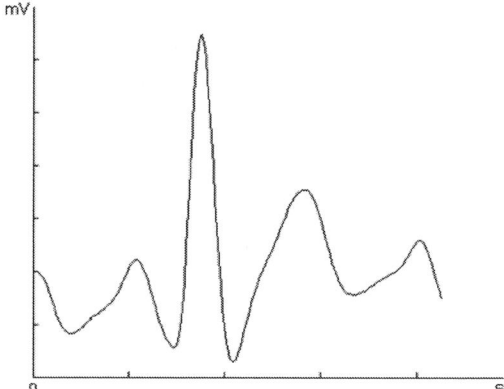

Figure 1.   Tipical electrocardiogram waveform.

This approach required a data signal bank in which each signal presented the main features of an electrocardiogram with angina.

Due to the presence of noise and DC components in the electrocardiograms, signal processing tools such as the third order Butterworth filter were used. The elimination of DC level allows more accuracy to the results.

The signal is them compared to a signal database and a correlation value between then is generated. The diagnosed cardiopathy corresponds to the signal from the database that shows the highest direct correlation with the signal under analysis.

II. Fuzzy Clustering

The fuzzy clustering process for data pruning [15] allows the system to process a smaller number of samples to describe the main features of the signal. Therefore, the required processing to generate the signal diagnosis is faster and consequently the required hardware to process it can be simplified.

The clustering process consists of generating clusters, where each cluster has its own function that will determine the system output to be controlled, according to the corresponding weights of each cluster for a particular system input [15].

In this work, the clustering process is used to indicate the value and location of each cluster, without requiring the generation of functions or rules corresponding to each cluster.

978-1-61284-863-1/11 $26.00 © 2011 IEEE

Consider the set of N input-output data pairs where X is the n dimensional input vector $X = [x_1, x_2, ..., x_n]$ corresponding to the acquisition time of each electrocardiogram signal and vector $Y = [y_1, y_2, ..., y_n]$ of the electrocardiogram samples generated for each time given by vector X. Here, n corresponds to the nth sample.

The parameters ai and bi of the corresponding functions in each rule are obtained through expression (1) [16].

$$\theta' = [(X')^T.X']^{-1}.(X')^T.y \qquad (1)$$

where,

$$X' = [\Gamma_1.X_e, \Gamma_2.X_e...., \Gamma_M.X_e] \qquad (2)$$

According to expression (2), X_e is a matrix $X_e = [X,1]$, and the activation of each rule is provided by Γ_i, which is a diagonal matrix whose normalized degree u_{ki} is the diagonal element.

$$u_{ki} = \frac{A_i(x_k)}{\sum_{i=1}^{M} A_i(k)} \qquad (3)$$

where,

u_{ki} – is the normalized degree of participation of each input for rule R_i as given by expression (3) [16]:
A_i is a group of fuzzy antecessors of a given i–rule, given by expression (4) [16].

$$A_i(x) = \prod_{j=1}^{n} \mu_{ij}(x_i) \qquad (4)$$

where the membership degree of each rule, regarding to the input x_i, is given by μ_{ij}.

Once A_i is achieved, the normalized degree of antecessor for rule R_i can be obtained.

By running the algorithm for the reduction of number of clusters, it is obtained a vector v that provides the prototype of the most important prototypes of clusters. Vector v is given by expression (5) [16]:

$$v^{(l)} = \frac{\sum_{k=1}^{N} \left(u_{ki}^{(l-1)}\right)^m . z_k}{\sum_{k=1}^{N} \left(u_{ki}^{(l-1)}\right)^m}, \quad 1 \le i \le M \qquad (5)$$

where
Z_k – is the matrix in which each column represents the input output pair as $Z_k = [X_k, Y_k]$;
m – is a fuzziness parameter (m > 1);
M – is the number of rules;
N – is input-output data pairs;
l – number of interactions.

The membership functions can be generated according to the information from a person that knows the process, or can be used known standard format such as trapezoidal, triangular, Gaussian, sine and sigmoidal. We have chosen the Gaussian function due to its easy implementation.

According to the fuzzy clustering process, 20 clusters were generated, that describe the cluster prototypes to be processed.

Figure 2 shows the 20 clusters obtained by fuzzy clustering process for the signal of figure 1. They describe the main changes in the electrocardiogram signal. These samples were selected according to the fuzzy clustering process.

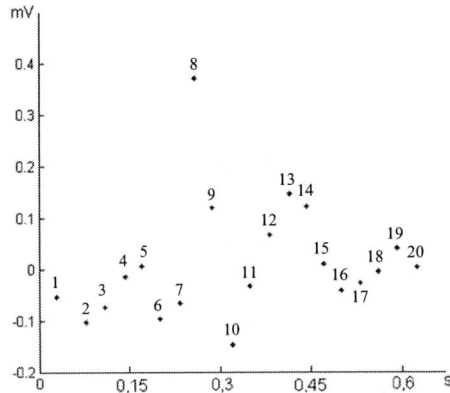

Figure 2. Cluster prototypes generated by the fuzzy clustering process.

III. CORRELATION

Many computer programs are used to obtain a diagnosis, such as Hidden Markov Models [1], Fuzzy classifiers [6], Artificial Neural Network and Rough Set Theory [7], Discrete Wavelet Transform [9]. Correlation was used in this work to reduce the processing required and to simplify the hardware used to implement it.

By using the fuzzy clustering, these samples will be reduced to a set of 20 samples, which are represented by the generated clusters.

Our system compares the 20 electrocardiogram samples from the data base, and 20 electrocardiograms samples of the signal under test. It is verified if exists a correlation between the signal under evaluation and the data bank. If a correlation is found, it means that the signals present similar variations, and thus the may have the same diagnosis.

The clusters generated by the fuzzy clustering process are compared with the clusters of signals from a database, whose diagnosis is known. This comparison is done by calculating the correlation among them.

The system will identify the diagnosis as the signal from the database that receives the highest correlation with the assessed signal.

The calculation of correlation between two signals can be obtained by expression (6) [16].

$$\rho = \frac{\sum x * y}{n * \sigma x * \sigma y} \qquad (6)$$

where,
ρ - is the correlation value;
x - is calculated according to equation (7).

$$x = X - MX \qquad (7)$$

X - –re points of the sampled signal;
MX - is the arithmetic mean of these sampled points, given as (8);

$$MX = \frac{\sum X}{n} \qquad (8)$$

y - is given by equation (9).

$$y = Y - MY \qquad (9)$$

Y - are the points of the signal pattern to be compared;
MY - is the arithmetic mean of these sampled points, give as (10);

$$MY = \frac{\sum Y}{n} \qquad (10)$$

n - is the number of points for X and Y;
σx - is the standard deviation of x;
σy - is the standard deviation of y.

According to the correlation calculation presented by equation (6), it can be observed that by using the fuzzy clustering process of the number of points to be processed in the correlation are smaller, thus the processing time becomes shorter.

IV. VALIDATION SYSTEM

In order to demonstrate the effectiveness of the techniques previously described, a system was created to validate the proposed theory. The validation system receives the samples of an electrocardiogram signal to be diagnosed. The signal is filtered and the most important features of the signal are obtained by clustering process. The diagnostics is obtained from the smaller set by correlation.

The proposed system was simulated on MATLAB®. The electrocardiogram signals used to create the database and to perform the tests were obtained from PhysioNet database [1].

After the simulation, the system was validated in an FPGA implementation on a XILINX Spartan®-3A Starter FPGA.

The Physionet Data presented in Figure 3 represents the electrocardiogram signal acquisition. Each signal is represented by 2,500 samples at a sampling frequency of 333 Hz.

It was calculated the arithmetic average of the signal in order and them subtracted from the signal on each sample in order to eliminate the DC offset.

In the second block, 20 samples are separated to be processed by the fuzzy clustering process. The input signal samples are are processed according to expression 1. The block outputs the 20 samples, obtained by the clustering system. In the third block, the 20 samples are correlated with the database signal samples, as described in section III. It outputs the generated correlation values.

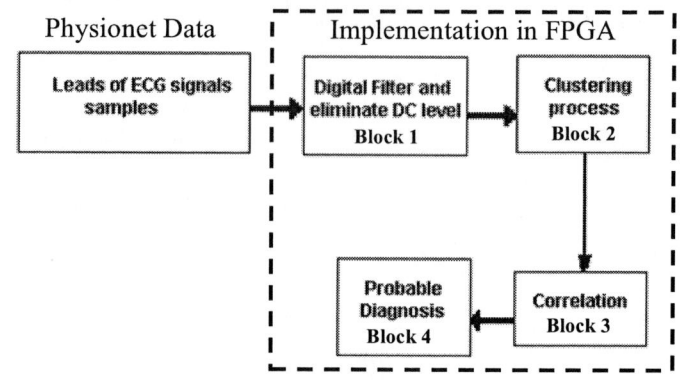

Figure 3. Describing the validation system.

In the final block, the correlation values from the previous block are compared, and the signal from the database that presents largest direct correlation is indicated as the probable diagnosis.

V. TESTS PERFORMED AND CONCLUSIONS

According to the literature, performance results are presented in terms of sensitivity (Se), positive predictivity (Pp) and accuracy (Acc).

The *Se* parameter indicates the percentage of correct diagnoses compared to diagnoses not detected. It can be obtained according to the equation (12) [17].

$$Se = \frac{Tp}{Tp + Fn} \qquad (12)$$

where *Tp* and *Fn* are the correct and undetected diagnoses, respectively.

The *Pp* parameter indicates the percentage of correct diagnoses in compared to wrong diagnoses. It is given by expression (13).

$$Pp = \frac{Tp}{Tp + Fp} \qquad (13)$$

where *Fp* indicates the wrong diagnoses.

The *Acc*–parameter indicates the accuracy of the system and can be obtained according to equation (14).

$$Acc = 1 - \frac{N_{err}}{N_{total}} \qquad (14)$$

where *Nerr* indicates the number of wrong diagnoses and N_{total} indicates the total number of diagnoses.

According to the presented validation system and the comparison parameters, tests were performed to verify its effectiveness.

Table 1 shows our work with other presented in the literature.

The system was validated using electrocardiogram signals of previously known medical diagnosis. The system compares the signal under test and signals of previously diagnosis data bank. After the processing described on Sections II and II, the system generates a diagnosis. The diagnosis is then compared to the medical diagnosis. The physician and the system diagnoses are used to generate the merit figures Pp, Se and Acc, as presented in Table I.

TABLE I.   COMPARATIVE FIGURES WITH OTHERS STUDIES

Ref.	Accuracy	Pp	Se
[1]	-	85 %	83 %
[2]	-	75 %	83 %
[3]	-	88 %	87 %
[4]	-	78 %	89 %
[5]	-	81 %	84 %
[6]	93 %	-	-
[9]	-	99 %	99 %
[15]	96 %	-	-
[This Work]	85 %	93 %	90 %

It can be observed from the table that our work is similar or superior to the other words. The proposed system allows the development of a fast and simple hardware implement since the use of fuzzy clustering reduces the number of samples to be processed.

The system presented is able to provide probable diagnoses with the same effectiveness of other processing systems. By using fuzzy clustering, processing is greatly reduced since the correlation is not conducted on all signal samples.

Since the calculations conducted require fewer samples, and consequently less memory, the system can be easily implemented in hardware, such as an FPGA with a soft processor. The system was implemented in a Xilinx Spartan®-3A Starter Kit with the Spartan-3A FPGA. It is a low cost board that runs at 50 MHz. It has 32 MB DDR2 SDRAM memory, I/O RS-232, serial port serial, 4 bottoms, 4 switches, LEDs, clock counter and JTAG USB download port [19]. Microblaze® is a XILINX 32-bit RISC soft processor Intellectual Property – IP [20]. The obtained results were the same as using MATLAB®.

The system was implemented in a Spartan-3A FPGA according to Section 5. The tests were conducted for many sets of samples, and it was observed that the number of clock cycles for 20 samples is approximately 9 times shorter than the number of clock cycles for 213 samples of the whole signal. Therefore, the fuzzy clustering process, used to reduce the number of samples to be processed, caused a reduction in the number of clock cycles in the hardware implementation.

Thus, there is an almost linear relationship between the number of clock cycles and samples to be processed.

ACKNOWLEDGMENT

The authors acknowledge CAPES, CNPq and FAPEMIG for their financial support.

REFERENCES

[1] Andreão R. V., "ST-Segment Using Ridden Markov Model Beat Segmentation: Aplication to Ischemia Detection," Institut National des Télecommunicatioons, France. Universidade Federal do Ceará, Fortaleza, Brazil. Computers in Cardiology 2004; 31:381 - 384.

[2] Vila J., Presedo J. et al. SUTIL: *Intelligent ischemia monitoring system.* Int J Med Inf 47(3): 193-214, 1997.

[3] Jager F., Moody G. B. and Mark R. G. Detection of transient ST segment episodes during ambulatory ECG monitoring. Comput. Biomed. Res. 31(5):305-22, 1998.

[4] Maglaveras N., Stamkopoulos T. et. Al. "An adaptive backpropagation neural network for real-time ischemia episodes detection: development and performance analysis using the European ST-T database". IEEE Trans Biomed Eng 45(7):193-214, 1998.

[5] Taddei A., Constantino G. et AL., A System for the Detection of Ischemic Episodes in Ambulatory ECG.   Computers   in Cardiology, Vienne, Autriche, 1995, pp. 705 – 708, 1995.

[6] Mrs. Anuradha, B.; Reddy, V. C. Veera, "Cardiac Arrhythmia Classification Using Fuzzy Classifiers," Journal of Theoretical and Applied Information Technology, 2005 – 2008.

[7] Setiawan, N.A.; Venkatachalam, P. A.; Hani, A. F. M., "Missing Data Estimation on Heart Disease Using Artificial Neural Network and Rough Set Theory," *IEEE Int Conf Intelligent and Advanced Syst, 07.*

[8] Souza, Camila B.; Andreão, Rodrigo V.; Segatto, Marcelo V., "Processamento de sinais de ECG para geração automática de alarmes," VI Workshop de Informática Médica – WIM2006.

[9] Zheng, Huabin; Wu, Jiankang, "Real-time QRS Detection Method, 2008 10th IEEE Intl. Conf. on e-Health Networking, Applications and Service

[10] Patil, Shantakumar B.; Dr Kumaraswamy Y. S., "Extraction of Significant Patterns from Heart Disease Warehouses for Heart Attack Prediction," IJCSNS – International Journal of Computer Science and Network Security, Vol. 9 No. 2, February 2009.

[11] Li SHI; Hui LI; Zhifu SUN; Wei LIU, "Research on Diagnosing Heart Disease Using Adaptive Network-based Fuzzy Interferences System," Proceedings of International Joint Conference on Neural Networks, Orlando, Floeida, USA, August 12-17, 2007.

[12] Zimmerman, TG; Syeda-Mahmood, T, "Automatic detection of Heart Disease from Twelve Channel Electrocardiogram Waveforms" IBM Almaden Reserarch Center, San Jose CA, USA, Computers in Cardiology, 2007.

[13] Stoco, Marcelo S.; Andreão, Rodrigo V.; Segatto, Marcelo V., "Detecçâo Automática de Batimentos Cardíacos Utilizando Transformada Wavelet," VI Workshop de Informática Médica – WIM2006.

[14] Chen, Ying-Hsiang and Yu,Sung-Nien, "Comparison of Different Wavelet Subband Features in the Classification of ECG Beats Using Probabilistic Neural Network",EMBS Annual International Conference New York City, USA, Aug 30-Sept 3, 2006.

[15] Afsar, Fayyaz A.; Akram, M. U.; Arif, M.; Khurshid, J., "A Pruned Fuzzy k-Nearest Neighbor Classifier with Application to Electrocardiogram Based Cardiac Arrhytmia Rcognition", Proceedings of the 12th IEEE International Multitopic Conference, Dec 23-24, 2008.

[16] M. Setnes, "Supervised Fuzzy for Extraction", IEEE Transactions on Fuzzy Systems, Vol. 8, N° 4, August 2000.

[17] Cintra, Evaldo Reno Faria; "Diagnóstico de cardiopatias baseado no reconhecimento de padrões pelo método de correlação," UNIFEI – Universidade Federal de Itajubá, 2006.

[18] Inan, Omer T.; Giovangrandi, Laurent and Kovacs, Gregory T. A., "Robust Neural-Network-Based Classification of Premature Ventricular Contractions Using Wavelet Transform and Timing Interval Features",IEEE Transactions on Biomedical Engineering, Vol. 53, no. 12, December 2006.

[19] A. Armato, E. Nardini, A. Lanatà, G. Valenza, C. Mancuso, E.P. Scilingo, De Rossi, "An FPGA based arrhythmia recognition system for wearable applications", Ninth International Conference on Intelligent Systems Design and Applications, 2009.

[20] Xilinx inc., "Spartan-3A/3AN FPGA Starter Kit Board User Guide",Jun, 2008.

Secret Sharing based Countermeasure for AES *S*-Box

Yi Wang, Zheng Yuan, Zhican Li and Renfa Li
Embedded Systems & Networking Laboratory, Hunan University
Hunan Provincial Key Laboratory of Network and Information Security, Hunan University
Changsha, China
e-mail:estellewy@hotmail.com

Abstract—Cryptographic devices are vulnerable to Differential Power Attack (DPA) in embedded systems. Masking methods are popularly used to defend against DPA by masking all intermediate data with random values. However, masking schemes on algorithm level are vulnerable to Higher-Order DPA (HODPA), while on gate level glitch attack is the biggest threaten attack. In this paper, we proposed a secret sharing based countermeasure for AES *S*-box, which can defend against both HODPA and glitch attack. The experimental results show that our proposed design takes up less hardware resources and achieves faster speed compared with the existing methods.

Keywords-secret sharing; AES; power analysis attack; glitch attack; FPGA

I. INTRODUCTION

Information security in embedded systems arouses more and more attention on it. Modern cryptography technologies are popularly used in embedded systems in order to secure sensitive information. Rijndael algorithm was adopted by National Institute of Standards and Technology (NIST) as the Advanced Encryption Standard (AES) in 2001 [1]. It has been successfully applied to many different embedded applications, such as smart cards, Automatic Teller Machine (ATM), Radio-Frequency Identification (RFID) and Virtual Private Networks (VPN).

Unfortunately, more and more attackers attack the secret information of the cryptographic algorithm by using non-invasive attacks. This kind of attack is called side channel attack (SCA). Power analysis attack is the most effective attacks among them, where differential power attack (DPA) and glitch attack are proved to be the easy schemes to retrieve the sensitive information.

DPA attack is able to recover secret information by comparing the differences between the sample power trace and the correct key power trace. Örs *et al.* [2] and Mangard *et al.* [3] successfully broke the AES by using DPA attack. Higher-Order differential power attack (HODPA) is a more powerful attack that exploits joint leakage information of several intermediate values to "crack" the secret information. Joye *et al.* [4] and Proutff *et al.* [5] analyzed 2ODPA attack in theory respectively. Waddle *et al.* proposed several different 2ODPA attacks to overcome the masked cryptographic algorithms [6].

Glitch attack is an efficient attack on gate level. It exploits the secret information by analyzing the temporary states of the output which is leaded by the different arrival of the input signals [7]. Mangrad proposed a theoretical analysis of glitch [7]. The schemes used in Golić [8] and Canright *et al.* [9] can be attacked successful by glitch attack.

Numerous efforts have been devoted to the development of efficient countermeasures for AES implementations against attacks during last years. Masking has the advantage of easy implementation and low cost, therefore it is a popularly used countermeasure. The strategy of masking is to randomize the intermediate values during the computation of cryptographic algorithm. Usually, masking can be applied to gate level and algorithmic level. On gate level, the key issue is how to make a secure AND gate which is the critical non-linear computation of AES. Golić *et al.* [10] and Fischer *et al.* [11] proposed masking schemes on gate level, but unfortunately, both of them cannot resist against glitch attack. On algorithmic level, it will not interfere the implementation on gate level, therefore glitch attack can be avoided. Unfortunately it still cannot resist against HODPA. Baek *et al.* proposed masking algorithms for multipliers over finite fields, but they cannot resist against HODPA [12].

Secret sharing was introduced firstly by A. Shamir [13] and G. Blakley [14] in 1979 separately. It is an important technology for protecting sensitive data. Secret sharing is popular used in key management, secure multiparty computation, and group signature. And it also can be applied to gate level masking scheme [16][17].

In this paper, we proposed a secret sharing based countermeasure for AES *S*-box, which has the ability to defend against both HODPA and glitch attack. We also ported the proposed method to Xilinx Virtex-5 FPGA platform. The detailed comparisons are given to show that our proposed method is faster and takes up less hardware resources compared with the existing methods. We also prove that our proposed countermeasure has the ability to resist against HODPA and glitch attack.

The remainder of this paper is as follows: Section II introduces the definitions of secret sharing, AES and *S*-box. Our proposed countermeasure is described in Section III. Section IV gives the security analysis of the proposed method. The experimental results are given in Section V. Section VI draws the conclusion.

This work is supported by "Chinese National Science Foundation" (No.60873074 and No.60673061); "Changsha Science Technology Scheme" (No.K1003028-11) and "the Fundamental Research Funds for Chinese Central Universities".

II. Preliminaries

A. Secret Sharing

A secret sharing scheme involves a dealer and a group of participants, and the dealer distributes a secret among these participants by allocating each of them a share. Only a sufficient number of participants can reconstruct the secret by combining their shares together.

There has an extensive investigation on secret sharing as it can separate the risk and tolerate the intrusion. Wolkerstorfer proposed a secret sharing hardware scheme to improve the privacy of network monitoring [15]. Nikova *et al.* proposed a secret sharing based implementation of the multiplicative inverse in the finite field $GF(16)$ using 5 shares [16]. Nikova *et al.* proposed a secret sharing based implementation of the block cipher Noekeon using 3 shares [17].

B. Advanced Encryption Standard

AES is a symmetric encryption algorithm [1]. It supports block size of 128-bit and key sizes of 128, 192 and 256 bits. The encryption process starts with the first key addition, followed by a number of round functions which depends on the key size. In the encryption, the round function is composed of four transformations: ShiftRows cyclically shifts to left the bytes in the last three rows of the state, with different offsets; SubBytes is the non-linear byte substitution and operates independently on each byte of the state; the MixColumns that multiplies modulo x^4+1 the columns of the state by the polynomial $\{03\}x^3+\{01\}x^2+\{01\}x+\{02\}$; and finally the AddRoundKey adds a round key to the state All the needed round keys are generated by a key schedule which takes the secret key and expands it as specified in the standard.

C. AES S-Box

The security of AES relies on the design of *S*-box. *S*-box is an invertible substitution table, and it consists of two transformations [1]:

1) Multiplicative inverse: the input bytes x of *S*-box should take a multiplicative inverse in the finite field $GF(2^8)$, and the inverse of x is x^{-1}.

2) Affine transformation: the input bytes should take a affine transformation which can be expressed in matrix form as:

$$S(x) = \begin{bmatrix} 1 & 0 & 0 & 0 & 1 & 1 & 1 & 1 \\ 1 & 1 & 0 & 0 & 0 & 1 & 1 & 1 \\ 1 & 1 & 1 & 0 & 0 & 0 & 1 & 1 \\ 1 & 1 & 1 & 1 & 0 & 0 & 0 & 1 \\ 1 & 1 & 1 & 1 & 1 & 0 & 0 & 0 \\ 0 & 1 & 1 & 1 & 1 & 1 & 0 & 0 \\ 0 & 0 & 1 & 1 & 1 & 1 & 1 & 0 \\ 0 & 0 & 0 & 1 & 1 & 1 & 1 & 1 \end{bmatrix} \bullet x^{-1} + \begin{bmatrix} 1 \\ 1 \\ 0 \\ 0 \\ 0 \\ 1 \\ 1 \\ 0 \end{bmatrix} \quad (1)$$

x^{-1} is the 8-bit input of the affine transformation. $S(x)$ is the 8-bit output.

III. Proposed Serect Sharing Based Countermeasure

In this section, we propose a secret sharing based countermeasure for AES *S*-box which can resist against HODPA and glitch attack.

The 8-bit *S*-box of AES composes of a multiplicative inverse operation and an affine transformation. In the proposed countermeasure, the multiplicative inversion operation is the same as in section II, then we define a linear function $n=L(m)$ of affine transformation according to (1). The linear function $n=L(m)$ can be expressed as:

$$\begin{aligned} s &= a+e+f+g+h+1; & t &= a+b+f+g+h+1; \\ u &= a+b+c+g+h; & v &= a+b+c+d+h; \\ w &= a+b+c+d+e; & x &= b+c+d+e+f+1; \\ y &= c+d+e+f+g+1; & z &= d+e+f+g+h. \end{aligned} \quad (2)$$

m is the input of the affine transformation, and each bit is represented by $[h, g, f, e, d, c, b, a]$. n is the output of *S*-box, each bit is represented by $[z, y, x, w, v, u, t, s]$.

In order to apply the share functions to (2), we need to build the sharing of 8 bits inputs and outputs with 16 Boolean functions. Nikova *et al.* had proven the theorem that the minimum number of shares ρ required to implement a product of φ variables should satisfy: $\rho \geq 1+\varphi$ [16]. Therefore, 2 shares should be needed for function $n=L(m)$. We define the secret dividing functions: $n_1=\pounds_1(m_2, p_2, r)$, $n_2=\pounds_2(m_1, p_1, r)$ which are represented as follows:

$$n_1=\pounds_1(m_2, p_2, r): \begin{aligned} s_1 &= a_2+e_2+f_2+g_2+h_2+p_2+r; \\ t_1 &= a_2+b_2+f_2+g_2+h_2+p_2+r; \\ u_1 &= a_2+b_2+c_2+g_2+h_2+r; \\ v_1 &= a_2+b_2+c_2+d_2+h_2+r; \\ w_1 &= a_2+b_2+c_2+d_2+e_2+r; \\ x_1 &= b_2+c_2+d_2+e_2+f_2+p_2+r; \\ y_1 &= c_2+d_2+e_2+f_2+g_2+p_2+r; \\ z_1 &= d_2+e_2+f_2+g_2+h_2+r; \end{aligned}$$

$$n_2=\pounds_2(m_1, p_1, r): \begin{aligned} s_2 &= a_1+e_1+f_1+g_1+h_1+p_1+r; \\ t_2 &= a_1+b_1+f_1+g_1+h_1+p_1+r; \\ u_2 &= a_1+b_1+c_1+g_1+h_1+r; \\ v_2 &= a_1+b_1+c_1+d_1+h_1+r; \\ w_2 &= a_1+b_1+c_1+d_1+e_1+r; \\ x_2 &= b_1+c_1+d_1+e_1+f_1+p_1+r; \\ y_2 &= c_1+d_1+e_1+f_1+g_1+p_1+r; \\ z_2 &= d_1+e_1+f_1+g_1+h_1+r. \end{aligned} \quad (3)$$

Where n_1, n_2 are the 2 shares of output n which satisfy $n_1+n_2=n$, while m_1, m_2 are the shares of input m which satisfy $m_1+m_2=m$. p is a fixed value 1, which satisfies $p_1+p_2=p=1$. r is a random value in the proposed secret dividing function.

IV. Security Analysis

In this section, we prove that our proposed countermeasure is theoretical secure.

A. Against HODPA

HODPA is an efficient attack to modern cryptographic algorithms. It can retrieve the secret information by exploiting the joint leakage information of several intermediate values [7].

978-1-61284-863-1/11 $26.00 © 2011 IEEE

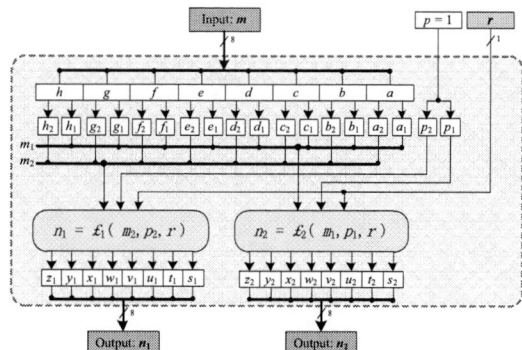

Figure 1. Secret dividing function

In the proposed countermeasure, because the input byte of the proposed S-box is divided into two shares, it is difficult for the attacker to retrieve the real intermediate values. Furthermore, the proposed secret dividing function has the characteristic of non-completeness, that is to say, $£_1$ is independent of m_1 and $£_2$ is independent of m_2. The attacker can only get two independent intermediate values by applying HODPA, which cannot recover the original intermediate values. Therefore, our proposed method can defend against HODPA.

B. Against Glitch Attack

Usually, the arrival time of the input signals are different, glitch attack can retrieve the secret information through the above differences by analyzing the temporary states of the output signals [7].

In our proposed method, the input byte of S-box is divided into two shares, therefore, one share is asynchronous with the other one, and the attacker cannot get the glitch information from the input byte. Moreover, a random value r is introduced in the secret dividing function, which randomizing the input byte and making it difficult for glitch attack.

V. IMPLEMENTATION AND RESULTS

In this section, we have implemented the proposed design using Hardware Description Language (HDL) and then synthesized our design using Xilinx ISE 12.1 and ported the design to Virtex-5 FPGA.

Fig. 2 shows the architecture of our proposed design. In Fig. 2, affine transformation of AES which are substituted by the proposed secret dividing function as described in section III. MUX and DE_MUX are controlled by C_1 and C_2 respectively, which determine whether share n_1 or share n_2 to be processed. During the procedure of key, we also divide the subkey into two shares after the key expanding operation.

Table I gives the comparison results of our proposed countermeasure with the existing masked S-box designs. From table I, it is obvious that our design has much better performance. On the same target platform of Virtex-4, our proposed secret sharing based design takes up 30% less area,

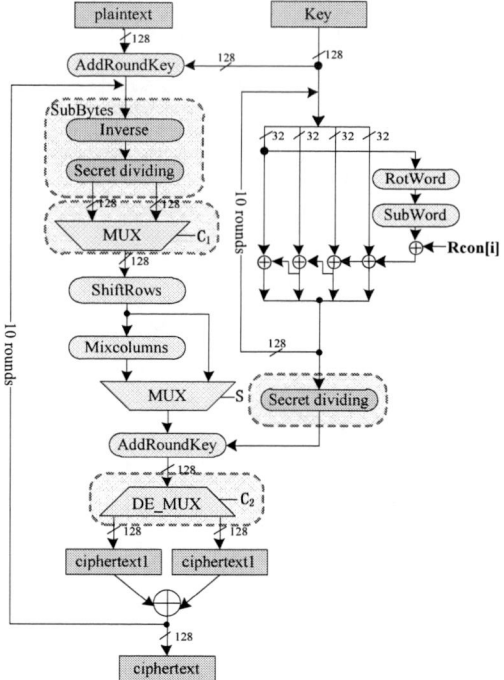

Figure 2. The architecture of secret sharing based countermeasure

and achieves the 1.83 times faster than Kamoun's one (Kamoun's design is the fastest among the exiting methods) [18].

Table II shows the experimental results of the proposed secret sharing based AES encryption and the existing methods. From table II, our design takes up the smallest area, which is 65% less than Kamoun's design (Kamoun's design is the smallest among the existing methods) [18]. Besides, our design achieves the fastest speed, which is 1.99 times faster than Kamoun's design (Kamoun's design is the fastest among the existing methods). The throughput of our proposed design is the best one, which is 1.76 times larger than Zheng' design (Zheng's design is the best among the existing methods) [23].

VI. CONCLUSION

In this paper, in order to resist against HODPA and glitch attack, we proposed a secret sharing based countermeasure for AES S-box. We proved that our proposed countermeasure has the ability to resist against both HODPA and glitch attack. The proposed method is based on two shares technology, therefore, the modified S-box could achieve less area, higher speed and higher throughput among the existing designs. When applying this method to AES implementation, the experimental results show that our proposed secret sharing based countermeasure for AES takes up the least hardware resources and achieve the fastest speed among the existing methods.

978-1-61284-863-1/11 $26.00 © 2011 IEEE

TABLE I. COMPARISON BETWEEN THE PROPOSED S-BOX AND THE EXISTING WORKS

	Platform	Technology	Area	Speed	DPA Resistant	HODPA Resistant	Glitch Resistant
Baek[12]	0.18μm CMOS	masking	954 gates	31.1 ns delay	Yes	No	No
Kamoun[18]	Xilinx Virtex-4 (LX25FF676)	masking	100 slices	16.67 ns delay	Yes	No	No
This Work	Xilinx Virtex-4 (XC4VLX200)	Secret sharing	70 slices	9.110 ns delay	Yes	Yes	Yes
This Work	Xilinx Virtex-5 (XC5VLX200)	Secret sharing	13 slices / 360 gates*	5.599 ns delay	Yes	Yes	Yes

*: the author estimate gate equivalents based on FPGA structure

TABLE II. COMPARISON BETWEEN THE PROPOSED AES AND THE EXISTING WORKS

	Platform	Technology	Area	Speed	Throughput	DPA Resistant	HODPA Resistant	Glitch Resistant
Baek[12]	0.18μm CMOS	masking	25,600 gates	15 MHz	11.9 Mbps	Yes	No	No
Trichina[20]	0.18μm CMOS	masking	20,506 gates	5 MHz	4 Mbps	Yes	No	No
zhao[22]	0.25μm CMOS	masking	48,000 gates	70 MHz	380 Mbps	Yes	No	No
zheng[23]	0.18μm CMOS	masking	49,000 gates	100 MHz	900 Mbps	Yes	No	No
Matsumoto[19]	Xilinx Virtex-2 (LC2VP30)	masking	3,017 Slices	44 MHz	512 Mbps	Yes	No	No
		secret sharing	10,619 Slices	63.7 MHz	337 Mbps	Yes	Yes	Yes
Kamoun[18]	Xilinx Virtex-4 (LX25FF676)	masking	2,281 Slices	137 MHz	-	Yes	No	No
This Work	Xilinx Virtex-5 (XC5VLX220)	secret sharing	795 Slices / 13,657 gates*	272.9 MHz	1,588 Mbps	Yes	Yes	Yes

*: the author estimate gate equivalents based on FPGA structure; -: not supported

REFERENCES

[1] National Institute of Standards and Technology. Advanced Encryption Standard (AES), FIPS-197, 2001.

[2] S. B. Örs, F. Gürkaynak, E. Oswald, and B. Preneel, "Power-analysis attack on an ASIC AES implementation," in ITCC 2004, IEEE Press, Apr. 2004, pp. 546-566.

[3] S. Mangard, N. Pramstaller and E. Oswald, "Successfully attacking AES hardware implementations," in CHES 2005, vol. 3659, Springer-Verlag, 2005, pp. 157-17.

[4] M. Joye, P. Paillier and B. Schoenmakers, "On second-order differential power analysis," In CHES 2005, vol. 3659, Springer-Verlag, 2005, pp. 293-308.

[5] E. Prouff, M. Rivain and R. Bevan, "Statistical analysis of second order differential power analysis," IEEE Transactions on Computers, vol. 58, 2009, pp. 799-811.

[6] J. Waddle and D. Wagner, "Towards efficient second-order power analysis," In CHES 2004, vol. 3156, Springer-Verlag, 2004, pp. 1–15.

[7] S. Mangard, E. Oswald and T. Popp. Power analysis attacks: revealing the secrets of smart cards. New York: Spinger-Verlag, 2007, pp. 3-33.

[8] J. D. Golić, "Techniques for random masking in hardware," IEEE Transactions on Circuits and Systems, vol. 54, Feb. 2007, Pages: 291-300.

[9] D. Canright and L. Batina, "A very compact "perfectly masked" S-box for AES," ACNS 2008, vol. 5037, Springer-Verlag, 2008, pp. 446-459.

[10] J. D. Golić and R. Menicocci, "Universal masking on logic gate level," IET Electronics Letters, vol. 40, May. 2004, pp. 526-528.

[11] W. Fischer and B. M. Gammel, "Masking at gate level in the presence of glitches, " in CHES 2005, vol. 3659, Springer-Verlag, 2005, pp. 187-200.

[12] Y.-J. Baek and M.-J. Noh, "DPA-resistant finite field multipliers and secure AES design," in Information Security Practice and Experience, vol. 3903, Springer-Verlag, 2006, pp. 1-12.

[13] A. Shamir, "How to share a secret," Communications of the ACM, vol. 22, Nov. 1979, pp. 612-613.

[14] G. R. Blakley, "Safeguarding cryptographic keys," Proceedings of the National Computer Conference, Jun. 1979, pp. 313-317.

[15] J. Wolkerstorfer, "Secret-Sharing Hardware Improves the Privacy of Network Monitoring," In DPM 2010 and SETOP 2010, vol. 6514, Springer-Verlag, May. 2010, pp. 51-63.

[16] S. Nikova, C. Rechberger, V. Rijmen, "Threshold Implementations Against Side-Channel Attacks and Glitches," Information and Communications Security, vol. 4307, Heidelberg: Springer-Verlag, 2006, pp. 529-545.

[17] S. Nikova, V. Rijmen and M. Schläffer, "Secure Hardware Implementation of Non-Linear Function of Glitches," In ICISC 2008, vol. 5461, Springer-Verlag, Dec. 2008, pp. 218-234.

[18] N. Kamoun, L. Bossuet, and A. Ghazel, "SRAM-FPGA implementation of masked S-Box based DPA countermeasure for AES," In IDT 2008, IEEE Press, Dec. 2008, pp. 74 – 77, doi: 10.1109/IDT.2008.4802469.

[19] T. Matsumoto, H. Mimura, D. Suzuki, "Complementary logics vs masked logics: which countermeasure is a better selection?," Proc. IEEE Symp. Circuit Theory and Design (ECCTD 09), IEEE Press, Aug. 2009, pp. 399 - 402. doi: 10.1109/ECCTD.2009.5274989.

[20] Trichina, E. and T. Korkishko, "Secure AES hardware module for resource constrained devices," in Security in Ad-hoc and Sensor Networks, vol. 3313, 2005, Springer-Verlag, pp. 215-229. doi: 10.1007/978-3-540-30496-8_18.

[21] E. Trichina, "Combinational logic design for AES subbyte transformation on masked data ," Cryptology ePrint Archive (http://eprint.iacr.org/) , Report 2003/236, 2003.

[22] J. Zhao, J. Han, X.Y. Zeng, J. Chen, "VLSI implementation of an AES algorithm resistant to Differential Power Analysis attack," In ASICON 2007, IEEE Press, Oct. 2007, pp. 838-841.

[23] X. Zheng, Y. Zhang, "Design and Implementation of a DPA Resistant AES Coprocessor," In WiCOM 2008, IEEE Press, Oct. 2008, pp. 1-4.

FPGA based Optimized SHA-3 Finalist in Reconfigurable Hardware

Qian Song, Yi Wang, Zhican Li, Quan Zhou,Wufei Wu, Demin Han, Wenlong Xu, Zuo Chen and Renfa Li

Embedded Systems & Networking Laboratory, Hunan University
Hunan Provincial Key Laboratory of Network and Information Security, Hunan University
e-mail:estellewy@hotmail.com

Abstract—A hash function is well-defined procedure to convert large, uncertain long message into fixed small integers. Secure Hash Algorithm (SHA) is an one-way message digest algorithm which is usually used in cryptographic applications such as authentication, digital signature and data integrity. In this paper, we proposed the reconfigurable structure for SHA-3 finalist BLAKE, Grøstl, JH, Keccak and Skein, separately. The proposed reconfigurable Grøstl, JH and Keccak could support different digested sizes. And Skein and BLAKE optimized three different modes using one single hardware core. The experimental results showed that our proposed structure could support different parameters of SHA-3 finalist with comparable performance among the existing works when ported to Xilinx Virtex-5 FPGA platform.

Keywords-SHA-3, FPGA, reconfigurable

I. Introduction

Secure Hash Algorithm (SHA) is a data encryption algorithm which is widely used in Automatic Teller Machine (ATM), Radio-Frequency Identification (RFID) and Virtual Private Networks (VPN) [1]. However, it has been announced that SHA-0, SHA-1 and SHA-2 might be attacked [2-5]. Therefore, National Institutes of Standards and Technology (NIST) officially announced to call SHA-3, which can be regarded as a new-secure hash algorithm. Till now, SHA-3 finalists are five hash algorithms: BLAKE, Grøstl, JH, Keccak and Skein [6].

The BLAKE [7] is processing with Hash Iterative Framework (HAIFA) iteration mode [8], whose compression function is built on the CHACHA core algorithm which is one of the fastest stream ciphers [9]. Aumasson *et al.* realized 1G (G is an encryption function), 4G and 8G mode of BLAKE separately, the maximum throughput achieved 3103Mb/s [7]. Kobayashi *et al.* [10] realized the 4G mode of BLAKE-32 with 2676Mb/s throughput using 1660 slices on Side-channel Attack Standard Evaluation Board II (SAEBO-GII) [11] platform. The similar design proposed by Homsirikamol *et al.* achieved 119 MHz when ported to Altera Stratix III FPGA platform [12]. Grøstl [13] is mainly composed of Message Digest (MD) iteration and Advanced Encryption Standard (AES) compression function. Baldwin *et al.* realized Grøstl-256 and Grøstl-512 separately which achieved throughput of 3242Mb/s and 3619 Mb/s on Xilinx xc5vlx330. They also proposed a reconfigurable structure of Grøstl-256 and Grøstl-512 by processing *P* and *Q* permutations in parallel and *S*-box in Block Random-Access Memory (BRAM), which achieved 7310Mb/s throughput [14]. Jungk and Reith proposed the reconfigurable structures of Grøstl-224 and Grøstl-256 [15], Grøstl-384 and

Grøstl-512 [16], and the structure shared *P* and *Q* permutations with *S*-Box generating on-the-fly. Their designs took up 6136 slices and 8308 slices separately. A pipelined structure for Grøstl-256 was proposed by Homsirikamol *et al.*, which achieved 7885Mb/s throughput and it is the best among the existing designs [11]. JH [17] proposed a new compression function structure, which used a large block cipher with constant key to construct a compression function. JH-256 had been realized by Baldwin *et al.* and Matsuo *et al.* separately, which achieved throughput of 1941Mb/s and 2639Mb/s using 1291 slices and 2661 slices separately on Xilinx Virtex-5 FPGA platform [14][18]. Homsirikamol *et al.* realized JH-256, which achieved 5516Mb/s throughput using 1018 slices on Xilinx Virtex-5 FPGA platform [11]. Keccak [19] is based on sponge construct [20] which can achieve higher frequencies by a shorter critical path. Homsirikamol *et al.* optimized Keccak-256 on Virtex-5 platform and Keccak-512 on Altera Stratix III FPGA platform [11]. The proposed Keccak-256 took up 1217 slices with throughput of 12817Mb/s, and the proposed Keccak-512 took up 4213 ALUTs with throughput of 12393Mb/s. Baldwin *et al.* realized Keccak-256 which achieved 8518Mb/s throughput using 1117 slices on Xilinx xc5vlx330 [14]. Skein is mainly composed of Unique Block Iteration (UBI) and Threefish functions. UBI transforms an input with the random length into an output with the fixed length. Threefish is the compression function which determining the size of Skein's state space [21]. Baldwin *et al.* proposed a four-rolled structure for Skein-512 which achieved 1945Mb/s throughput on Xilinx xc5vlx330 [15]. Tillich *et al.* proposed an eight-rolled architecture for Skein-256 and Skein-512 on Xilinx xc5vlx110, and their designs achieved 1751Mb/s and 3535Mb/s throughput separately [22]. The similar structure of Skein-256 and Skein-512 are proposed by Tillich *et al.*, and the maximum throughput could reach 1762Mb/s and 2501Mb/s in an UMC 0.18 μm CMOS standard cell technology [23].

In this paper, we firstly detailed the different characteristics of SHA-3 finalist. Then, we proposed the reconfigurable structures for SHA-3 finalist to support different message lengths. A detailed comparison of the proposed designs with the existing designs is given to show that the proposed reconfigurable designs took up less area compared with implementing the design separately.

II. Preliminary of SHA-3 Finalist

The SHA-3 finalists have different structures and different compression functions. NIST summarized the characteristics of these algorithms as shown in Table I [6]:

This work is supported by "Chinese National Science Foundation" (No.60873074 and No.60673061); "Changsha Science Technology Scheme" (No.K1003028-11) and "the Fundamental Research Funds for Chinese Central Universities".

TABLE I.   STRUCTURE AND CHARCTERASTICS

Algorithm	Designer	Structure	Compression Function
BLAKE[7]	Aumasson	HAIFA	LAKE,CHACHA
Grøstl[13]	Knudsen	Wide-pipe MD	AES permutations
JH[17]	Hongjun wu	Wide-pipe MD	AES
Keccak[19]	Daemen	Sponge	Iterated permutation
Skein[21]	Schneier	MD,UBI	Threefish

From Table I, these new candidate algorithms are designed with a variety of structures including HAIFA, wide-pipe MD, MD, UBI and Sponge. The compression functions include LAKE (a hash function with wide-pipe structure [24]), CHACHA, AES, Threefish and iterated permutation. Aumasson *et al.* [7] proposed BLAKE with HAIFA structure which mixed salt and counter in compression function in order to encourage the use of randomized hashing to overcome the weakness of iterative structure. Grøstl [13] employed Wide-pipe MD structure where the size of the internal state is significantly larger than the size of the output, and the compression function is mainly consist of P and Q permutations. JH [17] is an algorithm which has the same structure of Wide-pipe MD, and the compression function is an efficient differential propagation which is a byte-oriented Substitution-Permutation Network (SPN) as AES. Keccak [19] applies a Hermetic Sponge Strategy (HSS) with three steps: absorbing input data, squeezing data in a state and outputting data [20]. In the algorithm, Keccak-1600 is chosen from a set of 7 permutations as a candidate for SHA-3 competition. Skein [21] uses a new structure of UBI and Threefish. UBI structure includes three types: configuration UBI, message processing UBI and outputting UBI. Each UBI has a corresponding Threefish processing unit.

III. PROPOSED RECONFIRUBALE ARCHITECTURE FOR GRØSTL, SKEIN, JH, KECCAK AND BLAKE

A. Grøstl

According to the different output lengths, Grøstl can be divided into Grøstl-224, Grøstl-256, Grøstl-384 and Grøstl-512. We proposed a reconfigurable structure of Grøstl as shown in Fig. 1.

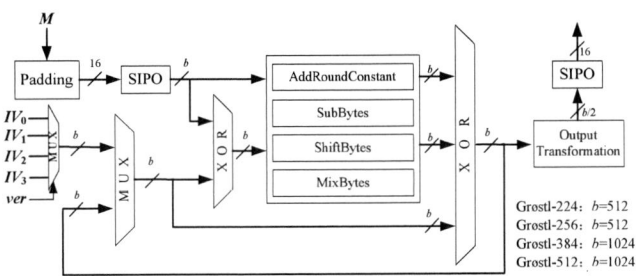

Figure 1.   The proposed reconfigurable structure of Grøstl

In Fig. 1 M represents the message and the signal *ver* is to select the parameters among 224, 256, 384 and 512.

B. Skein

According to the state space, Skein has three different versions, Skein-256, Skein-512 and Skein-1024. MIX function is a non-linear mixing function of Threefish which has iteration, four-rolled and eight-rolled modes. We proposed the scalable architecture for Skein-512 to support all three modes. Fig. 2 shows the proposed structure.

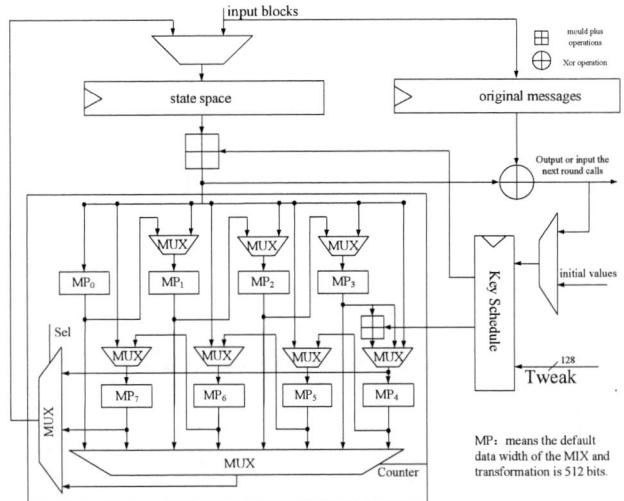

Figure 2.   The proposed scalable structure of Skein

Each MP_n ($n=0,1,2,...7$) represents a set of MIX transformations, and MUX unit represents Multiplexer which is controlled by *Sel* and *Counter* signals.

C. JH

JH can be divided into JH-224, JH-256, JH-384 and JH-512 according to the different output lengths. We proposed a reconfigurable structure of JH as shown in Fig. 3.

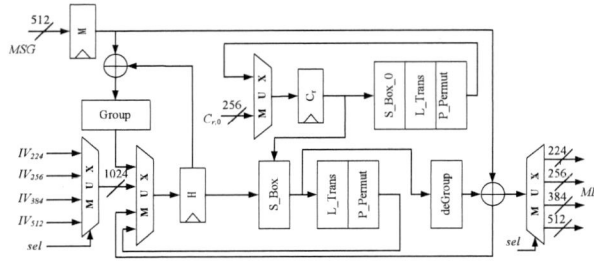

Figure 3.   The proposed reconfigurable structure of JH

In Fig. 3 *MSG* represents the input message block, $C_{r,0}$ is an initial vector of the round constant, *sel* signal is to select the initial values of JH-224, JH-256, JH-384 and JH-512.

D. Keccak

We proposed a reconfigurable structure for Keccak-224, Keccak-256, Keccak-384 and Keccak-512. Fig. 4 shows the proposed structure.

In Fig. 4 M represents the input message. The module R is a round function which iterates 18 times for each permutation. The *sel* signal is to select the different lengths of output messages.

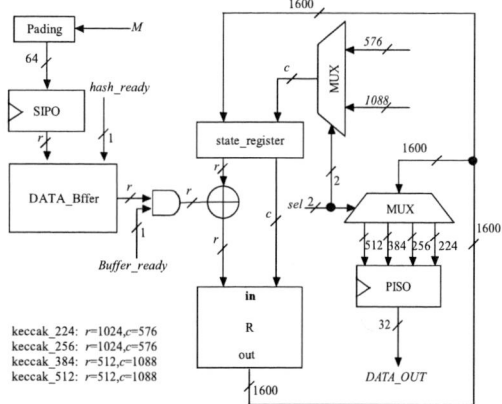

Figure 4. The proposed reconfigurable structure of Keccak

keccak_224: $r=1024, c=576$
keccak_256: $r=1024, c=576$
keccak_384: $r=512, c=1088$
keccak_512: $r=512, c=1088$

E. BLAKE

BLAKE has four different versions, BLAKE-28, BLAKE-32, BLAKE-48, and BLAKE-64. There have three different G function modes: 1G, 4G and 8G. We proposed the scalable structure for BLAKE-32 to support three different G modes. Fig. 5 shows the proposed structure of BLAKE.

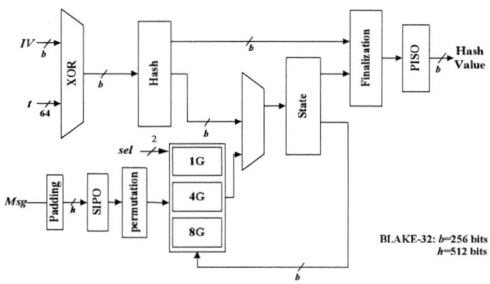

BLAKE-32: $b=256$ bits
$h=512$ bits

Figure 5. The proposed scalable structure of BLAKE

In Fig. 5 *Msg* represents the input message and the signal *sel* is to select the modes among 1G, 4G and 8G.

IV. COMPARISONS

We implemented Grøstl, Skein, JH, Keccak and BLAKE using Verilog hardware description language, and synthesizing with Xilinx ISE 12.1 tools. Table II shows the experimental results of the proposed reconfigurable structure and implemented individually on Xilinx Virtex-5 platform. From table II, it is obvious that the proposed reconfigurable structure of Grøstl takes up 1802 slices less compared with the total area of implementing for Grøstl-256 and Grøstl-512 individually. The proposed Skein takes up 1728 slices less compared with the total area of supporting three modes. The proposed JH takes up 1800 slices less compared with the total area of implementing for JH-256 and JH-512 individually and the proposed Keccak takes up 1423 slices less compared with the total area of implementing for Keccak-256 and Keccak-512 individually. The proposed BLAKE-32 takes up 2947 slices less compared with the total area of supporting three modes.

Table III shows the comparison between our reconfiguration designs and the existing methods. The Throughput/Area aspect of the proposed Grøstl is 1.4 times better than Kobayashi's design and 2.4 times better than Baldwin's design. Although Matsuo and Homsirikamol achieved 1.7 and 1.8 times better than our design in the aspect of Throughput/Area, but their design only can support one parameter. The throughput of the proposed Skein is 1.6 times larger than Baldwin's design, but it is 1.1 times less than Tillich's design. However, our scalable structure of Skein could support iteration, four-rolled and eight-rolled modes. The proposed JH achieved 1.6 times faster than Matsuo's design, but it is 0.8 times slower compared with Homsirikamol's design. The throughput of the proposed Keccak achieved 1.6 times larger than Baldwin's and Matsuo's designs, and also achieved 1.1 times larger than Homsirikamol's design. The proposed BLAKE achieved 6.8% faster than Baldwin's design.

TABLE II. THE RESULTS OF THE PROPOSED STRUCTURES

Algorithm	Platform	Area	Throughput	Frequency
Grøstl	256	xc5vlx220	2145slices	5548Mb/s	238.4MHz
	512	xc5vlx220	3936slices	8137Mb/s	238.4MHz
	R	xc5vlx220	4279slices	7623Mb/s	223.32MHz
	D	-	1802slices	-	-
Skein	512-1	xc5vlx30	1284slices	1214Mb/s	175.5 MHz
	512-4	xc5vlx30	1458slices	3074Mb/s	120.1 MHz
	512-8	xc5vlx30	1561slices	3645Mb/s	71.2 MHz
	R	xc5vlx30	2539slices	3139Mb/s	61.3MHz
	D	-	1728slices	-	-
JH	256	xc5vlx220	1452slices	4962Mb/s	378MHz
	512	xc5vlx220	1823slices	4962Mb/s	378MHz
	R	xc5vlx220	1475slices	4319Mb/s	329MHz
	D	-	1800slices	-	-
Keccak	256	xc5vlx220	1375slices	14438Mb/s	282MHz
	512	xc5vlx220	1446slices	7066Mb/s	276MHz
	R	xc5vlx220	1698slices	13414Mb/s	262MHz
	D	-	1423slices	-	-
BLAKE	32-1	xc5vlx220	1624slices	849Mb/s	136MHz
	32-4	xc5vlx220	1742slices	2828Mb/s	122MHz
	32-8	xc5vlx220	1730slices	1805Mb/s	78MHz
	R	xc5vlx220	2149slices	2927Mb/s	126MHz
	D	-	2947slices	-	-

R: the reconfigurable architecture of Grøstl, Skein, JH, Keccak and BLAKE.

D: the area difference between the reconfigurable architecture and the individual designs.

V. CONCLUSION

In this paper, we detail the design features and the existing methods of SHA-3 finalist. In order to improve the flexibility of hardware implementation, we proposed the new reconfigurable structures for Grøstl, JH and Keccak, which could support four different parameters of 224, 256, 384 and 512. We proposed the scalable structure for Skein, which could support iteration, four-rolled and eight-rolled modes computation. Similarly, we proposed the scalable structure for BLAKE, which could support 1G, 4G and 8G modes. The experimental results showed that our proposed designs take up smaller area compared with implementation individually. Moreover, we provide flexibility for the area-constraint applications.

TABLE III.   COMPARISON WITH THE EXISTING DESIGNS

Algorithm		Platform	Data path	Area	Throughput	Clock Frequency	Throughput/Area
Grøstl	Homsirikamol[12]	Xilinx Virtex-5	512-bit	1597slices	7885Mb/s	323.4MHz	4.94
		Xilinx Virtex-5	1024-bit	3188slices	10314Mb/s	292.1MHz	3.24
	Kobayashi[10]	Virtex-5 xc5vlx30	512-bit	4057slices	5171Mb/s	101MHz	1.27
	Baldwin[14]	Virtex-5 xc5vlx330	512-bit	2391slices	3242Mb/s	101.3MHz	1.36
		Virtex-5 xc5vlx330	1024-bit	4845slices	3619Mb/s	123.4MHz	0.75
	Matsuo[18]	Virtex-5 xc5vlx30	512-bit	2616slices	7885Mb/s	154MHz	3.01
Skein	Matsuo[18]	Virtex-5 xc5vlx30	256-bit-4	854slices	1402Mb/s	115MHz	1.64
	Kobayashi[10]	Virtex-5 xc5vlx30	256-bit-4	854slices	1482Mb/s	115MHz	1.74
	Tillich[22]	Virtex-5 xc5vlx110	256-bit -8	937slices	1751Mb/s	68.4MHz	1.87
		Virtex-5 xc5vlx110	512-bit -8	1632slices	3535Mb/s	69MHz	2.17
	Baldwin[14]	Virtex-5 xc5vlx330	512-bit -4	1786slices	1945Mb/s	83.65MHz	1.09
	Homsirikamol[12]	Xilinx Virtex-5	512-bit -4	1716slices	3209Mb/s	119.1MHz	1.87
JH	Homsirikamol[12]	Xilinx Virtex-5	256-bit	1018 slices	5416Mb/s	380.8MHz	5.32
		Xilinx Virtex-5	512-bit	1104slices	5610Mb/s	394.5MHz	5.08
	Matsuo[18]	Virtex-5 xc5vlx30	256-bit	2661slices	2639Mb/s	201MHz	0.99
	Baldwin[14]	Virtex-5 xc5vlx330	256-bit	1291slices	1941Mb/s	250.13MHz	1.50
Keccak	Homsirikamol[12]	Xilinx Virtex-5	1024-bit	1272slices	12817Mb/s	282.7MHz	10.08
	Baldwin[14]	Virtex-5 xc5vlx330	1024-bit	1117slices	8518Mb/s	189MHz	7.63
	Matsuo[18]	Virtex-5 xc5vlx30	512-bit	1117slices	8190Mb/s	189MHz	7.33
BLAKE	Aumasson[7]	Virtex-5 xc5vlx220	512-bit-1	390slices	575Mb/s	91MHz	1.47
		Virtex-5 xc5vlx220	512-bit-4	1217slices	2438Mb/s	100MHz	2.00
		Virtex-5 xc5vlx220	512-bit-8	1694slices	3103Mb/s	67MHz	1.83
	Kobayashi[10]	Virtex-5 xc5vlx220	512-bit-4	1660slices	2676Mb/s	115MHz	1.61
	Homsirikamol[12]	Virtex-5 xc5vlx220	512-bit-4	1851slices	2611Mb/s	117MHz	1.41
	Baldwin[14]	Virtex-5 xc5vlx220	512-bit-4	1118slices	1079Mb/s	118MHz	0.97
Grøstl	This paper	Virtex-5 xc5vlx220	1024-bit	4279slices	7623Mb/s	223.32MHz	1.78
Skein		Virtex-5 xc5vlx30	512-bit	2539slices	3139Mb/s	61.3MHz	1.24
JH		Virtex-5 xc5vlx30	512-bit	1475slices	4319Mb/s	329MHz	2.93
Keccak		Virtex-5 xc5vlx220	1024-bit	1698slices	13414Mb/s	262MHz	7.9
BLAKE		Virtex-5 xc5vlx220	512-bit	2149slices	2927Mb/s	126 MHz	1.36

REFERENCES

[1] National Institute of Standards and Technology, FIPS 180: "Secure Hash Standard," FIPS, 1993.

[2] X. Y. Wang and H. B. Yu, "How to Break MD5 and Other Hash Functions," EUROCRYPT'05, Lecture Notes in Computer Science Vol. 3494, Springer, 2007, pp.19-35.

[3] F. Chabaud, and A. Joux, "Differential Collisions in SHA-0," CRYPTO'98, vol. 1462, Springer-Verlag, 1998, pp. 56-71.

[4] X. Y. Wang, Y. L. Yin, and H. B. Yu, "Finding Collisions in the Full SHA-1," CRYPTO 2005, Springer-Verlag, 2005, pp.17-36.

[5] S. K. Sanadhya, and P. Sarkar, "New Local Collosions for the SHA-2 Hash Family," ICISC 2007, vol. 4817, Springer-Verlag, 2007, pp. 193-205.

[6] National Institute of Standards and Technology: "CRYPTOGRAPHIC HASH ALGORITHM COMPETITION," Gaithersburg: 2007[2011], http://csrc.nist.gov/groups/ST/hash/sha-3/index.html

[7] J. P. Aumasson, L. Henzen, and W. Meier, "SHA-3 proposal BLAKE," version 1.3, 2008, Available online at http://131002.net/blake/blake.pdf,

[8] E. Biham and O. Dunkelman, "A framework for iterative hash functions-HAIFA," Cryptology ePrint Archive, Report 2007/278, 2007, http://eprint.iacr.org/

[9] D. J. Bernstein, "ChaCha, a variant of Salsa20," January 2008, http://cr.yp.to/chacha.html .

[10] K. Kobayashi, J. Ikegami, and S. Matsuo, "Evaluation of hardware performance for the SHA-3 candidates using SASEBO-GII." Cryptology ePrint Archive, Report 2010/010, 2010, http://eprint.iacr.org/.

[11] National Institute of Advanced Industrial Science and Technology (AIST), Research Center for Information Security (RCIS), "Side-channel Attack Standard Evaluation Board (SASEBO)," http://www.rcis.aist.go.jp/special/SASEBO/SASEBO-GII-ja.html.

[12] E. Homsirikamol, M. Rogawski and K. Gaj, "Comparing Hardware Performance of Fourteen Round Two SHA-3 Candidates Using FPGAs," Cryptology ePrint Archive, Report 2010/445, 2010.

[13] P. Gauravaram, L. R. Knudsen, and K. Matusiewicz, "Grøstl–a SHA-3 candidate," October 2008, http://www.grøstl.info/Grøstl1.pdf,.

[14] B. Baldwin, N. Hanley, and M. Hamilton, "FPGA implementations of the round two SHA-3 candidates," FPL 2010 , 2010, pp. 400-407.

[15] B. Jungk, S. Reith, and J. Apfelbeck, "On Optimized FPGA Implementations of the SHA-3 Candidate Grøstl," IACR Eprint report 2009/206, Available online at http://eprint.iacr.org/2009/206.pdf.

[16] B. Jungk, and S. Reith, "On FPGA-based implementation of Grøstl," IACR Eprint report 2010/260, http://eprint.iacr.org/2010/260.pdf.

[17] H. J. Wu, "SHA-3 proposed JH," 2008, Available online at http://icsd.i2r.a-star.edu.sg/staff/hongjun/jh/index.html.

[18] S. Matsuo, M. Knezevic, and P. Schaumont, "How Can We Conduct 'Fair and Consistent' Hardware Evaluation for SHA-3 Candidate?" Second SHA-3 Candidate Conference, 2010, Available online at http://csrc.nist.gov/groups/ST/hash/sha-3/Round2/Aug2010/documents/papers/MATSUO_SHA-3_Criteria_Hardware_revised.pdf.

[19] G. Bertoni, J. Daemen, and M. Peeters, "Keccak sponge function family main ducument," 2008, http://keccak.noekeon.org/

[20] G. Bertoni, J. Daemen, and M. Peeters, "Sponge Functions," ECRYPT 2007,http://www.csrc.nist.gov/pki/HashWorkshop/Public_Comments/2007_May.html .

[21] N. Ferguson, S. Lucks, and B. Schneier, "The Skein hash function family," 2009, http://eprint.iacr.org/.

[22] S. Tillich, "Hardware Implementation of the SHA-3 Candidate Skein," IACR Eprint report 2009/159, http://eprint.iacr.org/ 2009/159.pdf.

[23] S. Tillich, M. Feldhofer, and M. Kirschbaum, "High-Speed Hardware Implementations of BLAKE, Blue Midnight Wish, CubeHash, ECHO, Fugue, Grøstl, Hamsi, JH, Keccak, Luffa, Shabal, SHAvite-3, SIMD, and Skein," IACR Eprint report 2009/510, http://eprint.iacr.org/2009/510.pdf.

[24] J. P. Aumasson, W. Meier, and R. C. W. Phan, "The hash function family LAKE," in Fast Software Encryption 2008, vol. 5086, Springer-Verlag, 2008, pp. 36-53.

A Code Reuse Method for Many-Core Coarse-Grained Reconfigurable Architecture Function Library Development

Shuo Li, Guo Chen, Ahmed Hemani
Department of Electronic Systems
School of Information and Communication Technology
Royal Institute of Technology
Stockholm, Sweden
Email: shuol, guoc, hemani@kth.se

Abstract—In this paper[1], a code reuse method is proposed to enhance the efficiency of the function library development of many core coarse-grained reconfigurable architecture. The method focuses on developing and using the precompiled ReConfigurable Functions (RCFs) in the function library. By applying this method on the RCF development, functions are objectified like classes in any objective-oriented programming language. Using a function is to instantiate a selected RCF. Similar functions can be instantiated from the same RCF. Thus, the total number of RCFs to be compiled is reduced and the global programming efficiency is increased and the labor requirement for application development is reduced.

Index Terms—Code Generation, Reconfigurable Architecture, Dynamically Reconfigurable Resource Array, DRRA

I. INTRODUCTION

Reconfigurable computing combines the flexibility of general processors and high efficiency of ASIC [1]. By using Network-on-Chip (NoC) as the interconnect backbone, massive processor integration in a reconfigurable computing architecture is enabled [2]. As the processing power will be dramatically increased due to the notable increasing of processors, the complexity of application development on the reconfigurable computing architecture will also be dramatically increased. One possible solution is to use optimally compiled function implementations as building blocks to reduce the application development time.

However, codeing and compiling the functions takes lots of time when the function library is large. In the method proposed in this paper, ReConfigurable Functions (RCFs) are used to reduce the workload of the function library development. Each RCF can have multiple instances according to the instantiation configurations. Instead of coding and compiling multiple similar functions, only one RCF is coded and compiled. The function library development time is then reduced.

During the compiler development of Dynamically Reconfigurable Resource Array (DRRA), which is a many core coarse-grained reconfigurable architecture developed in our

[1]This work is part of the CREST: Coarse Grain Reconfigurable Embedded Systems Technologies project (2010-01453) funded by Vinnova, Sweden

group, the mentioned function library development problem was addressed. The DRRA function library [3] is huge so that manually coding all functions is infeasible. Therefore, the code reuse method proposed in this paper is developed to enhance the efficiency of the function library development.

This code reuse method focuses on the RCF development and the code generation. In each RCF, code reuse is an inherent property and to generate proper programs is then become automatic and efficient.

The rest parts of this paper are organized as follows. Section II lists a group of related work. Section III gives a general description of DRRA. Section IV gives a detailed elaboration of the RCF model in DRRA function library. Section V describes the procedure to generate desired DRRA assembly program based on the code template provided by a RCF. Section VI gives the conclusion of our work.

II. RELATED WORK

As mentioned in Section I, reconfigurable computing combines the flexibility of general processors and high efficiency of ASIC. It is also becomes a notable topic in High-Performance Computing (HPC) [4]. Therefore, programming a reconfigurable architecture is already discussed in many papers.

In [5], the needs of compiler and application support for reconfigurable SoCs are addressed. The DRRA compiler is built up to provide a dedicated compiler. The DRRA function library is built to provide optimally compiled functions for application development.

In [6], high-level programming of coarse-grained reconfigurable architecture is discussed. Occam-pi was used as the programming language. A programming language called ARMLang is proposed in [7]. It is for regular processor arrays in particular reconfigurable meshes. In [8], Reconfigurable Computing C (RCC) is proposed. It is a subset of the ANSI-C. Hence, programmers do not have to learn a new language. In DRRA, Simulink is used as the application development language since (1) Simulink implicitly provides data/control

978-1-61284-863-1/11 $26.00 © 2011 IEEE

dependency as well as synchronization information and parallelism specification, (2) the simulation and debugging is visualized, (3) it is already well developed and algorithm designers know it well.

III. DRRA ARCHITECTURE

A. DRRA Overview

Dynamically Reconfigurable Resource Array (DRRA) is a coarse-grained reconfigurable architecture. All DRRA resources are integrated as a regular and seamlessly connected fabric on the logic die [9], as depicted in Fig. 1. The atomic resource in the DRRA architecture is the DRRA cell, which includes one morphable Data Path Unit (mDPU), one register file, one sequencer and two switchboxes. DRRA cells are connected by a seamless, sliding window, circuit switched interconnect fabric.

MDPU is 16-bit morphable data path unit with four 16-bit input ports corresponding to two complex numbers and two 16-bit output ports corresponding to one complex number. Depends on the configuration, one mDPU is able to operate on several modes at different time, for instance a multiply-accumulator (MAC) with internal and external accumulation, or a simple 16-bit adder.

Each register file contains 64 16-bit registers. It has two 16-bit input ports and two 16-bit output ports. It also has a DSP style address generation unit, which provides vectorized, circular buffer and bit-reverse addressing modes based on the configuration.

The switchbox is used to configure the circuit switching interconnect fabric. In the current DRRA architecture, the sliding window size is three hops. Thus, one DRRA cell can only communicate with another DRRA cell, who is less than three hops away.

The sequencer controls the mDPU, the register file and the switchbox within the same DRRA cell. To program the DRRA fabric is actually to program the sequencer. The program memory in each sequencer can store maximally 256 instructions. By using hierarchical control and interrupts, it is possible for a sequencer in one DRRA cell to control the sequencer in another DRRA cell.

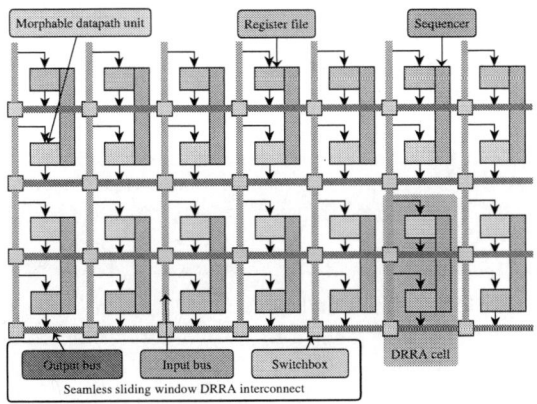

Fig. 1. DRRA Physical Layer Fabric

B. Sequencer Instructions

The instruction set of the sequencer has 16 instructions, which are differentiated by their instruction codes. These instructions define how and when the mDPU, the register file and the switchbox act in a DRRA cell. The details of all the instructions is beyond the scope of this paper. Therefore, only mDPU, register file, switchbox, delay, branch and hierarchical control instructions are briefly explained in this subsection.

The mDPU instruction defines the mDPU operating mode and the corresponding parameters. Taking the MAC mode as an example, the mDPU needs to know the number of clock cycles before it has to clear the internal accumulator. This information is provided in the instruction with the mode assignment.

The register file instruction defines the addressing mode, start and end addresses, write/read, etc. For example, a register file could work in circular buffer addressing mode, writing data one by one to address m to n.

The switchbox instruction defines the switch matrix inside the switchbox. For instance, the switchbox instruction could configure the interconnect in such a way that input port 0 of the mDPU is connected to the output port A of the register file in another DRRA cell. Note that the maximum communication distance is three hops.

These three instructions defines how the DRRA cell act. Delay, branch and hierarchical control instructions define when the DRRA cell act.

As the name suggest, delay instruction is just a multi-cycle no-operation instruction. This instruction is mostly used for inter-cell synchronization when the actions of the involved DRRA cells take predictable times.

Branch instruction defines several ways for the sequencer to make branching decision. For example, the next sequencer instruction to be executed can be based on one of the output of the mDPU. If it is equal to zero, instruction A will be executed otherwise instruction B will be executed. The limitation of the branch instruction is that it can only branch in two decisions. Case/if-elsif statement cannot be handled by one branch instruction but multiple instructions. Branching is one of the sources of the unpredictable action time of a DRRA cell.

The Hierarchical Control (HC) instruction defines the HC output of a DRRA cell and based on which value from which DRRA cell the current DRRA cell should perform which action. For example, DRRA cell A sends value X as its HC signal to the fabric. DRRA cell B waits for value X from DRRA cell A to execute instruction i. And DRRA cell C checks HC signal of DRRA cell A. If the value is X, executes instruction j otherwise executes instruction k. HC waiting and conditional branching is used for inter-cell synchronization no matter their action times are predictable or not.

It is possible to use these instructions to program the sequencer to control the mDPU, the register file and the switchbox in one DRRA cell and several other sequencers via HC. From the programmer's perspective, each DRRA cell can be considered as an individual programmable logic with interconnections to communicate with other DRRA cells within the sliding window.

IV. RECONFIGURABLE FUNCTION

From the programmer's perspective, a RCF in the DRRA function library is an abstract function consists of several function properties and function template. It has to be instantiated before being used in an application. Function configuration defines how to instantiate a reconfigurable function.

Function properties contain all function level properties for using the current function. For example, if FIR is a reconfigurable function. An example function property could be the function latency, which is the time between the arrival of the first input data and the birth of the first output data. In this example, the function latency is related to the number of taps of the FIR. The number of taps is a function property while the relationship between them is not a function property since to use the function, this relationship is totally unnecessary. Figure 2.a illustrates some example function properties. Note that all function properties are assigned with actual values only after the function is instantiated.

Function template contains all information below the function level. Code template and code generation method are included. For example, the relationship between the function latency and the number of taps is contained in the function template. Figure 2.b illustrated some example entries contained in the function template. The details of code template and code generation method are described in Section V together with the code generation process.

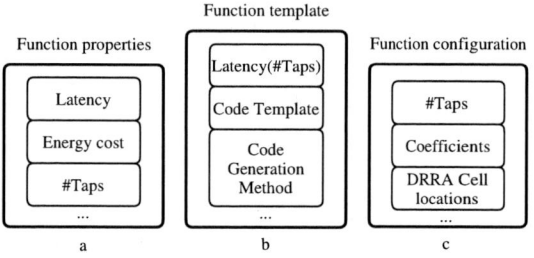

Fig. 2. Reconfigurable function

Function configuration consists of definitions of the tunable parameters defined in the function properties. For instance, the number of taps and the coefficients are included in the function configuration. Figure 2.c illustrated some example entries contained in the function configuration. By modifying the function configuration, various kinds of FIR in DRRA fabric could be generated. The work for the programmer is only to provide the configuration. The code generation is then automatically done.

V. CODE GENERATION

In this section, the code generate process is discussed. Figure 3 shows the flow of generating sequencer programs in an abstract view. The left part of Figure 3 is the RCF. As mentioned in Section IV, function properties and function

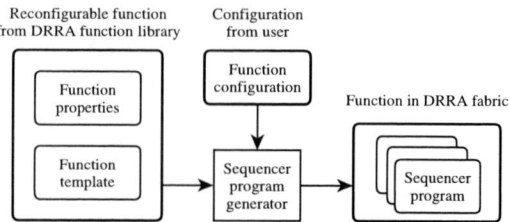

Fig. 3. Reconfigurable Function to Sequencer Programs

template are included. The sequencer program generator takes the RCF and the user defined function configuration as inputs and outputs the corresponding sequencer programs in the DRRA fabric.

A. Code Template

Code template consists of (1) global constants, (2) individual constants and (3) generalized sequencer programs. Global constants are used to assign function properties and generate all sequencer programs. Individual constants are used for generating the proper sequencer program for each individual DRRA cell. Global and individual constants provide what should be changed to generate programs. Generalized sequencer programs are the incomplete sequencer programs which contains global and individual constant names. Sequencer programs are completed by replacing global and individual constant names with their values. The following example code segment is part of a code template of a controller-worker RCF.

```
0  CtrlRow  :  0
1  CtrlCol  :  2
2  FirstJob  :  14
3
4  SelfRow  :  0
5  SelfCol  :  0
6  Ctrl  :  4
7
8  CONNECT  RegFile  CtrlRow  CtrlCol  A  to
     mDPU  SelfRow  SelfCol  0;
9  CONNECT  RegFile  CtrlRow  CtrlCol  B  to
     mDPU  SelfRow  SelfCol  1;
10 HC  wait  Ctrl  FirstJob  37;
```

The first part of this code segment defines three global constants: CtrlRow, CtrlCol and FirstJob. Their default values are 0, 2 and 14, respectively. The row index of the controller is CtrlRow. The column index of the controller is CtrlCol. The first job indicator is FirstJob. In the second part, three individual constants are defined. Their names are SelfRow, SelfCol and Ctrl. The default values of them are 0, 0 and 4, respectively. The row index of one worker is SelfRow. The column index of one worker is SelfCol and the controller is the Ctrlıth DRRA cell to the worker. In the third part, the incomplete sequencer program for one worker is listed. Line 8 connects port A of the controller's register file to input port

978-1-61284-863-1/11 $26.00 © 2011 IEEE   514

0 of the worker's mDPU. Line 9 connects port B to input port 1. The HC statement at line 10 specifies the worker should wait for the HC signal from the controller to be the value of FirstJob to execute the 37th sequencer instruction.

All constants should have correct values according to the function configuration after function instantiation. The values in the template are the default values and the relationships among all constants are fulfilled by the default values. By default, the controller is at (0, 2) (row, column) and the worker is at (0, 0). Therefore, in the HC statement at line 10, Ctrl should be 4 since the controller is the fourth DRRA cell from the worker's perspective. The numbering method is illustrated in Figure 4. Each block stands for one DRRA cell. The DRRA fabric is a N x 2 matrix of DRRA cells. The worker is cell 0 and the controller is cell 4.

Worker 0	2	Controller 4	...
1	3	5	...

Fig. 4. DRRA cell numbering

B. Code Generation Methods

As mentioned in the last subsection, constant names should be replaced by proper values. The values are obtained by the code generation methods. Take the Ctrl as the example, assume controller is at (CtrlRow, CtrlCol), the worker is at (SelfRow, SelfCol), the maxmium hop count is m and the number of DRRA cells per column is c. Ctrl = $Index$(CtrlRow, CtrlCol, SelfRow, SelfCol, m, c). The method $Index$ is shown in the following C# code.

```
0 public static int Index(int SelfRow, int
      SelfCol, int CtrlRow, int CtrlCol,
      int m, int c) {
1   int Col = CtrlCol - SelfCol + (SelfCol
      > MaxHopCount ? m : SelfCol);
2   int Row = CtrlRow - SelfRow + (SelfRow
      > MaxHopCount ? m: SelfRow);
3   return Col * c + Row; }
```

C. Generate the Codes

The sequencer programs are obtained by replacing constant names by constant values. Following the example code template in the last subsection, if one worker is required at (0, 1), SelfRow = 0, SelfCol = 1. In addition, according to the code generation method, Ctrl = 4. The global constants remain since only the worker location is changed. The generated code segment is as follows.

```
0 CONNECT RegFile 0 2 A to mDPU 0 1 0;
1 CONNECT RegFile 0 2 B to mDPU 0 1 1;
2 HC wait 4 14 37 0;
```

If we want more workers, we simply duplex the incomplete worker sequencer program and replace the constant names

with proper constant values. For example, if another worker is at (0, 4), following code segment will be generated for it. Ctrl is 2 instead of 4.

```
0 CONNECT RegFile 0 2 A to mDPU 0 4 0;
1 CONNECT RegFile 0 2 B to mDPU 0 4 1;
2 HC wait 2 14 37 0;
```

The code generation for DRRA functions become an automatic process. From the programmer perspective, a DRRA function can be used in an application as a predefined class. The programmer is not required to know any DRRA assembly details.

VI. CONCLUSIONS

By using the proposed method, programmers are able to focus more on algorithm and architecture optimization of the fabric than the physical layer details. In another hand, DRRA function development and DRRA application development can be separated and carried on in parallel. Therefore, the efficiency of DRRA application development can be greatly increased by applying this method. The required knowledge of DRRA fabric physical layer is minimized during application development. It is clear that this code reuse method can be easily extend to any reconfigurable architecture since this method only provides a framework and a guide line.

A variant of this code reuse method is included in the Simulink to DRRA compiler, which is under development. From the DRRA application perspective, function level global constants become the "individual" constants and application level constants become the "global" constants.

REFERENCES

[1] M. Iqbal, U. Awan, and S. Khan, "Reconfigurable computing technology used for modern scientific applications," in *Education Technology and Computer (ICETC), 2010 2nd International Conference on*, vol. 5, june 2010, pp. V5–36 –V5–41.

[2] S. Borkar, "Thousand core chips: a technology perspective," in *DAC '07: Proceedings of the 44th annual Design Automation Conference.* New York, NY, USA: ACM, 2007, pp. 746–749.

[3] O. Malik, A. Hemani, and M. Shami, "A library development framework for a coarse grain reconfigurable architecture," in *VLSI Design (VLSI Design), 2011 24th International Conference on*, jan. 2011, pp. 153 – 158.

[4] T. El-Ghazawi, "Is high-performance, reconfigurable computing the next supercomputing paradigm?" in *SC 2006 Conference, Proceedings of the ACM/IEEE*, nov. 2006, p. xv.

[5] A. Olugbon, T. Arslan, I. Lindsay, and S. MacDougall, "Providing compilers and application program support for reconfigurable socs: Radical but overdue," in *System-on-Chip, 2005. Proceedings. 2005 International Symposium on*, nov. 2005, pp. 54 –57.

[6] Z. ul Abdin, "High-level programming of coarse-grained reconfigurable architectures," in *Field Programmable Logic and Applications, 2009. FPL 2009. International Conference on*, 31 2009-sept. 2 2009, pp. 713 –714.

[7] H. Giefers and M. Platzner, "Armlang: A language and compiler for programming reconfigurable mesh many-cores," in *Parallel Distributed Processing, 2009. IPDPS 2009. IEEE International Symposium on*, may 2009, pp. 1 –8.

[8] F. Qi, X. Zhang, S. Wang, and X. Mao, "Rcc: A new programming language for reconfigurable computing," in *High Performance Computing and Communications, 2009. HPCC '09. 11th IEEE International Conference on*, june 2009, pp. 688 –693.

[9] M. Shami and A. Hemani, "An improved self-reconfigurable interconnection scheme for a coarse grain reconfigurable architecture," in *NORCHIP, 2010*, nov. 2010, pp. 1 –6.

Charge Collection Probability: Normal-Collector Configuration

Chee Chin Tan, Vincent K. S. Ong and K. Radhakrishnan
School of Electrical and Electronic Engineering
Nanyang Technological University
Singapore 639798
e-mail: tanc0184@e.ntu.edu.sg, vo@pmail.ntu.edu.sg, eradha@ntu.edu.sg

Abstract— The charge collection probability is the basis in the study of induced current generated when the semiconductor sample is subjected to some external excitation. In this paper, we present an analytical expression for the charge collection probability of the normal-collector configuration, with finite dimensions and surface recombination at the free surfaces. An excellent agreement has been found between the charge collection probability profiles computed using the newly derived analytical expression and those obtained using a device simulator. The analytical expression was then used to study the effects of the various physical parameters on the charge collection probability. This new analytical expression is expected to enhance our understanding of the charge collection process.

Keywords-charge-carrier processes; semiconductor diodes; semiconductor device modelling; simulation

I. INTRODUCTION

When a semiconductor sample is subjected to external excitation, i.e., high energy electron or photon beam, electron-hole pairs (EHPs) are generated in a region known as the generation volume. The EHPs tend to diffuse away from the generation volume due to the existence of the carrier concentration gradient in the generation volume and the bulk semiconductor. These diffused EHPs annihilate themselves in the bulk through recombination. However, when these diffused EHPs encounter some form of built-in electric field at the charge collecting junction, i.e., the p-n junction, the electrons and holes are separated, preventing them from further recombination. This separation of the EHPs contributes to an induced current detected at the external circuitry [1] and is known as charge collection [2]. The charge collection plays a significant role in the functionality and performance of semiconductor photonic devices, e.g., solar cell, and characterization techniques, e.g., electron-beam-induced current (EBIC).

The normalized induced current $I_N (x', z')$ detected at the external circuitry is given as [3, 4]

$$I_N(x',z') = \iint Q(x,z)g(x-x',z-z')dxdz. \qquad (1)$$

The generation volume function, $g(x-x')(z-z')$, gives the spatial distribution of the generated EHPs centered at (x', z'). $Q(x, z)$ is the charge collection probability which is defined as the spatially-dependent probability that the generated EHPs will be collected at the charge collecting junction [5]. The

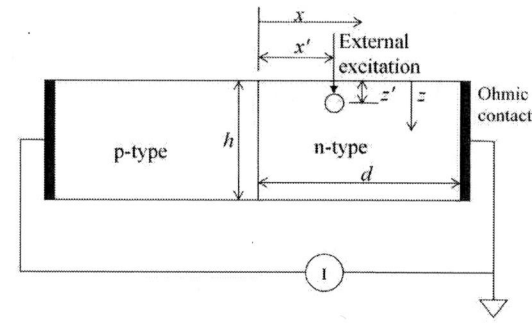

Figure 1. The normal-collector configuration

charge collection probability is one of the important parameters to study and to determine the generated induced current as a result of external excitation using (1).

Fig. 1 shows the normal-collector configuration, one of the commonly found configurations in semiconductor devices and their characterization techniques. In this configuration, the external excitation is injected on the surface normal to the charge collecting junction. The analytical expressions for the charge collection probability of a semi-infinite normal-collector configuration, i.e., $h = \infty$, can be found in literature [3, 6-8]. These analytical expressions have greatly contributed to the development and study of semiconductor materials and devices as well as their characterization techniques [4, 9-11].

However, when the normal-collector configuration has finite dimensions, i.e., both sample thickness h and sample width d are comparable to the diffusion length L, the use of the aforementioned analytical expressions become invalid. In the study of the charge collection of the normal-collector configuration with finite dimensions, the effect of the surface recombination at the bottom surface must be taken into consideration. Therefore, it is of great interest to derive a new analytical expression for the study of the charge collection for the normal-collector configuration with finite dimensions.

In this paper, we present a newly derived analytical expression for the charge collection probability of the normal-collector configuration with finite sample thickness h and sample width d. By making use the newly derived analytical expression, we studied the effects of various physical parameters on the charge collection probability. With this newly derived analytical expression and the studies done in this

978-1-61284-863-1/11 $26.00 © 2011 IEEE

paper, our understanding on the charge collection processes of the normal-collector configuration is enhanced.

II. ANALYTICAL EXPRESSION

It has been shown in [12, 13] that the charge collection probability satisfies

$$\frac{\partial^2 Q(x',z')}{\partial x'^2} + \frac{\partial^2 Q(x',z')}{\partial z'^2} - \lambda^2 Q(x',z') = 0 \qquad (2)$$

where $Q(x', z')$ is the charge collection probability at (x', z') and λ is the reciprocal of the diffusion length, i.e., $\lambda = 1/L$.

The boundary conditions for a given configuration are determined as follows [12, 13]. The Q at the charge collecting junction is equal to unity whereas the Q at the ohmic contact is equal to zero. The Q at all other surfaces, i.e., the free surface, satisfies

$$-\frac{\partial Q}{\partial \mathbf{n}} = sQ \qquad (3)$$

where $\mathbf{n}$ is the vector normal outwards from the free surface, s is the reduced surface recombination velocity at that surface, i.e., $s = v_s/D$, where v_s is the surface recombination velocity and D is the diffusion coefficient. Hence, the boundary conditions for the normal-collector configuration shown in Fig. 1 are

$$Q = 1, \qquad \text{for } x' = 0 \text{ and } 0 < z' < h$$
$$Q = 0, \qquad \text{for } x' = d \text{ and } 0 < z' < h$$
$$\frac{\partial Q}{\partial z'} = s_T Q, \qquad \text{for } z' = 0 \text{ and } 0 < x' < d \qquad (4)$$
$$\frac{\partial Q}{\partial z'} = -s_B Q, \quad \text{for } z' = h \text{ and } 0 < x' < d.$$

where s_T and s_B are the reduced surface recombination velocities at the top and the bottom surfaces respectively. This means that s_T and s_B are equal to v_{sT}/D and v_{sB}/D respectively where v_{sT} and v_{sB} are the surface recombination velocities at the top and bottom surfaces respectively.

The partial differential equation (2) with the boundary conditions (4) can be solved by using the Green's function method. The detailed derivation of the Green's function is available upon request. The Green's functions for a point generation source located at (x', z') are

$$G_I\left(x, z|x', z'\right) = \sum_{n=1}^{\infty} E_n \sinh\left(\mu_n x\right) \sinh\left[\mu_n\left(d - x'\right)\right]$$
$$\times \left[\cos\left(p_n z\right) + \frac{s_T}{p_n} \sin\left(p_n z\right)\right] \qquad (5)$$

$$G_{II}\left(x, z|x', z'\right) = \sum_{n=1}^{\infty} E_n \sinh\left[\mu_n\left(d - x\right)\right] \sinh\left(\mu_n x'\right)$$
$$\times \left[\cos\left(p_n z\right) + \frac{s_T}{p_n} \sin\left(p_n z\right)\right] \qquad (6)$$

where

$$E_n = \frac{\cos\left(p_n z'\right) + \dfrac{s_T}{p_n} \sin\left(p_n z'\right)}{\displaystyle\int_0^h \left(\cos\left(p_n z\right) + \dfrac{s_T}{p_n}\sin\left(p_n z\right)\right)^2 dz \times \mu_n \sinh\left(\mu_n d\right)}, \qquad (7)$$

$$\mu_n = \left(\lambda^2 + p_n^2\right)^{1/2}, \qquad (8)$$

and p_n is the positive roots of the transcendental equation

$$\tan\left(p_n h\right) = \frac{\left(s_T + s_B\right) p_n}{p_n^2 - s_T s_B}. \qquad (9)$$

$G_I(x, z| x', z')$ and $G_I(x, z| x', z')$ represent the Green's functions for region I which lies in $0 \leq x \leq x'$ and region II which lies in $x' \leq x \leq d$ respectively.

The charge collection probability of the normal-collector configuration can be determined from the Green's function and is given as [3]

$$Q(x', z') = \int_0^h \left.\frac{\partial G(x, z, | x', z')}{\partial x}\right|_{x=0} dz. \qquad (10)$$

Substituting (5) into (10) gives the charge collection probability at the point generation source location (x', z') as

$$Q\left(x', z'\right) = \sum_{n=1}^{\infty} Q_n \qquad (11)$$

$$Q_n = \frac{\cos\left(p_n z'\right) + \dfrac{s_T}{p_n}\sin\left(p_n z'\right)}{\displaystyle\int_0^h\left(\cos\left(p_n z\right) + \dfrac{s_T}{p_n}\sin\left(p_n z\right)\right)^2 dz \times \sinh\left(\mu_n d\right)}$$
$$\times\left[\frac{\sin\left(p_n h\right)}{p_n} - \frac{s_T}{p_n^2}\left(\cos\left(p_n h\right) - 1\right)\right]\sinh\left[\mu_n\left(d - x'\right)\right] \qquad (12)$$

By making use of (11) and (12), one can easily compute the charge collection probability of the normal-collector configuration as it involves only elementary functions and basic mathematical operations only. The newly derived analytical expression is found to be more general than the previous analytical expressions [3, 6-8] as it has taken into consideration the effects of the sample thickness h, sample width d and the surface recombination velocities at the free surfaces.

III. COMPUTATION AND SIMULATION

Given the values of the physical parameters, the values of p_n can be determined by solving (9) using the Mathematica software with a function known as *RootSearch* that is downloadable from the Wolfram Library Archive. The charge collection probability of the normal-collector configuration is then computed using (11) and (12) with the use any numerical

978-1-61284-863-1/11 $26.00 © 2011 IEEE   517

computation software, e.g., Matlab. In this paper, the infinite series of the analytical expression was approximated using 20 000 terms. This number of terms provides sufficient accuracy for the purpose at hand without consuming too much computational time and resources.

In order to verify the correctness of the analytical expression discussed in Section II, we compared the charge collection probability computed using the analytical expression with that obtained using the device simulator, i.e., MEDICI. The MEDICI simulations are similar to that in [14, 15] except that a different configuration is being used. In the MEDICI simulation, we first constructed the normal-collector configuration with the designed dimensions with a fine grid of 0.1 μm. The material used was Silicon and a uniform doping concentration of 10^{18} cm^{-3} was set at both the p- and n- regions. The value of the diffusion length was set by setting the value of the carrier lifetime appropriately. The point generation source was approximated by a square with sides of 0.2 μm.

IV. RESULTS AND DISCUSSION

Figs. 2-5 show the charge collection probability profiles of the normal-collector configuration computed using the analytical expression discussed in section II and those obtained from MEDICI simulations. An excellent match between the analytical and simulated results was found. This validates the correctness of the derived analytical expression.

The presence of surface recombination causes the generated charge carriers to recombine at the surface. This, in turn, reduces the number of charge carriers reaching the junction and eventually decreases the value of the charge collection probability. The effect of surface recombination at the top surface on the charge collection probability can be observed in Fig. 2. From this figure, we can see that the charge collection varies rapidly with v_{sT}, i.e., the charge collection probability decreases when v_{sT} increases, within the range of 10^3 cms^{-1} < v_{sT} < 10^6 cms^{-1} as compared to that outside this range. This means that it is possible to approximate v_{sT} to zero when v_{sT} < 10^3 cms^{-1} and v_{sT} to infinity when v_{sT} > 10^6 cms^{-1} since a small change in v_{sT} outside the range 10^3 cms^{-1} < v_{sT} < 10^6 cms^{-1} does not significantly affect the value of the charge collection probability.

The effect of the surface recombination at the bottom surface on the charge collection probability is shown in Fig. 3. Similar to the effect of the surface recombination at the top surface, the presence of the surface recombination at the bottom surface reduces the charge collection probability. However, since the generation volume is located further away from the bottom surface, the effect of the surface recombination at the bottom surface on the charge collection probability is not as strong as the effect of the surface recombination at the top surface.

Fig. 4 shows how the charge collection probability varies with the depth of the generation volume z'. It can be seen that the charge collection probability first increases with z' until it reaches a peak value and then decreases with z'. This observation can be explained as follows. When generation volume is further away from the top surface, i.e., when z' increases, it becomes less influenced by the surface

recombination at the top surface, i.e., there is less carrier recombination at the top surface. This causes an increase in the charge collection probability. Further increase in z' causes the generation volume to be nearer to the bottom surface. The nearer the generation volume is to the bottom surface, the more carriers will recombine at the bottom surface, and hence causes the charge collection probability to decrease with z'.

Another interesting observation drawn from Fig. 4 is that the charge collection probability is symmetrical along the line $z = h/2$ when the surface recombination at the top and bottom surfaces are of equal strength. This is observed in Fig. 4(a). This is because in this case, the normal-collector configuration is symmetrical in both geometric and boundary conditions. On the other hand, when the strengths of the surface recombination at the top and bottom surfaces are not equal, the peak of the charge collection probability along z axis shifts toward the surface with a lower surface recombination velocity. This indicates that the surface with a larger surface recombination velocity will have a greater impact on the charge collection than the other surface.

Reducing the sample thickness h will, in effect, bring the bottom surface closer to the generation volume. This causes more recombination to occur at both surfaces. This is because

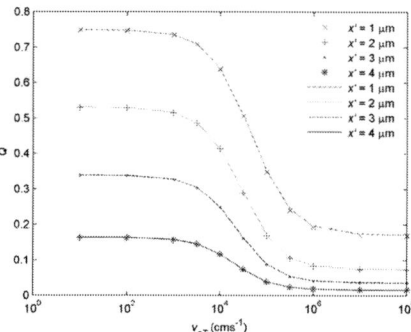

Figure 2. The plots of charge collection probability versus v_{sT} for different values of x'. The lines are computed using the derived analytical expression and the points are computed using MEDICI. The simulation parameters are d = 5 μm, h = 5 μm, L = 5 μm, z' = 0.3 μm and v_{sB} =10^4 cms^{-1}.

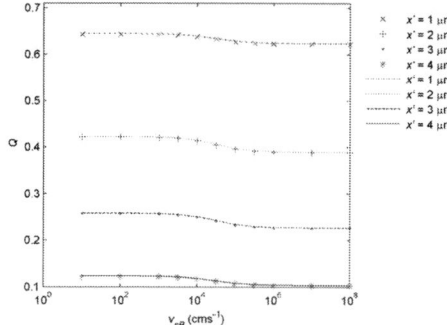

Figure 3. The plots of charge collection probability versus v_{sB} for different values of x'. The lines are computed using the derived analytical expression and the points are computed using MEDICI. The simulation parameters are d = 5 μm, h = 5 μm, L = 5 μm, z' = 0.3 μm and v_{sT} =10^4 cms^{-1}.

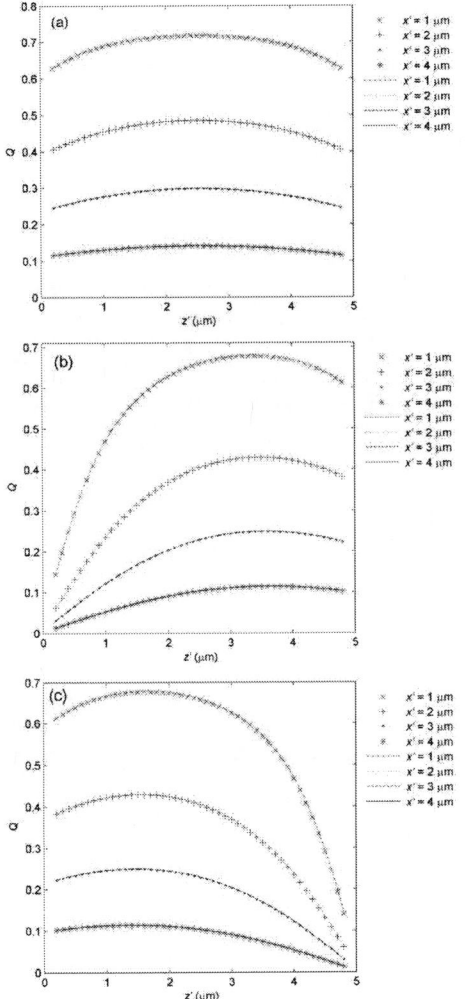

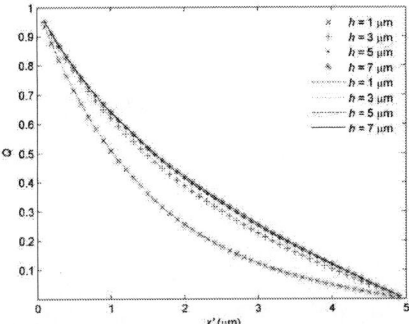

Figure 5. The plots of charge collection probability versus x' for different values of h. The lines are computed using the derived analytical expression and the points are computed using MEDICI. The simulation parameters are $d = 5$ μm, $L = 5$ μm, $z = 0.3$μm, $v_{sT} = 10^4$ cms^{-1} and $v_{sB} = 10^4$ cms^{-1}.

physical parameters, i.e., h, d, z' and the surface recombination at the free surfaces was also included in this paper.

Figure 4. The plots of charge collection probability versus z' for different values of x'. The lines are computed using the derived analytical expression and the points are computed using MEDICI. The simulation parameters are $d = 5$ μm, $h = 5$ μm, $L = 5$ μm, (a) $v_{sT} = 10^4$ cms^{-1} and $v_{sB} = 10^4$ cms^{-1}, (b) $v_{sT} = 10^6$ cms^{-1} and $v_{sB} = 10^4$ cms^{-1} (c) $v_{sT} = 10^4$ cms^{-1} and $v_{sB} = 10^6$ cms^{-1}.

the charge carriers that are partially reflected from one surface will reach the other surface and recombine. This will in turn decrease the value of the charge collection probability. Decreasing the value of h also causes the charge collection profile to concave upwards. These effects are as seen in Fig. 5.

V. CONCLUSION

The analytical expression for charge collection probability of finite dimensions normal-collector configuration was presented. An excellent agreement was found between the charge collection probability profiles computed using the newly derived analytical expression and those obtained using a device simulator. The study of the effects of the various

REFERENCE

[1] H. J. Leamy, "Charge collection scanning electron microscopy," Journal of Applied Physics, vol. 53, no. 6, pp. 51-80, 1982.

[2] D. B. Holt, Quantitative Scanning Electron Microscopy. London, U.K: Academic, 1974.

[3] C. Donolato, "On the analysis of diffusion length measurements by SEM," Solid-State Electronics, vol. 25, no. 11, pp. 1077-1081, 1982.

[4] O. Kurniawan and V. K. S. Ong, "Choice of generation volume models for electron beam induced current computation," IEEE Transactions on Electron Devices, vol. 56, no. 5, pp. 1094-1099, 2009.

[5] A.-A. S. Al-Omar, "The collection probability and spectral response in isotype heterolayers of tandem solar cells," Solid-State Electronics, vol. 50, no. 9-10, pp. 1656-1666, 2006.

[6] F. Berz and H. K. Kuiken, "Theory of life time measurements with the scanning electron microscope: steady state," Solid-State Electronics, vol. 19, no. 6, pp. 437-445, 1976.

[7] K. L. Luke and O. von Roos, "An EBIC equation for solar cells," Solid-State Electronics, vol. 26, no. 9, pp. 901-906, 1983.

[8] O. von Roos and K. L. Luke, "Analysis of the interaction of an electron beam with back surface field solar cells," Journal of Applied Physics, vol. 54, no. 7, pp. 3938-3942, 1983.

[9] K. L. Luke, "Determination of diffusion length in samples of diffusion-length size or smaller and with arbitrary top and back surface recombination velocities," Journal of Applied Physics, vol. 90, no. 7, pp. 3413-3418, 2001.

[10] O. Kurniawan and V. K. S. Ong, "An analysis of the factors affecting the alpha parameter used for extracting surface recombination velocity in EBIC measurements," Solid-State Electronics, vol. 50, no. 3, pp. 345-354, 2006.

[11] K. L. Luke, "The evaluation of surface recombination velocity from normal-collector geometry electron-beam-induced current line scans," Journal of Applied Physics, vol. 75, no. 3, pp. 1623-1631, 1994.

[12] C. Donolato, "A reciprocity theorem for charge collection," Applied Physics Letters, vol. 46, no. 3, pp. 270-272, 1985.

[13] C. Donolato, "An alternative proof of the generalized reciprocity theorem for charge collection," Journal of Applied Physics, vol. 66, no. 9, pp. 4524-4525, 1989.

[14] O. Kurniawan and V. K. S. Ong, "Charge collection from within a collecting junction well," IEEE Transactions on Electron Devices, vol. 55, no. 5, pp. 1220-1228, 2008.

[15] C. C. Tan and V. K. S. Ong, "An analytical expression for charge collection probability from within a U-shaped junction well," IEEE Transactions on Electron Devices, vol. 57, no. 11, pp. 3068-3073, 2010.

978-1-61284-863-1/11 $26.00 © 2011 IEEE

A Study of the Effect of Shallow Trench Isolation Technology on MOSFET DC Characteristic

Xia Fang, Lingling Sun, Senior Member, IEEE, Jun Liu, Huang Wang
Key Laboratory for RF Circuits and Systems of Ministry of Education
Hangzhou Dianzi University
Hangzhou 310037, China
Sunll@hdu.edu.cn

Abstract—An investigation of the effect of Shallow trench isolation (STI) technology on DC characteristic is presented. STI parameter of SA/SB impact on device characteristic is mainly considered, and discovered the small width device has different variation trench. Through comparing Agilent ICCAP simulation data with experimental data wafers fabricated using SMIC 0.13μm technology find that BSIM4 STI model is very well to fit all size devices except small width device. We design eight active lengths (i.e. SA=0.4μm, 0.8μm, 1.2μm, 1.6μm, 2.0μm, 2.4μm, 2.8μm and 3.2μm, and SA=SB) both PMOS and NMOS. Dependence of threshold voltage (V_{th}) on SA/SB is shown.

Keywords- Shallow trench isolation (STI), compact model, BSIM4, V_{th}

I. INTRODUCTION

STI technology originates in the 1980s. However, the high cost and immature STI technology enable it accepted until recent years. Indeed, today it has become the current mainstream technology to isolate the influence among devices in the large scale integrated circuit. STI technology overcomes the limitation of local oxidation on silicon (LOCOS) technology, e.g., severe bird's beak effect, LOCOS causing narrow width effect, not smooth surface and so on.

Although the STI has good isolation performance and flat surface, it is more or less impact on the device operating characteristic due to STI-induced stress. The result of non-uniform distribution of stress in channel affects the MOSFET characteristics and hence changes the circuit behavior. In fact, with the continuous shrinking of feature size, STI stress effect become significant and compact modeling efforts become more and more challenging. More empirical equations and parameters should be added for making the compact model more accurate. In BSIM4[1] model, it provides two different mechanisms within the influence of stress effect on the device characteristics. There are mobility-related and V_{th}-related. But it is not enough to representation the effect of STI-induced stress. For example, the effect of STI width (halo doping, Source/Drain contact space and so on) is not considered. Besides, there have been many reports about improving STI technology to reduce STI-induced stress[2][3]. In [2] study, it uses STI-wall-oxide nitridation and STI gap-fill-oxide densifying in pure N2 ambient to reduce the STI stress. In a way, this approach can reduce the active–area layout dependence of V_{th} and I_{ds}. And in [3], it uses nitric oxide (NO)-annealed wall oxide to suppress the inverse narrow width effect (this means that the narrower the width is, the smaller V_{th} is)[4]. However, the technology is just as far as possible to reduce the stress, and unable to remove it. So building an accurate STI stress model is still necessary for circuit design.

As shown in Fig. 1, the STI structure is around active area (AA). This paper presents the SA/SB (as shown in Fig. 2) impact on V_{th}, and the technology adopts SMIC 0.13μm. Generally, the effect is different between NMOS and PMOS[5]. In Fig. 2, SA and SB respectively are the distance between OD edges to poly from one and the other side. The article can be divided into four parts. The first part gives an introduction about STI. In the second section, we present the stress model of BSIM4. For third section, measurement and simulation results are discussed. Lastly, the fourth part makes a conclusion.

II. STI MODEL INTRODUCE

Currently, for enhancing device mobility, strain technology (e.g. embedded SiGe technology, stress memorization technique, dual stress liner and the parasitic stress form STI) is taken. But STI-induce stress is an intrinsic stress source and not intentionally built up for enhancing device performance. It is cased by the difference thermal expansion coefficient between Si and SiO_2, and a compressive state of stress develops as a peak at the AA-STI interface (as shown in Fig. 1). Fig. 1 shows the result of X-stress (in the direction of channel length, i.e. Sxx) distribution by using Sentaurus to simulate.

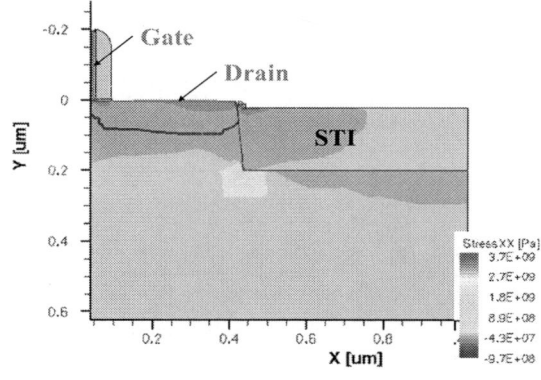

Figure 1. Sxx stress distribution by the TCAD software of Sentaurus

---

This work was supported by the 973 Program of China under Grant 2010CB327403.

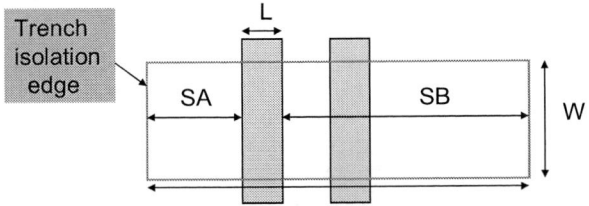

LOD=SA+SB+L OD: gate Oxide Definition

Figure 2. Shows the typical layout of a MOSFET

For STI compact model, there have a lot of efforts been devoted. Dunga et al. [6] proposes a modeling method through finding an equivalent stress level (as (1)) in the channel to account for the mobility enhancement, but the result isn't adequate for V_{th} shift. That is because mobility and V_{th} haven't the same equivalent stress level. Mobility-related is induced by the band structure modification, and V_{th}-related is a result of doping profile variation. Lately, Chi-Chao Wang et al. [7] gives an approach of layout decomposition that is to partition layout into a set of simple patterns for efficient model extraction. In BSIM4, the model includes STI influence on not only mobility and saturation velocity, but also the threshold voltage and other important second-order effects. The main model equations are written by (2)-(4).

$$\frac{\triangle \mu}{\mu} \infty S_{AVG}(L) \qquad (1)$$

$$\mu_{eff} = \frac{1+\rho_{ueff}(SA,SB)}{1+\rho_{ueff}(SA_{ref}+SB_{ref})}\mu_{effo} \qquad (2)$$

$$\upsilon_{sattemp} = \frac{1+KVSAT \cdot \rho_{\mu eff}(SA,SB)}{1+KVSAT \cdot \rho_{\mu eff}(SA_{ref},SB_{ref})}\upsilon_{sattempo} \qquad (3)$$

$$VTH0 = VTH0_{original} + \frac{KVTH0}{Kstress\_vth0}\cdot(Inv\_sa+Inv\_sb$$
$$-Inv\_sa_{ref}-Inv\_sb_{ref}) \qquad (4)$$

III. DISCUSSION MEASUREMENT AND SIMULATION RESULT

Because of the difference effect of STI-induced stress between NMOS and PMOS, the measure data both include NMOS and PMOS. The devices of NMOS or PMOS have large (L=5μm, W=5μm), small (L=0.13μm, W=0.3μm), narrow (L=5μm, W=0.13μm) and short (L=5μm, W=0.3μm). And for each feature size, SA/SB is set from 0.4μm to 3.2μm with step 0.4μm. There are a total of 64 devices (as illustrated in the table 1).

TABLE I.   ILLUSTRATED THE DESIGN MEASURE STRUCTURE FOR MODELING

SA/SB (NMOS/PMOS)	Large	Small	Narrow	Short
0.4μm	√	√	√	√
0.8μm	√	√	√	√
1.2μm	√	√	√	√
1.6μm	√	√	√	√
2.0μm	√	√	√	√
2.4μm	√	√	√	√

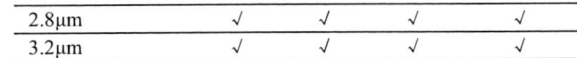

The device DC data is gotten by using Agilent4156C to measure. Then according to BSIM4 model extraction process, we make simulation data to fit measure data by using Agilent ICCAP software. ADS simulator is called. The Fig. 3 shows a part of device fitting result.

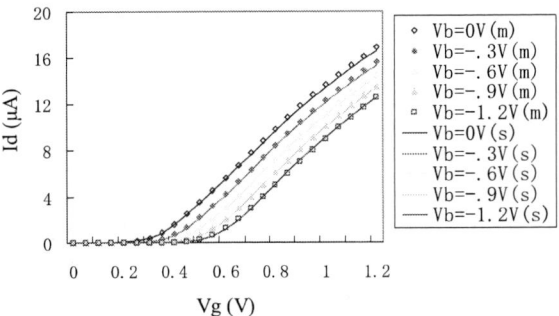

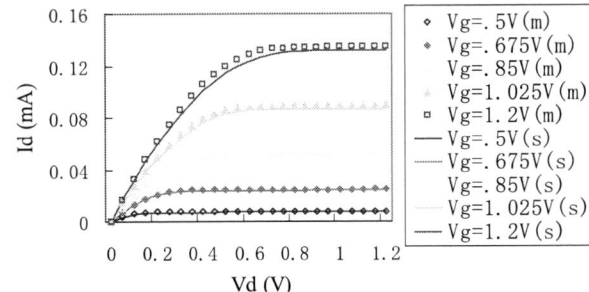

(a)

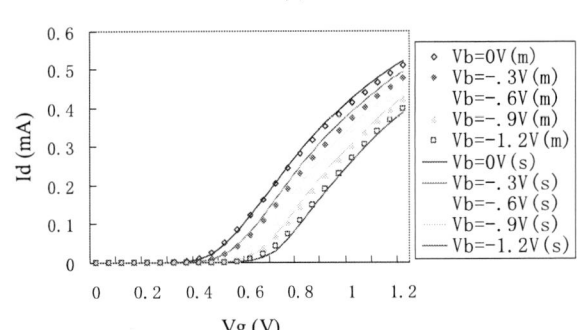

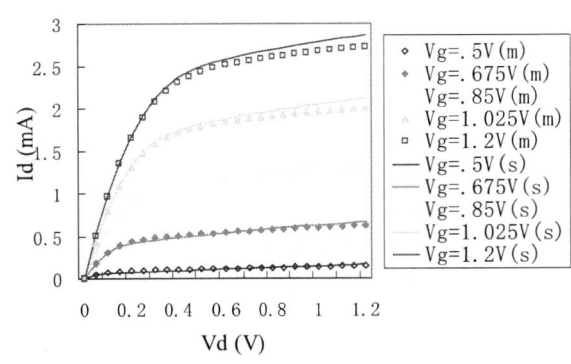

(b)

978-1-61284-863-1/11 $26.00 © 2011 IEEE

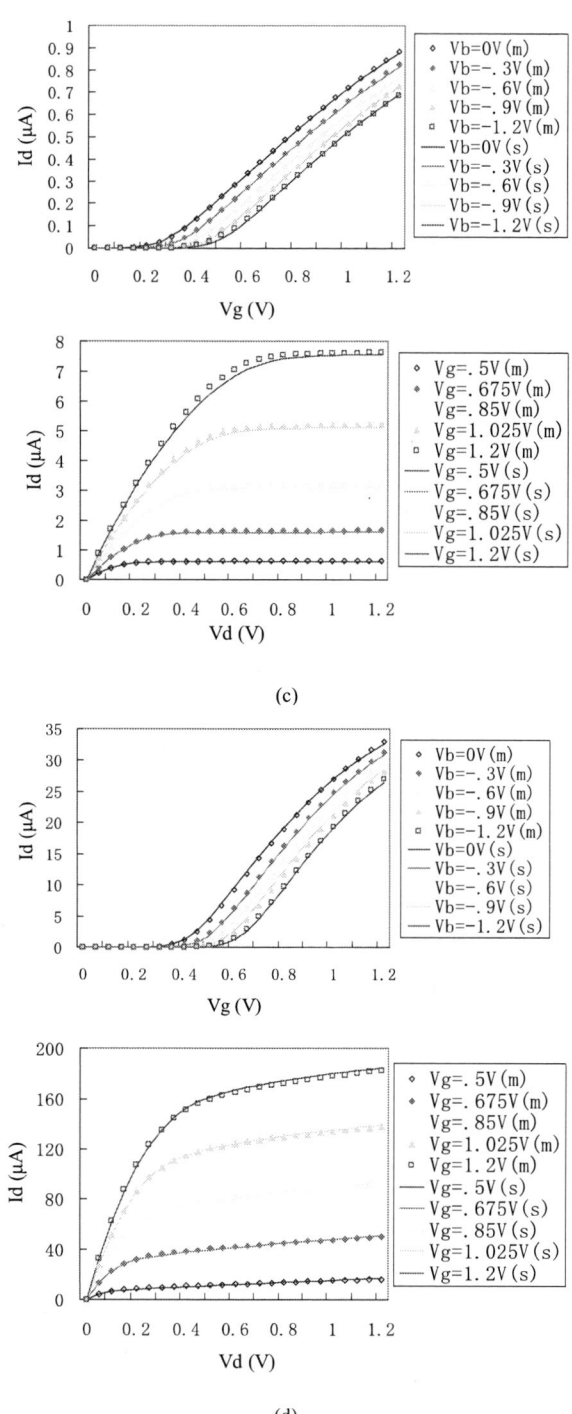

Figure 3. NMOS (SA/SB=3.2μm) DC characteristic fitting result, $I_dV_g(V_d=0.05V)$, $I_dV_d(V_b=0V)$, (a) Large, (b) Narrow, (c) Short, and (d) Small.

As we known, the STI stress parameters SA/SB influence on V_{th} greatly since the technology notes scaled down. According to transfer curves of the well fitting result, extracted V_{th} values are shown in Fig. 4 for NMOS. (a), (b), (c) and (d) are respectively large, narrow, short and small devices. It indicates V_{th} decreases with increasing SA/SB. In other words, the V_{th} of NMOS increases with enhancing STI-induced stress.

However, (c) and (d) measure data haven't shown the trench obviously, that due to W is so much smaller that Y-stress (in the direction of perpendicular to the channel plane, i.e. Syy) plays a more major role than Sxx stress. However, the BSIM4 modeling couldn't show this variation very well. The peak and valley points of Fig. 4 (d) illustrate that Syy and Sxx play a different role in different regions. That means when Syy is bigger than Sxx, V_{th} will increase with SA/SB increasing for V_{th} increasing with device width[8], and vice versa, V_{th} will decrease. In addition, PMOS has the same variation trench with NMOS, just as [2] and [7]. But PMOS is more independent of SA/SB (as shown in Fig. 5.). For PMOS, there is an abnormal point in Fig. 5 (b) below one which should come from measure process.

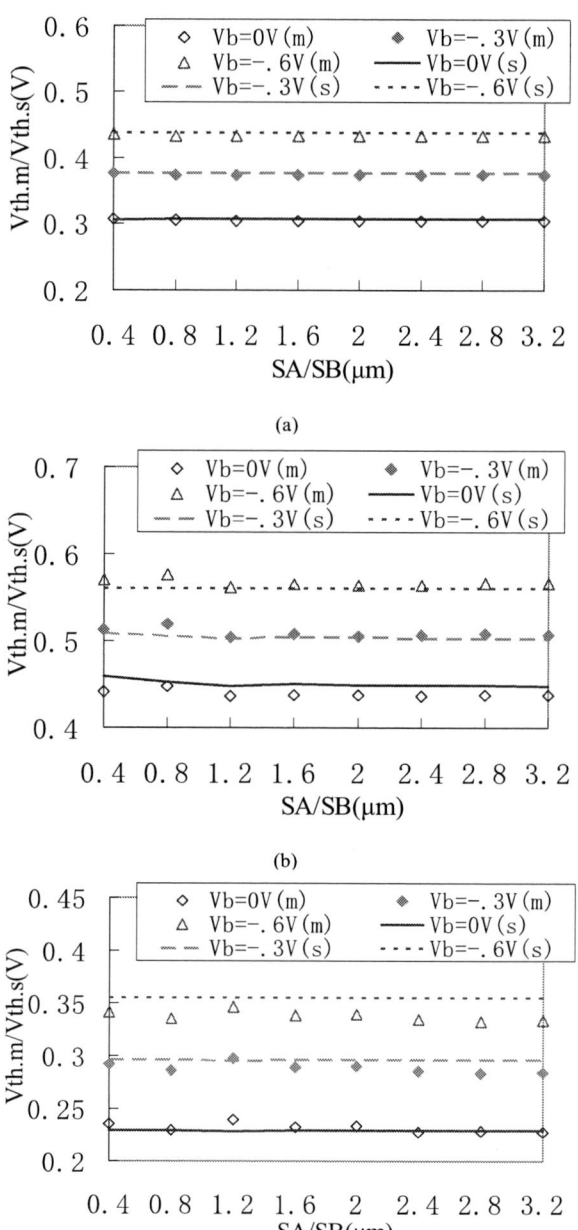

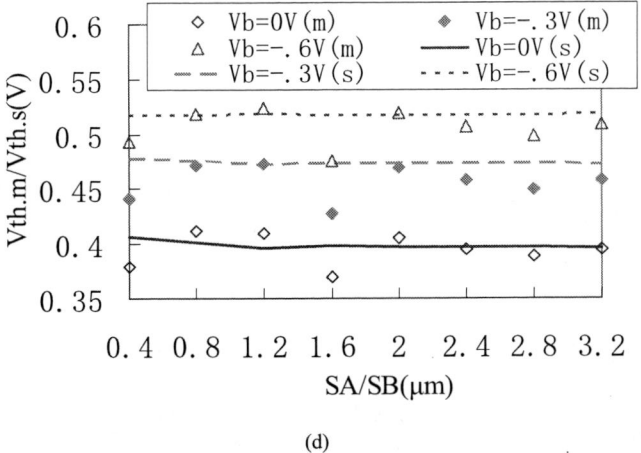

(d)

Figure 4. STI induced V_{th} changes with dependent on SA/SB for NMOS. (a) Large, (b) Narrow, (c) Short, and (d) Small.

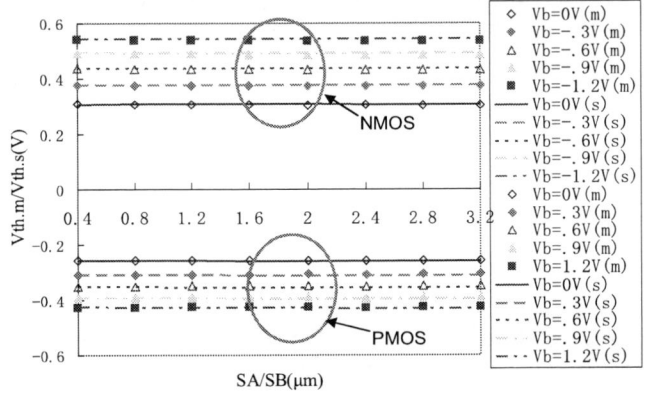

(a)

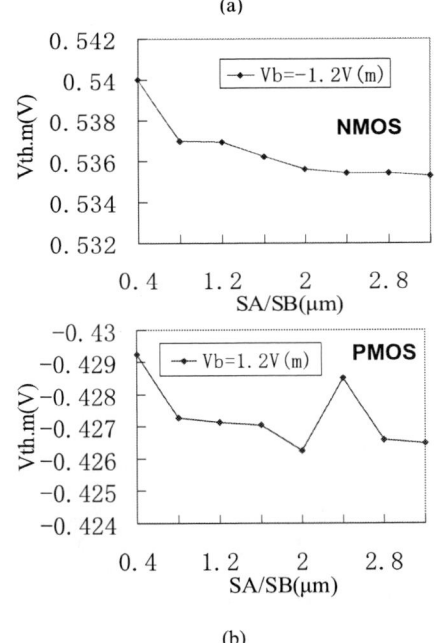

(b)

Figure 5. STI induced V_{th} changes with dependent on SA/SB for both large NMOS and PMOS. (a) all of the body bias, (b) the figure of above is NMOS measure date (V_b=-1.2V), and below one is for PMOS (V_b=1.2V).

Note: for all figures, (s) and (m) represent simulated (i.e. model) and measured result, respectively.

IV. CONCLUSION

In this paper, both measurement and simulation data of STI-induced stress impact on DC characteristic is presented. We use BSIM4 model to fit the group of measure data. Then through ICCAP function gets V_{th} versus SA/SB curves, and analysis the difference between NMOS and PMOS.

REFERENCES

[1] Mohan V. Dunga, et al., BSIM4.6.0 MOSFET Model User's Manual, University of California Berkeley, 2006.

[2] Masafumi Miyamoto, et al., "Impact of Reducing STI-Induced Stress on Layout Dependence of MOSFET Characteristics," IEEE Trans. Electron Devices, vol. 51, pp. 440-443, March 2004.

[3] Jongoh Kim, et al., "A Shallow Trench Isolation Using Nitric Oxide (NO)-Annealed Wall Oxide to Suppress Inverse Narrow Width Effect," IEEE Electron Device Letters, vol.21, no.12, pp. 575- 577, Dec 2000.

[4] Wang Xinzhu, Xu Qiuxia, Qian He, Shen Zuocheng and OuWen, "Shallow Trench Isolation Process for Deep Sub-Micron Technologies", Chinese Journal of Semiconductors, Vol. 23, No. 9, pp. 323-329, Mar 2002.

[5] P. B. Y. Tan, A. V. Kordesch and O. Sidek, "CMOS shallow trench isolation x-stress effect on channel width for 130nm technology," ICSICT, vol., no., pp.478-480, Oct. 2006

[6] M. V. Dunga, et al., "Modeling advanced FET technology in a compact model," TED, vol. 53, No. 9, pp.1971-1978, Sept. 2006.

[7] Chi-Chao Wang, Wei Zhao, Frank Liu, Min Chen, and Yu Cao, "Modeling of layout-dependent stress effect in CMOS design," ICCAD, vol., no., pp.513-520, Nov. 2009.

[8] Albert Victor Kordesch, Othman Sidek, Philip Beow Yew Tan, "Compact Modeling of Mechanical STI y-Stress Effect," ICSICT 8th International Conference on, vol., no., pp.1450-1452, Oct. 2006.

Low Frequency Noise Investigation of AlGaN/GaNOn Silicon Schottky Diode

Yihu Li[1,2], Yong-Zhong Xiong[2], S. Arulkumaran[3], Lei Wang[2], Wang Ling Goh[1], Geok Ing Ng[1], Shane Todd[2], Patrick Lo[2]

[1]Nanyang Technological University, Singapore

[2]Instiute of Microelectronics, A*STAR (Agency for Science, Technology, and Research), 117685, Singapore

[3] Temasek Laboratories@NTU, Nanyang Technological University, Singapore

Abstract— This paper presents the low frequency noise investigation of the AlGaN/GaN on silicon Schottky diode. The noise power spectral density of various sizes of the AlGaN/GaN Schottky diodes under different biasing voltages has been measured. The $1/f$ behaviour is observed before the corner frequency and the effect of the multi-finger contact is also been determined. With increasing number of fingers, higher $1/f$ noise is observed. However, if the frequency is increased further, the noise power density is noted to decay more rapidly. This therefore implies that a low thermal noise floor has been achieved.

Keywords— $1/f$ noise, power spectral density, Schottky diode, corner frequency, thermal noise floor

I. INTRODUCTION

For several decades, Schottky contacts are widely used as microwave detectors and mixer diodes [1]-[3]. Low frequency noise is an important parameter since it often defines the sensitivity or detection limit. Low frequency noise investigation and minimization are key issues in electronic devices.

The hetero-junction devices have become popular over the year, especially since the operation frequency has continued to augment. Gallium Nitride (GaN) based electronics become a promising solution for high temperature, high power and high frequency applications due to its wide direct band gap [4]-[6]. Among several approaches with different substrates, e.g. Sapphire or SiC, for cost effective purpose, GaN on Silicon is a potential solution and can be f using the compatible CMOS process.

Schottky diodes based on GaN on silicon process appear to be an optimal candidate for high frequency detection application. As a result, the low frequency noise investigation of a GaN on silicon Schottky contact is extremely important, calling for immediate attention. In this paper, the low frequency noise of n-GaN on silicon Schottky contacts is characterized under different DC biases. The effect of the contact size is also demonstrated. Furthermore, the influence of multi-finger structure too is extracted, in order to bring about a noise minimization design.

II. DEVICE STRUCTURE AND MEASUREMENT SETUP

The Aluminium Gallium Nitride/Gallium Nitride on silicon (GaN on silicon) devices are fabricated in-house at the

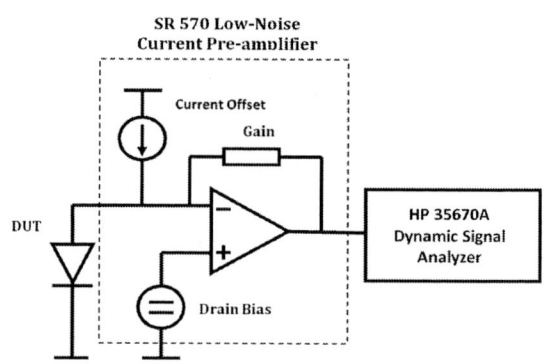

Fig. 1 Low frequency noise measurement setup.

Nanyang Technological University (NTU) [7]-[9]. The lengths of the Schottky contact are fixed at 0.25 µm and the widths are varied to control the sizing as well as the number of fingers. The measurement setup is depicted in Fig. 1. The entire noise measurement setup is contained within a shielded room. SR 570 Low-Noise current preamplifier is used to provide the DC supply and acts also as a low frequency noise current amplifier. An external multimeter is deployed to verify the biasing voltage applied to the Schottky diode. For each biasing point, the gain is adjusted to ensure that the output falls within suitable detection range of the HP 35670A Dynamic Signal Analyser. Instead of the line power source, Tthe internal chargeable battery is deployed for the current amplifier. This

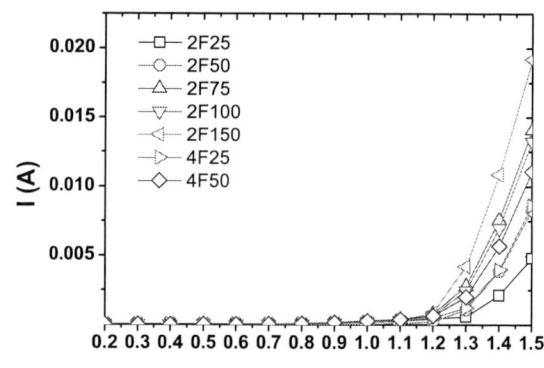

Fig. 2 I-V characteristics of various Schottky diodes.

helps to reduce the AC power source frequency coupling. Finally, the current offset is set to zero.

III. MEASUREMENT RESULTS

Several sets of measurements have been taken for measuring the low frequency noise performance of GaN on silicon Schottky diode, including the investigation of the effect of the contact size, multi-finger contact and the determination of the thermal noise floor. The noise power spectral densities are taken recorded with biasing voltage ranging from 0.1 V to 1.2 V, and in step of 0.1 V for each device. The temperature of the measuring environment is maintained at 292 K.

A. Schottky diode with different unit width

In this section, a series of devices with the same number of fingers were tested. The contact widths are: 25 μm, 50 μm, 75 μm, 100 μm and 150 μm. The number of fingers selected is two.

The I-V characteristics of all the Schottky diodes measured are shown in Fig. 2. It can be seen that before the diodes are turned on, the sub-threshold current is not proportional to their contact areas. This may be because the current transport

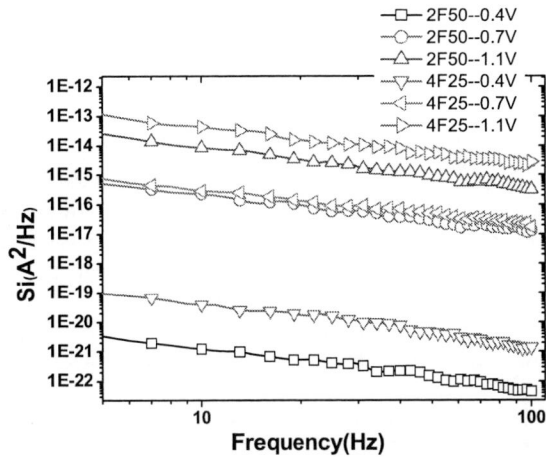

Fig. 5 Power spectral densities of 4-finger 25-μm and 2-finger 50-μm contacts at: $V_b = 0.4$ V, 0.7 V and 1.1 V.

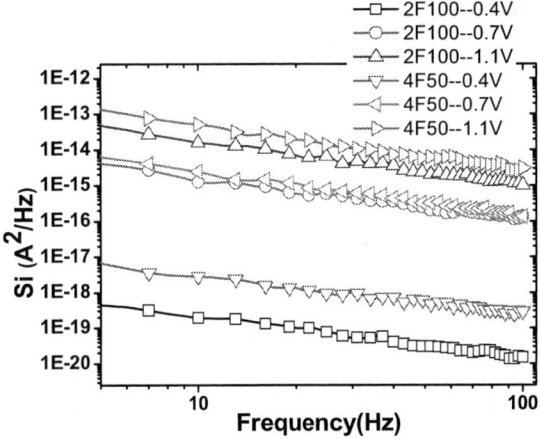

Fig. 6 Power spectral densities of 4-finger 50-μm and 2-finger 100-μm contacts at: $V_b = 0.4$ V, 0.7 V and 1.1 V.

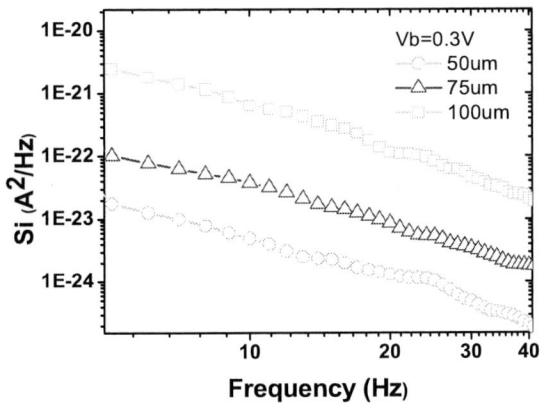

Fig. 3 Power spectral densities at $V_b = 0.3$ V.

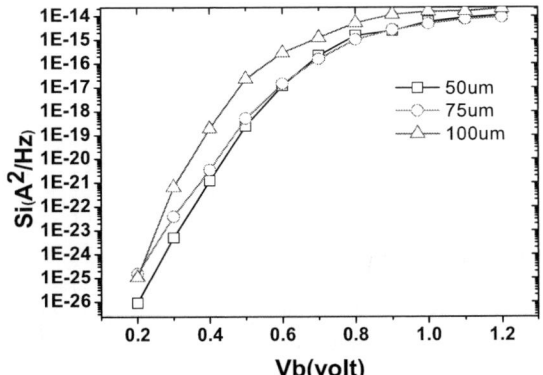

Fig. 4 Power densities at $f = 10$ Hz at different biasing voltages.

mechanism of Schottky diodes is not dominated by thermionic emission in at low voltages. At $V_b = 1.1$ V and beyond, the Schottky diodes enter the thermionic emission region, but after 1.3 V, the exponentially increase of the current is limited by the series resistance.

The low frequency noise is measured from 1 Hz to 100 Hz, in step of 1 Hz, and with an average number of 100. Fig. 3 illustrates the plot of the noise power spectral density from 5 Hz to 40 Hz at $V_b = 0.3$V and in log scale. In this frequency range, the typical $1/f^\gamma$ behaviour with γ close to 1 is observed. As expected, larger contact area implies higher noise density. An alternative plot of the noise power spectral density of the same group of Schottky contacts is given in Fig. 4. The frequency is fixed at 10 Hz and the biasing voltage varies from 0.2 V to 1.2 V. It can be observed that the noise power density approximately increases linearly with the biasing voltage until $V_b = 0.7$ V, and thereafter, the traces saturate after the diode is switched on. That also follows the characteristics of I-V curve, where an alternative current transportation mechanism starts to dominate when the biasing voltage is close to the "on"

978-1-61284-863-1/11 $26.00 © 2011 IEEE

voltage. Comparing with other devices [10], [11], GaN on silicon Schottky diodes are able to achieve better performance on sub threshold current and noise.

B. Multi-finger structure

Many researchers have proved that the RF performance of Schottky diode detector/mixer can be improved by dividing the contact area into multi-fingers. [12] The cut-off frequency is indeed increased by decreasing the series resistance. However, the noise performance is another disturbing issue that render special attention.

To compare the low frequency noise performance of multi-finger structures, two groups of measurement have been taken. In Fig. 5, the noise power spectral densities from 5 Hz to 100 Hz is plotted, a 2-finger 50-μm and a 4-finger 25-μm contacts are included. Three biasing voltages are selected: 0.4 V, 0.7 V and 1.1 V. At each biasing condition, the noise power plot has its respective $1/f$ behaviour. Unfortunately, the multi-finger structure shows a higher level of low frequency noise at each biasing voltage. Another pair of contacts comparing the 2-finger 100-μm and 4-finger 50-μm contacts is also characterised (see Fig. 6) and the results again led to the same conclusion. Since the low frequency noise usually limits the sensitivity of a detector, the increase of the noise will counter the benefit of higher cut-off frequency as the number of contact fingers is increased.

C. Thermal noise floor determination

In this section, other than the $1/f$ noise, the thermal noise floor and corner frequency will be determined and evaluated. Te effect of the multi-finger structure will also be discussed.

In order to balance the resolution and simulation time, higher frequency noise is evaluated separately from 100 Hz to 1.6 kHz and also 1.6 kHz to 20 kHz. Moreover, the pre-amplifier can adjust the gain to the suitable level for detection to improve the accuracy of the measurement. At $V_b = 0.2$ V, the corner frequency of the 2-finger 50-μm contact is around 724 Hz, and the thermal noise floor is as low as 1.67×10^{-30} A^2/Hz indicated in Fig. 7(a). Different from last section, the 4-finger 25-μm contact is noted to offer better performance. The corner frequency is around 580 Hz and the thermal noise floor is even lower, at 7.7×10^{-32} A^2/Hz. This is reasonable because the thermal noise consists of the shot noise of the Schottky barrier and thermal noise of the resultant series resistance. As 4-finger contact structure has less series resistance than 2 fingers', it will have lower thermal noise floor.

A non-linear behaviour has been observed at close to 100 Hz for the 4-finger 25-μm contact. This is due to the AC power source coupling. To remove this effect, a spectral analyser with DC battery may be used instead. As the biasing voltage is raised, the corner frequency is also increased; the level of thermal noise floor therefore also advances. For $V_b = 0.4$ V, one can observe a thermal noise floor at a corner frequency of around 7.3 kHz and 8.6 kHz for the 4-finger 25-μm and 2-finger 50-μm contacts, respectively.

In Fig. 7(b), with $V_b = 0.5$ V and above, the noise spectral density exhibits a $1/f$ behaviour until 20 kHz, which is within our frequency measurement limit. The corner frequency cannot be detected in this case. If the upper limit of the frequency is extended, the corner frequency and the thermal noise floor will appear. According to Luo's model [13], the current noise of a Schottky contact is a result of electron mobility fluctuation, and it can be expressed as

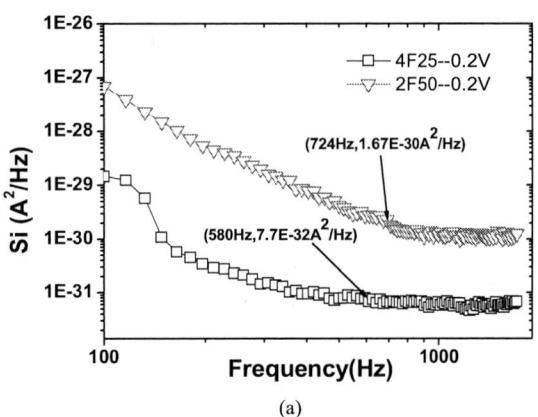

$$S_i^{1/f}(f) = \alpha_H \frac{I}{f} \frac{q^3}{3} \left(\frac{v_r}{v_d}\right)^2 \left[\frac{N_A(V_D - U)}{q\varepsilon\pi KTm^*}\right]^{1/2} \quad (1)$$

where v_r stands for the electron recombination velocity and v_d is the drift velocity of electrons. m^* denotes the effective mass of the electron, V_D is the built-in potential of the barrier, and the parameter α_H is the Hooge constant. Other samples stand for their usual meanings. The noise power spectral density is proportional to the current flowing through the Schottky barrier. At higher biasing voltage, the $1/f$ noise is more dominant over the thermal noise. Hence, the corner frequency is shifted to a higher value.

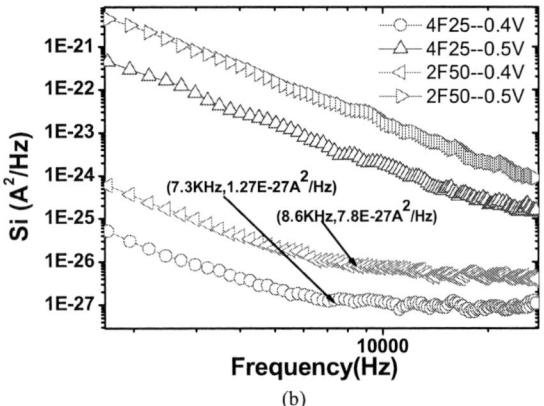

IV. CONCLUSIONS

In this paper, a low frequency noise investigation of GaN on silicon Schottky diode has been presented. The noise power spectral density from 1 Hz to 100 Hz has been investigated for

Fig. 7 (a) Thermal noise floor of 2-finger 50-μm and 4-finger 25-μm contacts spreading across 100 Hz to 1.6 kHz. (b) Thermal noise floor for 2-finger 50-μm and 4-finger 25-μm contacts, from 1.6 to 20 kHz.

various sizes of Schottky contacts. The biasing voltage is swept from 0.2 V to 1.2 V. As expected, the noise power density increases with the contact size and biasing voltage. $1/f$ behaviour has been observed before the corner frequency. As the frequency increases further, a thermal noise floor can be detected, consisting of shot noise and resistor thermal noise. GaN on silicon Schottky diode has a good performance on both $1/f$ noise and thermal noise. The effect of multi-finger contact has been extracted. Although the cut-off frequency can be boosted by engaging multi-finger contact, its $1/f$ noise performance will worsen.

ACKNOWLEDGMENT

The authors would like to thank the assistance provided by staff of the Institute of Microelectronics, Agency for Science Technology and Research (A*STAR), Singapore. The authors are especiallygrateful to the Temasek Laboratories@NTU for supporting the fabrication of the GaN on silicon devices. This work is part of the project entitled, "Device Characterization and Process-Design-Kit (PDK) Establishment", funded by A*STAR, Singapore. Grant Number: 102 169 0131

REFERENCES

[1] G. Hellings, J. Join, A. Lorenz, P. Malinowski and R. Mertens, "ALGaN schottky diodes for detector application in UV Wavelength range", *IEEE Trans. Electron Devices*, vol. 56, no. 11, Nov 2009.

[2] L. Liu, J. L. Hesler, H, Xu, A. W. Lichtenberger and R. M. WeikleII, "A broadband quasi-optical Tera-Hertz detector utilizing a zero bias schottky diode", *IEEE Microwave Wireless Components Lett*, vol. 20, no. 9, Sep 2009.

[3] J. H. Oh, S. W Moon, D. S. Kang and S. D. Kim, "High-performance 94-GHx single balanced mixer using disk-shaped GaAs schottky diode", *IEEE Electron Device Lett*, vol 30. No. 3, Mar, 2009.

[4] S. Aslam, R. E. Vest, D. Franz, F. Yan, and Y. Zhao, "Large area GaN, Schottky photodiode with low leakage current", *Electron. Lett.*, vol. 40, no. 17, pp. 1080–1082, Aug 2004.

[5] E. Monroy, T. Palacios, O. Hainaut, F. Omnes, and J.-F. Hochedez, "Assessment of GaN metal–semiconductor photodiodes for high-energy ultraviolet photodetection", *Appl. Phys. Lett.*, vol. 80, no. 17, pp. 3198–3200, Apr 2002.

[6] M.J. Shin & R.J. Trew, "GaN MESFETs for high-power and high temperature microwave applications", *Electron. Lett.*, vol. 31, no. 6, Mar 1995.

[7] Z. H. Liu, G. I. Ng, S. Arulkumaran, Y. K. T. Maung, K. L. Teo, S. C. Foo, and. V. Sahmuganathan, "Improved 2-D electron gas transport characteristics in AlGaN/GaN MIS HEMT with atomic layer-deposited Al_2O_3 as gate insulator", *Appl. Phys. Lett.*, 95, 223501 (2009).

[8] Z. H. Liu, S. Arulkumaran, and G. I. Ng, "Temperature dependence of ohmic contact characteristics in AlGaN/GaN HEMT from -50 to 200C", *Appl. Phys. Lett.*, 94, 142105 (2009).

[9] S. Arulkumaran, S. L. Selvaraj, T. Egawa, and G. I. Ng, "Sheet carrier density enhancement by Si_3N_4 passivation on non-polar a-plane (1120) sapphire grown AlGaN/GaN heterostructures", *Appl. Phys. Lett.*, 92, 092116 (2008).

[10] R. Singh & D. Kanjilal, "Temperature dependence of 1/f noise in Pd/n-GaAs Schottky barrier diode", *J. Applied Physics*, vol. 91, no. 1, Jan 2002.

[11] C. Wei & Y. Z. Xiong, X. Zhou "Investigation of low- frequency noise in N-channel FinFETs from weak to strong inversion", *IEEE Trans. Electron Devices*, vol. 56, no. 11, Nov 2009.

[12] S. Sankaran & Kenneth. K. O, "Schottky Barrier Diodes for Millimeter Wave Detection in a Foundry CMOS Process," *IEEE Electron Device Lett.*, vol. 26, no. 7 Jul 2005.

[13] M. Y. Luo, G. Bosman, A. van der Ziel and L. L. Hench, "Theory and experiments of 1/f noise in Schottky-barrier diodes operating in the thermionic-emission mode", *IEEE Trans. Electron Devices*, vol. 35, no. 8 Aug 1988.

Optimization of Vertical Silicon Nanowire based Solar Cell using 3D TCAD Simulation

Jitendra Kumar
Centre of Nanotechnology,
Indian Institute of Technology Roorkee
Uttarakhand, India 247667

S. K. Manhas[*], Dharmendra Singh
Dept of Electronics & Computer Engineering
Indian Institute of Technology Roorkee

Uttarakhand, India 247667
[*]e-mail: samanfec@iitr.ernet.in

Ramesh Vaddi
VIRTUS, IC Design Centre of Excellence
School of Electrical and Electronic Engineering, Nanyang
Technological University 50 Nanyang Avenue, Singapore

Abstract— **Solar cells based on nanowire (NW) array has shown promising potential for the low cost photovoltaic because of light absorption and charge carrier transport in this structure are in orthogonal direction to each other. In this study, we report the effect of variation of doping and defect densities on vertical NW solar cell bench-marked with standard planar structure using 3D-TCAD simulation. The performance of NW and planar structure for different amount of defect densities in the structures is investigated. We show that performance of NW solar cell continuously increases with wire doping. The results show that for increased efficiency, a high p-core and n-shell doping densities (~10^{19} cm^{-3}) are needed. This is attributed to radial structure of NW and increased field assisted charge separation. It is found that for same amount of illuminated area, NW structure has ~25% higher conversion efficiency. Further it is found that NW radial structure can tolerate defect density as high as 10^{18} cm^{-3}, with 82% higher conversion efficiency than planar structure. Our results have significant importance for design of vertical NW based solar cells and applications.**

Keywords- Nanowire Solar cell, Cell efficiency, Built-in field, Defect density

I. INTRODUCTION

Recently nanostructures like NW, nanopartical and quantum dot have shown to have enormous characteristics that can be exploited by using them as an active element in the solar cells [1]. Specially, NWs are expected to play a very important element in next generation solar cells. The radial geometry of NW solar cell has the advantage of orthogonalizing the solar radiation incident and charge carrier collection along with higher electric field. Because of this radial geometry it is considered to have very high defect density tolerance without losing significant conversion efficiency [2]. The use of NW in solar cell has been reported like in dye-sensitized solar cell [3], polymer based solar cell [4] and hybrid organic and inorganic solar cell [5]. Some novel structures have also been reported using single-crystalline n-CdS nanopillars, embedded in polycrystalline thin films of p-CdTe, to enable high absorption of light and efficient collection of the charge carriers [6].

In addition, silicon NWs have also been used to change the reflective properties of silicon solar cells [7]. In order to fabricate high quality, defect and contamination free silicon NWs solar cells, VLS techniques has been widely used [8], [9].

But very few studies have been done on studying the parameters that affect and guide the performance of NW solar cell. For fully utilizing the potential of NW based solar cells, we need to understand and optimize the parameters that affect its performance. In this work, we have carried 3D-TCAD simulation study and optimization of vertical pn-junction silicon NW solar cell for its doping and defect density tolerance.

The results were compared with planar p-n junction structure of same dimensions. Our results shows that radial p-n structure has 26% higher peak electric field compared to its planar counterpart. In addition to radial geometry, the larger E-field in NW junction enhances conversion efficiency compared to planar structure. For a radial structure without any defect density we get 31% enhanced conversion efficiency over planar pn junction cell. Our results shows that even with the defect density as high as 10^{18} cm^{-3} radial structure has 82% better conversion efficiency than planar cell with same defect density.

II. DEVICE STRUCTURE

The co-axial pn junction NW structure studied in this work is shown in Fig. 1 along with its 2D cross section view. In this structure, exterior "shell" (thickness 95nm) of NW is n-type, interior "core" is p-type (radius 95nm) and length of NW is 5 μm. Aluminum metal contacts are made on p-core at the bottom and transparent ohmic contact on the top surface of exterior n-shell. A 20 nm thick silicon nitrite (Si$_3$N$_4$) layer is used on outer surface for passivation and antireflective coating effect. A planar pn solar cell of same dimensions with top n$^+$ layer thickness of 95nm and bottom p layer of thickness 5μm is used for reference.

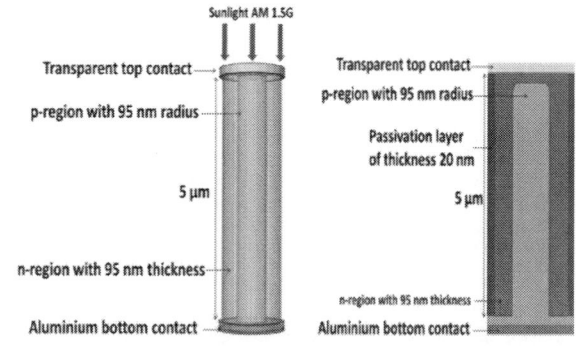

Fig.1. Structure of the co-axial nanowire solar cell.

Standard A.M. 1.5G solar spectrum is used with light incident normally on the top surface. In device simulation of implemented structure all important recombination statistics namely Shockley-Read-Hall recombination, Auger recombination, and surface recombination are incorporated. TCAD calculates optical generation rate in solar cell using raytracing model [10] and couples it with electrical simulation solving Poisson's and continuity equations.

To optimize NW solar cell structure we have evaluated short circuit current (I_{sc}), open circuit voltage (V_{oc}) and conversion efficiency with varying doping and defect densities. The conversion efficiency of the cell is extracted using relation

$$\eta = P_{out}/P_{in} = (V_{oc} \times I_{sc} \times FF)/P_{in} \qquad (1)$$

Where P_{out} is output power, P_{in} is incident power and FF is fill factor. Total power incident is calculated by multiplying power density of A.M. 1.5G global solar spectrum (0.1W/cm^2) with solar cell area on which solar illumination is incident.

III. RESULTS AND DISCUSSION

Fig. 2(a) and 2(b), respectively, shows the comparison of simulated electric field for radial and planar structures with varying n-layer doping density. We note that for low doping peak electric field at p/n-interface in radial structure is lower than that p/n-interface in planar structure. However, for higher doping (10^{19} cm^{-3}), the peak field for radial structure is comparable to planar structure. This enhancement of E-field in radial structure shows very good cell performance gain in terms of conversion efficiency over its planar cell as will be discussed in this paper.

The NW structure is optimized firstly by keeping p-layer doping constant and varying n-layer doping. Fig. 3(a-c) respectively shows the effect of n-layer doping variation on I_{sc}, V_{oc} and efficiency for radial and planar structure.

As doping density in n-layer increases recombination centers are increased, which results in decrease in I_{sc} for planar structure. But for radial structure it increases continuously because of its radial geometry and which give rise to higher electric field. These results show that increased E-field, as also seen in Fig. 2(a), is dominant over the dopant trap induced recombination current, and wire doping is a valuable parameter in increasing cell efficiency. Further we see from Fig. 3(b) that as doping density of n-layer is increased, due to increased built-in potential, V_{oc} increases. But since NW structure has lager surface area than planar structure, there is more surface recombination in NW structure than in planar case. This leads to the radial structure to having smaller V_{oc} than that of planar cell. From Fig. 3(c) best performance, for NW structure with n-layer doping density of 10^{19} cm^{-3} is obtained.

Fig. 4(a) and 4(b), respectively, shows the comparison of simulated electric field for radial and planar structures with varying p-layer doping density. We note that as doping increase peak electric field at p/n-interface increases for both radial and planar structure. However, for higher doping (10^{19} cm^{-3}), the peak field for radial structure is 26% higher than that of planar structure.

The NW structure is optimized secondly by keeping p-layer doping constant and varying n-layer doping.

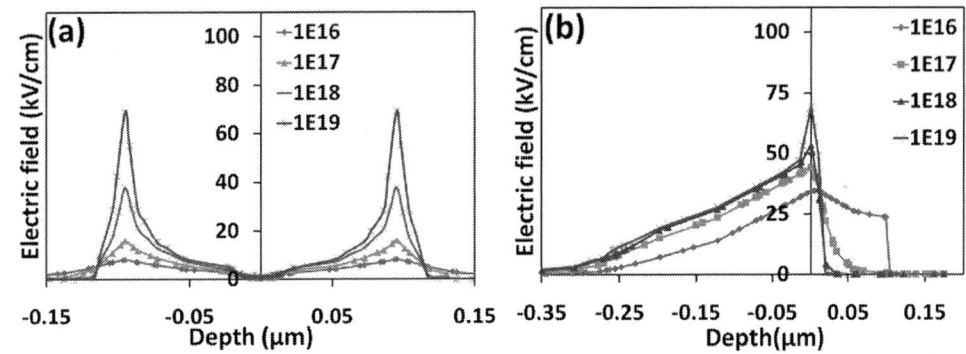

Fig.2. Efield variation with different amount of n-layer doping density (a) radial structure and (b) planar structure.

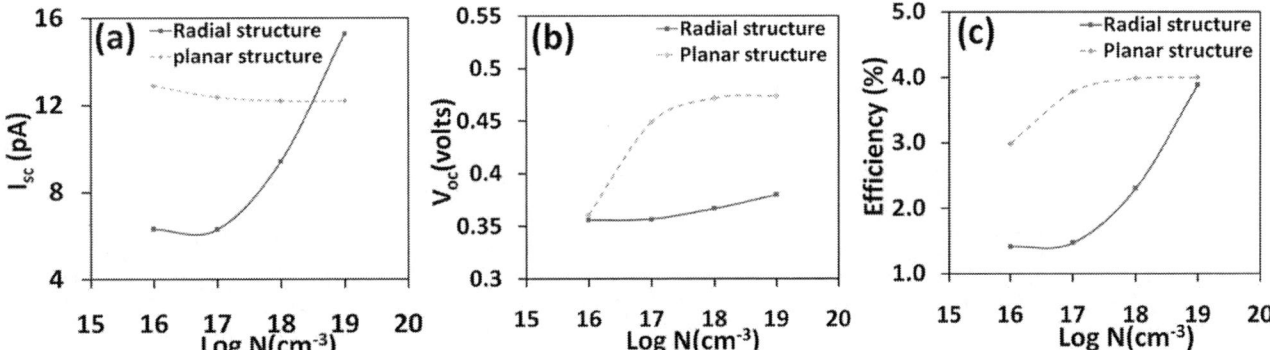

Fig. 3. Characteristics of radial and planar structures for different amount of doping in n-layer (a) I_{sc}, (b) V_{oc} and (c) Efficiency.

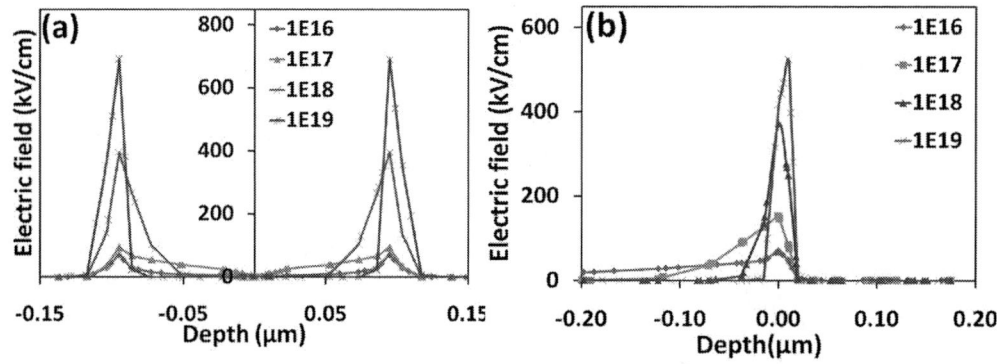

Fig.4. Efield variation with different amount of p-layer doping density (a) radial structure and (b) planar structure.

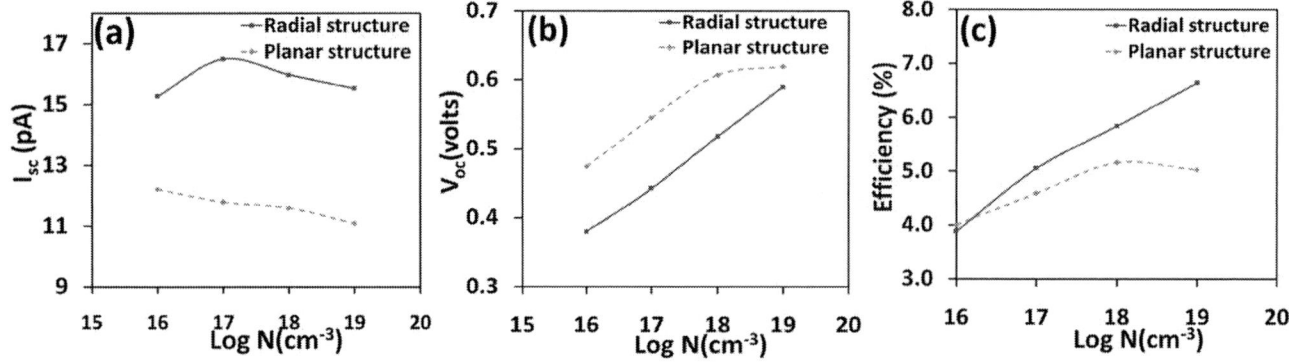

Fig. 5. Characteristics of radial and planar structures for different amount of doping in p-layer (a) I_{sc}, (b) V_{oc} and (c) Efficiency.

Fig. 5(a-c) shows the effect of p-layer doping variation on I_{sc}, V_{oc} and efficiency, respectively, for radial and planar structure. As doping density in p-layer increases diffusion length of minority carriers in p-layer decreases. This leads to reduction in I_{sc} of both radial and planar structure seen in Fig. 5(a).

With increase in doping density in p-layer, minority carrier concentration decreases, which in turns decreases reverse leakage (dark) current. This reduction in leakage current will increase V_{oc}. From simulation results the optimized value of p-core and n-shell doping density are 10^{19} cm^{-3} and 10^{19} cm^{-3} respectively.

At above optimized values of p-core and n-shell doping radial structure has 1.36 times higher conversion efficiency over planar structure as shown by Fig. 5 (c).

Although V_{oc} of NW structure is low as compared to planar structure, but since NW structure has higher I_{sc} than planar structure, as a result overall conversion efficiency of NW structure is higher than that of planar structure.

One of the most important effects of having junction in radial direction is that even with low quality material it gives sufficiently high efficiency.

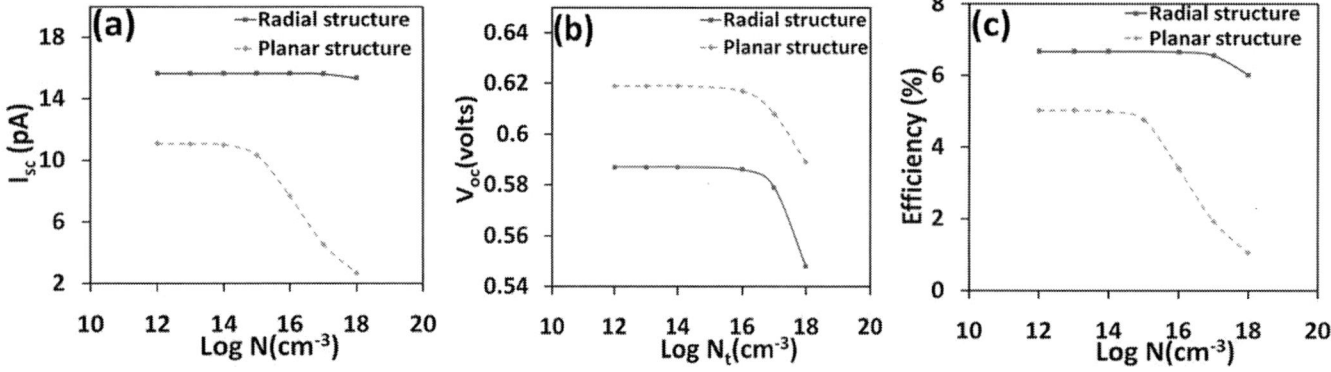

Fig. 6. Characteristics of radial and planar structures for different amount of defect density (a) I_{sc}, (b) V_{oc} and (c) Efficiency.

To compare the performance of radial and planar structures with different amount of defect density present in it, we have intentionally introduced donor type defects at intrinsic energy level in silicon of both radial and planar structures.

Fig. 6(a-c) shows the effect of defect density on I_{sc}, V_{oc} and efficiency for radial and planar structures. We note that as trap density increases minority carrier diffusion length will reduce. Since NW structure has junction in radial direction, hence this reduction in diffusion length will not affect radial junction solar cell performance and I_{sc} remains constant even with high as high as 10^{18} cm^{-3} amount of defect density.

However, for the case of planar case I_{sc} is reduced. Radial structure has higher I_{sc} than planar structure because of its radial geometry which give rise to higher electric field. In case of V_{oc}, since increasing trap density increases recombination centers, this in turns increases recombination current resulting in decrease in V_{oc} as shown by Fig. 6 (b). At low defect density of about 10^{12} cm^{-3} radial structure show 25% better conversion efficiency, while at high defect density as high as 10^{18} cm^{-3} the radial structure shows 82% better conversion efficiency as compared to its planar structure.

IV. CONCLUSION

Using 3D-TCAD simulation, we have optimized doping density and studied the effect of electric field in radial pn-junction structure and its effect on cell performance. It is seen that the performance of cell is strongly enhanced by increased electric field. We find that the best output in terms of conversion efficiency for p and n layer doping density of 10^{19} cm^{-3}. In addition, we studied the effect of defect density present in NW on solar cell performance. It is found that even with high amount of defect density, radial structure has higher conversion efficiency than its planar counterpart. This study serves as useful guideline for designing and developing NW based solar cell and related applications.

REFERENCES

[1] Allon I. Hochbaum, Peidong Yang, "Semiconductor Nanowires for Energy Conversion," Chem. Rev.,vol. 110, pp. 527-547, 2010.

[2] B. M. Kayes, H. A. Atwater, and N. S. Lewis, "Comparison of Device Physics Principles of Planar and Radial p-n Junction Nanorods Solar Cells," Journal of Applied Physics, vol. 97, pp. 114302-1-114302-11, May 2005.

[3] M. Law, L. E. Greene, J. C. Johnson, R. Saykally and P. Yang Lewis, "Nanowire Dye-Sensitized Solar cells," Nano Letters, vol. 4, pp. 455-459, 2005.

[4] Christoph J. Brabec, N. Serdar Sariciftci and Jan C. Hummelen, "Plastic Solar Cells," Journal of Advanced functional Materials, vol. 11, pp. 15-26, 2001.

[5] Li-Min Chen, Ziruo Hong, Gang Li, and Yang Yang, "Recent Progress in Polymer Solar Cells: Manipulation of Polymer:Fullerene Morphology and the Formation of Efficient Inverted Polymer Solar Cells," Journal of Advanced functional Materials, vol. 21, pp. 1434-1449, 2009.

[6] R. Kapadia, Z. Fan and A. Javey, "Design Constraints and Guidelines for CdS/CdTe nanopillars based photovoltaics," Applied Physics Letters, vol. 96, pp. 103116-1-103116-3, 2010.

[7] E. Garnett, P. Yang, "Light Trapping in Silicon Nanowire Solar cells," Nano Letters, vol. 10, pp. 1082-1087, 2010.

[8] Oki Gunawan, Supratik Guha, "Characteristics of Vapor-Liquid-Solid Grown Silicon Nanowire Solar cells," Elsevier Solar Energy Materials & Solar cells, vol. 93, pp. 1388-1393, 2009.

[9] Cheng Yung Kuo, Chie Gau, Bau Tong Dai, "Photovoltaic Characteristics of Silicon Nanowire Arrays Synthesized By Vapor-Liquid-Solid Process," Elsevier Solar Energy Materials & Solar cell, vol. 95, pp. 154-157, January 2011.

[10] Synopsys (2010), Sentauraus TCAD user manual, 2010.

Amorphous Carbon Step Coverage Improvement applied to Advanced Hard Mask for Lithographic Application

Zitu-Tin Lin, Chun-Chi Chen, Hung-Ju Chien, Hiroshi Matsuo

Module Technology Division
Powerchip Technology Corporation
Hsinchu, Taiwan
{ttlin, nckujim}@powerchip.com

Abstract—**Poor deposited step coverage (S/C) of carbon hard mask (HM) at alignment and overlay masks can cause worse overlay signal and further rework performance. A much better step coverage film can be obtained using C_2H_2 as precursor owing to superior sticking coefficient (higher C/H ratio). This C_2H_2 base amorphous (a-C) film property and etch selectivity are also compatible with other precursors at the same time. As to the rework ability, it improved a lot and the pattern is still clear even after five times of rework.**

Keywords—a-C; step coverage; etch selectivity; rework

I. INTRODUCTION

As the pattern size scaled down to sub-100nm, new technologies such as DET (double etching technology) and immersion lithography are emerging, because etching process become more difficult and marginal by applying thin photo resist (PR). The footing problem is a general issue with the smaller pattern size that can cause undesired side effects, such as pattern collapse, inaccurate line-width, and nonideal profile after etching process. In this paper, we tried to use a dual layer stack composing of a-C and MLR (SiON + SiO_2) as being HM and anti-reflectance coating (ARC) for solving above-mentioned issues. Commonly, the film S/C and etch selectivity between these dual layer are two concerned problems. We observed that using C_3H_6 base precursor for a-C deposition, the S/C of film would be deteriorated even though the good etch selectivity (>10) compared with SiON. However, it can be effectively improved while another C_2H_2 base a-C application in the different size of alignment and overlay marks.

II. EXPERIMENTAL

In this study we prepare three kinds of different size of alignment and overlay marks, and the shape is as below. Type1: line type, size: [W:2.0/H:0.7/L:24 μ m]; Type2: line type, size: [W:0.8/H:0.7/L:24 μ m]; and Type3: hole type, size: [W:0.4/H:0.7/L:0.4 μ m]. The schematic scheme is as Fig 1. Two various types of structure will be examined for profile influence on film S/C performance (Fig. 2). As to Plasma Enhance Chemical Vapor Deposition (PECVD) of a-C at

300°C, we carried out three kinds of precursor, including C_2H_2, C_3H_6 and C_7H_8 base by NVLS AHM, and the reaction equation is as in (1). In addition, the reaction formula for MLR is displayed in (2), where the N_2O treatment performed is advantageous to obtain preferable n and k of film properties for following lithographic process. As for S/C performance and overlay mark recognition, which are judged by Scanning Electron Microscopy (SEM) and Optical Microscopy (OM).

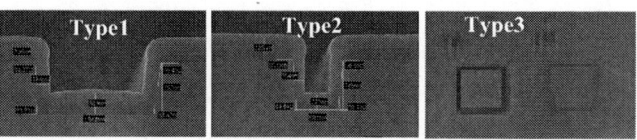

Figure 1. The schematic scheme of different size of alignment and overlay mark. Type1: line type, size: [W:2.0/H:0.7/L:24 μ m]; Type2: line type, size: [W:0.8/H:0.7/L:24 μ m]; and Type3: hole type, size: [W:0.4/L:0.4 μ m]

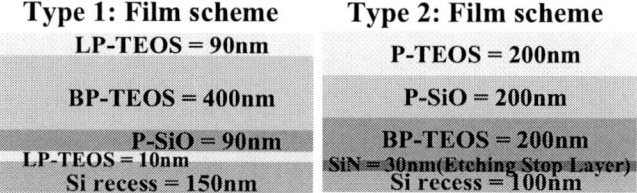

Figure 2. The film scheme of two types structure. Type1: Si recess 150nm, LP-TEOS 10nm, P-SiO 90m, BP-TEOS 400nm, and LP-TEOS 90nm. Type2: Si recess 100nm, LP-TEOS 10nm, SiN 30nm, BP-TEOS 200m, P-SiO 200nm, P-TEOS 200nm. The SiN layer is applied for stop layer of etching, where it can get flatter pre-structure at the bottom of alignment and overlay mark.

$$C_2H_2/C_3H_6/C_7H_8 + He + H_2 \overset{RF/\triangle}{\rightarrow} C_x + H \tag{1}$$

$$SiH_4 + N_2O + N_2/He \overset{RF/\triangle}{\rightarrow} SiON + N_2 + H_2 + H_2O$$

$$Si(SiON \text{ surface}) + N_2O \overset{RF/\triangle}{\rightarrow} SiO_x + N_2 + H_2 + H_2O$$

$$N_2O \overset{RF/\triangle}{\rightarrow} N_2 + O_2 + O^* \tag{2}$$

978-1-61284-863-1/11 $26.00 © 2011 IEEE

III. RESULTS AND DISCUSSION

As Fig. 3 is shown, it is noted that flatter pre-structure has significantly helpful to a-C film deposition process (type 2 film scheme). However for the weak corners, that being liable to cause undesired problem in the subsequent process while unsuitable precursor utilization. The worse result of film S/C for C_3H_6 base carbon is illustrated, but C_2H_2 one exhibits much better S/C performance (Fig. 4). For the former, it can be observed the a-C film is almost discontinuous and makes it be difficult for MLR to cover carbon entirely. Moreover, the lager width has, the better S/C performance acquired (Type1 > Type2 > Type3) (not shown). Comparing Fig. 3 (b) and (d), C_3H_6 base a-C poor deposited S/C on top and bottom corners will be easily pulled out through MLR process during N_2O plasma treatment condition or extremely PR rework (by plasma Asher) where the O radicals are dissociated from above process and then attack a-C film [1]. The schematic diagram is shown in Fig. 5.

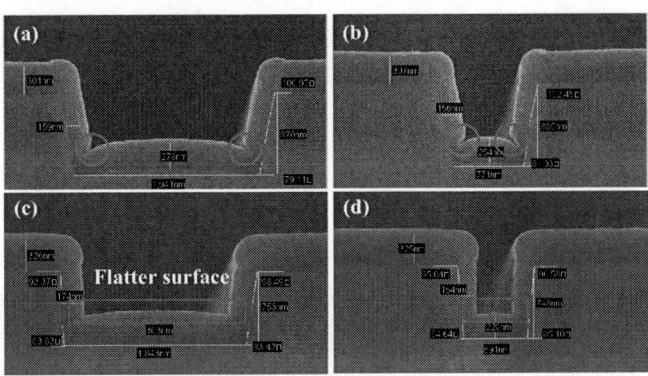

Figure 3. The SEM Cross section of step coverage performance on various type of structure. (a), (b): Type 1, (c), (d): Type 2. Conditions: C_2H_2 base a-C deposition.

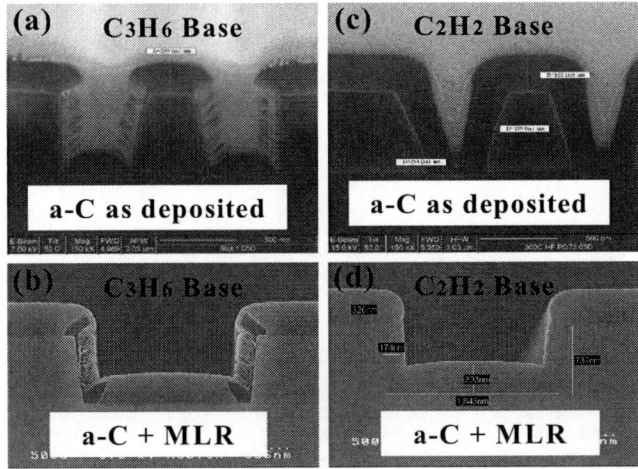

Figure 4. The SEM Cross section of step coverage performance on Type 2 structure. C_3H_6 base (a) a-C as deposited (b) a-C + MLR, C_2H_2 base (c) as deposited (d) a-C + MLR.

This incomplete hard mask can cause worse lithography alignment or rework performance. Nevertheless, applying

C_2H_2 base as the precursor, it demonstrated a preferable film S/C performance instead. On the other hand, another C_7H_8 base a-C is also taken into consideration for S/C performance investigation (Table 1). In general, film surface coverage has a definitely linear relationship with sticking efficiency, and higher sticking coefficient means more opportunity for C radical to adsorb on the film surface [2-4]. In previous experiment [5], it was indicated that a clear correlation exists between the ratio of C/H in hydrocarbon species and the sticking efficiency for the reaction behavior on hydrogenated diamond: the sticking efficiency enhances with C/H ratio increases. In our case, C_2H_2 precursor brings about much superior S/C performance on alignment and overlay marks owing to higher C/H ratio as compared to C_3H_6 base. Furthermore, it can improve lithographic application. Additionally, lower C/H ratio also probably can increase plasma reaction between C and H ions at sidewall, and result in discontinuous film.

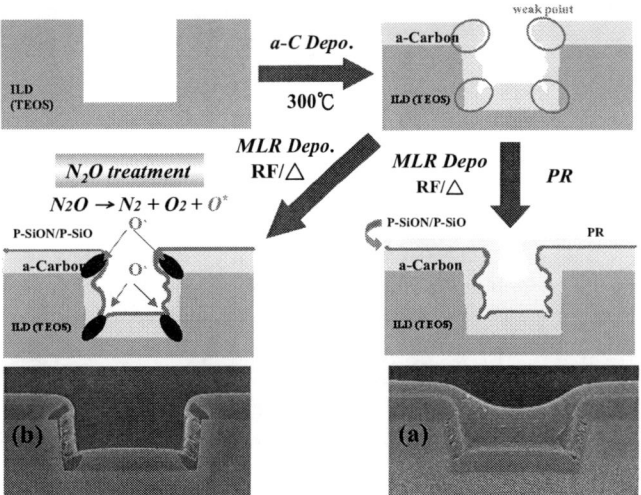

Figure 5. The schematic diagram of process flow and cross sectional SEM photographs of step coverage for comparison of N_2O treatment in the MLR deposition process. (a) Type1+ a-C+ MLR (No N_2O treatment) + PR, (b) Type2+ a-C+ MLR (N_2O treatment).

Table 1. The comparison table of different precursor of a-C film in terms of film behavior.

Precursor of a-C	C/H ratio	Step coverage	Sidewall
C2H2	1.000	50%	Smooth
C7H8	0.876	50%	
C3H6	0.500	~ 0%	Rough

As to rework ability, C_2H_2 base a-C exhibits obviously marvelous performance even after five times rework by plasma Asher, the alignment mark is still clear, however the C_3H_6 base a-C failed immediately afterwards one time oppositely. What is more film property (thickness U%, n, and, k value) is comparable with one another as well (Fig.6). In fact, except for high C/H ratio precursor application, such as using lower temperature O_2 plasma Asher or other gas (ex: $O_2/CHF_3/Ar$) Asher [6] instead for reworking process can also improve and even reach the similar excellent rework achievements.

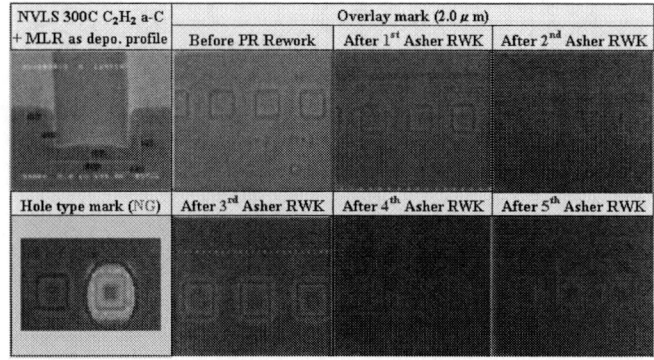

Figure 6. Rework ability for NVLS 300C C_2H_2 base a-C Film property: thickness: 289.5nm, U%: 1.09%, RI (193nm): 1.617, k (193nm): 0.431.

IV. CONCLUSION

The poor step coverage of a-C film at alignment and overlay marks occurred as using C_3H_6 base precursor as HM where carbon can be uprooted by MLR due to O radicals attack or by plasma Asher process during photo resist rework. However, apparent improvement on alignment accuracy and overlay control can be achieved by higher C/H ratio of C_2H_2 base a-C film utilization. In addition, this higher sticking coefficient of C_2H_2 base amorphous carbon film can provide a much wider margin for rework process and keep complete alignment shape and overlay mark for lithography.

REFERENCES

[1]   M. A. Goldman, D. B. Graves, G. A. Antonelli, S. P. Behera, and J. A. Kelber, "Oxygen radical and plasma damage of low-*k* organosilicate glass materials: Diffusion-controlled mechanism for carbon depletion," Journal of Applied Physics, vol. 106, no. 1, pp. 013311-013311-7, July 2009.

[2]   N. Sugiyama, N. Hirashita, T. Mizuno, Y. Moriyama, and S. Takagi, "Analysis of growth rate during Si epitaxy by hydrogen coverage model," Materials science in semiconductor processing, vol. 8, no. 1-3, pp. 11-14, Feb. 2005.

[3]   W. C. Hsin, D. S. Tsai, and Y. Shimogaki, "Surface reaction probabilities of silicon hydride radicals in SiH4/H2 thermal chemical vapor deposition," Industrial & Engineering Chemistry Research, vol. 41, no. 9, pp. 2129-2135, April 2002.

[4]   Achim von Keudell, " Formation of polymer-like hydrocarbon films from radical beams of methyl and atomic2hydrogen," Thin Solid Films, vol. 402, no. 1-2, pp. 1-37, Jan. 2002.

[5]   M. Eckert, E. Neyts, and A. Bogaerts, "Molecular Dynamics Simulations of the Sticking and Etch Behavior of Various Growth Species of (Ultra)Nanocrystalline Diamond Films," Chemical Vapor Deposition, vol. 14, no. 7-8, pp. 213-223, July 2008.

[6]   J. Hong, J. S. Jeon, Y. B. Kim, G. J. Min, and T. H. Ahn, "Novel technique to enhance etch selectivity of carbon antireflective coating over photoresist based on O₂/CHF₃/Ar gas chemistry," Journal of Vacuum Science & Technology A, vol. 19, no. 4, pp. 1379-1383, Nov. 2001.

Wideband Receiver for Software Defined Radio in GHz range using Standard 40nm CMOS technology

[1]F. Yang, [1]Y. Liu, [1]X. L. Zhang

[1]State Key Laboratory of Electronic Thin Films and Integrated Devices, University of Electronic Science and Technology
P. R. China 610054
Yliu1975@uestc.edu.cn

[2]Z. H. Lu, [2]X. P. Yu, [2]W. M. Lim, [3]C. H. Hu

[2]School of Electrical and Electronic Engineering
Nanyang Technological University, Singapore, 639798
[3]Marvell Semiconductor, Santa Clara, CA 95054, USA
zhlu@pmail.ntu.edu.sg

Abstract—**In this paper, a low power wideband receiver for Software-Defined Radio (SDR) is presented. In SDR design, the receiver should be able to receive signals over a broad-bandwidth so that a variety of different communication standards can be covered. In the proposed design, a 0.1-6 GHz analog front-end for SDR is demonstrated. Implemented in a standard 40nm CMOS technology, the proposed receiver is realized without inductors and able to work between 0.1-6 GHz while consuming less than 4.55mA current from a single 1.1-V supply voltage.**

Keywords- Low power receiver; 40nm CMOS; wideband; Software-Defined Radio;

I. INTRODUCTION

With substantial increase of the wireless communication market, the software-defined radio (SDR) technology attracts wide attention. In order to meet stringent market requirements, low-cost flexible communication systems capable of high data-rate and increased functionality are mostly desired. The concept of SDR, which can provide the multi wireless standards with a single hardware platform, has drawn great interest of people in the academic and industry. Extensive works have been dedicated to design receivers at radio frequency (RF) range in recent years. Several prototypes have been successfully developed to demonstrate the possibility of high data-rate at a short range. Up to now, SDR have been reported in many research works [1-3].

The programmability and configurability of SDRs are prospective advantages over traditional RF devices. Traditional RF devices either do not have or have a limited capability in programmability and configurability. With SDR, software modules define the baseband and protocol elements and provide an environment for easy application development. Without introducing new hardware, an SDR can control working parameters such as the frequency range, modulation type, bandwidth, maximum output power, on/off sensing and network protocols by programming the software. This enables a single wireless system to be reprogrammed for using different modulation, coding, and access protocols. This great flexibility of SDR provides opportunity for solving interoperability problems between the many different standards, implementing new standards, and minimizing the amount of hardware necessary to perform communications across these different standards. SDR also allows more efficient use of the spectrum by facilitating spectrum sharing and allowing equipment to be reprogrammed to more efficient modulation types.

For SDR receiver, it is desirable to achieve the multi-mode and multi-standard operation to satisfy requirements such as broadband, low cost, low power consumption and compatibility with digital signal processor (DSP). As has been demonstrated earlier, a direct up/down conversion architecture is the most suitable architecture to build an SDR [4]. In this paper, a 0.1-6GHz receiver is proposed to overcome above-mentioned challenges. The block diagram of the proposed SDR receiver is shown in Fig. 1. The RF signals are received from antenna, amplified by low noise amplifier (LNA) and then down-converted. The down-converted signals are amplified by intermediate frequency amplifier (IF AMP) and variable gain amplifier (VGA). Next the IF signals are sent to the analog digital converter (ADC). The digitalized signals are processed by DSP and all of the subsequent processing is implemented in software. In this work, we design the receiver front-end including LNA, mixer, IF amplifier and variable gain amplifier. External local oscillator (LO) from 100 MHz to 6 GHz will be used in the measurement.

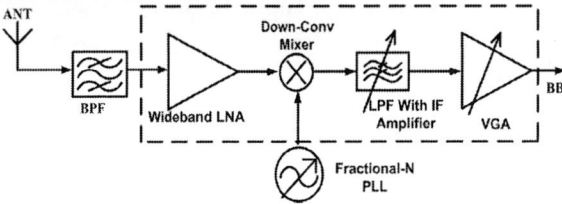

Figure 1. Block diagram of the proposed receiver for SDR.

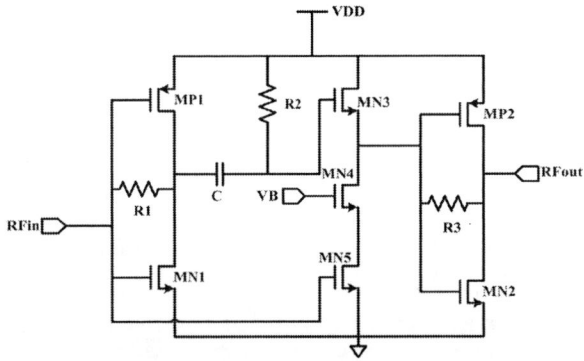

Figure 2. The schematic of the wideband inductorless LNA.

The proposed SDR receiver in this paper is realized in 40 nm CMOS technology. The considerations of an SDR analog

978-1-61284-863-1/11 $26.00 © 2011 IEEE

receiver front-end are described in Section II followed by a detailed circuit analysis to pave the way for the proposed receiver design in section III. Detailed circuit performance and simulation results are shown in Section IV. Section V is the conclusion.

II. DESIGN CONSIDERATIONS

The design of RF receivers needs to consider many key aspects including noise, sensitivity, linearity, power consumption, etc. These performances are defined by wireless standards for communications. The design requirements are well defined for each wireless communication standards.

Firstly, silicon based SDR wireless system is mainly used in medium distance. The typical signal power can be as low as -60dBm. Secondly, as the receiver should be universal, the working band should be broad, for example, from 0.1G to 6G. It should be able to obtain data-rate from kbps to tens of Mbps. Simple modulation schemes, e.g. ASK, OOK and complex modulation schemes such as OFDM can be applied in the receiver system. On the other hand, the requirement of linearity in the band range should be well controlled as well.

As shown in Figure 1, taking into account the image rejection, the system should be designed I and Q paths. Figure 1 only demonstrates one path. In this system, we design a down-conversion receiver and the output of IF signal is 1Mbps. The minimum input power is less than -70dBm. The small signal gain of the whole system must be higher than 40 dB, and the power consumption is lower than 5mW.

For this design, a wideband from 100 MHz to 6 GHz LNA becomes a great challenge due to the influence of the parasitic capacitors. Low noise figure and high AC signal gain is essential to the subsequent circuits because the level of noise mainly depends on the first stage of receiver. The down-conversion mixer receives two path signals: one comes from the output of LNA and the other is from external LO signal. LPF with IF amplifier and VGA play a role in rejecting high harmonics and enhance the demodulated signal, VGA also can decrease the amplitude of signal when the input signal is very strong.

III. CIRCUITS DESIGN

In this section，Design of circuit blocks of the receiver in Fig. 1 is described.

A. Wideband inductorless LNA

A wideband LNA is commonly used in SDR receivers to cover the whole bandwidth. Numerous wideband LNA topologies have been reported in literatures. Two techniques including active feedback [8] and thermal noise cancelling [9] have been used in this work for the inductorless wideband LNA design.

The schematic of wideband LNA is shown in Figure 2 with 50 ohm input impendence. The input NMOS (MN1) and PMOS (MP1) are short channel devices for the purpose of providing the input impendence and reducing the load capacitor. The gain of the first stage is not enough since the finite output resistance and thus it is necessary to increase it in the last stage (including MP2, MN2, and R3). The middle stage

(MN3, MN4, MN5, R2, and C) is used for thermal noise cancelling technology [4]. The simulation results are shown in figure 3-5. As shown in Fig. 3, the noise figure is less than 2.8dB in the overall band. Fig. 4 shows S11 is smaller than -11 dB and S21 is larger than 20 dB. Fig. 5 shows the IIP3 is larger than -10dBm over the whole bandwidth range (0.1-6 GHz).

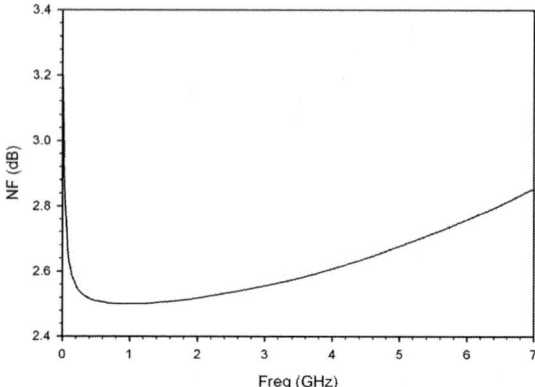

Figure 3.   Noise figure of the proposed LNA.

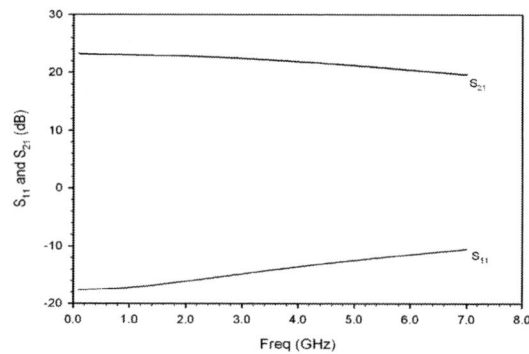

Figure 4.   S11 and S21 for the proposed LNA as functions of frequency.

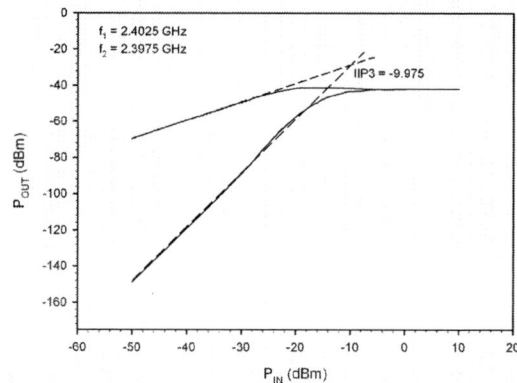

Figure 5.   Two-tone IIP3 of LNA.

B. Passive Mixer And IF Amplifier

Compared with active mixers [10], passive mixers [11] do not have static power consumption at the cost conversion gain.

The most commonly used CMOS passive mixer architecture employs a "current input and current output" approach, with a transimpedance amplifier (TIA) output stage to provide the low impedance at the mixer output [12]. The schematic of mixer in present study is showed in Figure 6. With the bandpass filter and the IF amplifier, they pure the IF signal spectrum and provide the signal gain, respectively. In this paper the passive mixer is designed with a low noise figure which is lower than 5 dB. Its conversion gain is about -5dB, and the gain of IF amplifier is 25 dB. The 1MHz signal is produced after mixing, and then amplified by IF amplifier.

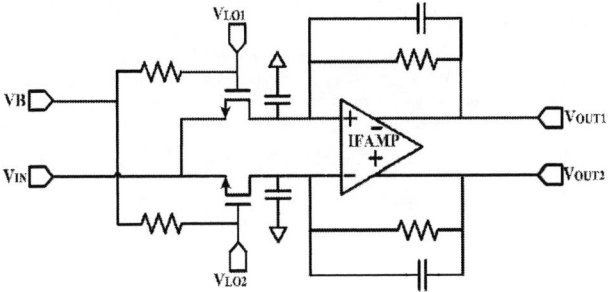

Figure 6.   The schematic of passive mixer and IF amplifier

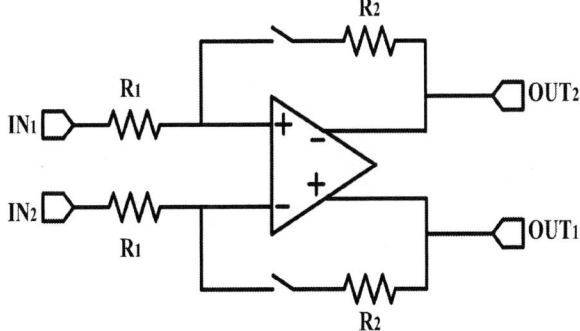

Figure 7.   Simplified schematic of VGA

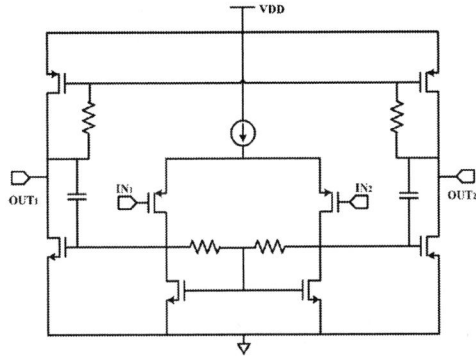

Figure 8.   The schematic of the amplifier in VGA

C.   VGA

Variable gain amplifier is widely used in RF receivers to improve the dynamic range of the overall system. There are two approaches used to realize VGAs depending on the control signal [13]. VGA controlled by analog signals take the advantage of variable transconductance and the gain can be controlled continuously.   A digitally controlled VGA would simplify the circuit and the gain varies as a discrete function of the control signal. The simplified schematic of digitally controlled VGA is shown in Figure 7. The gain is equaled to R_2 /R_1. According to requirement, the number of resistors would be selected to control the gain range. In this work we design a VGA gain ranges from -18 to 6 dB with 6 dB per step. Common-mode feedback circuit is also included in Fig. 8.

IV.   SIMULATION RESULTS

In this section, the system simulation results are presented and analysis is made. The noise figure of the receiver is shown in Figure 9. It ranges from 3.0 to 3.8 dB over the bandwidth range. Figure 10 shows the input and output data of the proposed demodulator, the input signal has a data rate of 1 Mbps with 100 MHz to 6 GHz carrier, the demodulator can work properly. The input amplitude of signal is 1 mV. Since the LNA is optimized at 1GHz, maximum gain of the receiver appears at 1GHz, while the signal power is minimized at 6 GHz input signal. The gain of the system can attain 40 dB or above. Figure 11 shows the layout of the wideband (0.1-6 GHz) receiver. The silicon area of the receiver is bondpad-limited. The effective area is 0.05 mm^2. The signal input by utilizing a ground-signal-ground (GSG) structure and the measurement of differential output by a ground-signal-signal-ground (GSSG) structure. A series of 7-pin eye pass pads are used for DC biasing. The proposed system has a wide operating frequency range from 100 MHz to 6 GHz with an input signal ranging from -70 to -30dBm. The input signal has a data-rate of 1 Mbps with 100 MHz - 6 GHz carrier. The total current consumption can be as low as 4.55mA.

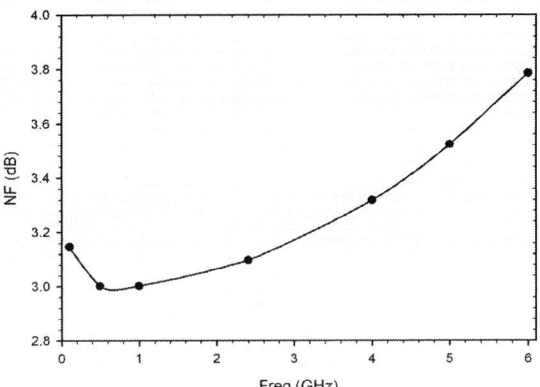

Figure 9.   Noise figure of the system

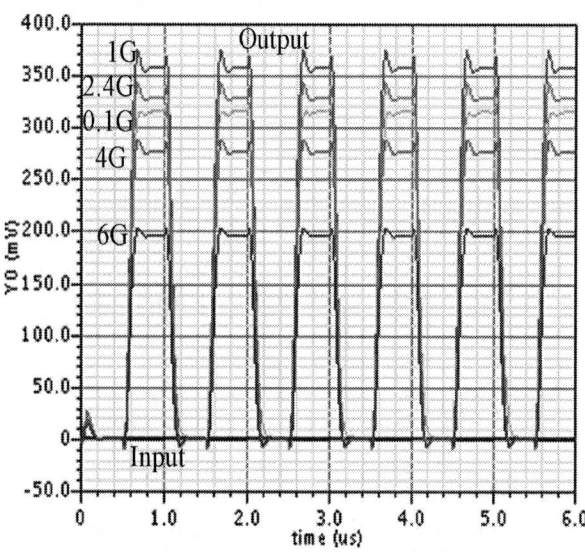

Figure 10. Output of the demodulator at different carrier frequency

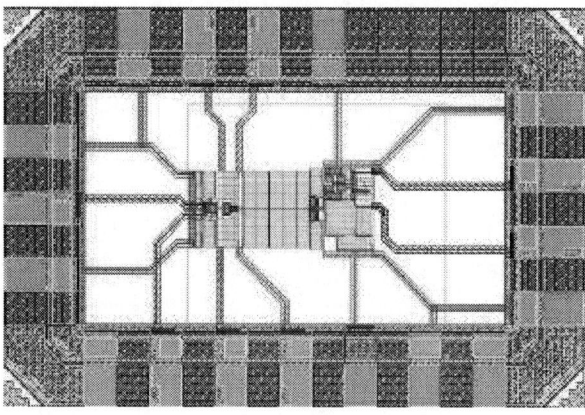

Figure 11. Layout of the receiver

V. CONCLUSION

This paper proposes a SDR receiver front-end in a standard 40 nm CMOS technology. The receiver is based on a 1 MHz-IF architecture that is able to cover the 0.1 to 6 GHz frequency range. The DC power consumption of the core circuit is less than 5 mW from a 1.1V supply. Additionally, energy scalability can be further achieved by balancing power and performance.

ACKNOWLEDGMENT

This work has been financially supported by NSFC under project No. 60806040, the Fundamental Research Funds for the Central Universities under project No.ZYGX2009X006, and the Young Scholar Fund of Sichuan under project No. 2011JQ0002.

REFERENCES

[1] Ingels, M.; Giannini, V.; Borremans, J.; Mandal, G.; Debaillie, B.; Van Wesemael, P.; Sano, T.; Yamamoto, T.; Hauspie, D.; Van Driessche, J.; Craninckx, J.; "A 5 mm^2 40 nm LP CMOS Transceiver for a Software-Defined Radio Platform,"IEEE Journal of Solid-State Circuits, vol. 45,pp: 2794 - 2806 , Oct. 2010

[2] Borremans, J.; Vengattaramane, K.; Giannini, V.; Debaillie, B.; Van Thillo, W.; Craninckx, J.; "A 86 MHz–12 GHz Digital-Intensive PLL for Software-Defined Radios, Using a 6 fJ/Step TDC in 40 nm Digital CMOS, "IEEE Journal of Solid-State Circuits, vol. 45, pp: 2116 – 2129, Sept. 2010

[3] Shun-Te Wang; Wu, J.-L.C.; Chun-Yen Hsu; Wen-Chun Ni; "Software downloading in reconfigurable networks of open wireless architecture using SDR technology," Communications Magazine, IEEE, vol. 44 , pp: 128 – 134, Oct. 2006

[4] Giannini, V.; Nuzzo, P.; Soens, C.; Vengattaramane, K.; Ryckaert, J.; Goffioul, M.; Debaillie, B.; Borremans, J.; Van Driessche, J.; Craninckx, J.; Ingels, M.; "A 2-mm^2 0.1–5 GHz Software-Defined Radio Receiver in 45-nm Digital CMOS,"IEEE Journal of Solid-State Circuits, vol. 44, pp:3486 - 3498, Dec. 2009

[5] J. Borremans, P. Wambacq, G. Van der Plas, Y. Rolain, and M. Kuijk, "A bondpad-size narrowband LNA for digital CMOS," in IEEE Radio Frequency Integrated Circuits (RFIC) Symp., pp. 677–680, Jun.2007

[6] S. C. Blaakmeer, E. A. M. Klumperink, D. M. W. Leenaerts, and B.Nauta, "A wideband noise-canceling CMOS LNA exploiting a transformer," in Proc. IEEE Radio Frequency Integrated Circuits (RFIC) Symp., Jun. 2006

[7] Donggu Im; Ilku Nam; Hong-Teuk Kim; Kwyro Lee; "A Wideband CMOS Low Noise Amplifier Employing Noise and IM2 Distortion Cancellation for a Digital TV Tuner," IEEE Journal of Solid-State Circuits, vol. 44 , pp: 686 - 698, Feb. 2009

[8] Borremans, J.; Wambacq, P.; Soens, C.; Rolain, Y.; Kuijk, M.; "Low-Area Active-Feedback Low-Noise Amplifier Design in Scaled Digital CMOS", IEEE Journal of Solid-State Circuits, vol. 43, pp: 2422 – 2433, Nov. 2008

[9] Bruccoleri, F.; Klumperink, E.A.M.; Nauta, B.; "Wideband CMOS low-noise amplifier exploiting thermal noise canceling,"IEEE Journal of Solid-State Circuits, vol. 39,no.2, pp: 275 - 282, Feb. 2004

[10] Do, A.V.; Boon, C.C.; Manh Anh Do; Kiat Seng Yeo; Cabuk, A.;"A Weak-Inversion Low-Power Active Mixer for 2.4 GHz ISM Band Applications," Microwave and Wireless Components Letters, IEEE, vol. 19 , pp: 719 – 721, Oct. 2009

[11] Sining Zhou; Chang, M.-C.F.; "A CMOS passive mixer with low flicker noise for low-power direct-conversion receiver", IEEE Journal of Solid-State Circuits, vol.40, pp: 1084 - 1093, May. 2005

[12] J. Zhan, B. R. Carlton, and S. S. Taylor, "Low-cost direct conversion RF front-ends in deep submicron CMOS," in Proc. IEEE RFIC Symp., pp. 203–206, Jul. 2007

[13] Quoc-Hoang Duong; Quan Le; Chang-Wan Kim; Sang-Gug Lee; "A 95-dB linear low-power variable gain amplifier," IEEE Circuits and Systems, vol. 53, pp: 1648 – 1657, Aug. 2006

A $\Delta\Sigma$ Fractional-N PLL with Fast Auto-Frequency Calibration for CMMB Tuners

Jing Jin, Xiaoming Liu, Peng Qin, and Jianjun Zhou

Center for Analog/RF Integrated Circuits (CARFIC), School of Microelectronics
Shanghai Jiao Tong University, Shanghai 200240, China
jinjng@ic.sjtu.edu.cn, zhoujianjun@sjtu.edu.cn

Abstract—A $\Delta\Sigma$ fractional-N Phase-Locked Loop (PLL) with fast Auto-Frequency Calibration (AFC) is presented in this paper. The proposed PLL employs a dual-core Voltage Controlled Oscillator (VCO), a programmable Frequency Divider (FD), a Phase/Frequency Detector (PFD), and a replica biased Charge Pump (CP) with static offset current to compensate fractional spur, etc. This PLL is intended for a China Mobile Multimedia Broadcasting (CMMB) TV tuner. The design is implemented in a 0.18-μm CMOS process and occupies an area of 2.1 mm×0.92 mm including the on-chip third-order loop filter and the digital circuits. The simulated power consumption is 10 mA at a 1.8 V supply voltage. The simulation result of the phase noise satisfies the CMMB specification. Also, with the help of a fast AFC scheme, the proposed PLL achieves fast settling performance. The settling time is less than 50 μs in typical case including both coarse tuning and fine tuning.

Keywords-Fractional-N PLL, AFC, Phase Noise.

I. INTRODUCTION

The $\Delta\Sigma$ fractional-N Phase-Locked Loops (PLLs) are widely used in wireless communication systems. The $\Delta\Sigma$ modulator generates a pseudo-random bit sequence to dither the instantaneous division ratio and its time-average value equals to the required fractional division ratio.

China Mobile Multimedia Broadcasting (CMMB) system is expected to be widely used in China, and it operates at the UHF band of 470-798 MHz. For the zero-IF receiver architecture, the VCO operates at twice of the Local Oscillator (LO) frequency and the quadrature LO signals are generated by a divide-by-2 circuit.

The proposed $\Delta\Sigma$ fractional-N PLL employs a gain compensated wideband Voltage Controlled Oscillator (VCO) along with an efficient Auto-Frequency Calibration (AFC) scheme to ensure wide tuning range, relatively constant loop parameters and fast settling. Also, a programmable Frequency Divider (FD) composed of nine divide-by-2/3 (/2/3) cells and the division range extension logic is designed. Finally, fractional spur performance is improved by using a replica biased CP with static offset current, generating static phase offsets between two inputs of PFD.

The contents of this paper are organized as follows. Section II introduces the phase noise specifications, the frequency plan and the overall PLL architecture. Section III describes

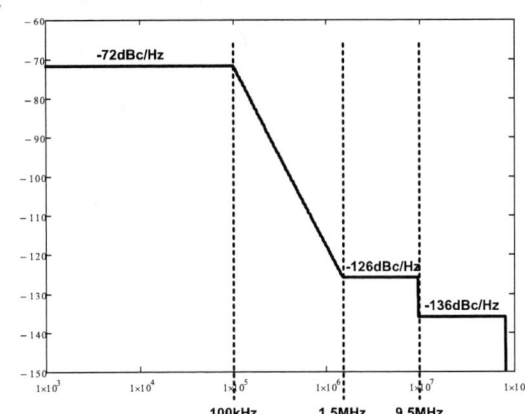

Figure 1   SSB phase noise mask for CMMB.

implementation details of each building block. Section IV explains the AFC scheme in this design. Finally, simulation results are presented in section V, followed by the conclusions in section VI.

II. PLL SPECIFICATIONS

A. Phase Noise Requirement

Phase noise specification derives from both the presence of the adjacent channel interference and total integrated phase noise requirement. The CMMB channel bandwidth is 8 MHz, and the system noise bandwidth is 7.512 MHz [1]. The integrated phase noise requirement is less than -33dBc referred to DVB-H, and the Phase Alternating Line (PAL) 1.5 MHz away from CMMB channel is the most critical adjacent channel interference [2]. The most critical required Carrier-to-Noise Ratio (CNR) is 14.3 dB [1]. According to above requirements, the overall Single Side Band (SSB) CMMB phase noise mask is shown in Fig. 1.

B. Frequency Plan

The LO frequency range is from 474 MHz to 794 MHz for CMMB, so the frequency range of the PLL output should be doubled, i.e. 948 MHz to 1.588 GHz. Considering the process, supply voltage and temperature (PVT) variation and the compatibility for DVB-H, a 300MHz margin is added to the

---

This research was supported by Chinese National Major Science and Technology Projects Program (No. 2009ZX01031-002-005).

978-1-61284-863-1/11 $26.00 © 2011 IEEE

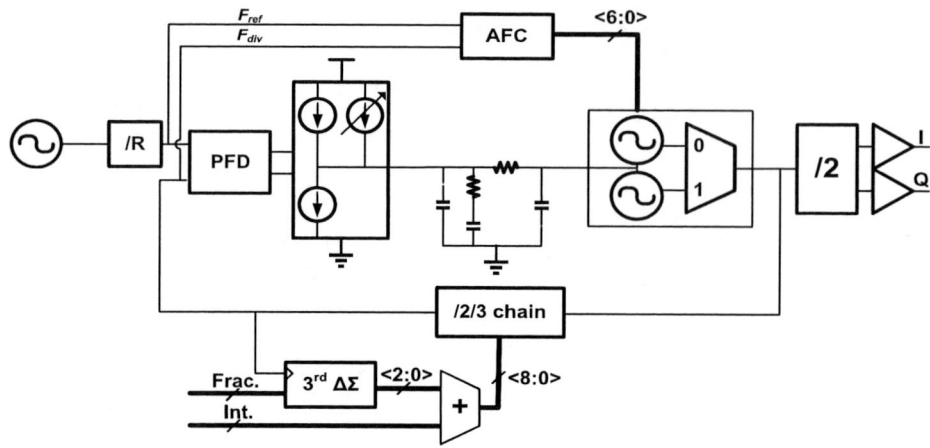

Figure 2   Proposed PLL architecture.

upper frequency, so the target frequency range is 900-1900MHz.

C. Overall PLL Architecture

The proposed PLL architecture is shown in Fig.2. The reference frequency can accommodate from 5 MHz to 40 MHz. Conventional PFD is used to drive the replica biased CP. Assisted with an efficient AFC and the gain-compensated dual-core VCO, the PLL can achieve a wide tuning range, good noise performance, relatively constant loop parameters, and fast settling performance. The programmable FD is employed to generate feedback signal for PFD.

III.   THE PROPOSE PLL AND ITS BUILDING BLOCKS

A. Gain-Compensated VCO

According to the frequency plan, the tuning range of VCO is 900-1900MHz. A dual-core VCO with switched-capacitor array is designed to ensure both tuning range and frequency overlap of the adjacent turning curves. Considering the cross-coupled VCO structure, four MOS-FETs are adopted to generate negative transconductance to compensate the loss in LC tank for oscillation. Tail resistor bank instead of bias transistor is used to adjust the current of the VCO and achieve low phase noise. The tail resistor does not contribute the flicker noise, and it also reduces the parasitic capacitance at the virtual ground node of NMOS-FET pair.

One of the serious problems related to the binary weighted switched-capacitor wideband VCO is the unacceptable large variation of tuning sensitivity (K_{VCO}). Such variation not only causes loop parameter variation but also possible errors in the selection of digital code of the switched-capacitor array, in which case the PLL would be unlocked.

In order to reduce the K_{VCO} variation, the varactor size is changed according to the digital control signal of the capacitor bank, and the number of subsections trades off the implementation complexity [3]. In this design, two subsections are divided for both the high core and low core VCOs.

Compared with the conventional binary weighted VCO with fixed varactor, the varactor size adaption greatly reduces

the K_{VCO} variation. According to the simulation results, the K_{VCO} variation of conventional binary weighted VCO is more than 200%, and that of the gain compensated dual-core VCO in this design is less than 100% while the implementation of the additional circuit almost costs nothing. Single core topology of the proposed VCO is shown in Fig. 3.

B. Programmable Frequency Divider

The programmable FD is mainly composed of nine divide-by-2/3 (/2/3) cells and the division range extension logic. To optimize the power consumption, MOS Current Mode Logic (MCML) /2/3 cells are only used in high frequency part. In low frequency part, CMOS static logic is adopted. The frequency division ratio, N, is controlled by the 9-bit division ratio control input, $p<8:0>$, and N can be expressed as follows:

$$N = 2^5 \cdot \prod_{i=6}^{8} \overline{p_i} + 2^8 \cdot p_8 + 2^7 \cdot p_7 + \ldots + 2^1 \cdot p_1 + 2^0 \cdot p_0 \quad (1)$$

According to (1), the continuous division ratio is from 32 to 511 stepped by 1.

C. Charge Pump with Static Current Offset

The CP based on the conventional structure employs two op-amps to eliminate charge sharing and current mismatch problems.

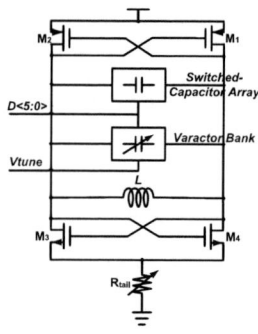

Figure 3   Single core of the proposed dual-core VCO

978-1-61284-863-1/11 $26.00 © 2011 IEEE   540

Some circuit techniques are also used to further improve the performance of the CP. Long channel transistor is used in current source to enlarge the output impedance of the CP. Also, transition gate with half-sized dummy device is adopted to implement switches to compensate both channel charge injection and clock feed-through.

The nonlinearity of PFD and CP is one the major sources of fractional spur, so additional static offset current which improves the linearity of the PFD and CP is integrated [4].

D. ΔΣ Modulator

A multi-bit third-order ΔΣ Modulator [5] is adopted in the proposed design. The noise transfer function is

$$H(z) = \frac{(1-z^{-1})^3}{1-z^{-1}+0.5z^{-2}-0.1z^{-3}} \qquad (2)$$

Compared with MASH-111 structure, the multi-bit ΔΣ Modulator extends input range, which helps to reduce the non-ideal effects near the integer boundary. Also, the narrower spread output bit pattern makes the synthesizer less sensitive to the substrate noise coupling.

IV. AUTO FREQUENCY CALIBRATION

The AFC refers to the procedure to find an optimal VCO sub-band tuning curve from the multiple curves, and the selected VCO sub-band is the closest to the target frequency. The time used in the AFC period always limits the PLL settling performance. The time complexity of linear search algorithm is N, and for binary search algorithm it is log_2N while N is the number of tuning curves.

A. AFC Fundamentals

The basic idea of the AFC in this design is shown in Fig. 4. Two counters, R CNTR and V CNTR, count the reference f_{ref} and feedback signal f_{div}, respectively, and the digital control logic block, CTRL, compares the results of the two counters. The maximum counter number required for the AFC to obtain the target tuning curve is [6]

$$CNTR_{MAX} = \left\lceil \frac{4 \cdot N \cdot f_{ref}}{(f_{step} / code)_{min}} \right\rceil \qquad (3)$$

where N is the division ratio; f_{ref} is the reference frequency; the denominator is the minimum value of the amount of VCO frequency change per unit capacitor bank code change. In this design the calculated maximum number is no more than 800, so a 10-bit counter is enough for both R CNTR and V CNTR.

B. AFC Design

Conventionally, the AFC adopts binary search algorithm. It would not stop searching until one of the two counters reaches $CNTR_{MAX}$, and then the AFC would give the result by checking the value of the other counter. However, the time consumed in conventional AFC procedure is quite long, and more efficient

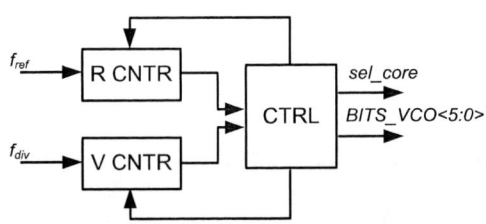

Figure 4   Block diagram of the AFC.

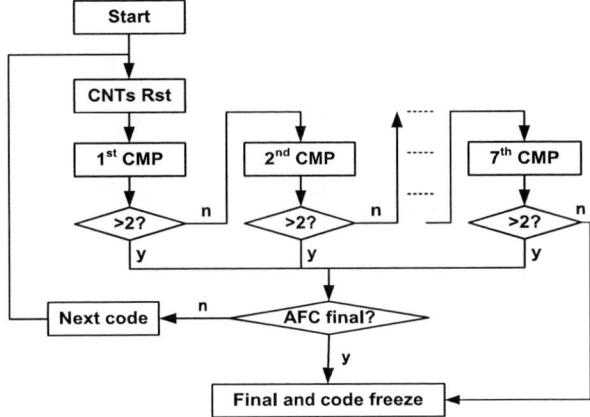

Figure 5   The AFC flow chart.

AFC algorithm is required. In this design, although the maximum comparison interval maintains the same, $CNTR_{MAX}$, different comparison interval is inserted. The simplified AFC flow chart is shown in Fig. 5 [7]. Several check points allow the AFC to quit the current counting and give the result much earlier than the final check especially when the frequency difference of the two signals is large. In the comparison of each code, seven check points are set before the counters reach $CNTR_{MAX}$.

The realization of the AFC algorithm needs careful consideration, because f_{ref} and f_{div} are asynchronous clocks. Meta-stability, which cannot be predicted, will occur in some situations. To eliminate this problem, f_{ref} is used as system clock. At each check point(R CNTR reaches the threshold number), if the number of V CNRT is 2 larger or less than that of R CNTR, AFC will turn to next code comparison; if not, two counters will go on operating till next check point. In the realization of the AFC, the difference between two counters is not directly checked by comparing the counter numbers but by checking the counter flag signals in flip-flop chains which avoid errors caused by the meta-stability problem in digital circuit.

V. SIMULATION RESULTS

The layout of the proposed ΔΣ fractional-N PLL is shown in Fig. 6, including dual-core VCO, programmable FD, PFD, CP, digital circuits, and input/output (I/O) pads. VCO is placed in the left corner and far away from the digital circuits for isolation.

The post-layout simulation output frequency range of the dual-core VCO is from 920MHz to 2.0GHz. The maximum

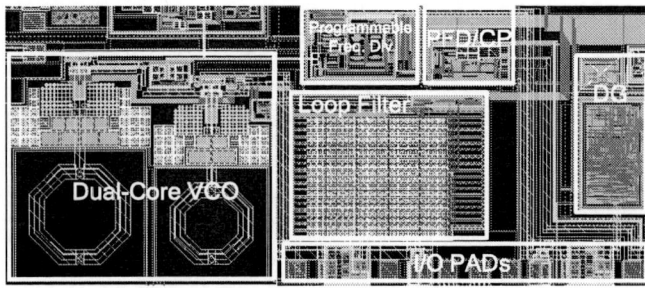

Figure 6   Layout of proposed design.

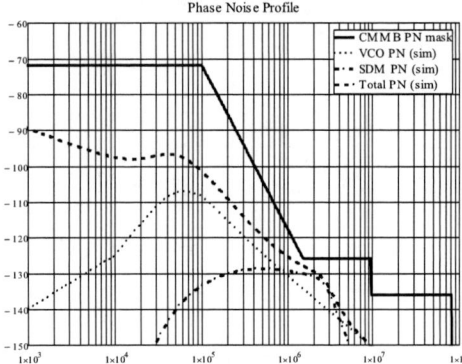

Figure 7   Phase noise performance.

K_{VCO} is 38MHz/V at 2.0GHz and the minimum K_{VCO} is 13MHz/V at 920MHz. The overall VCO tuning range reaches 74% and K_{VCO} variation is less than 100%.

Fig. 7 shows phase noise simulation results of the proposed PLL with typical loop parameters. The output phase noise is satisfied with the CMMB specification. The output phase noise of VCO and $\Delta\Sigma$ Modulator are also demonstrated in Fig. 7. The Quantization Noise (QN) is high-pass filtered by the $\Delta\Sigma$ modulation, while its noise transfer function demonstrates low-pass characteristic. The total QN-induced phase noise degradation presents a little hump in the output of the PLL as shown in Fig. 7.

Fig. 8 demonstrates PLL settling behavior including AFC period (coarse tuning) and fine tuning, while Fig. 8(a) is for the conventional binary search AFC algorithm and Fig. 8(b) is for the AFC algorithm in this design. *vco_bits* depicts the decimal value of capacity array control signal and *en_closeloop* shows whether the PLL is closed or not. The PLL opens when AFC operates. VCO control signal *vtune* is a fixed value, normally one-half of the supply voltage in AFC period, and after that period, *vtune* changes according to the PLL feedback mechanism. In Fig. 8(a), the period of each search is fixed; however, in the Fig. 8(b) AFC, only in the final search, two counters count to the maximum required number, and the comparison period of other comparisons is much less than that of the final one.

VI.   CONCLUSION

A fully integrated $\Delta\Sigma$ fractional-*N* Phase-Locked Loop with fast auto-frequency calibration for the CMMB tuner application is presented in this paper. The simulated output

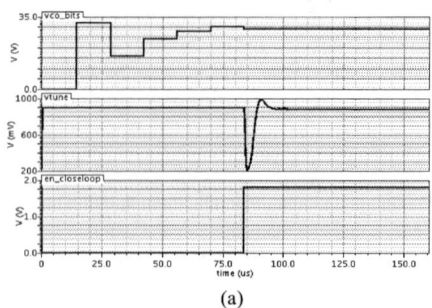

(a)

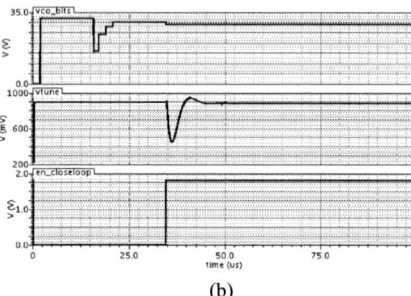

(b)

Figure 8   PLL settling behavior with (a) conventional AFC; (b) proposed AFC.

phase noise in typical case is -100 dBc/Hz, -125 dBc/Hz and -150 dBc/Hz at 100 kHz, 1 MHz and 10 MHz offset respectively, which perfectly satisfies CMMB phase noise requirements. VCO outputs from 920 MHz to 2.0 GHz with less than 100% tuning gain variation and has little effect on loop stability. The AFC scheme in this design achieves less than 50 μs settling time while the conventional AFC needs about 100 μs according to Fig. 8.

The design is implemented in a 0.18-μm CMOS process. The analog part including I/O pads occupies 1.7 mm×0.92 mm chip area and the digital part (AFC module and $\Delta\Sigma$ modulator) occupies 0.52 mm×0.23 mm. The total power consumption is 18 mW with a 1.8 V power supply.

REFERENCES

[1]   GY/T 220.1-2006 China Mobile Multimedia Broadcasting Part 1.

[2]   P. Antoine, P. Bauser, H. Beaulaton, et al., "A Direct-Conversion Receiver for DVB-H," *IEEE J. Solid-State Circuits*, vol. 40, pp. 2536-2546, Dec. 2005.

[3]   J. Kim, J. Shin, S. Kim, et al., "A Wide-Band CMOS LC VCO with Linearized Coarse Tuning Characteristics," *IEEE Trans. Circuits Syst. II, Exp. Briefs*, vol. 55, pp. 399-403, May 2008.

[4]   T. A. D. Riley, N. M. Filiol, Du Qinghong, et al., "Techniques for In-Band Phase Noise Reduction in $\Delta\Sigma$ Synthesizers," *IEEE Trans. Circuits Syst. II, Analog Digit. Signal Process.*, vol. 50, pp. 794-803, Nov. 2003.

[5]   W. Rhee, B. S. Song and A. Ali, "A 1.1-GHz CMOS Fractional-N Frequency Synthesizer with a 3-b Third-Order $\Delta\Sigma$ modulator," *IEEE J. Solid-State Circuits*, vol. 35, pp. 1453-1460, Oct. 2000.

[6]   J. Shin and H. Shin, "A Fast and High-Precision VCO Frequency Calibration Technique for Wideband $\Delta\Sigma$ Fractional-N Frequency Synthesizers," *IEEE Trans. Circuits Syst. I, Reg. Papers*, vol. 57, pp. 1573-1582, July 2010.

[7]   M. Marutani, H. Anbutsu, M. Kondo, et al., "An 18mW 90 to 770MHz Synthesizer with Agile Auto-Tuning for Digital TV-Tuners," *IEEE ISSCC Dig. Tech. Papers*, pp. 681-690, Feb. 2006.

A 1mW 5GHz Current Reuse CMOS VCO with Low Phase Noise and Balanced Differential Outputs

Wenrong Ying, Peng Qin, Jing Jin, and Tingting Mo*

Center for Analog/RF Integrate Circuits (CARFIC), School of Micro-Electronics
Shanghai Jiao Tong University, Shanghai 200240, China
Email: *motingting@ic.sjtu.edu.cn

Abstract—**This paper presents a low-power current reuse LC voltage controlled oscillator (VCO) with low phase noise and balanced differential outputs. A detailed analysis on different current control schemes, including various biasing techniques and the switching active core technique is conducted. A new current reuse VCO with novel current control scheme is proposed to achieve optimal phase noise and well balanced differential outputs at less power consumption. The proposed 5GHz VCO, simulated in a 0.18μm CMOS process, exhibits a phase noise of −117.7dBc/Hz at 1MHz offset while drawing only 0.88 mA current from a 1.2 V supply.**

Keywords-VCO; low power; current reuse

I. INTRODUCTION

Voltage controlled oscillator (VCO) is an essential building block in wireless transceivers. The design considerations for a VCO mainly include oscillation frequency, phase noise, tuning range, and power consumption. Among all the building blocks in wireless transceivers, VCO is one of the most power hungry components. Therefore, low power VCO is critical to the realization of low power wireless transceivers. Existing techniques for low power VCO design usually employ either transformer feedback (TF) structure [1], [2], [3], or current reuse topology [4], [5]. TF-VCO can achieve low phase noise under very low supply voltage, even below transistors' threshold voltage. Unfortunately, the inevitable transformer of TF-VCO occupies a quite large chip area and is unattractive to low cost integration. The current reuse VCO as first proposed in [4] is able to save half of the current consumption compared to conventional complementary cross-coupled VCOs. However, a current reuse VCO is fundamentally an asymmetrical circuit which is unable to produce balanced VCO outputs. Unbalanced VCO outputs generate extra noise/distortions in the circuits taking VCO signals, such as VCO buffers and VCO dividers, and thus deteriorate the system performance.

In this paper, several current control schemes used in current reuse VCOs to better balance the VCO outputs, including single bottom resistor biasing [4], external [5] resistor biasing, switching active core technique [6], and

newly proposed internal resistor biasing are carefully examined. Simulations show the internal resistor biasing scheme has better phase noise performance at low frequency offset, while the single bottom resistor biasing works better at high frequency offset. Based on this observation, a novel current control biasing scheme for current reuse VCO to achieve balanced differential outputs and phase noise optimization is proposed. The validation VCO, designed and simulated in a 0.18um CMOS process, exhibits the best power and phase noise performance compared with the state of the art low power CMOS VCO designs, yet achieving well balanced VCO outputs.

The rest of this paper is organized as follows. Section II describes different VCO topologies and their performances are summarized and compared. In Part A, the current reuse VCO is compared with the conventional cross coupled VCO. The mechanism of the low power feature of the current reuse VCO is explained. In Part B, detailed analysis of various biasing schemes and the switching active core technique for current control in the VCO is discussed. Section III proposes a novel VCO topology and it is designed and simulated to validate its superior performance. Section IV concludes this paper.

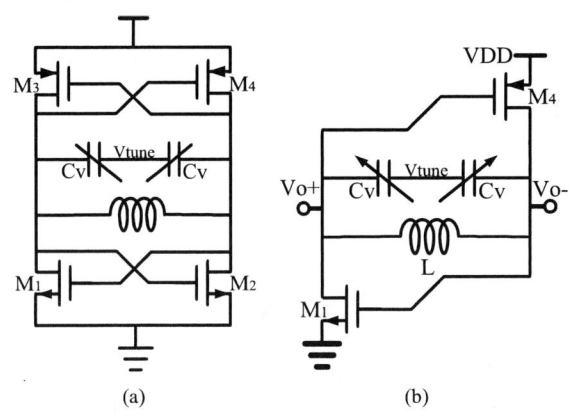

Figure 1.   (a) Conventional N-P Cross-coupled VCO (b) Current reuse VCO

This research was supported in part by Chinese National High Technology Research and Development Program (No. 2009AA011608) and in part by Chinese National Science Foundation Project (No.60801012).

II. VCO TOPOLOGY COMPARISON

A. Current reuse VCO versus conventional VCO

The conventional N-P cross coupled differential LC-VCO, as shown in Fig. 1(a), uses four active devices M1~M4 to provide negative conductance for LC tank loss compensation. During the first half period of oscillation, transistors M1 and M4 are on and transistors M2 and M3 are off, the current flows through the LC tank from right to left. During the second half period, M2 and M3 are on while M1 and M4 are off, and the current flows from left to right through LC tank.

Fig. 1(b) shows the current reuse VCO introduced in [4], in which the negative conductance is provided by a NMOS and a PMOS. It is interesting to note that the current reuse VCO in Fig. 1(b) takes the form of half of the conventional VCO in Fig. 1(a), but switches the transistors in a different way. During the first half period of oscillation, the two VCOs operate in a similar way. However, during the second half period, the current reuse VCO draws no current from the supply and as a consequence, the supply current in the current reuse VCO is about half of that in the conventional VCO. In fact, the current reuse VCO not only saves power but also improves the phase noise since it involves 2 transistors only and the parasitic capacitance is reduced by half compared with the conventional complementary cross-coupled VCO.

However, the major problem suffered by the current reuse VCO as shown in Fig. 1(b) is the voltage overdrive above the supply voltage during the transistor off period, which results in the unbalanced VCO outputs. Unbalanced VCO outputs could generate many problems, including deteriorated noise performance in circuits seeing the VCO signal. For example, if a VCO buffer sees the unbalanced VCO signal, the common-mode biasing noise could be up-converted to VCO frequency through nonlinearities. Therefore, techniques to improve the balance of differential outputs in current reuse VCOs are highly desirable.

B. Current control schemes for balanced VCO outputs

The mechanism for the unbalanced outputs in the current reuse VCO described above is that the output voltage is limited by the power supply voltage during transistor-on period, and limited only by the LC tank during transistor-off period. Therefore, a current control scheme forcing the VCO to operate in the current limited regime, where the output voltage is limited by the current during both periods, could potentially improve the output voltage balance.

A straightforward current control scheme is adding negative feedback biasing resistors in the VCO. As shown in Fig. 2, there are 3 different ways of adding the biasing resistors, i.e., bottom resistor biasing (BRB) [4] where a single biasing resistor is added between the source of NMOS and ground as shown in Fig. 2(a), external resistor biasing (ERB) [5] where 2 symmetrical biasing resistors are added outside the NMOS-PMOS pair as shown in Fig. 2(b), and internal resistor biasing (IRB) where 2 symmetrical biasing resistors are added inside the NMOS-PMOS pair as shown in Fig. 2(c). According to the authors' knowledge, this is first time proposed.

The BRB leads to unwanted 2nd –order distortion at the VCO outputs because of the body effect of NMOS caused by the bottom resistor. The ERB eliminates the body effect problem but deteriorates the quality factor of the LC tank, resulting in start-up difficulty and phase noise degradation. On the other hand, our proposed IRB technique eliminates both the body effect and the tank quality factor deterioration problems because the inserted biasing resistors have no effect on the negative resistance generated by the active devices. In addition, the IRB may reduce MOSFET flicker noise up-conversion from the active devices [7].

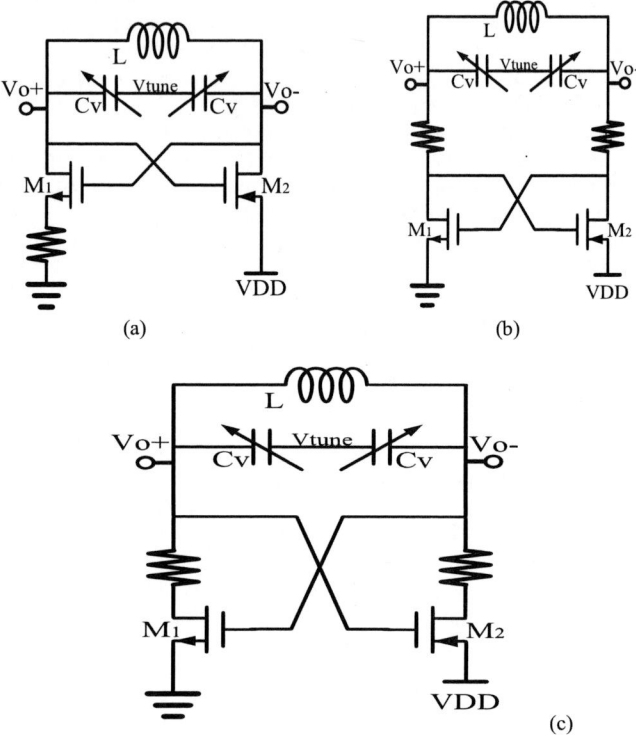

Figure 2.   Current control schemes: (a) the bottom resistor biasing (BRB); (b) the external resistor biasing (ERB); (c) the proposed internal resistor biasing (IRB)

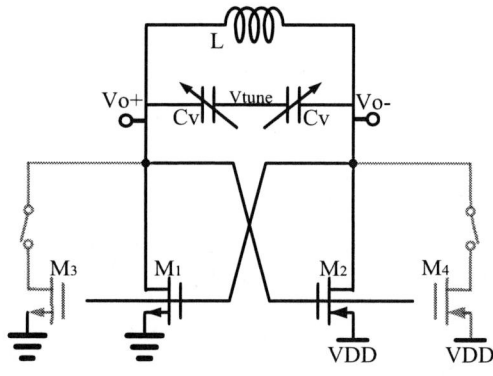

Figure 3.   Current control scheme with switching active core

Alternative way to operate the oscillator in current limited regime to improve the output balance is using smaller active transistors. Using smaller active transistors also results in less parasitic capacitance and thus better phase noise performance. However, smaller active transistors may generate insufficient negative conductance and lead to start-up problem in the oscillator. Therefore, the switching active transistor core technique [6] can be utilized. As shown in Fig. 3, two large transistors M3 and M4 are added in parallel with the original small transistors M1 and M2 to form the switching active core in the current reuse VCO. During normal operation, the active core of the VCO consists of only the small M1 and M2 to take advantage of their small parasitic capacitance and current limited operation. The large M3 and M4 will be switched in only when larger negative conductance is needed to start up the oscillation.

To have a fair comparison of all the current control schemes, four VCO topologies with different current control schemes, as depicted in Fig. 2 and Fig. 3, have been designed and simulated in a 0.18um CMOS. A conventional complementary LC VCO, as shown in Fig. 1(a), is also developed as the benchmark. For each VCO topology, phase noise performance and power consumption are optimized to achieve the best figure of merit (FoM), defined as the following:

$$FoM = L(\Delta f) - 20\log(\frac{f_o}{\Delta f}) + 10\log(\frac{P_{diss}}{1mW}) \quad (1)$$

where $L(\Delta f)$ is the VCO output phase noise, Δf is the offset frequency, f_o is the oscillation frequency, and P_{diss} is power dissipation in mW.

The performance summary of all five VCO topologies is shown in Fig. 4 and Table I. As can be seen, all of the current reuse VCOs have a better FoM than the benchmark conventional VCO does. At low offset frequency (e.g., 10kHz), the best performance is achieved by the VCO using the proposed IRB current control scheme. At high offset frequency (e.g., 1MHz), the VCO with BRB current control scheme scores the best, while the VCO with proposed IRB current control scheme comes a close 2nd-best.

TABLE I. VCO PERFORMANCE COMPARISON

Performance @5GHz, 1.2 V supply	1a	2a	2b	2c	3 S-off	3 S-on
Current (mA)	1.8	0.95	1.1	0.97	1.1	1.7
PN @ 10kHz offset (dBc/Hz)	-66.9	-74.2	-72.2	-76.1	-75.7	-62.9
@ 100kHz offset (dBc/Hz)	-93.5	-96.8	-94.9	-96.8	-96.7	-91.6
@ 1MHz offset (dBc/Hz)	-115.3	-117.2	-115.3	-116.9	-116.7	-115.7
Peak-to-peak Voltage Range (mV)	1154	961	910	926	871	921
FoM (dBc/Hz) 10kHz offset	-178	-188	-186	-190	-189	-174
FoM (dBc/Hz) 1MHz offset	-186.9	-191.3	-188.8	-190.9	-190.2	-187.3

1a is referred to conventional LC VCO as shown in Fig.1a; 2a, 2b and 2c are referred to VCO topologies depicted in Fig.2a,b,c; 3S-on and 3S-off are referred to the switching active core with switches on and off.

III. PROPOSED TOPOLOGY AND SIMULATION RESULTS

According to the above analysis, it is interesting to notice that the bottom biasing topology has better noise performance at high frequency offset while the IRB one has better noise performance at low frequency offset. Based on this point, it's easy to speculate that a combination these two methods may produce best noise performance at both low and high frequency offset spot. So finally, we propose the hybrid topology as in Fig. 5.

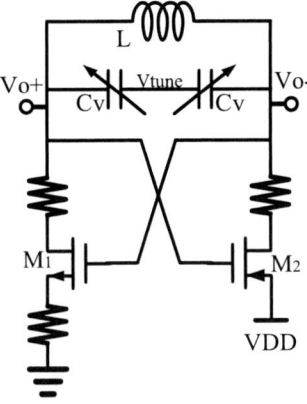

Figure 5. Hybrid topology of current reuse VCO

To validate our speculation, a 5GHz VCO with proposed structure is designed again in 0.18um CMOS technique. From Fig. 6, we can see the new design achieves phase noise of -75.7dBc/Hz, -97.5dBc/Hz, and -117.7dBc/Hz at 10kHz, 100kHz, and 1MHz offset respectively. Its current consumption is 0.88mA current from 1.2V supply. Compared these figures with what we got in the previous designs, the

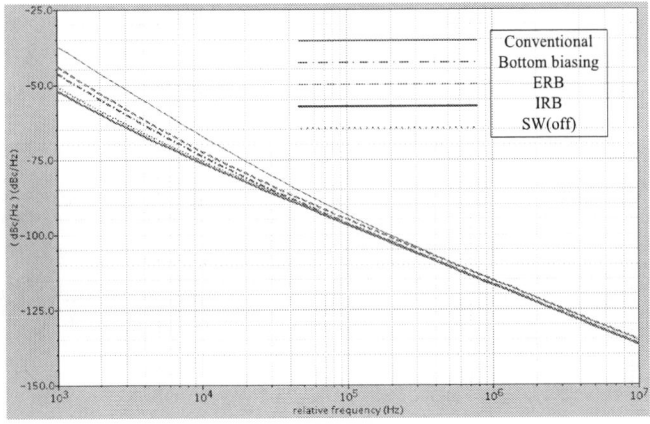

Figure 4. Phase noise comparison among VCOs in Fig. 2 and Fig. 3

978-1-61284-863-1/11 $26.00 © 2011 IEEE

newly proposed one owes the lowest power consumption and best phase noise at both 100kHz and 1MHz offset, which proves the effectiveness of the combination of the two biasing schemes. Table II summarize the whole performance of the proposed VCO, including FoM, and the one at 1MHz offset is the best.

We also take the symmetry of output waveform into account. Fig. 7(a) and (b) are the output waveforms of the bottom resistor biasing VCO and the proposed hybrid biasing VCO, respectively, where the solid waveforms are generated from negative output node (NMOS drain) and the dashed are from positive output node. The output of BRB VCO is 172-1044mV at negative output node and 160-1028mV at positive output node, with 12-16mV voltage shift. The output of the proposed VCO is 93-1053mV at negative output node and 89-1050mV at positive output node, with 3-4mV voltage shift, which is reduced by 3 times compared to the former.

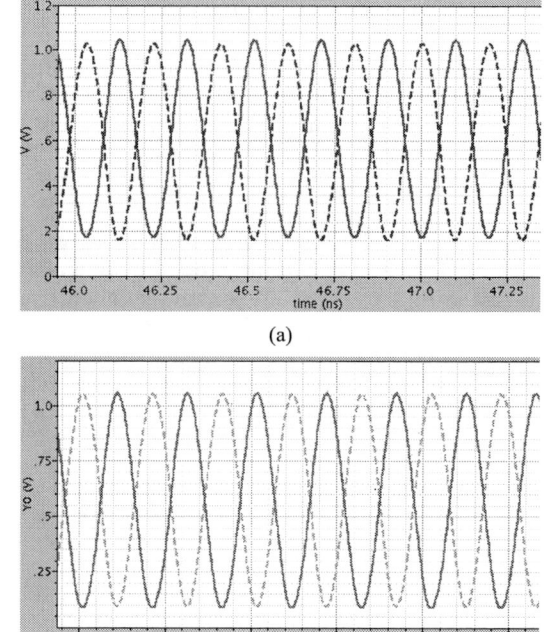

(a)

(b)

Figure 7. (a) The output voltage waveform of BRB VCO (b) The output voltage waveform of proposed VCO

TABLE II.   PERFORMANCE OF PROPOSED VCO

Performance@5GHz Under 1.2V Supply	
Current Consumption (mA)	0.88
Phase Noise @ 10kHz offset (dBc/Hz)	-75.7
@ 100kHz offset (dBc/Hz)	-97.5
@ 1MHz offset (dBc/Hz)	-117.7
Peak-to-peak Voltage Range (mV)	960
FoM (dBc/Hz)10kHz offset	-189.5
FoM (dBc/Hz)1MHz offset	-191.8

IV.   CONCLUSION

In this paper, the operation of current reuse VCO is analyzed in detail. Advantages and disadvantages of each topologies of current reuse VCOs are reviewed and a modified topology is newly proposed. Simulation results show that the proposed VCO achieves best phase noise performance while the power consumption is just 1mW.

REFERENCES

[1] K. Kwok and H. C. Luong, "Ultra-Low-Voltage High-Performance CMOS VCOs Using Transformer Feedback," *IEEE J. Solid-State Circuits*, vol. 40, no. 3, pp. 652–660, March. 2005.

[2] C. Lin, J. Kuo, K. Lin and H. Wang, "A 24GHz Low Power VCO With Transformer Feedback," *RFIC Symposium, Dig.* pp. 75–79, Nov. 2009.

[3] C. Hsieh, K. Kao and K. Lin, "An Ultra-Low-Power CMOS Complementary VCO Using Three-Coil Transformer Feedback," *RFIC Symposium, Dig.* pp. 91–94, Nov. 2009.

[4] S. Yun, S. Shin, H. Choi and S. Lee, "A 1mW Current-Reuse CMOS Differential LC-VCO with Low Phase Noise," *ISSCC Dig. Tech. Papers*, pp. 540–541, 2005.

[5] Z. Wang, H. S. Savci and N. S. Dogan, " 1-V Ultra-Low-Power CMOS LC VCO for UHF Quadrature Signal Generation," *ISCAS Dig.* pp. 4022–4025, May. 2006.

[6] D. Hauspie, E. Park, and J. Craninckx, "Wideband VCO With Simultaneous Switching of Frequency Band, Active Core, and Varactor Size," *IEEE J. Solid-State Circuits*, vol. 42, no. 7, pp. 1472–1480, July. 2007.

[7] S.Levantino, M. Zanuso, C. Samori and A. Lacaita, "Suppression of flicker noise upconversion in a 65nm CMOS VCO in the 3.0-to-3.6GHz Band," *ISSCC Dig. Tech. Papers*, pp. 50–51, 2010.

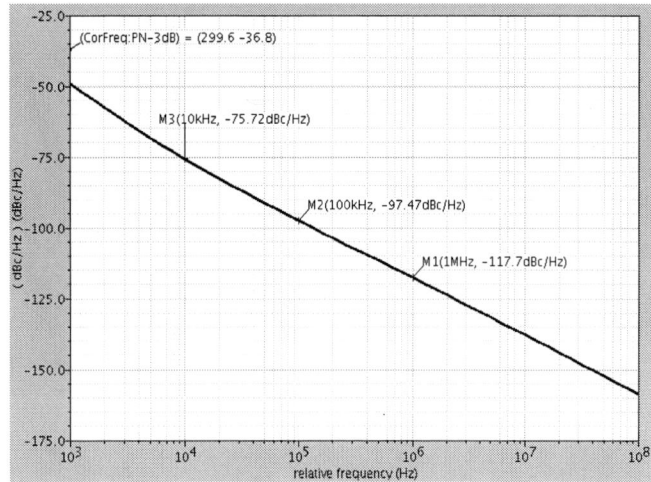

Figure 6.   Phase noise performance of the proposed current reuse VCO

A Curvature Compensated Bandgap Reference with low Drift and low Noise

Junmin JIANG, *Student Member, IEEE*, Zhihua NING and Lenian HE

Institute of Very Large Integrated Circuits

Zhejiang University

Hangzhou 310027, China

Email: zjujjm@gmail.com,{ningzh, helenian}@vlsi.zju.edu.cn

Abstract—A low temperature drift and low noise curvature-compensated CMOS bandgap reference is proposed. A dual-differential input pair amplifier was employed to add compensation with a high-order term of $T \ln T$ to the traditional 1st-order compensated bandgap. To reduce the offset of amplifier and output noise of bandgap reference, input differential pairs are large-sized PMOS. With a low dropout regulator stage, the voltage reference was increased to 3V. The voltage reference's temperature curvature is further corrected by 8-bit resistor trimming network. To avoid degrading the precision, noise performance is properly taken into account. The chip was fabricated by using TSMC's 0.35μm CMOS process. The temperature coefficient was measured only to be 2.1 ppm/°C over -40~125 °C after trimming. The output noise is 42 μVrms from 0.1Hz to 10Hz.

Index Terms—Voltage Reference, Bandgap, Curvature-compensation, Low drift, Low noise, Dual-differential input pair amplifier

I. INTRODUCTION

In high precision systems such as singal processing systems, power converters and RF circuits, the best performance totally depends on the accuracy of reference. It is the reason why voltage reference is the vital block in electronic systems. CMOS bandgap reference is a popular implementation of voltage references. The reference is required to be temperature independent and stable with variation of process. For instance, data converters require a voltage reference with less than 1 LSB error to guarantee the accuracy. For data converters whose operating temperature ranges from -40 ~ 125°C, 1 LSB error is 240 ppm for 12-bit and 12 ppm for 16-bit.

In order to reduce the temperature drift of voltage references, several creative methods for bandgap reference have been proposed[1]-[4]. A resistor ratio between a high-resistive poly-silicon resistor and a diffused resistor was used to provide high-order compensation[1]. However, the resistor's temperature cofficient varies quiet a lot with process variation. It's possible to hold the value of TC to tolerances of better than 250 ppm/°C for a poly resistor. Therefore, in [1], a complex trimming circuit is required in order to obtain a high precision reference for the design. Six temperature measurements are needed during one trimming operation[1]. Atrash[2] uses switching techniques to reduce offsets of the op-amp and current source devices. Still, the design only use 1st-order compensation and its drift is larger than 10 ppm/°C. Also a

large number of switches leads larger die area. Malcovati[3] uses V_{EB} linearization technique to compensate high-order terms of bandgap reference with current mode structure. It needs trimming circuits of 1st-order and high-order separately, increasing the complexity of trimming operation.

Voltage reference based on ΔV_{gs} was proposed[4]. The reference only uses MOSFETS in CMOS technology, which improves the simplicity of circuit design. However, V_{REF} in[4] depends on process parameters, thus, the magnitude might vary[5]. It was suggested that among various device parameters in semiconductor technologies, the characteristics of bipolar transistors have proven to be the most repoducible and well-defined quantities that can provide positive and negative TCs[6]. Therefore, bipolar operation still forms the core of this voltage reference.

For high precision applications, the noise generated by integreted devices and from power supply will degenerate the performance of voltage reference, thereby affect the static and dynamic performance of certain systems. For data converters, the parameters of Signal-to-Noise Distortion Ratio (SNDR) and Spurious-Free Dynamic Range (SFDR) are sensitive to noise. Therefore, if noise is not properly taken into account, the performance of systems that can be achieved might be misevaluated.

This paper proposes a new structure of bandgap reference. Reference is compensated with 1st-order and high-order terms to reduce temperature drift. Trimming with a switched resistor network is also used to minimize the process deviation. In our design considerations, noise is an important factor. The impact of noise is presented and optimization for decreasing noise is proposed. The measurement results and conclusion are given finally.

II. PROPOSED BANDGAP REFERENCE

Fig. 1 shows the circuits of bandgap reference proposed by this paper, which contains three parts: bandgap core, low dropout regulator (LDO) and curvature compensation circuit. The bandgap core generates proportional-to-absolute-temperature (PTAT) currents I_1 and I_2 flowing through Q_1 and Q_2, respectively, while the curvature compensation circuit produces a complementary-to-absolute-temperature (CTAT) current I_3 which flows through Q_3. Therefore, ΔV_{EB} the differential value of emitter-base voltage between Q_1 and

978-1-61284-863-1/11 $26.00 © 2011 IEEE

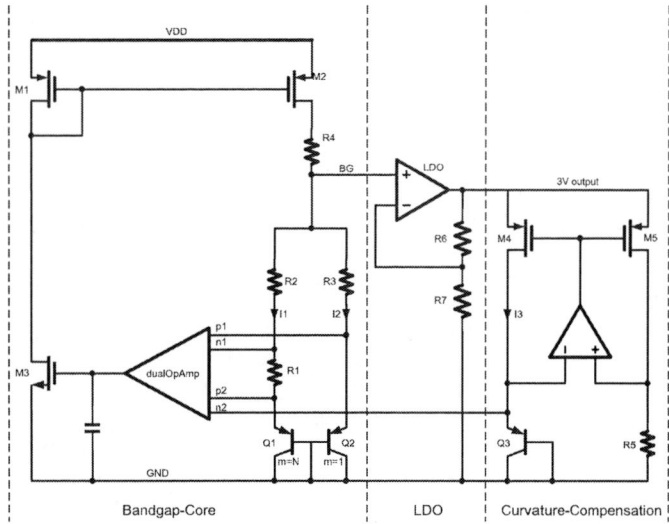

Fig. 1. Proposed bandgap reference

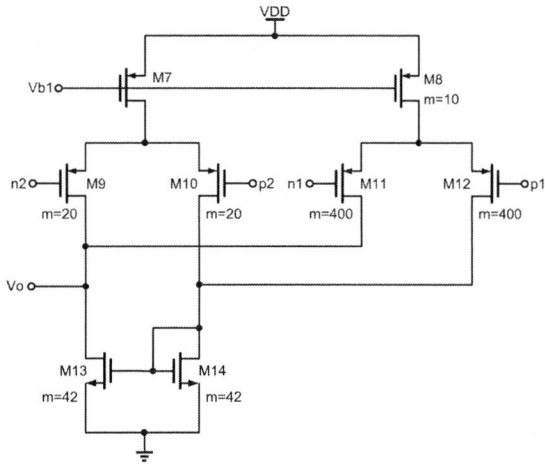

Fig. 2. Structure of dual-pair op amp

Q_3 is a high-order term associated with temperature, which was added to traditional 1st-order temperature compensated bandgap by dual differential-pair amplifier in bandgap core. A LDO is also integrated for driving ability and provides a stable power supply for curvature compensation circuit.

The structure of the dual-pair amplifier is shown in Fig. 2. The size of input differential transistors is very large, which would reduce the amplifier's offset and 1/f noise. Moreover, the input differential pairs are working in weak inversion region to keep a low quiescent current.

When PMOS works in weak inversion region, its drain current can be given as[6]:

$$I_{DS} = \frac{W}{L} I_0 e^{\frac{V_{SG}}{\xi V_T}} = \frac{W}{L} I_0 e^{n V_{SG}} \quad (1)$$

Where ξ is the process parameter and usually $1 < \xi < 3$, $n = \frac{1}{\xi V_T}$. I_0 is the drain current value when $V_{SG} = 0$. $\frac{W}{L}$ is the width and length ratio of PMOS.

Thus, the drain current of M9 to M12 can be expressed as:

$$I_{n1} = 400 I_0 \left(\frac{W}{L}\right) e^{n V_{SGn1}} \quad (2)$$

$$I_{p1} = 400 I_0 \left(\frac{W}{L}\right) e^{n V_{SGp1}} \quad (3)$$

$$I_{n2} = 20 I_0 \left(\frac{W}{L}\right) e^{n V_{SGn2}} \quad (4)$$

$$I_{p2} = 20 I_0 \left(\frac{W}{L}\right) e^{n V_{SGp2}} \quad (5)$$

As shown in Fig. 2, the load current mirror consists of MN13 and MN14. The multiplier of MN13 and MN14 are equal. So when MN13 and MN14 working in balanced status, their I_{DS} should be same. Therefore, we can obtain:

$$I_{n1} + I_{n2} = I_{p1} + I_{p2} \quad (6)$$

Two differential pairs' tail current mirror's ratio is 10. Thus, we can obtain:

$$I_{n1} + I_{p1} = 10(I_{n2} + I_{p2}) \quad (7)$$

From the schematic as seen in Fig. 1, the amplifier's input nodes voltages are:

$$V_{G,n1} = V_{EB1} + I_1 R_1 \quad (8)$$
$$V_{G,p1} = V_{EB2} \quad (9)$$
$$V_{G,n2} = V_{EB3} \quad (10)$$
$$V_{G,p2} = V_{EB1} \quad (11)$$

Caculate from Eq. (2) to (11),the final reference output voltage is[7]:

$$V_{REF} = V_{EB1} + \left(1 + \frac{2R_3 + 3R_4}{R_1}\right)(V_{EB2} - V_{EB1})$$
$$+ \left(1 + \frac{2R_3 + 3R_4}{R_1} + \frac{R_4}{R_3}\right)\frac{\xi T \ln T}{10} \quad (12)$$

where $V_{EB2} - V_{EB1}$ is the PTAT current term, which can provide 1st-order compensation. $T \ln T$ is the high order term compensation to V_{EB}. Then Eq. (12) could be simplified as follows:

$$V_{REF} = V_{EB1} + \alpha T + \beta T \ln T \quad (13)$$

where,

$$\alpha = \left(1 + \frac{2R_3 + 3R_4}{R_1}\right)\frac{k}{q}\ln(2N) \quad (14)$$

$$\beta = \frac{1}{10}\left(1 + \frac{2R_1 + 3R_4}{R_1} + \frac{R_4}{R_3}\right)\xi \quad (15)$$

III. NOISE ANALYSIS

For high precision applications, the noise generated by integreted devices and from power supply will degenerate the performance of voltage reference, thereby affect the static and dynamic performance of certain systems. Therefore, if noise performance is not properly taken into account, the performance that can be achieved might be misevaluated. There is a tradeoff among TC performance, noise and power dissipation.

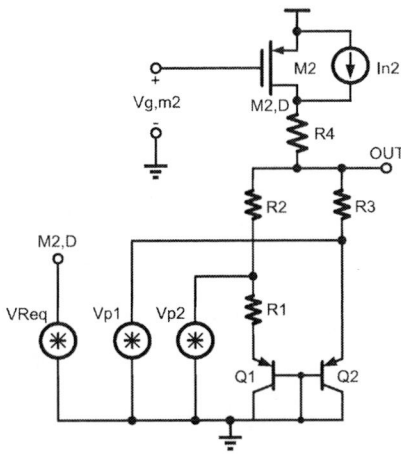

Fig. 3. Noise analysis model

TABLE I
NOISE CONTRIBUTION IN CIRCUIT

Device	Parameter	%	Device	Parameter	%
OP.MN13	Flick Noise	43.05	OP.MN14	Flick Noise	42.85
OP.MP7	Flick Noise	7.34	OP.MP12	Flick Noise	1.73
OP.MP11	Flick Noise	1.73	OP.MP8	Flick Noise	0.77
OP.MP9	Flick Noise	0.22	OP.MP10	Flick Noise	0.22
R_{eq}	Thermal	0.06	OP.MN13	Thermal	0.03
OP.MN14	Thermal	0.03	Others		1.97

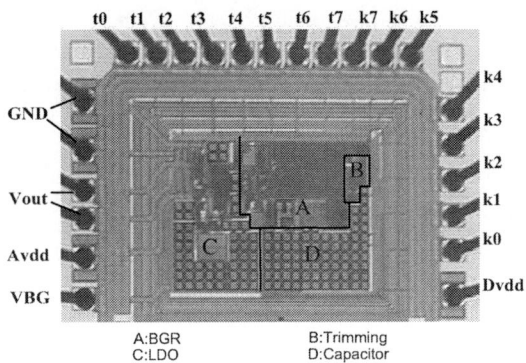

A:BGR   B:Trimming
C:LDO   D:Capacitor

Fig. 4. Chip mircophotograph

In order to get better TC performance, curvature compensation circuits and resistors' trimming circuits are added into design, which brings in more power dissipation and noise. In this paper, we try to evaluate the circuit's noise performance and present the method of optimization.

As shown in Fig. 1, the bandgap reference output is connected to current source which is decided by the gate voltage of M2, then all noise sources can be modeled by a noisy source at this point as $\bar{v}_{g,M2}$. Because of dual-pair input structure, op amp's input referred noise could be considered as two parts $\bar{v}_{p1}$ and $\bar{v}_{p2}$. Also MP12 and MP11's muliplier is far less than MP10 and MP9's, therefore, the noise from curvature compensation could be ignored. Then the main noise sources of this circuit are amplifier's input referred noise $\bar{v}_{p1}$ and $\bar{v}_{p2}$, current mirror noise of M2 $\bar{i}_{n,M2}$ and equivalent resistors' thermal noise $\bar{v}_{Req}$. The noise analysis model is built as shown in Fig. 3. Thus, we can obtain:

$$\bar{v}_{g,M2}^2 = \bar{v}_{p1}^2 \left(\frac{1}{g_{m2}} \frac{R_3 g_{m,Q2} + 1}{R_{eq}} \right)^2$$
$$+ \bar{v}_{p2}^2 \left(\frac{1}{g_{m2}} \frac{(R_1 + R_2) g_{m,Q1} + 1}{R_{eq}} \right)^2$$
$$+ \frac{4kT}{R_{eq} + R_4} \frac{1}{g_{m2}^2} + \frac{\bar{i}_{n,M2}^2}{g_{m2}^2} \quad (16)$$

where

$$R_{eq} = \left(R_1 + R_2 + \frac{1}{g_{m,Q1}} \right) // \left(R_3 + \frac{1}{g_{m,Q2}} \right) \quad (17)$$

g_{m2} is the transconductor of M2. $g_{m,Q1}$ and $g_{m,Q2}$ are the transconductor of PNP transistors of Q_1 and Q_2, respectively. Consequently, the output noise is:

$$\bar{v}_o^2 = \bar{v}_{g,M2}^2 (g_{m2} R_{eq})^2$$
$$= \bar{v}_{p1}^2 (R_3 g_{m,Q2} + 1)^2$$
$$+ \bar{v}_{p2}^2 [(R_1 + R_2) g_{m,Q1} + 1]^2$$
$$+ \frac{4kT R_{eq}^2}{R_{eq} + R_4} + \bar{i}_{n,M2}^2 R_{eq}^2 \quad (18)$$

In our design, $g_{m,Q1}$ and $g_{m,Q2}$ are constant in the circuit. To decrease the output noise $\bar{v}_o$, the resistors R_1 R_2 R_3 and R_{eq} should be decreased. It requires the current source to provide more current in the reference's output branch. Additionally, reducing resistance reduces the chip size. In our proposed bandgap reference in Fig 1, R_1 is $10.53k\Omega$, R_2 is $35.81k\Omega$, R_3 is $17.9k\Omega$. As seen in Eq. (12), V_{REF} depend on the ratio of R_1 R_3 and R_4. These resistors' absolute value could be decreased to get a better noise performance without affecting TC performance.

Table I is the simulation result of the output noise contribution. From Table I and Eq. (18), the output noise is mainly due to dual pair input amplifier's input referred noise $\bar{v}_{p1}$ and $\bar{v}_{p2}$. It is possible to see that the dual-pair amplifier is the major noise generator in bandgap reference, producing more than 97% of the noise in circuit. Because the input refered noise of dual-pair amplifier is amlified by the closed-loop gain.[8]

From the Table I, we can also see that most of the noise comes from 1st-stage of amplifier. In consideration of noise, we make input pair PMOS a very large size to reduce flick noise. As a consequence, the load devices of current mirror MN13, MN14 become the major noise sources. If the transistor operates as a constant current sources, it is required to minimize its transconductance to reduce current mirror load's flick noise[6]. Therefore, the transconductance of MN13 and MN14 should be decreased, as the tail current remains the same. But the amplifier's operation headroom will decrease correspondingly.

978-1-61284-863-1/11 $26.00 © 2011 IEEE

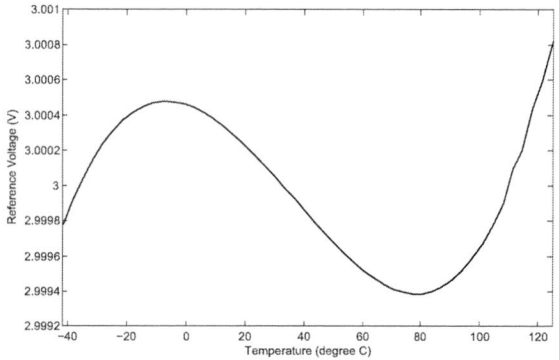

Fig. 5. TC Simulation result of voltage reference for temperature as function of -40 to 125°C

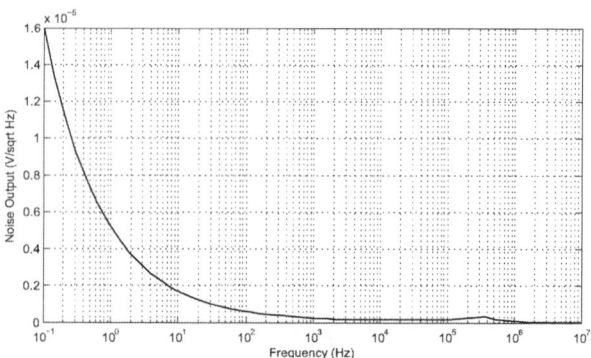

Fig. 7. Optimiezed output noise density of reference

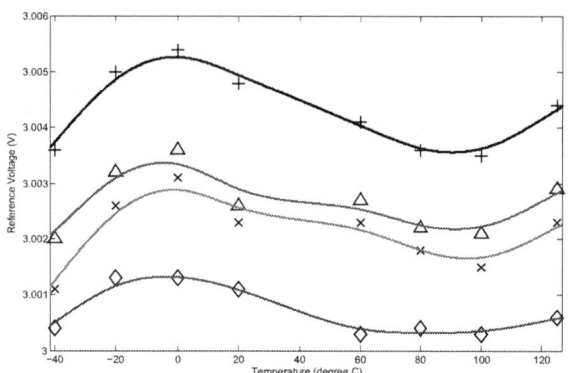

Fig. 6. TC Testing result of voltage reference for temperature as function of -40 to 125°C. + △ × and ◇ stand for four measured chips.

IV. MEASUREMENT RESULTS

The proposed voltage reference has been implemented by using TSMC's 0.35μm CMOS process. The chip micropho-tograph is show in Fig. 4. The chip area is 0.4mm². A B C and D in Fig. 6 indicate the area of bandgap-core, trimming cricuit, LDO and capacitor in the chip, respectively. 28 chips from different areas of the same Multi-Project Wafer (MPW) have been measured. Fig.5 is the simulation result of TC performance. Fig. 6 shows the measurement results of TC performance by setting trimming bits. Measurement results are connected by smoothing curves. The trend of curves meets the expectation. The temperature varition is from -40 to 125°C. When setting trimming bits samely, the chips' smallest temperature drift is only 2.1 ppm/°C. To achieve a more precise reference, each chip was trimmed separately. In this case the best value of TC is only 1.1 ppm/°C. All chips' TC are small than 7 ppm/°C.

Fig. 7 is the simulation result of output noise after opti-mization. The figure shows that the typical output noise is 16.01μVrms from 0.1Hz to 10Hz. The measurement result of noise was found to be 42μVrms from 0.1Hz to 10Hz.

V. CONCLUSIONS

A low temperature drift, low noise bandgap voltage ref-erence fabricated in TSMC's 0.35-μm CMOS process has been presented. The design considerations are minimizing the temperature drift and the noise. A dual pair input amplifier is used to add compensation with a high-order iterm of $T \ln T$ to the tranditional 1st-order compensated bandgap. The expression of output reference has benn calculated in detail. Noise analysis of bandgap circuit is made. To reduce the noise of the bandgap reference, input differential transistors of large size in amplifier are employed and they all work in weak inversion to reduce power dissipation. The best TC of reference is only $2.1ppm/°C$ over -40 to $125°C$. and the noise is $42\mu V$rms over the frequency of 0.1Hz to 10Hz. The performance make the reference very attractive for high precision application.

REFERENCES

[1] N. L. Ka, P. K. T. Mok, and Y. L. Chi, "A 2-V 23-μA 5.3-ppm/°C curvature-compensated CMOS bandgap voltage reference," *IEEE Journal of Solid-State Circuits*, vol. 38, no. 3, pp. 561– 564, 2003.

[2] A. H. Atrash and A. Aude, "A bandgap reference circuit utilizing switching to reduce offsets and a novel technique for leakage current compensation," in *The 2nd Annual IEEE Northeast Workshop on Circuits and Systems, 2004. NEWCAS 2004.*, ser. Circuits and Systems, 2004. NEWCAS 2004. The 2nd Annual IEEE Northeast Workshop on, 2004, pp. 297– 300.

[3] P. Malcovati, F. Maloberti, C. Fiocchi, and M. Pruzzi, "Curvature-compensated BiCMOS bandgap with 1-V supply voltage," *Solid-State Circuits, IEEE Journal of*, vol. 36, no. 7, pp. 1076–1081, 2001.

[4] N. L. Ka and P. K. T. Mok, "A CMOS voltage reference based on weighted ΔV_{GS} for CMOS low-dropout linear regulators," *IEEE Journal of Solid-State Circuits*, vol. 38, no. 1, pp. 146– 150, 2003.

[5] P. K. T. Mok and N. L. Ka, "Design considerations of recent advanced low-voltage low-temperature-coefficient CMOS bandgap voltage refer-ence ," in *Custom Integrated Circuits Conference, 2004. Proceedings of the IEEE 2004.*, ser. Custom Integrated Circuits Conference, 2004. Proceedings of the IEEE 2004, 2004, pp. 635 – 642.

[6] B. Razavi, *Design of Analog CMOS Integrated Circuits*. McGraw-Hill Education, 2000.

[7] Z. Ning and L. He, "A low drift curvature-compensated bandgap ref-erence with trimming resistive circuit," *Journal of Zhejiang University-SCIENCE*, 2011.

[8] D. Colombo, G. Wirth, S. Bampi, and C. Fayomi, "Impact of noise on trim circuits for bandgap voltage references," in *14th IEEE International Conference on Electronics, Circuits and Systems, 2007. ICECS 2007.*, ser. Electronics, Circuits and Systems, 2007. ICECS 2007. 14th IEEE International Conference on, 2007, pp. 775–778.

978-1-61284-863-1/11 $26.00 © 2011 IEEE

GSM/EDGE Power Amplifier Module with Improved Low-Power Efficiency

Jinbo Li, Tingting Mo*, and Feng Xu
Center for Analog/RF IC(CARFIC),School of Microelectronics
Shanghai Jiao Tong University
Shanghai, China
*Email: motingting@ic.sjtu.edu.cn

Abstract—**A dual mode 900MHz GSM/EDGE Power Amplifier module (PAM) is designed using IBM 0.35um SiGe BiCMOS technology. Bypass of the third stage is adopted for EDGE mode, leaving the final stage to work in Class E for high efficiency of GSM mode. Additionally, automatic level control (ALC) is employed to improve the power added efficiency (PAE) of low-power EDGE, along with the harmonic termination technique used in the inter-stage matching network for linearity requirement of EDGE. Simulation shows that the proposed PAM exhibits 29.5% PAE and -36.3dBc ACPR at 27dBm output for EDGE mode and 49% PAE at 35dBm output for GSM mode. After ALC, the average PAE of EDGE is boosted from 2.0% to 3.4%, which is 70% improved.**

Keywords- power amplifier module; SiGe BiCMOS; stage-bypass; automatic level control

I. INTRODUCTION

Since the first network deployment in 1992, GSM has become the most world-spread mobile system. Due to the fact that power amplifier stage dominates the power consumption in mobile handsets, the pursuit of high-performance GSM PA has never been ceased. Taking advantage of constant envelop of GSM signal, the peak power added efficiency (PAE) can reach 50 percent by using saturated power amplifier such as class E and F [1]. However, after introducing enhanced data rate for GSM evolution (EDGE), which adopts 8PSK modulation, to GSM system, challenges to satisfy linearity requirement without scarifying efficiency have appeared in the design of dual mode GSM/EDGE PA. The typical efficiency of a dual mode PA module for EDGE mode can be as low as 2 percent [2]. Research shows polar transmitter may be a possible way to mitigate the low efficiency of EDGE mode [3]. Unfortunately, this method brings high level circuit complexity.

In this paper, a dual mode GSM/EDGE PA module has been designed and a novel stage-bypass topology has been proposed. By using this method, different efficiency enhancement methods can be employed for different modes. As for GSM mode, three power stages are used and the final stage is biased in high efficiency class E mode, so the total PAE can reach as high as 49%. For EDGE mode, two power stages biasing in linear class AB are used and the 3rd stage is

bypassed. Generally speaking, there are two ways to improve the efficiency of a linear amplifier. One is the adaptive biasing [4], which boosts up gate/base voltage as the input signal increases; the other is automatic level control of supply voltage [5], which decreases the supply voltage when the output power level is low. The latter method is more effective in low and medium power range. By using it, an increase of the average efficiency of EDGE from 2.0% to 3.4% has been obtained.

However, biasing amplifier in class AB working mode cannot guarantee the linearity of EDGE. Harmonic and sub-harmonic terminations [6] have been employed in the input and inter-stage matching networks to further suppress the spurious output. About 3-4 dB improvement of adjacent channel power ratio (ACPR) is observed in EDGE mode.

Recently, SiGe HBT has become a competitive candidate to substitute relatively expansive GaAs technology for cellular handset power amplifiers [7]. This paper adopts IBM 0.35um SiGe BiCMOS for the power amplifiers, power detector, bias circuits and adaptive control logic design. Matching networks are also realized on chip, except the RF choke inductor at collector terminals. Simulation shows the proposed PAM exhibits 49% and 29.5% of PAE for GSM and EDGE at peak output power, respectively. And the average of PAE for EDGE can be increased by 70% after using ALC.

This paper is organized as follows. The proposed configuration of GSM/EDGE PAM is disclosed first. Detailed of PA core design, harmonic termination at input and inter-stage matching, and power detector circuit are discussed separately in section II. Simulations by ADS with dynamic link of Cadence are conducted to characterize the output power, peak PAE, average PAE and ACPR for GSM and EDGE mode respectively in section III. And at last, conclusion is made in section IV.

II. DESIGN OF GSM/EDGE POWER AMPLIFIER MODULE

A. Proposed topology

The overall topology of proposed dual mode GSM/EDGE PAM is depicted in Fig.1. There are three power stages for GSM mode, but only two for EDGE mode. The final stage is

---

This work has been supported by Chinese National Science Foundation Project **(No. 60801012)**

978-1-61284-863-1/11 $26.00 © 2011 IEEE

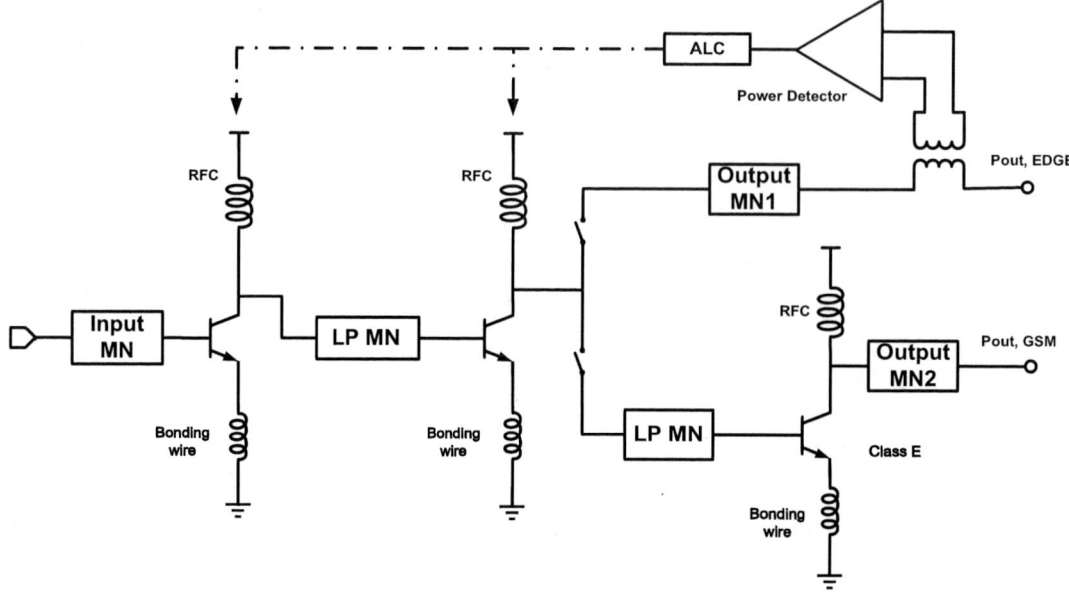

Figure 1. Proposed stage bypass topology of GSM/EDGE PA Module

bypassed due to the maximum output power of EDGE is 8 dB lower than that of GSM, which is approximate to the gain of the final stage. This stage-bypass configuration also set GSM free of choosing saturated power amplifier as the final stage to enhance the efficiency of GSM mode as well. Here, the 3rd stage is set as class E amplifier.

The ALC loop includes coupler, power detector, logic control unit and DC-DC converter. The input of the power detector circuit is from a 20dB coupler, and its voltage output is sent to the logic power level control, which adjust the collector voltage according to its output power level. For the average power level of EDGE mode, the collector voltage can be reduced from 3.5V to 2.4V, thus improved its efficiency. Details of the building blocks' design, such as power detector circuit, matching networks are discussed below.

B. PA core design

The geometry of the power transistors for the three stages has to be determined by the gain and output power they are going to handle. Load-pull and source-pull should be carefully carried on for each stage iteratively. The first two stages are biased at class AB mode, since EDGE adopts $3\pi/8$-shifted 8-PSK modulation, which needs a linear amplifier to preserve its envelope variation. Thanks to the bypass architect, the 3rd stage could be a saturated amplifier, here, we choose class E. Canon schematic of class E amplifier is employed, so details on its design is not covered in this paper. Just to mention that the output matching for GSM and EDGE has to be done separately.

Other design tricks include collect voltage and wire bonding, which are important to power amplifier design. SiGe technology has the shortcoming of lower breakdown voltage comparing to GaAs. The simulation shows that the third harmonic is detrimental to keeping Vce under BVCEO, thus

efforts should be made to suppress the 3rd-harmonic. As for inter-stage matching network, different matching structures are detailed in [8]. Low pass inter-stage matching is adopted here to suppress high frequency components, such as the 3rd harmonics as mentioned before.

Wire bonding is also in the consideration of this design. The bond wire model is employed in ADS, which is approximately 800pH of an 1mm gold wire bond with the loop height of 150um and radius of 12.7um (0.5mil).

C. Bias Network Analyse and Design

Bias circuits play an important role in PA design. General constant voltage bias through large inductor is simple, but large inductors occupy large chip area and cannot be easily integrated. In this paper, a beta helper circuit with a buffer as shown in Fig.2 is used to eliminate the large inductor for biasing. It could help to save the chip area as well as the cost.

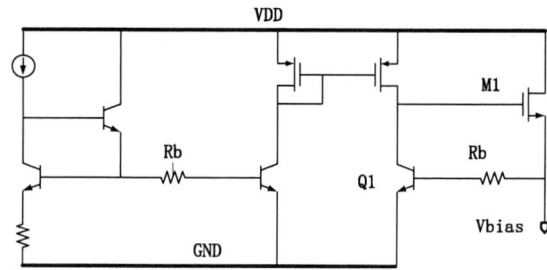

Figure 2. Constant voltage bias with buffer

It is straightforward that the output impedance of the buffer is low at low frequencies, but high at the RF frequency. Open loop amplifier with NMOS buffer stage as shunt-shunt feedback largely reduces output impedance. R_b serves to reduce current consumption and but limits bandwidth as well. To minimize the RF power loss due to the current leakage, the output impedance of the buffer should be high at the RF frequency so that the RF power leakage into the bias network is negligible [9]. On the other hand, the impedance should be low around DC frequency. Since GSM channel bandwidth is 200 kHz, the output impedance of buffer should be low at least within this bandwidth. Using the feedback theory, $Z_{bias}(\Delta\omega) \approx 1/g_{m1} \cdot (1 + g_{m\_Q1} \cdot r_{o\_Q1}/2)$ Since $g_{m\_Q1} \cdot r_{o\_Q1} >> 1$, $Z_{bias}(\Delta\omega) << 1/g_{m1}$. Based on simulation, $Z_{bias}(\Delta\omega) \approx 0$ within 200 kHz.

The bias circuit is also part of input or inter-stage matching network, it is worth to notice that this bias circuit provides a short circuit to DC and low-frequency components. Given the sub-harmonic component ($\Delta\omega$) contributes to both 3rd-nonlinearity and 5th-nonlinearity, it has been proven that the termination of DC or sub-harmonic components ($\Delta\omega = \omega_1 - \omega_2$, where ω_1 and ω_2 are two input frequencies) at the base of transistors is beneficial to the linearity of the PA [10].

D. Power Detector Design

The key building block of the ALC loop is the power detector. Usually, peak detector or power detector are used to track output for control. Since the fact that peak detector adopts large capacitor for longer holding time, digital modulation signal with higher PAR and unpredicted large interference makes the peak detector unable to catch up with the fast changes. Designers are inclined to employing power detectors for average power tracking instead. The proposed structure of CMOS power detector is depicted below in Fig. 3. M_1 and M_2 work in saturation mode. RF_p and RF_n are connected to differential RF input signal. The sum of current of M_1 and M_2 cancels 1st-order component and reflects the square of the input, which indicates the input power. DC bias current has been extracted by current mirror Q_1 and Q_2 for better logarithmic amplification. Since input frequency is very high, 1st-order RC low pass filter effectively reject ripples of the output.

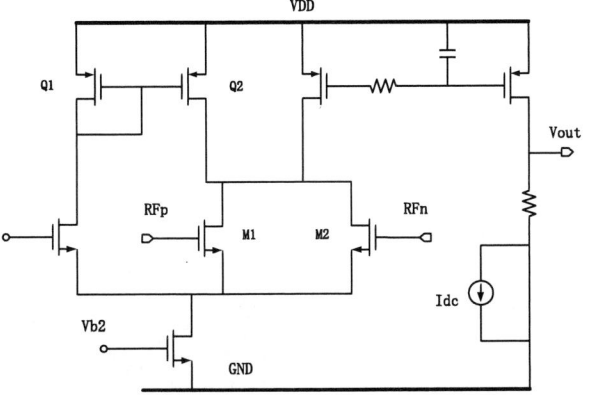

Figure 1. CMOS power detector working in saturation region

For better process-voltage-temperature (PVT) performance, V_{out} node should be calibrated through current array, which is

not covered here. The performance of this power detector circuit is offered in section III.

III. SIMULATIONS

This PAM design adopts IBM 0.35um SiGe BiCMOS technology, so the core amplifier can utilize the high performance SiGe HBT while the control circuit can use common CMOS technology.

First, the effect of with and without harmonic tuning on the output spectrum has been demonstrated in Fig.4. Around 3-4 dB improvement of ACPR is observed. It is proved that the harmonic tuning has the positive effect on the linearity of the power amplifier.

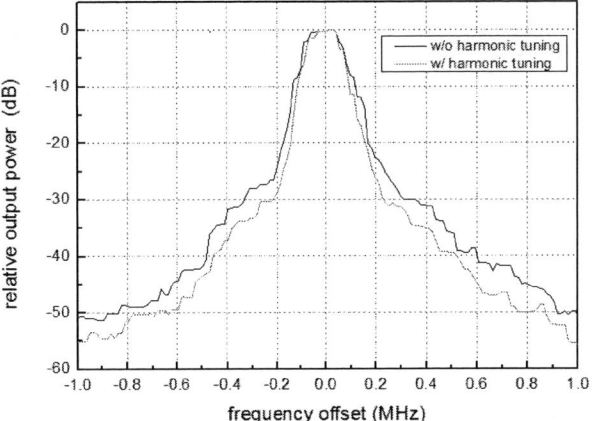

Figure 4. Output spectrum with and without harmonic tuning

Power detector circuit is also examined. Fig. 5 shows the output voltage versus inspected power. The curve is almost linear in the whole 35 dB of GSM and EDGE power dynamic range.

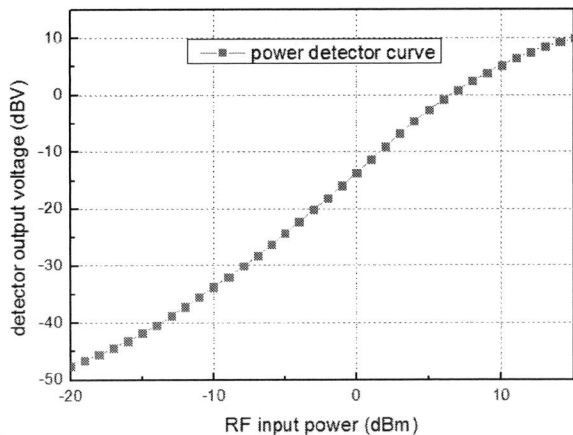

Figure 5. Output voltage versus inspected power of detector

Efficiency for GSM and EDGE mode are depicted in Fig. 6 and Fig. 7, respectively. Thanks for the high efficiency of class E operation, the 3-staged GSM PA exhibits 49% at 35dBm output. Fig. 7 tells efficiency improvement by automatic level control of supply voltage. Red line denotes PAE with supply voltage optimization, and blue denotes PAE without ALC. The peak PAE is untouched, but in the low power levels, the efficiencies are substantially enhanced. Combined with probability distribution function of EDGE mode, average PAE of output range of 3dBm to 27dBm increases from 2.0% to 3.4%, which is 70% improved.

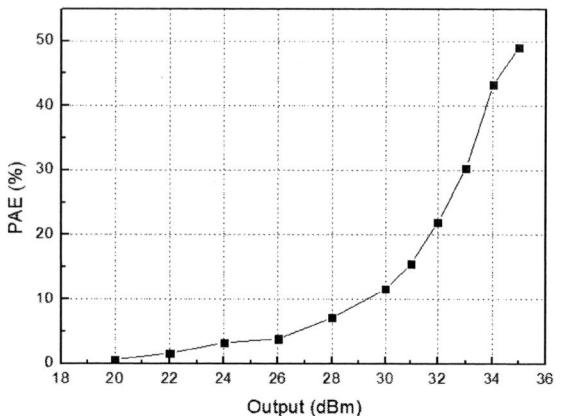

Figure 6. PAE of GSM versus output power

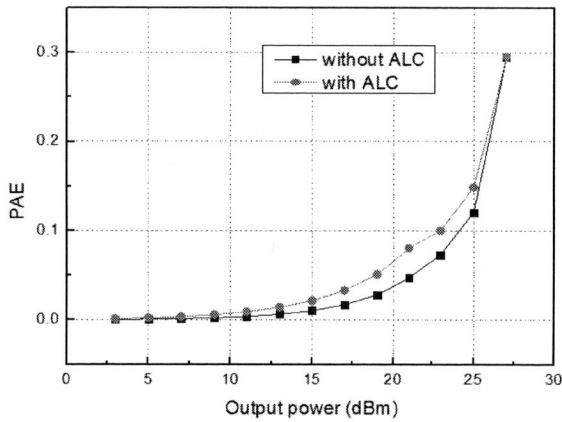

Figure 7. PAE of EDGE versus output power with and without ALC

The complete performance of the proposed PAM is summarized below in Table I. For EDGE mode, the two stage power amplifier achieves 27 dBm maximum output with a fixed power gain of 30 dB. ACPR at 200 KHz offset is measured as -36.3 dBc during maximum output. For GSM mode, the three stage power amplifier exhibits 49 percent PAE at the maximum output of 35 dBm. The low-power efficiency improvement of EDGE mode is also emphasized in this table.

TABLE I.   SUMMARY OF THE PERFORMANCE OF PROPOSED PAM

EDGE	Maximum Pout (dBm)	27
	Gain (dB)	30
	PAE@ Pout=27 dBm (%)	29.5
	PAE average without ALC (%)	2.0
	PAE average with ALC (%)	3.4
	ACPR@ 200kHz (dBc)	-36.3
GSM	Maximum Pout (dBm)	35
	PAE@ Pout=35 dBm (%)	49

IV.   CONCLUSION

In this paper, a dual mode GSM/EDGE PAM has been designed using IBM 0.35 SiGe BiCMOS technology. Stage-bypass topology has been introduced to separate high efficient design for GSM and linear design for EDGE. Class E operation is set at the final power stage for GSM efficiency boost. Harmonic tuning is adopted in input and inter-stage matching network for better linearity performance. Power detector and ALC circuit are employed to further enhance the PAE at low-power level for EDGE. Detailed circuits have been described and simulations have proved the effectiveness of our proposed method.

REFERENCES

[1]   F. H. Raab, "Class-F power amplifiers with maximally flat waveforms," Microwave Theory and Techniques, IEEE Transactions on, vol. 45, pp. 2007-2012, 1997.

[2]   "RF9810 Quad band GPRS/Linear EDGE+3.2V TD-SCDMA multi-mode transmit module" RFMD.com, Datasheet.

[3]   "The Polar Loop-TM transmitter for Quad-band GSM, GPRS, and EDGE Applications", White Paper, Skyworks Solutions Inc., May 16, 2005.

[4]   W. Kim, K. S. Yang, J. Han, J. Chang, and C. Lee, "An EDGE/GSM Quad-Band CMOS Power Amplifier", ISSCC Dig. Tech.Papers, pp. 429-431, Feb. 2011.

[5]   S. EGOLF, "Intelligent power management: a method to improve 2G/3G handset talk time", Microwave Journal, July, 2007.

[6]   J. Vuolevi and T. Rahkonen, "The effects of source impedance on the linearity of BTJ common-emitter amplifiers," in Circuits and Systems, 2000. Proceedings. ISCAS 2000 Geneva. vol.4, pp. 197-200, 2000.

[7]   M. Racanelli and P. Kempf, "SiGe BiCMOS technology for RF circuit applications," Electron Devices, IEEE Transactions on, vol. 52, pp. 1259-1270, 2005.

[8]   K. Mori, et al., "An L-band high efficiency and low distortion power amplifier module using an HPF/LPF combined interstage matching circuit", Microwave Symposium Digest., 2000 IEEE MTT-S International, vol.2, pp. 865-868, 2000.

[9]   D. Junxiong, et al., "A SiGe PA with dual dynamic bias control and memoryless digital predistortion for WCDMA handset applications", Solid-State Circuits, IEEE Journal of, vol. 41, pp. 1210-1221, 2006.

[10]   S. Watanabe, et al., "Simulation and experimental results of source harmonic tuning on linearity of power GaAs FET under class AB operation", Microwave Symposium Digest, 1996., IEEE MTT-S International, 1996, vol.3 pp. 1771-1774, 1996.

Comparative Study and Analysis of Noise Reduction Techniques for Front-End Amplifiers

Lei Liu[1,2], Xiaodan Zou[1], Wang Ling Goh[2], Minkyu Je[1]

[1]Institute of Microelectronics, A*STAR (Agency for Science, Technology and Research), Singapore
[2]Nanyang Technological University, Electrical & Electronics Engineering, Singapore
liul0021@ntu.edu.sg

Abstract—**This paper presents a comparative analysis of noise reduction techniques devised for front-end amplifiers (FEAs). Feedforward noise cancellation, signal-nulled noise feedback and current reuse techniques are briefly explained, analyzed, and compared. In order to gauge the effectiveness of the noise reduction technique in conjunction with power consumption, a noise efficiency factor (NEF) is used to analyze and compare the noise-power trade-off of different techniques reported. The analysis shows that the noise reduction technique based on the simple current reuse scheme provides the best noise-power trade-off, which is verified through circuit simulations.**

Index Terms—**front-end amplifier, low-noise amplifier, low-power amplifier, noise cancellation, noise reduction, noise efficiency factor.**

I. INTRODUCTION

Noise reduction has been one of the most important aspects of the front-end amplifier (FEA) design. In order to reduce noise from active devices, the most straightforward approach is to increase the operation current. However, the total amount of current consumption is preferred to be kept at the lowest, especially in battery powered systems such as portable electronics, and wearable and implantable biomedical devices [1]–[3]. Therefore, there is a strong trade-off between the noise performance and power consumption in the FEA design.

Recently, low-noise low-power FEAs with various noise improvement techniques have been reported [4, 5, 7]. Those noise reduction techniques are implemented by adding extra circuitry and hence the inevitable power consumption, aiming to cancel out a significant part of circuit noise. One representative example of such techniques is the feedforward noise cancellation scheme reported in [4]. In this structure, an additional amplification stage is used to cancel the noise generated by the input transistor. By biasing with sufficient amount of current, the noise from the additional amplification stage is minimized. This scheme breaks the tie between the noise performance and input matching in wideband RF FEA. Theoretically, it can achieve a very low noise at the cost of increased power consumption. Another example, which is based on the noise feedback scheme, is reported in [5]. In this approach, the overall output noise is suppressed by engaging two different feedback paths – one for signal and the other for noise. However, none of these device noise components can be completely cancelled. From our investigation, it is found that most of the reported FEA noise cancellation techniques require

additional circuitry and hence power consumption in order to suppress the noise. Nevertheless, the effectiveness of these noise reduction techniques has yet been evaluated considering the extra circuitry and power consumption required.

In this paper, various techniques reported for the noise performance improvement of AFEs are analyzed and compared based on a noise efficiency factor (NEF), which is commonly used to evaluate the noise-power trade-off of FEA designs. In Section II, various noise reduction techniques are introduced. Section III analyzes the well-known noise reduction techniques and compares the effectiveness of those techniques in terms of noise-power trade-off. In Section IV, the design result of a low-power low-noise neural recording amplifier is presented as an example to validate the analysis and discussion. Finally, a conclusion is drawn in Section V.

II. NOISE REDUCTION TECHNIQUES

In this section, common CMOS noise reduction techniques are reviewed in order to highlight their noise reduction efficiencies.

A. Feedforward Noise Reduction

These noise reduction techniques make use of the property that the noise at the output and at the feedback point are in phase, while the signals are out of phase. The noise can therefore cancel each other by negatively amplifying the noise at the feedback point and summing with the noise at the output. However, the signals are added constructively.

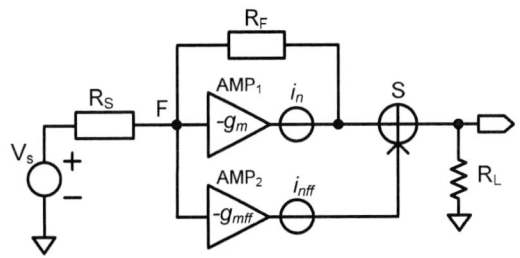

Figure 1. Feedforward noise reduction model

Fig. 1 presents the model of a well-known feedforward noise reduction technique [4]. The output noise is sensed at the AMP1 output and fed back to node F. The noise from the amplifier AMP_1 combines with this fed-back noise after amplification through the feedforward amplifier AMP_2. By doing so, the noise

---

This work was supported by the Science and Engineering Research Council of A*STAR (Agency for Science, Technology and Research), Singapore under Neurodevices program. The grant number for the project is 102-171-0162.

978-1-61284-863-1/11 $26.00 © 2011 IEEE

from AMP₁ can be totally cancelled when the gain of AMP2 denoted by A_2 and the feedback factor denoted by β are carefully set to satisfy that $\beta A_2 = 1$. Note that the noise from AMP₂ is not reduced but it can be suppressed by increasing the current consumption of AMP₂.

B. Signal-Nulled Noise Feedback

In the feedback noise reduction technique [5], as depicted in Fig. 2, the output noise is sensed at the summing node S, fed back to node F through the signal-nulled feedback network consisting of β_1 and β_2, amplified by AMP₂, and then added at node S in the opposite phase. While the noise is suppressed, the signal is not affected if β_1 and β_2 are designed such that node F becomes virtual AC ground. In this configuration, the noise is suppressed by the global feedback loop and hence the noise from AMP₂ is also reduced. This is in contrast to the feedforward noise reduction approach. However, the noise from either AMP₁ or AMP₂ cannot be completely eliminated. Consequently, for a better noise performance, the transconductances of the amplification devices in AMP₁ and AMP₂ need to be increased by consuming more power.

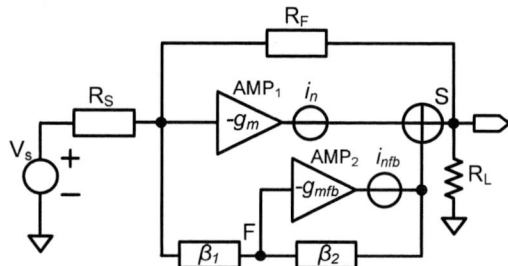

Figure. 2. Signal-nulled noise feedback model

C. Current Reuse

In the current reuse scheme, the multiple transistors are stacked in the same current branch to generate multiple transconductances from a single current branch. The resulting multiple transconductances are combined to provide a larger effective transconductance with the same amount of current consumed, when compared with a single device in a single branch. By obtaining a larger transconductance, a lower thermal noise can be achieved. One disadvantage of this scheme is that due to the limited voltage headroom, the signal swing range and linearity performance can be degraded unless careful attention is paid to the design.

III. NOISE-POWER TRADE-OFF ANALYSIS

By evaluating the additional power required for different noise reduction techniques, the noise-power trade-off of the techniques can be analyzed. A noise efficiency factor (NEF) which quantifies the noise-power trade-off is defined as [6]

$$NEF = V_{rms,in}\sqrt{\frac{2I_{total}}{4kT \cdot \pi \cdot U_T \cdot BW}} \qquad (1)$$

where $V_{rms,in}$ denotes the input referred noise, U_T the thermal voltage, I_{total} the total current consumption, and BW the bandwidth. In the following analysis, the same bandwidth and the same input transistor size are assumed. The noise from the feedback network is not included in the calculation because it can be avoided by using capacitive feedback network. It is also assumed that the thermal noise dominates.

A. Feedforward Noise Reduction

This technique is used to break the relation between the noise and input matching in wideband RF FEAs. Fig. 3 shows one representative example. By dissipating enough current at M₂, a good noise performance can be achieved at the expense of power consumption.

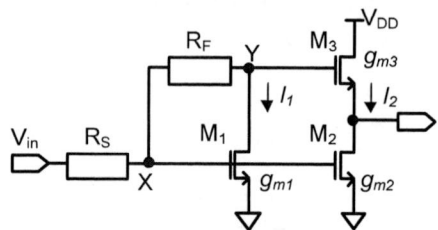

Figure. 3. Feedforward noise cancellation circuit [4]

The noise from M₁ is totally cancelled when the gain from the stage consisting of M₂ and M3 satisfies that

$$A_2 = \frac{g_{m2}}{g_{m3}} = 1 + \frac{R_F}{R_S} \qquad (2)$$

where g_{m2} and g_{m3} are the transconductances of transistors M₂ and M₃, respectively. The output noise is mainly contributed from the thermal noise of M₂, which can be expressed as

$$\overline{V_{no}^2} \approx \overline{V_{no,M_2}^2} = \frac{4kT\gamma g_{m2}}{g_{m3}^2} \qquad (3)$$

where T is the absolute temperature, and γ is the excess noise factor. The NEFs with and without feedforward noise cancellation are calculated as (4) and (5), and their ratio is derived in (6).

$$NEF_{w/o,NS} = K\sqrt{\frac{4kT\gamma}{g_{m1}}}\sqrt{I_1} \qquad (4)$$

$$NEF_{w,NS} \approx K\sqrt{\frac{4kT\gamma}{g_{m2}}}\sqrt{I_1 + I_2} \qquad (5)$$

$$\frac{NEF_{w,NS}}{NEF_{w/o,NS}} = \frac{1}{2}\sqrt{\frac{I_1 + I_2}{\sqrt{I_1 \cdot I_2}}} \qquad (6)$$

where $K = \sqrt{2/(4kT \cdot \pi \cdot U_T \cdot BW)}$. According to (6), the minimum NEF with the feedforward noise cancellation technique can be achieved when I_1 and I_2 are set to be equal.

$$\left.\frac{NEF_{w,NS}}{NEF_{w/o,NS}}\right|_{min} = \frac{\sqrt{2}}{2} \approx 0.71 \qquad (7)$$

From (7), we can conclude that the noise-power trade-off is improved by up to $\sqrt{2}/2$ times, compared to the circuit without

978-1-61284-863-1/11 $26.00 © 2011 IEEE

noise cancellation. Note that M_3 also generates noise, which is ignored in the analysis above.

B. Signal-Nulled Noise Feedback

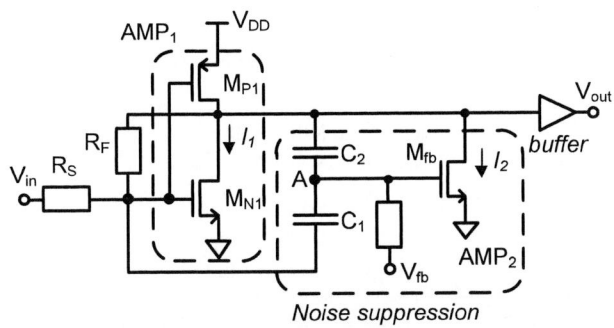

Figure 4. Signal-nulled noise feedback circuit [5]

The circuit using signal-nulled noise feedback technique is illustrated in Fig. 4. The noise sources are contributed by the two amplifiers, AMP$_1$ and AMP$_2$, as described in (8).

$$\overline{v_n^2} = 4kT\gamma(g_m + g_{mfb})R_{out}^2 \qquad (8)$$

where g_m and g_{mfb} are the transconductances of amplifier AMP$_1$ and AMP$_2$, respectively, and R_{out} is the total output impedance. In this configuration, the signal-nulled noise feedback is valid if the ratio of C_1/C_2 is equal to the voltage gain of AMP$_1$ and the node A does not carry any signal but noise. The output noise voltages with and without noise suppression are expressed in (9) and (10), respectively.

$$\overline{v_{no,w,NS}^2} = \frac{4kT\gamma(g_m + g_{mfb})\dfrac{(R_S + R_F)^2}{(1+g_m R_S)^2}}{(1 + g_{mfb}/g_m)^2} \qquad (9)$$

$$\overline{v_{no,w/o,NS}^2} = 4kT\gamma g_m \frac{(R_S + R_F)^2}{(1+g_m R_S)^2} \qquad (10)$$

The NEF improvement is evaluated in (11).

$$\frac{NEF_{w,NS}}{NEF_{w/o,NS}} = \frac{v_{no,w,NS}}{v_{no,w/o,NS}}\sqrt{\frac{I_1+I_2}{I_1}} \qquad (11)$$

By substituting (9) and (10) into (11),

$$\frac{NEF_{w,NS}}{NEF_{wo,NS}} = \sqrt{\frac{I_1+I_2}{\sqrt{I_1}+\sqrt{I_2}}\frac{1}{\sqrt{I_1}}} \; . \qquad (12)$$

Therefore, the maximum NEF improvement by a factor of 0.91 is achieved when $I_2 = (\sqrt{2}-1)^2 \cdot I_1$.

$$\left.\frac{NEF_{w,NS}}{NEF_{w/o,NS}}\right|_{min} = \sqrt{2\sqrt{2}-2} \approx 0.91 \qquad (13)$$

C. Current Reuse

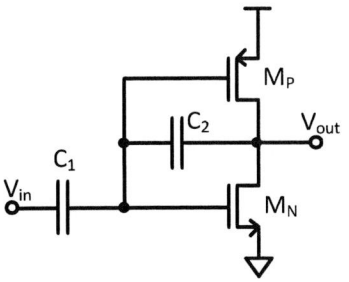

Figure 5. Noise reduction circuit employing current reuse technique

An example of amplifiers using the current reuse technique is depicted in Fig. 5. The output noise consists of the noise from both M$_P$ and M$_N$:

$$\overline{v_{no}^2} = 4kT\gamma(g_{mp} + g_{mn})R_{out}^2 \; . \qquad (14)$$

where g_{mp} and g_{mn} are the transconductances of M$_P$ and M$_N$, respectively. It is assumed that the M$_P$ and M$_N$ are matched so as to have the same transconductance. Hence, the noise-power trade-off of this circuit can be derived as

$$NEF_{w,NS} = \frac{\overline{v_{no}}}{g_{mp}R_{out} + g_{mn}R_{out}}\sqrt{I_{total}} \; . \qquad (15)$$

The NEF of the circuit using only NMOS or PMOS as the input transistor is given by

$$NEF_{w/o,NS} = \frac{4kT\gamma g_{mn}R_{out}^2}{g_{mn}R_{out}}\sqrt{I_{total}} \; . \qquad (16)$$

The ratio of NEFs with and without the technique applied is calculated as

$$\frac{NEF_{w,NS}}{NEF_{w/o,NS}} = \frac{\sqrt{2}}{2} \approx 0.71 \; . \qquad (17)$$

The NEF is improved by a factor of $\sqrt{2}/2$. Note that no other noise sources have been ignored during the analysis.

The results of the noise-power trade-off analysis performed for different noise reduction techniques are summarized in Table I. For the feedforward noise reduction technique, complete cancellation of the thermal noise from the input matching transistor is possible. However, the noise from the feedforward amplifier is not cancelled. This additional noise contribution can be minimized by increasing the current consumption of the feedforward amplifier, but at a penalty of degrading the current efficiency. For signal-nulled noise feedback technique, the noise from both main amplification and feedback amplification elements are suppressed, but none of them is completely cancelled. From Table I, we can conclude that among these noise reduction techniques, the current reuse technique is simple, but able to provide the maximum current efficiency, resulting in the best noise-power trade-off.

To further improve the noise-power efficiency, transistors operating in the weak inversion region are employed to maximize the current efficiency. However, this is often achieved at the expense of larger chip area. A combination of the noise reduction techniques discussed is also viable to improve the design.

TABLE I.   NOISE-POWER TRADE-OFF ANALYSIS SUMMARY

Noise reduction method	Minimum achievable NEF
Feed forward noise cancellation in [4]	$> \dfrac{\sqrt{2}}{2} NEF_{w/o,NS} \approx 0.71 \cdot NEF_{w/o,NS}$
Signal-nulled noise feedback in [5]	$\sqrt{2\sqrt{2}-2} NEF_{w/o,NS} \approx 0.91 \cdot NEF_{w/o,NS}$
Simple current reuse technique	$\dfrac{\sqrt{2}}{2} NEF_{w/o,NS} \approx 0.71 \cdot NEF_{w/o,NS}$

IV.   CIRCUIT IMPLEMENTATION

A low-power low-noise FEA with an aim to minimize the NEF was designed and implemented for use in the neural recording system. The designed circuit was fabricated in a 0.18-μm CMOS technology.

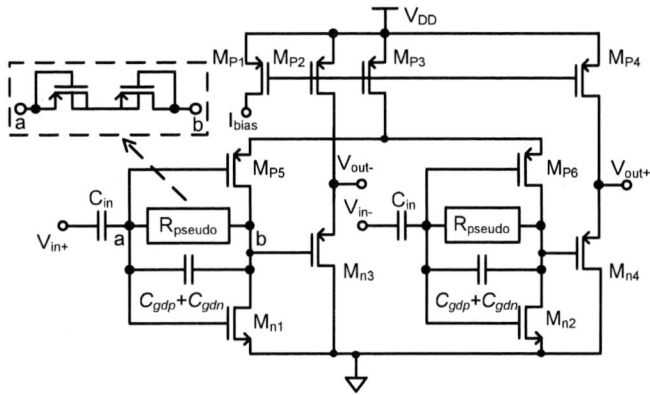

Figure 6.   Schematic of the neural recording amplifier with current reuse

The schematic of the neural recording amplifier is shown in Fig. 6, where the current reuse technique is employed. The input capacitor (C_{in}) and the parasitic gate-drain capacitance from both PMOS (C_{gdp}) and NMOS (C_{gdn}) input pairs form a closed-loop configuration. The gain of this amplifier is given by

$$A_v = \frac{C_{in}}{C_{gdp} + C_{gdn}} \quad . \tag{18}$$

Both PMOS and NMOS input pairs are biased in the weak inversion region to achieve better g_m/I efficiency. Furthermore, transistors that operate in weak inversion region give smaller thermal noise [9]. To suppress the flicker noise, the sizes of the input transistors were increased. In this design, the sizes of the input NMOS and PMOS transistors were designed to be 800 μm by 1 μm and 800 μm by 4 μm, respectively. The aspect ratio of the NMOS transistor is larger than that of the PMOS transistor because the NMOS generates approximately twice more flicker noise than the PMOS, given the same transistor size.

An output stage based on the PMOS source follower is used

to obtain a proper DC output level and improve the driving capability of the amplifier. The pseudo-resistors are implemented using two diode-connected PMOS transistors. Since the two PMOS transistors are connected in a balanced way, the variation of the resistance is smaller compared to the design in [8], when the voltage across the pseudo-resistor changes from -1 V to +1 V. Hence, the linearity of the amplifier is improved.

The simulation results are summarized in Table II. With the current reuse technique, the neural recording amplifier achieves an NEF of 1.88, which is much smaller than the theoretical lower limit of 2.9 derived in [8].

TABLE II.   SIMULATION RESULT SUMMARY

Parameters	Values
Gain	39.55 dB
Low cutoff frequency	0.1 Hz
Higher cutoff frequency	6.2 kHz
Input-referred noise	$3.8~\mu V_{rms}$
NEF	1.88
CMRR	> 38 dB
PSRR	> 100 dB
THD	0.66%

V.   CONCLUSION

A comparative study and analysis of various noise reduction techniques are presented. The power efficiencies of different noise reduction techniques are compared using the NEF. It is found that the noise reduction using current reuse technique has the best power efficiency among the methods discussed here. A neural recording FEA was also designed using the current reuse technique. The circuit simulation showed an excellent current efficiency, resulting in the NEF of 1.88.

REFERENCES

[1] C. Chestek, P. Samsukha, M. Tabib-Azar, R. Harrison, H. Chiel, and S. Garverick, "Wireless multi-channel sensor for Neurodynamic studies," in *Proc. IEEE Sensors Conf*, 2004, pp. 915–918.

[2] P. Mohseni, K. Najafi, S. J. Eliades, and X. Wang, "Wireless multichannel biopotential recording using an integrated FM telemetry circuit," *IEEE Trans. Neural Syst. Rehab. Eng.*, vol. 13, pp. 263–271, Sep. 2005.

[3] A. Nieder, "Miniature stereo radio transmitter for simultaneous recording of multiple single-neuron signals from behaving owls," *J. Neurosci. Methods*, vol. 101, pp. 157–164, 2000.

[4] F. Bruccoleri, E. A. M. Klumperink, and B. Nauta, "Wide-band CMOS low-noise amplifier exploiting thermal noise canceling," *IEEE J. Solid-State Circuits*, vol. 39, no. 2, pp. 275–282, 2004.

[5] L. Chin-Fu, *et al.*, "A noise-suppressed amplifier with a signal-nulled feedback for wideband applications," in *IEEE Asian Solid-State Circuits Conf. Dig. Tech. Papers*, Nov. 2008, pp. 453–456.

[6] M. S. J. Steyaert, and W. M. C. Sansen, "A micropower low-noise monolithic instrumentation amplifier for medical purposes," *IEEE J. Solid-State Circuits*, vol. 22, no. 6, pp. 1163–1168, 1987.

[7] C. F. Liao, and S. I. Liu, "A Broadband Noise-Canceling CMOS LNA for 3.1-10.6-GHz UWB Receivers," *IEEE J. Solid-State Circuits*, vol. 42, no. 2, pp. 329–339, 2007.

[8] R. R. Harrison, and C. Charles, "Low-Power Low-Noise CMOS Amplifier for Neural Recording Applications," *IEEE J. Solid-State Circuits*, vol. 38, no. 6, pp. 958–965, 2003.

[9] B. Gosselin, M. Sawan, and C. A. Chapman, "A Low-Power Integrated Bioamplifier With Active Low-Frequency Suppression", *IEEE Trans. Biomed. Eng.*, vol. 1, no. 3, pp. 184–192, Sept. 2007.

Temperature Insensitive Current Reference for the 6.27 MHz Oscillator

Wei-Bin Yang
Dept. of E. E.
Tamkang University
New Taipei City, Taiwan
robin@ee.tku.edu.tw

Zheng-Yi Huang
Dept. of E. E.
Tamkang University
New Taipei City, Taiwan
enjoyable_qq@hotmail.com

Ching-Tsan Cheng
Dept. of E. E.
Tamkang University
New Taipei City, Taiwan
t26213665@yahoo.com.tw

Yu-Lung Lo
Dept. of E. E.
National Kaohsiung Normal
University
Kaohsiung, Taiwan
yllo@nknu.edu.tw

Abstract—**This paper describes a circuit, which generates temperature-independent bias currents. In this paper, low- temperature coefficient reference is presented. The circuit is firstly employed to generate a current reference with temperature compensation, then combining the opposite characteristic curve current reference to minimize the variation of temperature. The proposed circuit has been design by a 0.18um CMOS technology process and using computer simulation to evaluate the thermal drift of the reference current. This current reference is used to provide a stable current for a current controlled oscillator(CCO). The proposed CCO achieves temperature coefficients of 22.3 ppm/℃ in the temperature range between -25 and 75℃.**

Keywords- *current reference, ring oscillator*

I. INTRODUCTION

Current and voltage references are indispensable circuit in analog, digital and power electronic systems. These should be designed stable as possible, the current and voltage references with high temperature immunity for proper operation. They are usually used to determine biasing points of sensitive analog circuits, for example amplifiers, oscillators, phase-locked loops (PLLs). Many high precision, temperature-independent reference circuits have been designed in the document over the last decades [1]-[3], many approaches have been made in order to design reliable current and voltage references in CMOS technology process.

Many CMOS current and voltage references consists of bipolar junction transistors (BJTs), which have been adapted to CMOS exploiting the parasitic lateral bipolar junction transistors in CMOS processes [4]-[5]. In addition, all-MOS voltage references have been proposed which use the thermal properties of MOS transistors worked in weak inversion region [6]or, recently the threshold voltage compensate the mobility temperature drift [7]-[11]. However, current references having MOS transistors operating in this region tend to have fairly large temperature coefficients. In these solutions, the circuits which want to obtain good temperature performance are complex and require a large area [12]-[14].

There is a design of current-controlled oscillator (CCO) in the following. To achieve the goal of compensating the frequency of CCO, an independent of temperature current reference is needed. When the proper current reference is brought in to the current-controlled oscillator, the frequency of CCO is compensated as following.

In this paper, the proposed circuit introduces a low-TC CMOS current reference utilizing MOS transistors operating in the weak inversion region and the current-controlled oscillator is compensated by the proper reference. The rest of this paper is organized as follows. Analysis for the proposed CMOS reference and current-controlled oscillator are both described in section II. Section III presents simulation results of several voltage and current references and oscillator to assess the performance. Finally, conclusions are given in Section IV.

II. CMOS CURRENT REFERENCE

A. MOSFETs in Weak Inversion Region

A model of the proposed CMOS reference can be used to describe the working of an n-channel MOS transistor in the weak inversion region [15]. Under The characteristic of an n-channel MOS transistor operating in the weak inversion region is similar to that of a BJT transistor and can be described as

$$I_D = I_{D0} S e^{q(V_{GS}-V_{TH})/nkT} \qquad (1)$$

where I_{DO} is the generation current, S is the geometrical shape factor of the transistor, q is the electron charge, n is a slope factor, k is the Boltzmann constant, T is the absolute temperature, V_{GS} is the gate-source voltage, and V_{th} is the threshold voltage of the transistor. From Eq. (1), the gate-source voltage of the MOSFET for a given drain current can be described as

$$V_{GS} = nV_T \ln \frac{I_D}{SI_D} + V_{TH} \qquad (2)$$

where V_T is the thermal voltage which is equal to kT/q. In this equation, the threshold voltage of the MOSFET can be described as [15]

$$V_{TH} = -\frac{kT}{q} \ln \frac{N_{D,poly}}{N_A} + \frac{2\sqrt{kTN_A\varepsilon_{si}\ln\frac{N_A}{n_i}-Q'_{SS}}}{C'_{OX}} \qquad (3)$$

where $N_{D,poly}$ is the doping concentration of donor atoms in the n+ poly gate and N_A is the doping concentration of acceptor atoms in the substrate, n_i is intrinsic carriers, ε_{si} is the relative dielectric constant of Silicon, Q'_{SS} is the surface-state charge, and $C'ox$ is the oxide capacitance per area.

978-1-61284-863-1/11 $26.00 © 2011 IEEE

Substituting Eq. (3) into Eq. (2) and taking the derivative of V_{GS} with respect to T, the temperature coefficient of V_{GS} can be written as

$$\frac{\partial V_{GS}}{\partial T} \approx n\frac{k}{q}\ln\frac{I_D}{SI_{D0}} - \frac{k}{q}\ln\frac{N_{D,poly}}{N_A}$$

$$= -\frac{k}{q}\ln\frac{N_{D,poly}(SI_{D0})^n}{N_A(I_D)^n} \qquad (4)$$

which indicates that the temperature coefficient of V_{GS} is a negative quantity.

B. Temperature Independent Current Reference

This section presents a temperature-compensated current reference, which is use an n-channel MOS transistor to operate in the weak inversion region [16]. In Fig.1, the circuit consists of a start-up circuit which is a PTAT current generator, a bandgap reference, and a zero-TC current replication circuit. The PTAT current generator generates a current proportional to absolute temperature, the value is given by

$$I_{PTAT} = \frac{nV_T}{R_1}\ln K \qquad (5)$$

where K is the size ratio of M_0 to M_1. The V_{ref} can be written as

$$V_{REF} = V_{GS,M9} + I_{PTC}R_2 \qquad (6)$$

The gate-source voltage of M_9 and the voltage drop across R_1 can be reduced by an n-channel MOS transistor work in the weak inversion region. The voltage drop across R_2 can be increased by a positive current. Therefore, the temperature compensation of V_{ref} is finished and the I_{ZTC} can be written as

$$I_{ZTC} = \frac{V_{REF}}{R_2} \qquad (7)$$

The current mirror comprising M_{11} and M_{12} copies I_{ZTC} to the output. The output reference current (I_{REF}) of the proposed current reference can be written as

$$I_{REF} = I_{ZTC}\frac{(W/L)_{12}}{(W/L)_{11}}$$

$$= \frac{1}{R_2 + R_3}(V_{GS,M9} + I_{PTAT}R_3)\frac{(W/L)_{12}}{(W/L)_{11}} \qquad (8)$$

C. The opposite Characteristic curve Temperature Independent Current Reference

This section design a temperature-compensated current reference [17], which has the opposite characteristic curve comparing with the literature [16]. In particular, the standard current reference in Fig. 2 In this circuit the diode-connected nMOS transistor M5 has been added. The KVL of this circuit structure can be expressed as

$$V_{GS1} + V_{GS5} - V_{GS2} - mR_1I = 0 \qquad (9)$$

gives

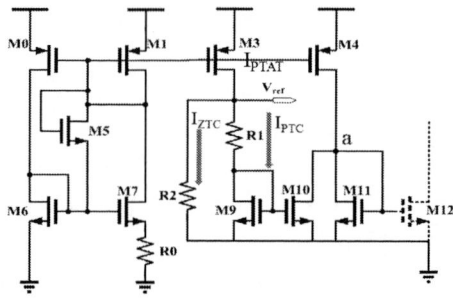

Fig. 1   Temperature compensation current reference

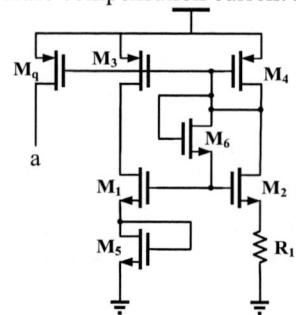

Fig. 2   The opposite characteristic curve temperature independent current reference

$$\sqrt{\frac{I}{\beta_{n0}}}\left(\frac{1}{\sqrt{\alpha_1}} + \frac{1}{\sqrt{\alpha_5}} - \sqrt{\frac{m}{\alpha_2}}\right) + V_{Tn} - mR_1I = 0 \qquad (10)$$

With reference to this circuit, the voltage drop across resistor R_1 is given by the sum of two terms with different temperature coefficients. One is related to the overdrive voltages of transistors M_2, M_3 and M_4 and has a positive temperature drift that is due to the negative drift of the mobility μ_n, while the other is the threshold voltage V_{Tn} whose temperature drift is related to different physical mechanisms [16]. Therefore, a reference current with a zero temperature coefficient can be obtained if the ratio and the size of these terms are properly chosen by design and temperature compensation is achievable. If the current ratio m is temperature independent, as in the standard MOS current mirror M_0-M_1, from Eq. 10, the temperature coefficient of the current can be expressed as

$$k_I = \frac{(2k_{VTn} + k_{u_n})V_{Tn} - (2k_{R1} + k_{u_n})mIR_1}{V_{Tn} + mIR_1} \qquad (11)$$

On the basis of (11), if

$$\frac{k_{u_n} + 2k_{V_{Tn}}}{k_{u_n} + 2k_{R_1}} > 0 \qquad (12)$$

k_I can be set to zero if

$$R_1 = \frac{V_{T_n}}{mI}\frac{k_{u_n} + 2k_{V_{Tn}}}{k_{u_n} + 2k_{R_1}} \qquad (13)$$

In conclusion, temperature compensation can be achieved and the opposite characteristic curve current reference comparing with the literature[16] can be obtained. The proposed circuit combined Fig. 1 and Fig. 2 at point A to minimize the temperature variation.

978-1-61284-863-1/11 $26.00 © 2011 IEEE

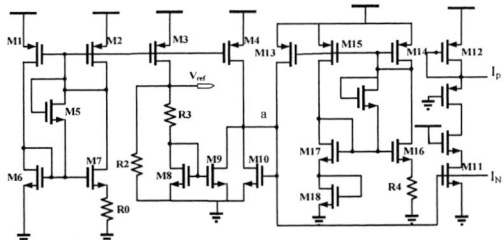

Fig. 3 Combination of the two architectures

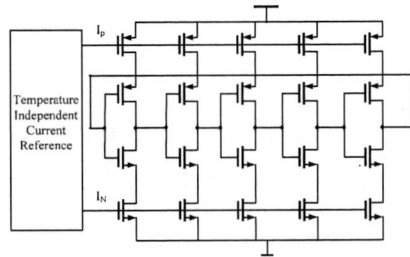

Fig. 4 Structure of CMOS ring oscillator with current-starved inverter stages

D. CURRENT CONTROLLED OSCILLATOR

The oscillation frequency of the ring oscillator, composed of N current-starved inverter stages, can be represented as[17]

$$f_{OSC} = \frac{1}{N \cdot \left(t_{PD\_rise} + t_{PD\_fall}\right)} = \frac{I_{source}}{N \cdot C_{load} \cdot V_{DD}} \quad (14)$$

Therefore, if I_{source} is stable with temperature drift, it can reduce the variation of oscillator's frequency obviously. Fig. 4 shows the structure of the ring oscillator, which is composed of an odd number of current-starved inverter stages with a temperature independent current reference.

III. SIMULATION RESULTS

The two kind of temperature-compensated current reference had been designed which have been presented in the previous Sections. The current references have been design -ed and simulated by HSPICE with reference to the models of the devices available in a 0.18 um CMOS technology.

The reference current versus temperature of temperature independent current reference is shown in Fig. 5 and the opposite characteristic curve current reference is shown in Fig. 6. It can be observed that the reference current, which has a nominal value of 1.315 uA at 25℃, has a residual temperature drift of about 0.96 nA in the temperature range between -25℃ and 75℃, in this range it shows a mean temperature drift of about 6.8 ppm/℃. In order to compensate the temperature drift of an oscillator itself, the temperature slope of proposed temperature independent current reference is slightly adjusted. The frequency of oscillator itself is proportional to temperature. Therefore the proposed current source is adjusted to complementary to temperature. Thus the currnet source is used to current controlled oscillator can achieve the oscillation frequency is 6.27 MHz and temperature drift of 14 kHz in range between -25℃ and 75℃. Based on this results, the temperature coefficients with the proposed oscillator is 22.3 ppm/℃. Thus the oscillation frequency has smaller temperature drift.

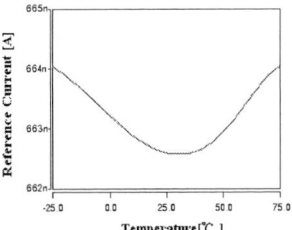

Fig. 5 Simulation plot of Fig. 1

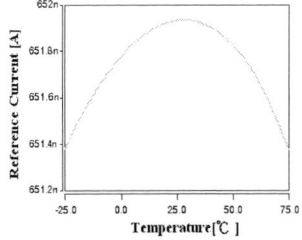

Fig. 6 Simulation plot of Fig.2

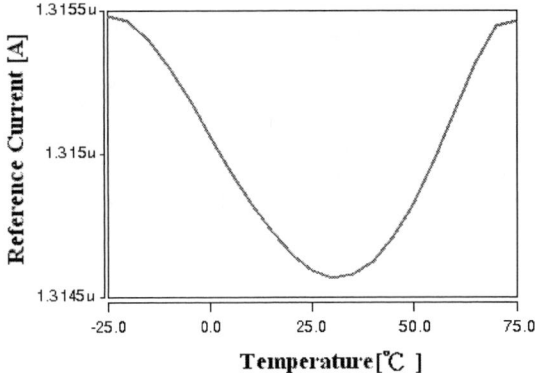

Fig. 7 Combination of the two architectures simulation plot

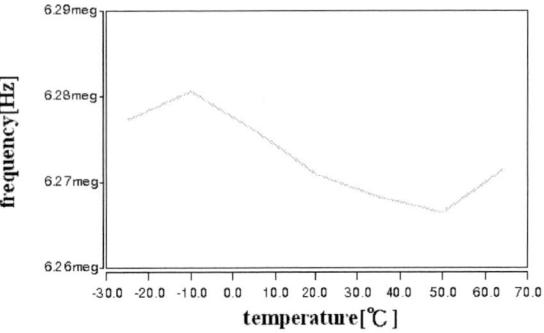

Fig. 8 The oscillation frequency versus temperature

IV. CONCLUSION

In this paper, the reference current source of combining two different temperature characteristic curve have been presented and analyzed. The temperature-compensated circuit, in particular, achieves a temperature drift of only 6.8 ppm/℃ in the temperature range between -25℃ and 75℃. The current controlled ring oscillator using this current source can oscillate to 6.27 MHz and the temperature coefficient 22.3 ppm/℃ is obtained. Both the current source and the current controlled ring oscillator have been designed by a 0.18 um CMOS technology process and their performances have been verified through computer simulations.

Acknowledgments

The authors would like to thank the National Chip Implementation Center and National Science Council, Taiwan, for fabricating this chip and supporting this work, respectively.

978-1-61284-863-1/11 $26.00 © 2011 IEEE   561

TABLE I
RING OSCILLATORS SPECIFICATION COMPARISON WITH REFERENCE

	Technology	Supply Voltage	Target Frequency	Temperature sensitivity	Power
This work	0.18 um	1.8 V	6.27 MHz	22.3 ppm/°C	32 uW
Y.-S. [18]	0.6 um	4 V	680 KHz	106 ppm/°C	0.4 mW
R. [19]	0.18 um	1.8 V	625 MHz	683 ppm/°C	0.59 mw
G. De [20]	0.35 um	1 V	80 KHz	824 ppm/°C	1.14 uW
K. R. [21]	0.13 um	3.3 V	1.25 GHz	340 ppm/°C	11 mW

V. REFERENCES

[1] R. J. Widlar, "New developments in IC voltage regulators, " *IEEE J. Solid-State Circuits*, vol. SC-6, no. 1, pp. 2–7, Feb. 1971.

[2] K. E. Kuijk, "A precision reference voltage source," *IEEE J. Solid-State Circuits*, vol. SC-8, no. 3, pp. 222–226, Jun. 1973.

[3] A. P. Brokaw, "A simple three-terminal IC bandgap reference," *IEEE J. Solid-State Circuits*, vol. SC-9, no. 6, pp. 388–393, Dec. 1974.

[4] Y. P. Tsividis and R. W. Ulmer, "A CMOS voltage reference, " *IEEE J.Solid-State Circuits*, vol. SC-13, no. 6, pp. 774–778, Dec. 1978.

[5] R. W. Ye and Y. P. Tsividis, "Bandgap voltage reference sources in CMOS technology," *Electron. Lett.*, vol. 18, pp. 24–25, Jan. 1982

[6] E. Vittoz and J. Fellrath, "CMOS analog integrated circuits based on weak inversion operation," *IEEE J. Solid-State Circuits*, vol. SC-12, no. 3, pp. 224–231, Jun. 1977.

[7] C.-H. Lee and H.-J. Park, "All-CMOS temperature-independent current reference," *Electon. Lett.*, vol. 32, pp. 1280–1281, Jul. 1996.

[8] W. M. Sansen, F. Op't Eynde, and M. Steyaert, "A CMOS temperature compensated current reference," *IEEE J. Solid-State Circuits*, vol. 23, no. 3, pp. 821–824, Jun. 1988.

[9] J. Georgiou and C. Toumazou, "A resistorless low current reference circuit for implantable devices," in *Proc. IEEE Int. Symp. Circuits and Systems 2002. ISCAS 2002*, vol. 3, pp. III-193–III-196.

[10] O. Cerid, S. Balkir, and G. Dundar, "Novel CMOS reference current generator," *Int. J. Electron.*, vol. 78, pp. 1113–1118, Jun. 1995.

[11] I. M. Filanovsky and A. Allam, "Mutual compensation of mobility and threshold voltage temperature effects with applications in CMOS," *IEEE Trans. Circuits Syst. I, Fundam. Theory Appl.*, vol. 48, no. 7, pp. 876–884, Jul. 2001.

[12] G. C. M. Meijer et al., "A new curvature corrected bandgap reference," *IEEE J. Solid-State Circuits*, vol. SC-17, no. 6, pp. 1139–1143, Dec.1982.

[13] B. Song and P. R. Gray, "A precision curvature-compensated CMOS bandgap reference, *IEEE J. Solid-State Circuits,* vol. SC-18, no. 6, pp. 634–643, Dec. 1983.

[14] M. Gunawan et al., "A curvature-corrected low-voltage bandgap reference,"*IEEE J. Solid-State Circuits*, vol. 28, no. 6, pp. 667–670, Jun.1993.

[15] R. J. Baker, "CMOS-Circuit Design, Layout and Simulation," 2nd ed., Piscataway, NJ: *IEEE Press,* 2005.

[16] Yoon-Suk Park; Hyoung-Rae Kim; Jae-Hyuk Oh; Yoon-Kyung Choi; Bai-Sun Kong;" Compact 0.7-V CMOS voltage current reference with 54/29-ppm/°C temperature coefficient," *SoC Design Conference (ISOCC)*, Page(s): 496 – 499,2009

[17] Karim Arabi and Bozena Kaminska, "Built-In Temperature Sensors for On-Line Thermal Monitoring of Microelectronics Structures", IEEE International Conference on VLSI in computers and processors, pp. 462-467, 1997

[18] Y.-S. Shyu and J.-C. Wu, "A process and temperature compensated ring oscillator," in *Proc. 1st IEEE Asia Pacific Conf.*, 1999, pp. 283–286.

[19] R. Vijayaraghavan, S. K. Islam, M. R. Haider, and L. Zuo, "Wideband injection-locked frequency divider based on a process and temperature compensated ring oscillator, " *IET Circuits, Devices & Syst.*, vol. 3, pp. 259–267, 2009.

[20] G. De Vita, F. Marraccini, and G. Iannaccone, "Low-voltage low-power CMOS oscillator with low temperature and process sensitivity," in *Proc. IEEE Int. Symp. Circuits and Systems (ISCAS 2007)*, 2007, pp. 2152–2155.

[21] K. R. Lakshmikumar, V. Mukundagiri, and S. L. J. Gierkink, "A process and temperature compensated two-stage ring oscillator, " in *Proc. IEEE Custom Integrated Circuits Conf. (CICC'07)*, 2007, pp. 691–694

A Novel Mode Switching Power Management System IC Design for Implantable Biomedical Instrumentations

Arnold C. Paglinawan*, Charmaine C. Paglinawan,
Glenn O. Avendaño
School of EECE, Mapúa Institute of Technology
Manila, Philippines
*ac_paglinawan@yahoo.com

Ying-Hsiang Wang, Wen-Yaw Chung
Dept. of Electronic Engineering, CYCU
Chung Li City. Taiwan, ROC

Abstract— A novel power management system with mode switching that widely applies to implantable biomedical devices is presented. Controlling the power transceiver output, monitor and simultaneously have special applications in the power path, are some of its major functions. Furthermore, a new temperature sensor and a temperature feedback regulated charger were designed for long term monitoring and small temperature variations. It also includes a system-on-enable (SOE) circuit and power feedback to increase system stability and accuracy. Moreover, constant power on the novel multi-level comparator was developed to reduce power and area consumption at the same time. The system also boasts a novel drug driver circuitry for implantable glucose biosensing applications. The whole system was fabricated in a TSMC 0.35μm 2P4M 3.3/5V CMOS technology it occupies a small area of 2.382 x 2.374 mm^2 (core size = 2.200 x 1.086 mm^2) and only dissipated 36.3mW of power. The transistors were designed in moderate inversion region to arrive at high performance and low power consumption. An in-depth analysis of the proposed system has been undertaken, and experimental results match the circuit simulations.

Keywords- Implantable biomedical system, Power management, Temperature sensor, Battery charger

I. INTRODUCTION

The current common biomedical implantable devices include pacemaker, left ventricular assist devices (LVAD), muscle stimulator, stimulating nervous system, cochlear implants, drug pumps, etc [1-3]. And there are basically two types of micro-power systems in such devices namely: 1) internal implantable battery and 2) wireless power transmission. The difference between these systems is whether to implant a battery in the human body or not. The first micro-power supply system is a combination of battery and microelectronic devices and packaging chamber to place inside the human body, when the batteries run out of energy then the patient will once again undergo a surgical operation to replace the battery. So in order to reduce the number of surgeries to replace the battery, the second type of micro-power system was developed. But using the wireless power transmission to replace the internal battery power supply will cause a long high-power energy transmission, resulting in skin abnormalities bringing much discomfort to the patient [1, 4].

With a leap forward in battery material technology, implantable devices have been further optimized due to the drastic reduction of size of the batteries, and at the same time, having rechargeable characteristics. This type of batteries (e.g. lithium-ion) can be combined with wireless power transmission now having dual characteristics of rechargeable energy storage implanted to the body [5]. Such a hybrid power system can reduce the case of battery replacement surgery and increase the time patients use this devices continuously, resulting in the reduction of high-power transmission continuously received by the body which causes discomfort [1, 4].

At present, consumer electronics power supplies with battery management are numerous but most of them in vitro and under conditions of room temperature. In contrast to consumer electronics, implantable devices emphasize different characteristics of the circuit, with special reference to noise, power, temperature, area, etc, does the need for a more stringent design consideration. At present, Medtronic, Bionics, ANS has been in the forefront of rechargeable power system which is applied to implantable biomedical devices [6, 7, 8, 9, 10]. The main parts of the power management system are the ff: 1) battery measurement system; 2) voltage regulator; 3) battery charger, 4) battery protection circuit, 5) watchdog device and 6) the use of programmable microcontrollers to be the interface logic control [6, 7, 8]. But in fact the optimal design of the implantable power management device is still in development, and the following must be taken into account: 1) the wireless transmission efficiency; 2) thermal effects; 3) charging efficiency; 4) battery-protection; 5) switching mode; 6) low power and 7) small area.

This proposed system has been designed for implantable systems. The following are the major features of the system: 1) includes a switch mode exchange interface; 2) avoided the use of microcontrollers as the core driver and control systems to reduce power and area consumption at the same time; 3) includes a self-timer driver circuit and charge pump, for long monitoring systems and applications such as high-voltage system, respectively; 4) adjustable charging waveform because the system used continuous temperature monitoring as a feedback control, so as to achieve low temperature and high efficiency charging effect of implant systems; 5) has a start-up circuit and power feedback regulation circuit for optimized power management system stability. Each and every block has been integrated in one chip for device reliability. Section 2 summarizes the full features of the system and the whole design methodology of the Power Management System Chip. The design implications of the power recovery block, temperature sensor block, battery charger block, mode interface and drug driver circuitry are tackled in Section 3. In order to verify the function of the Power Management System, the whole experimental methodology with the major results, together with the complete I/O configuration with its external components, are discussed in Section 4. In Section 5, the paper is finally concluded.

II. SYSTEM FEATURES AND DESIGN METHODOLOGY

A. System Architecture

Shown in Figure 1 is the system architecture of the implantable power management. It mainly contains three parts namely: 1) power supply system; 2) monitoring system and 3) battery charge system.

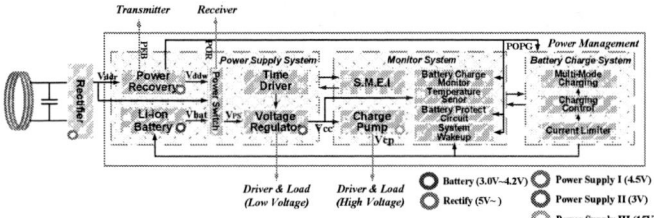

Figure 1. Implantable power management system architecture

After receiving the power from the outside source, this power is put into the power recovery circuit. And then converted to the internal system power supply (Vddw = 4.5V output voltage).

The power recovery circuit uses a low drop out voltage regulator circuit as its core circuit to improve power supply performance. The power recovery circuit mainly contains the ff: 1) over-voltage protection circuit, 2) voltage regulator circuits, and 3) power detection circuit and 4) Power feedback circuit. When the input voltage is too high (7.5V), using the over-voltage protection circuit inhibits the voltage at 7.5V, at the same time by using the power detection circuit feeds back the signal to external controller for proper power reduction. If the power coming to the circuit is appropriate and stable to start the system, the power recovery circuit will send a corresponding signal to the data receiver, that it is now ready to receive data. In order to prevent some data interference or any maloperation to happen, a delay signal will be used (Power on Reset, POR), to avoid signal maloperation situation happened.

Fig. 2 shows the Switch Mode Exchange Interface (SMEI) flowchart, it has six switching modes, respectively, 1) charge mode - it provides power to the internal battery and internal system power supply; 2) external power mode - will start if the internal battery does not need charging. It will directly use the external power to provide power for the whole system operation; 3) battery power mode - it will start if there is no external power received. It will then directly use the battery to provide whole system operation.; 4) protect mode - it will start

up in special circumstances, such as overheating, over-charging, etc.; 5) battery-saving mode - is used by the designer to whether it is time to start a fixed turn off feature, (example: turn off - when entering battery-saving mode, open – to enter power supply mode.); and 6) sleep mode - it will start if battery is fully discharged.

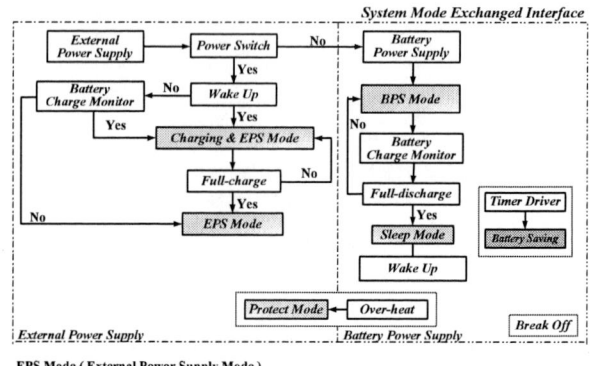

EPS Mode (External Power Supply Mode)
BPS Mode (Battery Power Supply Mode)

Figure 2. Flow chart of switch-mode power supply

III. CIRCUIT DESIGN AND CONSIDERATIONS

A. Power recovery circuit

When power is received externally from the body, it must be successfully restored through the circuits in order for the power to become a stable and reliable source. So the system for implantable devices must have specific specifications and functional considerations. The power recovery circuit, as shown in Figure 3, is composed of the ff: the start up circuits -, the power feedback – to give warning to the outside controller; the shunt voltage regulator (SVR) – for very large current; and the main voltage regulator circuit – that will increase the system stability and performance.

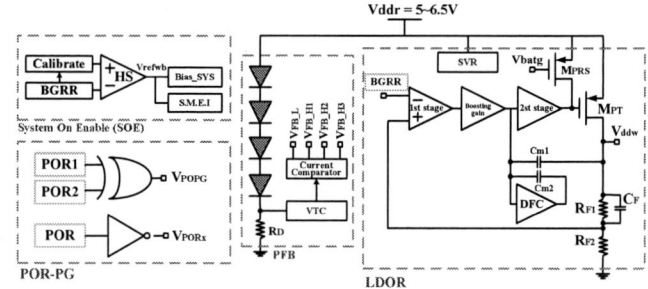

Figure 3. Power Recovery Circuit

The power recovery's start-up circuits are to be able to provide three output signals, namely: 1. The power signal to start the system (System on Enable, SOE); 2. The power to reset the starting signal (Power on Reset, POR) and 3. The power to start the pulse signal (Power on Pulse Generator, POPG). The purposes of the previous signals are: 1) for the sake of the implantable system can start to work after the receiver got the maximum voltage; 2) to avoid malfunction at low voltage. And also this system enable signal can operate at any circumstance including the internal battery's ability to

978-1-61284-863-1/11 $26.00 © 2011 IEEE

enter into sleep mode or protect mode. Lastly to be able to wake-up the system.

As mentioned above, in order to amend this error, a calibration circuit was designed. As shown in Figure 4 (a), the output signal of the BGR is connected to the calibration circuit and to the V- of the comparator. Then by using the transconductance characteristics of the transistor MPPS calibrate the V+ signals shown in Figure 4 (b). As clearly shown, this signal (i.e. calibrate signal (V+)) can bypass the early random voltage which is generated by the BGR to avoid the error in system's initialization.

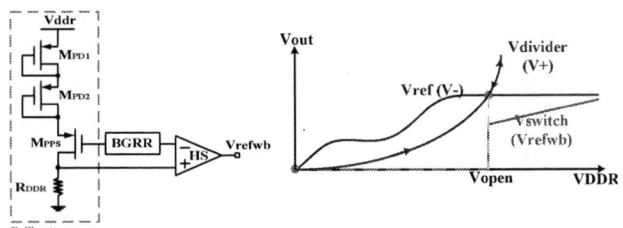

Figure 4. (a) Calibration circuit (b) Operating circumstances after calibration

B. Temperature Sensor Sub-System

Our body temperature does not change significantly under normal environmental conditions, the temperature sensor implanted into our body only needs to sense 35 °C ~ 42 °C, which is just a small range. And the changing range merely 6 °C ~ 8 °C (from 7°C ± 1° C). For this reason, the temperature sensor sub system will produce a precise signal output that corresponds to temperature. The structure of the temperature sensor sub system is illustrated in Fig. 5.

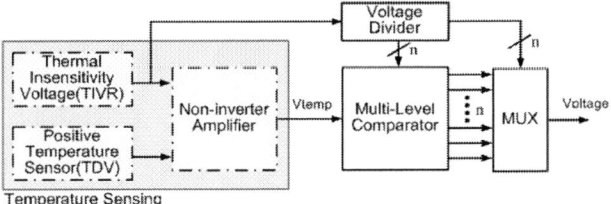

Figure 5. The structure of implantable temperature sensor

Normally, temperature sensors have problems regarding offset, especially in low voltage system. Nowadays, many researchers have work out this problem [2]-[4]. The way to solve this problem in our research is to first design two circuits: 1) a thermal insensitivity voltage reference (TIVR) and 2) a temperature-dependent voltage (TDV) circuit. Their output will serve as the inputs to the non-inverting amplifier. This non-inverting amplifier will then serve two purposes: 1) it will cancel the dc offset and 2) it will amplify the remaining small signal that we need (Fig. 5). The second part is that, we used a voltage divider across TIVR to make precise voltage levels which will serve as the reference levels for the multi-level comparator. Then, the output of the non-inverting amplifier, V_{temp}, will be compared to the precise voltage levels of the voltage divider, and be converted to a digital signal by using the multi-level comparator. And lastly, we utilized the same

precise voltage levels of the voltage divider and output of the multi-level comparator and used it as inputs to the multiplexer. The resulting output of the multiplexer will serve as our voltage temperature result.

IV. EXPERIMENTAL RESULTS AND DISCUSSIONS

The whole chip physical layout was done using Spring Soft's Laker™. It was verified with Mentor Graphics' Calibre™ and the post-layout simulation was done with Synopsys H-Spice™. A photomicrograph of the realized power management chip using full custom layout is shown in Fig. 6. The core die size is around 2.200 x 1.086 mm² and is packaged in a 48 pin DIP. The whole chip was fabricated using the TSMC 0.35μm 2P4M 3.3/5V CMOS technology.

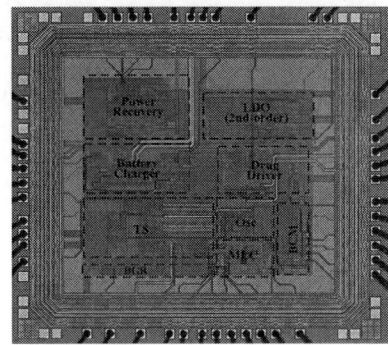

Figure 6. Microphotograph of the Fabricated Power Management Chip

NOTE:
TS – Temperature Sensor
BGR – Band Gap Reference
OSC – Oscillator
BCM – Battery Charge Meter
MLC – Multi-Level Comparator

In order to verify the performance of the power management system, computer simulations and actual test measurements were done. As shown in Figure 7, the result of all cases of switching mode system, divided into 15 sections (A ~ O). The following are the descriptions:

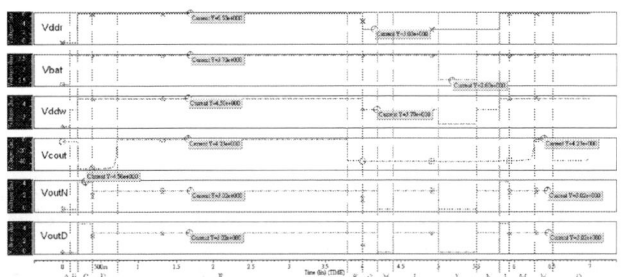

Figure 7. The Simulation Result of Switching Management Mode System

A. Initially there is no input power and battery supply is below 2.8V (over-discharge status). The Vddr (external power) can be seen as zero, resulting to Vddw (internal

978-1-61284-863-1/11 $26.00 © 2011 IEEE

power) to be at zero volts also. The V_{outN} (analog circuits power supply) and V_{outD} (digital circuits power supply) are at zero level, proving that there is no initial input power.

B. Even as the battery power is increased to 3.7V, the start-up circuit inside the chip still will not enable because all the other circuits (specifically the reference circuit) are still off. Therefore the whole system will still be in the lock out status and cannot activate (pre-implantation surgery battery load status).

C. The external power (Vddr = 6.5) was delivered and the battery power remained at 3.7V, Vddw increased steadily to 4.5V. Meanwhile the battery draws 35mA (charger's current specification) current from charger (not shown in the figure) but V_{cout} (battery charger's output) is still in the negative voltage (i.e. battery charger not activated). The V_{PORx} enable signal activates the reference circuits BGRR and BGRW resulting to the voltages V_{outN} and V_{outD} to be pulled to 4.5V.

D. After other circuits has been activated, the systems stability recovered, VoutN and VoutD output steadily goes down to the specified 3V, but the charger still does not activate successfully (i.e. delayed result).

E. Each part of the circuit at steady state, the Charger's V_{COUT} will be completely activated and be pulled up to 4.2V with supply current of 35mA.

F. As the charger's current is reduce to 100μA, the charger turns off.

G. The external power (Vddr) was pulled down to 3V (below the threshold voltage of 5V), thereby turning the power recovery circuit off resulting with Vddw decreased to the battery power supply of 3.7V.

H. To test the V_{timer}'s enabling performance, the second-order voltage regulator will be turn off, resulting to the voltages V_{outN} and V_{outD} to be pull down to the ground.

I. To test the V_{timer}'s disabling performance, the second-order voltage regulator must be turn on again, V_{outN} and V_{outD} will output 3V.

J. As battery power goes down to 2.6V (below the threshold of 3V), the whole system will enter the sleeping mode, V_{outN} and V_{outD} were pulled to ground potential, Vcout still in negative voltage.

K. As battery power goes back to 3.7V again, the whole system will recover to normal operation, V_{outN} and V_{outD} output of 3V, V_{ddw} remains at 3.7V.

L. As external power was returned to 6.5V, same as the initial state, V_{outN} and V_{outD} output is 4.5V, Vddw increased to 4.5V.

M. When each part of the system circuit recovers to the stable status, V_{POPG} has not enable the charger, so the charging function is still not active (i.e. V_{cout} negative voltage). (The draw current keeps on 100μA).

N. At this time V_{POPG} enabled the battery charger, Vcout output is at 4.2V. While the other circuits are still in stable status.

O. After the enabling of V_{POPG}, the current sensing circuit turns off the battery charger and pulls V_{cout} down to the negative voltage.

V. CONCLUSIONS

All aspects of power management and its considerations in implantable biomedical systems were developed. The whole system was fabricated in a single chip using the TSMC 0.35 μm 2P4M 3V/5V hybrid design process. The power management IC is subdivided into sub-systems namely: 1) power recovery system; 2) temperature sensor system; 3) the battery charging system; 4) application-specific system and 5) mode switching management systems. The power management interface, mainly located in the diode rectifier's output, allowed incoming signal range of 5 ~ 7.5V. The signal is sent back if higher than the 7.5 V limit using the power feedback circuit while the system reverts to the battery power if less than the % V limit. System-wide voltage regulator makes use of the damping-factor-control frequency compensation structure LDO regulator to avoid the use of large compensation capacitor.

The in vivo charger was stable even at changing temperatures making it suitable for implantable devices, but still need to consider the external wireless transmission for extended battery charging time on the influence on skin tissue abnormalities. With the mode switching management system, a microcontroller is not needed resulting to a significant power and size savings as compared to other general power management systems.

VI. ACKNOWLEDGEMENTS

The authors appreciate the technical support and chip fabrication from the National Chip Implementation Center (CIC) and TSMC and financial support from the National Science Council of the ROC.

REFERENCES

[1] O. Soykan, "Power sources for implantable medical devices," *Device Technology & Applications ELECTRONICS,* 2002.

[2] M. Sivaprakasam, W. Liu, M. S. Humayun, and J. D. Weiland, "A variable range bi-phasic current stimulus driver circuitry for an implantable retinal prosthetic device," *IEEE J SOLID-ST CIRC,* vol. 40, pp. 763-771, 2005

[3] J. Georgiou and C. Toumazou, "A 126-μW cochlear chip for a totally implantable system," *IEEE J SOLID-ST CIRC,* vol. 40, pp. 430-443, 2005.

[4] K. M. Alo, "Spinal Cord Stimulation for Complex Pain: Initial Experience with a Dual Electrode, Programmable, Internal Pulse Generator," *Pain Pract,* vol. 3, p. 31, 2003.

[5] Si, A. P. Hu, S. Malpas, and D. Budgett, "A Frequency Control Method for Regulating Wireless Power to Implantable Devices," *IEEE Transactions on Biomedical Circuits and Systems,* vol. 2, pp. 22-29, 2008.

[6] N. A. Torgerson and J. E. Riekels, "Battery recharge management for an implantable medical device," Google Patents, 2007.

[7] G. Echarri, R. Echarri, F. J. Barreras, and O. Jimenez, "Implantable power management system," Google Patents, 2001.

[8] P. M. Meadows, C. M. Woods, J. Chen, and H. Tsukamoto, "Implantable devices using rechargeable zero-volt technology lithium-ion batteries," Google Patents, 2007.

[9] T. Cameron, "Safety and efficacy of spinal cord stimulation for the treatment of chronic pain: a 20-year literature review," *J Neurosurg (Spine 3),* vol. 100, pp. 254-267, 2004.

[10] "International Research Foundation for RSD/CRP."

978-1-61284-863-1/11 $26.00 © 2011 IEEE

A 19-*n*W Sub-Bandgap Reference with 15ppm/°C Temperature Coefficient

Jia Hao Cheong, Minkyu Je, *Member, IEEE*

Integrated Circuits & Systems Laboratory
Institute of Microelectronics,
A*STAR (Agency for Science, Technology and Research), Singapore
cheongjh@ime.a-star.edu.sg

Abstract— **Driven by the increasing research interest in micropower devices such as implantable medical devices with battery operation and wirelessly powered wearable devices, voltage reference circuit which consumes power in the *n*W range has become important. In this paper, a 19-*n*W sub-bandgap reference circuit designed in 0.18-*μ*m CMOS process is presented. The sub-bandgap reference utilizes the difference of the gate-to-source voltage between a thick and a thin gate MOSFET to generate the temperature independent reference voltage output. Post-layout simulation shows that the sub-bandgap reference generates an output voltage of 201 *m*V with temperature coefficient of 15.4ppm/°C. It achieves a line regulation of 2%/V and a PSRR of -48dB at 10 *k*Hz frequency. The functionality of the circuit has been verified in measurement.**

Keywords- Bandgap, voltage reference, low power.

I. INTRODUCTION

Due to the rapid advancement in medical science in recent years, micro-power medical devices (e.g. implantable medical devices with battery operation and wearable devices with wirelessly powered mechanism) which consumes power from 1*μ*W to 10 *μ*W have attracted much research focus [1]. To obtain micro-power consumption, bandgap reference that is able to operate under low voltage and low power becomes important to provide more power headroom for other circuits [2].

In conventional bandgap reference design which uses bipolar junction transistors (BJT), the output voltage is normally greater than 1V (e.g. 1.25V) [3]. As the process goes into deep sub-micron and the component density increases as well as the low power, low voltage requirement for micro-power device, the supply voltage level below 1V is normally utilized. Hence, conventional bandgap reference voltage is not suitable [4-7]. As such, several sub-bandgap references have been developed in recent years for micro-power system.

Leung *et al.* developed a bandgap reference operating under 1V supply using resistor divider technique to reduce the input voltage [8]. Annema utilized dynamic threshold MOS to reduce the threshold voltage of the transistors in the bandgap reference [10]. Both techniques allow the bandgap reference to operate under 1V supply but they consume more than 1 *μ*W power. Ferreira *et al.* developed the voltage reference using

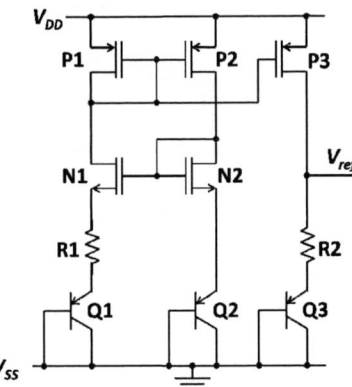

Fig. 1. Conventional bandgap reference using BJTs.

sub-threshold transistors but the power consumption is still limited by the resistors in the circuit [10].

Yan *et al.* developed a nano-watt sub-bandgap reference using the different temperature characteristics of a thick gate and thin gate MOSFET operating under subthreshold region [2]. In this paper, we applied similar concept in a resistor-less current reference circuit to achieve a sub-bandgap voltage reference with lower power consumption. Section II shows the subthreshold characteristic of MOSFETs. Section III described the proposed sub-bandgap reference circuit. Post-layout simulation results are shown in Section IV whereas the measurement results are shown in Section V. Section VI concludes the paper.

II. SUBTHRESHOLD OPERATION OF MOSFETs

Conventional bandgap voltage reference utilizes the base-emitter junction voltage (V_{BE}) of bipolar junction transistor (BJT) and a voltage that is proportional to absolute temperature (PTAT) to generate the temperature independent voltage as shown in Fig. 1. The base-emitter junction voltage of BJT is inversely proportional to the temperature and the thermal voltage (V_T) is directly proportional to the temperature. By summing them in appropriate proportion, a low temperature dependence reference voltage (V_{ref}) can be obtained as shown in (1).

$$V_{ref} = V_{EB3} + \frac{R_2}{R_1} V_T \ln(m) \qquad (1)$$

978-1-61284-863-1/11 $26.00 © 2011 IEEE

The resulting reference voltage is around 1.2V, which is the energy bandgap value of the silicon. Hence, it is not suitable for circuit working in sub-1V range.

In order to obtain reference voltage in sub-1V range, MOSFETs working under subthreshold region (also known as weak inversion) has been utilized [2], [10]. Under subthreshold operation, current flows when surface potential (ϕ_S) at the source end of the channel is less than twice the bulk Fermi potential (ϕ_B) [11]. The gate-source voltage (V_{GS}) of MOSFET in subthreshold operation has similar temperature characteristic as BJTs. MOSFETs operate in this region when V_{GS} is less than the threshold voltage (V_{TH}) and the drain current (I_D) has an exponential relation with V_{GS} as shown in (2)

$$I_D = I_0 e^{\left(\frac{V_{GS}-V_{TH}}{nV_T}\right)}\left(1-e^{-\frac{V_{DS}}{V_T}}\right) \qquad (2)$$

where $$I_0 = \mu \frac{W}{L}\sqrt{\frac{q\varepsilon_{si}N_{ch}}{2\varphi_S}}V_T^2 e^{\frac{(n-2)V_{BS}}{nV_T}} \qquad (3)$$

is a process dependent characteristic current [12], V_T is the thermal voltage which is 26 mV at room temperature, W is the effective width of the channel, L is the effective length of the channel, n is the subthreshold slope factor, and N_{ch} is the channel doping concentration [11], [13].

In weak inversion, MOSFET saturates when V_{DS} is greater than 3 times of V_T [15]. Assuming V_{BS} to be zero and V_{DS} is greater than $3V_T$, the drain current in subthreshold operation can be expressed as

$$I_D = I_0 e^{\left(\frac{V_{GS}-V_{TH}}{nV_T}\right)} \qquad (4)$$

The MOSFET operating under subthreshold region has similar temperature characteristic as BJT. Hence it can replace the BJT in the conventional bandgap reference to generate temperature low-dependent voltage.

III. PROPOSED VOLTAGE REFERENCE

A well known CMOS current reference circuit which uses

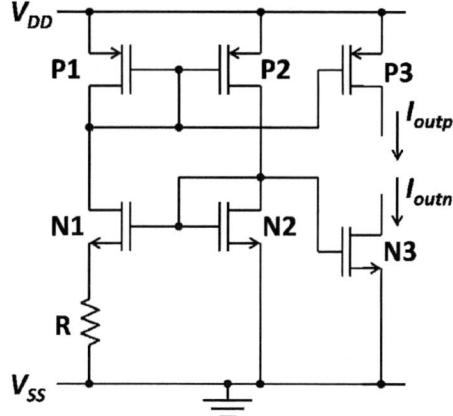

Fig. 2. Conventional CMOS current reference circuit.

only one resistor is shown in Fig. 2 [15]. PMOS, $P1$ and $P2$ act as current mirror. NMOS $N1$ and $N2$ are in subthreshold region. The NMOS $N1$ has larger W/L ratio compared to $N2$ and the current is defined by the V_{GS} difference of $N1$ and $N2$ as well as the resistor, R. The resistor is a drawback for some applications. If low current is required, large resistor would be needed, which will take up large area.

Based on the conventional design, Oguey et al. developed a resistor-less current reference by replacing the resistor, R, with an NMOS, $N5$, as shown in Fig. 3 [16]. The NMOS $N5$ is biased such that it is operating in the triode region. Additional PMOS, $P4$ and NMOS, $N4$ provide the bias voltage for $N5$.

In our sub-bandgap reference design, a thick-gate and a thin-gate NMOS are used to replace the $N1$ and $N2$ in the current reference circuit developed by Oguey et al. to generate the reference voltage as shown in Fig. 4.

The voltage reference circuit consists of two blocks – core circuit and start-up circuit. The core circuit is composed of transistors $N1$~$N4$ and $P1$~$P3$. Thick-gate transistors are utilized to achieve low current consumption with reasonable sizing of the transistors. Only one thin gate transistor, $N2$ is utilized to generate the temperature invariant reference voltage.

From (4), the gate-to-source voltage of a transistor under subthreshold operation can be expressed as

$$V_{GS} = V_{th} + nV_T \ln\left(\frac{I_D}{I_o}\cdot\frac{L}{W}\right) \qquad (5)$$

The output voltage, V_{out}, of the sub-bandgap reference circuit is equal to the difference between the gate-to-source voltages of the thick gate transistor, $N1$, and the thin gate transistor $N2$ [2]

$$V_{out} = V_{GS,tk} - V_{GS,tn} \qquad (6)$$

$$V_{out} = V_{th,tk} - V_{th,tn} + V_T \ln\left[\frac{I_{o,tn}\left(W_{tn}/L_{tn}\right)}{I_{o,tk}\left(W_{tk}/L_{tk}\right)}\right] \qquad (7)$$

The threshold voltage of NMOS is inversely proportional to the temperature and its derivative can be expressed as

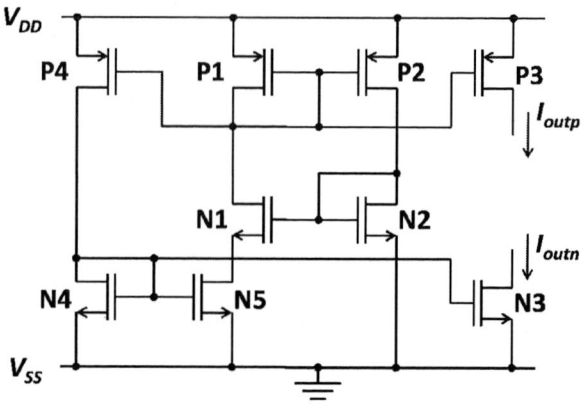

Fig. 3. Resistor-less current reference circuit.

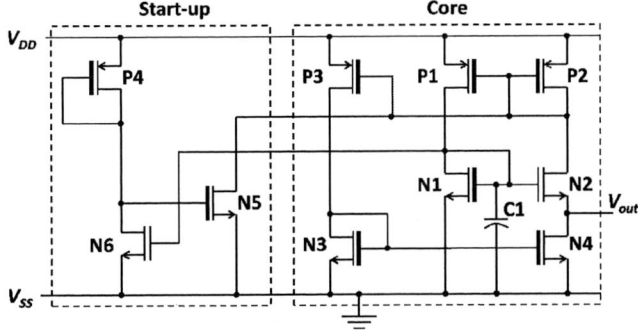

Fig. 4. Schematic of the proposed sub-bandgap reference circuit.

$$\frac{dV_{th}}{dT} = -\frac{1}{T}\left[\frac{E_g}{2q} - \phi_f\right]\left[2 + \frac{\gamma}{\sqrt{2\phi_f}}\right] \qquad (8)$$

On the other hand, the thermal voltage, V_T, is directly proportional to temperature and it is equal to kT/q. Therefore, by properly designing the size of the two transistors, $N1$ and $N2$, the output voltage can have better immunity on temperature's variation. In this design, $N2$ is 25 times larger than $N1$. The current flowing in each branch of the circuit is set to be 5.5 nA to achieve a total current consumption of 20 nA. A 1pF capacitor $C1$ is added to the circuit to improve the power-supply-rejection-ratio (PSRR) of the circuit.

The start-up circuit is composed by $P5$, $N6$ and $N7$. During start-up, $P5$ provides the current to charge up the gate voltage of $N6$. When $N6$ is turned on, it pulls down the gate voltage of $P1$, $P2$ and $P4$, so that the voltage reference circuit starts to operate. Under steady state, $N7$ will be turned on and pulled

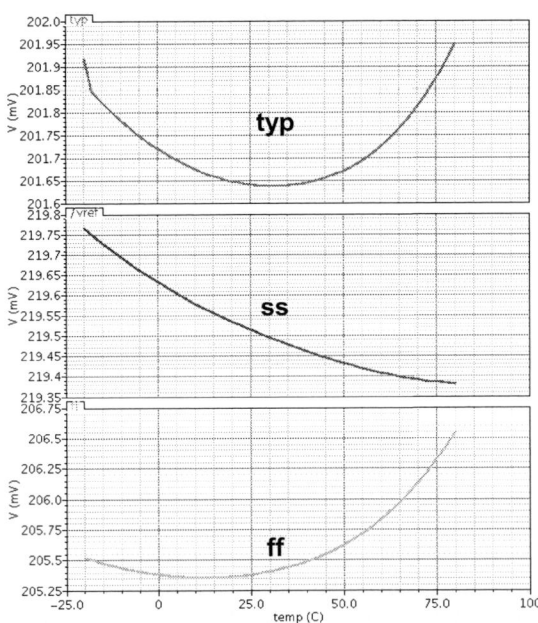

Fig. 5. Temperature variation post-layout simulation results of the sub-bandgap reference over typical, slow-slow and fast-fast corners.

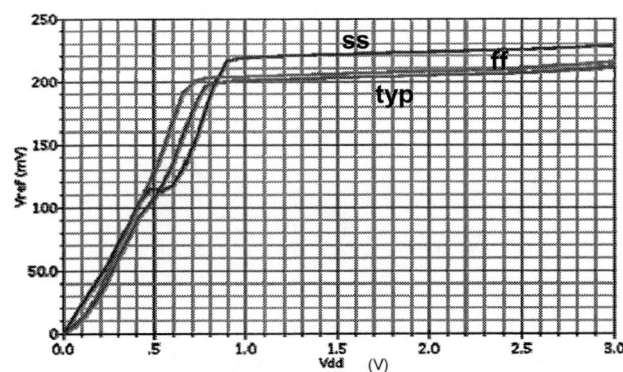

Fig. 6. Line regulation post-layout simulation result over typical, slow-slow and fast-fast corners.

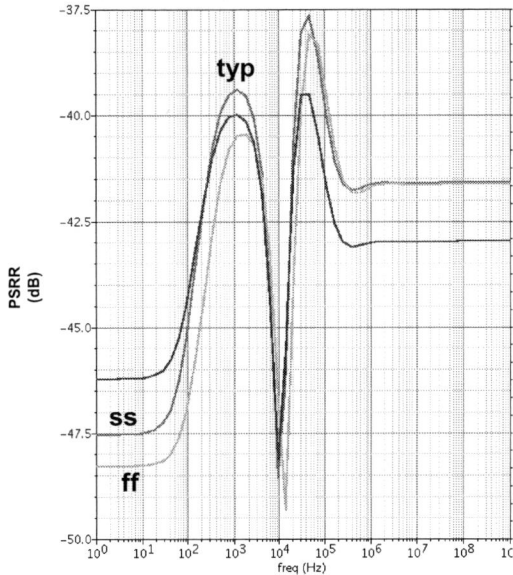

Fig. 7. PSRR post-layout simulation results of the sub-bandgap reference over typical, slow-slow and fast-fast corners.

down the gate voltage of $N6$, causing it to turn off.

IV. SIMULATION RESULTS

The sub-bandgap reference circuit was designed using CMOS 0.18μm process under 1.2V supply voltage. The simulations were done over 3 corners. The circuit consumes a total power of 19 nW under typical condition. The output voltage of the sub-bandgap reference is 201 mV under typical condition. The voltage variation across process is less than 9%. The post-layout simulation of the output voltage variation over temperature from -20°C to 80°C is shown in Fig. 5. The temperature coefficient for the typical simulation is 15.4ppm/°C.

The line regulation simulation result of the sub-bandgap reference was obtained by sweeping the DC supply voltage from 0V to 3V as shown in Fig. 6. The line regulation performance of the sub-bandgap reference is 2%/V under typical condition.

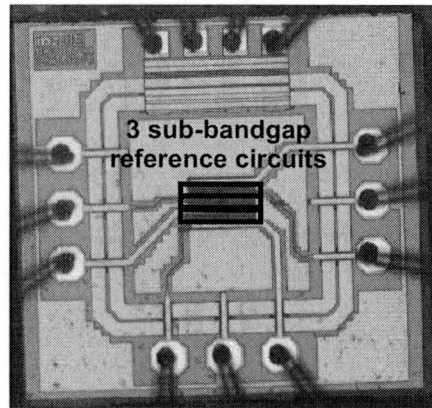

3 sub-bandgap reference circuits

Fig. 5. Die micrograph of the sub-bandgap reference circuit.

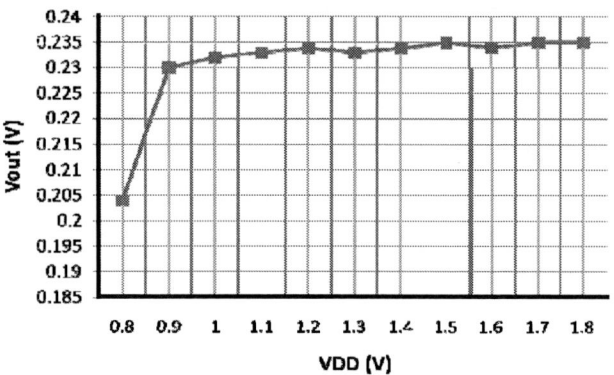

Fig. 6. Line regulation measurement result.

The power-supply rejection ratio (PSRR) simulation result of the sub-bandgap reference is shown in Fig. 7. The PSRR is less than -37.5dB across all frequency.

V. MEASUREMENT RESULTS

The proposed sub-bandgap reference circuit was fabricated with 0.18-μm CMOS process. The die micrograph is shown in Fig. 8. The sub-bandgap reference occupies an active area of 120μm × 170μm. The power consumption of the sub-bandgap reference circuit is 24 nW under 1.2V supply. The line regulation measurement result of the chip is presented in Fig. 9. The output voltage of the sub-bandgap reference is 232 mV and it stays constant over different supply voltage.

VI. CONSLUSION

A low power sub-bandgap reference circuit has been designed and fabricated with 0.18-μm CMOS process. The sub-bandgap reference utilizes the gate-to-source voltage differences of a thick and a thin gate MOSFETs in a current reference circuit structure to generate the temperature invariant reference voltage output. Post-layout simulation shows that the sub-bandgap reference circuit generates a reference voltage of 200 mV with temperature coefficient of 15ppm/°C while consuming a power of 19 nW. The functionality of the sub-bandgap reference circuit has also been verified in measurement. Table 1 shows the summary of the post-layout performance of the sub-bandgap reference circuit.

Table I
Post-layout performance summary

Supply voltage	1.2 V
Power	19.03nW
Reference voltage	201.64mV
Temperature Coefficient (-20~80°C)	15.4ppm/°C
PSRR @ 10kHz	-48.03dB
Line regulation	2%/V

REFERENCES

[1] N. Verma, D. C. Daly A. P. Chandrakasan, "Ultralow-power Electronics for Biomedical Applications," *Ann. Rev. Biomed. Eng.*, no. 10, pp. 247-274, 2008.

[2] W. Li and R. Liu W. Yan, "Nanopower CMOS sub-bandgap referece with 11ppm/°C temperature coefficient," *Electronics Letter*, vol. 45, no. 12, June 2009.

[3] B.-S. Song and P. R. Gray, "A Precision Curvature-Compensated CMOS Bandgap Reference," *IEEE. Journal of Solid-State Circuits*, vol. 34, no. 5, pp. 670-674, May 1999.

[4] L. MacEachern S. Miller, "A Nanowatt Bandgap Voltage Reference For Ultra-Low Power Applications," in *IEEE International Symposium on Circuits and Systems*, September 2006, pp. 645 - 648.

[5] S. J. Stratz C. J. B. Fayomi, "Novel Approach to Low-Voltage Low-Power Bandgap Reference Voltage in Standard CMOS Process," in *IEEE International Conference on Electronics, Circuits and Systems*, Dec 2006, pp. 208-211.

[6] E. K. F. Lee, "A Low Voltage CMOS Bandgap Reference Without Using An Opamp," in *IEEE International Symposium Circuits and Systems*, 2009, pp. 2533-2536.

[7] P. K. T. Mok and K. N. Leung, "Design Considerations of Recent Advanced Low-Voltage Low-Temperature-Coefficient CMOS Bandgap Voltage Reference," in *IEEE Custom Integrated Circuits Conference*, 2004, pp. 635-642.

[8] K. N. Leung and P. K. T. Mok, "A Sub-1-V 15-ppm/°C CMOS Bandgap Voltage Reference Without Requiring Low Threshold Voltage Device," *IEEE Journal of Solid-State Circuits*, vol. 37, no. 4, pp. 526-530, April 2002.

[9] A.-J. Annema, "Low Power Bandgap References Featuring DTMOSTs," *IEEE Journal of Solid-State Circuits*, vol. 34, no. 7, pp. 949-955, July 1999.

[10] L.H.C. Ferreira and T.C. Pimenta, "A CMOS Voltage Reference Based on Threshold Voltage for Ultra Low-Voltage and Ultra Low-Power," in *International Conference on Microelectronics*, 2005, pp. 10-12.

[11] M. Nishida and H. Ohyabu, "Temperature Dependence of MOSFET Characteristics in Weak Inversion," *IEEE Trans. Electron Devices* , vol. ED-24, no. 10, pp. 1245-1248, 1977.

[12] C. Hu and A. Niknejad, *Modeling and BSIM4.4.0, MOSFET Model - User's Manual.*: University of California, Berkeley, 2004.

[13] H. W. Li and D. E. Boyce R. J. Baker, *CMOS Circuit Design, Layout, and Simulation.*: IEEE Press, 1998.

[14] P. E. Allen and N. R. Strader R. L. Geiger, *VLSI- Design Techniques for Analog and Digital Circuits.*: McGraw-Hill, 1990.

[15] E. Vittoz and J. Fellrath, "CMOS analog circuits based on weak inversion operation," *IEEE J. Solid-State Circuits*, vol. SC-12, pp. 224-231, 1977.

[16] D. Aebischer H. J. Oguey, "CMOS Current Reference Without Resistance," *IEEE Journal of Solid-State Circuits*, vol. 32, no. 7, pp. 1132-1135, July 1997.

978-1-61284-863-1/11 $26.00 © 2011 IEEE

Double Regulated Voltage Supply for High Precision MEMS Accelerometers

Huey Jen Lim[1], Ravinder Pal Singh[1], Kevin Chai Tshun Chuan[1], David Nuttman[2] and Minkyu Je[1]

[1]Institute of Microelectronics, A*STAR (Agency for Science, Technology and Research)
11 Science Park Road, Singapore 117685
[2]Physical Logic Ltd, 31 Halekhi St. Bnei-Brak, 51200, Israel

limhj@ime.a-star.edu.sg

Abstract—**This work presents the power management block using a double regulated voltage supply design in order to meet the stringent requirement for high precision MEMS accelerometers ASIC. The architecture employs two stages of supply regulation using two bandgap circuits and two voltage regulators to achieve high PSRR (>100dB) and high line regulation (10µV/V) which are required for the ASIC to achieve high linearity and high dynamic range. The design is implemented in 6M1P TowerSemi 0.18µm CMOS process.**

Keywords-Low drop-out regulator (LDO), bandgap voltage reference, power-supply rejection ratio (PSRR)

I. INTRODUCTION

MEMS accelerometer has gained tremendous visibility in the industry over the past decade and has been widely applied in areas of motion sensing such as Inertial Measuring Units (IMUs) in navigation, airbag deployment in automobile, smartphones etc. The read-out circuit of a high precision capacitive sensing MEMS accelerometer is a key-challenge to achieving high linearity and high dynamic range for the whole system [1]. With very stringent specifications of the MEMS driver ASIC, a power management block design with high PSRR, line regulation, load regulation and low noise is required.

The study of power management techniques has increased spectacularly within the last few years corresponding to a vast increase in the use of portable, handheld battery operated devices. A power management system may contain several subsystems including linear regulators, switching regulators, and control logic. The control logic changes the attributes of each subsystem; turning the outputs on and off as well as changing the output voltage levels, to optimize the power consumption of the device. The nature of the linear regulator makes it appropriate for use in this application. Linear voltage regulator with high rejection to power supply variations is necessary for accurate or critical analog signal processing blocks inside this high-precision sensor system. In the implementation of a linear regulator, it has several advantages over a switching regulator in terms of simplicity, cost, switching noise, and electromagnetic compatibility.

II. CONVENTIONAL LINEAR VOLTAGE REGULATOR

A. Operational Overview of Voltage Regulator

A voltage regulator is a constant voltage source that adjusts its internal resistance to any occurring changes of load resistance to provide a constant voltage at the regulator's output. Designing a stable LDO for a wide range of load conditions, while achieving high power-supply rejection ratio (PSRR), low drop-out voltage, and low quiescent current, is the main target using current state-of-the-art CMOS technologies. A conventional CMOS voltage regulator implementation is shown in Figure 1. It comprises of a driving pass transistor, an op-amp, a bandgap reference circuit, a feedback resistor network (R1 and R2) to obtain the desired regulated output and a loading capacitor. The op-amp, which senses the difference between the bandgap reference voltage and the fraction of regulator output voltage, establishes a feedback loop to provide a regulated output. A large external output loading capacitor (Cload) creates the dominant pole to ensure stability and it provides an instantaneous charge source during fast load transient.

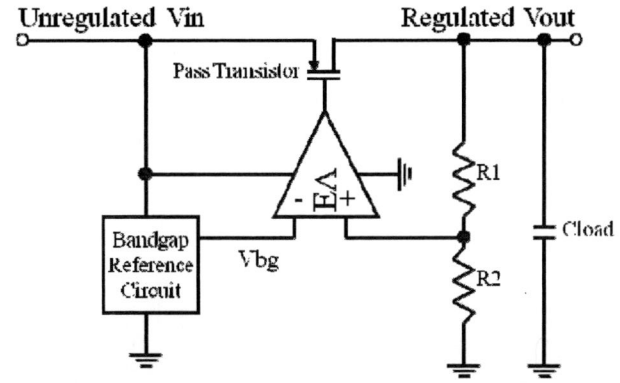

Figure 1.   Conventional linear voltage regulator

B. PSRR of Voltage Regulator

PSRR is the reverse gain of the linear regulator's output ripple over the input ripple at a particular frequency, i.e., its ability to eliminate output ripple caused by input variations. High PSRR values are desirable over the frequency range critical to the system application. The op-amp in conjunction with the transistor feedback loop provides immunity against supply noise, because it serves as a negative feedback structure with respect to the supply path. At dc, high PSR will be obtained from a large loop gain from the negative feedback path. Due to the reduction in loop gain at high frequencies, spurious signals can easily penetrate into the circuit. [2]

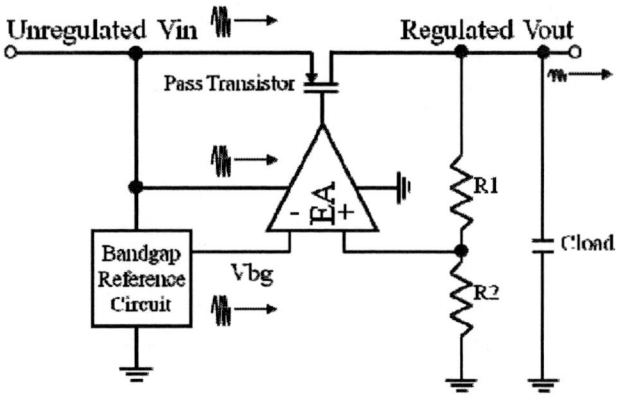

Figure 2.   Sources of ripple pass through regulator

The finite PSRR of the conventional regulator is due to various paths that could couple input ripples to the output as shown in Figure 2. The input ripple passes through the pass transistor due to its finite conductance as well as the regulated loop. The finite PSRR of the error amplifier and the bandgap circuit also will pass the input ripple to the output. [3]

III.   PROPOSED DOUBLE STAGE REGULATION ARCHITECTURE

A.   Proposed Architecture

In order to achieve the high PSRR and line regulation of the power supply, a double regulated technique is adopted. The proposed architecture is shown in Figure 3. The unregulated 3.3V supply input voltage is first regulated to an intermediate voltage of 2.5V and then further regulated to the output voltage of 1.8V in the second stage. The regulated 1.8V output voltage is the power source for all circuit processing blocks inside the ASIC.

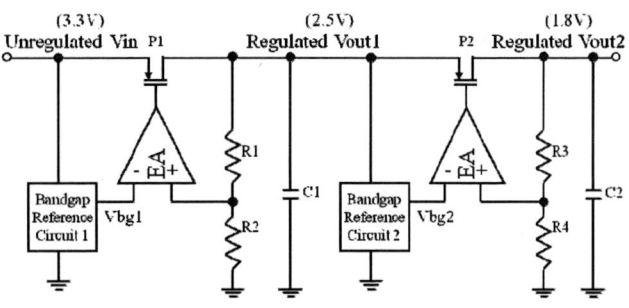

Figure 3.   Proposed double stage regulation architecture

The advantages of double stage regulation are as below:

- The higher the open-loop gain of the error amplifier, the better the PSRR of the regulator. By cascading two regulators in series, the open-loop gain of the whole system is almost doubled and higher PSRR is obtained.

- The PSRR of the bandgap is a critical contributor to the PSRR of the regulator, up to the bandgap's roll-off frequency. In this double regulation design, the PSRR of the bandgap circuit at the second stage, which is powered by the first stage output, is greatly improved. Thus, the limitation due to this factor is very much relaxed.

- The ripple that pass through the first stage will be regulated again in the second stage again and thus further reduce the input ripple to the output.

B.   Bandgap Reference Voltage Circuits

Two bandgap reference voltage circuits with different characteristics are used for the two stages of regulation. The circuits are shown in Figure 4.

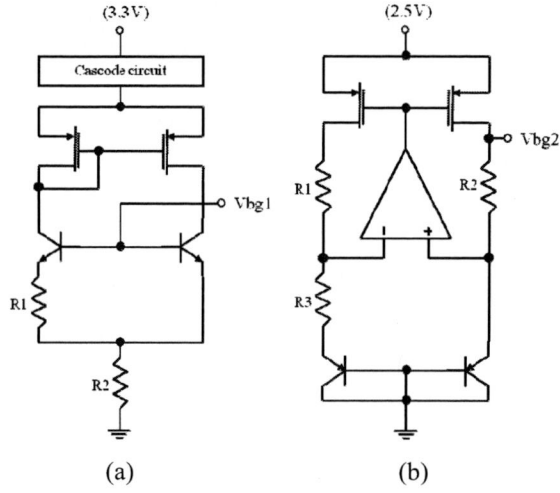

Figure 4.   Schematics of (a) conventional Brokaw bandgap circuit for 1st stage and (b) conventional op-amp based bandgap circuit for 2nd stage

Since, the bandgap output ripple gets amplified by the error amplifier, which in turn propagates the amplified ripple through the pass element to the output, a bandgap reference voltage circuit with high PSRR is desired in the first stage of regulation. A conventional Brokaw bandgap circuit [4] is used as shown in Figure 4(a). Cascode structure is adopted in the biasing currents design for the circuit to achieve higher PSRR.

For the second stage regulation, a bandgap reference voltage that could sustain low voltage operation is required. A conventional op-amp based bandgap circuit as shown in Figure 4(b) is used [5]. The circuit can function under supply voltage as low as the bandgap output voltage. The PSRR performance of the second bandgap reference circuit is not that critical as compared with the first stage and will be further discussed in Section IV.

C. Error Amplifier

The folded-cascode operational amplifier is adopted for the error amplifier design in order to achieve high loop gain. Close loop simulations are carried out for regulators in both the stages to study the stability of the whole system. Both the error amplifiers in the two stages are power-up by 3.3V in order to provide high gain and voltage overdrive signal for the pass transistor.

IV. SIMULATION RESULTS

The system is simulated, optimized and verified against all specifications, including load regulation, line regulation, temperature dependence, PSRR and transient response. The simulation results of PSRR and line regulation discussed in detailed below are unique for this double stage regulation.

A. PSRR Simulation

The PSRR of the system is simulated from frequency 1Hz to 1GHz as shown in Figure 5. Four nodes are monitored during the simulations, i.e. PSRR_Vbg1 (first stage bandgap voltage), PSRR_INT (first stage regulation output), PSRR_Vbg2 (second stage bandgap voltage) and PSRR_AVDD (second stage regulation output). With the cascode biasing circuit for the bandgap circuit in the first stage to isolate the fluctuation of unregulated input power supply, a high PSRR of 107dB at DC can be achieved. The PSRR of the intermediate output voltage of the first stage is thus mostly determined by the open loop gain of the error amplifier and a PSRR of 76dB at DC is obtained from simulation. With the bandgap circuit at second stage being powered by the first stage output, an even higher PSRR of 134dB at DC is achieved. The big improvement in PSRR of bandgap in second stage has greatly reduced the limiting factor as discussed earlier and made room for the output of second stage to achieve higher PSRR. A simulated PSRR result of >100dB below 2.8kHz could be achieved for the second stage output.

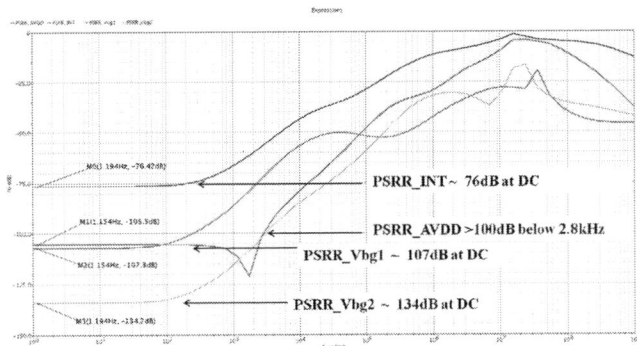

Figure 5. PSRR simulation result

B. Line Regulation Simulation

Another important parameter which is crucial to the linearity of the analog front end sensing circuit is the line regulation of the power supply. The system is tested with input power supply varies from 3.0V to 3.6V and the corresponding output variation is measured. The same four nodes are monitored during simulation.

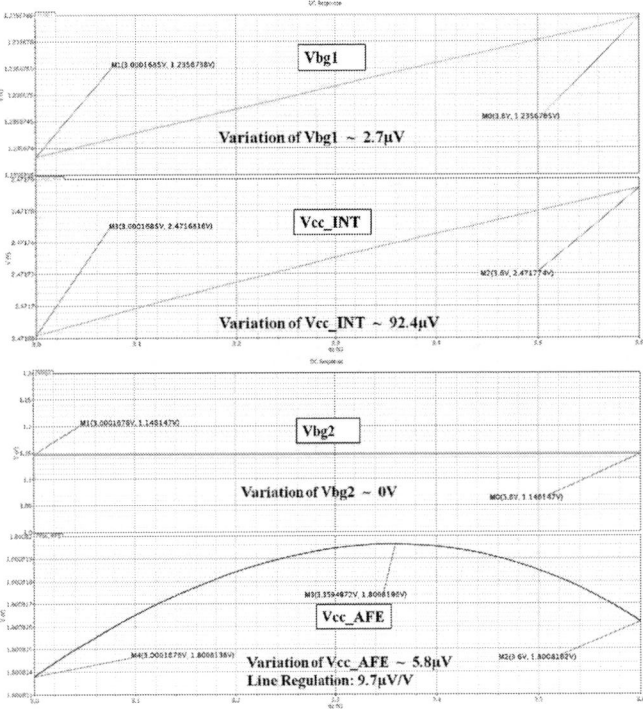

Figure 6. Line regulation simulation result

The first and second curves of Figure 6 show that variations of 2.7µV and 92.4µV are observed for the bandgap and the regulator output voltages of first stage respectively when subjected to input supply voltage changes. With the bandgap circuit of second stage powered by the first stage output, the voltage does not change during simulation and enable the second stage regulator output to achieve a variation of 5.8µV from the simulation.

C. Overall Design Specifications

The overall simulated design parameters for the power management system using the double regulated voltage supply technique is shown in Table I.

TABLE I.   OVERALL DESIGN SPECIFICATIONS

Parameter			Design Value	Unit
Name	Symbol	Formula/Comment		
Supply voltage	V_{supply}		3.3 ± 0.3	V
Supply noise	-		10	mV_{rms}
Current consumption	I_{PU}		1.9	mA
Regulated voltage output	V_{core}		1.8	V
Current drivability	$I_{load,max}$	<100mV drop from 1.8V.	182	mA
Power supply rejection ratio	$PSRR_{PM}$		>100dB @ 2.8KHz	dB
Line regulation	-		9.7	µV/V
Load regulation	-	When I_{load} = 40 + 5mA.	0.045	mV
Temp. coefficient	-		56	ppm/°C
Output noise	-		105	$µV_{rms}$

978-1-61284-863-1/11 $26.00 © 2011 IEEE   573

V. MEASUREMENT RESULTS

The ASIC has been fabricated and functionally tested. PSRR of around 85dB at DC is obtained during measurement as shown in Figure 7. The line regulation measurement result is shown in Figure 8.

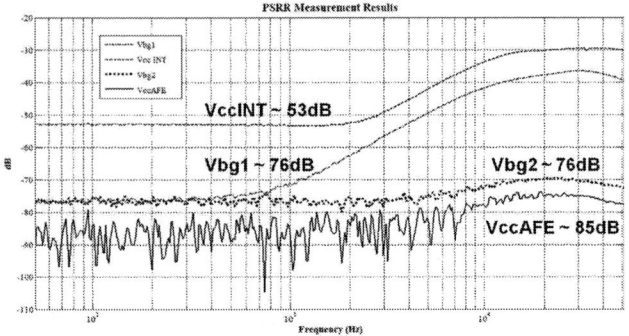

Figure 7.   PSRR measurement results

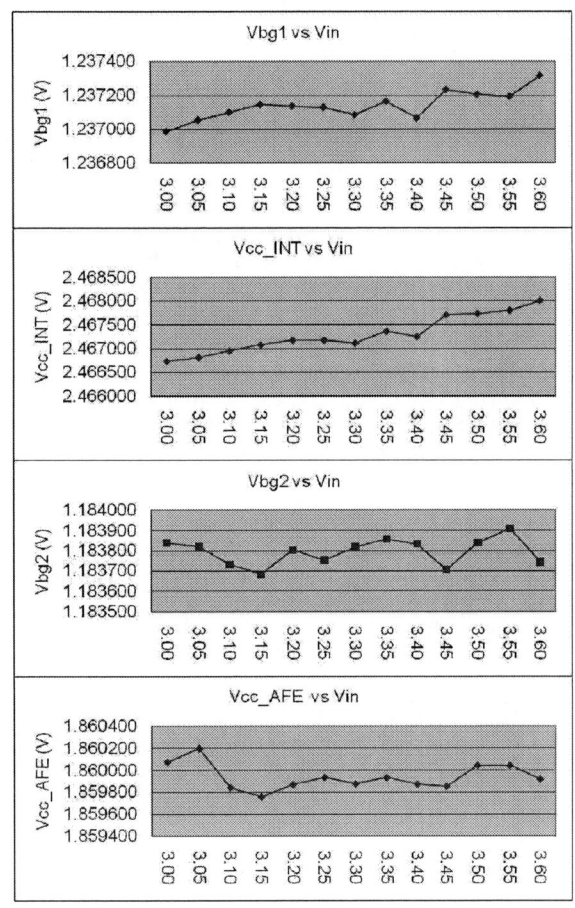

Figure 8.   Line regulation measurement results

The chip micrograph of the MEMS accelerometer ASIC is shown in Figure 9. The full ASIC is functionally tested and verified with the power management block.

Figure 9.   Chip micrograph of ASIC

VI. CONCLUSION

A double regulated voltage supply technique is used in the power management design. Simulation results show that high PSRR (> 100dB over 2.8kHz) as well as high line regulation (10μV/V) are achieved. Measurement results show that this highly regulated power source can meet the stringent requirements of the high precision MEMS accelerometer design.

ACKNOWLEDGMENT

The author would like to thank Mr. Goh Kim Seng for layout support.

REFERENCES

[1] K.T.C. Chai, D. Han, R.P. Singh, D.D. Pham, C.Y. Pang, J.W. Luo, D. Nuttman and M. Je, "118-dB dynamic range, continuous-time, opened-loop capacitance to voltage converter readout for capacitive MEMS accelerometer," IEEE Asian Solid-State Circuits Conference, pp. 1-4, Nov 2010

[2] G.T. Ong and P.K. Chan, "A low quiescent biased regulator with high PSR dedicated to micropower sensor circuits," IEEE Sensors Journal, Vol. 10, No. 7, pp. 1266-1275, July 2010

[3] M. El-Nozahi, A. Amer, J. Torres, K. Entesari and E. Sanchez-Sinencio, "High PSR low drop-out regulator with feed-forward ripple cancellation technique," IEEE Journal of Solid-State Circuits, Vol. 45, No. 3, pp. 565-577, March 2010

[4] A.P. Brokaw, "A simple three-terminal IC bandgap reference," IEEE Journal of Solid-State Circuits, Vol. SC-9, No. 6, pp. 388-393, December 1974

[5] B. Razavi, Design of Analog CMOS Integrated Circuits, New York: McGraw-Hill, 2001

978-1-61284-863-1/11 $26.00 © 2011 IEEE

Design and Optimzation of High Precision CMOS Voltage Reference Using Taguchi Orthogonal Array Technique

Hande Vinayak, *Student Member, IEEE*, Maryam Shojaei Baghini, *Senior Member, IEEE*, and Prakash Apte.
Department of Electrical Engineering,
Indian Institute of Technology(IIT)-Bombay, Mumbai, India.
Email: hande@ee.iitb.ac.in, mshojaei@ee.iitb.ac.in, apte@ee.iitb.ac.in

Abstract—**A CMOS voltage reference, which is based on the weighted compensation of thermal voltage and threshold voltage temperature variations is presented. Subthreshold NMOS transistors and resistive divider configuration are used to achieve reference voltage with low temperature coefficient. Taguchi orthogonal array technique is presented to optimize the circuit to attain precise reference voltage with high PSRR. The proposed voltage reference circuit is analyzed theoretically and compared with other methods. The circuit is designed and simulated in standard 180nm mixed mode CMOS technology. The minimum supply voltage is 1.2 V. A temperature coefficient of 3.6 ppm/°C is achieved with line sensitivity of 0.01%/V. Moreover, PSRR at 100 Hz and 1 MHZ is -100.8 dB and -31.2 dB respectively.**

I. INTRODUCTION

Voltage reference with low sensitivity to temperature and voltage supply is frequently needed in analog and mixed signal circuits, such as ADCs, DACs, Memories, Oscillators, etc. Sub-bandgap references less than 1V are quite common among low voltage applications. Different techniques for low voltage reference generation have been reported. Banba and Leunge used a current mode method to scale down the bandgap voltage [1][2] and Jiang has proposed a TIA (Trans Impedance Amplifier) to serve the same purpose [3]. But, temperature coefficient (TC) of these bandgap voltage reference i.e. without curvature compensation is usually greater than 30ppm/°C. Hence to improve the precision, several higher order curvature compensation methods have been put forward. Ka Leung et al. has designed a second order curvature compensation based on resistor combinations [4]. Malcovati et al. came up with a new technique for generation of non-linear current to compensate BJT's non-linear behavior [5]. Rincon-Mora et al. and Hong-Yi et al. have presented a piecewise-linear correction method to achieve low TC [7][8]. But, the high-order temperature compensation techniques usually suffer from various performance trade-offs, such as simple topology but low accuracy, or high accuracy but complicated compensation control scheme. Vita et al. proposed various techniques to improve TC but PSRR at low frequency is quite low [9][10].

In this paper, optimized low TC and high PSRR voltage reference circuit is presented. Thermal and threshold voltage variation with temperature mutually cancel each other's effect to reduce temperature dependency on reference voltage. Moreover, the reference voltage is logarithmically dependent on bias current, therefore it improves the supply voltage insensitivity. The presented reference voltage circuit is optimized for low TC with Taguchi orthogonal array based technique. Orthogonal array based optimization method achieves low TC, which is comparable with that obtained by any of the curvature based compensation techniques, without any additional circuit overhead.

The rest of the paper is organized as follows. Section II explains circuit configuration of proposed voltage reference. Section III discusses the temperature compensation scheme. Section IV explains supply voltage sensitivity to reference voltage. Section V overviews the Taguchi's optimization technique. Simulation results and comparison are shown in section VI and VII respectively. Section VIII discusses briefly about future developments for proposed voltage reference to achieve process insensitivity. Finally, conclusion is given in section IX.

II. CIRCUIT CONFIGURATION

The proposed voltage reference circuit is shown in Fig. 1. This circuit is comprised of start-up circuit, current generator and resistively divided active load, highlighted in Fig. 1. Generated current I_0 is almost independent of supply voltage. This supply invariant current is then mirrored and injected into active load.

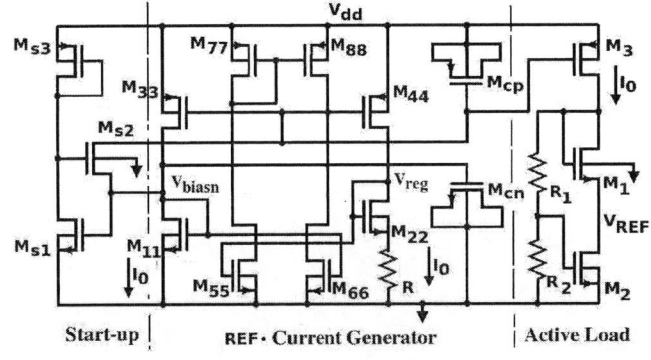

Fig. 1. Proposed CMOS voltage reference

A low TC reference voltage is achieved by mutual cancellaion of temperature dependencies of threshold and thermal voltage. The Taguchi optimization method used to find the effect of temperature on individual components statistically studied and adjusted in such a manner to achieve curvature compensation without any additional component.

A. Supply independent current generator

Current generator section of Fig. 1, shows a modified constant-gm bias circuit. Basic constant-gm bias circuit constitutes of devices M_{11} to M_{44} along with R. To reduce sensitivity of reference voltage to supply voltage, M_{55} to M_{88} devices are added to form a tail transistor-less differential amplifier [11].

The PMOS transistors form a current mirror, while the NMOS devices simply mirror the current in M_{11} when $V_{biasn} = V_{reg}$. If $V_{biasn} \neq V_{reg}$, then the imbalance causes the amplifier output to swing up or down providing the desired action. To make I_0 stable with respect to supply voltage, capacitors M_{cp} and M_{cn} are added.

The saturation and subthreshold drain current expressions for NMOS device are

$$I_{sat} = (1/2)K_n(V_{GS} - V_{TH})^2 \qquad (1)$$

$$I_{sub} = K_n(V_T)^2 exp\left[\frac{(V_{GS} - V_{TH})}{\eta.V_T}\right]\left[1 - exp\left(\frac{-V_{DS}}{V_T}\right)\right] \qquad (2)$$

where, $K_n = \mu_n.Cox.(W/L)_n$ and μ is the carrier mobility in the channel, V_{TH} is threshold voltage of device, η is subthreshold slope, which is assumed to be 1 for simplifying further analysis and V_T is thermal voltage [9]. If $V_{DS} \geqslant 3V_T$, the subthreshold current can be approximated as,

$$I = K_n(V_T)^2 exp\left[\frac{(V_{GS} - V_{TH})}{\eta.V_T}\right] \qquad (3)$$

Considering that the current flowing through devices M_{33} and M_{44} is equal i.e. I_0, M_{11} to M_{44} are biased in saturation region. The gain of differential amplifier M_{55} to M_{88} is sufficient to satisfy $V_{biasn} = V_{reg}$. Therefore, KVL for loop M_{11}-M_{22}-R results in the following expression,

$$I_0 = \frac{2}{R^2.K_{11}}.\left[1 - (1/\sqrt{M})\right] = \frac{2x}{R^2.K_{11}} \qquad (4)$$

where, M is scaling factor (i.e. Area of $M_{11} = M *$ Area of M_{22}) and x is $\left[1 - (1/\sqrt{M})\right]$.

The current I_0 is supply voltage independent and shows a PTAT (Proportional To Absolute Temperature) behavior.

III. TEMPERATURE COMPENSATION

Temperature dependence of the threshold voltage is given by

$$V_{TH} = V_{TH0} + \alpha(T - T_0) \qquad (5)$$

where, V_{TH0} is the threshold voltage at 0 K and α is the TC of threshold voltage. The value of threshold voltage coefficients are α_n = -0.75mV/°C and α_p = -0.98mV/°C for NMOS and PMOS transistors respectively, in standard 180nm mixed mode CMOS process. Therefore, magnitude of threshold voltage linearly decreases with temperature.

A resistively divided active load is used to generate a very low TC voltage reference. The supply voltage invariant I_0 is injected into two NMOS transistors, M_1 and M_2. Transistors M_1 and M_2 are biased in subthreshold region. The output voltage is given by

$$
\begin{aligned}
V_{REF} &= \left(1 + \frac{R_1}{R_2}\right)V_{GS2} - V_{GS1} \\
&= \left(1 + \frac{R_1}{R_2}\right)\left[\eta.V_T ln\left(\frac{I_0}{K_2.V_T^2}\right) + V_{TH2}\right] \\
&\quad - \left[\eta.V_T ln\left(\frac{I_0}{K_1.V_T^2}\right) + V_{TH1}\right] \\
&= \eta.V_T ln\left(\frac{A_1}{A_2}\right) + \frac{R_1}{R_2}V_{TH2} + \\
&\quad + \frac{R_1}{R_2}\eta.V_T ln\left(\frac{2x}{R^2 K_{11} K_2 V_T^2}\right) \qquad (6)
\end{aligned}
$$

where, $A_x = (W/L)_x$.

Differentiating (6) with respect to temperature,

$$
\begin{aligned}
\frac{dV_{REF}}{dT} &\approx \frac{dV_T}{dT}ln\left(\frac{A_1}{A_2}\right) + \frac{R_1}{R_2}\frac{dV_{TH2}}{dT} \\
&\quad + \eta\frac{R_1}{R_2}\frac{dV_T}{dT}\left[ln(R^2) - 2.ln(\mu_n) - 2.ln(Cox)\right] \\
&\quad - \eta\frac{R_1}{R_2}\frac{dV_T}{dT}\left[2 + ln(A_{11}) + ln(A_2) + ln(V_T^2)\right] \\
&\quad - \eta\frac{R_1}{R_2}\frac{dV_T}{dT}\left[2.ln(Cox) + ln(A_{11}) + ln(2x)\right] \\
&\quad - \eta\frac{R_1}{R_2}\frac{dV_T}{dT}\left[ln(A_2) + ln(V_T^2)\right] \qquad (7)
\end{aligned}
$$

where, V_T is 26mV, μ_n is $310cm^2/V.sec$ and Cox is $8.369e^{-7}F/cm^2$ for standard 180nm mixed mode CMOS process.

As $|V_T| << |\mu_n|$, effect of mobility variation due to temperature is highly reduced. Therefore, mobility related term is neglected in expression (7). After designing the constant-gm bias circuit, values of A_{11}, x and R are fixed. Therefore, overall effect of 3rd to 6th terms of expression (7) can be made nearly zero by selecting an appropriate value of A_2. Hence, simplified expression becomes,

$$
\begin{aligned}
\frac{dV_{REF}}{dT} &\approx \frac{dV_T}{dT}ln\left(\frac{A_1}{A_2}\right) + \frac{R_1}{R_2}\frac{dV_{TH2}}{dT} \\
&= \frac{k_B}{q}ln\left(\frac{(W/L)_1}{(W/L)_2}\right) - \alpha_n\frac{R_1}{R_2} \qquad (8)
\end{aligned}
$$

where, k_B is Boltzman constant.

Therefore, if equation (8) is satisfied, zero TC V_{REF} is obtained for any temperature. Once the dimensions of transistor M_1 are fixed, the ratio of resistors R_1 and R_2 is derived to achive low TC. Ratio of resistance is smaller than unity.

IV. SENSITIVITY TO SUPPLY VOLTAGE

The bias current I_0 varies due to channel length modulation effect. One way to reduce this effect is to use cascode

978-1-61284-863-1/11 $26.00 © 2011 IEEE

configuration. But penalty is supply voltage increment. The change in I_0 causes a change in reference voltage.

$$V_{REF} = \left(1 + \frac{R_1}{R_2}\right) \eta . V_T ln \left(\frac{I_0}{K_2 . V_T^2}\right) + \frac{R_1}{R_2} V_{TH2}$$
$$- \eta . V_T ln \left(\frac{I_0}{K_1 . V_T^2}\right) + \Delta V_{TH}$$

$$(9)$$

The expression (9) which is derived from (6), shows that reference voltage logarithmically depends on I_0. Therefore, reference voltage is weakly dependent on bias current. In addition, a variation of I_0 varies V_{GS1} and V_{GS2} in same direction, even though with not same magnitude; hence equation (6) indicates reference voltage is derived from difference of these two voltages. Therefore, this phenomenon further reduces the dependence of V_{REF} on I_0.

V. TAGUCHI OPTIMIZATION TECHNIQUE

Taguchi optimization technique is predominantly used in industrial engineering, in which the experiments are planned for designing of a product efficiently and reliably. Based on these experiments, the effects of various factors on design are calculated. Therefore, by controlling dominant factor, product is optimized to achieve better quality.

After incorporating temperature dependence factors in expressions (7) and (8), it is quite complicated to decide dimensions of devices which can give low TC V_{REF}. Therefore, the presented method of optimization helps to find out device dimensions to achieve precise reference voltage.

A. Orthogonal array

A matrix experiment consists of a set of experiments. Each experiment is performed by changing the setting of the various product or process parameters which has to study. After conducting matrix experiments, the data from all experiments in the set taken together is analyzed to determine the effects of various parameters. Matrix experiments are conducted using special matrices, called 'orthogonal arrays'[12]. Benefits of orthogonal arrays are,

1) Conclusions are valid over the entire region spanned by the control factors and their settings.
2) Large saving in the experimental effort.
3) Analysis is easy.

The optimizing parameter used for the analysis is low TC reference voltage value and PSRR. For optimizing proposed voltage reference, the controlling factors are width of device M_2, length of resistor R_1, width and length of device M_3. Table I shows the values at different levels of all 4 control factors.

The orthogonal array used for optimization is $L_{27}(3^{13})$. It means, this array can have 13 factors of 3 levels each. For optimizing reference voltage, only 4 factors are taken into account. A part of $L_{27}(3^{13})$ Taguchi orthogonal array is shown in TABLE II.

TABLE I
CONTROL FACTORS AND THEIR LEVELS

Factor	Level 1	Level 2	Level 3
W_2 (μm)	1	2	3
L_3 (μm)	20	30	40
L_{R_1} (μm)	330	350	370
W_3 (μm)	50	70	90

TABLE II
PART OF L_{27} TAGUCHI ORTHOGONAL ARRAY

Expt. Num.	1 W_2 (μm)	2 L_3 (μm)	3 L_{R_1} (μm)	4 W_3 (μm)	.	.	12	13
1	1	1	1	1	.	.	1	1
2	1	1	1	1	.	.	2	2
.	.	.	.	.	.	.	.	.
27	3	3	2	1	.	.	3	2

TABLE III
SIMULATION DATA OF EXPERIMENTS MENTIONED IN TABLE II

Exp. \ Temp.	0°C	25°C	50°C	75°C	100°C
1	396.82	398.66	400.89	403.62	406.75
2	396.82	398.66	400.89	403.62	406.75
.	.	.	.	.	.
27	303.39	304.55	304.73	304.46	304.00

B. Best Settings

The signal to noise ratio provides a measure for the impact of noise i.e. temperature, in proposed design, on performance. The larger the SNR, the more robust the product is against noise. As shown in factor effect plot in Fig. 2, factors W_2, L_3, L_{R1} and W_3 give highest SNR at $2\mu m$, $30\mu m$, $370\mu m$ and $50\mu m$ respectively. Highest possible SNR indicates the best setting of factors. The combination of these factors, at above mentioned sizes, gives lower TC and high PSRR reference voltage.

VI. SIMULATION RESULTS

The proposed voltage reference has been designed in standard 180nm mixed mode CMOS process and simulated by Spectre using foundry models. Simulation results show that the proposed and optimized voltage reference generates a mean reference voltage approximately 356.85mV with variations of

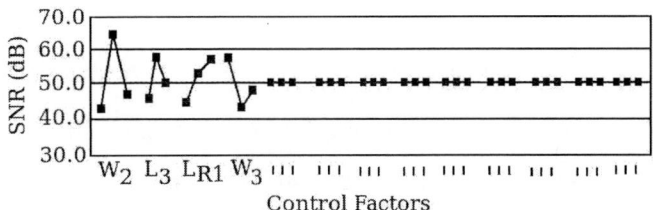

Fig. 2. Factor effect plot

0.08mV at room temperature, when supply voltage varies from 1.2V to 2V, as shown in Fig. 3.

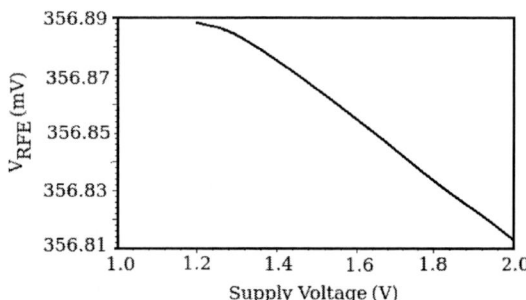

Fig. 3. Output reference voltage Vs supply voltage

Fig. 4 shows, the reference voltages' dependency on temperature. The TC achieved is 3.6 ppm/°C. As can be seen in Fig. 5, the power supply rejection ratio, without any filtering capacitor, is -100.74 dB and -31.22 dB at $100Hz$ and $1MHz$ respectively. At 27°C, the current drawn is 6.75uA from a supply voltage of 1.2V.

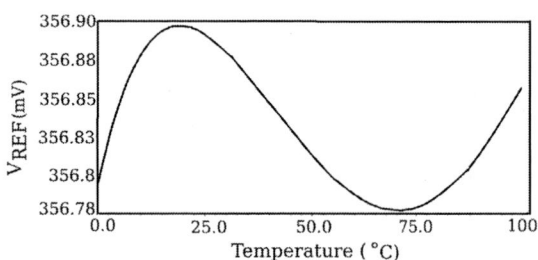

Fig. 4. Low TC reference voltage

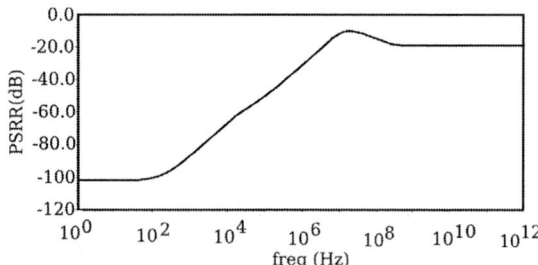

Fig. 5. PSRR at room temperature for supply voltage 1.2V

VII. COMPARISON

A comparison with other voltage references reported in literature is summarized in Table IV. Only standard CMOS process implementations are taken into consideration for comparison.

VIII. DISCUSSION AND FUTURE DEVELOPMENT

The optimization conditions for the proposed voltage reference are obtained with Taguchi optimization technique. The Taguchi optimization method also gives flexibility to optimize a circuit for process variations. Orthogonal arrays can be used

TABLE IV
COMPARISON

	This work	Magnelli et al.[14]	Vita et al.[10]	Ueno et al.[13]	Leung et al.[6]
Technology (µm)	0.18	0.18	0.35	0.35	0.6
Supply Voltage (V)	1.2-2	0.45-2	0.9-4	1.4-3	1.4-3
V_{REF} (mV)	356.84	263.5	670	745±25	309±19
TC (ppm/°C)	3.58 [0:100]	142 [0:125]	10 [0:80]	7 [-20:80]	36.9 [0:100]
Line Sensitivity(%/V)	0.01	0.44	0.27	0.002	0.083
PSRR (vdd) ≤ 100 Hz (dB) @ 1 MHz(dB)	1.2 V -100.74 -31.22	0.45 V -45 -12.2	0.9 V -47 -22@10K	2 V -45 -	1.4 V -47 -18
Die area(mm^2)	-	0.043	0.045	0.055	0.055

with more number of control factors to increase degree of freedom. Process corners can be considered as noise factors to achieve process insensitivity for the proposed circuit.

IX. CONCLUSION

A 3.6 ppm/°C voltage reference with supply voltage 1.2 V has been presented. The proposed voltage reference has been simulated with standard 180nm mixed mode CMOS process. The conditions of optimizations are studied and stated with Taguchi based optimization technique. Simulation results show that the proposed voltage reference circuit provides a precise reference voltage. Moreover, low-line sensitivity and high PSRR are achieved.

REFERENCES

[1] Banba et. al.,"A CMOS bandgap reference circuit with sub-1-V operation," *IEEE JSSCC*, pp. 670-673, May 1999.
[2] K. N. Leung et al., "A Sub-1-V 15-ppm/°C CMOS Bandgap Voltage Reference without Requiring Low Threshold Voltage Device," *IEEE JSSCC*, pp. 526-530, April 2002.
[3] Y.Jiang et al.," A Design of Low-Voltage Bandgap Reference Using Transimpedance Amplifier, *TCAS-II*, pp. 552-555, June 2000.
[4] K. N. Leung et al.," A 2-V 23-J.1A 5.3-ppm/°C curvature-compensated CMOS bandgap reference, IEEE JSSCC, pp. 561-564, March 2003.
[5] Malcovati et. al.,"Curvaure compensated BICOM bandgap with 1V supply voltage," *IEEE JSSCC*, pp. 1076-1081, July 2001.
[6] K.Leung et al.,"A CMOS voltage reference based on weighted ΔV_{GS} for CMOS low-dropout linear regulators," *IEEE JSSCC*, Jan. 2003.
[7] G.Rincon-Mora et.al.,"A 1.1 V Current-Mode and Piecewise-Linear Curvature Corrected Bandgap Reference," *IEEE JSSCC*, pp. 1551-1554, Oct. 1998.
[8] Hong-Yi et.al.,"Piecewise-Linear Curvature Compensated Bandgap voltage reference," *IEEE 15th ICECS*, pp.308-311, 2008.
[9] Vita et al.,"An ultra-low-power, temperature compensated voltage reference generator ," *Proc. of IEEE CICC, CA*, pp. 751-754, 2005.
[10] Vita et al.,"A sub-1-V, 10 ppm/°C, nanopower voltage reference generator ," *IEEE JSSCC*, pp. 15361542, Jul. 2007.
[11] R. Baker,"CMOS Circuit Design, Layout, and Simulation," *IEEE Press*, second edition, 2005.
[12] Madhav Phadke,"Quality Engineering using Robust Design," *Pearson Education*, 2008.
[13] K. Ueno et. al.,"A 300 nW, 15 ppm/°C, 20 ppm/V CMOS voltage reference circuit consisting of subthreshold MOSFETs," *IEEE JSSCC*, pp. 20472054, Jul. 2009.
[14] Luca Magnelli et. al.,"A 2.6 nW, 0.45 V Temperature-Compensated Subthreshold CMOS Voltage Reference," *IEEE JSSCC*, pp. 465-474, Feb. 2011.

A 2.72GOPS/11mW low power reconfigurable accelerator with a highly parallel datapath consisting of combinatorial circuits in 65nm CMOS

N.Ozaki, Y, Yasuda, Y,Saito, D.Ikebuchi, M.Kimura, H.Amano
H.Nakamura, K.Usami, M.Namiki, M.Kondo
Department of Information and Computer Science, Keio University
3-14-1 Hiyoshi, Yokohama, 223-8522 Japan (sld@am.ics.keio.ac.jp)
Department of Technology, Shibaura Institute of Technology
3-9-14 Shibaura, Minato-ku, Tokyo, 108-8548 Japan

Abstract—CMA (Cool Mega-Array) is a high energy-efficiency reconfigurable accelerator for battery-driven mobile devices. It consists of a large processing element (PE) array without memory elements for mapping the data-flow graph of the application being executed, a small simple programmable micro-controller for data management, and a data memory. Unlike traditional coarse grained reconfigurable processors in which each PE provides registers and context memory, a CMA rduces power consumption by doing away with that for switching of hardware context and storing intermediate data in registers and their clock distribution. Although the data-flow graph mapped on the PE array is static during execution, various application programs can be implemented by making the best use of flexible data management instructions in the micro-controller. When the delay time of the PE array is shorter than the data handling time taken by the micro-controller, the supply voltage for the PE array is scaled to reduce the power consumption without degrading the performance. In contrast, when the delay time of the PE array is longer, wave pipelining is applied to enhance performance of the PE array.

A prototype CMA chip (CMA-1) with 8×8 PE array with 24-bit data width was fabricated on the basis of 2.1×4.2-mm 65-nm CMOS technology, and achieves sustained performance of 2.5-GOPS/11.2-mW. This energy efficiency is comparable to that of the most-energy-efficient accelerators that have been reported.

I. INTRODUCTION

These days, as the mobile device spreads, performance gain , lower power consumption and short span of development period of embeded systems requiered.

To meet these requires ,Dynamically Reconfigurable Processor Array(DRPA) come to be paid attention in place of ASIC, as off-load engine[8], [4], [3], [1]. DRPA can realize high power efficiency with low operating frequency . Since datapaths required for computation are dynamically formed on an array of Processing Elements (PEs). But, these Dynamically Reconfigurable Architectures have the overhead of dynamic reconfiguration. The consuming power, consisting of dynamic and leakage power can be suppressed by lowering the supply voltage. Especially, the dynamic power is decreased squire order by lowering supply voltage. We originally developed Dynamically Reconfigurable Prosessor called MuCCRA. According to our investigation, MuCCRA consumes 20% of

power to dynamically reconfiguration. and 15% of power to standby.

Here, we propose a highly energy efficent architecture called Cool Mega-Array (CMA) . CMA has a large reconfigarable PE array which integrates many PEs but no registars,and hig speed micro-controller to control data communication between PE array and data-memory. This array forms a large combinatorial circuit unlike other typical accelerators.

CMA has advantages as follows:

- Realize applications by one time configuration
  CMA has a large PE array with combinatiorial circuit. To change the datapath on PE array , large power is required. Therefore, CMA prepare a lagere size PE array, realize applications at one time configuration. It become possible for PE array to eliminate registers by making PE array combinatiorial circuit, Becouse PE don't have to memorize variable on the way. CMA never change its configuration at executing mode. Instead, we prepare a high speed micro-controller in memory module to compensate its lack of flexibility . A micro-controller has high programmability and control the communication between data-memories and PE array. Howezer, micro-controller occupies small area of the chip and drive low power.
- the clocktree is restricted in the area of micro-controller
  At executing mode, CMA don't change its configuration , so clocktree for PE array which accupies large area on chip is gated.
- the supply voltage to PE array can be lowered even sub-threshold level Power consumption is proportional to the squair of supply voltage. So it is effective to save dinamically power consumption to lower the power supply to PE array.

In the present work, a CMA prototype (CMA-1) based on 65-nm CMOS technology which achieved 2.5 GOPS with an 11.2-mW power budget was fabricated. In the following section, the concept of CMA is introduced, and then an examples of implementation of CMA-1 is presented and evaluated.

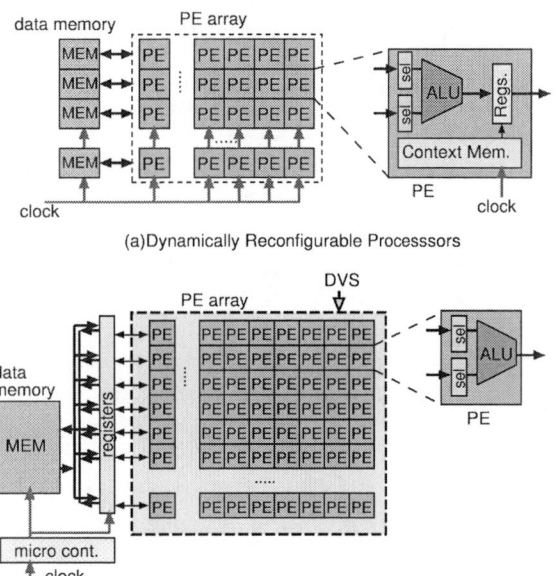

(a)Dynamically Reconfigurable Processsors

(b) Cool Mega-Array

Fig. 1. (a)Dynamically Reconfigurable Processors and (b)CMA architecture

II. CONCEPT OF CMA

The CMA architecture is mainly targeted at multi-media processing in battery-driven embedded systems. In such processing, a fixed amount of data must be processed in a certain time, but there is no advantage in reducing the processing time to less than that required. Minimizing the amount of energy used to process a fixed amount of data is important since it affects battery lifetime. The objective is thus to design an architecture for executing a fixed number of computations in the required time with the minimum amount of energy. The energy used is the product of power ($P = C \cdot V^2 \cdot f$) and time (T), thus, it is related to the square of supply voltage V. However, logic circuit delay increases drastically as V approaches threshold voltage Vth:

$$D = \beta \cdot \frac{C \cdot V}{(V - Vth)^\alpha} \quad (1)$$

where $alpha$ is about 1.6 [9]. Fortunately, most stream processing involves a large amount of parallelism, so the required performance can be achieved by parallel processing using many PEs. One key concept of the CMA architecture is reducing the supply voltage of the PE array as much as possible while still achieving the required computation time. Another key concept is reducing energy usage other than that required for computation as much as possible. The data must be transferred from the data memory to the computational module, where it is computed. The computation results must be written back to the data memory. This process is indispensable, so a certain amount of energy consumption is needed here. However, all other energy consumption, such as that for reducing costs and improving performance, should be eliminated.

Figure 1 shows the fundamental concept of the CMA architecture. In a typical CGDRP (Figure 1(a)), intermediate data is stored in registers(regs.) in each PE for pipeline processing. To achieve flexible execution ,the PE operations and interconnections are dynamically reconfigured. In some machines, the hardware context stored in the context memory (context mem.) in each PE is switched every clock cycle [3], [7], [4]. The clock is distributed to all PEs using a clock tree. The addresses of distributed memory modules (MEMs) are generated by PEs in the PE array as well as its access management.

In contrast, in the CMA architecture, the PE array consists of combinatorial circuits without registers for storing intermediate results and context memory. The supply voltage of the PE array can be scaled without worrying about the effects on the setup time or on the clock skew of the registers. There is no clock tree for distributing the clock in the PE array: as a result, the power consumption of the PE array can be much reduced.

Only the input and output data for the PE array are stored in registers. The micro-controller reads data from the data memory (MEM) and distributes data to the register attached to the input of the PE array. It also collects the results from the register attached to the output of the PE array and writes them back to the data memory. It flexibly manages the data transfer between the memory and registers by using mapping registers and vector operations. The above structure make it possible to implement various application programs without the need for power consuming dynamic reconfiguration in the PE array.

Since the computation in the PE array and data management by the micro-controller are done in a pipelined manner, their execution speeds must be balanced. If the computation delay is longer than the data management delay, the voltage supplied to the PE array can be reduced. The total power needed for computation can thus be reduced without degrading performance. On the other hand, if the data-management delay is longer than the computation delay, wave pipelining in the PE array can be used. The delay time for achieving wave pipelining can be also controlled by changing the voltage supplied to the PE array. There are three drawbacks to the CMA architecture. First, a glitch in the large combinatorial circuits can cause a large increase in the power used for switching. Second, using parallel processing only in the PE array can result in larger leakage current than in the case of using both parallel and pipelined processing between PEs.Third, the range of application programs that can be handled is limited because of dynamic reconfiguration of the PE array is not supported.

To evaluate the benefits of the CMA architecture and the the effect of the three above-mentioned drawbacks, two prototype CMA chips were constructed and evaluated.

III. FIRST PROTOTYPE, CMA-1

A. Overview of CMA-1 chip

The prototype CMA-1 was implemented by using a Fujitsu e-shuttle 65-nm 12-layer CMOS process in a 4.2 × 2.1-mm chip. A high-Vth library (CS202SZ) with middle-class delay was used for balancing leakage current against performance. The design was described in Verilog-HDL and synthesized

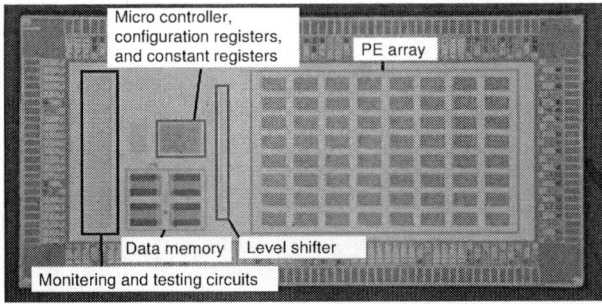

Fig. 2. Chip Photograph of CMA-1

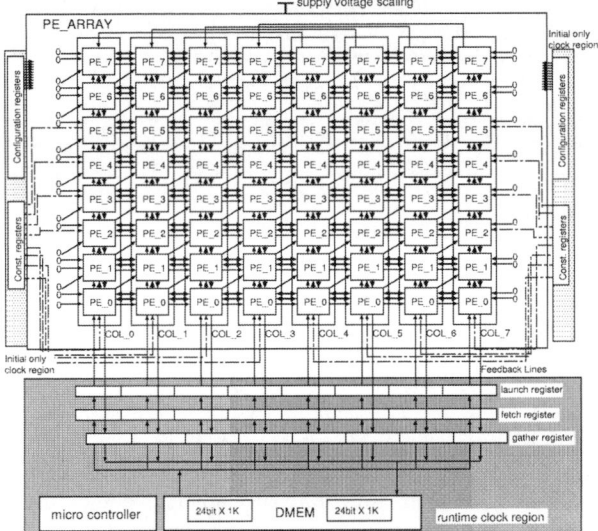

Fig. 3. Block Diagram of CMA-1

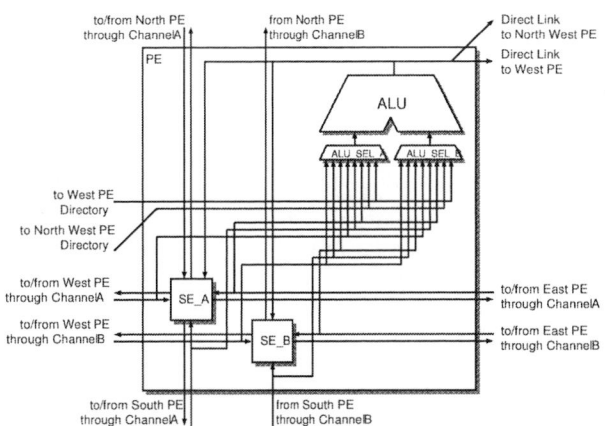

Fig. 4. PE structure

B. PE array

The PE array contains large combinatorial circuits that can operate at a supply voltage ranging from 0.5 to 1.2 V. It has an 8 × 8 structure and a 24-bit data width, which is commonly used for multi-media applications. Although the array is rather small, it is still large enough for implementing practical application programs. The PEs are connected by two sets of island-style global interconnection networks (thick arrows in the figure) and local direct interconnection links (thin arrows in the figure). The mapping tool tries to use the direct interconnection links for local communication first: It then maps the island-style global network for distant data communication. In each switching element (SE) in a PE, two lanes of global interconnection are exchangeable. A feedback line (from north to south) for transferring the result is placed in each PE. To avoid a forming feedback loop with the combinatorial circuits, once data is on the feedback line, it is directly forwarded to the gather register attached to outputs of the PE array. To prevent unnecessary data propagation on the networks, which increases the number of glitches on the data lines, the data can be forced to zero by appropriately setting the configuration data of SEs.

C. PE

As shown in Figure 4, a PE consists of a 24-bit arithmetic logic unit (ALU) and two input multiplexers (ALU_SELs). The ALU supports addition, subtraction, multiplication, and other simple operations, while the two ALU_SELs apply shift operations to the input data. The ALU consists of several functional units MULT, ADD/SUB, SHIFT, and other logic operation units. Operand isolation is applied to each unit. That is, the input data are forced to zero if the configuration data indicate that the corresponding functional unit is not being used. The main role of the ALU_SELs is selecting the input data from the two sets of global networks and from the direct links.

D. Configuration registers and constant registers

As in the case of other coarse-grain reconfigurable devices, the operation of each PE and the interconnections between

using the Synopsys Design Compiler (2007.12-SP3). The layout was done using the Astro (2007.03-SP3). The micro-controller operates at a 210-MHz clock rate. Since a PLL is not used to reduce chip power consumption, the maximum clock frequency is limited by the electric characteristics of the I/O pad. The main problem with using a large PE array is high leakage current. However, a preliminary estimation showed that as a result of eliminating context memory, registers and the clock tree, the area occupied by the PE array is about 25% that of a DRPA [4] based on the same process. Consequently, the PE array leakage current is about 1 to 2 mW. This range is tolerable, so techniques for reducing the leakage current (for example, using power gating for unused PEs) are unnecessary.

A chip photograph of CMA-1 is shown in Figure 2. The PE array occupies about 60% of the chip area, and the micro-controller (with constant registers and configuration registers) occupy about 5% of the chip area. To reduce the supply voltage of the PE array only, the power domain is divided by placing dividers at four locations in the I/O-filler, and level-shifter cells designed from scratch are inserted in the signal wires between the PE array and the micro-controller.

CMA-1 consists of a PE array, a micro-controller, a data memory (DMEM), and registers (Figure 3).

978-1-61284-863-1/11 $26.00 © 2011 IEEE   581

them are controlled by configuration data prepared before execution. In CMA-1, configuration registers are located outside the PE array(Figure 3), and all configuration signals are sent from them. The delay caused by long wires from outside the PE array does not degrade operational speed because the configuration data are not changed during execution. Constant data used in the application are also supplied from the constant registers (Const. registers in Figure 3). Although the clock must be delivered to these registers, it is gated during the execution at the root (initial only clock region). Thus, the clock tree is enabled only when the configuration data are loaded before execution.

The configuration multicast method "RoMultiC" [11] is used for loading the configuration data. A two-dimensional bitmap is assigned to each PE and the configuration data is multicasted to the configuration registers that have column and row bits set to 1. This reduces both the time and energy needed to load the configuration data as well as the total amount of configuration data.

E. Micro-controller

1) Three-step pipeline execution: The micro-controller is a programmable controller for reading the data from the 24-bit data memory, delivering them to the PE array, collecting the results from the PE array, and writing them into the data memory. It has 16 general-purpose registers. Dedicated 14-bit micro-operations are stored in a small instruction memory with 128 depth. Three registers, launch register (LR), fetch register (FR), and gather register (GR), each with eight 24-bit fields corresponding to the eight input/output of the PE array, are provided for three-step pipeline execution of the PE array computations. (1)The micro-controller reads a 24-bit data from the data memory (DMEM) and sends it to each field of the each FR each clock cycle (Distribution). This step takes eight clock cycles if all inputs of the PE array are used. The contents of the FR are then sent to LR (scatter). (2)Since the LR is directly connected to the inputs of the PE array, the combinatorial computation in the PE array immediately begins (computation). After a certain delay depending on the speed of the computations, the results are sent to the GR (gather), where they are stored. The timing of this storage depends on the instructions from the micro-controller. (3)The data in each field of the GR are written to the data memory one field per clock cycle (collection). To enable these three steps to be executed in a pipelined manner, a dual-port memory is used. The time of each step depends on the number of input data, output data and the computational complexity of the application program.

If the data management and computation are well balanced, the pipeline flows ideally. However, if the computation time of the PE array is shorter than the data alignment time of the micro-controller , the supply voltage can be reduced and energy is saved. If the alignment time is less than the computation time owing to insufficient performance of the PE array, a wave-pipeline technique can be applied by launching the next data before collecting the data currently being computed.

2) Data manipulation operations: The micro-controller provides several data manipulation istructions for flexible data management. General-purpose registers R0-R7 can be used as

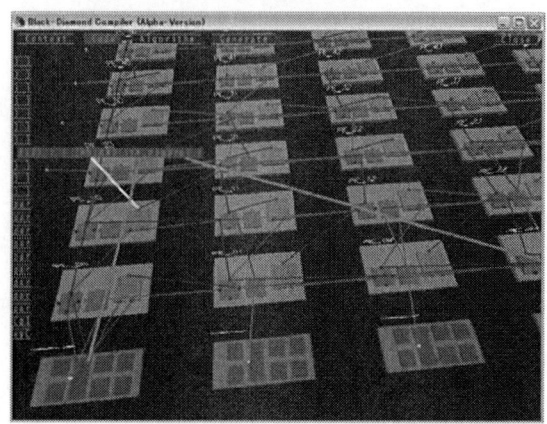

Fig. 5. GUI of black diamond

pointers of the data memory. LD_ADDRd, Rs (Load_Add instruction) reads the data memory indexed by register Rd and sends it to the d-th field of the FR by executing $Rd = Rd+Rs$ in a clock cycle. Thus, the pointer can be updated to the next element of the matrix in the data memory without additional instructions. LD_ADDI (Load_Add_immediate instruction) uses immediate data instead of Rs. A stride vector access with arbitrary stride can be easily implemented with them.

The micro-controller also provides block-transfer instructions with a bit-map vector. $LDVEC\ Rd, Rb$ (Load_VECTOR instruction reads eight data at most from the data memory continuously from the address stored in Rd. Bit-map register, Rb which indicates whether the data are to be sent to the corresponding field of the FR, is provided. It can thus be used when all inputs of the PE array are not used. The starting address stored in Rd is automatically incremented in accordance with the number of data transferred. Once the micro-controller starts the block-transfer operation, it can execute succeeding instructions during the block-transfer. After sending the data in the FR according to these instructions, by executing $LANCH$ instruction, the contents of FR is moved to LR and the computation in the PE array starts.

3) Data memory: Unlike typical pipelined parallel accelerators, which use power-consuming multi-bank data RAM modules (Figure 1), two 25-bit × 1024-depth dual-port RAM modules are used in DMEM. One RAM module is connected with the micro-controller during execution, while another is used for I/O data transfer. After the computation is finished, they can be switched within a clock cycle are provided to form a double-buffer mechanism. The time for I/O data transfer is thus hidden by overlapping it with the computation time.

IV. PROGRAMMING ENVIRONMENT

The "Black Diamond" retargetable compiler for dynamically reconfigurable processors [10], which uses a C-like programming language, was customized for CMA-1 and CMA-2. It compiles the described program and generates the configuration data for the PE array. Since a simple first-fit algorithm is used for mapping the target datapath, automatic mapping is often difficult when the utilization ratio of the PE array

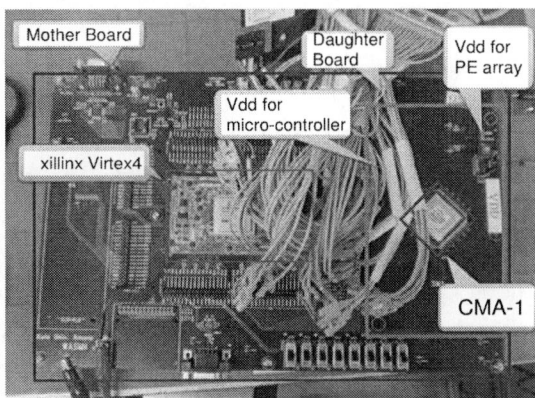

Fig. 6. Evaluation Board

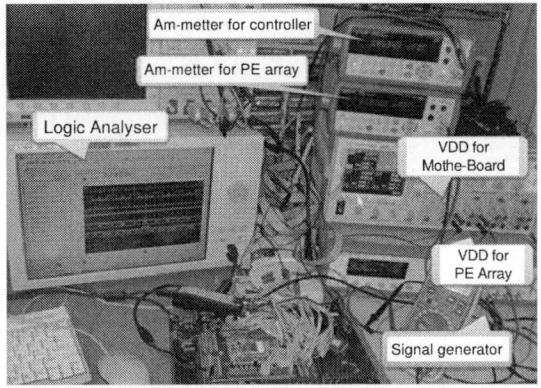

Fig. 7. Evaluation Emvironment

TABLE I
APPLICATIONS PROGRAMS DEVELOPED

af:	24-bit α blender
sf:	24-bit sepia filter,
alpha:	8-bit α blender
sepia:	8-bit sepia filter,
sepia & alpha	two filters
8bit	RGB data are compressed into 24 bits
DCT	discrete cosine transform for JPEG coder
edge:	edge filter
gray:	24-bit gray scale,
sad	sum of absolute differences
ssd	sum of squared differences
satd	sum of absolute transformed differences

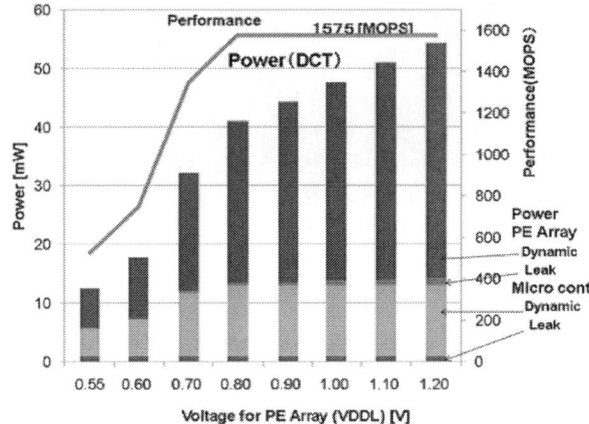

Fig. 8. Power/Performance vs.Voltage for PE Array. Micro cont. include all parts exept the PE array.

exceeds 60%. In such cases, the GUI shown in Figure 5 is used for manual allocation. Operand isolation is used for the unused PEs.

The micro-code of the micro-controller is described separately in simple assembly language. The timing for storing the results in the GR and the supply voltage for the PE array are fixed in accordance with the results of a delay computation, which is described later.

V. EVALUATION ENVIRONMENT

We developed original mother board and daughter board to evaluate CMA perfonrmance. Figure 6 is overview of boards, and figure 7 shows our evaluatio emvironment (Agilent Technologies DC Power Analizer and Puls Pattern Generator).

VI. EVALUATION

A. Applications used in evaluation

Twelve kernel-application programs for multi-media processing, shown in Table I, were developed for evaluation of CMA-1 by using the programming environment described in Section IV. The sepia & alpha programs implement two separate applications applied in the order.

B. Breakdown of power consumption

Fig8 shows the supply voltage for the PE array (VDDL) versus consuming power and performance when 2-Dimensional

DCT (Discrete Cosine Transform) for JPEG coder runs on the prototype chip. By making the best use of flexibility of memory access alignment, 64 data are transformed with 8 times data launching to the PE array in 72 clock cycles. In this application, the execution in the PE array is faster than preparing 8 data in the scatter registers. Thus, the performance is saturated with 1.0V VDDL, and the optimal energy efficiency is achieved with 0.7V VDDL. It is often reported in reconfigurable devices that large combinatorial circuits suffer a large crowbar current by glitches [4]. However, the power in the PE array is reduced almost related to $VDDL^2$ with the following reasons:

1) the propagation of data is suppressed as possible by setting proper configuration data for unused PE and switching elements
2) using the high threshold level transistors, the crowbar current can be suppressed as well as the leakage current.

C. Performance-and-power of CMA-1

The performance in units of power of the 12 application programs at various power supply levels is shown in Figure 9. Since the slack in delay for the PE array computation is used for saving energy, the energy efficiency for the programs with a small degree of parallelism is not degraded substantially. Energy efficiency can thus be achieved for a wide variety

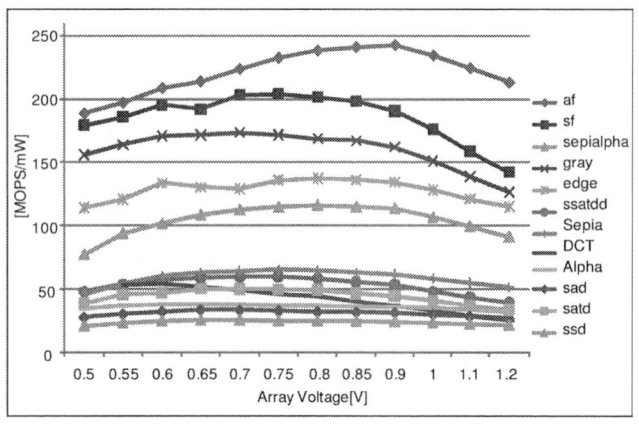

Fig. 9. Performance versus Array Voltage Applications:

of applications programs by controlling VDDL. When the voltage supplied to the PE array was reduced below 0.65 V, the energy efficiency was degraded because of the rapid increase in delay time. In this region, wave pipelining is efficient. Usually, the micro controller launches the next input data after writing the results of the previous computation in the PE array, so it must wait for the results for a longer delay time. However, if the time for data distribution is shorter than that for computation in the PE array, it can launch the next data during the computation.Wave pipelining was applied in the low-voltage region (0.5-0.65 V) for the application programs with long delay paths: 24-bit α blender (af), 24 -bit sepia filter (sepia), edge filter (edge), and gray scale (gray).

Reconfigurable devices with large combinatorial circuits have been reported to suffer from a large short-circuit current due to glitches [6]. However, the reduction in power consumed by PE array is almost proportional to $VDDL^2$ in the case of CMA-1. This is because data propagation was suppressed as much as possible by setting appropriate configuration data for the unused 'PEs and SEs by making the best use of the operand isolation mechanism of CMA-1.

The energy efficiency was maximized when VDDL was from 0.9 to 1.2 V, depending on the applicaton programs. When simple image processing was applied to input RGB data compressed into 24 bits (8 bits in Figure 9), 2.72-GOPS sustained performance was achieved with at 11.2 mW (243MOPS/mW). Compared with a 0.357GOS/13.4mW (24.9MOS/mW) dynamically reconfigurable processor[12] with the same die-size and developed with the same process technology, CMA-1 achieved about ten-times better energy efficiency.

This energy efficiency is comparable to that of the most-energy-efficient accelerators that have been reported. Compared with a 41.5-GOPS/ 0.775-W (54.8-MOPS/mW) dynamically reconfigurable accelerator [13] and a 3.2-GOPS/50-mW VLIW accelerator(64-MOPS/mW) [2], the CMA-1 has better energy efficiency. Although its energy efficiency is worse compared with that of a 494-GOPS/W SIMD accelerator [5], the reported evaluation result for that accelerator was the peak energy-consumption ratio only for the array part when the sub-threhold voltage is used, while the sustained performance for actual application programs running on an actual chip is shown here.

VII. SUMMARY

A "cool mega-array"(CMA), a high-energy-efficiency accelerator for embedded systems was devised from the concept up, and two CMA(CMA-1 and -2)chips wer fabricated and evaluated. CMA-1 is based on 65-nm CMOS technology in a 2.1 × 4.2-mm chip and achieves high energy efficiency; namely 2.72GOPS sustained performance with maximum power consumption of 11.2mW by using a combination of a PE array with a large combinatorial circuits and a small micro-controller with a sequentially accessed RAM module.

VIII. ACKNOWLEDGEMENT

This research was performed by Japan Science and Technology Agency [JST] of Core Research for Evolutional Science and Technology [CREST] as part of 「Technical improvement and itegration technology that aims at supper-energy saving of information systems」 of 「Super-low electric power of next generation efficient system LSI research by reformative power supply control」 program. And, CMA-1 chip was developed by the cooperation of STARC, e-shuttle Canpany and Fujitsu Japan ,through The University of Tokyo VLSI Design and Education Center.

REFERENCES

[1] C. Ebeling, D. C. Cronquist, and P. Franklin. RaPiD -Reconfigurable Pipelined Datapath. *International Workshop on Field-Programmable Logic and Applications(FPL04)*, Springer-Verlag:126–135, 1996.

[2] F.Clermidy, et.al. A 477mW NoC-Based Digital Baseband for MIMO 4G SDR. *ISSCC Dig. Tech. Papers*, pages 278–279, 2010.

[3] F.J.Veradas, M.Scheppler, W.Moffat, B.Mei. Custom Implementation of the Coarse-Grained Reconfigurable ADRES architecture for multimedia Purposes. *Proc. of International Conference on Field Programmable Logic and Applications (FPL05)*, pages 106–111, 2005.

[4] H.Amano, Y.Hasegawa, S.Tsutsumi, T.Nakamura, T.Nishimura, V.Tanbunheng, A.Parimala, T.Sano, M.Kato. MuCCRA chips: Configurable dynamically-reconfigurable processors. In Proc. of the ASSCC '07. IEEE Asian, Nov. 2007.

[5] H.Kaul, and et.al. A 300mV 494GOPS/W Reconfigurable Dual-Supply 4-Way SIMD Vector Processing Accelerator in 45nm CMOS. *ISSCC Dig. Tech. Papers*, pages 259–260, 2009.

[6] L.Cheng, et.al. Glitch Map: An FPGA Technology Mapper for Low Power Considering Glitches. *Proc. of DAC 2007*, pages 318–323, 2007.

[7] M. Motomura. STP Engine, a C-based Programmable HW Core featuring Massively Parallel and Reconfigurable PE Array:its Architecture, Tool, and System Implications. *Prof. of CoolChips XII.*, 2009.

[8] M. Motomura. Stp engine, a c-based programmable hw core featuring massively parallel and reconfigurable pe array:its architecture, tool, and system implications. *Prof. of CoolChips XII.*, 2009.

[9] T. Sakurai and A. R. Newton. Alpha-Power Law MOSFET Model and its Applications to CMOS Inverter Delay and Other Formulas. *IEEE Journal of Solid-State Circuits*, 25(2):584–594, Apr. 1990.

[10] V. Tunbunheng and H. Amano. Black-Diamond: a Retargetable Compiler Using Graph with Configuration Bits for Dynamically Reconfigurable Architectures. *Proc. of The 14th SASIMI*, pages 412–419, 2007.

[11] V.Tunbunheng, M.Suzuki, H.Amano. RoMultiC: Fast and Simple Configuration Data Multicasting Scheme for Coarse Grain Reconfigurable Devices. In Proc. of IEEE FPT, pages 129–136, 2005.

[12] Y.Saito and T.Sano and M.Kato and and V.Tunbunheng and Y.Yasuda and H.Amano. A Real Chip Evaluation of MuCCRA-3: A Low Power Dynamically Reconfigurable Processor Array. In *Proc. of Int'l Conf. on Engineering of Reconfigurable Systems and Algorithms (ERSA)*, 2009.

[13] Y.Tuyama, et.al. A 45nm 37.3GOPS/W Heterogeneous Multi-Core SoC. In *ISSCC Dig. Tech. Papers*, pages 100–101, 2010.

978-1-61284-863-1/11 $26.00 © 2011 IEEE

A New Source of Secure Pseudorandom Numbers Exploiting IMCGs Implemented in an FPGA

Mieczyslaw Jessa
Faculty of Electronics and Telecommunications
Poznan University of Technology
Poznan, Poland
mjessa@et.put.poznan.pl

Michal Jaworski
Faculty of Electronics and Telecommunications
Poznan University of Technology
Poznan, Poland
mj_2000@o2.pl

Abstract—**In this paper, we propose a new secure pseudorandom number generator that can be integrated with an arbitrary cryptographic system in the same field programmable gate array (FPGA). The described design uses the Improved Multiplicative Congruential Generators (IMCGs) defined recently. The specific properties of FPGAs and the IMCGs enable the generation of random numbers in only three cycles of the system clock. The output bit rate for eight IMCGs is 2.604 Gbit/s. It can be significantly increased using more IMCGs. The proposed generator can also produce secure bits as a specialized chip, which further increases the output bit rate.**

Keywords–pseudorandom number generators; FPGA; security; cryptography

I. INTRODUCTION

One of the most important elements of contemporary ciphers are pseudorandom number generators (PRNGs). The necessary but not sufficient condition for the security of PRNGs is very good statistical properties of the number sequences produced by these generators. Secure PRNGs must additionally ensure forward and backward unpredictability. Forward unpredictability means that subsequent numbers cannot be predicted from previously produced numbers. Backward unpredictability ensures that the seed cannot be found in a feasible time period from numbers produced by the PRNG. The structure of the PRNG is known and the only secret element is the key, which usually is the seed of the generator. During the last decades, many secure PRNGs were proposed. The most well-known are the RSA pseudorandom bit generators, Blum-Blum-Shub pseudorandom generators, nonlinear feedback shift registers (NLFSRs) and nonlinear combined LFSRs [1]. The described generators are slow, and the statistical properties of the output sequences often require additional improvement, known as postprocessing. Consequently, new solutions of secure PRNG and new implementations in FPGAs of new or known secure PRNGs is a very active research field.

In 2010, an enhancement of the well-known multiplicative congruential generator (MCG) was proposed [2]. The generator was named the Improved Multiplicative Congruential Generator (IMCG). The generator is an *l*-bit pseudorandom generator that can produce sequences with extremely long periods and very good statistical properties. Such sequences are obtained for a wide set of values of the

parameter characterizing the MCG. This property and significantly longer periods distinguish the proposed construction from the traditional MCG. The IMCG is easily implementable with integrated circuits, e.g., in FPGAs, and generates a new number in only three cycles of the system clock. The price is a delay in producing the first output bit by the IMCG. Implementing many IMCGs in the same FPGA significantly increases the speed of producing pseudorandom numbers. The basic disadvantage of the IMCG is the lack of security. In this paper, we propose a new secure PRNG that uses many IMCGs implemented in the same FPGA. The proposed generator requires only three cycles of the system clock to produce a new number, i.e., it needs the same number of slopes as a single IMCG or MCG. By increasing the number of IMCGs implemented in a reconfigurable circuit, we can achieve bit rates from a single Gbit/s to tens Gbit/s.

Section II contains information about the method of producing secure sequences with eight IMCGs. The security of the proposed generator, its implementation in Virtex-5 and the statistical properties of generated sequences are also presented. Section III contains basic conclusions and directions for future work.

II. A FAMILY OF SECURE PRNGS BASED ON IMCGS

A. The Improved Multiplicative Congruential Generator

The IMCG consists of three different functional blocks: the traditional Multiplicative Congruential Generator (MCG), the Combined Multiplicative Congruential Generator (CMCG) and the shuffling algorithm of Bays and Durham (Algorithm B) (Fig. 1). The key element of the IMCG is the CMCG. It improves the statistical properties of the numbers produced by the traditional MCG described by the formula [3]

$$p_n = \left(a \cdot p_{n-1} \right) \mod M, \quad n = 1, 2, \ldots \quad (1)$$

Improvements mean that sequences at the output of the CMCG pass all statistical tests described in "A Statistical Test Suit for Random and Pseudorandom Number Generators for Cryptographic Applications" [4]. The pseudocode describing the CMCG was published in [2]. Algorithm CMCG uses an additional table T with L cells. Algorithm B produces sequences with extremely long periods without degrading the statistical properties of the sequences produced by the CMCG. It was proposed by Bays and Durham to improve the quality of

978-1-61284-863-1/11 $26.00 © 2011 IEEE

the number sequences produced by MCGs [5],[6]. It can also be used to shuffle sequences produced by the CMCG.

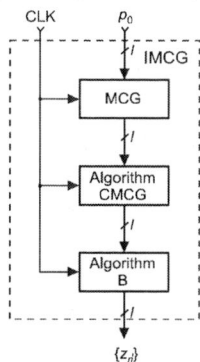

Figure 1. The structure of the IMCG [2].

Algorithm B requires an additional table T' with L' cells and initialization. The basic advantage of Algorithm B is the large period m_z of the output sequence obtained for L' $>>1$ [5], [6]:

$$m_z = O(m_u L'!)^{0.5}, \qquad (2)$$

where m_u is the period of the output sequence of the CMCG. Period (2) is achieved for

$$L' << m_u << L'!. \qquad (3)$$

If m_u does not satisfy (3), the period of $\{z_n\}$ either does not change or increases slightly [6].

The IMCG efficiently improves the quality of the maximal length sequences (m-sequences) produced by the MCG for a wide set of values of the multiplier a. Shorter sequences are not considered. Table I lists the values of the multiplier a that yield sequences that pass all NIST 800-22 statistical tests for exemplary values of M, a, R, L, and L'.

TABLE I.   THE VALUES OF MULTIPLIER a YIELDING SEQUENCES THAT PASS ALL NIST 800-22 TESTS FOR GENERATOR IMCG; $l = 31$, $M = 2^{31} - 1$, $p_0 = 1$, $\alpha = 16$, $L' = L = 256$, $R = 8$

No.	a	No.	a	No.	a	No.	a
1	7	6	103	11	181	16	313
2	11	7	127	12	263	17	317
3	31	8	131	13	269	18	331
4	53	9	139	14	277	19	353
5	73	10	149	15	307	20	457

During the tests, we applied two approaches proposed by NIST: (1) we examined the proportion R_β of sequences that pass the statistical test, and (2) we examined the distribution of P-values computed by the software; i.e., the value of P_T [4]. We also assumed the standard set of parameters proposed by NIST in v. 1.8. The significance level was $\beta = 0.01$. For each a, 1000 sequences of length 10^6 bits were tested. Sequences produced by the MCG failed at least one test for all a values considered.

The implementation of an IMCG described in [2] uses an array of 256 31-bit elements in a distributed RAM area. This approach resulted in significant utilization of the FPGA area, and further optimization is required for many IMCGs. Therefore, we decided to decrease the size of table L from 256

cells to 64 cells. Because the number of XOR-ed elements is critical for the quality of the sequences, we leaved the same R, i.e., $R=8$. In general, R can be larger, but it should not be smaller than 8. Because $L \geq \alpha \cdot R$ [2], we decreased the value of the parameter α from 16 to 8. Such values of α, R and L degrade the statistical properties of streams produced by IMCGs, but the sequences produced by the proposed PRNG still pass all NIST 800-22 tests for many a. Now, the set of available values of a is less numerous (see Table II).

TABLE II.   THE VALUES OF MULTIPLIER a YIELDING SEQUENCES THAT PASS ALL NIST 800-22 TESTS FOR GENERATOR IMCG; $l = 31$, $M = 2^{31} - 1$, $p_0 = 1$, $\alpha = 8$, $L = 64$, $L' = 256$, $R = 8$

No.	a	No.	a	No.	a	No.	a
1	---	6	103	11	---	16	313
2	11	7	---	12	263	17	317
3	31	8	131	13	269	18	331
4	53	9	139	14	277	19	---
5	73	10	149	15	---	20	457

B.   The Proposed Secure PRNG

To construct a secure PRNG, we need $K = 4j$ IMCGs, where $j > 1$. The generators are divided into four groups: A, B C, and D. Generators $A_1,...,A_j$, $B_1,...,B_j$, $C_1,...,C_j$, and $D_1,...,D_j$ produce numbers $p_{A_1}(k),...,p_{A_j}(k)$, $p_{B_1}(k),...,p_{B_j}(k)$, $p_{C_1}(k),...,p_{C_j}(k)$, and $p_{D_1}(k),...,p_{D_j}(k)$, respectively, with $k = 1,2,...$. The all numbers are encoded by l=31 bits. To simplify notation, we used index k instead n' introduced in [2]. The most significant bits of numbers $p_{A_1}(k),...,p_{A_j}(k)$ and $p_{B_1}(k),...,p_{B_j}(k)$ are added modulo 2 and yield symbol $s_0(k)$. The same operation is performed on the most significant bits of numbers $p_{C_1}(k),...,p_{C_j}(k)$ and $p_{D_1}(k),...,p_{D_j}(k)$. The result is symbol $s_1(k)$ (Fig. 2). Both symbols are encoded by one bit.

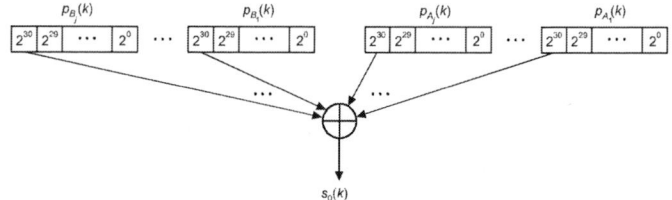

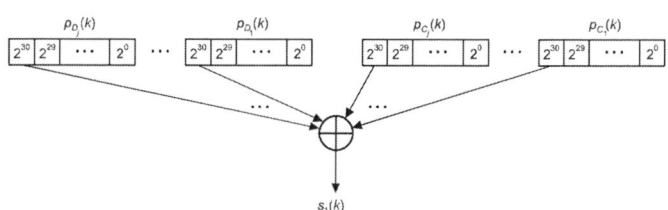

Figure 2. Producing two one-bit symbols $s_0(k)$ and $s_1(k)$ by the K IMCGs

The combination of both symbols indicates numbers for which binary representations are combined XOR. The output is j numbers $y_1(k),...,y_j(k)$. The rule is the following:

$$s_1(k) = 0 \wedge s_0(k) = 0 \Rightarrow \begin{cases} y_1(k) = p_{A_1}(k) \oplus p_{C_1}(k) \\ y_2(k) = p_{A_2}(k) \oplus p_{C_2}(k) \\ \vdots \\ y_j(k) = p_{A_j}(k) \oplus p_{C_j}(k) \end{cases}, \quad (4)$$

$$s_1(k) = 0 \wedge s_0(k) = 1 \Rightarrow \begin{cases} y_1(k) = p_{B_1}(k) \oplus p_{C_1}(k) \\ y_2(k) = p_{B_2}(k) \oplus p_{C_2}(k) \\ \vdots \\ y_j(k) = p_{B_j}(k) \oplus p_{C_j}(k) \end{cases}, \quad (5)$$

$$s_1(k) = 1 \wedge s_0(k) = 0 \Rightarrow \begin{cases} y_1(k) = p_{A_1}(k) \oplus p_{D_1}(k) \\ y_2(k) = p_{A_2}(k) \oplus p_{D_2}(k) \\ \vdots \\ y_j(k) = p_{A_j}(k) \oplus p_{D_j}(k) \end{cases}, \quad (6)$$

$$s_1(k) = 1 \wedge s_0(k) = 1 \Rightarrow \begin{cases} y_1(k) = p_{B_1}(k) \oplus p_{D_1}(k) \\ y_2(k) = p_{B_2}(k) \oplus p_{D_2}(k) \\ \vdots \\ y_j(k) = p_{B_j}(k) \oplus p_{D_j}(k) \end{cases}. \quad (7)$$

The K seeds of the IMCGs form the secret key

$$p_0 = \{ p_{A_0}(0), ..., p_{A_j}(0), p_{B_0}(0), ..., p_{B_j}(0), \\ p_{C_0}(0), ..., p_{C_j}(0), p_{D_0}(0), ..., p_{D_j}(0) \} . \quad (8)$$

The other parameters of IMCGs are known. The output of the generator is a fixed-point real number $y(k) = 0.y_1(k)y_2(k)...y_j(k)$ from the unit interval, i.e., $y(k) \in [0,1)$ or an integer number $z(k) = y_1(k)y_2(k)...y_j(k)$. The values of $z(k)$ belong to interval $[0, 2^{lj} - 1]$.

C. Security of the Proposed PRNG

The security of the proposed generator is based on two mechanisms: the random choice of numbers (generators) with binary representations that are combined XOR and discarding a large number of bits produced by K IMCGs. Because IMCGs produce uniformly distributed random numbers, any subsequence of digits of a random number can be used to form other uniform random numbers [6]. Thus, sequences $\{s_0(k)\}$ and $\{s_1(k)\}$ are also random, and zero and one are equally probable. Because no knowledge about the values of symbols $s_0(k)$ and $s_1(k)$ is available, the adversary cannot determine which of the numbers produced by the IMCGs are bitwise XOR-ed. Knowing only the result $y_i(k)$, $i=1,...,j$, the adversary cannot find the proper values for $p_{A_1}(k),...,p_{A_j}(k)$, $p_{B_1}(k),...,p_{B_j}(k)$, $p_{C_1}(k),...,p_{C_j}(k)$, $p_{D_1}(k),...,p_{D_j}(k)$ or the generators that provided the numbers for computing $y_i(k)$. Consequently, the adversary cannot concentrate his/her attack on the chosen generators. He/she must instead take into account the output of all generators for all iterations. It requires

studying the values of l-bit numbers produced by K independent generators. This process is equivalent to analyzing the value of a Kl-bit number, where only a part of the bits (unknown to the adversary) is used to compute $y_1(k),...,y_j(k)$. For l=31 and K=8 IMCGs, there are 2^{248} possibilities. Therefore the computation of the secret key is not possible in a feasible amount of time (backward unpredictability).

Although the numbers that provide a given $y_1(k),...,y_j(k)$ are unknown to the adversary, he/she can use any periodic behavior for predicting the next set of values. The IMCGs produce K independent numbers per iteration. The periods of the K sequences determined by subsequent iterations are enormous and the numbers are uniformly distributed in the interval $[0, 2^l - 1]$. Although the symbols $s_0(k)$ and $s_1(k)$ are equally probable and unknown to the adversary, the periods of sequences $\{s_0(k)\}$ and $\{s_1(k)\}$ may be significantly shorter. If they are shorter than the required number of numbers $y(k)$ or $z(k)$, it can simplify an attack. In this case, the adversary can focus on a smaller number of IMCGs. Because no theory is available to determine the shortest periods of $\{s_0(k)\}$ and $\{s_1(k)\}$, we must use simulations. During simulation, we check if the periods of sequences $\{s_0(k)\}$ and $\{s_1(k)\}$ are greater than or equal to the number of numbers we are going to produce. For example, we can assume that the PRNG should provide at least $N=10^{10}$ numbers, which corresponds to approximately 300 Gbits of encrypted binary data for a single secret key. Because the proposed IMCGs improve the properties of the maximal length sequences produced by the MCGs, it is sufficient to check only the periods for one initial point, i.e., for a single secret key. As long as the number of numbers used in a given application is not greater than the assumed N, no regular behavior is observed in the mechanism choosing the IMCGs. Because numbers produced by these generators are independent, uniformly distributed and they do not reveal regular behavior, we expect that the subsequent numbers in the output sequence cannot be predicted in a feasible amount of time from previously generated numbers (forward unpredictability).

D. An Implementation of the Proposed Generator in an FPGA

The method of producing $y_i(k)$, $i=1,...,j$ assumes the use of K Improved Multiplicative Congruential Generators, where K is a multiple of 4. Because K=4 does not ensure security, the smallest K for practical applications is 8. Table III contains the values of a assigned to generators $A_1,...,A_j$, $B_1,...,B_j$, $C_1,...,C_j$, and $D_1,...,D_j$. They were chosen from Table II.

TABLE III.   THE VALUES OF a ASSIGNED TO EIGHT IMCGS ($j = 2$)

A_1	A_2	B_1	B_2	C_1	C_2	D_1	D_2
a=11	a=31	a=53	a=73	a=103	a=131	a=139	a=149

Different assignments are possible, but the values of a assigned to the generators should be relatively prime. If this condition is not satisfied, the same numbers can be repeated regularly in the sequences produced by different IMCGs. In the experiments, we assumed that all the IMCGs started from the same seed, equal to 1, which yielded the key $p_0 = \{1,1,1,1,1,1,1,1\}$.

In the first step, we generated in parallel pseudorandom numbers using eight multiplicative congruential generators with values of parameter a from Table III. Implementation of the MCGs is based on a three-state machine (FSM). We generate one pseudorandom number for each MCG every three clock cycles. The MCGs were programmed in Verilog, and the only difference between the generators was the method used to multiply by a. We do not use a hardware multiplication block or a specialized FPGA. We multiply by shifting and adding/subtracting the relevant components. For example, multiplying a number p by 11 can be written as $(2^3 p + 2p + p)$. For a=31, we obtain $(2^5 p - p)$. The number produced by the MCG is provided to the next block, which implements the CMCG algorithm. The eight blocks work in parallel with the same clock and are all based on an FSM. The operations are synchronized with the previous and next steps of the proposed algorithm.

The next part of the hardware shuffles the values according to Algorithm B. In this block, there is no need to read data simultaneously. Therefore, Block RAM can be used, and the value for L' can be large. In our case L'= 256. This block, similar to the two previous blocks, was also designed as a three-state machine. This approach to the designs of the MCG, CMCG and Algorithm B ensures that the IMCG can provide a new pseudorandom number every three clock cycles.

From the eight simultaneously generated numbers coming from the corresponding eight generators, we next select two pairs and combine modulo 2 bits of numbers from each pair. This step of the algorithm is implemented using two 2 to 1 multiplexers (62-bit output) and one 62-bit modulo 2 operation. The basic results for the Virtex-5 – XC5VLX220T used in our design are the following (synthesized with Xilinx ISE 12.3 using default settings for synthesis and route; mapping has option "Optimization Strategy (Cover Mode)" set to Speed):

Place and Route Report (PAR):

Device Utilization Summary:

Number of BUFGs	1 out of 32	3%
Number of External IOBs	316 out of 680	46%
Number of LOCed IOBs	0 out of 316	0%
Number of OLOGICs	248 out of 800	31%
Number of RAMB18X2SDPs	8 out of 212	3%
Number of Slices	15036 out of 34560	43%
Number of Slice Registers	18054 out of 138240	13%
Number used as Flip Flops	18054	
Number used as Latches	0	
Number used as LatchThrus	0	
Number of Slice LUTS	41911 out of 138240	30%
Number of Slice LUT-Flip Flop pairs	48331 out of 138240	34%

Minimum period: 7.911 ns (maximum frequency 126.326 MHz)

Assuming a clock frequency of 126 MHz, the output bit rate is 2.604 Gbit/s. Choosing 12 IMCGs, we obtain an output rate of 3.906 Gbit/s. For 16 IMCGs, the output rate is 5.208 Gbit/s, and it continues to increase with increasing K.

E. Statistical Test Results

The results of the NIST 800-22 statistical tests obtained for K=8 are shown in Table IV. The minimum passing value for the standard set of parameters was approximately 0.9805. The minimum P_T value was 0.0001. An asterisk * denotes that this test consists of several subtests and that the worst result is shown. For tests marked with **, the minimum passing value for the standard set of parameters was approximately 0.9777. The results for K=12 and K=16 are very similar.

TABLE IV. THE RESULTS OF NIST 800-22 TESTS FOR 1000 SEQUENCES OF LENGTH 10^6 BITS; $l = 31$, $M = 2^{31} - 1$, $\alpha = 8$, $R = 8$ $L = 64$, $L' = 256$

Type of test	R_β	P_T
Frequency	0.9920	0.35373
Block Frequency	0.9920	0.09485
Cumulative Sums*	0.9940	0.13806
Runs	0.9910	0.32520
Longest Run of Ones	0.9900	0.27984
Rank	0.9930	0.71567
Spectral DFT	0.9890	0.06282
Non-overlapping Temp.*	0.9820	0.00862
Overlapping Templates	0.9920	0.38040
Universal	0.9880	0.46923
Approximate Entropy	0.9880	0.05328
Random Excursions*	0.9865	0.05823
Random Exc. Var.**	0.9831	0.00133
Serial*	0.9880	0.29251
Linear Complexity	0.9860	0.03781

III. CONCLUSIONS

In this paper, we proposed a new secure pseudorandom generator that can be entirely integrated in a single FPGA with an arbitrary digital encryption/decryption system. The described generator uses K Improved Multiplicative Congruential Generators defined recently. The values of K are a multiple of 4. In practice they are limited only by the available resources of the FPGA used in the design. The open research problem is the analytical determination of the periods of the sequences $\{s_0(k)\}$ and $\{s_1(k)\}$.

REFERENCES

[1] A. J. Menezes, P. C. van Oorschot and S. C. Vanstone, Handbook of Applied Cryptography. Boca Raton: CRC, 1997.

[2] M. Jessa, M. Jaworski, "High-Speed FPGA-Based Pseudorandom Generators with Extremely Long Periods," in *Proc. Int. Conf. on ReConFigurable Computing and FPGAs, ReConFig'10*, Dec. 2010, pp. 286-291.

[3] G. S. Fishman, Discrete-Event Simulation, New York: Springer, 2001.

[4] A. Rukhin, J. Soto, J. Nechvatal, M. Smid, E. Barker, S. Leigh, M. Levenson, M. Vangel, D. Banks, A. Heckert, J. Dray, S. Vo, A statistical test suite for random and pseudorandom number generators for cryptographic applications, NIST special publication 800-22, *National Institute of Standards and Technology*, 2001, Revised 2010, USA, Available at: http://csrc.nist.gov/rng/

[5] C. Bays and S. D. Durham, "Improving a poor random number generator," *ACM Trans. on Mathematical Software*, vol. 2, 1976, pp. 59–64.

[6] J. E. Gentle, Random Number Generation and Monte Carlo Methods, New York: Springer, 2003.

A Reed-Solomon Architecture for Soft-Core Implementation

Thullyo D. C. R. Ferreira, Luís H. C. Ferreira, Robson L. Moreno and Tales C. Pimenta

Federal University of Itajubá, Itajubá, 37500-903, Brazil (e-mail: thullyod@gmail.com; luis@unifei.edu.br).

Abstract—**This work presents a Reed-Solomon architecture for soft-core implementation, which is able to provide a robust communication in noisy links. It offers low power consumption by using a software oriented implementation in PicoBlaze that also reduces the used area of the logical device in FPGA. The implementation improves the BER performance and thus increase the communication range (up to 50%) and can be used in applications such as remote monitoring Wireless Sensor Network (up to 15% in noisy indoors) and medical applications (up to 31% in indoors with $BER = 10^{-7}$).**

Index Terms—**Reed-Solomon channel coding, soft-core, wireless sensor network, medical application, bit error rate, power reduction, PicoBlaze.**

I. Introduction

In 1948, Claude E. Shannon published his work on codification theory and proved the existence of codes that can minimize the probability of erroneous decoding. One of the basic postulates is that information can be treated like a measurable physical quantity, such as density or mass [1]. In 1960 Irving S. Reed and Gustave Solomon developed and published the Reed-Solomon - RS code [2], so powerful that can be found today in many applications, from compact disc players to deep-space communications [3]. The RS codes provide powerful burst error correction capability, thus benefiting real time applications and real time transmissions. The RS is an Error Correction Code - ECC that allows the recovery of a certain amount of error during data transmission without having to resend the data itself, thus increasing the communication links and providing power gain [4].

Two applications of ECC are highlighted in this article. The first is remote monitoring using Wireless Sensor Network - WSN for medical, security, surveillance, and industrial applications [5]. The second is biomedical implantable electronic devices that require power and data to be transferred by inductive coupling [6]. The main constraints of such systems are reliability, area, timing and efficiency, however the main issue is power consumption [7]. The ECC is used to reduce the power consumption in those applications.

The RS codes reduce the amount of energy per bit during the transmission of the information while maintaining the same Bit Error Rate - BER. This energy save allows the increase of communication range. Nevertheless it is important to observe whether the power savings can be achieved at the expense of power consumption in the encoding/decoding stages. This article proposes an implementation using a simple soft-core processor, like PicoBlaze by Xilinx®, to reduce power consumption in these stages since a simple soft-core

requires a small number of logic components, thus reducing the system power consumption.

The article focuses on a software implementation of an RS coding on a simple 8-bit microcontroller (PicoBlaze). Even though the soft-core processor has a simple architecture, it is capable of running RS codes, thus implementing the encoder and decoder. It was used the Peterson-Gorenstein-Zierler - PGZ [8] algorithm as the error-locator polynomial since it provides the simplest implementation for error-correction capability less than 4. The system was designed for applications in remote monitoring field and implantable medical devices.

II. RS Code and PGZ Review

The RS code is one of the most popular and it has been a long-time industry-standard ECC that has found numerous applications in communication systems, such as the NASA space communication RS(255, 233), digital storage on CD and DVD, HDTV transmissions and wireless communications. It is widely used because it offers great capacity of correcting both random and burst errors [4].

Reed-Solomon is a non-linear non-binary cyclic block code, usually denoted as $RS(n,k)$ where n indicates the total number of symbols of m bits in a coded block and k indicates the number of symbols in the original message. The RS code is capable of correction t errors where $t = (n - k)/2$ [3].

The RS code was developed based on abstract algebra and uses finite Field theory, known as Galois Field - GF. The GF is implemented according to a primitive polynomial. The polynomial must be primitive, so that all of its roots are primitive elements [9].

A. Encoding

Let polynomial $m(x)$ denote the message to be transmitted or stored. The message $m(x)$ is systematically encoded, thus generating the codeword parity symbols, $p(x)$, concatenated to the initial message and expressed as

$$c(x) = p(x) + x^{n-k}m(x). \tag{1}$$

The codeword is generated using a generator polynomial $g(x)$ given by Eq. (2) [9], so the roots of the generator polynomial must also be the roots of the codeword. The polynomial $p(x)$ is the remainder of the division between $x^{n-k}m(x)$ and $g(x)$.

$$g(x) = \prod_{i=b}^{b+2t-1} (x + \alpha^i), \tag{2}$$

978-1-61284-863-1/11 $26.00 © 2011 IEEE

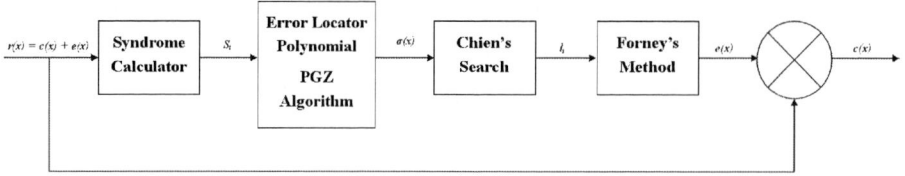

Fig. 1. RS Decoder Flow

where b is an initial number and α is a element of the field.

B. Decoding

The data received, $r(x)$, may be corrupted by the channel noise during the transmission and can be represented as

$$r(x) = c(x) + e(x), \qquad (3)$$

where $e(x)$ represents the error pattern. The RS decoder attempts to correct the errors by means of polynomial operations [7]. The *Syndrome Calculator* is the first block as shown in Fig. 1. It evaluates if data correction processing is necessary, by checking whether the data received contains errors ($e(x) \neq 0$).

The $2t$ syndrome values, S_i, are calculated by evaluating the received polynomial $r(x)$ at all the roots of the generator polynomial, α_i. The syndrome values are used to calculate the Error Locator Polynomial, $\sigma(x)$, as indicated by the flow diagram of Fig. 1. The equation of the calculate of the syndrome values can be written as [8]

$$S_i = r(\alpha^i) = \sum_{j=0}^{n-1} r_j(\alpha^i)^j \qquad 1 \leq i \leq 2t. \qquad (4)$$

The Syndrome polynomial can be expressed by

$$S(x) = \sum_{i=0}^{2t-1} S_{i+1}x^i. \qquad (5)$$

Assuming that the data received has ν errors at the locations $x^{j_1}, x^{j_2}, ..., x^{j_\nu}$. So the Error polynomial, $e(x)$, can be written as

$$e(x) = e_{j_1}x^{j_1} + e_{j_2}x^{j_2} + ... + e_{j_\nu}x^{j_\nu}. \qquad (6)$$

To correct the corrupted codeword each value of error e_{j_l} and its location x^{j_l} where $l = 1, 2, ..., \nu$, must be determined.

If $S(x) \neq 0$, it means that de data received contains errors. Therefore it is necessary to find the error locations. We have used the PGZ algorithm to obtain the error locator polynomial, $\sigma(x)$, that includes two main steps. Solving *Newton Identity* is the first step [8]:

$$\begin{bmatrix} S_2 & S_3 & ... & S_{t+1} \\ S_3 & S_4 & ... & S_{t+2} \\ \vdots & \vdots & \ddots & \vdots \\ S_{t+1} & S_{t+2} & ... & S_{2t} \end{bmatrix} \begin{bmatrix} \sigma_{t-1} \\ \sigma_{t-2} \\ \vdots \\ \sigma_0 \end{bmatrix} = \begin{bmatrix} -S_1 \\ -S_2 \\ \vdots \\ -S_t \end{bmatrix}. \qquad (7)$$

Then it is used syndrome values to find the σ values in Eq. (7). The *Error Locator Polynomial* is defined as

$$\sigma(x) = \sigma_0 + \sigma_1 x + ... + \sigma_{t-1}x^{t-1} + x^t. \qquad (8)$$

The second step is to solve the *Key equation*

$$\sigma(x)S(x) = -\omega(x) + \mu x^{2t}, \qquad (9)$$

where the *Error value polynomial* is defined as

$$\omega(x) = \omega_0 + \omega_1 x + ... + \omega_{t-1}x^{t-1}. \qquad (10)$$

In order to avoid malfunction of the PGZ algorithm, it was implemented the technique presented in [10]. After forming the error locator polynomial of Eq. (8) and the error value polynomial of Eq. (10)it is used the Chien's search and the Forney's method to form the error polynomial of Eq. (6). By using Eq. (3), the correct codeword can be obtained from

$$U(x) = r(x) + e(x) = c(x). \qquad (11)$$

The described algorithm was implemented in software using a simple soft-core processor (PicoBlaze) to reduce size and power consumption, as it will be presented next.

III. SOFT-CORE IMPLEMENTATION

A *soft-core processor* is a microprocessor fully described in Hardware Description Language - HDL, and can be synthesized in programmable devices, such as Field-Programmable Gate Array - FPGA. A soft-core processor implemented in FPGA is flexible because its parameters can be changed at any time by reprogramming the device [11].

The PicoBlaze is a cost-free 8-bit microcontroller architecture. It was chosen to implement the RS coding proposed in this paper due to its small size: approximately 130 slices (2%) of a Spartan-3E XC3S500E, where each slice is comprised of two Look-Up Tables - LUT and two D Flip-Flops. It runs at 113.666 MHz or 56.833 Million Instructions per Second - MIPS [12].

It was implemented the algorithm described in Section II, but the amount of elements needed depends on its variables n and k. Since the PicoBlaze is a small microcontroller, the memory available to store of these elements is limited to 64 locations, thus reducing the number of possible implementations. According to Table I, the maximum value for n is 15, since any higher value requires more memory than available by the PicoBlaze. It was used the rate k/n, presented in the

978-1-61284-863-1/11 $26.00 © 2011 IEEE

TABLE I
RATES OF CODES FOR VARIOUS RS CODES (MINIMUM MEMORY = $2 * 2^m + n$

	Minimum Memory	Rate of the code
RS(7, 3)	23	0.4
RS(7, 5)	23	0.7
RS(15, 7)	47	0.5
RS(15, 9)	47	0.6
RS(15, 11)	47	0.7
RS(15, 13)	47	0.9
RS(31, 23)	95	0.7

third column of Table I, as a comparison between the available codes. It was chosen the value 0.7, so that both $RS(7,5)$ and $RS(15,11)$ can be implemented and compared. They have the same area and the same *rate of code*, changing only the number of clock cycles.

By using the available tools, both codes were simulated and the approximate numbers of clock cycles are shown in the Table II. This table shows that at the same *code rate* and same area, the number of clock cycles is much lower for the $RS(7,5)$.

TABLE II
APPROXIMATE CLOCK CYCLES

	RS(7, 5)	RS(15, 11)
Encoding	1080	3420
Decoding	2380	13450
Total	**3460**	**16870**

Table III shows the decoding process latency for both codes. The latency of the second implementation is almost six times the latency of the first one.

TABLE III
RS(7, 5) AND RS(15, 11) COMPARISON

	RS(7, 5)	RS(15, 11)
Latency	$21\mu s$	$118.3\mu s$
Rate	0.714	0.733

IV. RS CODING GAIN

The RS symbol-error decoding probability, P_E, in terms of the channel symbol-error probability, p, can be described using the discrete probability distribution called *Bernoulli trial* and can be described as [3]

$$P_E \approx \frac{1}{2^m - 1} \sum_{j=t+1}^{2^m-1} j \binom{2^m - 1}{j} p^j (1-p)^{2^m-1-j}. \quad (12)$$

The bit-error probability P_B can be up bounded by the symbol-error probability for specific modulation types. For Multiple Frequency-Shift Keying - MFSK modulation with $M = 2^m$, the relationship between P_B and P_E is [13]

$$\frac{P_B}{P_E} = \frac{2^{m-1}}{2^m - 1}. \quad (13)$$

Figure 2 shows P_B versus E_b/N_0 for both uncoded and coded system using a specific modulation over an Additive White Gaussian Noise - AWGN channel.

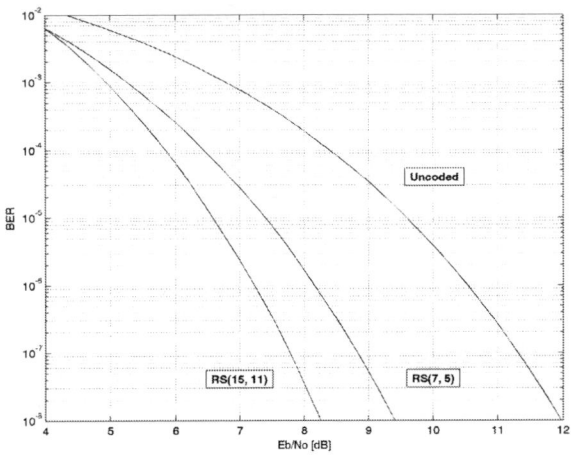

Fig. 2. BER versus E_b/N_0 performance of $RS(7,5)$ and $RS(15,11)$, t-error correcting RS coding systems with specific modulation over an AWGN channel

The coding gain of the RS code is usually calculated by setting the desired BER of both uncoded and coded systems for a specific modulation and then measuring the difference between their Signal-to-Noise Ratios - SNR [7]. For the two applications mentioned above was chosen two values of BER, 10^{-3} and 10^{-7}. It was chosen BER equals 10^{-3} for a channel coding WSN of a remote patient monitoring [7]. This high value considers a very noisy channel. It was also chosen BER equals 10^{-7} for an implantable medical device using a Binary Phase Shift Keying- BPSK modulation [6]. Table IV shows their gains, by observing the values of Fig. 2.

TABLE IV
GAINS USING RS CODES AT $BER = 10^{-3}$ AND $BER = 10^{-7}$

BER	10^{-3}	10^{-7}
RS(7, 5)	1.5484[dB]	2.4919[dB]
RS(15, 11)	1.8750[dB]	3.5444[dB]

The required transmission power P_{TX} necessary to achieve a minimum SNR power S/N on the receiver at a distance d can be modelled by [14]

$$P_{TX}[W] = \eta_u \frac{E_b}{N_0} N \left(\frac{4\pi}{\lambda}\right)^2 d^n, \qquad (14)$$

where η_u is the spectral efficiency, N is the Signal Noise, λ is the transmitted wavelength and n is the path loss exponent ($n = 2$ is for free space, $n = 3$ is for indoors, and $n = 4$ is for indoors with many obstacles).

Assuming that the same channel and transmitter, then Eq. 14 can be simplified to

$$\frac{\left(\frac{E_b}{N_0}\right)_U}{\left(\frac{E_b}{N_0}\right)_{RS}} = \left(\frac{d_{RS}}{d_U}\right)^n, \qquad (15)$$

where d_{RS} is the distance between the transmitter and receiver using RS coding and d_U is the distance without using it.

Eq. (15) and the values shown in Table IV can used to obtain the distance d_{RS} as a function of d_U. Table V shows the those gains for the both BER at all environments, as a function of d_U.

TABLE V
RS GAIN FOR $RS(7,5)$ AND $RS(15,11)$

BER	10^{-3}			10^{-7}		
n	2	3	4	2	3	4
RS(7,5)	19.51%	12.62%	9.32%	33.23%	21.08%	15.42%
RS(15,11)	24.09%	15.48%	11.40%	50.39%	31.26%	22.63%
Difference	4.58%	2.86%	2.08%	17.16%	10.18%	7.21%

Therefore, by using the same *rate of the code*, it is possible to increase the communication range up to 15.48% in a noise indoor environment with $RS(15,11)$. For implantable medical devices, it is possible to increase the communication range up to 31.26% in an indoor environment by using the same RS coding. It could mean the difference of placing the implantable device in a more suitable location while having the reader in a more convenient position.

According to the last line of Table V, the gain difference between the two implementations is not significant for a $BER = 10^{-3}$, in addition to an almost six times higher latency, as shown in Table III. Nevertheless, the gain difference between the two implementations using ($BER = 10^{-7}$) justifies its implementation, mainly for implantable devices.

V. CONCLUSION AND FUTURE WORK

The RS coding implemented in this work offers energy savings in terms of BER. It is possible to reduce the power consumption of the transmitter to increase the battery lifespan or increase the communication range while using the same transmitter. The power reduction obtained with this implementation can be used for low levels of BER (such as implantable medical devices) or for higher levels of BER (such as WSN).

The computational time could be further decreased by using a microcontroller to implement the algebraic operations over Finite Fields in a single instruction. In that case, the number of clock cycles needed to execute the decoding would be drastically reduced. We are working on the implementation of a customized soft-core for those algebraic operations and a compiler to generate the HDL customized to the number of bits in each symbol, thus further decreasing the number of logic components.

ACKNOWLEDGEMENTS

The authors acknowledge CAPES, CNPq and FAPEMIG for their financial support.

REFERENCES

[1] C. E. Shannon, "A mathematical theory of communication," *Bell system Technical Journal*, vol. 27, pp. 379–423, 1948.

[2] I. S. Reed and G. Solomon, "Polynomial codes over certain finite fields," *Journal of the Society for Industrial and Applied Mathematics*, vol. 8, pp. 300–304, 1960.

[3] B. Sklar, *Digital Communications: Fundamentals and Applications*, 2nd ed. Prentice Hall, 2001.

[4] R. H. Morelos Zaragoza, *The Art of Error Correcting Coding*. Wiley, Sep. 2006.

[5] K. Römer and F. Mattern, "The design space of wireless sensor networks," *IEEE Wireless Communications*, vol. 11, no. 6, pp. 54–61, December 2004.

[6] Y. Hu and M. Sawan, "A fully integrated low-power bpsk demodulator for implantable medical devices," *Circuits and Systems I: Regular Papers, IEEE Transactions on*, vol. 52, no. 12, pp. 2552 – 2562, dec. 2005.

[7] R. McSweeney, C. Spagnol, E. Popovici, and L. Giancardi, "Implementation of source and channel coding for power reduction in medical application wireless sensor network," in *Sensor Technologies and Applications, 2009. SENSORCOMM '09. Third International Conference on*, june 2009, pp. 271 –276.

[8] S.-F. Wang, H.-Y. Hsu, and A.-Y. Wu, "A very low-cost multi-mode reed-solomon decoder based on peterson-gorenstein-zierler algorithm," in *Signal Processing Systems, 2001 IEEE Workshop on*, 2001, pp. 37 –48.

[9] T. Barbosa, R. Moreno, T. Pereira, and L. Ferreira, "Fpga implementation of a reed-solomon codec for otn g.709 standard with reduced decoder area," in *Wireless Communications Networking and Mobile Computing (WiCOM), 2010 6th International Conference on*, sept. 2010, pp. 1 –4.

[10] A. Dur, "Avoiding decoder malfunction in the peterson-gorenstein-zierler decoder," *Information Theory, IEEE Transactions on*, vol. 39, no. 2, pp. 640 –643, mar 1993.

[11] F. Plavec, "Soft-core processor design," Master's thesis, University of Toronto, 2004.

[12] Picoblaze documentation Xilinx corporation web site. [Online]. Available: http://www.xilinx.com/

[13] J. Odenwalder, "Error control coding handbook," Linkabit Corporation, Tech. Rep., 1976.

[14] S. L. Howard, C. Schlegel, and K. Iniewski, "Error control coding in low-power wireless sensor networks: When is ecc energy-efficient?" *EURASIP Journal on Wireless Communications and Networking*, pp. 74 812: 1 – 14, 2006.

Debugging Methodology for A Synthesizable Testbench FPGA Emulator

A.W. Ruan H.C. Huang C.Q. Li Z.J. Song Y. B. Liao W.Tang

State Key Laboratory of Electronic Thin Films and Integrated Devices, University of Electronic Science and Technology of China
Chengdu, 610054, China
Phone: +86-28-83201942, Email: *ruanaiwu@uestc.edu.cn*

Abstract—Logic simulation provides SOC verification with full controllability and observability, but it suffers from very slow simulation speed for complex design. Using hardware emulation such as FPGA can have higher simulation speed. However, FPGA emulation approach has some limitations, i.e. unsynthesizable testbench and poor visibility for debugging. We address these problems by presenting a testbench synthesis engine as well as providing internal nodes probing on DUT. The proposed testbench synthesis engine is built by hardware constructs in terms of Verilog IEEE Simulation Model to correspond with testbench. Internal nodes are hardware-wired to DUT top-level during compilation, then sampled continuously by a sample logic into on-chip storage device (e.g. Block RAM, SDRAM and etc). Thus full observability can be achieved without stopping of DUT clock. Our experiment shows that, compared with a similar method in [13], simulation time is independent of number of probing nodes.

Index Terms—FPGA, Emulation, Debugging, Testbench, Synthesis

I. Introduction

Increasingly complex and sophisticated VLSI design, coupled with shrinking design cycles, requires shorter verification time and efficient debug method. In the meanwhile, logic simulation still plays an important role for digital circuits. Simulation-based verification is characterized by two inherently conflicting targets: the signal visibility and simulation performance. Achieving a proper trade-off between these two targets is of significant importance. Even though HDL simulator is the most widely used verification platform at the RTL and gate level, its major drawback is the low performance in verifying complex SOCs. Simulation acceleration increase performance by evaluating the HDL constructs in parallel. Unfortunately, the bottleneck of the communication between software and hardware limits the maximum simulation speed.

Comparing with simulation acceleration, FPGA emulation has no constant connection to the workstation during execution and the hardware platform receives no input from the workstation. By eliminating the connection to the workstation, the hardware platform now runs at its full speed and does not need to wait for any communication. However, although FPGA emulation has long provided the highest performance when compared with all other verification approaches, it has also suffered from severe drawbacks. One of drawbacks is that designers have to restrict their coding style or transform a huge unsynthesizable testbench into synthesizable one, as the existing emulation platform cannot support unsynthesizable HDL coding style. On the other hand, poor observability is presented due to internal nodes of Design under Test (DUT) hidden from designer when DUT is mapped to FPGA emulator.

In this paper, debugging methodology for a FPGA emulator with testbench synthesis engine is presented. The testbench synthesis engine is implemented by building hardware constructs in terms of Verilog IEEE Simulation Model [1] to correspond with testbench. In this way, the engine can support all Verilog syntaxes. On the other hand, internal nodes are hardware-wired to DUT top-level during compilation, then sampled continuously by a sample logic into on-chip storage device (e.g. Block RAM, SDRAM and etc). When on-chip cache is full-filled, nodes value are then uploaded to host PC and displayed in waveform to designer for debugging. Thus full observability can be achieved.

The paper is organized as follows. Section II describes some related works for FPGA debugging. Section III is an introduction to the testbench synthesis engine for a FPGA emulator on which our debugging method is implemented. We propose our debugging method in Section IV and Section V shows the experiment result. In Section VI, we conclude our work.

II. Related Work

Several approaches of FPGA debugging have been considered by researchers. We will go through these different FPGA debug methods in this section.

Readback and scan chain logics are often used to acquire internal nodes data from FPGA [2-3]. But they have common drawbacks that reading all data from FPGA costs a long time, and they are manipulated in gate level. RTL scan insertion is described in [4], which allows debug in RTL level. Scan chain method often causes large area overhead.

In modern FPGA debug method, Embedded Logic Analyzer (ELA) is widely used [5-6]. ELA samples internal signals if only a predefined condition is met, so controllability is added to FPGA while design still executes at or close to normal speed. However, any modification to selected signals or trigger conditions often needs design recompilation. To avoid recompilation, a bitstream instrumentation method for ELA is described in [7]. But it requires ELA designed to be bitstream-modifiable, which lacks flexibility. Recently, ELA debug in high-level design, such as for Matlab Simulink, is discussed by researchers in [8].

978-1-61284-863-1/11 $26.00 © 2011 IEEE

Another FPGA debug method known as simulation reconstruction was developed long time ago and has been evolved recently [9]. A snapshot method for FPGA debug is proposed to record the behavior of FPGA and replay it in HDL simulator. Only a minimum of FPGA internal nodes are sampled and all other missing nodes are reconstructed in software simulator by some algorithm. But the reconstruction process in software simulator needs a long time if the design is large, and some extra issues should be considered to ensure the correctness of the reconstructed value.

Besides checking waveform, [10] proposes a hardware debugging method that continuously records transitions of selected signals and makes them into state-transition diagram for debug. And [11] introduces method to synthesize System Verilog Assertions and bring assertion-based verification into FPGA debug.

There are many commercial tools that provide powerful automated features for FPGA debug. The drawbacks of Xilinx's ChipScope and Altera's SignalTap are already examined by many researchers. Synopsys Synplicity brings an advanced ELA debug tool named Identify RTL Debugger [12]. Compared with ChipScope and SignalTap, Identify provides RTL-level signal selection and trigger condition setting. Since ELA is inserted at RTL level and synthesized together with the design, FPGA execution speed is barely affected by the insertion of debug logics. Identify allows designers to set complex and nested trigger conditions, which enhance its debug capability. Besides, external RAM can be used instead of the Block RAM of FPGA, so designer can choose more signals to observe. But like other ELAs, it cannot achieve full visibility and user often does not know how to set the trigger condition to find design errors.

To overcome these limitations, a new RTL debugging methodology for FPGA-based cosimulation system is proposed in [13]. It introduces an internal-node probing method that within FPGA, which can reach 100% observability on iProve cosimulation platform. However, this method can be applied only to synchronous DUT circuit due to its cycle based synchronization, so any timing-related information will be lost when applied to gate-level simulation. It also needs a post process to generated VCD waveform for DUT internal nodes after simulation, so run-time debug is not allowed. Signals in testbench and DUT are displayed in separate waveforms, which make debugging inconvenient.

III. TESTBENCH SYNTHESIS ENGINE

The synthesizable testbench mode is a static emulation mode, that we map the entire design into FPGA emulator for high-performance verification. Fig.1 shows the block diagram of the proposed synthesizable tetbench emulator. Host PC can only access to the emulator for infrequent debugging purpose, such as displaying a message, loading content of a file to emulator from Host PC and etc. Details of the testbench synthesis engine are introduced in [15].

The flow of synthesizable testbench verification is shown in Fig.2. The testbench portion of user's design is behavioral

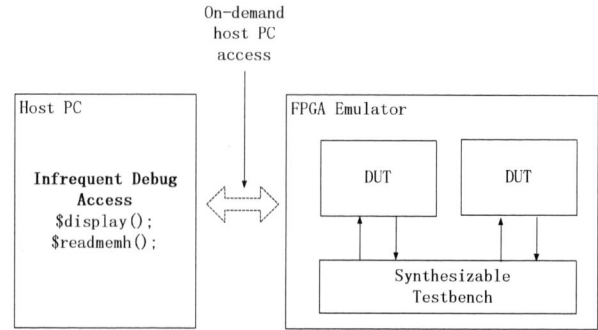

Fig.1. Proposed synthesizable testbench FPGA emulator

synthesized by an in-house developed behavioral compiler. After behavioral synthesis, synthesizable format of the testbench is generated. The synthesizable testbench is then simulated by a behavioral simulator within FPGA. The Behavioral simulator is an event-accurate Verilog simulator implemented by FPGA resources. It is the key element for simulating behavioral testbench using FPGA emulator. Synthesizable DUT portion is directly synthesized and mapped to FPGA emulator. DUT module and synthesizable testbench is then connected together through ports to form the entire design.

It can be seen that in synthesizable testbench verification mode, software logic simulator is no longer needed, and the entire design is mapped to FPGA emulator for fast execution.

In order to simulate Verilog behavioral testbench in FPGA emulator, we design a behavioral simulator in FPGA, which is a synthesizable simulation engine following the simulation

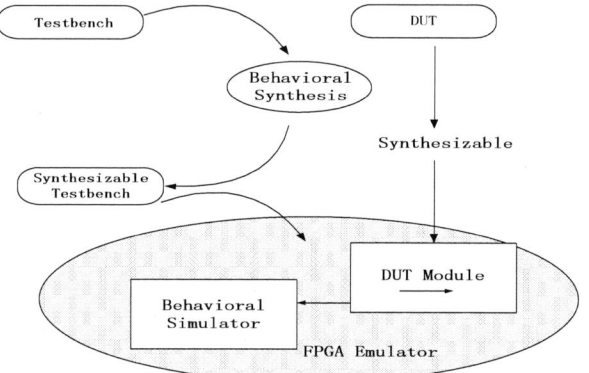

Fig.2. Synthesizable testbench verification flow

scheduling semantics that defined by IEEE Verilog Standard. The behavioral simulator can be used to simulate Verilog language in event-accurate style, so any form of behavioral testbench can be simulated in FPGA with the same functionality as it has in a software HDL simulator.

The behavioral simulator is built using completely synthesizable logics, making it possible to be mapped into FPGA emulator for fast execution. The basic architecture of the proposed behavioral simulator is depicted in Fig.3.

As shown in the figure, user's design, including both testbench and DUT, is compiled and converted into a structure that consists of several simulation elements and DUT modules.

978-1-61284-863-1/11 $26.00 © 2011 IEEE   594

A simulation element is generated from a single Verilog module using our proposed behavioral synthesis algorithm, representing a testbench module. And DUT keeps its original logic structure and connects to simulation elements through its ports. In the behavioral simulator architecture, a simulation scheduler is presented to manage simulation time. All simulation elements are connected to this unique simulation scheduler, so they can be synchronized to the same simulation time, but execute in parallel within each simulation cycle.

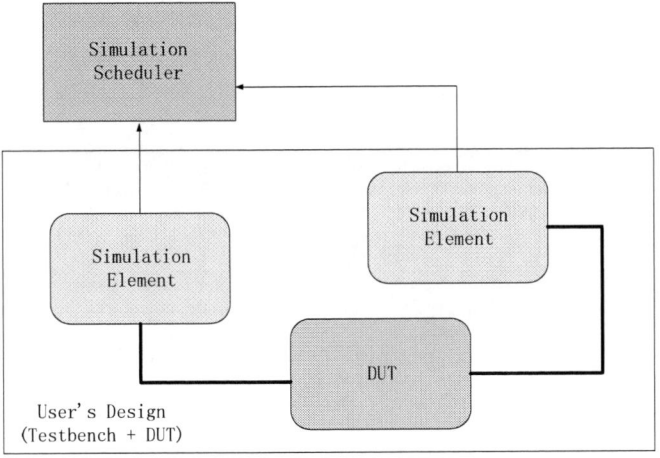

Fig.3. Architecture of the proposed behavioral simulator

IV. DEBUGGING DESIGN FLOW

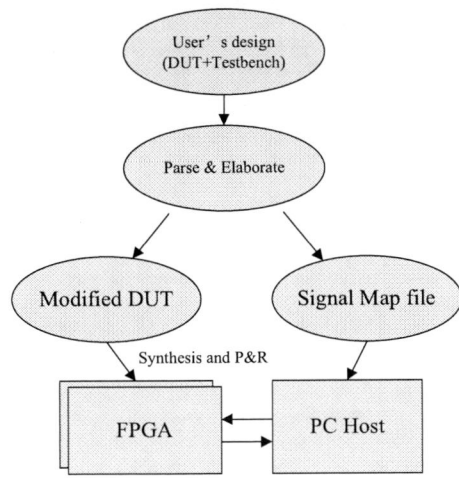

Fig.4. design flow

Fig.4. shows the overall design flow for the proposed debugging methodology. At first, user's design is parsed and elaborated by our Verilog compiler. Then DUT will be modified. At the same time, a signal map file is generated to record selected internal nodes in orders. Finally, synthesis tools take the modified DUT RTL codes and do the synthesis and P&R in FPGA. Signal map file is read by PC host for later debug control.

In order to seamlessly connect to proposed system, we design the debug system and place it in the middle of host PC and emulation platform as Fig.5 shows. As shown in the figure, the proposed debug system is targeted at the DUT, since it is where bugs commonly occur. And it connects with PC host through PCI/USB/JTAG.

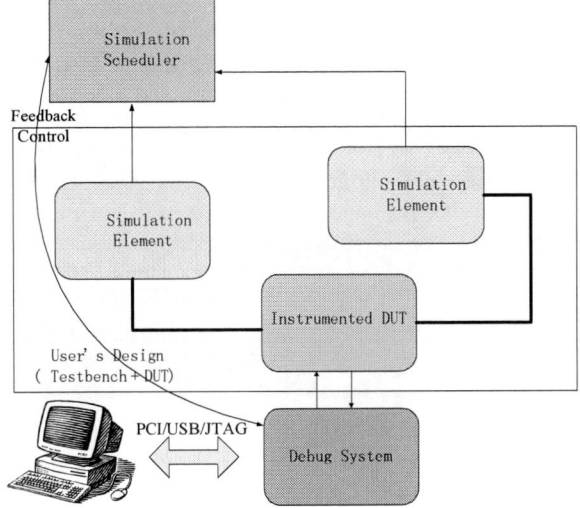

Fig.5. Block diagram of behavioral emulator with debug system

Debug is carried out around an instrumented DUT. The instrumented DUT is generated by an instrumentation compiler, which parse the original design and insert debug logics into it, such as scan-chain, sample logics and etc.

The debug system also communicates with Simulation Scheduler, to know where to capture and update the debug information. All debug information is sampled from DUT and updated to Host PC at correct time. System user can also control the debug system from Host PC, through a communication channel such as PCI, USB, JTAG and etc.

Verilog source files are parsed and elaborated by our Verilog compiler into internal C++ database, which provides rich API functions to process the parsed design. We use lex and yacc to parse the design according to the latest Verilog IEEE 1364-2005 standard.

We modify the design by calling API functions provided by Verilog compiler. In software side, DUT wrapper module is used to replace the original DUT module. Dangling signals are added in the wrapper module to hold the position for the probing nodes of DUT. Verilog VPI program would read node value from FPGA and place it directly to these dangling signals at the end of each simulation cycle.

The hardware side of ELA is in the FPGA and the software side is in the host PC. And they communicate through communication channel like PCI, USB, JTAG etc. The ELA starts in the trigger period by continuously checking for the trigger condition. At the same time, the sample logic continuously samples the selected nodes of DUT. If the trigger condition comes true, the trigger block send a 'start triggering' signals to sample logic and let sample logic store the sample values to the block RAM. If the block RAM comes

978-1-61284-863-1/11 $26.00 © 2011 IEEE

full, it will send data to the host PC through communication channel.

V. EXPERIMENT

We tested the proposed debugging method on an H.264/AVC Baseline Decoder IP, which is freely available from OPENCORES.ORG [14]. Experiment result of a cosimulation based RTL debugging method described in [13] is compared. The generated hardware side code is compiled using Quartus II Version 8.1 Build 163, and mapped to Altera Stratix II EP2S60F1020C4 FPGA device. Compilation is carried out on a computer composed of the 2GHz Intel Core 2T7250 processor and 2GB memory. In TABLE I, we make comparison of consumed FPGA resource in our method with that in [13]. The resource consumption is shown in the percentage of total available FPGA resource. Our method consumes more FPGA resource due to ELA in FPGA.

TABLE I   FPGA RESOURCE CONSUMPTION COMPARISON

Number of probing nodes	FPGA resource consumption	
	METHOD IN [13]	Proposed method
0	16%	16%
5,000	32%	63%
10,000	48%	82%
15,000	60%	90%

TABLE II DEBUGGING SPEED COMPARISON

Number of probing nodes	Time used for [13]	Time used for proposed method
0	6.73 seconds	0.11 seconds
2000	3.42 minutes	0.11 seconds
5000	18.48 minutes	0.11 seconds
15,000	Nil	0.11 seconds

We then perform simulation speed comparison for the DUT cell with different number of probing nodes. The communication channel we use is 33MHz PCI bus and the PFGA runs at 30MHz. With the increase of the total probing nodes, cosimulation speed of [13] drops dramatically since the data traffic between the hardware/software modules becomes very large. We can see that cosimulation time growth exponentially with the increase of probing nodes. And when the number exceeds 5000, cosimulation time becomes unbearable, which takes hours to finish. On the contrary, simulation time for our method is independent of number of probing nodes, as illustrated in TABLE II.

VI. CONCLUSION

In this paper, a debugging method for a synthesizable testbench emulator is proposed, which provide both high performance and good observability. However, we need study alternatives to reduce more FPGA resource consumed with the increase of probing nodes.

ACKNOWLEDGMENT

The authors would like to thank support from New Century Excellent Talents in University (NCET-09-0265) by Education Ministry, China and Program for Sichuan Province Science Foundation for Youths (No. 2010JQ0002), respectively.

REFERENCES

[1] IEEE Standard for Verilog Hardware Description Language, IEEE Std. 1364^{TM}-2005, Chapter 11.

[2] Virtex FPGA Series Configuration and Readback. Application Note XAPP138, Xilinx San Jose CA, March 2005

[3] T.Wheeler, P. Graham, B. Nelson and B. Hutchings. "Using design-level scan to improve FPGA design observability and controllability for functional verification", 11th International Conference on Field Programmable Logic and Applications, pp.483-492, Aug.2001.

[4] Y. Huang, C.C. Tsai and N. Mukherjee, "On RTL Scan Design", IEEE International Test Conference, pp.728-737, 2001

[5] G. Knittel, S. Mayer and C. Rothlaender, "Integrating Logic Analyzer Functionality into VHDL Designs", IEEE International Conference on Reconfigurable Computing and FPGAs, pp.127-132, Dec. 2008

[6] I. Mavroidis and I. Papaefstathiou, "Accelerating emulation and providing full chip observability and controllability", Design & Test of Computers, IEEE, pp.84-94, Dec. 2009

[7] P. Graham, B. Nelson and B. Hutchings. "Instrumenting Bitstreams for Debugging FPGA Circuits", 9th IEEE Symposium on Field-Programmable Custom Computing Machines, pp.41-50, April 2001

[8] K. Camera and R.W. Brodersen, "An integrated debugging environment for FPGA computing platforms", 16th International ACM/SIGDA Symposium on Field Programmable Logic and Applications, pp.311-316, Sept. 2008

[9] C.L. Chuang and W.H. Cheng, "Hybrid approach to faster functional verification with full visibility", Design & Test of Computers, IEEE, pp.154-162, April. 2007

[10] N. Ohba and K. Takano, "Hardware debugging method based on signal transition and transactions", IEEE Asia and South Pacific Design Automation Conference, pp.454-459, Jan. 2006

[11] S. Das, R. Mohanty, P. Dasgupta and P.P. Chakrabarti, "Synthesis of System Verilog Assertions", IEEE Design, Automation and Test in Europe, March. 2006

[12] Synopsys Inc, Synplicity Indentify RTL Debugger homepage, http://www.synopsys.com/Tools/Verification/HardwareAssistedVerification/Confirma/IdentifyPro/Pages/default.aspx

[13] S. Yang, H. Shim, W. Yang, and C.-M. Kyung, "A new RTL debugging methodology in FPGA-based verification platform", in Proc. Asia-Pacific Conf. Advanced Syst. Intgr. Circuits, pp.180-183, Aug. 2004

[14] OPENCORES.ORG, H.264/AVC Baseline Decoder Project, http://www.opencores.org/project,nova

[15] H.C. Huang, A.W. Ruan, et al., "A New Event-Driven Testbench Synthesis Engine for FPGA Emulation", accepted by IEEE 9th International Conference on ASIC, Oct. 2011.

A novel methodology for hardware acceleration and emulation

Y.B. Liao, C.Q. Li, H.C. Huang, C.Y. Xiang, A.W. Ruan, W.Tang

State Key Laboratory of Electronic Films and Integrated Devices, University of Electronic Science and Technology of China,
Chengdu 610054, China
Email: lyb@uestc.edu.cn

Abstract –The communication overhead between the software emulator in the workstation and the FPGA emulator is the speed bottleneck of the hardware acceleration platform. This paper presents a vector mode based hardware/software co-emulation methodology, which leverages the pipeline structure to transmit, receive and buffer data. This methodology reduces the communication overhead by carrying out a parallel mechanism in that while user's design is under test in the emulator, signal data are transmitting in the channel simultaneously, thus increasing the speed of hardware acceleration and emulation. The results of two experiments show that the acceleration factor is 747 and 157, respectively, compared with the traditional methodology.

Keywords – hardware acceleration, hardware/software co-emulation, instruction, direct memory access.

I   INTRODUCTION

Currently, with the rapid development of integrated circuits, verification of SOC chips has become a great challenge due to its integration and complexity. Circuit simulation is a traditional methodology of SOC verification. Simulators running on computers process data sequentially, whereas signals propagate simultaneously along numerous paths in the actual circuits. In the SOC era, traditional software-based simulation methodology cannot meet their verification needs. Therefore, FPGA-based hardware acceleration technologies are requested in SOC verification. The classic methodology of hardware acceleration downloads the DUT (Device under Test) to the FPGA, while part of RTL codes and testbench is still run on the simulator in the workstation. Research found that the speed bottleneck of this methodology is mostly caused by the ping-pong mode of data transmission between workstation software and the FPGA emulator, thus resulting in that channel transmission time takes too much proportion of total time [1][2]. In order to overcome the emulation channel transmission limits between workstations and FPGA, solutions are given by published papers [3-9].

Transaction-based methodology [3][4] packs information in multiple clock cycles to a transaction and sends it as a message to the hardware side, thus reducing both channel frequency and quantity of data transmission and leading to

better transmission efficiency. However, to achieve a high degree of transaction-level abstraction, the user would have to spend more time to prepare the transactor. Furthermore, debugging the transactor itself is also a time-consuming work.

Vericity proposed behavioral synthesis technique for eCelerator[5]. Part of the testbench which is used frequently is converted into synthesizable format and downloaded to FPGA with DUT. By adopting cache structure for transactions, it can reduce the amount of data exchanged through the channel, and therefore hardware acceleration is realized. The difficulty to use this tool lies in the need of splitting the testbench reasonably. Partition of software and hardware decides the effects of the emulation and acceleration.

Renate Henftling and etc proposed a re-use-centric architecture [6] structured with memories, micro-sequence units and protocol units. It produces synthesizable testbench and achieves behavioral-level testbench. But it will take too much FPGA hardware resources.

Kim et al proposed writing parts of the testbench with feedback loops into the downloadable format [7], increasing buffer stages in the software emulator and use of direct memory access (DMA) to transmit data. This approach not only reduced the amount of data transmitted on channels but also increased the parallelism of hardware and software emulation. However, the difficulty is to find out loops in complex testbench. In addition to that signals observed between input and output will be out of sync in the waveform window from the software simulator, making it hard to correctly observe and compare the emulation results.

Writing complex synthesizable testbench and compiling the modified testbench every time is very time-consuming. Choi and etc studied the way to change a testbench into the form of instructions which can be downloaded and executed in the hardware side [8]. Instructions will be pre-stored in the memory, and then the emulation process is to decode and execute the instructions in the memory. When the testbench changes, the only thing needed to be done is modifying the instructions instead of making any change on the hardware side, therefore avoiding the re-synthesis. Whereas the article only mentioned how to generate random test data rather than generate the instruction sequence with the same functions as the testbench. Testing staff have to write up instructions sequence on their own.

Coming to realize the difficulties of rewriting the testbench using high-level language into synthesizable form, Iakovos Mavroidis and etc presented a methodology to completely transfer the testbench to the FPGA verification board [9]. This methodology will split the testbench into C form and synthesizable form. While part of C form will run in the embedded CPU, synthesizable part will be download to FPGA with DUT. The article emphasizes that the direct exchanging of data part of the testbench should be written synthesizable, thereby improving the speed by reducing data exchange between hardware and software. But effective division of the testbench and transmission to the hardware side is quite complex and tedious work.

The methodology of hardware acceleration proposed by this paper intercepts data for once from the emulation process of a traditional platform as testbench and utilizes DMA channel to speed up data transfer, as well as increasing reasonable data caching mechanism, finally which reduces the ratio of channel transmission time in the entire emulation time, achieving accelerating emulation.

II TRADITONAL SPEED BOTTLENECK OF HARDWARE ACCELERATION AND SOLUTION

Take synchronized clock mechanism as example: the total time tsim for the simulator to run a clock change in the platform can be expressed as:

$$t_{sim} = t_{sw} + t_{channel} + t_{hw} \qquad (1)$$

t_{sim} is the time consumed by software side, $t_{channel}$ is the time consumed by data transmission channel, t_{hw} is the time consumed by hardware side. In traditional platform [10] $t_{channel}$ takes about 63% of t_{sim}, so $t_{channel}$ is the key factor to analyze. $t_{channel}$ consists of VPI call time t_{vpi}, drive start time t_{drive}, bus access time t_{bus} and the time to transfer data in a clock change for the emulation t_{clk}.

In events based and clock synchronization based mechanism, $t_{channel}$ can be expressed as:

$$t_{channel} = t_{vpi} + t_{drive} + t_{bus} + t_{clk} \qquad (2)$$

Experiments show that in expression (1) the time $(t_{vpi}+t_{drive}+t_{bus})$ has to be spend in the data transfer of each clock changes, and t_{clk} can be neglected in comparison. To shorten $t_{channel}$ here are two ways: The first methodology is to rewrite some of the $t_{estbench}$ into synthesizable format and download it to the hardware side, reducing the amount of data exchange between software and hardware; the second methodology is the use of DMA to achieve channel data transmission. This paper chooses the second methodology, and the $t_{channel}$ expression can be rewritten as:

$$t_{channel} = (t_{vpi} + t_{drive} + t_{bus} + Nt_{clk}) / N = \frac{t_{vpi} + t_{drive} + t_{bus}}{N} + t_{clk} \qquad (3)$$

In (3) N represents the changing times of clock during each DMA read and write. Obviously the greater N is, the smaller $t_{channel}$ is compared to (2).

III IMPLEMENTATION OF OUR HARDWARE ACCELERATION METHODOLOGY

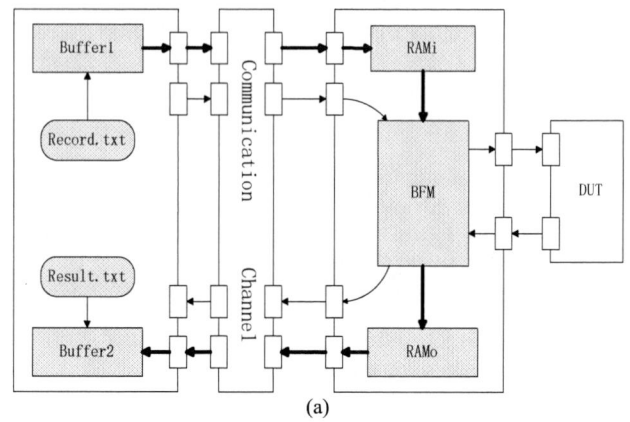

Figure 1. Framework and data pipeline of hardware acceleration platform

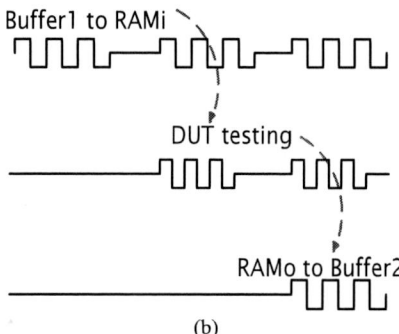

Figure 1(a) shows the way to transmit instructions as stimulus to the hardware side block by block for hardware acceleration. The software side intercepts data into instruction and caches them into memory, and then transfers them to the input memory RAMi of the hardware side through the DMA. Then the BFM (Bus Function Module) in the hardware side executes instructions to send stimulus to DUT. After the direct comparison in the hardware side, emulation results are stored into output memory RAMo, and then the software side will call the DMA function to read them back. Figure 1(b) shows the pipeline manner of data transmission between hardware side and software side. The total emulation time is decreased due to the parallel structure which reduces the awaiting time of the DUT and communication overhead.

A. Generation of testbench

Traditional co-simulation calls a lot of low-level API functions in the VPI of the software side [8], some of which are used to read and write the channel. We can get the needed emulation stimulus by recording the operation process of those functions and turning them into instruction format. The data in the format shown in figure 2 is the instruction to be transmitted to the cache in the RAMi of hardware side. Operation code is opcode to be executed. Offset is the offset in the instruction address in the RAM. Data means operand. Operation codes format are as follows:

Operation code (16bit)	Offset (16bit)	Data (32bit)

Figure 2.   Instruction format

NOP: empty instruction
WRITE: write the data in the RAMi to the corresponding port in address {base, address}.
READ: read the data in the address {base, offset} to the register rdata of the BFM, and then compare with data. If they are not equal, write rdata, as well as the output times, into the RAM for storing output values
BSST: assign the address to the segment register.
JUMP: jump to the address {base, address} in RAMi and execute the instruction inside.
RESTART: re-execute the instruction in the RAMi.
PAUSE: BFM suspend.

B. The design of hardware and software buffer

Traditional hardware acceleration methodology usually transfers emulation data to the hardware side first, then send stimulus to DUT by transactor. This will spend time in both transmitting data and sending stimulus respectively. In this paper we utilized the intercepted Record.txt file and extracted the data inside by C/C++ program in the software side. Then we turned it into instructions and buffing it in the computer memory so that the DMA function can transmit this large amount of data in one time. To parallel the processes of buffering data in the software, transmitting data by DMA and executing the instructions in the hardware side, we divided the RAMi in the hardware side as shown in figure 3.

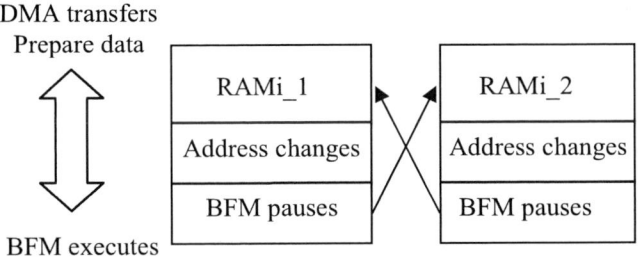

Figure 3.   Divisions of RAMi in the process of emulation

In Figure 3 RAMi is divided into two parts: RAMi_1 and RAMi_2. When the software side is buffering the data to be sent to RAMi_2 and calling DMA to send data to RAMi_1, the BFM of the hardware side is executing the instructions in the RAMi_2. Once there are no more instructions to execute,

BFM will stop. After DMA reloads new data into RAMi_1, BFM can restart execution in RAMi_1, and the software side starts buffering data to transmit to RAMi_1 later and calling DMA to send data to RAMi_2 at the same time. In this mechanism the function of this two RAMs make a conversion, the whole emulation process is being carried out in this way sequentially. The two processes of buffering data in the software and transmitting data by DMA are combined and parallel with the process of instruction execution of BFM. Furthermore, the DUT states can be observed by giving orders to BFM to stop by the software side or the instructions. Actually, this division also solved the limitation problem of the RAM space.

IV EXPERIMENT AND RESULTS

This experiment used C/C++ language to write software as data buffers in software side, and executed in Inter (R) Pentium (R) 2.40GHz.The hardware side used PCI as channels, on-board 33MHz clock and Altera Stratix II FPGA. Experiment mainly tests DUT's clock port's actual emulation frequency CPS (Cycle Per Second) in the hardware side. The DUTs are AES (Advanced Encryption Standard) [11] and ALU (Arithmetic Logical Unit) [12]. The frequencies of DUT emulation clock port and on-board clock are compared using the counter in the hardware side.

TABLE I.   . COMPARISON OF THE ACCELERATION FACTORS BETWEEN NEW AND TRADITIONAL HARDWARE ACCELERATOR

DMA transfer length	1024	2048	4096	6144	8000
AES speedup factor	593	702	747	739	720
ALU speedup factor	115	135	153	157	156

Table 1 shows that using stimulus file intercepted in the traditional hardware acceleration platform which is synchronization based, the emulation speed of these two DUT with new methodology are respectively 747 and 157 times faster than traditional methodology. And the longer the data length is, the faster the DMA transmission is. And as BFM in the hardware side executes at a fixed speed, the parallelism of them is decided by the data length, so different data transmission length will results in different acceleration effects. In this experiment, with the DMA transfer length increases, the emulation speed also increased. However, the comparison between the emulation speed of 6144 32bits data and 8000 32bit data via DMA indicates an increase in the emulation speed. This is because with the increase of the length of DMA transfer, the software side and hardware side will need more time to buffer data, so this methodology does not need to take up too much RAM space. In addition, the experiment seems to have different accelerating effects for different DUT, ALU acceleration effect was much better than AES. The causes of the differences are the DUT port quantities and the complexity of the testbench, which both will cost more time on the software side of traditional acceleration platform. However, the acceleration of the proposed methodology is not subject to the effects of testbench, so the advantage of the proposed acceleration methodology is more obvious if traditional hardware acceleration platform consumed more time on the software side.

V SUMMARY AND FUTURE WORK

Traditional hardware acceleration platform has been working out the problem that the communication time between software side and hardware side is too large in the proportion of the total time. Some existing solutions are not easy to observe the emulation results [1], some need complex users design in the hardware side [5][6][8][9]. The proposed methodology is simple in both software and hardware design while emulation accelerated significantly, and in the hardware side the emulation results are compared, allowing users to observe the emulation results directly in the software side with the number of errors and the corresponding location through the resulting files.

If only emulate once, the proposed methodology must run a traditional hardware acceleration platform at least once to intercept the data, which on the contrary increased the emulation time. But for a single DUT, interception for once is enough for multiple emulations, so this methodology is more suitable for the situation where requires multiple emulation debugging. The disadvantage of our methodology is the need to intercept in the traditional hardware acceleration platform for an original testbench, and only suitable for a single DUT. The future work will focuses on how to generate test bench in software side quickly and multi-DUT emulation methodology.

ACKNOWLEDGMENT

The authors would like to thank support from New Century Excellent Talents in University (NCET-09-0265) by Education Ministry, China and Program for Sichuan Province Science Foundation for Youths (No. 2010JQ0002), respectively.

REFERENCES

[1] Young-Il Kim, Wooseung Yang, Young-Su Kwon, et al. Communication-Efficient Hardware Acceleration for Fast Functional Simulation. Design Automation Conference, San Diego, USA, 2004: 293-298.

[2] Iakovos Mavroidis, Ioannis Mavroidis, Ioannis Papaefstathiou. Accelerating Emulation and Providing Full Chip Observability and Controllability. Design & Test of Computers, IEEE, 2009, 26(6): 84-94

[3] Young-Il Kim, Ki-Yong Ahn, Heejun Shim et al. Automatic Generation of Software/Hardware Co-Emulation Interface for Transaction- Level Communication. VLSI-TSA-DAT, Hsinchu, Taiwan, 2005: 196-199

[4] SohaHassoun, MuraliKudlugi, DuainePryor, et al. A Transaction-Based Unified Architecture for Simulation and Emulation. Very Large Scale Integration (VLSI) Systems, IEEE, 2005, 13(2): 278-287

[5] Verisity, eCeleretor Testbench Acceleration, http://www.verisity.com/products/ecelerator.html

[6] Renate Henftling, Andreas Zinn, Matthias Bauer, et al. Re-Use-Centric Architecture for a Fully Accelerated Testbench Environment. DAC 2003, ACM, 2003: 372-375

[7] Young-Il Kim, Chong-Min Kyung. TPartition: Testbench Partitioning for Hardware- Accelerated Functional Verification. Design & Test of Computers, IEEE, 2004, 21(6): 484-493

[8] Ho-seok Choi, Seung-beom Lee, Sin-chong Park. Instruction Based Testbench Architecture. Banff, Canada, 2005: 329-333

[9] Iakovos Mavroidis, Ioannis Papaefstathiou. Accelerating Hardware Simulation: Testbench Code Emulation. ICECE Technology, Dhaka, Bangladesh, 2008: 129-136

[10] A.W. Ruan, Y.B. Liao, P. Li, et al. An Improved Data Communication Mechanism for A SOC Hardware/Software Co-Emulation Environment. ICCCAS 2009, Milpitas, USA, 2009: 1029 – 1032

[11] OPENCORES.ORG, AES (Rijndael) IP Core, http://www.opencores.org/

[12] OPENCORES.ORG, ecpu_alu, http://www.opencores.org/

Ultra-low Power High Efficient Rectifiers with 3T/4T Double-gate MOSFETs for RFID Applications

Ramesh Vaddi and Tony T. Kim

VIRTUS, IC Design Centre of Excellence
School of Electrical and Electronic Engineering, Nanyang Technological University
50 Nanyang Avenue, Singapore
RVADDI@ntu.edu.sg,THKIM@ntu.edu.sg

Abstract— Recently, multi-gate MOSFETs such as double-gate MOSFETs have been identified as inevitable inclusion for future nano-scale circuit design. This paper explores the scope of tied-gate (3T), independent gate (4T), symmetric and asymmetric features of double-gate MOSFETs (DGMOSFETs) for ultra-low power and high efficient rectifiers for RFID applications. Various widely used rectifier topologies such as simple conventional rectifier, self-V_{th} cancellation (SVC) rectifier and differential drive rectifier etc, have been designed to investigate the better candidate for DGMOSFET technology. Analysis reveals that 3T differential drive rectifier topology shows the maximum power conversion efficiency (PCE) and higher DC output voltage level generation. Second part of the work further explores the effects of 3T/4T and symmetric/asymmetric features of DGMOSFETs on the performance of differential drive rectifier. Among the various DGMOSFET configurations for RFID rectifiers, symmetric tied-gate DGMOSFETs have the best power conversion efficiency and the lowest power consumption.

Key words— Double-gate MOSFETs, Ultra-low power, RFID rectifiers.

I.   INTRODUCTION

Long-range passive microwave transponders (tags) for RFID systems do not have an on-board battery and therefore must draw the power required for their operation from the electromagnetic field transmitted by the reader. The main concern in designing the passive RFID tag is to consume as low power as possible. Decreasing the power consumption decreases the RF power needed from the reader and thus increases the operation range of the RFID tag. Rectifier circuits for the RFID tags are usually designed either by Schottky diodes [1][2] or CMOS transistors [3-7]. Schottky diode rectifiers generally achieve a better performance due to the small turn-on voltage of the diodes. However, Schottky diodes are not readily supported in all CMOS technologies and therefore are not suited for CMOS implementations [3]. On the other hand, in CMOS technology the voltage drop of a diode-connected transistor is about the threshold voltage (V_{th}) of the device. This voltage drop can be reduced either by including a fixed voltage source between the drain and gate of diode connected transistors to reduce their turn-on voltage (static V_{th} cancelation) [5][6] or in more advanced

technologies by using zero or low-V_{th} devices [7]. Both approaches increase the reverse leakage of transistors when the transistors must be turned off and thus reduce the efficiency of the rectifiers. In [3], a four-transistor-cell CMOS rectifier is proposed which outperforms CMOS diode-based rectifiers, even when static V_{th}-cancelation technique is used to reduce the turn-on voltage of the diodes. One drawback of the four transistor cell rectifier is that it does not perform well when the received RF power is weak (e.g, a few μW).

Much of the work has been done on using CMOS front-ends and CMOS technology for RFID. Very few have considered double gate MOSFETs for ultra-low power RFID tag design. The usefulness of non-classical underlap channel architecture to enhance both gain and bandwidth of an OTA, alleviating gain-bandwidth trade-off associated with analog design, has been demonstrated in [8]. A. Kumar et al, [9] explores the application of independently driven double-gate MOSFETs for low-power low voltage analog integrated circuit design. In [10], P.Freitas et al, investigate new capabilities brought on by independently driven double gate CMOS transistors for analog baseband design. Since the gates are disconnected, the corresponding channels are coupled resulting in a dynamic threshold voltage tuning. This operation mode is exploited to create new analog functions and low-voltage circuits. A current mirror is redesigned using of independently driven double-gate MOSFETs and shown that this structure performs an efficient differential function relating to the potentials applied to the back gates. N. Mohankumar et al, study the influence of both channel and gate engineering on the analog and RF performances of double-gate MOSFETs for system-on-chip applications [11]. The gate engineering technique used here is the dual metal gate technology, and the channel engineering technique is the conventional halo doping process.

However, not many have explored the effect of various configurations of DGMOSFETs for ultra low power rectifier design for RFID applications. The rest of paper is organized as follows. Various DGMOSFET configurations and rectifier topologies implemented with 3T/4T DGMOSFETs are described in section II. Section III presents the simulation results and discussion on results. Finally, conclusions are presented in section IV.

978-1-61284-863-1/11 $26.00 © 2011 IEEE

II. THEORETICAL BACKGROUND

A. Various Double-gate MOSFET (DGMOSFET) Configurations

DGMOSFETs can have either a three-terminal (3T) configuration, where both the gates are shorted, or a four-terminal (4T) configuration, where the back-gate bias is fixed and the front gate acts as a control electrode. If the two gates in the DGMOSFETs are tied (3T), an identical voltage can be applied to both the gates. Conversely, when the two gates are independent (4T), different voltages can be applied.

The implementation of symmetry and asymmetry configurations in the DGMOSFETs can be realized in a number of ways. For example, it can be implemented by applying different gate voltages to the front and back gates, by assigning variations in oxide thickness at the front and back gates. An alternative way is varying gate material with different work functions. In this work, the asymmetric nature of the DGMOSFETs is brought by taking the variation in the gate oxide thickness for the front and back gates. The asymmetric feature of the DGMOSFETs offers more flexibility and freedom to circuit designers in the device control point of view.

In the following studies, we will primarily consider the four DGMOSFET structures that are shown in Fig. 1: 3T symmetric DGMOSFET (3TSDG) device ($V_{fg} = V_{bg} = V_{dd}$ and $T_{fox} = T_{box} = 1.4nm$), 3T asymmetric DGMOSFET (3TADG) device ($V_{fg} = V_{bg} = V_{dd}$ and $T_{fox} \neq T_{box}$), 4T symmetric DGMOSFET (4TSDG) device ($V_{fg} = V_{dd}$, $V_{bg} = 0$ and $T_{fox} = T_{box} = 1.4nm$), and 4T asymmetric DGMOSFET (4TADG) device ($V_{fg} = V_{dd}$, $V_{bg} = 0$ and $T_{fox} \neq T_{box}$).

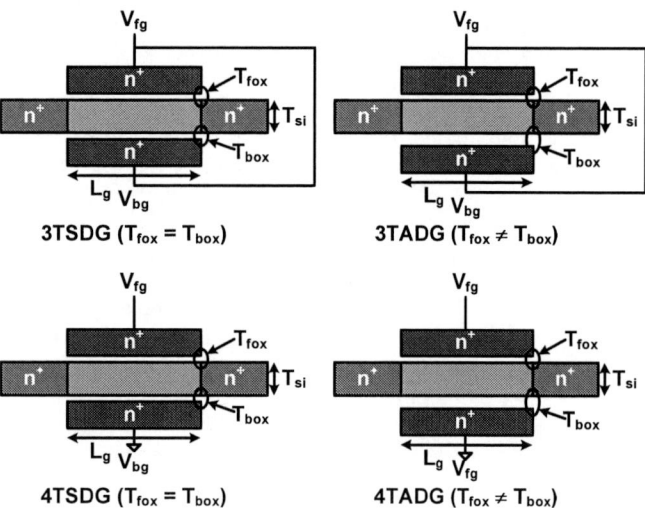

Fig. 1. Schematic of various configurations of DGMOSFETs.

B. Various RFID Rectifier Topology Implemenations with 3T/4T DGMOSFETs

As a RFID tag is a passive system, DC voltage must be generated to bias the circuits of the tag, which is done by a rectifier. The rectifier converts received RF signal into DC voltage. The main challenge in designing the RFID rectifier is to generate the required DC power using the low voltage amplitude of the input RF signal with acceptable power conversion efficiency. Power conversion efficiency (PCE) of a rectifier is defined by the output power divided by the input power. The PCE of a rectifier circuit is affected by various parameters such as circuit topologies, diode-device parameters, input RF signal frequency and amplitude, and output loading conditions. Since the input RF signal of RFIDs in long-range operations is extremely small, low turn-on voltage is the most important factor for the diode devices. Schottky diodes have been extensively utilized in previous rectifiers with a multi-stage configuration, because of its low turn-on voltage. The rectifier circuit using the Schottky diode achieves a high PCE, but it is not compatible with conventional CMOS technology and requires costly fabrication processing.

Four different rectifier configurations with 3T and 4T DGMOSFETs are illustrated in Fig. 2. Fig. 2 (a) shows a simple rectifier circuit widely used in RFID implemented with 3T DGMOSFETs. Diode-connected n-channel and p-channel MOSFETs are connected in series and the internal node is connected to RF input terminal through the coupling capacitor (C_C). The PCE for a rectifier using a diode-connected MOSFET is generally worse than that of the Schottky diode based one due to its higher threshold voltage (V_{th}), but when V_{th} cancellation techniques are utilized, the PCE can be improved dramatically. In order to reduce the effective turn-on voltage for achieving larger PCE, several V_{th} cancellation schemes have been proposed [5][12-14]. One uses a switched-capacitor technique to generate DC gate bias voltage from an external power supply [12] and others generate DC gate bias voltage from the output voltage of the rectifiers themselves [5][13][14]. Fig. 2 (b) shows a self-V_{th}-cancellation (SVC) rectifier circuit implemented with 3T DGMOSFETs. It is the same as the conventional rectifier circuit described in Fig. 2 (a), except that the gate electrodes of the n-MOS transistor and p-MOS transistor are connected to the output terminal and the ground terminal, respectively. This connection boosts the gate-source voltages of the n-MOS and p-MOS transistors as much as the output DC voltage. In other words, the threshold voltages of MOS transistors are equivalently decreased by the same amount to the output DC voltage. However, it has been found that "static" V_{th} cancellation schemes have limitation in achieving small ON-resistance and small reverse leakage current at the same time. Differential drive rectifier is an "active" V_{th} cancellation scheme in which V_{th} can be minimized in a forward bias condition and be increased in a reverse bias condition automatically by a cross coupled differential circuit configuration [4]. Fig.2 (c) shows a unit stage of the differential- drive rectifier circuit with the 4T configuration of DGMOSFETs and Fig. 2 (d) shows the same circuit implemented with 3T configuration of DGMOSFETs. The circuits have a cross-coupled differential configuration with a bridge structure. The similar units can be cascaded to produce larger PCE and DC output voltage levels required by other functional blocks.

978-1-61284-863-1/11 $26.00 © 2011 IEEE

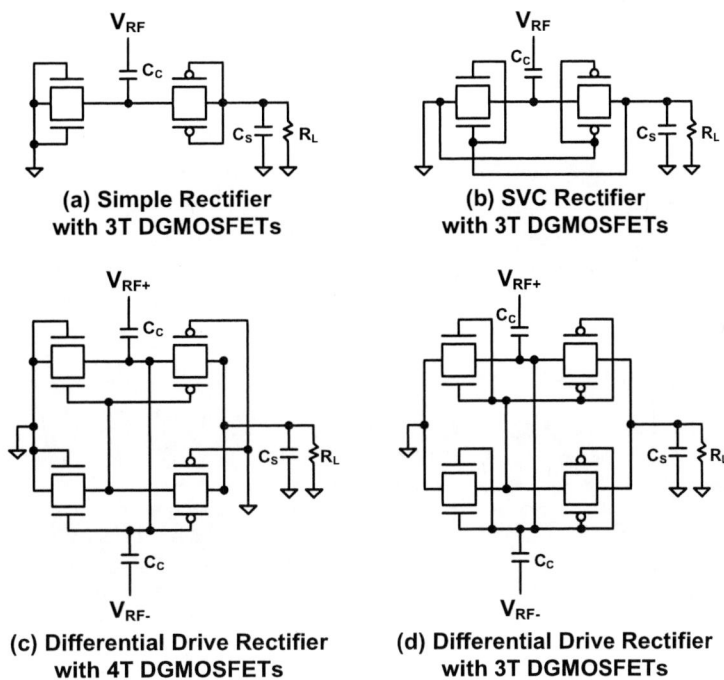

(a) Simple Rectifier
with 3T DGMOSFETs

(b) SVC Rectifier
with 3T DGMOSFETs

(c) Differential Drive Rectifier
with 4T DGMOSFETs

(d) Differential Drive Rectifier
with 3T DGMOSFETs

Fig. 2. Various rectifier configurations with DGMOSFETs.

III. RESULTS AND DISCUSSION

The various DGMOSFET rectifier topologies mentioned above are designed and simulated using HSPICE to analyze the best topology which suits for ultra low power RFID design with highest possible efficiency using DGMOSFET technology. The papameters of the passive devices are as follows: $C_C = C_S = 10pf$, $R_L = 10k\Omega$, and $f = 100MHz$. The W/L ratio for the minimum sized devices and the upsized devices are 2 and 10, respectively. Comparisons of the DC output voltages and the PCEs by the considered rectifier topoligies implemented with the minimum sized 3T DGMOSFETS with variations in input RF levels are presented in Fig. 3. The simple and SVC rectifier topologies demonstrate similar output voltage levels, but the output voltage levels of them are substantially lower than that of the differential drive rectifier. As the RF input signal amplitude changes from 0.1V to 0.9V (for a power generation of 0.5-365μW), the DC output power of the differential drive rectifier changes from 0.5 to 139μW while that of the SVC rectifier changes from 1.6pW to 1μW. As a result, larger power conversion efficiency values can be achieved from the differential drive rectifier as presented in Fig. 4. For the same changes in the input voltage levels, the PCEs is in the ranges of 42-38% for the differential rectifier, 0.04-2.1% for the SVC rectifier and 0.2-2.1% for the simple rectifier respectively.

The power consumption comparisons of the above rectifier topologies with minimum sized 3T DGMOSFETs are shown in Fig. 5. The differential rectifier consumes larger power than the simple and SVC rectifier topologies. Fig. 6 demonstrates the impact of sizing on the PCEs of the

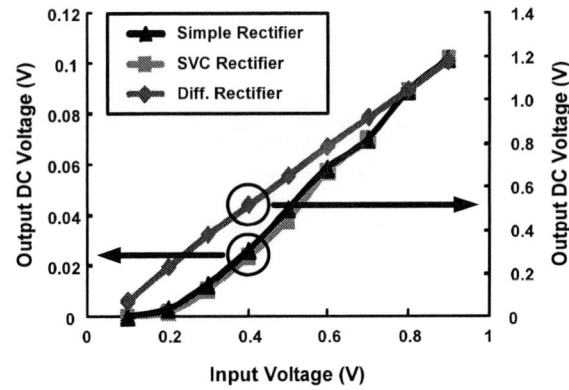

Fig. 3. Comparison of DC output voltages in various rectifier topoligies with minimum sized 3T DGMOSFETs sweeping RF input voltage.

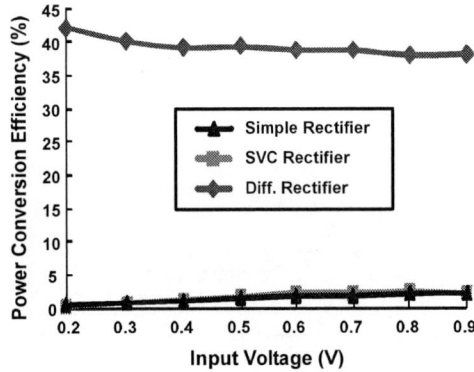

Fig. 4. Comparison of PCE in various rectifier topoligies with minimum sized 3T DGMOSFETs sweeping RF input voltage.

978-1-61284-863-1/11 $26.00 © 2011 IEEE

differential drive rectifier. Overall, 4T configuration has lower PCE than 3T configuration, but at very small RF input voltage levels, the 4T configuration shows higher PCE than the 3T one. Fig.7 shows the PCE comparisons of the differential drive rectifier implemented with the 3T/4T, symmetric/asymmetric features of the DGMOSFETs. Overall, the 3T structures have larger DC output levels and PCE than the 4T structure. The effect of the symmetric and asymmetric oxide thickness on the PCE is insignificant except at the input voltage of 0.1V where the asymmetric feature shows a larger PCE value.

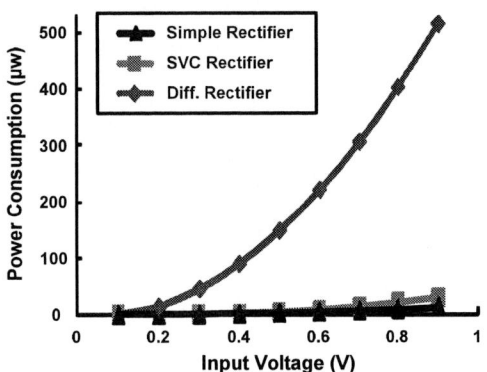

Fig. 5. Comparison of power consumption in various rectifier topoligies with minimum sized 3T DGMOSFETs sweeping RF input voltage.

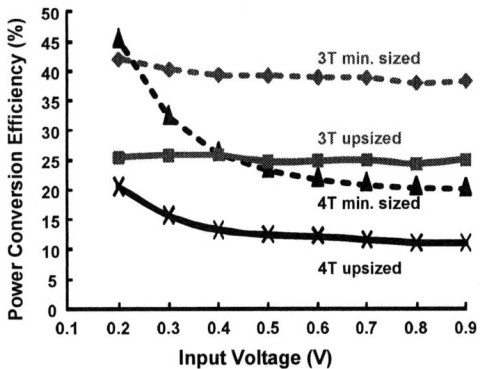

Fig. 6. Comparison of PCE of the differential drive rectifier with minimum and up-sized 3T/4T DGMOSFETs sweeping RF input voltage.

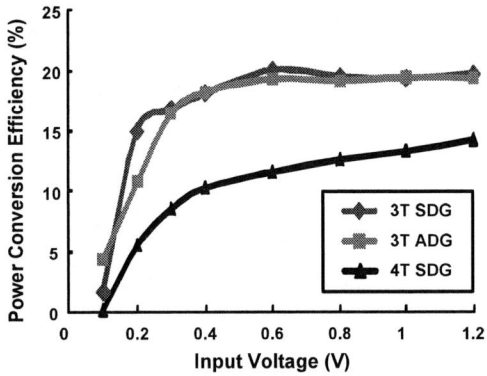

Fig. 7. Comparison of PCE of the differential drive rectifier with 3T/4T and symmetric/asymmetric DGMOSFETs sweeping RF input voltage.

IV. CONCLUSIONS

Various rectifier topologies are explored with DGMOSFET technology to suggest a better candidate for the RFID design. The simulation results have shown that the differential drive rectifier with the minimum-sized 3T DGMOSFETs is the most promising candidate for ultra-low power and efficient RFID rectifier design due to the higher power conversion efficiency ranging from 38% to 42%. In the differential rectifier, the 3T rectifier topology generates higher PCE and DC output voltage than the 4T one. Finally, the asymmetric gate oxide thicness has negligible impact on PCE when the input voltage level is high. However, the PCE becomes sensitive to the configuration of the gate oxide thickness when the input voltage is low.

REFERENCES

[1] U.Karthaus, and M. Fischer, "Fully integrated passive UHF RFID transponder IC with 16.7µW minimum RF input power," *IEEE Journal of Solid-State Circuits*, vol. 38, no. 10, pp. 1602-8, September 2003.

[2] R.E.Barnett, Jin Liu, S.Lazar, "A RF to DC voltage conversion model for multi-stage rectifiers in UHF RFID transponders," IEEE Journal of Solid-State Circuits, vol. 44, no. 2, pp. 354-70, February, 2009.

[3] S. Mandal, and R. Sarpeshkar, "Low-Power CMOS Rectifier Design for RFID Applications," *IEEE Transactions on Circuits and Systems I: Regular Papers*, vol. 54, no. 6, pp. 1177-1188, July, 2007.

[4] A. Sasaki, K. Kotani, and T. Ito, "Differential-drive CMOS rectifier for UHF RFIDs with 66% PCE at −12 dBm input," IEEE Asian Solid-State Circuits Conference (A-SSCC), pp.105-108, November,2008.

[5] H. Nakamoto, D. Yamazaki, T. Yamamoto, H. Kurata, S. Yamada, K. Mukiada, T. Ninomiya, T. Ohkawa, S. Masui, and K. Gotoh, "A Passive UHF RF Identification CMOS Tag IC Using Ferroelectric RAM in 0.35-µm Technology," *IEEE Journal of Solid-State Circuits*, vol. 42, no. 1, pp. 101-110, January, 2007.

[6] W. Che, Y. Yang, C. Xu, N. Yan, X. Tan, Q. Li, H. Min, and J. Tan, "Analysis, design and implementation of semi-passive Gen2 tag," IEEE International Conference on RFID, pp. 15-19, April, 2009.

[7] J. Yi, W.-H. Ki, and C.-Y. Tsui, "Analysis and Design Strategy of UHF Micro-Power CMOS Rectifiers for Micro-Sensor and RFID Applications," *IEEE Transactions on Circuits and Systems I: Regular Papers*, vol. 54, no. 1, pp. 153-166, January, 2007.

[8] A. Kranti and G. Alastair Armstrong, " Nonclassical Channel Design in MOSFETs for Improving OTA Gain-Bandwidth Trade-Off ", IEEE Trans. Circuits and Syst.-I, Vol. 57, No. 12,pp. 2010.

[9] A. Kumar, and S.Tiwari, "A Power -Performance Adaptive Low Voltage Analog Circuit Design Using Independently Controlled Double Gate CMOS Technology "Proc. of . International Symposium on Circuits and Systems, 2004, pp. I-197 -I-200 Vol.1.

[10] P.Freitas, G.Billiot, H.Lapuyade, and J.B.Begueret, " Analog Design Considerations For Independently Driven Double Gate MOSfets And Their Application in a Low-Voltage OTA" ,14th IEEE International Conference on Electronics, Circuits and Systems, ICECS, 2007 , pp: 198 – 201.

[11] N. Mohankumar, Binit Syamal, and C.K. Sarkar," Influence of Channel and Gate Engineering on the Analog and RF Performance of DG MOSFETs" IEEE Trans. Ele. Dev., Vol. 57, N0. 4,pp,82-826, Apr. 2010.

[12] T. Umeda, H.Yoshida, S. Sekine, Y. Fujita, T. Suzuki, and S. Otaka, "A 950-MHz rectifier circuit for sensor network tags with 10-m distance," *IEEE J. Solid-State Circuits*, vol. 41, no. 1, pp. 35–41, Jan. 2006.

[13] K. Kotani ,and T. Ito, "High efficiency CMOS rectifier circuit with self- Vth-cancellation and power regulation functions for UHF RFIDs," in *Proc. IEEE ASSCC*, Nov. 2007, pp. 119–122.

[14] K. Kotani ,and T. Ito, "Self-Vth-cancellation high-efficiency CMOS rectifier circuit for UHF RFIDs," *IEICE Trans. Electron.*, vol. E92-C, no. 1, pp. 153–160, Jan. 2009.

Low-Power 4-Bit Flash ADC For Digitally Controlled DC-DC Converter

Guolei Yu and Liter Siek
VIRTUS – IC Design Center of Excellence
Nanyang Technological University
Singapore 639798
Emails: YU0003EI@e.ntu.edu.sg; ELSIEK@ntu.edu.sg

Abstract—**A low-power flash analog-to-digital converter (ADC) using logic gates as comparators is presented. The circuit is designed by using 0.35µm CMOS technology. A conversion delay of 12.55nS has been achieved at 200 Mega sample per second (MSPS). The static current consumptions at zero error bit from the power supply and sampled input are 63µA and 93µA respectively. The high speed and low power characteristic make the ADC structure ideal for the high frequency digitally controlled DC-DC converter applications.**

Keywords—**flash ADC; DC-DC converter; digital control**

I. INTRODUCTION

Digitally controlled DC-DC converter has emerged as a potential alternative to the traditional analog DC-DC converter due to its scalable design, process portability, flexible control scheme as well as low sensitivity to the variation of process parameters, voltage supply, ambient temperature, and external component values [1].

The industry adapts the digitally controlled DC-DC conversion technology rather slowly due to the chip size, performance and design topology unfamiliarity. Digitally controlled DC-DC converters cannot achieve the same DC accuracy as the analog counterpart due to the sampling of the analog-to-digital converter (ADC). The small signal bandwidth of the digitally controlled DC-DC converter also limits its dynamic response to load current transient, input voltage transient and reference voltage tracking.

The major control blocks in the digitally controlled DC-DC converter system include the ADC, the discrete-time compensator, the digital pulse-width modulator (DPWM) and the driver. The output voltage of the DC-DC converter is sampled and quantized by an ADC. A discrete-time compensator computes the duty-cycle control signal based on the output of the ADC. DPWM converts the duty-cycle control signal to pulse-width modulator (PWM) signal. The output power transistor is driven on or off, according to the PWM signal, by a driver stage which in its simplest form is an inverter chain with gradually increased sizes.

The ADC conversion time (including sample and hold circuit delay time), digital compensator computation time and DPWM delay time limit the obtainable small signal bandwidth of the digitally controlled DC-DC converter, which limits the

response of the digital DC-DC converter in the event of a sharp voltage change during load transient. Due to the ability of the digital control in switching into different complex mathematical computation, it is therefore well suited for multimode system. For example, during the normal operation when the load current does not change much, the control is by a linear Proportional-Integral-Derivative (PID) controller. When there is a load transient, the system changes to non-linear control in order to obtain the optimal response. One of such non-linear control method is the charge balance control (also known as time optimal control) [1]. For this kind of control method, it is important to determine the voltage profile during the load transient in the shortest possible time. To address this issue, multi-sampling or sampling at higher frequency can be used with the penalty of high switching loss. Another approach is to use asynchronous ADC to detect the output voltage change in continuous time. When the output voltage changes due to a sudden load current transient, and the quantized output voltage error exceeds a certain preset threshold, the transient control loop will bypass the normal PID control loop and bring the output voltage back to the nominal value in the shortest possible time. To be able to do this, a low power asynchronous ADC is required [2].

This paper will describe the design of an asynchronous ADC scheme for the digitally controlled DC-DC converters by using logic gates and simple logic circuits. The new ADC structure is able to scale with the process advancement and it can function at low supply voltage. The simulation result shows the functionality of the proposed ADC structure.

II. ANALYSIS OF THE ADC STRUCTURE

The requirement of the ADC in a digitally controlled DC-DC converter is analyzed as following:

1) Accuracy: For a typical DC-DC converter, a DC accuracy of ±1% and a transient accuracy of ±10% are usually required. That is to say, the LSB of the ADC should be less than 2% of the required output voltage. And the ADC should be able to resolve the DC voltage from 90% to 110% of the required output voltage in order for the controller to properly compensate the transient response. In another word, a window ADC, rather than an ADC for the entire range, is needed.

978-1-61284-863-1/11 $26.00 © 2011 IEEE

The required number of bits (NOB) for the ADC:

$$NOB = \log_2 \frac{20\%}{2\%} = 3.3 < 4 \qquad (1)$$

Hence, a 4-bit ADC is sufficient. For this design, a 4-bit ADC will be targeted to demonstrate the concept.

2) *Conversion time:* For clocked ADC like SAR or pipelined ADCs, from the time the signal is available at the input to when the digital code is produced, a finite time is needed, which is called conversion time. The conversion time should be just a small fraction of the switching period for robust response in the event of sudden output change.

3) *Power consumption:* Low power operation is desired to improve the overall system efficiency.

4) *Digital process compatibility:* The ADC should not depend extensively on the matching of analog components; else changing technology node requires extensive simulation and rework in layout.

5) *Asynchronous operation:* This is to enable asynchronous control for fast load transient recovery without waiting for a complete clock cycle to recover.

The above criteria rules out the traditional type of ADC like SAR, two-step or those ADCs that require more than one clock cycle to compute. To achieve high speed, flash type ADC is preferred. But traditional flash ADCs need large number of matched comparators, which are not so suitable for this application [3]. One solution to this problem is to use inverters as analog comparators in the flash ADC, which is referred to as Threshold Inverter Quantization (TIQ) method [4].

Inverters are basic building blocks in the digital system. The voltage transfer characteristic of the inverter closely resembles that of an analog comparator. The switching point of the inverter, V_{SP}, is determined by the relative sizes of the PMOS and NMOS transistors. If the sizes are chosen correctly, the switching points of the inverters can be used to quantize the input voltage of the inverter, as in (2), where μ, w, l and V_{th} are the mobility, width, length and threshold voltage of NMOS and PMOS respectively, while VDD is the power supply voltage to the inverter.

$$V_{SP} = \frac{\sqrt{\frac{u_n \cdot w_n \cdot l_p}{u_p \cdot w_p \cdot l_n}} \cdot V_{th,n} + (VDD - V_{th,p})}{\sqrt{\frac{u_n \cdot w_n \cdot l_p}{u_p \cdot w_p \cdot l_n}} + 1} \qquad (2)$$

The TIQ ADC is low power and high speed. However, the drawback is that the threshold voltage changes with power supply voltage, temperature and process, and also there is no correlation between the NMOS and PMOS threshold voltages. Using an accurately pre-regulated power supply for the inverters will improve the accuracy of the switching point. However, this approach will add significant complexity to the whole system. The variation of the switching point is not acceptable for DC-DC converter IC because the output of the DC-DC converter IC is required to be tightly regulated even

with voltage or temperature variations. The TIQ comparators' internal threshold cannot be easily trimmed because it involves the sizing adjustment of the MOS transistors.

To address the above concern, a new flash ADC which is suitable for the digitally controlled DC-DC converter application is proposed as shown in Fig. 1.

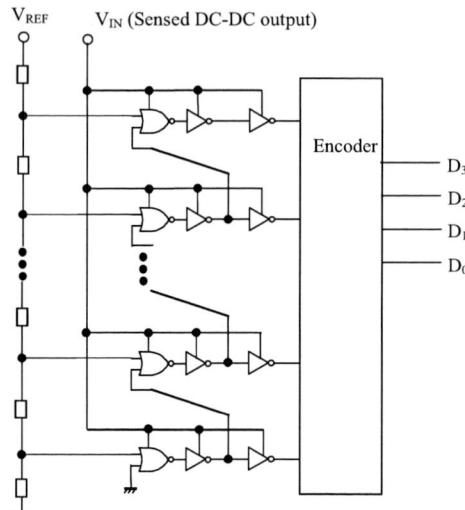

Figure 1. Flash ADC structure using NOR gate as comparator.

In the proposed structure, the power supply node of the logic gate is used as one input of the comparator, which is connected to the output voltage of the DC-DC converter, while the scaled reference voltage is connected to the usual input of the logic gate. In this configuration, the logic gate functions as a comparator to determine the error voltage between the reference voltage and the sensed DC-DC converter output. The dynamic and static current consumption of the logic gates are drawn directly from the DC-DC converter output. This eliminates the need of a pre-regulated voltage buffer as in the TIQ case. Moreover, as the buck converter configuration is most widely used for portable applications, drawing current from the output of the DC-DC converter reduces the effective current consumption by the input to output ratio. For example, if the DC-DC converter is converting a 3.3V input voltage to 1.8V output, the effective current drawn by the logic gate comparators is reduced by about 1.83 times.

III. BLOCK CIRCUIT DESIGN

A. Comparator

The output of DC-DC converter varies within a narrow range, which is possible to operate a few adjacent logic gates at shoot-through condition when both PMOS transistor and NMOS transistor are on. To reduce the current consumption, the NOR gate is used here instead of inverters. If the sensed voltage is lower than the voltage required to make the output of certain NOR gate to be high, it will be lower than the voltage

978-1-61284-863-1/11 $26.00 © 2011 IEEE

required to make the output of all the NOR gate above this NOR gate to be high. Hence, the output voltages of all the NOR gates above this NOR gate is set to be low. By doing this, the static current consumption is reduced to one NOR comparator only. This method also effectively removes the bubble error.

Since the logic gates use the minimum length in the particular process, the tolerance of the switching point due to process variation needs to be considered. In this design, the switching point is fixed by the reference voltage ladder, so re-arranging (2) for the quantization levels: V_{IN}:

$$V_{IN}[n] = (k+1) \cdot V_{SP}[n] - k \cdot V_{th,n} + V_{th,p}, \quad \text{where } k = \sqrt{\frac{u_n \cdot w_n \cdot l_p}{u_p \cdot w_p \cdot l_n}} \quad (3)$$

The relative tolerance between the resistors is usually quite small and ignored here. The offset is mainly attributed to the tolerance in the threshold voltage of NMOS and PMOS transistors, which is determined by the size of the transistors as well as the process parameters as shown in (4).

$$\sigma_{vth} = \frac{X_{VT}}{\sqrt{w \times l}} \quad (4)$$

where X_{VT} is process dependent parameter corresponding to 1 sigma (σ) variation for the threshold voltage. From Eqn. (3) and (4), the 1σ variation of the quantization level is:

$$\sigma_{VDD} = k \cdot \sigma_{vth,n} + \sigma_{vth,p} \quad (5)$$

For the process we used in this design, the 1σ variation of the quantization level is 33.7mV which is about 1.5LSB. Simple digital calibration technique as in [5] can be utilized to trim the variation in the quantization level to be less than 1/3 LSB.

B. Boost stage

The boost stage consists of two inverters. The main function is to make the sharp transition between the logic zero and the logic one level. One thing to take note is that the logic high level is the same as the sensed DC-DC converter output voltage level. The output of the boost stages are in the thermometer code format.

C. Encoder

The encoder converts the thermometer code to binary logic output. It mainly consists of zero/one detector and a ROM structure. The ROM structure also functions as a level shifter to convert the logic high level from the sensed DC-DC output voltage level to the power supply voltage of the compensator block for subsequent duty cycle calculation.

IV. SIMULATION RESULT

The simulation result is as shown in Table 1.

Fig. 2(a) shows the ADC output voltage when the input to the ADC is swept from 0V to 3.3V. It can be seen that when the input voltage is below the "window" range, all the bits of the ADC output are of logic low, while when the input range is

above the "window" range, all the bits are of logic high. This is the desired operation for the window ADC. Fig. 2(b) zooms into the range where the input voltage is within the conversion window, which shows the proper operation of the ADC.

Table 1: Simulation result summary

Number of Bits	4
Technology	CMOS, 0.35um
Supply voltage	3.3V
Output voltage	1.8V
LSB voltage	22.5mV
INL/DNL @200MSPS	0.49/0.29 LSB
"Window" range	1.618V~1.936V
Static current from the power supply at zero error	63uA
Static current consumption from the input at zero error	93uA
Maximum conversion delay @200MSPS	12.55nS

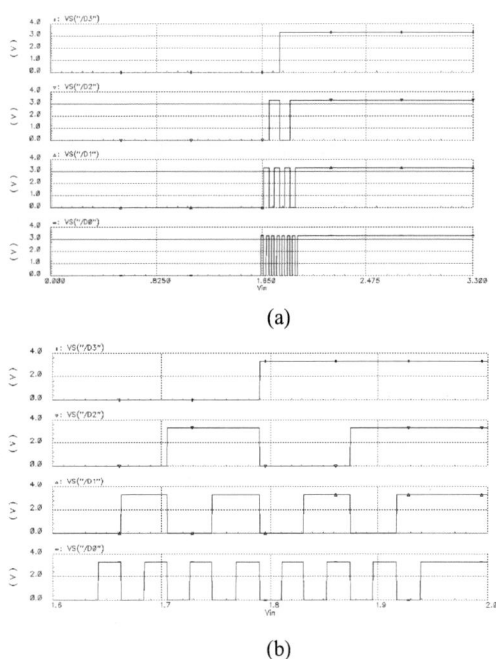

(a)

(b)

Figure 2. (a) DC sweep of the input voltage from 0 to 3.3V, showing the proper operation of the "window" ADC. (b) DC sweep of the input voltage across the "window".

Fig. 3 shows the transient result at the sampling rate of 200 Mega-Sample-Per-Second (MSPS). It can be seen that the ADC functions properly without missing code. The INL and DNL are both less than 0.5LSB.

978-1-61284-863-1/11 $26.00 © 2011 IEEE

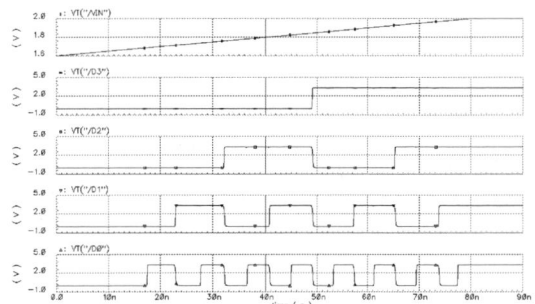

Figure 3. Simulation result of the 4-bit ADC operating at the speed of 200 MSPS

V. LAYOUT DESIGN

The layout of the ADC is as shown in Fig. 4.

The total active area is measured to be 0.034mm². The reference generator block, which consists of the poly resistor dividers to provide the reference for each comparator, takes about 0.009mm². The comparator and boost stage takes about 0.012mm². "0" detector blocks takes about 0.01mm². The encoder block takes about 0.002mm². The rest of the area is occupied by wirings.

The proposed flash ADC has been taped out by using VIS 0.35um 2-Poly-4-Metal CMOS process. Currently, the preparation for evaluation is being carried out. The actual measurement result will be compared with the simulation result when the prototype chips return.

VI. SUMMARY

In this paper, a new flash ADC structure has been proposed which is suitable for the digitally controlled DC-DC converter application. The new ADC uses mainly logic gates and simple encoder structure, and consumes 63uA at 1.8V V_{out} when the output voltage is in regulation. The flash structure enables asynchronous conversion, which will facilitate the implementation of fast transient recovery algorithm in the digitally controlled DC-DC converter.

ACKNOWLEDGMENT

This work is supported by Panasonic Semiconductor Asia. Guolei Yu would like to thank Panasonic Semiconductor Asia for sponsoring his PhD study in NTU.

REFERENCES

[1] Liu. Y F, Meyer. E, Liu. X., "Recent Developments in Digital Control Strategies for DC/DC Switching Power Converters", IEEE Transactions on Power electronics, Nov 2009, pp. 2567-2577.

[2] Z. Zhao and A. Prodic, "Continuous-time digital controller for high-frequency DC–DC converters," *IEEE Trans. Power Electron.*, vol. 23, no. 2, pp. 564–573, Mar. 2008.

[3] A. V. Peterchev, J. Xiao, and S. R. Sanders, "Architecture and IC implementation of a digital VRM controller," *IEEE Trans. Power Electron.*, vol. 18, no. 1, pt. 2, pp. 356–364, Jan. 2003.

[4] S. Saggini, E. Orietti, P. Mattavelli, A. Pizzutelli, A. Bianco, "Fully-digital hysteretic voltage-mode control for dc-dc converters based on asynchronous sampling," in IEEE APEC 2008, pp. 503-509.

[5] CY Chen, M. Q. Le and K. Y. Kim, " A Low Power 6-bit Flash ADC with reference voltage and common-mode calibration," in IEEE Journal of Solid-state Circuits, vol. 44, no.4, pp. 1041-1046, Apr. 2009.

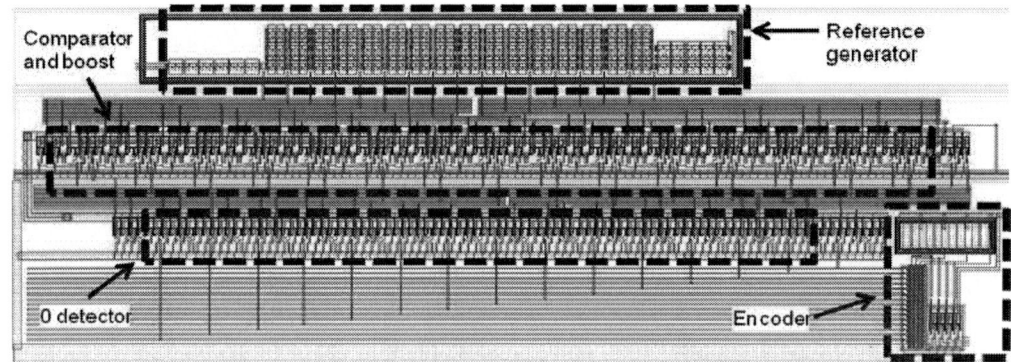

Figure 4.   Layout of the proposed ADC

978-1-61284-863-1/11 $26.00 © 2011 IEEE

Synthesizable Verification IP to Stress Test System-On-Chip Emulation and Prototyping Platforms

Subramanian Shiva Shankar
Lantiq Asia Pacific
Singapore
ShivaShankar.Subramanian@lantiq.com

Jayaratnam Siva Shankar
Lantiq Asia Pacific
Singapore
SivaShankar.Jayaratnam@lantiq.com

Abstract—One of the biggest challenges today with Pre-silicon System-on-Chip verification is to stress out the SoC to uncover as many corner case design issues by injecting heavy real time data traffic into the system. The inherent efficiency and the performance of the Emulation and FPGA prototyping systems make them the ideal platforms to run these tests. A typical solution is to inject data traffic through protocol exercisers with proprietary hardware (vendor specific slow down solutions) which can bridge the emulated DUT with a real time device or use software API's with transaction based SCE-MI communication infrastructure. The need for a complex input output interface makes the former difficult to be used with all emulators / FPGA prototyping systems while SCE-MI communication infrastructure being protocol specific is a disadvantage. So, a synthesizable verification architecture compliant with SCE-MI 2.0 infrastructure through which the protocol specific traffic is injected through industry standard interfaces .i.e. PIPE (PCIe), UTMI (USB), MII (Ethernet) based on user configured stimuli has been designed and implemented. Being synthesizable, the verification environment can run in both emulation and prototyping platforms effectively stress testing the complete system.

Keywords - Stress Testing, SCE-MI, PIPE, UTMI

I. INTRODUCTION

As we proceed to the modern era, the complexity of the System on Chip (SoC) designs has increased considerably which also poses serious challenges to verify them. One of the biggest challenges today with the system level verification is to fully stress out the system during the emulation or prototyping stages to uncover as many corner cases as possible by injecting as heavy data traffic as possible. Moreover, a verification environment at this level will also enhance Hardware Software co-verification as software development has become a substantial part of the design cycle. [5] Therefore, this paper proposes a reconfigurable synthesizable verification architecture in which the protocol specific traffic is injected based on preconfigured user input which can be integrated with any design independent of emulation or prototyping system. The verification environment is also complaint with SCE-MI 2.0 communication infrastructure.

Previously, traffic injection in emulation / prototyping stages has been done in two ways. The first is by using a real

device like Peripheral Component Interface Express (PCIe) wireless card along with proprietary hardware which is available from the corresponding Electronic Design Automation (EDA) companies. The proprietary hardware / vendor specific slow down solutions acts as an intermediate system controlling the data transfer between the two devices i.e. the emulated Device Under Test (DUT) sees a protocol compliant system running at emulated system speed and a real world device sees a system running at real speed. But it needs a complex input output interface which is not available in all emulators / prototyping platforms. Moreover, if multiple instantiations of a particular module is available in the system, then multiple hardware would be needed to stress test the system concurrently.

The second is by injecting traffic through software API that is transferred to the emulator / Field Programmable Gate Array (FPGA) prototype through Standard Co-Emulation Modelling Interface (SCE-MI) 2.0 communication infrastructure. SCE-MI 2.0 standard defines a multi-channel communication interface which allows communication between emulated DUT with software Bus Functional Model. [1] But the SCE-MI 2.0 communication infrastructures, which are available from the EDA companies, are protocol specific which is a huge disadvantage.

II. FUNDAMENTAL THEORY

High level verification languages such as e, vera and now systemverilog are useless for developing models, transactors and testbenches to run in emulators or FPGA prototypes as none of these languages are synthesizable. [2] So building synthesizable testbenches would be a good solution to run system level tests in emulation or prototyping environments. Transactorizing the synthesizable testbench will enable existing software testbenches to be integrated with high-level systemC / C / C++ environments. [2] Most of the bus architectures that get the inputs from external endpoints, like Universal Serial Bus (USB) and PCIe have a clear distinction between protocol specific higher layers and the lower level physical layers. The interface between the higher level layers and the lower level layers is governed by a parallel data interface with some control signals which effectively assist the

978-1-61284-863-1/11 $26.00 © 2011 IEEE

movement of data through the interface. [4] So traffic injection through these interfaces, adhering to the protocol makes the stress testing process effective and debugging also becomes easy. Moreover, injection of data through these standard interfaces is the only possible way to test the complete system and those modules which do not directly communicate with the inputs / outputs of the SoC.

III. ARCHITECTURE OF THE VERIFICATION ENVIRONMENT

The verification architecture is split into 3 modules namely the Transmit (TX), Receive (RX) and the protocol specific higher layers. All three modules are implemented as finite state machines. The TX and the RX layers directly communicate with the respective transmit and the receive interfaces of the DUT through the corresponding standard interfaces such as USB Transceiver Macrocell Interface (UTMI) for USB, PHY Interface for PCIe (PIPE) for PCIe. There are two different functions for the TX and the RX layers. The first is to perform the initial handshaking in order to prepare the verification environment for the higher level transactions. The second would be to encode the control signals of the standard interface to a single bit enable signal so that the higher layers function independent of the physical layer control signals. The encoding of the control interface signals to one signal is done to make the communication interface common and simple across various standard interfaces used by different protocols.

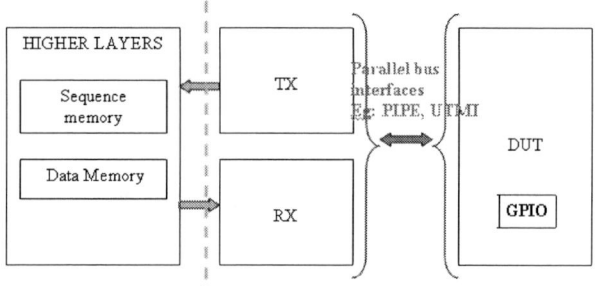

Figure 1-Top Level Architecture of the Verification Environment with Memory Model

The protocol specific higher layer is a state machine which can imitate the top level transaction layers of the protocol along with some behavioural memory models to store the data. The actual data that is communicated to and from the DUT is stored in the data memories. Data is written into the memory in case of write transactions while read back in case of read transactions. The transaction sequences can be programmed through the sequence memory through which the user can control the sequence of the packets being exchanged. Various top-level errors can be easily injected by programming them through the sequence memory. Currently the data for the sequence memory is generated manually. However, the contents can be generated through high level scripts. Automatic reply mode has been designed for the higher layers, which if enabled automatically sends the responses without the need of filling the sequence memory. Figure 1 shows the

top level architecture of the verification environment with the memory model.

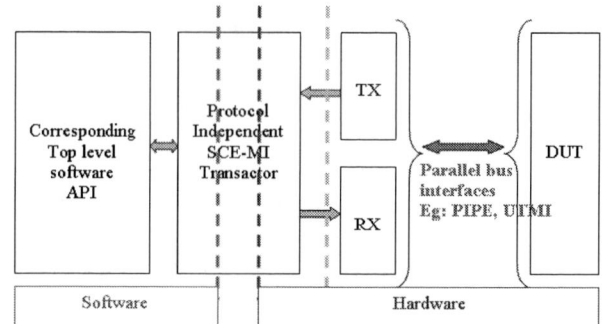

Figure 2 – Top Level Architecture if the Verification Environment with SCE-MI

Figure 2 shows the verification environment with the SCE-MI communication infrastructure. In this case the memory models which mimic the function of an endpoint is removed and connected to a software model through protocol independent SCE-MI 2.0 communication infrastructure. The system is initially designed with the behavioral memory models mimicking the higher layers to test the TX / RX interfaces and will be hooked up to the protocol independent SCE-MI once its tested.

IV. PROTOCOL INDEPENDENT SCEMI 2.0 COMMUNICATION INFRASTRUCTURE

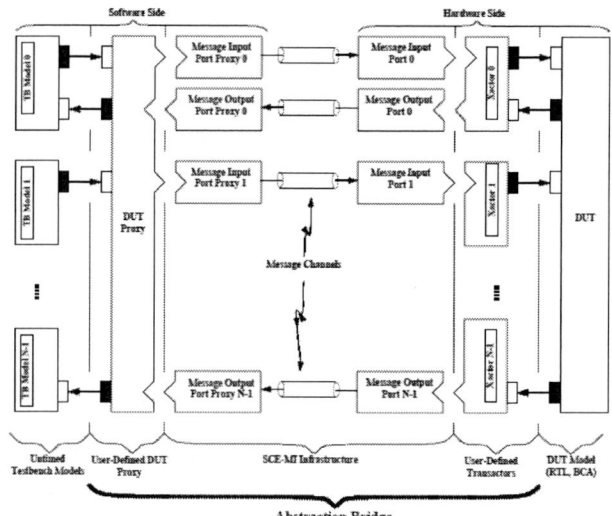

Figure 3 – SCE-MI 2.0 Multi Channel Abstraction Bridge

Figure 3 shows the top-level architecture of SCE-MI 2.0 multi channel Abstraction Bridge. The most important components of the bridge which takes care of the communication between the hardware and the software are the Message Input and the Output ports and their corresponding proxies. The message input and output ports are found in the hardware side and the corresponding proxies are found in the software side. The main aim of a SCE-MI 2.0 communication

978-1-61284-863-1/11 $26.00 © 2011 IEEE   610

infrastructure being protocol independent is that the underlying infrastructure should be able to support any protocol with any packet size. This can be done in various ways, but the implementation should not affect the throughput of the system by adding too much additional overhead. So it is necessary to design a system with minimum overhead to achieve the planned stress. Therefore, the solution proposed is to identify the packet size and append it before the actual data, which will be communicated, so that the message ports and port proxies will know the size of the data to be transmitted or received. This is also the reason for which the interface between the TX/RX and the higher layers is kept constant for various protocols as this interface will communicate with the message input/output port once the higher layers are removed.

In this way, the proposed SCE-MI 2.0 infrastructure can be reused for any protocol by identifying the packet size before being communicated to the software. Once the packet size is identified, the data can be sent through the data bus setting the corresponding enable signal to logic high.

V.   PIPE ADAPTER

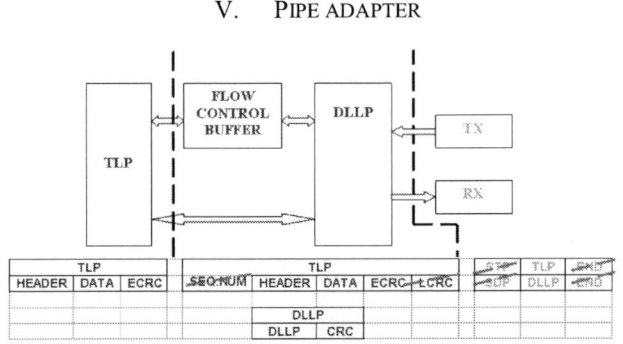

Figure 4 – Architecture of PIPE Adapter

PIPE Adapter is a verification environment which can effectively inject and receive PCIe traffic through the PIPE Interface which was designed based on the verification architecture proposed in chapter 3. The architecture of the environment has been shown in figure 4. The abstraction between the lower level layers and the higher layers is demarcated using the red dashed line in figure 4. Since PCIe is a multi layered protocol, the data link layer functionalities are also handled in the hardware layer itself, as it is hardware intensive. So if the current environment is integrated with SCE-MI 2.0 communication infrastructure only the transaction layer functionalities will be taken care by the software layer. The flow control buffer is available temporarily for the data to be transferred between the memory model and the DUT.

The TX and RX layers take care of receiving and sending of both the Transaction Layer Packet (TLP) and the Data Link Layer Packet (DLLP) by adding the framing. The DLLP layer takes care of the data link layer functionalities such as sending and receiving flow control update packets, acknowledgment packets and can handle link level data integrity of the TLPs. The TLP layer is the module, which takes care of transaction

layer packets. This has multiple memory models in it, which can store and send the data. The sequence memory has a limit of 32 distinct transactions of maximum size after which the memory rolls over. The input to the sequence memory will be the header of a TLP packet. Since the TLP header can either be 12 bytes or 16 bytes, the width of the sequence memory has been set to 16 bytes. For example, if a memory read packet with 16-byte header is sent by the root complex, the PIPE Adapter can either be programmed to send a completion packet with data which refers to a successful completion or a completion without data to induce an error by programming the sequence memory. In this way error injection becomes very easy through programming of the sequence memory. PIPE adapter can support all types of PCIE transactions including Memory, configuration, IO and Interrupt packets.

VI.   UTMI ADAPTER

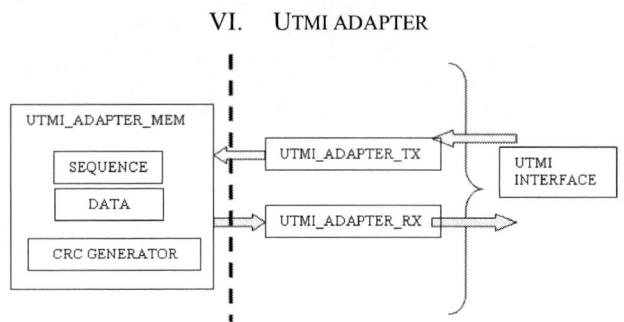

Figure 5 – Architecture of UTMI Adapter

UTMI Adapter is a verification environment to stress test the system by injecting USB traffic through the UTMI interface designed based on the verification architecture proposed in chapter 3. The TX and the RX layers carry on the initial handshaking process (chirping) and then send the data to the higher layers. UTMI adapter can only support bulk transactions. Figure 5 shows the architecture of UTMI adapter.

Table 1 – Encoded PID's used in Sequence Memory

Packet Type	PID	Encoded PID
DATA0	8'b11000011	3'b000
DATA1	8'b01001011	3'b001
ACK	8'b11010010	3'b010
NAK	8'b01011010	3'b011
STALL	8'b00011110	3'b100
NYET	8'b10010110	3'b101
PING	8'b10110100	3'b110

The higher layer of the UTMI Adapter has a Cyclic Redundancy Check (CRC) generator and a CRC checker for the outgoing and incoming packets respectively. The higher layer (UTMI_ADAPTER_MEM) has a behavioural memory model, the sequence memory through which the transaction sequences are programmed. 32 distinct transaction sequences can be programmed through the sequence memory. Similarly the data memory has a memory limit of 16K bytes. The packet types shown in Table 1 are supported by UTMI Adapter and they can be easily identified through the Packet ID (PID) defined according to USB 2.0 specification. Since the PID's are 8 bits and are unique for each packet, it would be too big to be programmed into the sequence memory. So they were encoded into 3 bit encoded PID's and they were programmed into the sequence memory. Table 1 shows the encoded PID's. Errors can be easily injected as six responses can be programmed for each sequence. For example, for a bulk out transaction, even if the data received has a positive CRC, a negative acknowledgement can be programmed through the sequence memory.

VII. RESULTS

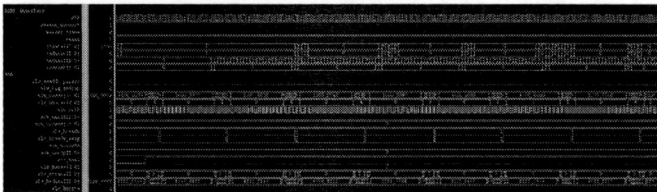

Figure 6 – Memory Write Transactions at the PIPE & AHB Interfaces

In both the cases i.e. UTMI Adapter and PIPE Adapter, the verification environments were integrated with the SoC through the standard interfaces and enormous amount of stress was created at the respective standard interfaces, in turn stress testing the respective host controllers, Advanced High-performance Bus (AHB) Bridge, Crossbar and the Double Data Rate memory (DDR) modules in the SoC. Figure 6 shows the stress levels created in the PIPE Interface and the corresponding AHB bridge when continuous memory write transactions were injected through the interface. The UTMI Adapter was integrated with the SoC which had 2 host controllers. Figure 7 shows the stress levels created in the UTMI interface by injecting bulk out transactions through the first host controller and bulk in transactions through the other host controller. Similarly figure 8 shows the stress levels in the AHB bridges of the corresponding host controllers in the above specified testcase. In both the cases a particular sequence of data was written into the verification environment and the same was read back to verify the data integrity.

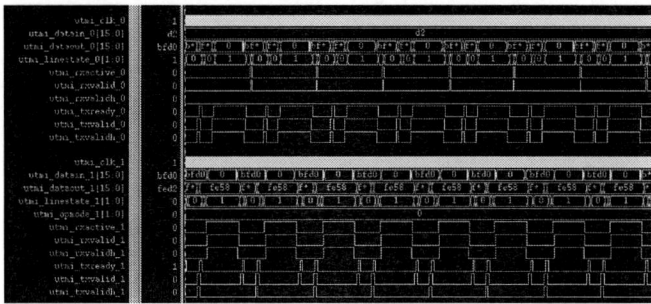

Figure 7 – Stress created at the UTMI Interface

The environment proposed above can be integrated and reused easily across various projects as the data injection happens through the standard interfaces, in turn reducing the

cost and precious time. The architecture also takes care of data integrity which makes it efficient for stress testing and can also inject packets with least inter packet delay as the physical level layers are bypassed. Moreover, the provision to integrate with SCE-MI 2.0 communication infrastructure makes the solution versatile and futuristic.

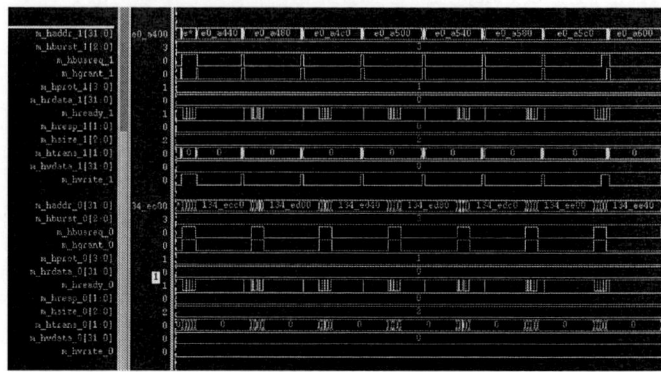

Figure 8 – Stress created at the AHB bridge Interface

VIII. CONCLUSION

The advantage of a synthesizable / SCE-MI supported verification environment arises when there are more than one host's which are available in the SoC and both of them needs to be stress tested. Having such an environment makes things easier as the verification environment is inside the emulated system. Moreover, the complexity of modern system on chip architectures is increasing exponentially and there is a need to capture corner case design issues and improve verification coverage. Capturing the corner case issues needs accurate testcases which can only be generated through powerful verification environments. The proposed model can also be used for pre silicon driver development where the endpoints can be replaced by hardware models proposed above and driver development can be done even before is silicon is available. The proposed solution can also be used to undertake performance studies on the existing system and identify bottlenecks for improvement. Thus the verification environment proposed above is an alternative idea which can be implemented in both emulation systems and FPGA prototypes without any external needs and also without compromising on the controllability of the testcases from the user perspective.

IX. REFERENCES

[1] SCE-MI Reference Manual, Version 2.0, Brian Baily et. al. Accellera, March 2007

[2] "Delivering Synthesizable Verification IP for Test Benches", Bluespec Inc., 2009

[3] PCI Express Base Specification Revision 3.0, Chamath Abhayagunawardhana et al. ,PCI-SIG, August 2010

[4] PHY Interface for PCI Express Architecture, Jeff Morris, Andy Martwick, Brad Hosler, Matthew Myer, Jim Chaoate, Intel Corporation, June 2003

[5] Hardware Emulation and FPGA Prototyping for Embedded Systems Design , Lauro Rizatti, Eve, December 2010

978-1-61284-863-1/11 $26.00 © 2011 IEEE

1-bit heuristic adaptive quantizer (HAQ) for on chip image compression in CMOS image sensors

Michael Barrow[a], Amine Bermak[a], Shoushun Chen[b]

[a]ECE Department, Hong Kong University of Science and Technology [b]EEE Department, Nanyang Technological University
Email: mjbarrow@alumni.ust.hk, eebermak@ust.hk, eechenss@ntu.edu.sg

Abstract—This paper presents an algorithm for implementing a single bit adaptive quantizer based on fast boundary adaptation rule (FBAR). The peak signal to noise ratio (PSNR) gain and performance gain of the algorithm over prior designs is found to be larger than that displayed by prior art compared with a reference FBAR implementation. A maximum increase of *1.44db* was seen. In addition, the new design facilitates an improved bits per pixel ratio (bpp) when integrated with the QTD compressor utilized in previous prototypes. The presented algorithm is hardware friendly and designed for low power implementation, with simulation results also showing an improvement of relative energy cost over previous work. Experimental evidence for image sizes ranging from 64x64 pixels to 512x512 pixels and the heuristic adaptive quantizer (HAQ) algorithm are detailed in this paper.

I. INTRODUCTION

Image sensors are increasingly prevalent in today's electronic system designs with the most widespread application of CMOS image sensors is in portable devices [1]. In 2010 camera phones alone accounted for 62% of total CMOS image sensor sales [2]. With the ever increasing image processing demands of new mobile handsets, and the myriad of emerging novel applications, extensive processing is required in today's image sensor application.

Compression facilitates low energy manipulation of a captured image on the sensor and also reduces the energy required to transfer image data to different parts of a hypothetical device utilizing the sensor. This allows for lower power sensors and more extensive processing on chip.

Compression is therefore an instrumental technique used to meet the identified demands of low power and extensive processing in emerging CMOS image sensor designs.

Today's system design space features numerous well known compression algorithms and standards such as JPEG, SPHIT etc. which utilize the compression techniques of a discrete cosine transform (DCT)[3] or discrete wavelet transform (DWT)[4]. However these compression methods require extensive memory and processing which lies beyond the power and area budget for the typical portable application.

Recently there have been several proposed CMOS image sensor designs featuring on chip compression [5], [6], [7] which feature low power consumption and on sensor image compression.

This work contributes a hardware friendly "heuristic adaptive quantizer" (HAQ) architecture for sensors in order they better target emerging CMOS image sensor applications.

II. FBAR THEORETICAL BASIS

The architecture described in [5] provides quantization research contributions based on predictive and adaptive manipulation of quantizer thresholds. Quantization introduces distortion which must be minimized in an optimum fashion. This design and later evolutions [6], [7] feature sub-optimal implementations of a boundary adaption rule or BAR quantizer.

There are several methods of minimising distortion in an N-point scalar quantizer. Distortion is commonly expressed as a product of rth power of possible threshold differences between an input signal and the probability that this delta could happen. According to the literature the most common method used to minimize distortion is the "Holder norm and its rth power"[8]

$$\Delta(x, Q(x)) = D_r \equiv \sum_{i=1}^{N} \int_{x_{i-1}}^{x_i} |x - y_i|^r p(x) dx \qquad [8] \quad (1)$$

The most commonly used powers are; r=1 (mean absolute error) r=2 (mean square error).

For the powers of r=1 or r=2 one may use the Lloyd I algorithm [8] however the reviewed sensors use the FBAR rule defined by (2) as the basis for quantizer threshold adaptation because it has comparable accuracy but far faster convergence then Lloyd I.

$$\Delta x_j = \eta \left(\sum_{k=j+1}^{N} \delta_k^r \frac{\mathbb{1}_{R_k}}{N - j} - \sum_{k=1}^{j} \delta_k^r \frac{\mathbb{1}_{R_k}}{j} \right) j = 1, \dots, N-1$$
$$[8] \quad (2)$$

In order to reduce complexity (number of computations) and required storage, prior art implements FBAR with $r = 0$ when adapting threshold levels. This has the advantage of reducing the required computations for threshold adaption to $O(\eta/(N(N-1)))$ [8] where N is the number of Quantization regions. In the reviewed literature only one bit quantizers are implemented, so only two input comparisons are required to adapt quantizer thresholds. Such an implementation, henceforth referred to as "FBAR$_0$" is expressed as follows:

$$\Delta x = \eta \left(\mathbb{1}_{R_2} - \mathbb{1}_{R_1} \right) \qquad [7] \quad (3)$$

The cost of this simplification is that when $r = 0$ it is not possible to minimise distortion, so SNR is reduced. This is because input variance is not used to compand the quantization levels around commonly occurring signal levels

978-1-61284-863-1/11 $26.00 © 2011 IEEE

within the pixel array (or η is highly unlikely to be optimum). Because the variance terms are removed in FBAR$_0$, prior art has implemented a heuristic rule to re-introduce a measure of it [5]. This rule will henceforth be referred to as the "Λ rule".

III. DESCRIPTION OF FBAR$_0$ HEURISTIC RULES

Prior art features evolutionary improvement in performance, characterized by two distinct image sensor architectures [5], [7]. The first, henceforth referred to as prototype one ([5]) focused on low power and sacrificed SNR for simplified architecture. The second generation henceforth referred to as prototype two ([7]) reduced power by removing a large amount of registers.

In consideration of the implemented FBAR$_0$ versus the general form (see eq.2), this work contributes an improved algorithm and improved architecture focussed on improving PSNR with minimal power cost. Given the interrelation, dependence and hardware overlap of the sensor system blocks described in prior art [5], [6], [7], this work focused on the single bit quantizer block. This benefits in being both most isolated in terms of its functionality and the main source of noise. In consideration of spatial continuity in images, prior art implements a heuristic for fast convergence of very large pixel intensity gradients (increasing η), but not one for very small gradients. We see the case for both large and small gradients by considering the example image in fig 1:

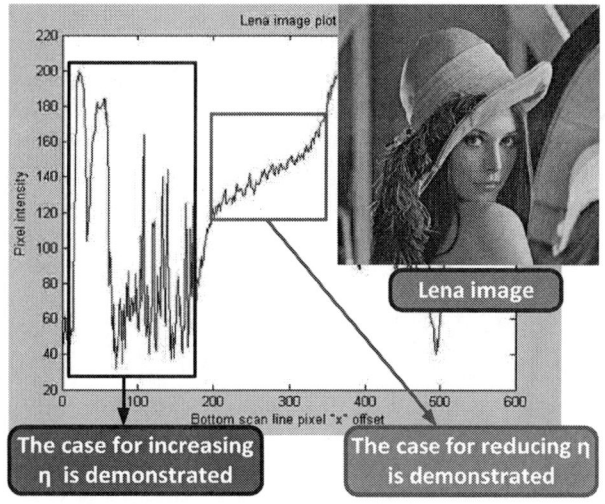

Fig. 1. a scan line of the "Lena" image [9] demonstrating the hypothesis

Sharp contrast between the shoulder and hair of the Lena image corresponds to a large gradient. This could benefit from the heuristic in Prior art (see [7]). However the minimal contrast in the region of Lena's back conversely corresponds to minimal intensity gradient. Previous work [6] employs the hilbert curve when scanning an image pixel array, so Lena from fig 1 is presented to the prior art quantizer in a spatially contiguous manner, compounding the case for a heuristic rule targeting such regions.

To summarize, the hypothesis explored in this contribution is that PSNR could be improved in the general case if:

- The original rule to increase η when the local gradient is large is preserved
- A new rule is used to reduce η when the local gradient is small is introduced
- Both rules are applied in a complimentary way

The theoretical foundation of this new rule is identical to that of the Λ rule in the cited literature, in that introducing a measure of local variance to FBAR$_0$ can help speed convergence, however it operates on the opposite extreme of image intensity variation.

IV. PROPOSED IMPLEMENTATION

A new η augmentation heuristic rule, henceforth referred to as the "γ rule" was identified for a one bit quantizer. In addition a modified FBAR$_0$ implementation was derived that includes both γ and Λ rules. The result is a newly contributed algorithm of

$$FBAR_0 = f(\eta, \Lambda, \gamma, N)$$

where N must be 2 (for a one bit quantizer). An implementation, known as the "HAQ algorithm" used to characterize the γ heuristic is presented below in fig2.

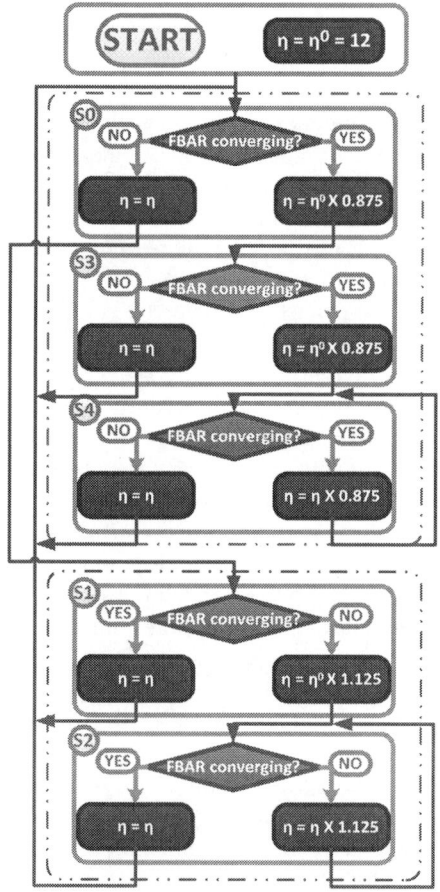

Fig. 2. Flow chart of the HAQ algorithm

Convergence of FBAR$_0$ is measured over two samples and is the previous quantizer output bit xored with the current output bit

TABLE I
SIMULATED PSNR (DB) AND BIT-PER-PIXEL (BPP) FOR 10 TEST IMAGES. ALL IMAGES WERE SCANNED USING A HILBERT CURVE AND SOURCED FROM A SCIENTIFIC REPOSITORY [9]. MEAN PSNR AND BPP WERE OBTAINED BY AVERAGING DATA FROM SWEEPING η_0 FROM 1 TO 50. "R"($psnr/bpp$) IS THE QUALITY TO COMPRESSION RATIO AND MEASURE OF OVERALL PERFORMANCE.

Operation modes	size of test images											
	64x64			128x128			256x256			512x512		
	PSNR	BPP	R	PSNR	BPP	R	PSNR	BPP	R	PSNR	BPP	R
FBAR$_0$	19.89	0.83	23.96	21.34	0.81	23.35	22.07	0.80	27.59	18.74	0.76	24.66
FBAR$_0$+Λ	19.74	0.79	24.99	21.23	0.74	28.69	21.94	0.73	30.06	18.50	0.67	27.61
HAQ	20.67	0.76	27.20	22.47	0.71	31.65	23.34	0.70	33.5	22.23	0.61	36.44

TABLE II
POST PLACE/ROUTE SIMULATED POWER AND DELAY FIGURES OF THE HAQ QUANTIZATION ARCHITECTURE AND TWO REFERENCE ARCHITECTURES IMPLEMENTED IN TSMC 0.18μm TECHNOLOGY. ENERGY AND LATENCY COST ARE CALCULATED RELATIVE TO THE PREVIOUS HEURISTIC IMPLEMENTATION FOR EACH DESIGN. MEASUREMENTS WERE OBTAINED WITH IDENTICAL HDL SYNTHESIS AND PLACE/ROUTE TOOL SETTINGS

Architecture	Heuristic	Power(μW)	Period(nS)	Energy(nJ/τ)	Energy Cost(nJ/τ)%	Latency Cost%
FBAR$_0$	1	7.281	6.370	0.0464	0%	0%
FBAR$_0$+Λ	2	7.687	9.920	0.0763	64%	56%
HAQ	3	8.309	12.270	0.1019	34%	23%

A. Algorithm key contribution

It is of critical importance to highlight the "Hold off" in this algorithm. The γ rule reduces η in a less aggressive manner then the Λ rule increases it, as shown by the inclusion of step **S3**. Although the Λ rule is preserved in its original form from prior art, it is applied less aggressively as shown by the large number of steps that leave η unaltered. These steps embody the aforementioned "Hold off". Either rule has a second chance at being applied before η is radically altered by the other. This enforces a complementary behaviour between the two and prevents rapid oscillation between applying both of them on consecutive input samples. Experimental implementations without hold off only exhibited a marginal PSNR improvement.

Fig 2A demonstrates the application of the Λ rule in the contributed algorithm. It is seen that after the quantization threshold has risen once the rule is in the "hold off state" **S3**. After a second consecutive rise, the heuristic will be applied **S4** this continues for consecutive samples that the quantizer threshold is below the input signal. The heuristic is also applicable to consecutively lower input signals.

Fig 2B demonstrates the application of the γ rule in the contributed algorithm. If FBAR$_0$ begins to converge, i.e. the quantizer threshold must be increased and reduced in consecutive cycles, the γ rule limits threshold change to $\eta_0 \times 0.875$ This is seen in **S1**. The rule is then in the "hold off state". If FBAR$_0$ (3) continues to converge, the heuristic will be applied to all consecutive oscillating samples and the algorithm stays in **S2**.

B. Hardware architectural implementation

The HAQ algorithm was implemented as a finite state machine (FSM). The optimized "γ" and "Λ" can be applied in a mutually exclusive fashion by using the same "η" adjusted right three bits ($\eta \gg = 3$) and either added (for "Λ") or subtracted (for "γ"). This simplicity results in a low power and latency increase relative to prior art. (see table II).

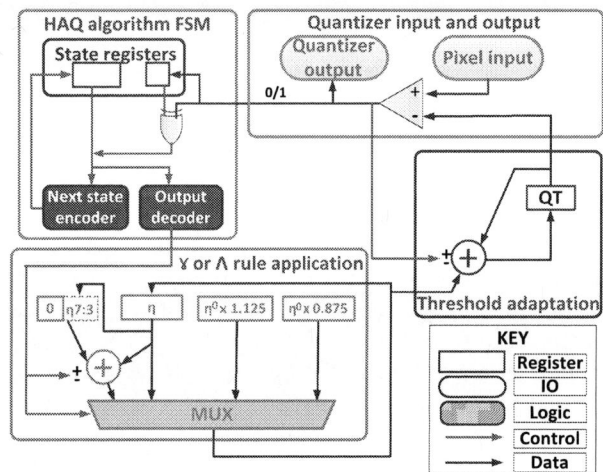

Fig. 3. "HAQ architecture". "QT" is the quantization threshold register. All data busses are eight bit except the comparator output bus, which is one bit. The "State registers" are segmented into a 3-bit block required for storing the HAQ algorithms state, and a 1-bit block to indicate FBAR$_0$ convergence.

Fig 3 demonstrates the HAQ algorithm implemented for low hardware overhead with respect to prior art [7]. The critical path from the heuristic rule to the quantizer threshold or "QT" register is only extended by the additional time needed for data to propagate through the widened MUX. On the control path, additional delay is incurred by the FSM output decoder logic or next state logic. In terms of additional register count, three additional one bit registers are required in the algorithm FSM.

V. EXPERIMENTAL RESULTS

Comparative experiments with prior art using test images were carried out to asses the fitness of HAQ architecture as a replacement. Output data was fed into a QTD compressor as described in prior art ([5]) and the design was also synthesised with TSMC 0.18μm technology (See fig 5). A variety of image sizes were tested to determine how well the design scales.

A. PSNR improvement

It was found that the HAQ architecture offered an additional gain in PSNR of up to $+1.144db$ over "FBAR$_0$ architecture" and up to $+0.764db$ over "FBAR$_0$+Λ architecture". It was observed that the average increase in PSNR over FBAR$_0$+Λ architecture is larger than the increase seen in when the Λ rule is introduced to augment standard FBAR$_0$ (see table I).

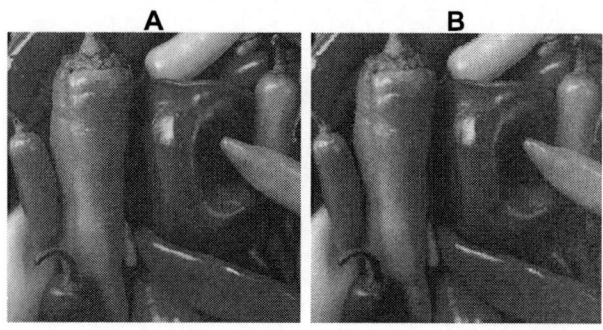

Fig. 4. "Peppers2" image [9] and quantized "Peppers2" image.

Fig 4A shows an original image. Fig 4B shows an image quantized with the HAQ algorithm. The quantized image, has a PSNR of $33.03db$ and was encoded at $0.77bpp$ using QTD compression. η_0 was 8.

In addition to PSNR improvement, compression ratio was found to improve by $0.04bpp$ on average compared with Λ rule prior art. Table I shows the HAQ algorithms performance improvement with respect to prior art (see [6]). "R" is the combined merit of PSNR and bpp ($PSNR/bpp$) and is the performance measure. This data demonstrated that the HAQ algorithm provides a gain in performance much larger then that of prior art compared with standard FBAR$_0$ in all resolutions considered in prior art. The data also demonstrated the trend of improved performance at higher resolution observed in [6], however HAQ performance scales better with an average four times improvement at the highest resolution. Average performance results are significant in demonstrating an improvement, regardless of input image and "η_0" (see fig 2) which may not be changed during field use of an image sensor.

B. Relative power and latency cost

It was found through hardware synthesis that the gain in PSNR and bpp from adding the "γ" rule incurred a low power and latency cost, relative to prior art implementing only the Λ rule. In summary of table II, HAQ architecture incurs a 23% latency cost respective to 56% for previous architecture. Energy per cycle cost also exhibits an improvement of a 34% increase compared with 64%.

VI. CONCLUSION

The hypothesis that FBAR$_0$ could be improved with a heuristic to adapt the convergence based on detecting small signal variance was shown to be true in a single bit quantizer. An FBAR$_0$ heuristic is shown to improve the performance of an existing single bit adaptive η quantizer. The HAQ algorithm

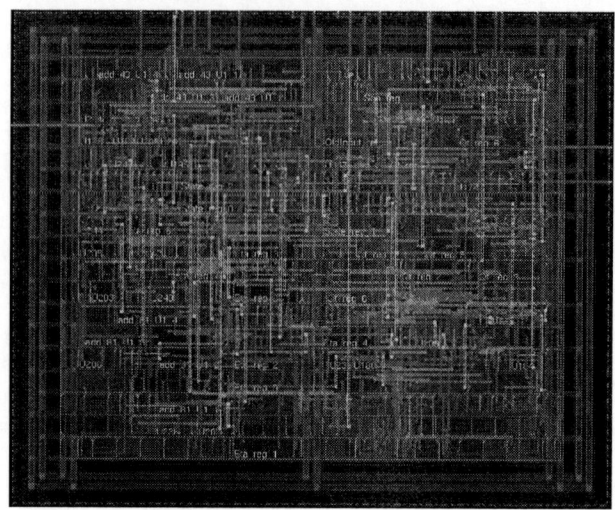

Fig. 5. Layout of HAQ architecture used in characterizing hardware performance. The architecture was synthesised with TSMC $0.18\mu m$ technology

is contributed as an improvement over prior art heuristic quantizers with PSNR demonstrated as having up to a $1.144db$ increase. Both PSNR and Performance were found improve by a larger margin than prior designs compared to a standard FBAR$_0$. Improvements make good use of the hardware design space as the added energy cost is around half that of prior art. In summary, designs able to leverage HAQ will better target today's low power CMOS image sensor applications.

ACKNOWLEDGMENT

The support of Wang Yang is acknowledged

REFERENCES

[1] El Gamal, A., "Trends in CMOS image sensor technology and design," *Electron Devices Meeting, 2002. IEDM '02. Digest. International*, pp.805-808, December 2002.

[2] Lineback, R., IC Insights Inc, Scottsdale, AZ, USA., "O-S-D Report 2011'", 2011. [Online], Available: http://www.icinsights.com/data/articles/documents/256.pdf"

[3] Kawahito et al., "CMOS Image Sensor with Analog 2-D DCT-Based Compression Circuits," *IEEE Journal of Solid-State Circuits*, Vol. 32, No.12, pp.2029-2039, December 1997.

[4] Vishwanath, M. and Owens, R.M. and Irwin, M.J., "VLSI architectures for the discrete wavelet transform," *Circuits and Systems II: Analog and Digital Signal Processing, IEEE Transactions on*, pp.305-316, May 1995.

[5] Shoushun, C., Bermak, A. Yan, W. Martinez, D., "Adaptive-Quantization Digital Image Sensor for Low-Power Image Compression", *Circuits and Systems I: Regular Papers, IEEE Transactions on*, Vol. 54, Number 1, pp. 13-25, Jan 2007

[6] Shoushun Chen and Bermak, A. and Yan Wang., "A CMOS Image Sensor With On-Chip Image Compression Based on Predictive Boundary Adaptation and Memoryless QTD Algorithm", *Very Large Scale Integration (VLSI) Systems, IEEE Transactions on*, Vol. 19, Number 4, pp. 538-547, April 2011.

[7] Chen Shoushun and Bermak, A. and Wang Yan and Martinez, D., "A CMOS Image Sensor with combined adaptive-quantization and QTD-based on-chip compression processor", *Custom Integrated Circuits Conference, 2006. CICC '06. IEEE*, pp. 329-332, Sept 2006.

[8] Martinez, D. Van Hulle, M. M., "Generalized Boundary Adaptation Rule for Minimizing rth Power Law Distortion in High Resolution Quantization", *Neural Netw.*, Vol. 8, Number 6, pp. 891-900, 1995

[9] Signal and Image Processing Institute, University of Southern California, Los Angeles, CA, "The USC-SIPI image database", 2011. [Online], Available: http://sipi.usc.edu/database/inex.html

0.5-V High-Speed Circuit Designs for Nanoscale SoCs
–Challenges and Solutions-

Kiyoo Itoh, Akira Kotabe, Dai Hisamoto, Ryuta Tsuchiya, and Riichiro Takemura

Central Research Laboratory, Hitachi, Ltd.,
Kokubunji, Tokyo 185-8601, Japan,
kiyoo.itoh.pt@hitachi.com

Abstract—Some solutions are proposed and evaluated by simulation after the challenges facing the creation of 0.5-V nanoscale SoCs are clarified. First, the repair techniques and nanoscale FD-MOSFETs are discussed in terms of their V_t-variation. Second, 0.5-V dual-V_{DD} dual-V_t logic circuits with gate-source reverse-biasing schemes are proposed. Third, a boosted word-voltage six-transistor (6-T) SRAM cell is evaluated with a 25-nm planar FD-SOI MOSFET and then a FinFET, revealing that the FinFET drastically improves the voltage margin and speed of the 6-T cell. Finally, the feasibility of a 0.5-V 25-nm SoC comprising a 1-Gb SRAM and 160-Mgate logic block is studied. We conclude that an SoC like this with a competitive speed while reducing the power to about one-tenth that of a conventional 1-V 32-nm CMOS LSI is possible, if the above-described devices and circuits are used and the within-wafer V_t-variations are stringently controlled and/or compensated for.

Keywords: 0.5-V 25-nm 6-T SRAM cell, FD-MOSFETs, repair, boosted word voltage, worst design

I. INTRODUCTION

Reducing the power consumption of SoCs (used in PCs, mobile devices, and even medical equipment) as well as 3D through-silicon-via (TSV)-chip stacks is becoming an increasingly important task to complete. The most effective and straightforward way to reduce the power consumption is to reduce the operating voltage V_{DD}. For memory-rich high-speed SoCs, such as an SoC with a 1-Gb SRAM and a 160-Mgate random logic block (Fig. 1), this reduction has been extremely difficult. This is because the reduction is prevented by two unscalable MOS parameters, namely, the MOS threshold voltage (V_t) and the variation in V_t [1, 2]. Reducing V_t, which is necessary to maintain the necessary level of speed under voltage scaling, intolerably increases the subthreshold current (leakage). The V_t-variation, which rapidly increases with the device scaling, particularly reduces the voltage margin of the six-transistor SRAM cell (6-T cell) [3]. Consequently, the V_{DD} of SoC products tends to become saturated at about 1 V, which is the so-called 1-V wall [1, 2]. Thus, innovative circuits and devices to cope with the leakage and variation issues are strongly needed to breach this wall and open the door to the 0.5-V nanoscale era. To the best of our knowledge, there have been few papers consistently and systematically describing 0.5-V circuits and devices for nanoscale SoCs.

In this paper, the importance of the suppression of V_t-variations for low-voltage high-speed SoCs is stressed first. Second, the concepts of the gate-source reverse biasing of low-

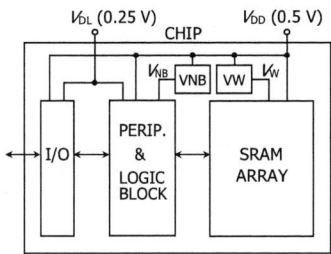

Figure 1. SoC incorporated with 1-Gb SRAM and a 160-Mgate logic block. VW/VNB; V_W/V_{NB} generators, PERIP.; peripheral circuits of SRAM array, $V_{NB} = -0.25$ V, and $V_W > V_{DD}$.

V_t MOSFETs and the applications to static logic circuits are described in terms of the low-V_{DD} operations. Third, a boosted word-line scheme to widen the voltage margin of a 6-T SRAM cell is proposed and evaluated by simulation and then discussed. Several repair techniques and a 25-nm planar FD-SOI MOSFET [4] and FinFET [5, 6] are used for the simulation. Finally, the possibility of a 0.5-V high-speed SoC (Fig. 1) is examined. The V_t used in this paper is defined as the sum of the constant current V_t (nA/μm) and 0.3 V.

II. SUPPRESSION OF V_t VARIATIONS

It is well known that the V_t-variation consists of two components, the within-wafer (WIW) systematic V_t-variation mainly caused by the short channel effect (SCE) and the local V_t variation (ΔV_t) coming from the random dopant fluctuation (RDF) in a MOS channel. The WIW V_t-variation varies the average V_t (V_{t0}) of each SoC in a wafer, degrading the voltage margin and speed. Moreover, it particularly increases the leakage of an SRAM cell array. The ΔV_t is most problematic for the SRAM cell array [3] due to the large variations in speed and voltage margin of each cell. The problem stems from the inherent features of the cell array, in which each of a huge number of tiny cells is randomly activated with static and ratio operations. The ΔV_t is of less concern for the logic block. This is because in practice, the minimum operating V_{DD} (V_{min}) of the cell array is higher, even after using the repair techniques, than that of the logic block when taking at least the averaging effect of RDF [1, 2] into account. Since the V_{DD} for the SoC for successful operation must be higher than the V_{min} of the SRAM, the speed variations of the logic block at the V_{DD} are mitigated, thanks to the lower V_{min}, than those of the SRAM cell array. In any event, to be more exact, the maximum ΔV_t (ΔV_{tmax}) in a chip must be reduced to ensure properly operations of the SoC

978-1-61284-863-1/11 $26.00 © 2011 IEEE

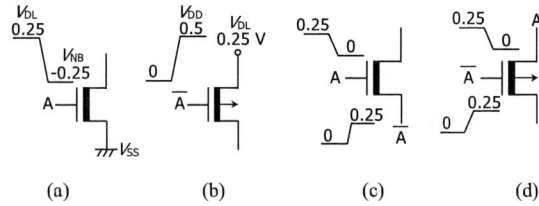

Figure 2. G-S offset driving of (a) nMOSFET and (b) pMOSFET, and G-S differential driving of (c) nMOSFET and (d) pMOSFET. Each MOSFET has a low V_t ($V_{tl} = 0$).

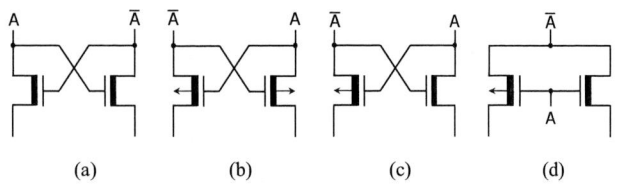

Figure 3. G-S DIFF of a pair of (a) nMOSFETs, (b) pMOSFETs, and (c) and (c) n/p MOSFETs. Each MOSFET has a low V_t ($V_{tl} = 0$). A, $\overline{A}$: 0.25 V or 0, as in Figure 2.

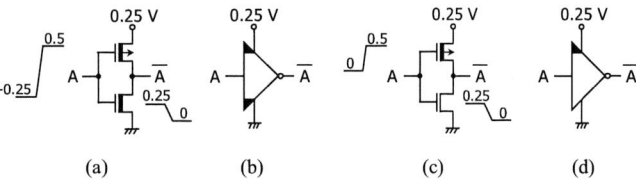

Figure 4. Inverter with (a) low-V_t (V_{tl}) n/p MOSFETs and (b) its logic symbol, and with (c) high-V_t (V_{th}) nMOSFET and low-V_t (V_{tl}) pMOSFET and (d) its logic symbol. $V_{th}/V_{tl} = 0.25/0$ V.

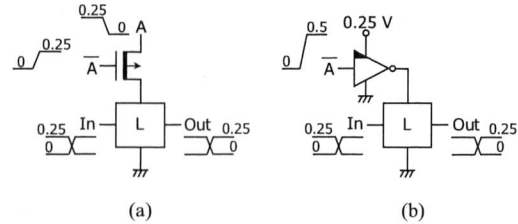

Figure 5. (a) pMOS power switch and (b) inverter power switch. L: logic block and $V_{th}/V_{tl} = 0.25/0$ V.

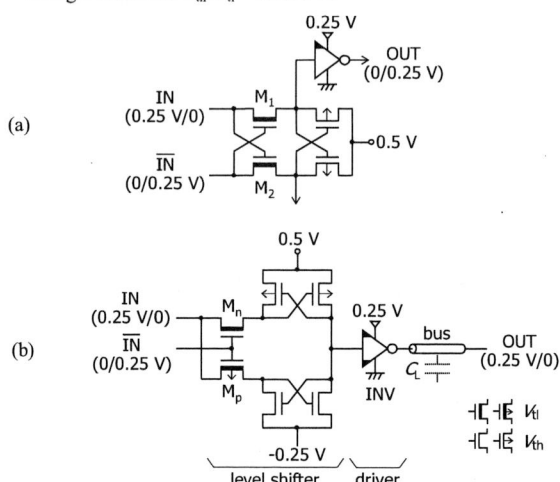

Figure 6. (a) Level shifter and (b) bus driver [3]. $V_{th}/V_{tl} = 0.25/0$ V and $C_L = 3$pF.

even for the worst case design. ΔV_{tmax} is given as $m\sigma(V_t)$, where $\sigma(V_t) = A_{vt}(LW)^{-0.5}$. m, $\sigma(V_t)$, A_{vt}, and LW are constant depending on the circuit count in the block, the standard deviation of the V_t distribution, the Pelgrom constant, and the MOSFET size [1, 2], respectively. Thus, FD-MOSFETs (e.g., planar FD-SOI MOSFETs and FinFET) and the repair techniques (redundancy and error checking and correction (ECC)) [7] are indispensable for reducing ΔV_{tmax}. For example, a planar ultra-thin buried-oxide 25-nm FD-MOSFET (SOTB [4]), in which the SOI, BOX, and gate oxide are 10 nm, 5 nm, and 1 nm thick, respectively, are estimated to achieve an A_{vt} of 0.5 mV·μm and an S-factor of 91 mV/decade. If such a SOTB and repair techniques for eliminating excessive ΔV_t MOSFETs are combined, the estimated ΔV_{tmax} of a 25-22-nm SoC (Fig. 1) is 48 mV for the average MOSFET (i.e., $8F^2$) in a 160-Mgate logic block, and 63 mV for the transfer MOSFET (i.e., the smallest MOSFET of $1.5F^2$) in 1-Gb SRAM cells [2].

III. DUAL-VDD (0.5/0.25 V) STATIC LOGIC CIRCUITS

Low-power high-speed logic circuits are created by reducing the V_{DD} while maintaining the leakage, gate-over drive (e.g., $V_{GS} - V_{t0} = V_{DD} - V_{t0}$) of a MOSFET and the speed variation, which is approximately given as $\Delta V_{tmax}/(V_{DD}-V_{t0})$ [1, 2]. For a given ΔV_{tmax}, the reduction in V_{t0} meets the requirements if the large resulting leakage is sufficiently reduced. A reduction to 0 V, if possible, is thus most desirable.

The ΔV_{tmax} reduction further reduces the speed variation. One of the most effective ways is to use the gate(G)-source(S) reverse biasing scheme of a low-V_t MOSFET [1, 2]. There are two schemes for this, the G-S offset driving (G-S OFST) and the G-S differential driving (G-S DIFF) [7], which are conceptually shown in Fig. 2. The G-S OFST usually requires dual-V_{DD} circuit configurations (i.e., V_{DL}/V_{NB} (< 0) for nMOS and V_{DD}/V_{DL} (<V_{DD}) for pMOS). Here, a negative supply V_{NB} is internally generated by a well-known on-chip generator. MOSFETs can be cut off during the inactive period if the minimum tolerable V_t for leakage is 0.25 V, and $V_{DD}/V_{DL} = $ 0.5/0.25 V, highV_t (i.e., V_{th})/lowV_t (i.e., V_{tl}) = 0.25/0 V, and V_{NB} = − 0.25 V. In the inactive period, they are cut off using a G-S reverse bias of 0.25 V, while during the active period, they provide a large amount of current with a gate-over drive of 0.25 V. For the G-S DIFF, the G-S is self-reverse biased during the inactive periods, so it is cut off as long as V_{DL} is higher than 0.25 V. This driving thus enables the lowest possible voltage operation at 0.25 V with a gate-over drive of 0.25 V. The G-S DIFF of a pair of MOSFETs, as shown Fig. 3, creates unique ultra-low voltage (i.e., 0.25 V) circuits. One of the two MOSFETs in circuit (a) or circuit (b) turns on, depending on the polarity of the differential input (A and /A) voltage of 0.25 V. For circuit (c), both the pMOS and nMOS are activated or inactivated, depending on the input. For circuit (d), either the pMOS or nMOS is activated, depending on the input. Figure 4 shows the applications of a G-S OFST to a low-V_t CMOS inverter. A large input-voltage swing reduces the amount of leakage while increasing the on current due to using low V_ts.

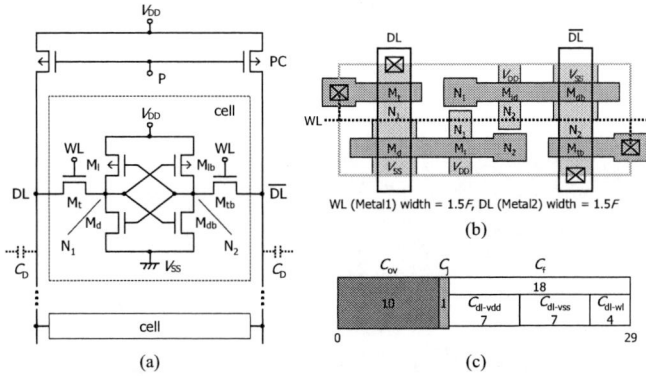

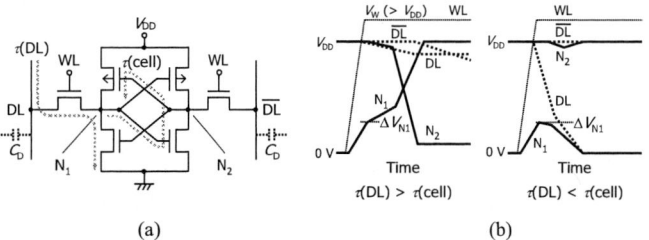

Figure 7. (a) 6-T SRAM cell, (b) cell layout and (c) components of C_D per cell (aF). PC; precharge circuit, P; precharge, and C_D; data-line capacitance. Cell size = 19F x 7F, F = 25 nm, L = 25 nm, and $W(M_l)/W(M_t)/W(M_d)$ = 45/50/70 nm.

Figure 5 depicts the applications to a power switch on a logic block (L). The G-S DIFF of the V_{tl}-pMOS power switch (Fig. 5(a)) allows the whole block to operate at 0.25-V. An inverter using a G-S OFST (Fig. 5(b)) works as a power switch despite the need for a high 0.5-V power supply. Figure 6(a) shows a level shifter [3] with a G-S DIFF (Fig. 3(a)) and G-S OFST (Fig. 4(c)). A fast level shifting from V_{DL} to V_{DD} is possible. If necessary, the high level is further shifted by the inverter shown in Fig. 4(c). Figure 6(b) shows a bus driver [3] consisting of a dual-V_t level shifter and a V_{tl} CMOS inverter (INV) to drive a large bus capacitance (C_L) of 3 pF. The level shifter consists of a V_{tl}-CMOS receiver (M_n, M_p) and a 0.75-V power-supply (= V_{DD} + V_{NB}) V_{th}-cross-coupled pMOS and nMOS circuit. For the bus itself, the power and speed are improved by a low-voltage swing of 0.25 V on the bus. For the preceding circuit, they are also improved by the G-S DIFF applied to the receiver (i.e., Fig. 3(d)), and a G-S OFST to the inverter (i.e., Fig. 4(a)). According to the inputs (IN, /IN), either the pMOS or nMOS circuit in the receiver is quickly activated without incurring any leakage, so the inverter is over-driven at a large voltage swing of 0.75 V. Thus, the inverter quickly drives C_L without leakage.

The simulation results when using the 25-nm SOTB at $V_{DD}/V_{DL}/V_{NB}$ = 0.5/0.25/–0.25 V and a V_{th}/V_{tl} = 0.3/0.05 V revealed that a four-stage inverter chain with MOSFET power switches [3], similar to that in Fig. 5(a), reduces the power to one-fifth that of the conventional 0.5-V V_{th}-CMOS circuits for a given speed. Moreover, the 3-pF-load bus driver in Fig. 6(b) reduces the power to one-fourth that of the conventional 0.5-V V_{th}-CMOS bus driver [3], while reducing the necessary total channel width to half.

IV. 0.5-V 6-T SRAM CELLS

In this section, the feasibility of a 0.5-V low-power high-speed 25-nm 1-Gb SRAM is discussed, while taking both V_{t0} and ΔV_t issues into account. The subthreshold leakage must be reduced to achieve such an SRAM by increasing the average V_t (V_{t0}) of the cell array. For example, if a V_{t0} of 0.28 V at 25°C is used, which is an approximately estimated V_{t0} for designing a 1-V high-speed 32-nm 291-Mb SRAM [8], a hypothetical 1-V 32-nm 1-Gb SRAM may consume a leakage of 10 A at 100°C using this V_{t0} [3], and the leakage power is thus 10 W. If V_{t0} is

increased to 0.36 V to reduce it to 2 A, and V_{DD} is reduced to 0.5 V in the 25-nm generation, the power is reduced to 1 W. An increase in V_{t0} and a decrease in V_{DD} for achieving such a low-power 1-Gb SRAM, however, inevitably degrade the voltage margin and speed of the cells. A boosted word-voltage scheme [3] for the 6-T cell shown in Fig. 7 was recently proposed to tackle these problems. This scheme simply boosts the word-line voltage V_W to a level higher than V_{DD} for both the read and write operations with a fixed pulth width, while leaving the other voltages (i.e., the power supply of the cells and pre-charge voltage of data lines) at V_{DD}. Combined with the repair techniques and FD-SOI MOSFETs for smaller ΔV_{tmax}, as discussed previously, this may solve the problems. Note that the increase in power by the raised V_W is marginal because only one word line is activated. Although this scheme is disregarded based on past design knowledge, the advantage is conceptually understandable if the dynamic behaviors [9] of cell node N_1 are considered, as shown in Fig. 8. For example, even if the raised N_1 voltage (ΔV_{N1}), when V_W is applied, is high enough to turn on M_{db}, the cell will not necessarily flip. If ΔV_{N1} is slowly discharged in accordance with the DL discharging, the feedback loop of the cell comprising M_{db} and M_l is activated within the word-pulse width, so the cell flips. If ΔV_{N1} is rapidly discharged, however, the feedback loop within the word-pulse width is deactivated, and thus, the cell does not flip. Consequently, the margin is closely related to the relationship between the DL-discharging time τ(DL), the feedback-loop delay τ(cell), and the word-pulse width. Therefore, the magnitude of the data-line capacitance C_D is crucial. Figure 7(c) depicts the data-line capacitance per cell C_d. The C_d consists of two components. One is the MOS-structure dependent capacitance, that is, the sum of the gate-drain overlap capacitance C_{ov} and the junction capacitance C_j of the transfer MOSFET, when the drain is connected to the data line. The other is the MOS-structure independent capacitance C_f, that is, the sum of the DL-WL capacitance C_{dl-wl}, the DL-V_{DD} capacitance C_{dl-vdd}, and the DL-V_{SS} capacitance C_{dl-vss}. In any event, despite the importance, the dependences of the voltage margin on C_D and the MOS structures, especially for FinFET, remain unknown.

A. Voltage Margin of SOTB 6-T Cell

The degradations of the margin and speed strongly depend on the ΔV_{tmax}s of the MOSFETs in the cell (Fig. 7). Each MOSFET has its own ΔV_{tmax}, depending on the size. The δ_1 is ΔV_{tmax} for a transfer MOS (M_t) and load MOS (M_l), both of which are almost the same size, while the δ_2 is that for driver

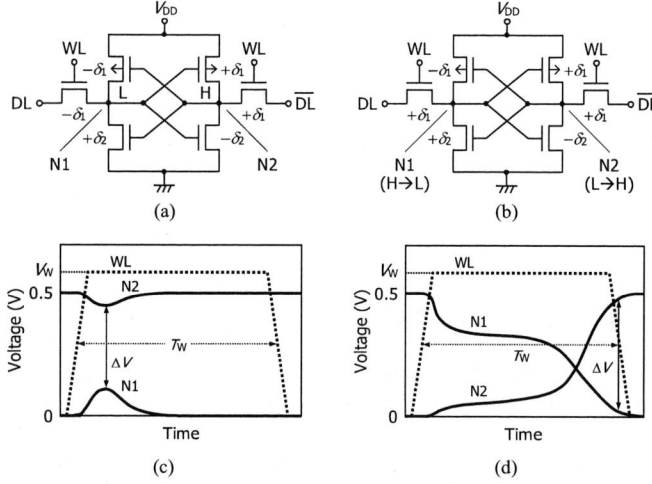

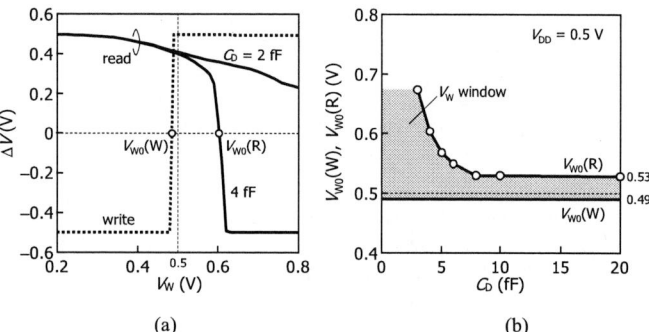

Figure 9. Worst combination in ΔV_{tmax} (i.e., δ_1 and δ_2) for (a) read and (b) write, and corresponding schematic waveforms for (c) read and (b) write [3]. ΔV: dynamic margin.

Figure 10. (a) ΔV vs. V_W for various C_{DS}, and (b) $V_{w0}(W)$ and $V_{w0}(R)$ vs. C_D. $V_{DD} = 0.5$ V, $V_{t0} = 0.36$ V, $\delta_1/\delta_2 = 63/51$ mV, $T_W = 1$ ns and 100°C.

MOS (M_d). They are estimated to be 63 mV and 51 mV for the 1-Gb SRAM, respectively, for a 25-nm SOTB after repair [3]. The worst combination of δ for a read operation [3] is developed when the raised N_1 voltage by a M_t and M_d ratio operation is the highest with the worst set of δ_1 and δ_2 in a cell (Fig. 9(a)). This combination minimizes the voltage margin (i.e., the voltage difference, ΔV, between the two storage nodes, N_1 and N_2). The worst combination for a write operation [3] is shown in Fig. 9(b). This combination makes it most difficult for the cell to flip N_1 from "H" to "L" when "L" is written from DL, and thus minimizes ΔV. Figure 10 depicts ΔV versus V_W for the worst δ combinations, and $V_{w0}(W)$ and $V_{w0}(R)$ versus C_D at $V_{DD} = 0.5$ V, T_W (i.e.,V_W pulse width) = 1 ns, and 100°C. Here, $V_{w0}(W)$ and $V_{w0}(R)$ are the critical V_Ws for successful write and read operations, each of which is defined as the V_W necessary for $\Delta V = 0$ in Fig. 10(a). The acceptable V_W for a successful read and write is thus between $V_{w0}(W)$ and $V_{w0}(R)$. Note that $V_{w0}(W)$ (= 0.49 V) close to V_{DD} (= 0.5 V) calls for a word boosting. Indeed, the word boosting is indispensable, taking another simulation result of a $V_{w0}(W)$ of 0.53 V at −10°C into account. Even so, the window is saturated to be as small as 0.04 V at a larger C_D, making the V_W setting in the window difficult. The window must be wide enough to ensure low cost, stable, and high-speed operations even under V_{DD} and

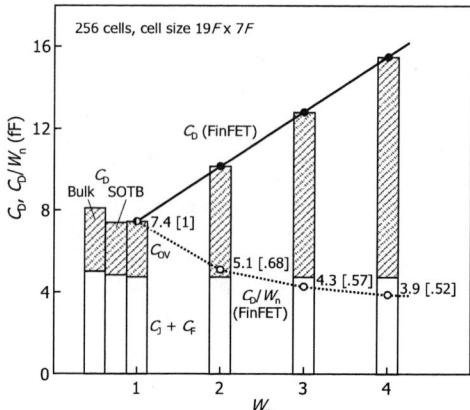

Figure 11. Data-line capacitance C_D vs. W_n.

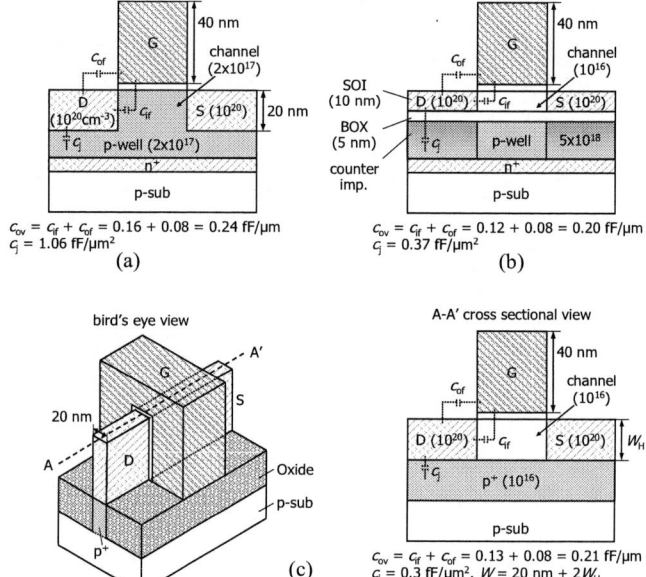

Figure 12. Structures of (a) planer bulk, (b) SOTB, and (c) bulk FinFET.

thus V_W fluctuations. If the V_W window becomes wider for a larger C_D, more memory cells can be connected to one data line using a much easier V_W setting, resulting in a smaller chip size with a smaller area overhead at the data-line divisions. Moreover, it allows V_W to be set higher, resulting in a faster speed.

B. Voltage Margin and Speed of FinFET 6-T Cell

The V_W window is drastically widened by the FinFET [5, 6]. The effectively reduced C_D and inherently smaller ΔV_{tmax} thanks to the wider channel width are responsible for the widened window as follows. Figure 11 depicts the total data-line capacitance C_D when 256 FinFET cells are connected to one data line. The C_D is the sum of the total gate-drain overlap capacitance C_{OV} and total junction capacitance C_J, and the total MOS-strucure independent capacitance, that is, a fixed capacitance C_F, as mentioned previously. Here, the Fin-Fin capacitance is negligible for the cell layout. W_n is the ratio of the channel widths to the original when the channel widths of

978-1-61284-863-1/11 $26.00 © 2011 IEEE

the respective MOSFETs in Fig. 7 are simultaneously widened at the same rate by increasing the height. Therefore, a W_n of four implies that the respective channel widths are simultaneously quadrupled for a given layout MOSFET footprint. As a reference, the C_{DS} of a conventional bulk MOSFET and the SOTB [4] are added at $W_n = 1$ in Fig. 11. Widening the channel width is impossible for a given cell size in such planar MOSFETs, which is different from that in a FinFET. Thus, they exist only at $W_n = 1$. The corresponding MOS structures [4, 10] are shown in Fig. 12. The capacitance components of the MOSFET c_{ov}, c_{if}, c_{of}, and c_j are the gate-drain capacitance, inner fringe capacitance, and outer fring capacitance per length, and the junction capacitance per area, respectively. These capacitances were estimated by using the data from previous papers [4, 11-14] and the results from our device simulation. For a FinFET, C_D increases with W_n due to the increased C_{OV}. However, C_D/W_n, which is the effective C_D of FinFET with $W_n = 1$, significantly decreases due to C_D being occupied by a fixed capacitance C_F. Consequently, it decreases with a larger W_n than those of planar MOSFETs with $W_n = 1$, exemplified by 0.52 for $W_n = 4$. Therefore, a higher $V_{W0}(R)$ and a higher speed are expected with an increasing W_n. Moreover, the δ_1 and δ_2 of a FinFET are reducible from those (i.e., 63/51 mV) for $W_n = 1$ according to the square root of LW. The V_{W0} window is thus expected to widen due to the further increased $V_{W0}(R)$ and decreased $V_{W0}(W)$. A reduction in C_{OV} further enhances such advantages of the FinFET.

Figure 13(a) shows the $V_{W0}(R)$ and $V_{W0}(W)$ versus the W_n at $V_{DD} = 0.5$ V for a FinFET. The larger the W_n, the wider the V_W window due to the higher $V_{W0}(R)$ and lower $V_{W0}(W)$, as expected. Note the rapid increase in $V_{W0}(R)$. This is because the feedback speed τ(cell) of the cross-coupled MOSFETs is kept almost constant despite increasing the W_n due to their loads being dominated by the corresponding gate capacitances, while the wider channel widths of M_t and M_d shorten τ(DL). The reductions in δ_1 and δ_2 are also responsible for the increase in $V_{W0}(R)$ and decrease in $V_{W0}(W)$. Figure 13(b) shows the signal developing speed τ_0 versus W_n for 256 cells connected to a data line at $V_{DD} = 0.5$ V. The worst (i.e., slowest) δ combination for the speed, which differs from that for the voltage margin, is used. It is developed when the V_t of each MOSFET has the highest V_t, namely, $V_t(M_t, M_{tb}) = V_{t0} + \delta_1$, $V_t(M_l, M_{lb}) = -(V_{t0} + \delta_1)$ and $V_t(M_d, M_{db}) = V_{t0} + \delta_2$. V_W is set to the intermediate of $V_{W0}(W)$ and $V_{W0}(R)$, as shown with the dotted line in Fig. 13(a), except for $W_n = 4$, where it is equal to that for $W_n = 3$ to ensure device reliability. Here, τ_0 is the time from the intermediate of the V_W pulse to the differential signal of 100 mV. The τ_0s of SOTB at $V_{DD} = 0.7$ V and 1 V for $W_n = 1$ are also shown for reference. Obviously, the FinFET at $V_{DD} = 0.5$ V is faster than the 0.7-V SOTB, or as competitive as 1-V SOTB, if $W_n > 3$. In other words, a FinFET at least halves the power consumption for a given speed.

V. DISCUSSION

Based on our investigation, it turned out that a 0.5-V high-speed 25-nm SoC is feasible, if the gate-source reverse biasing logic circuits and the boosted word-voltage 6-T cell are used. In particular, the results for the SRAM cell are kept effective since the definition of $V_{W0}(R)$ and $V_{W0}(W)$ and the

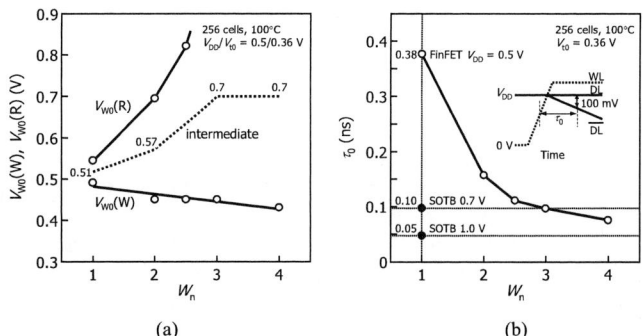

Figure 13. (a) $V_{W0}(W)$ and $V_{W0}(R)$ vs. W_n, and (b) speed vs. W_n.

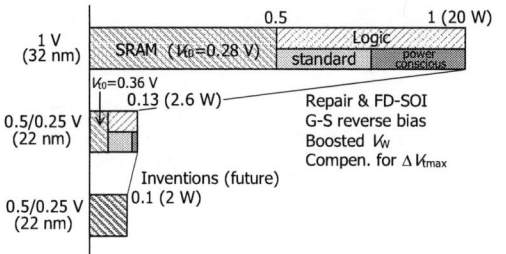

Figure 14. Estimated reduction in active power for a chip comprising a 1-Gb SRAM and a 160-Mgate logic block.

combinations of δ are justified as follows, although detailed analysis will be needed. The V_t-imbalances between the two CMOS inverters for a cross-coupled circuit in the cell may flip the cell according to the imbalances after turning off the word pulse, causing malfunctios. To avoid the flipping, ΔV must be higher than the difference in the logical threshold voltage between the two inverters, ΔV_c, which is approximately given as the sum of δ_1 and δ_2 (= 114 mV). $V_{W0}(R)$ and $V_{W0}(W)$ must thus be defined as the V_W at $\Delta V = \Delta V_c$, instead of the V_W at $\Delta V = 0$ V. Even so, $V_{W0}(R)$s and $V_{W0}(W)$s defined at the two different ΔVs are almost the same for practical C_{DS} (> 4fF) since ΔV is quite sensitive to V_W in the vicinity of $\Delta V = 0$ V, as seen in Fig. 10(a). The use of the $V_{W0}(R)$ and $V_{W0}(W)$ defined in the previous section is thus kept effective. It should be noted that, in practice, the cell stability can be further improved because combinations of δ_1 and δ_2 are statistically unrealistic. For example, the maximum difference in V_t value between a pair of MOSFETs (e.g., M_d and M_{db}) is likely to be $2^{0.5}\Delta V_{tmax}$ instead of the $2\Delta V_{tmax}$ shown in the figure. Since the voltage margin was wide sufficiently even with such unrealistic combinations, it is further improved with realistic and more relaxed ΔV_{tmax} combination.

The within-wafer (WIW) V_t-variation may affect the promising results described above due to intolerably wide variations in the speed and leakage of the chips. Thus, in addition to stringently controlling the fabrication processes and suppressing the SCE, compensating for the variations in V_{t0} using an on-chip substrate or well reverse-bias (V_{BB}) generator is indispensable at low V_{DD}s. Such on-chip V_{BB} generators have been proposed for an MPU [15] and a 0.6-V DRAM [16] using bulk CMOS processes. However, they may generate hazardous bulk junction currents caused by a positive V_{BB} bias. For the SOTB, even a difference in V_{t0} of 100 mV between chips can

978-1-61284-863-1/11 $26.00 © 2011 IEEE

be compensated for by varying V_{BB} by 0.5 V [17] without such junction currents because the BOX cuts the substrate current. For a FinFET, the compensation for V_{t0} variations using a V_{BB} control may be done for an ultra-thin BOX applied to an SOI FinFET [17], while for a bulk FinFET [10] it remains unknown. An on-chip low-dropout voltage down-converter for the V_{DD} [18] might assist in the V_{t0} compensation only if the compensation when using an on-chip V_{BB} generator is insufficient. As for advanced MOSFETs, we found that the FinFET is suitable for creating SRAM cells with excellent performances even at 0.5 V. This is also the case for the logic block, particularly ones in which the circuit loads are dominated by the interconnect capacitances, as in the data line in an SRAM cell array. Even for loads dominated by the gate capacitance, the speed variations are reduced due to the wider channel width, although the power increases if the channel width is unnecessarily wide. The advantages of a FinFET will be eventually further enhanced in the less-than-22-nm generations due to its wider channel width.

Figure 14 shows the reduction in active power of the SoC (Fig. 1) estimated on the assumption that a 1-V 32-nm CMOS SoC consumes 20 W, that the SRAM and the logic block consume the same power, and that half the power of the logic block comes from power-conscious circuits accepting dual-V_{DD} and dual-V_t circuits that reduce the power to one-fourth, as mentioned before. Since the V_{t0} of the array can be increased to 0.36 V and the V_{DD} can be reduced to 0.5 V using the above-described 0.5-V devices and circuits, the power of a 25-nm SoC is reduced to 2.6 W. It will be further reduced to less than one-tenth (< 2 W) by using gate-source reverse biasing circuits throughout the chip, along with other inventions expected in the near future. The level of speed may continue to be compatitive even at low V_{DD}s if a FinFET is used.

VI. CONCLUSION

After clarifying the challenges faced with 0.5-V nanoscale SoCs, we proposed some solutions and evaluated them by conducting simulations. First, we discussed repair techniques and nanoscale FD-MOSFETs in terms of their V_t-variation. Second, we proposed 0.5-V dual-V_{DD} dual-V_t logic circuits with gate-source reverse-biasing schemes. Third, a boosted word-voltage six-transistor (6-T) SRAM cell was evaluated using a 25-nm planar FD-SOI MOSFET and then a FinFET, revealing that the FinFET drastically improves the voltage margin and speed of the 6-T cell. Finally, the feasibility of a 0.5-V 25-nm SoC comprising a 1-Gb SRAM and 160-Mg logic block was studied. We concluded that such a competitive speed SoC while reducing the power to about one-tenth that of a conventional 1-V 32-nm CMOS LSI is possible, if the above-described devices and circuits are used and the within-wafer V_t-variations are stringently controlled and/or compensated for.

REFERENCES

[1] K. Itoh, "Adaptive circuits for the 0.5-V nanoscale CMOS era," ISSCC Dig., pp. 14-20, Feb. 2009.

[2] K. Itoh, M. Yamaoka, and T. Oshima, "Adaptive circuits for the 0.5-V nanoscale CMOS era," IEICE Trans. Vol. E93-C, No. 3, pp. 216-233, March 2010.

[3] A. Kotabe, K. Itoh, R. Takemura, R. Tsuchiya, and M. Horiguchi, "Device-Conscious Circuit Designs for 0.5-V High-Speed Memory-Rich Nanoscale CMOS LSIs," CICC Dig., 16-1, Sept. 2011.

[4] R. Tsuchiya, et al., "Controllable Inverter Delay and Suppressing Vth Fluctuation Technology in Silicon on Thin BOX Featuring Dual Back-Gate Bias Architecture," IEDM Dig., pp. 475-478, Dec. 2007.

[5] D. Hisamoto, T. Kaga, Y. Kawamoto, and E. Takeda, "A Fully Depleted Lean-channel Transistor (DELTA)–A novel vertical ultra thin SOI MOSFET–," IEDM Dig., pp. 833-836, Dec. 1989.

[6] D. Hisamoto, et al., "FinFET–A Self-Aligned Double-Gate MOSFET Scalable to 20 nm," IEEE Trans. Electron Devices, Vol. 47, No. 12, pp. 2320-2325, Dec. 2000.

[7] M. Horiguchi and K. Itoh, Nanoscale Memory Repair, Springer, Jan. 2011.

[8] Y. Wang, et al., "A 4.0 GHz 291 Mb Voltage-Scalable SRAM Design in a 32 nm High-k + Metal-Gate CMOS Technology with Integrated Power Management," IEEE J. Solid-State Circuits, Vol. 45, No. 1, pp. 103-110, Jan. 2010.

[9] M. Yamaoka, K. Osada, and T. Kawahara, "A Cell-activation-time Controlled SRAM for Low-voltage Operation in DVFS SoCs Using Dynamic Stability Analysis," ESSCIRC Dig., pp. 286-289, Sept. 2008.

[10] C. C. Wu, et al., "High Performance 22/20nm FinFET CMOS Devices with Advanced High-K/Metal Gate Scheme," IEDM Dig., pp. 600-603, Dec. 2010.

[11] T. Chiarella, et al., "Migrating from Planar to FinFET for Further CMOS Scaling: SOI or Bulk?" ESSDERC Dig., pp. 84-87, Sept. 2009.

[12] Y. Taur and T. Ning, Fundamentals of Modern VLSI Devices, Cambridge University Press, 1998.

[13] J. Zou, et al., "Predictive 3-D Modeling of Parasitic Gate Capacitance in Gate-all-Around Cylindrical Silicon Nanowire MOSFETs," IEEE Trans. Electron Devices, Vol. 58, No. 10, pp. 3379-3387, Oct. 2011.

[14] A. Yagishita, "FinFET SRAM Process Technology for hp32 nm node and beyond," ICICDT, pp. 1-4, May 2007.

[15] M. Miyazaki, et al., "A 1000-MIPS/W microprocessor using speed adaptive threshold-voltage CMOS with forward bias," ISSCC Dig., pp. 420-421, Feb. 2000.

[16] K. Hardee, et al., "A 0.6V 205MHz 19.5ns tRC 16Mb embedded DRAM," ISSCC Dig., pp. 200-201, Feb. 2004.

[17] K. Itoh, N. Sugii, D. Hisamoto, and R. Tsuchiya, "FD-SOI MOSFETs for the low-voltage nanoscale era," Int. SOI Conf. Dig., pp. 1-4, Oct. 2009.

[18] Y. Okuma et al., "0.5-V Input Digital Low-Dropout Regulator (LDO) with 98.7% Current Efficiency in 65 nm CMOS," IEICE Trans. Electron., vol. E94-C, No. 6, pp. 938-944, 2011.

IEEE
445 Hoes Lane
Piscataway, NJ 08854-4141

ISBN 978-1-61284-863-1